JAMMING AND RHEOLOGY

JAMMING AND RHEOLOGY

Constrained dynamics on microscopic and macroscopic scales

Edited by

Andrea J. Liu

Sidney R. Nagel

CRC Press
Taylor & Francis Group
Boca Raton London New York

CRC Press is an imprint of the
Taylor & Francis Group, an **informa** business

CRC Press
Taylor & Francis Group
6000 Broken Sound Parkway NW, Suite 300
Boca Raton, FL 33487-2742

First issued in paperback 2020

ISBN 13: 978-0-367-57879-4 (pbk)
ISBN 13: 978-0-7484-0879-5 (hbk)

Visit the Taylor & Francis Web site at
http://www.taylorandfrancis.com

and the CRC Press Web site at
http://www.crcpress.com

Typeset in Times by Deerpark Publishing Services

Publisher's Note
This book was prepared from camera-ready copy supplied by the editors.

British Library Cataloguing in Publication Data
A catalogue record for this book is available from the British Library

Library of Congress Cataloging in Publication Data
A catalog record for this book has been requested

We dedicate this volume to two pioneering theorists in the field of soft condensed matter, Shlomo Alexander and Tom Witten.

CONTENTS

ACKNOWLEDGEMENTS

Figure acknowledgements

"Jamming and the Statics of Granular Materials" (Dov Levine).
Figures 2 and 5 American Physical Society. Figure 4 American Association for the Advancement of Science.
"Rheology, and how to stop aging" (Jorge Kurchan)
Figure 1 American Physical Society and Elsevier Science. Figures 2 and 5 EDP Sciences and Elsevier Science.
"The Viscous Slowing Down of Supercooled Liquids and the Glass Transition: Phenomenology, Concepts, and Models" (Gilles Tarjus and Daniel Kivelson).
Figures 3 and 6 American Physical Society. Figure 5 EDP Sciences. Figures 7 and 10 Elsevier Science. Figure 8 American Institute of Physics. Figure 13 Academic Press. Reproduced with permission.
"Jamming, Friction and Unsteady Rheology" (Mark O. Robbins).
Figure 7 American Association for the Advancement of Science. Figure 8 American Chemical Society. Reproduced with permission.

Reprint acknowledgements

"Model for force fluctuations in bead packs," S. N. Coppersmith, C.-h. Liu, S. Majumdar, O. Narayan, and T. A. Witten, Phys. Rev. E 53, 4673–4685 (1996). Copyright 1996 by the American Physical Society.
"Force Distributions in Dense Two-Dimensional Granular Systems," F. Radjai, M. Jean, J.-J. Moreau and S. Roux, Phys. Rev. Lett. 77, 274–277 (1996). Copyright 1996 by the American Physical Society.
"Force Transmission in Granular Media," C. Thornton, KONA 15, 81–90 (1997). Copyright 1997 Hosokawa Micron International Inc. Reproduced with permission.
"Force Distribution in a Granular Medium," D. M. Mueth, H. M. Jaeger, and S. R. Nagel, Phys. Rev. E 57, 3164–3169 (1998). Copyright 1998 by the American Physical Society.
"Jamming and Static Stress Transmission in Granular Materials," M. E. Cates, J. P. Wittmer, J.-P. Bouchaud and P. Claudin, Chaos 9, 511–522 (1999). Copyright 1999, American Institute of Physics. Reproduced with permission.
"Statistical Mechanics of Stress Transmission in Disordered Granular Arrays," S. F. Edwards and D. V. Grinev, Phys. Rev. Lett. 82, 5397–5400 (1999). Copyright 1999 by the American Physical Society.
"Elastoplastic arching in two dimensional granular heaps," F. Cantelaube, A. K. Didwania and J. P. Goddard, in Physics of Dry Granular Media, ed. H. J. Herrmann, et al. (Kluwer Academic Publishers, Amsterdam, 1998), 123–128. Reproduced with kind permission of Kluwer Academic Publishers.
"Stress propagation through frictionless granular material," A. V. Tkachenko and T. A. Witten, Phys. Rev. E 60, 687–696 (1999). Copyright 1999 by the American Physical Society.
"Rigidity loss transition in a disordered 2D froth," F. Bolton and D. Weaire, Phys. Rev. Lett. 65, 3449–3451 (1990). Copyright 1990 by the American Physical Society.

"Osmotic pressure and viscoelastic shear moduli of concentrated emulsions," T. G. Mason, M.-D. Lacasse, G. S. Grest, and D. Levine, J. Bibette and D. A. Weitz, Phys. Rev. E 56, 3150–3166 (1997). Copyright by the American Physical Society.

"Universality and its Origins at the Amorphous Solidification Transition," W. Q. Peng, H. E. Castillo, P. M. Goldbart, and A. Zippelius, Phys. Rev. B 57, 839–847 (1998). Copyright by the American Physical Society.

"Anomalous viscous loss in emulsions," A. J. Liu, S. Ramaswamy, T. G. Mason, H. Gang, and D. A. Weitz, Phys. Rev. Lett. 76, 3017–3020 (1995). Copyright 1995 by the American Physical Society.

"Yielding and Rearrangements in Disordered Emulsions," P. Hebraud, F. Lequeux, J. P. Munch, and D. J. Pine, Phys. Rev. Lett. 78, 4657–4660 (1997). Copyright 1997 by the American Physical Society.

"Dynamics of viscoplastic deformation in amorphous solids," M. Falk and J. S. Langer, Phys. Rev. E 57, 7192–7205 (1998). Copyright 1998 by the American Physical Society.

"The Mode Coupling Theory of Structural Relaxations," W. Gotze and L. Sjogren, Transport Theory and Statistical Physics 24, 801–853 (1995). Reproduced with permission of Marcel Dekker, Inc.

"Light scattering spectroscopy of orthoterphenyl–idealized and extended mode coupling analysis," H. Z. Cummins, G. Li, W. M. Du, Y. H. Hwang, and G.Q. Shen, Prog. Theor. Phys. S126, 21–34 (1997). Copyright 1997, The Progress of Theoretical Physics. Reproduced with permission.

"Glass transition in colloidal hard spheres: mode-coupling theory analysis," W. Van Megen and S. M. Underwood, Phys. Rev. Lett. 70, 2766–2769 (1993). Copyright 1993 by the American Physical Society.

"Emulsion glasses: a dynamic light-scattering study," H. Gang, A. H. Krall, H. Z. Cummins, and D. A. Weitz, Phys. Rev. E. 59, 715–721 (1999). Copyright 1999 by the American Physical Society.

"Scaling concepts for the dynamics of viscous liquids near an ideal glassy state," T. R. Kirkpatrick, D. Thirumalai, and P. G. Wolynes, Phys. Rev. A 40, 1045–1054 (1989). Copyright 1989 by the American Physical Society.

"A Topographic View of Supercooled Liquids and Glass Formation," F. H. Stillinger, Science 267, 1935–1939 (1995). Copyright 1995 American Association for the Advancement of Science. Reprinted with permission.

"A thermodynamic theory of supercooled liquids," D. Kivelson, S. A. Kivelson, X. Zhao, Z. Nussinov, and G. Tarjus, Physica A219, 27–38 (1995). Reproduced with permission of Elsevier Science Ltd.

"Long-Lived Structures in Fragile Glass-Forming Liquids," A. I. Melcuk, R. A. Ramos, H. Gould, W. Klein, and R. D. Mountain, Phys. Rev. Lett. 75, 2522–2525 (1995). Copyright 1995 by the American Physical Society.

"On a Dynamical Model of Glasses," J. P. Bouchaud, A. Comtet, and C. Monthus, J. Phys. (France) I, 5, 1521–1526 (1995). Copyright 1995 by Springer-Verlag.

"High-Frequency Asymptotic Shape of the Primary Relaxation in Supercooled Liquids," R. L. Leheny and S. R. Nagel, Europhys. Lett. 39, 447–452 (1997). Reproduced with permission of EDP Sciences.

"Ergodicity Breaking Transition and High-Frequency Response in a Simple Free-Energy Landscape," M. Ignatiev and B. Chakraborty, Phys. Rev. E. 60, R21–R24 (1999). Copyright 1999 by the American Physical Society.

"Dynamic heterogeneity in ortho-terphenyl studied by multidimensional deuteron NMR," R. Bohmer, G. Hinze, G. Diezemann, B. Geil, H. Sillescu, Europhys. Lett. 36, 55–60 (1996). Reproduced with permission of EDP Sciences.

"How Long Do Regions of Different Dynamics Persist in Supercooled o-terphenyl" C. Y. Wang and M. D. Ediger, J. Phys. Chem. B 103, 4177–4184 (1999). Copyright 1996 American Chemical Society. Reprinted with permission.

"Origin of the difference in the temperature dependences of diffusion and structural relaxation in a supercooled liquid," D. N. Perera and P. Harrowell, Phys. Rev. Lett. 81, 120–123 (1998). Copyright 1998 by the American Physical Society.

"Dynamics of highly supercooled liquids: heterogeneity, rheology and diffusion," R. Yamamoto and A. Onuki, Phys. Rev. E 58, 3515–3529 (1998). Copyright 1998 by the American Physical Society.

"Liquid to Solidlike Transitions of Molecularly Thin Films Under Shear," M. L. Gee, P. M. McGuiggan, J. N. Israelachvili and A. M. Homola, J. Chem. Phys., 93, 1895–1906, (1990). Copyright 1990, American Institute of Physics. Reproduced with permission.

"Structure and Shear Response in Nanometer Thick Films," P. A. Thompson, M. O. Robbins and G. S.

Grest, Israel Journal of Chemistry 35, 93–106 (1995). Copyright 1995 Laser Pages Publishing Ltd. Reproduced with permission.

"Glasslike Transition of a Confined Simple Fluid," A. L. Demirel and S. Granick, Phys. Rev. Lett. 77, 2261–2264 (1996). Copyright 1996 by the American Physical Society.

"Nonlinear Bubble Dynamics in a Slowly Driven Foam," A. D. Gopal and D. J. Durian, Phys. Rev. Lett. 75, 2610–2613 (1995). Copyright 1995 by the American Physical Society.

"Kinetic Theory of Jamming in Hard-Sphere Startup Flows," R. S. Farr, J. R. Melrose and R. C. Ball, Phys. Rev. E 55, 7203–7211 (1997). Copyright 1997 by the American Physical Society.

"Friction in Granular Layers: Hysteresis and Precursors," S. Nasuno, A. Kudrolli and J. P. Gollub, Phys. Rev. Lett. 79, 949–952 (1997). Copyright 1997 by the American Physical Society.

"Kinematics of a two-dimensional granular Couette experiment at the transition to shearing," C. T. Veje, D. W. Howell, and R. P. Behringer, Phys. Rev. E 59, 739–745 (1999). Copyright 1999 by the American Physical Society.

"Theory of Powders," S. F. Edwards and R. B. S. Oakeshott, Physica A 157, 1080–1090 (1989). Reproduced with permission of Elsevier Science Ltd.

"Density Fluctuations in Vibrated Granular Materials," E. R. Nowak, J. B. Knight, E. Ben-Naim, H. M. Jaeger, and S. R. Nagel, Phys. Rev. E 57, 1971–1982 (1998). Copyright 1998 by the American Physical Society.

"Diffusing-Wave Spectroscopy of Dynamics in a Three-Dimensional Granular Flow," N. Menon and D. J. Durian, Science 275, 1920–1922 (1997). Copyright 1997 American Association for the Advancement of Science. Reprinted with permission.

"Rheology of Soft Glassy Materials," P. Sollich, F. Lequeux, P. Hebraud, and M. E. Cates, Phys. Rev. Lett. 78, 2020–2023 (1997). Copyright 1997 by the American Physical Society.

"Energy flow, partial equilibrium and effective temperatures in systems with slow dynamics," L. F. Cugliandolo, J. Kurchan and L. Peliti, Phys. Rev. E 55, 3898–3914 (1997). Copyright 1997 by the American Physical Society.

"Frequency-domain study of physical aging in a simple liquid," R. L. Leheny and S. R. Nagel, Phys. Rev. B 57, 5154–5162 (1998). Copyright 1998 by the American Physical Society.

"Sheared Foam as a Supercooled Liquid?" S. A. Langer and A. J. Liu, Europhysics Lett. 49, 68–74 (2000). Reproduced with permission of EDP Sciences.

"An Exact Solution of a One-Dimensional Asymmetric Exclusion Model with Open Boundaries," B. Derrida, E. Domany, and D. Mukamel, Journal of Statistical Physics, 69, Nos. 3–4, 667–687 (1992). Reproduced with permission of Plenum Publishing Corporation.

"Jamming and Kinetics of Deposition-Evaporation Systems and Associated Quantum Spin Models," M. Barma, M.D. Grynberg, and R. B. Stinchcombe, Phys. Rev. Lett. 70, 1033–1036 (1993). Copyright 1993 by the American Physical Society.

"Slow Relaxation in a Model with Many Conservation Laws Deposition and Evaporation of Trimers on a Line," M. Barma, D. Dhar, Phys. Rev. Lett., 73, 2135–2138 (1994). Copyright 1994 by the American Physical Society.

"Spontaneous Symmetry-Breaking in a One-Dimensional Driven Diffusive System," M. R. Evans, D. P. Foster, C. Godreche, and D. Mukamel, Phys. Rev. Lett. 74, 208–211 (1995). Copyright 1995 by the American Physical Society.

"Spontaneous jamming in one-dimensional systems", O. J. O'Loan, M. R. Evans, and M. E. Cates, Europhys. Lett. 42, 137–142 (1998). Reproduced with permission of EDP Sciences.

"Self-Organization and a Dynamic Transition in Traffic-Flow Models," O. Biham, A. A. Middleton, D. Levine, Phys. Rev. A 46, R6124–R6127 (1992). Copyright 1992 by the American Physical Society.

Jamming and Rheology: An Introduction

Andrea J. Liu

Dept. of Chemistry and Biochemistry, University of California, Los Angeles, CA 90095

Sidney R. Nagel

James Franck Institute, The University of Chicago, Chicago, IL 60637

Things get jammed. Traveling on a highway we get caught in traffic jams. We are frustrated at the whole-foods supermarket when grains and beans jam in the process of flowing out of the hopper into our bags. On a larger scale, factories using powders as the raw materials are plagued by the jamming that occurs as the powders become clogged in the conduits that were designed to have them flow smoothly from one side of the factory floor to the other.

In these systems, one can see the structural arrest of macroscopic particles. Vibrations are necessary to induce any motion in materials such as granular media, since thermal energy is insufficient to lift a particle by its own diameter. As the vibration intensity is lowered, these materials undergo a transition from a state reminiscent of a fluid or gas to one more akin to a solid with a yield stress.

In colloidal suspensions of smaller particles, such jamming also occurs when the pressure or density is sufficiently large. Above this transition, a colloidal suspension is a disordered solid with a yield stress. A dense suspension of deformable particles such as an emulsion or foam is an example of a class of soft amorphous solids known as Bingham or Herschel-Bulkeley plastics, which flow under high shear stress. When the applied shear stress is lowered below the yield stress, these systems stop flowing. Thus, foams and emulsions can unjam when either the pressure (or density) is lowered, or the shear stress is raised.

Even molecular systems can become stuck in a part of phase space that does not support macroscopic motion. The prototypical example is a liquid that ordinarily flows readily but that, as the temperature is lowered, turns into a rigid glass – an amorphous solid with a yield stress. One of the oldest unsolved problems in condensed matter physics is the nature of this glass transition [1]. As the temperature is lowered, the stress relaxation time increases dramatically. Rapidly increasing relaxation times are not unique to the glass transition or to jamming transitions; the relaxation time diverges near a second-order phase transition, as well. However, for such a transition, there is a structural length scale, the correlation length, which diverges. Because long length scale perturbations relax more slowly, the diverging correlation length gives rise to a diverging relaxation time. Near the glass transition, by contrast, no structural correlation length has yet been found and it is not clear if there is any length scale that can be used to characterize the growth of the glassy response.

Is there really a meaningful connection among these different jamming and glass transitions? This is still an open question, but at first glance the answer appears to be "No" because the jamming and glassy phenomena apparently have very different physical origins. In supercooled liquids, the increase in the relaxation time with decreasing temperature clearly has a thermal origin. In colloidal suspensions, increase in relaxation time with increasing density has an entropic origin. Finally, in athermal systems such as Bingham plastics (foams and emulsions) or granular systems of macroscopic solid particles, the increase in the relaxation time appears to have a purely kinetic origin.

In spite of these differences, however, there are striking similarities in the phenomenology of these different systems. For example, the dynamics of colloids near the colloidal glass transition have many characteristics in common with the dynamics of glass-forming liquids and the same theoretical framework has been used to analyze the dynamical data from experiments on all these different systems. [See the papers in the section: *Experimental Manifestations of Mode Coupling.*] In addition, the dynamics appear to become spatially inhomogeneous [see papers in the section: *Inhomogeneities near Jamming*] and temporally intermittent [see papers in the section: *Unsteady Rheology*] near the jamming or glass transition in all systems. The stress response in the jammed or glassy state also appears to be inhomogeneous in many different systems. [See papers in the section: *Elasticity and Inhomogeneous Response to Stress of Dense Amorphous Packings*].

Jamming looks so similar in so many different systems that one suspects that there should be a common conceptual framework with which to address this phenomenon. It appears that if we are to construct a theory of jamming we may learn a lot by studying what has been proposed to explain the glass transition. The converse is also certainly true that, since the problem of the glass transition has not been solved, progress in that field may benefit from the work that has recently been done to understand jamming. The caveat is, of course, that there is the danger that one may only be importing ignorance from one field to another. However, having said that, the rather obvious similarities of these two fields is so striking that one wonders why the connection has not been made before.

In that spirit, we have suggested [2] that one can draw a schematic phase diagram that includes both types of behavior. A sketch of how this diagram would look is shown in Figure 1. The choice of axes is dictated by the parameters that control the transition to jamming in the different systems, namely temperature, pressure and shear stress.

The ordinary phase diagram for the glass transition would be in the $T - 1/P$ plane. At high pressure (or equivalently high density) there is a transition between a supercooled liquid and a glass that occurs at a given temperature. As the density or pressure is lowered, the temperature at which the glass transition takes place normally decreases. This glass-transition line is represented by the curve separating the Liquid and the Glass in this plane. There is of course considerable debate about whether the transition is a true thermodynamic one occurring at a finite temperature or whether the complete arrest of the dynamics only occurs at $T = 0$. For the purposes of this discussion, the line separating the jammed (or glassy) phase from the liquid state corresponds to a relaxation time that has increased to a certain fixed value (such as 10^5 sec as is sometimes used to define the glass transition temperature, T_g, in supercooled liquids).

The ordinary phase diagram for the jamming transition would be in the horizontal, or $1/P - s$, plane. At fixed pressure, one must apply a shear stress higher than the yield stress in order for the system to flow. Thus, the yield stress as a function of pressure is the curve that separates the unjammed regime from the jammed regime. At high pressure, the yield stress is high; as the pressure decreases the yield stress also decreases. Note that it is impossible to tell whether the yield stress is truly nonzero or if it is only nonzero on the time scale of

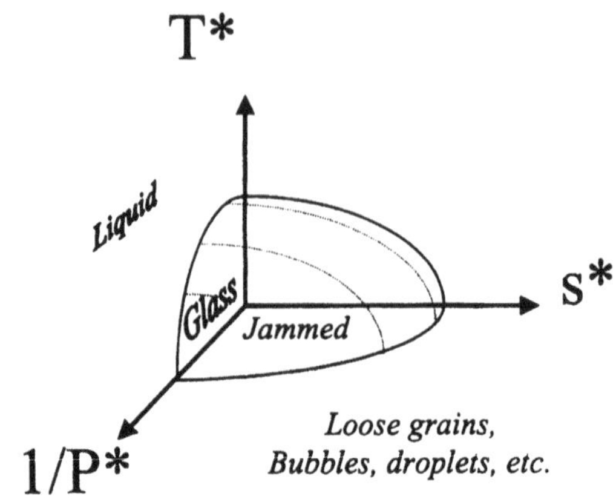

Fig. 1. Schematic jamming phase diagram. Here, the temperature, pressure and shear stress are scaled to be dimensionless. In the case where the particles have some modulus M and repel each other bv deforming each other, we have scaled shear stress by $s^* = s/M$. inverse pressure by $1/p^* = M/p$ and temperature by $T^* = kT/Ma^3$, where a is the size of the individual particles. (The pressure could in principle be replaced by the density as was originally done [2].) When the interactions are attractive, the axes must be scaled differently; for systems with Lennard-Jones interactions

$$V(r) = 4\varepsilon\left[\left(\frac{\sigma}{r}\right)^{12} - \left(\frac{\sigma}{r}\right)^{6}\right]$$

for example, we use the scalings $T^* = kT/\varepsilon$ and $s^* = s\sigma^3/\varepsilon$. In addition, we note that the pressure axis must be shifted because a negative pressure is required to achieve a low density, so $1/p^* = \varepsilon/\sigma^3(p+p_o)$.

a rheological measurement. The same caveat that holds for the glass transition – that the transition corresponds to a relaxation time that has reached some fixed value-therefore holds for the entire surface of the jammed phase.

The aim of this "phase diagram" is to unite the three axes and indicate how the introduction of a temperature might affect the jamming transition and how the introduction of a shear might lower the glass transition temperature. The jammed, or glassy, region of phase space is in the region close to the origin. It is important to realize that when we talk of temperature in the context of this phase diagram we mean something more general than the ordinary thermal energy of a system. In addition, we mean the kinetic energy that can be induced in a material via vibrations (such as is often done in granular materials) or via the random Brownian motion that is important for the diffusion of colloidal particles. Of course the shape of the phase diagram in Figure 1 is artificial and there is no reason at the outset without further data to believe that the surfaces should be concave rather than convex.

(Indeed recent data from David Weitz and his group indicate that the shape might be convex.) It is also very likely that the shape of the surface is not always simple. For a shear-thickening material like corn starch in water, one would expect reentrant behavior: there would be a liquid at low shear stress, a jammed state at the same density at high shear rates (high shear stresses), and a liquid again at even higher shear stress when the jammed state breaks up.

There are several benefits of having such a diagram. On the one hand, it suggests a number of different experiments. For example, there is virtually no study of how the introduction of a temperature (i.e., kinetic energy) can affect the possibility that a system will jam at a given stress. In a typical foam, the thermal energy is a million times too small to cause rearrangements of bubbles. However, if one were able to increase the thermal energy by a factor of a million without destroying the foam, the phase diagram predicts that the yield stress would decrease. Likewise, the introduction of vibrations in a granular material can lead to a lowered yield stress. In a similar vein, there is little known, especially in simple supercooled liquids, about how the introduction of a shear stress can influence the glass transition temperature. The phase diagram predicts that if one could apply a large shear stress to a supercooled liquid (comparable to the thermal energy at the glass transition divided by the molecular volume), the glass transition would shift to lower temperatures. Experiments that elucidate such effects will be important for gaining a better understanding of both the jamming and the glass transitions.

Another benefit of having such a phase diagram is that it suggests that the jamming and the glass transitions may be linked at an underlying level. That is, the same processes that are important for jamming a material under an applied stress may be important for arresting the dynamics of a liquid as the temperature is lowered.

There are several categories of questions that can be asked about jamming. (i) What is the nature of the jammed state? (ii) How does one get into the jammed state? (iii) Are there similarities between jamming in driven, athermal systems and jamming in quiescent thermal systems (such as the glass transition in molecular liquids)? In this volume we have collected a series of articles that have addressed what we think are some of the important issues in the fields of jamming and glassy dynamics. We have organized them loosely into topics, but many of the papers could be placed under more than one topic.

Fluctuations, temperature and the fluctuation-dissipation theorem

One question that appears over and over again in all out-of-equilibrium systems is whether there is a useful concept of temperature. [See the papers in the section: *Effective Temperature.*] This appears naturally in the subject of supercooled liquids where the term "fictive temperature" has long been used to describe the characteristic out-of-equilibrium thermal characteristics of the system. More recently, the question has been addressed as to whether there exists for these systems a fluctuation-dissipation theorem. The existence of such a theorem relies on there being a temperature. Generalizations of such theorems have been produced for supercooled systems although there have been no experimental confirmations of the predictions. There have been other attempts to define a "temperature" for athermal systems such as granular materials, foams and colloids. One question is whether the fluctuations in a driven steady-state system that is far from equilibrium can be used to define a temperature. Are all the temperatures that one defines in this way consistent with one another? Is there a zeroeth law of thermodynamics (i.e., is there a notion of equilibrium)? For static systems in some metastable state, can one define a quantity similar to temperature which describes the state of the system?

Force chains in jammed systems

When a system becomes jammed, it is held rigid due to propagation of forces from one boundary surface to another. One common feature of many jammed systems is that this propagation of forces is very heterogeneous. [See papers in the section: *Force Chains and Fluctuations.*] Thus, for example, in a granular system one can actually see the "force chains" by placing the stressed material between two crossed circular polarizers. Simulations of granular materials under a variety of loads also clearly show this heterogeneous force propagation. These fluctuations in the forces can be very large – as large as the average force itself – and a high fraction of the contacts between particles carry almost no force. Experimental measurements and numerous computer simulations seem to indicate that the distribution of the forces is robust – that is it varies only slightly between one system and preparation condition and another. Several questions naturally arise. Are these fluctuations reminiscent of a temperature? How does one model these fluctuations? Clearly there is disorder, but it is not clear how one puts disorder into a theory. In particular, it is not apparent how to model the disorder: should one assume complete randomness or should one build in correlations? (In a thermal system, by contrast, there is an ergodic assumption that all accessible states are equally possible.) Most of the studies of "force chains" have been on granular systems. Are similar force heterogeneities seen in other jammed systems such as foams and molecular glasses? Ideally, one could develop a comprehensive theory that would describe how the force distributions vary from one system to another. For example, why is the distribution of forces exponential at large forces in a granular material but Gaussian for a foam?

Force propagation in granular materials

In addition to asking how to treat force heterogeneities, one can ask how to treat the *average* propagation of forces in a granular material. [See the papers in the section: *Force Propagation in a Granular Pile.*] Is it appropriate to use conventional theories of elasticity or elasto-plasticity to treat these very hard but delicately balanced materials? Recently, in the literature, there has been an assault on this approach in which a different set of constitutive relations have been proposed to close the equations. These new equations are not based on elastic behavior but rather are based solely on relations between various components of the stress tensor in the material. One question that has been investigated at length is whether the new approaches allow one to explain experimental results that could not be understood by the more conventional theories. That is, are the two approaches really fundamentally different? Another question is what is the nature of the plastic region in the more conventional theories and what experimental implications does the plastic region have? In relating this topic to the previous one on force inhomogeneities, one is also led to ask whether plasticity is the consequence of a broad spectrum of force fluctuations. In particular, many observed force distributions show that there is a high fraction of particle pairs that have almost no force between them. If so many contacts are on the verge of breaking, does this imply that the assembly is particularly sensitive and that ordinary elastic response theory might fail? A technical issue is whether the equations that govern the static stress patterns are elliptic or hyperbolic. The more recent theories assert that the equations are hyperbolic so that stresses will propagate along lines like a wave. This has the consequence that the stress needs only to be specified along a (e.g., top) surface and the stress then propagates (downward) due to those applied forces. Is this scenario experimentally justified? The issue is still a subject of heated debate.

Friction in confined systems

When a fluid is confined in a narrow region, a small stress may not be sufficient to force the system to move: there is a yield stress. For systems driven just above the yield stress, there is "stick/slip" motion where the fluid flows in short bursts. [See the papers in the section: *Unsteady Rheology.*] As the force on the driving spring is increased still further, the motion becomes more and more smooth. Such behavior is seen in many other systems such as granular materials (in avalanches as well as in forced flow between plates), foams, and colloids, and goes by different names depending on the context. On the largest scale, this stick/slip motion is felt in earthquakes. The relation to jamming is obvious: the flow only stops because the particles get jammed and new force chains are set up that span the entire system. Questions that have been asked are: What is the nature of the rearrangement events during one of these processes? Can the stick/slip event be predicted from other external observations of the system? How do the stick/slip events change their character as the stress is increased? Recently, experiments and simulations have been able to observe the nature of the individual "avalanche" events. It appears that the size of these events does not diverge as the stress decreases towards the yield stress.

Glass transition

The nature of the glass transition is one of the oldest unsolved questions in condensed matter physics. [See the papers in the section: *Glass Transition: Signatures and Models.*] In a fundamental sense this is a jamming phenomenon: as the temperature of a liquid is lowered, it becomes more sluggish. This jamming is an equilibrium phenomenon and is caused by the decrease in thermal energy available for molecular rearrangements. Thus, as distinct from many of the cases discussed above, it is not caused by the kinetics of driving the system out of equilibrium. Despite this difference, are there similarities between these jamming phenomena? Certainly many of the questions asked are similar: What notion of temperature can be used to describe the system as soon as it has left equilibrium? Is there a fluctuation-dissipation theorem below the glass transition temperature? As the glass transition temperature is approached, does one become increasingly more sensitive to kinetic jamming due to shear or other forcing of the system?

Many experiments appear to indicate that supercooled liquids are very heterogeneous as are the other jammed systems (e.g., force chains, inelastic collapse). Several recent experiments indicate that the time scales for translational diffusion in liquids decouple from those important for rotational relaxation and viscosity. Are there different "temperatures" that are important as soon as the system begins to jam that govern the different properties? Recent experimental results have also indicated that structural glasses have a great deal more in common with spin glasses than had previously been suspected. There is a tail, extending to high frequencies, that appears in the susceptibility of both systems (as well as in plastic crystals). This ubiquitous feature is not understood although it may imply a definite correspondence between the physics of these systems.

Kinetic jamming and rheology

Glassy systems have a peculiar rheology. How does one account for a yield stress and how does one model the response of the system above the yield threshold? How does the shear modulus depend on frequency as the frequency is lowered? [See the papers in the section: *Elasticity and Inhomogeneous Response to Stress of Dense Amorphous Packings.*] Many soft systems (foams, colloids etc.) have a very non-Maxwellian frequency dependence with both real and imaginary parts of the modulus varying as a fractional power of the frequency as the frequency approaches zero. Molecular glass-forming liquids do not have this characteristic as far as we can tell. Can this anomalous frequency dependence be understood in terms of some simple model? Several attempts at construction such a model have been tried. Some of these focus on the microscopics at the level of the individual bubbles or particles and the forces constraining them. Can one relate the weak regions that occur in these systems to the nature of the force distributions that were discussed above or is the nature of the weak regions most accurately described by a geometric effect? Other theories operate on a more phenomenological, coarse-grained level; energy can flow between different cells according to a specified set of rules. Both kinds of theories can produce non-Maxwellian behavior. Another whole class of theories, is based on jamming due to essentially free-volume effects such as the random sequential deposition model. These models can be generalized to treat other dynamical systems such as traffic flow problems. [See the papers in the section: *Traffic*.]

The idea for this volume grew out of the program on Jamming and Rheology that was held at the Institute for Theoretical Physics at Santa Barbara in the Autumn of 1997. The conference that was held as part of the program had a somewhat unusual feature. Because the subject matter for the conference spanned many different fields, no one could be expected to be an expert in all relevant specialties. For that reason the organizers planned a series of tutorial talks in conjunction with the research talks in the different areas. These overviews were designed to give a survey of a particular area and indicate how the work in that subfield related to the topic of jamming in the wider context of the workshop. We have used that same idea in this volume. At the outset we have a series of overview papers, written specifically for this book, that were meant to provide a broader review of the field. The second part contains a series of reprints that cover many of the topics that we consider to be the most relevant in a discussion of jamming. These papers are divided into sections dealing with individual topics.

We thank all of the staff at the ITP for their generous hospitality. In particular we would like to thank David Gross, Director of the Institute, who first suggested this volume. We particularly thank Deborah Storm for helping us to get the overview talks transcribed. We also thank her and Daniel Hone for helping to arrange the details of the initial program. We thank Judy Sweeney at UCLA for all her help in putting this volume together and tracking down all the loose ends. We are extremely grateful to Tom Witten for his encouragement and support of the entire project from its inception. Without him, it is doubtful that there would have been a workshop at all. Finally we would like to thank our co-organizers of the workshop, Sam Edwards and Mark Robbins, who helped in so many ways to give substance and shape to this program.

References

[1] P.W. Anderson, Science **267**, 1615 (1995). "The deepest and most interesting unsolved problem in solid state theory is probably the theory of the nature of glass and the glass transition. This could be the next breakthrough in the coming decade. The solution of the problem of spin glass in the late 1970s had broad implications in unexpected fields like neural networks, computer algorithms, evolution, and computational complexity. The solution of the more important and puzzling glass problem may also have a substantial intellectual spin-off. Whether it will help make better glass is questionable."

[2] A. J. Liu and S. R. Nagel, "Jamming is not just cool any more," Nature **396**, 21-22 (1998).

OVERVIEWS

Jamming and the Statics of Granular Materials

Dov Levine
Department of Physics, The Technion, Haifa, 32000, Israel

The effect of jamming on the static properties of granular materials is discussed. Various models for stress propagation, such as elasto-plasticity, isostaticity, and fragility are examined critically. Analogy is made to (frictionless) emulsion rheology, and the effect of friction on the results is considered. The relevance of the specifics of the construction history is emphasized.

I. INTRODUCTION

The static properties of an aggregate of granular material are inexorably linked to the process by which it was formed. This, in itself, is not peculiar to granular materials; for example, rapidly solidified metallic alloys may well behave differently from their slowly cooled counterparts, and the behavior of a plastic depends on the extruding process used in its formation. This sensitivity implies that the system, although static, is far from thermodyamic equilibrium. In the cases of the plastic and the metal, solidification takes place as the system cools down due to contact with the external environment. In contrast, for a granular material, the cessation of motion of the grains arises from the dissipation of kinetic energy. If this dissipation is very rapid, the system is unable to explore much phase space, and it *jams*, arriving at a state which may well be special in ways which will be discussed in the following. A system which expends its energy slowly, so that it has ample time to "find" a favorable state (whatever that may be), should not be said to have jammed, and its properties may be quite different from those of a jammed state. As physicists, it is incumbent upon us to ask whether there are any statistical properties which typify the jammed state, and if so, what their physical consequences are. For granular materials, at stake is an understanding of stress propagation, fluctuations in the forces between grains, and the probability distribution of stress itself. Although much effort has been expended on these questions, hard results remain elusive. And this issue is not academic – the various possibilities are *fundamentally* different. There is one property, however, which derives directly from the jamming scenario: the critical importance of the specifics of the construction history on the physics of the material. Since a jammed system does not have the ability to explore its phase space, we would expect, at least with respect to some physical properties, that each system will be different (Of course, it may certainly be that other physical properties "self-average", that is, all similarly prepared systems will have the same values.). This picture is dramatically different from non-jamming systems, for example an ordinary fluid, for which the macroscopic constraints (*e.g.*, size and shape of container, or temperature) determine all the properties precisely, and no variation is found.

The prototypic granular materials are composed of very hard grains, which are large enough so that temperature cannot lead to any random grain motion – for all thermodynamic intents and purposes, the system is at $T = 0$. Each individual grain is an elastic solid: they repel one another when compressed (*i.e.* they resist the compression), but once separated they cease to interact (see below). This means that there is no attractive force tending to bring two grains to a certain equilibrium separation – in the jargon, we say that *granular materials can not support tensile stresses*. Typically, there is friction between grains. I shall consider the term "packing" to describe aggregates of macroscopic grains, with or without friction, with no attractive interactions. I will use the term "granular material" to mean a packing for which the individual grains are not significantly deformed from their equilibrium states. In this paper, I will not consider so-called "wet" granular materials, such as moist beach sand, for which there is a cohesive force between the grains owing to a liquid layer between them.

When sand, the prototype granular material, is at rest in a container, it has a volume fraction Φ (ratio of sand volume to total volume), which typically lies between .59 and .64 [1]. The former is generally referred to as *random loose packing*, and the latter as *random close packing*, or *dense random packing*. The as-poured state is usually at the lower density, and higher densities may be attained by compactifying, such as by gently tapping or otherwise agitating the container. Effectively, volume fractions higher than .64 are never observed. These densities are familiar in the context of hard-sphere colloidal suspensions and emulsions [2–6], where there is no friction. Indeed, a very slightly compressed emulsion, is, in the above sense, an idealised granular material in which the effects of friction and gravity are absent, and I will make extensive analogy to emulsions in this paper. For sand under gravity, the compactification is anomalously slow, and is characterized by a logarithmic dependence on time (as measured in number of taps or shakes) [1]. The slowness of the compacification serves to underscore the difficulty inherent in moving from state to state, and lends credence to the jamming hypothesis.

An important issue in the mechanics of granular materials is *static indeterminacy*. This is the statement that for a packing with a given topology [7], there are many nontrivial (*i.e.*, not simply all scaled up) sets of intergrain forces for which all the grains are in mechanical equilibrium. The reason for this is that the number of variables in the system, for example, the number of intergrain forces, may well be larger than the number of (mechanical) equilibrium conditions, being the equations of force and torque balance on each grain. Such ideas are not new, but they have been recently applied in the context of packings [8–13]. Briefly stated, the argument runs as follows: Any given grain in a d dimensional packing has, if it is not perfectly round, d translational degrees of freedom (these correspond to d equations of force balance) and $\frac{d(d-1)}{2}$ rotational degrees of freedom (corresponding to $\frac{d(d-1)}{2}$ equations of torque balance). Thus, for each grain, there are $d+\frac{d(d-1)}{2}$ equations constraining the possible forces between the grains [14]. If the forces are not simply central [15], but are nontrivial vectors (as is the case if there is friction), then, per grain, there are $d\bar{z}/2$ (scalar) components of intergrain force, where $\bar{z}$ is the average number of contacts that a grain has. Thus, in order for the system to be perfectly defined, that is, to have a unique equilibrium solution, the number of unknowns (the components of the forces) must equal the number of constraint equations, giving $\bar{z} = d+1$. A system which has exactly one such solution is called *isostatic.* If $\bar{z} < d+1$, then there are too few variables: the system is overconstrained, and in general there is no solution. If $\bar{z} > d+1$, there are fewer constraint equations than unknowns, and there is a multitude of solutions consistent with the packing. What this means is that, in the last case, for a given (underdetermined) packing, there are many different stress patterns which are consistent with the packing being in mechanical equilibrium. Note that we have tacitly assumed throughout that all the constraint equations are linearly independent, which is not always the case, though for a disordered packing of variously shaped grains this is probably true. Note too the interesting special case of grains which interact solely via normal forces. In this case, if the grains are obliquely shaped, then the critical coordination number is $\bar{z} = d(d+1)$, while if the grains are all spherical,then $\bar{z} = 2d$.

Another way to understand static indeterminacy is from the continuum equations defining the stress tensor $\sigma_{ij}(\mathbf{r})$. For static systems, we have d force balance equations

$$\partial_i \sigma_{ij} = f_j, \tag{1}$$

where f_i is the body force (*e.g.* gravity) acting on the system. These equations hold for *any* static system for which a continuum picture applies. On the other hand, there are $\frac{d(d+1)}{2}$ independent components of the stress tensor (by virtue of its being symmetric), leaving $\frac{d(d-1)}{2}$ components undetermined. One now needs to introduce this number of relations among the stress components in order to determine the state of stress of the system. The traditional way of doing this is to introduce a strain field (see below) and relating it to the stress. These *constitutive relations*, as they are traditionally called, are meant to reflect the physics of the system under examination, and may be derived, in principle, from microscopic considerations. Thus, gels, wood, plastics, and metals, all of which satisfy Eq. (1), each has its own distinctive constitutive relation.

As stated above, it is often useful to introduce a *displacement field*, u_i, (where i indexes the components of the vector $\mathbf{u}$) which gives the displacement of a small volume element of material relative to a reference state. In traditional *elasticity theory*, one begins from a reference state which is stress-free [16], and assumes that the free energy F of the system may be expanded to quadratic order in the strain field u_{ij} [17]. This is simply the generalization of the Hooke law for a simple spring, where the energy of a displacement x is given by kx^2, where k is the spring constant. In an elastic solid, k is replaced by a tensor relating stress and strain, reflecting the tensorial nature of the deformation. In principle, this tensor can vary in space, in which case it is a tensor field, which we shall call, for want of a better name, the *elasticity field* [18]. Since the stress and the strain are conjugate variables,

$$\sigma_{ij} = \left(\frac{\partial F}{\partial u_{ij}}\right)_T \tag{2}$$

where T is the temperature of the system, which for us, as stated earlier, is effectively zero. Eq. (2) represents $\frac{d(d+1)}{2}$ relations between σ_{ij} and u_{ij}, with σ_{ij} as unknown quantities. However, since u_{ij} is derived from a *vector* field u_i, being its derivative, we have $d+\frac{d(d+1)}{2}$ unknowns (that is, u_i and σ_{ij}), and the same number of governing equations, Eqs. (1) and (2). Provided that the applied stress is not so large as to impart irreversible (plastic) deformations, elasticity theory describes most solids extemely well [19]. However, for an isostatic system, there are no additional relations necessary, since the system is well-defined and fully determined by force balance alone, and so an isostatic system need not obey elasticity theory.

It has recently been proposed [20] that, for granular materials, in lieu of the aforementioned procedure of elasticity theory, there may exist local, linear relations between components of the stress tensor, which may be written (in two dimensions)

$$a_{jk}\sigma_{jk} = 0 \tag{3}$$

(additional relations are needed in three dimensions). In particular, for a sand*pile* constructed from a point source, it is argued that the principal axes of the stress tensor of a given grain [22] are frozen at the moment it stops moving, and that the addition of susbsequent grains will not affect this. This approach has been criticized as simplistic by Savage [23], but it has one interesting defining feature: Stress is propagated along linear paths (See Fig. 1) related to the characteristics of the differential equation for the stress obtained by supplementing Eq. (1) with the linear closure Eq. (3). This equation is *hyperbolic* in nature, like a wave equation, and as such requires specification of the stress on a *portion* of the boundary in order for a solution to exist. In a sense, such models are cousins of the simple stress propagation models [24–28] which envisage that a grain passes stress down (in the direction of gravity) to its neighbors beneath it. This stands in stark contrast to elastic models, where the differential equation for the stress field is *elliptic*, whose solution requires the specification of data (stress or displacement field) on the entire boundary of the material. When a weight is placed on a simple elastic material, the entire material deforms, not just the portion beneath the disturbance. We note here (see also [21]) that there is a precedent for such behavior: if an elastoplastic material is at yield (see below and next section), then stresses propagate in it according to a hyperbolic equation. The difference is that in the aforementioned proposal, the local, linear relation is posited whether the material is at yield or not.

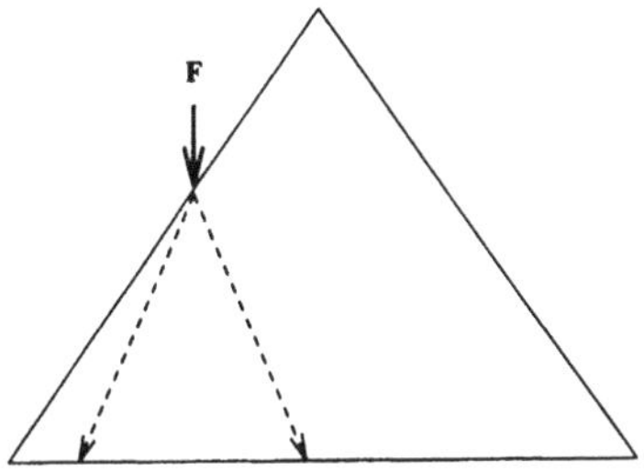

FIG. 1. Schematic depiction of force propagation in a sandpile with a linear closure as in Equation 3. Forces are propagated along the characteristics of a wave-like equation. To solve for the stress distribution, it is necessary to specify a boundary condition at the free surface (the top) of the pile.

The nature of the constitutive or closure relations, and the attendant issue of boundary conditions has fueled sharp debate, the principal question being whether one can or cannot, *in principle*, calculate the stresses at the base of a sandpile (or a differently shaped mass of granular material). This question arises because of the fascinating feature of many sandpiles [29,30] that directly beneath the apex, there is a pressure dip, contrary to normal intuition. One may now inquire whether this pressure distribution, at the base of the pile, can be calculated. For the hyperbolic models, there is no problem – the boundary condition is specified at the free surface of the pile, and this (supplemented by a scaling hypothesis for the stress field) fully determines the stress distribution at the base. For certain choices of the vector a_{ij} in Eq. (3), such a dip is found. On the other hand, if one treats the pile as an elastic medium (see the discussion in [31,23]) (analogous to, though different from, a pile-shaped piece of rubber or metal) then the elliptic nature of the equations for the stress field require us to specify boundary conditions everywhere, in particular at the base of the pile itself. If the stresses must be prescribed, then the meaning is that they cannot be calculated. One may instead specify the displacements at the base; clearly, this requires a reference state (see next section). It is unclear that such a reference state is well-defined; certainly it is not a stress-free state as is standard in elasticity theory. Moreover, for extremely rigid grains, specification of displacements may be ill-defined [32]. Whatever the resolution of this controversy, attempting to treat the sandpile, or any other granular system, as an elastoplastic material without regard to its construction history seems misguided. It is precisely during construction that the system develops stress paths, and in the elastic picture, this is when the elasticity field is set up. Without knowledge of this field, the elasticity theory is not defined. One way of incorporating the construction history is to solve the problem by adding layer upon layer of material, at each stage solving the elasto-plastic problem (say, by finite elements) assuming a constitutive relation. The way in which new material is added to the existing pile defines the construction history. Such a study has been carried out by Savage [23], who concludes that the dip in the pressure beneath the apex is not inevitable, but depends on the nature of the support upon which the pile sits. If the support is rigid, no dip is obtained, while if the table is allowed to sag a bit, a dip is apparant in the data. This debate should, as all scientific debates should, be settled experimentally, and careful experiments are essential.

The experimental results that do exist are fascinating. In a precise experiment on a granular material in a long, rather small diameter vertical cylinder [33,34] of diameter 3.8 cm (filled with 2 mm diameter glass balls), the apparant mass at the bottom of the column was measured as a function of the total mass of material in the cylinder. Since some of the weight is taken up by the sidewalls, the apparant mass M_E is different (typically less, since friction is usually mobilized upwards unless special measures are taken) from the total, actual mass M_T (which is the weight of each grain times the number of grains).

The simple, classical argument of Janssen [35,36] predicts an initial linear dependence of M_E for small M_T, which asymtotes to a value M_∞ for large M_T. If a weight is placed on the top of the column (not touching the cylinder walls), then the above result is only slightly modified, so as to give the correct pressure at the top layer of the granular material. In any case, the function $M_T(M_E)$ is monotonic increasing, as intuition would suggest. It is therefore a great surprise that for substantial overloads, an oscillatory dependence of M_E on M_T was observed, with a period which appears to be roughly the Janssen decay length. For (at least) one of the models [20] employing a linear closure of the form of Eq. 3, such oscillatory behavior is predicted, and derives from the fact that in this model, stress is "reflected" off the sidewalls, in wavelike fashion. It will be interesting to see whether elasto-plastic theories, incorporating construction histories mimicking the experiment, can reproduce this unusual dependence. It seems clear that these inriguing results, supplimented by further experiments, will go a long way in sorting out the various proposals.

II. ISOSTATICITY, FRAGILITY, AND ELASTICITY

The case of an isostatic system of spherical particles with central forces only is of particular interest since it is what is expected for an emulsion at critical volume fraction Φ_c. An emulsion consists of liquid drops (say oil) immersed in another liquid (say water); the drops being stabilized by a surfactant (say detergent). Φ_c is the volume fraction at which droplets touch in point contacts, with no droplet deformation. In the case of an emulsion whose volume fraction is slowly increased to Φ_c from the liquid state, there is ample time for the system to find a good local free energy minimum, though the global minimum (believed to be a close-packed crystal for a monodisperse emulsion) is usually too far away to be accessed. Under these conditions, the emulsion achieves random close packing (RCP) of spheres at $\Phi_c \approx .64$ [6]. Indeed, for a three dimensional emulsion at Φ_c, computer simulation [37] indicates that $\bar{z} = 6$.

What happens for granular materials? If the system consists of sand (for example) poured into a container in some particular fashion, then due to the rapid dissipation of kinetic energy, the density of the resulting packing is considerably lower than $\Phi_c = .64$. In this case, it is interesting to ask whether the packing achieved is *marginally stable*, that is, on the boundary separating stable from unstable in the space of states of the system. A marginally stable state may be thought of as the most unstable of stable states, and we may speculate that it is what occurs for as-poured sand since the system jams up, stopping at the first state [38] which can support whatever external load is present during the aggregate's formation. [39] A "better", more stable packing may be obtained by "annealing", by gentle agitation. Of course, the marginal state hypothesis is not necessarily correct, and in the following I discuss some possible descriptions of the statics of granular materials.

There are several interesting possibilities for the state of such as-poured granular material. First, it may be isostatic, having exactly the correct number of contact points so as to define the forces uniquely [9–11]. Or, it may be *fragile* [40,41], meaning that it is able to support only those external stresses which were imposed during the construction process, and no others–these others, even if infinitesimal, would generate irreversible rearrangements in the bulk. We might term this "strong fragility". A slightly relaxed definition of fragility might allow for the support of a certain subset of external stresses, perhaps close to those the system grew up with, but the application of other, different external stresses would cause the system to rearrange irreversibly. Another possibility might be called "weak fragility" – this would be the ability to sustain all infinitesimal applied stresses save one; that is, only if we push in a certain way will the system rearrange irreversibly. If the state generated was marginally stable, it would presumably fall into one of these categories. Another possibility, the traditional view [31,23], is that the the system is, in fact, an elasto-plastic medium. An elasto-plastic material is one which possesses regions which behave according to elasticity theory, until they are so stressed that they reach a "yield criterion", at which they deform irreversibly. The accepted yield criterion for granular materials is the *Coulomb criterion* [42,36], which states that yield is reached when the maximum ratio of shear stress to normal stress equals μ, the friction coefficient [43]. The basis for the elasto-plastic picture is that each individual *grain* is itself a solid body obeying the dictates of elasticity theory. This being said, it is clear that stress is propagated in a granular medium via grain deformation, and so why should the material *as a whole* be anything but elastic [23,44]? The reason that this is not so clear is that if the material is fragile or isostatic, it is unable to suffer certain applied stresses, regardless of their strength, without internal rearrangement. Since the grains, although elastic, are so hard, to a certain extent they may be expected to behave like ideal hard particles, and it is here that the ideas of fragility and isostaticity may enter.

Why should a granular material be isostatic? Consider again an emulsion, which at low volume fractions behaves like a viscous liquid, the viscosity enhanced by the presence of the droplets. As the volume fraction increases, the viscosity increases, diverging at Φ_c, when the system develops a shear modulus [5,45,46]. Just at this point, the droplets are undeformed spheres, with average coordination 6. If the volume fraction is increased further, the average coordination number increases (growing as

$\sqrt{\Phi - \Phi_c}$ [37]) as the droplets deform and squeeze closer to each other. This trend can be seen in Fig. 2, which gives a histogram of the percent $N(z_c)$ of droplets with z_c contacts. Thus, we conclude that *at* Φ_c, an emulsion is isostatic, but as the osmotic pressure is increased, the number of neighbors in contact with a given droplet rises, on average (and, of course, the contact region is no longer a point, but a finite area). For these higher than critical volume fractions, the emulsion appears to behave like an elastic solid, albeit one with highly (osmotic) pressure-dependent elastic moduli.

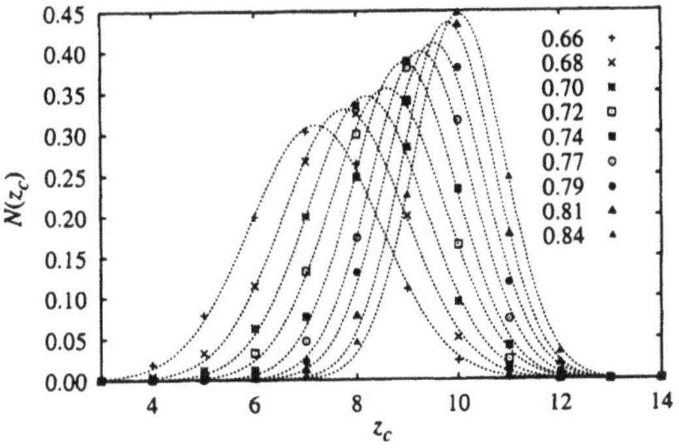

FIG. 2. The probability distribution of the coordination number for a 4913 soft sphere system. The curves are fits to Gaussian distributions. The system is uniformly compressed at the volume fractions indicated in the graph. From reference 46.

There are several differences between the emulsion discussed above and a sand-like granular material. The most obvious is that there is friction between the grains constituting the granular material, but not between emulsion drops, which interact via central forces. Also obvious is that sand grains are typically not spherical, while emulsion drops, up to Φ_c, are. Emulsion drops are easily deformed, while sand grains are extremely hard. For small deformations, sand grains, being elastic solids, deform according to the Hertzian interaction [16], which says that the energy required to bring two grains closer together by an amount ξ scales as $\xi^{5/2}$. On the other hand, when droplets are compressed together, the energy stored scales as ξ^{α}, where α is a number between 2.1 and 2.6, depending on the number of neighbors in contact with the drop [47,48]. Emulsions, owing to the deformability of the constituent droplets, are easily compressed, while a granular aggregate is hard, and does not compress appreciably even under large external stresses. However, as I stated in the beginning of this article, aside from the question of friction, to which we will return later, I believe that in the (very) small compression limit, where $\Phi \approx \Phi_c$, an emulsion is very like a standard granular material. As stated above, for almost any compression, until grains break or pulverize, there is essentially no change in the volume of a granular material, unless there are internal rearrangements. For an emulsion, if the external applied stress (the osmotic pressure) is small enough, the volume fraction will be very close to Φ_c, and the droplet deformations miniscule. Thus, a granular material is essentially always in a state like that of a near-critical emulsion, with the aforementioned caveat about friction. Another way to put this is to say that a system of hard spheres behaves like a critical volume fraction emulsion.

Let us examine a system composed of hard spheres, which by definition, interact only via an excluded volume interaction – no interpenetration is allowed. Moukarzel has posited that such a system must be isostatic [10,11], or, more precisely, that a system of spheres with repulsive central forces, in the limit of infinite rigidity (*i.e.* arbitrarily large bulk modulus) must be isostatic. The argument given relies on decomposing the stresses between spheres into two parts, one dependent on the external load, the other not. Moukarzel goes on to demonstrate that under certain conditions, if this division is valid, an isostatic packing is the only one consistent for arbitrarily rigid particles having no tensile stresses. The argument, though compelling, is not entirely airtight, since the existence of the decomposition is not obvious. Nonetheless, the result seems quite reasonable, and is consistent with what we have said about critical volume fraction emulsions, even though emulsion drops are certainly not very rigid. This is because what is really important is that the particles not deform, and whether this arises because they are exremely rigid or because the applied stress is small is immaterial. A blow-up of the emulsion's radial distribution function $g(r)$ for distances r close to one droplet diameter (D) is shown in Fig. 3 for various volume fractions. Note that as $\Phi \to \Phi_c$, the distribution becomes sharper and sharper, indicating that the drops are becoming spherical.

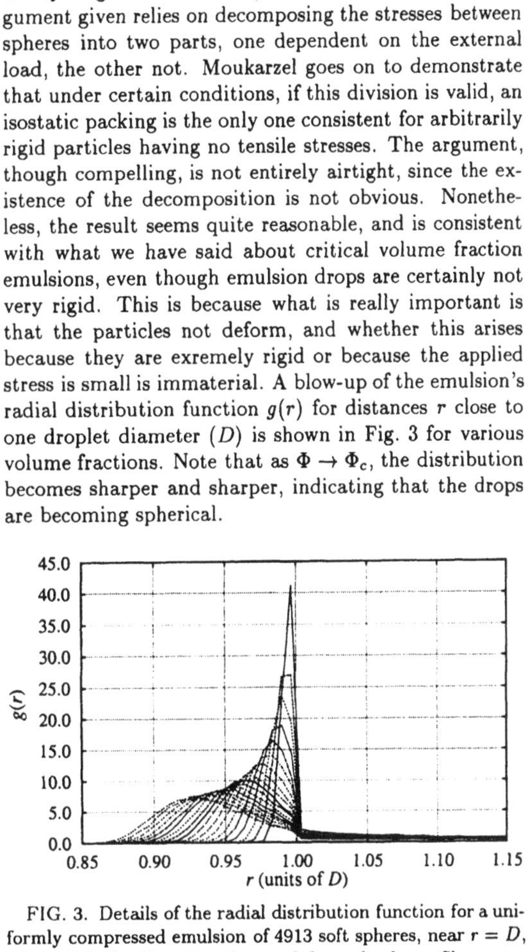

FIG. 3. Details of the radial distribution function for a uniformly compressed emulsion of 4913 soft spheres, near $r = D$, where D is the diameter of an undeformed sphere. Shown are 18 different volume fractions in the range $0.66 < \Phi < 0.85$ in steps of ≈ 0.01.

Edwards [49] also argues that a system of highly rigid grains should be isostatic, especially if the grains are not spherical. His argument may be expressed in a "table analogy": Suppose a three legged table is constructed from infinitely rigid materials. When placed on the floor (which need not be flat), it will stand, resting on all three legs, irrespective of their lengths. A four legged table similarly constructed would always have one leg in the air (perhaps only infinitesimally) unless the legs were machined with infinite precision. Likewise, Edwards argues, a grain will always find itself resting in exactly the critical number (the minimal number necessary for stability) of neighbors. As new grains are added, the precise neighbors in contact may change to account for the new distribution of forces in the system, but the grains always have the critical number of contacts.

To what extent may we idealize the system and say that sand grains are perfectly rigid? As mentioned earlier, two grains of sand (or any two simple elastic bodies), when compressed, deform as elastic bodies, according to the Hertz interaction. But if the stress applied to a granular material is not enormous, the deformations, though nonzero, are so small that it is tempting to try to bypass their use in calculations, if possible. This situation is not unlike that of a man of weight W standing on a cinderblock - the force required to drag the block increases by μW (μ being the coefficient of friction), but even a careful measurement of the block would reveal no change in its dimensions. Indeed, the very statement of the law of classical friction, that the frictional force is proportional to the weight of the body, emphasizes the point that considering the deformation of a hard solid body, as this pertains to friction, is a thankless task. Another way of saying this is to say that the frictional force of a solid body depends in a highly singular fashion on the (elastic) deformation of the body, but in a simple, linear way on the applied stress. However, although it may be *practically* hopeless to compute the response of a granular medium by considering the deformations of the individual grains, the purely theoretical question of whether they may be altogether neglected in calculations is less clear [23]: The theory of elasticity is not predicated on the size of the strain field, but with its existence and non-singular nature.

One difficulty of an elastic or elasto-plastic approach is the nature of the reference state with respect to which displacements are reckoned. As stated above, standard elasticity theory is based on the existence of a stress-free reference state, and this is not natural for granular materials. When a granular solid is created, it is by jamming from a poured, "liquid" state. Thereafter, it may be compactified. But in the jamming process, local stresses freeze in–this is one of the hallmarks of jamming. Since the Hertz law is non-Hookean, the elasticity field [50] depends strongly on these stresses, and in fact follow them–where there are large frozen stresses, the elasticity field will be large; where small, small frozen stresses. Since the specific pattern of frozen stresses is different for each granular system, due to differences in their preparation, the elastic reference state of each granular system will be different. This is seen in the large stress fluctuations which are a hallmark of granular materials [51]. Moreover, the elasticity of these materials is inherently non-linear; one way to see this is by noting that there is a sudden "switching on" of the elastic interaction when grains come into contact. Were the elasticity field somehow known everywhere, and provided that no new contacts were induced (nor prior contacts lost), then for a sufficiently small externally applied stress we might speculate that (regions of) the medium would respond elastically [23,31,32]. But there would be little resemblance between this elasticity and that of a homogeneous rubber block. In fact, it is not at all clear what it would behave more like–the rubber block or an idealized isostatic or fragile material. Thus, the construction history of the material is of real relevance to its properties.

One intriguing possibility addressing the ramifications of isostaticity is proposed by Edwards, Grinev, and Ball [9]. They encode the construction history in the *fabric tensor* [52,9], which describes the topology and geometry [53] of the packing. For a packing of grains with perfect friction (*i.e.*, $\mu = \infty$), they find a geometry-dependent correction to Eq. (1), which may manifest itself on length scales of the order of several grain sizes. Moreover, they obtain geometry-dependent relations involving the stress tensor and its spatial derivatives, which serve to close Eq. (1) (or its corrected form). Interestingly, in the mean field limit, in which certain quantities are replaced by their averages, they recover a linear relation between components of the stress tensor, reminiscent of closures like Eq. 3. In this sense, they provide a framework in which the linear closures may be understood, and shed light on their limitations and generalizations.

Another analysis based on an isostatic packing was performed by Tkachenko and Witten [13], who consider what they call a "sequential packing", in d dimensions. Such an aggregate is constructed by building the system up layer by layer, such that each grain is in contact with d neighbors below it from the previous layer, and d above it from the subsequent layer. Under the assumption that the beads are frictionless ($\mu = 0$), they show that the components of the stress tensor are linearly related. This result is similar to that of Edwards, Grinev, and Ball [9], with the important difference being the absence of friction. When taken together, the two results indicate that, at least in an average sense, *in an isostatic packing, with or without friction, a linear relation between stress components exists.* This should still be taken with a caveat, however, since the case of *finite* friction has yet to be calculated. This notwithstanding, a consequence of the linear relation is that stress propagates according to a wave-like equation, rather than an elliptic one as for elastic media.

One question which has yet to be addressed for these isostatic packings is the way in which they yield. For granular materials, it is generally accepted that the material yields when the Coulomb criterion is met. It is not clear what the yield criterion for isostatic packings is. In particular, it would be of interest to build isostatic piles of grains with finite μ, and to see how the angle of repose of the pile depends on μ. This should be feasible, and it would be of great interest to see whether a Coulomb-like criterion were to arise.

Does the presence of friction materially change the considerations above? The conclusion of de Gennes [44] and Tkachenko and Witten [13] is that it does. The argument given is that in order to determine what the forces between grains are in the presence of friction, we must know the history of the grains in question, in particular, how they deformed when they came into contact. This fact is well known to simulators, as first discussed by Cundall and Strack [54]. Thus, if the deformations are not merely a convenience, but are necessary, the material must behave like an elastic medium, at least in portions of the material. The descriptions involved in the two cases are somewhat different, but both require that after a grain has ceased to move, and is in contact with its neighbors, that it remain in contact with these same neighbors. Without this, it is not clear that one can define a reference state unambiguously. de Gennes calls this case "quasi-elastic". Tkachenko and Witten take their earlier packing, which was isostatic without friction, and ask what would result if there *were* friction. Their conclusion is that the frictional contacts would be built up by distortions of the grains, and that these distortions would minimize the elastic energy of deformation near the contacts.

While this view is persuasive, it is not unassailable, for three reasons: (1) Nothing constrains an old contact from breaking when additional grains are added subsequently; (2) Friction is dissipative, so it is not clear that minimization of elastic deformation energy is correct in principle; (3) In adding friction to a (previously) isostatic packing, one generates a system which is no longer isostatic, but manifestly underdetermined, and it is not at all clear that such a system can be created (with nonzero probability) in any experimental setting. Still, should it be demonstrated that any underdetermined packing obeys elasticity theory, it would represent a considerable advance in our understanding. This is unlikely, however, since it one might well consider an isostatic packing and add a small number of additional connections, leaving the vast majority of the system untouched. It does not seem likely that this would cause the properties of the system to change in such a drastic fashion. More likely, it seems to me, is that, if the additional bonds do not percolate throughout the system, there would be an isostatic backbone and passive "spectator" grains, which could be removed from the system without effecting any but local rearrangement. This suggests that if there is a causal connection between overdetermined packings and elasticity, that the packing must be everywhere overdetermined.

III. FORCE CHAINS AND STRESS DISTRIBUTIONS

One fascinating feature of granular materials is the extremely inhomogeneous way in which stress is distributed in the bulk. Dating back to Dantu [55], and most recently by Howell, *et al* [56], stress-induced birefringence has been used to visualize stress in granular materials. Certain materials, such as plexiglas, when stressed, become birefringent. When viewed between crossed polarizers, the stressed beads "light up", the others are dark. In this manner we may see whether the stress is concentrated along paths, or is distributed in a more-or-less uniform fashion. As seen in Fig. 4, the stress is distributed along a well-defined network of paths, the so-called *force chains*. This phenomenon is seen clearly in computer simulations employing static friction as well [54,52,57–59]. One of the most striking manifestations of this is the remarkable sensitivity to tiny perturbation of sound propagation in granular packings [51] - small changes to a "load bearing" path have an enormous effect on the signal received.

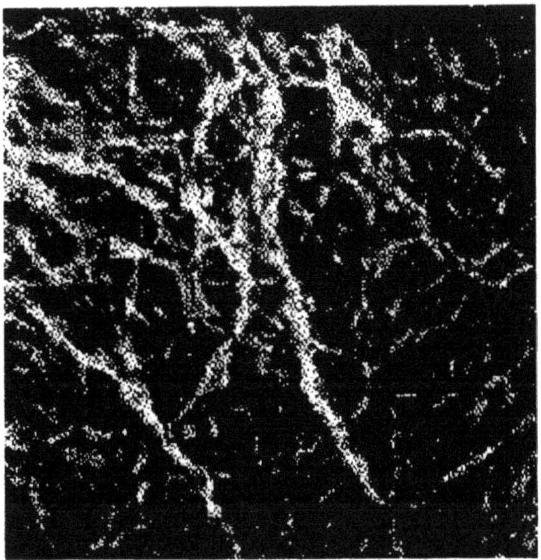

FIG. 4. A compressed packing of photoelastic grains viewed through crossed polarizers. The highly stressed "force chains" appear to "light up". From reference 21.

The disadvantage of stress-induced birefringence is

that it is difficult to quantify the extent that a given grain is stressed in a precise way. Thus, though we can see stress lines, we cannot get a quantitative sense of their nature. This difficulty is surmounted by performing experiments which measure the (normal) force which a particle exerts on the wall of its confining container. One such method [60] causes a mark to be made by lining the inside of the container with a layer of carbon paper. When a bead presses the carbon paper sufficiently hard, a record of its impression is left on (a paper lining) the wall; the greater the force, the larger the mark. A calibration is obtained by measuring the size of the marks made by applying known forces, and forces from 0.8 - 80 N may be measured. The disadvantage of this technique is that only forces between beads and container walls may be measured, not bulk intergrain forces. Thus, if there is a difference between the properties of the surface and the bulk, such as might arise if there is enhanced ordering at the walls, this could not be discerned.

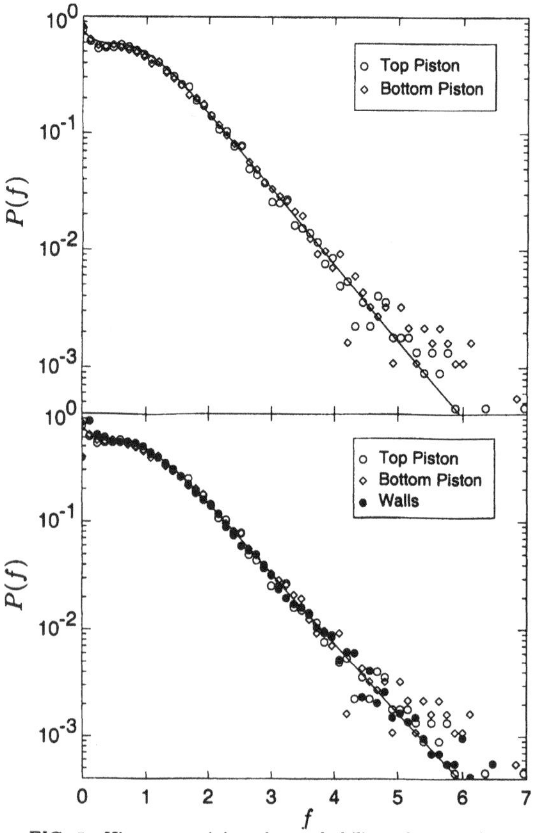

FIG. 5. Histogram giving the probability of measuring a (normalized) force f, for the experiment described in reference 60. The solid curve is the function of Eq.4. The figure is taken from reference 60.

Mueth, *et al* place glass beads in a open cylindrical container, and then press the material between two pistons. Carbon paper is affixed to the sides of the cylinders and to the top and bottom piston. These measurements yield a surprising result for the force distribution, that is, the probability $P(f)$ of measuring a normalized force f (Here, the actual force F is normalized by the mean force $< F >$, that is, $f = \frac{F}{<F>}$.). The experimental results are shown in Fig. 5. Also shown in this figure is a curve of the form

$$P(f) = a(1 - be^{-f^2})e^{-\beta f}. \tag{4}$$

For $a = 3, b = .75$, and $\beta = 1.5$ an excellent fit to the data is obtained. There are several features of significance in these data. First, there is an exponential decay at high f, that is, for forces greater than the mean force. Second, for forces smaller than the mean, the distribution is quite flat, with a slight upturn near $f = 0$. In a 2D computer simulation employing the contact dynamics technique, Radjai, *et al* [57] compressed a set of frictional disks and measured the same quantity, $P(f)$. They found that the data fit well to a curve composed of two pieces:

$$P(f) = \begin{cases} cf^{\alpha} & \text{f<1} \\ ce^{\beta(1-f)} & \text{f >1} \end{cases} \tag{5}$$

with c=.47, $\alpha = -0.3$, and $\beta = 1.4$. The two fitting forms, Eqs. 4 and 5 are practically indistinguishable over the range of the data.

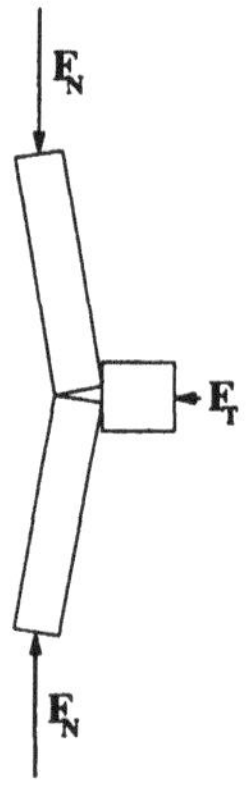

FIG. 6. The force F_T required to resist the buckling induced by the forces F_N at the ends of a force chain (shown as a rod) is small compared to F_N.

The interpretation of this arises from a later work of Radjai, *et al* [58], it too a simulation in which a set of two-dimensional frictional disks is compressed. This study indicates that the strong forces, those having $f > 1$, tend to point preferentially in the direction of compression, while the weaker forces, with $f < 1$, align preferentially in the orthogonal direction. This is because there are strong force chains which develop during, and which resist, the compression. In the absence of any other material, these force chains would become unstable, and Euler buckle, like a thin rod under compression [16,8]. In order to prevent this buckling, we may press in at the sides of the rod or force chain, as in Fig.6. However, the force F_T needed to prevent the buckling is very small compared to the compressive, destabilizing force F_N, as depicted in Fig.6. The small stabilizing forces in the material, analogous to F_T, are aligned perpendicular to the strong compressive stresses. This suggests an interesting picture, namely, that in a granular material, there is a network of load-bearing grains which span the system, and a set of (many) weakly stressed grains, which are necessary to prevent the load-bearing force chains from buckling and breaking. In the presence of a strong externally applied stress, the force chains are aligned, if the external compression is isotropic, so will the force chains be. This idea makes more precise our earlier discussion of spectator grains. It would be interesting to see whether the network of force chains is isostatic. This is not addressed in the simulation papers, perhaps due to the difficulty in defining this network precisely. However, one intriguing statistic given in reference [57] is the number of contacts in the various samples. For the samples listed, varying from 500 to 4025 particles, the mean coordination number varies as $2.93 < \bar{z} < 3.28$, not far from the isostatic value of $\bar{z} = 3$.

No comprehensive theory exists which predicts the force distribution over the entire measured range. It is instructive to note that the simplest of the stress propagation models, the so-called "q-model" [24,25], in which the vertical force on a grain is passed down probabilistically to the n grains supporting it from below, gives $P(f) \propto e^{-nf}$ for all f. This is analogous to the case of random elastic collisions of gas molecules, where the energy of a pair of particles is redivided between them after they collide; the randomness arising from the random angles of collision. In this case, the energy ultimately becomes distributed in a Boltzmann distribution.

It is interesting to compare these results to those found in simulations of emulsions. In a three-dimensional simulation of a compressed emulsion, Lacasse, *et al* [37] found that the distribution of forces between droplets was Gaussian for volume fractions larger than Φ_c, where the emulsion is expected to behave as an elastic material. The $P(f)$ distribution for deformable particles is related to the portion of the radial distribution function of the packing for values of r corresponding to compressed particles. This is what is shown in Fig.3 for the simulation of Lacasse, *et al* [46]. However, this issue too is controversial: In their 2D simulations of sheared emulsions, Langer and Liu [61] find that while the system is being sheared, and is flowing, $P(f)$ is Gaussian. However when they turn the shear off (a rapid quench), the system settles into a local minimum, and they observe an exponential fall-off at high forces, much like the frictional case discussed above. More work is needed to determine whether this descrepancy is due to the difference between two and three dimensions, or the way in which the states were prepared, or the rapidity of the quench.

It is important to note that the distribution of forces, $P(f)$, should not be taken as conclusively supportive of a particular model, be it isostatic, hyperbolic or elastic. This is because while it is true that a model of isotropic, homogeneous elasticity is unlikely to exhibit the $P(f)$ measured in experiment and simulation, if we allow for highly inhomogeneous elasticity, then we may build in any force distribution we wish by suitably choosing the elasticity field. Once again, it is a question of construction - The material self-organizes into a highly inhomogeneous state where force chains exist, which have an unusual $P(f)$.

This having been said, force chains appear to be the rule in an isostatic material. Moreover, Moukarzel [11] has argued that an isostatic network is highly susceptible to minute changes. He argues that for an overdetermined medium a small perturbation will lead to small particle displacements, since the extra contacts will be able to "soak up" the new stresses resulting from the perturbation. On the other hand, an isostatic network, which is fully determined, does not have this option - there is nothing to mollify the perturbation, and so the particles must rearrange so as to find a new stationary state.

IV. CONCLUSIONS

The salient feature of static granular packings is their enormous heterogeneity. This is borne out by the existence of force chains, and the unusual force distribution $P(f)$ with its exponential fall-off for large f, and its nearly flat form for small f. Such behavior is a feature of isostatic or fragile packings. During the construction process, the system self-organizes, and if the dissipation takes place fast enough, it will jam. The questions one may ask concern two related but distinct areas:

- What is the nature of the state which the system settles (or jams) into? Is it isostatic or is there a finite density of "extra" contacts? Or can we describe it by building up, layer by layer, at all stages allowing the material to deform in an elasto-plastic way, and thus describe the generation of an inhomogeneous elasticity field?

- How will the system, as constructed, respond to an external pertubation in the form of a small added stress? Will it respond elastically, by compressing and expanding in places, in accordance with the elasticity field generated during construction? Or will it respond as a fragile system, necessarily rearranging unless the added preturbation is carefully chosen? Or will it rearrage under practically any and all added perturbations, as an isostatic packing has been conjectured to do?

These questions, and the debate they inspire, reflects the fundamental nature of research on static properties of granular materials. The jamming metaphor has provided interesting avenues of approach; it is up to experiment to ultimately decide.

ACKNOWLEDGMENTS

It is a pleasure to thank M.E. Cates, S.F. Edwards, P.G. de Gennes, J.D. Goddard, M.-D. Lacasse, A. J. Liu, S.R. Nagel, A. Tkachenko, D.A. Weitz, and T.A. Witten for stimulating and insightful conversations. I would like to thank the Institute for Theoretical Physics, U.C.S.B. for providing an atmosphere which facilitated many of these discussions. Finally, I would like to acknowledge a great debt of gratitude to Shlomo Alexander, to whose memory I humbly dedicate this article.

Support from the Israel Science Foundation, grant 211/97, is gratefully acknowledged.

---

[1] E.R. Nowack, J.B. Knight, M. Povinelli, H.M. Jaeger, and S.R. Nagel, *Powder Technology* **94**, 79 (1997).

[2] P.N. Pusey and W. van Megen, Phys. Rev. Lett. **59**, 2083 (1987).

[3] J.C. van der Werff, C.G. de Kruif, C. Blom and J. Mellema *Phys .Rev.* /bf A39, 795 (1989).

[4] T. Shikata and D.S. Pearson, *J. Rheol.* 38, 601 (1994).

[5] T.G. Mason, J. Bibette, and D.A. Weitz, Phys. Rev. Lett. **75**, 2051 (1995).

[6] T.G. Mason and D.A. Weitz, Phys. Rev. Lett. **75**, 2770 (1995).

[7] A packing is described by a connectivity matrix telling which neighbors a given grain is in contact with; this is often called the topology of the packing.

[8] S. Alexander, *Physics Reports* **296**, 65 (1998).

[9] S.F. Edwards, D.V. Grinev, and R.C. Ball, *preprint.*

[10] C.F. Moukarzel, Phys. Rev. Lett. **81**, 1634 (1998).

[11] C.F. Moukarzel, *cond-mat* /9807004.

[12] R.C. Ball and D.V. Grinev, *cond-mat* /9810124.

[13] A.V. Tkachenko and T.A. Witten, *cond-mat* /9811171.

[14] To be precise, one should subtract out the $d + \frac{d(d-1)}{2}$ degrees of freedom of the aggregate as a whole–translations and rotations *en masse*. This may be done if desired, but for a system of many particles, or one for which these degrees of freedom are constrained by the boundary, it is not necessary.

[15] Strictly speaking, by "central force" I mean here a force whose direction is the normal to the plane tangent to both grains at the point of contact. In this case, the force is specified by one scalar component. By "vector force" I mean a force with components in the tangent plane as well as along the normal; this is defined by d scalar components in d dimensions.

[16] L.D.Landau and E.M. Lifshitz, "Theory of Elasticity", (Pergamon Press, Oxford, UK) (1970).

[17] u_{ij} is the symmetrized spatial derivative of the displacement vector field u_i, and is defined by

$$u_{ij} = \frac{1}{2}\left(\frac{\partial u_i}{\partial x_j} + \frac{\partial u_j}{\partial x_i} + \frac{\partial u_l}{\partial x_i}\frac{\partial u_l}{\partial x_j}\right).$$

For most purposes, we may neglect the second order strain term $\frac{\partial u_l}{\partial x_i}\frac{\partial u_l}{\partial x_j}$, but for thin shells [16] and tenuous, fragile networks [8], for example, it is important.

[18] The elasticity field is ordinarily a fourth rank tensor which is spatially dependent. In a homogeneous elastic body, however, it is constant in space, and, often, most of the components are zero, leaving us with what are commonly called the "elastic constants" of the material.

[19] The material behaves elastically until stresses exceed some yield criterion. See, for example, L.E. Malvern, *Introduction to the Mechanics of Continuous Media*, Prentice-Hall, New Jersey (1969). For granular materials, the accepted yield criterion is due to C.A. Coulomb (*Mém. Math. Phys* (Paris) **X**, 161 (1785)), and is a generalization of simple static friction between two solid bodies.

[20] J.P. Wittmer, M.E. Cates, and P. Claudin, *J. Phys. I France* **7**, 39 (1997).

[21] Note that in critical state soil mechanics, where it is assumed that the material is everywhere critical (*i.e.*, at yield), no strain field is employed; instead, the yield condition, a relation of the form of Eq. 3, is used to complete the set of equations.

[22] This is an idealization–the stress tensor of the granular material is really defined for a small volume of material, not an individual grain.

[23] S.B. Savage, in *Powders and Grains 97*, R.P. Behringer and J.T. Jenkins, eds., Balkema, Rotterdam (1997).

[24] C.-h. Liu, S.R. Nagel, D.A. Sheeter, S.N. Coppersmith, S. Majumdar, O. Narayan, and T.A. Witten, *Science* **269**, 513 (1995).

[25] S.N. Coppersmith, C.-h. Liu, S. Majumdar, O. Narayan, and T.A. Witten, Phys. Rev. E **53**, 4673 (1996).

[26] J.E.S. Socolar, Phys. Rev. E **57**, 3204 (1998).

[27] P. Claudin, J.-P. Bouchaud, M.E. Cates, and J.P. Wittmer, Phys. Rev. E **57**, 4441 (1998).

[28] C. Eloy and E. Clement, *J. Phys. I* (France) **7**, 1541 (1997).

[29] J. Smid and J. Novosad, *Proceedings of the 1981 Powtech Conference; Ind. Chem. Eng. Symp.* **63** D3V 1-12 (1981).

[30] R. Brockbank, J.M. Huntley, and R. C. Ball, *J. Phys II* (France) **7**, 1521 (1997).

[31] F. Cantelaube and J.D. Goddard, in *Powders and Grains 97*, R.P. Behringer and J.T. Jenkins, eds., Balkema, Rotterdam (1997).

[32] J. P. Bouchaud, P. Claudin, M.E. Cates, and J.P. Wittmer, in *Physics of Dry Granular Media*, H.J. Herrmann, J.P. Hovi, and S. Luding, eds., NATO ASI 1997. *cond-mat* /9711135.

[33] L. Vanel and E. Clément, to appear in *Europhys. Journal* **B**.

[34] L. Vanel, P. Claudin, J.-P. Bouchaud, M.E. Cates, E. Clément, and J.P. Wittmer, *cond-mat* /9904094.

[35] H.A. Janssen, *Z. Vereins Deutsch Ing.* **39**, 1045 (1895).

[36] R. Nedderman, *Statics and Kinematics of Granular Materials*, Cambridge University Press (1992).

[37] M.-D. Lacasse, D. Levine, and G.S. Grest, to be published.

[38] This is an idealization, of course. But the message remains as long as the construction of the aggregate proceeds via the stoppage of individual grains, (or small groups), without subsequent wholesale collective reconstruction

[39] Of course, a grain of sand does not actually stop at the first place it can–it rolls, slides, and jumps until it slows down and can be stopped. But this is still very far from searching for an optimal place. Additionally, it is not clear that the state arrived at by having individual particles lose energy and stop is actually marginally stable, since marginal stability is a collective property of the entire system. This being said, a state which is constructed by having each constituent grain stop without optimization is likely to be very special–the prototype example being a DLA (diffusion limited aggregation) cluster.

[40] M.E. Cates, J.P. Wittmer, J.-P. Bouchaud, and P. Claudin, Phys. Rev. Lett. **81**, 1841 (1998).

[41] R.S. Farr, J.R. Melrose, and R.C. Ball, Phys. Rev. E **55**, 7203 (1997).

[42] V.V. Sokolovskii, *Statics of Granular Materials*, Pergamon, Oxford (1965).

[43] The Coulomb criterion, like any yield criterion, must be independent of the basis in which the tensors are represented. In two dimensions, for example, the criterion may be expressed in the following basis-independent way:

$$\frac{\sigma_1}{\sigma_2} = \frac{1 + sin(\phi)}{1 - sin(\phi)},$$

where $\phi = arctan(\mu)$, and σ_1, σ_2 are the principal stresses – *i.e.*, the eigenvalues of the stress tensor.

[44] P.-G. de Gennes, preprint; P. Evesque and P.-G. de Gennes, preprint.

[45] M.-D. Lacasse, G.S. Grest, D. Levine, T.G. Mason, and D.A. Weitz, Phys. Rev. Lett. **76**, 3448 (1996).

[46] T.G. Mason, M.-D. Lacasse, G.S. Grest, D. Levine, J. Bibette, and D.A. Weitz, Phys. Rev. E **56** (1997), 3150-3166.

[47] M.-D. Lacasse, G. Grest, and D. Levine, Phys. Rev. E **54**, 5436 (1996).

[48] Strictly speaking, these effective power laws hold for non-infinitesimal compressions; see reference [47].

[49] S.F. Edwards, private communication.

[50] I will leave off the inaccurate and misleading word "constant", thus, *elasticity field* instead of *elastic* constant *field*. See also [18].

[51] C.-h. Liu and S.R. Nagel, Phys. Rev. Lett. **68**, 2301 (1992).

[52] C. Thornton and D.J. Barnes, *Acta Mechanica* **64**, 45 (1986).

[53] The geometry of a packing involves not only which grains are in contact with which, but also the distance between them.

[54] P.A. Cundall and O.D.L. Strack, *Geotechnique* **29**, 47 (1979).

[55] P. Dantu, in *Proc. 4th Int. Conf. on Soil Mech, and Foundation Eng.*, vol. 1, 144 (1957).

[56] D. Howell, R.P. Behringer, C. Veje, Phys. Rev. Lett. **82**, 5241-5244 (1999); C.T. Veje, D.W. Howell, R.P. Behringer, Phys. Rev. E **49**, 739-745 (1999).

[57] F. Radjai, M. Jean, J.J. Moreau, and S. Roux, Phys. Rev. Lett. **77**, 274 (1996).

[58] F. Radjai, D. Wolf, M. Jean, and J.J. Moreau, Phys. Rev. Lett. **80**, 61 (1998).

[59] F. Radjai, D. Wolf, S. Roux, M. Jean, and J.J. Moreau, in *Powders and Grains 97*, R.P. Behringer and J.T. Jenkins, eds., Balkema, Rotterdam (1997).

[60] D.M. Mueth, H.M. Jaeger, and S.R. Nagel, Phys. Rev. E **57**, 3164 (1998), and references therin.

[61] A. Liu and S. Langer, to be published.

The viscous slowing down of supercooled liquids and the glass transition: phenomenology, concepts, and models

G. TARJUS

Laboratoire de Physique Théorique des Liquides, Université Pierre et Marie Curie, 4 Place Jussieu, 75252 Paris cedex 05, France

D. KIVELSON

Department of Chemistry and Biochemistry, University of California, Los Angeles, CA 90095, USA

Abstract

The viscous slowing down of supercooled liquids that leads to glass formation can be considered as a classical, and is assuredly a thoroughly studied, example of a "jamming process". In this review, we stress the distinctive features characterizing the phenomenon. We also discuss the main theoretical approaches, with an emphasis on the concepts (free volume, dynamic freezing and mode-coupling approximations, configurational entropy and energy landscape, frustration) that could be useful in other areas of physics where jamming processes are encountered.

Introduction

When cooling a liquid, usually under isobaric $P = 1$ atm conditions, one can often bypass crystallization, thereby obtaining a supercooled liquid that is metastable relative to the crystal. When the temperature is further lowered, the viscosity of the liquid, as well as the relaxation times associated with the primary (α) relaxation of all kinds of structural, dielectric, macro- and micro-scopic observables, increase rapidly, until a temperature is reached at which the liquid can no longer flow and equilibrate in the time scale of the experiment. The system effectively appears as a rigid amorphous material and is then called a glass. Glass formation thus results from the strong viscous slowing down of a liquid with decreasing temperature [1], a slowing down that can be considered as a classical example of a "jamming process". A characteristic of this process that is unanimously recognized, a unanimity rare in this otherwise quite open and controversial field, is that it is a dynamic effect. The so-called "glass transition" is not a *bona fide* thermodynamic phase transition, but represents a crossover below which a liquid falls out of equilibrium on the experimental time scale. The transition temperature, T_g, depends on this time scale, set either by the observation time (corresponding, for instance, to a relaxation time of 10^2 or 10^3 sec or a viscosity of 10^{13} Poise) and/or by the cooling rate (in a typical differential scanning calorimetry measurement, 10 K per minute). The dependence on cooling rate is, however, weak—a difference of a few K in T_g for an order-of-magnitude change of the rate. This is so because on further lowering of the temperature, the viscosity and α-relaxation times rapidly become enormous and out of reach of any experimental technique.

In the following, we shall focus on the jamming process occurring in the supercooled liquid state.[1,2] Both the crystal[2], which is the stable phase below the melting point T_m but can be ignored in discussing the glass transition, and the glassy state itself will be excluded from the present discussion. By appropriately eliminating the crystal (experimentally as well as theoretically), metastable supercooled liquids can be treated by equilibrium thermodynamics, statistical mechanics, and conventional linear-response formalisms. Glasses on the other hand, although mechanically stable, are out-of-equilibrium states; especially near T_g, they display nonlinear responses and relaxations

[1] Although not as widely used, there are other ways of generating glassy structures, such as vapor deposition, in situ polymerization or chemical reactions.[3]

[2] Note that for several liquids, such as m-fluoroaniline and dibutylphthalate at atmospheric pressure and atactic polymers, crystallization has never been observed.

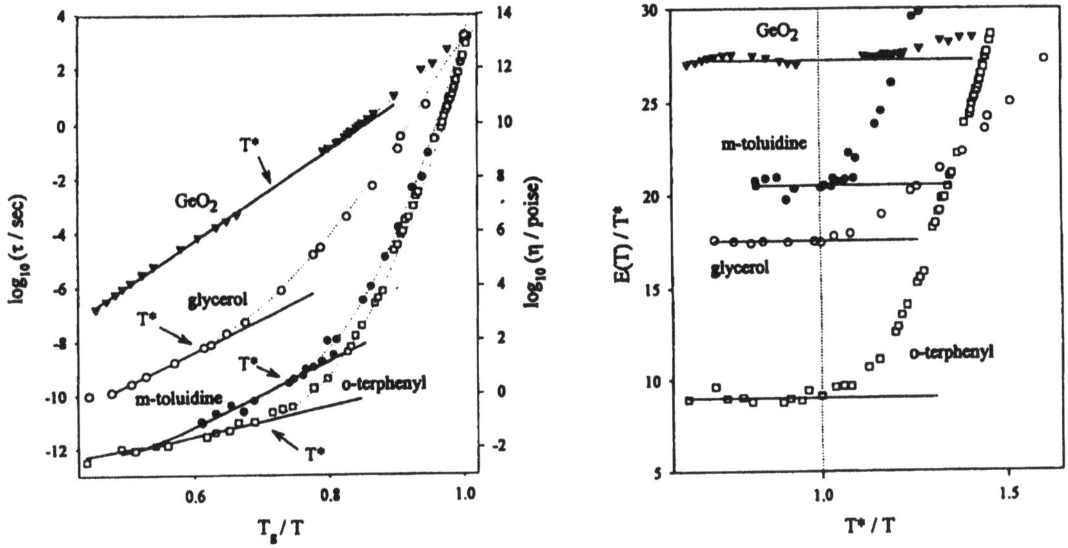

Figure 1. Super-Arrhenius T-dependence of the viscosity η and α-relaxation times τ_α in glass-forming liquids. **a)** Logarithm (base 10) of η and τ_α versus reduced inverse temperature T_g/T for several liquids. For GeO_2, a system forming a network of strong intermolecular bonds, the variation is almost linear, whereas the other liquids (glycerol, m-toluidine, and ortho-terphenyl) are characterized, below some temperature T^*, by a strong departure from linear dependence: the behavior is nearly Arrhenius in the former case and super-Arrhenius in the latter. (Data taken from references cited in Ref. (8) and from C. Alba-Simionesco, private communication.) **b)** Effective activation free energy $E(T)$, obtained from data shown in **a)**, as a function of inverse temperature. Both $E(T)$ and T are divided by the crossover temperature T^* shown in **a)**.

known as aging or annealing, and their properties depend on their history of preparation.[4,5] These phenomena will not be considered in this review.

I. Salient phenomenology

The distinctive feature of glass-forming liquids is the dramatic, continuous increase of viscosity and α-relaxation times with decreasing temperature. This sort of jamming is observed in a large variety of substances: covalently bonded systems like SiO_2, hydrogen-bonded liquids, ionic mixtures, polymers, colloidal suspensions, molecular van der Waals liquids, etc. The emphasis will be placed on those liquids (the vast majority) that do not form 2- or 3-dimensional networks of strong bonds because they show the most striking behavior when passing from the high-temperature liquid phase to the deeply supercooled and very viscous regime.

1. Strong, super-Arrhenius T-dependence of viscosity and α-relaxation times

The viscosity η and α-relaxation times τ_α can change by 15 orders of magnitude for a mere decrease of temperature by a factor two[3]. Such a dramatic variation is conveniently represented on a logarithmic plot of η or τ_α versus $1/T$, i.e., an Arrhenius plot: see Fig. 1a. A system like GeO_2, an example of a network-forming system, is characterized by an almost linear variation, which indicates an Arrhenius temperature dependence. For all other liquids on the figure there is a marked upward curvature, which represents a faster-than-Arrhenius, or super-Arrhenius, temperature dependence and is often described by an empirical Vogel-Fulcher-Tammann (VFT) formula (also called Williams-Landel-Ferry formula in the context of polymers studies[6]),

$$\tau_\alpha = \tau_0 \exp\left(D\frac{T_0}{T - T_0} \right) \tag{1}$$

where τ_0, D and $T_0 < T_g$ are adjustable parameters. On the basis of such Arrhenius plots, with the temperature scaled by T_g, Angell proposed the now

[3] The (shear) viscosity η can be related to a time characteristic of α-relaxation, the average shear stress relaxation time τ_S, by $\eta = G_\infty \tau_S$, where G_∞ is the infinite-frequency shear modulus; G_∞ is typically of the order of 10^{10}-10^{11} erg cm^{-3}, so that a viscosity of 10^{13} Poise roughly corresponds to a time of 10^2 or 10^3 sec.

standard classification of glass-forming liquids into *strong* (Arrhenius-like) and *fragile* (super-Arrhenius) systems; [7] in Eq. 1, the smaller the value of D, the more fragile the liquid. There are, of course, alternative fitting formulas that have been used, some which do not imply a singularity at a nonzero temperature as does the expression in Eq. 1. [8]

A different way of representing the phenomenon is to plot the effective activation free energy for α-relaxation, $E(T)$, obtained from

$$\tau_\alpha = \tau_{\alpha,\infty} \exp\left(\frac{E(T)}{k_B T}\right) \tag{2}$$

where k_B is the Boltzmann constant and $\tau_{\alpha,\infty}$ is a high-T relaxation time, or from a similar equation for the viscosity. This is illustrated in Fig. 1b, where the temperature has been scaled for each liquid to a temperature T^* above which the dependence is roughly Arrhenius-like. Although the determination of T^* is subject to some uncertainty, [8,9] the procedure emphasizes the crossover from Arrhenius-like to super-Arrhenius behavior that is typical of and quite distinct in most supercooled liquids. The appreciable size of the effective activation free energies $E(T)$, namely, 40 $k_B T_g$ at the glass transition, is indicative of thermally activated dynamics. Such a large effective activation free energy for weakly bonded fragile molecular liquids such as orthoterphenyl is an intriguing feature of the phenomenology. Another peculiar property of $E(T)$ for fragile systems is that it increases significantly between T^* and T_g (a factor of 3, i.e., a factor of 5 or 6 in units of the thermal energy $k_B T$ for fragile liquids). Such a variation is not commonly encountered. For instance, in the field of critical phenomena, the slowing down of dynamics that occurs when approaching the critical point is usually characterized by a power law growth of the relaxation time; in terms of effective activation free energy, this corresponds to a logarithmic growth and it is slower than the variation described by the VFT formula, Eq. 1. Unusually strong slowing down, with exponentially growing times similar to Eq. 1, is found in some disordered systems like the random field Ising model and it is known as "activated dynamic scaling". [10]

2. *Nonexponential relaxations*

In an "ordinary" liquid above the melting point, relaxation functions are usually well described, after some transient time, by a simple exponential decay. Deviations are observed, but they are neither systematic nor very marked. The situation changes at lower temperatures, and the α-relaxation is no longer characterized by an exponential decay. A better representation is provided by a "stretched exponential" (or Kohlrausch-Williams-Watts function),

$$f_\alpha(t) \propto \exp\left[-\left(\frac{t}{\tau_\alpha}\right)^\beta\right] \tag{3}$$

where β ($0 < \beta \leq 1$) is the stretching parameter; the smaller β, the more "stretched" the relaxation. Although not unambiguously established, the degree of departure from exponential behavior, or stretching, appears to increase (i.e., β decreases) with decreasing temperature.

Alternatively, in frequency space, the spectrum of the imaginary part of the suceptibility, which is characterized by a peak at a frequency

$$\omega_\alpha \propto 1/\tau_\alpha$$

tends to be broader (when plotted as a function of log ω) than the simple Lorentzian or Debye spectrum that is just the Fourier transform of the time-derivative of an exponential relaxation function (see Fig. 2). Fitting formulas related to Eq. 3, formulas like the Cole-Davidson function for frequency-dependent susceptibilities,

$$\left(1 - i(\omega/\omega_\alpha)\right)^{-\beta'}$$

are used to fit the spectroscopic data, but similar

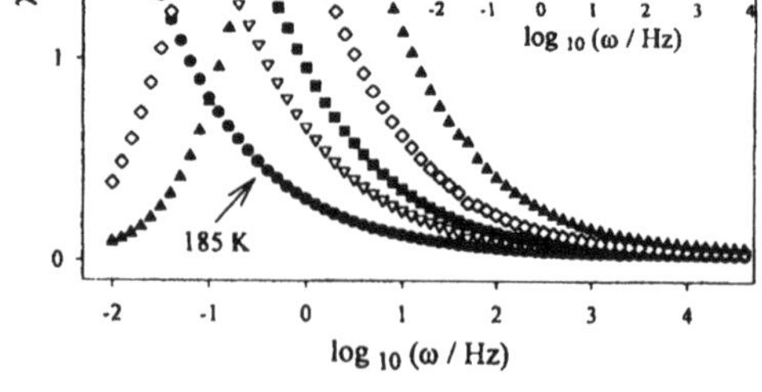

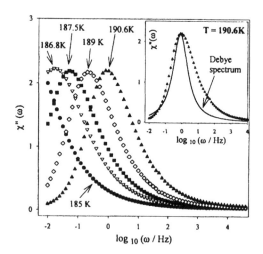

Figure 2. Imaginary part of the dielectric susceptibility χ'' of liquid m-toluidine versus $\log_{10}(\omega)$ for several temperatures close to T_g (T_g = 183.5 K). The inset shows that the α peak is broader than a Debye spectrum that would correspond to a purely exponential relaxation in time. (Data from C. Tschirwitz, E. Rössler, and C. Alba-Simionesco, private communication.)

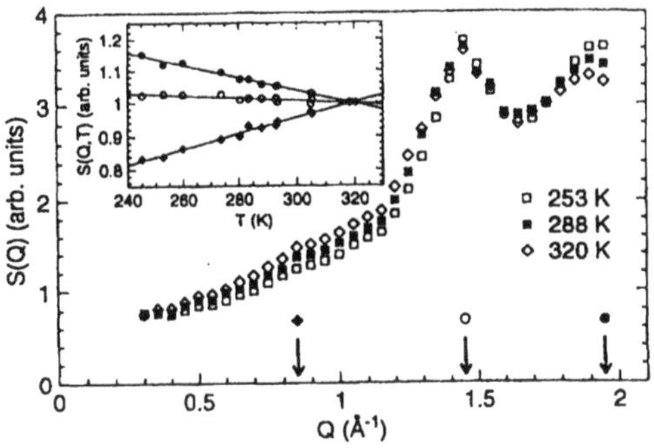

Figure 3. Static structure factor S(Q) of liquid (deuterated) ortho-terphenyl at several temperatures from just below melting ($T_m = 329$ K) to just above the glass transition ($T_g = 243$ K). The inset shows the weak variation with temperature of $S(Q)$ for three values of Q indicated by the symbols above the Q-axis. (From Ref. 13.)

trends are observed: the α peak, as observed for instance in the imaginary part of the dielectric susceptibility as a function of log ω, broadens as the temperature is lowered towards T_g,[11] which indicates increasing departure from Debye/exponential behavior. Except for network-forming systems, the stretching of the α relaxation is significant (β is typically between 0.3 and 0.6 for fragile liquids at T_g). However, and this point may not have been given enough attention, the stretching, or broadening in frequency space, is relatively small when compared to the extremely rapid variation with temperature of the α-relaxation time itself. This is to be contrasted for instance with the activated critical slowing discussed above. There, the power law growth of the activation free energy when the temperature is decreased toward the critical point comes with a more striking stretching of the relaxation function that occurs on a logarithmic scale: in this case, in place of a stretched exponential behavior as in Eq. 3, $\ln(f(t))$ goes as some power of $(1/\ln(t))$.[12]

3. No marked changes in structural quantities

It is tempting to associate the huge increase in α-relaxation times and viscosity with the growth of a *structural* correlation length. However, no such growth has been detected so far in supercooled liquids. Quite to the contrary, the variation of structure in liquids and glasses, as measured in neutron and X-ray diffraction experiments, appear rather bland[13,14] (see Fig. 3). The ordinary, high- temperature liquid has only short-range order whose signature in the static structure factor $S(Q)$ is a broad peak (or a split peak for some molecular systems as illustrated in Fig. 3) at a wave vector Q that roughly corresponds in real space to some typical mean distance between neighboring molecules. As the temperature is lowered and the supercooled regime is entered, there are small, continuous variations of the structure factor that mostly reflect the change in density (typically, a 5% change between T_m and T_g) and, possibly, some adjustments in the local arrangements of the molecules. There is no sign, however, of a significantly growing correlation length, nor of the appearance of a supermolecular length.

In network-forming and H-bonded systems, an additional "pre-peak" is sometimes detectable at wave vectors somewhat lower than that of the main peak, but it is attributed to specific effects induced by the strongly directional intermolecular bonds and not to a length scale that would correlate with the viscous slowing down.[15]

In contrast to this lack of structural signature for the existence of an increasing super-molecular correlation length with decreasing temperature, there is significant evidence, as discussed below, that corresponding "dynamical" correlation length do exist.

4. Rapid entropy decrease and Kauzmann paradox

The absence of marked changes in the structure, *at least at the level of two-particle density correlations*, or of a strong increase in any directly measured static susceptibility is a puzzling feature of the jamming process associated with glass formation. The only static quantity that shows behavior that might be relevant is the entropy.

Below the melting point, T_m, the heat capacity $C_P(T)$ of a supercooled liquid is larger than that of the corresponding crystal. (At T_g, the C_P of the liquid drops to a value that is characteristic of the glass and is close to the C_P of the crystal, but this is a consequence of the system no longer being properly equilibrated.) As a result of this "excess" heat capacity, the entropy difference between the liquid and the crystal decreases with temperature, typically by a factor of 3 between T_m and T_g for fragile liquids. The effect is illustrated in Fig. 4 and leads to the famous Kauzmann paradox:[16] if the entropy difference is extrapolated to temperatures below T_g, its extrapolated value vanishes at some nonzero temperature T_K, which results in the unpleasant feature that the entropy of the liquid becomes equal to that of the crystal (even more unpleasant: if the extrapolation is carried to still lower temperatures, the entropy of the liquid becomes negative, which violates the third law of thermodynamics). The

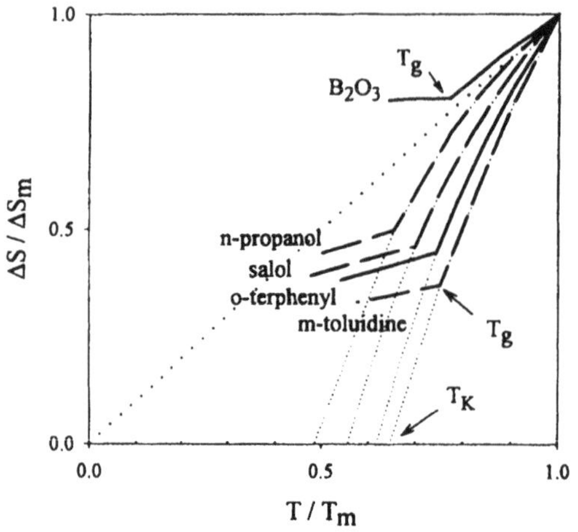

Figure 4. Kauzmann's representation of the "entropy paradox": entropy difference ΔS between the liquid and the crystal (normalized by its value ΔS_m at the melting point) versus reduced temperature T/T_m. The break in the slopes of the full lines signals the glass transition at T_g. The dashed lines indicate an extrapolation of the entropy difference curves below T_g. Except for the strong, network-forming liquid B_2O_3, the extrapolated entropy difference vanishes at a nonzero temperature T_K. (Data from Refs. 16, 18, and from H. Fujimori and C. Alba-Simionesco, private communication.)

paradox is that this *extrapolated* entropy crisis is avoided for a purely dynamic reason, the intervention of the glass transition: what would occur if one were able to keep the supercooled liquid equilibrated down to temperatures below T_g? There are certainly many ways to answer the question. The paradox could be resolved by the existence between T_g and T_K of an intrinsic limit of metastability of the liquid[16] or of a second-order phase transition[4] (a speculation that gains additional credibility with the observation that the VFT temperature T_0 at which the extrapolated viscosity and α-relaxation times diverge (see Eq. 1) is often found close to T_K[17,18]). Even more simply, one might find that the extrapolation breaks down above T_K and that the entropy-difference curve levels off and goes smoothly to zero at zero K, in much the same way as it does in the Debye theory of crystals. These are of course all speculations, but it remains that the rapid decrease of the entropy of the supercooled liquid relative to that of the crystal represents an intriguing aspect of the phenomenology of fragile glass-formers.

[4]Note that a low-T first-order transition does not resolve the paradox because it can be supercooled.

5. *Two-step relaxation and secondary processes*

As we stressed before, the salient features related to glass formation concern the long-time (low-frequency) primary or α relaxation. As the α-relaxation time increases with decreasing temperature, so too does the window between this time and typical microscopic, picosecond or sub-picosecond times. When the relaxation function is plotted against the logarithm of the time, one then observes what is sometimes called a "two-step relaxation". An illustration is given in Fig. 5 by the dynamic structure factor of the fragile ionic glass-former $Ca_{0.4}K_{0.6}$ $(NO_3)_{1.4}$ obtained by neutron techniques.[19] At high temperature, the relaxation function is essentially a one-step process. However, as the liquid becomes more viscous, the relaxation proceeds in two steps separated by a plateau. Although the terminology is far from being universally accepted, the approach to the plateau from the short-time side is often referred to as β or fast-β relaxation. If one is to fit the long-time part by a stretched exponential (Eq. 3), there is a large range of "mesoscopic" times that is not adequately described and that widens as the temperature is lowered. Power law functions of time are often used to reproduce the relaxation function in this mesoscopic range.

This two-step relaxation feature is common to all fragile liquids. In addition, there may also appear additional secondary processes, detected first by Johari and Goldstein in dielectric spectroscopy.[20] Such secondary processes, whose presence and strength strongly vary from one liquid to another, have characteristic frequencies that are intermediate between those of the α and fast-β relaxations. They are denoted Johari-Goldstein-β,

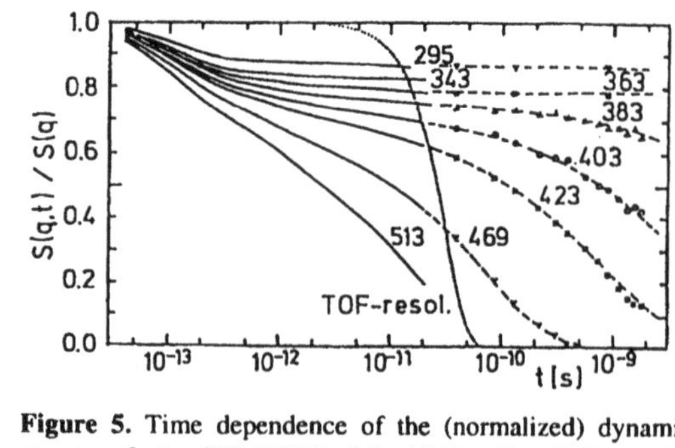

Figure 5. Time dependence of the (normalized) dynamic structure factor $S(Q,t)/S(Q)$ of liquid $Ca_{0.4}K_{0.6}$ $(NO_3)_{1.4}$ (for $Q \approx 1.9Å^{-1}$) at various temperatures. The continuous lines are obtained by Fourier transforming neutron time-of-flight data and the symbols represent neutron spin-echo results. When T decreases, two steps separated by a plateau appear in the relaxation. (From Ref. 19.)

slow-β, or simply β processes.[2,21] To make the description more complete, one should also mention the so-called "boson peak" that may be present on the high-frequency side (~10^2–10^3 GHz) of light and neutron scattering (or absorption) spectra.[2,22] Here we do not discuss either the slow-β processes or the boson peak.

II. A selection of questions

After this brief review of the salient aspects of the phenomenology of supercooled liquids as they get glassy, we discuss in more detail a number of questions, whose answers give justification or put constraints on the theoretical picture one can build to explain the viscous slowing down.

1. How universal is the behavior of glass-forming liquids ?

Universality is a key concept in physics and it has proven to be central in the field of critical phenomena. By the standards of critical phenomena studies, the observed behavior of glass-forming liquids is not universal, the main reason being that no singularity is detected experimentally (or approached asymptotically close), as stressed above. However, if one is willing to take a broader view of the notion of universality, one can find considerable generality or "universality" in the properties, those mostly associated with long-time and low-frequency phenomena, that characterize the approach to the glass transition. For instance, the super-Arrhenius T-dependence of the viscosity and α-relaxation times and the nonexponential character of the relaxation function are observed for virtually all glass-formers, be they polymeric, H-bonded, ionic, van der Waals, etc., with the exception of a minority of strong network-forming systems; and, for a given liquid, these properties are found by a large variety of experimental techniques, such as dielectric relaxation, light and neutron scattering, NMR, viscosity measurements, specific heat spectroscopy, volume and enthalpy relaxation, optical probe methods.

The presence of an underlying "universality" is supported by the fact that experimental data covering a wide range of temperatures and a great diversity of substances can be collapsed onto master curves with only a small number of species-dependent adjustable parameters. A good example is provided by the scaling plot of the frequency-dependent dielectric susceptibility proposed by Nagel and coworkers.[11] As shown in Fig. 6, the data taken over a 13 decade range of frequencies for many different liquids can be placed with good accuracy onto a single curve after scaling with only three parameters associated with the α-peak position, width and intensity. The master-curve for the temperature dependence of the effective activation free energy for viscosity and α relaxation put forward by Kivelson *et al.*[8] is another example. It is nonetheless fair to say that the fits resulting from these various scaling procedures are far from perfect, which leaves room for debate and conflicting interpretations. One can also ask the question whether the universality holds only up to the implied high-frequency cut-off of the susceptibility scaling curve of Nagel *et al.* (i.e., whether one should be focusing on slow behavior) or whether it extends higher in light of the fact that similarities have also been observed in the high-frequency susceptibilities.[23]

2. Is the α-relaxation homogeneous or heterogeneous ?

The observation stressed above that the α-relaxation is nonexponential in the supercooled liquid range and its representation by a stretched exponential as in Eq. 2 can be formally interpreted in terms of a superposition of exponentially decaying functions with a distribution of relaxation

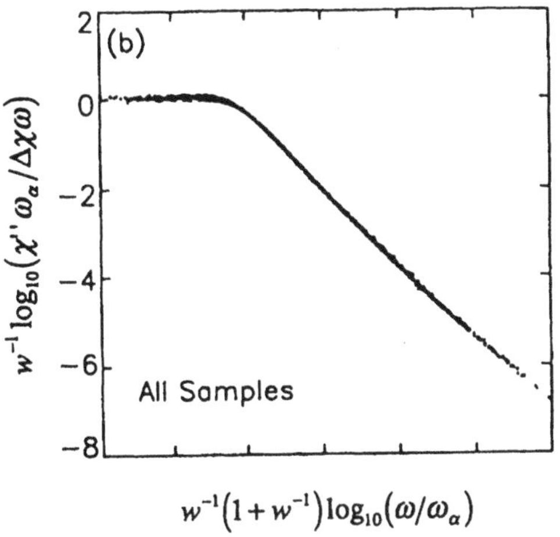

Figure 6. Scaling plot for the imaginary part χ'' of the dielectric susceptibility of several glass-forming liquids (glycerol, salol, propylene glycol, dibutyl-phtalate, α-phenyl-o-cresol, ortho-terphenyl, ortho-phenylphenol). Experimental data similar to those shown in Fig. 2, but for 13 decades of frequency, are collapsed for all temperatures and all liquids onto the master-curve $w^{-1}\log_{10}(\chi''\ \omega_\alpha/\Delta\chi\ \omega)$ vs $w^{-1}(1 + w^{-1})\log_{10}(\omega/\omega_\alpha)$; ω_α is the α-peak position, w is a shape factor that characterizes the deviation from Debye behavior, and Δχ is the static susceptibility. (From Ref. 11.)

times; but, this *per se* does not guarantee that the dynamics be "heterogeneous", in the sense that relaxation of the molecules differs from one environment to another with the environment life time being longer than the relaxation time. An alternative explanation can be offered within a "homogeneous" picture in which relaxation of the molecules everywhere in the liquid is intrinsically nonexponential.

In recent years, there has been mounting evidence that heterogeneities, sufficiently long-lived to be relevant to the α-relaxation and to be at least partly responsible for its nonexponential feature, do exist in supercooled liquids.[24] The heterogeneous character of the slow dynamics has been demonstrated in several experiments: multi-dimensional NMR,[25] photobleaching probe rotation measurements,[26] nonresonant dielectric hole burning.[27] These techniques involve the selection of a subensemble of molecules in the sample that is characterized by a fairly narrow distribution of relaxation times (and in general a relaxation slower than average) and the further monitoring of the gradual return to the equilibrium situation. Additional evidence of the spatially heterogeneous nature of the dynamics in fragile supercooled liquids is provided by the so-called breakdown of the Stokes-Einstein relation between the translational diffusion constant and the viscosity, and the concomitant "decoupling" between rotational and translational time scales:[28,29] see Fig.7.

The size of the heterogeneities is not directly observable in the above mentioned experiments,

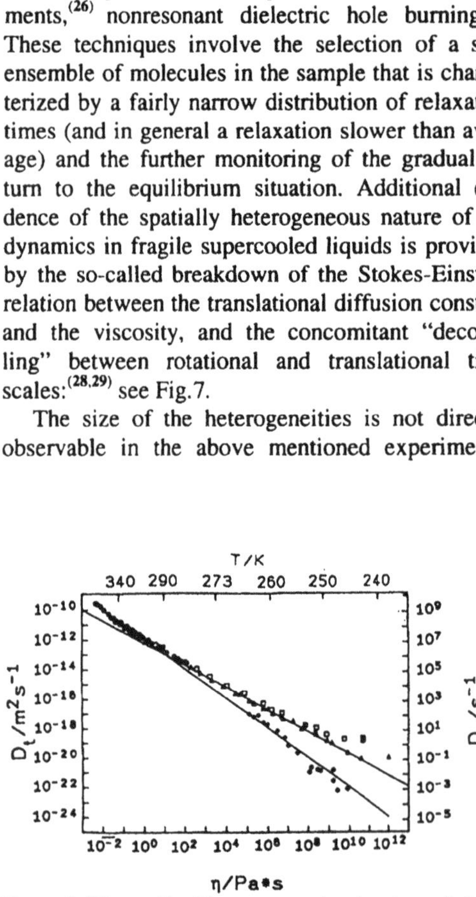

Figure 7. "Decoupling" between rotational and translational time scales: logarithm (base 10) of the rotational (D_R) and translational (D) diffusion coefficients for ortho-terphenyl as a function of temperature and of the logarithm (base 10) of the reduced inverse viscosity. D_R (triangles) follows the viscosity at all temperatures whereas D (squares, crosses and dots) departs from viscosity (and from D_r) below $T \approx 290$ K, as indicated by a value of ξ less than 1. (From Ref. 28).

but various estimates, obtained, e.g., from optical studies of the rotational relaxation of probes of varying size,[30] NMR measurements,[31] the study of excess light scattering,[32] and the influence of a well-defined 3-dimensional confinement[33] lead to a typical length of several nanometers in different fragile liquids near T_g. One should recall that these signatures are all dynamical and that no signature at such a length scale has been detected so far in small-angle neutron and X-ray diffraction data. If the heterogeneous character of the α-relaxation appear reasonably well established, at least for deeply supercooled fragile liquids, several points concerning the lifetime, the size, and the nature of the heterogeneities need still be settled.

3. Is density or temperature the dominant control variable?

The phenomenon of viscous slowing down and glass formation as it is studied most of the time (and described in the preceding sections) takes place under isobaric $P = 1$ atm conditions. As a consequence, when the temperature is lowered, there is also an increase of the density of the liquid. This increase is small (a typical variation of 5% between T_m and T_g), but it could still have a major influence on the dynamics. Actually, there are theoretical models of jamming, such as those based on free volume concepts and hard sphere systems, that attribute the spectacular increase of viscosity and α-relaxation times of fragile glass-formers (almost) entirely to the density changes. It is thus important to evaluate the role of density and temperature in driving the jamming process that leads to the glass transition *at 1 atm*.

Basic models and theories are usually formulated in terms of either density or temperature as control variable, but experiments are carried out with pressure and temperature as external control variables. The data must be converted, when enough experimental results are available, in order to analyze the influence of density at constant temperature and that of temperature at constant density, for a range of density and temperature that is characteristic of the phenomenon *at 1 atm*. Extant analyses[34] are far from exhaustive. However, as illustrated in Fig. 8, the characteristic super-Arrhenius T-dependence of the viscosity, η, and α-relaxation times, τ_α, appears predominantly due to the variation of temperature and not to that of density. This conclusion is confirmed by a comparative study of the contributions induced by density variation (at constant temperature) and by temperature variations (at constant density) to the rate of change

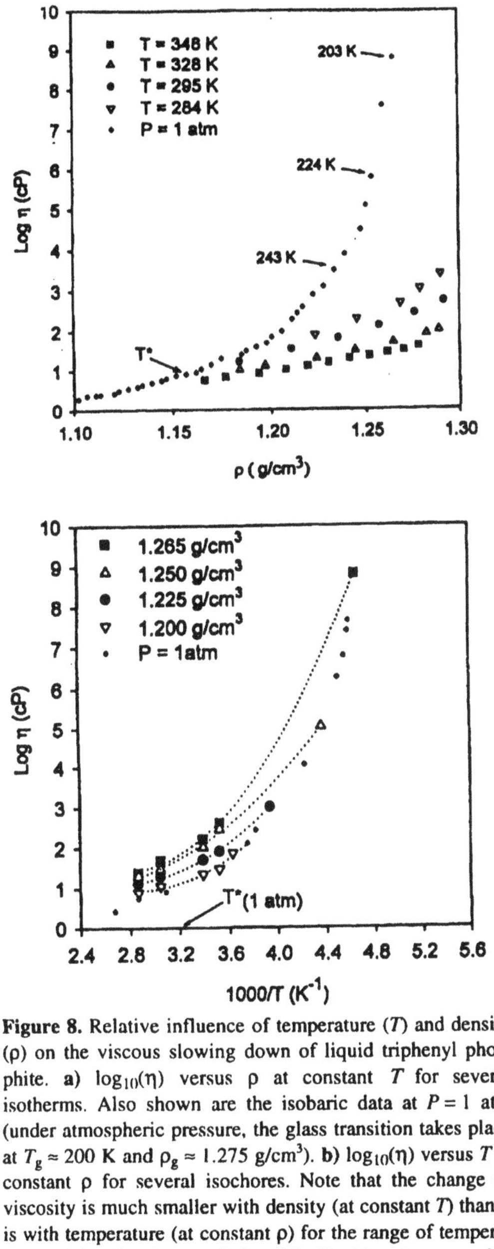

Figure 8. Relative influence of temperature (T) and density (ρ) on the viscous slowing down of liquid triphenyl phosphite. **a)** $\log_{10}(\eta)$ versus ρ at constant T for several isotherms. Also shown are the isobaric data at $P = 1$ atm (under atmospheric pressure, the glass transition takes place at $T_g \approx 200$ K and $\rho_g \approx 1.275$ g/cm^3). **b)** $\log_{10}(\eta)$ versus T at constant ρ for several isochores. Note that the change of viscosity is much smaller with density (at constant T) than it is with temperature (at constant ρ) for the range of temperature and density characteristic of the liquid and supercooled liquid phases at atmospheric pressure. (From Ref. 34.)

of η and τ_α at constant (low) pressure in the viscous liquid regime of several molecular and polymeric glass-formers.[34]

How general is the above result? Temperature appears to be the dominant variable controlling the viscous super-Arrhenius slowing down of supercooled liquids at low pressure, but this may not be the case at much higher pressure (although not much data are presently available to confirm this point), and it is most likely not true for describing the concentration-driven congestion of dynamics in colloidal suspensions. In the absence of a "super-universal" picture of the jamming associated with glass formation, we shall restrict ourselves, as we have implicitly done above, to the consideration of supercooled liquids at 1 atm.

4. *What are the relevant characteristic temperatures?*

There is no unbiased way of presenting the phenomenology of glass-forming liquids. Choices must be made about the emphasis put on the different aspects, about the best graphic representations, and about the way in which one analyzes experimental data. To make sense out of the wealth of observations and measurements, it is natural to look for characteristic temperatures about which to organize and scale the data. However, since no singularity is directly detected, the selection of one or several relevant temperatures is far from straightforward. The temperatures that can be easily determined experimentally are the boiling point, T_b, the melting point, T_m, and the glass transition temperature(s), T_g. Unfortunately, the former two are generally considered as irrelevant to the jamming phenomenon, and the latter has an operational rather than a fundamental nature (see the introduction).

Several other candidates have been suggested, that can be split into two groups. First, there are "extrapolation temperatures" *below* T_g, i.e., temperatures dynamically inaccessible to supercooled liquids, at which extrapolated behavior diverges or becomes singular. This is the case of the VFT temperature T_0 (see Eq. 1) and the Kauzmann temperature T_K (see section II-4 and Fig. 4). In the second group are "crossover temperatures" *above* T_g at which a new phenomenon seems to appear (decoupling of rotations and translations, emergence of a secondary β process, etc.), a crossover of behavior or a change of α-relaxation mechanism seems to take place (passage from Arrhenius to super-Arrhenius T-dependence, arrest of the relaxation mechanisms described by the mode-coupling theory, putative emergence of activated barrier crossing processes, etc.). A variety of such temperatures for the fragile glass-former OTP are shown in Fig. 9: the location of the different characteristic temperatures is illustrated on an Arrhenius plot of the viscosity. It is interesting to note that most putative crossover behaviors occur in the region of strong curvature where $\eta \sim 1$–10^2 Poise or $\tau_\alpha \sim 10^{-10}$–$10^{-8}$

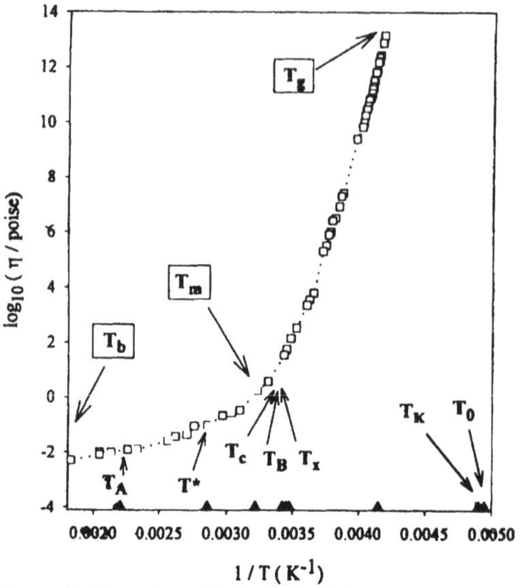

Figure 9. Illustration of the various choices of characteristic temperature for describing the viscous slowing down of liquid ortho-terphenyl. $\log_{10}(\eta)$ is plotted versus inverse temperature. "Extrapolation" temperatures: T_0 (VFT, see Eq. 1) = 200–202 K[18,35], T_K (Kauzmann, see I-4) = 204 K[18]. "Crossover" temperatures: T^* (see Fig. 1 and III-5) = 350 K[8], T_c (MCT, see III-2) = 276–290 K[38], T_A = 455 K[36], T_B = 290 K[37], T_X = 289 K[37]. Also shown are the experimentally measured boiling (T_b = 610 K), melting (T_m = 311 K), and glass transition (T_g = 246 K) temperatures. (Viscosity data from Ref. 38.)

sec, while, on the other hand, the temperatures obtained by extrapolation of data to low T lie fairly close to each other, some 40 K below T_g.

5. What can be learned from computer simulations?

Computer simulation studies,[39] in particular those based on Molecular Dynamics algorithms, have proven extremely valuable in investigating the structure and dynamics of simple, ordinary liquids. Their contribution to the understanding of the glass transition is unfortunately limited, the main reason being the restricted range of lengths and times that are accessible to Molecular Dynamics simulations: typical simulations on atomistic models consider 10^3–10^4 atoms and can follow relaxations for less than 10^{-8} sec (when expressing the elementary time step in terms of parameters characteristic of simple liquids). As a result, the viscous, deeply supercooled regime of real glass-forming liquids, where strong super-Arrhenius behavior, heterogeneous dynamics, and other significant features associated with the jamming process develop, is out of reach, as is the laboratory glass transition that occurs on a time scale of 10^2 or 10^3 sec. Simple liquid models do form "glasses" on the observation, i.e., simulation, time with many of the attributes of the laboratory glass transition: abrupt change in the thermodynamic coefficients, dependence on the cooling rate, aging effects, etc.[5] However, these supercooled simulation liquids are not truly fragile (the T-dependence of the α-relaxation time shows only small departure from Arrhenius behavior) and are not deeply supercooled so that one's ability to extract insights into the deeply supercooled fragile liquids is questionable.

Computer simulations can be useful in studying the moderately supercooled liquid region, where one can observe the onset of viscous slowing down (see for instance Fig. 9). The major interest of such studies is that static and dynamic quantities that are not experimentally accessible can be investigated, such as multi-body (beyond two-particle) correlations involving a variety of variables and microscopic mechanisms for transport and relaxation. They also allow for testing in detail theoretical predictions made in the relevant window of times and lengths (e.g., those of the mode-coupling theory) and for analyzing properties associated with configurational or phase space (see below).

In addition to the much studied one- or two-component systems of spheres with spherically symmetric interaction potentials,[39] the models investigated in computer simulations can be divided into two main groups: on one hand, more realistic microscopic models for molecular glass-formers that attempt to describe species-specific effects;[40] on the other hand, more schematic systems, coarse-grained representations,[41] lower-dimensional systems[42] or toy-models,[43] that bear less detailed resemblance with real glass-formers, but can be studied on much longer time scales and with bigger system sizes.

III. Theoretical approaches

There is a large number of theories, models, or simply empirical formulae that attempt to reproduce pieces of the phenomenology of supercooled

[5] However, these model glasses are effectively high-temperature structures, since they correspond to liquid configurations that are kinetically arrested at a temperature at which the primary relaxation time is of the order only of nanoseconds. Even with systems specially designed to avoid crystallization, such as binary Lennard-Jones mixtures, the cooling rates to prepare glasses (10^8 to 10^9 K/sec) are orders of magnitude higher than those commonly used in experiments.

liquids. There are fewer approaches, however, that address the question of why and how the viscous slowing down leading to the glass transition, with its salient characteristics described in the preceding sections, occurs in liquids as they are cooled. In the following, we shall briefly review the main theoretical approaches, with an emphasis on the concepts and methods that may prove useful in other areas of physics where some sort of jamming process is also encountered[6]. More specifically, we shall discuss phenomenological models based on free volume and configurational entropy, the description of a purely dynamic arrest resulting from mode-coupling approximations, ideas relying on the consideration of the topographic properties of the configurational space (energy and free-energy landscapes) or on the analogy with generalized spin glass models, and approaches centered on the concept of frustration.

1. Free volume

Free-volume models rest on the assumption that molecular transport in viscous fluids occurs only when voids having a volume large enough to accommodate a molecule form by the redistribution of some "free volume", where this latter is loosely defined as some surplus volume that is not taken up by the molecules. In the standard presentation by Cohen and Turnbull,[46] a molecule in a dense fluid is mostly confined to a cage formed by its nearest neighbors. The local free volume, v_f, is roughly that part of a cage space which exceeds that taken by a molecule. It is assumed that between two events contributing to molecular transport, a reshuffling of free volume among the cages occurs at no cost of energy. Assuming also that the local free volumes are statistically uncorrelated, one derives a probability distribution, $P(v_f)$, which is exponential,

$$P(v_f) \propto \exp\left(-\gamma \frac{v_f}{\overline{v_f}}\right) \tag{4}$$

where $\overline{v_f}$ is the average free volume per molecule and γ is a constant of order 1. Since the limiting mechanism for the diffusion of a molecule is the occurrence of a void, i.e., a local free volume v_f larger than some critical value, v_0, that is approximately equal to the molecular volume, the diffusion constant D is given by the probability of finding a free volume equal to v_0; this leads to an expression for D, and by extension for the viscosity η, which is similar to the formula first

$$\eta \propto \frac{1}{D} \propto \exp\left(\gamma \frac{v_0}{\overline{v_f}}\right) \tag{5}$$

proposed by Doolittle.[47]

In the Cohen-Turnbull formulation, the average free volume per molecule is given by $\overline{v_f} = v - v_0$, where $v = 1/\rho$ is the average total volume per molecule. The free-volume concept, in zeroth order, relies on a hard-sphere picture in which thermal activation plays no role. For application to real liquids, temperature enters through the fact that molecules, or molecular segments in the case of polymers, are not truly "hard" and that, consequently, the constant-pressure volume is temperature-dependent,

$$\overline{v_f(T)} \propto \alpha_P (T - T_0) \tag{6}$$

where α_P is the coefficient of isobaric expansivity and T_0 is the temperature at which all free volume is consumed, i.e. $v = v_0$. Inserting the above equation in the Doolittle formula, Eq. 4, gives the VFT expression, Eq. 1. An unanswered, but fundamental question associated with Eq. 6 is why the free volume should be consumed at a nonzero temperature, T_0? An extended version of the free-volume approach has been developed by Cohen and Grest, in which the cages or "cells" are divided into two groups, liquid-like and solid-like, and concepts from percolation theory are included to describe the dependence upon the fraction of liquid-like cells.[48] (See also the model for molecular diffusivity in fluids of long rod molecules by Edwards and Vilgis.[49])

The main criticisms of the free volume models are (i) that the concept of free volume is ill-defined, which results in a variety of interpretations and difficulty in finding a proper operational procedure even for simple model systems, and (ii) that the pressure dependence of the viscosity (and α-relaxation times) is not adequately reproduced. This latter feature has been emphasized in many studies,[17,34,50] and it is a consequence of the observation made above (see II-3) that the viscous slowing down of glass-forming liquids at 1 atm and more generally at low pressure is primarily controlled by temperature and not by density or volume. Glass formation in supercooled liquids does not predominantly results from the drainage of free volume, but rather from thermally activated processes.

---

[6] Models addressing more specific questions, such as the "coupling model"[44] or the "continuous time random walk" approach,[45] are discussed in the reviews cited in Ref. (1).

2. *Mode-coupling approximations*

The theory of glass-forming liquids that has had the highest visibility for more than a decade is the mode coupling theory.[51] It predicts a dynamic arrest of the liquid structural relaxation without any significant change in the static properties. All structural quantities are assumed to behave smoothly and jamming results from a nonlinear feedback mechanism that affects the relaxation of the density fluctuations. Formally, the theory involves an analysis of a set of nonlinear integro-differential equations describing the evolution of pair correlation functions of wave-vector- and time-dependent fluctuations that characterize the liquid. These equations have the form of generalized Langevin equations, and they can be derived by using the Zwanzig-Mori projection-operator formalism. The equation for the quantity of prime interest in the theory, the (normalized) correlation function of the density fluctuations,

$$\phi_Q(t) = \left\langle \rho_Q(t)\rho_Q(0)^* \right\rangle \Big/ \left\langle \left|\rho_Q(0)\right|^2 \right\rangle \tag{7}$$

where

$$\rho_Q(t) = \sum_j \exp(i\mathbf{Q}\cdot\mathbf{r}_j)$$

and $\mathbf{r}_j$ denotes the position of the *j*th particle, can be written as

$$\frac{d^2}{dt^2}\phi_Q(t) + \Omega_Q^2\phi_Q(t) + \int_0^t dt' m_Q(t-t')\frac{d}{dt'}\phi_Q(t') = 0 \tag{8}$$

where Ω_Q is a microscopic frequency obtainable from the static structure factor,

$$S(Q) \propto \left\langle \left|\rho_Q(0)\right|^2 \right\rangle$$

and $m_Q(t)$ is the time-dependent memory function that is formally related to the correlation function of a *Q*-dependent random force. The above equation being exact, the crux of the mode-coupling approach consists in formulating an approximate expression for $m_Q(t)$. The mode-coupling scheme has been implemented for liquids both in the frame of the kinetic theory of fluids[51] and that of the fluctuating nonlinear hydrodynamics.[52] It essentially boils down to approximating the memory function $m_Q(t)$ as the sum of a bare contribution coming from the fast relaxing variables and a mode-coupling contribution coming from the slowly decaying bilinear density modes,

$$m_Q(t) = \gamma\delta(t) + \sum_{Q'} V_{QQ'}\phi_{Q'}(t)\phi_{|Q-Q'|}(t) \tag{9}$$

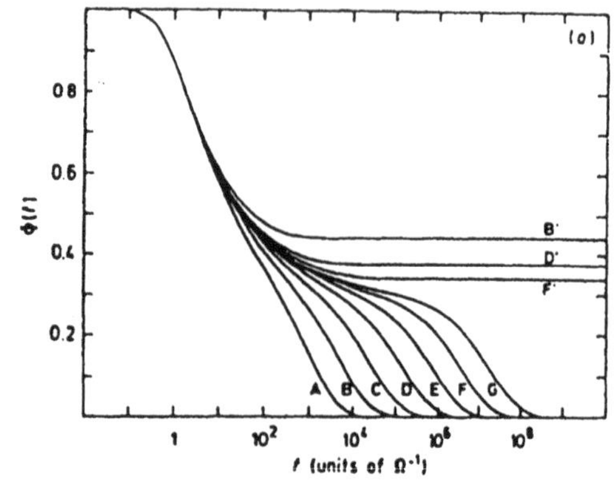

Figure 10. Mode-coupling scenario of kinetic freezing and appearance of a 2-step relaxation: time dependence of the (normalized) density-density correlation function for a schematic mode-coupling model. The curves from A to G correspond to the approach toward the dynamic singularity from the ergodic state. The other curves correspond to the nonergodic state. (From Ref. 51.)

where the vertices $V_{QQ'}$ can be expressed in terms of the static structure factor. The self-consistent solution of the resulting set of nonlinear equations predicts a slowing of the relaxation that is attributed, within a purely homogeneous picture (see II-2), to a cage effect and to the feedback mechanism above mentioned. This solution exhibits a dynamic arrest at a critical point, T_c, which represents a transition from an ergodic to a nonergodic state with no concomitant singularity in the thermodynamics and structure of the system.

The main achievements of the mode-coupling approach are the predicted anomalous increase in relaxation time and the appearance of a two-step relaxation process with decreasing temperature, as indeed observed in real fragile glass-formers (compare Fig. 10 to Fig. 7) and in molecular dynamics simulations.[39] Early on, however, it was realized, both from empirical fits to experimental data and from comparison to simulation data on model systems, that the dynamic arrest at T_c did not describe the observed glass transition at T_g, nor the transition to an "ideal glass" at a temperature below T_g. This is illustrated in Fig. 11. Thus, the T_c was interpreted as a temperature above T_g. The singularity at T_c is avoided because of the breakdown of the simple mode-coupling approximation, Eq. 9, and the T_c of what is called the "idealized" mode-coupling theory is taken as a crossover below which additional relaxation mechanisms, such as activated processes, presumably take over. Unfortunately, beyond some empirical introduction,[51,52] activated processes are not theoretically described by mode-coupling ap-

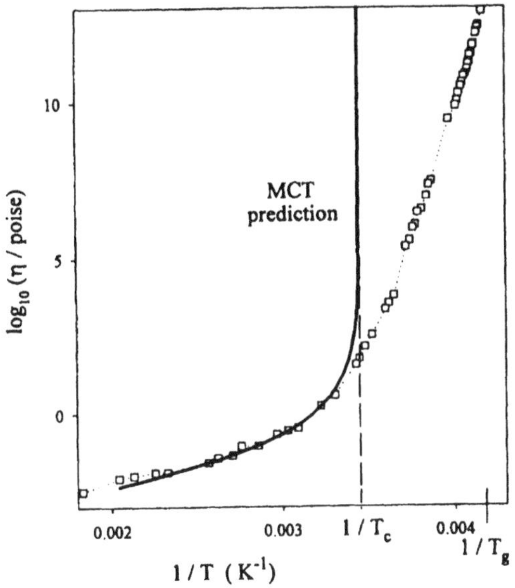

Figure 11. Breakdown of the "idealized" mode-coupling theory illustrated on $\log_{10}(\eta)$ versus $1/T$ for liquid ortho-terphenyl (see caption to Fig. 9). The continuous line is the mode-coupling fit to the experimental data. The predictions break down below a point at which the viscosity is of the order of 10 Poise; the dynamic singularity is not observed, and T_c is interpreted as a crossover temperature.

proaches, and so the theory of α relaxation has not been extended to temperatures below T_c. To draw once again a parallel with critical phenomena (where a singularity occurs at T_c in the instance by Kawasaki,[53] are known to describe quite well the standard critical slowing down, but not structure and the thermodynamics of the system), mode-coupling approximations, as formulated for the activated dynamic scaling such as that observed in the random field Ising model (see section I-1). This failure is related to the underlying nature of the approximation that corresponds to a one-loop self-consistent resummation scheme in a perturbative treatment[52,54] (see also below in III-4 the parallel with spin glass models).

Mode-coupling approaches can thus describe at best the dynamics of moderately supercooled liquids[7] (see Fig. 11). Because of the many detailed predictions it makes in this regime, the mode-coupling theory has stimulated the use and the development of experimental techniques, such as neutron and depolarized light scattering, and molecular dynamics simulations that are able to probe the early stage of the viscous slowing down; but, the very fact that the predicted dynamic singularity is not observed makes it difficult to reach any clear-cut conclusion about the quantitative adequacy of the theory, and this has led to much debate in recent years.[55]

3. Configurational entropy and (free) energy landscape

The existence of a crossover temperature in the moderately supercooled liquid region where α-relaxation times are of the order of 10^{-9} sec (hence in the same region as the T_c predicted by the mode-coupling theory) was advocated 30 years ago by Goldstein.[56] Goldstein argued that below this crossover flow is dominated by potential energy barriers that are high compared to thermal energies and slow relaxation occurs as a result of thermally activated processes taking the system from one minimum of the potential energy hypersurface to another. The idea that molecular transport in viscous liquids approaching the glass transition could be best described by invoking motion of the representative state point of the system on the potential energy hypersurface had also been suggested by Gibbs.[57] In his view, the slowing down of relaxations with decreasing temperature is related to a decrease of the number of available minima and to the increasing difficulty for the system to find such minima. The viscous slowing down would thus result from the decrease of some "configurational entropy" that is a measure of the number of accessible minima. These two concepts, potential energy hypersurface, also denoted "energy landscape", and "configurational entropy", have gained a renewed interest in recent years, boosted by the analogy with the situation encountered in several generalized spin glass models (see below).

The Adam-Gibbs approach[58] represents a phenomenological attempt to relate the α-relaxation time of a glass-forming liquid to the "configurational entropy". In the picture, α-relaxation takes place by increasingly cooperative rearrangements of groups of molecules. Any such group, called a "cooperatively rearranging region", is assumed to relax independently of the others. It is a kind of a long-lived heterogeneity. Molecular motion is activated and the effective activated free energy is equal to the typical energy barrier per molecule, which is taken as independent of temperature, times the number of molecules that are necessary to form a cooperatively rearranging region whose size permits a transition from one configuration to a new one independently of the environment. This latter number goes as the inverse of the configurational entropy per molecule, $S_C(T)/N$, where N is the total

[7] It is possible that they are also applicable to the fast-β relaxations even below T_c.[51]

number of molecules in the sample. Since S_C decreases with decreasing temperature, the reasoning leads to an effective activation free energy that grows with decreasing temperature, i.e., to a super-Arrhenius behavior,

$$\tau_\alpha = \tau_0 \exp\left(\frac{C}{TS_C(T)}\right) \tag{10}$$

where C is proportional to N times the typical energy barrier per molecule. If the configurational entropy vanishes at a nonzero temperature, an assumption somewhat analogous to that in Eq. 6 for the free volume model, but one that is inherent for instance in the Gibbs-di Marzio approximate mean field treatment of a lattice model of linear polymeric chains,[59] then the α-relaxation times diverge at this same nonzero temperature. In particular if the configurational entropy is identified as the entropy difference between the supercooled liquid and the crystal[8], the Adam-Gibbs theory allows one to correlate the extrapolated divergence of the α-relaxation times with the Kauzmann paradox (see I-4): the Kauzmann temperature T_K would then signal a singularity both in the dynamics and in the thermodynamics of a supercooled liquid[9]. Note also that by using a hyperbolic temperature dependence to fit the experimental data on the heat capacity difference between the liquid and the crystal, $\Delta C_P(T) = K/T$, and using this formula to extrapolate the configurational entropy down to the Kauzmann temperature, one converts Eq. 10 to a VFT formula,

$$\tau_\alpha = \tau_0 \exp\left(\frac{CT_K}{K(T - T_K)}\right) \tag{11}$$

with the VFT temperature T_0 equal to the Kauzmann temperature T_K. When comparing to experimental data, the configurational–entropy based expressions provide a good description at least over a restricted temperature range, but the resulting estimates for the critical number of molecules composing a cooperatively rearranging region is often

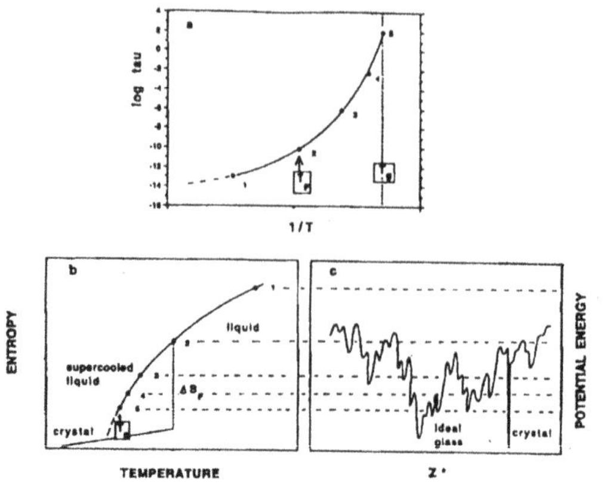

Figure 12. Illustration of the putative relation between the rapid increase of α-relaxation time (**a**), the decrease of the entropy difference between the liquid and the crystal (**b**), and the characteristic energy level on the (schematic) potential energy landscape (**c**) for a fragile glass-former at various temperatures. The ideal glass level corresponds to the Kauzmann temperature (see I-4) and to an extrapolated divergence of the relaxation time; Goldstein's crossover (see III-3) takes place somewhat below point 2. (From Ref. 17.)

is often found to be unphysically small (only a few molecules at T_g)[1]

Building upon the early suggestion made by Goldstein, Angell[62] and others proposed that the apparent passage with decreasing temperature from flow dynamics described by a mode-coupling approach to activated dynamics such as pictured by the configurational-entropy theory of Adam and Gibbs could be rationalized by considering the physics of exploration of the energy landscape: see Fig. 12.

The energy landscape is the potential energy in configurational space. It can be envisaged as an incredibly complex, multi-dimensional ($3N$ dimensions for a system of N particles) set of hills, valleys, basins, saddle-points, and passage-ways around the hills. At constant volume and constant number of particles, this landscape is independent of temperature. However, the fraction of space that is statistically accessible to the representative state point of the system decreases with decreasing temperature, and the system becomes constrained to deeper and deeper wells. (Recall that below the melting point the deepest energy minima corresponding to the crystalline part of configurational space must be excluded when studying the supercooled liquid.) At low enough temperature, when the representative point of the supercooled liquid is mostly found in fairly deep and narrow wells, it seems reasonable to define a "configurational en-

[8] This phenomenological choice for the entropy of configuration has been criticized by Goldstein who showed for several glass-formers that only half of the entropy difference between the liquid and the crystal comes from strictly "configurational" sources; the remainder comes mostly from changes in vibrational anharmonicity or differences in the number of molecular groups able to engage in local motions.[60]

[9] A recent careful, but conjectural analysis of dielectric relaxation data suggests that these data are consistent with the existence of a critical point, both structural and dynamical, at the approximate T_0 specified by the VTF expression in Eq. 1.[61]

tropy" that is proportional to the logarithm of the number of minima that are accessible at a given temperature. The liquid configurations corresponding to these accessible minima have been called "inherent structures" and Stillinger and coworkers have devised a gradient-descent mapping procedure to find the inherent structures and study their properties in computer simulations.[63] Interestingly, Stillinger has also shown, with fairly general arguments, that if one is to use the above defined notion of configurational entropy, an "ideal glass transition" of the type commonly associated with the Kauzmann paradox, i.e., one characterized by the vanishing of the configurational entropy at a nonzero temperature, cannot occur for systems of limited molecular weight and short-range interactions.[64]

It may be more fruitful to investigate in place of the potential energy landscape a free-energy landscape. Such a landscape can only be defined if one is able to construct a free-energy functional by a suitable coarse-graining procedure, as can be done for instance in the case of mean-field spin glass models (see below). A free-energy landscape is temperature-dependent, and it is important to note that the "configurational entropy", also called "complexity",[65] that one can define from the logarithm of the number of accessible free-energy minima differs from the "configurational entropy" computed from the potential energy landscape. In particular, the behavior of the complexity is not restricted by the Stillinger arguments given above.

The "landscape paradigm" is very appealing in rationalizing many observations on liquids and glasses and, more generally, in establishing a framework to describe qualitatively slow dynamics in complex systems that span a wide range of scientific fields.[66] It has been used to motivate, in addition to the Adam-Gibbs theory and other phenomenological approaches like the soft-potential model,[67] simple stochastic models of transport based on master equations.[68] Nevertheless, it has not so far offered a way for elucidating the *physical* mechanism that is responsible for the distinctive features of the viscous slowing down of supercooled liquids.

4. Analogy with generalized spin glass models

If one takes seriously the observation that the extrapolated temperature dependence of both the viscosity (and α-relaxation times) and the "configurational" entropy (taken as the difference of entropy between the liquid and the crystal) become divergent or singular at essentially the same temperature $T_0 \approx T_K$ (see Fig. 9), one is naturally led to postulate the existence at this temperature of an underlying thermodynamic transition, usually referred to as the "ideal glass transition". Looking for analogies with phase transitions in spin glasses is then appealing. However, the kind of dynamic activated scaling that would be required to describe the slowing down of relaxations when approaching the ideal glass transition (see I-1) is not found in the most studied Ising spin glasses.[69] Kirkpatrick, Thirumalai, and Wolynes argued that generalized spin glass models, such as Potts glasses and random p-spin systems, would be better candidates.[70,71] The random p-spin model, for instance, is defined by the following hamiltonian:

$$H = \sum_{i_1 < i_2 < \cdots < i_p} J_{i_1 i_2 \ldots i_p} \sigma_{i_1} \sigma_{i_2} .. \sigma_{i_p} \tag{12}$$

where the σ_i's are Ising variables and the couplings $J_{i_1 i_2 .. i_p}$'s are quenched independent random variables that can take positive and negative values according to a given probability distribution. The behavior of these systems, at least when solved in the mean-field limit where the interactions between spins have infinite range, bears many similarities with the theoretical description of glass-forming liquids outlined above. Indeed, mean-field Potts glasses (with a number of states strictly larger than 4) and mean-field p-spin models (with $p \geq 3$) have essentially the following characteristics:

(i) At high temperature, the system is in a fully disordered (paramagnetic) state. At a temperature T_D, there appears an exponentially large number of metastable "glassy" states whose overall contribution to the partition function is equal to that of the paramagnetic minimum. The free energy and all other static equilibrium quantities are fully regular at T_D, but the dynamics have a singularity of the exact same type as that found in the mode-coupling theory of liquids. At T_D, the system is trapped in one of the metastable free-energy minima, and ergocity is broken.

(ii) Below T_D, a peculiar situation occurs. The partition function has contributions from, both, the paramagnetic state and the exponentially large number of "glassy" (free-energy) minima, the logarithm of which defines the configurational entropy or complexity. This latter decreases as the temperature is lowered.

(iii) At a nonzero temperature $T_S < T_D$, the configurational entropy vanishes. The system undergoes a *bona fide* thermodynamic transition to a spin-glass phase. The transition has been termed "random first order"[70,71] because it is second-order in the usual thermodynamic sense (with, e.g., no latent heat), but shows a discontinuous jump in the order parameter. (Technically, within the replica formalism, it corresponds to a one-step replica symmetry breaking with a discontinuous jump of the Edwards-Anderson order parameter.[69–71])

These mean-field systems are the simplest, analytically tractable models found so far that display a high-temperature mode-coupling dynamic singularity, a nontrivial free-energy landscape, and a low-temperature ideal (spin) glass transition with an "entropy crisis"[10]. Analyzing them sheds light on the mode-coupling approximation, whose validity for fluid systems is otherwise hard to assess. The mode-coupling approximation becomes exact in the mean-field limit, because the barriers separating the metastable minima diverge (in the thermodynamic limit) at and below T_D as a result of the assumed infinite range of the interactions. One expects that in a finite-range model, provided the same type of free-energy landscape is still encountered, barriers are large but finite and ergodicity is restored by thermally activated processes. Accordingly, the dynamic transition is smeared out, and the activated relaxation mechanisms that take over must be described in a nonperturbative way, as suggested for instance by Kirkpatrick, Thirumalai, and Wolynes[71] in their dynamic scaling approach based on entropic droplets.

An advantadge of an analogy between glass-forming liquids and generalized spin glasses[11] is that the powerful tools that have been developed in the theory of spin-glass models to characterize the order parameter and the properties associated with the existence of a large number of metastable glassy states, among which the replica formalism,[69] can be used *mutatis mutandis* to study liquids and glasses.[73] However, to make the analogy really successful, one must still find a short-range model (even more convincing would be a model without quenched disorder) that actually displays dynamic activated scaling and a random first-order transition and make progress in describing the slow relaxation.

5. *Intrinsic frustration without randomness*

Spin glasses, and related systems like orientational glasses, vortex glasses and vulcanized matter,[74] owe their fascinating behavior to two main ingredients: *randomness*, namely the presence of an externally imposed quenched disorder, and *frustration*, which expresses the impossibility of simultaneously minimizing all the interaction terms in the energy function of the system. Liquids and glasses (sometimes called "structural glasses" to stress the difference with spin glasses) have no quenched randomness, but frustration has been suggested as a key feature to explain the phenomena associated with glass formation.[75–79] Frustration in this context is attributed to a competition between a short-range tendency for the extension of a locally preferred order and global constraints that preclude the periodic tiling of the whole space with the local structure.

The best studied example of such an intrinsic frustration concerns single-component systems of spherical particles interacting with simple pair potentials. What is usually called "geometric" or "topological" frustration can be more easily understood by comparing the situations encountered in 2 and 3 dimensions[76] (see Fig. 13). In 2 dimensions, the arrangement of disks that is locally preferred, in the sense that it maximizes the density and minimizes the energy, is a hexagon of 6 disks around a central one, and this hexagonal structure can be extended to the whole space to form a triangular lattice. In 3 dimensions, as was shown long ago by Frank,[80] the locally preferred cluster of spheres is an icosahedron; however, the 5-fold rotational symmetry characteristic of icosahedral order is incompatible with translational symmetry, and formation of a periodic icosahedral crystal is forbidden. Geometric or topological frustration is thus absent in the 2-dimensional case but present in the

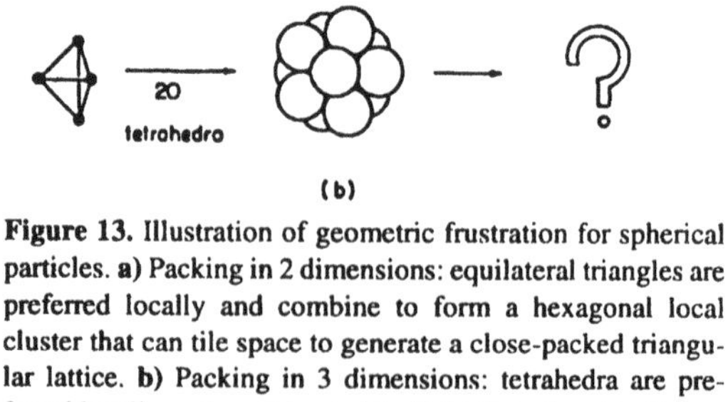

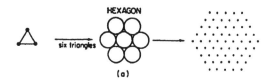

Figure 13. Illustration of geometric frustration for spherical particles. **a)** Packing in 2 dimensions: equilateral triangles are preferred locally and combine to form a hexagonal local cluster that can tile space to generate a close-packed triangular lattice. **b)** Packing in 3 dimensions: tetrahedra are preferred locally and combine (with slight distortions) to form a regular icosahedral cluster; however, the 5-fold symmetry axes of the icosahedron preclude a simple icosahedral space-filling lattice. (From Ref. 76.)

[10] They also show aging phenomena, as discussed in Ref. (5).
[11] See also the frustrated percolation model.[72]

3-dimensional case. A consequence of this, for instance, is that crystallization is continuous, or weakly first-order, in 2 dimensions (with some subtleties related to ordering in 2 dimensions[(81)]) whereas it is strongly first-order in 3 dimensions and accompanied by the breaking of the local icosahedral structure to make the face-centered-cubic or hexagonal-close-packed order that allows to tile space periodically. (In contrast, aligned cubes in 3 dimensions have no frustration and undergo a continuous freezing transition to a crystalline state.[(82)]). The geometric frustration that affects spheres in 3-dimensional Euclidean space can be relieved in curved space with a specially tuned curvature; the creation of topological defects (disclination lines) can then be viewed as the result of forcing the ideal icosahedral ordering into "flat" space. This picture of sphere packing disrupted by frustration has been further developed in models for simple atomic systems and metallic glasses,[(75,76)] and the slowing down of relaxations has been tentatively attributed to the topological constraints that hinder the kinetics of the entangled defect lines;[(76)] however, the treatment remains only qualitative and incomplete.

A significant difficulty in applying the concept of geometric or topological frustration to supercooled liquids is that real fragile glass-formers are in general either mixtures or single-component systems of nonspherical molecules with a variety of shapes, all of which obscures the detailed mechanisms and constraints that are responsible for the frustration. Attempts have been made to get around this problem by proposing a more coarse-grained description of frustration[12].[(77-79)]

In Stillinger's "tear and repair" mechanism for relaxation and shear flow[(78)] and in the more recently introduced "frustration-limited domain theory",[(79)] frustration is described as the source of a strain free energy that opposes the spatial extension of the locally preferred structure and grows superextensively with system size. It results in the breaking up of the liquid into domains, whose size and growth with decreasing temperature are limited by frustration, the weaker the frustration the larger the domains. The super-Arrhenius temperature dependence of the viscosity and α-relaxation times and the heterogeneous nature of the dynamics are attributed to these domains (see also Ref. (77)). Progress has been made along these lines, by making use of a scaling approach based on the concept of avoided critical behavior[13].[(79,84)] However, the putative order variable characterizing the locally preferred structure of the liquid has not yet been properly identified, and, as in the case of the generalized spin-glass models discussed above, one must still give convincing evidence that the 3-dimensional statistical-mechanical frustrated models that have been suggested as minimal theoretical descriptions do show the expected activated dynamics.

[12] In addition, several "toy models" possessing frutration, but no quenched disorder have been studied by computer simulation: see for instance Ref. (83).

[13] This approach differs from both those based upon spin-glass analogies and those in which the slow kinetics are attributed to frustration-induced entangled defect lines in that these others scale about a low-temperature characteristic point signifying ultimate slowing down, whereas in the frustration-limited domain theory the scaling is carried out about a high-temperature characteristic point signifying the initiation of anomolously slow dynamics. For the same reason, it also differs from the domain (or cluster) picture that has been proposed on the basis of an analogy between a supercooled liquid approaching the glass transition and a mean-field model with purely repulsive interactions near its spinodal.[(85)]

IV. Conclusion

The viscous slowing down of supercooled liquids that leads to glass formation can be considered as a classical and thoroughly studied example of a "jamming process". In this review, we have stressed the distinctive features characterizing the phenomenon: strong, super-Arrhenius temperature dependence of the viscosity and the α-relaxation times, nonexponential and heterogeneous character of the α relaxation, absence of marked changes in the structural (static) quantities, rapid decrease of the liquid entropy relative to that of the crystal, appearance of a sequence of steps (or regimes) in the relaxation functions. These features are common to most glass-forming liquids (with the exception of systems forming 2- and 3-dimensional networks of strong intermolecular bonds).

We have also discussed the main theoretical approaches that have been proposed to describe the origin and the nature of the viscous slowing down and of the glass transition. We have emphasized the concepts, such as free volume, dynamic freezing and mode-coupling approximations, configurational entrop y and (free) energy landscape, and frustration, that could be useful in other areas of physics where jamming processes are encountered.

References

1. Other aspects of supercooled liquids and glasses are presented in previous reviews: J. Jäckle, Rep. Prog. Phys. **49**, 171 (1986); C. A. Angell, J. Non-Cryst. Solids **131-133**, 13 (1991); M. D. Ediger, C. A. Angell, and S. R. Nagel, J. Phys. Chem. **100**, 13200 (1996).
2. More specific and detailed presentations can be found for instance in the *Proceedings of the International Meetings on "Relaxations in Complex Systems"*: (i) K. L. Ngai and G. B. Wright Eds., Office of Naval Research, Arlington (1985); (ii) K. L. Ngai and G. B. Wright Eds., J. Non-Cryst. Solids **131-133** (1991); (iii) K. L. Ngai, E. Riande, and G. B. Wright Eds., J. Non-Cryst. Solids **172-174** (1994); (iv) K. L. Ngai Ed., J. Non-Cryst. Solids **235-238** (1998).
3. C. A. Angell, Science **267**, 1924 (1995).
4. L. C. E. Struik, *Physical Aging in Amorphous Polymers and Other Materials* (Elsevier, Amsterdam, 1978). G. W. Scherer, *Relaxation in Glasses and Composites* (John Wiley, New York, 1986). I. M. Hodge, J. Non-Cryst. Solids **169**, 211 (1994).
5. J. P. Bouchaud, L. Cugliandolo, J. Kurchan, and M. Mezard, in *Spin Glasses and Random Fields*, A. P. Young Ed. (World Scientific, Singapore, 1998), pg. 161. J. Kurchan, this volume.
6. J. D. Ferry, *Viscoelastic Properties of Polymers* (John Wiley, New York, 3rd Edition, 1980).
7. C. A. Angell, in *Relaxation in Complex Systems*, K. L. Ngai and G. B. Wright Eds. (Office of Naval Research, Arlington, 1985), pg. 3.
8. D. Kivelson, G. Tarjus, X-L Zhao, and S. A. Kivelson, Phys. Rev. E **53**, 751 (1996).
9. F. Stickel, E. W. Fisher, and R. Richert, J. Chem. Phys. **102**, 6251 (1995); *ibid.* **104**, 2043 (1996).
10. D. S. Fisher, G. M. Grinstein, and A. Khurana, Physics Today Dec. 1988, 56 (1988).
11. P.K. Dixon, L. Wu, S.R. Nagel, B.D. Williams and J.P. Carini, Phys. Rev. Lett. **65**, 1108 (1990). L. Wu, P. K. Dixon, S. R. Nagel, B. D. Williams, and J. P. Carini, J. Non-Cryst. Solids **131-133**, 32 (1991). S. R. Nagel, in *"Phase Transitions and Relaxation in Systems with Competing Energy Scales"*, T. Riste and D. Sherrington Eds. (Kluwer Academic, Netherlands, 1993), pg. 259.
12. J. Villain, J. Phys. (Paris) **46**, 1843 (1985); D. S. Fisher, Phys. Rev. Lett. **56**, 416 (1986). A. T. Ogielski and D. A. Huse, *Phys.* Rev. Lett. **56**, 1298 (1986).
13. A. Tölle, H. Schober, J. Wuttke, and F. Fujara, Phys. Rev. E. **56**, 809 (1997).
14. See also: B. Frick, D. Richter, and Cl. Richter, Europhys. Lett. **9**, 557 (1989). E. Kartini, M. F. Collins, B. Collier, F. Mezei, and E. C. Swensson, Phys. Rev. B **54**, 6292 (1996). R. Leheny, N. Menon, S. R. Nagel, K. Volin, D. L. Price, and P. Thiyagarjan, J. Chem. Phys. **105**, 7783 (1996).
15. S. R. Elliot, Nature **354**, 445 (1991); J. Phys. Condens. Matter **4**, 7661 (1992). D. Morineau, C. Alba-Simionesco, M.-C. Bellissent-Funel, M.-F. Lauthie, Europhys. Lett. **43**, 195 (1998).
16. W. Kauzmann, Chem. Rev. **43**, 219 (1948).
17. C. A. Angell, J. Res. Natl. Inst. Stand. Technol. **102**, 171 (1997).
18. R. Richert and C. A. Angell, J. Chem. Phys. **108**, 9016 (1998).
19. W. Knaak, F. Mezei, and B. Farago, Europhys. Lett. **7**, 529 (1988).
20. G. P. Johari and M. Goldstein, J. Chem. Phys. **53**, 2372 (1970).
21. E. Rössler, Phys. Rev. Lett. **69**, 1620 (1992). E. Rössler, U. Warschewske, P. Eiermann, A. P. Sokolov, D. Quitmann, J. Non-Cryst. Solids, **172-174**, 113 (1994). E. Bartsch, F. Fujara, B. Geil, M. Kiebel, W. Petry, W. Schnauss, H. Sillescu, and J. Wuttke, Physica A **201**, 223 (1993).
22. E. Rössler, V. N. Novikov, and A.P. Sokolov, Phase Transitions **63**, 201 (1997), and references therein.
23. U. Schneider, P. Lukenheimer, R. Brand, and A. Loidl, J. Non-Cryst. Solids **235-237**, 173 (1998).
24. See for instance the review by H. Sillescu, J. Non-Cryst. Solids, in press (1999).
25. K. Schmidt-Rohr and H.W. Spiess, Phys. Rev. Lett. **66**, 3020 (1991). A. Heuer, M. Wilhelm, H. Zimmermann, and H. W. Spiess, Phys. Rev. Lett.**75**, 2851 (1995). R. Böhmer, G. Hinze, G. Diezemann, B. Geil, and H. Sillescu, Europhys. Lett. **36**, 55 (1996). R. Böhmer, G. Diezemann, G. Hinze, and H. Sillescu, J. Chem. Phys. **108**, 890 (1998);
26. M. T. Cicerone and M. D. Ediger, J. Chem. Phys. **103**, 5684 (1995); C.-Y. Wang and M. D. Ediger, J. Phys. Chem. B , in press (1999).
27. B. Schiener, R. Böhmer, A Loidl, and R. V. Chamberlin, Science **274**, 752 (1996); B. Schiener, R. V. Chamberlin, G. Diezemann, and R. Böhmer, J. Chem. Phys. **107**, 7746 (1997).
28. F. Fujara, B. Geil, H. Sillescu and G. Fleischer, Z. Phys. B-Condensed Matter **88**, 195 (1992). I. Chang, F. Fujara, G. Heuberger, T. Mangel, and H. Sillescu, J. Non-Cryst. Solids **172-174**, 248 (1994).
29. M. T. Cicerone and M. D. Ediger, J. Chem. Phys. **104**, 7210 (1996). F. Blackburn, C.-Y. Wang, and M. D. Ediger, J. Phys. Chem. **100**, 18249 (1996).
30. M. T. Cicerone, F. R. Blackburn, and M. D. Ediger, J. Chem. Phys. **102**, 471 (1995). M. T. Cicerone and M. D. Ediger, J. Non-Cryst. Solids **235-238**, (1998).
31. U. Tracht, M. Wilhelm, A. Heuer, H. Feng, K. Schmidt-Rohr, and H.W. Spiess, Phys. Rev. Lett., (1998).
32. C. T. Moynihan and J. Schroeder, J. Non-Cryst. Solids **160**, 52 (1993).
33. G. Barut, P. Pissis, R. Pelster, and G. Nimtz, Phys. Rev. Lett.**80**, 3543 (1998).
34. M. L. Ferrer, C. Lawrence, B. G. Demirjian, D. Kivelson, C. Alba-Simionesco, and G. Tarjus, J. Chem. Phys. **109**, 8010 (1998), and references therein.
35. W. T. Laughlin and D. R. Uhlmann, J. Phys. Chem. **76**, 2317 (1972). M. Cukierman, J. W. Lane, and and D. R. Uhlmann, J. Chem. Phys. **59**, 3639 (1973). R. G. Greet and D. Turnbull, J. Chem. Phys. **46**, 1243 (1967).
36. C. Hansen, F. Stickel, T. Berger, R. Richert, and E. W. Fischer, J. Chem. Phys. **107**, 1086 (1997)

37. R. Rössler and A. P. Sokolov, Chem. Geol. **128**, 143 (1996); E. Rössler, K.-U. Hess, and V. N. Novikov, J. Non-Cryst. Solids **223**, 207 (1998).
38. W. Petry, E. Bartsch, F. Fujara, M. Kiebel, H. Sillescu, and B. Farago, Z. Phys.B **83**, 175 (1991).
39. See for instance the various reviews: J. L. Barrat and M. L. Klein, Annu. Rev. Phys. Chem. **42**, 23 (1991). J. P. Hansen, Physica A **201**, 138 (1993). W. Kob, J. Phys. Condens. Matter, in press (1999); as well as the special issue: *"Glasses and the glass transition. Challenges in Materials Theory and Simulation"*, S. Glotzer Ed. (Computational Materials Science **4**, 1995).
40. P. Sindzingre and M. L. Klein, J. Chem. Phys. **96**, 4681 (1992). P. H. Poole, F. Sciortino, U. Essmann, and H. E. Stanley, Nature **360**, 324 (1992). L. J. Lewis and G. Wahnstrom, Phys. Rev. E **50**, 3865 (1994). M. Wilson and P. A. Madden, Phys. Rev. Lett. **72**, 3033 (1994). J. Horbach, W. Kob, and K. Binder, Phil. Mag. B **77**, 297 (1998).
41. K. Binder, J. Baschnagel, W. Paul, H. P. Wittmann, and W. Wolfgardt, Computational Materials Science **4**, 309 (1995), and references therein.
42. T. Muranaka and Y. Hiwatari, Phys. Rev. E **51**, R2735 (1995); D. N. Perera and P. Harrowell, Phys. Rev. E **52**, 1694 (1995); R. Yamamoto and A. Onuki, J. Phys. Soc. Jpn. **66**, 2545 (1997). A. I. Mel'cuk, R. A. Ramos, H. Gould, W. Klein, and R. D. Mountain, Phys. Rev. Lett. **75**, 2522 (1995).
43. G. H. Fredrickson and H. C. Andersen, Phys. Rev. Lett. **53**, 1244 (1984). T. A. Weber and F. H. Stillinger, Phys. Rev. B **36**, 7043 (1986). F. Ritort, Phys. Rev. Lett. **75**, 1190 (1995). M. Foley and P. Harrowell, J. Chem. Phys. **98**, 5069 (1993). See also J. Jäckle, Prog. Theor. Phys. Suppl. **126**, 53 (1997).
44. K. L. Ngai and R. W. Rendell, in *Supercooled Liquids: Advances and Novel Applications*, J Fourkas, D. Kivelson, U. Mohanty, and K. A. Nelson Eds. (ACS Symposium Series 676, Washington, 1997), pg. 45.
45. J. T. Bendler and M. F. Schesinger, in *"Relaxations in Complex Systems"* K. L. Ngai and G. B. Wright Eds. (Office of Naval Research, Arlington, 1985), pg. 261.
46. M. H. Cohen and D. J. Turnbull, J. Chem. Phys. **31**, 1164 (1959); D. Turnbull and M. H. Cohen, *ibid.* **34**, 120 (1961), **52**, 3038 (1970).
47. A. K. Doolittle, J. Appl. Phys. **22**, 1471 (1951).
48. G. S. Grest and M. H. Cohen, Adv. Chem. Phys. **48**, 455 (1981).
49. S. F. Edwards and Th. Vilgis, Phys. Scripta T **13**, 7 (1986).
50. M. Goldstein, J. Phys. Chem. **77**, 667 (1973). J. P. Johari and E. Whalley, Faraday Symp. Chem. Soc. **6**, 23 (1973). S. A. Brawer, *Relaxation in Viscous Liquids and Glasses* (American Ceramic Society, Colombus, 1985). D. M. Colucci, G. B. McKenna, J. J. Filliben, A. Lee, D. B. Curliss, K. B. Bowman, and J. D. Russel, J. Polym. Sci. B: Polym. Phys. **351**, 1561 (1997).
51. W. Götze, in *"Liquids, Freezing, and the Glass Transition"*, J. P. Hansen, D. Levesque, and J. Zinn-Justin Eds. (North Holland, Amsterdam, 1991), pg. 287. W. Götze and L. Sjögren, Rep. Prog. Phys. **55**, 241 (1992). W. Götze, J. Phys.: Cond. Mat. **11**, A1 (1999).
52. S. P. Das and G. F. Mazenko, Phys. Rev. A **34**, 2265 (1986). B. Kim and G.F. Mazenko, Adv. Chem. Phys. **78**, 129 (1980).
53. K. Kawasaki, in *Phase Transitions and Critical Phenomena*, C. Domb and M. S. Green Eds; (Academic Press, New York, 1976), vol. 5a.
54. J. P. Bouchaud, L. Cugliandolo, J. Kurchan, and M. Mezard, Physica A **226**, 243 (1996). C. Z.-W. Liu and I. Oppenheim, Physica A **235**, 369 (1997).
55. For a sample of viewpoints and contributions, see: G. Li, W. M. Du, X.K. Chen, H. Z. Cummins, and N. J. Tao, Phys. Rev. A **45**, 3867 (1992); G. Li, W. M. Du, A. Sakai, and H. Z. Cummins, Phys. Rev. A **46**, 3343 (1992); H. Z. Cummins, W. M. Du,M. Fuchs, W. Götze, S. Hildebrand, G. Li, and N. J. Tao, Phys. Rev. E **47**, 4223 (1993). X. C. Zeng, D. Kivelson, and G. Tarjus, Phys. Rev. E **50**, 1711 (1994); P. K. Dixon, N. Menon, and S. R. Nagel, Phys. Rev. E **50**, 1717 (1994). A. P. Sokolov, W. Steffen, and E. Rossler, Phys. Rev. E **52**, 5105 (1995). F. Mezei and M. Russina, J. Phys.: Cond. Mat. **11**, A341 (1999). J. Gapinski, W. Steffen, A. Patkowski, A. P. Sokolov, A. Kisliuk, U. Buchenau, M. Russina, F. Mezei, and A. Schober, preprint (1998).
56. M. Goldstein, J. Chem. Phys. **51**, 3728 (1969).
57. J. H. Gibbs, in *Modern Aspects of the Vitreous State*, J. D. Mackenzie Ed. (Butterworths, London, 1960), vol. 2, pg. 152.
58. G. Adam and J. H. Gibbs, J. Chem. Phys. **43**, 139 (1965).
59. J. H. Gibbs and E. A. Di Marzio, J. Chem. Phys. **28**, 373 (1958).
60. M. Goldstein, J. Chem. Phys. **64**, 4767 (1976).
61. N. Menon and S. R. Nagel, Phys. Rev. Lett. **74**, 1230 (1995).
62. C. A. Angell, J. Phys. Chem. Solids **49**, 863 (1988).
63. F. H. Stillinger and T. A. Weber, Science **225**, 983 (1983); F. H. Stillinger, *ibid.* **267**, 1935 (1995); S. Sastry, P. G. Debenedetti, and F. H. Stillinger, Nature **393**, 554 (1998). See also R. J. Speedy and P. G. Debenedetti, Mol. Phys. **88**, 1293 (1996). T. Keyes, J. Chem. Phys. **101**, 5081 (1992).
64. F. H. Stillinger, J. Chem. Phys. **88**, 7818 (1988).
65. R. G. Palmer, Adv. Phys. **31**, 669 (1982).
66. C. A. Angell, Nature **393**, 521 (1998).
67. U. Zürcher and T. Keyes, in *Supercooled Liquids: Advances and Novel Applications*, J Fourkas et al. Eds. (ACS Symposium Series 676, Washington, 1997), pg. 82.
68. S. A. Brawer, J. Chem. Phys. **81**, 954 (1984); H. Bässler, Phys. Rev. Lett. **58**, 767 (1987); J. C. Dyre, Phys. Rev. Lett. **58**, 792 (1987); U. Mohanty, I. Oppenheim, and C. H. Taubes, Science **266**, 425 (1994); G. Diezemann, J. Chem. Phys. **107** (1997).
69. See for instance: K. Binder and A. P. Young, Rev. Mod. Phys. **58**, 801 (1986); *Spin Glass Theory and Beyond*, M. Mezard, G. Parisi, and M. A. Virasoro Eds. (World Scientific, Singapore, 1987); *Spin Glasses and Random Fields*, A. P. Young Ed. (World Scientific, Singapore, 1998).
70. T. R. Kirkpatrick and P. G. Wolynes, Phys. Rev. A **35**, 3072 (1987); Phys. Rev. B **36**, 8552 (1987); T. R. Kirk-

Kirkpatrick and D. Thirumalai, Phys. Rev. Lett. **58**, 2091 (1987); J. Phys. A **22**, L149 (1989).

71. T. R. Kirkpatrick, D. Thirumalai, and P. G. Wolynes, Phys. Rev. A **40**, 1045 (1989).72. M. Nicodemi and A. Coniglio, J. Phys. A **30**, L187 (1997); A. Coniglio, A. de Candia, and M. Nicodemi, J. Phys.: Cond. Mat. **11**, A167 (1999).73. G. Parisi, in *Supercooled Liquids: Advances and Novel Applications*, J Fourkas, D. Kivelson, U. Mohanty, and K. A. Nelson Eds. (ACS Symposium Series 676, Washington, 1997), pg. 110. R. Monasson, Phys. Rev. Lett. **75**, 2847 (1995). S. Franz and G. Parisi, J. Physique (Paris) I **5**, 1401 (1995). M. Mezard and G. Parisi, J. Phys. A: Math. Gen. **29**, 6515 (1996); cond-mat/9812180 (1998).74. See the articles collected in *Spin Glasses and Random Fields*, A. P. Young Ed. (World Scientific, Singapore, 1998).75. M. Kleman and J. F. Sadoc, J. Phys. Lett. (Paris) **40**, L569 (1979). J. P. Sethna, Phys. Rev. Lett. **51**, 2198 (1983). S. Sachdev and D. R. Nelson, Phys. Rev. Lett. **53**, 1947 (1984); Phys. Rev. B **32**, 1480 (1985). S. Sachdev, Phys. Rev. B **33**, 6395 (1986).76. D. R. Nelson and F. Spaepen, Solid State Phys. **42**, 1 (1989).77. J. P. Sethna, J. D. Shore, and M. Huang, Phys. Rev. B **44**, 4943 (1991).78. F. H. Stillinger, J. Chem. Phys. **89**, 6461 (1988); F.H. Stillinger and J. A. Hodgdon, Phys. Rev. E **50**, 2064 (1994).79. D. Kivelson, S. A. Kivelson, X.-L. Zhao, Z. Nussinov, and G. Tarjus, Physica A **219**, 27 (1995); G. Tarjus and D. Kivelson, J. Chem. Phys. **103**, 3071 (1995); D. Kivelson and G. Tarjus, Phil. Mag. B **77**, 245 (1998).80. F.C. Frank, Proc. Royal Soc. London **215A**, 43 (1952).81. K. J. Standburg, Rev. Mod. Phys. **60**,161 (1988).82. E. A. Jagla, Phys. Rev. E **58**, 4701 (1998).83. L. Gu and B. Chakraborty, Mat. Res. Symp. **455**, 229 (1997). B. Kim and S. J. Lee, Phys. Rev. Lett. **79**, 3709 (1997). S. J. Lee and B. Kim, cond-mat/9901077.84. L. Chayes, V. J. Emery, S.A. Kivelson, Z. Nussinov, and G. Tarjus, Physica A **225**, 129 (1996).85. W. Klein, H. Gould, R. A. Ramos, I. Clejan, and A. I. Mel'cuk, Physica A **205**, 738 (1994).

Jamming in colloidal dispersions: hard-sphere suspensions, emulsions and foams

Douglas J. Durian
Department of Physics and Astronomy
University of California, Los Angeles, CA 90095

Andrea J. Liu
Department of Chemistry and Biochemistry
University of California, Los Angeles, CA 90095

Colloidal objects such as solid particles or droplets of liquid or gas can be dispersed in a liquid. The resulting suspension properties have long been studied and exploited for applications. Here we consider jamming behavior, not as a function of temperature as is normally done for molecular liquids, but rather as a function of particle concentration and of shear stress. Hard sphere colloidal particles, at low volume fractions, are free to diffuse and explore configuration space using thermal energy. As the volume fraction increases, the space available for motion decreases and the system becomes jammed. Soft droplets of gas (liquid) tightly packed together in a foam (emulsion) are jammed, too. By application of external shear forces, the droplets can rearrange and explore configuration space and hence unjam. In this article we review progress in understanding the nature of these jamming transitions, as well as the jammed states themselves. Many observations can be summarized by phase diagrams of jamming behavior as a function of particle concentration, applied force or temperature.

I. INTRODUCTION

Jamming behavior in colloidal dispersions is rich because one can approach the jammed state by different routes, depending on the nature of the suspended particles. We will consider suspensions of hard spheres and of deformable spheres (emulsions and foams). Small colloidal particles of silica or plastic, for instance, are widely available and find many uses [1,2]. Much of colloid science has focused on the interaction between particles, with a view towards either preventing or enhancing aggregation or gravitational sedimentation. Sedimentation can be prevented by thermal energy for small enough spheres, depending on the density difference with the suspending liquid. Aggregation can be prevented, even in the presence of long-range Van der Waals attractions, by engineering a suitable short-range repulsion. Examples include surface molecules that dissociate and leave a net charge (the so-called double-layer force), or else grafted polymers that resist compression entropically (steric stabilization). Thus, it is possible to make systems of particles that move with thermal energy and interact with each other mainly through excluded volume interactions, *i.e.* that behave as ideal hard spheres. As the volume fraction of particles is raised toward random close packing, $\Phi_c \approx 0.64$ in three dimensions [3] (largely independent of polydispersity), particle motion becomes increasingly restricted and the suspension viscosity increases. Eventually, all motion is arrested, and the viscosity is infinite; the system is jammed.

This and other facets of jamming can be studied with suspensions of soft deformable spheres, such as foams and emulsions. An emulsion is a suspension of liquid drops in an immiscible liquid; for example, oil droplets suspended in water [4]. A foam is a suspension of gas bubbles in liquid [5,6]. In both systems, surfactants are typically used to stabilize the interfaces against rupture. At low packing fractions, the droplets in an emulsion and the bubbles in a foam are spherical, but if the packing fraction exceeds Φ_c, then the droplets or bubbles deform from the spherical shape. Above Φ_c, emulsions and foams are jammed; the droplets or bubbles are packed into a specific disordered configuration, and thermal energy is insufficient to cause rearrangements. As a result, these systems can support some shear elastically, like a solid. When subjected to shear forces, the closely-packed bubbles or droplets distort in shape. The distortion increases the surface area, and hence the interfacial energy of a bubble; this is the origin of the elastic restoring force. Once the shear stress exceeds the yield stress, however, the bubbles can hop around each other, allowing the foam or emulsion to flow. The hopping process will be referred to as a rearrangement event.

In this article, we will discuss two different ways in which one can approach the jammed state in a colloidal dispersion, as well as the properties of the jammed state itself. In the first way, one starts with a dilute colloidal dispersion and gradually increases the packing fraction Φ, giving the system enough time to find a local minimum but not the global minimum, which, for a monodisperse system at sufficiently high Φ, is a crystalline packing. As Φ increases towards Φ_c, the viscosity of the disordered suspension increases rapidly; this is a signature of jamming. This route to the jammed state is discussed in Section II and is typically studied using hard-sphere suspensions.

Another way to approach the jammed state is to start with a sheared dispersion above Φ_c. It is impossible to achieve packing fractions above Φ_c in a hard-sphere suspension without partially ordering the system. However, because emulsion droplets and foam bubbles are deformable, they can easily be packed above Φ_c. In this regime, the suspension has a shear modulus whose magnitude is set by the energy per unit volume required to deform the colloids; for an emulsion or foam, this characteristic energy density is the Laplace pressure, the interfacial tension divided by the characteristic size of a droplet or bubble. The system has a shear modulus because the tightly-packed colloids deform under shear, leading to an increase in the free energy. However, if a shear stress larger than the yield stress σ_y is applied, the colloids will change their relative positions to avoid deforming too far. These rearrangements cause the suspension to flow at high applied shear stresses, and give rise to a well-defined viscosity. Now consider what happens as the applied shear stress (or the strain rate) is decreased. The viscosity increases, and eventually diverges as the shear stress approaches the yield stress from above. This second route to the jammed state is discussed in Section III.

II. APPROACHING JAMMING BY INCREASING THE PACKING DENSITY: THE COLLOIDAL GLASS TRANSITION

The jamming that occurs as one increases the packing fraction towards Φ_c has been studied most extensively in hard-sphere suspensions. A monodisperse suspension of spheres crystallizes at $\Phi = 0.49$ [7]. Above this packing fraction, the system can be prepared in a metastable fluid state and measurements can be made on time scales short compared to the crystallization time. As Φ increases in the metastable fluid state, several properties show dramatic changes similar to those observed in the glass transition of a supercooled liquid.

Most work on the hard-sphere glass transition has focused on two main questions. First, is there true structural arrest at a packing fraction below the random-close-packing value, Φ_c? It is clear that structural arrest must occur by Φ_c, but it would interesting if the onset of structural arrest were below Φ_c. Second, to what extent is the glass transition in hard-sphere systems similar to the glass transition in atomic or molecular systems (supercooled liquids)? In one important respect, the two systems are very different; in hard-sphere suspensions, the glass transition is controlled by free volume, or entropy, while in supercooled liquids, the glass transition is controlled by energy barriers. Nonetheless, as we shall see, the two systems appear to share many common features. This may be due to the fact that both systems are jamming (*i.e.* spontaneously restricting themselves to a small part of phase space).

A. Where is the colloidal glass transition?

A very detailed set of experiments on the hard-sphere glass transition were carried out using light scattering techniques [8–14]. The decay of thermal fluctuations in Fourier components of the number density, $n(q)$, was measured by dynamic light scattering in terms of the intermediate scattering function $f(q,t) = \langle n(q,t)n(-q,0)\rangle / \langle n(q,0)n(-q,0)\rangle$. The decay of this function exhibits two distinct, well-separated relaxation times, the fast β-relaxation time and a slower α-relaxation time (see the Overview on the glass transition for a discussion of these two time scales in the context of supercooled liquids). At low packing fractions, both relaxation times are short and $f(q,t)$ decays rapidly to zero with time t. As Φ increases, however, the intermediate scattering function develops a plateau that extends to longer and longer times; in other words, the α-relaxation time increases more than the β-relaxation time. At a packing fraction of $\Phi \approx 0.58$ [9,10], the α-relaxation time exceeds the laboratory time scale of 1000 s. At the same packing fraction, the mechanism of crystallization changes; at lower packing fractions, small crystallites appear by homogeneous nucleation, but for $\Phi \geq 0.58$, the samples develop larger, asymmetric crystallites whose shape and orientation depend on the shear history of the sample [8,10]. Finally, $f(q,t)$ can be fitted to ideal mode-coupling theory [15,16] to yield a β-relaxation time with a maximum and an α-relaxation time appears to diverge at $\Phi \approx 0.58$ (see Fig. 3 of Ref. [12]). Based on the coincidence of these three observations (the α-relaxation time exceeding 1000 s, the change in nucleation mechanism and the fits to mode-coupling theory), the colloidal glass transition was identified occurring at $\Phi \approx 0.57$ or 0.58 [10,14].

Similar dynamic light scattering experiments have been repeated on emulsions [17]. A marked increase is also observed in the ratio of the α-relaxation time to the β-relaxation time between $\Phi = 0.57$ and $\Phi = 0.58$, and the glass transition has also been placed around $\Phi = 0.58$ based on fits to mode coupling theory. However, instead of fitting to ideal mode-coupling theory, it was necessary to fit to extended mode-coupling theory (see Ref. [27] for a short review). In the ideal theory, the α-relaxation time diverges at the ideal glass transition ($\Phi = 0.58$). The extended theory, however, allows for finite α-relaxation times even beyond the ideal glass transition. For emulsions over the range $0.54 \leq \Phi \leq 0.62$, the intermediate scattering function always decays at long times, contrary to the prediction of ideal mode-coupling theory, but good fits to the data can be obtained using extended mode-coupling theory [17]. The relaxation observed at

$\Phi > 0.58$ was attributed to to mechanisms involving deformation of droplets; perhaps thermal noise can allow deformable droplets to rearrange, leading to activated hopping [17].

Rheological measurements also seem to be well-described by ideal mode-coupling theory. The complex dynamic shear modulus for concentrated hard-sphere suspensions has been measured [18] for packing fractions up to $\Phi = 0.56$. The storage modulus starts to develop a plateau at high frequency at $G'(\omega) \approx kT/a^3$, where a is the particle radius. When the data are fit to ideal mode-coupling theory [15] there appears to be a glass transition at $\Phi \approx 0.58$ [18]. A similar plateau has been observed in experiments on supercooled liquids [19]; there, the plateau value is $G'(\omega) \approx kT/V$, where V is the molecular volume.

The suspension viscosity also shows signs of an approach to a glass transition [20–23]. The most recent low-shear viscosity measurements [24] cover the concentration range from $0 < \Phi \leq 0.56$; over this range, the viscosity increases by a factor of roughly 3000. The analysis shows that the position of the glass transition depends on the functional form used for the fit. If the data in the range $0.48 \leq \Phi \leq 0.56$ are fit to a power law that diverges at $\Phi_g = 0.58$, the fit is reasonable with an exponent of 2.84 ± 0.08, in good agreement with the prediction of mode coupling theory (between 2 and 3 [15]). However, the Vogel-Fulcher form provides a much better fit, with a predicted divergence at $\Phi_g = 0.64$.

In summary, mode-coupling analyses of three different experiments, namely the decay of number density fluctuations, the complex dynamic shear modulus and the low-shear viscosity, all suggest that structural arrest occurs at $\Phi = 0.58$. Moreover, the nature of crystallites that nucleate in the suspension changes at around $\Phi = 0.58$. On the other hand, when the viscosity measurements are analyzed in terms of free-volume theory with a Vogel-Fulcher form, the predicted glass transition is at $\Phi = 0.64$, consistent with the random-close-packing density.

In spite of the fact that idealized mode-coupling theory appears to describe colloidal systems quite well, there is some experimental evidence that there is no true structural arrest at $\Phi = 0.58$, contrary to the prediction of the theory. Recent extensive dynamic light scattering measurements over an extended time range [14] show that the intermediate scattering function decays at long times even at $\Phi = 0.583$, above the colloidal glass transition. It is argued that the system is in fact nonergodic at $\Phi = 0.583$, even though the relaxation time is finite, because the relaxation time depends on how long the system was equilibrated before the measurement was made (the waiting time). However, the longest waiting time is only about an order of magnitude longer than the relaxation time, so it is possible that the waiting time is simply not long enough, and that the α-relaxation time is in fact finite.

Finally, optical microscopy studies [25,26] show that the long-time self-diffusion coefficient appears to be nonzero, even at $\Phi = 0.60$, well above the colloidal glass transition, indicating that the system is in fact still ergodic there.

Given this uncertainty as to whether true structural arrest occurs below Φ_c, it makes sense simply to adopt a practical dynamical definition of where the glass transition is. This must be based on a laboratory time scale, such as 1000 s. We will therefore refer to systems whose relaxation times exceed 1000 s as jammed.

B. Comparison to glass transition of supercooled liquids

The phenomenology of the colloidal glass transition is strikingly similar to that of glass-forming liquids, despite the fact that the underlying physics controlling the two glass transitions (free volume vs. energy barriers) appears to be different.

In comparing glass transitions of hard spheres to atomic and molecular liquids, it is important to keep in mind that the accessible dynamic range is much larger for supercooled liquids. For supercooled liquids, the glass transition temperature is often defined as the temperature at which the relaxation time exceeds 1000 s. Thus, the definition of the colloidal glass transition [13] appears consistent with the definition used for supercooled liquids. However, the relaxation time scales for colloids and molecular liquids far from the glass transition differ by many orders of magnitude; the characteristic relaxation time for a molecular liquid is of order 10^{-11} s, while the same time scale for a colloid in a dilute suspension is roughly 1 ms. In order to have a consistent definition of the glass transition in both systems, the *ratio* of relaxation times near the transition to the characteristic relaxation times far away from the transition should be comparable. If this definition were employed, then the colloidal glass transition would be identified as the packing fraction at which the relaxation time exceeds 10^{11} s. This is, of course, impossible to measure in practice. However, we note that this implies that the colloidal glass transition is much further from the onset of true structural arrest (wherever it may be) than the glass transition measured for supercooled liquids.

It has been argued [13] that activated hopping processes should be suppressed in colloidal suspensions because the particles are overdamped by the solvent. As a result, ideal mode-coupling theory might describe the colloidal glass transition well, even though it clearly fails to describe supercooled liquids [27]. However, the recent evidence that the relaxation times are finite even above the colloidal glass transition, discussed in Section II.A above, suggests that ideal mode-coupling theory also fails to describe the colloidal glass transition. Because the dynamic range is necessarily more limited for colloids, it is difficult to resolve this question.

One important feature of the glass transition in supercooled liquids is that no significant structural signature has been found. All measured structural quantities do not appear to change much in spite of the dramatic increase in the relaxation time near the transition [27]. In particular, there is no discernible diverging length scale that might account for the diverging time scale. The same is true of colloidal systems near the glass transition [10,11]. The static structure factor shows no significant change with packing fraction and shows that there is only short-ranged order.

Finally, there is now a body of evidence for spatial heterogeneities in the dynamics in supercooled liquids [27]. There is some evidence for kinetic heterogeneities in colloidal suspensions as well. Optical microscopy studies [25,26] show that particles can move relatively quickly for short periods of time when they are not caged by neighboring particles, but can then be trapped for long times inside cages. Deviations from Gaussian behavior in the squared-displacement distribution are also observed (see Fig. 7 of Ref. [25]). Such deviations have been shown to be related to kinetic heterogeneities in simulations of bidisperse Lennard-Jones systems [29].

C. Crystallization kinetics: a puzzle

One aspect of the colloidal glass transition is that samples can crystallize completely up to approximately $\Phi = 0.61$, even though the colloidal glass transition has been identified to be at $\Phi = 0.58$. It is difficult to imagine how a disordered suspension could crystallize completely without some preceding structural relaxation. If the system can overcome the barriers to crystallization, it is not clear why it cannot overcome barriers to structural relaxation as well.

There are two more observations that make the crystallization kinetics even more puzzling. First, a glassy sample at $\Phi = 0.619$, which did not crystallize at all on earth over a period of more than a year, was found to crystallize in microgravity in a matter of days [30]. Second, density-matched samples at $\Phi = 0.631 \pm 0.004$ were observed to crystallize in a matter of days [31]. The behavior of a sample that was centrifuged to mimic the effects of gravity was similar [32]. The dependence on the magnitude of gravity (or an applied stress from centrifugation) indicates that packing fraction may not be the only important variable in determining the onset of jamming in colloidal suspensions. Here we suggest that the ratio of the thermal energy, kT, to the gravitational potential energy to raise a particle by its diameter, $\delta\rho g r^4$ (where $\delta\rho$ is the density difference between the particles and solvent and r is the characteristic particle size), may also be an important parameter. For the particles used in the microgravity experiment, this ratio is $kT/\delta\rho g r^4 \approx 60$ on earth. The effect of decreasing the particle size or decreasing gravity may be similar to the effect of raising the thermal energy. This would shift the system further from the jamming transition, which could explain the structural relaxation that apparently occurs when the system crystallizes.

III. JAMMING ABOVE Φ_C

In order to study the properties of the jammed state at $\Phi > \Phi_c$, we turn to systems such as foams or emulsions, which are simply dispersions of easily-deformable bubbles of a gas or droplets of a liquid [4–6]. At low packing fractions, the droplets or bubbles do not touch and are spherical in order to minimize their individual surface areas. In this regime, the materials behave very similarly to the hard sphere dispersions discussed above. Above Φ_c, however, the droplets or bubbles form flattened facets at the point of contact. The increase in interfacial area costs surface-tension energy, and gives rise to a repulsive force between neighbors. Bubbles throughout the entire system accordingly adjust their positions and shapes in order to minimize total interfacial area. The resulting bubble shapes can range from spherical to nearly-polyhedral as Φ ranges from Φ_c to 1. It is important to note that the resulting bubble configuration is perfectly static, frozen in some local minimum close to the initial conditions. The typical energy barrier to rearrange bubble configurations is of order $\Sigma(\gamma_y r)^2$. where Σ is the interfacial tension, γ_y is the yield strain (of order a few percent for three-dimensional foams or emulsions) and r is the typical bubble size. The thermal energy is roughly $10^5 - 10^7$ times smaller for typical bubble sizes (microns or larger). As a result, the bubbles cannot rearrange and explore phase space in search of a global minimum. This is a classic hallmark of the jammed state.

From a jamming point of view, emulsions and foams are almost identical. Two obvious differences are the differences in the compressibilities and viscosities of liquid droplets and gas bubbles. However, as long as the energy cost to deform a bubble is small compared to the cost to compress a bubble, the bubble will deform at constant volume. The cost to deform a bubble is measured by the Laplace pressure, namely Σ/r, the interfacial tension divided by the characteristic bubble size. For gas bubbles bigger than approximately 1μm, the Laplace pressure is small compared to atmospheric pressure and the bubbles are effectively incompressible. The viscosity difference between gas bubbles and liquid droplets is probably also not important compared to the properties of the surfactant monolayer at the interface, which largely determines the boundary conditions for flow inside and outside the bubbles.

The main differences between emulsions and foams are practical ones. First, emulsion bubbles can be made smaller and can be density matched with the suspending liquid, so that thermal motion could be more important at volume fractions near Φ_c. Second, both systems are not in equilibrium; given time, they phase separate at a macroscopic scale into two phases. However, the kinetics of phase separation are much slower for emulsions than for foams, and are negligible on the time scales for rheological measurements. In emulsions, the droplet size distribution is therefore independent of time, while in foams, it coarsens with time. On one hand, it is an advantage to work with a system with a fixed size distribution; in foams, the microscopic dynamics are more complicated because the coarsening process itself leads to rearrangement events. However, the advantage of the coarsening process is that it leads to a size distribution that scales with time, and is therefore both polydisperse and reproducible. Since emulsions do not coarsen significantly, the only reproducible size distribution is a monodisperse one. The disadvantage of a monodisperse distribution is that the emulsion will crystallize under shear; only a sufficiently polydisperse system will remain disordered under shear. In order to study jamming, it is crucial that the system remains disordered, since a crystalline packing exhibits completely different dynamics. For example, when a periodic emulsion is subjected to shear, the rearrangement events occur simultaneously for every droplet at periodic intervals in time. In a disordered packing, on the other hand, the rearrangement events are localized and intermittent. Because the jamming properties of emulsion and foams are so similar, we will refer to both liquid droplets and gas bubbles as "bubbles."

Below we discuss progress in understanding the jammed state of foams and emulsions, as well as the unjammed, sheared state. To begin, we briefly note the general approaches that have been employed. Experimentally, efforts have focused on measurements of macroscopic rheology (relations between time-dependent stress and strain) [33], and measurements of the underlying microscopic bubble motion. Since the bubble interfaces strongly scatter light, optical imaging is restricted to surface bubbles or index-matched emulsions. Fortunately, a wide variety of information can be extracted from light that has been multiply-scattered by the material. The technique of diffusing-wave spectroscopy (DWS) was developed and applied originally to dilute colloidal suspensions of hard spheres [34]. It quantifies Brownian motion via the mean-squared displacement of the spheres as a function of delay time. DWS has since been applied to the study of bubble rearrangments in foams caused not by thermal motion, but rather by the coarsening process (diffusion of gas from small to larger bubbles) [35,36]. DWS can also give information on the size and rate of rearrangements induced by steady [37,38] or oscillatory [39,40] shear. To obtain such information directly, one may alternatively view quasi-two dimensional systems such as foam-like phases within a monolayer of lipid at an air-water interface [41]. As an important variation, DWS can also gives rheological information based on thermal motion [42].

Theoretically, since jammed foams and emulsions are disordered, the key tool continues to be computer simulations. One approach is to minimize the area of the films in between bubbles by brute force; this was pioneered for two-dimensional dry foams $\Phi = 1$ [43], and was only recently extended to wetter foams [44,45] and to three-dimensional systems [46]. An equivalent approach is to employ a large-Q Potts model where bubbles are represented by domains of like spin [47]. These faithfully model microscopic structure and surface tension forces, but cannot account for viscous forces arising from flow of liquid between bubbles [48]. Dissipative dynamics in a random foam were first modeled for a strictly two-dimensional dry system by considering the forces acting on the vertex at which three films meet [49]. The only model capable of treating dynamics in three-dimensional wet foams is the bubble model of Durian [50,51], in which the foam is constructed from spherical bubbles that can overlap. Bubble motion is then set by the competition between two pairwise-additive interactions between neighboring bubbles, a harmonic repulsion proportional to overlap that accounts for bubble deformation and surface tension, and a force proportional to the velocity difference between neighboring bubbles that accounts for the viscous drag. Though based on a caricature of the film-scale behavior, this captures the essential features of the collective bubble-scale and packing behavior of real foams [50–54].

A. The jammed state

Without application of external forces, jammed systems of foams and emulsions are basically static (except for tiny capillary-wave like excitation of the interfaces, and slow evolution by gas diffusion or drainage). The simplest probe of the collective response of the jammed state is to observe the response to oscillatory shear. If the perturbation is sufficiently small, then the bubbles distort without altering their packing configuration. This is resisted by a restoring force due to the increase in interfacial area, which grows in proportion to the strain and may thus be characterized by a shear modulus, G. If the material were not jammed, the bubbles could move around one another and rearrange in order to relax stress. Thus, a nonzero value of G indicates jamming.

Just as the approach to jamming in hard spheres was studied by the divergence of viscosity as $\Phi \rightarrow \Phi_c^-$, the degree of jamming in foams and emulsions may be studied by the way G vanishes as $\Phi \rightarrow \Phi_c^+$. It has long been known that the scale for G is set by the ratio Σ/r

of surface tension to bubble size, and that G decreases with increasing liquid content [55,56]. However, the form of G vs Φ has remained controversial until experiments with monodisperse emulsions eliminated irreproducibilities [57]. The behavior is roughly $G \propto \Phi(\Phi - \Phi_c)$, which has been explained using simulations of the bubble model augmented with anharmonicities in the bubble-bubble repulsions [52]. Identical behavior is found in polydisperse foams [58], which can now be made reproducibly as well [59].

One can also study the response of the jammed state to small-amplitude oscillatory shear as a function of frequency ω. This response is characterized by the complex dynamic shear modulus, $G^*(\omega)$, defined as the ratio of the oscillatory stress $\sigma(t)$ to the imposed oscillatory strain $\gamma(t) = \gamma_o \exp(i\omega t)$ [60]. Results are typically displayed separately for the real and imaginary parts, $G'(\omega)$ and $G''(\omega)$, which characterize energy storage and dissipation, respectively. In this notation, the shear modulus is $G = \lim_{\omega \to 0} G'(\omega)$. Typical solids follow the so-called Kelvin model, $G^*(\omega) = G + i\eta\omega$, at low frequencies. Jammed foams and emulsions, however, have strikingly different dissipation behavior. Instead of being proportional to ω, as required by causality, $G''(\omega)$ is found to be roughly constant at the lowest accessible frequencies. This has been seen in foams [56,61], emulsions [62], and other soft glassy materials as well [63,64]. An extremely slow activated process, coarsening, or some other aging process could lead to stress relaxation at very long times and hence be responsible for the observed behavior in $G''(\omega)$ [64]. We note that the behavior of $G''(\omega)$ implies that $G'(\omega)$ might decrease at very low ω, below the range accessible to experiments. Thus, it is possible that the system does not actually have a nonzero shear modulus or yield stress at very long time scales. In this sense, a foam may only be jammed at sufficiently short time scales, just as a molecular liquid may only be jammed on experimentally-accessible time scales. Note that in gas-liquid foams, the stress does indeed relax to zero following a step strain, due to coarsening, but with a relaxation time that is unexpectedly much longer than the time between coarsening-induced rearrangements [61]. In emulsions, it is more surprising that the shear modulus might vanish at long time scales since coarsening is essentially suppressed.

There is another feature in $G^*(\omega)$ that seems to be characteristic of the jammed state. Namely, the leading dependence of $G'(\omega)$ and $G''(\omega)$ at higher frequencies is $\sqrt{(\omega)}$. This is found in both emulsions [62] and foams [65,61], and has been explained in terms of non-affine motion of the bubbles in response to the imposed shear. Since the bubble-packing is disordered, stresses cannot be distributed uniformly throughout the medium. Rather, there will be regions where tightly-packed bubbles distort to support stress, and other "weak" regions where bubbles shift their positions to avoid distortion. This inhomogeneous response to an applied stress is reminiscent of force chains in granular media, though here the stresses are presumably more evenly distributed because the spheres are highly deformable [66]. Assuming that weak regions have locally anisotropic elastic behavior, but that the sample as a whole is isotropic, yields a $\sqrt{i\omega}$ contribution to $G^*(\omega)$ [62]. At much higher frequencies, of order $\Sigma/\eta r$, where η is the viscosity of the continuous phase, the leading dependence is $G^*(\omega) \sim \omega$ as viscous forces become dominant and uniform throughout the medium.

B. Jamming vs. oscillatory shear flow

So far the discussion has concerned linear mechanical properties of the jammed state, where bubbles deform slightly in response to shear but do not alter their nearest neighbors. If the amplitude of the imposed oscillatory shear strain is increased, the response can become nonlinear. First, and least interesting, the deformation at fixed packing configuration may become anharmonic just like a spring stretched too far. Second, and more important to jamming, rearrangements may be induced in the packing configuration. The effect of both is to cause stress to grow less than linearly with strain. Experimentally, the linear regime is only attained at asymptotically small strains, so the strain at which nonlinearities become important cannot be precisely defined. For very large strains, the stress appear to grows as strain raised to a power less than one; the intersection of this and the linear behavior, on a log-log scale, is a convenient way to define the "yield strain", γ_y [67]. For very dry emulsions [67] and foams [58], the resulting yield strain is about twenty percent; near Φ_c it drops to a few percent. These numbers coincide well with multiple-light scattering measurements of changes in the microscopic bubble dynamics caused by application of a step-strain [37]. One can also define the yield stress, σ_y, as the minimum applied shear stress required for the foam to induce rearrangements. The yield stress is approximately given by the product of the shear modulus and the yield strain. Thus, it also vanishes as the packing fraction is decreased towards Φ_c from above [67].

What is the nature of the rearrangements induced by large-amplitude shear? This is one of the central questions, because shear "unjams" the material by allowing bubbles to move around each other and explore phase space - just as thermal energy allows molecules to move and explore phase space in a liquid. Multiple-light scattering techniques have used to observe motion in foams under steady shear [37,68,38], discused further in the next sub-section, and are applicable to oscillatory shear as well [39,40]. If the bubble motion is periodic, the DWS correlation function will also be periodic, decaying from one to zero in the time it takes for bubbles to move about a wavelength, and exhibiting an identical peak or "echo" at multiples of the oscillation period. This indeed is found for emulsions strained at small amplitudes, where the packing configuration is fixed [39]. At larger amplitudes, the echo height can be reduced due to bubble rearrangements induced by the applied shear. This gives a measure of yielding based on microscopic information, and is in good agreement with macroscopic mechanical measurements. This also happens for foams, where successive peaks already decay in height due to bubble rearrangements induced by coarsening of the size distribution via diffusion of gas from small to large bubbles [40]. A very interesting feature observed in emulsions, where coarsening is absent, is that the only decay in echo height is from time zero to the first echo; all subsequent echos have the same height [39]. The suggested explanation is that some fraction of the bubbles are trapped in specific packing configurations, but the rest of the bubbles are liquefied ("unjammed") by application of shear, and follow different random trajectories in each oscillation period. This is strongly reminicent of kinetic heterogeneities observed in supercooled liquids. Here, as the strain amplitude increases, the fraction of unjammed bubbles increases and the echo height accordingly decreases.

C. Jamming vs. steay shear flow

Large scale flows of foams and emulsions can be smooth and homogeneous, as is familiar from common experience with shaving cream and mayonnaise. The microscopic basis for such macroscopic deformations is the cumulative effect of a series of local bubble rearrangements from one packing configuration to another. With time, neighboring bubbles eventually wander far apart. The faster the shear rate, the more frequent and smooth the rearrangements, and hence the more liquid-like the material. In this final subsection, we will consider the approach to the jammed state as a function of imposed shear rate.

The traditional macroscopic characterization of foam or emulsion flow is by measurement of stress vs strain rate. This is not straightforward to accomplish because care must be taken to prevent slip at the walls or fracture within the sample. Data at low shear rates are typically analyzed using the Herschel-Bulkeley form:

$$\sigma = \sigma_y + \sigma_y(\dot{\gamma}T)^x \tag{1}$$

where σ_y is the yield stress, below which there is no flow, and T is a time scale and $x \leq 1$. The value of x appears to depend on packing fraction [67] but satisfies $x \leq 1/2$; the value $x = 1$ corresponds to a Bingham plastic [33]. According to the bubble model, $x \approx 1/3$ [53]. At very high shear rates (barring breakup or rupture of bubbles), the system eventually behaves like an ordinary viscous liquid, with $\sigma \approx \eta_\infty \dot{\gamma}$. The viscosity of the material depends on shear rate through the relation $\eta = \sigma/\dot{\gamma}$, and increases with decreasing shear rate. As the shear rate decreases towards zero, the viscosity diverges approximately as $\sigma_y/\dot{\gamma}$ and the system jams.

What is the nature of the bubble rearrangement dynamics during shear flow, and how does it depend on strain rate? At high enough strain rates, where $\dot{\gamma}T \gg 1$, successive rearrangements can merge together and produce flow that is smooth and laminar all the way down to the bubble scale. This is seen experimentally by the shape of the DWS correlation function [68]. It can also be seen in movies generated by simulations of the bubble model [54]. In this regime, the duration τ_d of the rearrangement events is longer than the time between events, which is given by the ratio $\gamma_y/\dot{\gamma}$ of yield strain to strain rate [68]. Thus, for strain rates greater than $\dot{\gamma}_x = \gamma_y/\tau_d$, bubbles are always rearranging and viscous forces dominate. According to this picture, the time T that appears in the Herschel-Bulkeley relation scales as τ_d/γ_y. If the system behaves as a Bingham plastic ($x = 1$ in Eq. 1), then the viscosity in this regime should be a constant, with $\mu_p = \sigma_y/\dot{\gamma}_x$, which is approximately equal to $G\tau_d$. The importance of the duration τ_d of rearrangements for foam rheology was first seen through a change in the morphology of viscous fingering patterns at high strain rates [69]. A more precise means of characterizing the crossover macroscopically is by the linear response to small perturbations [61]. Indeed, the relaxation of transient stress following a step-strain superposed upon steady shear flow shows liquid-like behavior for strain rates greater than about the same $\dot{\gamma}_x = \gamma_y/\tau_d$. Simulations of the bubble model give an even more complete picture of the high shear rate regime [71]. At high strain rates, $\dot{\gamma} \gg \dot{\gamma}_x$, the distribution of bubble velocities is Gaussian, the velocity autocorrelation function decays exponentially with time and bubbles wander diffusively around the average flow. All this is consistent with the behavior of an ordinary liquid.

At lower strain rates, the behavior is markedly different. For $\dot{\gamma} \ll \dot{\gamma}_x$, the duration of a rearrangement event is short compared to the time between events. As a result, the flow is characterized by intermittent rearrangements of bubbles. At very low shear, where the viscosity is approximately $\sigma_y/\dot{\gamma}$, viscous stresses are unimportant and surface tension forces sum to zero for all bubbles (except those undergoing rearrangement). This is called the quasistatic regime. Here, the rearrangements are sudden stick/slip-like events that briefly punctuate smooth deformation whenever the packing becomes unstable. The rate of decay of the DWS correlation function gives the rate of these events as being proportional to the strain rate, as expected for a quasistatic process. More interestingly, the shape of the DWS correlation function shows that these are localized events, involving some small number of bubbles that change nearest neighbors [37,38,68].

Direct imaging of monolayer foams are also reveal rearrangements as restricted to a few bubbles [41]. Analogous behavior has been seen for bond breaking in simulations of supercooled liquids under shear [70]. These rearrangements, as mentioned before, are reminiscent of kinetic heterogeneities in that some small regions are active while the rest of the system remains jammed. Many simulations have been performed in the quasistatic or low shear rate regime, and are in agreement with this picture [50,51,53,54,47,45]. An exception is the vertex model [49], where rearrangements involving a broad distribution of bubbles, up to the system size, were found and cited as evidence for self-organized criticality.

When the flow consists of intermittent, localized rearrangement events, the bubble velocity fluctuations are very different from those in an ordinary liquid. According to simulations of the bubble model, the velocity distribution becomes Lorentzian rather than Gaussian, the velocity autocorrelation function exhibits stretched exponential decay, and the bubble wandering is subdiffusive (at least at short times) [71]. This behavior is strongly reminiscent of supercooled liquids. At high temperatures/strain-rates, the behavior is that of an ordinary liquid; at lower temperatures/strain-rates, there is a crossover to a nearly-jammed state with unusual localized dynamics; at zero temperature/strain rates, the material is completely jammed.

The analogy to a supercooled liquid can be made more explicit by studying the viscosity as a function of decreasing strain rate. Jamming in a supercooled liquid is characterized by an extremely rapid increase of viscosity with decreasing temperature, often characterized by the Vogel-Fulcher form, $\eta = \eta_\infty \exp[A/(T - T_0)]$. In simulations of the bubble model, the viscosity increases with decreasing shear rate, but the shear rate cannot be directly compared to a temperature, because these are two completely different variables. Instead, a quantity Γ can be defined that characterizes stress fluctuations in the sheared system, and that corresponds to the temperature in a non-driven, equilibrium system via a linear response relation [72]. The relation between η and Γ turns out to be the same as the relation between η and T in a supercooled liquid. This is evidence that a common mechanism may underlie jamming in these two very different systems.

IV. DISCUSSION

It has been suggested [73] that the jamming behavior of supercooled liquids, colloidal dispersions and granular systems might be unified by a "jamming phase diagram," depicted schematically in Fig. 1. These systems all jam at sufficiently high densities, low temperatures, and low applied stresses (loads), and can unjam as the density is lowered, the temperature is raised, or the applied stress is increased. Parts of the proposed phase diagram are speculative, such as the curve in the temperature-load plane. However, certain cuts of the three-dimensional diagram represent the jamming behavior that we have discussed for colloidal suspensions.

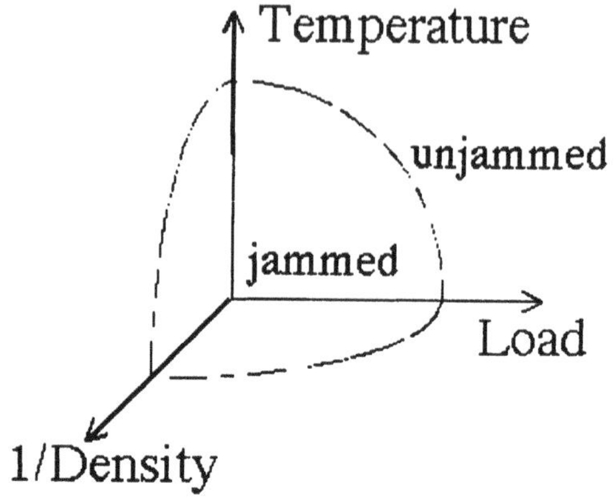

FIG. 1. A possible phase diagram for jamming proposed by Liu and Nagel in Ref. 71. The jammed region, near the origin, is enclosed by the depicted surface.

In Section II.C, we speculated that the puzzling crystallization kinetics observed in colloidal suspensions above the colloidal glass transition might be explained by introducing a second important parameter, namely $kT/\delta\rho g r^4$, in addition to the packing fraction Φ, to describe the jamming behavior. If this is true, then a colloidal suspension might be described by the two-dimensional cut at zero load shown in Fig. 2(a), in the temperature-density plane. According to this cut, there is a packing fraction, denoted here by Φ_j, below which the system is always unjammed, even at $kT/\delta\rho g r^4 = 0$. If we assume that $\Phi_g \approx 0.58$ marks the jamming transition for the colloidal particles studied by van Megen and coworkers at room temperature under earth's gravity, then $\Phi_j < \Phi_g$. Above Φ_j, a system that is jammed on earth might unjam in a microgravity environment; this would explain the results of a recent shuttle experiment [30] as well as terrestrial experiments on density-matched samples [31]. Note that for hard spheres, the value of $kT/\delta\rho g r^4$ required to unjam the system must diverge at Φ_{max} (namely, the random-close-packing density Φ_c). This is shown by the solid line in Fig. 2(a). On the other hand, for deformable spheres, the value of $kT/\delta\rho g r^4$ required to unjam the system could be finite up to Φ_c. This possibility is shown by the dashed line in Fig. 2(a). This is a fundamental difference between hard and deformable particles.

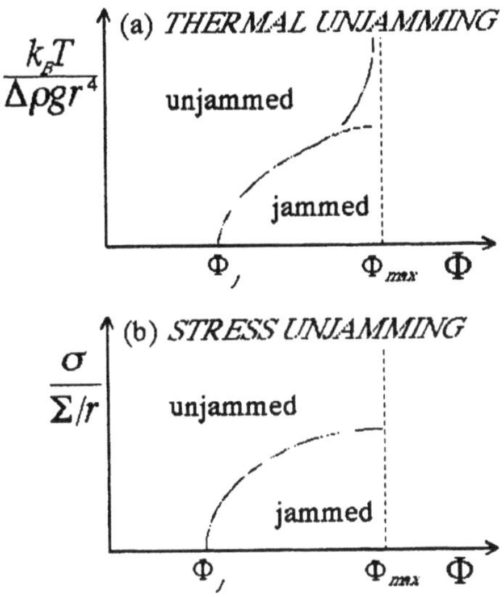

FIG. 2. (a) A two-dimensional slice of the jamming phase diagram that might represent the behavior of suspensions above the colloidal glass transition and below $\Phi_{max} = \Phi_c$ (where $\Phi_c \approx 0.64$ corresponds to random-close-packing, where the spheres are in actual contact). For hard particles, the temperature required to unjam the system must diverge at $\Phi_{max} = \Phi_c$ (solid line), but for deformable particles, the unjamming temperature can be finite (dashed line). (b) The two-dimensional slice at $T = 0$ of the jamming phase diagram in Fig. 1 represented by foams or emulsions. Because the bubbles are deformable, $\Phi_{max} = 1$. The curve separating the jammed and unjammed regimes is simply the yield stress as a function of volume fraction.

In section III.C, we discussed the jamming of a foam or emulsion with decreasing strain rate. This process could equally well be expressed in terms of decreasing shear stress, since shear stress and shear strain rate are conjugate variables related by Eq. 1. Thus, the systems are jammed for applied stresses lower than the yield stress σ_y, and unjam when the applied stress exceeds σ_y. Note that the yield stress decreases as the packing fraction decreases towards Φ_c. If there is indeed a colloidal glass transition that lies below Φ_c at Φ_j for an athermal system, then the yield stress should vanish at Φ_j. Thus, a foam or emulsion can be described by the two-dimensional slice of Fig. 1 shown in Fig. 2(b), where we have plotted the applied stress (load) necessary to unjam the system, σ_y, as a function of packing fraction Φ. Because bubbles are deformable, the maximum value of Φ is $\Phi_{max} = 1$. Note that the slice of Fig. 1 shown here corresponds to $T = 0$; the appropriate dimensionless measure of temperature for bubbles above close-packing is $kT/\Sigma(\gamma_y r)^2$, where Σ is the interfacial tension, γ_y is the yield strain and r is the typical bubble size.

It is possible that hard spheres might show similar behavior to Fig. 2(b) between Φ_j and $\Phi_{max} = \Phi_c$; to our knowledge, the behavior under applied shear of a hard-sphere suspension above the glass transition has not yet been studied. If this is indeed the case, then such suspensions might be used to study the temperature-load plane shown in Fig. 1. The jamming phase diagram raises several interesting questions that remain to be explored for colloidal suspensions.

ACKNOWLEDGMENTS

We thank Paul Chaikin, Daniel Kivelson, Sidney Nagel and David Weitz for stimulating discussions. The support of this work by the National Science Foundation through grants CHE-9624090 (AJL) and DRM-9623567 (DJD), as well as by NASA through grant NAG3-1419 (DJD), is gratefully acknowledged.

[1] W. B. Russel, D.A. Saville, and W. R. Schowalter, *Colloidal dispersions* (Cambridge University Press, New York, 1989).

[2] T. G. M. Van de Ven, *Colloidal Hydrodynamics* (Academic Press, London, 1989).

[3] D. J. Cumberland and R. J. Crawford, *The packing of particles* (Elsevier, New York, 1987).

[4] P. Becher, *Emulsions; theory and practice* (Reinhold, New York, 1965).

[5] R. K. Prud'homme and S. A. Khan, eds., *Foams: Theory, Measurement, and Application.* Surfactant Science Series **57**, (Marcel Dekker, NY, 1996).

[6] D. J. Durian and D. A. Weitz, "Foams," in Kirk-Othmer Encyclopedia of Chemical Technology, 4 ed., edited by J.I. Kroschwitz (Wiley, New York, 1994), Vol. 11, pp. 783-805.

[7] W. G. Hoover and F. H. Ree, J. Chem. Phys. **49**, 3609 (1968).

[8] P. N. Pusey and W. van Megen, Nature, **320**, 340 (1986).

[9] P. N. Pusey and W. van Megen, Phys. Rev. Lett. **59**, 2083 (1987).

[10] W. van Megen and P. N. Pusey, Phys. Rev. A **43**, 5429 (1991).

[11] I. Snook, W. van Megen, P. Pusey, Phys. Rev. A **43**, 6900 (1991).

[12] W. van Megen and S. M. Underwood, Phys. Rev. E **47**, 248 (1993); Phys. Rev. Lett. **70**, 2766 (1993).

[13] W. van Megen and S. M. Underwood, Phys. Rev. E **49**, 4206 (1994).

[14] W. van Megen, T. C. Mortensen, S. R. Williams and J. Müller, Phys. Rev. E **58**, 6073 (1998).

[15] W. Götze and L. Sjögren, Rep. Prog. Phys. **55**, 241 (1992).

[16] W. Götze and L. Sjögren, Phys. Rev. A **43**, 5442 (1991).

[17] H. Gang, A. H. Krall, H. Z. Cummins and D. A. Weitz, Phys. Rev. E **59**, 715 (1999).
[18] T. G. Mason and D. A. Weitz, Phys. Rev. Lett. **75**, 2770 (1995).
[19] N. Menon, S. R. Nagel and D. C. Venerus, Phys. Rev. Lett. **73**, 963 (1994).
[20] C. G. de Kruif, E. M. F. van Iersel and A. Vrij, J. Chem. Phys. **83**, 4717 (1985).
[21] G. N. Choi and I. M. Krieger, J. Coll. Int. Sci. **113**, 101 (1986).
[22] L. Marshall and C. F. Zukoski, J. Phys. Chem. **94**, 1164 (1990).
[23] D. A. R. Jones, B. Leary and D. V. Boger, J. Coll. Int. Sci. **147**, 479 (1991).
[24] Z.-D. Cheng, Ph. D. dissertation (Princeton University, 1998).
[25] A. Kasper, E. Bartsch and H. Sillescu, Langmuir **14**, 5004 (1998).
[26] E. Bartsch, Curr. Op. Coll. Int. Sci. **3**, 577 (1998).
[27] G. Tarjus and D. Kivelson, overview on glass transition.
[28] I. Snook, W. van Megen and P. N. Pusey, Phys. Rev. A **43**, 6900 (1991).
[29] M. M. Hurley and P. Harrowell, J. Chem. Phys. **105**, 10521 (1996).
[30] J. Zhu, M. Li, R. Rogers, W. Meyer, R. H. Ottewill, STS-73 Shuttle Crew, W. B. Russel and P. M. Chaikin, Nature **387**, 883 (1997).
[31] W. K. Kegel, Langmuir (in press).
[32] A sample that was centrifuged for a long time so that the density difference between the upper and lower parts of the sample was significant (ranging from approximately $\Phi = 0.627$ at the upper end to $\Phi = 0.635 \pm 0.006$ at the lower end) did not crystallize at the lower end over a period of about a year. However, this failure to crystallize was attributed in Ref. 21 to the fact that the density at the lower end was extremely close to the random-close-packing limit. Thus, the results of the density-matching experiment appear consistent with the results of the microgravity shuttle experiment of Ref. 20.
[33] A. M. Kraynik, Ann. Rev. Fluid Mech. **20**, 325 (1988).
[34] D. A. Weitz and D. J. Pine, in *Dynamic Light Scattering: The Method and some Applications*, edited by W. Brown (Claredon, Oxford, 1993), p. 652; G. Maret, Current Opinion in Colloid & Interface Science **2**, 251 (1997).
[35] D. J. Durian, D. A. Weitz, and D. J. Pine, Science **252**, 686 (1991).
[36] D. J. Durian, D. A. Weitz, and D. J. Pine, Phys. Rev. A **44**, R7902 (1991).
[37] A. D. Gopal and D. J. Durian, Phys. Rev. Lett. **75**, 2610 (1995).
[38] J. C. Earnshaw and A. H. Jaafar, Phys. Rev. E **49**, 5408 (1994); J. C. Earnshaw and M. Wilson, J. Phys.: Condens. Matter **7**, L49 (1995); J. C. Earnshaw and M. Wilson, J. Phys. II **6**, 713 (1996).
[39] P. Hebraud, F. Lequeux, J. P. Munch, and D. J. Pine, Phys. Rev. Lett. **78**, 4657 (1997).
[40] R. Hohler, S. Cohen-Addad, and H. Hoballah, Phys. Rev. Lett. **79**, 1154 (1997).
[41] M. Dennin and C. M. Knobler, Phys. Rev. Lett. **78**, 2485 (1997).
[42] T. G. Mason and D. A. Weitz, Phys. Rev. Lett. **74**, 1250 (1995).
[43] D. Weaire and J. P. Kermode, Phil. Mag. B **48**, 245 (1983), and Phil. Mag. B **50**, 379 (1984).
[44] F. Bolton and D. Weaire, Phys. Rev. Lett. **65**, 3449 (1990).
[45] S. Hutzler, D. Weaire, and F. Bolton, Phil. Mag. B **71**, 277 (1995).
[46] A. M. Kraynik and D. A. Reinelt, J. Coll. I. Sci **181**, 511 (1996).
[47] Y. Jiang, P. J. Swart, A. Saxena, M. Asipauskas, and J. A. Glazier, Phys. Rev. E, preprint (1998).
[48] D. M. A. Buzza, C. Y. D. Lu, and M. E. Cates, J. de Phys. II **5**, 37 (1995).
[49] T. Okuzono and K. Kawasaki, Phys. Rev. E **51**, 1246 (1995).
[50] D. J. Durian, Phys. Rev. Lett. **75**, 4780 (1995).
[51] D. J. Durian, Phys. Rev. E **55**, 1739 (1997).
[52] M. D. Lacasse, G. S. Grest, D. Levine, T. G. Mason and D. A. Weitz, Phys. Rev. Lett. **76**, 3448 (1996); M. D. Lacasse, G. S. Grest and D. Levine, Phys. Rev. E **54**, 5436 (1996). T. G. Mason, M. D. Lacasse, G. S. Grest, D. Levine, J. Bibette, and D. A. Weitz, Phys. Rev. E **56**, 3150 (1997).
[53] S. A. Langer and A. J. Liu, J. Phys. Chem. B **101**, 8667 (1997).
[54] S. Tewari, D. Schiemann, D. J. Durian, C. M. Knobler, S. A. Langer, and A. J. Liu, preprint (1999).
[55] H. M. Princen, J. Col. I. Sci. **91**, 160 (1983); H. M. Princen, J. Coll. I. Sci. **105**, 150 (1985); and H. M. Princen and A. D. Kiss, J. Coll. I. Sci. **112**, 427 (1986).
[56] S. A. Khan, C. A. Schnepper, and R. C. Armstrong, J. Rheology **32**, 69 (1989).
[57] T. G. Mason, J. Bibette, and D. A. Weitz, Phys. Rev. Lett. **75**, 2051 (1995).
[58] A. Saint-Jalmes, D. J. Durian, preprint submitted to J. Rheology (1999).
[59] A. Saint-Jalmes, M. U. Vera, and D. J. Durian, preprint submitted to EJP-B (1999).
[60] C. W. Macosko, *Rheology Principles, Measurements, and Applications* (VCH Publishers, Inc., New York, 1994); H. A. Barnes, J. F. Hutton, and K. Walters, *An introduction to rheology* (Elsevier, New York, 1989); J. D. Ferry, *Viscoelastic properties of polymers* (Wiley, New York, 1980).
[61] A. D. Gopal and D. J. Durian, preprint (1999).
[62] A. J. Liu, S. Ramaswamy, T. G. Mason, H. Gang, and D. A. Weitz, Phys. Rev. Lett. **76**, 3017 (1996).
[63] P. Sollich, F. Lequeux, P. Hebraud, and M. E. Cates, Phys. Rev. Lett. **78**, 2020 (1997).
[64] P. Sollich, Phys. Rev. E **58**, 738 (1998).
[65] S. Cohen-Addad, H. Hoballah, and R. Hohler, Phys. Rev. E **57**, 6897 (1998).
[66] S. A. Langer and A. J. Liu, J. Phys. Chem. B **101**, 8667 (1997).
[67] T. G. Mason, J. Bibette, and D. A. Weitz, J. Coll. I. Sci. **179**, 439 (1996).
[68] A. D. Gopal and D. J. Durian, J. Coll. I. Sci. **to appear** (1999).
[69] S. S. Park and D. J. Durian, Phys. Rev. Lett. **72**, 3347 (1994).
[70] R. Yamamoto and A. Onuki, Phys. Rev. E **58**, 3515 (1998).

[71] I. Ono, S. A. Langer, and A. J. Liu, preprint (1999).
[72] S. A. Langer and A. J. Liu, preprint (1999).
[73] A. J. Liu and S. R. Nagel, Nature (News & Views editorial), **396**, 21 (Nov. 5, 1998).

Jamming, Friction and Unsteady Rheology

Mark O. Robbins
Department of Physics and Astronomy, The Johns Hopkins University, Baltimore, MD 21218
(July 12, 1999)

I. INTRODUCTION

The very existence of friction is intimately related to jamming at the interface between two solids. Static friction is the force needed to overcome jamming and initiate sliding motion. Kinetic friction is the force that impedes sliding once it begins, and reflects energy dissipation due to sliding. In many cases the sliding is jerky due to periodic jamming and unjamming. This type of stick-slip motion can involve molecular scale or macroscopic slips between each jamming event. The intermittent nature of the motion can be important in determining how much energy is dissipated during sliding, and thus the kinetic friction.

In this overview chapter I will describe some of the many ways in which jamming produces friction on scales ranging from macroscopic to atomic. The goal will be to point to the commonality between processes in friction and the other manifestations of jamming discussed in this book. More complete treatments of friction can be found in a number of books [1–6], including the seminal book by Bowden and Tabor [1] and the collection of articles in "Fundamentals of Friction [2]."

The chapter begins with a brief review of the way friction is measured, the types of results that are observed, and what is known about the geometry of the contacts between two surfaces. Then the results of simple models of ideal, flat crystals are described. These reveal how the jamming that produces static friction is intimately related to the existence of metastable states, in this case due to the deformability of elastic solids. However, this type of metastability does not appear general enough to explain many examples of static friction.

Real surfaces are far from flat and chemically pure. Based on analogies to charge-density-wave conductors, flux lattices and other elastic systems, it seems natural that disorder such as surface roughness or chemical heterogeneity could lead to jamming. A brief review of studies of this effect is presented, and it is shown that pinning by disorder is exponentially weak in our three dimensional world.

A layer of some other material also normally separates two surfaces. This may be a thick lubricant film, dust, grease, wear debris produced by sliding, or a molecularly thin layer of hydrocarbon, water or other material adsorbed from the air. These additional objects between the two surfaces are called "third bodies" by tribologists and are known to affect the measured friction. Three of the papers following this overview chapter examine the behavior of a simple class of third bodies: Molecularly thin films of various fluids between ideal crystals. As the film thickness decreases to a few times a characteristic molecular diameter, these films undergo a jamming transition that is very similar to bulk glass transitions. Jammed systems exhibit a shear response that is typical of solid friction, and it is argued that jamming of third bodies provides a simple and general explanation for the prevalence of friction between the objects around us.

The chapter concludes by considering the unsteady stick-slip motion that often arises when systems become unjammed by sufficiently large stress. Similar unsteady motion occurs in all of the systems described in the following reprints [7–13]. Different types of stick-slip motion are identified, and some of its origins are explored.

II. FRICTION MEASUREMENTS

The basic geometry of a friction measurement is shown in Figure 1. A solid is pushed down onto another solid with a normal load L. One then determines the relation between the lateral velocity v and lateral force F, which is called the friction. In essence this is just a rheology measurement like those made on fluids, granular media, or foams. The difference is that one generally does not know the area, thickness or constituents of the interfacial region where shear occurs. Thus it is difficult to map F and v into the usual rheological variables: shear stress $\tau = F/A$ and shear rate $\dot{\gamma} \equiv \partial v_x / \partial z$. Progress has been made in developing well characterized systems, as discussed below. However, even in the best cases, one does not know how shear is distributed within the interfacial region.

Most macroscopic systems exhibit a force/velocity relation that is similar to the rheology of plastic materials (Fig. 2(a)) [1–3]. The velocity remains equal to zero until a threshold force called the static friction F_s is exceeded. Once sliding has begun, a usually smaller kinetic friction force F_k acts in the direction opposite to the motion. In many cases F_k is relatively independent of velocity for reasons that will be discussed below.

One might expect a very different force/velocity relation for lubricated systems. If the lubricant was Newtonian and its thickness h was constant, F would rise linearly with velocity, and there would be no static friction (Fig. 2(b)). However, if the velocity is kept at zero for a long enough interval, the applied load pushes any *fluid* lubricant out of the contacts. One then observes static friction and a velocity independent kinetic friction at very low velocities (Fig. 2(a)). As the velocity increases, hydrodynamic lift causes the lubricant layer to thicken, and the friction drops sharply. Eventually the lubricant thickness begins to saturate and the force rises. This force/velocity relation is called a Stribeck curve [14,15]. The large velocity region is well described by elasto-hydrodynamics [15,16], a continuum theory that includes the elastic deformation of the bounding solids as well as the hydrodynamics of the fluid between them. The small velocity region is very similar to what one finds for non-lubricated systems and presumably reflects the same jamming.

The two basic laws for static and kinetic friction that are still taught and used today were written down by Amontons exactly three hundred years ago and were known to da Vinci even earlier [4]. Amontons' first law is that friction is proportional to load, and the second is that friction is independent of the apparent macroscopic area of contact between the two solids. As a result the friction between two bodies is normally reported as a friction coefficient μ defined as the ratio of force to load: $\mu = F/L$. Different coefficients of friction are typically measured for static and kinetic friction, and denoted by μ_s and μ_k, respectively. In some systems μ varies with load or surface area. We will see that these failures of Amontons' laws can help to understand their successes in many other systems.

As noted above, friction is a measure of rheology in the interfacial region, and it is not surprising that measured friction coefficients depend strongly on the material at the interface between the solids. To obtain reproducible results one must carefully control the surface conditions, including the ambient humidity. In some rare cases the two solids are in direct atomic contact. However, this is usually true only for experiments in ultra-high vacuum or when wear has exposed momentarily clean fracture surfaces. In most cases there are other objects between the two surfaces that are called "third bodies" by tribologists [17]. These range from sand grains, to dust, to wear debris, to thick lubricant films, down to single monolayers of grease, water or airborne hydrocarbons. In Amontons' original experiments a layer of grease was applied to the contacting surfaces. Most machines attempt to maintain a thick lubricant layer between the surfaces in order to minimize friction and wear. Theoretical treatments of friction have typically ignored third bodies, or used continuum mechanics to treat thick lubricant films. In part this may be due to the lack of precise information about the material at the interface. The ability to measure and control material in the contact is one of the great advances in experimental techniques over the last decade.

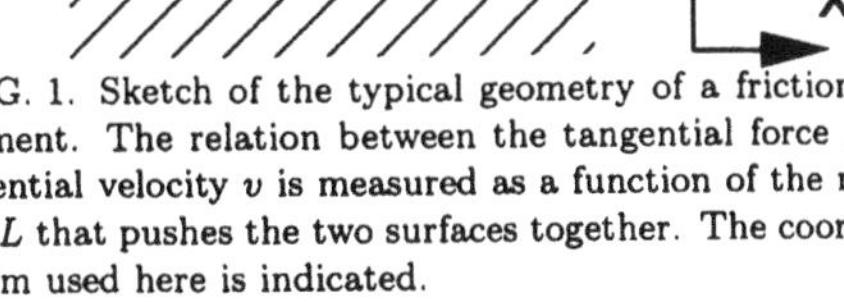

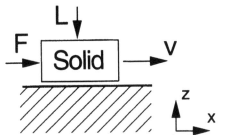

FIG. 1. Sketch of the typical geometry of a friction measurement. The relation between the tangential force F and tangential velocity v is measured as a function of the normal load L that pushes the two surfaces together. The coordinate system used here is indicated.

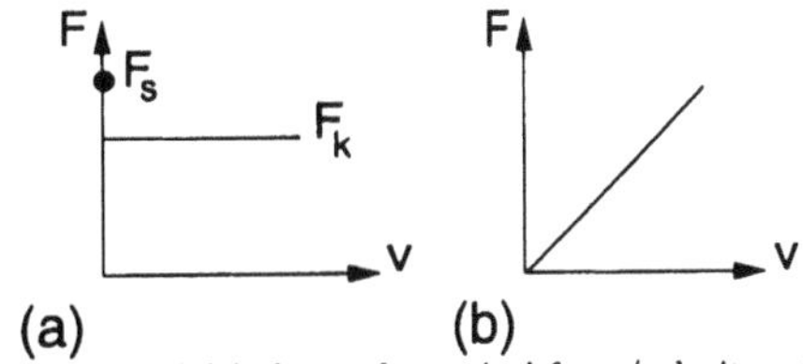

FIG. 2. Panel (a) shows the typical force/velocity relation for a friction measurement. There is no motion until the static friction F_s is exceeded. Once motion starts, the kinetic friction F_k resists motion. Panel(b) shows the force/velocity relation predicted for a Newtonian lubricant at constant film thickness or for sliding between two incommensurate solids.

III. SIMPLE MODELS OF JAMMING

The existence of static friction implies that the surfaces and any material in the interfacial region must jam. The interfacial region becomes trapped in a local potential energy minimum. When an external force is applied to the top wall, it moves away from the minimum until the derivative of the potential energy balances the external force. The static friction is the maximum force the potential can resist, i.e. the maximum slope of the potential.

The simplest picture for jamming is that macroscopic peaks and valleys on the two surfaces interlock and prevent lateral motion as shown in Fig. 3(a). This was the basis of some of the oldest attempts to explain Amontons' laws. Parent and Euler [4] considered surfaces with no microscopic resistance to sliding, but roughened so that they make an angle θ relative to the average surface plane. A static friction $F_s = L \tan\theta$ must be overcome before the top surface can move up the ramp formed by the bottom surface and begin to slide. There is no dependence on surface area, and the predicted coefficient of friction $\mu_s = \tan\theta$ covers all possible values.

One might expect that Parent and Euler's model would apply to the system of glass spheres considered by Nasuno et al. in one of the following reprints [10]. Here glass spheres were glued to the bottom of a glass plate. The plate was then placed on a pack of spheres of the same size. One expects that the spheres on the plate will burrow between spheres in the pack when a load is applied. In order to move the plate, the spheres must be

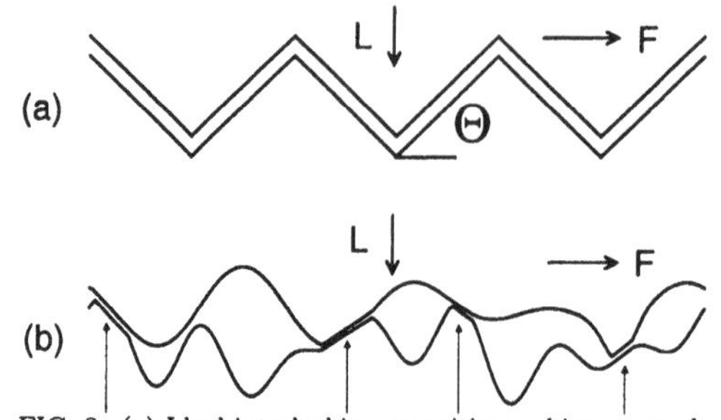

FIG. 3. (a) Ideal interlocking asperities making an angle θ relative to the average surface. (b) Contacts between random surfaces (arrows) occur at the peaks of isolated asperities. For every ramp that goes up there is another going down and the net force is zero. The vertical scale has been exaggerated and the angle in each contact would be orders of magnitude smaller [18].

lifted out of the holes they created. The average angle required was worked out by Belidor in 1737 [4] and would give $\mu_s = 0.35$. However, glass beads can never slide past each other without microscopic friction as assumed by Parent, Euler and Belidor. In order to get a good model of friction in granular systems one must follow Coulomb [4] and include a microscopic friction coefficient as well as the mean slope. One then returns to the question of determining the origin of friction between two surfaces.

Parent and Euler's picture of friction due to interlocking peaks, or asperities as they are more frequently called, can not explain many experimental trends and observations [1]. One is that decreasing θ by smoothing surfaces can actually increase static friction. Indeed magnetic hard disks are purposefully roughened to reduce μ_s. Another is that experiments show that μ depends on the materials in contact more strongly than their surface morphology. For example, a single monolayer of fatty acid can reduce the friction by an order of magnitude without changing the surface roughness. Finally, Parent and Euler's model would give vanishing kinetic friction because the force applied to move up the ramps would be exactly balanced by the force exerted by the ramps on the way back down [19].

Over the last 50 years, a variety of experimental techniques [1,3,20,21] have shown that contacts between most macroscopic surfaces look like Fig. 3(b), where the roughness is exaggerated by magnifying the vertical scale. Molecules from opposing surfaces only contact where two peaks meet. The random spacing between peaks prevents the intimate mating envisioned in Fig. 3(a). Except for very soft materials, like rubber, the total area of intimate molecular contact, A_{real}, is much less than the apparent geometrical area of the surfaces A_{app}. Elastic deformation of the solids leads to relatively flat contact regions with small slopes and diameters of order one to ten micrometers. Parent and Euler's picture fails both because typical values of tan θ are much smaller than μ_s and because the effect of the contacts averages to zero: For every contact with an upward slope there is another sloping down.

The implications of these experiments is that static friction must result from jamming within the areas of intimate molecular contact. The diameters of these contacts are much larger than molecular diameters (1 to 10μm vs. $\sim$1 nm), and are usually much larger than the thickness h of the interfacial region where shear or jamming occurs. One thus expects that friction results from jamming within many independent regions of dimension h within the contact. These regions will add constructively to the potential barrier that determines the static friction, leading to a friction that is linear in the real area of contact. Of course the local potential barriers will depend on the local pressure p which will decrease with increasing A_{real}. We can summarize this by saying that a piece of contact area dA contributes a static friction $dF_s = \tau_s(p)\, dA$ where τ_s is the yield stress at the local pressure.

The conclusions of the above paragraph might seem to conflict with Amontons' second law which says that the apparent area of contact does not influence the friction. However, the real and apparent areas are known to be very different. Bowden and Tabor [1] suggested a simple phenomenological theory for τ_s that not only yields Amontons' laws for a wide range of systems but can also explain experiments where Amontons' laws fail [7,22–25]. They assumed a simple linear relation between τ_s and p

$$\tau_s(p) = \tau_0 + \alpha p. \tag{3.1}$$

Then integrating the force over the entire contact area A_{real} gives:

$$F_s = \tau_0 * A_{real} + \alpha * L, \tag{3.2}$$

where L is the total normal load and the result is independent of the distribution of pressures in the contacts. Dividing by the total load gives the coefficient of static friction:

$$\mu_s = \alpha + \tau_0/\overline{p}, \tag{3.3}$$

where $\overline{p} = L/A_{\text{real}}$ is the average pressure over all contacts. This expression for the coefficient of friction will be independent of load and macroscopic area if $\overline{p}$ is much larger than τ_0 or if $\overline{p}$ is constant. As we now discuss, the latter holds for simple models of ideally plastic [1] or elastic [26,27] surfaces.

Plastic materials will deform to increase the contact area when the pressure exceeds the hardness of the material. Given the small values of $A_{\text{real}}/A_{\text{app}}$ observed in many systems, it seems likely that this failure condition may be met at typical values of the normal load [1,20]. The real area of contact at a given load will then grow until p decreases to the hardness H, and the coefficient of friction will have the constant value:

$$\mu_s = \alpha + \tau_0/H. \tag{3.4}$$

Typical values of τ_0 are usually small compared to H and thus α will normally dominate the friction coefficient [28].

The argument for elastic surfaces is more complicated and the reader is referred to the original paper by Greenwood and Williamson [26] and a more recent paper by Volmer and Natterman [27]. The basic starting point is that real surfaces often exhibit power law noise spectra characteristic of self-affine fractals. The typical height variation δh over a lateral distance δl scales as $\delta h \sim \delta l^{0.5}$. The peaks on such surfaces have a broad distribution of heights. Increases in load will push the surfaces together, creating new contacts and expanding pre-existing contacts. The distribution of heights is such that the fraction of contacts that have a given area remains nearly unchanged, but the total number of each size increases linearly with load. Since the total area is proportional to load, the pressure remains constant. The constant value must be smaller than the hardness or plastic deformation will occur.

At large enough loads the real area of contact must become equal to the apparent area. Materials such as steel are so stiff and hard that the machines applying the load would be likely to fail before this condition was met. However, the rubber on tires and shoes is much more easily deformed. Experiments on polymers show that Amontons' laws fail when $A_{real} = A_{app}$, but the linear relation assumed by Bowden and Tabor (Eq. 3.1) is still followed [23]. The same linear relation holds for other situations where $A_{real} = A_{app}$, including experiments on MoS_2 and mica [7,22,24,25]. It even applies to the case of tape, where P is zero, but a large force per unit area τ_0 must be overcome to initiate sliding. Thus Eq. 3.1 is capable of describing a wide range of systems, encompassing both the failures and successes of Amontons' laws.

If static friction is just the rheology of a jammed system at the interface, one might expect that a relation like Eq. 3.1 would describe the yield stress of bulk systems. This expectation is born out by experiments on a variety of inorganic [29] and organic [30,31] solids that show a linear increase in bulk yield stress with hydrostatic pressure. The yield stress of granular materials is also known to increase with pressure, although this is normally attributed to the increased friction at contacts between grains due to the increased load [10,32–34]

To summarize this section, a wide range of experimental evidence indicates that static friction results from jamming of material at the interface between two solids. If the yield stress of this material increases linearly with pressure, then Amontons' laws and many exceptions to them can be understood. A yield stress that increases linearly with pressure is found in bulk experiments, but it is not clear what produces this type of jamming in friction.

IV. CRYSTALLINE SURFACES IN DIRECT CONTACT

Theoretical studies of friction have historically focused on the case illustrated in Fig. 4(a) – two clean crystalline surfaces in direct contact [35–42]. The crystals are generally assumed to deform elastically, while sliding leads to plastic deformation at the interface. This implicitly assumes that the interactions across the interface are weaker than the interactions within the crystals, although the models are often extended beyond this regime.

Even the idealized case of Fig. 4(a) is difficult to analyze analytically. Figure 4 (b) and (c) show two even simpler one-dimensional (1D) models that have been used to understand whether and how two crystals can jam. Both models replace the bottom solid by a periodic potential, and retain only the bottom layer of atoms from the top wall. In the Tomlinson model (Fig. 4(b)), the remaining atoms are coupled to the center of mass of the top wall by springs, and coupling between neighboring atoms is ignored [35,36,43]. In the Frenkel-Kontorova model (Fig. 4(c)), the atoms are coupled to nearest-neighbors by springs, and the coupling to the atoms above is ignored [37–41]. Many hybrids and variants of these models have also been considered [42], but the qualitative features of the results are universal.

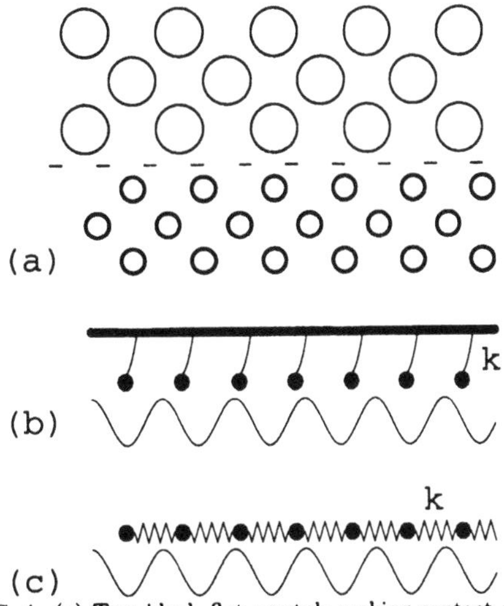

FIG. 4. (a) Two ideal, flat crystals making contact at the plane indicated by the dashed line. The nearest-neighbor spacings in the bottom and top walls are a and b respectively. The Tomlinson model (b) and Frenkel-Kontorova model (c) replace the bottom surface by a periodic potential $F_0 \sin(2\pi x/a)$. The former model keeps elastic forces between atoms on the top surface and the center of mass of the top wall, and the latter includes springs between neighbors in the top wall.

The dimensionless quantities that characterize both simple 1D models are the ratio between the lattice constants of the two surfaces, $\eta = b/a$, and the ratio of the strength of the periodic potential to that of the springs. If we assume that the force from the potential is a simple sine wave, $F_0 \sin(2\pi x/a)$, then we can characterize the potential's strength by the ratio $\lambda \equiv 2\pi F_0/ak$ of the maximum derivative of the force to k. The models are not self-consistent if λ is much larger than one. In this limit the interactions at the interface are stronger than those in the bounding solids and the solids should yield before the interface. One should then consider models of sliding at an interface within the weaker solid.

The simplest case to consider is that of identical lattice constants, $\eta = 1$. In this case, all atoms go up and down over the potential in phase, just as in Fig. 3(a). The static friction per atom for both models is $F_s = F_0$, the maximum force from the periodic potential. The static friction is less than the maximum force in two or more dimensions because the atoms can move around the maxima in the potential and pass over a saddle point. This type of sliding occurs in internal shear of bulk fcc systems [44,45] and Hirano and Shinjo have noted this extra degree of freedom lowers the static friction relative to one-dimensional models [38].

The kinetic friction is difficult to treat within these simple models because it necessarily involves dissipation. In the Tomlinson model each atom is an independent harmonic oscillator with no ability to dissipate. There are more degrees of freedom in the Frenkel-Kontorova model, but the vibrations are all exact normal modes with no allowed energy transfer. What is usually done in both models is to add a phenomenological viscous friction force, $F_{drag} = -\Gamma v$, to the equations of motion for each atom. This might be thought of as the damping associated with coupling to external degrees of freedom in a Langevin model, but the thermal noise term that would be coupled to the damping is usually ignored. In the following we specialize to the overdamped case where the inertia of the atoms can be ignored. The effects of inertia are discussed briefly in the section on stick-slip motion.

Although the above approximation involves assuming friction in order to calculate friction, one can still get non-trivial behavior. Indeed this version of the Tomlinson model is mathematically identical to simple models for Josephson junctions [46], to the single-particle model of charge-density wave depinning [47], and to the equations of motion for a contact line on a periodic surface [48,49]. Fig. 5 shows the results for the Tomlinson model as a function of the dimensionless strength of the interface potential, λ. When the potential is much weaker than the springs, the atoms can not deviate significantly from the center of mass. As a result, they go up and down over the periodic potential at constant velocity and the force averages out to zero. The kinetic friction is just due to the drag force on each atom and rises linearly with velocity. The same result holds for all spring constants in the Frenkel-Kontorova model with equal lattice constants.

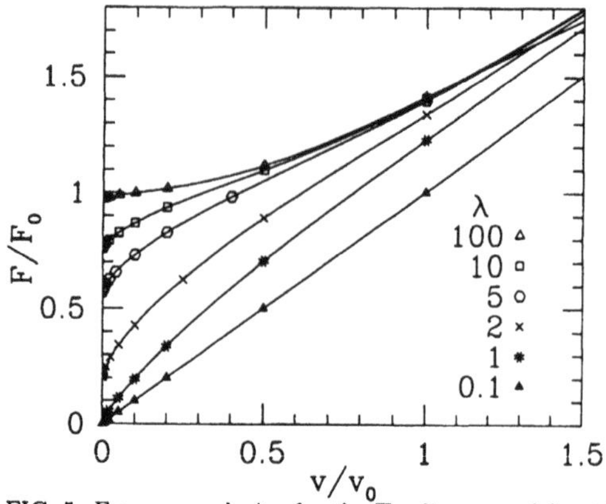

FIG. 5. Force vs. velocity for the Tomlinson model at the indicated values of λ. The force per atom F is normalized by the static friction F_0, and the velocity is normalized by $v_0 \equiv F_0/\Gamma$ where Γ is the phenomenological damping rate.

As the potential becomes stronger, the periodic force begins to contribute to the kinetic friction of the Tomlinson model. There is a transition at $\lambda = 1$, and at larger λ the kinetic friction remains finite in the limit of zero velocity. This constant approaches the static friction F_0 as $\lambda \to \infty$. The origin of this new low frequency behavior is the onset of multistability [36], a feature that is important in many jammed systems. The condition for metastability is that the spring and periodic forces balance:

$$k(x - x_{CM}) = F_0 \sin[2\pi x/a)], \qquad (4.1)$$

where x_{CM} is the center of mass position. As shown graphically in Fig. 6, there is only one solution for weak interfacial potentials ($\lambda < 1$), but there are an increasing number of metastable solutions as λ increases beyond unity. Once an atom is in a given metastable minimum it is trapped there until the center of mass moves far enough away that the second derivative of the potential vanishes and the local minimum disappears. The equations of motion then become unstable, and the atom pops forward very rapidly to the next minimum. This only reduces the force from F_0 by about ka. Thus the friction force remains constant as v goes to zero at a value that is of order the static friction and approaches it as k goes to zero.

The above discussion explains the kinetic friction entirely in terms of the spring extension, but kinetic friction is fundamentally due to dissipation. The existence of a finite low velocity limit for the friction implies that the rate of dissipation rises linearly with velocity, rather than exhibiting the quadratic dependence characteristic of viscous fluids. If the rate of dissipation is linear in velocity there is the same amount of energy dissipated for every

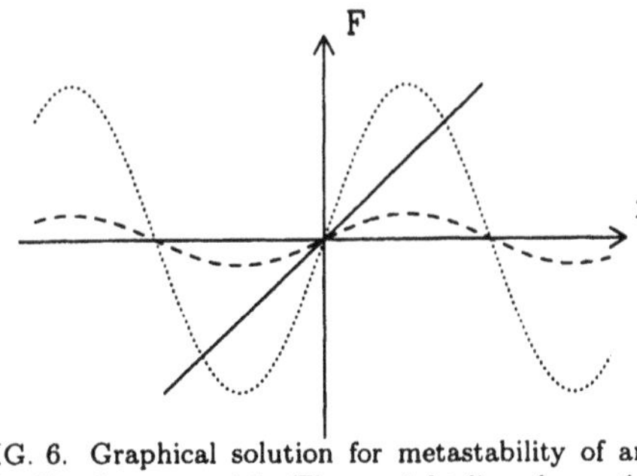

FIG. 6. Graphical solution for metastability of an atom in the Tomlinson model. The straight line shows the force from the spring for $x_{CM} = 0$. The dotted and dashed curves show periodic potentials with amplitudes corresponding to $\lambda = 3$ and 0.5, respectively. For $\lambda > 1$ there are multiple intersections with the spring force, indicating the existence of multiple metastable states.

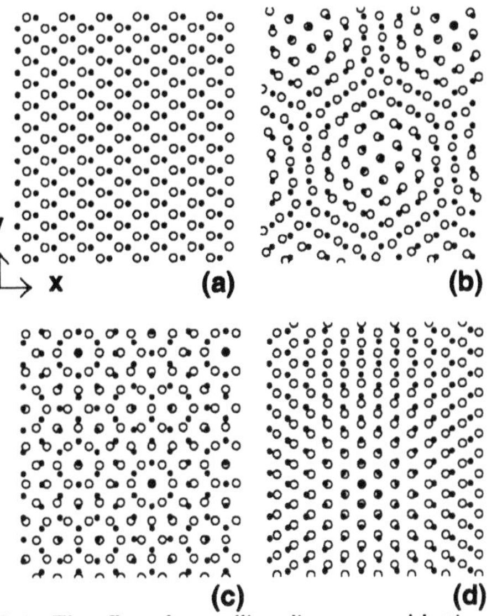

FIG. 7. The effect of crystalline alignment and lattice constant on commensurability is illustrated by projecting atoms from the bottom (filled circles) and top (open circles) surfaces into the plane of the walls. In (a)-(c) the two walls have the same structure and lattice constant but the top wall has been rotated by 0°, 8.2° or 90°, respectively. In (d) the walls are aligned, but the lattice constant of the top wall has been reduced by a small amount. Only case (a) is commensurate. The other cases are incommensurate and atoms from the two walls sample all possible relative positions with equal probability. Reproduced by permission from Ref. [51].

lattice constant advanced in the low velocity limit - how does this come about?

The only mechanism of dissipation is through the phenomenological damping force, which is proportional to the velocity of the atom. The velocity is essentially zero except in the rapid pops that occur as a state becomes unstable and the atom pops to the next metastable state. During this pop the atom's velocity is always of order $v_0 \equiv F_0/\Gamma$ - independent of the average velocity of the center of mass. Moreover, the time of the pop is nearly independent of v_{CM} and so the total energy dissipated is independent of v_{CM}. One can of course show that this dissipated energy is consistent with the limiting force determined from arguments based on the extension of the springs [48–50]. This basic idea that kinetic friction is due to dissipation during pops that remain rapid as $v_{CM} \to 0$ is very general, although the phenomenological damping used in the model is far from realistic.

The case of equal lattice constants is extremely unlikely, and becomes even more so when one considers three-dimensional objects with two-dimensional surfaces. Figure 7 shows that even identical crystals only match up when they are perfectly aligned. The most likely case is that the ratio of lattice constants is an irrational number. In this case the crystals are said to be incommensurate, while a rational ratio would correspond to a commensurate case. Incommensurate crystals share no common period, and thus lattice sites on the top wall lie at all possible positions relative to lattice sites on the bottom wall. This situation is necessarily translationally invariant, so there is no variation in the ground state with the position of the top wall.

One might assume that this translational invariance immediately implied that the static friction vanished for incommensurate systems. This is indeed the case in either the Tomlinson or Frenkel-Kontorova model if the potential is weak compared to the springs. In this case there is no multistability and atoms stay near their lattice sites. Since these sites sample all possible values of the force with equal probability, the average force, and hence the static friction, is zero. The situation changes when the interfacial potential becomes strong enough to introduce multistability for individual atoms. The Tomlinson case is the simplest. Although I have not seen it discussed for incommensurate systems, the mathematics is identical to Fisher's mean field theory of charge-density-wave pinning by random impurities [50]. If atoms start on their lattice sites and are then allowed to relax, they will reach exactly the same ground state energy for any center of mass position x_{CM} of the top wall. All that changes is that the role of individual atoms is interchanged. If there is multistability, then some or all of the atoms will remain in these metastable states as x_{CM} changes. Thus even though there is a solution for the new x_{CM} with the ground state energy, atoms are unable to cross barriers to reach this solution. The result is a net force that reflects the metastable minimum that the system is trapped in. The static friction becomes non-zero when the potential is strong enough to produce multistability ($\lambda > 1$), and its value is just equal to the kinetic friction in the limit $v \to 0$ (Fig. 5).

The physics of the Frenkel-Kontorova model is much richer, but has the same qualitative features [37–41]. The main difference is that the static friction and ground state depend strongly on η. For any given irrational value of η there is a threshold potential strength λ_c. For weaker potentials the static friction vanishes, and for stronger potentials metastability produces a finite static friction. The metastable states take the form of locally commensurate regions that are separated by domain walls where the two crystals are out of phase.

The one-dimensional Frenkel-Kontorova model has been analyzed in great detail [39–41]. Within the locally commensurate regions the ratio of the periods is a rational number p/q that is close to η. In order to get a pinning force near F_0, the springs must be weak enough to allow a metastable state with $p = q = 1$. As q increases, the pinning force decreases and the value of λ_c rises. The range of η where jamming occurs grows with increasing potential strength until it spans all values. At this point there is an infinite number of different metastable ground states that form a "Devil's staircase" as η varies [39,40].

The two-dimensional case that would represent the interface between three dimensional crystals has received less attention, but seems to have the same general behavior [38,52,53]. The value of η in such systems depends strongly on both the direction of sliding and the relative orientation of the crystalline axes of the two solids (Fig. 7). Thus, if this model described the metastability in real systems, one would expect strong variations in friction with alignment and sliding direction, including the absence of static friction in certain orientations [54]. Anisotropy is seen in certain systems [55–57], but the variations seen in typical experiments, where no effort is made to control orientation, are usually only of order 20%.

In order to get static friction at all orientations within the Frenkel- Kontorova model one would need to have a very strong interfacial potential compared to the cohesive potentials within each solid. As noted earlier, this is inconsistent with the assumption that sliding occurs at the interface rather than within the weaker solid. Strong interfacial interactions are most likely to occur between clean, reactive surfaces in ultrahigh vacuum. Experiments on clean metal surfaces do indeed show that the surfaces weld together at the interface and that sliding occurs through fracture within one of the solids. The friction forces are anomalously large, tend to depend on area, and material wears away at extraordinarily high rates.

In summary of this section, simple models predict that static friction between two flat crystals should only be observed in the unlikely case of commensurate walls, or when the walls are incommensurate and there is multistability. The condition for the latter is that the interfacial potential is large compared to interactions within the crystals. There have been relatively few experiments on flat, clean, crystalline surfaces, both because of the difficulty in making flat surfaces and because of the difficulty in removing chemical contamination. However, the experiments that have been done are consistent with the simple models described above.

One class of experiments measures the friction between an adsorbed layer and a substrate using a quartz crystal microbalance [58–60]. The substrate is atomically flat over large regions and the thickness of the adsorbed layer can be varied from a fraction of a monolayer up to several layers. When the layer is a fluid or incommensurate crystal there is no static friction. The friction varies linearly with velocity as in Fig. 2(b) and incommensurate crystals actually slide as much as an order of magnitude more easily than fluid layers of the same atom [58,60,61]. Because there is no metastability, the slope can be determined from the fluctuation-dissipation theorem. Theories based on dissipation through excitations of phonons [61–63] and electrons [64–66] have successfully explained the experimental results, although the relative importance of the two contributions has not been determined because of uncertainties in the interfacial potential [60].

Another experiment used a crystalline atomic-force-microscope tip that was rotated relative to a crystalline substrate. Static friction was observed when the crystals were aligned, and decreased below the threshold of detection when the crystals were rotated out of alignment [55]. Unfortunately the crystals were orders of magnitude smaller than typical contacts. Hirano and coworkers have also looked at the orientational dependence of the friction between macroscopic mica surfaces [57]. They found as much as an order of magnitude decrease in friction when the mica was rotated to become incommensurate. However, this large orientational anisotropy was only observed in vacuum. No anisotropy could be seen in ambient air.

A final set of experiments used MoS_2 lubricating two surfaces [67]. The MoS_2 forms plate-like crystals that slide over each other within the contact. They appear to be randomly oriented and thus incommensurate. In ultrahigh vacuum the measured friction coefficient was always lower than 0.002 and in many cases dropped below the experimental noise. In contrast, films exposed to air give friction coefficients of 0.01 to 0.05. This and the mica experiments provide a strong hint that contamination between surfaces is important to static friction as we discuss further below [51,54].

V. ROUGHNESS AND CHEMICAL HETEROGENEITY

There are many examples of physical systems where disorder leads to multistability and a pinning force like static friction. The most studied examples are probably charge-density-wave conduction [47,50,68] and magnetic flux lattices [69,70], but other examples include fluid invasion of porous media [71–73], motion of magnetic domain walls [74–76], and spreading on disordered surfaces [48,49,77–79]. In each case, one finds that disorder produces pinning below a critical spatial dimension and no pinning in higher dimensions. The pinning is typically exponentially weak at the critical dimension.

To determine the critical dimension and pinning force one uses scaling theory and compares the energy gained by conforming to the disorder to the energy cost of the elastic deformation [80]. Spreading [77–79] and friction [27,81,82] are different than other systems because the elastic object that is deformed has a higher dimensionality than the disordered region. In spreading, the contact line where the fluid interface intersects the disordered solid is a 1D "surface" of a 2D elastic interface. In friction, the 2D surface of a 3D elastic solid interacts with a disordered substrate. In both cases an elastic deformation with a wavevector q at the surface creates an elastic deformation over a distance of order $1/q$ into the bulk. The energy cost of deformations normally scales as q^2, but when integrated over the bulk, one factor of q is canceled and the energy density scales as q. This changes the critical dimension from the charge-density wave case.

Consider the general case of a d dimensional system with a $d-1$ dimensional surface that interacts with a random potential. The energy per unit surface area needed to deform a region of diameter l scales as $q \sim 1/l$. The energy density gained by conforming to the disorder in this region scales as $1/l^{(d-1)/2}$ if the disorder is uncorrelated. This is just the usual result that the fluctuation in the mean of n independent quantities scales as $1/\sqrt{n}$ with n proportional to the area of the deformed region. Comparing the elastic cost to the gain one finds that disorder always wins at lengths longer than some size l_p for $d < 3$, but not for $d > 3$. Thus contact lines are always pinned, and friction is a marginal case.

Volmer and Natterman [27] have considered the case of friction in detail. There is always pinning, even for weak disorder, but the pinning force is exponentially small. By making several assumptions one can obtain a static friction that satisfies Amontons' laws. However, it seems unlikely that these assumptions would be satisfied broadly enough to provide a general explanation of Amontons' laws. For example, one key assumption is that l_p is comparable to the diameter of individual contacts. Given the large size of contacts (10 μm), this would require very weak disorder. One can estimate τ_s using the fact that the pinning and elastic energies are roughly equal at l_p. The elastic energy is the cost of deforming by a lattice constant a over a wavelength of l_p, while the theoretical yield stress τ_y is the stress needed to deform by a over a wavelength of a lattice constant. Thus $\tau_s \sim \tau_y l_p/a \sim 10^{-4}\tau_y$ which is far smaller than experimental values. In the case of plastic materials it would imply $\mu < 10^{-4}$.

The basic reason that it is difficult for incommensurate potentials or disorder to pin crystals is that the atoms must move by distances of order a lattice constant in order to produce a metastable state. This type of plastic deformation can only occur when the forces are strong enough to tear the crystal apart. Stresses large enough to produce plastic deformation may occur in some cases, but are expected to lead to extremely high wear rates that would not be acceptable in practical applications. Replacing the crystals with elastic amorphous walls, does not resolve this problem. However, as we now discuss, introducing easily deformable "third bodies" between the solid walls allows them to jam together without producing excessive wear.

VI. JAMMING IN CONFINED FLUIDS

The Surface Force Apparatus (SFA) allows the mechanical properties of fluid films to be studied as a function of thickness over a range from hundreds of nanometers down to contact. The fluid is confined between two atomically flat surfaces. The most commonly used surfaces are mica, but silica, polymers, and mica coated with amorphous carbon, sapphire, or aluminum oxide have also been used. The surfaces are pressed together with a constant normal load and the separation between them is measured using optical interferometry. The fluid can then be sheared by translating one surface at a constant tangential velocity [7,83] or by oscillating it at a controlled amplitude and frequency in a tangential [8,84,85] or normal [86,87] direction. The steady sliding mode mimics a typical macroscopic friction measurement, while the oscillatory mode is more typical of bulk rheological measurements. Both modes reveal the same sequence of transitions in the behavior of thin films.

When the film thickness h is sufficiently thick, one observes the rheological behavior typical of bulk fluids [86,87]. Flow can be described by the bulk viscosity μ_B with the usual no-slip boundary condition - equal velocities of the fluid and bounding solid at their interface. The shear stress τ is just

$$\tau = \mu_B v/h \tag{6.1}$$

where v is the velocity difference between the two walls.

As the film thickness decreases below 100nm, the behavior begins to change [86,87]. Experiments can not determine whether this is due to changes in viscosity or in boundary condition, because the flow profile is not measured. However simulations indicate that changes in viscosity do not become significant until the thickness becomes much smaller [9,88–92]. The experiments are consistent with small deviations from the no-slip boundary condition that are also found in simulations. These change the effective width of the film from h to $h+2S$ in Eq. 6.1 where S is called the slip length and the factor of two comes from the presence of two interfaces. The slip length quantifies the degree to which the viscosity near the interfaces is different from the bulk value. In some cases the fluid slips more easily over the wall and S is positive. In other cases a layer of fluid solidifies on to the wall and S is negative. The displacement of the slip plane is comparable to the diameter of fluid molecules in most experiments, and several simulation studies have examined the factors that determine this boundary condition [90,92–96].

When the film thickness decreases below of order ten molecular diameters, the deviations from bulk behavior become dramatic [7,85]. The shear stress becomes far too large to interpret in terms of a bulk viscosity and slip length, because the width of the fluid region $h + 2S$ would have to be much smaller than an Angström. Experimental results are often expressed in terms of Eq. 6.1 with an effective viscosity $\mu_{eff} \equiv \tau h/v$ and effective shear rate $\dot{\gamma}_{eff} \equiv v/h$. The following reprints by Gee et al. [7] and Demirel and Granick [8] show that low velocity values of μ_{eff} rise many orders of magnitude beyond μ_B and that relaxation times determined from the shear rate dependence of μ_{eff} increase even more dramatically over bulk values. Both changes are strong evidence that the film is becoming jammed. By the time h reaches two or three molecular diameters, most films behave like solids.

In some cases the transition from liquid to solid behavior of the film appears to occur discontinuously, as if the film underwent a first-order freezing transition [97]. Simulations suggest that this is most likely to happen when the crystalline phase of the film is commensurate with the solid walls, and when the molecules have a relatively simple structure that facilitates order [7,9,98–100]. In most cases the transition looks like a continuous glass transition, but at temperatures and pressures that are far from the glass region in the bulk phase diagram. In fact, the molecules may not readily form glasses in the bulk. This suggests that one must treat the film thickness as an additional thermodynamic variable that may shift or alter phase boundaries. Cases where a single interface stabilizes a different phase than the bulk are central to the field of wetting. The presence of two interfaces separated by only a few nanometers leads to more pervasive phase changes [7].

In one of the following reprints [7], Gee et al. describe the "liquid to solid-like transitions" of five different liquids, cyclohexane, octamethylcyclotetrasiloxane (OMCTS), n-octane, n-tetradecane, and a branched isoparaffin 2-methyloctadecane. The liquids have different molecular structures that lead to different types of ordering in thin layers, and provide a reasonable sampling of the types of fluids that have been studied. For simple molecules one finds strong oscillations in the normal force on the walls as the film thickness decreases [89,92,101,102]. The period of the oscillations is a characteristic molecular dimension, and the oscillations reflect ordering of molecules into layers that are parallel to the bounding walls. The factors that control layering are discussed in detail in the reprint by Thompson et al. [9], references therein, and in some subsequent work [99].

For the five liquids studied by Gee et al., and many others studied later [25,103–106], the viscosity and relaxation times become too large to measure when the thickness has dropped to one to three molecular diameters. These thin films are jammed and resist shear like a solid. No motion occurs until a yield stress is exceeded, and the calculated yield stress is comparable to values for bulk solid phases of the molecules at higher pressure or lower temperature. The response of the walls to a lateral force becomes typical of static friction, and Gee et al. analyze their results in terms of equation 3.2 using S_c and C in place of τ_0 and α, respectively. The values of τ_0 are of order a few to 20MPa, and values of α range from 0.3 to 1.5. However, the latter are influenced by the change in film thickness at the relatively low pressures used (< 40 MPa) and would probably not be representative of the behavior in a typical mechanical device. Gee et al. also describe unsteady stick-slip motion of the walls for certain fluids [7,107]. This is discussed in the next section.

In another reprint, Demirel and Granick [8] examine the transition from liquid to solid in OMCTS, following an analogy to studies of bulk glass transitions [108,109]. They measure the frequency dependence of the real and imaginary parts of the elastic moduli of thin films. They find that all the data can be collapsed onto a universal curve using a generalization of time-temperature scaling [108,109], and conclude by comparing the spectrum of relaxation times to those of bulk glasses.

Figure 8 shows that simulation results for the viscosity of thin films can also be collapsed using Demirel and Granick's approach [110,111]. The simulation model used is described in detail in the reprint by Thompson et al. [9]. Data for different thicknesses, normal pressures, and interaction parameters taken from Figs. 9, 15 and 16 of this paper were scaled by the low shear rate viscosity μ_0. The shear rate was then scaled by the rate $\dot{\gamma}_c$ at which the viscosity dropped to $\mu_0/2$. Also shown on the plot (circles) are data for different temperatures that were obtained for longer chains in films that are thick enough to exhibit bulk behavior [110–112]. The data fit well on to the same curve, providing a strong indication that a similar glass transition occurs whether thickness, normal pressure, or temperature is varied. In all cases the high shear rate data is characterized by a power law shear thinning with an exponent near -2/3. Similar exponents are observed in experimentally measured viscosities [105].

As in the case of bulk glasses, the issue of whether there is a true glass transition at finite temperature is controversial. The simulations seem to show a clear change in behavior, and the divergence of the relaxation time $\dot{\gamma}_c^{-1}$ and μ_0 can be fit to a free volume theory [9,111,113]. In the glassy state the response shows a constant yield stress which would correspond to a slope of -1 on Fig. 8. There is no case where we see a smooth crossover to this slope from the value of -0.69 indicated by the dashed line in Fig. 8, but our data is in a very different frequency range than experiments and spans fewer decades. Demirel and Granick get a reasonable collapse of data at all thicknesses [8]. This would indicate that the relaxation time never truly diverges. However, there

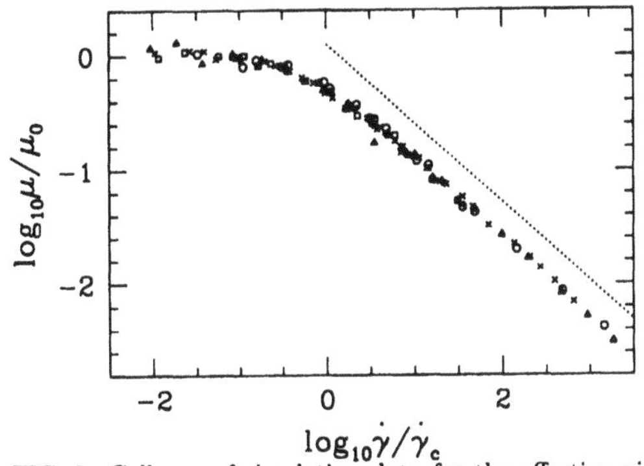

FIG. 8. Collapse of simulation data for the effective viscosity vs. shear rate as a glass transition is approached by decreasing temperature (circles), increasing normal pressure at fixed number of fluid layers (triangles) or decreasing film thickness at fixed pressure with two different sets of interaction potentials (squares and crosses). The dashed line has a slope of -0.69. Reproduced by permission from Ref. [111].

is some ambiguity in their collapse of data for 1 and 2 layers, since the moduli are nearly independent of rate at these thicknesses.

Whether the relaxation time and viscosity truly diverge or not, they clearly become large compared to measurement times. Thin films of some molecules support stress over at least several days in the SFA [7,24]. The pressures in most macroscopic friction measurements are roughly two orders of magnitude higher, making the relaxation times even longer. Thus jamming of thin fluid layers between asperities provides a possible explanation for the prevalence of static friction in macroscopic experiments. This has been explored in a recent paper [51] using the wall geometries shown in Fig. 7. The shear stress was found to follow Bowden and Tabor's phenomenological law (Eq. 3.1) up to the largest pressures studied (~ 1.5 GPa). In contrast to the case of bare walls, where the friction can depend strongly on the degree of incommensurability [37–41], τ was nearly independent of the orientation of the walls and the direction of sliding. The results were also independent of other factors that are not controlled in the experiments, such as the length of the molecules and their density.

VII. STICK-SLIP MOTION

Any jammed system can be unjammed by the application of a sufficiently large shear stress. The subsequent dynamics depend on many factors, including the types of metastable state in the system, the times needed to transform between states, and the mechanical properties of the device that imposes the stress. At high rates or stresses, systems usually slide smoothly. At low rates the motion often becomes intermittent, with the system alternately jamming and unjamming. Everyday examples of such stick-slip motion include the squeak of hinges and the music of violins.

The alternation between jammed and unjammed states of the system reflects changes in the way energy is stored. While the system is jammed, elastic energy is pumped in to the system by the driving device. When the system unjams, this elastic energy is released into kinetic energy, and eventually dissipated as heat. The system then jams once more, begins to store elastic energy, and the process continues. Both elastic and kinetic energy can be stored in all the mechanical elements that drive the system. The whole coupled assembly must be included in any analysis of the dynamics.

The simplest type of intermittent motion is illustrated by the multistable regime ($\lambda > 1$) of the Tomlinson model (Fig. 4(b)). Consider an atom that starts at rest in a metastable state and is pulled by a wall moving with constant center of mass velocity v. As the wall moves forward, the force from the spring increases. The atom moves gradually up the local metastable potential well, storing more and more energy as the force increases. When the local well becomes unstable, the atom pops rapidly to the next potential well. In the process, potential energy is converted to kinetic energy and dissipated by the damping term. As noted in Section IV, this dissipation approaches a constant value as $v_{\rm CM} \rightarrow 0$ because the characteristic atomic velocity during each pop remains large in this limit ($v_{max} \sim v_0 = F_0/\Gamma$). As $v_{\rm CM}$ increases, the motion becomes smoother and smoother, and for $v_{\rm CM} > v_0$ there is no period where the atoms are jammed.

This type of intermittent motion is very similar to the atomic-scale stick-slip observed with atomic force microscopes (AFMs) [114]. Even though the tip usually contains several atoms, its potential energy varies with the periodicity of the underlying surface. The effective spring constant reflects the entire mechanical system that imposes stress, including the internal stiffness of the tip, the cantilever that moves it, and the piezoelectric devices that displace the cantilever. In most cases, k is nearly equal to the cantilever stiffness, and the force is measured from the cantilever deflection. As the tip scans over the surface one often sees oscillations of the force that reflect sticking and slipping with the periodicity of an atomic spacing [114,115]. This indicates that $\lambda > 1$ and an estimate of λ can be obtained by reversing the direction of the scan and seeing how far one must move before popping in the opposite direction.

The above examples of stick-slip motion involve a simple ratcheting over a surface potential through a regular series of hops between neighboring metastable states. The slip distance is determined entirely by the periodicity of the surface potential. The granular media [10,12] and foams [11] discussed in the following reprints have a much richer potential energy landscape due to their many internal degrees of freedom. As a result, stick-slip motion between neighboring metastable states can involve a complicated sequence of slips of varying length. For example, Nasuno et al.'s study of stick-slip in bead packs [10] reveals an erratic series of slip distances that are much smaller than the bead diameter at low velocity (Fig. 2(b) of Ref. [10]). This implies that their system has metastable states that are separated by very small wall displacements. Presumably a small cluster of beads becomes unstable and rearranges at each slip event. This shifts stress to the remaining beads, causing them to move slightly within their potential wells. The net displacement will be much less than a bead diameter if the fraction of the beads involved in the rearrangement is small. Such local rearrangements are studied in the reprints by Veje et al. [12] and Gopal and Durian [11] who consider sand and foam, respectively.

Many examples of stick-slip involve a rather different type of motion that can lead to intermittency and chaos [116,117]. Instead of jumping between neighboring metastable states, the system slips for very long distances before sticking. For example, Gee et al. [7] observe slip distances of many microns in their studies of confined films. This distance is much larger than any characteristic periodicity in the potential, and varies with velocity, load, and the mass and stiffness of the SFA. The fact that the SFA does not jam after moving by a lattice constant indicates that sliding has changed the state of the system in some manner, so that it can continue sliding even at forces less than the yield stress.

One simple property that depends on past history is the amount of stored kinetic energy. This can provide enough inertia to carry a system over potential energy barriers even when the stress is below the yield stress. Inertia is readily included in the Tomlinson model and has been thoroughly studied in the mathematically equivalent case of an underdamped Josephson junction [46]. One finds a hysteretic response function like that sketched in Fig. 9. Both static and moving steady-states exist over a range of forces between F_{min} and the static friction F_s. Inertia leads to similar hysteresis in the dynamics of a sand grain along a sand pile [118].

Hysteresis in the Tomlinson model can produce stick-slip motion if the top wall has its own mass M and is connected through a much weaker external spring k' [119] to a stage that moves at a constant velocity u less than u_{min} (Fig. 9). The time-dependence looks much like that in the more complicated system shown in Fig. 10. While the wall is pinned, the force exerted by the external spring k' rises at a constant rate $dF/dt = k'u$. When the force exceeds the static friction, the wall starts to move at a velocity greater than u_{min}. As it catches up with the moving stage, the force drops. When it falls to F_{min}, the wall stops, and the process repeats.

The systems studied in the following reprints have many additional structural degrees of freedom that can change with time to produce hysteresis. Such changes lead to complex multivalued force/velocity curves [20,117,120,121] like that in Fig. 3(b) of Ref. [10] that are often modelled with "rate-state" equations. The changes in the system with sliding or sticking time are lumped into a single phenomenological state variable whose value depends on the sliding history [20,117,120,121]. Unfortunately it is difficult to determine the nature of the state variable without making microscopic structural measurements of the system. This can be difficult in experiments, but is relatively easily done in simulations.

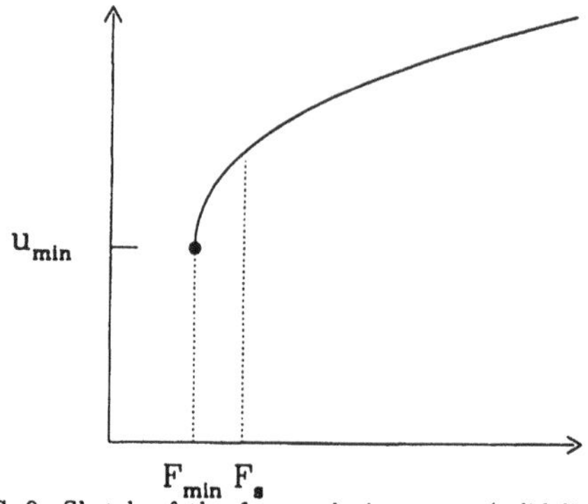

FIG. 9. Sketch of the force-velocity curve (solid line) for the underdamped Tomlinson model. The system is bistable over the range of forces between F_{min} and the static friction F_s (dotted lines). The size of this region depends on the degree of underdamping. Sliding velocities less than u_{min} are unstable.

One structural variable that is found to change in both experiments and simulations is the density. The degree of dilation is known to play a critical role in the yield and dynamics of granular media [122], and its role is discussed in the following reprints by Nasuno et al. [10] and Veje et al. [12]. Simulations of stick-slip motion in confined fluid films (Fig. 10) also showed dilation during slip [107], and dilation was subsequently measured in experiments on these systems [123]. In all cases, work must be done to expand the volume against the imposed external pressure (or gravity). Once this is done there is more room for shear to occur. The system may be able to keep sliding as long as it takes more time for the volume to contract, than for the system to traverse between metastable states.

Simulations of stick-slip in confined films show that other types of structural change can occur. In cases where confinement induces a crystalline structure in the film, the film may undergo periodic melting and crystallization transitions as it goes from static to sliding states and then back [107,124]. An example of this behavior is shown in Fig. 10. Glassy films [107,111] and sand [125] can undergo periodic "melting" and jamming transitions. As in equilibrium, the structural differences between glass and fluid states are small. However, there are strong changes in the self-diffusion and other dynamic properties when the film goes from the static glassy to sliding fluid state. These melting and freezing transitions are induced by shear and not by the negligible changes in temperature. Shear-melting transitions at constant temperature have been observed in both theoretical and experimental studies of bulk colloidal systems [44,45].

In the cases just described, the entire film transforms to a new state, and shear occurs throughout the film. Another type of behavior is also observed. In some systems shear is confined to a single plane - either a wall/film interface, or a plane within the film [110,111]. There is always some dilation at the shear plane to facilitate sliding. In some cases there is also in-plane ordering of the film to enable it to slide more easily over the wall [110,111]. This ordering remains after sliding stops, and provides a mechanism for the long-term memory seen in some experiments [7,103,126].

The dynamics of the jamming and unjamming transitions during stick-slip motion are crucial in determining the range of velocities where it is observed, the shape of the stick-slip events, and whether stick-slip disappears in a continuous or discontinuous manner [6,104,107,116,121]. Current models are limited to energy balance arguments [107] or phenomenological models of the nucleation and growth of jammed regions [6,104,121]. Microscopic models and detailed experimental data on the jamming and unjamming process are lacking. Kinetic models for other systems, like that in the reprint by Farr et al. [13], may help to understand these complex processes.

VIII. SUMMARY AND CONCLUSIONS

The preceding sections have described the key role jamming plays in many aspects of friction. At the same time they have attempted to draw parallels to other manifestations of jamming in this book.

The ubiquitous observation of static friction implies that jamming occurs at the interfaces between almost all objects. Jamming of crystals, pinning by disorder, and jamming of third bodies between the surfaces were discussed. The latter mechanism seems to provide the most robust explanation of static friction and Amontons' laws [51,54]. Third bodies are known to be present on almost all interfaces, whether in the form of air born hydrocarbons, wear debris or dust. Their interactions are dominated by hard core repulsions at typical pressures, and one may expect progress in the understanding of jamming in granular media, bulk glasses, and other systems to be directly relevant to our understanding of friction.

Two types of transition to a jammed state were described. The first is the solidification of fluid films as the thickness decreases. Both experiment and theory indicate that this transition is closely related to equilibrium glass transitions in many systems [7-9,111], although crystallization may also occur [97,100,107]. The second type of transition occurs during stick-slip motion where the system alternates between jammed and unjammed states. These non-equilibrium transitions present an even greater theoretical challenge, because of the extra dynamic degrees of freedom and the lack of general principles for determining the stable state of non-equilibrium systems.

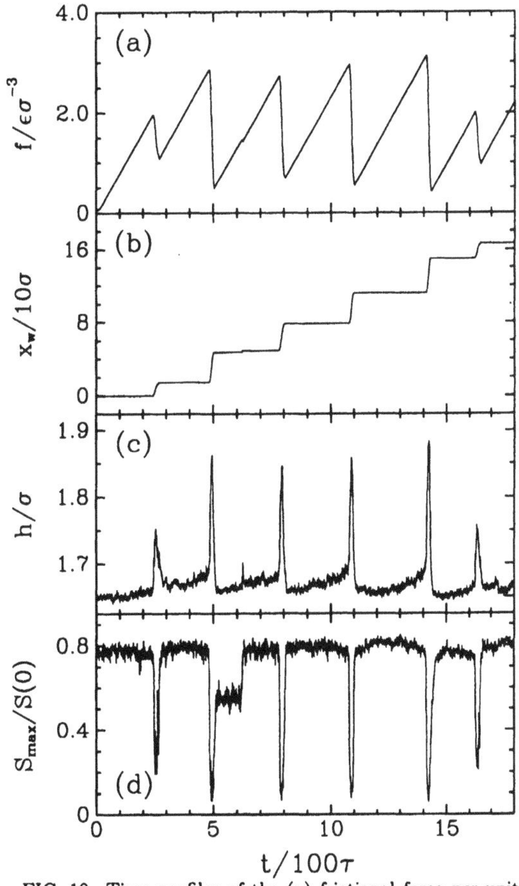

FIG. 10. Time profiles of the (a) frictional force per unit area f, (b) displacement of the top wall x_w, (c) wall spacing h, and (d) Debye-Waller factor $S_{max}/S(0)$ during stick-slip motion of spherical molecules that form two crystalline layers in the static state. Note that the system dilates (c) during each slip event. The coinciding drop in Debye-Waller factor shows a dramatic decrease from a typical crystalline value to a characteristic value for a fluid. Quantities are normalized by ϵ, σ and τ which are the characteristic energy, length and time scales of the Lennard-Jones potential.

ACKNOWLEDGMENTS

I thank my collaborators, A. R. C. Baljon, M. Cieplak G. S. Grest, G. He, R. Mountain, M. Müser, E. D. Smith, and P. A. Thomspon, for their help and insight. I would also like to thank the many experimentalists, especially S. Granick, J. Klein, J. Krim, J. N. Israelacvhili and P. M. McGuiggan, whose new techniques have changed the types of questions we can ask about friction. Finally, I gratefully acknowledge support from the National Science Foundation through Grant No. DMR-9634131 and through the Institute for Theoretical Physics.

[1] F. P. Bowden and D. Tabor, *The Friction and Lubrication of Solids*, (Clarendon Press, Oxford, 1986).

[2] *Fundamentals of Friction*, I. L. Singer and H. M. Pollock, Eds. (Elsevier, Amsterdam, 1992).

[3] E. Rabinowicz, *Friction and Wear of Materials* (J. Wiley, New York, 1965).

[4] D. Dowson, *History of Tribology* (Longman, New York, 1979).

[5] *Physics of Sliding Friction*, B. N. J. Persson and E. Tosatti, Eds. (Kluwer Academic, Dordrecht, 1996).

[6] B. N. J. Persson, *Sliding Friction* (Springer-Verlag, Berlin, 1998).

[7] M. L. Gee, P. M. McGuiggan, J. N. Israelachvili and A. M. Homola, J. Chem. Phys. **93**, 1895 (1990).

[8] A. L. Demirel and S. Granick, Phys. Rev. Lett. **77**, 2261 (1996).

[9] P. A. Thompson, M. O. Robbins, and G. S. Grest, Israel J. of Chem. **35**, 93 (1995).

[10] S. Nasuno, A. Kudrolli and J. P. Gollub, Phys. Rev. Lett. **79**, 949 (1997).

[11] A. D. Gopal and D. J. Durian, Phys. Rev. Lett. **75**, 2610-2613 (1995).

[12] C. T. Veje, D. W. Howell, and R. P. Behringer, Phys. Rev. E **59**, 739 (1999).

[13] R. S. Farr, J. R. Melrose and R. C. Ball, Phys. Rev. E**55**, 7203 (1997).

[14] R. Stribeck, Z. Ver. dt. Ing. **46**, 1341 (1902); *ibid* 1432 (1902); *ibid* 1463 (1902).

[15] D. Dowson in [2], pp. 325-349.

[16] D. Dowson and G. R. Higginson, *Elastohydrodynamic Lubrication* (Pergamon Press, Oxford, 1966).

[17] Y. Berthier, M. Brendel and M. Godet, STLE Tribology Trans. **32**, 490 (1989). Y. Berthier, L. Vincent and M. Godet, Wear **125**, 25 (1988).

[18] See for example J. A. Greenwood in Ref. [2], pp. 37-56.

[19] This was first pointed out by John Leslie [4].

[20] J. H. Dieterich and B. D. Kilgore, Pure Appl. Geophys. **143**, 283 (1994); Tectonophysics **256**, 219 (1996).

[21] P. Berthoud and T. Baumberger, Proc. R. Soc. A**454**, 1615 (1998).

[22] I. L. Singer in [2], pp. 237-261.

[23] B. J. Briscoe, D. J. Tabor, J. Adhesion **9**, 145 (1978). B. J. Briscoe, D. C. B. Evans, Proc. R. Soc. (London) **A380**, 389 (1982).

[24] A. M. Homola, J. N. Israelachvili, P. M. McGuiggan, and M. L. Gee, Wear **136**, 65 (1990).

[25] J. V. Alsten and S. Granick, Langmuir **6**, 876 (1990). A. L. Demirel and S. Granick, J. Chem. Phys. **109**, 1 (1998).

[26] J. A. Greenwood and J. B. P. Williamson, Proc. Roy. Soc. (London) **295**, 300 (1966).

[27] A. Volmer and T. Natterman, Z. Phys. B **104**, 363 (1997).

[28] See for example [22,23,1].

[29] P. W. Bridgeman, Proc. Amer. Acad. Arts Sci. **71**, 387 (1936).

[30] J. Boyd and B. P. Robertson, Trans. ASME **67**, 51 (1945).

[31] B. J. Briscoe and B. Scruton and F. R. Willis, Proc. Roy. Soc. A **333**, 99, (1973).

[32] M. J. Adams in [2] pp. 183-208.

[33] J.-C. Géminard, W. Losert and J. P. Gollub, Phys. Rev. E**59**, 5881 (1999).

[34] S. Nasuno, A. Kudrolli, A. Bak and J. P. Gollub, Phys. Rev. E **58**, 2161 (1998).

[35] G. A. Tomlinson, Phil. Mag. **7**, 905 (1929).

[36] G. M. McClelland in *Adhesion and Friction*, M. Grunze, H. J. Kreuzer, Eds. (Springer Verlag,Berlin, 1990), Vol. 17; p. 1. G. M. McClelland and J. N. Glosli in [2] pp. 405-425.

[37] Y. I. Frenkel and T. Kontorova, Zh. Eksp. Teor. Fiz. **8**, 1340 (1938).

[38] M. Hirano and K. Shinjo, Phys. Rev. B**41**, 11837 (1984); Wear **168**, 121 (1983). K. Shinjo and M. Hirano, Surf. Sci. **283**, 473 (1993).

[39] A review of the Frenkel-Kontorova model and the systems it describes in different dimensions is given in P. Bak, Rep. Progr. Phys. **45**, 587 (1982).

[40] S. Aubry, Physica D**7**, 240 (1983).

[41] An extensive review of the one-dimensional Frenkel-Kontorova model with $\eta = 1$ can be found in: O. M. Braun and Y. S. Kivshar, Phys. Rep. **306**, 1 (1998).

[42] See for example the combined Tomlinson and Frenkel-Kontorova of M. Weiss and F. J. Elmer, Phys. Rev. B**53**, 7539 (1996).

[43] J. N. Glosli and G. M. McClelland, Phys. Rev. Lett. **70**, 453 (1993).

[44] M. J. Stevens and M. O. Robbins, Phys. Rev. E**48**, 3778 (1993).

[45] B. J. Ackerson, J. B. Hayter, N. A. Clark, and L. Cotter, J. Chem. Phys. **84**, 2344 (1986) and references therein.

[46] See for example D. E. McCumbers, J. App. Phys. **39**, 3113 (1993).

[47] G. Grüner, A. Zawadowski and P. M. Chaikin, Phys. Rev. Lett. **46**, 511 (1981).

[48] E. Raphael and P. G. deGennes, J. Chem. Phys. **90**, 7577 (1989).

[49] J. F. Joanny and M. O. Robbins, J. Chem. Phys. **92**, 3206 (1990).

[50] D. S. Fisher, Phys. Rev. B**31** 1396 (1985).

[51] G. He, M. Müser and M. O. Robbins, Science **284**, 1650 (1999).
[52] M. R. Sorensen, K. W. Jacobsen, and P. Stoltze, Phys Rev. B **53**, 2101-2112 (1996).
[53] J. A. Snyman and H. C. Snyman, Surf. Sci. **105**, 357 (1981), and references therein.
[54] M. O. Robbins and E. D. Smith, Langmuir **12**, 4543 (1996).
[55] M. Hirano, K. Shinjo, R. Kaneko and Y. Murata, Phys. Rev. Lett. **78**, 1448 (1997).
[56] M. Liley, D. Gourdon, D. Stamou, U. Meseth, T. M. Fischer, C. Lautz, H. Stahlberg, H. Vogel, N. A. Burnham, and C. Duschl, Science **280**, 273 (1998).
[57] M. Hirano, K. Shinjo, R. Kaneko and Y. Murato, Phys. Rev. Lett. **67**, 2642 (1991). This paper also cites earlier work on anisotropy in friction.
[58] J. Krim, D. H. Solina and R. Chiarello, Phys. Rev. Lett. **66**, 181 (1991). C. Mak and J. Krim, Phys. Rev. **B58**, 5157 (1998).
[59] J. Krim, Sci. Am. **275**, 74 (1996).
[60] M. O. Robbins and J. Krim, MRS Bull. **23(6)**, 23 (1998).
[61] M. Cieplak, E. D. Smith and M. O. Robbins, Science **265**, 1209 (1994).
[62] E. D. Smith, M. O. Robbins and M. Cieplak, Phys. Rev. B **54**, 8252 (1996).
[63] M. S. Tomassone, J. B. Sokoloff, A. Widom and J. Krim, Phys. Rev. Lett. **79**, 4798 (1997).
[64] B. N. J. Persson, Phys. Rev. B**44**, 3277 (1991).
[65] J. B. Sokoloff, Phys. Rev. B**52**, 5318 (1995).
[66] M. S. Tomassone and A. Widom, Phys. Rev. B **56**, 4938 (1997).
[67] J. M. Martin, C. Donnet, Th. Le Mogne, and Th. Epicier Phys. Rev. B **48**, 10583 (1993).
[68] G. Grüner, Rev. Mod. Phys. **60**, 1129 (1988).
[69] A. I. Larkin and Y. N. Ovchinnikov, J. Low Temp. Phys. **34**, 409 (1979).
[70] G. Blatter, M. V. Fiegel'man, V. B. Geshkenbein, A. I. Larkin and V. M. Vinokur, Rev. Mod. Phys. **66**, 1125 (1994).
[71] J. P. Stokes, A.P. Kushnick and M. O. Robbins, Phys. Rev. Lett. **60**, 1386 (1988).
[72] N. Martys, M. O. Robbins and M. Cieplak, Phys. Rev. **B44**, 12294 (1991).
[73] C. S. Nolle, B. Koiller, N. Martys and M. O. Robbins, Phys. Rev. Lett. **71**, 2074 (1993).
[74] R. Bruinsma and G. Aeppli, Phys. Rev. Lett. **52**, 1547 (1984).
[75] J. Koplik and H. Levine, Phys. Rev. **B32**, 280 (1985).
[76] H. Ji and M. O. Robbins, Phys. Rev. B **46**, 14519 (1992).
[77] P. G. de Gennes, Rev. Mod. Phys. **57**, 827 (1985).
[78] J. F. Joanny and P. G. deGennes, J. Chem. Phys. **81**, 552 (1984).
[79] M. O. Robbins and J. F. Joanny, Europhys. Lett. **3**, 729 (1987).
[80] Y. Imry and S. Ma, Phys. Rev. Lett. **35**, 1399 (1975). G. Grinstein and S. Ma, Phys. Rev. B **28**, 2588 (1983).
[81] C. Caroli and P. Nozieres in Ref. [5], p. 27.
[82] B. N. J. Persson and E. Tosatti in Ref. [5], p. 179.
[83] J. N. Israelachvili, P. M. McGuiggan and A. M. Homola, Science **240**, 189 (1988).
[84] J. V. Alsten and S. Granick, Phys. Rev. Lett. **61**, 2570 (1988).
[85] S. Granick, Science **253**, 1374 (1991).
[86] D. Y. C. Chan and R. G. Horn, J. Chem. Phys. **83**, 5311 (1985).
[87] J. N. Israelachvili, J. Coll. Int. Sci. **110**, 263 (1986). J. N. Israelachvili and S. J. Kott, *ibid* **129**, 461 (1989).
[88] P. A. Thompson, G. S. Grest, M. O. Robbins, Phys. Rev. Lett. **68**, 3448 (1992).
[89] M. Schoen, C. L. Rhykerd, D. J. Diestler and J. H. Cushman, Science **245**, 1223 (1989). M. Schoen, J. H. Cushman, D. J. Diestler and C. L. Rhykerd, J. Chem. Phys. **88**, 394 (1988).
[90] L. Bocquet and J.-L. Barrat, Phys. Rev. E **49**, 3079 (1994); J. Phys. Condens. Matter **8** 9297 (1996).
[91] E. Manias, G. Hadziioannou, I. Bitsanis, G. ten Brinke, Europhys. Lett. **24**, 99 (1993). E. Manias, G. Hadziioannou, G. ten Brinke, J. Chem. Phys. **101**, 1721 (1994).
[92] J. Magda, M. Tirrell and H. T. Davis, J. Chem. Phys., 83 (1985) 1888.
[93] P. A. Thompson and M. O. Robbins, Phys. Rev. A**41**, 6830 (1990).
[94] J.-L. Barrat and L. Bocquet, Phys. Rev. Lett. **82**, 4671 (1999).
[95] P. A. Thompson and S. M. Troian, Nature **389**, 360 (1997).
[96] C. J. Mundy, S. Balasubramanian, K. Bagchi, J. I. Siepmann, and M. L. Klein, Faraday Discuss. **104**, 17 (1996).
[97] J. Klein and J. Kumacheva, Science **269**, 816 (1995).
[98] U. Landman, M. W. Ribarsky, J. Chem. Phys. **97**, 1937 (1992). T. K. Xia, J. Ouyang, M. W. Ribarsky, and U. Landman, Phys. Rev. Lett. **69**, 1967 (1995).
[99] J. Gao, W. D. Luedtke, and U. Landman, Science **270**, 605 (1995); J. Chem. Phys. **106**, 4309 (1997); J. Phys. Chem. **B101**, 4013 (1997).
[100] M. J. Stevens, M. Mondello, G. S. Grest, S. T. Cui, H. D. Cochran, and P. T. Cummings, J. Chem. Phys. **106**, 7300 (1997).
[101] R. G. Horn and J. N. Israelachvili, J. Chem. Phys. **75**, 1400 (1981).
[102] S. Toxvaerd, J. Chem. Phys., 74 (1981) 1998.
[103] H. Yoshizawa, Y.-L. Chen, and J. N. Israelachvili, J. Phys. Chem. **97**, 4128 (1993).
[104] H. Yoshizawa and J. N. Israelachvili, J. Phys. Chem. **97**, 11300 (1993).
[105] H.-W. Hu, G. A. Carson and S. Granick, Phys. Rev. Lett. **66**, 2758 (1991).
[106] G. Luengo, J. N. Israelachvili, and S. Granick, Wear **200**, 328 (1996).
[107] P. A. Thompson and M. O. Robbins, Science **250**, 792 (1990). M. O. Robbins and P. A. Thompson, Science **253**, 916 (1991).
[108] J. D. Ferry, *Viscoelastic Properties of Polymers* (Wiley, New York, 1980), 3rd. ed..
[109] W. Götze and Sjögren, Rep. Prog. Phys. **55**, 241 (1992).

[110] A. R. C. Baljon and M. O. Robbins, in *Micro/Nanotribology and its Applications*, B. Bhushan, Ed. (Kluwer, Amsterdam, 1997), p. 533.

[111] M. O. Robbins and A. R. C. Baljon, in *Microstructure and Microtribology of Polymer Surfaces*, V. V. Tsukruk and K.J. Wahl, Eds. (American Chemical Society, Washington DC, 2000), pp. 91-115.

[112] A. R. C. Baljon and M. O. Robbins, Science **271**, 482 (1996).

[113] G. S. Grest, M. H. Cohen, in it Advances in Chemical Physics, E. Prigogine, S. A. Rice, Eds. (Wiley, New York, 1981) p. 455.

[114] A review of AFM results can be found in R. W. Carpick and M. Salmeron, Chem. Rev. **97**, 1163 (1997).

[115] T. Gyalog, M. Bammerlin, R. Lüthi, E. Meyer and H. Thomas, Europhys. Lett. **31**, 269 (1995).

[116] F. Heslot, T. Baumberger, B. Perrin, B. Caroli and C. Caroli, Phys. Rev. **E49**, 4973 (1994).

[117] A. Ruina, J. Geophys. Res. **88**, 10359 (1983). J. C. Gu, J. R. Rice, A. L. Ruina and S. T. Tse, J. Mech. Phys. Sol. **32**, 167 (1984).

[118] H. M. Jaeger, C. H. Liu, S. R. Nagel and T. A. Witten, Europhys. Lett. **11**, 619 (1990).

[119] The values of M and k' must also give a characteristic frequency that is much lower than the frequency at which atoms ratchet over the periodic potential u/a.

[120] J. H. Dieterich, J. Geophys. Res. **84**, 2161 (1979).

[121] A. A. Batista and J. M. Carlson, Phys. Rev. E**57**, 4986 (1998).

[122] See for example, H. M. Jaeger, S. R. Nagel and R. P. Behringer, Rev. Mod. Phys. **68**, 1259 (1996)

[123] A. Dhinojwala and S. Granick, private communication.

[124] M. Lupowski and F. van Swol, J. Chem. Phys. **95**, 1995 (1991).

[125] P. A. Thompson and G. S. Grest, Phys. Rev. Lett., **67**, 1751 (1991).

[126] A. L. Demirel and S. Granick, Phys. Rev. Lett. **77**, 4330 (1996).

Nonequilibrium stochastic jamming models

Robin Stinchcombe

Theoretical Physics, Dept. of Physics, Oxford University, 1 Keble Road, Oxford OX1 3NP

1. Introduction

While granular materials, colloids, emulsions, molecular glasses, etc have distinguishing features, a wide variety of such systems also share certain properties such as jamming and slow dynamics. So it may be helpful to consider very simple models which can give such properties, both to obtain a minimal description and to seek any underlying unity or universality which may wait to be uncovered.

What are the simplest models for non-equilibrium behaviour in general, and in particular for such special collective effects as jamming and other non-equilibrium transitions, non-trivial steady states, sensitivity to preparation, ageing, dynamic slowing, and the like?

Clearly the models must involve many interacting constituents, as is the case for strongly collective behaviour in equilibrium statistical mechanics. There, the most important and simplest class of primitive collective models is lattice-based. This class comprises the Ising, XY, and Heisenberg models for magnestism, and for liquid crystals, etc. and lattice gases for the liquid-gas transition, together with other lattice models describing many other transitions, from geometric ones such as percolation (Potts models) to quantum ones (transverse Ising model, etc)

In the lattice gas the hard core interaction between particles is taken care of by excluding more than one particle from a site, having set up the lattice with the nearest neighbour site separation equal to particle diameter. The two possibilities for the state at each site makes the lattice gas equivalent to an Ising model in which spin up/down corresponds to particle/vacancy at a site. There are many other equivalences and relationships between the subclasses of equilibrium lattice model.

The lattice gas (and related models) has simple non-equilibrium generalisations [1,2,3] in which dynamics is added through prescribed stochastic or deterministic microscopic evolution rules. Such models, which still incorporate the hard core interactions through the single occupancy condition (one particle excluding another from a site), provide the primitive models for non-equilibrium behaviour, particularly in their stochastic versions. In order that the models should have non-trivial steady states, the (stochastic) evolution rules should not satisfy detailed balance conditions [4]. Then even the simplest such models show highly non-trivial non-equilibrium behaviour, such as jamming transitions (§§2, 3), sensitivity to preparation (§2), and the like. Ageing is also quite typical (§5), and does not even require the lack of detailed balance.

Such minimal lattice based stochastic models, and their continuum versions, are the principal subject of this overview. 'Driven' models (i.e. containing biased motion) is an important subclass (§3), which provides models for particulate flow, including traffic – e.g. for "laned" traffic in their one-dimensional versions. As such they exhibit jamming transitions ("jamming via exclusion") analogous to those occurring on highways subject to a large influx of vehicles. Two-dimensional lattice models have been used to mimic city traffic. Other models with deposition processes (§2) have jamming transitions analogous to 'poisoning' in catalysis. And so on.

It remains to be seen whether such models can adequately describe the jamming transitions and the glassy dynamics seen in e.g. compacted granular materials, colloidal systems, emulsions and glassy systems. Certainly they have been proposed, particularly for granular media and colloidal suspensions (§2, 3, 4, 6) in both discrete and continuum versions.

The models are specified in terms of the possible processes (configuration C • configuration C^1) and associated rates ($W(C,C^1)$). In the standard description [4] the Master Equation gives the rate of change of the probability $P(C)$ of configuration C as the sum, over configurations C^1, of $[P(C^1)W(C^1,C) - P(C)W(C,C^1)]$. From this it is easy to see why with detailed balance ($W(C,C^1)$ a ratio of Gibbs probabilities) the steady states are trivially the "Gibbs" states, i.e. having the equilibrium Boltzmann-Gibbs weight. Without detailed balance conditions on rates W the steady states are in general non-trivial, and thermodynamic descriptions do not in general apply. Non-Gibbsian rates are quite natural if e.g. geometric (e.g. caging) considerations decide them.

2. Deposition models, lattice- and continuum-based

The original lattice-based deposition-only models [5,6] were introduced to describe "random sequential absorption" of molecules on a surface. A simple version considers dimers depositing on empty nearest neighbour sites on a chain. Then, deposition occurs in unit time step on a randomly chosen bond (joining sites ℓ, $\ell + 1$, say) with probability ε (say), or not (with probability $1 - \varepsilon$) if the two sites ℓ, $\ell + 1$ are both empty; otherwise the dimer is excluded. Simulation, which is very easy for such processes, or a little thought, shows that the configuration evolves to a final completely jammed state having only isolated vacancies. The approach to the jammed state, e.g. as measured through the coverage density, is exponential in time, because each empty adjacent pair of sites fills at rate ε. This is true in any dimension, and actually regardless of lattice and molecular shape. The characteristic time and asymptotic coverage are less trivial to obtain, but are known exactly in certain cases [6].

Innocuously minor generalisations of the models can lead to very different results. The first of these generalisations considered here is the continuum car parking version (see e.g. [6]). This illustrates dramatically the very different long time slowing effects in continuum as opposed to lattice models, when excluded volume considerations enter directly. The continuum car parking model has been used [7] to mimic the late inverse logarithmic time dependence of the density observed in granular compaction experiments [8,9,10]. Shaking is assumed to cause particles in a given level to fall into spaces in the level below. This is seen as analogous to cars attempting to park at an already partly-filled kerb, and succeeding in doing so (with probability ε) if the space they attempt to enter can accommodate them. This is the one-dimensional continuum car parking model. If at time t the kerb coverage (ratio of filled to total space) is $c(t)$, the distribution of spaces x has Poisson form $p(x,c) \sim \exp[-x/\{a(1/c - 1)\}]$, where a is the 'car' length. The rate of decrease of the coverage ($-dc/dt$) is proportional to the probability that a chosen space has sufficient size to accommodate a car, i.e. to the integral of $p(x,c)$ from $x = a$ to ∞; which is about $p(a,c)$. That results in [7] $c \sim 1/\ell n\ t$). Here the excluded volume effects were crucial in slowing the deposition rate $\varepsilon\ p(a,c)$, which takes a Vogel-Fulcher dependence ($\varepsilon \exp\ [C/(1 - C)]$) on coverage, resulting in the anomalously slow symptotic time evolution. Such effects are missing from the simplest lattice-based versions, where individual vacancies are never less than a finite fraction of the molecular/car size. Additional rearrangement rules (as e.g. in [11]) can however produce Vogel-Fulcher rates in lattice based models.

A second generalisation of the lattice-based dimer deposition model [12,13] allows also for evaporation of dimers (made from any nearest neighbour pair of atoms on the surface). This could describe catalytic processes, or reduction of jamming in shaken granular media through exchanges either way between successive levels, etc. It turns out (from exact solution) that the asymptotic decay of auto correlations is power law in time. More interesting is the same lattice-based model but with k-mers ($k \geq 3$), rather than dimers ($k = 2$), depositing and evaporating, e.g. on a linear chain. In this model the number of jammed states is exponential in system size [13], due to exponentially-many macroscopic conserved variables. These divide the dynamic evolution into exponentially many unconnected sectors, making the evolution strongly non-ergodic and making the eventual steady state dependent on the initial state. The conservation laws result in a splitting of the "critical" (long time asymptotic) dynamics into exponentially many dynamic universality classes [14]. The same "dynamic diversity" type of phenomenon also occurs with e.g. dimer diffusion models [15]. Whether continuum models can have this type of behaviour is not known: the usual procedures for continuum limits do not seem to apply. It is not yet clear whether the strong non-ergodicity, exponentially many jammed states, and diverse dynamics of the lattice models will be inherited by any continuum versions, or whether these properties have related analogues in the behaviour of realistic exclusion systems (e.g. grains).

3. Jamming transitions in driven systems and traffic models

Driven lattice gases (having rules producing biased motion) have received much attention both in connection with non-equilibrium phase transitions and in attempts to clarify non-equilibrium states of flowing particles [2]. In connection with steady states, the normal associations of constant gradients with constant currents fail completely.

In the simplest driven stochastic model, fully biased hard core diffusion, alias "the asymmetric exclusion process" [3], the microscopic prescribed process on any chosen bond ℓ, $\ell + 1$ of, say, a linear chain is: particle at ℓ hops with probability p (or stays put with probability $1-p$) onto site $\ell + 1$ provided $\ell + 1$ was vacant. The collective effects make it impossible to force a steady particle current greater than about p/4 through the system. An associated steady state phase transition is seen [16, 17]. The steady density–velocity characteristics and dynamic shock waves of the model are not unlike those seen in highway traffic [18], for which it provides the simplest model. Various generalisations have been proposed as models for flow of grains and suspensions, etc.

An exact solution for the steady state phase transition in the primitive one-dimensional model has been obtained, first by a recursion method [17] and then by an elegant steady state operator algebra technique [19,20]. The operator algebra has been extended to the full dynamic situation [21], but not fully reduced except in a special case.

It turns out that a mean field approach provides an exact location of the phase transition (even though fluctuation effects are large in $d = 1$); and it gives a qualitatively correct understanding, as follows. Denoting particle density at site ℓ by $n(\ell)$, in a mean field approximation neglecting correlations the average particle current across bond ℓ, $\ell + 1$ is $J(\ell,\ell + 1)=p\ n\ (\ell)\ \{1-n(\ell + 1)\}$. Equating the difference of currents on adjacent bonds to the rate of change on $n(\ell)$ at the intervening site ℓ gives a (continuity) equation, which determines $n(\ell,t)$ for given boundary and initial conditions. Gradient expansions are adequate near the transition where the characteristic length $1/a$, given below, is long. They produce a continuum version of the continuity equation containing the divergence of the continuum current $J = n(1-n) - dn/d\ell$. This continuity equation is the noiseless Burgers equation [22]. It can be solved completely via exact linearisation using the Cole-Hopf transformation [23]. The general solution is a superposition of solutions. The corresponding steady state solution, from J = constant, takes different forms on either side of a transition at $J = J^*$, where $J^* = p/4$ is the critical current. For $J < J^*$, the profile is a steady shock of width $1/a$ where $a^2 = |J - J^*|$; specifically $n = \frac{1}{2} + a \tanh a\ (\ell - \text{constant})$. For $J > J^*$, the form is similar, but with a tan function. In a large system this latter form requires small a, as n is limited to the interval 0 to 1. The transition can be seen as boundary induced [17] if, in analogy with cars entering or leaving a highway, particles are injected at attempt rate α at the left hand boundary site $\ell = 0$ (subject to it being unoccupied); and ejected at rate β at the right boundary site $\ell = L$ (so $J = \alpha\ n(0) = \beta(1-n(L))$. That makes $J \sim J^*$ for α,β, both greater than ½, and makes $J < J^*$ otherwise. So the boundary effects are influencing the steady state throughout the system, however large it may be.

Fluctuation effects neglected in mean field theory are most severe in the lowest dimensions, and they normally prevent equilibrium phase transitions in $d = 1$. That is not the case here: the exact solutions [17,19] establish the transition in $d = 1$. Its occurrence is related to the fact (see §4) that the non-equilibrium models map into equilibrium quantum spin models, and the steady state transition maps into a conventional equilibrium quantum transition existing in one space and one time dimension.

The one dimensional asymmetric exclusion model also maps, via a density-height transformation, to an interface growth model [24]. In its continuum limit the height variable h, related to the particle density n in the Burgers equation by $n = dh/d\ell$, satisfies the Kardar-Parisi-Zhang equation [25]. All this applies in mean field theory, where the equations are noiseless, but also without making the mean field approximation: then the equations have noise terms, representing the effect of correlation fluctuations. The application of field theory to these noisy equations generates the correct critical exponents [26,25]. One of these is the dynamic critical exponent z (=3/2 in $d = 1$) relating the width w of the shock front or growing interface with time through $t \sim w^z$. In the mean field description the shock front does not broaden, which is clearly wrong. The result $z = 3/2$ is also provided by an exact treatment using the Bethe ansatz [27, 28]. Subsequent field theoretic treatments of the model are contained in [29,30,31]; the operator algebra results for steady state properties are reviewed in [20].

The asymmetric exclusion process provides a steady state jamming transition and crudely accounts for basic dynamical aspects of single-lane traffic behaviour, such as bunching regions travelling at different speeds from the local particle velocity. But it is too oversimplified for some effects and for more detailed comparisons. One generalisation, allowing for randomly chosen hopping rates [32], provides a Bose-Einstein type transition corresponding to traffic piling up behind slow vehicles. Other generalisations, allowing for different velocity states [33,34] produce more realistic descriptions. Primitive models for city traffic include the two dimensional asymmetric exclusion process and rather more successful models (see especially [35]) which can provide a transition to complete jamming as well as self-organisation effects.

A two-species generalisation of the asymmetric exclusion process has been introduced [36,37] which shows spontaneous symmetry breaking. Here, + and – particles attempt to move in opposite directions on an open chain. The dynamical rules are invariant under charge conjugation combined with space inversion, and when the steady state preserves the symmetry the currents of + and – particles are equal. There is a transition to a broken-symmetry state where the two currents become unequal. The symmetry-breaking is analogous to that encountered in heavy traffic conditions on two-way single-lane roads with passing places.

A further two-species model, in which a "bus route" interpretation the two species correspond to bus or passengers at a site, shows sharp crossovers and coarsening; one limit gives jamming via a symmetry-breaking phase transition [38,39]. This corresponds to clustering of buses or, in another interpretation in which the two species correspond to different mobility states, to clogging of a suspension forced along a pipe.

Away from the traffic context, further generalisations of the fully asymmetric exclusion process have been much investigated, and some of the generalisations are referred to in the next subsection.

4. Generalised non-equlilibrium particle models, mappings and methods

The partially asymmetric exclusion process (particles hop to vacant nearest neighbour sites to left or right with different probabilities p,q) has for $p \neq q$ a jamming transition and other properties very like the fully asymmetric case. The continuum version is again the Burgers equation. The case $p = q$ is symmetric hard core diffusion. This is of interest in many contexts, e.g. gas diffusion in solid. But it is relatively simple because of the uncoupling of the governing hierarchy of kinetic equations, and because of the corresponding simplified form of the operator algebra describing it [21]. With boundary injection the profiles are linear in space [40,21]. As mentioned in §2, symmetric or asymmetric diffusion of excluding unreconstituting dimers has the dynamic diversity associated with exponentially many macroscopic conserved quantities [15], so having far richer properties, not all properly understood, than the diffusing hard core particle model.

The largest source of generalisations of the asymmetric exclusion process comes from combining its hard core hopping with other processes. Examples involve combination with pair creation (same as dimer deposition) and or pair annihilation (evaporation), or k-particle annihilation or creation, or k particles transforming to j particles, etc. These models provide descriptions of surface reactions, catalysis etc [41]. Other processes which can be combined include aggregation and fragmentation. Such systems can exhibit dynamical [42] and steady slate [43] phase transitions, including shear-induced clustering.

Such models can be lattice-or continuum-based. For cases where the particle or vacancy density evolves towards zero (e.g. hard core hopping with pair annihilation) the late-time dependence is expected to be the same in continuum and lattice based versions, since effective spacings become much greater than lattice constants. The late time behaviour is related to coarsening and ageing, as discussed in §5.

Multispecies models are another source of generalisations and these provide further examples of phase transitions. A three species model with simple rules has been shown to exhibit phase separation on a cyclic chain [44,45].

Some such models are amenable to exact solutions or to exact mappings which can be very helpful. Single species models (particle or vacancy at each site) can be mapped to an equivalent spin ½ system (spin up or down at each site) – and there are obvious generalisations for multi-species systems. The time evolution involves moving particles to and from sites, corresponding to spin flips. Such moves are generated by an evolution operator exp[-Ht] where H is an appropriate quantum spin Hamiltonian, since it contains spin flip operators. Such mappings provide procedures for the understanding and solution of the non-equilibrium particle models via standard viewpoints and techniques for the equivalent quantum spin systems [12,13,27,28,46].

In some cases the quantum spin procedures provide exact solutions. An example [47,48,49] is one-dimensional asymmetric hardcore diffusion (rates p,q for right and left hopping) combined with pair creation and annihilation (rates ε, ε^1). Here the Jordan Wigner representation of the spin operators takes H into a fermion Hamiltonian. Further, when the rates are related by $p + q = \varepsilon + \varepsilon^1$ the Hamiltonian is of free fermion form, and can be diagonalised by a Bogoliubov-Valatin transformation. This provides a complete solution for all dynamical and steady state properties. By this means correlation functions [47,48,49] and even persistence functions [50,51] have been obtained. Exponential time decays occur when ε, ε^1 are both non zero because of a gap in the spectrum. When either $\varepsilon = 0$ or $\varepsilon^1 = 0$, density and auto correlation functions have power law long-time dependences, while the persistence time dependence is power law [50] or stretched exponential [51] depending on whether diffusion is symmetric ($p = q$) or not.

Another source of exact solutions is field theory. Since the original work of Forster, Nelson and Stephen [26], the field theoretic approaches have been developed [52,53] so that they now provide powerful direct approaches to the discrete particle non-equilibrium processes. This allows some processes to be exactly renormalised to all orders in perturbalion theory, for example the reaction $A + A \to A$ [54]. Other applications are contained in [55,56,57,58].

5. Coarsening, ageing, fluctuations and "glassy" behaviour

We now turn to ageing and coarsening in spin and non-equilibrium particle models. The standard simplest demonstration of ageing is a spin system evolving towards an ordered single-domain low temperature state via coarsening of domain configurations. Exactly analogous ageing occurs in many non-equilibrium systems, such as those with decay processes, and this can be seen directly or sometimes via equivalences.

For generality we consider coarsening in a d-dimensional exclusion model having symmetric diffusion together with k-particle annihilation. This includes the hard core particle model with symmetric diffusion and pair annihilation, but no creation (i.e. $p = q$; $\varepsilon = 0$, $\varepsilon^1 \neq 0$) as a special case ($k = 2$). In the late time regime, annihilation will have thinned out the particles until their separation is large and on average equal to $1/n(t)^{1/d}$, where n is the density. Then, provided the diffusion is efficient enough to smooth the density, mean field considerations of the following direct type will apply: at late times hard core effects and lattice versus continuum distinctions become immaterial and the characteristic rate for a particle to annihilate is $\varepsilon^1 n^k$. Equating this to the rate of decrease of n gives $n \sim t^{-1(k-1)}$ at long times in mean field theory. Correlation functions, which explicitly demonstrate characteristic ageing time-dependences, can be calculated similarly.

Any such mean field argument is strictly limited to dimensions sufficiently large to make fluctuation effects unimportant, e.g. as raised earlier in connection with possible suppression of a mean field transition. In the ageing argument we are more interested in whether fluctuations modify the critical exponent in the density-time relation. That can be decided by the following consistency check [59]. Due to k-particle annihilation alone, the power of t with which the particle spacing increases in d-dimensions is $1/[d(k-1)]$. The corresponding power for the increase of the diffusion length is ½. So diffusion can smooth the otherwise increasingly inhomogenous particle distribution at long times, as assumed in the mean field discussion, if d is greater than the 'upper critical dimension' $2/(k-1)$.

For the special 1d case with $k = 2$ (and $p = q$, $\varepsilon = 0$, $\varepsilon^1 \neq 0$) this criterion is clearly violated, so the mean field discussion for that particular non-equilibrium particle model gives an incorrect critical exponent. That can be confirmed by comparison with the exact free-fermion results [47,49] or with field theoretic approaches [58].

The 1d model is actually equivalent to the domain wall model in spin systems. There single spin-flip Glauber dynamics causes domain wall diffusion, and pair annihilation and creation of domain walls, at temperature-dependent rates set by detailed balance. At low temperatures the creation process can be neglected. Identifying the domain wall with the particles in the non-equilibrium model provides the equivalence. Actually the detailed balance condition corresponds to the free fermion condition $p + q = \varepsilon + \varepsilon^1$ in the particle system. Further discussion of the equivalence can be found in [60].

Long-time dependences of exponential power law, and stretched exponential character were referred to for specific cases mentioned above. Generalised models can have more extreme slow dynamics and even anomalously slow rates, analogous to Vogel-Fulcher or exponential inverse temperature squared behaviour (EITS) in glasses. One such model [61] has been proposed, and recently solved [62], which involves a lattice of spins evolving with kinetically constrained Glauber rules. A given spin can flip only if its left neighbour is up; in that case the flips from up to down or from down to up occur with rates 1 or $\varepsilon = \exp(-1/T)$ (simple activation). Since detailed balance applies, the steady states are simple equilibrium states corresponding to free spins in a unit downward field. So at high temperatures spins are equally likely to be up or down. After a quench from such a state to very low temperatures (where $\varepsilon << 1$) the state coarsens until inhibited by the need to utilise the slow process more and more. The spin model has a lattice-gas equivalent (in which the rate ε might again be set by activation, as we'll continue to take here, or by other considerations such as a local caging effect for example). Here spin up (down) corresponds to particle (vacancy), which we represent by 1(0). Using this equivalent lattice gas, the processes are: $11 \to 10$, rate 1; $10 \to 11$, rate ε. After the quench the slowly relaxing domains have 1 followed by one or more 0's. The more 0's there are following the 1, the more slow moves that are needed to relax the configuration towards equilibrium. For a large number d of 0's, the number of slow moves needed is $n(d) \sim b\ell nd$, where b is a constant of order 1. The characteristic rate to relax to a domain of equilibrium size $D - \exp(1/T)$ is therefore about $\exp[n(D)\ \ell n\ \varepsilon] \sim \exp[-b/T^2\}$. This is the EITS behaviour. The model also provides stretched exponential relaxation of the autocorrelation function, with temperature-dependent stretching exponent [62].

6. Discussions

The simple non-equilibrium stochastic models (lattice-based and continuum exclusion models, or equivalent spin ones) have been motivated by their incorporation of bare essentials (many consitutents, interaction of hard core type, simple dynamic rules) and their ability to mimic collective effects, and to incorporate geometric constraints, fluctuation behaviour, etc. In that sense they provide minimal models for non-equilibrium behaviour. Typical features they provide include non-Gibbsian steady states (§2), and dissipation, and non-linear effect. Among the representative examples discussed are ones which show

(i) asymptotic time dependences of power law, exponential, stretched exponential or inverse log type (§§2,3,4,5)
(ii) rates of Vogel Fulcher or exponential inverse temperature squared type (§§ 2,5)
(iii) primitive ageing due to coarsening (§§5)
(iv) partial or complete jamming, and other non-equilibrium phase transitions such as symmetry breaking and phase separation (§§2,3,4).
(v) strong non-ergodicity related to exponentially many disconnected dynamic sectors, and steady states decided by initial conditions (§§2).

It is remarkable that they exhibit such a wide range of phenomena, given their very limited ingredients: all had no mass, momentum, force, friction ingredients (so without generalisation they could not exhibit phenomena such as inelastic collapse governed by such ingredients [63,10])

It could however appear that the particular microscopic dynamic processes incorporated in some of the models considered are contrived so that they will give rise to features such as (ii), (iv), (v). So one should consider what justifies these prescribed dynamic rules; how are they connected to the real microscopic physics, and in particular how relevant are they for rheology, glassy dynamics and for specific systems such as granular media, colloids, emulsions and molecular glasses. The aim is for relevance, but maintaining simplicity.

A related question concerns universality. Should there be a universal account of e.g. jamming or of glassy dynamics, and if so can simple non-equilibrium models provide it. Probably the models themselves (e.g. those referred to in §§2,3) are useful in already showing that there are different forms of jamming (partial or complete, arrived at with or without re-organization, etc) if that were not already obvious. These distinctions seem to appear in real traffic situations (including the obvious differences between city and highway jams), in related flow problems, and in catalytic poisoning etc, where surface mobility matters; these are systems for which the simple models seem quite relevant. So any universal view of jamming would have to allow for a range of universality classes or cross-over situations.

The subtle distinctions between the behaviour of granular media, colloids, emulsions and molecular glasses (as discussed elsewhere in this volume) suggest that the aim for simple models can only provide descriptions of limited value.

Nevertheless, it seems worthwhile to seek a more robust and generic origin for e.g. glassy behaviour within the simple models, based on the real physics of particular classes of system. And in particular, rules incorporating excluded volume concepts, as for example reviewed in [64] would seem an appropriate starting point for granular systems, and possibly some other classes. How distilled the description can be while still accounting for more subtle effects is not yet known; so far no models seem capable of accounting for e.g. the noise spectrum in highly compacted shaken granular systems in the "glassy" regime where the spectrum is becoming very broad [8].

Acknowledgements

It is a pleasure to acknowledge the stimulus and free exchange of ides provided by colleagues at the 1997 ITP Jamming Workshop, and to thank the Institute Director and the workshop organisers for the opportunity to participate. Partial support from EPSRC grants GR/K97783 and GR/MO4426 is also gratefully acknowledged.

References

1. "*Non Equilibruim Statistical Mechanics in One Dimension*" ed V. Privman (Cambridge University Press, to be published) and references therein.
2. B. Schmittmann and R K P Zia, *Statisitcal Mechanics of Driven Diffusive Systems in Phase Transitions and Critical Phenomena*, C Domb and J L Lebowitz, eds (Academic, New York), (1995).
3. T M Liggett "*Interacting Particle systems*" (Springer-Verlag, New York, 1983)
4. N G van Kampen, "*Stochastic Processes in Physics and Chemistry*" (North-Holland, Amsterdam, 1992), 2nd Edn.
5. P J Flory, *J Am. Chem. Soc* 61 (1939) 1518, for review of the area, see.
6. J W Evans, *Rev. Mod. Phys.* 65 (1993) 1281
7. E. Ben-Naim, J B Knight, and E R Nowak, Report No. cond-mat/9603150; J Chem. Phys. **100**, 6778 (1994)
8. J B Knight, C G Frandrich, C Ning Lau, H M Jaeger and S R Nagel, Phys. Rev. E **51**, 3957 (1995)
9. H M Jaeger and S R Nagel, Science **255**, 1523 (1992)
10. H M Jaeger, S R Nagel, and R P Behringer, Phys. Today **49**, No. 4, 32 (1996)
11. Caglioti E, Loreto V, Herrmann H J, Nicodemi M, *Physical Review Letters*, 1997, Vol. 79, No. 8, pp. 1575-1578
12. M Barma, M D Grynberg, R B Stinchcombe, *Phys. Rev. Lett* 70, 1033 (1993)
13. R B Stinchcombe, M D Grynberg, M Barma, *Phys Rev* E47, 4018 (1993)
14. M Barma and D Dhar, *Phys. Rev. Lett* 73, 2135 (1994)
15. M Barma and D Dhar, Proceedings of the 19th IUPAP Conference on Statistical Physics, Xiamen, China, 1995 (World Scientific, Singapore) ed. B-L Hao (1996)
16. J Krug, *Phys. Rev. Lett* 67, 1882 (1991)
17. B Derrida, E Domany and D Mukhamel, *J. Stat. Phys.* 69, 667 (1992)
18. B S Kerner and H Rehborn, *Phys. Rev.* E53, R1297 (1996)
19. B Derrida, M R Evans, V Hakim, V Pasquier, *J. Phys.* A26, 1493 (1993)
20. Derrida B *Physics Reports - Review Section of Physics Letters*, 1998, Vol. 301, No.1-3, pp.65-83
21. R B Stinchcombe and G M Schutz, *Phys. Rev. Lett* 75, 140 (1995)
22. J M Burgers "*The Non-linear Diffusion Equation*" (Riedel, Boston 1974).
23. E Hopf, *Commun. Pure Appl. Math 3*, 201, (1950); J D Cole, *Quart: Appl. Math* 9, 225 (1951)
24. Meakin P, Ramanlal P, Sander L M, Ball R C, *Physical Review A*, 1986, Vol 34, 5091
25. M Karda, G Parisi and Y-C Zhang, *Phys. Rev. Lett* 56, 889 (1986)
26. D Forster, D Nelson and M J Stephen, *Phys. Rev.* A16, 732 (1977).
27. L-H Gws and H Spohn *Phys. Rev. Lett* 68, 725 1992)
28. Gwa L H, Spohn H, *Physical Review A* 1992, vol. 46, 844
29. T Hwa and E Frey, *Phys. Rev.* A44, R7873 (1991)
30. E Frey and U C Tauber, *Phys. Rev.* E50, 1024 (1994)
31. E Frey, U C Tauber and T Hwa, *Phys. Rev.* E53, 4424 (1996)
32. Evans M R, *Journal of Physics A – Mathematical and General*, 1997, Vol. 30, 5669
33. Nagel K, Schreckenberg M, *Journal De Physique I*, 1992, Vol. 2, 2221
34. Sasvari M, Kertesz J, *Physical Review E*, 1997, Vol. 56, 4104
35. Biham O, Middleton A A, Levine D, *Physical Review A*, 1992, Vol. 46, No. 10, R6124
36. M R Evans, D P Foster, C Godrèche and D Mukamel, *Phys. Rev. Lett.* 74, 208 (1995)
37. Evans M R, Foster D P, Godrèche C, Mukamel D, *Journal of Statistical Physics*, 1995, Vol. 80, 69.
38. O'Loan O J, Evans M R, Cates M E *Physical Review E*, 1998, vol. 58, A, 1404
39. O'Loan O J, Evans M R, Cates M E *Europhysics Letters*, 1998, Vol. 42, 137
40. H Spohn, in Statistical Physics and Dynamical Systems: Rigorous Results (Birkhauser Boston, Boston, 1985)
41. Ziff R M, Gulari E, Barshad Y, *Physical Review Letters*, 1986, Vol. 56, 2553
42. S N Majumdar, S Krishnamurthy, & M Barma *Physical Review letters* 81, 3691 (1998)
43. O'Loan O J, Evans M R, Cates, M E, *Physica A*, 1998, Vol. 258, 109.
44. M R Evans, Y Kafri, H M Koduvely and D Mukamel *Phys. Rev. Lett.* 80, 425 (1998).
45. Evans M R, Kafri Y, Koduvely H M, Mukamel D *Physical Review E*, 1998, Vol. 58, A, 2764.
46. F C Alcaraz, M Droz, M Henkel and V Rittenberg, *Ann. Phys.* (N.Y.) 230, 250 (1994)
47. M D Grynberg and R B Stinchcombe, *Phys. Rev. Lett* 74, 1242 (1995)
48. Grynberg M D, Stinchcombe R B, *Physical Review E*, 1995, Vol. 52, A, 6013
49. Grynberg M D, Stinchcombe R B, *Phys. Rev. Lett* 76, 851 (1996)
50. B Derrida, V Hakim and R Zeitak *Physical Review Letters* 77, 2871 (1996)
51. M J Stephen and R B Stinchcombe *Journal of Physics A*, to be published (1999)
52. M Doi, *J. Phys.*, A: Math. Gen. 9 1465, 1479, (1976)
53. L Peliti, *J. Physique* 46 1469, (1985)
54. Peliti L, *Journal of Physics A – Mathematical and General*, 1986, Vol. 19, L365
55. B P Lee, *J. Phys.* A: Math. Gen. 27, 2633 (1994)
56. B P Lee and J Cardy, *Phys. Rev.* E50, R3287 (1994)
57. B P Lee and J Cardy *J.Stat. Phys.* 80, 971, (1995)
58. M Howard and J Cardy: *J.Phys.* A Math. Gen. 28 3599 (1995)
59. K Kang and S Redner, *Phys. Rev.* A32, 435 (1985)
60. Stinchcombe R B, Santos J E, Grynberg M D *Journal of Physics A – Mathematical and General*, 1998, Vol. 31, 541
61. J Jäckle and S Eisinger, Z. Phys. B 84, 115 (1991).
62. P Sollich and M R Evans – to be published
63. S McNamara and W R Young, *Phys. Fluids A* 4, 496 (1992); Phys. Rev. E50, R28 (1994)
64. Herrmann H J, *Physica A*, 1999, Vol.263, 51

Rheology, and how to stop aging[a]

Jorge Kurchan

Laboratoire de Physique Théorique de l'École Normale Supérieure de Lyon, 46 Allée d'Italie, F-69364, Lyon Cedex-07, France

Recent analytical developments in glass theory are equally relevant to the understanding of anomalous rheology, with characteristic features such as the Reynolds dilatancy and the driving-power dependence of the viscosity arising naturally. A notion of effective temperature based on the fluctuation-dissipation relation can be introduced in the limit of small driving power. Within mean-field, the analogue of the Edwards compactivity can be computed, and it coincides with this effective temperature. The approach does not invoke any particular geometry for the constituents of the fluid, provided it has glassy behaviour.

1. Aging and Rheology

Systems which have properties that depend on the time since preparation t_w are said to 'age'. The simplest example is phase-separation of two inmiscible fluids. The fluids form domains whose size keep increasing with time, and if the system is infinite this process never stops. Another simple case is that of a dry foam, with the average bubble size growing like a power of time.

In many aging systems such as inorganic glasses, plastics, gels and spin-glasses, we do not have (or we do not know) a simple way to visualize 'what is growing', but we can still measure quantities that depend on the waiting time t_w. The measurements fall basically into two categories: two-time correlations (mean-squared displacement of particles or polymers, spin autocorrelations, etc); and responses (contraction of a sample at time $t_w + \tau$ following a pressure application since time t_w, magnetization evolution after a field has been turned on at t_w, etc.).

Figures 1 and 2 show the characteristic aging curves obtained: the typical relaxation time τ^{rel} does not immediately become infinite, but grows together with the waiting time. If a system has a very long but finite equilibration time t^{equil} (like, for example, a supercooled liquid just above the glass transition), τ^{rel} grows with the age until it levels off at $t_w \sim t^{equil}$.

If we inject power into any of these aging systems, depending on the form of the drive there is the possibility of stabilizing the age of the sample in a power-dependent level: the younger the larger the power input. The phase-separation example is very clear: this is what we do when we shake a mixture of vinegar and oil to make the oil droplets smaller.

The viscosity of certain gelling systems is known to increase with time, and can hence be taken as a measure of their age. When shear forces are applied to these systems their viscosity stabilises to a shear-rate dependent value[1].

Another intriguing case is that of gently, regularly tapped sand[2]. Aging under these circumstances means compactification, which makes the mobility of grains decrease with time[3]. The Reynolds effect, the fact that sand swells when sheared, is yet another example of rejuvenation and stabilization of aging through power injection.

In fact, the relation between rheology and aging in glasses is much like the relation between driven and decaying turbulence. Just like with turbulence, the driven, stationary situation is in many senses simpler.

In the last years there has been a development of analytical ideas originated in spin-glass theory. In section 2, I briefly review them as applied to purely relaxational structural glasses (there is by now quite a large literature on this case, see[4], and references therein). In section 3, I describe the implications for rheology of the same theoretical ideas. This line of research has been much less explored so far, and there are a couple of questions (Reynolds dilatancy, Edwards compactivity) that have not, to my knowledge, been discussed yet within this context.

---

[a] Proceedings of 'Jamming and Rheology: Constrained Dynamics on Microscopic and Macroscopic Scales' ITP, Santa Barbara, 1997. The original talk can be found in http://www.itp.ucsb.edu/online/jamming2.

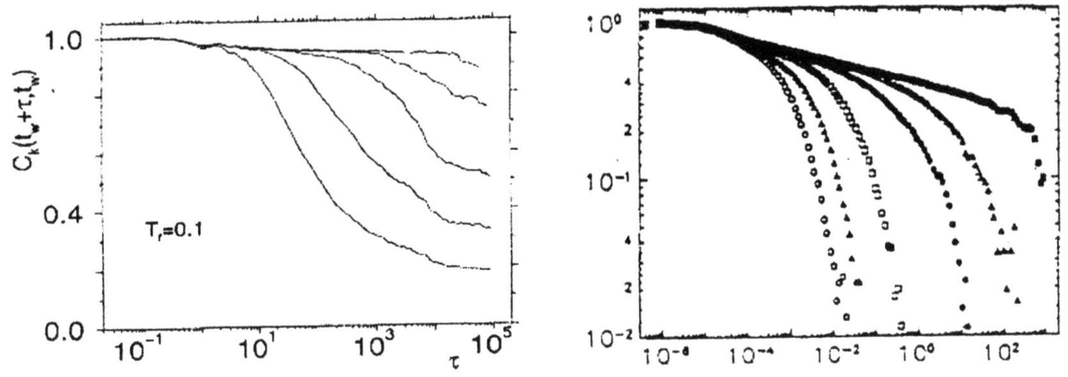

Figure 1: Autocorrelation decay for different waiting times. Left: Lennard-Jones binary mixture, molecular dynamics simulation [28] (waiting times from 10 to 39810). Right: Light scattering data for laponite gels [40] (waiting times of 11 to 100 hours). (See also [41] for similar curves for polymer melt models, [44] for spin-glass simulations, and [42,43] for polymers in random media).

2. (Relaxing) Glasses

The considerations so far have been generic to systems with slow dynamics. It turns out that these systems fall into two classes[4]. The first class includes spin-glasses and ferromagnetic domain growth, and is characterized by having a sharp transition temperature (or density) below which slow dynamical properties appear. Systems of the second class, comprising window glasses, plastics and gels, have instead a crossover range in temperature or density in terms of which some characteristic time (typically related to the viscosity) grows rapidly. In what follows I shall only deal with the latter, although some of the discussion also applies to the former class.

2.1. The Kirkpatrick-Thirumalai-Wolynes realization of the Adam-Gibbs scenario

In a series of papers[5], Kirkpatrick, Thirumalai and Wolynes pointed out that the essential features of structural glasses can be seen (albeit in a rather caricatural way) in microscopic models within an approximation that, depending on the context, is called 'direct interaction', 'mode-coupling' or 'mean-field'[7] and is exact for a family of fully connected disordered models[4]. In fact, their discussion can be extended without real complications to the whole family of approximations consisting in 'closing' the problem by reexpressing everything in terms of one and two-point correlations.

For ease of presentation, let us run the argument

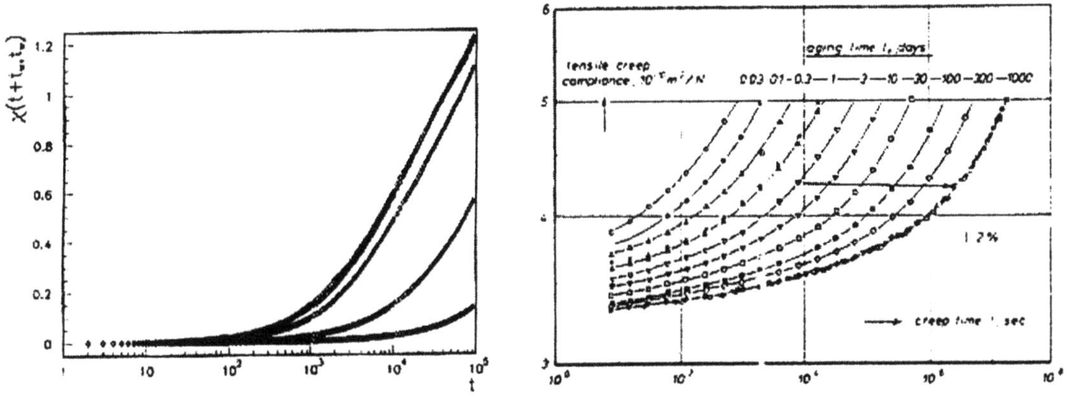

Figure 2: Response to perturbations applied after different waiting times. Left: tagged particle response in a kinetic glass model (t_w = 10 to 10^5) [39]. Right: aging experiments in plastic (PVC) [36]. (See also also [37] for dielectric susceptibility measurements in glycerol, [44] for spin-glass simulations, and [43] for polymers in random media.)

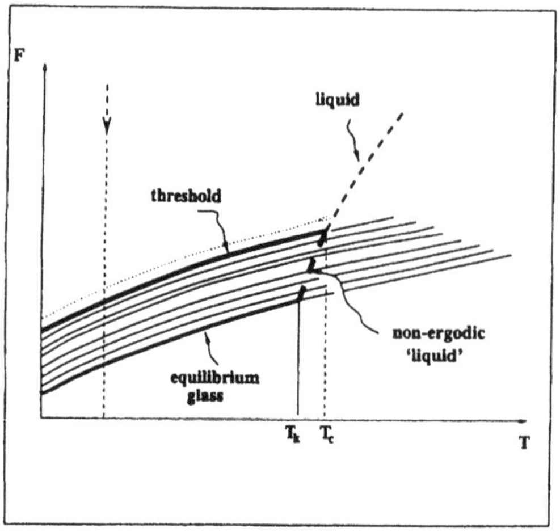

Figure 3: A sketch the temperature-dependence of metastable states.

on the mean-field p-spin spherical glass[8], a model that is simple and has all the required ingredients except an ordered crystalline state (which we shall ignore completely as it is not relevant for the glass problem). The model has quenched disorder, although this is not essential, see Refs.[9]. It involves a single mode (i.e. there is no space dependence), the extension to many coupled spatial modes is straightforward[16].

This model has exponentially many stable states at low temperatures. Fig. 3 shows a sketch of the free energies of these states in terms of temperature[10]. At given temperature, the density of states per unit of free energy increases exponentially with the free energy, up until the level labeled 'threshold'. At free energies just above the threshold, the landscape has no more minima, just saddles.

The thin dashed line labeled 'liquid' corresponds to the solution that dominates at all temperatures above T_c (in the language of replica theory[11] it is replica symmetric). KTW noted that the dynamical equations for the correlations just above T_c are exactly the simplest[12] mode-coupling equations[13] for a liquid, with 'mode coupling transition temperature' $= t_c$. As t_c is approached from above, the dynamics becomes slower and slower, the typical (*'alpha'*) relaxation time τ^{rel} diverges.

On the other hand, an equilibrium calculation gives a transition at T_k, not at T_c. T_k is for this model the Kauzmann temperature: at temperatures below T_k we have an 'equilibrium glass' (Fig. 3) phase with the Gibbs distribution split between many states that have the lowest free energy density and are separated by infinite barriers. (The replica solution has a one-step breaking.)

The situation in the intermediate regime between T_c and T_k is more subtle. An equilibrium calculation in this range gives a solution for the Gibbs distribution that is the continuation of, and in every aspect similar to the liquid phase (the thick dashed line labeled 'non-ergodic liquid'). However, if we start from a temperature T equilibrium configuration (chosen with the Gibbs probability distribution) at $T_k < T < T_c$ and study its dynamics we find that ergodicity is broken: the system never goes on to explore all the configurations belonging to the equilibrium state, but stays in a neighbourhood of the initial point. In other words, as we cross T_c, even if nothing spectacular happens from the point of view of Gibbs measure, the (single) state becomes fractured in exponentially many ergodic components.

It was well known that in realistic systems there was no true freezing at the mode-coupling transition temperature T_c. KTW pointed out that in a finite-dimensional system ergodicity would be restored between T_k and T_c by activated processes, and in fact one would only observe a crossover where the dynamics becomes slower than the experimental time at a certain (*so defined*) T_g intermediate between T_k and T_c. This will become more clear when we discuss the glassy dynamics in the next subsection.

2.2. Out of equilibrium relaxation. Effective temperatures

Below T_g, there is clearly a problem with the description above: the dynamics starting from an equilibrium configuration is invoked, although when ergodicity is broken the system could not have arrived in that configuration in the first place. In order to understand the glass phenomenon such as it happens in nature, we have to follow the dynamics of a system that undergoes a quench to low temperatures (or high pressures), and see what happens.

Within the present approximation, one has a set of two exact coupled equations for the autocorrelation $C(t,t')$ and the response function $R(t,t')$, with temperature entering as a parameter. At temperatures $T > T_c$ (the liquid phase), we find that after a short time, correlation and response functions depend only on time-differences and satisfy the fluctuation-dissipation theorem (FDT):

$$TR(t-t') = \frac{\partial C(t-t')}{\partial t'} \tag{1}$$

indicating that the system is in equilibrium. Using (1) the two equations collapse into a single one for the correlation: these are the usual mode-coupling equations for the autocorrelations of the liquid.

If we now start decreasing slowly the temperature, we still manage to equilibrate at every step with a timescale τ^{rel} that will diverge at T_c. Hence, no matter how slowly we cool, there will come a temperature such that the cooling rate is too fast compared to τ^{rel}. The system then falls out of equilibrium: this is signalled by the fact[14] that the correlation and response start depending on two times $C(t,t') \neq C(t-t')$ and $R(t,t') \neq R(t-t')$ and the fluctuation-dissipation relation is violated: $TR(t,t') \neq \partial C(t,t')/\partial t'$. Even if we stabilize the temperature at a certain $T < T_c$, neither stationarity nor FDT are achieved, no matter how long we wait. The system is *aging*: it keeps forever memory of the time since it crossed the transition. The curves of autocorrelation relaxation $C(t_w + \tau, t_w)$ and the corresponding ones for the integrated susceptibility $\chi(t_w + \tau, t_w)$ defined as:

$$\chi(t_w + \tau, t_w) = \int_{t_w}^{t_w+\tau} R(t_w + \tau, s)\mathrm{d}s \qquad (2)$$

show the characteristic waiting-time dependence we see in Figs. 1 and 2.

Furthermore, at each step we can compute the energy density, and find that *it is just above the threshold line* of Fig. 3. In order to understand this, bear in mind that the threshold level is such that below it there are minima in free energy, while above it only saddles: in other words, *it is the level above which the phase-space becomes connected* (hence the name 'threshold' in Ref. [14]). A system that relaxes at constant temperature, approches more and more the threshold level, thus seeing a lanscape that becomes less and less well connected, and this is why the dynamics slows down as time passes.

A parametric $\chi(t,t_w)$ vs. $C(t,t_w)$ plot[15] would give, if FDT were satisfied, a straight line with gradient $-1/T$. As mentioned above, FDT is violated in the aging regime, and we obtain a form like Fig. 4. Remarkably, for long times the plot tends to two straight lines, one with the usual gradient $-1/T$, and another with gradient (say) $-1/T_{eff}$. Thus, we have defined the effective temperature T_{eff} as the factor that enters in the fluctuation-dissipation ratio.

2.3. *Beyond the simplest description*

As mentioned above, the mode-coupling transition at T_c cannot be a true one in a real life. In this context it is easy to see why: a situation with free-energy density above the equilibrium one cannot last forever, as there will always be nucleation processes in a finite-dimensional system allowing the free-energy to decrease[16].

Thus, a more realistic glass model will be able, with time, to penetrate a certain amount below the threshold level – how much depends on the thermal history. If we are at a temperature below but close to T_c, the system might even go down to the thick dashed line in Fig. 3, and equilibrate after a long time t^{equil}. Indeed, we can paraphrase Kauzmann's original argument: if we could cool slow enough, we could ideally follow all the thick dashed line, until we met the 'true' thermodynamic transition at T_k.

If real systems unlike the mean-field case relax below the threshold, to what extent do the aging features encountered in mean-field survive? The characteristic aging curves of Figs. 1 and 2 belong to realistic systems, and show a situation that is very similar to mean-field from the qualitative point of view. A stronger test is the existence of an effective temperature as in Fig. 4. There has been quite a lot of numerical activity to check this in finite dimensional systems, with encouraging results[27,28] (see Fig. 5). Several experiments are also now under way.

For the moment there is no real theory taking into account activated processes below T_g, so we have to content ourselves with learning what we can from the mean-field scenario, but always bearing in mind where are its deficiencies. This is the approach we describe next for the rheological case.

3. Anomalous rheology

In order to study the rheology of these models, we have to couple them to forces that can do work on them. The simplest case is to add a force field that does not derive from a global potential[17,21,18] or is time-dependent[19]. (The first studies of this kind[20] were motivated by mean-field models of neural

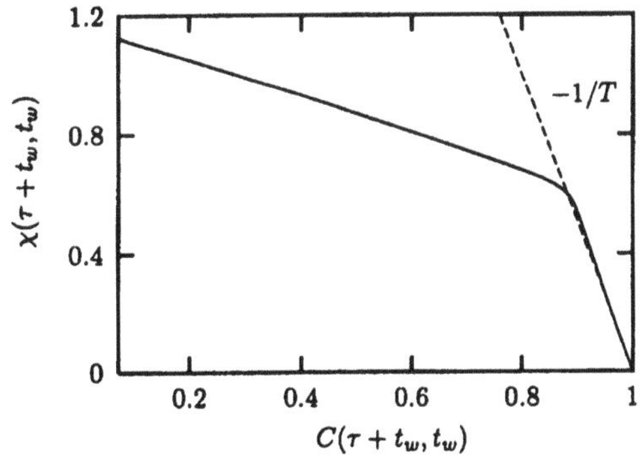

Figure 4: A fluctuation-dissipation plot (see [15,22]). The straight line to the left defines the effective temperature.

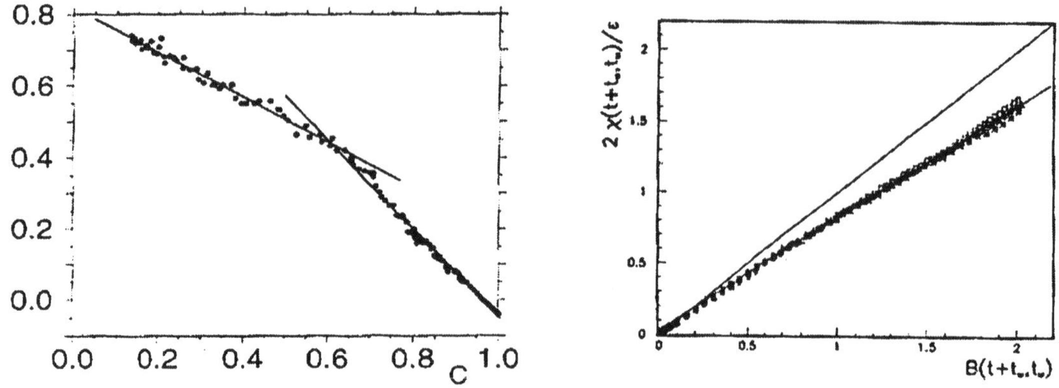

Figure 5: Fluctuation-dissipation plots. Left: Lennard-Jones glass, molecular dynamics [28]. (See [27] a similar curve for Monte Carlo simulation). Right: lattice gas [39]; the thin line is the FDT value. See [38] for the corresponding curves for ising systems, [43] for polymers in random media and [29] for models of granular media.

networks, the forcing terms were added in order to destroy the undesired glass phase).

3.1. Threshold level and Reynolds dilatancy

If we cool the system in the presence of a small drive down to a subcritical temperature, we can calculate the evolution of the energy. It turns out[21] that in the stationary regime obtained, the system remains 'surfing' above the threshold level, the closer the smaller the driving power: it costs an arbitrarily small power input to keep the system 'just above the threshold (this is, as we shall see, a mean-field peculiarity). In agreement with the discussion at the beginning, aging in this situation is interrupted, the correlation and response functions follow a typical aging pattern up to a time τ^{rel}, after which they become stationary (i.e. $C(t,t') = C(t-t')$, $R(t,t') = R(t-t')$). The typical correlation τ^{rel} time depends on the strength of the non-conservative forces, and is larger the smaller these forces: hence a less driven system is more viscous, and it behaves as if it were stationarized in an older age.

On the other hand, we can ask what would happen had we prepared the system in any low-lying sub-threshold state, (although this *cannot* be done within mean-field by just cooling!). For small driving forces the system remains trapped, the correlations do not decay beyond a certain small value, and the system is solid. Only by applying large driving forces can we make the system escape the deep state, and it will then resettle just above the threshold.

Hence, the system becomes free to move by raising its energy density (or volume, in the presence of gravity or pressure), the Reynolds dilatancy effect. Note that the crucial factor, the existence of a threshold level, has not been put by hand – it arises naturally from a microscopic calculation.

3.2. Effective macroscopic temperatures. Edwards compactivity

When we drive a system which would age, we can make it stationary even with a very small amount of power input. We may then ask if the situation so obtained is in some sense near equilibrium. In order to test this, we make a χ vs. C plot as in Fig. 4. Remarkably, in the limit of small driving the curve looks exactly like Fig. 4, not at all what we would have obtained close to equilibrium.

We have mentioned before that one can define an effective temperature T_{eff} through the inverse gradient of the line to the left of Fig. 4. This corresponds to the fluctuation-dissipation ratio *associated with the slow motion*. The importance of this temperature is that one can show[22] that:

- A thermometer tuned to respond only to the lower frequencies will measure exactly T_{eff}.
- The effective temperatures of two different systems become the same when the systems are coupled strongly enough[b].
- The effective temperature is *macroscopic*: it stays non-zero in the limit of zero thermal bath temperature (that is, in the limit in which Boltzmann's constant is negligible).

On the other hand, Edwards and collaborators[24] introduced a definition of 'packing entropy' of a granular medium as the logarithm of the number of packings at given volume, and from it defined a 'compactivity' (playing the role of a temperature) as the inverse of the derivative of this entropy with

[b] If the coupling is too weak, something different[23] happens.

respect to the volume. Within the present context, Edwards' 'temperature' is, at zero bath temperature, just the inverse of the logarithmic derivative of the number of stable minima of the energy with respect to the energy. (A generalization for non-zero bath temperature can be readily made substiting 'energy' by 'TAP-free energy'). Using the known[25,26] density of minima for this model, one can readily compute Edwards' temperature evaluating the energy at the threshold level, as is appropiate for weakly driven systems. *One finds that it coincides with the fluctuation-dissipation T_{eff} temperature defined above*[c]. Hence, Edwards' temperature inherits the 'zero-th law' properties of T_{eff} we have described.

Let us point out a suggestion that this mean-field like scenario already gives us. We have mentioned before that the plot of Fig. 4 only becomes two stright lines in the limit of small drive (or, in an aging problem, for large times). Hence, we have a concept of a single, well-defined T_{eff}, with the properties described above, *only in this limit*. It is then plausible the compactivity concept might itself only be relevant for weakly driven granular media, (and this quite apart from the question of its validity beyond mean-field).

An interesting question is to what extent the particular form of driving affects the stationary measure attained. This dependence should be weak, or at least controllable, if 'ergodic' arguments (as in Edwards compactivity) are to be useful. In the mean-field case one can easily check that observables that are not correlated with the driving force take the same values independently of the form of these forces *provided they are weak*[21]. It would be very interesting to know to what extent this is a mean-field peculiarity.

3.3. Beyond the simplest description

The mean-field calculations are useful because they sometimes give results that were not expected *a priori*, and that may carry through to realistic systems: when the mean-field dynamical calculation showed the existence of an FDT temperature, this was looked for and found numerically in realistic aging models[27,28]. One can expect that the same will happen for realistic driven systems (see Ref. [29]).

In order to understand experiments and simulations it is however necessary to be aware of the limitations of the preceding discussion, and to have an at least qualitative idea of what new elements activated processes bring in. As we have seen above, in finite dimensions nucleation allows for penetration below the threshold level[16]. This means that in order to mantain a stationary situation 'surfing' above the threshold, we need some finite (although small) energy input, unlike the mean-field situation where this could be done with arbitrarily small drive. Furthermore, the possibility of getting trapped and untrapped below the threshold can produce intermittency (see [21]) and hysteresis effects.

The greatest challenge now is the inclusion of activated, non-perturbative processes in the analytic treatment.

3.4. Relation with other approaches

Let us conclude by very briefly mentioning three other approaches to the rheology of soft glasses that are potentially related to the one discussed here.

A possible strategy[30,3] is to study finite dimensional systems whose infinite-dimensional version falls within the solvable category described above and that reproduce many of the qualitative features of soft glasses. Note that the scope there is not so much to find new microscopic models of glasses (polymer models, Lennard-Jones particles, etc. are already themselves perfectly good candidates), but rather models which 'interpolate' between reality and some solvable limit.

Another quite different approach is the 'soft glassy rheology' (SGR) model developed by Sollich et al.[31]. It is built upon Bouchaud's (purely relaxational) trap model[32], supplementing it with driving forces plus – and this is essential – a *macroscopic effective temperature* representing the noise generated by the interactions. A very interesting question that arises immediately is whether the FDT-related effective temperature we discussed above is in any sense a microscopic derivation of the noise temperature in the SGR model. The question merits further consideration, but one should note that in spite of the fact that the trap models were originally inspired in mean-field statics, their dynamics is distinct and not directly related to mean-field dynamics (see [33,4] for a lengthy discussion of this point).

Another recent approach is due to Hébraud and Lequeux[34], where in particular they discuss the amplitude dependence of the response to an oscillating strain. Work is in progress[35] to study this feature within the present scenario.

[c] This is directly checked by calculating the derivative of (2.17) in [26], and comparing with the value of T_{eff} in [14]

4. Conclusion

Compared to other, more mesoscopic analytical approaches to anomalous rheology, the present one has the disadvantage that the mechanisms involved are not as explicit. It has, however, the merit of being able to give unforseen results (threshold level, FDT violations, two transition temperatures, etc.), and this because there is quite a large distance between what goes into the model (microscopic Hamiltonian and dynamics) and what comes out of it (macroscopic correlations and responses).

Acknowledgements

I wish to thank Leticia Cugliandolo and Remi Monasson for discussions, and very specially Daniel Bonn for many conversations on the experimental results.

References

1. G. Chaveteau, R. Tabary, M. Renard and A. Omari, SPE 50752 in *SPE International symposium on Oilfield Chemistry, Houston, Texas* (1999).
2. J.B. Knight, C. G. Fandrich, C. Ning Lau, H. M. Jaeger and S. R. Nagel; *Phys. Rev.* **E51**, 3957 (1995).
3. M. Nicodemi and A. Coniglio, cond-mat/9803148.
4. For a review on glassy dynamics with or without quenched disorder, see J-P Bouchaud, L. F. Cugliandolo, J. Kurchan and M. Mézard cond-mat/9702070; in *Spin-glasses and random fields*, A. P. Young ed. (World Scientific, Singapore).
5. T. R. Kirkpatrick and D. Thirumalai, *Phys. Rev.* B **36**, 5388 (1987), *ibid* **37**, 5342 (1988), *Phys. Rev.* A **37**, 4439 (1988). D. Thirumalai and T. R. Kirkpatrick, *Phys. Rev.* B **38**, 4881 (1988). T. R. Kirkpatrick and D. Thirumalai, *J. Phys.* **A22**, L149 (1989). T. R. Kirkpatrick and P. Wolynes, *Phys. Rev.* A **35**, 3072 (1987). T. R. Kirkpatrick and P. Wolynes, *Phys. Rev.* B **36**, 8552 (1987). T. R. Kirkpatrick, D. Thirumalai, P. G. Wolynes, *Phys. Rev. A* **40**, 1045 (1989).
6. S. Franz and J. Hertz, *Phys. Rev. Lett.* **74**, 2114 (1995).
7. J.P. Bouchaud, L. F. Cugliandolo, J. Kurchan and M. Mézard, *Physica* A **226**, 243 (1996).
8. A Crisanti and H-J Sommers, *Z. Phys* B **87**, 341 (1992).
9. For a sample of models without quenched disorder see: J.P. Bouchaud and M. Mézard, *J. Phys.* I (France) **4**, 1109 (1994); E. Marinari, G. Parisi, and F. Ritort, *J. Phys.* A **27**, 7615 (1994) pp. 7615 and 7647; L. F. Cugliandolo, J. Kurchan, G. Parisi and F.Ritort, *Phys. Rev. Lett.* **74**, 1012 (1995); L. F. Cugliandolo, J. Kurchan, R. Monasson and G. Parisi, *J. Phys.* A **29**, 1347 (1996); P. Chandra, L. B. Ioffe and D. Sherrington, *Phys. Rev. Lett.* **75**, 713 (1995); P. Chandra, M. V. Feigel'man and L. B. Ioffe, *Phys. Rev. Lett.* **76**, 4805 (1996) and Ref.[4].
10. J. Kurchan, G. Parisi and M. A. Virasoro, *J. Phys.* I (France), **3**, 1819 (1993).
11. M. Mézard, G. Parisi and M.A. Virasoro, *Spin Glass Theory and Beyond*, (World Scientific, Singapore, 1987); K. H. Fischer and J. A. Hertz, *Spin Glasses*, (Cambridge Univ. Press, 1991).
12. E. Leutheusser, *Phys. Rev.* A **29**, 2765 (1984).
13. W. Götze, in *Liquids, freezing and glass transition*, eds. JP Hansen, D. Levesque, J. Zinn-Justin Editors, Les Houches 1989 (North Holland). W. Götze and L. Sjögren, *Rep. Prog. Phys.* **55**, 241 (1992).
14. L. F. Cugliandolo and J. Kurchan, *Phys. Rev. Lett.* **71**, 173 (1993); *Phil. Mag.* B **71**, 501 (1995).
15. L. Cugliandolo and J. Kurchan, *J. Phys.* A **27**, 5749 (1994).
16. In what follows, whenever reference is made to 'finite dimensions' I mean a theory with essentially finite dimensional effects taken into account. Variational methods, perturbative resummation schemes and closure approximations can be made involving many spatial modes, but this is still mean field-ish from the present point of view.
17. H. Horner, *Z. Phys.* **B100**, 243 (1996).
18. F. Thalmann, Thesis University of Grenoble (1998), and cond-mat 9807010.
19. H. Horner, *Z. Phys.* B **57**, 29 (1984); *ibid.*, 39 (1984).
20. J. A. Hertz, G. Grinstein and S. Solla, in: J. L. van Hemmen, I. Morgernstern (eds.), *Proceedings of the Heidelberg Colloquium on Glassy Dynamics and Optimization, 1986* (Berlin: Springer Verlag, 1987). G. Parisi, J. Phys. **A19**, L675 (1986); A. Crisanti and H. Sompolinsky; Phys. Rev. **A36**, 4922 (1987).
21. L. F. Cugliandolo, J. Kurchan, P. Le Doussal and L. Peliti; *Phys. Rev. Lett.* **78**, 350 (1997).
22. L. F. Cugliandolo, J. Kurchan and L. Peliti, *Phys. Rev.* **E55**, 3898 (1997).
23. L. F. Cugliandolo and J. Kurchan, cond-mat/9807226.
24. S.F. Edwards and A. Mehta, *J. Phys. France* **50** (1989) 2489; A. Mehta and S.F. Edwards, *Physica* **A 157** (1989) 1091; R. B. S. Oakshot and S.F. Edwards, *Physica* **A 189** (1992) 188; C. C. Mounfield and S.F. Edwards, *Physica* **A 210** (1994) pp. 279, 290 and 301; R. Monasson, O. Pouliquen, *Physica* **A 236**, 395 (1997).
25. A. Crisanti and H. J. Sommers, *J. Phys.* (France) **I 5** (1995) 805.
26. A. Cavagna, I. Giardina and G. Parisi, Phys. Rev. **B57**, 11251 (1998).
27. G. Parisi, *Phys. Rev. Lett.* **79**, 3660 (1997).
28. W. Kob and J-L. Barrat, *Phys. Rev. Lett.* **78**, 4581 (1997); cond-mat 9804168; J-L. Barrat and W. Kob; cond-mat 9806027.
29. M. Nicodemi, cond-mat/9809346.
30. A. Coniglio and H.J. Herrmann, *Physica* **A225**, 1 (1996); M. Nicodemi, A. Coniglio, H.J. Herrmann, *Phys. Rev.* E **55**, 3962 (1997); *J. Phys.* **A 30**, L379 (1997); J. Arenzon, cond-mat/9806328.
31. P. Sollich, F. Lequeux, P. Hebraud, M. Cates; *Phys. Rev. Lett.* **78**, 2020 (1997); P. Sollich, cond-mat 9712001.
32. J.P. Bouchaud, *J. Phys. I* (France) **2**, 1705 (1992).
33. J. Kurchan and L. Laloux, *J. Phys.* **A29**, 1929 (1996).
34. P. Hébraud and F. Lequeux, cond-mat 9805373.
35. J-L Barrat, L. Berthier and J. Kurchan, work in progress.

36. L. C. E. Struick, *Physical Aging in Amorphous Polymers and Other Materials* (Elsevier, Houston, 1978).
37. R. L. Leheny and R. S. Nagel,*Phys. Rev.* **B57** 5154 (1998).
38. S. Franz and H. Rieger, *J. Stat. Phys* **79** 749 (1995); G. Parisi, F. Ricci-Tersenghi and J. J. Ruiz-Lorenzo,*Phys. Rev.* **E 57** 13617 (1998).
39. M. Sellitto; cond-mat/9804168.
40. M. Kroon, G. H. Wegdam and R. Sprick,*Phys. Rev.* **E54** (1996) 6541; D. Bonn, H. Tanaka, G. Wegdam, H. Kellay and J. Meunier,*Europhys. Lett.*, to be published.
41. W. Paul and J. Baschnagel, in *Monte Carlo Dynamics Simulations in Polymer Science*, K. Binder Eds. (Oxford University Press, 1995) 307.
42. A. Barrat; *Phys. Rev.* **E 55**, 5651 (1997).
43. H. Yoshino, cond-mat/9802283.
44. H. Rieger, *J. Phys.* **A 26**, L615 (1993).

What exactly happens when two systems with different effective temperatures are coupled (in particular if the coupling is very weak) is discussed further in:

Leticia F. Cugliandolo, Jorge Kurchan, Physica A263, 242 (1999).

The proof of the fact that FDT-temperature corresponds to a true temperature is made more properly in:

'A scenario for the dynamics in the small intropy production limit' Leticia F. Cugliandolo, Jorge Kurchan cond-mat/9911086, to appear in "Frontiers in Magnetism", special issue of the Journal of the Physical Society of Japan.

Two further works dealing with granular matter:

'Emergence of macroscopic temperatures in systems that are not thermodynamical microscopically: towards a thermodynamical description of slow granular rheology' J. Kurchan cond-mat/ 9909306.

'Edwards measures for powders and glasses' A. Barrat, J. Kurchan, V. Loreto, M. Sellitto cond-mat/0006140.

Statistical Physics of the Jamming Transition: The Search for Simple Models

S. F. Edwards and D. V. Grinev*

Polymers and Colloids group, Cavendish Laboratory, University of Cambridge, Madingley Road, Cambridge, CB3 0HE, UK

February 22, 1999

Abstract

We investigate universal features of the jamming transition in granular materials, colloids and glasses. We show that the jamming transition in these systems has common features: slowing of response to external perturbation, and the onset of structural heterogeneities.

1. Introduction

Many studies in physics concern the onset of instabilities into turbulence or some form of chaotic break up, the simplest being laminar flow into turbulent flow, but the collapse of civil engineering structures is much studied. There are not many studies of the reverse phenomenon, where some chaotic system becomes more regular, or even stops altogether. The jamming phenomenon exists in quantum as well as classical systems, but in this article we will discuss only the latter, and indeed systems which are readily accessible to intuitive understanding.

The approach of this article will be to list the most common jamming systems, and see if there exist simple physical or mathematical systems which lead to interpretable experiments and soluble theories. In this paper, we consider mostly our own work, or studies that we know of in detail. The word "jamming" derives from systems coming to rest, and an archetypical problem is in the flow of granular or colloidal systems. We will try to answer the following questions:

- What is the jammed state?
- How does the jamming transition occur?
- What are the common features of the jamming transition in granular materials, colloidal suspensions and glasses?

In this paper, we confine ourselves to the discussion of the jamming transition in granular materials, colloids and glasses. We will not comment on other systems that exhibit the jamming transition, which are many: foams, vortices in superconductors, field lines in turbulent plasma etc. All these systems throw light on one another and exhibit universal behaviour. We will demonstrate that the jamming transition in these systems can be characterized by the slowing of response to external perturbations, and the onset of structural heterogeneities.

2. Granular Media

The jamming transition in dry granular media is a very common phenomenon which can be observed in everyday life. Stirring jammed sugar with a spoon can be hard but manageable. A jam within a silo can cause its failure. Nonetheless, such an ubiquity does not make the problem less difficult. In this section we will not even attempt to analyze the dynamical problem of a granular flow coming to rest, and the onset of jamming. Instead, we will try to characterize the static jammed state of a granular material. The difficulty of a theoretical analysis lies in the fact that dense granular media can expand upon shearing, a well known Reynolds dilatancy, that can be viewed as a counterpart of jamming [15,25]. Consider a cylindrical vertical pipe in the following cases (see Figure 1):

* Electronic address: dg218@phy.cam.ac.uk

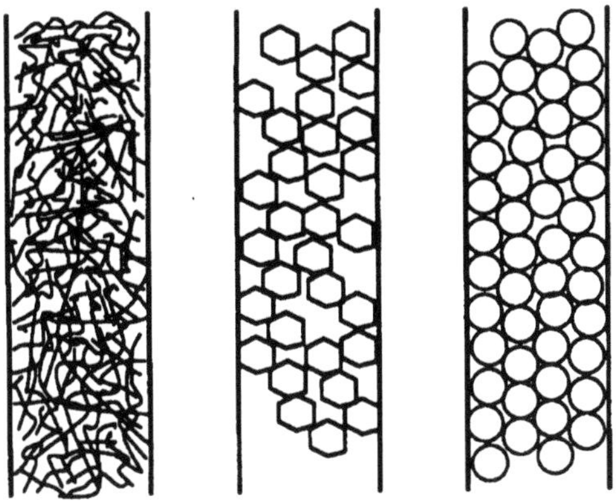

Figure 1. Types of granular materials.

- Rough walls, filled with pieces of twisted wire. This wire will entangle and not flow at all, indeed one does not need the pipe.
- Rough walls, with rough, approximately spherical particles. These can flow under certain conditions, but can also jam; it is a well studied problem in chemical engineering.
- Smooth walls, smooth spheres. This is difficult, perhaps impossible, to jam.

If one asks for the simplest granular material in the above list, it must be the third one if we model the granular material as an assembly of discrete rigid particles whose interactions with their neighbours are localized at point contacts. Therefore, the description of the network of intergranular contacts is essential for the understanding of the stress transmission and the onset of the jammed state in granular assemblies. The random geometry of a static granular packing can be visualized as a network of intergranular contacts. For any aggregate of rigid particles the transmission of stress from one point to another can occur only via the intergranular contacts. Therefore, the contact network determines the network of intergranular forces. In general, the contact network can have a coordination number varying within the system and different for every particular packing. It follows then that the network of intergranular forces is indeterminate i.e. the number of unknown forces is larger than the number of Newton's equations of mechanical equilibrium. Thus, in order for the network of intergranular forces to be perfectly defined, the number of equations must equal the number of unknowns. This can be achieved by choosing the contact network with a certain fixed coordination number. In this case the system of equations for intergranular forces has a unique solution and the complete system of equations for the stress tensor can be derived. This is the simplest model of a granular material. If this cannot be solved, one will be left with empiricism. This is from the point of view of physics; it is not a good philosophy for engineers. The geometric specification of our system is as follows: we will need z contact point vectors $\vec{\mathcal{R}}^{\alpha\beta}$, centroid of a grain α, $\vec{R}^{\alpha}$, $\vec{r}^{\alpha\beta}$ the vector from $\vec{R}^{\alpha}$ to $\vec{\mathcal{R}}^{\alpha\beta}$, and the distance $\mathcal{R}$ between grains α and β, $\vec{R}^{\alpha\beta}$ (see Figure 2).

Grain α exerts a force on grain β at a point $\vec{\mathcal{R}}^{\alpha\beta} = \vec{R}^{\alpha} + \vec{r}^{\alpha\beta}$. The contact is a point in a plane whose normal is $\vec{n}^{\alpha\beta}$. The vector $\vec{R}^{\alpha}$ is defined by:

$$\vec{R}^{\alpha} = \frac{\sum_{\beta} \vec{\mathcal{R}}^{\alpha\beta}}{z} \tag{1}$$

so that $\vec{R}^{\alpha}$ is the centroid of contacts, and hence the relation:

$$\sum_{\beta} \vec{r}^{\alpha\beta} = 0$$

We note that z is the number of contacts per grain, and $\sum_{\beta}$ means summation over the nearest neighbours. We define the distance between grains α and β

$$\vec{R}^{\alpha\beta} = \vec{r}^{\beta\alpha} - \vec{r}^{\alpha\beta} \tag{2}$$

Hence $\vec{R}^{\alpha}$, $\vec{r}^{\alpha\beta}$ and $\vec{n}^{\alpha\beta}$ are geometrical properties of the aggregate under consideration and the other shape specifications do not enter the equa-

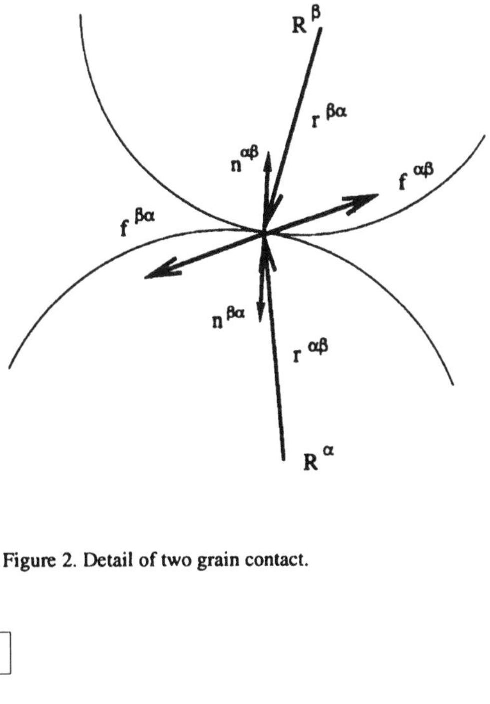

Figure 2. Detail of two grain contact.

tions. In a static array, Newton's equations of intergranular force and torque balance are satisfied. Balance of force around the grain requires

$$\sum_{\beta} f_i^{\alpha\beta} = g_i^{\alpha} \tag{3}$$

$$f_i^{\alpha\beta} + f_i^{\beta\alpha} = 0 \tag{4}$$

where $\vec{g}^{\alpha}$ is the external force acting on grain α.

The equation of torque balance is

$$\sum_{\beta} \varepsilon_{ijkl} f_k^{\alpha\beta} r_l^{\alpha\beta} = C_i^{\alpha} \tag{5}$$

Friction is assumed to be infinite [17]. It can be verified that, for the intergranular forces in the static array to be determined by these equations, the coordination number $z = 3$ in 2-D and $z = 4$ in 3-D is required. The microscopic version of stress analysis is to determine all of the intergranular forces, given the applied force and torque loadings on each grain, and geometric specification of a granular array. The number of unknowns per grain is $zd/2$. The required force and torque equations give $d + (d(d-1))/2$ constraints. The system of equations for the intergranular forces is complete when the coordination number is $z_m = d + 1$. In addition, the configuration of contacts and normals is only acceptable if all the forces are compressive. If tensile forces are allowed, then we would be studying a sandstone (an interesting problem, but not a subject for a study of jamming). Many investigators do not believe that z will be this small, and would invoke the kind of arguments used in bridgework, where the same problem arises when extra spares lead to too few equations for solution. However the simplest assumption is to assume that if the geometry gives more than $z_m/2$ contacts, some will contain no force. At all events we can restrict ourselves to systems where there really are $z_m/2$ contacts per grain The main question is: can one observe the jammed state in this simple model where the static stress state can be determined? We define the tensorial force moment:

$$S_{ij}^{\alpha} = \sum_{\beta} f_i^{\alpha\beta} r_j^{\alpha\beta} \tag{6}$$

which is the microscopic analogue of the stress tensor. With $C_i^{\alpha} = 0$, S_i^{α} will be symmetric. To obtain the macroscopic stress tensor from the tensorial force moment, we coarse-grain, i.e. average it over an ensemble of configurations:

$$\sigma_{ij}(\vec{r}) = \left\langle \sum_{\alpha=1}^{N} S_{ij}^{\alpha} \delta(\vec{r} - \vec{R}^{\alpha}) \right\rangle \tag{7}$$

The number of equations required equals the number of independent components of a symmetric stress tensor $\sigma_{ij} = \sigma_{ji}$, and is $d\,(d + 1)/2$. At the same time, the number of equations available is d. These are vector equations of the stress equilibrium $\partial\sigma_{ij}/\partial x_j = g_i$ which have their origin in Newton's second law. Therefore we have to find $d(d-1)/2$ equations, which possess the information from Newton's third law, to complete and solve the system of equations which governs the transmission of stress in a granular array.

Given the set of equations (3–5) we can write the probability functional for the integranular force $f_{ij}^{\alpha\beta}$ as

$$P\{f_i^{\alpha\beta}\} = \mathcal{N}\delta\left(\sum_{\beta} f_i^{\alpha\beta} - g_i^{\alpha}\right) \times\delta\left(\sum_{\beta} \varepsilon_{ikl} f_k^{\alpha\beta} r_l^{\alpha\beta}\right) \times\delta\left(f_i^{\alpha\beta} + f_i^{\beta\alpha}\right) \tag{8}$$

where the normalization, $\mathcal{N}$, which is a function of a configuration, is

$$\mathcal{N}^{-1} = \int \prod_{\alpha,\beta} P\{f_i^{\alpha\beta}\} \mathcal{D}^{\alpha\beta} \tag{9}$$

The probability of finding the tensorial force moment S_{ij}^{α} on grain α is

$$P\{S_{ij}^{\alpha}\} = \int \prod_{\alpha,\beta} \delta\left(S_{ij}^{\alpha} - \sum_{\beta} f_i^{\alpha\beta} r_j^{\alpha\beta}\right) P\{f_i^{\alpha\beta}\} \mathcal{D}^{\alpha\beta} \tag{10}$$

where $\int \mathcal{D} f^{\alpha\beta}$ implies integration over all functions $f^{\alpha\beta}$, since all the constraints on $f^{\alpha\beta}$ have been experienced. We assume that the $z = d + 1$ condition means that the integral exists.

It has been shown [18,19,20] that the fundamental equations of stress equilibrium take the form

$$\nabla_j \sigma_{ij} + \nabla_j \nabla_k \nabla_m K_{ijkl} \sigma_{lm} + \cdots = g_i \tag{11}$$

$$P_{ijk}\sigma_{jk} + \nabla_j T_{ijkl}\sigma_{kl} + \nabla_j \nabla_l U_{ijkl}\sigma_{km} + \cdots = 0 \tag{12}$$

However, there are difficulties with the averaging procedure [18] which is highly non-trivial. The simplest mean-field approximation gives the equation $\sigma_{11} = \sigma_{22}$ for the case of an isotropic and homogeneous disordered array [18]. Though this equation gives the diagonal stress tensor $\sigma_{ij} = p\delta_{ij}$, it is not rotationally invariant. The only linear, algebraic and rotationally invariant equation is $\mathrm{Tr}\,\sigma_{ij} = 0$. However, this equation cannot be accepted, for a stable granular aggregate does not support tensile stresses. We believe that, at least in the simplest case, the fundamental equation for the microscopic stress tensor should be linear and algebraic (because of the linearity of Newtons' second law for intergranular forces). In this paper we offer an alternative way which is considered below.

The leading terms of the system of equations (11,12) arise from the system of discrete linear equations for S_{ij}^{α} [18,19]

$$\sum_{\beta} S_{ij}^{\alpha} M_{jk}^{\alpha} R_k^{\alpha\beta} - S_{ij}^{\beta} M_{jk}^{\beta} R_k^{\beta\alpha} = g_i^{\alpha} \quad (13)$$

$$S_{11}^{\alpha} - S_{22}^{\alpha} = 2S_{12}^{\alpha} \tan\theta \quad (14)$$

and if $\tan\theta^{\alpha}$ has an average value $\tan\phi$

$$\sigma_{11} - \sigma_{22} = 2\sigma_{12} \tan\phi \quad (15)$$

which is known as the Fixed Principal Axes equation [21,22,23,24], and has been used with notable effect to solve the problem of the stress distribution in sandpiles. For a homogeneous and isotropic system, the averaging process gives the stress tensor $\sigma_{ij} = p\delta_{ij}$ which is simply hydrostatic pressure, as is to be expected. Rotating S_{ij}^{α} by some arbitrary angle θ

$$\begin{pmatrix} A & B \\ C & D \end{pmatrix} \quad (16)$$

where

$$A = S_{11}^{\alpha} \cos^2\theta - 2S_{12}^{\alpha} \sin\theta\cos\theta + S_{22}^{\alpha} \sin^2\theta$$

$$B = (S_{11}^{\alpha} - S_{22}^{\alpha}) \sin\theta\cos\theta + S_{12}^{\alpha}(\cos^2\theta - \sin^2\theta)$$

$$C = (S_{11}^{\alpha} - S_{22}^{\alpha}) \sin\theta\cos\theta + S_{12}^{\alpha}(\cos^2\theta - \sin^2\theta)$$

$$D = S_{11}^{\alpha} \sin^2\theta + 2S_{12}^{\alpha} \sin\theta\cos\theta + S_{22}^{\alpha} \cos^2\theta$$

one can easily that (14) constraints the off-diagonal components of S_{ij}^{α} to be zero.

The system of equations (13,14) is solved by Fourier transformation and the macroscopic stress tensor is obtained by averaging over the angle θ

$$\sigma_{11}(k) = \langle S_{11}(k) \rangle_{\theta} = \frac{g_1(k^3 + 3j^2k) + g_2 j(j^2 - k^2)}{(k^2 + j^2)^2} \quad (17)$$

$$\sigma_{22}(k) = \langle S_{22}(k) \rangle_{\theta} = \frac{g_2(j^3 + 3k^2 j) + g_1 k(j^2 - k^2)}{(k^2 + j^2)^2} \quad (18)$$

Figure 3. Jammed pipe.

$$\sigma_{12}(k) = \langle S_{12}(k) \rangle_{\theta} = \frac{(g_1 j - g_2 k)(j^2 - k^2)}{(k^2 + j^2)^2} \quad (19)$$

By doing the inverse Fourier transformation one can see that the macroscopic stress tensor is diagonal. There must also be constraints on the permitted configurations (due to the absence of tensile forces) which are not so easily expressed, for they affect each grain in the form

$$S_{ik}^{\alpha} M_{kl}^{\alpha} R_l^{\alpha\beta} n_i^{\alpha\beta} > 0 \quad (20)$$

which has not yet been put into continuum equations other than $\mathrm{Det}\,\sigma > 0$ and $\mathrm{Tr}\,\sigma > 0$. However, this condition must be crucial, for without it jamming becomes a very ubiquitous phenomenon provided the density of grains is such that they are all in contact. This example shows the utility of simple models. Apply this type of grain in a vertical pipe (the grain is a thick tile, see Figure 3).

This system is clearly jammed, but the problem is how it can get into this configuration. Another limiting case is to think of a sphere with many spines pointing in the radial direction. These spines mean that spheres can only approach or retreat along the direction joining their centres, and the only other motion permitted to a group of two or more is that of rigid rotation of the group. As soon as a line (in 2-D) or shell (3-D) of these objects occurs, they jam; so they always jam. Again, although this is a possible material, most materials fall into the classes above. So there are trivial jamming problems, but it is proving difficult to produce an effective analytic theory for intermediate materials. It is natural then to use computer simulations, and there is a notable literature in existence [25]. However, rather than comment on this litera-

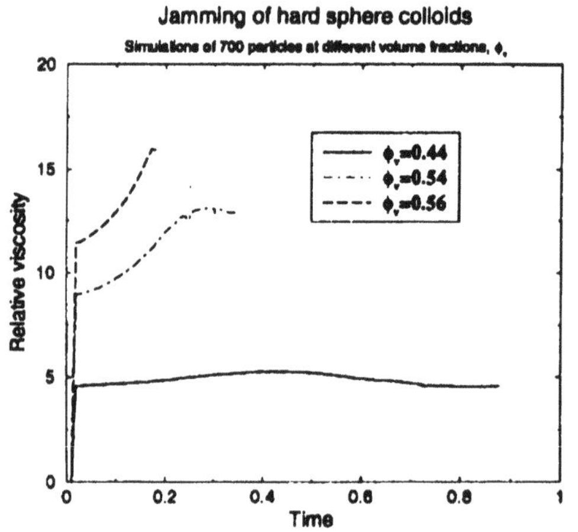

Figure 4. Relative viscosity of hard sphere colloids versus time, for various volume fractions.

ture for granular materials we move to the related problem of colloidal flow which offers a natural route from grains to glasses.

3. Colloidal Suspensions

The simplest model of a colloid which exhibits the jamming transition is the sheared suspension of hard monodisperse spheres, interacting hydrodynamically through a Newtonian solvent of viscosity ,η_0. Such a system at equilibrium is characterized by its volume fraction ϕ_v. The behaviour of this system at equilibrium is well known [33]. With increasing ϕ_v, the system crystallizes, with phase coexistence occurring between $\phi_v \approx 0.50$ and $\phi_v \approx 0.55$. If simple shear is applied, we need one other parameter, which is the Peclet number

$$\mathrm{Pe} = \frac{6\pi\dot{\gamma}\eta_0 d^3}{k_B T} \tag{21}$$

where d is the particle diameter, $\dot{\gamma}$ the shear rate and T the temperature. The Peclet number gives the ratio of shear forces to Brownian forces. Pairwise hydrodynamic interactions between the neighbouring particles can be divided into squeeze terms along the line of centres and terms arising from relative shear motion. To leading order the squeeze hydrodynamic force on a particle is given by the well-known Reynolds lubrication formula

$$f_i = -\sum_j \frac{3\pi\eta_0}{8h_{ij}}\{(v_i - v_j)\cdot n_{ij}\}n_{ij} + O\left(\ln\frac{2}{h_{ij}}\right) \tag{22}$$

where the sum is over nearest-neighbour particles j, h_{ij} is the gap between the neighbour surfaces with the unit of distance the particle diameter, n_{ij} is the unit vector along the line of centres i to j and v_i, v_j are the particle centre velocities.

In the absence of other interactions, Brownian forces are left to control the approach of particles. If conservative (steric or charge) forces are present, these control the gaps between particles, and dominate over the Brownian forces at high shear rates. When one studies the shear flow of such a system, it exhibits thickening effects: a rise of viscosity with increasing shear rate (for a review, see [34]). At volume fractions approaching random close packing, $\phi_{RCP} \approx 0.64$, discontinuous thickening with a large jump in viscosity occurs at a critical Pe. However, at a lower ϕ_v and lower Pe a more continuous rise can be observed (see Figure 4). Hence, the jamming transition in this simple system can be either continuous or discontinuous, which depends sensitively on the volume fraction.

This happens in various experimental systems but in particular, in those whose particles, by polymer coating or surface charges, do not flocculate (if they want to stick together then jamming is not an obvious phenomenon). Colloids with repulsive interactions exhibit shear thinning i.e. the viscosity decreases as the shear rate increases. The presence of aggregating forces can greatly alter the shear thinning. We will discuss the regime of shear thickening (which we call later the jamming transition) in the simplest model, gradually incorporating various interactions. The physical picture can be obtained by combining theory and computer simulations [35,36,37,38,39,40,41]. The simulation technique for particles under quasi-static motion determined by a balance of conservative and dissipative forces has been proposed in [35]. The motion of N colloidal particles, immersed within a hydrodynamic medium, is governed by an equation of quasi-static force balance

$$-\mathcal{R}(\vec{X})\cdot\vec{V} + \vec{F}_C(\vec{X}) + \vec{F}_B(t) = \vec{0} \tag{23}$$

where $\vec{X}$ represents the $6N$ particle position coordinates and orientations, $\mathcal{R}$ is a $6N \times 6N$ resistance matrix and $\vec{V} = d\vec{X}/dt$ is the particle velocity. The terms F_C and F_B represent conservative and Brownian forces. The effects of inertia are ignored i.e. the Reynolds number is small

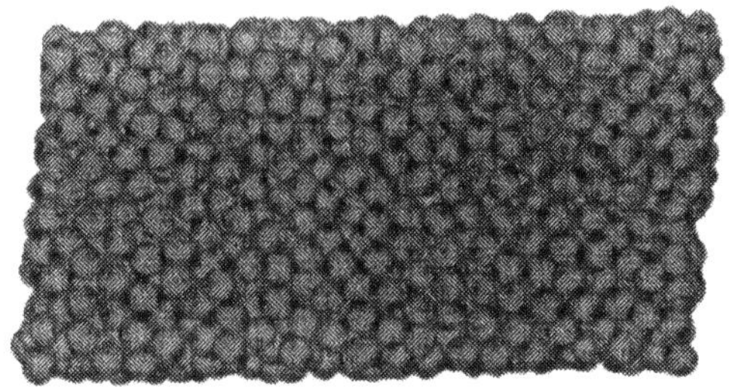

Figure 5. Snapshot from a 3-D simulation of 2000 hard spheres at $\phi_v = 0.56$. All particles are shown.

$$Re = \frac{\rho_s \dot{\gamma} d^2}{\eta_0} << 1 \tag{24}$$

Bearing in mind that for a typical colloid Pe/Re ~ 10^7, it follows that it is possible to achieve Re << 1 and Pe >> 1 simultaneously. Equation (23) can be solved on a computer [35,36]. Although simulation shows exactly what is happening in the sense one can see every sphere flowing its path (see Figures 5, 6), it is still not agreed as to whether the phenomenon of thickening is an order-disorder transition [42,43] or whether it is due to the development of clusters of particles along the compression axis [44,45].

A microscopic kinetic theory for the origin of the increased bulk viscosity at high ϕ_v and Pe = ∞ has been recently proposed [38,40]. It attributes the viscous enhancement to the presence of hydrodynamic clustering, i.e. incompressible groups of particles which lie near the compression axis. The rigidity of clusters is provided by divergent lubrication drag coefficients. The theory predicts a critical ϕ_v above which jamming occurs. This model gives a flow-jam phase transition at any strain and may be of more general applicability (e.g. colloids with conservative and Brownian forces). Assuming the cluster length to be additive on collision, one can obtain the standard Smoluchowski aggregation equation

$$\frac{dX_k}{dt} = \frac{1}{2}\sum_{i,j=1}^{\infty}\left[K_{ij}X_iX_j - X_{i+j}\right]\left(\delta_{i+j,k} - \delta_{i,k} - \delta_{j,k}\right) \tag{25}$$

This equation relies upon a mean field approximation, i.e. it is assumed that each cluster is embedded in an average flow, composed of the other clusters, "spectator" particles and solvent. Eqn (25) governs the evolution of the concentrations X_K of clusters of size k monomers as a function of time, given that clusters can collide, aggregate and break up at rates dependent on their sizes and specified by the aggregation kernel K_{ij}. The mathematical procedure of solving Equation (25) has been reported in [38,40]. We discuss the generic structure of the theory. The rheology of this system results from a competition between a binary aggregation process and single cluster breakup. The paradigm of hydrodynamic clustering modelled in terms of aggregation-breakup laws provides the qualitative physical picture of the jamming transition. At low ϕ_v breakup dominates. The system achieves a steady state with a population of clusters whose average size increases and a viscosity which increases as the volume fraction rises. In the vicinity of some volume fraction ϕ_c, average cluster size diverges, and so does the viscosity

$$\eta \propto \eta_0\left(1 - \frac{\phi_v}{\phi_c}\right)^{-2} \tag{26}$$

Above ϕ_c the flow is transient and dominated by aggregation. The average cluster size diverges after a certain strain, which falls as ϕ_v is further increased. The system undergoes the jamming transition and can never reach a steady state. This model predicts $\phi_c \approx 0.49$, although experiments suggest a divergence of bulk viscosity at a higher volume fraction (which is still less than ϕ_{RCP}). This difference is due to the presence of conservative interactions in the real hard core colloid. In conclusion, we briefly discuss the jamming behaviour of long flexible chain polymers. A good way to understand this is to plot viscosity against the molecular weight of the polymer in the molten state. The molecular weight is equivalent to the length of the chain. The short chains slide past each other in many ways, giving a linear relation between length and shear viscosity, but an entanglement crisis occurs at a critical molecular weight M_c,

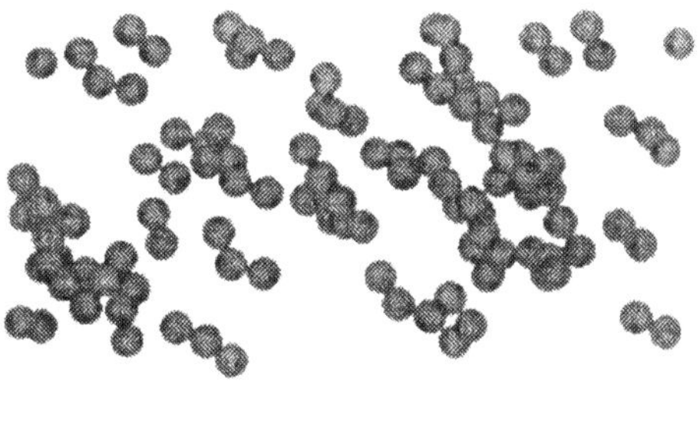

Figure 6. Snapshot from a 3-D simulation of 2000 hard spheres at $\phi_v = 0.56$. Only particles who have separations $h < 10^{-4}d$ are shown. The flow direction is left to right.

when the polymer can only wriggle in Brownian motion up and down and "out" of a tube formed by its neighbours in reptation. Whether there is sharp transition is not established in a difficult slow experiment, but all agree that the dependence of the viscosity jumps to $M^{3.4}$, a colossal change which at first sight is just like a jamming, but very slowly the melt flows flows due to reptation. Polymers very easily form glasses and it only takes a fall in temperature to make the melt solidify, although the specific heat shows that although reptation (a slow translatory motion) has ceased, there is plenty of other movement taking place, but movement which averages over time to zero. A further fall in temperature destroys this motion, and one enters a state like a conventional glass.

4. Glasses

Studies of glasses seem to fall into two camps. Those taking continuum field representation of the material and solving under statistical thermodynamics conditions, principally mode-coupling methods, and alternatively the translation, into comparatively simple equations and simple physical models. The former has given intuitive thinkers a hard time because it is not obvious in those cases, such as the polymeric glasses discussed below, where one should have a picture of what goes, what indeed is going on. The modelling approach is weak; for example, in an assembly of packed spheres where an enormous number of motions is possible, but strong in polymer glasses where the motion is obvious and the jamming of motion is the cessation of the centre of mass diffusion. The mode-coupling theory [46] of the glass transition for simple liquids has an ergodicity to non-ergodicity transition. It starts with an equation of motion for the density correlation function, which is an integro-differential Langevin equation, with a non-linear memory kernel. This non-linear memory term governs the transition to the non-ergodic state, and can be characterized by measurements of the dynamic structure factor. We refer the reader to the literature [46] for details of this approach. The glass transition is not limited to special types of materials. Every class of material can be transformed in an amorphous solid if the experimental parameters are adjusted to the dynamics of the system. Consider therefore, two extreme examples. Consider the system consisting of spherical molecules such as rare gases. The hopping time of the spheres is very short, and the dynamics extremely fast. Nevertheless, it has been shown that such fluids undergo a glass transition from the super cooled melt, if one cools the system with a quenching rate of $q \propto 10^{12}$ K s^{-1}, when the molecules jam effectively to rest. On the other hand, consider polystyrene which consists of very large molecules. The dynamics of such a system is much more complicated in comparison to spheres because there are many degrees of freedom. It is most significant that the centre of mass diffusion of a single molecule is small. We want to stress characteristics of the glass transition in general. The most significant points we want to discuss are:

- The divergence of the transport or inverse transport properties, such as viscosity, inverse diffusion coefficient, and relaxation times.
- The extreme broad relaxation phenomena of the stress, modulus etc.
- The quantitative definition of the term cooperativity.
- Influence of external parameters on T_g.

It has been recognised that the relaxation time follows the Vogel–Fulcher Law (VF):

$$\tau \propto e^{A/(T-T_0)} \tag{27}$$

This law has many names. In polymer physics it is often called the Williams–Landel–Ferry (WLF) law [47]. The law is not valid over the whole temperature range (see Figure 7). The divergence given by (27) comes at T_0, where T_0 is a temperature below the freezing temperature T_g, and the empirical

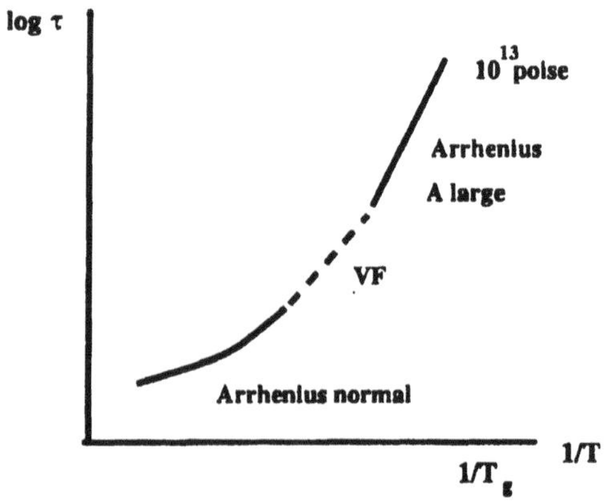

Figure 7. Vogel-Fulcher Law.

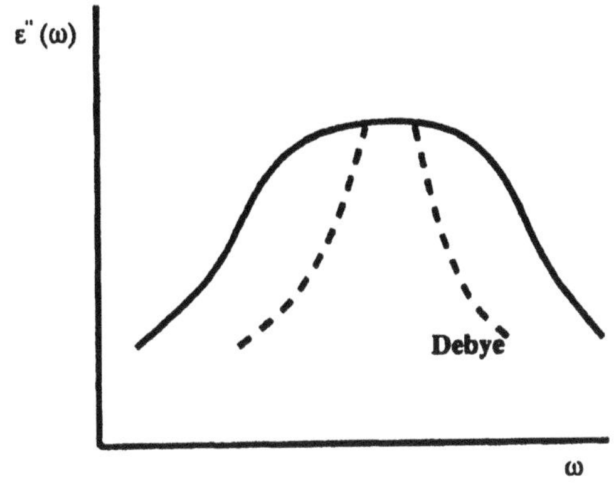

Figure 8. Kohlrausch-Williams-Watts Law.

rule is given by:

$$T_0 \sim T_g - (20 \sim 30)^0 \tag{28}$$

The physical meaning of T_g is still unclear, and we want to try to clarify this point. The VF law is much stronger than the critical slowing down in phase transition phenomena where, by scaling arguments, the relaxation time is given by

$$\tau \sim |T - T_0|^{-\nu z} \tag{29}$$

where ν is the correlation length exponent i.e. $\xi \sim (T - T_0)^{-\nu}$ and z the dynamical exponent

$$\tau \sim \xi^z \tag{30}$$

There have been attempts to fit data for freezing transitions (glass transition, spin glass transition) by Eqn. (29), and it turned out that νz is very high $\nu z \sim 10,\ldots,20$ which seems very unnatural. This again indicates a physical significance to the VF law. Another peculiar point lies in the relaxation properties. An empirical law was found long ago by Kohlrausch and in the early seventies recovered by Williams and Watts in their studies of the broadening of relaxation processes [48]. For example, measurement of the dielectric constant shows a much larger half width in the imaginary part compared to the Debye process. Empirically this is described by the Kohlrausch–Williams–Watts (KWW) law (see Figure 8)

$$\phi(t) \sim e^{-(t/\tau)^\beta} \tag{31}$$

where $\phi(t)$ can be any quantity which relaxes, i.e.

$$\phi(t) = \frac{\varepsilon(t) - \varepsilon(\infty)}{\varepsilon(0) - \varepsilon(\infty)} \tag{32}$$

However, there has been no convincing explanation for a unique value for β. We doubt that there is more to the physics of (31) than a wide relaxation spectrum due to different physical processes, and a more relevant question is how both the KWW and VF laws link together in general, and what is the relationship to cooperativity. It is believed that the glass transition exhibits a large amount of cooperative motion as the system is close to T_g. This might be indicated as well by the VF law, which is the crossover to the Arrhenius behaviour right at T_g but with an extremely high activation energy (see Figure 7). This activation energy is so high that it could hardly be attributed to only one molecule, and the phenomenological interpretation is that there are cooperative regions of some linear size diverging at some temperature

$$\xi \sim (T - T_g)^{power} \tag{33}$$

Another quite general question is the state of time-temperature superposition principle. This says that if one measures a physical quantity, $D(t)$, at some temperature T, and if the measurement is repeated at some temperature T_1, the quantity $D(t)$ can be resolved by a "shift factor" a_T, i.e. $D(T,t)$ is not a function of two variables but only of a combination of both:

$$D(T,t) = D\left(\frac{t}{a_T}\right) \tag{34}$$

where a_T is often given by

$$a_T = \frac{C_1 + T_g}{T - T_g + C_2} \tag{35}$$

near T_g, and

$$a_T = \frac{\delta E}{T} - \frac{\delta E}{T^*} \tag{36}$$

far from T_g. Hence (35) is of the Volger-Fulcher form.

Polymeric glasses offer two challenges:

- There is a clear intuitive picture of what is happening. The "tube" closes at its ends, or contracts at "entanglement" points.
- The molecular weight offers, as with viscosity, a new degree of freedom, and hence new laws emerge. Any theory of glass must encompass VF, KWW, and the experiments we now describe.

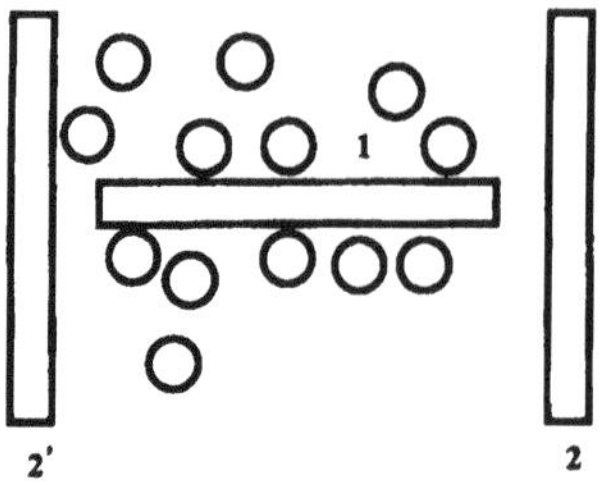

Figure 9. "Jammed" rod.

For polymers, the glass transition temperature depends on the molecular weight, and an empirical rule is given by Flory and Fox:

$$T_g(L) = T_g(\infty) - \frac{\text{constant}}{L} \tag{37}$$

where L is the length of the molecules i.e. the molecular weight. The agreement of (37) with experiments is not particularly good, but it gives an estimate of $T_g(L)$. Hence (37) tells us that there is not a very significant dependence on the molecular weight, unless L is small. This point should be investigated using the knowledge of polymer dynamics in melts which has recently emerged. Another empirical law we want to discuss is the mixing rule in plastification. Mixing two glass forming polymers together, the new T_g is often given by

$$\frac{1}{T_g(\text{mix})} = \frac{\phi_1}{T_{g_1}} + \frac{\phi_2}{T_{g_2}} \tag{38}$$

to zeroth order. The ϕ_i is the volume fraction of the i'th polymer. We have discussed so far the most important experimental results. Clearly there is a need for a solvable model in the framework of which the VF and KWW laws can be derived. Any fundamental theory of glass transition should relate the mobility of a molecule within the cage to mobilities of molecules forming the cage. These mobilities of surrounding molecules are coupled with those of their neighbours *etc.*, and therefore in general there is no small parameter that can justify the decoupling approximation. The small molecule model of a glass transition involves cages, but polymers are simpler because they have a tube, and a tube of such a model can be modelled as a straight tube, indeed restrictions along the tube can be included, but are not here for simplicity. The reason why we take rods was the advantage of simple geometry, no internal degrees of freedom, and very slow dynamics, so that we have no problems with high quenching rates, so we do not have to worry about thermodynamics. It is obvious that the small width to the length ratio and the sufficiently large number of neighbouring rods justify the decoupling procedure. If the solution of the rods is dense, the concentration $c \leq d^2L$ (d is the diameter of the rod, L its length and c the concentration) and severe constraints are acting on the rods. For example, they cannot move rotationally, and they can only make progress along their length. Such a solution has been called entangled. Suppose now we have such a solution of highly entangled rods. A rod can slide between the entangling rods until it meets rods which block it (see Figure 9).

The motion of a rod will then be like a particle diffusing along a line but meeting gates which open and close randomly through thermal fluctuations. If no barriers we present the probability $P(x,t)$ of finding the test rod at x (which is the coordinate of the rod down the tube) and time t satisfies the simple diffusion equation

$$\left(\frac{\partial}{\partial t} - D_0\frac{\partial^2}{\partial x^2}\right)P(x,t) = 0 \tag{39}$$

This has the solution

$$P(x,t) = \int_{-\infty}^{\infty} G_0(x,x';t,t')P_0(x',t')\mathrm{d}x' \tag{40}$$

where $P_0(x',t')$ is the initial probability function and $G_0(x,x';t,t')$ is the standard Green function of the diffusion equation. Suppose now , a reflecting barrier is placed at position R at time t_R and removed at some time t_Q. Then using the method of images to find the Green function it is straightforward to calculate $P(x,t)$ [49]. The same method can be applied when there are many barriers appearing and disappearing along the path of the rod. After neglecting correlations between barriers one can see that the rod (1) can only diffuse if

$$D = D_0(1-\alpha) \tag{41}$$

where $\alpha \sim \varepsilon(cDL^2)^{3/2}$, cDL^2 being the Onsager number. Eqn. (41) gives $D = 0$ if $\alpha = 1$ i.e. $\varepsilon cDL^2 = 1$. This is the jammed state. The result can be modified to include cooperativity. The complete solution has been obtained in [49,50] by mapping the problem onto the self-avoiding walk problem. The VF law appears when summed over n (where n is the number of rods that loops are made out of)

$$D \sim D_0\exp\left(-\frac{\alpha_1^2}{1-\alpha_2}\right) \tag{42}$$

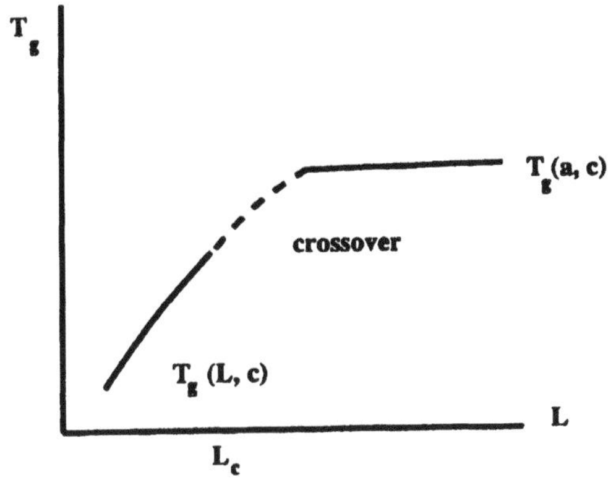

Figure 10. T_g as a function of L.

where the parameter α_1 contains generalised constants and a minimum loop size, and $\alpha_2 = \alpha - \alpha_1$. It can be shown that the number of rods moving cooperatively in the loops are given by

$$\bar{n} \sim (1-\alpha_2)^{-2} \tag{43}$$

and the size of the loop is therefore

$$\xi^2 \sim \bar{n}L^2 \sim \frac{L^2}{(1-\alpha_2)^2} \tag{44}$$

Thus ξ diverges if $\alpha_2 \to 1$, as the phenomenological interpretation requires. So far we have used α to denote the expression cdL^2, a combination of constants well known in liquid crystal theory since the work of Onsager. However, its role in our theory is much more general; for example, d can be temperature dependent through the fact that Van der Waals forces appear in the form $e^{-E/nT}$, whereas hardcore forces do not contain T. Thus, at the level of a model one can regard $\alpha = 1$ at $T = T_g$. Until one studies a detailed physical case, there is no purpose in false verisimilitude. Thus the above model now says that the VF law is a direct consequence of cooperativity. Turning now to the relaxation behaviour, it can be shown that for the rod model, the stress relaxation follows, to first order, the law

$$\sigma(t) \sim e^{-(t/\tau)^{1/2}} \tag{45}$$

Cooperativity as indicated by (42) does not change (45) drastically, giving rise to logarithms

$$\sigma(t) \sim e^{-(t/\tau)^{1/2}}\left(1+\text{const.}\log\frac{t}{\tau}\right) \tag{46}$$

This leads us to the conclusion that cooperativity is responsible for the VF behaviour, but not for the relaxation phenomena. Further study of this model allows a derivation of (37,38). A final interesting feature of the tube model is that the tube itself must be a random walk of step length a (in the rheology literature this is called the primitive path) and the "rod" of the preceding discussion is of length a for a very long polymer, and L for a long, but not very long molecule. The freezing of large scale motion is the freezing of motion on the scale of the primitive path step length, and the long term reptative motion of the whole chain is not a vital constitutive of the glass temperature. Hence T_g is a function of a and the density of the material, and also additional parameters (e.g. chain stiffness), i.e. $T_g = T_g(a(c,L),\ldots)$. But a is a function of the density itself, so that the effect of a diluent acts on both a and c, and we expect a different concentration dependence of T_g above M_c. T_g dependence on L is roughly given by Figure 10.

This simple model has been modified for arrays of rigid rods with fixed centres of rotation [51]. The rods attached to the sites of the cubic lattice can rotate freely but cannot cross each other. It is important to note that the glass transition in this system is decoupled from the structural transition (nematic ordering) and the only parameter is the ratio of the length of the rods to the distance between the centers of rotation. The Monte-Carlo study in 2-D shows that with increasing the parameter of the model a sharp crossover to infinite relaxation times can be observed. In 3-D the simulation gives a real transition to a completely frozen state at some critical length L_c.

4.1. An Important Analogy

Recent crucial experiments on granular materials [52,53] show that external vibrations lead to a slow approach of the packing density to a final steady-state value. Depending on the initial conditions and the magnitude of the vibration acceleration, the system can either reversibly move between steady-state densities, or can become irreversibly trapped into metastable states; that is, the rate of compaction and the final density depend sensitively on the history of vibration intensities that the system experiences (see Figure 11).

The function which has been found to fit the ensemble averaged density $\rho(t)$ better than other functional forms, is [53]:

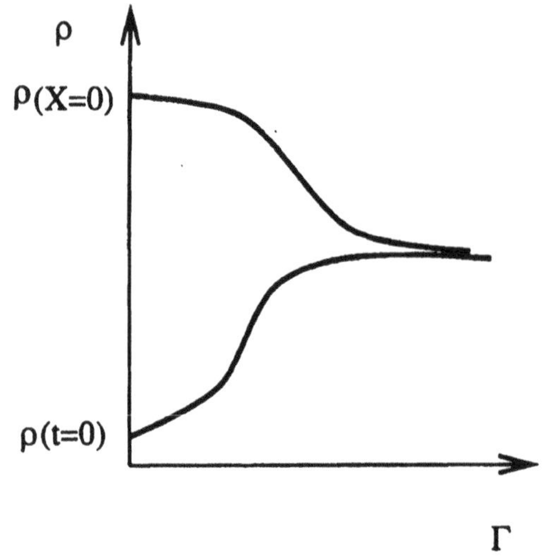

Figure 11. Dependence of the steady-state packing density on the tapping history (Nowak et al.). Experimental values of density packing fraction are in the following correspondence with model parameters: $\rho(X=0) = 1/v_0 \approx 0.64$, $\rho(t=0) = \rho_0 = 1/v_1 \approx 0.58$ and $\rho(X=\infty) = 2/(v_0+v_1) \approx 0.62$. The vibration intensity is parametrized by $\Gamma = a/g$.

$$\rho(t) = \rho_f - \frac{\Delta\rho_f}{1 + B\log\left(1 + \frac{t}{\tau}\right)} \tag{47}$$

where the parameters ρ_f, $\Delta\rho_f$ and τ depend only on Γ. This is an analogue of a glass forming material where the lower curve corresponds to the situation where the quenching speed inflicted on the glass is faster than the speed at which the glass relaxes back to equilibrium. The special feature of the granular material is that it has no power of its own to proceed to the equilibrium state, so that every aspect can be studied without having to worry about the fact that the true glass is always seeking equilibrium. It is worth considering a simple model for the density response to external vibrations [54]. If we assume that all configurations of a given volume are equally probable, it is possible to develop the formalism [55,56,57,58,59,60] analogous to conventional statistical mechanics. We introduce the volume function W (the analogue of a Hamiltonian) which depends on the coordinates of the grains, and their orientations. Averaging over all possible configurations of the grains in real space gives us a configurational statistical ensemble, which describes the random packing of grains. An analog of the microcanonical probability distribution is:

$$P = e^{-(S/\lambda)}\delta(V - W) \tag{48}$$

We can define the analogue of temperature as:

$$X = \frac{\partial V}{\partial S} \tag{49}$$

This fundamental parameter is called compactivity [55]. It characterizes the packing of a granular material, and may be interpreted as being characteristic of the number of ways it is possible to arrange the grains within the system into a volume ΔV, such that the disorder is ΔS. Consequently, the two limits of X are 0 and ∞, corresponding to the most and least compact stable arrangements. This is clearly a valid parameter for sufficiently dense powders, since one can in principle calculate the configurational entropy of an arrangement of grains, and therefore derive the compactivity from the basic definition. We will use the canonical probability distribution

$$P = e^{(Y-W)/\lambda X} \tag{50}$$

where λ is a constant which gives the entropy the dimension of volume. We call Y the effective volume; it is the analogue of the free energy:

$$e^{-Y/\lambda X} = \int e^{-W(\mu)/\lambda X} d(\text{all}) \tag{51}$$

$$V = Y - X\frac{\partial Y}{\partial X} \tag{52}$$

Examples of volume functions for particular systems can be found elsewhere [57,58,59,60]. We consider the rigid grains powder dominated by friction deposited in a container which will be shaken or tapped (in order to consider the simplest case, we ignore other possible interactions, e.g. cohesion, and do not distinguish between the grain-grain interactions in the bulk and those on the boundaries). We assume that most of the particles in the bulk do not acquire any non ephemeral kinetic energy, i.e. the change of a certain configuration occurs due to continuous and cooperative rearrangement of a free volume between the neighbouring grains. The simplest volume function is

$$W = v_0 + (v_1 - v_0)(\mu_1^2 + \mu_2^2) \tag{53}$$

where two degrees of freedom μ_1 and μ_2 define the "interaction" of a grain with its nearest neighbours. If we assume that all grains in the bulk experience the external vibration as a random force, with zero correlation time, then the process of compaction can be seen as the Ornstein-Uhlenbeck process for the degrees of freedom μ_i, $i = 1,2$ [61]. Therefore we write the Langevin equation:

$$\frac{d\mu_i}{dt} + \frac{1}{\nu}\frac{\partial W}{\partial \mu_i} = \sqrt{D} f_i(t) \tag{54}$$

where $\langle f_i(t) f_j(t')\rangle = 2\delta_{ij}\delta(t - t')$, and ν characterizes the frictional resistance imposed on the grain by its nearest neighbours. The term $f_i(t)$ on the RHS of (54) represents the random force generated by a tap. The derivation of this gives the analogue of the Einstein relation that $\nu=(\lambda X)/D$. If we identify f with the amplitude of the force a used in the tapping, the natural way to make this dimensionless is to write the "diffusion" coefficient as:

$$D = \left(\frac{a}{g}\right)^2 \frac{\nu\omega^2}{v} \tag{55}$$

that is we have a simplest guess for a fluctuation-dissipation relation:

$$\lambda X = \left(\frac{a}{g}\right)^2 \frac{\nu^2\omega^2}{v} \tag{56}$$

where v is the volume of a grain, ω the frequency of tapping, and g the gravitational acceleration. The standard treatment of the Langevin equation (\ref{langeq}) is to use it to derive the Fokker-Planck equation:

$$\frac{\partial P}{\partial t} = \left(D_{ij} \frac{\partial^2}{\partial \mu_i \partial \mu_j} + \gamma_{ij} \frac{\partial}{\partial \mu_i} \right) P = 0 \tag{57}$$

where $D_{ij} = D\delta_{ij}$ and $\gamma_{ij} = \gamma\delta_{ij}$. This equation can be solved explicitly [54,61]. We can calculate the volume of the system as a function of time and compactivity, $V(X,t)$. Though this is a simple model, it is too crude to give a quantitative agreement with experimental data; however, it gives a clear physical picture of what is happening. It is possible to imagine an initial state where all the grains are improbably placed, i.e. where each grain has its maximum volume v_1. So if one could put together a powder where the grains were placed in a high volume configuration, it will just sit there until shaken; when shaken, it will find its way to the distribution (50). It is possible to identify physical states of the powder with characteristic values of volume in our model. The value $V = Nv_1$ corresponds to the "deposited" powder, i.e. the powder is put into the most unstable condition possible, but friction holds it. When $V = Nv_0$ the powder is shaken into closest packing. The intermediate value of $V = (v_0 + v_1)/2$ corresponds to the minimum density of the reversible curve. Thus we can offer an interpretation of three values of density presented in the experimental data [52]. Our theory gives three points $\rho(X = 0)$, $\rho(X = \infty)$ and $\rho(t = 0)$ which are in the ratio: v_0^{-1}, $2/(v_0 + v_1)$, v_1^{-1} and these are in reasonable agreement with experimental data: $\rho(X = 0) = 1/v_0 \approx 0.64$, $\rho_0 = 1/v_1 \approx 0.58$ and $\rho(X = \infty) = 2/(v_0 + v_1) \approx 0.62$. Another important issue is the validity of the compactivity concept for a "fluffy", but still mechanically stable, granular array, e.g. for those composed of spheres with $\rho \leq 0.58$. In our theory, $\rho(X = \infty)$ corresponds to the beginning of the reversible branch (see Figure 11). We foresee that granular materials will throw much light onto glassy behaviour in the future, for questions like the pressure fluctuations in the jammed material are accessible in granular materials, and not entirely in glasses.

5. Discussion

In Section 2 we have given the analysis of stress for the case of granular arrays with the fixed values of the coordination number. However, a realistic granular packing can have a fluctuating coordination number which need not necessarily be 3 in 2-D or 4 in 3-D. How one can extend the formalism of Section 2 to make it capable of dealing with arbitrary packings? The simplest (however sensible) idea is to assume that in 3-D the major force is transmitted only through *four* contacts (we call them active contacts). The rest of the contacts transmit only an infinitesimal stress and can be christened as passive. A considerable experimental evidence for this conjecture does exist. Photoelastic visualization experiments [5,26] show that stresses in static granular media concentrate along certain well-defined paths. A disproportionally large amount of force is transmitted via these stress paths within the material. Computer simulations [31,32,29,30] also confirm the existence of well-defined stress-bearing paths. At the macroscopic scale, the most obvious characteristic of a granular packing is its density. As it has been shown in Section 4 it is natural to introduce the volume function W and compactivity $X = \partial V/\partial S$. The statistical ensemble now includes now the set of topological configurations with different microscopic force patterns. Therefore, in order to have a complete set of physical variables, one should combine the volume probability density functional (48) with the stress probability density functional (10). The joint probability distribution functional can be written in the form

$$P\left\{V \mid S_{ij}^{\alpha}\right\} = e^{-S/\lambda}\delta(V - W)$$
$$\times\prod_{\alpha,\beta}\delta\left(S_{ij}^{\alpha} - \sum_{\beta} f_i^{\alpha\beta} r_j^{\alpha\beta}\right)\Theta(\vec{f}^{\alpha\beta}\cdot\vec{n}^{g\alpha\beta})P\left\{\vec{f}^{\alpha\beta}\right\} \tag{58}$$

We speculate that this mathematical object is necessary for the analysis of the stress distribution in granular aggregates with an arbitrary coordination number. However, this is an extremely difficult problem which involves a mathematical description of force chains i.e. the mesoscopic clusters of grains carrying disproportionally large stresses and surrounded by the sea of spectator particles. An explicit theoretical analysis of this problem will be a subject of future research.

The aim of this paper has been to show that simple models can capture the basic physics of the jamming transition and provide insight into universal features of this phenomenon. We have christened the "jamming transition" as diverse physical phenomena which take place in granular materials, colloids and glasses. However, we argue that the existence of similar features like the slowing of response and the appearance of heterogeneous correlated structures gives hope that there are universal physical laws that govern jamming in various systems. Due to the extreme complexity of this problem, the simplest models should be considered and solved at first. This would help to capture basic mechanisms of these phenomena, and establish a theoretical framework which will incorporate higher complexities and details, and become a predictive tool.

Acknowledgements

We acknowledge financial support from Leverhulme Foundation (S. F. E.) and Shell (Amsterdam) (D. V. G.). The authors have benefited from many conversations with Professors Robin Ball, Mike Cates, Joe Goddard, Sidney Nagel and Tom Witten. We are most grateful to Dr John Melrose and Alan Catherall for providing the diagrams and many illuminating discussions on the jamming transition in colloids. We thank the Institute for Theoretical Physics (University of California in Santa-Barbara) for hosting the "Jamming and Rheology" programme, and for warm hospitality.

References

1. J. Feda, *Mechanics of Particulate Materials: The Principles*, Elsevier, Amsterdam, 1982.
2. A. Mehta (ed.), *Granular Matter: An Interdisciplinary Approach*. Springer-Verlag, New-York, 1993.
3. D. Bideau and A Hansen (eds.), *Disorder and Granular Media*, Elsevier, Amsterdam, 1993.
4. E. Guyon *et al.*, Nonlocal and nonlinear problems in the mechanics of disordered systems: application to granular media and rigidity problems, *Reports on Progress in Physics*, **53**, 373 (1990).
5. H. M. Jaeger, S. R. Nagel, and R. P. Behringer. Granular solids, liquids, and gases, *Review of Modern Physics*, **68**, 1259, (1996).
6. H. J. Herrmann, J.-P. Hovi and S. Luding (eds.), *Pysics of Dry Granular Media*, Kluwer, Dordrecht, 1998.
7. B. J. Briscoe and M. J. Adams (eds.), *Tribology in Particulate Technology*, Adam Hilger, Bristol, 1987.
8. A. Bose, *Advances in particulate materials*, Boston, Butterworth-Heinemann, 1995.
9. N. A. Fleck and A. C. F. Cocks (eds.), *Mechanics of Granular and Porous Materials*, Kluwer, Dodrecht, 1997.
10. K. L. Johnson, *Contact Mechanics*, Cambridge University Press, Cambridge, 1987.
11. A.N. Schofield and C. P. Roth. *Critical State Soil Mechanics*. McGraw Hill, 1968.
12. D. M. Wood. *Soil Behaviour and Critical State Soil Mechanics*. Cambrudge University Press, Cambridge, 1990.
13. V. V. Sokolovskii, *Statics of Granular Materials*, Pergamon, Oxford, 1965.
14. R. M. Nedderman, *Statics and Kinematics of Granular Materials*, Cambridge University Press, Cambridge, 1992.
15. O. Reynolds, On the dilatancy of media composed of rigid particles in contact, *Phylosophical Magazine*, **20**, 469 (1885).
16. M. E. Harr, *Mechnics of Particulate Media: A Probabilistic Approach*, McGraw-Hil, New York, 1977.
17. F. P. Bowden, D. Tabor, *The friction and lubrication of solids*, Clarendon Press, Oxford, 1986.
18. S. F. Edwards and D. V. Grinev, The Statistical-Mechanical Theory of Stress Transmission in Granular Materials, *Physica A*, **263**, 545 (1999).
19. S. F. Edwards, D. V. Grinev and R. C. Ball, Statistical Mechanics of Stress Transmission in Disordered Granular Arrays, *Physical Review Leters*, submitted.
20. R. C. Ball and D. V. Grinev, The Stress Transmission Universality Classes of Rigid Grain Powders, *Physical Review Letters*, submitted.
21. J-P. Bouchaud, M. E. Cates, P. Claudin, Stress distribution in granular media and nonlinear wave equation, *Journal de Physique I (France)*, **5**, 639, (1995).
22. J.P. Wittmer, P. Clauidn, M. E. Cates, J-P. Bouchaud, An explanation for the central stress minimum in sand piles, *Nature*, **382**, 336, (1996)

23. J. P. Wittmer, P. Claudin, M. E. Cates, Stress propagation and arching in static sandpiles, *Journal de Physique I (France)*, **7**, 39, (1997).
24. P. Claudin, J. P Bouchaud, M. E. Cates, J. P. Wittmer, Models of stress fluctuations in granular media, *Physical Review E*, **57**, 4441, (1998).
25. J. D. Goddard, Continuum modeling of granular assemblies, *Pysics of Dry Granular Media*, 1, H. J. Herrmann, J.-P. Hovi and S. Luding (eds.), Kluwer, Dordrecht, 1998.
26. C. H. Liu *et al.*, Force fluctuations in bead packs, *Science* **269**, 513 (1995),
27. D. M. Mueth, H. M. Jaeger and S. R. Nagel, Force distribution in a granular medium, *Physical Review E*, **57**, 3164 (1998).
28. B. Miller, C. O'Hern and R. P. Behringer, Stress fluctuations for continuously sheared granular materials, *Physical Review Letters* **77**, 3110 (1996).
29. F. Radjai *et al.*, Bimodal character of stress transmission in granular packings, *Physical Review Letters*, **80**, 61 (1998).
30. F. Radjai *et al.*, Force distributions in dense two-dimensional granular systems , *Physical Review Letters*, **77**, 274 (1996).
31. C. Thornton, Computer-simulated experiments on particulate materials, *Tribology in Particulate Technology*, B. J. Briscoe and M. J. Adams (eds.), Adam Hilger, Bristol 1987.
32. C. Thornton, Force transmission in granular media, *KONA Powder and Particle*, **15**, 81 (1997).
33. W. B. Russel, D. A. Saville and W. R. Schowalter, *Colloidal Dispersions*, Cambridge University Press, Cambridge, 1989.
34. H. A. Barnes, Shear thickening (dilatancy) in suspensions of nonaggregating solid particles dispersed in Newtonian fluids, *Journal of Rheology*, **33**, 329 (1989).
35. J. R. Melrose and R. C. Ball, A simulation technique for many spheres in quasi-static motion under frame-invariant pair drag and Brownian forces, *Physica A*, **247**, 444 (1997).
36. J. R. Melrose and R. C. Ball, The pathological behaviour of sheared hard spheres with hydrodynamic interactions, *Europhysics Letters*, **32**, 535 (1995).
37. J. R. Melrose, J. H. van Vliet and R. C. Ball, Continuous shear thickening and colloid surfaces, *Physical Review Letters*, **77**, 4660 (1996).
38. R. S. Farr, J. R. Melrose and R. C. Ball, Kinetic theory of jamming in hard-sphere startup flows, *Physical Review E*, **55**, 7203 (1997).
39. R. S. Farr, L. E. Silbert, J. R. Melrose and R. C. Ball, Power law shear thinning in concentrated colloids with weakly aggregating forces: a kinetic theory, *Journal of Chemical Physics*, submitted.
40. R. S. Farr, R. C. Ball, J. R. Melrose, Hydrodynamic clustering in the rheology of hard sphere colloids: a flow-jam phase transition, *Physical Review Letters*, submitted.
41. A.A. Catherall, J. R. Melrose and R. C. Ball, Shear thickening and order-disorder effects in concentrated colloids at high shear rates, *Journal of Rheology*, submitted.
42. R. L. Hoffman, Discontinuous and dilatant viscosity behavior in concentrated suspensions. I. Observation of flow instability *Transactions of Society of Rheology*, **16**, 155 (1972).
43. R. L. Hoffman, Explanations for the cause of shear thickening in concentrated dispersions, *Journal of Rheology*, **42**, 111 (1998).
44. J. F. Brady, G. Bossis, The rheology of concentrated suspensions of spheres in simple shear flow by numerical simulation *Journal of Fluid Mechanics*, **155**, 105 (1985).
45. R. J. Butera, M. S. Wolfe, J. Bender and N. J. Wagner Formation of a highly ordered colloidal microstructure upon flow cessation from high shear rates, *Physical Review Letters*, **77**, 2117 (1996).
46. W. Götze, L. Sjögren, Relaxation processes in supercooled liquids, *Reports on Progress in Physics*, **55**, 241 1992.
47. J. D. Ferry, *Viscoelastic properties of polymers*, John Wiley, New York, 1980.
48. G. Williams, D. C. Watts, *Journal of the Chemical Society, Faraday Transactions*, **68** 80 (1970).
49. S. F. Edwards, K. E. Evans, Dynamics of highly entangled rod-like molecules, *Journal of the Chemical Society, Faraday Transactions* 2, **78**, 113 (1982).
50. S. F. Edwards, T. Vilgis, The dynamics of the glass transition, *Physica Scripta*, **T13**, 7 (1986).
51. S. Obukhov, D. Kobzev, D. Perchak and M. Rubinstein, Topologically induced glass transition in freely rotating rods, *Journal de Physique I*, **7**, 563 (1997).
52. E. R. Nowak, J. B. Knight, M. L. Povinelli, H. M. Jaeger and S. R. Nagel, Reversibility and irreversibility in the packing of vibrated granular material, *Powder Technology*, **94**, 79 (1997).
53. E. R. Nowak, J. B. Knight, E. BenNaim, H. M. Jaeger and S. R. Nagel, Density fluctuations in vibrated granular materials, *Physical Review E*, **57**, 1971 (1998).
54. S. F. Edwards and D. V. Grinev, Statistical Mechanics of Vibration-Induced Compaction of Powders, *Physical Review E*, **58**, 4758 (1998).
55. S. F. Edwards and R. B. S. Oakeshott, Theory of powders *Physica A*, **157**, 1080 (1989).
56. S. F. Edwards, The role of entropy in the specification of a powder, *Granular Matter: An Interdisciplinary Approach*, A. Mehta (ed.), 121, Springer-Verlag, New-York, 1993.
57. C. C. Mounfield and S. F. Edwards, The statistical mechanics of granular systems composed of elongated grains, *Physica A*, **210**, 279 (1994).
58. S. F. Edwards and C. C. Mounfield, The statistical mechanics of granular systems composed of spheres and elongated grains, *Physica A*, **210**, 290 (1994).
59. C. C. Mounfield and S. F. Edwards, The statistical mechanics of granular systems composed of irregularly shaped grains, *Physica A*, **210**, 301 (1994).
60. A. Higgins and S. F. Edwards, A theoretical approach to the dynamics of granular materials, *Physica A*, **189**, 127 (1992).
61. H. Risken, *The Fokker-Planck Equation*, Springer-Verlag, New-York, 1990.

Part 1

FORCE CHAINS AND FLUCTUATIONS

Model for force fluctuations in bead packs

S. N. Coppersmith,[1,2] C.-h. Liu,[3,4] S. Majumdar,[5] O. Narayan,[6,7] and T. A. Witten[8]
[1]AT&T Bell Laboratories, Murray Hill, New Jersey 07974
[2]James Franck Institute, University of Chicago, 5640 Ellis Avenue, Chicago, Illinois 60637*
[3]Exxon Research & Engineering Company, Route 22 East, Annandale, New Jersey 08801
[4]Xerox Webster Research Center, 800 Phillips Road, Webster, New York 14580*
[5]Department of Physics, Yale University, New Haven, Connecticut 06511
[6]Department of Physics, Harvard University, Cambridge, Massachusetts 02138
[7]Department of Phyiscs, University of California, Santa Cruz, California 95064*
[8]James Franck Institute, University of Chicago, 5640 Ellis Avenue, Chicago, Illinois 60637
(Received 27 November 1995)

We study theoretically the complex network of forces that is responsible for the static structure and properties of granular materials. We present detailed calculations for a model in which the fluctuations in the force distribution arise because of variations in the contact angles and the constraints imposed by the force balance on each bead of the pile. We compare our results for the force distribution function for this model, including exact results for certain contact angle probability distributions, with numerical simulations of force distributions in random sphere packings. This model reproduces many aspects of the force distribution observed both in experiment and in numerical simulations of sphere packings. Our model is closely related to some that have been studied in the context of self-organized criticality. We present evidence that in the force distribution context, "critical" power-law force distributions occur only when a parameter (hidden in other interpretations) is tuned. Our numerical, mean field, and exact results all indicate that for almost all contact distributions the distribution of forces decays exponentially at large forces. [S1063-651X(96)07005-5]

PACS number(s): 02.50.Ey, 81.05.Rm

I. INTRODUCTION

Disordered geometric packings of granular materials [1] have fascinated researchers for many years [2]. Such studies, with their applicability to the geometry of glass-forming systems, initially were concerned with categorizing the void shapes and densities. More recently, partly in recognition of the ubiquity of granular materials and their importance to a wide variety of technological processes, interest has focused on how the forces supporting the grains are distributed. Visualizations of two-dimensional granular systems [3] demonstrate weight concentration into "force chains." It is natural to expect that similar concentrations of forces will occur in three dimensions. The distinctive forces in bead packs also give rise to distinctive boundary-layer flow [4] and novel sound-propagation properties [5].

Reference [6] presents experiments, simulations, and theory characterizing the inhomogeneous forces that occur in stationary three-dimensional bead packs, focusing particularly on the relative abundance of forces that are much larger than the average. If the bead pack were a perfect lattice, then, at any given depth, no forces would be greater than some definite multiple of the average force. At the other extreme, if the network of force-bearing contacts were fractal [7], then fluctuations in the forces (characterized, say, by their variance) would become arbitrarily large compared to the average force at a given depth, as the system size is increased. Reference [6] shows that the forces in bead packs are intermediate between these two extremes. The forces are unbounded, but the number of large forces falls off exponentially with the force. The fluctuations remain roughly the same as the average force, regardless of how large the bead pack becomes. A simple model was introduced to understand the results of the experiments and simulations.

This paper presents the detailed analysis of the model introduced in Ref. [6]. The model yields force distributions which agree quantitatively with those obtained in numerical simulations of sphere packings. Generic distributions of contacts lead to force distributions which decay exponentially at large forces, though a special distribution exists for which the force distribution is a power law. We discuss the relationship of this model to other related systems and present the analysis leading to the results that are quoted in Ref. [6] without derivation.

The paper is organized as follows. Section II defines the model, discusses several limiting cases that have been discussed previously in other contexts, and then presents our analysis of the force distribution expected in the context of force chains in bead packs. Special emphasis is placed on one particular contact distribution, the "uniform" distribution, which is the most random distribution consistent with the constraint of force balance. We first present a mean field solution for this model, and then show that this mean field solution is exact. We also obtain exact results for a countable set of nongeneric distributions as well as the mean field and numerical results for other contact distributions. Evidence is presented that almost all contact distributions lead to exponentially decaying force distributions. Section III discusses numerical simulations of sphere packings, which we analyze to obtain contact probability distributions to be used in the q model. We show that the force distribution predicted by the

*Present address.

model with this contact distribution agrees quantitatively with the force distribution in the simulation. The Appendix presents some mathematical identities concerning the uniform q distribution which are used in the text.

II. THE q MODEL

A. Definition of the model

Here we introduce the model, which assumes that the dominant physical mechanism leading to force chains is the inhomogeneity of the packing causing an unequal distribution of the weights on the beads supporting a given grain. Spatial correlations in these fractions as well as variations in the coordination numbers of the grains are ignored. We consider a regular lattice of sites, each with a particle of mass unity. Each site i in layer D is connected to exactly N sites j in layer $D+1$. Only the vertical components of the forces are considered explicitly (it is assumed that the effects of the horizontal forces can be absorbed in the random variables q_{ij} defined below). A fraction q_{ij} of the total weight supported by particle i in layer D is transmitted to particle j in layer $D+1$. Thus, the weight supported by the particle in layer D at the ith site, $w(D,i)$, satisfies the stochastic equation

$$w(D+1,j)=1+\sum_i q_{ij}(D)w(D,i). \tag{2.1}$$

We take the fractions $q_{ij}(D)$ to be random variables, independent except for the constraint $\Sigma_j q_{ij}=1$, which enforces the condition of force balance on each particle. We assume that the probability of realizing a given assortment of q's at each site i is given by a distribution function $p(q_{i1},\ldots,q_{iN})=\{\Pi_j f(q_{ij})\}\delta(\Sigma_j q_{ij}-1)$. We define the induced distribution $\eta(q)$ as

$$\eta(q)=\prod_{j\neq k}\int dq_{ij}\,p(q_{i1},\ldots,q_{ik}=q,\ldots,q_{iN}). \tag{2.2}$$

Because $p(q_{i1},\ldots,q_{iN})$ is a probability distribution and $\Sigma_{j=1}^{N} q_{ij}=1$, the induced distribution must satisfy the conditions $\int_0^1 dq\ \eta(q)=1$, $\int_0^1 dq\ q\,\eta(q)=1/N$.

In this paper we focus on the force distribution $Q_D(w)$, which is the probability that a site at depth D is subject to vertical force w. We obtain $Q_D(w)$ for different distributions of q's. We will also consider the force distribution $P_D(v)$ for the normalized weight variable $v=w/D$. For $\eta(q)=\delta(q-1/N)$, where each particle distributes the vertical force acting on it equally among all its neighbors, the force distribution at a given depth is homogeneous: $Q_D(w)=\delta(w-D)$, or $P_D(v)=\delta(v-1)$. At the other extreme, there is a "critical" limit, when q can only take on the values 1 or 0, so that weight is transmitted to a single underlying particle. For this, as discussed in the next section, the force distribution obeys a scaling form and decays as a power law at large forces, $Q(w)\propto w^{-c}$, where $c(N\geqslant 3)=\frac{3}{2}$ and $c(N=2)=\frac{4}{3}$. We demonstrate that this power law does not occur when q can take on values other than 1 and 0, as is the case for real packings. Generic continuous distributions of q's lead to a distribution of weights that, normalized to the mean, is independent of depth at large D and which decays exponentially at large weights. We solve the model exactly for a countable infinite set of q distributions, and present mean field and numerical results for other distributions of q's.

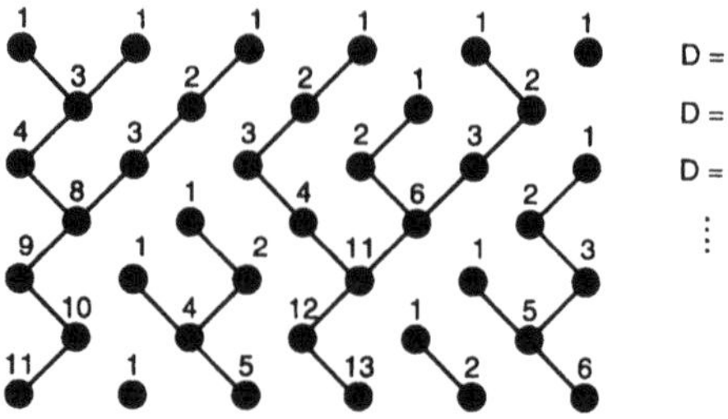

FIG. 1. Schematic diagram showing the paths of weight support for a two-dimensional system in the $q_{0,1}$ limit where each site transmits its weight to exactly one neighbor below. The numbers at each site are the values of $w(i,D)$.

B. q model for the "critical" case

We first consider the case where each particle transmits its weight to exactly one neighbor in the layer below, so that the variable q is restricted to taking on only the values 0 and 1. We denote this (singular) limiting case of our model by the "$q_{0,1}$ limit." Figure 1 shows the paths of weight support for a two-dimensional system in this limit. The solid lines correspond to bonds for which $q=1$. The paths of weight support of particles in the top row are coalescing random walks. Since a random walk of length D has typical transverse excursion of $D^{1/2}$, for the two-dimensional case the maximum weight supported by an individual grain at depth D scales as $D^{3/2}$ [8]. Because $D^{3/2}\gg D$, the mean weight supported at depth D, it is plausible that in the $q_{0,1}$ limit the model yields a broad weight distribution.

The defining equations of the $q_{0,1}$ limit of our model are known to be identical to those of Scheidegger's model of river networks [9] and a model of aggregation with injection [10,11]; the model is also equivalent to that of directed Abelian sandpiles [12–14]. (The number of neighbors below a particle, N, corresponds to the dimensionality d in these models.) The last equivalence follows [13,14] if we define $G_0(\vec{X}_1;\vec{X}_0)$ as the probability that the weight of site $\vec{X}_1$ is supported by site $\vec{X}_0$ in the same row or below it. The conditional probability that $\vec{X}_1$ is supported by $\vec{X}_0$, given that l of the N neighboring particles in the row below are supported by $\vec{X}_0$, is l/N. Thus,

$$G_0(\vec{X}_1;\vec{X}_0)=\frac{1}{N}\sum_{i=1}^{N} G_0(\vec{X}_1-\vec{e}_i;\vec{X}_0)+\delta_{\vec{X}_1,\vec{X}_0}, \tag{2.3}$$

where $\{\vec{X}_1-\vec{e}_i\}$ are the neighbors of $\vec{X}_1$ in the row below it, and the δ-function term follows because each particle must support its own weight. Similarly, the probability that two sites $\vec{X}_1$ and $\vec{X}_2$ in the same row are supported by $\vec{X}_0$ satisfies

$$G(\vec{X}_1,\vec{X}_2;\vec{X}_0)=\frac{1}{N^2}\sum_i\sum_j G(\vec{X}_1-\vec{e}_i,\vec{X}_2-\vec{e}_j;\vec{X}_0) \tag{2.4}$$

for $\vec{X}_1 \neq \vec{X}_2$. These equations are precisely those that describe the behavior of the correlations of the avalanches in the directed Abelian sandpile [12,15]. In this model, an integer "height" variable $z(\vec{X})$ is assigned each site $\vec{X}$ on a lattice. The dynamics are defined by the rule that if any $z(\vec{X})$ exceeds a critical value z_c, then the variables at m nearest neighbor sites along a preferred direction increase by 1, while $z(\vec{X})$ decreases by m. In this context $G_0(\vec{X}_1;\vec{X}_0)$ is identified with the probability that adding a particle at $\vec{X}_0$ creates an avalanche that topples over the site $\vec{X}_1$. Higher order correlations are mapped similarly. The distribution of weights in our model is mapped to the distribution of avalanche sizes.

All these models [9–13] have been studied as examples of self-organized criticality [16] because they lead to power-law correlations without an obvious tuning parameter. However, in the context of our model, the $q_{0,1}$ limit is a singular one, where the probability of $q \neq \{0,1\}$ has been tuned to zero. As we shall show in this paper, generic distributions $\eta(q)$, for which the probability that $q \neq \{0,1\}$ is nonzero (no matter how small), yield completely different results, with the distribution of weights decaying exponentially at large weights. With hindsight, we identify the probability for a river to split in the river network model [9], and the probability for a colloidal particle to fragment in the aggregation model [10,13] as hidden parameters that were tuned to zero. The corresponding parameter for directed Abelian sandpiles is less obvious.

The equivalence of our model in the $q_{0,1}$ limit to the models discussed above can be exploited to obtain some results for the distribution of weights. Recalling that the dimensionality d in these models corresponds to our N, we know that the weight distribution function at a depth D, $Q_D(w)$, has a scaling form for all N:

$$Q_D(w) = D^{-a} g(w/D^b), \qquad (2.5)$$

where $g(x) \to x^{-c}$ as $x \to 0$ [with a cutoff at w of $O(1)$].

The normalization constraints, $\int_0^\infty dw\, Q_D(w) = 1$ and $\int_0^\infty dw\, w Q_D(w) = D$, yield the conditions

$$a = bc, \quad 1 + a = 2b, \qquad (2.6)$$

so that there is only one free exponent. For $d=2$, the random walk argument at the beginning of this section suggests that $b = \frac{3}{2}$ [8] which agrees with the exact result [11]. For $d>2$, random walks are less likely to coalesce, and this argument breaks down. In mean field theory one obtains the analytic result $b=2$ [10], and exact analytic results for directed Abelian sandpiles in all dimensions [12] show that mean field theory is valid for $d \geq 3$ (with logarithmic corrections in $d=3$), and confirm the result $b = \frac{3}{2}$ for $d=2$. (Our exponent b can be identified with $\alpha+1$ of Ref. [12].)

As $D \to \infty$, the argument of the scaling function g in Eq. (2.5) is small for any finite w. Thus, in the $q_{0,1}$ limit of our model, the distribution of weights, $Q(w)$, is independent of D as $D \to \infty$, and is of a power-law form, and hence is infinitely broad.

C. q model away from criticality

The rest of this paper concerns probability distributions of the q's that do not have the property that q takes on only the values 1 and 0. We argue that all such distributions lead to force distributions that differ qualitatively from those described in the previous section. The $q_{0,1}$ limit is the only one that yields a power-law force distribution; other distributions lead to a much faster, typically exponential, decay. In addition, for other q distributions, the distribution for the *normalized* weight $v = w/D$, $P_D(v)$ converges to a fixed distribution $P(v)$ as $D \to \infty$. In contrast, in the $q_{0,1}$ limit, the quantity $Q_D(w)$ converges to a fixed function. In this section we present evidence for these assertions via both numerical simulations and mean field analysis.

1. Numerical simulations

Our numerical investigations all indicate that that for all q distributions except for the $q_{0,1}$ limit, the normalized force

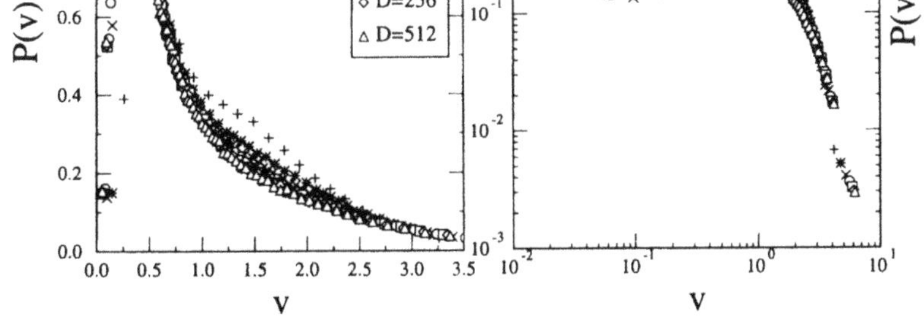

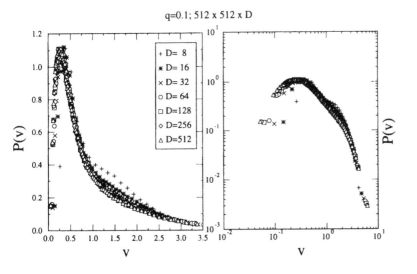

FIG. 2. Linear-linear and log-log plots of the normalized weight distribution function $P_D(v)$ vs v for a three-dimensional system on fcc lattice ($N=3$), for the q distribution defined in Eq. (2.7) with $q_0=0.1$. The distribution $P_D(v)$ appears to become independent of D as D becomes large, and decays faster than a power law at large v.

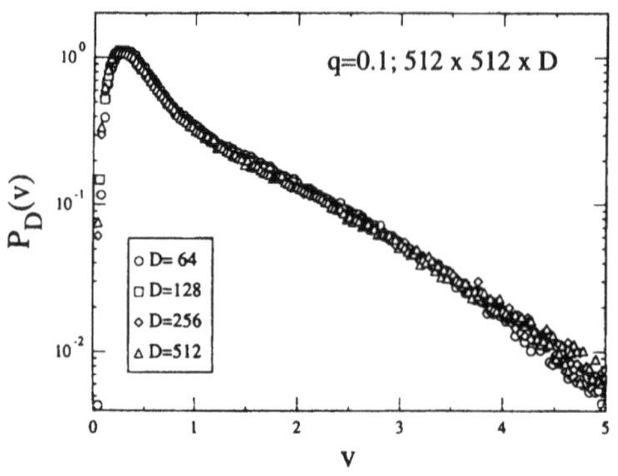

FIG. 3. Semilog plot of the normalized weight distribution function $P_D(v)$ vs v for a three-dimensional system on a fcc lattice ($N=3$), for the q distribution defined in Eq. (2.7) with $q_0=0.1$. The behavior of $P_D(v)$ at large v is consistent with exponential decay.

distribution $P_D(v)$ becomes independent of D as $D\rightarrow\infty$. To illustrate typical behavior, we consider the specific q distribution consisting of $N-1$ bonds emanating down from each site with value $q=q_0<1/(N-1)$ and one bond with $q=1-(N-1)q_0$, which has the induced distribution

$$\eta_{q_0}(q)=\frac{1}{N}\,\delta(q-[1-\{N-1\}q_0])+\frac{N-1}{N}\,\delta(q-q_0). \quad (2.7)$$

Figure 2 displays the normalized force distribution $P_D(v)$ versus v for several different depths D in a three-dimensional fcc system ($N=3$) of dimension $512\times512\times D$, with $q_0=0.1$. Periodic boundary conditions are imposed in the transverse directions. As D becomes large, $P_D(v)$ converges to a function independent of D which decays faster than a power law. Figure 3 is a semilog plot of $P_D(v)$ versus v for several values of D, showing that the decay of $P(v)$ at large D is roughly exponential. To see that this behavior is qualitatively different from that of the $q_{0,1}$ limit, in Fig. 4 we display numerical results for $P_D(v)$ versus v for a system which is identical except that $q_0=0$. In contrast to the $q=0.1$ case, $P_D(v)$ decays as a power law at large v. Also, $P_D(v)$ shows no signs of becoming independent of D as $D\rightarrow\infty$. This is consistent with the result in the previous section that $Q_D(w)$ becomes independent of D at large D.

2. *Mean field theory*

The technique of the mean field analysis for a general q distribution is a generalization of that used for the $q_{0,1}$ case [10]. The weight supported by a given site at depth D, $w_i(D)$, depends not only on the weight supported by the sites at depth $D-1$ but on the values of q for the relevant bonds:

$$w_i(D)=\sum_j q_{ij}w_j(D-1)+1. \quad (2.8)$$

In general the values of w at neighboring sites in layer D are not independent; the mean field approximation consists of ignoring these correlations.

As discussed above, when q is allowed to take on values other than 0 and 1, it is useful to study the force distribution function as a function of the *normalized* weight at a given depth, $v=w/D$. In terms of the normalized weight variable v, the mean field approximation leads to a recursive equation for the weight distribution function $P_D(v)$:

$$P_D(v)=\prod_{j=1}^{N}\left\{\int_0^1 dq_j\eta(q_j)\int_0^\infty dv_jP_{D-1}(v_j)\right\}\times\delta\left(\sum_{j=1}^{N}[(D-1)/D]v_jq_j-(v-1/D)\right). \quad (2.9)$$

The quantity $\eta(q)$ is defined in Eq. (2.2). The constraint that the q's emanating downward from a site must sum to unity enters only through the definition of $\eta(q)$ because there is no restriction on the q's for the *ancestors* of a site. The only approximation here is the neglect of possible correlations between the values of v among the ancestors.

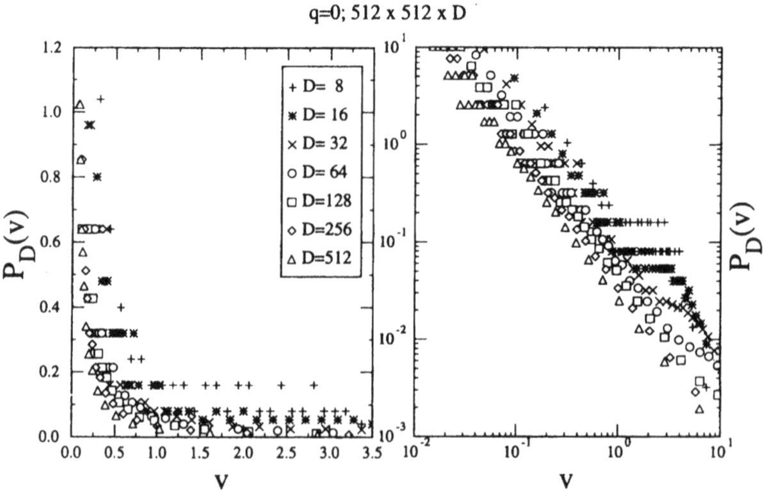

FIG. 4. Linear-linear and log-log plots of the normalized weight distribution function $P_D(v)$ vs v for a three-dimensional system on a fcc lattice ($N=3$), for the q distribution defined in Eq. (2.7) with $q_0=0$. For this special case, the distribution $P_D(v)$ does not become independent of D as D becomes large. The asymptotic decay of $P_D(v)$ at large v is a power law.

By Laplace transforming, one finds that $\tilde{P}_D(s)$, the Laplace transform of the distribution function of the normalized weight $P_D(v)$, obeys

$$\tilde{P}_D(s)=e^{-s/D}\left[\int_0^1 dq\ \eta(q)\tilde{P}_{D-1}[sq(D-1)/D]\right]^N. \quad (2.10)$$

Since as $D\rightarrow\infty$ the distribution $P_D(v)$ becomes independent of D [17,18], Eq. (2.10) then becomes

$$\tilde{P}(s)=\left[\int_0^1 dq\ \eta(q)\tilde{P}(sq)\right]^N. \quad (2.11)$$

First we show that the weight distribution $P(v)$ decays faster than any power of v for all q distributions except those that only take on the values 0,1. We expand the Laplace transform $\tilde{P}(s)$ in powers of s, $\tilde{P}(s)=1+\Sigma_{j=1}^{\infty}P_{ij}s^j$, and plug into Eq. (2.11), obtaining

$$1+P_1s+\sum_{j=2}^{\infty}P_js^j=\left[1+P_1s/N+\sum_{j=2}^{\infty}P_js^j\langle q^j\rangle\right]^N. \quad (2.12)$$

Here, s is the Laplace transform variable, $\langle q^j\rangle=\int_0^1 dq\ q^j\eta(q)$, and we have used $\langle q\rangle=1/N$. Equating the coefficients of s^j on the left and right hand sides of the equation, we obtain a linear equation for P_j:

$$P_j[N\langle q^j\rangle-1]=G(P_{j-1},P_{j-2},\ldots,P_1), \quad (2.13)$$

where G is some complicated polynomial. This can be iterated to obtain P_j for successively higher values of j.

Since Eq. (2.13) is linear in P_j, P_j can diverge only if its coefficient $[N\langle q^j\rangle-1]$ is zero. If q can take on only the values 0 and 1, then $\langle q^j\rangle=\langle q\rangle$ and $[N\langle q^j\rangle-1]=0$ for all $j>1$. However, for *any* other distribution of q's restricted to the interval [0, 1], the distribution for q^j is shifted towards the origin compared to the distribution for q^k, whenever $j>k$. Since $\langle q^j\rangle<\langle q\rangle=1/N$ for all $j>2$, Eq. (2.13) has a nonzero coefficient for P_j for all $j>2$, which means that all moments $\langle v^j\rangle$ of $P(v)$ are finite. (For the special case of $j=1$, the equation is degenerate; P_1 is set by the normalization of v.) If $\ln P(v)$ were to behave asymptotically as $-a\ln v$, then $\langle v^j\rangle$ would diverge for all $j>a-1$. Hence $P(v)$ must fall off faster than any power of v and $(d\ln P(v)/d\ln v)\rightarrow-\infty$ as $v\rightarrow\infty$.

D. Weight distributions away from criticality: Mean field results

Now we consider the distribution of weights for noncritical distributions of q's. Motivated by the geometrical disorder present in granular materials, we focus especially on continuous distributions. First we calculate this distribution within a mean field approximation for the simplest possible continuous distribution, $f(q_{ij})$=const, or $\rho(q_{i1},\ldots,q_{iN})=(N-1)!\delta(\Sigma_jq_{ij}-1)$ (the uniform q distribution). We show that within mean field theory, all "typical" continuous q distributions lead to a force distribution that decays exponentially at large weights. We will show later that the mean field solution is *exact* for a countable set of q distributions, including the uniform q distribution.

1. Mean field theory for the uniform distribution

One example of a q distribution that can lead to an exponentially decaying distribution of weights is the "uniform" distribution of q's, for which the probability of obtaining the values $q_{i1},\ldots,q_{iN}$ is $\rho(q_{i1},\ldots,q_{iN})=(N-1)!\delta(\Sigma_jq_{ij}-1)$. We show in the Appendix that this distribution induces $\eta_u(q)=(N-1)(1-q)^{N-2}$. Thus, for this q distribution in the limit $D\rightarrow\infty$ the mean field force distribution is the solution to the self-consistent equation:

$$\tilde{P}(s)=\left[\int_0^1 dq(N-1)(1-q)^{N-2}\tilde{P}(sq)\right]^N. \quad (2.14)$$

First consider $N=2$. For this case $\eta(q)=1$, so Eq. (2.14) becomes

$$\tilde{P}(s)=\left[\int_0^1 dq\ \tilde{P}(sq)\right]^2. \quad (2.15)$$

Letting $\tilde{V}(s)=[\tilde{P}(s)]^{1/2}$ and $u=qs$, one obtains

$$s\tilde{V}(s)=\int_0^s du\ \tilde{V}^2(u). \quad (2.16)$$

Differentiating with respect to s yields

$$\tilde{V}(s)+s\frac{d\tilde{V}}{ds}=\tilde{V}^2(s), \quad (2.17)$$

which can be integrated to yield

$$\tilde{V}(s)=\frac{1}{1-Cs}. \quad (2.18)$$

The constant of integration C is determined by the definition of the mean, $\int_0^{\infty}dv\ vP(v)=-d\tilde{P}/ds|_{s=0}=1$. Thus, $C=d\tilde{V}/ds|_{s=0}=-\frac{1}{2}$. Hence one finds $\tilde{P}(s)=4/(s+2)^2$ and $P(v)=4ve^{-2v}$.

This method can be generalized for all N. Defining $\tilde{V}_N(s)=[\tilde{P}_N(s)]^{1/N}$, inserting in Eq. (2.14), and differentiating $N-1$ times, one finds that $\tilde{V}_N(s)$ obeys the differential equation:

$$\frac{d^{N-1}}{ds^{N-1}}[s^{N-1}\tilde{V}_N(s)]=(N-1)!\tilde{V}_N^N(s). \quad (2.19)$$

A solution to this equation is $\tilde{V}_N(s)=C/(s+C)$, where C is any constant. This can be shown by induction: Assume that

$$\frac{d^{N-2}}{ds^{N-2}}\left(s^{N-2}\frac{C}{s+C}\right)=(N-2)!\left(\frac{C}{s+C}\right)^{N-1}. \quad (2.20)$$

Then

$$\frac{d^{N-1}}{ds^{N-1}}\left(s^{N-1}\frac{C}{s+C}\right)=\frac{d^{N-1}}{ds^{N-1}}\left((s+C-C)s^{N-2}\frac{C}{s+C}\right) \tag{2.21}$$

$$=-C\frac{d}{ds}\left[(N-2)!\left(\frac{C}{s+C}\right)^{N-1}\right] \tag{2.22}$$

$$=(N-1)!\left(\frac{C}{s+C}\right)^{N}. \tag{2.23}$$

Since direct substitution can be used to show that the identity holds for $N=2$, it holds for all N.

The condition $d\widetilde{V}/ds=-1/N$ is satisfied when $C=N$. Hence one finds the weight distribution:

$$P_N(v)=\frac{N^N}{(N-1)!}v^{N-1}e^{-Nv}. \tag{2.24}$$

The question of uniqueness of this solution is discussed below.

2. Mean field asymptotic force distribution for generic continuous q distributions

We now show that, within mean field theory, generic continuous q distributions lead to weight distribution functions $P(v)$ for the normalized weight v which have the asymptotic forms $P(v)\propto v^{N-1}e^{-Nv}$ as $v\to\infty$ and $P(v)\propto v^{N-1}$ as $v\to 0$.

We consider q distributions of the form $p(q_{i1},\ldots,q_{iN})=\{\Pi_j f(q_{ij})\}\delta(\Sigma_j q_{ij}-1)$ [the uniform distribution is $f(q_{ij})=\text{const}$]. If $f(q_{ij})$ has a nonzero limit as $q_{ij}\to 0$, and does not have a δ-function contribution at $q_{ij}=0$, then phase space restrictions imply that the induced distribution $\eta(q)\sim(1-q)^{N-2}$ for $q\to 1$. This is because if a site receives a fraction q of the weight from one of its predecessors, then the fractions received by all the *other* successors of that predecessor, $\{q_2\cdots q_N\}$, must add up to $1-q$. For q close to 1, this gives a phase-space volume of the order of $(1-q)^{N-2}$.

To determine the large v asymptotics of $P(v)$, we use the result of Sec. II C 2, that $P(v)$ must fall off faster than any power of v. We write the $D\to\infty$ limit of Eq. (2.9) as

$$P(v)=\left\{\prod_{j=1}^{N}\int_0^\infty dv_j F(v_j)\right\}\delta\left(v-\sum_j v_j\right), \tag{2.25}$$

$$F(v_j)=\int_0^1 dq_j P(v_j/q_j)\eta(q_j)/q_j. \tag{2.26}$$

Since $P(v)$ decays quickly (in particular, faster than $1/v$), the apparent singularity near $q=0$ in Eq. (2.26) is not really there. The integral is dominated by $q\approx 1$. This follows because

$$P(v/q)=P(v)\exp\left[-\frac{\partial \ln P(v/q)}{\partial q}\bigg|_{q=1}(1-q)+\cdots\right]$$

$$=P(v)\exp\left[\frac{v}{q^2}\frac{\partial \ln P(v/q)}{\partial (v/q)}\bigg|_{q=1}(1-q)+\cdots\right]$$

$$=P(v)\exp\left[\frac{\partial \ln P(u)}{\partial \ln u}\bigg|_{u=v}(1-q)+\cdots\right]. \tag{2.27}$$

Since $\partial\ln P(u)/\partial\ln u\to-\infty$ as $u\to\infty$, this expression becomes very small as $1-q$ increases. Thus, for large v, since $\eta(q)\sim(1-q)^{N-2}$ for $q\approx 1$,

$$F(v)\sim P(v)\bigg/\left[\left|\frac{\partial \ln P(v)}{\partial \ln v}\right|\right]^{N-1}. \tag{2.28}$$

Already it is clear that $P(v)$ for any generic q distribution has the same large-v asymptotics as the uniform distribution, since the asymptotics are determined entirely by the phase-space restrictions on $\eta(q)$ for $q\approx 1$. This decay also can be demonstrated explicitly by assuming faster and slower decays and showing inconsistency with Eq. (2.25). If $P(v)$ were to decay faster than exponentially, then the convolution in Eq. (2.25) would be dominated by the region where all the v_j's are roughly equal. But since $[P(v/N)]^N \gg P(v)$, Eq. (2.25) cannot be satisfied. On the other hand, if $P(v)$ were to decay slower than exponentially, then the convolution would be dominated by the region where one of the v_i's is $\approx v$ and the others are $O(1)$. Equation (2.25) would then imply

$$P(v)\sim P(v)\bigg/\left[\left|\frac{\partial \ln P(v)}{\partial \ln v}\right|\right]^{N-1}. \tag{2.29}$$

Since the expression in square brackets diverges with v, this is not possible either. Thus one must have $P(v)=h(v)\exp[-av]$, where $h(v)$ varies more slowly than an exponential. Equation (2.25) then implies

$$h(v)=\left\{\prod_{j=1}^{N}\int_0^\infty dv_j h(v_j)/v^{N-1}\right\}\delta\left(v-\sum_j v_j\right). \tag{2.30}$$

This is satisfied by $h(v)\sim v^{N-1}$, so that

$$P(v)\sim v^{N-1}\exp[-av] \tag{2.31}$$

for $v\to\infty$.

Hence we have shown that for generic continuous q distributions, within mean field theory $P(v)\to v^{N-1}\exp(-av)$ as $v\to\infty$.

3. Mean field theory for singular q distributions

We have shown that all q distributions which satisfy the condition $\int_q^1 dq\,\eta(q)\sim(1-q)^{N-1}$ as $q\to 1$ have a weight distribution within mean field theory that is of the form $P(v)\sim v^{N-1}\exp[-av]$ for large v. This condition on $\eta(q)$ is satisfied under fairly general assumptions: one requires (1) that the probability density for any q_{ij} in Eq. (2.8) have a nonzero $q_{ij}\to 0$ limit and (2) that it not have a δ-function contribution at $q_{ij}=0$. However, as we shall see below, to compare the results of the q model to molecular dynamics simulations and to experiments on real bead packs, it is useful to consider the case where there is a finite probability for some of the q_{ij}'s to be zero, which implies that the induced distribution $\eta(q)$ has a δ function at $q=0$ (and in some cases at $q=1$) [19]. Such a choice for $\eta(q)$ is also useful in examining the crossover from the critical $q_{0,1}$ limit to the smooth q distributions considered in the preceding section. We will see that q distributions of this type lead to force distributions $P(v)$ that decay exponentially, though with different power laws multiplying the exponential than for continuous q distributions.

We first note that, when $\eta(q)$ has a finite weight at $q=1$, it is impossible for $\tilde{P}(s)$ to diverge at any s. The solutions of the form $\tilde{P}(s)\sim 1/(q+s/s_0)^N$ obtained in Sec. II D 1 were possible because, in Eq. (2.11), the integral over q reduces the singularity, which is compensated by the exponentiation. With a finite weight at $q=1$, close to a divergence at s_0 one would have $\tilde{P}(s)\propto[\tilde{P}(s)]^N$, which would be impossible as $s\to s_0$.

It is instructive to consider first a simplified version of such singular q distributions. Let us consider the case of $N=2$, and assume that $\eta(q)$ has the form

$$\eta(q)=\tfrac{1}{2}(1-\theta)\{\delta(q)+\delta(1-q)\}+\theta\delta(q-\tfrac{1}{2}), \tag{2.32}$$

with $0<\theta<1$. This $\eta(q)$ satisfies the conditions $\int dq\ \eta(q)=1$ and $\int dq\ q\,\eta(q)=\frac{1}{2}$ for all θ. Equation (2.11) then simplifies to

$$\tilde{P}(s)=[\tfrac{1}{2}(1-\theta)+\tfrac{1}{2}(1-\theta)\tilde{P}(s)+\theta\tilde{P}(s/2)]^2, \tag{2.33}$$

where we have used the fact that $\tilde{P}(0)=1$. Equation (2.33) can be solved as follows: for small s, we know that $\tilde{P}(s)=1-s+O(s^2)$ [the coefficient of the linear term being fixed by the normalization condition $\int dv\ v P(v)=1$]. Starting with a small negative value of s, where $\tilde{P}(s)$ is approximated as $1-s$, Eq. (2.33) can be iterated to find $\tilde{P}(2^n s)$ for $n=1,2,\ldots$ [the correct root of the quadratic equation is chosen by requiring $\tilde{P}(s)=1$ for $\tilde{P}(s/2)=1$]. Eventually the result of this iteration scheme is complex rather than real, signifying that s is in a region where $\tilde{P}(s)$ has a branch cut. It is easiest to find the origin s_0 of this branch cut by adjusting $\tilde{P}(s_0/2)$ so that Eq. (2.33) has a double root for $\tilde{P}(s_0)$, and then iterating *backwards* to obtain $\tilde{P}(s_0/2^n)$. As $n\to\infty$, by matching on to the requirement that $\tilde{P}(s)=1-s$ for $s\to 0$, one can obtain s_0. It is clear from Eq. (2.33) that in the vicinity of s_0 $\tilde{P}(s)$ is of the form $\tilde{P}(s_0)+\alpha\sqrt{s-s_0}$. This yields

$$P(v)\sim v^{-3/2}\exp[-s_0 v] \quad \text{for } v\to\infty. \tag{2.34}$$

Although the power-law prefactor is different from that in Eq. (2.31), there is still an exponential decay.

We now consider possible changes to Eq. (2.34) from choosing $\eta(q)$ of a more complicated form than Eq. (2.32). For any $\eta(q)$ of the form

$$\eta(q)=\sum_{i=0}^{n} c_i\delta(q-\lambda^i)+\left(1-\sum c_i\right)\delta(q)\delta(q) \tag{2.35}$$

with $0<\lambda<1$, one can use the method outlined above to find that $\tilde{P}(s)$ has a square-root branch cut at some s_0. This answer is not affected by making n large, so long as c_0 remains nonzero. As $n\to\infty$, with all the c_i's for $i>0$ tending to zero, we can approach arbitrary continuous distributions for $\eta(q)$ with δ functions at $q=0$ and $q=1$.

For $N>2$, Eq. (2.33) is changed to a higher order equation. This, however, does not generically change the results above. Even for higher order equations, the degeneracy of the roots generally occurs only pairwise, so that close to the point of degeneracy the singularly ranging roots still have a square-root singularity. It will, however, be possible to find *nongeneric* choices for $\eta(q)$ that could result in an asymptotic form $P(v)\sim v^{-(1+1/m)}\exp[-av]$ with $N\geq m\geq 2$.

E. Beyond mean field theory

1. Proof that mean field theory is exact

In this section we prove that the mean field solution presented in the preceding section is an *exact* solution of the model with the uniform q distribution for any N.

In general, the mean field theory presented above is not exact because it does not account for the fact that two neighboring sites in row $D+1$ both derive a fraction of their weight from the same site in row D. Suppose a site j in row $D+1$ has $w(j)$ much larger than the average value. Then it is likely that the weight supported by an ancestor $w(i)$ in row D is larger than average also. Because this ancestor transmits its weight to a neighboring site in row $D+1$ as well, there is a "correlation" effect that creates a greater likelihood that in a given layer sites supporting large weight are close together. On the other hand, there is a "anticorrelation" effect arising because $\Sigma_j q_{ij}=1$; if a large fraction of the weight from site i is transmitted to site j, then small fractions are transmitted to the other "offspring" sites. When the q's are chosen from the uniform distribution, these "correlation" and "anticorrelation" effects cancel exactly.

The result that the mean field correlation functions are exact for the uniform distribution of q's can be understood by considering the system in terms of weights on bonds. Each bond $\{ij\}$ corresponds to a particle with "energy" $E_{ij}=v_i q_{ij}$. Moving down by one layer corresponds to having groups of N particles colliding at each site and emerging with different energies, subject to the constraint that the total energy of all N particles colliding at each site is unchanged by the collision.

For the "uniform" q distribution, each collision takes N particles of energies $e_{\alpha_1},\ldots,e_{\alpha_N}$ and changes their energies to $E_{\alpha_1},\ldots,E_{\alpha_N}$, subject only to the constraint that $\Sigma e_\alpha=\Sigma E_\alpha$. If we start with a "microcanonical" ansatz for the phase-space density, i.e., that it is uniform over the space $\Sigma E_\alpha=E$, then it is preserved by the collisions. Hence, the microcanonical density is the correct one for this system.

With a microcanonical density for a large collection of particles, the density for any finite subgroup is canonical (in the thermodynamic limit) [20]. Thus, we have shown for this case that the distribution of "bond forces" is exponential, which is the most random distribution consistent with the constraint that the sum of the forces is fixed [20,21].

Note that this argument does not hold for q distributions other than the uniform one. For instance, in the $q_{0,1}$ limit, each collision takes all the energy of the group and gives it to one of the colliding particles. Thus, even if we start with the microcanonical distribution, it breaks down at the very first step. For general q distributions, the phase-space density is not separable, i.e., mean field theory is not exact and there are spatial correlations within each layer.

The explicit algebraic proof proceeds by constructing exact recursion relations for the correlation functions describing the weight distribution in the model in row $D+1$ in terms those for row D, and showing that the mean field correlation functions are invariant under this recursion. We ignore the weight added in each row because we are looking for the fixed distribution very far down the pile.

Let $P_D(u_i)$ be the probability that site i in row D supports weight u_i, $P_D(u_{i_1},u_{i_2})$ be the probability that sites i_1 and i_2 support weight u_{i_1} and u_{i_2}, respectively, and $P_D(u_{i_1},u_{i_2},\ldots,u_{i_n})$ be the normalized joint distribution describing the probability that sites $i_1,i_2,\ldots,i_n$ support weights $u_{i_1},\ldots,u_{i_n}$, respectively. The mean field joint probability distributions are given by the mean field $P(u)$ and

$$P(u_1,u_2,\ldots,u_n)=P(u_1)P(u_2)\cdots P(u_n). \tag{2.36}$$

Consider the M-point correlation function in row $D+1$ that is obtained when all the correlation functions in row D are the the mean field ones. Let $\{u_i\}$ be the weights in row D and $\{v_i\}$ be the weights in row $D+1$. Consider a cluster of sites $j=1,\ldots,M$ in row $D+1$, with ancestors in row D at sites $i=1,\ldots,p$. (The labels do not imply any particular spatial relation of the sites.) The q's describing the bonds emanating from ancestor i are q_{il}, where $l=1,\ldots,N$. We define $\eta_{il}(j)$ to be 1 if sites i and j are connected by bond il and zero otherwise. The M-point correlation function in row $D+1$, $P_{D+1}(v_1,\ldots,v_M)$, must obey

$$P_{D+1}(v_1,\ldots,v_M)=\prod_{i=1}^{p}\left\{\int_0^1 dq_{i1}\cdots\int_0^1 dq_{iN}(N-1)!\times\delta\left(1-\sum_{k=1}^{N}q_{ik}\right)\int_0^\infty du_i P_D(u_i)\right\}\times\prod_{j=1}^{M}\delta\left(v_j-\sum_{i=1}^{p}\sum_{l=1}^{N}\eta_{il}(j)q_{il}u_i\right). \tag{2.37}$$

We define the general Laplace transform

$$\tilde{P}(s_1,\ldots,s_n)=\int_0^\infty dv_1\cdots\int_0^\infty dv_n P(v_1,\ldots,v_n)\times e^{-s_1v_1-\cdots-s_nv_n}. \tag{2.38}$$

Laplace transforming Eq. (2.37), one obtains

$$\tilde{P}_{D+1}(s_1,\ldots,s_M)=\prod_{i=1}^{p}\int_0^1 dq_{i1}\cdots\int_0^1 dq_{iN}(N-1)!\times\delta\left(1-\sum_{k=1}^{N}q_{ik}\right)\times\tilde{P}_D\left(\sum_{j=1}^{M}\sum_{l=1}^{N}\eta_{il}(j)q_{il}s_j\right). \tag{2.39}$$

For $\tilde{P}_D(x)=(1+x/N)^{-N}$, one can use the condition $\Sigma_{l=1}^{N}q_{il}=1$ to write

$$\tilde{P}_{D+1}(s_1,\ldots,s_M)=\prod_{i=1}^{p}\int_0^1 dq_{i1}\cdots\int_0^1 dq_{iN}(N-1)!\times\delta\left(1-\sum_{k=1}^{N}q_{ik}\right)\times\left[\sum_{l=1}^{N}q_{il}\left(1+\sum_{j=1}^{M}\eta_{il}(j)s_j/N\right)\right]^{-N}. \tag{2.40}$$

Using the identity [22]

$$\prod_{n=1}^{N}(a_n)^{-1}=(N-1)!\times\int_0^1 dx_1\cdots\int_0^1 dx_N\frac{\delta(1-x_1-\cdots-x_N)}{(a_1x_1+\cdots+a_Nx_N)^N} \tag{2.41}$$

with $a_n=1+\Sigma_{j=1}^{M}\eta_{in}(j)s_j/N$, one finds

$$\tilde{P}_{D+1}(s_1,\ldots,s_M)=\prod_{i=1}^{p}\prod_{n=1}^{N}\frac{1}{1+\Sigma_{j=1}^{M}\eta_{in}(j)s_j/N}. \tag{2.42}$$

If a given bond $\{in\}$ connects to no sites in the descendant cluster, then every term in the sum in the denominator of Eq. (2.42) is zero, and the $\{in\}$th term in the product is unity. If the bond connects to a site in the descendant cluster, then $\eta_{in}(j)$ is unity for exactly one j. Each site j in the descendant cluster is connected to exactly N antecedents in row D, so

$$\tilde{P}_{D+1}(s_1,\ldots,s_M)=\prod_{j=1}^{M}\frac{1}{(1+s_j/N)^N}. \tag{2.43}$$

Thus, the mean field correlation functions are preserved from row to row for this q distribution.

2. Other q distributions

We have identified a countable set of q distributions for which mean field theory is exact, those of the form $f(q_{ij})=q^r$, for all integer r (the uniform distribution is $r=0$). The resulting force distribution $P_r(v)\propto v^{r+N-1}e^{-Nrv}$ has Laplace transform $\tilde{P}_r(s)=[1+s/(Nr)]^{-Nr}$. The demonstration that this solution is exact follows precisely the same line of reasoning as for the $r=0$ case presented in the preceding section, utilizing the identity [22]

$$\prod_{n=1}^{N}(a_n)^{-r}=\frac{\Gamma(Nr)}{[\Gamma(r)]^N}\int_0^1 dx_1x_1^{r-1}\cdots\int_0^1 dx_Nx_N^{r-1}\times\frac{\delta(1-x_1-\cdots-x_N)}{(a_1x_1+\cdots+a_Nx_N)^{Nr}}. \qquad (2.44)$$

In terms of the particle collision picture discussed at the beginning of Sec. II E 1, a general value of r corresponds to the particles having an energy which is the sum of $r+1$ components (which may be viewed as as spatial coordinates in some underlying space), each one of which is conserved individually in a collision. The microcanonical phase-space density, uniform in the $N(r+1)$-dimensional space, is preserved by the collisions, and yields the $P_r(v)$ stated here.

The result that mean field theory yields an exact solution of the model holds only for a very limited class of q distributions. For general q distributions, the phase-space density is not separable, i.e., mean field theory is not exact, and there are spatial correlations within each layer. For example, Fig. 5 shows $P(v)$ for a two-dimensional system with $N=2$ and the q distribution where the two bonds emanating from each take on the values q_0 and $1-q_0$, with $q_0=0.1$. In the model, a site (i,D) is connected to sites $(i,D+1)$ and $[i+1\ (\text{mod}L),D+1]$; in the mean field calculation, site (i,d) is connected to sites $(p_1(i),D+1)$ and $(p_2(i),D+1)$, where $\vec{p}_1$ and $\vec{p}_2$ are permutations of $(1,\ldots,L)$. This method of simulating mean field theory destroys the spatial correlations between ancestor sites, while ensuring that every site has exactly two ancestors and two descendants. The numerical data were obtained by averaging $P(v)$ for rows 10 001–20 000 in a system of transverse extent $L=20\,000$. This figure demonstrates explicitly that the mean field force distribution $P(v)$ is not exact for this distribution. However, the deviations of the mean field theory from the direct simulation are extremely small for $v\gtrsim 0.1$, so mean field theory provides an accurate quantitative estimate for $P(v)$ over a large range of v.

3. Uniqueness of the steady state distribution

In this section we show that our results (numerical and analytical) for the force distribution do not depend on either the boundary conditions imposed at the top of the system or on the specific realization of randomness a particular system might have.

Consider a system of finite transverse extent L, in which weights $\{w(i)\}$ are put on the particles in the top row. The weight then propagates downwards according to Eq. (2.8). If we now consider the same system with a different loading on the top row, $\{w(i)+\delta w(i)\}$, then since Eq. (2.8) is linear in w, the difference between the two solutions satisfies the homogeneous equation

$$\delta w(D+1,j)=\sum_i q_{i,j}(D)\,\delta w(D,i). \qquad (2.45)$$

Summing up both sides, we see that $\Sigma_j\delta w(D+1,j)=\Sigma_i\delta w(D,i)$, which means that the total excess weight placed on the top of the system propagates downwards unaltered. Such a change only affects the normalization of our distributions. Thus, if we are interested in the normalized distribution $P_D(v)$, we can without loss of generality consider perturbations $\{\delta w(D,i)\}$ satisfying the constraint $\Sigma_i\delta w(D,i)=0$.

Equation (2.45) can be viewed as a two stage process: (1) each $\delta w(D,i)$ splits into N parts, $q_{i,j}(D)\delta w(D,i)$, and (2) the N fragments $q_{i,j}(D)\delta w(D,i)$, with i running over the neighbors of j in the row above it, combine to give $\delta w(D+1,j)$. The important thing is that all the $q_{i,j}$'s are positive. Thus if we define the total difference between the two configurations as $\Delta(D)=\Sigma_i|\delta w(D,i)|$, then because all the q's are positive and $\Sigma_jq_{ij}=1$, Δ is unchanged in the first step, while in the second step it can either stay constant or decrease (depending on whether the signs of the fragments are

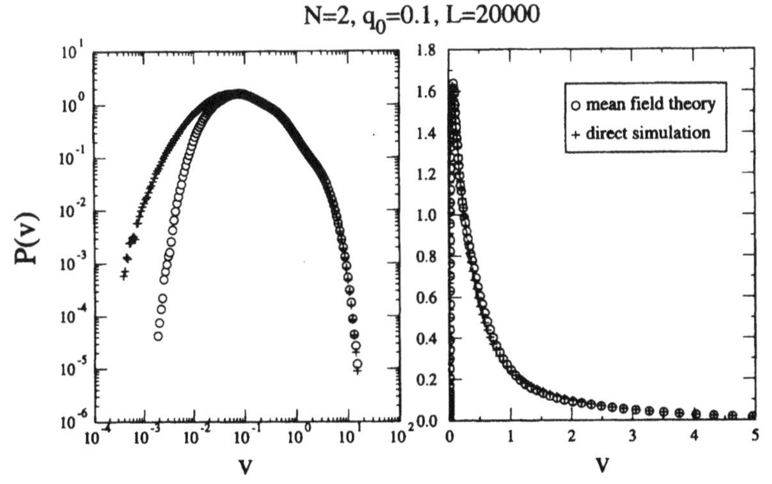

FIG. 5. Comparison of force distribution $P(v)$ versus v for simulation of mean field theory and of the original model equations (2.1) for a system with the q distribution Eq. (2.7) with $N=2$ and $q_0=0.1$. Both data sets were obtained by averaging the bottom 10 000 rows of a $20\,000\times 20\,000$ system. Mean field theory does not yield the exact $P(v)$ for this q distribution. Nonetheless, it provides an accurate quantitative estimate for $P(v)$ over a broad range of v.

the same or different). Further, while for any *particular* value of D it is possible for $\Delta(D)$ to be equal to $\Delta(D+1)$, the only way in which $\Delta(D)$ can remain unchanged as D increases is if all the positive δw's are segregated from all the negative ones. Even if this is the case in the top row, this becomes increasingly unlikely as D is increased. In fact, if the minimum distance between positive and negative δw is l in the top row, and if there are no δ functions in $\eta(q)$, then $\Delta(D+1)$ *must* be less than $\Delta(D)$ for $D>l$. Thus for a system of finite transverse extent, the distribution of weights at the bottom of the system is independent of the loading on the top row in the limit that the height of the system is infinite. For the case when all dimensions of the system are made infinite the situation is trickier; due to the conservation of $\Sigma\,\delta w$ under the evolution of Eq. (2.45) discussed above, if one were to make δw positive on one side of the top row and negative on the other half, for a system of transverse extent L it would require a height $O(L^2)$ for the effects of this perturbation to "diffuse" away. For generic loading at the top, however, we do not expect such an anomalous concentration of fluctuations into only the longest wavelength modes of the system, and $\Delta(D)$ should decay with D even if all dimensions of the system are enlarged.

We have seen that the distribution of weight at the bottom of any infinite system is independent of the details of how forces are distributed at the top, at least in the limit when the height of the system is taken to infinity before its transverse dimensions. This is true for each system individually, and is therefore also true for the full ensemble of systems with different realizations of randomness (the choice of q_{ij}'s), so that the solutions we have obtained so far for quantities like $P(v)$ are unique. For any particular system, however, the weights on the different sites at the bottom *do* depend on the q_{ij}'s; in fact, with all the q_{ij}'s specified, the weights on the different sites are completely determined. Even for a single system, however, statistics can be obtained by measuring quantities across all the sites in the bottom row; for a system of infinite transverse size the measurements then lead to distributions. At least for the "uniform" q distribution, any quantity like, say, $P(v)$, is the same, whether obtained by averaging over sites in a single system or for a single site over the entire ensemble. This is because, as we have seen, the ensemble averaged distribution of $\{v_1,\ldots,v_L\}$ across the bottom row is of the form $P(v_1,\ldots,v_L) = P(v_1)P(v_2)\cdots P(v_L)$. For any single system chosen randomly from the ensemble, this is the probability density that the normalized weights in the bottom row take on the specific values $\{v_1,v_2,\ldots,v_L\}$. The probability that l of these L sites will have v_i greater than some v_0 is then

$$\binom{L}{l}\left(\int_{v_0}^{\infty}P(v)dv\right)^{l}\left(1-\int_{v_0}^{\infty}P(v)dv\right)^{L-l}; \quad (2.46)$$

as $L\to\infty$, l/L is sharply peaked around $\int_{v_0}^{\infty}P(v)dv$, so that the site averaged result is the same as the ensemble average. We expect this to be the case even for more general q distributions, for which the ensemble averaged $P(v_1,v_2,\ldots,v_L)$ does not have a product form, so long as the transverse correlation lengths are finite.

III. NUMERICAL SIMULATIONS OF SPHERE PACKINGS

We now discuss the relevance of the q model to granular materials. Although we have shown that the q model yields an exponentially decaying force distribution independent of the details of the q distribution, to make quantitative comparison of this model to granular systems, we must know the q distribution for a granular material. To make this comparison, we have performed molecular dynamics simulations of three-dimensional sphere packings, analyzed the contact distributions to estimate the distribution of q's, and then calculated the force distribution in the sphere packing and compared it to that predicted by the q model. Our simulations yield results for the contact force distributions that are consistent with previous work [23–26]; the new ingredient here is that the geometry of the packing is characterized simultaneously, allowing testing of the statistical assumptions underlying the q model.

Our simulation consists of 500 spherical beads of weight and diameter unity in a uniform gravitational field with gravitational constant $g=1$, interacting via a central force F of the Hertzian form, $F=F_0(\delta r)^{3/2}$. Here, F_0 is the force constant, chosen so that a sphere has a deformation of $\delta r=0.001$ when subjected to its own weight, and δr is the deformation of each bead at the contact. The box containing the beads had a fixed bottom, and lateral dimensions of 5.5 $\times$5.5. In each simulation, the spheres are initially placed in a loose rectangular lattice with lattice constants of $1\times1\times1.5$ and have random initial velocities uniformly distributed in the range $-V_{\max}<V_{x,y,z}<V_{\max}$, where $V_{\max}=50$ is large enough to yield significantly different packings from run to run. By freezing the motion of the beads whenever the total kinetic energy of the system reaches a maximum, the kinetic energy of the system is reduced and eventually the spheres all settle to the bottom of the box. Starting with a flat bottom, the regularity of layerlike packing reduces as the height increases. A rough bottom was obtained by selecting the beads with height between H and $H+1$ (typically, $H\sim10$) and this rough bottom was used for the next simulation. Within a few iterations, the statistical properties of the rough bottom becomes independent of its initial configuration; this configuration of spheres at the bottom of the box is then fixed and used as a boundary condition for subsequent packing simulations.

In our packings, a sphere can have up to six contacts on its bottom half. However, on average, the three strongest vertical forces at these contacts sustain over 98% of the load; three or fewer particles supported at least 90% of the weight for over 92% of the particles. Therefore, comparison with the q model with $N=3$ is reasonable.

We estimate the q distribution for the sphere simulation by calculating the fractions of the total vertical force supported by each of the three strongest contacts [27]. To display our results for the q distribution for the simulation of hard spheres, we define the variables $\alpha_1=(q_3-q_2)/\sqrt{3}$, $\alpha_2=q_1$ [28]. Because $\Sigma_{j=1}^{3}q_j=1$, the possible values of the α's can all be represented as points in the interior of an equilateral triangle, where the values of q are the perpendicular distances to each side of the triangle. Moreover, for the uniform q distribution, the density of points in the triangle is constant. If one orders the q's so that $q_3>q_2>q_1$, then in terms of the α variables, all the points lie in the triangle shown in Fig. 6, which is bounded by the lines $\alpha_1>0$, $\alpha_2>0$, and $\sqrt{3}\alpha_1+\alpha_2<1$. As Fig. 6 demonstrates, there is some deviation from the uniform q distribution because a nonzero fraction of the particles have $q_1=\alpha_2=0$. A reasonable description of the numerically observed particle contact distribution is obtained by taking each particle and assigning with probability p, l, and u into "point," "line," and "uniform" pieces. In the point piece one of the q's has value unity, and the other two are zero. In the line piece one of the q's is set to zero, and the other two are determined as in the $N=2$ uniform distribution. Finally, the particles in the uniform piece have their q's determined exactly as in the $N=3$ uniform distribution. Our numerical data for the spheres are consistent with the values $p=0.017\pm0.0023$, $l=0.1635\pm0.007$, $u=1-l-p=0.8195\pm0.007$.

We now discuss our results for the distribution of vertical forces. First, we calculated the force distribution at several different depths D. Our numerical data indicate that if one considers the normalized force $v=w/D$, the force distribution $P(v)$ indeed becomes independent of depth for $D\gtrsim5$, and it decays exponentially at large v. The data were obtained by making a histogram of the vertical force exerted by spheres in horizontal slices of width $\Delta D=1$. The scales are set by the normalization requirements $\int_0^\infty dv\ P(v)=1$, $\int_0^\infty dv\ vP(v)=1$.

We now compare the results of these molecular dynamics simulations to results from the q model. Figure 7 shows $P(v)$ calculated via numerical simulation of the q model Eq. (2.1) with $f(q)=1$ at depth $D=1024$ on a periodically continued fcc lattice of side 1024, with $N=3$. Within our approximation of placing the grains on a uniform lattice, the reasonable choice for N is the dimensionality d of the system: For $d=3$ the grains are approximated as being in triangular lattice layers, with each layer staggered relative to the next, so that each grain has three neighbors. As expected, since the mean field distribution is exact for the uniform case, there is excellent agreement with Eq. (2.24). On the same graph we show $P(v)$ obtained in the sphere simulation described above. Both the sphere simulation and the q model exhibit a $P(v)$ that decays exponentially at large v. The quantitative agreement between the two is surprisingly good considering the "arching" [1] in the sphere simulation, as reflected in the line and point pieces of the q distribution for the spheres. To examine the effects of arching on the results, we examined the force distribution resulting from the "q model" with the three-piece q distribution, which more closely approximates that of the sphere simulation. Figure 8 shows the numerically calculated $P(v)$ for the q model with the three-piece distribution with $p=0.017$ and $l=0.1635$, together with the solution for the uniform distribution and the numerical data from the sphere simulation. Changing the q distribution has little effect on $P(v)$; to the extent that there is a change, it appears to improve the already good agreement between the q model and the sphere simulation.

Thus, our simulations indicate that our sphere packings are reasonably well-described (at the ~15% level) by the uniform q distribution. Deviations from this q distribution are observed; accounting for them improves the already good agreement between the q model and the simulations.

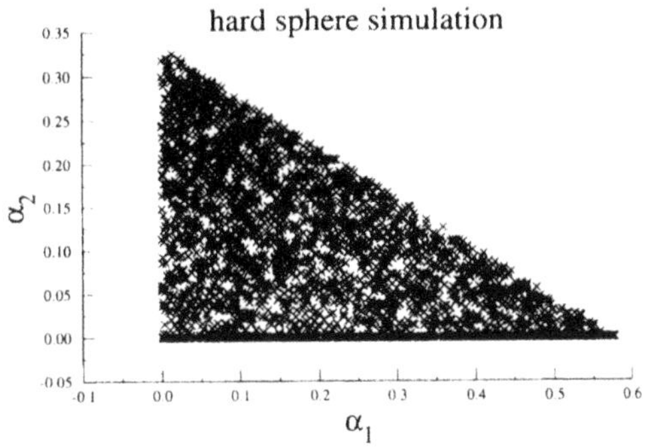

FIG. 6. Scatter plot of contact variables α_1, α_2 (defined in the text) obtained from the sphere simulation described in the text. The graph has 3229 points. On this plot, the uniform q distribution would have a uniform density of points. The "arching" in the simulation is reflected in the fact that a nonzero fraction of points have $\alpha_2=0$.

IV. DISCUSSION

This paper presents a statistical model for the force inhomogeneities in static bead packs and compares the results to numerical simulations of disordered sphere packings. The irregularities of the packing are described probabilistically, in terms of spatially uncorrelated random variables. Although there is a special q distribution for the q model that leads to a force distribution that decays as a power law at large forces, we have presented evidence that the force distribution decays exponentially at large forces for almost all q distributions. We obtain exact results for all the multipoint force correlation functions at a given depth for a countable set of q distributions, including one that is "generic" (the "uniform" distribution). The force distribution function for the uniform case agrees quantitatively with that obtained for the sphere simulation. Our numerical calculations demonstrate that a modified distribution of q's which more closely approximates that observed for the sphere simulation improves the already good agreement between the force distribution predicted by the q model using the uniform q distribution and simulations of spheres. Thus, this model appears to contain some essential features of the force inhomogeneities in granular solids.

Neither our simulations nor the q model of Eq. (2.1) captures all features of real bead packs. In our simulations, we have included only central forces and have ignored friction; the q model ignores the vector nature of the forces, assuming that only the component along the direction of gravity plays a vital role. The qualitative consistency between the results obtained using the different methods as well as with experiment [6] provides some indication that the effects that we have neglected do not determine the main qualitative features of the force distribution at large v.

Several avenues for future investigations are evident. It should be straightforward to extend the analysis of the model to calculate longitudinal (along the direction of gravity) correlations of the forces. It is not obvious how to measure these correlations experimentally, but comparison to sphere simulations is clearly possible and would provide further tests of

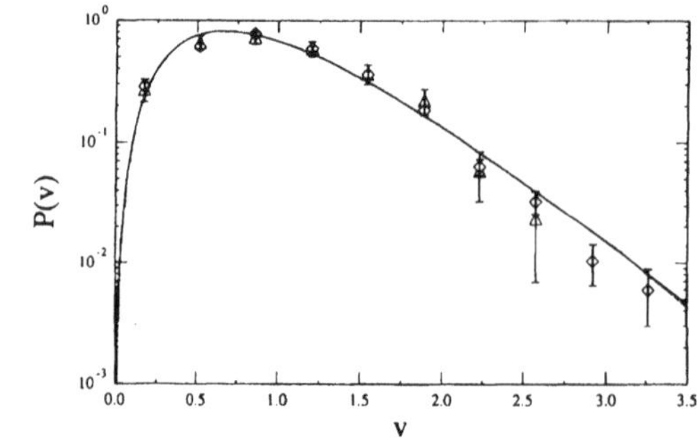

FIG. 7. The distribution of forces $P(v)$ as a function of normalized weight $v=w/D$ at a given depth D. Dashed line: $P(v)$ at $D=1024$ obtained via numerical simulation of the model Eq. (2.1) with $f(q)=1$ on a periodically continued fcc lattice of transverse extent 1024. Solid line: $P_u(v)$ obtained from the analytic mean field solution, Eq. (2.24). The points are $P(v)$ obtained in the sphere simulation described in the text at depth $D=10$ (triangles) and averaged over depths $D=6$ through $D=13$ (diamonds). There are no adjustable parameters; the scales are set by the normalization requirements $\int_0^\infty dv\ P(v)=1$, $\int_0^\infty dv\ vP(v)=1$.

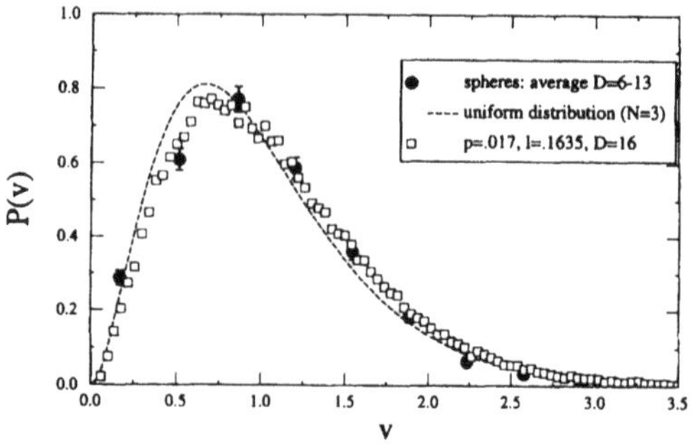

FIG. 8. The distribution of forces $P(v)$ as a function of normalized weight $v=w/D$ at a given depth D. Dashed line: $P_u(v)$ obtained from the analytic mean field solution for q model with $N=3$, Eq. (2.24). Solid circles: $P(v)$ obtained in the sphere simulation averaged over depths $D=6$ through $D=13$. Open squares: $P(v)$ at $D=16$ obtained via numerical simulation of the model Eq. (2.1) with the three-part q distribution described in the text, with parameter values given on the graph, on a periodically continued fcc lattice of transverse extent 256. This figure demonstrates that using the measured q distribution instead of the uniform q distribution improves the already good agreement between the q model and the sphere simulation.

the statistical model. Similarly, the theory makes clearcut predictions for the multipoint correlation functions, which can be tested both by experiment and by simulations. The model can be generalized to apply to a broader variety of situations by including vector forces as well as incorporating boundary effects. Most interestingly, we plan to investigate whether the statistical theory developed here can be extended to provide new insight into the complex dynamical effects exhibited by granular materials [1].

In summary, we have presented and analyzed a statistical model for force inhomogeneities in stationary bead packs. The model, which predicts that force inhomogeneities decay exponentially at large forces for almost all contact distributions, agrees well with numerical simulations of sphere packings as well as experiment [6].

ACKNOWLEDGMENTS

We thank S. Nagel for many useful conversations as well as collaboration and A. Pavlovitch for discussions and for providing us with Ref. [26] prior to publication. O.N. acknowledges support from Harvard University and the Institute for Theoretical Physics at Santa Barbara (NSF Grant No. PHY89-04035). This work was supported in part by the MRSEC Program of the National Science Foundation under Contract Nos. DMR-8819860 and DMR-9400379.

APPENDIX: UNIFORM q DISTRIBUTION

Here we consider the "uniform" q distribution, which is the simplest q distribution consistent with the restriction that $\Sigma_{1=1}^{N} q_i=1$. It is obtained by choosing each of $q_1, q_2, \ldots, q_{N-1}$ independently from a uniform distribution between 0 and 1, setting $q_N=1-\Sigma_{j=1}^{N-1} q_j$, and then keeping only those sets where $q_N \geq 0$. Here we show that $\eta_u(q)=1/(N-1)(1-q)^{N-2}$ for this distribution.

For $N=2$, if one chooses q_1 between 0 and 1, then $q_2=1-q_1$ must also be between 0 and 1, so that $\eta_u(q)=1$. When $N=3$, configuration will be retained only if $q_1+q_2+q_3 \leq 1$. Therefore, the probability of obtaining a value of q is given by

$$\eta_u(q)=M\int_0^1 dq_1\int_0^1 dq_2\delta(1-q_1-q_2-q)$$

$$=M\int_0^{1-q} dq_1=M(1-q), \quad \text{(A1)}$$

where M is a normalization constant. Since $\int_0^1 dq\ \eta(q)=1$, one immediately finds $M=2=(N-1)$.

For general N, $\eta_u(q)$ can be written:

$$\eta_u(q)=V(q)\Big/\int_0^1 V(q)dq, \quad \text{(A2)}$$

where

$$V(q)=\int_0^1 dq_2\int_0^{1-q_2} dq_3\cdots\int_0^{1-\Sigma_{i=1}^{N-1} q_i} dq_n \times\delta\left(1-q-\sum_{i=2}^{N} q_i\right). \quad \text{(A3)}$$

Using the identity

$$\int_0^{1-\Sigma_{j=1}^{n-k} q_m}\left(1-\sum_{m=1}^{n-k+1} q_m\right)^{k-1} dq_{n-k+1} =\frac{1}{k}\left(1-\sum_{m=1}^{n-k} q_m\right)^k, \quad \text{(A4)}$$

one can show that

$$\eta_N(q)=(N-1)(1-q)^{N-2}. \quad \text{(A5)}$$

[1] See, e.g., H. M. Jaeger, J. B. Knight, C.-h. Liu, and S. R. Nagel, Mat. Res. Soc. Bull. **19**, 25 (1994); H. M. Jaeger and S. R. Nagel, Science **255**, 1523 (1992); *Disorder and Granular Media*, edited by D. Bideau and A. Hansen (North-Holland, Amsterdam, 1993); *Powder and Grains 93*, edited by C. Thornton (A. A. Balkema, Rotterdam, 1993); *Granular Matter*, edited by A. Mehta (Springer, Berlin, 1993).

[2] See, e.g., J. D. Bernal, Proc. R. Soc. London A **280**, 299 (1964).

[3] P. Dantu, Ann. Ponts Chaussees IV, 144 (1967); A. Drescher and G. de Josselin de Jong, J. Mech. Phys. Solids **20**, 337 (1972); T. Travers, M. Ammi, D. Bideau, A. Gervois, J. C. Messager, and J. P. Troadec, Europhys. Lett. **4**, 329 (1987).

[4] S. B. Savage, Adv. Appl. Mech. **24**, 289 (1984); R. M. Nedderman, U. Tuzun, S. B. Savage, and G. T. Houlsby, Chem. Eng. Sci. **37**, 1597 (1982); C. S. Campbell, Annu. Rev. Fluid Mech. **22**, 57 (1990).

[5] C.-h. Liu and S. R. Nagel, Phys. Rev. Lett. **68**, 2301 (1992); Phys. Rev. B **48**, 15 646 (1993); C.-h. Liu, *ibid.* **50**, 782 (1994).

[6] C.-h. Liu *et al.*, Science **269**, 513 (1995).

[7] See, e.g., B. Mandelbrot, *The Fractal Geometry of Nature* (Freeman, San Francisco, 1983).

[8] H. Takayasu and I. Nishikawa, in *Proceedings of the 1st International Symposium for Science on Form*, edited by S. Ishizaka (KTK, Tokyo, 1986), p. 15.

[9] A. E. Scheidegger, Bull. Intl. Acco. Sci. Hydrol. **12**, 15 (1967).

[10] H. Takayasu, I. Nishikawa, and H. Tasaki, Phys. Rev. A **37**, 3110 (1988).

[11] G. Huber, Physica **170A**, 463 (1991).

[12] D. Dhar and R. Ramaswamy, Phys. Rev. Lett. **63**, 1659 (1989).

[13] S. N. Majumdar and C. Sire, Phys. Rev. Lett. **71**, 3729 (1993).

[14] D. Dhar (unpublished).

[15] The use of the term "sandpile" in this context is rather unfortunate, since we are exploiting a mathematical equivalence and do not suggest that the dynamical properties of models of SOC are directly related to the static properties of bead packs.

[16] P. Bak *et al.*, Phys. Rev. Lett. **59**, 381 (1987).

[17] The claim that $P_D(v)$ becomes independent of D as $D\rightarrow\infty$ is supported by the results of numerical simulations as well as by the fact that we obtain a consistent analytic solution for the distribution $P_{D\rightarrow\infty}(v)$.

[18] Note that in the $q_{0,1}$ limit, Eq. (2.11) has a singular solution, $\tilde{P}(s)=1$, that disagrees with the exact results. This is because [as can be seen by doing a similar mean field analysis in the $q_{0,1}$ limit in terms of $Q_D(w)$] the terms proportional to $1/D$ dropped in going from Eq. (2.10) to Eq. (2.11) are important for the critical case. This is reasonable physically; for a critical system, although the injection process becomes less and less important in terms of the normalized variables as $D\rightarrow\infty$, the "relaxation time" [the number of levels one must propagate downwards before a perturbation $\delta P(v)$ decays away] diverges, so that $P(v)$ never settles to an equilibrium form, and the injection cannot be ignored. No such complications are expected away from criticality, where the relaxation time should be finite.

[19] Some $p(q_1,\ldots,q_N)$ that are not of product form induce $\eta(q)$'s which have δ functions at $q=0$ and not at $q=1$ (but not vice versa).

[20] See, e.g., A. Katz, *Principles of Statistical Mechanics: The Information Theory Approach* (Freeman, San Francisco, 1967).

[21] L. Rothenburg, Ph.D. thesis, Carleton University, Ottowa, Canada, 1980.

[22] J. Zinn-Justin, *Quantum Field Theory and Critical Phenomena* (Clarendon, Oxford, 1990), p. 214 [Eq. (9.39)]; P. Ramond, *Field Theory: A Modern Primer* (Benjamin-Cummings, Reading, MA, 1981), p. 157 [Eq. (4.8)].

[23] P. A. Cundall, J. T. Jenkins, and I. Ishibashi, in *Powders and Grains*, edited by J. Biarez and R. Gourves (Balkema, Rotterdam, 1989), pp. 319–322.

[24] J. T. Jenkins, P. A. Cundall, and I. Ishibashi, in *Powders and Grains* (Ref. [23]), pp. 257–274.

[25] K. Bagi, in *Powders and Grains* (Ref. [1]), pp. 117–121.

[26] A. Pavlovitch (unpublished).

[27] For beads with more than three contacts, say, f_1, f_2, f_3, and f_4, only the three strongest contacts were picked and renormalized to derive the q's, namely, $q_1=f_1/(f_1+f_2+f_3)$, etc.

[28] The q's are related to the α's via $q_1=\alpha_2$, $q_2=\frac{1}{2}(1-\sqrt{3}\alpha_1-\alpha_2)$, $q_3=\frac{1}{2}(1+\sqrt{3}\alpha_1-\alpha_2)$.

Force Distributions in Dense Two-Dimensional Granular Systems

Farhang Radjai,[1] Michel Jean,[2] Jean-Jacques Moreau,[2] and Stéphane Roux[3]
[1]*HLRZ, Forschungszentrum, 52425 Jülich, Germany*
[2]*LMGC, Université de Montpellier II, Place Eugène Bataillon, 34095 Montpellier Cedex 05, France*
[3]*LPMMH-ESPCI, 10 rue Vauquelin, 75231 Paris Cedex 05, France*
(Received 27 March 1996)

Relying on contact dynamics simulations, we study the statistical distribution of contact forces inside a confined packing of circular rigid disks with solid friction. We find the following: (1) The number of normal and tangential forces lower than their respective mean value decays as a power law. (2) The number of normal and tangential forces higher than their respective mean value decays exponentially. (3) The ratio of friction to normal force is uniformly distributed and is uncorrelated with normal force. (4) When normalized with respect to their mean values, these distributions are independent of sample size and particle size distribution. [S0031-9007(96)00589-3]

PACS numbers: 46.10.+z, 83.70.Fn

Despite the highly uniform density of a random packing of noncohesive particles, photoelastic visualizations provide a striking evidence of the heterogeneous distribution of contact forces on a scale definitely larger than the typical particle size [1–3]. A quantitative characterization of these distributions is relevant both to mechanical processing (compression, compaction, flow, grinding) and fundamental understanding (mesoscopic scales, instability thresholds) of granular media [4–7].

This Letter is concerned with a numerical study of this problem in confined two-dimensional packings at static equilibrium. We are interested in the statistical distributions P_N and P_T of normal forces and (absolute values of) friction forces N and T, independently of contact orientations. We also study the distribution P_η of the dimensionless variable $\eta = T/N$, which is a measure of friction "mobilization" within the Coulomb range $[0, \mu]$, where μ is the coefficient of friction between disks. Scaling with sample size and relation among the three distributions will be considered too.

Numerical results will be presented here for four samples of 500, 1200, 4025, and 1024 particles, referred to as samples A, B, C, and D, respectively. Particle radii are uniformly distributed between 3.8 and 7.5 mm in samples A and B, and between 1.5 and 7.5 mm in sample C. Sample D contains 192 particles of radius 1.6 mm, 320 particles of radius 1.05 mm, and 512 particles of radius 0.65 mm. Particles are contained in a rectangular frame composed of one planar base, two immobile walls, and one horizontal plane (the lid) free to move vertically and on which a downward force of 6600 N is applied. The acceleration of gravity is set to zero in order to avoid force gradients in the sample. Particle-particle and particle-base coefficients of friction are 0.2 and 0.5, respectively. All other coefficients of friction are zero.

For this investigation, we have relied on the contact dynamics (CD) approach to the dynamics of perfectly rigid particles with unilateral contacts. Since particles cannot interpenetrate, the allowed configurations of the system, characterized by a set of inequalities, define a region in the configuration space presenting a large number of edges and corners. Moreover, the basic Coulomb's law of friction, relevant to most of the granular media of interest, is a *nonsmooth* law in the sense that friction force and sliding velocity at a contact are not related together as a function. Finally, in the case of collisions velocity jumps occur, so that the evolution is not globally governed by differential equations in the classical sense.

The most commonly used algorithms are based on regularization schemes. In this way, impenetrability is approximated by a steep repulsive potential and Coulomb's law by a viscous friction law, to which smooth computational methods can be applied. The dominant feature of the CD method is that the conditions of *perfect rigidity* and *exact Coulombian friction* are implemented, with no resort to any regularization. At a given step of evolution, all kinematic constraints implied by lasting interparticular contacts and the possible rolling of some particles over others are *simultaneously* taken into account, together with the equations of dynamics, in order to determine all contact forces in the system. The method is thus able to deal properly with the *nonlocal* character of the momentum transfers—resulting from the perfect rigidity of particles in contact.

Detailed descriptions of the CD method can be found in the literature [8–10]. In relation with the present investigation, we would just like to underline the point that dynamics is an essential ingredient of this approach. It is well known that a granular system at static equilibrium is *hyperstatic;* i.e., for given boundary conditions there is a continuous set of possible contact forces. This is due both to the absence of an internal displacement field (because of perfect rigidity) and to the nonsmooth character of the friction law [11]. In the CD method, the force network at static equilibrium is determined through the *dynamic processes* from which it relaxed. In other words, as in real granular systems, the static values of forces are reached

 0031-9007/96/77(2)/274(4)$10.00

asymptotically as the kinematic energy of the system is dissipated in friction and collisions.

Of course, this does not mean that the *statistical* distribution of forces is necessarily dependent on the preparation process. The most probable force distribution may well result from the generic disorder of granular systems [3]. However, density is a major control parameter of the mechanical properties of granular materials, and only in the steady state, reached after enough shear-induced volume change, it acquires a rather well-defined value for a given confining pressure [12]. That is why we applied the same procedure to prepare all samples in the same state: Filling the box with particles under gravity, shearing by moving the base horizontally (dilation occurs then), stopping shear and applying the confining load on top of the sample, and, finally, setting the gravity to zero and allowing the system to relax to equilibrium under the load. Although the algorithm is quite efficient compared to other available techniques, the whole procedure requires hundreds of CPU hours on a fast Unix workstation (Sparc 20) for each sample.

Figure 1 shows the network of normal forces in sample D. One can observe both large contact-to-contact fluctuations and a subnetwork of "force chains" that seem to carry a significant portion of the applied external stress. Forces range from 0.003 to 1127 N, i.e., a range of 6 orders of magnitude, which clearly requires a scaling analysis. The mean normal force is $\langle N \rangle = 249$ N and more than 60% of contacts carry a force lower than the mean.

Figure 2 displays semilogarithmic plots of probability distributions P_N of normal forces in the four samples. Forces are normalized with respect to their mean in each sample. The normalized distributions coincide over almost the whole range, and the data for forces larger than the mean are well fitted by an exponential decay. In order to see the behavior at low forces, we have shown in Fig. 3 the normalized log-log plots of the distribution of the logarithm of the forces. The data for forces lower than the mean have a power-law distribution. We conclude that the normalized distribution of normal forces is independent of our sample sizes and can be approximated by a power-law decay with a crossover to an exponential cutoff,

$$P_N \propto \begin{cases} (N/\langle N \rangle)^{\alpha}, & N < \langle N \rangle, \\ e^{\beta(1-N/\langle N \rangle)}, & N > \langle N \rangle. \end{cases} \tag{1}$$

We find $\alpha = -0.3$ and $\beta = 1.4$. It is important to notice the collapse of normalized data on the same distribution in spite of the fact that the size dispersity of particles is not the same in all samples. The mean values seem, however, to depend on size dispersity since they do not scale with system size as shown in Table I. On the other hand, the lack of statistics at low normal forces in sample D as compared to sample B, giving rise to the fluctuations observed in Fig. 3, suggests that the "branching process" generating low forces from the high applied force on the system is more efficient in systems with a continuous distribution of particle sizes.

The semilogarithmic and log-log plots of the probability distributions of the T are displayed in Figs. 4 and 5. The data are normalized with respect to the mean $\langle T \rangle$ in each sample, and, as we see, they nicely collapse on the same distribution. This is again essentially a power-law decay with a crossover to an exponential cutoff,

$$P_T \propto \begin{cases} (T/\langle T \rangle)^{\alpha'}, & T < \langle T \rangle, \\ e^{\beta'(1-T/\langle T \rangle)}, & T > \langle T \rangle. \end{cases} \tag{2}$$

We find $\alpha' = -0.5$ and $\beta' = 1$.

We also studied the probability distribution P_η of $\eta = T/N$. This is a uniform distribution except for a small peak at $\eta = \mu$. The uniformity of this distribution may

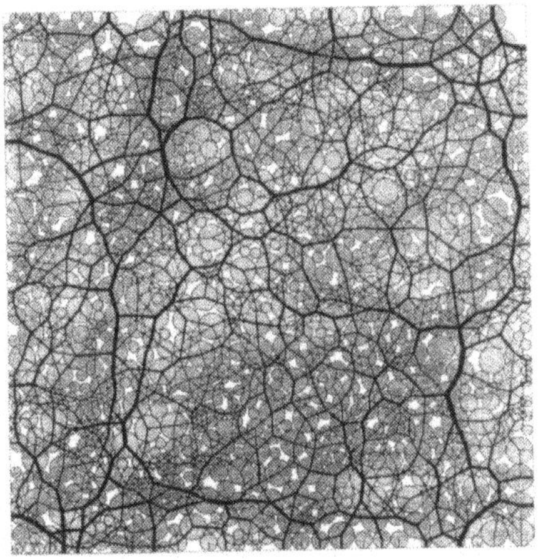

FIG. 1. Network of normal forces in sample D; see Table I. Forces are encoded as the widths of intercenter connecting segments.

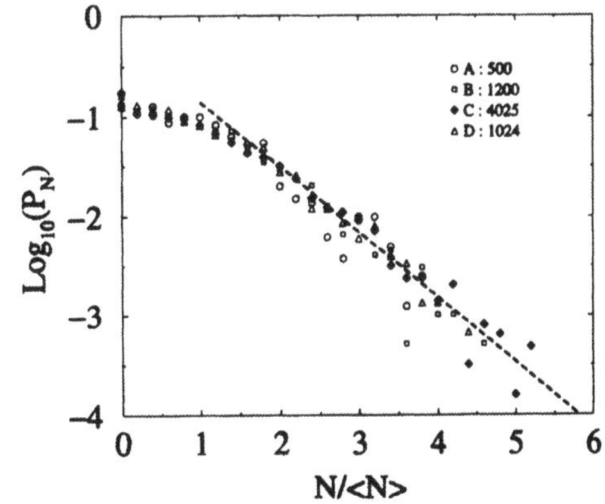

FIG. 2. Semilogarithmic plots of the probability distributions of normalized normal forces $N/\langle N \rangle$.

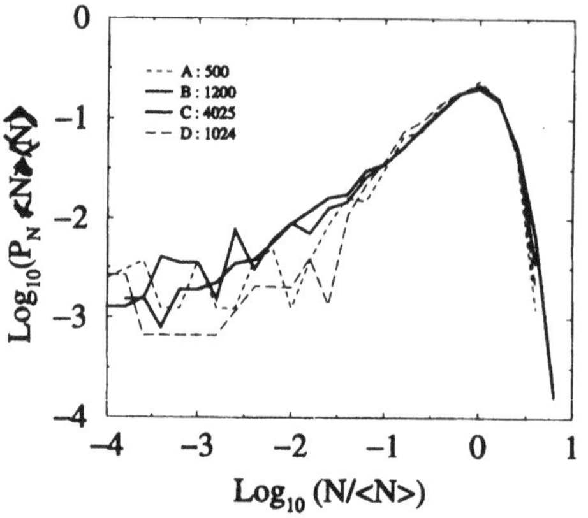

FIG. 3. Log-log plots of the probability distributions of normalized normal forces $N/\langle N\rangle$.

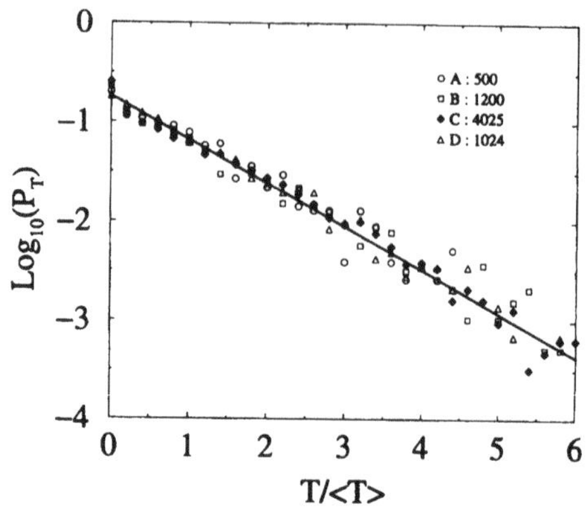

FIG. 4. Semilogarithmic plots of the probability distributions of normalized friction forces $T/\langle T\rangle$.

be attributed to the random structure of the contact network. On the other hand, it is likely that the rather weak singularity at $\eta = \mu$ is a "signature" of the dynamics of preparation. Indeed, only at *sliding* contacts is the friction force fully mobilized. If a granular assembly relaxes asymptotically towards static equilibrium, then the set of the last sliding contacts at the equilibrium threshold might remain fully mobilized. We checked that when the system is sheared by the motion of the basal plane, the peak at $\eta = \mu$ can rise to 50% of contacts, whereas the distribution remains uniform within $[0, \mu[$. Finally, we note that the normal forces for which $\eta = \mu$ are much smaller than the average, so that the peak may well result also from an imperfect relaxation.

Another important result regarding friction mobilization is the statistical independence of η with respect to N. Whatever the value of N, friction is indifferently mobilized within the Coulomb range $[0, \mu N]$ (apart from the above discussed small peak). Such an assumption allows one to relate in a simple way P_N to P_T. Let $P(N,T)$ be the joint probability distribution of normal and friction forces. Since η is statistically independent of N, we may write $P(N,T)$ as a product of P_N and P_η times the Jacobian of the transformation $(N,T) \to (N,\eta)$,

$$P(N,T)\,dN\,dT = \frac{1}{N} P_N(N) P_\eta(\eta)\,dN\,dT\,. \quad (3)$$

TABLE I. Number of particles p, number of contacts c, width L, mean normal force $\langle N\rangle$, and mean friction force $\langle T\rangle$ in our samples A, B, C, and D.

Sample	p	c	L (mm)	$\langle N\rangle$ (N)	$\langle T\rangle$ (N)
A	500	806	260	592	51
B	1200	1969	389	213	18
C	4025	6293	620	219	21
D	1024	1498	65	249	23

One may check that integration of the two members of Eq. (3) with respect to N and T over $[0, +\infty]$, with the substitution $T = \eta N$ in the right-hand side together with the constraint $\eta \in [0, \mu]$, implies a normalized P_η over the Coulomb range $[0, \mu]$. Introducing the uniform distribution $P_\eta(\eta) = \mathbf{1}([0,\mu])$ in Eq. (3) and integrating with respect to N over $[0, +\infty]$ with the substitution $N = T/\eta$ in the right side, we get the following relation between P_N and P_T:

$$P_T(T) = \frac{1}{\mu} \int_{1/\mu}^{+\infty} P_N(xT)\,\frac{dx}{x}\,. \quad (4)$$

This equation implies that the initial power law of the two distributions P_N and P_T should be the same: $\alpha = \alpha'$. Moreover, an exponential upper cutoff of normal forces yields an exponential-integral cutoff for friction forces, i.e., essentially an exponential decay times a slowly varying function. Going back to Figs. 2–5, we see that such refinements are out of reach within the

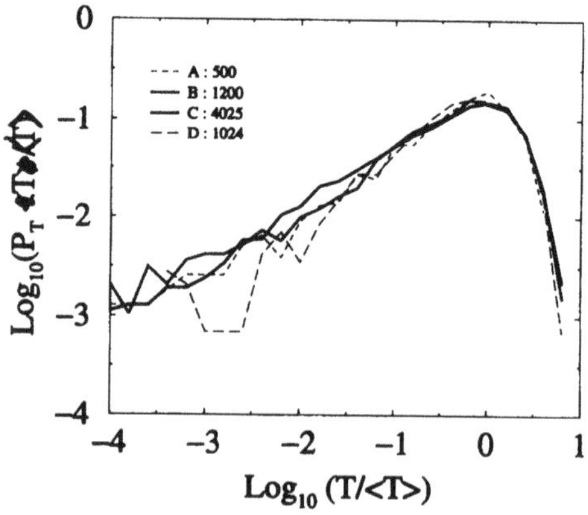

FIG. 5. Log-log plots of the probability distributions of normalized friction forces $T/\langle T\rangle$.

statistical precision. On one hand, the cutoff may well be an exponential-integral function. On the other hand, the equality of exponents is consistent with the *rough* determination of these exponents.

Equation (4) can, however, be directly checked from the data. In Fig. 6 we have plotted both P_T and the probability distribution obtained from P_N via Eq. (4) for sample C. They are almost the same with a very good precision, although we assumed a uniform distribution of η with no additional peak on the edge $\eta = \mu$. This validation of Eq. (4) is also an indirect check of the statistical independence of η with respect to N.

Finally, integration of Eq. (4) with respect to T yields the following relation between the mean values:

$$\langle T \rangle = \frac{\mu}{2} \langle N \rangle. \tag{5}$$

This relation is approximately satisfied for our samples, as can be seen in Table I.

In view of these findings, we would like to underline some salient aspects of the problem. One important point concerns the scale of statistical homogeneity of granular systems. Despite local force fluctuations, the present study shows that for a sample as small as 1200 particles the force distributions are clearly defined over several decades. An increase in sample size does nothing but improve statistics. Hence, as far as stress is concerned, the linear scale of statistical homogeneity in a 2D assembly is a few tens of particle diameters. This is what comes out also from the study of anisotropy in angular distributions of contact forces [13,14]. This observation is crucial for a continuum approach to the mechanics of granular media, needed in most of the usual technological problems.

Another point is that only the exponential tail of the distribution of *normal* forces, comprising nearly 40% of contacts in our simulations, has been observed in experiments [3]. Weaker forces are technically difficult to measure, and their distribution has not been observed. The exponential tail has also been obtained through the usual simulation methods [15], and, what is more, a recent theoretical model provides plausible statistical arguments in favor of it [3]. This statistical model is likely to apply only to the subnetwork of force chains, which carries in effect most of the applied external load and in our simulations belongs to the exponential tail. The characteristic force at this scale is essentially imposed by the external load and the ratio of the system size to the largest particle size. On the other hand, the power-law decay of weak forces, if confirmed by other investigators, indicates the self-similar nature of *weak* contacts that do not belong to the subnetwork of large forces. Indeed, such contacts do not *feel* the external load, and hence their distribution can give rise to a power law through a self-similar branching process with no intrinsic scale. This observation also suggests that the exponents α and α' depend on the interparticular friction coefficient μ.

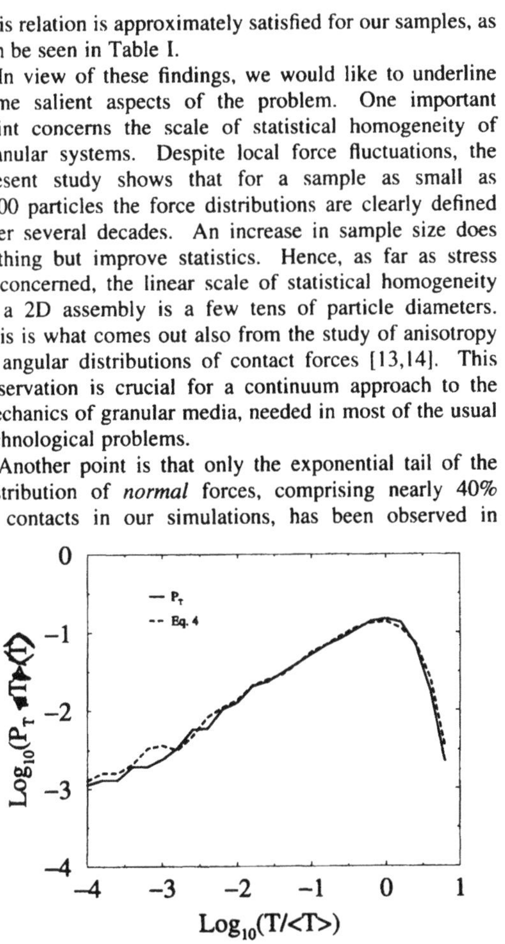

FIG. 6. Log-log plots of the distribution P_T of normalized friction forces and the one obtained by applying Eq. (4) to P_N in sample C.

We gratefully acknowledge many fruitful conversations with D. E. Wolf. This work has been supported by the Groupement de Recherche "Physique des Milieux Hétérogènes Complexes" of the CNRS.

[1] P. Dantu, in *Proceedings of the 4th International Conference on Soil Mechanics and Foundations Engineering* (Butterworths Scientific Publications, London, 1957).

[2] T. Travers, D. Bideau, A. Gervois, and J.C. Messager, J. Phys. A **19**, L1033–1038 (1986).

[3] C.H. Liu, S.R. Nagel, D.A. Schecter, S.N. Coppersmith, S. Majumdar, O. Narayan, and T.A. Witten, Science **269**, 513 (1995).

[4] E. Guyon, S. Roux, A. Hansen, D. Bideau, J.P. Troadec, and H. Crapo, Rep. Prog. Phys. **53**, 373–419 (1990).

[5] H.M. Jaeger and S.R. Nagel, Science **255**, 1523 (1992).

[6] *Disorder and Granular Media,* edited by D. Bideau and A. Hansen (Elsevier, North-Holland, Amsterdam, 1993).

[7] D.E. Wolf, in *Computational Physics: Selected Methods—Simple Exercises—Serious Applications,* edited by K.H. Hoffmann and M. Schreiber (Springer, Heidelberg, 1996).

[8] J.J. Moreau, Eur. J. Mech. A, Solids **13**, 93–114 (1994).

[9] M. Jean, in *Mechanics of Geometrical Interfaces,* edited by A.P.S. Selvadurai and M.J. Boulon (Elsevier Science B. V., Amsterdam, 1995), pp. 463–486.

[10] F. Radjai and S. Roux, Phys. Rev. E **51**, 6177 (1995).

[11] F. Radjai, L. Brendel, and S. Roux, Phys. Rev. E (to be published).

[12] A.N. Schofield and C.P. Wroth, *Critical State Soil Mechanics* (McGraw-Hill, London, 1968).

[13] L. Rothenburg and R.J. Bathurst, Géotechnique **39**, 601–614 (1989).

[14] B. Cambou, in *Powders and Grains 93,* edited by C. Thornton (Balkema, Rotterdam, 1993).

[15] K. Bagi, in *Powders and Grains 93* (Ref. [14]).

Review

Force Transmission in Granular Media[†]

C. Thornton
School of Engineering, Aston University, U.K.

Abstract

Powders and other granular media are composed of discrete particles and force transmission through a system of particles can only occur via the interparticle contacts. The distribution of contact orientations affects the magnitude of the contact forces which are not uniformly distributed. Since the contact stiffness and contact area are functions of the contact force, the manner in which the contact forces are distributed determines the shear strength, compressibility and conductivity characteristics of granular media. The paper reviews the current understanding of force transmission in granular media as gleaned from both physical experiments and numerical simulations and discusses the mechanical implications.

1. Introduction

Powders, and other granular media composed of discrete particles, exhibit very complex non-linear, hysteretic, stress-strain behaviour which is both stress level and stress path dependent. The complexity arises from the fact that the ensemble macroscopic response is a function of the spatial, size and shape distributions of the constituent particles and the distributions of other internal variables associated with the interactions between contiguous particles. The macroscopic state of stress is a function of the distribution of contact forces, Thornton and Barnes [1], and the ensemble moduli are related to the distribution of contact stiffnesses, Thornton [2]. The distribution of contact forces is also relevant to the fracture and crushing of constituent particles. The contact stiffnesses are functions of the contact areas and the distribution of contact areas controls the conductivity of granular media. Since the contact areas and contact stiffnesses are themselves functions of the contact forces, it is of scientific and industrial interest to understand how the intrinsic properties of the constituent particles affects the force transmission in granular media.

For any assemblage of discrete particles subjected to external loading, the transmission of force from one boundary to another can only occur via the interparticle contacts. Intuitively, therefore, we expect that the distribution of contacts will determine the distribution of forces within the system of particles and that the forces will not necessarily be distributed uniformly. Direct observations of stress distribution in photoelastic studies of two-dimensional arrays of discs have been reported by a number of researchers [3-7]. Konishi et al [8] used the same photoelastic technique to examine arrays of oval shaped particles. Photoelastic investigations of dynamic stress wave propagation in disc assemblies have been reported by Rossmanith and Shukla [9], Shukla [10] and dynamic photoelastic studies of regular arrays of elliptical discs have been performed by Zhu et al [11].

In all static photoelastic studies of disc assemblies it has been observed that the load is largely transmitted by relatively rigid, heavily stressed chains of particles forming a relatively sparse network of larger than average contact forces. Groups of particles separating the strong force chains are only lightly loaded. The implication is that, in a random system of particles, the applied load will search for the shortest and most direct transmission path, and the less ziz-zag the chosen pathway the higher the proportion of load that will be transmitted. The existence of strong force chains in three-dimensional packings of spheres has recently been demonstrated by Liu et al [12].

A more efficient alternative to photoelastic studies of force propagation in granular media is to perform numerical simulations of particle systems subjected to external loading regimes. A well established computational technique, which we have used at Aston since 1980, was developed by Cundall and Strack [13] and is commonly known as the Distinct (or Discrete) Element Method. In their paper, Cundall and Strack [13] simulated a simple shear test reported by Oda and Konishi [7] and demonstrated

*Aston Triangle, Birmingham B4 7ET, UK
[†]Received 19 June, 1997

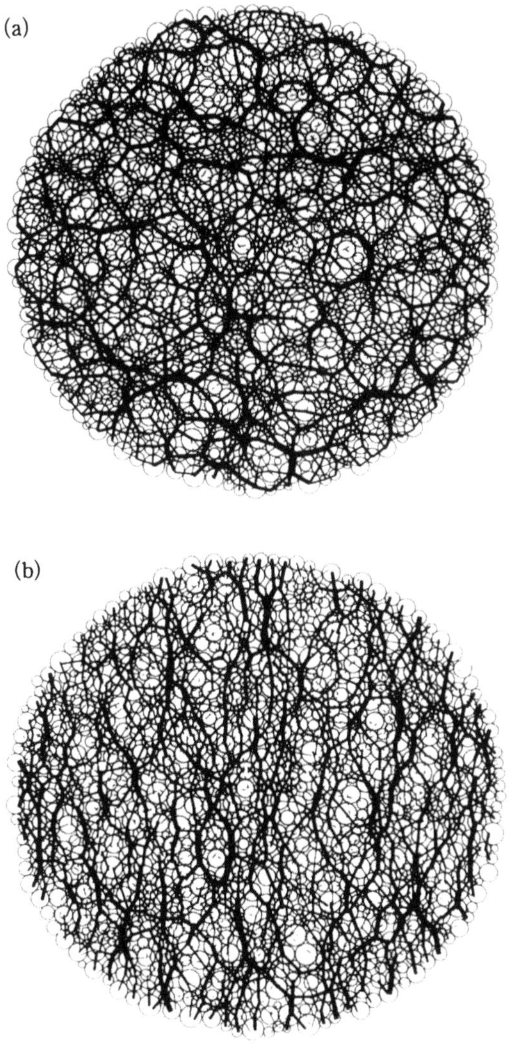

Fig. 1 Contact force transmision in a 2D array of polydisperse discs (a) isotropic stress state (b) deviatoric stress state.

good qualitative agreement obtained between the numerically simulated force transmission patterns with those obtained in the photoelastic experiment [7]. Sadd et al [14] have examined wave propagation in granular media using both photoelastic experiments and distinct element simulations. They demonstrated that, by using non-linear hysteretic normal and tangential interaction laws (similar to those used in the Aston version of the TRUBAL code), the relative errors between the experimental and simulated data were less than 15%.

2. Granular Dynamics

The computational technique may also be described as Granular Dynamics since the interactions between discrete particles are modelled as a dynamic process and the evolution of the system is advanced using an explicit finite difference scheme. The method is similar to Molecular Dynamics but with the difference that particles only interact when in contact and the contact interaction rules are more complex.

Application of the program involves cyclic calculations. At any time t, interparticle force increments are calculated at all contacts from the relative velocities of the contacting particles using incremental force-displacement rules. The interparticle forces are updated and, from the new out-of-balance force and moment of each particle, new particle accelerations (both linear and rotational) are obtained using Newton's second law. Numerical integration of the accelerations, using a small timestep Δt, provides new velocities which are then numerically integrated to give displacement increments from which the new particle positions are obtained. Having obtained new positions and velocities for all the particles, the program repeats the cycle of updating contact forces and particle locations. Checks are incorporated to identify new contacts and contacts that no longer exist.

In the original 2D simulation code BALL, linear springs and dashpots were used to model the interactions between contiguous particles [1, 13] and, in order to simulate quasi-static deformation, it was necessary to incorporate mass proportional damping terms into the equations of motion in order to dissipate sufficient energy. In the Aston version of the 3D simulation code TRUBAL, the interactions between contiguous particles are modelled by algorithms based on theoretical contact mechanics. Details of the contact mechanics theories used are provided by Thornton and Yin [15], Thornton [16]. Simulations of quasi-static deformation are performed using a representative volume element, with periodic boundaries, subjected to uniform strain fields and the bulk mechanical properties are calculated directly using statistical mechanics formulations [1, 2]. Mass proportional damping is no longer used in the TRUBAL code. Instead, the particle density is scaled up by a factor of 10^{12} in order to provide sufficient inertial damping to permit quasi-static simulations to be performed within a reasonable timescale.

3. Visualisation of force transmission pathways

Thornton and Barnes [1] reported two-dimensional simulations of quasi-static shear deformation of a compact polydisperse system of 1000 discs using the BALL code. The particles were initially randomly generated within a prescribed circular area such that there were no interparticle contacts. The boundary of the assembly was located by linking the centres of the peripheral particles to provide a convex boundary which enclosed all the disc centres. Deformation of the system was achieved by applying strain or stress control to the boundary particles, the effects of which then propagated through the system as the system evolved with time. In this way the system was subjected to isotropic compression followed by shear deformation during which the mean stress $(\sigma_1+\sigma_2)/2$ was maintained constant.

Figure 1a shows the force transmission pattern at the start of the shear stage when both the structure and the applied stress field are isotropic. The orientation of the contact forces is shown by the straight lines and the magnitude of each force is indicated by the thickness of the line, scaled to the current maximum force. It is clear from the figure that the force distribution is random. Macroscopically the orientation and distribution of the contact forces are both homogeneous and isotropic (i.e. there appears to be no regional preference regarding concentrations and no preferred contact force orientation). Microscopically, however, the distribution is inhomogeneous. Even when the stress state is isotropic, some contacts transmit forces several times those of others. Large forces tend to be transmitted along chains of particles which form the boundaries of enclosed, approximately circular, regions inside which there are relatively unloaded particles.

When a principal stress difference $(\sigma_1-\sigma_2)$ is applied to shear the system of particles, there is a rapid change in the force distribution as shown in **Figure 1b**. Macroscopically the distribution of forces becomes more and more anisotropic. The circular patterns initially formed by the large force chains rapidly adjust to form the boundaries of enclosed regions which are elongated in the direction of the compressive principal strain. The obliquity of contact forces along high force chains is low whereas the ratio of tangential to normal contact force is much higher in the unloaded regions. This indicates that, as confirmed from examining the simulation data, sliding is more likely to occur at contacts that carry small forces. Further discussion of the simulations in terms of the deformation mechanisms associated with the force transmission patterns shown in **Figure 1** are provided in [1].

Using the 3D simulation code TRUBAL, quasi-static deformation tests have been simulated in a periodic cell on polydisperse systems of 8000 elastic spheres. **Figure 2** illustrates the contact force transmission through such a system when subjected to an axisymmetric stress state $(\sigma_1>\sigma_2=\sigma_3)$. For clarity, sections through the mid-third of the periodic cell are shown. **Figure 2a** shows a view in the direction of σ_1. Although large forces are evident there appears to be no directional bias. This is due to the fact that both the structure and the state of stress are isotropic in the plane orthogonal to the viewing direction. However, when the force transmission is viewed in the direction of σ_3, as shown in **Figure 2b**, it is clear that strong force chains exist which tend to align themselves in the σ_1 direction.

4. Distribution of contact forces in terms of magnitude

It is necessary to characterise in a quantitative manner the force transmission behaviour described above. In this section, we consider the variation in magnitude only and leave any considerations of orientation until the next section.

Liu et al [12] used a layer of carbon paper on the inside surface of a cylindrical container to examine the distribution of forces on the base when the container was filled with beads. Their results indicated that the number of contacts carrying a given force decreased exponentially as the magnitude of the force increased. Similar exponential distributions of the contact force magnitude have been obtained in 3D simulations of isotropic compression of small assemblies of binary mixtures of 432 elastic spheres [17, 18].

A simple theoretical model was proposed [12] for the probability distribution of the magnitude of the normal contact forces. The model assumes that the inhomogeneity of the particle arrangement causes an unequal distribution of force transmission between contiguous particles which leads to chains of particles transmitting larger than average forces, as observed in photoelastic disc assemblies. Adopting a stochastic analysis, a mean field solution for the probability distribution P was obtained which may be written in the form

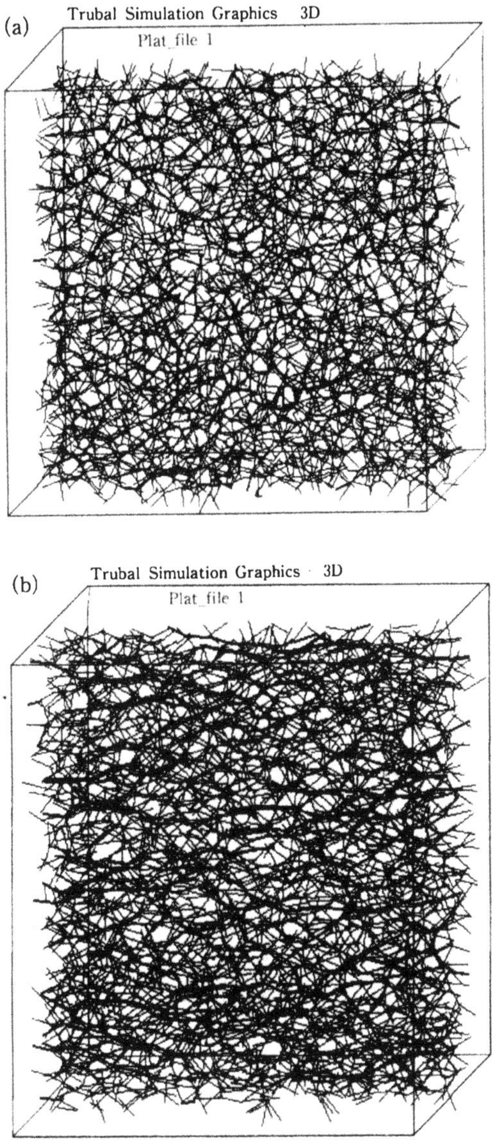

Fig. 2 Contact force transmission in a 3D array of polydisperse spheres (spheres not shown) during axisymmetric compression (a) view in direction of σ_1 (b) view in direction of σ_3.

$$P = \frac{k^k}{(k-1)!}\left(\frac{N}{\langle N\rangle}\right)^{k-1} exp\left[-k\left(\frac{N}{\langle N\rangle}\right)\right] \quad (1)$$

where $\langle N\rangle$ is the average normal contact force. **Figure 3** illustrates how the distribution changes from an exponential distribution to an almost Gaussian distribution as the parameter k is varied from 1 to 12.

Using a different simulation technique from that described in the previous section, Radjai et al [19]

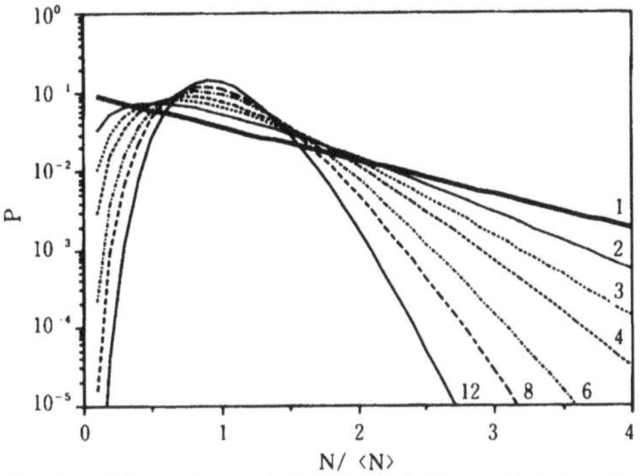

Fig. 3 Theoretical model for the probability distribution P of the normalised contact normal force magnitudes (N/⟨N⟩) given by equation (1) for values of k = 1 to k = 12.

have examined force networks in two-dimensional systems of rigid spheres. In a rigid sphere array the contact force is not a function of the relative displacement of the two spheres forming the contact but is the result of the geometrical configuration of the whole system and the boundary conditions. The simulation technique has been termed the Contact Dynamics Method and is described by Radjai and Roux [20]. From the probability distributions obtained, Radjai et al [19] found that although the larger than average normal contact forces exhibited an exponential decay, the smaller than average normal contact forces had a power law distribution. They suggested that the probability P_N of the normal forces N takes the form

$$P_N \propto (N/\langle N\rangle)^{\alpha} \quad N < \langle N\rangle \quad (2)$$

$$P_N \propto exp[\beta(1 - N/\langle N\rangle)] \quad (3)$$

Similar expressions were obtained for the probability P_T of the tangential forces T.

We have performed three dimensional simulations of two polydisperse systems of 8000 elastic spheres subjected to isotropic compression in a periodic cell. The only difference in the particle specifications was that one system was composed of "hard" spheres and the other of "soft" spheres. The particles in the hard sphere system were attributed with a Young's Modulus, E = 70 GPa; a Young's Modulus, E = 70 MPa was specified for the particles in the soft sphere system. Both systems were isotropically compressed to stress levels of 50 kPa, 100 kPa, 200 kPa and 400 kPa.

Figure 4 shows that the contact force distributions obtained for the hard sphere system are insensitive to stress level. Superimposed on the figure is the theoretical model given by (1) with k = 3. It can be seen that this simple one parameter model well describes the distribution of the larger than average

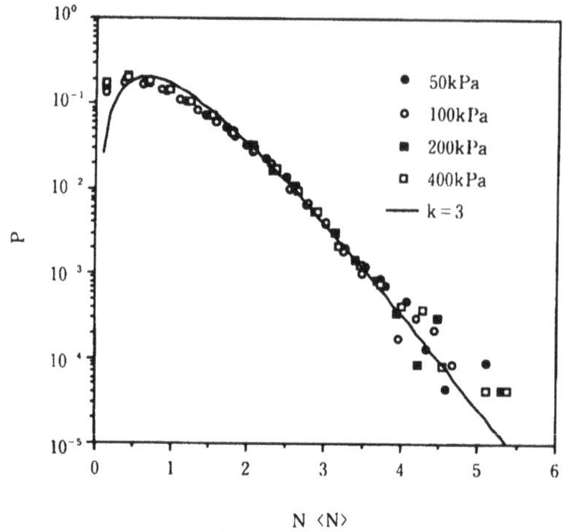

Fig. 4 Contact force distributions obtained at different stress levels for a polydisperse system of hard sheres compared with equation (1) using k=3.

contact forces. In **Figure 5** the data is replotted to compare with the alternative correlations given by (2) and (3). The fit to the data shown in **Figure 5a** was obtained with $\beta=2.15$ compared to a value of $\beta=1.4$ obtained in the 2D simulations of rigid discs reported in [19]. However, it can be seen that the exponential distribution is only reasonable for contact forces which are more than twice the magnitude of the average contact force. **Figure 5b** shows that the smaller than average forces tend to follow a power law distribution as suggested by (2) with $\alpha=-0.16$. Exactly the same exponent was obtained for the soft particle system and the value can be compared with the value of $\alpha=-0.3$ reported in [19].

The contact force distributions obtained for the soft particle system, **Figure 6**, clearly show that the distribution of the larger than average forces depends on the stress level and cannot be represented by the simple exponential relationship provided by (3). In **Figure 7**, a comparison of the results of the simulations with the predictions given by (1) for k=3 and k=4 suggests that, for soft particle systems, k increases with stress level. Although a fit to the larger than average forces in any one data set may be obtained by adjusting k, the distribution of the smaller than average forces tend to obey a power law which is independent of the elastic modulus of the particles and the stress level. Consequently, from the results of our own simulations, we conclude that there is no simple universal rule to completely characterise the contact force networks in granular media.

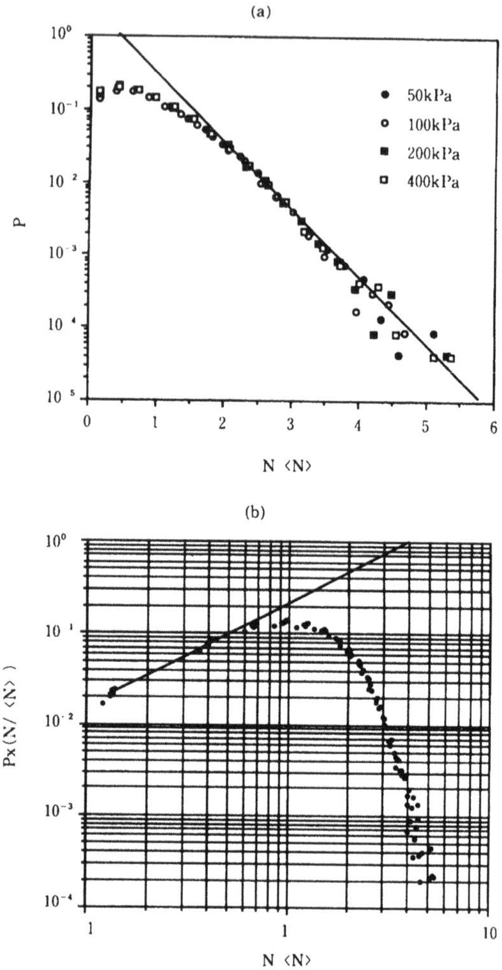

Fig. 5 Contact force distributions obtained at different stress levels for a polydisperse system of hard spheres (a) comparison with equation (3), $\beta=2.15$ (b) comparison with equation (2), $\alpha=-0.16$.

5. Distribution of contact forces in terms of magnitude and orientation

Granular media exhibit anisotropic characteristics. Consequently, it is necessary to take into account the contact orientations in order to provide descriptions of the macroscopic stress, modulus and conductive properties. Here we consider only the stress state.

For a large system of particles occupying a volume V the ensemble average stress tensor can be defined in terms of the interparticle contact forces P [1] by the equation

$$\sigma_{ij}=\frac{1}{V}\sum_{1}^{2M}x_iP_j=\frac{2M}{V}\langle x_iP_j\rangle \qquad (4)$$

where ⟨.⟩ denotes statistical average, $i=1,3$; $j=1$,

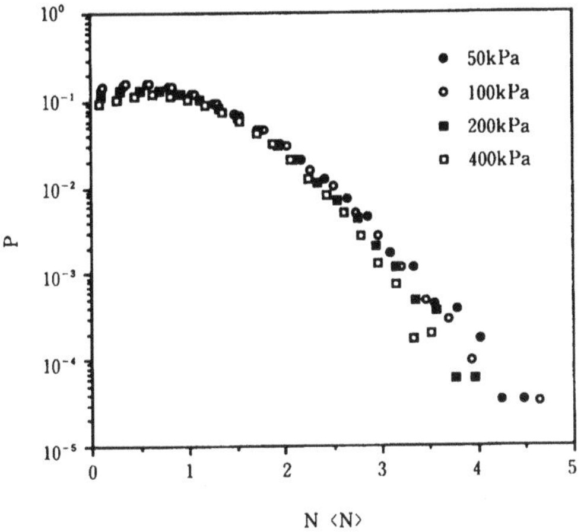

Fig. 6 Contact force distributions obtained at different stress levels for a polydisperse system of soft spheres.

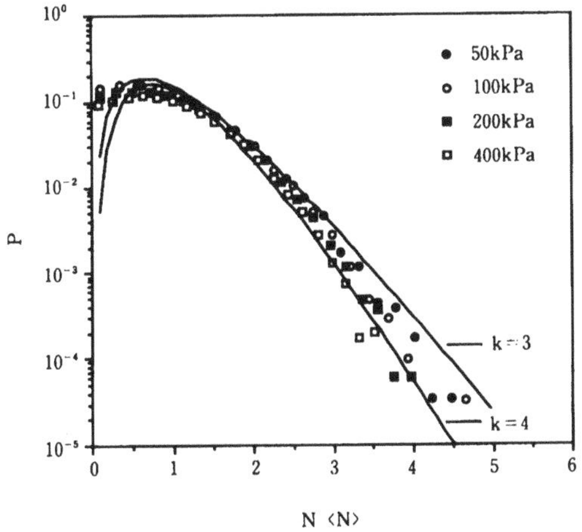

Fig. 7 Comparison of simulation data for the soft sphere system shown in figure 6 with equation (1) for k=3 and k=4.

3 and x_i defines the coordinates of the contact referenced to the particle centroid. The summation is performed over the M contacts in the system, each contact being counted twice.

For discs or spheres, the contact forces may be partitioned into their normal and tangential components (N,T) so that $P_i = Nn_i + Tt_i$ where n_i are the direction cosines of the unit contact normal vector and the vector t_i defines the direction of the relative tangential velocity at the contact. Also, the coordinates may be expressed in terms of the particle radius $x_i = Rn_i$ and (4) may be rewritten in terms of the normal and tangential force contributions to the stress tensor

$$\sigma_{ij} = \sigma_{ij}^N + \sigma_{ij}^T \tag{5}$$

where

$$\sigma_{ij}^N = \frac{2M}{V}\langle RNn_in_j\rangle \text{ and } \sigma_{ij}^T = \frac{2M}{V}\langle RTn_it_j\rangle \tag{6}$$

Since $n_in_i = 1$ and $n_it_i = 0$, (6) may also be written as

$$\sigma_{ij} = \sigma_{kk}\left[\left\langle\frac{RNn_in_j}{RN}\right\rangle + \left\langle\frac{RTn_it_j}{RN}\right\rangle\right] = \sigma_{kk}(N_{ij} + T_{ij}) \tag{7}$$

where

$$\sigma_{kk} = \frac{2M}{V}\langle RN\rangle \tag{8}$$

It follows from (7) that the anisotropy of the normal force contribution σ_{ij}^N is defined by the tensor N_{ij} which characterises the orientational distribution of contact normal forces. We may approximate the actual distribution with a continuous distribution by introducing a probability density function $P(\vec{n})$

$$N_{ij} = \left\langle\frac{RNn_in_j}{RN}\right\rangle \Rightarrow \frac{1}{4\pi}\int_\Omega P(\vec{n})\, n_in_j d\Omega \tag{9}$$

which satisfies the conditions

$$\int_\Omega P(\vec{n})\, d\Omega = 1 \text{ and } P(\vec{n}) = P(-\vec{n}) \tag{10}$$

The probability density function may be represented by a Fourier series expansion expressed in terms of even rank tensors [21]

$$P(\vec{n}) = P_0 + P_{ij}f_{ij} + \cdots\cdots \text{ where } f_{ij} = n_in_j - \delta_{ij}/3 \tag{11}$$

and

$$P_0 = \frac{1}{4\pi}\int_\Omega P(\vec{n})\, d\Omega = \frac{1}{4\pi} \tag{12}$$

$$P_{ij} = \frac{15}{8\pi}\int_\Omega P(\vec{n})\, f_{ij} d\Omega = \frac{15}{8\pi}(N_{ij} - \delta_{ij}/3) \tag{13}$$

Hence, we see that the orientational distribution of the normal contact forces may be represented by a tensorial Fourier series whose coefficients are functions of the deviatoric components of the normal force contribution to the stress tensor. The Fourier approximations to the normal contact force distribution may be further illustrated by polar diagrams, as shown in **Figure** 8.

Figure 8a shows the Fourier approximation to the normal contact force distribution during axisymmetric compression ($\sigma_1 > \sigma_2 = \sigma_3$). The distribution is clearly anisotropic and reflects the fact that the large contact forces tend to align in the σ_1 direction, as seen in **Figure** 2, and that contact separation tends to occur in directions orthogonal to the direction of σ_1. The 3D shape may be described as a peanut. For the case of axisymmetric extension ($\sigma_1 = \sigma_2 > \sigma_3$), the corresponding shape resembles a doughnut, as illustrated in **Figure 8b**. For general stress states ($\sigma_1 > \sigma_2 > \sigma_3$), there is a smooth continuous change in the shape of the normal contact force distribution, between the two extreme cases shown in **Figure** 8, as the intermedi-

(a)

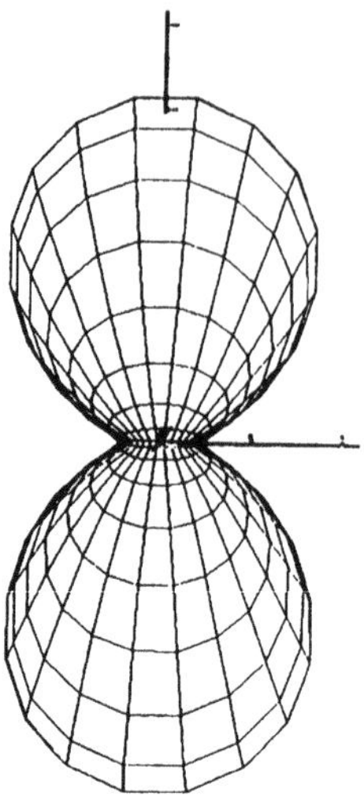

(b)

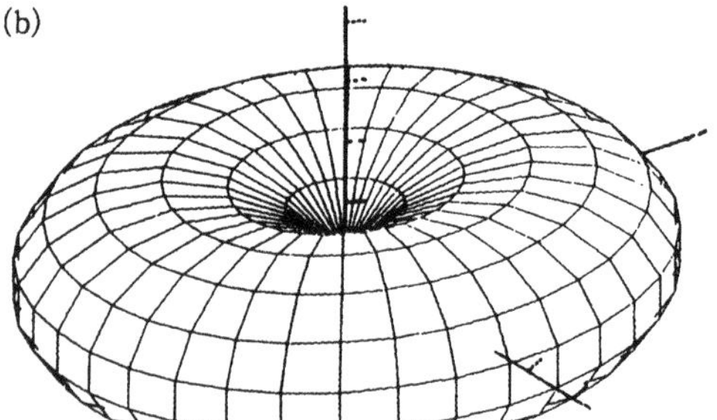

Fig. 8 Probability density distributions of contact normal forces represented by equation (11) for (a) axisymmetric compression and (b) axisymmetric extension.

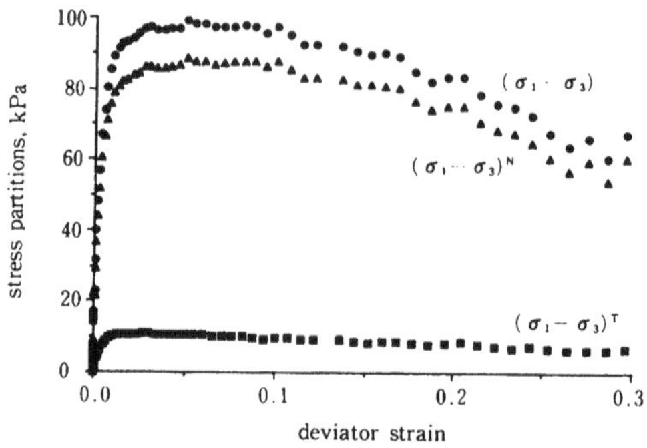

Fig. 9 Evolution of the deviator stress $(\sigma_1 - \sigma_3)$, the normal contact force contribution to the deviator stress $(\sigma_1 - \sigma_3)^N$ and the tangential contact force contribution to the deviator stress $(\sigma_1 - \sigma_3)^T$ during axisymmetric compression of a dense polydisperse system of elastic spheres.

ate principal stress σ_2 varies.

6. Force transmission during shear deformation

When subjected to an isotropic state of stress the tangential contact force contribution to the stress tensor is zero. Although tangential contact forces exist, the distribution of positive and negative shear directions is self-cancelling in the statistical averaging. During shear, the tangential contact forces contribute only to the deviatoric part of the stress tensor ; the normal contact forces contribute to both the isotropic and deviatoric parts.

Thornton and Sun [22, 23] reported numerical simulations of two polydisperse systems of 3620 elastic spheres, one dense the other loose, in which the mean stress, $(\sigma_1+\sigma_2+\sigma_3)/3=100$ kPa, was maintained constant during shear. **Figure 9** shows the evolution of the deviator stress $(\sigma_1-\sigma_3)$, the normal contact force contribution to the deviator stress $(\sigma_1-\sigma_3)^N$, and the tangential contact force contribution to the deviator stress $(\sigma_1-\sigma_3)^T$ during axisymmetric compression $(\sigma_2=\sigma_3)$ of the dense system. It is clear from the figure that the mobilised deviator stress is primarily a function of the normal force contribution and that the tangential force contribution is small. The results shown are for a coefficient of interparticle friction $\mu=0.3$. Increasing the interparticle friction increases the tangential force contribution but also increases the normal force contribution which remains the major contribution to the stress tensor. Similar trends were observed for the loose system. Results of recent, unpublished, numerical simulations indicate that this is also true for systems of non-spherical particles.

During shear, the distribution of contact orientations becomes anisotropic primarily due to contact separation in directions approximately orthogonal to the direction of the major principal stress and consequently, in dense systems, the total number of contacts decreases. The average number of contacts per particle, the coordination number, is normally defined by Z=2M/V. However, in all simulations it is found that there are always a number of particles which have only one or no contacts and, hence, such particles do not contribute to the force transmission through the system. Consequently, we define a mechanical coordination number

$$Z_m = \frac{2M - N_1}{N - N_0 - N_1} \tag{14}$$

where N_1 and N_0 are the number of particles with only one or no contacts respectively. **Figure 10** shows the evolution of the mechanical coordination number during axisymmetric compression for both dense and loose systems. It can be seen that after a

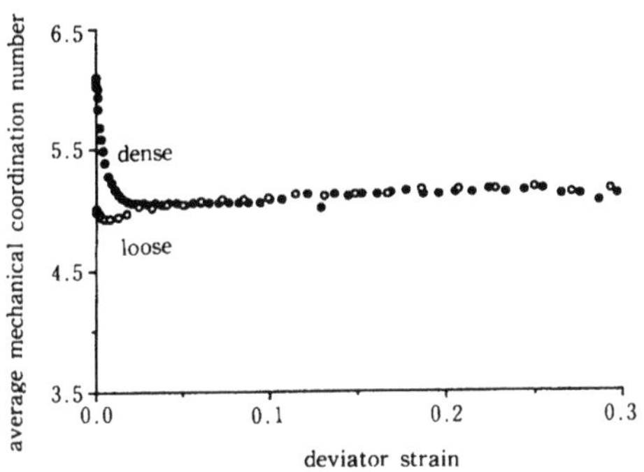

Fig. 10 Evolution of the mechanical coordination number, defined by equation (14), during axisymmetric compression of dense (●) and loose (o) polydisperse systems of elastic spheres.

small amount of straining the mechanical coordination number becomes essentially constant and independent of the initial packing density, irrespective of whether the system is expanding or contracting. This reflects an underlying stability requirement and corresponds, in statistical physics terminology, to a percolation threshold. If the coefficent of interparticle friction is increased, the critical mechanical coordination number decreases.

In other simulations which we have carried out we have found that the critical mechanical coordination number can momentarily drop below the critical value [24]. Since we impose a uniform strain field in periodic cell simulations, the critical coordination number is regained but fluctuations continue to occur. These fluctuations are coincident with fluctuations in the stress-strain curve and with transient jumps in the fluctuating kinetic energy density (granular temperature) which may be of one or two orders of magnitude. This phenomena indicates local transient instabilities in the columns of particles transmitting the larger than average contact forces. In the presence of real boundaries, we anticipate that strain localisation may occur when the mechanical coordination number falls below the critical value.

7. Concluding remarks

It has been demonstrated that, in granular media, load is transmitted primarily along relatively rigid, heavily stressed chains of particles forming a relatively sparse network of larger than average contact forces. Drawing an analogy with percolation theory, the heavily stressed chains of particles form the backbone to the percolating cluster. The distribution of the magnitude of the larger than average contact forces is exponential but for general states of stress it is necessary to account for the orientations of the contact forces by adopting a tensor representation of the force network. Nevertheless, it is the rapid development of strong force chains aligned in the direction of the major axis of compression that results in the deviator stress being primarily related to the normal contact force contribution to the stress tensor.

From our numerical simulations we observe that the system of particles acts in a way similar to that of a highly redundant space structure. There is a multiplicity of pathways along which force transmission may be achieved in order to establish a stable stress state. In order to achieve a stable state of stress it is not necessary that all the possible pathways are used. Hence, the system does not use all the potential pathways but naturally optimises the selection to match the loading direction. If the loading direction is suddenly changed then there is a rapid reselection of a more appropriate and efficient set of pathways. In terms of statistical physics, this is a percolation problem and the system can be considered to be self-organising. This self-organised optimisation of the force transmission is reflected in a critical value of the mechanical coordination number which corresponds to a percolation threshold. If the mechanical coordination number falls below the critical value then instabilities occur.

Numerical simulations have shown that increasing the interparticle friction does not produce a sigificant increase in the amount of energy dissipated but decreases the percentage of sliding contacts and reduces the value of the critical mechanical coordination number. This demonstrates that friction is, primarily, a kinematic constraint which permits reorientation of force transmission paths without undue particle rearrangement. Enhanced friction at the contacts increases the stability of the system and reduces the number of contacts required to achieve a stable configuration. When sliding does occur it tends to be at the relatively unloaded contacts in the regions between the chains of particles carrying the larger than average contact forces. Consequently the tangential contact force contribution to the stress tensor is small and the shear strength mobilised is primarily due to the development of the strong force network rather than due to work done in sliding particles over each other as implied in traditional soil mechanics.

Finally, it should be said that the above is true for "hard" particle systems but "soft" particle systems, and mixtures of hard and soft particles, may behave differently. Consider an equivalent space lattice which is composed of "branches" joining the centres of contacting spheres. Even when each branch has the same stiffness the effective stiffness depends on the direction of the applied loading. If we consider a hard sphere system, the space lattice consists, in effect, of a random array of stiff springs. Under load, branches which are favourably orientated will pick up the load with very little reduction in the branch length. Consequently, the load is not shared with other less favourably oriented branches. This leads to the type of force transmission networks described in Section 3. In contrast, the space lattice representing the soft particle system consists of a random array of highly compressible springs. Even though it will be the most favourably orientated branches which will initially pick up the load, the reduction in branch length as the applied loading is increased will permit load transfer to other less favourably orientated branches. Hence, in soft sphere systems, the force transmission is more uniformly distributed. However, we have observed that at very low stress levels the force transmission through soft particle systems is exponential, indicating that the force transmission also depends on stress level. At sufficiently high stress levels it is anticipated that hard particles will deform plastically and the force distribution will become more uniform.

References

1) Thornton C and Barnes D J : Computer simulated deformation of compact granular assemblies. Acta Mechanica 64 (1986) 45-61.
2) Thornton C : On the relationship between the modulus of particulate media and the surface energy of the constituent particles. J. Phys.D : Appl. Phys. 26 (1993) 1587-1591.
3) Wakabayshi T : Photoelastic method for determination of stress in powdered mass. In Proc. 7th Japan Nat. Cong. Appl. Mech. (1957) 153-158.
4) Dantu P : A contribution to the mechanical and geometrical study of non-cohesive masses. In Proc. 4th Int. Conf. on Soil Mechanics & Foundation Engineering (1957) 144-148.
5) de Josselin de Jong G and Verruijt A : Étude photoélastique d'un empilement de disques. Cahiers du Groupe Francais de Rheologie 2 (1969) 73-86.
6) Drescher A and de Josselin de Jong G : Photoelastic verification of a mechanical model for the flow of a granular material. J. Mech. Phys. Solids 20 (1972) 337-351.
7) Oda M and Konishi J : Microscopic deformation mechanism of granular material in simple shear. Soils and Foundations 14 (1974) 25-38.
8) Konishi J, Oda M and Nemat-Nasser S : Inherent anisotropy and shear strength of assembly of oval cross-sectional rods. In Deformation and Failure of Granular Materials (P A Vermeer and H J Luger, Eds.) Balkema (1982) 403-412.
9) Rossmanith H P and Shukla A : Photoelastic investigation of dynamic load transfer in granular media. Acta Mechanica 42 (1982) 211-225.
10) Shukla A : Dynamic photoelastic studies of wave propagation in granular media. Optics and Lasers in Engineering 14 (1991) 165-184.
11) Zhu Y, Shukla A and Sadd M H : The effect of microstructural fabric on dynamic load transfer in two dimensional assemblies of elliptical particles. J. Mech. Phys. Solids 44 (1996) 1283-1303.
12) Liu C-h, Nagel S R, Schecter D A, Coppersmith S N, Majumdar S, Narayan O and Witten T A : Force fluctuations in bead packs. Science 269 (1995) 513-515.
13) Cundall P A and Strack O D L : A discrete numerical model for granular assemblies. Géotechnique 29 (1979) 47-65.
14) Sadd M H, Tai Q M and Shukla A : Contact law effects on wave propagation in particulate material using distinct element modeling. Int. J. Non-Linear Mechanics 28 (1993) 251-265.
15) Thornton C and Yin KK : Impact of elastic spheres with and without adhesion. Powder Technology 65 (1991) 153-166.
16) Thornton C : Coefficient of restitution for collinear collisions of elastic-perfectly plastic spheres. J. Appl. Mech. 64 (1997) 383-386.
17) Cundall P A, Jenkins J T and Ishibashi I : Evolution of elastic moduli in a deforming granular material. In Powders & Grains (J Biarez and R Gourvès, eds.) Balkema (1989) 319-322.
18) Bagi K : On the definition of stress and strain in granular assemblies through the relation between micro- and macro-level characteristics. In Powders & Grains 93 (C Thornton, Ed.) Balkema (1993) 117-121.
19) Radjai F, Jean M, Moreau J-J and Roux S : Force distributions in dense two-dimensional granular systems. Phys. Rev. Lett. 77 (1996) 274-277.
20) Radjai F and Roux S : Friction-induced self-organization of an array of particles. Phys. Rev. E 51 (1995) 6177.
21) Onat E T and Leckie F A : Representation of mechanical behavior in the presence of changing internal structure. J. Appl. Mech. 110 (1988) 1-10.
22) Thornton C and Sun G : Axisymmetric compression of 3D polydisperse systems of spheres. In Powders & Grains 93 (C Thornton, Ed.) Balkema (1993) 129-134.
23) Thornton C and Sun G : Numerical simulation of general 3D quasi-static shear deformation of granular media. In Numerical Methods in Geotechnical Engineering (I M Smith, Ed.) Balkema (1994) 143-148.
24) Thornton C : Micromechanics of elastic sphere assemblies during 3D shear. In Proc. Workshop on Mechanics

and Statistical Physics of Particulate Materials, Institute for Mechanics and Materials Report No. 94-9, University of California San Diego, 64-67.

Author's short biography

Colin Thornton

Colin Thornton graduated in Civil Engineering from Aston University where he also obtained his Ph.D and, since 1971, has lectured on soil mechanics and geotechnical engineering. He is a member of the International Society for Soil Mechanics and Foundation Engineering's Technical Committee on the Mechanics of Granular Materials, the Engineering and Physical Sciences Research Council's Process Engineering College and the IChemE Particle Technology Subject Group Commitee. In 1993, he organised the 2nd International Conference on the Micromechanics of Granular Media and edited the conference proceeding, Powders & Grains 93 (Balkema, Rotterdam). Since 1980 his reseach has focussed on the micromechanics of granular media and computer simulations of particle systems which have been applied to quasi-static shear deformation, hopper flow, particle/wall collisions and the coalescence, attrition and fracture of agglomerates.

PHYSICAL REVIEW E VOLUME 57, NUMBER 3 MARCH 1998

Force distribution in a granular medium

Daniel M. Mueth, Heinrich M. Jaeger, and Sidney R. Nagel
The James Franck Institute and Department of Physics, The University of Chicago, 5640 South Ellis Avenue, Chicago, Illinois 60637
(Received 18 August 1997)

We report on systematic measurements of the distribution of normal forces exerted by granular material under uniaxial compression onto the interior surfaces of a confining vessel. Our experiments on three-dimensional, random packings of monodisperse glass beads show that this distribution is nearly uniform for forces below the mean force and decays exponentially for forces greater than the mean. The shape of the distribution and the value of the exponential decay constant are unaffected by changes in the system preparation history or in the boundary conditions. An empirical functional form for the distribution is proposed that provides an excellent fit over the whole force range measured and is also consistent with recent computer simulation data. [S1063-651X(98)02603-8]

PACS number(s): 81.05.Rm, 46.10.+z, 05.40.+j, 83.70.Fn

INTRODUCTION

Granular materials have a rich set of unusual behavior which prevents them from being simply categorized as either solids or fluids [1]. Even the most simple granular system, a static assembly of noncohesive, spherical particles in contact, holds a number of surprises. Particles within this system are under stress, supporting the weight of the material above them in addition to any applied load. The interparticle contact forces crucially determine the bulk properties of the assembly, from its load bearing capability [2,3] to sound transmission [4–6] or shock propagation [7,8]. Only in a crystal of identical, perfect spheres is there uniform load sharing between particles. In any real material the slightest amount of disorder, due to variations in the particle sizes as well as imperfections in their packing arrangement, is amplified by the inherently nonlinear nature of interparticle friction forces and the particles' nearly hard-sphere interaction. As a result, stresses are transmitted through the material along "force chains" that make up a ramified network of particle contacts and involve only a fraction of all particles [9–11].

Force chains and spatially inhomogeneous stress distributions are characteristic of granular materials. A number of experiments on two-dimensional (2D) and 3D compression cells have imaged force chains by exploiting stress-induced birefringence [9–16]. While these experiments have given qualitative information about the spatial arrangement of the stress paths inside the granular assembly, the quantitative determination of contact forces in three dimensional bead packs is difficult with this method. Along the confining walls of the assembly, however, individual force values from all contacting particles can be obtained. Liu *et al.*'s experiments [10] showed that the spatial probability distribution $P(F)$ for finding a normal force of magnitude F against a wall decays exponentially for forces larger than the mean $\bar{F}$. This result is remarkable because, compared to a Gaussian distribution, it implies a significantly higher probability of finding large force values $F \gg \bar{F}$.

A number of fundamental questions remain, however. While several model calculations [10,19], computer simulations [20–24], as well as experiments on shear cells [25] and 2D arrays of rods [11] have corroborated the exponential tail for $P(F)$ in the limit of large F, other functional forms so far have not been ruled out [26]. Furthermore, there has been no consensus with regard to the shape of the distribution for forces smaller than the mean. The original "q model" by Coppersmith *et al.* [19] and Liu *et al.* [10] predicted power law behavior with $P(F) \propto F^{\alpha}$ and $\alpha \approx 2$ for small F, while recent simulations by Radjai *et al.* [20–22] and Luding [23] found $\alpha \leqslant 0$. So far, experiments have lacked the range or sensitivity required for a firm conclusion. The roles of packing structure and history, identified in much recent work as important factors in determining stresses in granular media, have not yet been explored experimentally in this system. Finally, the existence of correlations between forces remains unclear. Shear cell data by Miller *et al.* [25] have been interpreted as an indication of correlations between forces against the cell bottom surface.

In this paper we present results from a set of systematic experiments designed to address these issues. We have refined the carbon paper method [10,17,18] for determining the force of each bead against the constraining surface and are now able to measure force values accurately over two orders of magnitude. With this improvement we are able to ascertain the existence of the exponential behavior and to obtain close bounds on its decay constant in the regime $F > \bar{F}$. For $F < \bar{F}$ we find that $P(F)$ flattens out and approaches a constant value. In addition, our experiments investigated the effects of the packing history. We studied both the influence of the boundary conditions posed by the vertical container walls on the distributions of forces $P(F)$ as well as the spatial correlations in the arrangement of beads due to crystallization near a wall during system preparation. None of these variations on the experiment are found to influence $P(F)$ significantly. Finally, we have also measured the lateral correlations between forces on different beads and find that no correlations exist.

EXPERIMENTAL METHOD

The granular medium studied was a disordered 3D pack of 55 000 soda lime glass spheres with diameter $d = 3.5$

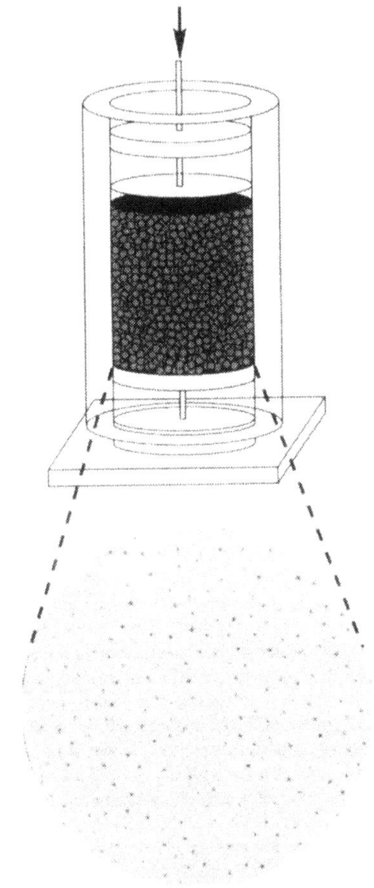

FIG. 1. Sketch of the apparatus used for experiments with "floating walls." The lower piston is fixed and the cylinder is supported by friction with the bead pack. A load is applied to the upper piston and the beads press the carbon paper into white paper, leaving marks which are used to determine the contact forces. A detail of the obtained raw data is shown in the photograph (field of view: 76 mm across).

± 0.2 mm. The beads were confined in an acrylic cylinder of 140 mm inner diameter. The top and bottom surfaces were provided by close-fitting pistons made from 2.5 cm thick acrylic disks rigidly fixed to steel rods. The height of the bead pack could be varied, but experiments described in this paper were performed with a height of 140 mm. Once the cell was filled with beads, a load, typically 7600 N, was applied to the upper piston using a pneumatic press while the lower piston was held fixed. In most experimental runs, the outside cylinder wall was not connected to either piston so that the cylinder was supported only by friction with the bead pack (see Fig. 1). We shall refer to this as the "floating wall" method. The system could also be prepared with the bottom piston rigidly attached to the cylinder wall, which we shall refer to as the "fixed wall" method. To estimate the bead-bead and bead-wall static friction coefficients, we glued beads to a plate resting on another glass or acrylic plate and inclined the plates until sliding occurred. We found the static coefficient of friction to be close to 0.2 for both glass-glass and glass-acrylic contacts.

As the beads were loaded into the cell, they naturally tended to order into a 2D polycrystal along the lower piston. The beads against the upper piston, by contrast, were irregularly packed. We were able to enhance ordering on the lower piston by carefully loading the system, or disturb it by placing irregularly shaped objects against the surface which were later removed. For some experiments, the cell was inverted during or after loading with beads. By varying the experiment in these ways, we probed the effect of system history on the distribution of forces.

Contact forces were measured using a carbon paper technique [18,17,10]. With this method, all constraining surfaces of the system were lined with a layer of carbon paper covering a blank sheet of paper. For the blank sheet we used color copier paper, which is smoother, thicker, and has a more uniform appearance than standard copier paper. Beads pressed the carbon onto the paper in the contact region and left marks whose darkness and area depended on the force on each bead. After the load had been applied to the bead packing, the system was carefully disassembled and the marks on the paper surface were digitized on a flatbed scanner for analysis. A region from a typical data set taken from the area over one of the pistons is shown in Fig. 1. Each experiment yielded approximately 3800 data points over the interior cylinder wall and between 800 and 1100 points for each of the piston surfaces, depending on how the system was prepared. The position of each mark was identified and the thresholded area and integrated darkness were calculated. At the scan resolution used, marks ranged from several pixels to several hundred pixels in area.

The force was determined by interpolating the measured area and darkness on calibration curves that were obtained by pressing a single bead with a variable, known force onto the carbon paper. This was achieved by slowly lowering a known mass through a spring onto a single bead. The spring was essential as it greatly reduced the otherwise large impulse which occurs when a bead makes contact with the carbon paper and quickly comes to rest. Both area and darkness of the mark left on the copier paper were found to increase monotonically with the normal component of the force exerted by each bead, as seen in Fig. 2. Note that the only requirement is that these curves are monotonic; we do not assume any particular functional relationship. With this carbon paper technique, we were able to measure forces between 0.8 and 80 N with an error of less than 15%. We ensure that the beads do not slide relative to the carbon paper during an experiment by measuring the eccentricity of each mark. We find that the eccentricities ϵ are narrowly distributed with a mean of 0.1, corresponding to a ratio of major to minor axis $a/b = 1/\sqrt{1-\epsilon^2}$ of 1.005 for both piston surfaces and container walls.

We find that for less than approximately 0.8 N, little or no mark is left on the copier paper. A consequence, visible in Fig. 1, is that there are regions where there may have been one or more contacts with normal force less than 0.8 N, or alternatively, which may have had no bead in contact with the surface. This ambiguity presents a problem for the precise determination of the mean force $\bar{F}$. To estimate the

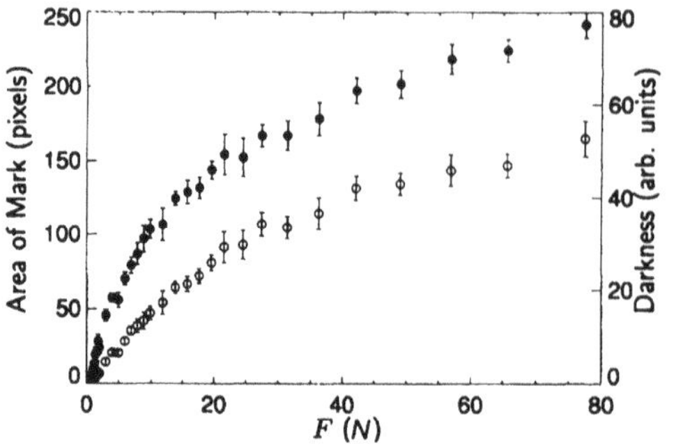

FIG. 2. Calibration curves for the conversion of pressure mark size or intensity to normal force. The solid circles represent the mark area and the open circles its integrated darkness.

number of contacts below our resolution, we could fill the voids with the maximum possible number of additional beads, using a simple computer routine. However, this overestimates the number of actual contacts with the carbon paper. Instead, we used the following method. The average number of beads touching a piston surface was measured by placing double-sided tape on the piston and lowering it onto the pack. The tape was sufficiently sticky that the weight of a single bead would affix it to the tape. Subtracting the average number of contacts with $F>0.8$ N from this number, we found that 6.4% of the beads on the lower piston and 4.3% of the beads on the upper piston have $F<0.8$ N. The upper piston had fewer points below 0.8 N because the total number of beads in contact with that piston was typically smaller than on the bottom, raising the mean force and decreasing the fraction of beads with $F<0.8$ N. The weight supported by the walls was calculated by subtracting the net weight on the two pistons. For experiments performed with floating walls, we verified that the pistons had equal net force (since the effects of gravity and the weight of the walls can be neglected with respect to the applied force).

RESULTS

While we conducted experiments with both fixed walls and floating walls, most experiments were performed with the walls floating to reduce asymmetry. In this configuration the cylindrical wall of the system was suspended solely by friction with the bead pack. Since the applied load was much greater than the weight of the system, any remaining asymmetry between the top and bottom of the system must have come primarily from system preparation, and not from gravity. In Fig. 3 we show the resulting force distributions $P(f)$ (where $f \equiv F/\bar{F}$ is the normalized force) for all system surfaces, averaged over fourteen experimental runs performed under identical, floating wall conditions. We find that, within experimental error, the distributions $P(f)$ for the upper and lower piston surfaces are identical and, in fact, independent of floating or fixed wall conditions. Note that the lowest bin contains forces from 0 to roughly 1 N which includes both measured forces as well as an estimated number of undetectable contacts, giving it a greater uncertainty than other bins. For forces greater than the mean ($f>1$), the probability of a bead having a certain force decays exponentially, $P(f) \propto e^{-\beta f}$, with $\beta = 1.5 \pm 0.1$.

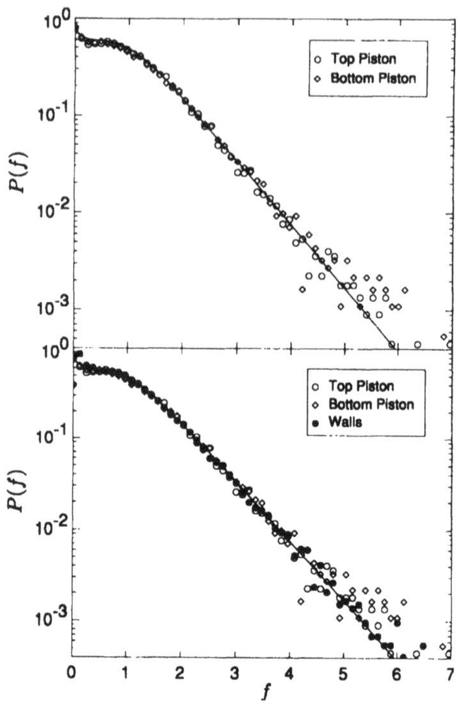

FIG. 3. The distribution $P(f)$ of normalized forces f against the top piston (open circles), the bottom piston (diamonds), and the walls (solid circles). The upper panel shows $P(f)$ for the pistons, averaged over fourteen identical experiments. The curve drawn is a fitting function as explained in the text [Eq. (1)]. The lower panel shows the same data, but with data from the walls included as well.

Also shown in Fig. 3 is a curve corresponding to the functional form

$$P(f) = a(1 - be^{-f^2})e^{-\beta f}. \qquad (1)$$

An excellent fit to the data is obtained for $a=3$, $b=0.75$, and $\beta = 1.5$. This functional form captures the exponential tail at large f, the flattening out of the distribution near $f \approx 1$, and the slight increase in $P(f)$ as f decreases towards zero.

For the mean force against the side wall we observe a dependence with the depth z into the pile from the top piston which strongly depends on the boundary conditions (Fig. 4). For fixed wall boundary conditions (solid symbols) the angle-averaged wall force $\bar{F}_w(z)$ is greatest near the upper piston, decaying with increasing depth into the pile. On the other hand, for floating wall conditions (open symbols) $\bar{F}_w(z)$ stays roughly constant. Using $\bar{F}_w(z)$ we compute the set of normalized forces $f_{w,i} \equiv F_{w,i}/\bar{F}_w(z_i)$ exerted by individual beads i against the side walls. We find that the probability distribution $P(f_w)$ is independent of z within our experimental resolution and is practically identical to that found on the upper and lower piston surfaces, with a decay constant $\beta_w = 1.5 \pm 0.2$ for the regime $f_w > 1$. This distribu-

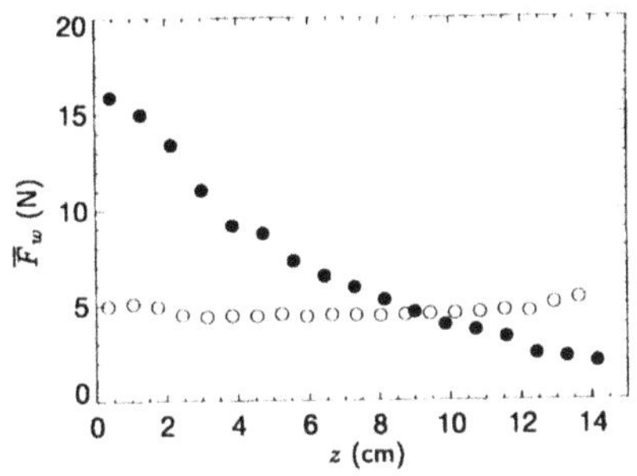

FIG. 4. The mean normal force $\bar{F}_w(z)$, measured along the wall at height z below the top surface of the packing, for fixed wall (solid circles) and floating wall (open circles) boundary conditions.

tion is shown in the lower panel of Fig. 3 by the solid symbols. Since along the walls we were unable to determine directly the number of contacts with force less than 0.8 N, we estimated it to be 4.3%, based on our result for the disordered piston. The uncertainty in β_w is predominantly due to the uncertainty in this estimate. Note that within the resolution of our measurements, the probability distributions in Fig. 3 are the same for all surfaces.

In contrast to observations reported previously [10,27], we observe that the mean force on any portion of the piston is independent of position. The radial dependence of the mean force against the pistons found previously [10] was an artifact of the compression method, and does not occur if the load is applied using a pneumatic press with carefully aligned pistons.

The first few layers of monodisperse beads coming into contact with the lower piston tend to order in a hexagonal packing while farther into the system a random packing is observed. To probe the effect of boundary-induced crystallization, the degree of bead ordering was varied in some experiments. We used the measured positions of the marks left on the copier paper to compute the radial distribution function

$$g(r)=\frac{1}{Nn_0\pi r}\sum_{i=1}^{N}\sum_{j=i+1}^{N}\delta(r_{ij}-r) \tag{2}$$

where n_0 is the average density of points, N is the total number of points, and r_{ij} is the distance between the centers of marks i and j. If filled from the bottom up without container inversion, the packing structure over the lower piston surface clearly exhibits a larger degree of crystalline order than that touching the top piston surface, as seen in Figs. 5(a) and 5(b). Vertical lines are drawn to indicate peaks expected in $g(r)$ for a 2D hexagonal packing. The radial distribution function for the lower piston in an experiment where ordering along this piston is disturbed is shown in Fig. 5(c). Despite the significant differences in degree of ordering evident from Figs. 5(a)–5(c), no significant effect on $P(f)$ was observed.

Since beads generally move downward as the cell is loaded, friction forces tend to be oriented upward. The process of adding beads to fill the cell, therefore, breaks the symmetry of the system by building an overall directionality into the force network. With different packing histories, however, such as inverting the system once or more during or after loading, we systematically disrupted this directionality. Again no measurable effect on $P(f)$ was found.

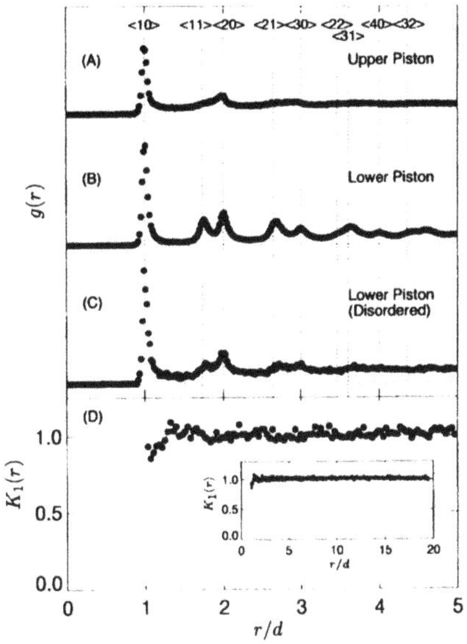

FIG. 5. Pair distribution function $g(r)$ for (a) upper piston, (b) lower piston, and (c) lower piston with disrupted ordering. The horizontal axis gives the distance r between any two points, normalized by the bead diameter d. Vertical lines indicate the distances between points separated by hexagonal lattice translation vectors and are labeled by the vector indices. (d) Force pair correlation function $K_1(r)$ for the bottom piston. The inset shows $K_1(r)$ out to 20 bead diameters, a distance equal to the radius of the cell and half its height.

Our experiments also allowed for a direct calculation of correlations between normal forces impinging on a given container surface. We computed the lateral force-force pair correlations

$$K_n(r)=\frac{\sum_{i=1}^{N}\sum_{j=i+1}^{N}\delta(r_{ij}-r)f_i^n f_j^n}{\sum_{i=1}^{N}\sum_{j=i+1}^{N}\delta(r_{ij}-r)} \tag{3}$$

over both piston surfaces and the walls. As an example, Fig. 5(d) shows the first order correlation $K_1(r)$ for the lower piston in experiments where ordering was not disrupted [corresponding to $g(r)$ in Fig. 5(b)]. The featureless shape of $K_1(r)$ is characteristic of all cases examined ($n \in \{1,2,3\}$) and indicates no evidence for force correlations.

DISCUSSION

The key features of the data in Fig. 3 are the nearly constant value of the probability distribution for $f<1$ and the exponential decay of $P(f)$ for larger forces. No comprehensive theory exists at present that would predict this overall shape for $P(f)$. The exponential decay for forces above the mean is predicted by the scalar q model as a consequence of a force randomization throughout the packing [10,19]. In this mean field model the net weight on a given particle is divided randomly between N nearest neighbors below it, each of which carries a fraction of the load. Only one scalar quantity is conserved, namely the sum of all force components along the vertical axis. Randomization has an effect analogous to the role played by collisions in an ideal gas [10,19]. The result is a strictly exponential distribution $P(f) \propto e^{-Nf}$ for the normal forces across the contact between any two beads.

The calculations for the original q model were done for an infinite system without walls. If one assumes that each particle at a container boundary has N neighbors in the bulk and a single contact with the wall, then the net force transmitted against the wall is a superposition of N independent contact forces on each bead, so that the probability distribution for the net wall force is modified by a prefactor f^{N-1}, much in the way a phase-space argument gives rise to the power law prefactor in the Maxwell-Boltzmann distribution. Thus, the original q model predicts a nonmonotonic behavior for $P(f)$ with vanishing probability as $f \rightarrow 0$. Such a "dip" at small force values has also been found in recent simulations by Eloy and Clement [26]. It is, however, in contrast to the data in Fig. 3 and to recent simulation results on 2D and 3D random packings by Radjai and co-workers [20–22]. These simulations indicated that the distribution of normal contact forces anywhere, and at any orientation, in the packing did not differ from that found for the subset of beads along the walls. In fact, for both normal and tangential contact forces inside and along the surfaces of the packings, Radjai *et al.* observed distributions that were well described by

$$P(f) \propto \begin{cases} f^{-\alpha}, & f<1 \\ e^{-\beta f}, & f>1, \end{cases} \tag{4}$$

with α close to zero and positive and $1.0<\beta<1.9$, depending on which quantity was being computed, the dimension of the system, and the friction coefficient. While we were unable to measure experimentally forces below about $f \approx 0.1$, the simulation data by Radjai and co-workers extends to $f \approx 0.0001$. Power law behavior with $\alpha>0$ in Eq. (4), if indeed correct, would lead to a divergence in $P(f)$ as $f \rightarrow 0$. However, we observe that our empirical function, Eq. (1), which does not diverge, provides a fit essentially indistinguishable from a power law $f^{-\alpha}$ over the range $0.001<f<1$ as long as α is positive and close to zero. We can thus equally well fit the simulation data for normal forces in Refs. [20–22], over its full range, with Eq. (1). For the case of 3D simulations and friction coefficients close to 0.2, this is possible using the same coefficients as for the experimental data in Fig. 3.

We point out that the fitting function in Eq. (1) is purely empirical. In particular, we do not have a model that would predict the $(1-be^{-f^2})$ prefactor of the main exponential. It may be possible to think of this prefactor, in some type of modified q model, as arising from considerations similar to phase-space arguments. The fact that it clearly differs from the usual f^N dependence expected for N independent vector components would then point to the existence of correlations between the contact forces on each bead. Such correlations obviously exist, in the form of constraints; yet how these constraints conspire to give rise to a specific functional form for $P(f)$ as in Eq. (1) remains unclear. Eloy and Clement [26] have attempted to take into account some of the correlations that might apply to forces acting locally on a given bead. Using a modified q model they include the possibility of a bias in the distribution of q's, leading to a screening of small contact forces by larger ones. The resulting $P(f)$, nevertheless, still tends to zero as $f \rightarrow 0$.

Finally, we note that a "dip" in $P(f)$ for small forces can always be introduced by averaging our data over areas large enough to contain several pressure marks. Data by Miller *et al.* [25] on shear cells, using stress transducers of various sizes, similarly show an increasingly pronounced "dip" for the larger transducers. They did not, however, observe the pronounced narrowing of the distribution that is expected in the limit of sufficiently large areas and attributed this to possible force correlations. Our data for the force pair correlations in Fig. 5 indicate that no simple correlations exist between forces within the plane of any of the confining walls. This result is in accordance with the q model [28].

CONCLUSION

We have found that the distribution of forces, shown in Fig. 3, is a robust property of static granular media under uniaxial compression. Its shape turns out to be identical, within experimental uncertainties, for all interior container surfaces and furthermore appears to be unaffected by changes in the boundary conditions or in the preparation history of the system. The exponential decay for forces above the mean emerges as a key characteristic of the force distribution. The exponential tail of the distribution can be understood on the basis of a scalar model (q model), where it emerges as a result of a randomization process that occurs as forces are transmitted through the bulk of the bead pack. The consequences of the vector nature of the contact forces on the distribution, however, remain unclear. A second key aspect of the measured distribution is the absence of either a "dip" or a power law divergence for small forces; instead, our data is most consistently fit by a functional form that approaches a finite value as $f \rightarrow 0$. This empirical fitting form, Eq. (1), provides an excellent fit over the full range of forces for our experimental data, as well as for simulation results on 3D packings obtained by Radjai *et al.* and for simulations performed by Thornton.

ACKNOWLEDGMENTS

We would like to thank Sue Coppersmith, John Crocker, David Grier, Hans Herrmann, Chu-heng Liu, Onuttom Narayan, Farhang Radjai, David Schecter, and Tom Witten for many useful discussions. This work was supported by the NSF under Grant No. CTS-9710991 and by the MRSEC Program of the NSF under Grant No. DMR-9400379.

[1] H. M. Jaeger, S. R. Nagel, and R. P. Behringer, Rev. Mod. Phys. **68**, 1259 (1996).
[2] T. Travers, M. Ammi, D. Bideau, A. Gervois, J.-C. Messager, and J.-P. Troadec, Europhys. Lett. **4**, 329 (1987).
[3] E. Guyon, S. Roux, A. Hansen, D. Bideau, J.-P. Troadec, and H. Crapo, Rep. Prog. Phys. **53**, 373 (1990).
[4] C. Liu and S. R. Nagel, Phys. Rev. Lett. **68**, 2301 (1992).
[5] M. Leibig, Phys. Rev. E **49**, 1647 (1994).
[6] S. Melin, Phys. Rev. E **49**, 2353 (1994).
[7] A. V. Potapov and C. S. Campbell, Phys. Rev. Lett. **77**, 4760 (1996).
[8] R. S. Sinkovits and S. Sen, Phys. Rev. Lett. **74**, 2686 (1995).
[9] M. Ammi, D. Bideau, and J. P. Troadec, J. Phys. D **20**, 424 (1987).
[10] C. Liu, S. R. Nagel, D. A. Schecter, S. N. Coppersmith, S. Majumdar, O. Narayan, and T. A. Witten, Science **269**, 513 (1995).
[11] G. W. Baxter, in *Powders and Grains 97*, edited by R. P. Behringer and J. T. Jenkins (Balkema, Rotterdam, 1997), pp. 345–348.
[12] P. Dantu, in *Proceedings of the 4th International Conference On Soil Mechanics and Foundation Engineering* London, 1957 (Butterworths, London, 1958), Vol. 1, pp. 144–148.
[13] P. Dantu, Ann. Ponts Chauss. **IV**, 193 (1967).
[14] T. Wakabayashi, in *Proceedings of the 9th Japan National Congress for Applied Mechanics*, Tokyo, 1959 (Science Council of Japan, Tokyo, 1960), pp. 133–140.
[15] T. Travers, M. Ammi, D. Bideau, A. Gervois, J.-C. Messager, and J.-P. Troadec, J. Phys. (France) **49**, 939 (1988).
[16] D. Howell and R. P. Behringer, in *Powders and Grains 97* (Ref. [11]), pp. 337–340.
[17] F. Delyon, D. Dufresne, and Y.-E. Lévy, Ann. Ponts Chauss. **53–54**, 22 (1990).
[18] D. Dufresne, F. Delyon, and Y.-E. Lévy, J. Sci. LPC **2**, 209 (1994).
[19] S. N. Coppersmith, C. Liu, S. Majumdar, O. Narayan, and T. A. Witten, Phys. Rev. E **53**, 4673 (1996).
[20] F. Radjai, M. Jean, J. J. Moreau, and S. Roux, Phys. Rev. Lett. **77**, 274 (1996).
[21] F. Radjai, D. Wolf, M. Jean, and J. J. Moreau (unpublished).
[22] F. Radjai, D. E. Wolf, S. Roux, M. Jean, and J. J. Moreau, in *Powders and Grains 97* [11], pp. 211–214.
[23] S. Luding, Phys. Rev. E **55**, 4720 (1997).
[24] C. Thornton (unpublished).
[25] B. Miller, C. O'Hern, and R. P. Behringer, Phys. Rev. Lett. **77**, 3110 (1996).
[26] C. Eloy and E. Clément, J. Phys. (France) I **7**, 1541 (1997).
[27] H. Kuno and I. Kuri, Powder Technol. **34**, 87 (1983).
[28] O. Narayan (private communication).

Part 2

FORCE PROPAGATION IN A GRANULAR PILE

CHAOS VOLUME 9, NUMBER 3 SEPTEMBER 1999

Jamming and static stress transmission in granular materials

M. E. Cates
Department of Physics and Astronomy, University of Edinburgh, JCMB King's Buildings, Mayfield Road, Edinburgh EH9 3JZ, Great Britain

J. P. Wittmer[a)]
Départment de Physique des Matériaux, Université Lyon I & CNRS, 69622 Villeurbanne Cedex, France

J.-P. Bouchaud and P. Claudin
Service de Physique de l'Etat Condensé, CEA, Ormes des Merisiers, 91191 Gif-sur-Yvette Cedex, France

(Received 11 January 1999; accepted for publication 15 April 1999)

We have recently developed some simple continuum models of static granular media which display "fragile" behavior: They predict that the medium is unable to support certain types of infinitesimal load (which we call "incompatible" loads) without plastic rearrangement. We argue that a fragile description may be appropriate when the mechanical integrity of the medium arises adaptively, in response to a load, through an internal jamming process. We hypothesize that a network of force chains (or "granular skeleton") evolves until it can just support the applied load, at which point it comes to rest; it then remains intact so long as no incompatible load is applied. Our fragile models exhibits unusual mechanical responses involving hyperbolic equations for stress propagation along fixed characteristics through the material. These characteristics represent force chains; their arrangement expressly depends on the construction history. Thus, for example, we predict a large difference in the stress pattern beneath two conical piles of sand, one poured from a point source and one created by sieving. © *1999 American Institute of Physics.* [S1054-1500(99)00303-1]

Granular materials are microscopically heterogenous. Despite this, it is natural to search for continuum models that can describe their static and dynamic behavior. The problem of granular statics implicitly requires knowledge of the construction history of the medium. At the microscopic scale, the construction history determines exactly where every grain is, and how it has been deformed from its original shape. Given this information, the microscopic forces follow from the local contact mechanics. But such a microscopic description of granular materials is, in practical terms, impossible and unlikely to be a useful guide to their macroscopic behavior. For example it is often assumed that if elasticity governs the local contact mechanics, the continuum behavior of the assembly as a whole must be elastic. This may be unjustified: the physics of the granular assembly involves additional, strongly nonlinear physics—namely, whether each contact is actually present or not. If, as we believe, the contact network is an adaptive structure that has organized itself to support the specific load applied during construction, it may obey continuum equations quite unlike those of conventional elastic or elastoplastic media.

I. INTRODUCTION

In this paper we consider assemblies of cohesionless rough particles, whose rigidity is sufficient that individual particle deformations remain always small. Such assemblies are sometimes argued to be governed by the continuum mechanics of a Hookean elastic solid (perhaps with a very high modulus). But this implicitly assumes that each granular contact is capable equally of supporting tensile as compressive loads. For a cohesionless medium this is certainly untrue: Cohesionless granular assemblies are therefore not elastic.[1] The question is not one of principle, but of degree—how important is the prohibition of tensile forces? This is not completely clear; some would argue[2] that it represents a negligible effect and that an elastic model remains basically sound, so long as the mean stresses in the material remain compressive everywhere. However, a fully elastic granular assembly would be one in which grains were, effectively, *glued* permanently to their neighbors on first contact. Because the packing is microscopically disordered, it is possible that, during subsequent loading *a significant fraction* of such contacts would become tensile, even if the load being applied remains everywhere compressive *on average*. If so, the absence of tensile forces is a major, even dominant, factor.

Notice that the absence of tensile forces is a distinct physical effect from the one addressed by most elastoplastic continuum theories of granular media (see, e.g., Ref. 3). These are like elastic models, but they allow for the fact that the ratio of shear and normal forces at a contact cannot exceed a fixed value determined by a coefficient of static friction; this is usually assumed to translate to a similar condition on the bulk stress components acting across any plane. The resulting plasticity is similar to that arising in metals, for example, and not related to the prohibition of tensile forces: It applies equally for cohesive contacts. Of course, in applying such theories to cohesionless media one should assume that the mean stresses are everywhere compressive: How-

[a)]Electronic mail: jwittmer@dpm.univ-lyon1.fr

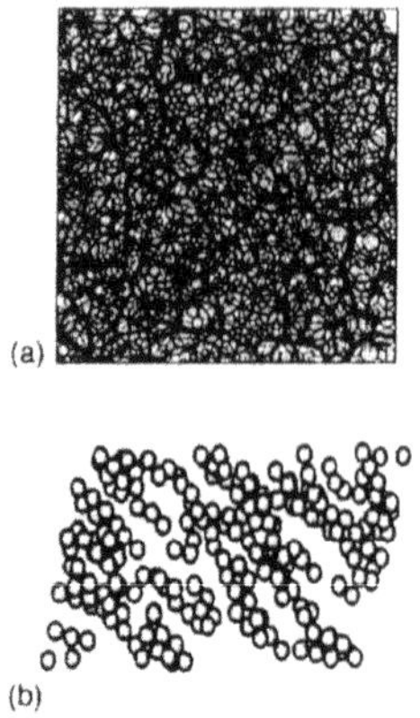

FIG. 1. (a) Granular skeleton in two-dimensional frictional hard spheres by F. Radjai *et al.* (Ref. 6) (Figure courtesy of F. Radjai.) (b) The jamming transition in a sheared colloid. The data are from a computer simulation of a hard sphere colloidal suspension at $\phi=0.54$ which has been strained to $\gamma=0.22$. Shown in the figure are *only* those spheres which have come into very close contact ($\leqslant 10^{-5}$ radius) with at least one neighbor. As seen from the figure, the contact geometry is strongly anisotropic and suggests the formation of "force chains" running top left to bottom right. (The simulation is by J. Melrose, Cavendish Laboratory; the figure is courtesy of him.)

ever, as emphasized above, this constraint *does not* ensure that individual contact forces are *all* compressive as is actually required.

These considerations suggest a physically very different picture of granular media, already well developed and respected in the sphere of discrete modeling.[4–6] In this picture, *nonlinear* physics is dominant, and the contact network of grains is always liable to reorganize as loads are applied: It is an "adaptive structure."[5] The contact network defines a loadbearing granular skeleton shown in Fig. 1(a): This is often thought of as a network of "force chains," or roughly linear chains of strong particle contacts, alongside which the other grains play a relatively minor role in the transmission of stress.[5,6] If these ideas are true, the continuum mechanics of the material has to be thought about afresh. Since this widely accepted picture of force chains implies a microscopically heterogeneous character of the contact network in the material, it is not necessarily obvious that a continuum description of it is possible at all. However, we have in recent years developed continuum models for granular materials which, we now believe, do capture some of the physics of force chains, and of their geometrical dependence on the construction history. This interpretation, which has evolved significantly beyond the empiricism of our early work,[7–9] is developed below.

The proposal that granular assemblies under gravity cannot properly be described by the ideas of conventional elastoplasticity has been opprobriously dismissed in some quarters: We stand accused of ignoring all that is "long and widely known] among geotechnical engineers.[10] However, we are not the first to put forward such a subversive proposal. Indeed workers such as Trollope[11] and Harr[12] have long ago developed ideas of force transfer rules among discrete particles, not unrelated to our own approaches, which yield continuum equations quite unlike those of elastoplasticity.[13–16] More recently, dynamical *hypoplastic* continuum models have been developed[17] which, as explained by Gudehus[18] describe an "anelastic behavior without [the] elastic range, flow conditions and flow rules of elastoplasticity." Our own models, though not explicitly dynamic, are similarly anelastic in a specific manner that we describe as fragile.

In Sec. II below, we describe a generic "jamming" mechanism for the construction of a granular skeleton that, we argue, points toward fragile mechanical behavior. This scenario is related, but not identical, to several other current ideas in the literature on granular media.[1,6,17,19–25] These include the emergence of rigidity by successive buckling of force chains[21] and the concept of mechanical percolation.[17] In particular there is a strong link between fragile media and *isostatic* models of granular assemblies.[1,26] In an isostatic network, the requirement of force balance at the nodes is enough to determine all the forces acting, so these can be calculated without reference to a strain or displacement variable. Isostatic networks require a mean coordination number with a specific critical value ($z=2d$ with d the dimension of space). In this sense, isostatic contact networks are "exceptional," and may appear remote from real granular materials.

However, it is increasingly clear[23,25] that almost all disordered packings of *frictionless* spheres actually approach an isostatic state in the rigid particle limit. Since friction is ignored, there is still a missing link between this result and the physics of real granular media—a link provided by the concept of force chains, as we show below (Sec. II B). More generally, the idea that the granular skeleton could engineer itself to maintain an isostatic or fragile state is closely connected with the concepts of self-organized criticality (SOC)[22] (see also Ref. 27). The concepts provide a generic mechanism whereby an overdamped dynamical system under external forcing can come to rest at a marginally stable (critical) state. In the SOC scenario, this state is characterized by hierarchical (fractal) correlations and large noise effects [compare Fig. 1(a)]. In this article we ignore these complications and describe only our minimalist, noise-free models of the granular skeleton; these represent, in effect, regular arrays of force chains. The effect of noise on the resulting continuum equations is of great interest, but these require a separate discussion, which is made elsewhere.[13,14]

II. COLLOIDS, JAMMING, AND FRAGILE MATTER

A. Colloids

We start by describing a simple model of jamming in a colloid, sheared between parallel plates.[16] This is the simplest plausible scenario in which an adaptive skeleton forms in response to an applied load; we believe it sheds much light on the related problem of dry granular media as discussed in Sec. III below. We will first use it to illustrate some general ideas on the relationship between jamming and fragility.

Consider a concentrated colloidal suspension of hard particles, confined between parallel plates at fixed separation, to which a shear stress is applied [Figs. 1(b) and 2(a)]. Above a certain threshold of stress, this system exhibits en-

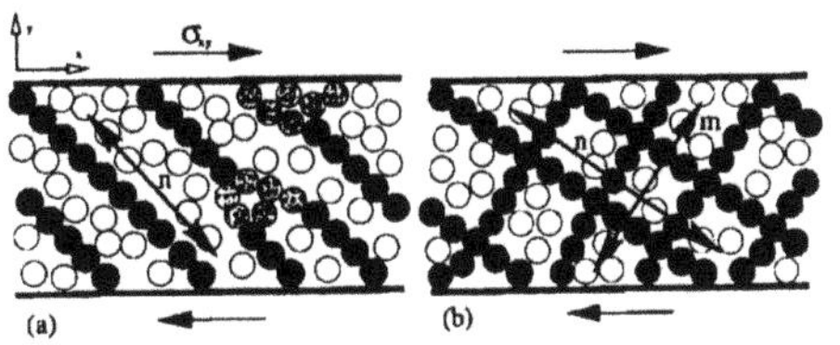

FIG. 2. (a) A jammed colloid (schematic). Black: force chains; gray: Other force-bearing particles; white: Spectators. (b) Idealized rectangular network of force chains.

ters a regime of strong shear thickening; see, e.g., Ref. 28. The effect can be observed in the kitchen, by stirring a concentrated suspension of corn-starch with a spoon. In fact, computer simulations suggest that, at least under certain idealized conditions, the material will jam completely and cease to flow, no matter how long the stress is maintained.[29] In these simulations, jamming apparently occurs because the particles form force chains[4] along the compressional direction [Fig. 1(b)]. Even for spherical particles the lubrication films cannot prevent direct interparticle contacts; once these arise, an array or network of force chains can indeed support the shear stress indefinitely. (We ignore Brownian motion, here and below, as do the simulations; this could cause the jammed state to have finite lifetime.)

To model the jammed state, we start from a simple idealization of a force chain: A linear string of at least three rigid particles in point contact. Crucially, this chain can only support loads *along its own axis* [Fig. 3(a)]: Successive contacts must be collinear, with the forces along the line of contacts, to prevent torques on particles within the chain.[19] Note that neither friction at the contacts, nor particle asphericity, can change this "longitudinal force" rule. (Particle deformability, however, does matter; see Sec. III C below.)

As a minimal model of the jammed colloid, we therefore, take an assembly of such force chains, characterized by a unique director (a headless unit vector) **n**, in a sea of "spectator" particles, and incompressible solvent. This is obviously oversimplified, for we ignore completely any interactions between chains, the deflections caused by weak interactions with the spectator particles, and the fact that there must be some spread in the orientation of the force chains themselves. Nonetheless, with these assumptions, in static equilibrium and with no body forces acting, the pressure tensor p_{ij} (defined as $p_{ij}=-\sigma_{ij}$, with σ_{ij} the usual stress tensor) must obey

$$p_{ij}=P\delta_{ij}+\Lambda n_i n_j. \tag{1}$$

Here P is an isotropic fluid pressure, and $\Lambda(>0)$ a compressive stress carried by the force chains.

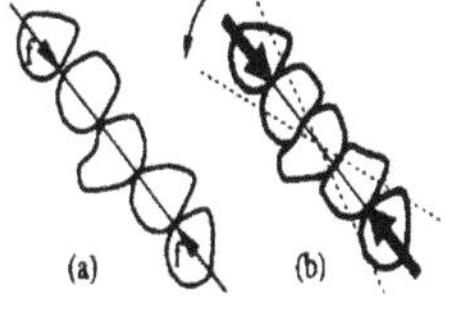

FIG. 3. (a) A force chain of hard particles (any shape) can statically support only longitudinal compression. Note that neither friction at the contacts, nor particle asphericity, can change this "longitudinal force" rule. (b) Finite deformability allows small transverse loads to arise.

B. Jamming and fragile matter

Equation (1) defines a material that is mechanically very unusual. It permits static equilibrium only so long as the applied compression is along **n**; while this remains true, incremental loads (an increase or decrease in stresses at fixed compression axis of the stress tensor) can be accommodated reversibly, by what is (at the particle contact scale) an elastic mechanism. But the material is certainly not an ordinary elastic body, for if instead one tries to shear the sample in a slightly different direction (causing a rotation of the principal stress axes) static equilibrium cannot be maintained without changing the director **n**. Now, **n** describes the orientation of a set of force chains that pick their ways through a dense sea of spectator particles. Accordingly **n** cannot simply rotate; instead, the existing force chains must be abandoned and new ones created with a slightly different orientation. This entails dissipative, plastic, reorganization, as the particles start to move but then re-jam in a configuration that can support the new load. The entire contact network has to reconstruct itself to adapt to the new load conditions; within the model, this is true even if the compression direction is rotated only by an infinitesimal amount.

Our model jammed colloid is thus an idealized example of fragile matter: It can statically support applied shear stresses (within some range), but only by virtue of a self-organized internal structure, whose mechanical properties have evolved directly in response to the load itself. Its incremental response can be elastic only to *compatible* loads; *incompatible* loads (in this case, those of a different compression axis), even if small, will cause finite, plastic reorganizations. The inability to elastically support *some* infinitesimal loads is our chosen technical definition of the term fragile.[16]

We argue that jamming may lead *generically* to mechanical fragility, at least in systems with overdamped internal dynamics. Such a system is likely to arrests as soon as it can support the external load; since the load is only just supported, one expects the state to be only marginally stable. Any incompatible perturbations then force rearrangement; this will leave the system in a newly jammed but (by the same argument) equally fragile state. This scenario is related, but not identical, to several other ideas in the literature.[17,19–23] The link between jamming and fragility is schematically illustrated in Fig. 4.

Now consider again the idealized jammed colloid of [Fig. 2(a)]. So far we allowed for an external stress field (imposed by the plates) but no body forces. What body forces can the system support *without* plastic rotation of the director? Various models are possible. One is to assume that Eq. (1) continues to apply, with $P(\mathbf{r})$ and $\Lambda(\mathbf{r})$ now varying in space. If P is a simple fluid pressure, a localized body

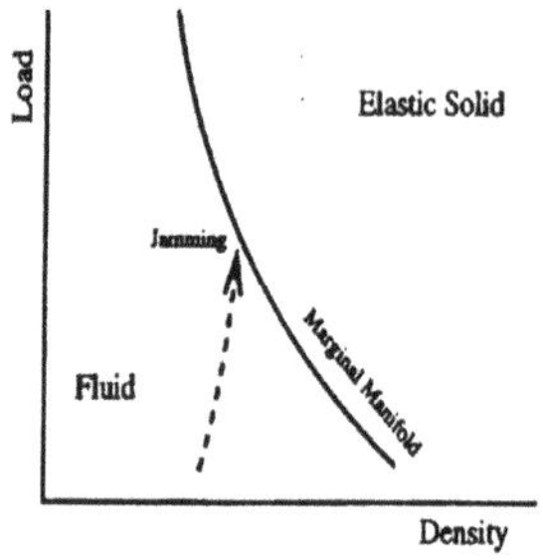

FIG. 4. Schematic "phase diagram" of a jamming system. If, as load or density is increased (dashed arrow), the granular skeleton arrests on first being able to support the applied load, it can remain indefinitely on the "marginal manifold" separating conventional solids from liquidlike packings. Incompatible loads will move the system to another point on the same manifold.

force can be supported only if it acts along **n**. Thus (as in a bulk fluid) no static Green function exists for a general body force. [Note that, since Eq. (1) is already written as a continuum equation, such a Green function would describe the response to a load that is localized in space but nonetheless acts on many particles in some mesoscopic neighborhood.] For example, if the particles in Fig. 2(a) were to become subject to a gravitational force along y, then the existing force chains could not sustain this but would reorganize. Applying the longitudinal force rule, the new shape is easily found to be a catenary, as realized by Hooke,[30] and emphasized by Edwards.[19] On the other hand, a general body force can be supported, in three dimensions, if there are several different orientations of force chain, possibly forming a network or granular skeleton.[4–6,17] A minimal model for this is

$$p_{ij}=\Lambda_1 n_i n_j+\Lambda_2 m_i m_j+\Lambda_3 l_i l_j, \qquad (2)$$

with **n**,**m**,**l** directors along three nonparallel populations of force chains; the Λ's are compressive pressures acting along these. Body forces cause $\Lambda_{1,2,3}$ to vary in space.

We can thus distinguish two levels of fragility, according to whether incompatible loads include localized body forces [*bulk* fragility, e.g., Eq. (1)], or are limited to forces acting at the boundary [*boundary* fragility, e.g., Eq. (2)]. In disordered systems one should also distinguish between macro-fragile responses involving changes in the *mean* orientation of force chains, and the micro-fragile responses of individual contacts. We expect micro-fragility in granular materials (see Ref. 27), although the models discussed here, which exclude randomness, are only macro-fragile; in practice the distinction may become blurred. In any case, these various types of fragility should not be associated too strongly with minimal models such as Eqs. (1) and (2). It is clear that many granular skeletons have a complex network structure where many more than three directions of force chains exist. Such a network may nonetheless be fragile.

Fragility in fact requires any connected granular skeleton of force chains, obeying the longitudinal force rule (LFR), to have a mean coordination number $z=2d$ with d dimension of space [e.g., Fig. 2(b) in two dimensions]. This coordination number describes the skeleton, rather than the medium as a whole; but otherwise it is the same rule as applies for packings of *frictionless* hard spheres. These also obey the LFR—not because of force chains, but because there is no friction. Regular packings of frictionless spheres (which show isostatic mechanics) have been studied in detail recently;[1,13,20] and Moukarzel has argued that disordered frictionless packings of hard spheres are also generically fragile[23] (see also Ref. 25). These arguments appear to depend only on the LFR and the absence of tensile forces, so they should, if correct, equally apply to any granular skeleton that is made of force chains of three or more completely rigid particles.

C. Fixed principal axis model

Returning to the simple model of Eq. (2), the chosen values of the three directors (two in $d=2$) clearly should depend on how the system came to be jammed (its "construction history"). If it jammed in response to a constant external stress, switched on suddenly at some earlier time, one can argue that the history is specified purely by the stress tensor itself. In this case, if one director points along the major compression axis then by symmetry any others should lie at right angles to it [Fig. 2(b)]. Applying a similar argument to the intermediate axis leads to the ansatz that all three directors lie along principal stress axes; this is perhaps the simplest model in $d=3$. One version of this argument links force chains with the fabric tensor,[17] which is then taken coaxial with the stress.[6] (The fabric tensor is the second moment of the orientational distribution function for contacts and/or interparticle forces.)

With the ansatz of perpendicular directors as just described, Eq. (2) becomes a "fixed principle axes" (FPA) model.[8,9,13] Although grossly oversimplified, this leads to nontrivial predictions for the jammed state in the colloidal problem, such as a constant ratio of the shear and normal stresses when these are varied in the jammed regime. Such constancy is indeed reported by Laun[28] in the regime of strong shear thickening; see Ref. 16.

III. GRANULAR MATERIALS

We believe that these simple ideas on jamming and fragility in colloids are equally relevant to cohesionless, dry granular media constructed from hard frictional particles. For although the formation of dry granular aggregates under gravity is not normally described in terms of jamming, it is a closely related process. Indeed, the filling of silos and the motion of a piston in a cylinder of grains both exhibit jamming and stick-slip phenomena associated with force chains; see Ref. 31. And, just as in a jammed colloid, the mechanical integrity of a sandpile entirely disappears as soon as the load (in this case gravity) is removed.

In the granular context, a model like Eq. (2) can be interpreted by saying that a fragile granular skeleton of force chains is laid down at the time when particles are first buried at a free surface (see Fig. 5); so long as subsequent loadings are compatible with this structure, the skeleton will remain intact—if not grain by grain, then at least in its average prop-

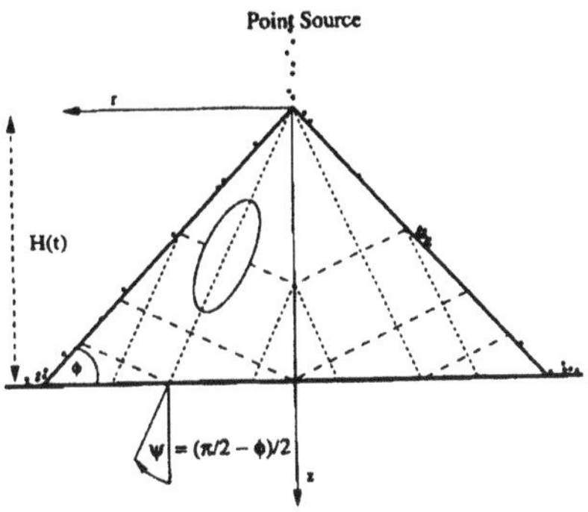

FIG. 5. Definition of normal construction history of a pile. The grains fall down from the point source on the pile and roll down the slopes, which are at the repose angle ϕ. The height of this pile is $H(t)$. Sketch of the geometry of the FPA model: The stress ellipsoid has fixed inclination angle Ψ; its ellipticity varies from zero at the center of the pile to a maximum in the outer region. The outward and inward stress propagation characteristics are indicated by short-dashed and long-dashed lines; these are at right angles and coincident with the principal axed of the stress ellipsoid.

erties. If in addition the skeleton is rectilinear (perpendicular directors) this forces the principal axes to maintain forever the orientation they had close to the free surface (FPA model). However, we do not insist on this last property and other models, which correspond to an oblique family of directors in Eq. (2), will be described below.[9,15,16]

In what follows we review in more detail the nature of our fragile models and the role they might play within a continuum mechanical description of granular media. We will mainly be concerned with the *standard sandpile*, which we define to be a conical pile, constructed by pouring cohesionless hard grains from a point source onto a perfectly rough, rigid support as shown in Fig. 5. We assume that this construction leads to a series of shallow surface avalanches whereby all grains have come to rest, at the point of burial, very close to the free surface of the pile. (Very different conditions may apply for *wedges* of sand; see Ref. 15.) An alternative history is the *sieved pile*, which is a cone created by sieving a series of concentric discs one on top of the other. In the standard sandpile, it is well known that the vertical normal stress has a minimum, not a maximum, beneath the apex.[32,33] A striking feature of our modeling approach is that it not only accounts for this "stress-dip" reasonably well, but predicts that it should be entirely absent for a sieved pile. This proposal is currently being subject to careful experimental verification.[34]

A. Continuum modeling of granular media

The equations of stress continuity express the fact that, in static equilibrium, the forces acting on a small element of material must balance. For a conical pile of sand we have, in $d=3$ dimensions

$$\partial_r\sigma_{rr}+\partial_z\sigma_{zr}=\beta(\sigma_{\chi\chi}-\sigma_{rr})/r, \quad (3)$$

$$\partial_r\sigma_{rz}+\partial_z\sigma_{zz}=g-\beta\sigma_{rz}/r, \quad (4)$$

$$\partial_\chi\sigma_{ij}=0, \quad (5)$$

where $\beta=1$. Here z,r, and χ are cylindrical polar coordinates, with z the downward vertical. We take $r=0$ as a symmetry axis, so that $\sigma_{r\chi}=\sigma_{z\chi}=0$; g is the force of gravity per unit volume; σ_{ij} is the usual stress tensor which is symmetric in i,j. The equations for $d=2$ are obtained by setting $\beta=0$ in Eqs. (3) and (4) and suppressing Eq. (5). These describe a wedge of constant cross section and infinite extent in the third dimension.

The Coulomb law states that, at any point in a cohesionless granular medium, the shear force acting across any plane must be smaller in magnitude than $\tan\phi$ times the compressive normal force. Here ϕ is the angle of friction, a material parameter which, in simple models, is equal to the angle of repose. We accept this here, while noting that (i) ϕ in principle depends on the texture (or fabric) of the medium and hence on its construction history; (ii) for a highly anisotropic packing, the existence of an orientation-independent ϕ is questionable; (iii) the identification of ϕ with the repose angle ignores some complications such as the Bagnold hysteresis effect (which may in turn be coupled to density changes). Setting these to one side, we note that the Coulomb law is an *inequality*: Therefore, when combined with stress continuity, it cannot lead to closed equations for the granular stresses. To close the system of equations, further assumptions are clearly required. One choice is to assume that the material is an elastic continuum wherever it does not violate the Coulomb condition. (This is the simplest possible type of elastoplastic model.) A second choice it to treat the Coulomb condition as though it were an *equality*. This is the basis of the so-called "rigid plastic" approach to granular media. We return to both of these modeling schemes after first describing our own approach.

1. Constitutive relations among stresses

We view cohesionless granular matter as assembly of rigid particles held up by friction. The static indeterminacy of frictional forces can, we argue, then be circumvented by assuming the existence of some local *constitutive relations* (c.r.'s) among components of the stress tensor.[7–9] The c.r.'s among stress components are taken to encode the network of contacts in the granular packing geometry; they therefore, depend explicitly on its construction history. The task is then to postulate and/or justify physically suitable c.r.'s among stresses, of which only one (the *primary* c.r.) is required for systems with two dimensional symmetry, such as a wedge of sand; for a three dimensional symmetric assembly (the conical sandpile) a secondary c.r. is also needed.

The above nomenclature has caused confusion to some commentators on our work. In solid mechanics the term "constitutive relation" normally refers to a material-dependent equation relating stress and strain. In fluid mechanics one has instead equations relating stress and (in the general case of a viscoelastic fluid) strain-rate history. Instead, our models of granular media entail equations relating stress components to one another, in a manner that we take to depend on the construction history of the material. Clearly such equations are intended to describe constitutive properties of the medium: They relate its state of stress to other discernable features of its physical state. We see no alternative to the term constitutive relations for such equations.

In the simplest case, which is the FPA model[8,9] one hypothesizes that, in each material element, the orientation of the stress ellipsoid became "locked" into the material texture at the time when it last came to rest, and does not change in subsequent compatible loadings. This is a bold, simplifying assumption, and it may indeed be far too simple, but it exemplifies the idea of having a local rule for stress propagation that depends explicitly on construction history. At first sight the idea of locking in the principal axes seems to contradict the conception of an adaptive granular skeleton which can rearrange in response to small incremental loads. Remember though that this locking in only applies for compatible loads—those which the existing skeleton can support. Any incompatible load will cause reorganization. We, therefore, require that any incompatible loads are specified in defining the construction history of the material.

For the standard sandpile geometry (see Fig. 5), where the material comes to rest on a surface slip plane, such loads do not in fact arise after material is buried. The FPA constitutive hypothesis then leads to the following primary c.r. among stresses:

$$\sigma_{rr}=\sigma_{zz}-2\tan\phi\sigma_{zr}, \tag{6}$$

where ϕ is the angle of repose. Eq. (6) is algebraically specific to the case of a standard sandpile created from a point source by a series of avalanches along the free surface. The conceptual basis of the FPA model is not so narrow: Indeed, we applied it already to jammed colloids in Sec. II. More generally the FPA model is arguably the simplest possible choice for a history-dependent c.r. among stresses; but this does not mean it will be sensible in all geometries.

A consequence of Eq. (6) for a standard sandpile, is that the major principal axis everywhere bisects the angle between the vertical and the free surface. It should be noted that in cartesian coordinates, the FPA model reads

$$\sigma_{xx}=\sigma_{zz}-2\,\mathrm{sign}(x)\tan\phi\sigma_{xz}, \tag{7}$$

where $x=\pm r$ is horizontal. From Eq. (7), the FPA constitutive relation is seen to be discontinuous on the central axis of the pile: the local texture of the packing has a singularity on the central axis which is reflected in the stress propagation rules of the model. (This is physically admissible since the centerline separates material which has avalanched to the left from material which has avalanched to the right.) The paradoxical requirement, on the centerline, that the principal axes are fixed simultaneously in two different directions has a simple resolution: The stress tensor turns out to be isotropic there (see Fig. 5). The constitutive singularity leads to an "arching" effect for the standard sandpile, as previously put forward by Edwards and Oakeshott[19] and others.[11,35]

The FPA model is one of a larger class of OSL (for "oriented stress linearity") models in which the primary constitutive relation (in the sandpile geometry) is, in Cartesians

$$\sigma_{xx}=\eta\sigma_{zz}+\mu\,\mathrm{sign}(x)\sigma_{xz}, \tag{8}$$

with η,μ constants. Note that the boundary condition, that the free surface of a pile at its angle of repose ϕ is a slip

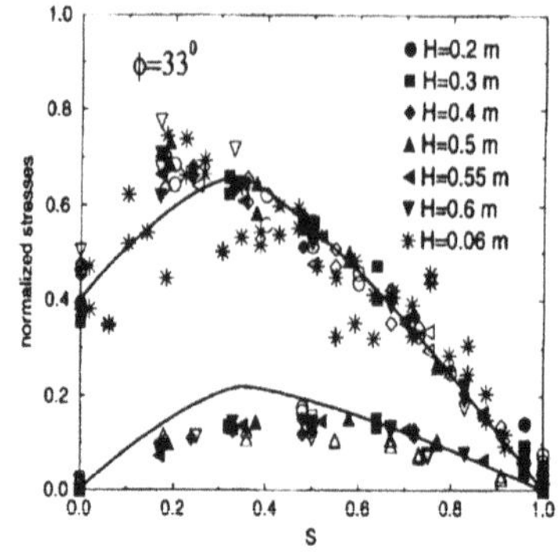

FIG. 6. Comparison of FPA model using a uniaxial secondary closure (Refs. 8 and 9) with scaled experimental data of Smid and Novosad (Ref. 32) and (*) that of Brockbank *et al.* (Ref. 33) which was averaged over three piles. Upper and lower curves denote normal and shear stresses. The data is used to calculate the total weight of the pile which is then used as a scale factor for stresses. The horizontal coordinate $S=r\tan(\phi)/H$ is scaled by the pile height H.

plane, yields one equation linking η and μ to ϕ; thus, for a sandpile geometry, the OSL scheme represents a one-parameter family of primary c.r.'s. The OSL models were developed[9] to explain experimental data on the stress distribution beneath a standard sandpile.[32,33,36] With a plausible choice of secondary c.r. (of which several were tried, with only minor differences resulting), the experimental data (Fig. 6) is found to support models in the OSL family with η close, but perhaps not exactly equal, to unity (the FPA value). This is remarkable, in view of the radical simplicity of the assumptions made.

As explained by Wittmer *et al.*,[8,9] the OSL models, combined with stress continuity [Eq. (8)] yield hyperbolic equations having *fixed characteristic rays* for stress propagation. In fact they are wave equations;[7,9] moreover they are essentially equivalent to Eq. (2), with (in general) an oblique triplet of directors (these become mutually orthogonal only in the case of FPA). The constitutive property that OSL models describe is that these characteristic rays (and not, in general, the principal axes) have orientations that are locked in on burial of an element, and do not change when a further compatible load is applied. As demonstrated already in Sec. II, there is every reason to identify such characteristics, in the continuum model, with the mean orientations of force chains in the underlying material.

Note that unless the OSL parameter is chosen so that $\mu=0$, a constitutive singularity on the central axis, as mentioned above for the FPA case, remains. (The characteristics are asymmetrically disposed about the vertical axis, and invert discontinuously at the centerline $x=0$.) The case $\mu=0$ corresponds to one studied earlier by Bouchaud *et al.*,[7] and of the OSL family it is the only candidate for describing a sieved pile, in which the construction history clearly cannot lead to a constitutive singularity at the axis of the pile. The resulting Bouchaud, Cates, and Claudin (BCC) model could be called a "local Janssen model" in that it assumes *local* proportionality of horizontal and vertical compressive stresses—an assumption which, when applied *globally* to average stresses in a silo, was first made by Janssen.[37] The BCC model predicts a smooth maximum, not a dip, in the pressure beneath the apex of a pile. This is what we expect, therefore, in the case of a sieved pile.[9]

2. Rigid-plastic models

A more traditional, but related, approach is one based on the (Mohr–Coulomb) rigid-plastic model.[38] To find so-called limit state solutions in this model, one postulates that the Coulomb condition is everywhere obeyed *with equality*.[39] That is, through every point in the material there passes some plane across which the shear force is exactly tan ϕ times the normal force. By assuming this, the model allows closure (modulo a sign ambiguity discussed below) of the equations for the stress without invocation of an elastic strain field.

This limit-state analysis of the rigid plastic model is equivalent to assuming a "constitutive relation" (sometimes called "incipient failure everywhere"[9])

$$\sigma_{rr}=\sigma_{zz}\frac{1}{\cos^2\phi}[\sin^2\phi+1 \pm 2\sin\phi\sqrt{1-(\cot\phi\,\sigma_{zr}/\sigma_{zz})^2}], \tag{9}$$

whereas the Coulomb inequality requires only that σ_{rr} lies between the two values ($\pm$) on the right. It is a simple exercise to show that for a sandpile at its repose angle, only one solution of the resulting equations exists in which the sign choice is everywhere the same. This requires the negative root (conventionally referred to as an "active" solution) and it shows a hump, not a dip, in the vertical normal stress beneath the apex. Savage,[10] however, draws attention to a "passive" solution, having a pronounced dip beneath the apex.[39] The passive solution actually contains a pair of matching planes between an inner region where the positive root of Eq. (9) is taken, and an outer region where the negative is chosen. In fact (see Ref. 15) there are more than one such matched solutions, corresponding to different types of discontinuity in the stress (or its gradient) at the matching plane and/or the pile center. Moreover, there is no physical principle that limits the number of matching surfaces; by adding extra ones, a very wide variety of results might be achieved.

It is interesting to compare the mathematics, and physics, of Eq. (9) with that of the OSL models introduced above. The rigid-plastic model yields a local c.r. among stresses; like OSL the resulting equations are hyperbolic. It also exhibits fragility: Because a yield plane passes through every material point, certain incremental loads will cause reorganization. Therefore, anyone who defends the rigid-plastic model as a cogent description of sandpiles cannot reasonably object to these same features in our own models. Conversely, we cannot object in principle to a model in which a Coulomb yield plane passes through every material point. However, we still see no reason why it should be a good model;[9] in particular we cannot see how to make a link between the characteristic rays in this model (which are always load dependent) and the underlying geometry of the contact network in the medium. In contrast, this link arises naturally in the OSL framework.

3. Elastoplastic models

The simplest elastoplastic models assume a material in which a perfectly elastic behavior at small strains is conjoined onto perfect plasticity (the Coulomb condition with equality) at larger ones. In such an approach to the standard sandpile, an inner elastic region connects onto an outer plastic one. In the inner elastic region the stresses obey the Navier equations, which follow from those of Hookean elasticity by elimination of the strain variables. The corresponding strain field is usually not discussed, but tacitly treated as infinitesimal: The high modulus limit is taken. It has been argued that, for a sandpile on a rigid support FPA-like solutions can be found within a purely elastoplastic model, at least in two dimensions.[40] However, since these show a cusp in the vertical stress on the centerline, they imply an infinitesimal displacement field incompatible with a continuous elasticity across the midline.[3] On the other hand, it is possible to obtain a stress dip, in an elastoplastic model, by assuming that the supporting base is not rigid but subject to basal sag.[3] This explanation cannot explain the data of Huntley[33] which involves an indentable (rather than sagging) base.[15] Moreover, it would predict a similar dip for a sieved pile, unlike or own models; experiments on this point are now available, and suggest that indeed no dip is seen in this case.[34]

Objections to the elastoplastic approach to modeling sandpiles can also be raised at a much more fundamental level.[15,16] Specifically, to make unambiguous predictions for the stresses in a sandpile, these models require boundary information which, at least for the simpler models, can be given no clear physical meaning or interpretation. We return to this point below.

B. Fragile vs elastoplastic descriptions

1. Problems defining an elastic strain

In the FPA model and its relatives, strain variables are not considered. No elastic modulus enters, and there is no intrinsic stress scale. The resulting predictions for a conical pile therefore obey what is usually called radial stress-field (RSF) scaling. Formally one has for the stresses at the base

$$\sigma_{ij}=ghs_{ij}(r/ch), \tag{10}$$

where h is the pile height, $c=\cot\alpha$ and s_{ij} a reduced stress: α is the angle between the free surface and the horizontal so that for a pile at repose, $\alpha=\phi$. This form of RSF scaling, which involves only the forces at the base,[41] might be called the "weak form" and is distinct from the "strong form" in which Eq. (10) is obeyed also with z (an arbitrary height from the apex) replacing h (the overall height of the pile). Our OSL models obey both forms; only the weak form has been tested directly by experiment but it is well-confirmed in many systems (Smid and Novosad,[32] Huntley[33]).

The observation of RSF scaling, to experimental accuracy, suggests that elastic effects *need not be considered ex-*

plicitly. This does not of itself rule out elastic or elastoplastic behavior which, at least in the limit of large modulus, can also yield equations for the stress from which the bulk strain fields, and hence also the modulus, cancel. (Note that it is tempting, but entirely wrong, to assume that a similar cancellation occurs at the *boundaries* of the material; we return to this below.) The cancellation of bulk strain fields in elastoplastic models disguises a serious problem in their application to the standard sandpile and related geometries.[15,16] The difficulty is this: There is no obvious definition of strain or displacement for such a construction history. To define a physical displacement or strain field, one requires a *reference state*. In (say) a triaxial strain test (see, e.g., Ref. 42) an initial state is made by some reproducible procedure, and strains measured from there. The elastic part is identifiable in principle, by removing the applied stresses (maintaining an isotropic pressure) and seeing how much the sample recovers its shape. In contrast, a pile constructed by pouring grains onto its apex is not made by a plastic and/or elastic deformation from some initial reference state of the same continuous medium. The problem of the missing reference state occurs whenever the solidity of the body itself arises purely because of the load applied. Thus, for the jammed colloid considered in Sec. II above, the unloaded state is simply a fluid. For the sandpile, it is grains floating freely in space. On cannot satisfactorily define an elastic strain with respect to either of these reference states.

A route to defining a strain variable does however exist,[2] so long as one ignores the fact that tensile forces are prohibited. In effect, one assumes that when grains of sand arrive at the free surface, each one forms permanent or "glued" elastic contacts with its neighbors;[16] this contact network can then, by assumption, elastically support arbitrary incremental loads. This is an admissible physical hypothesis, though contradictory to our own hypothesis of an adaptive, fragile granular skeleton. We do not yet know which hypothesis is more correct; the test of this lies in experiment. (It does not lie in a sociological comparison of how physicists and engineers approach their work, as offered by Savage.[3])

If the "glued pile" model is correct, then a strain variable is defined from the relative displacement that has occurred between adjoining particles since the moment they first were glued together.[2] However, the resulting displacement field, found by integrating the strain, is unlikely to be single valued. Put differently, if a glued assembly is created under gravity and then gravity is switched off, it will revert to a state in which there are residual elastic strains throughout the material, even though there is now no body force acting [Fig. 9(a)]. This is because the particle contact network was itself created under partially loaded conditions. Many elastic and elastoplastic calculations, such as all those reviewed by Savage,[3] entirely ignore the problem of quenched stresses, and therefore embody an implausible "floating model" of a sandpile shown in Fig. 9(b).

Note that these effects do not become small when the limit of a large modulus is taken; the quenched stresses remain of order the stress that was acting during formation, and can take both signs (tensile as well as compressive).[23] So, if one creates a glued pile under gravity and then slowly switches off the body force, tensile forces will arise long before g has gone to zero. In this sense, the response to gravity of a cohesionless pile is completely nonlinear. Correspondingly, in an unglued pile, no smooth deformation can connect the state of a pile created under gravity with an unloaded state of the same contact network: As the load is removed, such a pile will undergo large-scale reorganization.

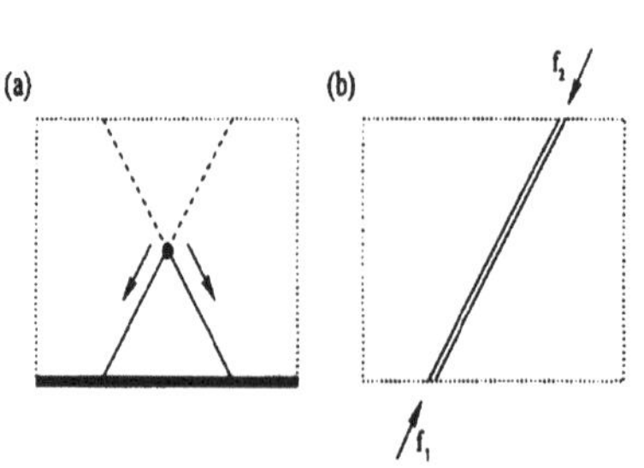

FIG. 7. (a) The response to a localized force is found by resolving it along characteristics through the point of application, propagating along those which do not cut a surface on which the relevant force component is specified. For a pile under gravity, propagation is only along the downward rays. (b) Admissible boundary conditions cannot specify separately the force component at both ends of the same characteristic. If these forces are unbalanced (after allowing for body forces), static equilibrium is lost.

2. Boundary conditions and determinacy in hyperbolic models

Models, such as OSL, that assume local constitutive equations among stresses provide hyperbolic differential equations for the stress field. Accordingly, if one specifies a zero-force boundary condition at the free (upper) surface of a granular aggregate on a rough rigid support, then any perturbation arising from a small extra body force (in two dimensions, for simplicity) propagates along two characteristics passing through this point. In the OSL models these characteristics are, moreover, straight lines. Therefore the force at the base can be found simply by summing contributions from all the body forces as propagated along two characteristic rays onto the support; the sandpile problem is, within the modeling approach by Bouchaud *et al.*[7] and Wittmer *et al.*,[8,9] mathematically well posed. There is no need to consider any elastic strain field and the paradoxes concerning its definition in cohesionless poured sand, discussed above, do not arise.

Note that in principle, one could have propagation also along the "backward" characteristics [see Fig. 7(a)]. This is forbidden since these cut the free surface; any such propagation can only arise in the presence of a nonzero surface force, in violation of the boundary conditions. Therefore, the fact that the propagation occurs only along downward characteristics is not related to the fact that gravity acts downward; it arises because we know already the forces acting at the free surface (they are zero). Suppose we had instead an inverse problem: A pile on a bed with some unspecified overload at the top surface, for which the forces acting at the base had been measured. In this case, the information from the known forces could be propagated along the *upward* characteristics to find the unknown overload. More generally, in OSL models, each characteristic ray will cut the surface of a (convex) patch of material at two points; the sum of the forces along the ray at the two ends must then be balanced by the longitudinal component of the body force integrated along the ray [see Fig. 7 (b)]. These models are thus "boundary fragile."

In three dimensions, the mathematical structure of these models is somewhat altered,[7] but the conclusions are basically unaffected. The propagation of stresses is governed by a Green's function which is the response to a localized overload; OSL models predict that for (say) sand in a horizontal bed, the maximum response at the base is not directly beneath a localized overload but on a ring of finite radius (proportional to the depth) with this as its axis.[7] (This could be difficult to test cleanly because of noise effects, but there are related consequences for stress–stress correlations which are discussed in Ref. 13 and 14.) On the other hand, for different geometries, such as sand in a bin, the stress propagation problem is not well-posed even with hyperbolic equations, unless something is known about the interaction between the medium and the sidewalls. But by assuming a constant ratio to shear and normal forces at the walls, further interesting predictions can be made, for example that the total weight increment measured at the base of a cylindrical silo, in response to an overload on the top, is a nonmonotonic function of the height of the fill.[43] These predictions represent clear signatures of hyperbolic stress propagation and, if confirmed experimentally, would be hard to explain by other means.

3. The problem of elastic indeterminacy

The well-posedness of the standard sandpile is not shared be models involving the elliptic equations for an elastic body. For such a material, the stresses throughout the body can be solved only if, at all points on the boundary, either the force distribution or a displacement field is specified.[44] Accordingly, once the zero-stress boundary condition is applied at the free surface, nothing can in principle be calculated unless either the forces or the displacements at the base are known (and the former amounts to specifying in advance the solution of the problem). The problem does not arise from any uncertainty about what to do mathematically: One should specify a displacement field at the base. Difficulties nonetheless arise if, as we argued above, no physical significance can be attributed to this displacement field for cohesionless poured sand.

To give a specific example, consider the static equilibrium of an elastic cone of finite modulus, which is placed in an unstressed state (without gravity) onto a completely rough, rigid surface; gravity is then switched on. This generates a pressure distribution with a smooth parabolic hump as in Fig. 8(a). (The roughness can crudely be represented by a set of pins.) Starting from any initial configuration, another can be generated by pulling and pushing parts of the body horizontally across the base (i.e., changing the displacements there); if this is rough, the new state will still be pinned and will achieve a new static equilibrium. This will generate a stress distribution, across the supporting surface and within the pile, that differs from the original one. Indeed, if a large enough modulus is now taken (at fixed forces), this procedure allows one to generate arbitrary differences in the stress

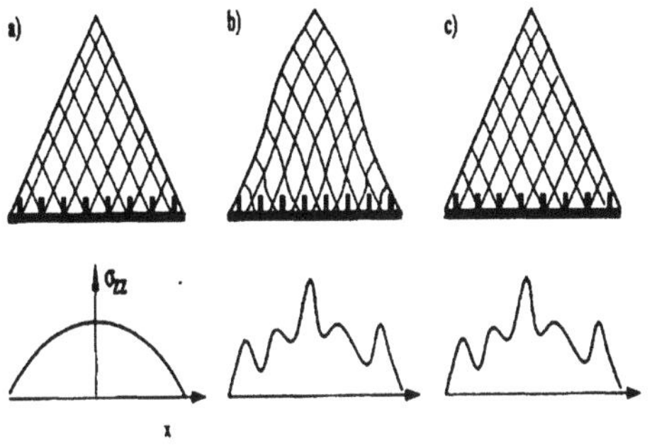

FIG. 8. Starting from an elastic cone or wedge on a rough support, any initial stress distribution can be converted to another by displacements with respect to the rough "pinning" surface (a)→(b). Taking the limit of a high modulus (b)→(c) at fixed surface forces, an arbitrary stress field remains, while recovering the initial shape of the cone and satisfying the free surface boundary conditions. This shows the physical character of "elastic indeterminacy" for an elastic or elastoplastic body on a rough support.

distribution while generating neither appreciable distortions in the shape of the cone, nor any forces at its free surface. This corresponds to a limit $Y\to\infty$, $\underline{u}\to 0$ at fixed $Y\underline{u}$ where Y is the modulus and $\underline{u}$ the displacement field at the base.

Analogous remarks apply to any simple elastoplastic theory of sandpiles, in which an elastic zone, in contact with part of the base, is attached at matching surfaces to a plastic zone. A natural presumption for the standard sandpile might be that $Y\underline{u}=0$ (that is, the basal displacements vanish before the high modulus limit is taken). This is consistent with the glued pile interpretation of elastic models—one assumes that glue also firmly attaches grains to the support as they arrive. However, the same interpretation, as shown above, also requires explicit consideration of quenched stresses (see Fig. 9). Note in any case that elastic and elastoplastic predictions

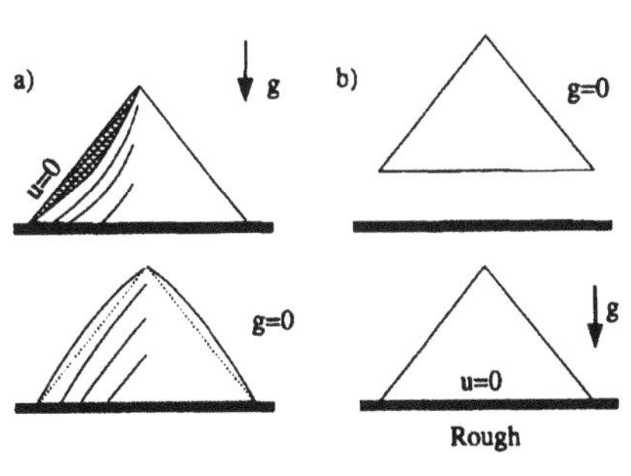

FIG. 9. (a) Quenched stresses in an *elastic* sandpile. Layers are added in a state of zero stress to a pre-existing, gravitationally loaded pile. Such a pile (if gravity is removed) will spring into a new shape, characterized by a nonzero internal stress field which includes tensile stresses. These require rearrangements if the grains are cohesionless; the response to gravity is intrinsically nonlinear. (b) The "floating" model of a sandpile. An unstrained, isotropic elastoplastic cone is brought into contact with a rough surface and gravity then switched on. This is the only construction history we can think of that completely avoids quenched stresses in the formation of the pile.

for the sandpile are indeterminate, in a rigorous mathematical sense, if the $Y \to \infty$ limit is taken before the basal displacements $\underline{u}$ have been specified.

Experiments (reviewed in detail in Cates *et al.*[15]) report that for sandpiles on a rough rigid support, the forces on the base can be measured reproducibly; and, although subject to statistical fluctuations on the scale of several grains, do not vary too much among piles constructed in the same way. In contrast, for any simple elastic or elastoplastic model that does not include a specification of the basal displacements, there is a very large indeterminacy in the predicted stress distribution, even after averaging over any statistical fluctuations. An elastoplastic modeller who believes that the experiments measure something well-defined is then obliged to explain why and how the basal displacements (even if infinitesimal) are fixed by the construction history. Note that basal sag[3,10] is *not* a candidate for the missing mechanism, since it does not resolve the elastic indeterminacy in these models; the latter arises primarily from the *roughness*, rather than the rigidity, of the support. An alternative view is that of Evesque,[45] who directly confronts the issue of elastic indeterminacy and seemingly concludes that the experimental results themselves *are and must be indeterminate*; he argues that the external forces acting on the base of a pile are effectively chosen at will, rather than actually measured, by the experimentalist (see also Ref. 46). To what extent this viewpoint is based on experiment, and to what extent on an implicit presumption in favor of elastoplastic theory, is to us unclear.

C. Crossover from fragile to elastic regimes

We have emphasized above the very different modeling assumptions of the fragile and elast(oplast)ic approaches to granular media. However, we have recently shown that hyperbolic fragile behavior can be recovered from an elastoplastic description by taking a strongly anisotropic limit.[16] Moreover, the crossover between elastic and hyperbolic behavior, at least for one simple model of the granular skeleton,[16] is controlled by the *deformability* of the granular particles. For simplicity in this section, we restrict attention to the FPA model.

The FPA model describes, by definition, a material in which the shear stress must vanish across a pair of orthogonal planes fixed in the medium—those normal to the (fixed) principal axes of the stress tensor. According to the Coulomb inequality (which we also assume) the shear stress must also be less than $\tan\phi$ times the normal stress, across planes oriented in all other directions. Clearly this combination of requirements can be viewed as a limiting case of an elastoplastic model with an anisotropic yield condition

$$|\sigma_{tn}| \leq \sigma_{nn} \tan\Phi(\theta), \tag{11}$$

where θ is the angle between the plane normal $\mathbf{n}$ and the vertical (say) and $\mathbf{t}\cdot\mathbf{n}=0$. An anisotropic yield condition should arise, in principle, in any material having a nontrivial fabric, arising from its construction history. The limiting choice corresponding to the FPA model for a sandpile is $\Phi(\theta)=0$ for $\theta=(\pi-2\phi)/4$ (this corresponds to planes where $\mathbf{n}$ lies parallel to the major principal axis), and $\Phi(\theta)=\phi$ otherwise. (There is no separate need to specify the second, orthogonal plane across which shear stresses vanish, since this is assured by the symmetry of the stress tensor.) By a similar argument, all other OSL models can also be cast in terms of an anisotropic yield condition, of the form $|\sigma_{tn}-\sigma_{nn}\tan\Psi(\theta)| \leq \sigma_{nn}\tan\Phi(\theta)$ where $\Phi(\theta)$ vanishes, and $\Phi(\theta)$ is finite for two values of θ. (This fixes a *nonzero* ratio of shear and normal stresses across certain special planes.)

At this purely phenomenological level there is no difficulty in connecting hyperbolic models smoothly onto *highly anisotropic* elastoplastic descriptions. Specifically, consider a medium having an orientation-dependent friction angle $\Phi(\theta)$ that does not actually vanish, but is instead very small ($\leq \epsilon$, say) in a narrow range of angles (say of order ϵ) around $\theta=(\pi-2\phi)/4$, and approaches ϕ elsewhere. (One interesting way to achieve the required yield anisotropy is to have a strong anisotropy in the *elastic* response, and then impose a uniform yield condition to the strains, rather than stresses.)

Such a material will have, in principle, mixed elliptic/hyperbolic equations of the usual elastoplastic type. The resulting elastic and plastic regions must nonetheless arrange themselves so as to obey the FPA model to within terms that vanish as $\epsilon \to 0$. If ϵ is small but finite, then for this elastoplastic model the results will depend on the basal boundary condition, but only through these higher order corrections to the leading (FPA) result. Thus, although elastoplastic models do suffer from elastic indeterminacy (they require a basal displacement field to be specified), the extent of the influence of the boundary condition on the solution depends on the model chosen. Strong enough (fabric-dependent) anisotropy, in an elastoplastic description, might so constrain the solution that it is *primarily the granular fabric* (hence the construction history) and only minimally the boundary conditions which actually determine the stresses in the body. For models such as that given above there is a well-defined limit where the indeterminacy is entirely lifted, hyperbolic equations are recovered, and it is quite proper to talk of local stress propagation "rules" determined by the construction history of the material. Our continuum modeling framework is based precisely on these assumptions.

The crossover just outlined can also be understood directly in terms of the micromechanics of force chains, at least within the simplified picture developed in Sec. II. We consider a regular lattice of force chains [see Fig. 2 (b)], for simplicity rectangular (the FPA case), which is fragile if the chains can support only longitudinal forces. As mentioned in Sec. III C, this is true so long as such paths consist of linear chains of rigid particles, meeting at frictional point contacts: The forces on all particles within each chain must then be collinear, to avoid torques. This imposes the (FPA) requirement that there are no shear forces across a pair of orthogonal planes normal to the force chains themselves. Suppose now a small degree of particle deformability is introduced.[16] This relaxes *slightly* the collinearity requirement, but only because the point contacts are now flattened [see Fig. 3 (b)]. The ratio ϵ of the maximum transverse load to the normal one will, therefore, vanish with (some power of) the mean compression. This yield criterion applies only across two special planes; failure across others is governed by some smooth yield requirement (such as the ordinary Coulomb condition: The ratio of the principal stresses lies between given limits). The granular skeleton just described, which was fragile in the limit of rigid grains, is now governed by a strongly anisotropic elastoplastic yield criterion of precisely the kind described above. This indicates how a packing of frictional, deformable rough particles, displaying broadly conventional elastoplastic features when the deformability is significant, can approach a fragile limit when the limit of a large modulus is taken. (It does not prove that *all* packings become fragile in this limit.) Conversely it shows how a packing that is basically fragile in its response to a gravitational load could nonetheless support very small incremental deformations, such as sound waves, by an elastic mechanism.

The question of whether sandpiles are better described as fragile, or as ordinarily elastoplastic, remains open experimentally. To some extent it may depend on the question being asked. However, we have argued, on various grounds, that in calculating the stresses in a pile under gravity a fragile description may lie closer to the true physics.

IV. CONCLUSIONS

The jammed state of colloids, if it indeed exists in the laboratory, has not yet been fully elucidated by experiment. It is interesting that even very simple models, such as Eq. (1), can lead to nontrivial and testable predictions (such as the constancy of certain measured stress ratios). Such models suggest an appealing conceptual link between jamming, force chains, and fragile matter.[15,16] However, further experiments are needed to establish the degree to which they are useful in describing real colloids.

For granular media, the existence of tenuous force-chain skeletons is clear;[4–6,17,24] the question is whether such skeletons are fragile. Several theoretical arguments have been given, above and elsewhere, to suggest that this may be the case, at least in the limit of rigid particles. Moreover, simulations show strong rearrangement under small changes of compression axis; the skeleton is indeed "self-organized."[5,6] Experiments also suggest cascades of rearrangement[27,31] in response to small disturbances. These findings are consistent with the fragile picture.

The standard sandpile (a conical pile formed by pouring onto a rough rigid support) has played a central role in our discussions. From the perspective of geotechnical engineering, the problem of calculating stresses in the humble sandpile may appear to be of only of marginal importance. The physicist's view is different: The sandpile is important, because it is one of the simplest problems in granular mechanics imaginable. It, therefore, provides a test-bed for existing models and, if these show shortcomings, may suggest ideas for improved physical theories of granular media.

Given the present state of the data, a conventional elastoplastic interpretation of the experimental results for sandpiles may remain tenable; more experiments are urgently required.[34] In the mean time, a desire to keep using tried-and-tested modeling strategies until these are demonstrably proven ineffective is quite understandable. We find it harder to accept the suggestion[10] that anyone who questions the general validity of traditional elastoplastic thinking is somehow uneducated.

In summary, we have discussed a new class of models for stress propagation in granular matter. These models assume local propagation rules for stresses which depend on the construction history of the material and which lead to hyperbolic differential equations for the stresses. As such, their physical basis is substantially different from that of conventional elastoplastic theory. Our approach predicts fragile behavior, in which stresses are supported by a granular skeleton of force chains that respond by finite internal rearrangement to certain types of infinitesimal load. Obviously, such models of granular matter might be incomplete in various ways. Specifically we have discussed a possible crossover to elastic behavior at very small incremental loads, and to conventional elastoplasticity at very high mean stresses (when significant particle deformations arise). However, we believe that our approach, by capturing at the continuum level at least some of the physics of force chains, may offer important insights that lie beyond the scope of previous continuum modeling strategies.

ACKNOWLEDGMENTS

The authors thank R. C. Ball, E. Clement, C. S. and M. J. Cowperthwaite, S. F. Edwards, P. Evesque, P.-G. de Gennes, G. Gudehus, J. Goddard, D. Levine, C. E. Lester, J. Melrose, S. Nagel, F. Radjai, J.-N. Roux, J. Socolar, C. Thornton, L. Vanel, and T. A. Witten for illuminating discussions. Work funded in part by EPSRC (UK) GR/K56223 and GR/K76733.

[1] E. Guyon, S. Roux, A. Hansen, D. Bideau, J.-P. Troadec, and H. Crapo, Rep. Prog. Phys. **53**, 373 (1990).
[2] For recent discussions see P.-G. de Gennes, Physica A **261**, 267 (1998); P. Evesque and P.-G. de Gennes, Compt. Rend. de l'Acad. des Sci. **326**, 761 (1998).
[3] S. B. Savage, in *Proceeding of Physics of Dry Granular Media, Cargese, France, October 1997*, edited by H. J. Herrmann, J. P. Hovi, and S. Luding (NATO Advanced Study Institute, Kluwer, 1998), pp. 25–96.
[4] P. Dantu, Ann. Ponts Chaussees **4**, 144 (1967).
[5] C. Thornton, in KONA Powder and Particle **15**, 103–108 (1997); C. Thornton and G. Sun, in *Numerical Methods in Geotechnical Engineering*, edited by I. Smith (Balkema, Rotterdam, 1994), pp. 143–148.
[6] F. Radjai, D. E. Wolf, M. Jean, and J. J. Moreau, Phys. Rev. Lett. **90**, 61 (1998); F. Radjai, S. Roux, and J. J. Moreau, *ibid.* **90** (to be published).
[7] J.-P. Bouchaud, M. E. Cates, and P. Claudin, J. Phys. I **5**, 639 (1995).
[8] J. P. Wittmer, P. Claudin, M. E. Cates, and J.-P. Bouchaud, Nature (London) **382**, 336 (1996).
[9] J. P. Wittmer, P. Claudin, and M. E. Cates, J. Phys. I **7**, 39 (1997).
[10] S. B. Savage, in *Proceedings of the 3rd International Conference on Powders and Grains, Durham NC 18-23 May 1997*, edited by R. P. Behringer and J. T. Jenkins (Balkema, Rotterdam, 1997), pp. 185–194.
[11] D. H. Trollope, in *Rock Mechanics in Engineering Practice*, edited by K. G. Stagg and O. C. Zienkiewicz (Wiley, New York, 1968), Chap. 9, pp. 275–320.
[12] M. E. Harr, *Mechanics of Particulate Media* (McGraw-Hill, New York, 1977).
[13] J.-P. Bouchaud, P. Claudin, M. E. Cates, and J. P. Wittmer, in *Physics of Dry Granular Media*, edited by H. J. Herrmann, J. P. Hovi, and S. Luding (NATO Advanced Study Institute, Kluwer, 1997), pp. 97–122.
[14] P. Claudin, J.-P. Bouchaud, M. E. Cates, and J. P. Wittmer, Phys. Rev. E **57**, 4441 (1998).

[15] M. E. Cates, J. P. Wittmer, J.-P. Bouchaud, and P. Claudin, Philos. Trans. R. Soc. London, Ser. A **356**, 2535 (1998).
[16] M. E. Cates, J. P. Wittmer, J.-P. Bouchaud, and P. Claudin, Phys. Rev. Lett. **81**, 1841 (1998).
[17] D. Kolymbas, Archive of Applied Mechanics **61**, 143 (1991); D. Kolymbas and W. Wu, in *Modern Approaches to Plasticity*, edited by D. Kolymbas (Elsevier, Rotterdam, 1985), pp. 213–223.
[18] G. Gudehus, in *Proceedings of the 3rd International Conference on Powders and Grains, Durham NC, USA 18–23 May 1997*, edited by R. P. Behringer and J. T. Jenkins (Balkema, Rotterdam, 1997), pp. 169–183.
[19] S. F. Edwards and R. B. S. Oakeshott, Physica D **38**, 88 (1989); S. F. Edwards and C. C. Mounfield, Physica A **226**, 1 (1996); **226**, 12 (1996) **226**, 25 (1996); S. F. Edwards, **249**, 226 (1998).
[20] R. C. Ball and S. F. Edwards (in preparation); S. F. Edwards and D. V. Grinero, Physica A **263**, 545–553 (1999).
[21] S. Alexander, Phys. Rep. **296**, 65 (1998); P. S. Cundall and O. D. L. Strack, in *Mechanics of Granular Materials: New Models and Constitutive Relations*, edited by J. T. Jenkins and M. Satake (Elsevier, Rotterdam, 1983).
[22] P. Bak, *How Nature Works; The Science of Self-Organized Criticality* (Oxford University Press, Oxford, 1997).
[23] C. Moukarzel, Phys. Rev. Lett. **81**, 1634 (1998); to appear in *Proceedings of Rigidity Theory and Applications, Traverse City MI, June 14–18 1998*, Fundamental Material Science Series, Plenum (cond-mat/9807004).
[24] C. H. Liu, S. R. Nagel, D. A. Schecter, S. N. Coppersmith, S. Majumdar, and T. A. Witten, Science **269**, 513 (1995).
[25] A. V. Thachenko and T. A. Witten, cond-mat/9811171.
[26] S. Ouaguenouni and J. N. Roux, Europhys. Lett. **39**, 117 (1997).
[27] P. Claudin and J.-P. Bouchaud, Phys. Rev. Lett. **78**, 231 (1997).
[28] H. M. Laun, J. Non-Newtonian Fluid Mech. **54**, 87 (1994).
[29] R. S. Farr, J. R. Melrose, and R. C. Ball, Phys. Rev. E **55**, 7203 (1997); R. C. Ball, J. R. Melrose, Adv. Colloid Interface Sci. **59**, 19 (1995); J. R. Melrose and R. C. Ball, Europhys. Lett. **32**, 535 (1995); see also E. Aharonov, D. Sparks, cond-mat/9812204.
[30] R. Hooke, *A Description of Helioscopes, and Some Other Instruments* (John Martin, London, 1676). See anagram on title page concerning "The true Mathematical and Mechanical form of all manner of Arches."; reproduced as Fig. 2.2 of Ref. 42.
[31] E. Kolb, T. Mazozi, J. Duran, and E. Clement (in preparation); P. Claudin and J.-P. Bouchaud, Granular Matter **1**, 71 (1998).
[32] J. Smid and J. Novosad, in *Proceedings of 1981 Powtech Conf., Ind. Chem. Eng. Symp.* **63**, D3V 1-12 (1981).
[33] R. Brockbank, J. M. Huntley, and R. C. Ball, J. Phys. II **7**, 1521 (1997).
[34] L. Vanel, D. W. Howell, E. Clément, and R. P. Behringer (in preparation).
[35] D. H. Trollope and B. C. Burman, Geotechnique **30**, 137 (1980).
[36] T. Jokati and R. Moriyama, J. Soc. Powder Technol. Jpn. **60**, 184 (1979).
[37] H. A. Janssen, Zeitschrift des Vereines Deutscher Ingenieure **39**, 1045 (1895).
[38] R. M. Nedderman, *Statics and Kinematics of Granular Materials* (Cambridge University Press, Cambridge, 1992).
[39] Savage's claim that active and passive solutions can be "generally regarded as bounds between which other states can exist, i.e., when the material is behaving in an elastic or elastoplastic manner" is disproved by counterexample in Ref. 15.
[40] F. Cantelaube and J. D. Goddard, in *Proceedings of the 3rd International Conference on Powders and Grains, Durham NC, USA 18–23 May 1997*, edited by R. P. Behringer and J. T. Jenkins (Balkema, Rotterdam, 1997), pp. 231–234.
[41] P. Evesque, J. Phys. I **7**, 1305 (1997).
[42] D. Muir Wood, *Soil Behaviour and Critical State Soil Mechanics* (Cambridge University Press, Cambridge, 1990).
[43] P. Claudin, Ph.D. thesis, Université de Paris XI 1999.
[44] L. D. Landau and E. M. Lifshitz, *Theory of Elasticity*, 3rd ed. (Oxford, Pergamon, 1986).
[45] P. Evesque, private communication 1996.
[46] P. Evesque and S. Boufellouh, in *Proceedings of the 3rd International Conference on Powders and Grains, Durham NC, USA 18–23 May 1997*, edited by R. P. Behringer and J. T. Jenkins (Balkema, Rotterdam, 1997), pp. 295–298.

Statistical Mechanics of Stress Transmission in Disordered Granular Arrays

S. F. Edwards and D. V. Grinev
Cavendish Laboratory, University of Cambridge, Madingley Road, Cambridge CB3 OHE, United Kingdom
(Received 28 September 1998; revised manuscript received 2 March 1999)

We give a statistical-mechanical theory of stress transmission in disordered arrays of rigid grains with perfect friction. Starting from the equations of microscopic force and torque balance we derive the fundamental equations of stress equilibrium. We illustrate the validity of our approach by solving the stress distribution of a homogeneous and isotropic array. [S0031-9007(99)09454-5]

PACS numbers: 83.70.Fn, 45.05.+x

Transmission of stress and statistics of force fluctuations in static granular arrays are fundamental, but unresolved, problems in physics [1,2]. Despite several theoretical attempts [3,4] and a vast engineering literature [5,6] the connectivity of granular media is still poorly understood at a fundamental level. In this Letter we propose a theory of stress transmission in disordered arrays of rigid cohesionless grains with perfect friction. A real granular aggregate (e.g., sand or soil) is a very complex object [6]. However, simple models are easier to comprehend, and extra complexities can always be incorporated subsequently. In our case the rigid grain paradigm provides a crucial starting point from which to appreciate the theoretical physics of the problem. We model the granular material as an assembly of discrete rigid particles whose interactions with their neighbors are localized at pointlike contacts. Therefore the description of the network of intergranular contacts is essential for the understanding of force transmission in granular assemblies. Grain α exerts a force on grain β at a point $\mathcal{R}^{\alpha\beta} = \mathbf{R}^{\alpha} + \mathbf{r}^{\alpha\beta}$. The contact is a point in a plane whose normal is $\mathbf{n}^{\alpha\beta}$. The vector $\mathbf{R}^{\alpha}$ is defined by

$$\mathbf{R}^{\alpha} = \frac{\sum_{\beta} \mathcal{R}^{\alpha\beta}}{z}, \tag{1}$$

so that $\mathbf{R}^{\alpha}$ is the centroid of contacts, and hence

$$\sum_{\beta} \mathbf{r}^{\alpha\beta} = 0, \qquad \mathbf{R}^{\alpha\beta} = \mathbf{r}^{\alpha\beta} - \mathbf{r}^{\beta\alpha}, \tag{2}$$

where z is the number of contacts per grain and $\sum_{\beta}$ means summation over the nearest neighbors. Hence $\mathbf{R}^{\alpha}$, $\mathbf{r}^{\alpha\beta}$, and $\mathbf{n}^{\alpha\beta}$ are geometrical properties of the aggregate under consideration and the other shape specifications do not enter. Friction is assumed to be infinite and the geometry is frozen after the deposition and cannot be changed by applying or removing an external force on the boundaries. In a static array Newton's equations of intergranular force and torque balance are satisfied. Balance of force around the grain α requires

$$\sum_{\beta} f_i^{\alpha\beta} = g_i^{\alpha}, \tag{3}$$

$$f_i^{\alpha\beta} + f_i^{\beta\alpha} = 0, \tag{4}$$

where $i = 1, 2, 3$ are Cartesian indices and $\mathbf{g}^{\alpha}$ is the external force acting on grain α. Further on in this Letter $\mathbf{g}$ is used also for the external forces at the boundaries.

The equation of torque balance is

$$\sum_{\beta} \epsilon_{ikl} f_k^{\alpha\beta} r_l^{\alpha\beta} = C_i^{\alpha}. \tag{5}$$

The centroid of the contact points need not coincide with the centroid of the forces, e.g., the center of mass of a solid grain, but we will assume it is so in order to keep the analysis simple so that we ensure that the macroscopic stress tensor is symmetric, at least on average. It can be verified that, for the intergranular forces in the static array to be determined by these equations, the coordination number $z = 3$ in 2D and $z = 4$ in 3D is required. In this paper we present the results for the 2D case only. The microscopic version of stress analysis is to determine all of the intergranular forces, given the applied force, torque loadings on each grain, and geometric specification of a granular array. The number of unknowns per grain is $zd/2$. Required force and torque equations give $d + \frac{d(d-1)}{2}$ constraints. The system of equations for the intergranular forces is complete when the coordination number is $z_m = d + 1$. Theory which confirms this observation has been proposed for periodic arrays of grains with perfect and zero friction [7]. It is clear that the coordination number z controls the connectivity of granular media. We will assume that z is indeed 3 in 2D, for this is surely the simplest situation, and one which is physically possible. The ultimate goal, however, is to determine the macroscopic stress tensor at every point of a granular array, given external loadings and geometric specification. The macroscopic state of stress is a function of the distribution of contact forces. For any aggregate of discrete grains subjected to external loading, the transmission of stress from one point to another can occur only via the intergranular contacts. Therefore it is clear that the network of contacts determines the distribution of stresses within the granular array. The network of contacts is determined by the deposition history of the sample and the external loading on the boundaries. We define the tensorial force moment

$$S_{ij}^{\alpha} = \sum_{\beta} f_i^{\alpha\beta} r_j^{\alpha\beta}, \tag{6}$$

0031-9007/99/82(26)/5397(4)$15.00

which is the microscopic analog of the stress tensor. With $C_i^\alpha = 0$, S_{ij}^α will be symmetric. Our goal is to find a complete system of equations for the macroscopic stress tensor σ_{ij}, which is supported by the given network of contacts in the state of mechanical equilibrium. Given an assembly of discrete grains which is represented by a very complex network of contacts, we associate a continuous medium to have continuously distributed properties. Such spatial smoothing or coarse-graining can be accomplished formally. To obtain the macroscopic stress tensor from the tensorial force moment to the macroscopic stress tensor we coarse-grain, i.e., average it over an ensemble of configurations,

$$\sigma_{ij}(\mathbf{r}) = \left\langle \sum_{\alpha=1}^{N} S_{ij}^\alpha \delta(\mathbf{r} - \mathbf{R}^\alpha) \right\rangle. \tag{7}$$

In the simplest cases of isotropic and homogeneous arrays this is not a problem. The difficulties appear when the array under consideration is anisotropic or inhomogeneous. Within the confines of this paper we explore only the simplest cases. The number of equations required equals the number of independent components of a symmetric stress tensor $\sigma_{ij} = \sigma_{ji}$ and is $\frac{d(d+1)}{2}$. At the same time, the number of equations available is d. These are vector equations of the stress equilibrium $\frac{\partial \sigma_{ij}}{\partial x_j} = g_i$ which have their origin in Newton's second law. Therefore we have to find $\frac{d(d-1)}{2}$ equations, which possess the information from Newton's third law, to complete and solve the system of equations which governs the transmission of stress in a granular array. Thus in 2D there is one missing equation, and we derive it in terms of the geometry of the system.

Given the set of Eqs. (3)–(5) we can write the probability functional for the intergranular force $f_i^{\alpha\beta}$ as

$$P\{f_i^{\alpha\beta}\} = \mathcal{N} \delta\left(\sum_\beta f_i^{\alpha\beta} - g_i^\alpha\right) \times \delta\left(\sum_\beta \epsilon_{ikl} f_k^{\alpha\beta} r_l^{\alpha\beta}\right) \times \delta(f_i^{\alpha\beta} + f_i^{\beta\alpha}), \tag{8}$$

where the normalization, $\mathcal{N}$, which is a function of a configuration, is

$$\mathcal{N}^{-1} = \int \prod_{\alpha,\beta} P\{f_i^{\alpha\beta}\} \mathcal{D} f^{\alpha\beta}. \tag{9}$$

The probability of finding the tensorial force moment S_{ij}^α on grain α is

$$P\{S_{ij}^\alpha\} = \int \prod_{\alpha,\beta} \delta\left(S_{ij}^\alpha - \sum_\beta f_i^{\alpha\beta} r_j^{\alpha\beta}\right) P\{f_i^{\alpha\beta}\} \mathcal{D} f^{\alpha\beta}, \tag{10}$$

where $\int \mathcal{D} f^{\alpha\beta}$ implies integration over all functions $f^{\alpha\beta}$, since all the constraints on $f^{\alpha\beta}$ have been experienced. We assume that the $z = d + 1$ condition means that the integral exists.

We exponentiate the delta functions and thus introduce the set of conjugate fields ζ_{ij}^α, γ_i^α, λ_i^α, and $\eta_i^{\alpha\beta}$.

$$P\{S_{ij}^\alpha\} = \int \prod e^{iA} \mathcal{D} f^{\alpha\beta} \mathcal{D} \zeta^\alpha \mathcal{D} \gamma^\alpha \mathcal{D} \lambda^\alpha \mathcal{D} \eta^{\alpha\beta}, \tag{11}$$

where A is

$$A = \sum_\alpha \zeta_{ij}^\alpha \left(S_{ij}^\alpha - \sum_\beta f_i^{\alpha\beta} r_j^{\alpha\beta}\right) + \gamma_i^\alpha \left(\sum_\beta f_i^{\alpha\beta} - g_i^\alpha\right) + \lambda_i^\alpha \left(\sum_\beta \epsilon_{ikl} f_k^{\alpha\beta} r_l^{\alpha\beta}\right) + \eta_i^{\alpha\beta}(f_i^{\alpha\beta} + f_i^{\beta\alpha}). \tag{12}$$

The λ^α field term gives the symmetry of S_{ij}^α. After integrating out the $f^{\alpha\beta}$ and $\eta^{\alpha\beta}$ fields we find the following linear equation for the conjugate fields:

$$\zeta_{ij}^\alpha r_j^{\alpha\beta} - \gamma_i^\alpha = \zeta_{ik}^\beta r_k^{\beta\alpha} - \gamma_i^\beta. \tag{13}$$

The idea of the conjugate fields method is to use these equations for the ζ field in the stress probability functional, in order to derive the complete system of equations for the stress tensor. The general solution of the above equation is a sum of the ζ^0 field which is the particular solution and depends on γ, and ζ^* which is the complementary function

$$\zeta_{ij}^\alpha = \zeta_{ij}^{\alpha 0} + \zeta_{ij}^{\alpha *}. \tag{14}$$

If we introduce the fabric tensor F_{ij}^α and its inverse M_{ij}^α:

$$F_{ij}^\alpha = \sum_\beta R_i^{\alpha\beta} R_j^{\alpha\beta}, \qquad M_{ij}^\alpha = (F^\alpha)_{ij}^{-1}, \tag{15}$$

we can rewrite Eq. (13) in the following form:

$$\zeta_{ij}^\alpha = M_{jl}^\alpha \sum_\beta R_l^{\alpha\beta}(\gamma_i^\alpha - \gamma_i^\beta) + M_{jl}^\alpha \sum_\beta R_l^{\alpha\beta} r_k^{\beta\alpha}(\zeta_{ik}^\beta - \zeta_{ik}^\alpha), \tag{16}$$

which permits an expansion based on the first two terms, i.e.,

$$\zeta_{ij}^\alpha \simeq M_{jl}^\alpha \sum_\beta R_l^{\alpha\beta}(\gamma_i^\alpha - \gamma_i^\beta) + M_{jl}^\alpha \sum_\beta R_l^{\alpha\beta} r_k^{\beta\alpha} M_{km}^\beta \sum_\delta R_m^{\beta\delta}(\gamma_i^\beta - \gamma_i^\delta) + \dots. \tag{17}$$

The stress-force equation.—The next step is to integrate out the γ field which gives us the stress-force equation

$$\sum_\beta M_{jl}^\alpha R_l^{\alpha\beta} S_{ij}^\alpha - \sum_\beta M_{jl}^\beta R_l^{\beta\alpha} S_{ij}^\beta = g_i^\alpha. \tag{18}$$

So by expanding β quantities about α quantities we reach

$$\nabla_j \sigma_{ij} + \nabla_j \nabla_k \nabla_m K_{ijkl} \sigma_{lm} + \ldots = g_i, \tag{19}$$

where $K_{ijkl} = \langle R_i^{\alpha\beta} R_j^{\alpha\beta} R_k^{\alpha\beta} R_l^{\alpha\beta} \rangle$ and gives a correction to the standard equation of stress equilibrium $\nabla_j \sigma_{ij} = g_i$ at the length scale which is small compared to the size of the system. These corrections correspond to the presence of the second, third, etc., nearest neighbors and topological correlations and must vanish in the $k \to 0$ limit.

The stress-geometry equation.—So far the well-known equations have been derived by using the information from Newton's second law. But we still have unused information from Newton's third law. By integrating out the ζ^* field we obtain the missing equations we are looking for. Let us consider that part of the Eq. (11) which contains the ζ^* field,

$$\int e^{i \sum_{\alpha=1}^{N} \zeta_{ij}^{\alpha *} S_{ij}^{\alpha}} \delta(\zeta_{ij}^{\alpha *} r_j^{\alpha\beta} - \zeta_{ij}^{\beta *} r_j^{\beta\alpha}) \prod_{\alpha=1}^{N} \mathcal{D} \zeta_{ij}^{\alpha *}, \tag{20}$$

and

$$\zeta_{ij}^{\alpha *} r_j^{\alpha\beta} - \zeta_{ij}^{\beta *} r_j^{\beta\alpha} = 0. \tag{21}$$

Counting the degrees of freedom in this equation we note that it can give only two (scalar) equations in 2D and three in 3D. Using $\mathbf{R}^{\alpha\beta}$ and $\mathbf{Q}^{\alpha\beta} = \mathbf{r}^{\alpha\beta} + \mathbf{r}^{\beta\alpha}$, we can get these equations by projecting the vector equation into

$$\zeta_{ij}^{\alpha *} \sum_\beta R_i^{\alpha\beta} R_j^{\alpha\beta} + \sum_\beta (\zeta_{ij}^{\alpha *} - \zeta_{ij}^{\beta *}) r_j^{\beta\alpha} R_i^{\alpha\beta} = 0, \tag{22}$$

$$\zeta_{ij}^{\alpha *} \sum_\beta Q_i^{\alpha\beta} R_j^{\alpha\beta} + \sum_\beta (\zeta_{ij}^{\alpha *} - \zeta_{ij}^{\beta *}) r_j^{\beta\alpha} Q_i^{\alpha\beta} = 0. \tag{23}$$

It should be emphasized that the system under consideration is disordered and therefore $\mathbf{Q}^{\alpha\beta} \neq 0$ (whereas for a honeycomb periodic array $\mathbf{Q}^{\alpha\beta} = 0$). Assuming as before that $\zeta_{ij}^{\alpha *} - \zeta_{ij}^{\beta *}$ gives rise to gradient terms we can exponentiate (22) and (23) by parametric variables ϕ^α and ψ^α,

$$\int e^{i \sum_{\alpha=1}^{N} \zeta_{ij}^{\alpha *} (S_{ij}^{\alpha} - \phi^\alpha F_{ij}^\alpha - \psi^\alpha G_{ij}^\alpha)} \prod_\alpha \mathcal{D} \zeta_{ij}^{\alpha *} \mathcal{D} \phi^\alpha \mathcal{D} \psi^\alpha, \tag{24}$$

where F_{ij}^α is given by (15) and

$$G_{ij}^\alpha = \frac{1}{2} \left(\sum_\beta Q_i^{\alpha\beta} R_j^{\alpha\beta} + Q_j^{\alpha\beta} R_i^{\alpha\beta} \right). \tag{25}$$

After integrating out the $\zeta^{\alpha *}$, ϕ^α, and ψ^α fields, we find the following equation for S_{ij}^α:

$$\begin{vmatrix} S_{11}^\alpha & F_{11}^\alpha & G_{11}^\alpha \\ S_{22}^\alpha & F_{22}^\alpha & G_{22}^\alpha \\ S_{12}^\alpha & F_{12}^\alpha & G_{12}^\alpha \end{vmatrix} = 0. \tag{26}$$

Note that although there are explicit forms generalizing (26) in 3D, these are more complex algebraically as a consequence of the higher coordination number. F_{ij}^α and G_{ij}^α will depend on configuration and averaging (26) is quite complex. The simplest array will have $\mathbf{Q}^{\alpha\beta}$ orthogonal to $\mathbf{R}^\alpha$, i.e., if $\mathbf{R}^{\alpha\beta} = (X^{\alpha\beta}, Y^{\alpha\beta})$, then $\mathbf{Q}^{\alpha\beta} = (Y^{\alpha\beta}, -X^{\alpha\beta})$. It follows that F_{ij}^α and G_{ij}^α can be written as

$$F_{ij}^\alpha = \begin{pmatrix} 1 & 0 \\ 0 & 1 \end{pmatrix}, \qquad G_{ij}^\alpha = \begin{pmatrix} \sin\theta^\alpha & \cos\theta^\alpha \\ \cos\theta^\alpha & -\sin\theta^\alpha \end{pmatrix}. \tag{27}$$

Then Eq. (26) can be rewritten as

$$S_{22}^\alpha - S_{11}^\alpha = 2 S_{12}^\alpha \tan\theta^\alpha. \tag{28}$$

Thus if we are given S_{12}^α, the probability of finding $S_{11}^\alpha - S_{22}^\alpha$ is

$$P\{S_{11}^\alpha - S_{22}^\alpha | S_{12}^\alpha\} = \frac{2}{\pi} \frac{|S_{12}^\alpha|}{(S_{11}^\alpha - S_{22}^\alpha)^2 + (S_{12}^\alpha)^2}. \tag{29}$$

Mathematically it is more convenient to introduce $(\xi^\alpha)^2 = (S_{11}^\alpha - S_{22}^\alpha)^2 + (S_{12}^\alpha)^2$ and determine the probability of finding $S_{11}^\alpha - S_{22}^\alpha$ given ξ^α.

$$P\{S_{11}^\alpha - S_{22}^\alpha | \xi^\alpha\} = \frac{1}{2\pi} \frac{1}{\sqrt{(\xi^\alpha)^2 - (S_{11}^\alpha - S_{22}^\alpha)^2}}. \tag{30}$$

The mean values of $S_{11}^\alpha - S_{22}^\alpha$ and S_{12}^α are zero; hence we predict, rather obviously, hydrostatic pressure. However, notice that we are able to predict the fluctuations away from hydrostatic pressure, and would do more on correlations if one could find a pathway to measure them.

Another approach to deal with the system of discrete equations [(18) and (26)] is to solve it for S_{ij}^α, and then average the solution. This way seems to be feasible at least for the simplest granular systems (e.g., isotropic or periodic arrays) and may provide deep insight into the origins of the non-Gaussian statistics of stress fluctuations [8]. In complex cases this can be accomplished in some approximation, or by using computer simulations. By applying Fourier transformation to (18) and (26) one can obtain $S_{ij}(\mathbf{k})$. The macroscopic stress tensor is obtained by averaging over the distribution of angles θ^α

$$i\sigma_{11}(\mathbf{k}) = \langle S_{11}(\mathbf{k}) \rangle_\theta = \frac{g_1(k_1^3 + 3k_2^2 k_1) + g_2(k_2^3 - k_1^2 k_2)}{|\mathbf{k}|^4}, \tag{31}$$

$$i\sigma_{22}(\mathbf{k}) = \langle S_{22}(\mathbf{k}) \rangle_\theta = \frac{g_2(k_2^3 + 3k_1^2 k_2) + g_1(k_1 k_2^2 - k_1^3)}{|\mathbf{k}|^4}, \tag{32}$$

$$i\sigma_{12}(\mathbf{k}) = \langle S_{12}(\mathbf{k})\rangle_\theta = \frac{(g_1k_2 - g_2k_1)(k_2^2 - k_1^2)}{|\mathbf{k}|^4}, \quad (33)$$

where $|\mathbf{k}|^2 = k_1^2 + k_2^2$ and $\sigma_{ij}(\mathbf{r}) = \int \sigma_{ij}(\mathbf{k})e^{i\mathbf{kr}}d^3\mathbf{k}$. By doing the inverse Fourier transformation one can see that the macroscopic stress tensor is diagonal. There must also be constraints on the permitted configurations (due to the absence of tensile forces) which are not so easily expressed, for they affect each grain in the form

$$S_{ik}^{\alpha}M_{kl}^{\alpha}R_l^{\alpha\beta}n_i^{\alpha\beta} > 0, \quad (34)$$

which has not yet been put into continuum equations other than $\mathrm{Det}\sigma > 0$ and $\mathrm{Tr}\sigma > 0$.

Discussion.—In this Letter we have derived the fundamental equations of stress equilibrium,

$$\nabla_j\sigma_{ij} + \nabla_j\nabla_k\nabla_m K_{ijkl}\sigma_{lm} + \ldots = g_i, \quad (35)$$

$$P_{ijk}\sigma_{jk} + \nabla_j T_{ijkl}\sigma_{kl} + \nabla_j\nabla_l U_{ijklm}\sigma_{km} + \ldots = 0. \quad (36)$$

In order to solve these equations one needs to know the geometric quantities K_{ijkl}, P_{ijk}, T_{ikl}, and U_{ijklm}. In practice details of the distribution of intergranular contacts are not known in advance, but should be obtained from the deposition history of the system or experimental measurements of two-body correlation functions.

If the system is strongly anisotropic (i.e., there exists a preferred direction characterized by some angle ϕ) and $\tan\theta^{\alpha}$ has an average value $\tan\phi$, then Eq. (26) becomes in the mean-field approximation

$$\sigma_{11} - \sigma_{22} = 2\sigma_{12}\tan\phi, \quad (37)$$

where ϕ is the angle of repose. It is known as the fixed principal axes equation [3], and has been used with notable effect to solve the problem of the stress distribution in sandpiles. Explicit mathematical expressions for the 3D case are more complex, and will be reported elsewhere. The issue of whether the derived system of Eqs. (35) and (36) is robust against the inclusion of real friction, softness of grains, etc., illuminates the existence of a whole array of fascinating theoretical and experimental problems. Other important issues which are not addressed in this Letter are that of stress fluctuations and the response of a granular aggregate to external perturbations. In general, cohesionless granular materials are quasistatic or "fragile" [9], which means that they cannot support certain types of infinitesimal changes in stress without configurational rearrangements.

In conclusion, our theory in its present form gives a simplified, but physical, picture of stress behavior in cohesionless granular media. Further development is needed to make it a predictive tool which could be able to match experimental findings.

We acknowledge financial support from Leverhulme Foundation (S. F. E.), Shell (Amsterdam), and Gonville & Caius College (Cambridge) (D. V. G.). Invaluable discussions with Professor Robin Ball are gratefully acknowledged.

[1] *Granular Matter: An Interdisciplinary Approach,* edited by A. Mehta (Springer-Verlag, New York, 1993); for review see, e.g., H. M. Jaeger, S. R. Nagel, and R. P. Behringer, Rev. Mod. Phys. **68**, 1259 (1996).

[2] C. H. Liu *et al.*, Science **269**, 513 (1995); B. Miller, C. O'Hern, and R. P. Behringer, Phys. Rev. Lett. **77**, 3110 (1996).

[3] J. P. Wittmer, P. Claudin, and M. E. Cates, J. Phys. I (France) **7**, 39 (1997).

[4] S. N. Coppersmith *et al.*, Phys. Rev. E **53**, 4673 (1996).

[5] R. M. Nedderman, *Statics and Kinematics of Granular Materials* (Cambridge University Press, Cambridge, 1992).

[6] D. M. Wood, *Soil Behaviour and Critical State Soil Mechanics* (Cambridge University Press, Cambridge, 1990).

[7] R. C. Ball and D. V. Grinev, cond-mat/9810124.

[8] D. M. Mueth, H. M. Jaeger, and S. R. Nagel, Phys. Rev. E **57**, 3164 (1998).

[9] M. E. Cates, J. P. Wittmer, J. P. Bouchaud, and P. Claudin, Phys. Rev. Lett. **81**, 1841 (1998).

ELASTOPLASTIC ARCHING IN TWO DIMENSIONAL GRANULAR HEAPS

F. CANTELAUBE, A. K. DIDWANIA AND J. D. GODDARD
Department of Applied Mechanics and Engineering Sciences
University of California, San Diego, La Jolla, Ca 92093-0411, USA

Abstract. This paper is a continuation of our previous work [1] on the modeling of the stress distribution in deep two-dimensional granular heaps or "wedges". We amend the previous treatment of symmetric heaps and consider asymmetric heaps or "berms", with one face at the angle of repose. We find continuous families of solutions and asymmetric pressures under symmetric heaps.

1. Introduction - Basic Equations

Several distinct models have been put forth to explain the dip in pressure profile measured at the bottom of granular heaps [1–3]. In Ref. [1], a standard elastoplastic model was employed to treat the symmetric two-dimensional (2D) heap with both faces at the angle of repose. A continuous one-parameter family of solutions was found, but we have since concluded that only three are admissible.

In this paper we present briefly some further calculations based on the same elastoplastic model [1]. After a brief description of the basic equations, we discuss the admissible symmetric solutions for the symmetric heap and consider general solutions for an asymmetric heap.

With cartesian axes x and y oriented, respectively, in the vertical and horizontal directions, we consider a 2D heap or "wedge" defined by $\theta = \tan^{-1}(y/x) \in [\beta', \beta]$ (see Fig. 1). With left face inclined at the angle of repose and right face no more steeply, this interval is given by:

$$\beta' = \phi_\mu - \frac{\pi}{2} < 0, \text{ and } \quad -\beta' \leq \beta \leq \frac{\pi}{2}, \tag{1}$$

H.J. Herrmann et al. (eds.), Physics of Dry Granular Media, 123–128.

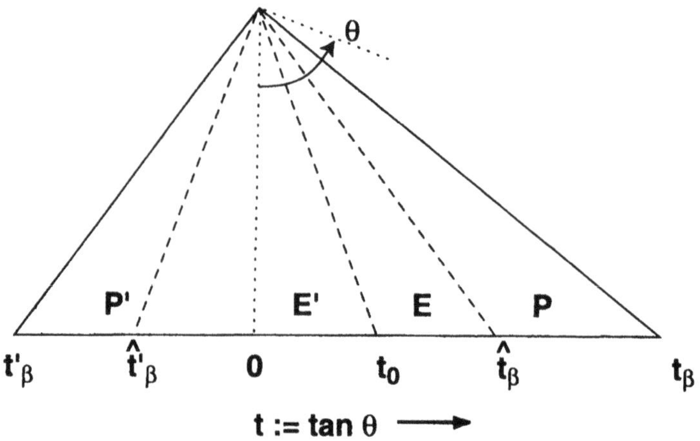

Figure 1. Plastic and elastic zones in the heap.

where ϕ_μ is the is friction angle. With $t = \tan\theta$, we let

$$t_\beta = \tan\beta,\ \hat{t}_\beta = \tan\hat{\beta},\ t_0 = \tan\beta_0,\ t'_\beta = \tan\beta',\ \hat{t}'_\beta = \tan\hat{\beta}'. \quad (2)$$

We identify plastic zones P and P' located near the free surfaces, where $t'_\beta \le t \le \hat{t}'_\beta$ and $\hat{t}_\beta \le t \le t_\beta$, and the core, which consists of two elastic zones E and E' separated by the line $t = t_0$ (see Fig. 1).

As previously discussed in Ref. [1], the basic equations are the equilibrium equations

$$\begin{aligned} \sigma_{xx,x} + \sigma_{xy,y} &= -1, \\ \sigma_{xy,x} + \sigma_{yy,y} &= 0, \end{aligned} \quad (3)$$

where subscripts preceded by commas denote partial derivatives. Identifying a Coulomb yield function as

$$\mathcal{F} = (\sigma_{xx} - \sigma_{yy})^2 + 4\sigma_{xy}^2 - (\sigma_{xx} + \sigma_{yy})^2 \sin^2\phi_\mu, \quad (4)$$

we have $\mathcal{F} < 0$ in the elastic zones, and $\mathcal{F} = 0$ in the plastic zones. In the elastic zones we also have a compatibility restriction on stresses given in Ref. [1].

1.1. ELASTIC ZONES

An admissible set of elastic solutions in the region E is

$$\left.\begin{aligned} \sigma_{xx}/x &= (a_1 - 1) + b_1 t, \\ \sigma_{yy}/x &= (a_2 - 1) + b_2 t, \\ \sigma_{xy}/x &= -b_2 - a_1 t, \end{aligned}\right\} \quad (5)$$

with constants $a_{1,2}$ and $b_{1,2}$. Similar solution holds also in E', involving constants $a'_{1,2}$ and $b'_{1,2}$, say.

1.2. PLASTIC ZONES

We now discuss the solutions in the plastic zone P, the equations for P' involving essentially the same forms. Since all the stress components must vanish at the free surface, suitable linear form for the plastic solutions of Eq. (3), corresponding to the "simple" solutions of [4] (denoted as "FPA" by others [3]), are

$$\begin{aligned}\sigma_{xx}/x &= a_{11}(t/t_\beta - 1),\\ \sigma_{yy}/x &= a_{22}(t/t_\beta - 1),\\ \sigma_{xy}/x &= a_{12}(t/t_\beta - 1),\end{aligned} \tag{6}$$

where the a_{ij} are constants which, by Eqs. (3), obey

$$\begin{aligned}-a_{11} + a_{12}t_\beta^{-1} &= -1,\\ -a_{12} + a_{22}t_\beta^{-1} &= 0.\end{aligned} \tag{7}$$

Eqs. (7) together with $\mathcal{F} = 0$ give

$$\frac{a_{11}}{a_{22}} = \frac{1 + \sin^2\phi_\mu \pm \frac{2}{\sin\beta}\sqrt{\sin(\phi_\mu - \frac{\pi}{2} + \beta)\sin(\phi_\mu + \frac{\pi}{2} - \beta)}}{\cos^2\phi_\mu}, \tag{8}$$

which is valid for $\frac{\pi}{2} - \phi_\mu \leq \beta \leq \frac{\pi}{2} + \phi_\mu$, the lower limit representing $-\beta'$ [see Eq. (1)]. When we set $\beta = -\beta'$, the radical in Eq. (8) vanishes and Eqs. (7-8) yield the simple solution for a symmetric heap [1]

$$\left.\begin{aligned}a_{11} &= 1 + \cos^2\beta',\\ a_{22} &= \sin^2\beta',\\ a_{12} &= -\tfrac{1}{2}\sin 2\beta'.\end{aligned}\right\} \tag{9}$$

1.3. STRESS MATCHING

Matching of all stresses at the boundary between zones P and E, $t = \hat{t}_\beta$, yields

$$\left.\begin{aligned}a_{11}(\hat{t}_\beta/t_\beta - 1) &= (a_1 - 1) + b_1\hat{t}_\beta\\ a_{22}(\hat{t}_\beta/t_\beta - 1) &= (a_2 - 1) + b_2\hat{t}_\beta\\ a_{12}(\hat{t}_\beta/t_\beta - 1) &= -b_2 - a_1\hat{t}_\beta\end{aligned}\right\} \tag{10}$$

with similar matching between P' and E'. Matching the normal and shear stresses at $t = t_0$ gives

$$\left.\begin{aligned}(a_1 - 1) + b_1 t_0 &= (a'_1 - 1) + b'_1 t_0,\\ (a_2 - 1) + b_2 t_0 &= (a'_2 - 1) + b'_2 t_0,\\ -b_2 - a_1 t_0 &= -b'_2 - a'_1 t_0.\end{aligned}\right\} \tag{11}$$

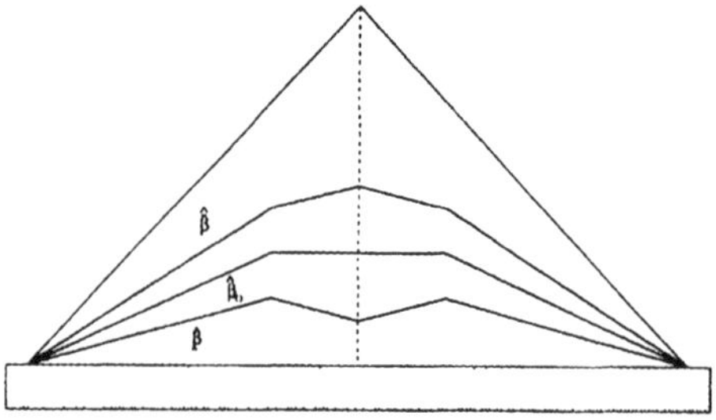

Figure 2. Schematic representation of the three symmetric pressure profiles in a 2D symmetric heap.

However, a discontinuity in the tensile stress parallel to $t = t_0$ may occur. In such a case the elastic compatibility is violated, and we must impose the yield condition $\mathcal{F} = 0$. This point was overlooked in Ref. [1], leading to a one-parameter family of solutions with elastic singularity on the axis of a symmetric heap, a situation which we now rectify.

2. Symmetric Solutions for the Symmetric Heap

For the symmetric heap with both faces at the angle of repose, we have

$$t'_\beta = -t_\beta, \;\; \hat{t}'_\beta = -\hat{t}_\beta, \;\; t_0 = 0, \tag{12}$$

with

$$\beta = \frac{\pi}{2} - \phi_\mu. \tag{13}$$

In this case, Eqs. (12) and (13) simplify Eqs. (10) and (11), which yields exactly three values of $\hat{\beta}$ corresponding to three distinct stress distributions. A schematic representation of the vertical pressure profiles is shown in Fig. 2.

We denote by $\hat{\beta}_0$, $\underline{\hat{\beta}}$ and $\overline{\hat{\beta}}$ those values of $\hat{\beta}$, for which the normal stress distributions exhibit a plateau, an arch (*i.e.* a pressure dip under the apex) or peak (with maximum under the apex), respectively. The plateau is the only one with a purely elastic core, the others possess a singularity at $t = 0$ and represent the extremal states discussed in [1]. We note here that, in addition to these symmetric solutions, there exists asymmetric pressure solutions for the symmetric heap which we do not discuss in this article.

3. Asymmetric Solutions

We can also calculate the stress distribution for asymmetric heaps, which are decribed by $\beta > -\beta'$ in Eq. (1), but now we find continuous families

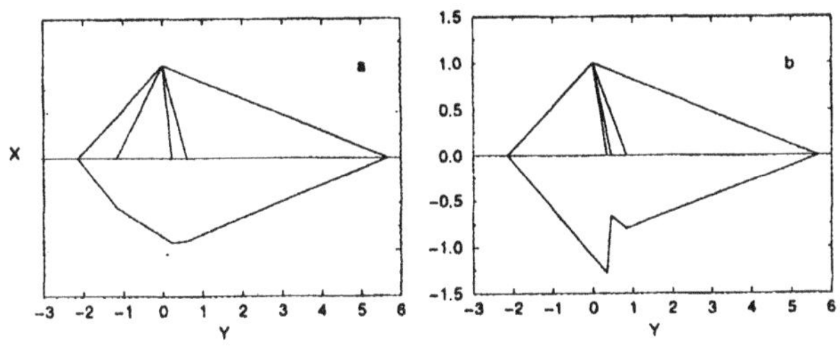

Figure 3. Peaked (a) and arched (b) normal stress distributions at the bottom of a 2D heap with $\hat{\beta}' = -65°$ and $\hat{\beta} = 80°$

of solutions. For example, we show in Fig. 3 two possible normal stress distributions at the bottom of a 2D heap with angle $\beta = 80°$. The top half of the figure displays the heap shape and the zones P', E', E, P, and the bottom half shows the corresponding stress distribution. In Fig. 3 (a) the angles are $\hat{\beta}' = -49°$, $\beta_0 = 12.3°$ and $\hat{\beta} = 31°$, while in Fig. 3 (b) we have $\hat{\beta}' = -19°$, $\beta_0 = 24°$, $\hat{\beta} = 40°$. Figs. 3 (a,b) represent peaked and arched distributions, respectively. In addition to these two types of solutions we also find plateaus. A more detailed discussion of the problem will be given in a later publication [5].

4. Conclusions

The standard elastoplastic model appears fully capable of describing arching in granular heaps, without hypotheses as to new mechanisms of the "stress propagation" in granular media [3]. Although we could adopt a more complex (*e.g.* anisotropic) elastoplastic model, the simple model considered here already exhibits a multiplicity of solutions that cannot be resolved in purely static models.

Acknowledgements. Partial support from the U.S. National Aeronautics and Space Administration (Grant NAG3-1888), the U.S. Air Force Office of Scientific Research (Grant F49620-96-1-0246), and the National Science Foundation (Grant CTS-9510121) is gratefully acknowledged.

References

1. Cantelaube F. and J. D. Goddard, 1997. In R. B. Behringer and J. T. Jenkins, editors, *Powders & Grains*, (Balkema, Rotterdam, 1997).
2. Savage, S. 1997. In R. B. Behringer and J. T. Jenkins, editors, *Powders & Grains*, (Balkema, Rotterdam, 1997)
3. Wittmer J., M. E. Cates and P. Claudin 1997. *J. Physique I France*, **7**, 39-80.
4. Sokolovskii V. V. *Statics of Soil Media*, (Butterworths Scientific Publication, 1960).
5. Cantelaube F., J. D. Goddard and A. Didwania 1997. *in preparation*.

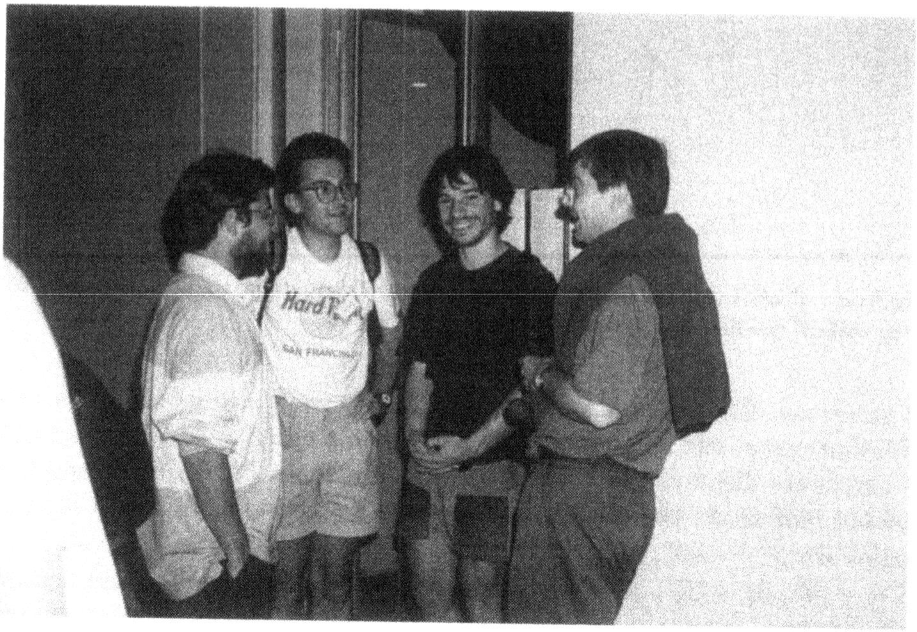

Loic Vanel (left), Thomas Boutreux, Philippe Claudin, and Michael Cates

Stress propagation through frictionless granular material

Alexei V. Tkachenko and Thomas A. Witten
The James Franck Institute, The University of Chicago, Chicago, Illinois 60637
(Received 16 November 1998)

We examine the network of forces to be expected in a static assembly of hard, frictionless spherical beads of random sizes, such as a colloidal glass. Such an assembly is minimally connected: the ratio of constraint equations to contact forces approaches unity for a large assembly. However, the bead positions in a finite subregion of the assembly are underdetermined. Thus to maintain equilibrium, half of the exterior contact forces are determined by the other half. We argue that the transmission of force may be regarded as unidirectional, in contrast to the transmission of force in an elastic material. Specializing to sequentially deposited beads, we show that forces on a given buried bead can be uniquely specified in terms of forces involving more recently added beads. We derive equations for the transmission of stress averaged over scales much larger than a single bead. This derivation requires the ansatz that statistical fluctuations of the forces are independent of fluctuations of the contact geometry. Under this ansatz, the $d(d+1)/2$-component stress field can be expressed in terms of a d-component vector field. The procedure may be generalized to nonsequential packings. In two dimensions, the stress propagates according to a wave equation, as postulated in recent work elsewhere. We demonstrate similar wavelike propagation in higher dimensions, assuming that the packing geometry has uniaxial symmetry. In macroscopic granular materials we argue that our approach may be useful even though grains have friction and are not packed sequentially. [S1063-651X(99)02007-3]

PACS number(s): 45.05.+x, 83.70.Fn

I. INTRODUCTION

The nature of the forces within a static pile of grains has proven more subtle than one might expect. Such a pile is an assembly of many hard, spheroidal bodies that maintain their positions via a balance of gravitational forces and contact forces with their neighbors [1,2]. On the one hand, determining these forces is a prosaic equilibrium problem. Since the number of grains is large, the long-established notions of continuum solid mechanics appear applicable. On the other hand, a pile of grains or beads is not a solid. The forces between beads are more problematic than those between the atoms of a conventional solid. These latter forces are smoothly varying on the scale of the separation and they arise from a potential energy that includes attraction. The forces on a grain are different. First, they vary sharply with interparticle distance, and there is no attraction. Second, the frictional part of a contact force is not determined by the macroscopic positions of the grains. Rather, it depends on how each contact was formed. The resulting macroscopic behavior of the pile is also clearly different from that of a conventional solid. Arbitrarily slight forces can disturb the pile, so that the notion of stable equilibrium is suspect. Despite these complexities, we expect the mechanics of a granular pile to be universal. Hard, round grains appear to form piles of the same nature independent of their composition or detailed shape. We are led to think of these as nondeformable objects that exert normal forces of constraint and transverse forces limited by Coulomb's static friction limit.

Recently, a puzzling discovery has underlined the subtlety of the forces in a conical heap of poured sand [3]. The supporting force under the center, where the pile is deepest, is not maximal. Instead, the maximal force occurs along a circle lying between the edge and the center of the pile. From this circle the force decreases to a *minimum* at the center. In order to explain this puzzling "central dip," a number of inventive approaches have been taken. Some [4] seek to account for the minimum qualitatively by viewing the pile as a stack of concentric "wigwams," whose sloping sides support the load. Others [5] have shown that the central dip is compatible with the conventional continuum mechanics, in which the pile is viewed as a central elastic zone flanked by an outer plastic zone which is at the Coulomb static friction limit. A third group [6–8] has argued that granular material requires a new constitutive law, a homogeneous, local, linear constraint on the stress arising from the packing geometry. We shall call it a "null stress" law. Their proposed law gives a continuum mechanics as simple as that of a liquid or a solid, yet different from either. The hallmark of this law is the hyperbolic equations governing the transmission of forces. Hyperbolic equations, such as the wave equation, obey causality. The wave at the present is unaffected by the future. In the null-stress picture, the vertical direction plays the role of time. Accordingly, forces at a given point in the pile are only influenced by forces above that point. Force *propagates* as in a traveling wave. The transmission is *unidirectional*, in contrast with conventional continuum elasticity. The equations of continuum mechanics are elliptical. According to these, a contact within the pile should be influenced by all the forces above or below it, as sketched in Fig. 1.

A separate approach has given indirect evidence for the unidirectional transmission of forces. Coppersmith's [9] heuristic q model aims to account for the point-to-point variability of the contact forces. It imposes a unidirectional prescription for determining the downward forces from one grain in terms of the forces acting on it from above. Both this model and refinements [10,11] of it yield an exponential falloff of probability for large forces. This exponential falloff agrees well with measured distributions [9,12,13]. This exponential

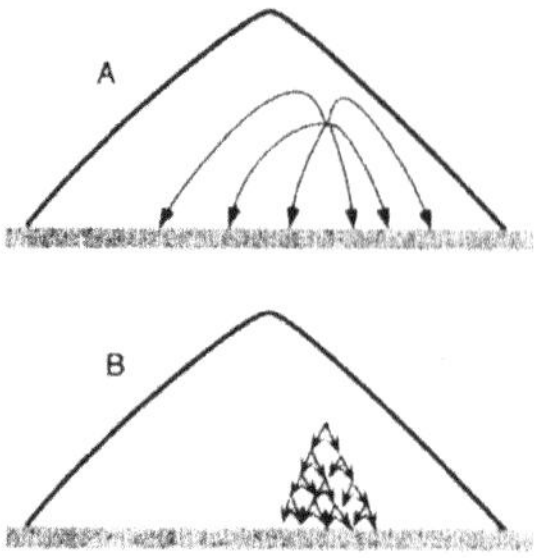

FIG. 1. Contrast between elastic (elliptical) and unidirectional (hyperbolic) force propagation in a sand-pile shaped object. (a) Elastic response to an imposed local downward force. Pictured lines of force represent the current of vertical force, i.e., the stress contracted into a downward unit vector. Near the source, this field is symmetric about a horizontal plane through the object [32]; part of the force is transmitted through points above the source. (b) Unidirectional response to an imposed local downward force. The imposed force is transmitted to neighbors below the source, and is further transmitted to neighbors below these. No force is transmitted to points above the source.

falloff contrasts with the Gaussian falloff expected for a heterogeneous elastic solid.

In this paper we grapple with the relationship between the conventional elastic view and the newer null-stress picture. Our work builds on several recent studies of the relationship between the connectivity of a structure and the force transmission in it. Alexander's recent review [14] has explored the nature of unconstrained degrees of freedom in a minimally connected or isostatic [1] network. Ball and Edwards [15] have explored force transmission through minimally connected networks, assuming a fixed coordination number for all particles. They have shown that such lattices can have constitutive equations of the null-stress form. Our main aim here is to broaden the class of systems that must show unidirectional force transmission, as required by the null-stress picture.

We present the discussion of the problem of stress transmission in a granular packing on several levels of generality. In Secs. II–IV we focus on properties of the system of frictionless, spherical beads. Possible experimental realizations are hard-sphere colloidal dispersions [16,17] or weakly deformed droplet emulsions [18]. First, in Sec. II we use general counting arguments to show that such a packing is minimally coupled. We then relate this fact to the inadequacy of elastic description for such a system. For a small subsystem within the pile a counting of equations and unknowns shows that approximately half of the surface forces transmitted from outside the subsystem are redundant. In equilibrium these cannot be independently specified, but have a fixed relation to the other half of the forces. This requirement for balance amongst many forces implies the existence of soft modes—infinitesimal deformations with no restoring force. In the continuum limit, the soft modes impose conditions on the stress field of the null-stress form.

In order to obtain the particular form of such macroscopic description, we limit further discussion to so-called sequential packing specified in Sec. III. In Sec. IV we give a microscopic prescription for determining the contact forces in a unidirectional way, and develop a Green's-function formalism for determining the forces in the d-dimensional sequential packing. This picture allows one to decouple the geometric features of the packing from the pattern of transmitted forces. In Sec. IV we also explore the macroscopic consequences of these force laws, leading to an expression for the stress tensor with d variables rather than the $d(d+1)/2$ variables of a general stress tensor. In two dimensions, our formalism places a constraint on the stress, whose form agrees with the null stress law of Wittmer *et al.* [6]. In higher dimensions, the constraints on the stress also lead to a unidirectional equation for the transmission of stress in the form of a wave equation. In Sec. VI we consider the relevance of our findings for real granular piles, which are not sequentially packed and which have friction. Friction can alter the transmission of forces qualitatively, restoring the elastic transmission of stress.

II. UNDERDETERMINATION WITHIN A FRICTIONLESS PACK

In this section we consider how the constraints inherent in the packing of impenetrable, frictionless beads determine the contact forces between the beads. Forces in frictionless packs have been studied by simulation [19–22]. Theoretical properties of the forces have been established for simplified systems [15,23,24]. We begin with a summary of the well-recognized enumeration of equations and unknowns, as discussed, e.g., by Alexander [14] and by the Cambridge group [15]. We then consider the role of external forces acting on a subregion of the pack, and show that a subset of these suffices to determine the others.

For definiteness we consider a system of rigid spherical beads whose sizes are chosen randomly from some continuous distribution. We suppose that the diameters of all the spheres are specified and the topology of their packing is given. That is, all the pairs of contacting beads are specified. The topology can be characterized by the average coordination number, i.e., the average number of nearest neighbors, $\bar{Z}=2N_c/M$ (here N_c is the total number of contacts, and M is the number of beads). The necessary condition for the packing of a given topology to be realizable in a space of dimensionality d is that the coordinates of the bead centers, $\mathbf{x}_\alpha$ satisfy the following equations, one for each pair of contacting beads α and β:

$$(\mathbf{x}_\alpha-\mathbf{x}_\beta)^2=(R_\alpha+R_\beta)^2. \quad (1)$$

Here R_α, R_β are the radii of the beads. There are N_c such equations (one for each contact), and Md variables (d for each bead). The number of equations should not exceed the number of variables; otherwise, the coordinates $\mathbf{x}_\alpha$ are overdetermined. Thus, the coordination number of a geometrically realizable packing should not exceed the critical value of $2d$: $\bar{Z}\leqslant 2d$. We assume all the equations imposed by the topological constraints to be independent. If they were not independent, they would become so upon infinitesimal variation of the bead sizes. For instance, the hexagonal packing in two dimensions has the coordination number 6 which is higher than the critical value, $\langle Z\rangle=4$; but the extra contacts are eliminated by an infinitesimal variation of the bead diameters. In other words, the creation of a contact network with coordination number higher than $2d$ occurs with probability zero in an ensemble of spheres with a continuous distribution of diameters. We shall ignore such zero-measure situations henceforth.

The above consideration gives the upper limit on the average coordination number $\bar{Z}$. The lower limit can be obtained from the analysis of mechanical stability of the packing: it gives a complementary inequality: $\bar{Z} \geq 2d$. We will consider a packing to be mechanically stable if there is a nonzero measure set of external forces which can be balanced by interbead ones. The packing of frictionless spheres is always characterized by $\langle Z \rangle = 2d$, as we now show. Stability requires that the net force on each bead be zero; there are Md such equations. The forces in these Md equations are the N_c contact forces. The Md equilibrium conditions determine the magnitudes of the N_c contact forces. (Their directions are determined by the geometry of the packing.) The number of equilibrium equations Md should not exceed the number of force variables N_c; otherwise these forces would be overdetermined. Thus $Md \leq N_c$, or $\bar{Z} \geq 2d$. To avoid both overdetermined co-ordinates and overdetermined forces, we must thus have $\bar{Z} = 2d$.

Similar counting arguments have been discussed previously [14,23]. A subset of them has been applied to granular packs with friction [15]. Here we emphasize a further feature of a frictionless bead pack that has not been well appreciated: the coordinates and forces within a subregion of a large bead pack are necessarily *underdetermined*. Quantifying this indeterminacy will play an important role in our reasoning below. To exhibit the indeterminacy, we consider some compact region within the packing, containing M' beads. This unbiased selection of beads must have the same average coordination number $\bar{Z}$ as the system as a whole: $\bar{Z}' = 2d$. Let N_{ext} be the number of contacts of this sub-system with external beads, and N_{int} be the number of the internal contacts. The average coordination number $\bar{Z}'$ can be expressed $\bar{Z}' = (N_{\text{ext}} + 2N_{\text{int}})/M'$ (any internal contact should be counted twice). Since there are $M'd$ equations of force balance for these beads, one is able to determine all $N_{\text{ext}} + N_{\text{int}}$ contact forces in the system, whenever $M'd = N_{\text{ext}} + N_{\text{int}}$. Evidently, if the forces on the N_{ext} contacts are not specified, the internal forces cannot be computed: the system is underdetermined. The number of external forces N_0 required is given by $N_0 = M'd - N_{\text{int}}$. This N_0 may be related to the average coordination number $\bar{Z}'$:

$$N_0 = M'\left[d - \frac{\bar{Z}'}{2}\right] + \frac{N_{\text{ext}}}{2}. \tag{2}$$

We now observe that the quantity in $[\cdots]$ vanishes on average. This is because the average of $\bar{Z}'$ for any subset of particles is the same as the overall average. There is no systematic change of $\bar{Z}'$ with M'. Thus if one-half (on average) of mutually independent external forces is known (let us call them "incoming" ones), the analysis of force balance in the region enables one to determine all the remaining forces, including the other half of external ones ("outgoing"). We are free to choose the incoming contacts at will, provided these give independent constraint equations.

This observation supports the unidirectional, propagating stress picture, discussed in the Introduction. Indeed, one can apply the above arguments to the slabs of the packing created by cutting it with horizontal surfaces. In a given slab of material, we choose the forces from the slab above as the incoming forces. According to the preceding argument, these should determine the outgoing forces transmitted to the slab beneath. This must be true provided that the constraints from the upper slab are independent.

Such force transmission contrasts with that of a solid body, as emphasized in the Introduction. If a given set of forces is applied to the top of a slab of elastic solid, the forces on the bottom are free to vary, provided the total force and torque on the slab are zero. Yet in our bead pack, which appears equally solid, we have just concluded that stability conditions determine all the bottom forces individually. In deducing this peculiar behavior, we did not exclude tensile forces; we may replace all the contacts by stiff springs that can exert strong positive or negative force, without altering our reasoning. In this sense our result is different from the recent similar result of Moukarzel [23]. The origin of the peculiar behavior lies in the minimal connectivity of the beads.

In a subregion of the minimal network, the constraints can be satisfied with no internal forces. Moreover, numerous (roughly $N_{\text{ext}}/2$) small external displacements can be applied to the subregion without generating proportional restoring forces. We call these motions with no restoring force "soft modes." If we replace these external displacements with external forces and require no motion, compensating forces must be applied elsewhere to prevent motion of the soft modes. If the applied forces perturb all the soft modes, there must be one compensating force for each applied force to prevent them from moving—on average $N_{\text{ext}}/2$ of them. The subregion is "transparent" to external forces, allowing them to propagate individually through the region.

This transparent behavior would be lost if further springs were added to the minimal network, increasing $\bar{Z}$. Then the forces on a subregion would be determined even without external contacts. The addition of external displacements would deform the springs, and produce proportional restoring forces. There would be no soft modes, and no transparency to external forces.

A simple square lattice of springs provides a concrete example of the multiple soft modes predicted above. Its elastic energy has the form

$$H = K\int dx\,dy[(u^{xx})^2 + (u^{yy})^2]. \tag{3}$$

This functional does not depend on u^{xy}, thus there are shear deformations $(u^{xx} = u^{yy} = 0)$ which cost no elastic energy. This means that the stress field should be orthogonal to any such mode, i.e.,

$$\sigma^{ij} u_o^{ij} = 0, \tag{4}$$

where $u_o^{xx} = u_o^{yy} = 0$, and u_o^{xy} is an arbitrary function of $(x;y)$. The above equation implies that $\sigma^{xy} = 0$, i.e., the principal axes of the stress tensor are fixed and directed along x and y. This provides a necessary closure for the standard macroscopic equation of force balance,

$$\partial^i \sigma^{ij} = f^j_{\text{ext}}; \tag{5}$$

here $\mathbf{f}_{\text{ext}}$ is an external force. Since $\sigma^{xy}=0$ the two unknown components of the stress field, σ^{xx} and σ^{yy} propagate independently along the corresponding characteristics, $x=\text{const}$ and $y=\text{const}$:

$$\partial^x \sigma^{xx} = f^x_{\text{ext}}, \tag{6}$$

$$\partial^y \sigma^{yy} = f^y_{\text{ext}}. \tag{7}$$

The propagation of the solution along characteristics is a property of hyperbolic problems such as wave equation. The above equations without external force imply that each component of the stress tensor $\hat{\sigma}$ satisfies a wave equation of the form

$$\left(\frac{\partial^2}{\partial t^2} - \frac{\partial^2}{\partial s^2}\right)\hat{\sigma} = 0, \tag{8}$$

where $t \equiv x+y$ and $s \equiv x-y$. Thus, the fact that the original elastic energy has the soft modes results in hyperbolic, rather than elliptic equations for the stress field. One has now to specify the surface forces (or displacements) at a single noncharacteristic surface—a line not parallel to x or y— in order to determine the stress field in all the sample.

A frictionless granular packing behaves similar to this example: they both are minimally coupled, they both have soft modes, they both have unidirectional propagation. In both examples only the surface of the sample stabilizes the soft modes. The above consideration of regular lattice can be easily extended to the case of arbitrary angle between the characteristic directions x and y. Instead of starting with a square lattice, we could have applied a uniform x-y shear, altering the angle between the horizontal and vertical springs. The reasoning above holds for this lattice just as for the original square one.

The nature of the soft modes in a disordered bead pack is less obvious than in this lattice example. We have not proven, for instance, that all the forces acting on the top of a slab correspond to independent soft modes, which determine the forces at the bottom. Otherwise stated, we have not shown that the soft modes seen in the microscopic displacements have continuum counterparts in the displacement field of the region. However, the following construction, as in the lattice example above, suggests that the soft modes survive in the continuum limit.

To construct the pack, we place circular disks one at a time into a two-dimensional vertical channel of width L. (Such sequential packings will figure prominently in the next section.) Since the disks are of different sizes, the packing will be disordered. We place each successive disk at the lowest available point until the packed disks have reached a height of order L, as shown in Fig. 2. We now construct a second packing, starting from a channel of slightly greater width $L+\delta$. We reproduce the packing constructed in the first channel as far as possible. We use an identical sequence of disks and place each at the lowest point, as before. There must be a nonvanishing range of δ for which the contact topology is identical. The motion of the wall over this range is thus a soft mode. As the side wall moves, the top surface will move by some amount ϵ, proportional to δ. Now, instead of holding the side wall fixed, we exert a given force f^x on it. Likewise, we place a lid on the top, remove gravity, and exert another force f^y. Evidently unless $f^x/f^y=\epsilon/\delta$, a motion of the soft mode would result in work, and the system would move. Thus f^y plus the condition of no motion determines f^x. This condition translates into an imposed proportionality between the stresses σ^{yy} and σ^{xx}, as in the lattice example above. The soft modes have continuum consequences.

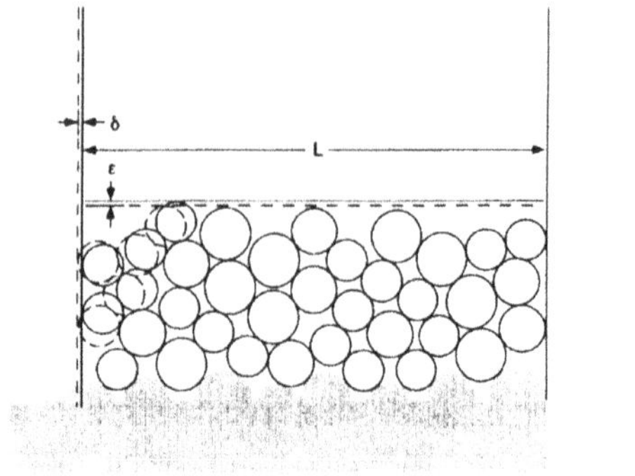

FIG. 2. A sequential packing in a channel. When the left wall is displaced outward by a small amount δ, the beads shift to the positions shown by dashed lines, and the top of the pile shifts by an amount ϵ.

III. SEQUENTIAL PACKING UNDER GRAVITY

In the previous section we have shown that a packing of frictionless spherical beads is an anomalous solid from the point of view of classical elastic theory. The fact that the average coordination number in such a packing is exactly $2d$ for the infinite system supports unidirectional, propagating stress. Now we elaborate this concept in more detail, by deriving particular laws for microscopic and macroscopic force transfer adopting a particular packing procedure. We suppose that the beads are deposited one by one in the presence of gravity. The weight of any new sphere added to the existing packing must be balanced by the reactions of the supporting beads. This is possible only if the number of such supporting contacts is equal to the dimensionality d. Any larger number of contacts requires a specific relationship between the sizes and coordinates of the supporting beads, and thus occurs with vanishing probability. As a result, the eventual packing has an average coordination number $2d$, similar to any stable, frictionless pack. In addition, it has a further property: a partial timelike ordering. Namely, among any two contacting beads there is always one which has found its place earlier than the other (the supporting one), and any bead has exactly d such supporting neighbors. Note that the supporting bead is not necessarily situated below the supported one in the geometrical sense. The discussed ordering is topological rather than spatial.

One could expect that although any bead has exactly d supporters at the moment of deposition, this may change later. Specifically, adding an extra bead to the packing may result in the violation of positivity of some contact force in the bulk [7]. This will lead to a rearrangement of the network. For the moment we assume that the topology of the sequential packing is preserved in the final state of the system, and return to the effect of rearrangements in Sec. V.

The partial ordering of the sequential packing considerably simplifies the calculation of the force distribution. Indeed, any force applied to a bead can be uniquely decomposed into the d forces on the supporting contacts. This means that the force balance enables us to determine all the "outcoming" (downward) forces if the "incoming" ones are known. Therefore, there is a simple unidirectional procedure of determination of all the forces in the system. Below, we use this observation to construct a theory of stress propagation on the macroscopic scale.

IV. MEAN-FIELD STRESS

We will characterize any interbead contact in a sequential packing with a unit vector directed from the center of supported bead α toward the supporting one β,

$$\mathbf{n}_{\alpha\beta}=\frac{\mathbf{x}_\beta-\mathbf{x}_\alpha}{|\mathbf{x}_\beta-\mathbf{x}_\alpha|}. \tag{9}$$

The stress distribution in the frictionless packing is given if a non-negative magnitude of the force acting along any of the above contact unit vector is specified. We denote such scalar contact force as $f_{\alpha\beta}$.

The total force to be transmitted from some bead α to its supporting neighbors is the sum of all the incoming and external (e.g., gravitational) forces:

$$\mathbf{F}_\alpha=(\mathbf{f}_{\rm ext})_\alpha+\sum_{\beta(\rightarrow\alpha)}\mathbf{n}_{\beta\alpha}f_{\beta\alpha}. \tag{10}$$

Here $\beta(\rightarrow\alpha)$ denotes all the beads supported by α. Since there are exactly d supporting contacts for any bead in a sequential packing, the above force can be uniquely decomposed onto the corresponding d components, directed along the outcoming vectors $\mathbf{n}_{\alpha\gamma}$. This gives the values of the outcoming forces. The f's may be compactly expressed in terms of a generalized scalar product $\langle\cdots|\cdots\rangle_\alpha$:

$$f_{\alpha\gamma}=\langle\mathbf{F}_\alpha|\mathbf{n}_{\alpha\gamma}\rangle_\alpha. \tag{11}$$

The scalar product $\langle\cdots|\cdots\rangle_\alpha$ is defined such that $\langle\mathbf{n}_{\alpha\gamma}|\mathbf{n}_{\alpha\gamma'}\rangle_\alpha=\delta_{\gamma\gamma'}$ (all the Greek indices count beads, not spatial dimensions). In general, it does not coincide with the conventional scalar product. If now some force is applied to certain bead in the packing, the above projective procedure allows one to determine the response of the system, i.e., the change of the contact forces between all the beads below the given one. In other words one can follow how the perturbation propagates downward. Since the equations of mechanical equilibrium are linear, and beads are assumed to be rigid enough to preserve their sizes, the response of the system to the applied force is also linear. This linearity can be destroyed only by violating the condition of positivity of the contact forces, which implies the rearrangement of the packing. While the topology (and geometry) of the network is preserved, one can introduce the Green's function to describe the response of the system to the applied forces. Namely, force $\mathbf{f}_\lambda$ applied to certain bead λ results in the following additional force acting on another bead μ (lying below λ):

$$\mathbf{f}_\mu=\hat{\mathbf{G}}_{\mu\lambda}\cdot\mathbf{f}_\lambda. \tag{12}$$

Here $\hat{\mathbf{G}}_{\lambda\mu}$ is a tensor Green function, which can be calculated as the superposition of all the projection sequences (i.e., trajectories), which lead from λ to μ.

The stress field σ^{ij} in the system of frictionless spherical beads can be introduced in the following way [25]:

$$\sigma^{ij}(\mathbf{x})=\sum_\alpha\sum_{\beta(\leftarrow\alpha)}f_{\alpha\beta}n^i_{\alpha\beta}n^j_{\alpha\beta}R_{\alpha\beta}\delta(\mathbf{x}_\alpha-\mathbf{x}). \tag{13}$$

Here $R_{\alpha\beta}=|\mathbf{x}_\alpha-\mathbf{x}_\beta|$. As we have just shown, the magnitude of the force $f_{\alpha\beta}$ transmitted along the contact unit vector $\mathbf{n}_{\alpha\beta}$ can be expressed as an appropriate projection of the total force F_α acting on the bead α from above. This allows one to express the stress tensor in terms of the vector field F_α:

$$\sigma^{ij}(\mathbf{x})=\sum_\alpha\sum_{\beta(\leftarrow\alpha)}\langle F_\alpha|\mathbf{n}_{\alpha\beta}\rangle_\alpha n^i_{\alpha\beta}n^j_{\alpha\beta}R_{\alpha\beta}\delta(\mathbf{x}_\alpha-\mathbf{x}). \tag{14}$$

In order to obtain the continuous macroscopic description of the system, one has to perform the averaging of the stress field over a region much larger than a bead. At this stage we make a mean-field approximation for the force $\mathbf{F}_\alpha$ acting on a given bead from above: we replace $\mathbf{F}_\alpha$ by its average $\bar{\mathbf{F}}$ over the region. To be valid, this assumption requires that

$$\sum_{\alpha\beta}\langle(\mathbf{F}_\alpha-\bar{\mathbf{F}})|\mathbf{n}_{\alpha\beta}\rangle_\alpha\mathbf{n}^i_{\alpha\beta}\mathbf{n}^j_{\alpha\beta}R_{\alpha\beta}\ll\sum_{\alpha\beta}\langle\bar{\mathbf{F}}|\mathbf{n}_{\alpha\beta}\rangle_\alpha\mathbf{n}^i_{\alpha\beta}\mathbf{n}^j_{\alpha\beta}R_{\alpha\beta}. \tag{15}$$

For certain simple geometries, the mean-field approximation is exact. One example is the simple square lattice treated in Sec. II. In any regular lattice with one bead per unit cell, all the $\mathbf{F}_\alpha$'s must be equal under any uniform applied stress. Thus replacing $\mathbf{F}_\alpha$ by its average changes nothing. If this lattice is distorted by displacing its soft modes, the $\mathbf{F}_\alpha$ are no longer equal and the validity of the mean-field approximation can be tested. Figure 3 shows a periodic distortion with fourbeads per unit cell. For example, under an applied vertical force, the bottom forces oscillate to the left and right. Nevertheless, the stress crossing the bottom row, similar to that crossing the row above it, is the average force times the length. One may verify that the $\mathbf{F}_\alpha$ may also be replaced by its average when the applied force is horizontal. Though the mean-field approximation is exact in these cases, it is clearly not exact in all. In the lattice of Fig. 3 the mean field ap-

FIG. 3. A buckled square lattice illustrating the propagation of inhomogeneous forces. Bottom row of sites has alternating wide and narrow spacing. Arrows indicate the unequal forces on these sites.

proximation may be inexact if one considers a region not equal to a whole number of unit cells.

A disordered packing may be viewed as a superposition of periodic soft modes such as those of Fig. 3. Each such mode produces fluctuating forces, such as those of the example. But after averaging over an integer number of unit cells, the stress may depend on only the average force $\overline{\mathbf{F}}$. A disordered packing need not have a fixed coordination number as our example does. This is another possible source of departure from the mean-field result.

Now, it becomes an easy task to perform a local averaging of Eq. (14) for the stress field in the vicinity of a given point $\mathbf{x}$, replacing $\mathbf{F}_\alpha$ by its average

$$\overline{\sigma^{ij}}(\mathbf{x}) = \rho \overline{F}^k(\mathbf{x}) \tau^{kij}(\mathbf{x}). \tag{16}$$

Here ρ is the bead density, $\overline{\mathbf{F}}(\mathbf{x})$ is the force $\mathbf{F}_\alpha$ averaged over the beads α in the vicinity of the point $\mathbf{x}$, and the third-order tensor $\hat{\tau}$ characterizes the local geometry of the packing:

$$\tau^{kij}(\mathbf{x}) = \overline{|\mathbf{n}_{\alpha\beta}\rangle^k_\alpha n^i_{\alpha\beta} n^j_{\alpha\beta} R_{\alpha\beta}}. \tag{17}$$

This equation is similar in spirit to one derived by Edwards for the case of a $d+1$ coordinated packing of spheres with friction [26]. Our geometric tensor τ plays a role analogous to that of the fabric tensor in that treatment.

The stress field satisfies the force balance equation, Eq. (5). Since this is a vector equation, it normally fails to give a complete description of the tensor stress field. In our case, however, the stress field has been expressed in terms of the vector field $\mathbf{F}$. This creates a necessary closure for the force balance equation. It is important to note that the proposed macroscopic formalism is complete for a system of arbitrary dimensionality: there is a single vector equation and a single vector variable. We now discuss the application of the above macroscopic formalism in two special cases. First we consider the equations of stress propagation in two dimensions. Then we discuss a packing of arbitrary dimensionality but with uniaxial symmetry. It is assumed to have no preferred direction other than that of gravity.

A. Two-dimensional packing

In two dimensions, according to Eq. (16), the stress tensor $\hat{\sigma}$ can be written as a linear combination of two τ tensors

$$\hat{\sigma} = F_1 \hat{\sigma}_1 + F_2 \hat{\sigma}_2, \tag{18}$$

where $[\hat{\sigma}_1]^{ij} = \tau^{1ij}$ and $[\hat{\sigma}_2]^{ij} = \tau^{2ij}$. Since the $\hat{\sigma}_1$ and $\hat{\sigma}_2$ are properties of the medium and are presumed known, the problem of finding the stress profile $\hat{\sigma}(x)$ becomes that of finding F_1 and F_2 under a given external load. Rather than determining these F's directly, we may view Eq. (18) as a constraint on $\hat{\sigma}$. The form (18) constrains $\hat{\sigma}$ to lie in a subspace of the three-dimensional space of stress components $\vec{\sigma} \equiv (\sigma^{xx}, \sigma^{yy}, \sigma^{xy})$. It must lie in the two-dimensional subspace spanned by $\vec{\sigma}_1$ and $\vec{\sigma}_2$. This constraint amounts to one linear constraint on the components of σ, of the form

$$\sigma^{ij} u^{ij} = 0, \tag{19}$$

where the $\hat{u}$ tensor is determined by $\hat{\sigma}_1$ and $\hat{\sigma}_2$. Specifically, $\hat{u}$ may be found by observing that the determinant of the vectors $\vec{\sigma}$, $\vec{\sigma}_1$, $\vec{\sigma}_2$ must vanish. Expanding the determinant by minors to obtain the coefficients of the σ^{ij}, one finds

$$\hat{u} = \begin{pmatrix} \begin{vmatrix} \sigma_1^{yy} & \sigma_2^{yy} \\ \sigma_1^{xy} & \sigma_2^{xy} \end{vmatrix} & \begin{vmatrix} \sigma_1^{xx} & \sigma_2^{xx} \\ \sigma_1^{yy} & \sigma_2^{yy} \end{vmatrix} \\ \begin{vmatrix} \sigma_1^{xx} & \sigma_2^{xx} \\ \sigma_1^{xy} & \sigma_2^{xy} \end{vmatrix} & \begin{vmatrix} \sigma_1^{xy} & \sigma_2^{xy} \\ \sigma_1^{xx} & \sigma_2^{xx} \end{vmatrix} \end{pmatrix}. \tag{20}$$

Equation (19) has the same "null-stress" form as that introduced by Wittmer *et al.* [6], whose original arguments were based on a qualitative analysis of the problem. By an appropriate choice of the local coordinates (ξ, η), the $\hat{u}$ tensor can be transformed into coordinates such that $u^{\xi\xi} = u^{\eta\eta} = 0$. Then the null stress condition becomes $\sigma^{\xi\eta} = \sigma^{\eta\xi} = 0$. This implies that, according to force balance equation (5), the nonzero diagonal components of the stress tensor "propagate" independently along the corresponding characteristics, $\xi = \text{const}$ and $\eta = \text{const}$:

$$\partial^\xi \sigma^{\xi\xi} = f^\xi_{\text{ext}}, \quad \partial^\eta \sigma^{\eta\eta} = f^\eta_{\text{ext}}. \tag{21}$$

Our microscopic approach gives an alternative foundation for the null-stress condition, Eq. (19), and allows one to relate the tensor $\hat{u}$ in this equation to the local geometry of the packing. Our general formalism is not limited to the two-dimensional case, and in this sense, is a generalization of the null-stress approach.

B. Axially symmetric packing

Generally, there are two preferred directions in the sequential packing: that of the gravitational force, $\mathbf{g}$, and that of the growth surface $\mathbf{n}$. In the case when these two directions coincide, the form of the third-order tensor $\hat{\tau}$, Eq. (17), should be consistent with the axial symmetry associated with the single preferred direction, $\mathbf{n}$. Since τ^{kij} is symmetric with respect to $i \leftrightarrow j$ permutation, it can be only a linear combination of three tensors: $n^k n^i n^j$, $n^k \delta^{ij}$, and $\delta^{ki} n^j + \delta^{kj} n^i$, for general spatial dimension d.

Let σ^{ij} be the stress tensor in the d-dimensional space $(i,j = 0, \ldots, d-1$, and index 0 corresponds to the vertical direction). From the point of view of rotation around the vertical axis the stress splits into scalar σ^{00}, $(d-1)$-dimensional vector σ^{0a} $(a=1,\ldots,d-1)$ $(d-1)$-dimensional tensor σ^{ab}. According to our constitutive Eq. (16), the stress should be linear in vector **F**, which itself splits into a scalar F^0 and a vector F^a with respect to horizontal rotations. Since the material tensor τ is by hypothesis axially symmetric, the only way that the "scalar" σ^{00} may depend on **F** is to be proportional to "scalar" F^0. Likewise, the only way "tensor" σ^{ab} can be linear in **F** is to be proportional to $\delta^{ab}F^0$. Therefore, in the axially symmetric case

$$\sigma^{ab}=\lambda\,\delta^{ab}\sigma^{00}, \tag{22}$$

where the constant λ is, e.g., τ^{011}/τ^{000}. This constitutive equation allows one to convert the force balance equation (5) to the following form:

$$\partial^0\sigma^{00}+\partial^a\sigma^{a0}=f^0_{\text{ext}},\quad \partial^0\sigma^{a0}+\lambda\partial^a\sigma^{00}=f^a_{\text{ext}}. \tag{23}$$

In the case of no external force, we may take ∂^0 of the first equation and combine with the second to yield a wave equation for σ^{00}. Evidently σ^{ab}, being a fixed multiple of σ^{00}, obeys the same equation. Similar manipulation yields the same wave equation for σ^{0a} and σ^{a0}. Thus every component of stress satisfies the wave equation with vertical direction playing the role of time and $\sqrt{\lambda}$ being the propagation velocity.

V. DISCUSSION

In this section we consider how well our model should describe real systems of rigid, packed units. As stated above, our model is most relevant for emulsions or dense colloidal suspensions, whose elementary units are well described as frictionless spheres. Under very weak compression the forces between such units match our model assumptions. However, our artificial procedure of sequential packing bears no obvious resemblance to the arrangements in real suspensions. We argue below that our model may well have value even when the packing is not sequential. More broadly we may consider the connection between our frictionless model and real granular materials with friction. The qualitative effect of adding friction to our sequential packing is to add constraints so that the network of contacts is no longer minimally connected. Thus the basis for a null-stress description of the force transmission is compromised. We argue below that friction should cause forces to propagate as in an elastic medium, not via null-stress behavior.

A. Sequential packing

We first consider the consequences of our sequential packing assumption. One consequence is that each bead has exactly d supporting contacts. These lead successively to earlier particles, forming a treelike connectivity from supported beads to supporters. Although the counting arguments of Sec. II show that the propagating stress approach should be applicable to a wide class of frictionless systems, the continuum description of Sec. IV depends strongly on the assumed sequential order. Now, most packings are not sequential, and even when beads are deposited in sequence, they may undergo rearrangements that alter the network of connections. However, it is possible to modify our arguments to take account of such rearrangements. Our reasoning depends on the existence of d supporting contacts for each bead. Further, every sequence of supporting contacts starting at a given bead must reach the boundary of the system without passing through the starting bead: there must be no closed loops in the sequence.

Even in a nonsequential packing we may define a network of supporting contacts. First we define a downward direction. Then, for any given bead in the pack, we *define* the supporting contacts to be the d lowest contacts. With probability 1, each bead has at least d contacts. Otherwise it is not stable. Typically a supporting bead lies lower than the given bead. Thus the typical sequence of supporting contacts leads predominantly downward, away from the given bead, and returns only rarely to the original height. A return to the original *bead* must be even more rare. One may estimate the probability that there is a loop path of supporting contacts under simple assumptions about the packing. As an example we suppose the contacts on a given bead to be randomly distributed amongst the 12 sites of a randomly oriented close-packed lattice. We further imagine that these sites are chosen independently for each bead, with at least one below the horizontal. Then the paths are biased random walks with a mean steplength of 0.51 diameters and a root-mean-square steplength of about 1.2 times the mean. The probability of a net upward displacement of 1 or more diameter is about 1%. It appears that our neglect of loop paths is not unreasonable.

B. Friction

The introduction of friction strongly affects most of our arguments. Friction creates transverse as well as normal forces at the contacts. The problem is to determine positions and orientations of the beads that lead to balanced forces and torques on each. If the contact network is minimally connected, the forces can be determined without reference to deformations of the particle. But if the network has additional constraints, it is impossible to satisfy these without considering deformation. This is no less true if the beads are presumed very rigid. We first give an example to show that in a generic packing the deformability alters the force distribution substantially. We then give a prescription for defining the deformation and hence the contact forces unambiguously.

In our example we imagine a two-dimensional sequential packing and focus on a particular bead, labeled 0, as pictured in Fig. 4. We presume that the beads are deposited gently, so that each contact forms without tangential force. Thus when the bead is deposited, it is minimally connected: its weight determines the two (normal) supporting forces, labeled 1 and 2. Thenceforth no slipping is allowed at the contact. Later during the construction of the pack bead 0 must support the force from some subsequent bead. This new force is normal, since it too arises from an initial contact. But the new force creates tangential forces on the supporting contacts 1 and 2. To gauge their magnitude, we first suppose that there is no friction at contacts 1 and 2, while the supporting beads remain immobile. Then the added force F leads to a compres-

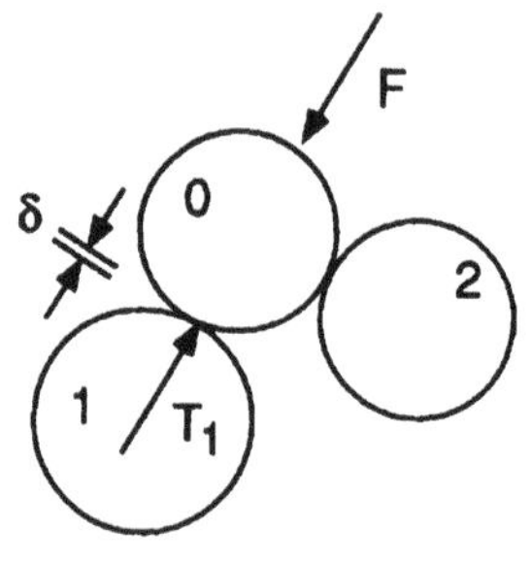

FIG. 4. The effect of friction on a triad of beads. In the absence of friction, the applied force F is transmitted entirely to contact 1, causing a displacement δ. This would result in a sliding displacement of contact 2 by an amount δ. With friction, contact 2 cannot slide; it must deform the contact region by an amount of order δ. Thus the applied force F is shared between contacts 1 and 2. The force is distributed so as to minimize the total elastic energy at contacts 1 and 2.

sion. We denote the compression of the contact 1 as δ. With no friction, the contact 2 would undergo a slipping displacement by an amount of order δ. Friction forbids this slipping and decrees deformation of the contact instead. The original displacement there would create an elastic restoring force of the same order as the original F. Thus the imposition of friction creates new forces whose strength is comparable to those without friction. The frictional forces are not negligible, even if the beads are rigid. Increasing the rigidity lessens the displacements δ associated with the given force F, but it does not alter the ratio of frictional to normal forces. Neither are the frictional forces large compared to the normal forces. Thus a coefficient of friction μ of order unity should be sufficient to generate enough frictional force to prevent slipping of a substantial fraction of the contacts.

The contact forces T_1 and T_2 cannot be determined by force balance alone, as they could in the frictionless case. Now the actual contact forces are those which minimize the elastic energy of deformation near the two contacts. This argument holds not just for spheres but for general rounded objects.

Though the new tangential forces complicate the determination of the forces, the determination need not be ambiguous. We illustrate this point for a sequential packing on a bumpy surface with perfect friction. We choose a placement of the successive beads so that no contact rearrangements occur. If only a few beads have been deposited in the container, the forces are clearly well determined. Further, if the forces are presumed well determined up to the Mth bead, they remain so upon addition of the $(M+1)$st bead. We presume as before that the new bead exerts only normal forces on its immediate supporters. Each supporter thus experiences a well-defined force, as shown in Sec. II. But by hypothesis, these supporting beads are part of a well-connected, solid object, whose contacts may be regarded as fastened together. Thus the displacements and rotations of each bead are a well-defined function of any small applied load. Once the $(M+1)$st bead has been added, its supporting contacts also support tangential force, so that it responds to future loads as part of the elastic body.

We conclude that a sequential packing with perfect friction, under conditions that prevent contact rearrangements, transmits forces as a solid body. Small changes in stress $\delta\sigma^{ij}$ in a region give rise to proportional changes in the strain $\delta\gamma^{k\ell}$. This proportionality is summarized by an elasticity tensor $K^{ijk\ell}$: $\delta\sigma^{ij}=K^{ijk\ell}\delta\gamma^{k\ell}$. The elastic tensor K should depend in general on how the pack was formed; thus it may well be anisotropic.

This elastic picture is compromised when the limitations of friction are taken into account. As new beads are added, underlying contacts such as contacts 1 and 2 of Fig. 4 may slip if the tangential force becomes too great. Each slipping contact relaxes so as to satisfy a fixed relationship between its normal force N and its tangential force T: viz., $|T| =\mu|N|$. If μ were very small, virtually all the contacts would slip until their tangential force was nearly zero. Then the amount of stress associated with the redundant constraints must become small and the corresponding elastic moduli must become weak. Moreover, as μ approaches 0, the material on any given scale must become difficult to distinguish from a frictionless material with unidirectional stress propagation. Still, redundant constraints remain on the average and thus the ultimate behavior at large length scales (for a given μ) must be elastic, provided the material remains homogeneous.

C. Force-generated contacts

Throughout the discussion of frictionless packs we have ignored geometric configurations with probability zero, such as beads with redundant contacts. Such contacts occur in a close-packed lattice of identical disks, for example. Though such configurations are arbitrarily rare in principle, they may nevertheless play a role in real bead packs. Real bead packs have a finite compressibility; compression of beads can create redundant contacts. Thus for example a close-packed lattice of identical spheres has six contacts per bead, but if there is a slight variability in size, the number of contacts drops to 4. The remaining two beads adjacent to a given bead do not quite touch. These remaining beads can be made to touch again if sufficient compressive stress is applied. Such stress-induced redundant contacts must occur in a real bead with some nonzero density under any nonzero load. These extra contacts serve to stabilize the pack, removing the indeterminate forces discussed in Sec. II. To estimate the importance of this effect, we consider a large bead pack compressed by a small factor γ. This overall strain compresses a typical contact by a factor of order γ as well. The number of new contacts may be inferred from the pair correlation function $g(r)$. Data on this $g(r)$ is available for some computer-generated sequential packings of identical spheres of radius R [27]. These data show that $g(r)$ has a finite value near 1 at $r=2R$. Thus the number of additional contacts per bead that form under a slight compression by an amount δr is given by $\delta\bar{Z}=6\phi g(2R)\delta r/R\simeq4\gamma$. Here $\phi\simeq0.6$ is the volume fraction of the beads. These extra contacts impose constraints that reduce the number of undetermined boundary forces in a compact region containing M' beads and roughly $M'^{2/3}$ surface beads. The remaining number of undetermined boundary forces now averages $\frac{1}{2}N_{\text{ext}}-M'\delta\bar{Z}$. The first term is of order $M'^{2/3}$, and must thus be smaller than the second term for $M'^{1/3} \simeq (\delta \bar{Z})^{-1}$. For M' larger than this amount, there are no further undetermined forces and the region becomes mechanically stable. Moukarzel [23] reaches a similar conclusion by a somewhat different route.

If the pack is compressed by a factor of γ, stability occurs for $M'^{1/3} \gtrsim 1/\gamma$—a region roughly $1/\gamma$ beads across. In a typical experiment [28] the contact compression γ is 10^{-4} or less, and the system size is far smaller than 10^4 beads. Thus compression-induced stability should be a minor effect here. Still, this compression-induced stability might well play a significant role for large and highly compressed bead packs such as compressed emulsions [20]. In some of the large packs of Ref. [3], compression-induced stability may also be important.

D. Experimental evidence

We have argued above that undeformed, frictionless beads should show unidirectional, propagating forces while beads with friction should show elastic spreading of forces. The most direct test of these contrasting behaviors is to measure the response to a local perturbing force [7]. Thus, e.g., if the pile of Fig. 1 is a null-stress medium, the local perturbing force should propagate along a light cone and should thus be concentrated in a ringlike region at the bottom [8]. By contrast, if the pile is an elastic medium the perturbing force should be spread in a broad pattern at the bottom, with a maximum directly under the applied force. Existing experimental information seems inadequate to test either prediction, but experiments to measure such responses are in progress [29].

As noted above, emulsions and colloids are good realizations of the frictionless case. The contacts in such systems are typically organized by hydrostatic pressure or by flow, rather than by gravity. Still, our general arguments for unidirectional propagation should apply. Extensive mechanical measurements of these systems have been made [16,18]. The shear modulus study of Weitz and Mason [18] illustrates the issues. The study spans the range from liquidlike behavior at low volume fractions to solidlike behavior at high volume fractions. In between these two regimes should lie a state where the emulsion droplets are well connected but little deformed. The emulsion in this state should show unidirectional force transmission. It is not clear how this should affect the measured apparent moduli.

Other indirect information about force propagation comes from the load distribution of a granular pack on its container, such as the celebrated central dip under a conical heap of sand [3]. These data amply show that the mechanical properties of a pack depend on how it was constructed. Theories postulating null-stress behavior have successfully explained these data [6]. But conventional elastoplastic theories have also proved capable of producing a central dip [5]. An anisotropic elastic tensor may also be capable of explaining the central dip.

Another source of information is the statistical distribution of individual contact forces within the pack or at its boundaries. The measured forces become exponentially rare for strong forces [9,12]. Such exponential falloff is predicted by Coppersmith's "q model" [9], which postulates unidirectional force propagation. Still, it is not clear whether this exponential falloff is a distinguishing feature of unidirectional propagation. A disordered elastic material might well show a similar exponential distribution.

Computer simulations should also be able to test our predictions. Recent simulations [20,30,31] have focussed on stress-induced restructuring of the force-bearing contact network. We are not aware of a simulation study of the transmission of a local perturbing force. Such a perturbation study seems quite feasible and would be a valuable test. We have performed a simple simulation to test the mean-field description of stress in frictionless packs. Preliminary results agree well with the predictions. An account of our simulations will be published separately.

VI. CONCLUSION

In this study we have aimed to understand how force is transmitted in granular media, whether via elastic response or via unidirectional propagation. We have identified a class of disordered systems that ought to show unidirectional propagation. Namely, we have shown that in a general case a system of frictionless rigid particles must be isostatic, or minimally connected. That is, all the interparticle forces can in principle be determined from the force balance equations. This contrasts with statically undetermined, elastic systems, in which the forces cannot be determined without self-consistently finding the displacements induced by those forces. Our general equation-counting arguments suggest that isostatic property of the frictionless packing results in the unidirectional propagation of the macroscopic stress.

We were able to demonstrate this unidirectional propagation explicltly by specializing to the case of sequential packing. Here the stress obeys a condition of the previously postulated null-stress form [6]; our system provides a microscopic justification for the null-stress postulate. Further, we could determine the numerical coefficients entering the null-stress law from statistical averages of the contact angles by using a mean field hypothesis (decoupling ansatz). We have devised a numerical simulation to test the adequacy of the sequential packing assumption and the mean-field hypothesis. The results will be reported elsewhere.

If we add friction in order to describe macroscopic granular packs more accurately, the packing of rigid particles no longer needs to be isostatic, and the system is expected to revert to elastic behavior. This elasticity does not arise from softness of the beads or from a peculiar choice of contact network. It arises because contacts that provide only minimal constraints when created can provide redundant constraints upon further loading.

We expect our formalism to be useful in understanding experimental granular systems. It is most applicable to dense colloidal suspensions, where static friction is normally negligible. Here we expect null-stress behavior to emerge at scales large enough that the suspension may be considered uniform. We further expect that our mean-field methods will be useful in estimating the coefficients in the null-stress laws. In macroscopic granular packs our formalism is less applicable because these packs have friction. Still, this friction may be small enough in many situations that our picture remains useful. Then our microscopic justification may account for the practical success of the null-stress postulate [6] for these systems.

ACKNOWLEDGMENTS

The authors are grateful to the Institute for Theoretical Physics in Santa Barbara for hospitality and support that enabled this research to be initiated. The authors thank the participants in the Institute's program on Jamming and Rheology for many valuable discussions. Many of the ideas reported here had their roots in the floppy network theory of the late Shlomo Alexander. This work was supported in part by the National Science Foundation under Grant Nos. PHY-94 07194, DMR-9528957, and DMR 94-00379.

[1] For a review, see E. Guyon, S. Roux, A. Hansen, D. Bideau, J.-P. Troadec, and H. Crapo, Rep. Prog. Phys. **53**, 373 (1990).

[2] *Friction, Arching, Contact Dynamics*, edited by D. E. Wolf and P. Grassberger (World Scientific, Singapore, 1997).

[3] T. Jotaki and R. Moriyama, J. Soc. Powder Technol. Jpn. **60**, 184 (1979); J. Smid and J. Novosad, in Proceedings of 1981 Powtech Conference [Ind. Chem. Eng. Symp. **63**, D3V 1 (1981)]; R. Brockbank, J. M. Huntley, and R. C. Ball, J. Phys. II **7**, 1521 (1997).

[4] S. F. Edwards and C. C. Mounfield, Physica A **226**, 1 (1996).

[5] J. D. Goddard and A. K. Didwania, Q. J. Mech. Appl. Math. **51**, 15 (1998); F. Cantelaube and J. D. Goddard, *Powders and Grains 97*, edited by R. P. Behringer and J. T. Jenkins (Balkema, Rotterdam, 1997), p. 185.

[6] J. P. Wittmer, P. Claudin, M. E. Cates, and J. P. Bouchaud, Nature (London) **382**, 336 (1996); J. P. Wittmer, M. E. Cates, and P. Claudin, J. Phys. I **7**, 39 (1997).

[7] P. Claudin and J. P. Bouchaud, Phys. Rev. Lett. **78**, 231 (1997); P. Claudin, J.-P. Bouchaud, M. E. Cates, and J. P. Wittmer, Phys. Rev. E **57**, 4441 (1998).

[8] M. E. Cates, J. P. Wittmer, J. P. Bouchaud, and P. Claudin, Phys. Rev. Lett. **81**, 1841 (1998).

[9] S. N. Coppersmith, C.-H. Liu, S. Majumdar, O. Narayan, and T. A. Witten, Phys. Rev. E **53**, 4673 (1996).

[10] J. E. S. Socolar, Phys. Rev. E **57**, 3204 (1998).

[11] E. Clément, C. Eloy, J. Rajchenbach, and J. Duran, in *Lectures on Stochastic Dynamics*, edited by T. Pöschel and L. Schimanski-Geier (Springer, Heidelberg, 1997).

[12] D. M. Mueth, H. M. Jaeger, and S. R. Nagel, Phys. Rev. E **57**, 3164 (1998).

[13] F. Radjai, M. Jean, J. J. Moreau, and S. Roux, Phys. Rev. Lett. **77**, 274 (1996).

[14] S. Alexander, Phys. Rep. **296**, 65 (1998).

[15] S. F. Edwards, Physica A **249**, 226 (1998).

[16] J. C. VanderWerff, C. G. Denkruif, C. Blom, and J. Mellema, Phys. Rev. A **39**, 795 (1989).

[17] H. Watanabe, M. L. Yao, K. Osaki, T. Shikata, H. Niwa, and Y. Morishima, Rheol. Acta **36**, 524 (1997); **37**, 1 (1998).

[18] T. G. Mason, J. Bibette, and D. A. Weitz, Phys. Rev. Lett. **75**, 2051 (1995).

[19] M. D. Lacasse, G. S. Grest, and D. Levine, Phys. Rev. E **54**, 5436 (1996).

[20] S. A. Langer and Andrea J. Liu, J. Phys. Chem. B **101**, 8667 (1997).

[21] S. Luding, Phys. Rev. E **55**, 4720 (1997).

[22] R. S. Farr, J. R. Melrose, and R. C. Ball, Phys. Rev. E **55**, 7203 (1997).

[23] C. F. Moukarzel, Phys. Rev. Lett. **81**, 1634 (1998); *Proceedings of Rigidity Theory and Applications, Traverse City, MI, June, 1998*, Fundamental Material Science Series (Plenum, New York, in press).

[24] G. Oron and H. J. Herrmann, Phys. Rev. E **58**, 2079 (1998).

[25] L. D. Landau and E. M. Lifshitz, *Theory of Elasticity*, 3rd ed. (Pergamon, Oxford, 1986), Sec. 2.

[26] R. C. Ball (private communication); R. C. Ball and S. F. Edwards (unpublished); D. V. Grinev and R. C. Ball (unpublished).

[27] R. Jullien, A. Pavlovitch, and P. Meakin, J. Phys. A **25**, 4103 (1992).

[28] C.-H. Liu and S. R. Nagel, Phys. Rev. B **48**, 15 646 (1993).

[29] S. Nagel (private communication); J.-P. Bouchaud (private communication).

[30] C. Thornton and J. Lanier, *Powders and Grains 97*, edited by R. P. Behringer and J. T. Jenkins (Balkema, Rotterdam, 1997), p. 223.

[31] B. Miller, C. O'Hern, and R. P. Behringer, Phys. Rev. Lett. **77**, 3110 (1996).

[32] J. N. Goodier, Philos. Mag. **22**, 678 (1936); H. M. Smallwood, J. Appl. Phys. **15**, 758 (1944); E. Guth, *ibid.* **16**, 20 (1945); Z. Hashin, *Proceedings of the 4th International Congress on Rheology*, edited by E. H. Lee (Interscience, New York, 1965), Vol. 3, p. 30.

Part 3

ELASTICITY AND INHOMOGENEOUS RESPONSE TO STRESS OF DENSE AMORPHOUS PACKINGS

Rigidity Loss Transition in a Disordered 2D Froth

F. Bolton and D. Weaire
Physics Department, Trinity College, Dublin 2, Ireland
(Received 5 October 1990)

Disordered two-dimensional soap froth has been simulated, with a gas fraction ϕ less than unity. Upon decreasing ϕ, it is found that the system loses its rigidity at $\phi_c = 0.84$. This value is identified with the dense random packing of hard disks. Results are presented for the variation of yield stress and shear modulus, the latter being tentatively related to elastic-network theory.

PACS numbers: 82.70.Rr, 68.90.+g

In this Letter we present the first results of simulations of the two-dimensional soap froth with gas fraction ϕ less than unity. We explore the nature of the transition by which this system loses its rigidity as ϕ is decreased. The question is simply: How does foam (which acts as an elastic solid under low stress) fall apart, as it must, if we steadily increase the liquid fraction?

A remarkably simple scenario emerges for this transition, linking it with two classic computational problems—random hard-disk packings and the rigidity of random elastic networks.

Until now, theories of disordered 2D froth structure and properties had the limited objective of understanding the case defined by $\phi = 1$, in which the cell walls (curved lines) meet at point vertices, rather than joining liquid Plateau borders, which are the consequence of $\phi < 1$. Very recently, various discrepancies[1,2] in the comparison of experimental results with theoretical expectations have forced these Plateau borders upon our attention, and it has been realized that even borders of quite modest size can be significant in their effects.[3,4] The theorems[4] and procedures which we used to demonstrate this are of limited applicability, being confined to small, three-sided Plateau borders. If ϕ is decreased to values considerably less than unity, we encounter borders with ever greater numbers of sides. These can only be analyzed by recourse to a fresh approach to the simulation problem, which takes account of them. We have developed the necessary program which can equilibrate large samples in a manner broadly similar to that of the previous work.[5]

With Plateau borders incorporated, the equilibrium conditions are as follows. The cell walls (which are still not given any finite width) join the Plateau borders smoothly; the two border edges have a common tangent with the adjoining cell wall. The border edges themselves have radii of curvature r which satisfy $\Delta p = \sigma r^{-1}$, where σ is surface tension, while the cell walls satisfy $\Delta p = 2\sigma r^{-1}$ as before.[5] The cell areas are fixed and hence the cell pressures are variables, but the pressure is put equal to a common value throughout the Plateau borders, which is adjusted to achieve a specified ϕ. Both liquid and gas are treated as incompressible.

The technical details of the program cannot be detailed here: Its success is self-evident from Fig. 1. This sample of 100 cells was created by the Voronoi procedure,[5] with periodic boundary conditions.

A sequence of structures was generated by reducing ϕ by intervals of 0.01, equilibrating the structure for each successive value. The starting structure at $\phi = 1$ was characterized by $\mu_2 = 1.44$ and $\mu_2^A = 0.15$. These are respectively the second moment of the distribution of numbers of cell sides and that of cell area normalized by division by the square root of mean cell area. The degree of disorder represented by this choice of structure is roughly typical of a mature 2D froth.[6]

Such a simulation allows us to address for the first time the loss of rigidity which must occur as ϕ is decreased, since the system must eventually consist of isolated bubbles.

How and when does this happen? Only an *ordered* hexagonal array of cells has previously been studied.[7] In this case, there is a sudden collapse at $\phi = 0.9069$, with no change in the shear elastic modulus up to that point. This has very limited bearing on the behavior of the typical disordered froth.

In various preliminary runs, we found it impossible to equilibrate structures below $\phi = 0.84$ and percolation of the Plateau borders was evident as the cause. Hence we estimate this to be the critical value ϕ_c for the disordered system. This critical density is indeed distinct from the value for the ordered froth, but it has a simple, and related, significance: it is the packing density for random hard disks. As Bideau and Troadec[8] have shown, there is a wide range of random mixtures of hard disks for which the packing fraction 0.84 ± 0.1 is obtained.

Such random hard-disk packings have coordination number close to $Z = 4$. Bideau and Troadec observed a slightly lower figure but Weaire[9] has argued that this is largely attributable to the effects of rigid wall boundary conditions. The identification of the limiting 2D froth with the dense random hard-disk packing is reinforced by the observation that the value $Z = 4$ is indeed approached as $\phi \rightarrow \phi_c$, as shown in Fig. 2. Note that Z is defined as the average number of neighbors with which a cell makes contact. As triangular Plateau borders

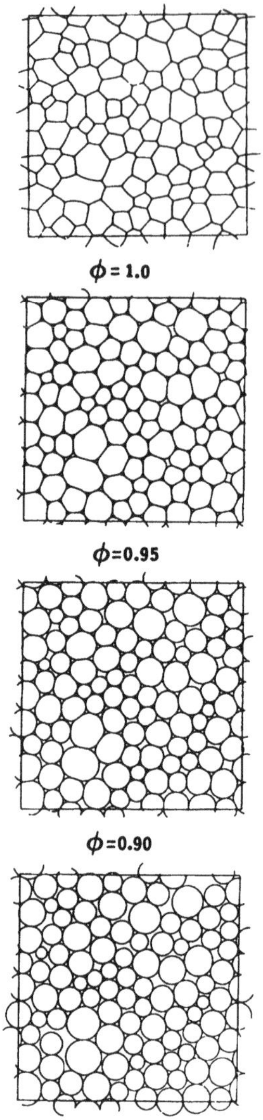

FIG. 1. A sequence of disordered simulated froth structures, with decreasing gas fraction ϕ.

coalesce to form many-sided borders, Z drops from its initial value $Z = 6$, which is fixed by Euler's theorem.

Figure 2 also shows the variation of energy, shear modulus, and yield stress.[10] The dimensionless quantities shown are defined in terms of the mean gas cell area $\bar{A}$, gas fraction ϕ, and surface-tension parameter (energy per unit-cell edge length) σ by

$$E^* = \bar{A}^{1/2} \phi^{-1} \sigma^{-1} E . \tag{1}$$

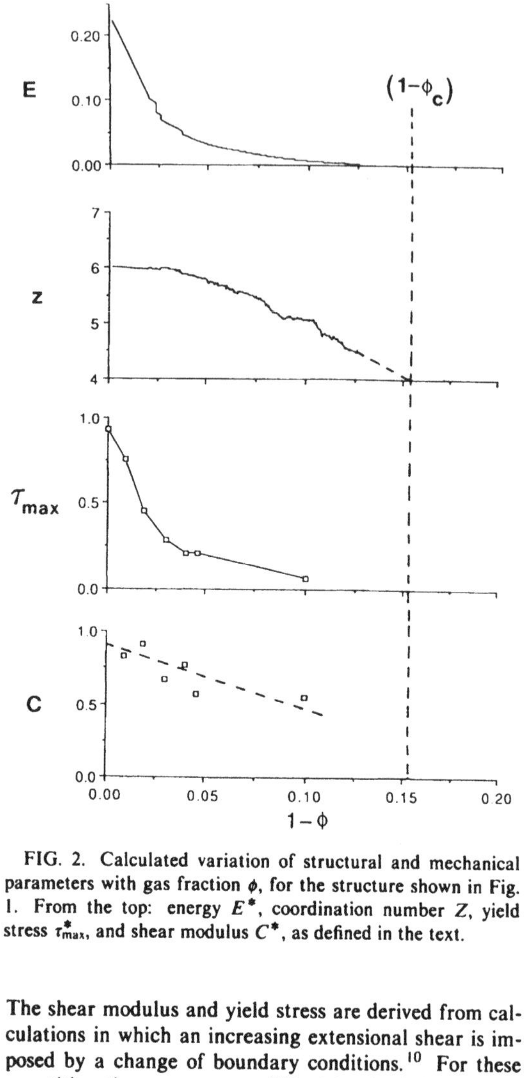

FIG. 2. Calculated variation of structural and mechanical parameters with gas fraction ϕ, for the structure shown in Fig. 1. From the top: energy E^*, coordination number Z, yield stress τ^*_{max}, and shear modulus C^*, as defined in the text.

The shear modulus and yield stress are derived from calculations in which an increasing extensional shear is imposed by a change of boundary conditions.[10] For these quantities there are some difficulties in following the variation into the critical region. In part this is due to the small size of our sample: We intend to undertake studies of much larger ones in the future.

Bearing in mind the variation of Z, this transition is highly reminiscent of that recently investigated for elastic networks, in which elastic springs are randomly severed until there is a loss of rigidity (rigidity percolation).[11] Loosely speaking, the contacts between the disks act as elastic springs, at least close to ϕ_c. These springs act only under compression so that they are effectively severed as a contact is lost, upon decreasing ϕ. Note that the critical value $Z = 4$ recurs in the elastic-network model, as the value at which the number of constraints defined by zero deformation of the springs equals the number of degrees of freedom. This strongly suggests that the prediction of the effective-medium theory,[11] which is also successful for elastic networks, that the shear constant goes to zero *linearly*, should describe this system well in the critical region (although not necessarily at the transition itself).

The yield stress[10] τ_{max} shows a dramatic initial drop with decreasing ϕ, reinforcing our assertion that Plateau border effects can be very large, even when the borders themselves are small. Indeed a simple analysis which treats the elastic modulus as constant and estimates the change in yield strain and hence stress, due to Plateau borders, suggests that the derivative of τ_{max} is infinite at $\phi=1$.

Note that although we have used the word *percolation* from time to time, the rigidity loss transition is *not* provoked by the growth of a percolating Plateau border, as one might suspect. The analogy with elastic-network models is particularly valuable in understanding this. One does not have to cut such a network in two in order to cause it to lose its rigidity.

Such results present a challenge to further experimentation and perhaps an opportunity to review some existing data to determine, for example, Z as a function of ϕ. Recent investigations include not only the prototypical 2D soap froth,[2] but also analogous structures in lipid monolayers.[12,13] Photographs of the latter system[12,13] provide a particularly interesting comparison with our Fig. 1. It may even be possible to perform two-dimensional rheological measurements in this case, an objective which seems unattainable for the soap froth itself.

In conclusion, we have established the broad features of this type of transition for the first time. Its intrinsic interest is reinforced by its relation to hard-disk packings and elastic networks. Ultimately it can offer us insights into the practical problem of three-dimensional foam rheology. Indeed, the entire scenario which we have established should carry over to the three-dimensional case, with minor modifications, such as critical coordination number $Z=6$.

Conversations and correspondence with A. Kraynik were helpful in delineating this problem. Research was funded by Eolas (Irish Science and Technology Agency).

[1] J. Stavans and J. A. Glazier, Phys. Rev. Lett. **62**, 1318 (1989).

[2] J. A. Glazier, Phys. Rev. A **40**, 7398 (1989).

[3] D. Weaire, Phys. Rev. Lett. **64**, 3202 (1990).

[4] F. Bolton and D. Weaire, Philos. Mag. (to be published).

[5] J. P. Kermode and D. Weaire, Comput. Phys. Commun. **60**, 75 (1990).

[6] J. A. Glazier, M. P. Anderson, and G. S. Grest, Philos. Mag. (to be published).

[7] A. Kraynik, Annu. Rev. Fluid Mech. **20**, 325 (1988).

[8] D. Bideau and J. P. Troadec, J. Phys. C **17**, L731 (1984).

[9] D. Weaire, Philos. Mag. B **51**, L19 (1985).

[10] T. L. Fu and D. Weaire, J. Rheol. **32**, 271 (1988).

[11] S. Feng, M. F. Thorpe, and E. Garboczi, Phys. Rev. B **31**, 276 (1985).

[12] K. J. Stine, S. A. Rauseo, B. G. Moore, J. A. Wise, and C. M. Knobler, Phys. Rev. A **41**, 6884 (1990).

[13] B. Berge, A. J. Simon, and A. Libchaber, Phys. Rev. A **41**, 6893 (1990).

Osmotic pressure and viscoelastic shear moduli of concentrated emulsions

T. G. Mason,[1] Martin-D. Lacasse,[2] Gary S. Grest,[2] Dov Levine,[3] J. Bibette,[4] and D. A. Weitz[5]
[1]Department of Chemical Engineering, Johns Hopkins University, Baltimore, Maryland 21218
[2]Corporate Research Science Laboratories, Exxon Research and Engineering Company, Annandale, New Jersey 08801
[3]Department of Physics, Technion, Haifa, 32000 Israel
[4]Centre de Recherche Paul Pascal, Avenue Dr. A. Schweitzer, F-33600 Pessac, France
[5]Department of Physics and Astronomy, University of Pennsylvania, Philadelphia, Pennsylvania 19104
(Received 24 February 1997)

We present an experimental study of the frequency ω dependence and volume fraction φ dependence of the complex shear modulus $G^*(\omega,\varphi)$ of monodisperse emulsions which have been concentrated by an osmotic pressure Π. At a given φ, the elastic storage modulus $G'(\omega)=\mathrm{Re}[G^*(\omega)]$ exhibits a low-frequency plateau G_p', dominating the dissipative loss modulus $G''(\omega)=\mathrm{Im}[G^*(\omega)]$ which exhibits a minimum. Above a critical packing fraction φ_c, we find that both $\Pi(\varphi)$ and $G_p'(\varphi)$ increase quasilinearly, scaling as $(\varphi-\varphi_c)^\mu$, where $\varphi_c \approx \varphi_c^{\mathrm{rcp}}$, the volume fraction of a random close packing of spheres, and μ is an exponent close to unity. To explain this result, we develop a model of disordered droplets which interact through an effective repulsive anharmonic potential, based on results obtained for a compressed droplet. A simulation based on this model yields a calculated static shear modulus G and osmotic pressure Π that are in excellent agreement with the experimental values of G_p' and Π. [S1063-651X(97)15008-5]

PACS number(s): 82.70.Kj, 81.40.Jj, 62.20.Dc

I. INTRODUCTION

An emulsion is an immiscible mixture of two fluids, one of which is dispersed in the continuous phase of the other, typically made by rupturing droplets down to colloidal sizes through mixing. To inhibit recombination, or coalescence, a surfactant which concentrates at the interfaces must be added to create a short-ranged interfacial repulsion between the droplets [1,2]. For an appropriate surfactant, a quantity much less than the mass of the liquids is often sufficient to make this interfacial repulsion strong enough to render the emulsion kinetically stable against coalescence and demixing for many years. This kinetic stability differentiates emulsions from thermodynamically stable microemulsions which form spontaneously without mixing when the proper proportions of certain fluids and surfactants are placed in contact.

Despite being comprised solely of fluids, emulsions consisting of highly concentrated droplets can possess a striking shear rigidity that is characteristic of a solid. The nature of this elasticity is unusual; it exists only because the repulsive droplets have been compressed by an external osmotic pressure Π, and thus concentrated to a sufficiently large droplet volume fraction φ, which permits the storage of interfacial elastic shear energy. For instance, if $\Pi(\varphi)$ approaches the characteristic Laplace pressure required to deform the droplets $(2\sigma/R)$, where σ is the interfacial tension and R is the undeformed droplet radius, the droplets pack together and deform, creating flat facets where neighboring droplets touch. As the osmotic pressure is raised even further, φ tends toward unity, and the emulsion resembles a biliquid foam. Provided the droplets are compressed by an osmotic pressure, additional energy can be stored by imposing shear deformations which create additional droplet surface area; this gives rise to the emulsion's elastic modulus. However, if a concentrated emulsion is diluted, so that the osmotic pressure drops well below the Laplace scale, the resulting emulsion of unpacked spherical droplets loses it shear rigidity and is dominantly viscous, like the suspending fluid. Thus emulsions are versatile materials whose rheological properties can range from viscous to elastic depending on the applied osmotic pressure, and therefore φ. It is precisely this broad range of rheological behavior that gives rise to many technological applications; low viscosity oils can be made effectively rigid if emulsified in water and osmotically compressed to high volume fractions, while high viscosity oils can be made to flow more readily if emulsified at dilute volume fractions in water.

Emulsions possess microscopic mechanisms for both elastic energy storage and viscous dissipation. They are viscoelastic, exhibiting a stress response to a dynamically applied shear strain that is partially liquidlike and partially solidlike. The energy storage and dissipation per unit volume can be represented by the frequency-dependent complex viscoelastic shear modulus $G^*(\omega,\varphi)$, which is defined only for perturbative shears in which the stress and strain are linearly proportional [3,4]. The real part $G'(\omega)=\mathrm{Re}[G^*(\omega)]$, or storage modulus, is the in-phase ratio of the stress with respect to an oscillatory strain, and reflects elastic mechanisms, whereas the imaginary part $G''(\omega)=\mathrm{Im}[G^*(\omega)]$, or loss modulus, is the out-of-phase ratio of the stress with respect to the strain and reflects dissipative mechanisms. Linearity and causality imply that $G'(\omega)$ and $G''(\omega)$ are interrelated by the Kramers-Kronig relations [3–5] indicating their inherent link to the dissipation of shear stress and strain fluctuations in an emulsion. Understanding $G^*(\omega,\varphi)$ for well-controlled emulsions over a wide range of φ would provide valuable insight into the importance of the elastic and dissipative mechanisms as the droplets become packed and deformed.

In this paper, we present experimental measurements of the osmotic pressure and complex shear modulus of monodisperse emulsions compressed to different volume fractions.

By using well-controlled emulsions consisting of droplets of a single size [6,7], our approach offers several advantages over previous rheological experiments [8,9,19] which were made using emulsions having a broad distribution of droplet sizes. Indeed, polydisperse emulsions are difficult to study because they contain droplets with many different Laplace pressures so that, at a fixed osmotic pressure, the large droplets may deform significantly while the small droplets remain essentially undeformed. Moreover, the droplet packing and deformation cannot be easily connected to φ because small droplets can fit into the interstices of larger packed droplets. By contrast, using monodisperse emulsions eliminates these inherent difficulties: all the droplets have the same Laplace pressure. Moreover, the volume fraction can be simply related to the packing of identical spheres, thus allowing for meaningful comparisons with theoretical predictions which have usually assumed that the emulsion is monodisperse and ordered.

The earliest calculations of $\Pi(\varphi)$ and $G(\varphi)$ for emulsions and foams [11–17] are based on perfectly ordered crystals of droplets. In such systems at a given volume fraction and applied shear strain, all droplets are compressed equally and deform affinely under the shear; thus all droplets have exactly the same shape. Describing the dependence of Π and G on φ then reduces to the "simpler" problem of solving for the interfacial shape of a single droplet within a unit cell. Nevertheless, calculating the exact shape and area of such a single droplet at all $\varphi > \varphi_c$ is a very difficult free-boundary problem that can only be solved analytically for simple cases [16], or numerically [16,17]. Real emulsions, however, exhibit a disordered droplet structure, and a comparison of experimental results to these theoretical predictions is inappropriate. In particular, the comparison of the φ dependence of the low-frequency plateau value of the storage modulus of disordered, monodisperse emulsions to the static shear modulus predicted by these studies has demonstrated the existence of significant discrepancies [18].

The origin of the elasticity of an emulsion arises from the packing of the droplets; forces act upon each droplet due to its neighboring droplets pushing on it to withstand the osmotic pressure. However, all these forces must balance to maintain mechanical equilibrium. Calculations of the elastic properties of such disordered packings are complicated by the many different droplet shapes and the necessity of maintaining mechanical equilibrium as the droplets press against one another in differing amounts. While a general theory of the elasticity of disordered packings may ultimately lead to a precise analytical description of emulsion elasticity, computer simulations including adequate interdroplet interactions and accounting for the complexity associated with disorder can provide insight into the origins of the φ-dependent shear modulus. In order to understand the effects introduced by disorder, we developed a model for compressed emulsions which includes a disordered structure as well as realistic droplet deformations [10]. In this model, we formulate an anharmonic potential for the repulsion between the packed droplets, based on numerical results obtained for individual droplets when confined within regular cells [16]. Numerical results for the osmotic pressure Π and the static shear modulus G obtained from this model are in excellent agreement with our experimental values of Π and the elasticity, as can

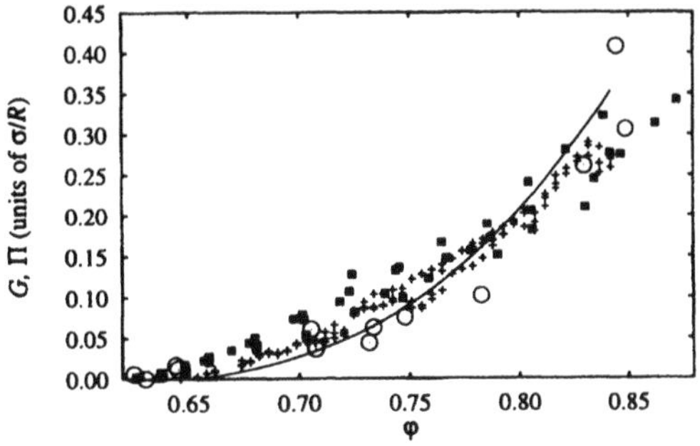

FIG. 1. The scaled shear modulus and osmotic pressure as a function of φ. The computed scaled static shear modulus $G/(\sigma/R)$ (+) and osmotic pressure $\Pi/(\sigma/R)$ (line), as obtained from the model presented in Sec. IV B 2, are compared with the experimental values of $G'_p(\varphi_{\rm eff})$ (■) and $\Pi(\varphi_{\rm eff})$ (○).

be shown from Fig. 1. We measure the frequency dependent storage modulus $G'(w,\varphi)$, and take the low-frequency plateau values $G'_p(\varphi)$ as the static shear modulus $G(\varphi)$. Our model of emulsions as disordered packings of repulsive elements is very general, and may also be applicable to other materials which become elastic under an applied osmotic compression, provided the potential between the elements is appropriately modified.

The structure of this paper is as follows. In Sec. II, we review the theoretical predictions for the osmotic pressure and shear rheology of emulsions. In Sec. III, the experimental aspects of this study are described; Sec. III A describes the emulsion preparation and the rheological measurement techniques; Sec. III B presents the results of our measurements; and Sec. III C compares our experimental observations to existing predictions and previous measurements. In order to understand the difference found between our results and the predictions existing for ordered arrays of droplets, in Sec. IV we present the results of numerical studies based on a model that can account for disorder. In Sec. IV A, we describe the details and the motivation of the model, while, in Sect. IV B we present and discuss the simulation results. A brief conclusion closes the paper.

II. THEORY

In order to understand the properties of packings of deformable spheres, it is useful first to review the packing of static, solid spheres. Their packing determines the critical volume fraction φ_c at which the onset of droplet deformation occurs and the coordination number z_c of nearest neighbors touching a given droplet. The highest volume fraction of monodisperse hard spheres is attained for ordered crystalline structures, including face-centered-cubic (fcc) and hexagonal close packing (hcp). These have $\varphi_c^{\rm cp} = \pi\sqrt{2}/6 \approx 0.74$ and $z_c^{\rm cp} = 12$. By randomly varying the stacking order of the planes, a random hexagonally close-packed (rhcp) structure can be made, but this does not alter either φ_c or z_c. Other ordered packings are less dense. For example, the body-centered-cubic (bcc) packing has $\varphi_c^{\rm bcc} = \pi\sqrt{3}/8 \approx 0.68$ and $z_c^{\rm bcc} = 8$, while the simple cubic (sc) packing has

$\varphi_c^{sc} = \pi/6 \approx 0.52$ and $z_c^{sc} = 6$. The strict definition of a packing excludes conditions of mechanical stability. However, under an interdroplet potential that is purely repulsive and spherically symmetric as the one found in emulsions, both bcc and sc are unstable against weak random mechanical agitations.

By contrast to ordered packings, mechanically stable disordered packings occur at significantly lower volume fractions. By shaking loosely packed macroscopic ball bearings [20], or through entropically driven Brownian motion for colloidal-sized particles, the packing density can be increased up to a reproducible limit termed random close packing (rcp) for which $\varphi_c^{rcp} \approx 0.64$ [21] (conjectured to be $2/\pi$ [22]) and at an average coordination number $\bar{z}_c^{rcp} \approx 6$ [10]. From experimental observation, this is the highest volume fraction at which disordered monodisperse hard spheres can be packed.

While increasing the volume fraction of a dilute colloidal system toward φ^{rcp}, the packing of spheres undergoes an ergodic to nonergodic transition, or a colloidal glass transition, at a value φ_g well below φ^{rcp}. Above φ_g, every sphere is confined into a local region by the cage formed by its neighbors; however, there remains some degree of local translational free volume within its cage. Despite this motion, the global configuration remains locked into a glassy structure, since the probability for a sphere to diffuse out of its cage over a reasonable time scale is essentially zero. Below φ_g, the system exhibits an ergodic behavior. The colloidal glass transition is well-described by mode-coupling theory (MCT), which assumes that the vibrational modes of the glassy structure at different wave vectors are inherently coupled [23]; it predicts that $\varphi_g \approx 0.58$. Light-scattering measurements [24] and mechanical rheological measurements [25] of disordered colloidal hard-sphere suspensions support this prediction for φ_g.

The behavior of an emulsion for $\varphi < \varphi_c$ is expected to be reminiscent of that of hard spheres; any elastic behavior is entropic in nature [25]. We emphasize, however, that the magnitude of this entropic elasticity is significantly lower than that controlled by surface tension, since $k_B T \ll \sigma R^2$; nevertheless, below φ_c it is measurable. As the volume fraction is increased further, one eventually reaches a volume fraction at which the droplets can no longer pack without deforming; for a disordered monodisperse emulsion, this occurs initially at $\varphi \sim \varphi_c^{rcp}$. Since the interactions between emulsion droplets are purely repulsive, work must be done against surface tension to compress and deform the droplets. This work is done through the application of an osmotic pressure and the resulting excess surface area of the droplets determines the equilibrium elastic energy stored at a fixed osmotic pressure. The additional excess surface area created by a perturbative shear deformation determines the static shear modulus $G(\varphi)$. Thus the elasticity and the osmotic pressure are both controlled by the surface tension of the droplets, or their Laplace pressure. Although Π and G represent fundamentally different properties, they both depend on the degree of droplet deformation and therefore φ. In principle, both can be determined if all the droplet shapes are known. These shapes depend upon the overall positional structure, or packing, of the droplets as they press against their neighbors in mechanical equilibrium, and also upon the detailed geometries of the individual contacts.

For an emulsion of oil in water stabilized by an ionic surfactant, each interdroplet contact is in reality a charged system of interfaces oil-surfactant-water-surfactant-oil, making the contact purely repulsive and thus stable against coalescence. The presence of a thin water layer between interacting droplets exists at all volume fractions, including φ near unity. The screened double-layer repulsion has a strength determined by the surface potential and a range characterized by the Debye length λ_D [26]. These depend on the interfacial concentration of the surfactant, the bulk ionic concentration in the aqueous continuous phase, and the temperature. Two droplets forced together will begin to deform before their interfaces actually touch due to the electrostatic repulsion; the droplet system minimizes its total free energy by reducing the energy due to electrostatic repulsion at the expense of creating some additional surface area by deforming the droplet interfaces. Thus, the droplets have an effective radius larger than their actual size, and consequently, they deform for φ below φ_c [27,28]. This electrostatic repulsion can be accounted for by using an effective volume fraction which incorporates a first-order correction for the film thickness h,

$$\varphi_{\text{eff}} \approx \varphi[1 + 3/2(h/R)], \tag{1}$$

for $h \ll R$ [27]; this φ_{eff} represents the actual phase volume fraction of packing, allowing us to account exclusively for the effects of the packing. Although this approximation assumes that the droplets are spherical, it is valid to within 10% even for nearly polyhedral droplets near $\varphi \approx 1$ [29].

A. Osmotic pressure

Just as the structure and interactions between atoms determine the pressure-volume equation of state for homogeneous solids, the structure and interactions (deformability) between droplets determines the osmotic equation of state $\Pi(\varphi)$ of dispersions of droplets. The osmotic equation of state for emulsions governs the (osmotic) compression of the droplets at fixed total droplet volume, allowing the free exchange of solvent with a reservoir [30]. As the droplets are compressed by the osmotic pressure, their total surface area $A(\varphi)$ increases above that of the undeformed droplets A_o, which is, for example, $4\pi N R^2$ for a monodisperse collection of N droplets of undeformed radius R. For any monodisperse emulsion in d dimensions, the osmotic pressure is obtained from

$$\Pi(\varphi)/(\sigma/R) = d\varphi^2 \frac{\partial}{\partial \varphi}\left[\frac{A(\varphi)}{A_o}\right]. \tag{2}$$

In this equation and what follows, we assume that the surface tension is constant. Below φ_c, the droplets are not compressed, so $A(\varphi)$ is constant ($A = A_o$) and any surface tension contribution to Π vanishes (an entropic contribution remains). By contrast, when the droplets are compressed above φ_c, their surface area increases as they press against neighboring droplets and deform, and Π increases.

The droplet response to compression has three characteristic regimes in three dimensions [16]. First, when the drop-

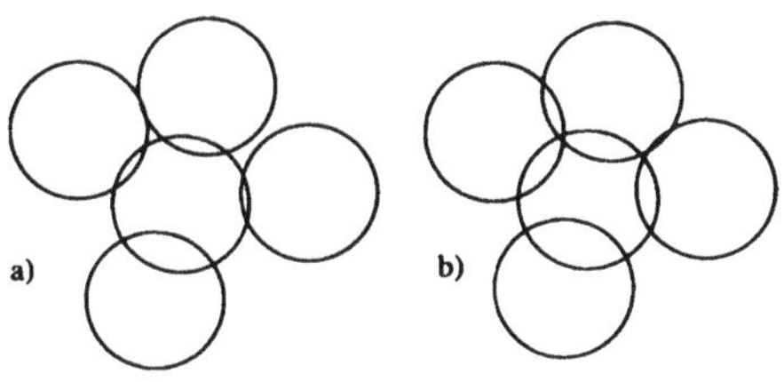

FIG. 2. Schematic representation of a uniform compression of a disordered cluster of droplets. In the initial state (a), the droplets are compressed to $\varphi > \varphi_c$ and are in mechanical equilibrium (surrounding droplets are not shown). The overlap of the circles is meant to indicate the degree of compression schematically, and the central force between each pair of droplets is some function of the overlap. In (b), the compression has been increased uniformly. Only for a Hookian force will the cluster remain in mechanical equilibrium.

lets are infinitesimally deformed, the interaction energy U between two droplets resulting from the increase of surface area is [13,16]

$$U(f)/(\sigma A_o) \sim \left(\frac{fR}{\sigma A_o}\right)^2 \left[a - \ln\left(\frac{fR}{\sigma A_o}\right)\right], \qquad (3)$$

where f is the interdroplet force and a is a dimensionless constant. The range of validity of this expression is rather 'narrow, just after contact, and this behavior is significant only at the onset of the response of ordered emulsions. In terms of interdroplet distance, to first order, this potential can be shown [10] to be equivalent to $U(\xi) \sim -\xi^2/\ln\xi$.

The second regime of the droplet elastic response spans the much broader range of deformation that follows contact, and is therefore the most important when considering the response of real (disordered) emulsions. In this regime, the response of the droplets is *anharmonic*, with a behavior that can be approximated by a power law, where the power depends on the coordination number [10,16]. The anharmonicity of the potential has profound consequences on the deformations: even under a uniform compression, it implies that there are nonaffine particle displacements. To illustrate this, consider a *disordered* system of monodisperse spherical objects interacting with repulsive forces only, and compress the system uniformly at some $\varphi > \varphi_c^{\mathrm{rcp}}$. If the system is in mechanical equilibrium, any droplet i with z_i neighbors will have z_i forces $\mathbf{f}_{ij}$, $j = 1, \ldots, z_i$ acting on it and such that $\Sigma_j \mathbf{f}_{ij} = \mathbf{0}$. Now increase the compression uniformly, which amounts to reducing the interdroplet distances r_{ij} between centers of droplet i and j by a constant factor $\lambda < 1$, as pictured in Fig. 2. For a Hookian force, $\mathbf{f}_{ij}(\lambda \mathbf{r}_{ij}) = \lambda \mathbf{f}_{ij}(\mathbf{r}_{ij})$, and thus mechanical equilibrium remains after the transformation. For any other force, the droplet must move in different amounts so as to achieve a new equilibrium state. Thus, even for a relatively benign case of uniform compression, a nonharmonic force will lead to nonaffine motion of the constituent particles. Here we implicitly assumed that the number of contacts before and after compression is the same, an assumption that is clearly not true for a disordered system. Thus systems with a Hookian interdroplet potential are very likely not to respond affinely, since the creation of any new contact will change the conditions for local mechanical equilibrium.

This anharmonic region is followed by a third regime, in which the droplet response to compression sharply rises, due to volume conservation effects. This is the regime near the biliquid foam limit where $\varphi \to 1$, so that most of the continuous-phase liquid has been extracted; there only remains thin veins for which the radii of curvature of the free surfaces are very small, reflecting large Laplace pressures [19].

To capture the essential predictions for the osmotic equation of state of an ordered emulsion, we first consider the response of an array of droplets to a uniform compression near contact, i.e., at $\varphi \gtrsim \varphi_c$. For this purpose, we introduce a dimensionless displacement ξ defined by

$$\xi = 1 - (\varphi_c/\varphi)^{1/d}. \qquad (4)$$

For an ordered array in d dimensions uniformly compressed at $\varphi > \varphi_c$, ξ is the dimensionless ratio of the perpendicular displacement of the facet toward the droplet center to the undeformed radius R. To obtain a rough estimate of $\Pi(\varphi)$, we make the following assumptions: the compression is assumed to be affine and the logarithmic term of Eq. (3) is ignored. Thus, for $\varphi \gtrsim \varphi_c$, $(\varphi - \varphi_c) \sim \xi$, and hence the energy can be approximated by $U(\varphi)/(\sigma A_o) \sim (\varphi - \varphi_c)^2$, reflecting a Hookian spring force, $f/(\sigma R) \sim \xi$. Using Eq. (2), one finds that the osmotic pressure in the weak compression limit is

$$\Pi(\varphi)/(\sigma/R) = B\varphi^2(\varphi - \varphi_c), \qquad (5)$$

where both constants B and φ_c depend on the geometry of the droplet packing. Since φ^2 varies little in the vicinity of φ_c, the dominant scaling of Π is linear with respect to the difference of φ above packing.

Since the derivation of Eq. (5) relies on the assumptions of affinity and harmonicity, which are true for two-dimensional (2D) systems but are false for real (3D) disordered emulsions, it is not surprising to realize the similarity between this linear scaling form and that of Princen [11,12] who derived it for an ordered 2D monodisperse system of deformable circles of constant area. When the logarithmic corrections are included in the derivation of $\Pi(\varphi)$ for 3D systems [13], the linear form of Eq. (5) is still dominant.

Equation (3) is only valid at infinitesimal compression; it is thus appropriate to consider a more representative interdroplet potential. In fact, Eq. (5) can be shown to be a special case of a more general approach (see the Appendix): if the response of the droplets is assumed to be a power law $U(\xi) \sim [(1-\xi)^{-3} - 1]^\alpha$ [16], then the osmotic pressure obeys

$$\Pi(\varphi)/(\sigma/R) \sim \varphi^2(\varphi - \varphi_c)^{\alpha - 1}. \qquad (6)$$

Since $\alpha \gtrsim 2$ for real droplets moderately compressed by six neighbors, Eq. (5) is somewhat recovered.

B. Static shear modulus

Similarly to the osmotic pressure, the static shear modulus $G(\varphi)$ is determined by the additional deformation of the droplets away from their equilibrium shapes due to a perturbative static strain [31]. Princen [11] analyzed an ordered (monodisperse) 2D array of deformable circles and showed that $G=0$ for $\varphi<\varphi_c$, and then discontinuously jumps to nearly the Laplace pressure at φ_c, reflecting the elasticity of the circles themselves. The existence of an exact solution in two dimensions is possible because the droplet surface (more exactly the perimeter) is parametrized by only one radius of curvature, and therefore the minimum free surface is always an arc of a circle. Three-dimensional problems are much more elaborate. However, using the potential of Eq. (3) and uniaxially straining an emulsion with a sc packing, Buzza and Cates predicted a sharp but continuous rise of the uniaxial static shear modulus $G(\varphi)$ at φ_c [14]. This behavior arises because of the logarithmic divergence in the droplet response at small compression: as two droplets begin to touch ($\xi=0$), the effective spring constant f/ξ increases continuously but very sharply (divergent derivative) from zero to a finite value. As a result, the static shear modulus G of *ordered* emulsions does not exhibit a discontinuity at φ_c, as it would for a harmonic potential, but rather shows a very sharp but continuous rise [14,16]. It is not clear however, that the characteristic onset [cf. Eq. (3)] of the force at infinitesimal compression is determinant in the φ dependence of the static shear modulus for *disordered* emulsions at $\varphi>\varphi_c$. Indeed, while a quasilinear scaling of Π similar to Eq. (5) was measured experimentally for polydisperse (disordered) emulsions [19] (for which it was assumed that $\varphi_c\approx 0.71$), as well as for monodisperse emulsions [18] (for which $\varphi_c\approx\varphi_c^{\rm rcp}$), no sharp rise in $G(\varphi)$ was observed; rather, for both cases, a smooth increase of $G(\varphi)$ was observed at the same φ_c [8,18].

At this point, we have the following picture: the theoretical predictions for the static shear modulus of *ordered* monodisperse emulsions are that $G(\varphi)$ should exhibit a sharp rise (most likely continuous) at φ_c and then continue to increase with $d^2G/d\varphi^2\lesssim 0$; by contrast, existing experimental data for *disordered* monodisperse emulsions [18], which are displayed in Fig. 1, show that the static shear modulus $G(\varphi)$ increases smoothly at φ_c (with $(dG/d\varphi)|_{\varphi_c}\approx 0$), followed by a region of slight positive curvature ($d^2G/d\varphi^2>0$).

To reconcile this difference, the effect of disorder and the interdroplet potential must be taken into account. The behavior of $G(\varphi)$ near φ_c has been investigated in simulations, which unfortunately have been restricted to two dimensions, with disorder introduced through polydispersity [32–35]. These simulations find a jump of the static shear modulus at φ_c and a negative second derivative. We shall investigate these issues in more detail below.

Under strong compression, for which $\varphi\approx 1$ and the highly deformed droplets are nearly polyhedral, the use of the undeformed droplet radius to characterize the Laplace pressure is inadequate. Instead, the Laplace pressure must be obtained from the curvatures at a point on the nonspherical droplet free interface, i.e., $\sigma(R_1^{-1}+R_2^{-1})$, where the R_i's represent the local radii of curvature. One or both these radii can become vanishingly small at the free edges and Plateau borders [7]. This implies that the osmotic pressure needed to remove all the water to create perfectly polyhedral droplets with sharp edges becomes very large, and that the slope of the surface area $A(\varphi)$ diverges as φ approaches unity, regardless of the emulsion's packing structure. For 2D systems, for example, the osmotic pressure is expected to diverge as $\Pi(\varphi)/(\sigma/R)\sim(1-\varphi)^{-1/2}$ [12,34]. By contrast, the static shear modulus clearly does not diverge in the same limit. Assuming that the emulsion can be treated as a biliquid foam, two conditions for mechanical equilibrium must be imposed, namely, that the films meet at equal angles of 120°, and that only four edges can meet at equal tetrahedral angles. For a random 3D isotropic system of flat interfaces, theoretical work suggests that $G(1)/(\sigma/R)\approx 0.55$ at $\varphi\approx 1$ [36]. Princen's measurements [8,19] of polydisperse emulsions qualitatively support these predictions; he reported a divergence of Π and $G(1)/(\sigma/R)\approx 0.5$.

C. Viscoelastic response

While much attention has been given to the static shear modulus, emulsions are in fact viscoelastic. Thus the shear modulus is in reality a function of frequency. Moreover, in addition to a storage, or elastic modulus, they also possess a loss, or viscous modulus. The loss modulus is typically significantly less than the storage modulus for most compressed emulsions, and, as such, has received relatively little attention. The effective viscosity $\eta_{\rm eff}$ of a highly compressed emulsion under low frequency shear of infinitesimal amplitude was predicted by Buzza, Lu, and Cates [15]. Within this theory, the contribution to the viscosity due to capillary flow of the water through the thin films between the droplets is $(R/h)\eta_w$, where η_w is the water viscosity, and the contribution from the surface dilational viscosity of the surfactant monolayer κ_B, as more surface area is created, is κ_B/R:

$$\eta_{\rm eff}=\eta_w\,\frac{R}{h}+\frac{\kappa_B}{R}. \tag{7}$$

Using $\kappa_B\approx 10^{-2}$ P cm, this suggests that the dilational contribution dominates with $\eta_{\rm eff}\approx 10^4\eta_w$ for micrometer-sized droplets with $h\gtrsim 1$ nm. Regardless of the dissipative mechanism or packing structure, this theory implies that the low-frequency behavior of the loss modulus for concentrated emulsions varies linearly with frequency as $G''(\omega)\sim\eta_{\rm eff}\omega$. However, the magnitude of the prefactor predicted theoretically is significantly smaller than that measured experimentally. A possible origin for this anomalous viscous loss for disordered concentrated emulsions was suggested by Liu *et al.* [37]. Their approach allowed for some local, randomly distributed weak regions (faults) within the packing having a zero shear modulus in one plane. By averaging over the random orientations of these planes, an anomalously large contribution to the loss modulus was found, with an unusual frequency dependence, $G''(\omega)\sim\omega^{1/2}$. Due to the Kramers-Kronig relations, a similar power law must contribute to $G'(\omega)$. Both contributions were observed using a light-scattering technique for measuring the high-frequency viscoelastic moduli of emulsions [38].

At volume fractions well below φ_c, emulsion droplets can deform only slightly during momentary collisions with neighbors as they undergo Brownian motion. Thus, at these volume fractions, an emulsion's osmotic pressure and rheological properties should resemble those of hard-sphere suspensions. Since the free volume available for translation V_f of each sphere vanishes as $\varphi \rightarrow \varphi_c$, the entropic energy density, proportional to k_BT/V_f, should diverge for hard spheres. This sets the scale for $\Pi(\varphi)$ and $G'(\omega,\varphi)$, which should also diverge for hard spheres. For emulsions however, a divergence at φ_c is precluded by the possibility of deformation of the droplets. Such entropic contributions arising from the effects of excluded volume have not been incorporated into previous theories of Π and G' for concentrated emulsions of colloidal droplets. Instead, these theories have assumed that the droplets are sufficiently large that entropic contributions to the free energy can be neglected, forcing Π and G' to be zero below φ_c and purely interfacial in origin above φ_c. A complete theory for the viscoelasticity of emulsions must account for the crossover from the entropically dominated regime below φ_c to the interfacially dominated regime above φ_c.

To address this behavior at a heuristic level, a model for $G'(\omega,\varphi)$ and $G''(\omega,\varphi)$ for concentrated emulsions near φ_c has been proposed [25], by analogy with a similar model for concentrated suspensions of hard spheres near the colloidal glass transition. As with hard spheres, we assume that emulsion droplets form a colloidal glass when concentrated to the glass transition volume fraction φ_g, although we allow for the possibility that the deformability of the droplets may slightly alter the observed value of φ_g compared to that of hard spheres. Below φ_g, on the liquid side of the glass transition, MCT makes quantitative predictions for the asymptotic behavior of the temporal (droplet) density autocorrelation function which exhibits the β-relaxation plateau [39]. A universal feature of MCT below φ_g is that the autocorrelation function of any microscopic variable coupled to density fluctuations has the same generic form in the β-relaxation regime, and, therefore, the form for the stress autocorrelation function is predicted to be the same [23]. The glassy contribution to the viscoelastic moduli at low frequencies can be obtained by Fourier transforming the stress autocorrelation function into the frequency domain, while respecting the Kramers-Kronig relations; the magnitude of $G^*(\omega,\varphi)$ is set by the thermodynamic derivative of the stress with respect to strain. Below φ_g, the generic MCT form for the density autocorrelation function leads to a frequency plateau in $G'(\omega,\varphi)$ which reflects entirely entropic energy storage, and a frequency minimum in $G''(\omega,\varphi)$, which reflects rearrangements of the spheres at low ω and internal cage motion at high ω. Above φ_g, the frequency plateau in $G'(\omega,\varphi)$ persists, but the rise in $G''(\omega,\varphi)$ toward low ω disappears as the structural frustration associated with nonergodicity prevents colloidal relaxations of the hard spheres. At higher frequencies, contributions to both G' and G'' proportional to $\omega^{1/2}$ arise from a diffusional boundary layer between the spheres [40,41], and the φ-dependent solvent viscosity [42] contribution to G' proportional to ω. The rheological model superposes the low-frequency MCT and high-frequency contributions to the moduli; this implicitly assumes a wide separation of time scales between each of these processes. This model has provided a successful interpretation of the measured frequency dependencies of the viscoelastic moduli for hard spheres, which include a plateau in $G'(\omega)$ and minimum in $G''(\omega)$ [25], and it may also serve as a basis for understanding the moduli of emulsions when the droplets are not strongly compressed.

III. EXPERIMENT

A. Methodology

To make model monodisperse emulsions suitable for our study, crude polydisperse emulsions of polydimethylsiloxane (PDMS) silicone oil droplets in water are fractionated using a procedure based on a droplet-size-dependent depletion attraction [6]. The surfactant is sodium dodecylsulfate (SDS) at a concentration of $C = 10$ mM, only slightly above the critical micelle concentration, making micelle-induced depletion attractions negligible [6], yet sufficiently large to guarantee good interfacial stability [43]. Our own observations with optical microscopy have confirmed that the droplets are stable against coalescence at all φ studied. We have measured the surface tension of the SDS solution in contact with silicone oil and find $\sigma = 9.8$ dyn/cm using a duNouy ring method. Our emulsions have a polydispersity that has been measured to be about 10% of the radius using angle-dependent dynamic light scattering from a dilute emulsion [7]. Light-scattering measurements of the angle-dependent intensity from concentrated emulsions with $\varphi > 0.6$ confirm that the droplet structure factor resembles that of a disordered glass; at lower φ, a liquidlike structure has been observed [7]. All measurements have been made at room temperature.

We determine the osmotic equation of state $\Pi(\varphi)$, for an emulsion having $R = 0.48$ μm by first setting the osmotic pressure to concentrate a dilute emulsion, waiting for equilibration of φ, and then measuring φ by weighing the emulsion before and after the water has been evaporated. To set Π over a large dynamic range, we use three different techniques in order of decreasing compression: polymer dialysis in which a hydrophilic polymer withdraws water from between the droplets thereby deforming them; centrifugation in which the density difference between the oil and water in the presence of an effective gravity is used to concentrate an initially uniform dispersion of droplets into a cream; and simple creaming in the much lower gravitational field of the earth.

In the polymer dialysis technique [44], a dilute emulsion is enclosed in a semipermeable cellulose bag and immersed in a reservoir of strongly hydrophilic dextran solution having a known osmotic pressure that increases with polymer content. The cellulose bag has a pore size which is much smaller than the droplet radii and radii of gyration of the polymer, so only water and SDS can be freely exchanged between the polymer solution and the emulsion. To prevent destabilization of the droplet interfaces by a loss of SDS from the emulsion, the SDS concentration in the polymer solution is also fixed at 10 mM. The polymer's affinity for water drives water out of the emulsion, thereby raising φ. The measured volume fraction remains constant after two weeks of equilibration; this implies that the emulsion's Π has been set to that of the polymer solution. We repeat this procedure at several different polymer concentrations to apply different Π.

Due to imprecision of the dialysis calibration for $\Pi < 10^4$ dyn/cm^2, we use centrifugation at different speeds to set Π at these lower values. We centrifuge a known amount of dilute emulsion, and then determine φ by skimming a small amount of the creamed emulsion off the top of the column and evaporating the water. After creaming, if all the droplets occupy a distance much less than that of the centrifuge's lever arm, the spatial gradient in the acceleration g can be neglected, and the osmotic pressure at the top can be determined: $\Pi = \ell \Delta\rho g \varphi_i$, where ℓ is the column height, φ_i is the initial volume fraction before centrifugation, and $\Delta\rho$ is the density mismatch between the oil droplets and water. This maximum osmotic pressure reflects the buoyant stress of all droplets below the exposed layer, independent of the spatial gradient in φ, since the total volume of droplets is known. For large Π, equilibration of φ typically takes several hours to one day. As the speed of the centrifuge is lowered to obtain very small Π, the equilibration time becomes many days, making centrifugation impractical. Thus, to achieve the lowest Π, we have allowed an emulsion to cream in the earth's gravity, and after an equilibration time of half a year, we have measured φ of the skimmed cream.

To investigate the dependence of the linear viscoelastic moduli on the droplet size, we measure $G'(\omega)$ and $G''(\omega)$ for four silicone oil-in-water emulsions having radii of $R = 0.25$, 0.37, 0.53, and 0.74 μm using a mechanical controlled-strain rheometer [45]. To set C and φ simultaneously, we first wash the purified emulsion with a SDS solution at $C = 10$ mM, and then we concentrate it to nearly $\varphi \approx 1$ by centrifugation. This highest φ is measured by evaporation of a sample removed from this reservoir. Lower φ are set by diluting samples with a 10-mM SDS solution to the total volume required by the rheometer geometry. All emulsions have been made with PDMS (viscosity $\eta_o = 12$ cP), except for the emulsion with $R = 0.53$ μm made with polyphenylmethylsiloxane (PPMS, $\eta_o = 235$ cP).

In order to measure $G^*(\omega, \varphi)$ at high volume fractions, we employ a cone and plate geometry, while for $\varphi \leqslant 0.60$, we use a double-wall Couette geometry with a larger surface area to increase the rheometer's stress sensitivity. Vigorous preshearing along an applied strain can reduce the measured stress as a result of emulsion fracturing, especially at high φ. Thus our measurements are performed directly after loading the sample. During loading, all emulsions are necessarily presheared perpendicular to the direction of the azimuthally applied strain as the two rheometer surfaces are moved into position; this preshear is radial for the cone and plate geometry, and axial for the double wall Couette geometry. A motor actuates a sinusoidal strain of amplitude γ at a frequency ω, and the magnitude of the stress $\tau(\omega)$, as well as its phase lag relative to the strain $\delta(\omega)$, are detected by a torque transducer. In the linear regime at small strains, the stress is also sinusoidal, and the storage modulus is $G'(\omega) = [\tau(\omega)\cos(\delta(\omega))]/\gamma$, while the loss modulus is $G''(\omega) = [\tau(\omega)\sin(\delta(\omega))]/\gamma$ [3]. By measuring the moduli of an emulsion using both geometries, we verified that the results are reproducible and independent of the geometry. We enclose the emulsion with a water-filled vapor trap to prevent any evaporation that may change φ; this can cause the elasticity of the emulsion to initially grow with time as it develops a skin layer having higher φ. Standing waves in the gap of either geometry are negligible over the range of frequencies and elasticities we probe.

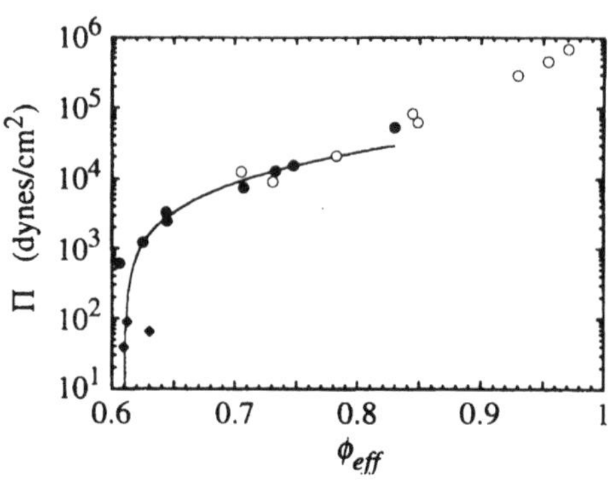

FIG. 3. The osmotic equation of state Π as a function of effective volume fraction, φ_{eff}, of a monodisperse emulsion having radius $R = 0.48$ μm measured using dextran dialysis (○), centrifugation (●), and creaming in the earth's gravitational field (solid diamonds). The solid line is a fit to $\Pi(\varphi_{\mathrm{eff}})$ using the nearly linear weak compression prediction of the single droplet model in Eq. (5) for $\varphi_{\mathrm{eff}} \leqslant 0.80$. The effective volume fraction accounts for the thin films of water and is only slightly different than the oil volume fraction φ (see text).

Some previous measurements of the elastic moduli of emulsions employed a geometry in which slip was purposely induced at the walls of the cell, necessitating a complex correction for its effects to ensure that the proper moduli were determined [8,9]. We follow a different procedure, and ensure that no slip whatsoever occurs along the rheometer walls. We roughen the metal walls of the cells to a length scale somewhat larger than the droplet diameter; this eliminates wall slip [46]. We sandblasted the cone and plate, creating a roughness depth ranging from about 5 to 500 μm, larger than the micrometer-sized droplets. We verified that the measured $G'(\omega)$ and $G''(\omega)$ are the same for larger roughnesses introduced by milling regular grooves of 100 μm or 1 mm in the surfaces. The absence of slip has also been confirmed by varying the gap between the surfaces, and verifying that the measured moduli do not depend on the gap thickness.

B. Experimental results

The osmotic equation of state for an emulsion with $R = 0.48$ μm measured using polymer dialysis (open circles), centrifugation (solid circles), and ordinary creaming (diamonds) is shown in Fig. 3. Near $\varphi_{\mathrm{eff}} \approx 0.6 \approx \varphi_c^{\mathrm{rcp}}$, the osmotic pressure rises sharply, by several orders of magnitude, although the exact nature of this rise is obscured by experimental uncertainty in the measurement of φ, which is accurate to approximately 2%. Good agreement between the centrifugation and dialysis methods can be seen as $\Pi(\varphi)$ continues to rise, albeit less rapidly, well above φ_c. Near $\varphi_{\mathrm{eff}} \approx 1$, the osmotic pressure begins to rise more sharply again, reflecting the resistance of the droplets against assuming polyhedral shapes with small radii of curvature near their edges. For such extreme osmotic compressions, the water films can rupture allowing droplets to fuse, making the emulsion unstable, and the onset of droplet coalescence limits the highest φ we are able to explore.

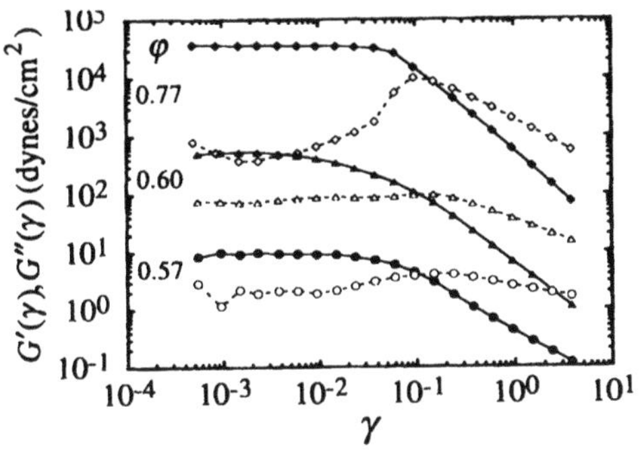

FIG. 4. The γ dependence of the storage, G', (solid symbols) and loss, G'' (open symbols) moduli of a monodisperse emulsion with $R \approx 0.53$ μm, for effective volume fractions of $\varphi_{eff} \approx 0.77$ ($\Diamond$), 0.60 ($\triangle$), and 0.57 ($\bigcirc$), measured at $\omega = 1$ rad/s.

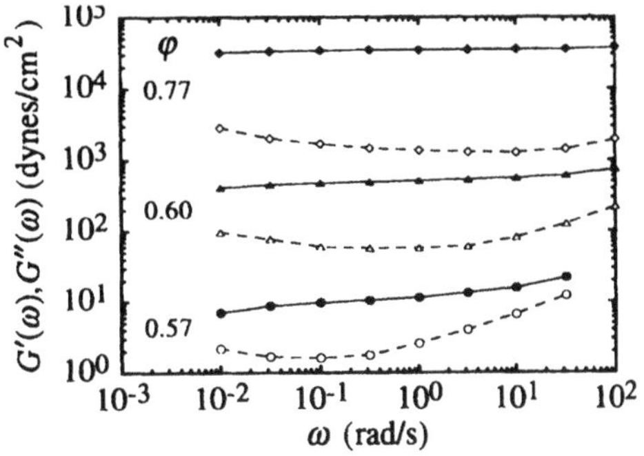

FIG. 5. The frequency dependence of the storage, G', (solid symbols) and loss, G'' (open symbols) moduli of a monodisperse emulsion with $R \approx 0.53$ μm, for $\varphi_{eff} = 0.77$ ($\Diamond$), 0.60 ($\triangle$), and 0.57 ($\bigcirc$). The results for the two larger φ_{eff} were obtained with $\gamma = 0.005$, while those for the lowest were obtained with $\gamma = 0.015$.

To measure $G^*(\omega)$, we first establish the strain regime where the emulsion's stress response is linear. We set the frequency of the applied strain to $\omega = 1$ rad/s and sweep from small to large strain amplitudes to determine the extent of the linear regime. The measured linearity of $G^*(\omega)$ is not noticeably influenced by ω, although its asymptotic magnitude as $\gamma \rightarrow 0$ may vary with ω. The strain dependencies of $G'(\gamma,\varphi)$ and $G''(\gamma,\varphi)$ for a series of volume fractions are shown in Fig. 4 for $R = 0.53$ μm and $\omega = 1$ rad/s. The moduli are independent of strain below $\gamma \approx 0.02$ for the two lowest φ, showing that the linear regime exists only at very small strains. For these low strain values, $G'(\gamma)$ is greater than $G''(\gamma)$, reflecting the emulsion's dominantly elastic nature. At larger strains however, there is a slight but gradual drop in the storage modulus while the loss modulus begins to rise noticeably, indicating the approach to nonlinear yielding behavior and plastic flow. At very large strains, beyond the yield strain marked by the onset of the drop in $G'(\gamma)$, we observe that the temporal stress wave form is not sinusoidal, but becomes flattened at the peaks [47]. Since this response is nonlinear, G' and G'' are not strictly defined here; they are only apparent properties which reflect the peak stress to strain ratio and the phase lag defined by the temporal zero crossing of the stress relative to the strain. At these high values of strain, the apparent G'' dominates the apparent G', reflecting the dominance of energy loss introduced by the nonlinear flow.

To explore the time scales for stress relaxation, we fix a small strain amplitude which lies within the linear regime where G' and G'' are independent of γ, and measure $G^*(\omega)$ as a function of frequency. Using this very low value of peak strain ensures that our spectra reflect the emulsion's true linear moduli and are not influenced by increased dissipation typical at larger γ, in the nonlinear regime. By sweeping ω from high to low, we obtain $G'(\omega,\varphi)$ and $G''(\omega,\varphi)$; these are shown in Fig. 5. At all φ, we observe a low-frequency regime in which $G'(\omega)$ is constant or depends slightly on frequency. At the highest φ, $G'(\omega)$ is essentially independent of ω. At the lowest φ however, a plateau is still observed, but over a narrower frequency range, with $G'(\omega)$ dropping at low frequencies and rising at high frequencies. The low-frequency drop presumably reflects the very slow relaxation of the glassy structure of the emulsion, while the high-frequency rise reflects the fact that the system is comprised solely of fluids, whose viscous behavior dominates at sufficiently high frequencies. In order to define an equivalent to the static shear modulus, we define the plateau value $G_p'(\varphi)$ of the storage shear modulus $G'(\omega,\varphi)$; this is well defined at high volume fractions, while at lower φ, it is defined by the inflection point in $G'(\omega)$. In both cases, it reflects the overall magnitude of the static shear modulus $G(\varphi)$. The measured low-frequency plateau modulus increases over three decades from low to high φ.

By contrast to the plateau behavior of the dominant storage modulus $G'(\omega)$, the smaller loss modulus $G''(\omega)$ exhibits a minimum at frequencies close to the inflection point in $G'(\omega)$. The magnitude of this minimum, $G_m''(\varphi)$, also increases over three decades from lowest to highest φ. The minimum is shallow at the highest φ, but becomes more pronounced at lower φ.

We investigate how the droplet size influences $G_p'(\varphi)$ by examining emulsions having radii $R = 0.25$, 0.37, 0.53, and 0.74 μm. For $\varphi \leq 0.52$, the loss modulus dominates the storage modulus, and therefore $G_p'(\varphi)$ or $G_m''(\varphi)$ cannot be de-

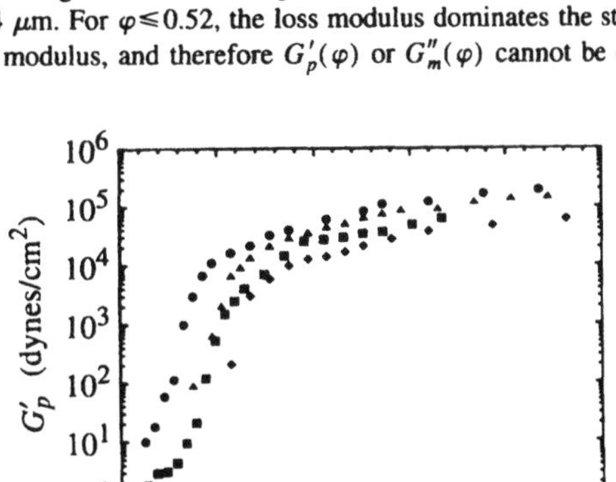

FIG. 6. The plateau storage modulus G_p' as a function of volume fraction for monodisperse emulsions having radii $R = 0.25$ μm (●), 0.37 μm ($\triangle$), 0.53 μm (■), and 0.74 μm ($\Diamond$).

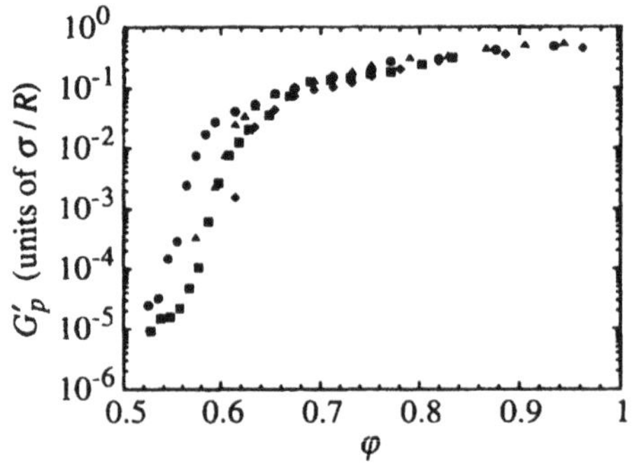

FIG. 7. The volume fraction dependence of the plateau storage modulus $G'_p(\varphi)$, scaled by (σ/R), for four monodisperse emulsions having radii $R=0.25$ μm (●), 0.37 μm (△), 0.53 μm (■), and 0.74 μm (◇).

fined. However, they are well defined at larger φ, and we plot $G'_p(\varphi)$ for different radii in Fig. 6. The plateau modulus for each emulsion rises many orders of magnitude around $\varphi \approx 0.60$. Emulsions comprised of smaller droplets have distinctly smaller φ at which the onset of the rise occurs. At high φ, where the droplets are strongly compressed, G'_p is larger for smaller droplets. By contrast with Π, the plateau modulus does not diverge as φ approaches unity.

To investigate the role of the interfacial deformation of the droplets on the emulsion elasticity, we scale $G'_p(\varphi)$ by (σ/R), and plot the results in Fig. 7. At high φ, this scaling collapses the data for different droplet sizes. However, at low φ there are large systematic deviations from this scaling. To reconcile these apparently different onset volume fractions, we must account for the electrostatic repulsion between the interfaces of droplets stabilized by ionic surfactants; this alters the φ dependences of G and Π. By using φ_{eff} [cf. Eq. (1)] instead of φ, we account for the thin water films stabilizing the charges between the droplets. These thin films will make the apparent packing size of each droplet larger. However, the thickness of the film will be determined by a balance between the screened electrostatic forces between droplets and the deformation of their interfaces. Thus the actual film thickness will be only weakly dependent on droplet size, but will make a relatively larger contribution for the packing of small droplets than for large droplets.

The film thickness itself depends on φ, but in some unknown fashion. Thus we linearly interpolate between a maximum film thickness, $h_{\max}$, at low φ, below rcp, where the droplets are not deformed, and a minimum film thickness $h_{\min}$, between the facets of the nearly polyhedral droplets at $\varphi_{\max}$ near $\varphi \approx 1$. Stable Newton black films of water at a similar electrolyte concentration have been observed with $h_{\min} \approx 50$ Å [48]. This is comparable to the calculated Debye length $\lambda_D \approx 30$ Å, for 10-m*M* SDS solution. Thus we assume that $h_{\min}=50$ Å; this makes a larger correction for the smaller droplets. To determine the maximum film thickness, we vary $h_{\max}$ until the scaled $G'_p(\varphi_{\mathrm{eff}})$ for all droplet sizes collapse onto one universal curve. We find that the film thickness for weak compression which gives the best collapse is $h_{\max}=175$ Å, and is the same for all droplet sizes, as shown in Fig. 8. This film thickness agrees with the mea-

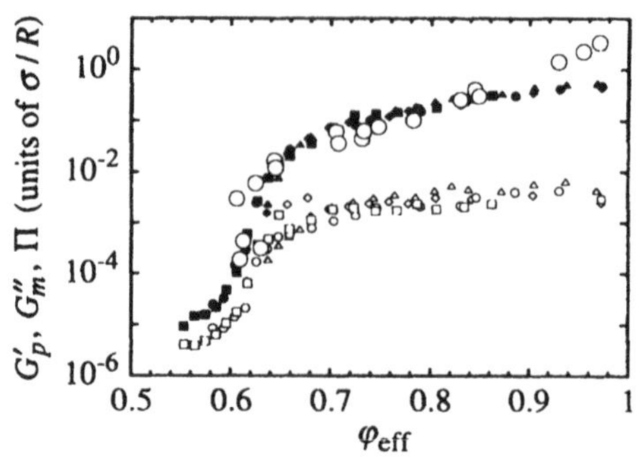

FIG. 8. The scaled plateau storage modulus $G'_p/(\sigma/R)$ (small solid symbols), and the scaled minimum of the loss modulus $G''_m/(\sigma/R)$ (small open symbols), as a function of φ_{eff} for monodisperse emulsions having radii $R=0.25$ μm (○), 0.37 μm (△), 0.53 μm (□), and 0.74 μm (◇). The (○) symbols are the measured values of the scaled osmotic pressure $\Pi/(\sigma/R)$. The maximum film thickness has been adjusted to $h_{\max}=175$ Å to give the best collapse of $G'_p/(\sigma/R)$.

sured separation between the surfaces of monodisperse ferrofluid emulsion droplets at the same SDS concentration [49], lending credence to its value. Near rcp, the film increases the volume fraction more for smaller droplets, about 5% for $R=0.25$ μm, and only 1% for $R=0.74$ μm.

The onset of a large elastic modulus now occurs near rcp, at $\varphi_{\mathrm{eff}} \approx \varphi_c^{\mathrm{rcp}}$, as expected. We note that this value is not a

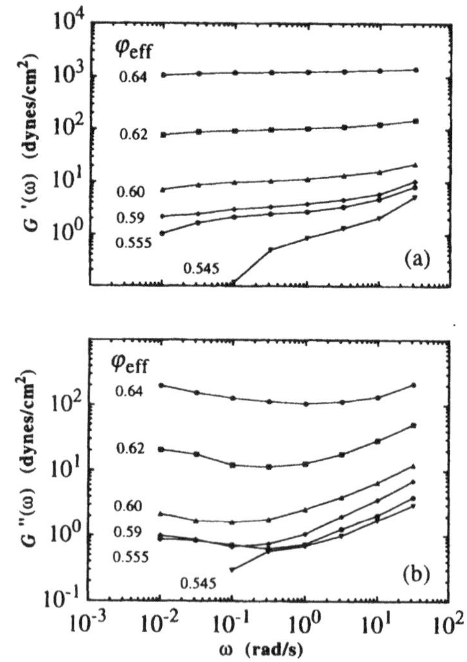

FIG. 9. The frequency dependence of (a) the storage modulus, $G'(\omega)$, and (b) the loss modulus, $G''(\omega)$, for a series of effective volume fractions below the critical packing volume fraction φ_c for $R \approx 0.53$ μm. The lines merely guide the eye.

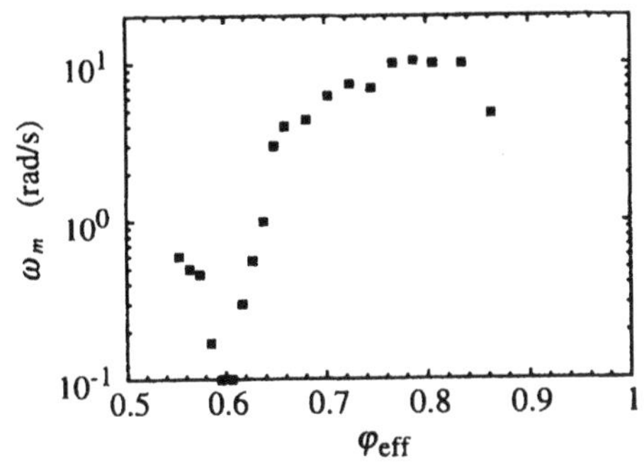

FIG. 10. The effective volume fraction dependence of the frequency where the minimum in the loss modulus occurs, ω_m, for $R=0.53$ μm.

result of our assumption for $h(\varphi)$, but is essentially determined by the measured elastic onset of the largest emulsion, because the 1% adjustment to its φ is very small. The excellent collapse of the data for G_p' and the agreements of $h_{\rm min}$ and $h_{\rm max}$ with independent observations strongly support our use of $\varphi_{\rm eff}$ to account for the electrostatic repulsion between the droplet interfaces.

We also plot the value of the minimum of the loss modulus G_m'' as a function of $\varphi_{\rm eff}$ for each of the droplet sizes in Fig. 8; they also collapse onto a single curve, although the collapse is not as good as for G_p'. At high $\varphi_{\rm eff}$, well above rcp, the elastic modulus is significantly larger than the loss modulus; however, even at lower $\varphi_{\rm eff}$, where the droplets are not deformed, the elastic modulus is still dominant, albeit by not as much. We show more details of this behavior in Fig. 9, where we plot the frequency dependence of both the storage $G'(\omega)$ and the loss $G''(\omega)$ moduli for a series of volume fractions for $R=0.53$ μm. The plateau in G' persists down to $\varphi_{\rm eff}\approx 0.56$ after dropping three orders of magnitude from random close packing; moreover G'' approaches G' as $\varphi_{\rm eff}$ decreases. By contrast, for $\varphi_{\rm eff}\approx 0.55$, $G'(\omega)$ does not exhibit a flat plateau, but instead has a low-frequency dropoff which appears within our measurable frequency range, while $G''(\omega)$ begins to dominate at low frequencies. Measurements at lower $\varphi_{\rm eff}$ are precluded by the stress sensitivity of our rheometer.

The frequency where the minimum in the loss modulus occurs, ω_m, indicates roughly where the contributions from the high- and low-frequency relaxations in the emulsion are equal. Its behavior is plotted as a function of $\varphi_{\rm eff}$ in Fig. 10 for the emulsion having $R=0.53$ μm. Evident is a pronounced dip from $\omega_m\approx 0.5$ rad/s near $\varphi_{\rm eff}=0.57$ to $\omega_m\approx 0.1$ rad/s near $\varphi_{\rm eff}=0.59$, and there is a rapid subsequent rise to nearly a constant value of $\omega_m\approx 10$ rad/s at higher $\varphi_{\rm eff}$. Above this cusp, the frequency of the minimum becomes relatively insensitive to volume fraction, saturating at higher $\varphi_{\rm eff}$.

We have repeated measurements of the frequency spectra, plateau moduli, and minimum in the loss moduli as a function of φ for monodisperse emulsions having a range of oil viscosities: $\eta_o\approx 12$, 235, and 1070 cP. In no case did we observe significant changes in either the magnitudes or frequency dependencies of G' and G''.

C. Discussion

The measured osmotic equation of state is in good agreement with the quasilinear scaling form proposed in Eq. (6). This is shown in Fig. 3 by the solid line, which is a fit to Eq. (5), choosing $\alpha=2$. With this choice, a effective critical value of $\varphi_c=0.60(2)$ is obtained, in reasonable agreement with $\varphi_c^{\rm rcp}$. A similar behavior has been reported for polydisperse emulsions [19], albeit with a considerably larger $\varphi_c\approx 0.72$. This increase in φ_c suggests that polydisperse emulsions can pack more efficiently to higher $\varphi_{\rm eff}$ because smaller droplets can fit in the interstices of larger droplets without deforming. Near $\varphi_{\rm eff}\approx 1$, our data show that Π begins to diverge, with $\Pi\approx\sigma/R$ at $\varphi_{\rm eff}\approx 0.9$, similar to the observed behavior of polydisperse emulsions [19]. This suggests that $\Pi(\varphi_{\rm eff})$ for highly compressed emulsions is relatively insensitive to the polydispersity, with the average droplet size setting the characteristic Laplace pressure scale.

Perhaps the most surprising result comes when we compare the normalized elastic modulus $G_p^{IH}(\varphi)/(\sigma/R)$, with the normalized osmotic pressure $\Pi(\varphi)/(\sigma/R)$. We find that their magnitudes are similar over a range of $\varphi_{\rm eff}$ above φ_c, as shown on a linear plot in Fig. 1. $G_p'(\varphi)$ tracks the osmotic pressure, rising nearly linearly with $\varphi_{\rm eff}$ above the critical volume fraction $\varphi_c\approx 0.64$ corresponding to rcp. The association of this rise with rcp is evidence that the macroscopic rheology is probing the elasticity of the packing of disordered droplets. The similarity between $\Pi(\varphi)$ and $G_p'(\varphi)$ over a large range of φ is reminiscent to the critical-state theory of soil mechanics, where the resistance to shear is proportional to the hydrostatic pressure with a proportionality constant increasing with (soil) packing density [50].

When the droplets are highly compressed, near $\varphi_{\rm eff}\approx 1$, the emulsion's elasticity resembles that of a dry foam and is determined by σ/R. For a disordered monodisperse foam, $G_p'(\varphi)$ is predicted to be $0.55\sigma/R$ [36]. As can be seen in Fig. 8, we find that $G_p'(\varphi)$ approaches $0.6\sigma/R$, in excellent agreement with this prediction. The absence of a divergence of $G_p'(\varphi)$ near $\varphi_{\rm eff}\approx 1$ indicates that volume preserving shear does not cause the local radii of curvature at the droplet's edges and Plateau borders to vanish; instead, the shear merely stretches the interfaces. By contrast, the measured Π does exhibit a pronounced increase in slope as $\varphi_{\rm eff}$ approaches unity. This supports its predicted divergence due to the vanishing radii of curvature as water is squeezed out, although an extensive test of the predicted power law, $\Pi\sim(1-\varphi)^{-1/2}$, is precluded by droplet coalescence.

The existence of a well-defined minimum in $G''(\omega)$ (cf. Fig. 5) at high φ contrasts with the monotonic rise found for a foam [51]. The minimum reflects viscous relaxations at both high and low frequencies. The high-frequency rise in $G''(\omega)$ presumably reflects the increasing importance of the molecular solvent viscosity, while the low-frequency rise reflects glasslike configurational rearrangements of the colloidal droplets. As the droplets become more highly concentrated with increasing $\varphi_{\rm eff}$, they cannot rearrange as easily, so this relaxation is pushed to very low frequencies, permitting the existence of a dominant plateau elasticity. This plateau in $G'(\omega)$ and the corresponding minimum in $G''(\omega)$ have been corroborated by recent dynamic light-scattering measurements of the viscoelastic moduli of concentrated emulsions [52].

The measured frequency dependencies of $G'(\omega,\varphi)$ and $G''(\omega,\varphi)$ for emulsions below φ_c (see Fig. 9) resemble those of glassy hard-sphere suspensions at similar φ [25]. For hard-sphere suspensions, φ is the thermodynamic variable which plays the role of temperature in a normal liquid-glass transition; as it is raised near the glass-transition volume fraction, $\varphi_g \approx 0.58$, a dominant frequency plateau $G'(\omega)$ and a minimum in $G''(\omega)$ have also been observed. These features are the consequences of the glassy relaxation of the droplet configurations, and the rheological behavior can be described using a model based on MCT [25]. This model also accounts for the frequency plateau $G'(\omega)$ and minimum in $G''(\omega)$. This similarity to hard spheres is reasonable, since the droplets are spherical at these φ. Viscosity measurements for dilute φ at the same surfactant concentration have shown that the surface elasticity of the surfactant prevents coupling of flows outside the droplets to their interior [53]. Above φ_g, a hard-sphere suspension loses its low-frequency relaxation and become nonergodic. This implies an ideal zero-frequency elastic modulus in the rheological model [no drop in G' nor rise in $G''(\omega)$ toward low ω]. However, our emulsion data do show evidence of a low-frequency relaxation even for volume fractions well above φ_g where the droplets are highly compressed. This is reflected by the increase in $G''(\omega)$ as ω decreases. This difference suggests that the deformability of the droplets allows a persistent relaxation for emulsions even above φ_g, unlike hard spheres.

Assuming that this minimum frequency is proportional to the β-scaling frequency in simple mode-coupling theory, which is expected to show a dip (cusp) at the glass transition volume fraction [23], we can identify $\varphi_g \approx 0.59$ for our emulsion from Fig. 10. This is similar to $\varphi_g \approx 0.58$ measured for hard-sphere suspensions [24,25]. These observations are evidence that emulsions first become solids at φ_g, although their elastic moduli are entropic in origin and weak compared to moduli dominated by droplet deformation above φ_c. The rise in $G''(\omega)$ toward low frequencies above the glass transition is indicative of structural relaxations that persist above φ_g; it may be possible to account for $\omega_m \approx 10$ rad/s at these volume fractions by using a modified MCT, which can account for additional relaxation due to the possibility of thermally induced deformations of the droplets.

It is surprising that the (σ/R) scaling for $G_p'(\varphi)$ also produces a reasonable collapse of the data for $G_m''(\varphi)$ (see Fig. 8), since this scaling is based on an elastic mechanism associated with energy storage, not dissipation. This observation suggests that the Laplace pressure also sets the scale for the loss modulus, just as it does for the storage modulus. This is consistent with a proposed model for the loss modulus [37]. As with $G_p'(\varphi)$, the magnitude of $G_m''(\varphi)$ increases dramatically near φ_c where the droplets begin to pack. In the dry foam limit, $\varphi_{\text{eff}} \rightarrow 1$, $G_m''(\varphi)$ approaches $4\times10^{-3}\sigma/R$, about two decades lower than $G_p'(\varphi)$. Since the minimum in $G''(\omega)$ cannot be described by the viscosity of Eq. (7) alone, a meaningful comparison with this theoretical model of dissipation is not possible. Instead, our observations suggest that the the Kramers-Kronig relations connecting $G''(\omega)$ to $G'(\omega)$ may lead to an understanding of the scaling of the minima in the loss modulus with volume fraction and droplet size.

The independence of our results on the internal viscosity of the droplets reflects the domination of the surface tension in the deformation of the droplets at our observation frequencies. To estimate the frequency above which this viscosity may be important, we compare the Laplace pressure with the maximum internal viscous stress possible during shear, $\eta_o \gamma\omega \approx \sigma/R$. Solving for the frequency, we find $\omega \approx (\sigma/R)/(\eta_o \gamma)$. For unity strain amplitude, $\sigma = 10$ dyn/cm, and $R = 1$ μm droplets, the frequency for our most viscous droplets ($\eta_o \approx 1000$ cP) would be $\omega \approx 10^4$ s^{-1}, well above our range of mechanical rheometer. This argument agrees with our observations that the behavior is independent of η_o at low frequencies, although it suggests that the spectra at either higher frequencies or for emulsions with much larger internal viscosities may be influenced by η_o.

IV. NUMERICAL STUDIES

A. Model and method

The difference between the theoretical predictions for the φ dependence of the static shear modulus G of ordered emulsions and the experimental data for $G_p'(\varphi)$ leaves us with several unanswered questions. The effects of disorder, the exact form of the potential, the existence of nonaffine relaxation processes all must be investigated in detail. Existing results provided by 2D simulations are not of great relevance, since most of these effects depend strongly on the underlying dimensionality.

The exact deformation of a single droplet under compression has been studied [16] with the help of Brakke's software [54], which triangulates (discretizes) the surfaces to be minimized under a given set of constraints. This procedure is very intensive computationally; thus only relatively small systems can be studied using this approach [17]. The results obtained for the compression of a single droplet, however, provide valuable physical insights on the increase of the droplet surface upon compression. Moreover, the knowledge of the response potential obtained for an individual droplet can in turn be used in a more coarse-grained model which can represent more droplets and thus include the effects of disorder.

A natural candidate for such a coarse-grained model is to represent a collection of N droplets by N pointlike particles confined in a "bulk" system obtained by imposing periodic boundary conditions. The resulting system has a reduced number of $3N$ degrees of freedom. While the total interfacial area of an N-droplet emulsion is essentially a function of $3N$ variables, to a good approximation, the interfacial area of an individual droplet can be described by a function of only the respective positions of all its interacting neighbors. A much cruder approximation, which should be valid at very small compression, consists in approximating the droplet's potential by a sum of two-body interactions and neglecting higher-order terms. While high-order interactions are necessary to account for volume conservation effects such as the divergence of the osmotic pressure at high φ, it is not clear how important they are for moderately compressed emulsions.

In an extensive numerical study [16] of the response of a single droplet to compression by various Wigner-Seitz cells, it was shown that, for moderately compressed emulsions, the interaction potential U can be approximated by a power law,

$$U(\xi)=k'\sigma R^2\xi^\alpha, \tag{8}$$

where k' is some constant, ξ is defined as in Eq. (4), and α is a power larger than 2. The striking point of this study is that high-order terms are important since it was demonstrated that α and k' depend on the number of interacting neighbors. A better fit over a wider range of the data was obtained by

$$U(\xi)=k\sigma R^2[(1-\xi)^{-3}-1]^\alpha, \tag{9}$$

which has the advantage of being reducible to terms in $(\varphi-\varphi_c)$ for ordered structures. For the sake of comparison, we shall use both potentials in the present study, with $k'=3^\alpha k$.

In order to reconcile computational tractability and the inclusion of three-body and higher interactions, we use the following approach for studying disordered structures: we construct disordered systems of hypothetical soft spheres that interact through a two-body, short-range, central-force repulsive pairwise potential represented by a power law [either Eq. (8) or (9)] with a form (coefficient and exponent) depending on the average coordination number of the system. While still mean field in nature, this potential is a definite improvement over simple two-body interaction potentials. Moreover, the anharmonicity of this potential implies that the system will deform nonaffinely, and this model enables us to measure these effects directly.

Under pairwise repulsive potentials, the particles can be thought of as soft compressible spheres, pushing one another and deforming when their separating distance is smaller than the sum of their undeformed radii. The total energy of the system is the sum of all the energy involved in the interacting pairs. The osmotic pressure is obtained from the virial [55].

With the help of this model, we can study the factors influencing the φ dependence of the static shear modulus G, and thus account for our experimental data. For this purpose, we separately investigate the effects of disorder, and the form of the interaction potential on the elastic response.

B. Numerical results and discussion

1. Ordered systems

It is instructive first to investigate the behavior of regular structures of compressible spheres responding to compression with the repulsive potentials introduced in Eqs. (8) and (9). Since ordered structures can be described by a single node, these problems can be solved analytically and, here, the elastic properties of ordered systems are derived for uniform compressions and uniaxial shear deformations. The deformations are always applied along the principal axes of the usual representative unit cells. The details of the calculations are presented in the Appendix.

For all structures, we define the displacement by [generalizing Eq. (4)]

$$\xi=1-r/R, \tag{10}$$

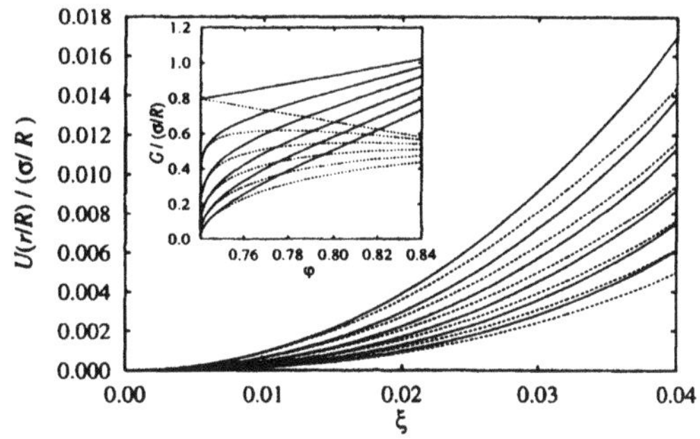

FIG. 11. Two interaction potentials for a maximum displacement corresponding to $\varphi=0.84$. The corresponding static shear modulus of a fcc lattice undergoing a uniaxial strain is shown in the inset. Dashed curves are for Eq. (8) while solid curves are for Eq. (9). Curves are for, top to bottom, $\alpha=2.0$, 2.1, 2.2, 2.3, 2.4, and 2.5.

where R is the undeformed soft sphere radius, and r is half the distance between the centers of two interacting spheres $(r<R)$. By dealing with ordered structures, the shear deformation always leads to the equilibrium (affine) configuration of the system, and, thus, the notion of "center of the droplet" still has a meaning.

A springlike potential only takes into account two-body interactions, and thus cannot capture the effects of volume conservation. It is instructive, however, to see how sensitive the shear modulus of ordered structures is to the form of the potential assumed by our hypothetical soft spheres. Figure 11 compares both potentials of Eqs. (8) and (9), with α ranging from 2.0 to 2.5 ($k=1$ has been kept constant in order to spread the curves). In the inset, the static shear modulus of a fcc lattice of hypothetical soft spheres interacting through these potentials and undergoing a uniaxial shear deformation is presented for the same values of α. Numerical surface calculations of the shape of a single droplet uniformly and moderately compressed in a fcc lattice have shown that the excess surface energy per contact can be fit by Eq. (9) with

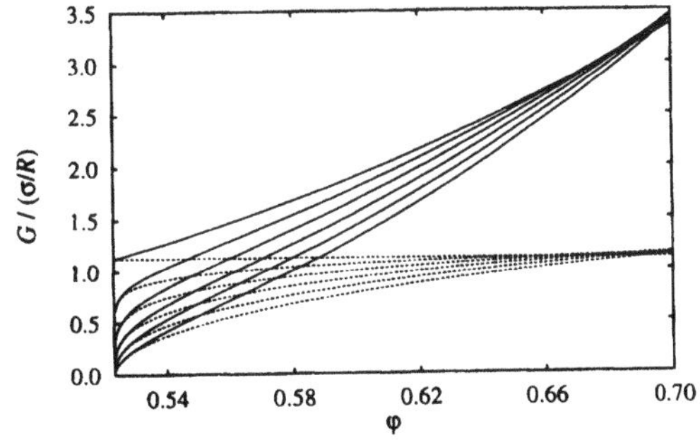

FIG. 12. The static shear modulus of an sc lattice undergoing a uniaxial deformation. The dashed curve are for Eq. (8), while the solid curves are for Eq. (9). Curves are for $\alpha=2.0$, 2.1, 2.2, 2.3, 2.4, and 2.5, from top to bottom.

an exponent $\alpha \approx 2.4$ [16]. The static shear modulus shown in this figure for $\alpha = 2.4$ compares very well with the true uniaxial shear modulus as obtained from numerical calculations of a single droplet uniaxially sheared in a fcc lattice [16].

For anharmonic potentials (i.e., $\alpha > 2$), G exhibits two characteristic features: a sharp onset at φ_c so that $(\partial G/\partial\varphi)|_{\varphi_c}$ diverges, followed by a region where $\partial^2 G/\partial\varphi^2 \lesssim 0$. This is in contrast with experimental data obtained for disordered emulsions, showing a vanishing first derivative at φ_c, followed by an increase of positive curvature. For harmonic potentials ($\alpha = 2$), G exhibits a discontinuity at φ_c [see the analytical form in Eqs. (A4) and (A6)].

Figure 12 shows the same quantities for a sc lattice. Note that for moderate compressions, the true response of a droplet in a sc lattice has an exponent close to 2.2. The static shear modulus obtained from our hypothetical soft spheres interacting with Eq. (9) (and $\alpha = 2.2$) compares very well with the true estimate of G, as obtained from numerical surface calculations [16] or from an expansion at small compression [14].

For a bcc structure, the lattice is unstable under a uniaxial shear along one of its principal axes as such a strain deformation gradually transforms a bcc lattice into an fcc lattice, thus continuously decreasing its energy.

One can also compute the value of the modulus for simple shear strains, as applied along the principal planes (e.g., [100]). The corresponding values of G are only positive for fcc and bcc lattices, since a sc lattice is unstable with respect to this deformation. Calculations of G for simple shear strains give results similar to those obtained for uniaxial deformations, and will therefore not be presented here.

Although a pairwise-potential model can only include radial compressive forces, we demonstrated that it can nevertheless reproduce the qualitative features of the shear moduli of ordered structures, in particular the φ dependence of G.

The calculations of the osmotic pressure are the same for all lattice structures, and are presented in the Appendix. In the vicinity of $\varphi \gtrsim \varphi_c$, both potentials show the same scaling, i.e.,

$$\Pi/(\sigma/R) \approx \frac{k\alpha z_c}{2a_{\text{cell}}}\left[\left(\frac{\varphi}{\varphi_c}\right)-1\right]^{\alpha-1}, \tag{11}$$

where z_c is the coordination number, and a_{cell} is a lattice specific constant (see the Appendix). The osmotic pressure thus contrasts with G, having a smooth rise $((\partial\Pi/\partial\varphi)|_{\varphi_c} \approx 0)$ at φ_c.

2. *Disordered systems*

In order to investigate the effects of disorder on the shear modulus, we numerically study disordered systems of N hypothetical soft spheres interacting through the same potentials Eqs. (8) and (9). The systems are cubic and have periodic boundary conditions (PBC's), with N ranging from 1000 to 4913. Smaller systems were also studied, but because of the combined effects of the softness of the potential and PBC's, these systems have a tendency to order spontaneously in fcc at high (osmotic) pressure.

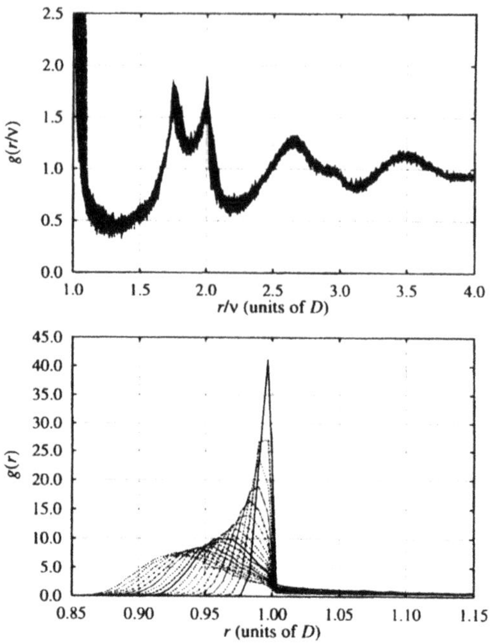

FIG. 13. The radial distribution function of a $N = 4913$ hypothetical soft-sphere system. In (a), the distance has been rescaled by the compression factor $\nu = (\varphi_c/\varphi)^{1/3}$. In (b), the details of $g(r)$ are shown near $r = D$. The curves are for 18 different volume fractions ranging from $\varphi = 0.66$ to 0.85 in steps of ~ 0.01.

The initial configurations are prepared at a volume fraction φ_i by randomly selecting the coordinates of the deformable droplets and then relaxing the system by slowly increasing the potential to its desired value. We find that we must choose $\varphi_i \gtrsim \varphi_c^{\text{rcp}}$ to avoid the slow relaxation observed at random close packing. The systems are then (uniformly) compressed and relaxed in small increments. The relaxation is done by minimizing the energy through a conjugate-gradient algorithm [56] modified to ensure convergence to the closest minimum. At the end of each relaxation, the energy is computed as well as the coordination number and the osmotic pressure. The system is compressed this way until it reaches $\varphi \approx 0.85$, at which value the procedure is reversed, and the shear modulus is computed at each value as φ is decreased.

To investigate the behavior of the packing, in Fig. 13(a) we plot the radial distribution function $g(r)$ of a system of 4913 soft spheres as it is uniformly compressed. The distances are measured in diameter (D) units, and are rescaled by the factor $\nu = (\varphi_c/\varphi)^{1/3} \leqslant 1$. Note how all the curves collapse for large r, showing that there are no large scale rearrangements: contact effects dominate. The radial distribution exhibits the two characteristic peaks of random close packings as discussed by Bernal [57] and Finney [58]: the first one at $r \approx 1.75D$ is related to different local geometries, while the second one is related to colineation of spheres as supported by the sharp drop at $r = 2.0D$ (representing an angle π between three osculatory spheres). These colinea-

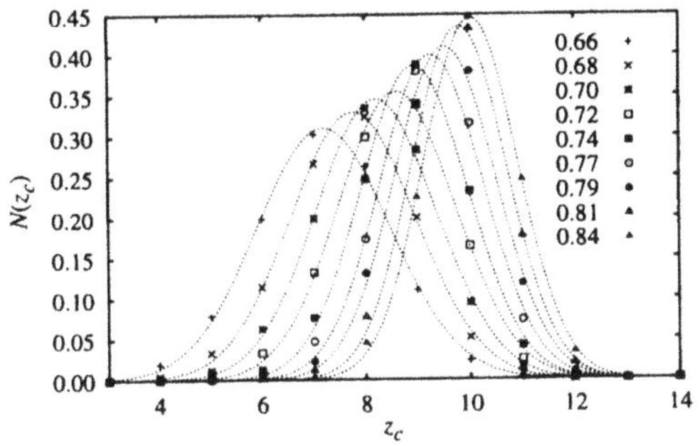

FIG. 14. The probability distribution of the coordination number for the $N=4913$ hypothetical soft sphere system. The curves are fits to Gaussian distributions. The system is uniformly compressed at the volume fractions indicated in the graph.

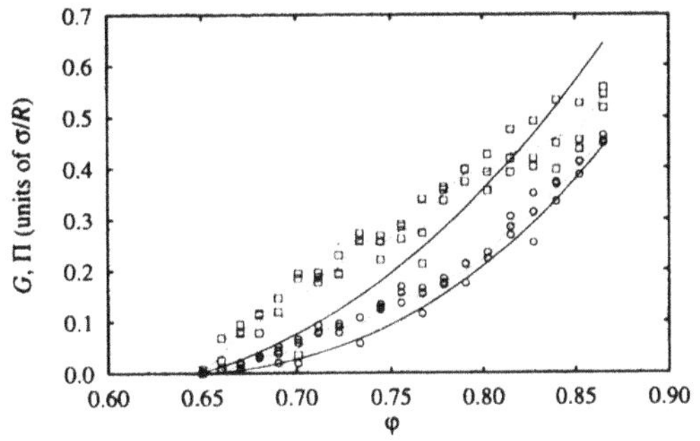

FIG. 15. The static shear modulus of a disordered system undergoing a uniaxial deformation. The potential is as Eq. (9) with $\alpha=2$ (top) and 2.5 (bottom). The symbols represent G (one for a uniaxial strain in each spatial direction) while the solid lines represent Π. (Dotted lines merely guide the eye).

tions, however, are affected by compression, as can be seen from the broadening of the peak: this broadening is due to aligned (angle of π) osculatory triplets that stay aligned and stretched even after compression. This is one indication that the relaxation is nonaffine.

Figure 13(b) shows an expanded view of the contact peak. It is sharp at $\varphi \gtrsim \varphi_c^{\rm rcp}$ and then broadens as the system is compressed, showing that a wide range of interacting contacts is taking place. The sharp cusp at $r=D$ found at moderate φ shows that some spheres can still "escape" from interactions with some of their neighbors. The part of the first peak at $r>D$ represents spheres about to touch: any small compression will bring these in contact, thus increasing the coordination number. This presence of such almost osculatory neighbors is also present in packings of hard spheres, making the evaluation of the coordination number rather difficult, and leading to an overestimate in most cases [59]. For soft compressible sphere systems, these "almost" neighbors play an important role in the elastic response. This effect is better seen from Fig. 14, which shows the probability distribution of the coordination number for different uniform compressions. This plot exhibits some interesting features of disordered systems. The absence of any node having $z_c=12$ at low compression is striking, showing that rcp has short-range order which favors smaller coordination numbers. The curves are well described by a Gaussian, although there seems to be some systematic skewness at the tails. The mean coordination number increases as the system is compressed, while its distribution appears to be narrower. For an emulsion, the increase of the coordination number plays an important role for two reasons: it increases the number of contacts, and it changes the response of the individual droplets. The first effect is captured by the present model, while the second can be taken care of by modifying the interaction potential as the coordination number changes.

The static shear modulus $G(\varphi)$ is obtained by gradually applying a uniaxial strain in small step increments, relaxing the system (always using the same conjugate-gradient method) at each small shear increment. The size of the shear increments has been tested for reliability: the same results are obtained after halving its value, showing that we are in a regime where G does not depend on the value of the shear strain step increment. We determined that a shear step increment of $\Delta\lambda=10^{-4}$ is an optimal value. For most cases, a maximum strain of $\gamma=0.002$ is sufficient to obtain good numerical data while minimizing the computation: our results demonstrate that with this value, we are in the linear, perturbative regime.

Figure 15 shows the shear modulus and the osmotic pressure of a disordered system of $N=3375$ hypothetical soft spheres interacting with a potential of the form of Eq. (9) with fixed values of $\alpha=2.0$ and 2.5. For both sets of curves, we use $k=1$, so that the amplitude is arbitrary and the two sets should not be directly compared. For $\alpha=2$, G and Π have different slopes at the onset, while for $\alpha=2.5$ the two curves are much closer. The curvature $(d^2G/d^2\varphi)$ is slightly negative for Eq. (9) and $\alpha=2$. Simulations of systems of soft spheres interacting with with Eq. (8) and $\alpha=2$ (i.e., a straight harmonic potential) have a negative curvature that is even more pronounced. The results for Eq. (9) and $\alpha=2.5$ definitely show a positive curvature at the onset, similar to the one observed experimentally. We thus see that the form of the potential has an effect on the qualitative response of the shear modulus of disordered emulsions.

For droplets in ordered lattices, the response to compression depends on the number of neighbors, with an exponent that varies approximately linearly from $\alpha \approx 2.0$ at $z_c=6$ (for sc) to $\alpha \approx 2.4$ at $z_c=12$ (for fcc), when fitting surface calculations results to Eq. (9) [16]. In view of determining the elastic response of disordered systems as precisely as possible, we use the following scheme for the potential: we use Eq. (9) with an exponent that varies according to $\alpha=2+0.4(\bar{z}_c-6)/6$, and a coefficient $k(\bar{z}_c)$ obtained from a cubic interpolation scheme between the values obtained for ordered lattices [16]. For a system a $N=1000$ soft spheres, we obtain the curve shown in Fig. 1, where the static shear modulus and the osmotic pressure are compared with the experimental values of the scaled plateau modulus $G_p'(\varphi_{\rm eff})$ and $\Pi(\varphi_{\rm eff})$. For a large range of φ, the agreement between the measurements and simulation results is excellent both in magnitude and overall shape. Such good agreement would not have been obtained without the realistic droplet potential and the disordered droplet structure inherent in the model.

These features of the simulation suggest that the resemblance between $G(\varphi)$ and $\Pi(\varphi)$ found experimentally for emulsions may be fortuitous, resulting from the combined effects of disorder and the particular response of the droplet to compression which can be obtained from purely geometric arguments. However, it may be possible that other heterogeneous dispersions of repulsive elements which interact through more general anharmonic potentials than Eq. (9) may also exhibit the same similarity between $G(\varphi)$ and $\Pi(\varphi)$. Unfortunately, our present knowledge of statistical geometry is not sufficient to either support or rule out such a conjecture.

V. CONCLUSION

Our measurements clearly demonstrate the similarity between the longitudinal $\Pi(\varphi)$, which maintains the static deformation of the droplets, and the transverse $G(\varphi)$ for monodisperse emulsions having a disordered, glassy structure. The monodispersity has enabled us to interpret the strong rise from the entropic to the Laplace scale in terms of packings of monodisperse spheres. In addition, it has allowed us to meaningfully compare the measurements with a three-dimensional model that incorporates both a realistic droplet repulsive potential and a disordered droplet positional structure. The excellent agreement of the simulation based on this model with the experimental results confirm that the origin of this similarity lies in these two essential features.

This central result provides a first insight into the elasticity of disordered packings of identical repulsive elements forced together under an applied osmotic pressure. By contrast to conventional homogeneous solids, in which the moduli of compression and shear are comparable, for disordered heterogeneous solids (e.g., emulsions) the osmotic pressure itself, not the osmotic modulus of compression, may be closer in magnitude to the shear modulus. This similarity may hold for other materials besides emulsions. For instance, our results should be directly applicable to foams comprised of gas bubbles, and they may also provide a guide to the viscoelastic behavior of concentrated microgel beads [60] and multilamellar vesicles [61,62]. The model we have introduced may even provide realistic predictions for their $G(\varphi)$ and $\Pi(\varphi)$, provided an average response of a bead or vesicle to deformation can be calculated.

Despite the success of our model for describing the static elastic modulus and osmotic pressure of compressed emulsions, it cannot predict the full ω dependence of the viscoelastic moduli, since it does not consider dissipative mechanisms. The exact nature of the slow glassy relaxation of the disordered droplet structure, indicated by the rise in $G''(\omega)$ toward small ω, remains obscure for $\varphi>\varphi_g$, especially when the droplets are strongly deformed well above φ_c. In addition to dissipative mechanisms, entropic contributions to $G'(\varphi,\omega)$ and $G''(\varphi,\omega)$ must likewise be included as Π falls well below the Laplace pressure scale, before a meaningful comparison with the experimental data can be made for $\varphi<\varphi_c$. There, the frequency dependence of the viscoelastic moduli exhibits the characteristic rheological features of a colloidal glass: a plateau in $G'(\omega)$, a minimum in $G''(\omega)$, and a frequency associated with minimum $\omega_m(\varphi)$ which exhibits a cusp at the glass transition volume fraction. For undeformed droplets below φ_c, there is a strong similarity between the emulsion's ω-dependent viscoelastic moduli and those of disordered hard spheres, indicating the importance of entropy and the influence of the colloidal phase behavior and glass transition on concentrated emulsion rheology.

This study of the rheology for repulsive monodisperse droplets provides a foundation for comparison with future studies which consider the role of interdroplet attractions and polydispersity. In the case of attractive emulsions, which may form very tenuous solid aggregates or gels of droplets at φ much less than φ_c, the simplicity of our interpretation of the rheology in terms of familiar packings such as rcp may be precluded. In fact, the thermodynamic concept of an osmotic pressure of attractive droplets at dilute φ may be completely different, since the aggregate may not be able to reversibly re-expand once it has been compressed. Likewise, the simple packing interpretation we have used in this study may become much more complicated for polydisperse emulsions. However, by contrast to the past approaches, the effects of polydispersity can now be precisely studied by combining different monodisperse emulsions to generate systematically controllable size distributions. We anticipate that the results of both of these studies will lead to important new results.

ACKNOWLEDGMENTS

We thank Shlomo Alexander, Paul Chaikin, Herman Cummins, Doug Durian, Eric Herbolzheimer, Andrea Liu, David Morse, Tom Witten, and Denis Weaire for many stimulating discussions and suggestions. We are grateful to the *Fonds FCAR du Québec* (M.D.L.), the U.S.-Israel Binational Science Foundation (D.L.), and the NSF (DMR96-31279, D.A.W.) for financial support.

APPENDIX CALCULATION OF G AND Π FOR ORDERED LATTICES

For the sake of demonstration, we derive the shear modulus of a fcc lattice of springlike droplets under a uniaxial strain. To represent the fcc lattice, we choose a unit cell of one node connected to its neighbors by the following 12 vectors:

$$\begin{gathered} 2r'(0,\pm 1,\pm 1)/\sqrt{2}, \\ 2r'(\pm 1,0,\pm 1)/\sqrt{2}, \\ 2r'(\pm 1,\pm 1,0)/\sqrt{2}, \end{gathered} \tag{A1}$$

where r' depends on the amount of uniform compression, which is imposed through a factor $\nu=(\varphi_c/\varphi)^{1/3}\leq 1$: $r'=\nu R$, where R is the equivalent of the undeformed droplet radius; more correctly in the present context, it is half the range of interaction of our springlike potential. In the present case, $\varphi_c=\varphi_c^{\mathrm{cp}}$ is understood although we use φ_c to simplify the notation. This potential is taken to be

$$U(r)=k\sigma R^2\left[\left(\frac{R}{r}\right)^3-1\right]^\alpha, \tag{A2}$$

for $r<R$, and zero otherwise, and k is some constant we define as unity. In order to determine the effect of the exponent α on the shear modulus G, we impose a uniaxial strain on our unit cell: the z direction is stretched by a factor $\lambda = 1+\epsilon$, while the perpendicular xy plane is compressed by a factor $\lambda^{-1/2}$. The volume of our unit cell is unchanged by this transformation and remains $2^{5/2}\nu^3R^3$. Applying a uniform compression and a uniaxial shear to the vectors of Eq. (A1), the excess energy density u of our model is

$$u(\lambda,\nu)/(\sigma/R)=\frac{k}{2\sqrt{2}}\,\nu^{-3}\left\{2\left[\left(\frac{\nu}{\sqrt{2}}\,(\lambda^{-1}+\lambda^2)^{1/2}\right)^{-3}-1\right]^{\alpha}+\left[\left(\frac{\nu}{\sqrt{2}}\,(2\lambda^{-1})^{1/2}\right)^{-3}-1\right]^{\alpha}\right\}. \quad \text{(A3)}$$

The shear modulus is obtained for small ϵ (expanding to second order), using $u(\epsilon,\varphi)-u(0,\varphi)=(3/2)G(\varphi)\epsilon^2$ [63],

$$G(\varphi)/(\sigma/R)=\frac{9k\alpha}{16\sqrt{2}}\,\frac{\varphi^2}{\varphi_c^{\alpha+1}}\,(\varphi-\varphi_c)^{\alpha-2}[\varphi(\alpha-2)+\varphi_c]. \quad \text{(A4)}$$

One can also define a simpler potential

$$U(r)=k'\sigma R^2\left[1-\frac{r}{R}\right]^{\alpha}, \quad \text{(A5)}$$

which is a simple harmonic spring when $\alpha=2$. In this case, we use $k'=3^{\alpha}k$ in order to perform a direct comparison with Eq. (A2). For the same uniaxial strain, the shear modulus of a fcc system of deformable spheres interacting with such a potential is found to be

$$G/(\sigma/R)=\frac{k'\alpha}{16\sqrt{2}}\,\frac{\varphi^{1-\alpha/3}}{(\varphi_c)^{2/3}}\,(\varphi^{1/3}-\varphi_c^{1/3})^{\alpha-2}\times[(\alpha+6)(\varphi_c)^{1/3}-7\varphi^{1/3}]. \quad \text{(A6)}$$

A similar procedure is used for determining the shear modulus G of other types of strains and lattices.

The calculation of the osmotic pressure is independent of the lattice structure, since a uniform compression yields the equilibrium configuration. The osmotic pressure is obtained (at $\epsilon=0$) from the energy density by

$$\Pi=\varphi\,\frac{\partial u}{\partial\varphi}-u. \quad \text{(A7)}$$

For Eq. (A5), one finds

$$\Pi/(\sigma/R)=\frac{k'\alpha z_c}{6a_{\text{cell}}}\left(\frac{\varphi}{\varphi_c}\right)^{2/3}\left[1-\left(\frac{\varphi_c}{\varphi}\right)^{1/3}\right]^{\alpha-1}, \quad \text{(A8)}$$

where z_c and φ_c depend on the lattice and $a_{\text{cell}}=8$, $32/3^{3/2}$, and $2^{5/2}$ for sc, bcc, and fcc, respectively.

For potential Eq. (A2), one finds a similar form, namely,

$$\Pi/(\sigma/R)=\frac{k\alpha z_c}{2a_{\text{cell}}}\,\frac{\varphi^2}{\varphi_c^{1+\alpha}}(\varphi-\varphi_c)^{\alpha-1}. \quad \text{(A9)}$$

Note that for a disordered structure the coordination number $z_c(\varphi)$ depends on the volume fraction.

[1] P. Becher, *Emulsions: Theory and Practice* (Reinhold, New York, 1965).
[2] F. Sebba, *Foams and Biliquid Foams—Aphrons* (Wiley, Chichester, 1987).
[3] R. B. Bird, R. C. Armstrong, and O. Hassager, *Dynamics of Polymeric Liquids* (Wiley, New York, 1977).
[4] J. D. Ferry, *Viscoelastic Properties of Polymers* (Wiley, New York, 1980).
[5] P. M. Chaikin and T. Lubensky, *Principles of Condensed Matter Physics* (Cambridge University Press, Cambridge, 1995).
[6] J. Bibette, J. Colloid Interface Sci. **147**, 474 (1991).
[7] T. G. Mason, A. H. Krall, H. Gang, J. Bibette, and D. A. Weitz, in *Encyclopedia of Emulsion Technology*, edited by P. Becher (Marcel Dekker, New York, 1996), Vol. 4, p. 299.
[8] H. M. Princen, J. Colloid Interface Sci. **112**, 427 (1986).
[9] H. M. Princen, J. Colloid Interface Sci. **105**, 150 (1985)
[10] H. M. Princen, Langmuir **3**, 36 (1987).
[11] H. M. Princen, J. Colloid Interface Sci. **91**, 160 (1983).
[12] H. M. Princen, Langmuir **2**, 519 (1986).
[13] D. C. Morse and T. A. Witten, Europhys. Lett. **22**, 549 (1993).
[14] D. M. A. Buzza and M. E. Cates, Langmuir **10**, 4503 (1994).
[15] D. M. A. Buzza, C.-Y. D. Lu, and M. E. Cates, J. Phys. (France) II **5**, 37 (1995).
[16] M.-D. Lacasse, G. S. Grest, and D. Levine, Phys. Rev. E **54**, 5436 (1996).
[17] A. M. Kraynik and D. A. Reinelt (unpublished).
[18] T. G. Mason and D. A. Weitz, Phys. Rev. Lett. **75**, 2051 (1995).
[19] M.-D. Lacasse, G. S. Grest, D. Levine, T. G. Mason, and D. A. Weitz, Phys. Rev. Lett. **76**, 3448 (1996).
[20] G. D. Scott and D. M. Kilgour, Br. J. Appl. Phys., J. Phys. D **2**, 863 (1969).
[21] J. G. Berryman, Phys. Rev. A **27**, 1053 (1983).
[22] A. Gamba, Nature (London) **256**, 521 (1975).
[23] W. Götze and L. Sjogren, Rep. Prog. Phys. **55**, 241 (1992).
[24] W. van Megen and S. M. Underwood, Phys. Rev. Lett. **70**, 2766 (1993); Phys. Rev. E **49**, 4206 (1994).
[25] T. G. Mason and D. A. Weitz, Phys. Rev. Lett. **75**, 2770 (1995).
[26] J. N. Israelachvili, *Intermolecular and Surface Forces* (Academic, London, 1992).
[27] H. M. Princen, M. P. Aronson, and J. C. Moser, J. Colloid Interface Sci. **75**, 246 (1980).
[28] D. M. A. Buzza and M. E. Cates, Langmuir **9**, 2264 (1993).
[29] K. J. Lissant, in *Emulsion and Emulsion Technology*, edited by K. J. Lissant (Marcel Dekker, New York, 1974), Vol. 1, p. 1.

[30] M. L. McGlashan, *Chemical Thermodynamics* (Academic, New York, 1979).
[31] L. D. Landau and E. M. Lifshitz, *Theory of Elasticity* (Pergamon, Oxford, 1986).
[32] A. M. Kraynik, Annu. Rev. Fluid Mech. **20**, 325 (1988).
[33] F. Bolton and D. Weaire, Phys. Rev. Lett. **65**, 3449 (1990).
[34] S. Hutzler and D. Weaire, J. Phys. Condens. Matter **7**, L657 (1995).
[35] D. Durian, Phys. Rev. Lett. **75**, 4780 (1995); Phys. Rev. E **55**, 1739 (1997).
[36] D. Stamenovic, J. Colloid Interface Sci. **145**, 255 (1991).
[37] A. J. Liu, S. Ramaswamy, T. G. Mason, H. Gang, and D. A. Weitz, Phys. Rev. Lett. **76**, 3017 (1996).
[38] T. G. Mason and D. A. Weitz, Phys. Rev. Lett. **74**, 1250 (1995).
[39] W. Götze and L. Sjogren, Phys. Rev. A **43**, 5442 (1991).
[40] I. M. de Schepper, H. E. Smorenburg, and E. G. D. Cohen, Phys. Rev. Lett. **70**, 2178 (1993).
[41] R. A. Lionberger and W. B. Russel, J. Rheol. **38**, 1885 (1994).
[42] A. J. C. Ladd, J. Phys. Chem. **93**, 3484 (1990).
[43] J. Bibette, Phys. Rev. Lett. **69**, 2439 (1992).
[44] V. A. Parsegian, R. P. Rand, N. L. Fuller, and D. C. Rau, Meth. Enzym. **127**, 400 (1986).
[45] R. W. Whorlow, *Rheological Techniques* (Horwood, Chichester, 1980).
[46] A. Yoshimura, R. K. Prud'homme, H. M. Princen, and A. D. Kiss, J. Rheol. **31**, 699 (1987).
[47] A. S. Yoshimura, and R. K. Prud'homme, Rheol. Acta **26**, 428 (1987).
[48] D. Exerowa, D. Kashchiev, and D. Platikanov, Adv. Colloid Interface Sci. **40**, 201 (1992).
[49] F. L. Calderon, T. Stora, O. M. Monval, P. Poulin, and J. Bibette, Phys. Rev. Lett. **72**, 2959 (1994).
[50] G. Y. Onoda and E. G. Liniger, Phys. Rev. Lett. **64**, 2727 (1990).
[51] S. A. Khan, C. A. Schnepper, and R. C. Armstrong, J. Rheol. **32**, 69 (1988).
[52] T. G. Mason, H. Gang, and D. A. Weitz, J. Opt. Soc. Am. (to be published).
[53] T. G. Mason, J. Bibette, and D. A. Weitz, Adv. Colloid Interface Sci. **179**, 439 (1996).
[54] K. Brakke, Exp. Math. **1**, 141 (1992).
[55] M. P. Allen and D. J. Tildesley, *Computer Simulation of Liquids* (Clarendon, Oxford, 1987).
[56] G. Forsythe, M. Malcolm, and C. Moler, *Computer Methods for Mathematical Computations* (Mir, Moscow, 1980).
[57] J. D. Bernal, Proc. R. Soc. London, Ser. A **280**, 299 (1964).
[58] J. L. Finney, Proc. R. Soc. London, Ser. A **319**, 479 (1970).
[59] J. D. Bernal and J. Mason, Nature (London) **188**, 910 (1960).
[60] E. Bartsch, M. Antonietti, W. Schupp, and H. Sillescu, J. Phys. Chem. **97**, 3950 (1992).
[61] D. Roux, F. Nallet, and O. Diat, Europhys. Lett. **24**, 53 (1993).
[62] P. Panizza, Ph.D. thesis, Université Bordeaux I, 1996.
[63] L. R. G. Treloar, Rep. Prog. Phys. **36**, 755 (1973).

Universality and its origins at the amorphous solidification transition

Weiqun Peng, Horacio E. Castillo, and Paul M. Goldbart
Department of Physics, University of Illinois at Urbana-Champaign, 1110 West Green Street, Urbana, Illinois 61801-3080

Annette Zippelius
Institut für Theoretische Physik, Universität Göttingen, D-37073 Göttingen, Germany
(Received 25 July 1997)

Systems undergoing an equilibrium phase transition from a liquid state to an amorphous solid state exhibit certain universal characteristics. Chief among these are the fraction of particles that are randomly localized and the scaling functions that describe the order parameter and (equivalently) the statistical distribution of localization lengths for these localized particles. The purpose of this paper is to discuss the origins and consequences of this universality, and in doing so, three themes are explored. First, a replica-Landau-type approach is formulated for the universality class of systems that are composed of extended objects connected by permanent random constraints and undergo amorphous solidification at a critical density of constraints. This formulation generalizes the cases of randomly cross-linked and end-linked macromolecular systems, discussed previously. The universal replica free energy is constructed, in terms of the replica order parameter appropriate to amorphous solidification, the value of the order parameter is obtained in the liquid and amorphous solid states, and the chief universal characteristics are determined. Second, the theory is reformulated in terms of the distribution of local static density fluctuations rather than the replica order parameter. It is shown that a suitable free energy can be constructed, depending on the distribution of static density fluctuations, and that this formulation yields precisely the same conclusions as the replica approach. Third, the universal predictions of the theory are compared with the results of extensive numerical simulations of randomly cross-linked macromolecular systems, due to Barsky and Plischke, and excellent agreement is found. [S0163-1829(98)04102-2]

I. INTRODUCTION

During the last decade there has been an ongoing effort to obtain an ever more detailed understanding of the behavior of randomly cross-linked macromolecular systems near the vulcanization transition.[1–4] This effort has been built from two ingredients: (i) the Deam-Edwards formulation of the statistical mechanics of polymer networks;[5] and (ii) concepts and techniques employed in the study of spin glasses.[6] As a result, a detailed mean-field theory for the vulcanization transition—an example of an amorphous solidification transition—has emerged, which makes the following predictions: (i) For densities of cross links smaller than a certain critical value (on the order of one cross-link per macromolecule) the system exhibits a liquid state in which all particles (in the context of macromolecules, monomers) are delocalized. (ii) At the critical cross-link density there is a continuous thermodynamic phase transition to an amorphous solid state, this state being characterized by the emergence of random static density fluctuations. (iii) In this state, a nonzero fraction of the particles have become localized around random positions and with random localization lengths (i.e., rms, displacements). (iv) The fraction of localized particles grows linearly with the excess cross-link density, as does the characteristic inverse square localization length. (v) When scaled by their mean value, the statistical distribution of localization lengths is universal for all near-critical cross-link densities, the form of this scaled distribution being uniquely determined by a certain integrodifferential equation. For a detailed review of these results, see Ref. 4; for an informal discussion, see Ref. 7.

In the course of the effort to understand the vulcanization transition for randomly cross-linked macromolecular systems, it has become clear that one can also employ similar approaches to study randomly *end-linked* macromolecular systems,[8] and also randomly cross-linked *manifolds* (i.e., higher dimensional objects);[9] in each case, a specific model has been studied. For example, in the original case of randomly cross-linked macromolecular systems, the macromolecules were modeled as flexible, with a short-ranged excluded-volume interaction, and the cross links were imposed at random arc-length locations. On the other hand, in the case of *end-linked* systems, although the excluded-volume interaction remained the same, the macromolecules were now modeled as either flexible or stiff, and the random linking was restricted to the ends of the macromolecules. Despite the differences between the unlinked systems and the styles of linking, in all cases identical critical behavior has been obtained in mean-field theory, right down to the precise form of the statistical distribution of scaled localization lengths.

Perhaps even more strikingly, in extensive numerical simulations of randomly cross-linked macromolecular systems, Barsky and Plischke[10,11] have employed an off-lattice model of macromolecules interacting via a Lennard-Jones potential. Yet again, an essentially identical picture has emerged for the transition to and properties of the amorphous solid state, despite the substantial differences between physical ingredients incorporated in the simulation and in the analytical theory.

In light of these observations, it is reasonable to ask whether one can find a common theoretical formulation of the amorphous solidification transition (of which the vulcanization transition is a prime example) that brings to the fore the emergent collective properties of all these systems that are model independent, and therefore provide useful predictions for a broad class of experimentally realizable systems. The purpose of this paper is to explain how this can be done. In fact, we approach the issue in two distinct (but related) ways, in terms of a replica order parameter and in terms of the distribution of random static density fluctuations, either of which can be invoked to characterize the emergent amorphous solid state.

The outline of this paper is as follows. In Sec. II we construct the universal replica Landau free energy of the amorphous solidification transition. In doing this, we review the replica order parameter for the amorphous solid state and discuss the constraints imposed on the replica Landau free energy by (a) symmetry considerations, (b) the smallness of the fraction of particles that are localized near the transition, and (c) the weakness of the localization near the transition. In Sec. III we invoke a physical hypothesis to solve the stationarity condition for the replica order parameter, thereby obtaining a mean-field theory of the transition. We exhibit the universal properties of this solution and, in particular, the scaling behavior of certain central physical quantities. In Sec. IV we describe an alternative approach to the amorphous solidification transition, in which we construct and analyze the Landau free energy expressed in terms of the distribution of static density fluctuations. Although we shall invoke the replica approach in the construction of this Landau free energy, its ultimate form does not refer to replicas. As we show, however, the physical content of this Landau theory is identical to that of the replica Landau theory addressed in Secs. II and III. In Sec. V we exhibit the predicted universality by examining the results of extensive numerical simulations of randomly cross-linked macromolecular networks, due to Barsky and Plischke. In Sec. VI we give some concluding remarks.

II. UNIVERSAL REPLICA FREE ENERGY FOR THE AMORPHOUS SOLIDIFICATION TRANSITION

We are concerned, then, with systems of extended objects, such as macromolecules, that undergo a transition to a state characterized by the presence of random static fluctuations in the particle density when subjected to a sufficient density of permanent random constraints (the character and statistics of which constraints preserve translational and rotational invariance). In such states, translational and rotational symmetry are spontaneously broken, but in a way that is hidden at the macroscopic level. We focus on the long-wavelength physics in the vicinity of this transition.

In the spirit of the standard Landau approach, we envisage that the replica technique has been invoked to incorporate the consequences of the permanent random constraints, and propose a phenomenological mean-field replica free energy, the $n\rightarrow 0$ limit of which gives the disorder-averaged free energy, in the form of a power series in the replica order parameter. We invoke symmetry arguments, along with the recognition that near to the transition the fraction of particles that are localized is small and their localization is weak. The control parameter ϵ is proportional to the amount by which the constraint density exceeds its value at the transition. As we shall see, the stationarity condition for this general, symmetry-inspired Landau free energy is satisfied by precisely the order-parameter hypothesis that exactly solves the stationarity conditions derived from semimicroscopic models of cross-linked and end-linked macromolecules. From the properties of this solution we recover the primary features of the liquid-amorphous solid transition.

In a system characterized by static random density fluctuations, one might naïvely be inclined to use the particle density as the order parameter. However, the disorder-averaged particle density cannot detect the transition between the liquid and the amorphous solid states, because it is uniform (and has the same value) in both states: a subtler order parameter is needed. As shown earlier, for the specific cases of randomly cross-linked[1,3,4] and end-linked[8] macromolecular systems, the appropriate order parameter is instead:

$$\Omega_{\mathbf{k}^1,\mathbf{k}^2,\ldots,\mathbf{k}^g}\equiv\left[\frac{1}{N}\sum_{i=1}^{N}\langle e^{i\mathbf{k}^1\cdot\mathbf{c}_i}\rangle_\chi\langle e^{i\mathbf{k}^2\cdot\mathbf{c}_i}\rangle_\chi\cdots\langle e^{i\mathbf{k}^g\cdot\mathbf{c}_i}\rangle_\chi\right], \quad (1)$$

where N is the total number of particles, $\mathbf{c}_i$ (with $i=1,\ldots,N$) is the position of particle i, the wave vectors $\mathbf{k}^1,\mathbf{k}^2,\ldots,\mathbf{k}^g$ are arbitrary, $\langle\cdots\rangle_\chi$ denotes a thermal average for a particular realization χ of the disorder, and $[\cdots]$ represents averaging over the disorder.

We make the Deam-Edwards assumption[5] that the statistics of the disorder is determined by the instantaneous correlations of the unconstrained system. (It is as if one took a snapshot of the system and, with some nonzero probability, added constraints only at those locations where two particles are in near contact.) This assumption leads to the need to work with the $n\rightarrow 0$ limit of systems of $n+1$, as opposed to n, replicas. The additional replica, labeled by $\alpha=0$, represents the degrees of freedom of the original system before adding the constraints, or, equivalently, describes the constraint distribution.

We assume, for the most part, that the free energy is invariant under the group S_{n+1} of permutations of the $n+1$ replicas. In this replica formalism, the replica order parameter turns out to be

$$\Omega_{\hat{k}}\equiv\left\langle\frac{1}{N}\sum_{i=1}^{N}\exp(i\hat{k}\cdot\hat{c}_i)\right\rangle_{n+1}^{\mathrm{P}}. \quad (2)$$

Here, hatted vectors denote replicated collections of vectors, viz., $\hat{v}\equiv\{\mathbf{v}^0,\mathbf{v}^1,\ldots,\mathbf{v}^n\}$, their scalar product being $\hat{v}\cdot\hat{w}\equiv\sum_{\alpha=0}^{n}\mathbf{v}^\alpha\cdot\mathbf{w}^\alpha$, and $\langle\cdots\rangle_{n+1}^{\mathrm{P}}$ denotes an average for an effective pure (i.e., disorder-free) system of $n+1$ coupled replicas of the original system. We use the terms *one-replica sector* and *higher-replica sector* to refer to replicated vectors with, respectively, exactly one and more than one replica α for which the corresponding vector $\mathbf{k}^\alpha$ is nonzero. In particular, the order parameter in the one-replica sector reduces to the disorder-averaged mean particle density, and plays only a minor role in what follows. The appearance of $n+1$ replicas in the order parameter allows one to probe the correlations between the density fluctuations in the constrained system and the density fluctuations in the unconstrained one.

We first study the transformation properties of the order parameter under translations and rotations, and then make use of the resulting information to determine the possible terms appearing in the replica free energy. Under independent translations of all the replicas, i.e., $\mathbf{c}_i^\alpha \to \mathbf{c}_i^\alpha + \mathbf{a}^\alpha$, the replica order parameter, Eq. (2), transforms as

$$\Omega_{\hat{k}} \to \Omega'_{\hat{k}} = e^{i\Sigma_{\alpha=0}^{n}\mathbf{k}^\alpha\cdot\mathbf{a}^\alpha}\Omega_{\hat{k}}. \tag{3}$$

Under independent rotations of the replicas, defined by $\hat{R}\hat{v} \equiv \{R^0\mathbf{v}^0, \ldots, R^n\mathbf{v}^n\}$ and $\mathbf{c}_i^\alpha \to R^\alpha \mathbf{c}_i^\alpha$, the order parameter transforms as

$$\Omega_{\hat{k}} \to \Omega'_{\hat{k}} = \Omega_{\hat{R}^{-1}\hat{k}}. \tag{4}$$

As discussed in detail in Ref. 4, because we are concerned with the transition between liquid and amorphous solid states, both of which have uniform macroscopic density, the one-replica sector order parameter is zero on both sides of the transition. This means that the sought free energy can be expressed in terms of contributions referring to the higher-replica sector order parameter, alone.

We express the free energy as an expansion in (integral) powers of the replica order parameter $\Omega_{\hat{k}}$, retaining the two lowest possible powers of $\Omega_{\hat{k}}$, which in this case are the square and the cube. We consider the case in which no external potential couples to the order parameter. Hence, there is no term linear in the order parameter. We make explicit use of translational symmetry, Eq. (3), and thus obtain the following expression for the replica free energy (per particle per space dimension) $\mathcal{F}_n(\{\Omega_{\hat{k}}\})$:[12]

$$nd\mathcal{F}_n(\{\Omega_{\hat{k}}\}) = \overline{\sum_{\hat{k}}} g_2(\hat{k})|\Omega_{\hat{k}}|^2 - \overline{\sum_{\hat{k}_1\hat{k}_2\hat{k}_3}} g_3(\hat{k}_1,\hat{k}_2,\hat{k}_3) \times \Omega_{\hat{k}_1}\Omega_{\hat{k}_2}\Omega_{\hat{k}_3}\delta_{\hat{k}_1+\hat{k}_2+\hat{k}_3,\hat{0}}. \tag{5}$$

Here, the symbol $\overline{\Sigma}$ denotes a sum over replicated vectors $\hat{k}$ lying in the higher-replica sector, and we have made explicit the fact that the right-hand side is linear in n (in the $n \to 0$ limit) by factoring n on the left-hand side. In a microscopic approach, the functions $g_2(\hat{k})$ and $g_3(\hat{k}_1,\hat{k}_2,\hat{k}_3)$ would be obtained in terms of the control parameter ϵ, together with density correlators of an uncross-linked liquid having interactions renormalized by the cross linking. Here, however, we will ignore the microscopic origins of g_2 and g_3, and instead use symmetry considerations and a long-wavelength expansion to determine only their general forms. In the saddle-point approximation, then, the disorder-averaged free energy f (per particle and space dimension) is given by[6]

$$f = \lim_{n\to 0} \min_{\{\Omega_{\hat{k}}\}} \mathcal{F}_n(\{\Omega_{\hat{k}}\}). \tag{6}$$

Bearing in mind the physical notion that near the transition any localization should occur only on long length scales, we examine the long-wavelength limit by also performing a low-order gradient expansion. In the term quadratic in the order parameter we keep only the leading and next-to-leading order terms in $\hat{k}$; in the cubic term in the order parameter we keep only the leading term in $\hat{k}$. Thus, the function g_3 in Eq. (5) is replaced by a constant and the function g_2 is expanded to quadratic order in $\hat{k}$. By analyticity and rotational invariance, g_2 can only depend on $\{\mathbf{k}^0,\ldots,\mathbf{k}^n\}$ via $\{|\mathbf{k}^0|^2,\ldots,|\mathbf{k}^n|^2\}$, and, in particular, terms linear in $\hat{k}$ are excluded. In addition, by the permutation symmetry among the replicas, each term $|\mathbf{k}^\alpha|^2$ must enter the expression for g_2 with a common prefactor, so that the dependence is in fact on $\hat{k}^2$. Thus, the replica free energy for long-wavelength density fluctuations has the general form:

$$nd\mathcal{F}_n(\{\Omega_{\hat{k}}\}) = \overline{\sum_{\hat{k}}}\left(-a\epsilon + \frac{b}{2}|\hat{k}|^2\right)|\Omega_{\hat{k}}|^2 - c\overline{\sum_{\hat{k}_1\hat{k}_2\hat{k}_3}}\Omega_{\hat{k}_1}\Omega_{\hat{k}_2}\Omega_{\hat{k}_3}\delta_{\hat{k}_1+\hat{k}_2+\hat{k}_3,\hat{0}}. \tag{7}$$

To streamline the presentation, we take advantage of the freedom to rescale $\mathcal{F}_n$, ϵ, and $\hat{k}$, thus setting to unity the parameters a, b, and c. Thus, the free energy becomes

$$nd\mathcal{F}_n(\{\Omega_{\hat{k}}\}) = \overline{\sum_{\hat{k}}}\left(-\epsilon + \frac{|\hat{k}|^2}{2}\right)|\Omega_{\hat{k}}|^2 - \overline{\sum_{\hat{k}_1\hat{k}_2\hat{k}_3}}\Omega_{\hat{k}_1}\Omega_{\hat{k}_2}\Omega_{\hat{k}_3}\delta_{\hat{k}_1+\hat{k}_2+\hat{k}_3,\hat{0}}. \tag{8}$$

By taking the first variation with respect to $\Omega_{-\hat{k}}$ we obtain the stationarity condition for the replica order parameter:

$$0 = nd\frac{\delta\mathcal{F}_n}{\delta\Omega_{-\hat{k}}} = 2\left(-\epsilon + \frac{|\hat{k}|^2}{2}\right)\Omega_{\hat{k}} - 3\overline{\sum_{\hat{k}_1\hat{k}_2}}\Omega_{\hat{k}_1}\Omega_{\hat{k}_2}\delta_{\hat{k}_1+\hat{k}_2,\hat{k}}. \tag{9}$$

This self-consistency condition applies for all values of $\hat{k}$ lying in the higher-replica sector.

III. UNIVERSAL PROPERTIES OF THE ORDER PARAMETER IN THE AMORPHOUS SOLID STATE

Generalizing what was done for cross-linked and end-linked macromolecular systems, we hypothesize that the particles have a probability q of being localized (also called the "gel fraction" in the context of vulcanization) and $1-q$ of being delocalized, and that the localized particles are characterized by a probability distribution $2\xi^{-3}p(\xi^{-2})$ for their localization lengths ξ. Such a characterization weaves in the physical notion that amorphous systems should show a spectrum of possibilities for the behavior of their constituents, and adopts the perspective that it is this spectrum that one should aim to calculate. This hypothesis translates into the following expression for the replica order parameter:[3,4]

$$\Omega_{\hat{k}} = (1-q)\delta_{\hat{k},\hat{0}} + q\delta^{(d)}_{\tilde{\mathbf{k}},\mathbf{0}}\int_0^\infty d\tau\, p(\tau)e^{-\hat{k}^2/2\tau}, \tag{10}$$

where we have used the notation $\tilde{\mathbf{k}} \equiv \Sigma_{\alpha=0}^{n}\mathbf{k}^\alpha$. The first term on the right-hand side term represents delocalized particles, and is invariant under independent translations of each replica [cf. Eq. (3)]. In more physical terms, this corresponds to the fact that not only the average particle density but the individual particle densities are translationally invariant. The second term represents particles that are localized, and is only invariant under common translations of the replicas (i.e., translations in which $\mathbf{a}^\alpha = \mathbf{a}$ for all α). This corresponds to the fact that the individual particle density for localized particles is not translationally invariant, so that translational invariance is broken microscopically, but the average density remains translationally invariant, i.e., the system still is macroscopically translationally invariant.

By inserting the hypothesis (10) into the stationarity condition (9), and taking the $n \to 0$ limit, we obtain

$$0 = \delta^{(d)}_{\mathbf{k},\mathbf{0}} \left\{ 2(3q^2 - \epsilon q + q\hat{k}^2/2) \int_0^\infty d\tau \, p(\tau) e^{-\hat{k}^2/2\tau} - 3q^2 \int_0^\infty d\tau_1 p(\tau_1) \int_0^\infty d\tau_2 p(\tau_2) e^{-\hat{k}^2/2(\tau_1+\tau_2)} \right\}. \quad (11)$$

In the limit $\hat{k}^2 \to 0$, the equation reduces to a condition for the localized fraction q, viz.,

$$0 = -2q\epsilon + 3q^2. \quad (12)$$

For negative or zero ϵ, corresponding to a constraint density less than or equal to its critical value, the only physical solution is $q=0$, corresponding to the liquid state. In this state, all particles are delocalized. For positive ϵ, corresponding to a constraint density in excess of the critical value, there are two solutions. One, unstable, is the continuation of the liquid state $q=0$; the other, stable, corresponds to a nonzero fraction,

$$q = \frac{2}{3}\epsilon \quad (13)$$

being localized. We identify this second state as the amorphous solid state. From the dependence of the localized fraction q on the control parameter ϵ and the form of the order parameter Eq. (10) we conclude that there is a continuous phase transition between the liquid and the amorphous solid states at $\epsilon = 0$, with localized fraction exponent $\beta = 1$ (i.e., the classical exponent[13]). It is worth mentioning that microscopic approaches go beyond this linear behavior near the transition, yielding a transcendental equation for $q(\epsilon)$, valid for all values of the control parameter ϵ; see Refs. 3, 4.

From Eq. (8) it is evident that the liquid state is locally stable (unstable) for negative (positive) ϵ: the eigenvalues of the resulting quadratic form are given by $\lambda(\hat{k}) = -\epsilon + \hat{k}^2/2$.

Now concentrating on the amorphous solid state, by inserting the value of the localized fraction into Eq. (11), we obtain the following integrodifferential equation for the probability distribution for the localization lengths:

$$\frac{\tau^2}{2}\frac{dp}{d\tau} = \left(\frac{\epsilon}{2} - \tau\right) p(\tau) - \frac{\epsilon}{2} \int_0^\tau d\tau_1 p(\tau_1) p(\tau - \tau_1). \quad (14)$$

All parameters can be seen to play an elementary role in this equation by expressing $p(\tau)$ in terms of a scaling function:

$$p(\tau) = (2/\epsilon)\pi(\theta); \quad \tau = (\epsilon/2)\theta. \quad (15)$$

Thus, the universal function $\pi(\theta)$ satisfies

$$\frac{\theta^2}{2}\frac{d\pi}{d\theta} = (1-\theta)\pi(\theta) - \int_0^\theta d\theta' \pi(\theta')\pi(\theta - \theta'). \quad (16)$$

Solving this equation, together with the normalization condition $1 = \int_0^\infty d\theta \pi(\theta)$, one finds the scaling function shown in Refs. 3,4. The function $\pi(\theta)$ has a peak at $\theta \simeq 1$ of width of order unity, and decays rapidly both as $\theta \to 0$ and $\theta \to \infty$. By combining these features of $\pi(\theta)$ with the scaling transformation (15) we conclude that the typical localization length scales as $\epsilon^{-1/2}$ near the transition. The order parameter also has a scaling form near the transition:

$$\Omega_{\hat{k}} = [1 - (2\epsilon/3)]\delta_{\hat{k},\hat{0}} + (2\epsilon/3)\delta^{(d)}_{\mathbf{k},\mathbf{0}}\, \omega(\sqrt{2\hat{k}^2/\epsilon}),$$

$$\omega(k) = \int_0^\infty d\theta \pi(\theta) e^{-k^2/2\theta}. \quad (17)$$

Equation (16) and the normalization condition on $\pi(\theta)$ are precisely the conditions that determine the scaling function for the cross-linked and end-linked cases; they are discussed extensively, together with the properties of the resulting distribution of localization lengths and order parameter, in Refs. 3, 4, and 8.

As discussed in this section, the localized fraction $q(\epsilon)$ and the scaled distribution of inverse square localization lengths $\pi(\theta)$ are universal near the transition. We now discuss this issue in more detail.

First, let us focus at the mean-field level. Recall the mean-field theory of ferromagnetism[14] and, in particular, the exponent β, which characterizes the vanishing of the magnetization density order parameter (from the ferromagnetic state) as a function of the temperature at zero applied magnetic field. The exponent β takes on the value of 1/2, regardless of the details of the mean-field theory used to compute it. The functions $q(\epsilon)$, $\pi(\theta)$, and $\omega(k)$ are universal in the same sense. The case of $q(\epsilon)$ is on essentially the same, standard, footing as that of the magnetization density. What is not standard, however, is that describing the (equilibrium) order parameter is a universal scaling *function*, $\omega(k)$ [or, equivalently, $\pi(\theta)$] that is not a simple power law. This feature arises because the usual presence of fields carrying *internal* indices, such as Cartesian vector indices in the case of ferromagnetism, is replaced here by the external continuous replicated wave-vector variable $\hat{k}$. There are two facets to this universal scaling behavior of the order parameter. First, for systems differing in their microscopic details and their constraint densities there is the possibility of collapsing the distribution of localization lengths on to a single function, solely by rescaling the independent variable. Second, there is a definite prediction for the dependence of this rescaling on the control parameter ϵ.

Now, going beyond the mean-field level, in the context of vulcanization de Gennes[15] has shown that the width of the critical region, in which fluctuations dominate and mean-field theory fails, vanishes in the limit of very long macromolecules in space dimension $d=3$ or higher. Thus, one may anticipate that for extended objects the mean-field theory discussed here will be valid, except in an exceedingly narrow region around the transition. Nevertheless, if—as is usually the case—the effective Hamiltonian governing the fluctuations is the Landau free energy then the universality discussed here is expected to extend, *mutatis mutandis*, into the critical regime.

That the amorphous solid state given by Eq. (17) is stable with respect to small perturbations (i.e., is locally stable) can be shown by detailed analysis. Moreover, as we shall see in Sec. V, it yields predictions that are in excellent agreement with subsequent computer simulations. However, there is, in principle, no guarantee that this state is globally stable (i.e., that no states with lower free energy exist).

Up to this point we have assumed that the free energy is invariant under interchange of all replicas, including the one representing the constraint distribution ($\alpha=0$), with any of the remaining n, i.e., that the free energy is symmetric under the group S_{n+1} of permutations of all $n+1$ replicas. This need not be the case, in general, as the system can be changed, (e.g., by changing the temperature) after the constraints have been imposed. In this latter case, the free energy retains the usual S_n symmetry under permutations of replicas $\alpha=1,\dots,n$. The argument we have developed can be reproduced for this more general case with only a minor change: in the free energy, we can no longer invoke S_{n+1} symmetry to argue that all of the $|\mathbf{k}^{\alpha}|^2$ must enter the expression for g_2 with a common prefactor. Instead, we only have permutation symmetry among replicas $\alpha=1,\dots,n$ and, therefore, the prefactor b for all of these replicas must be the same, but now the prefactor b_0 for replica $\alpha=0$ can be different. This amounts to making the replacement

$$\hat{k}^2 \to \overline{k^2} \equiv b_0 b^{-1} |\mathbf{k}^0|^2 + \sum_{\alpha=1}^{n} |\mathbf{k}^{\alpha}|^2 \tag{18}$$

in the free energy. Both the rest of the derivation and the results are unchanged, except that $\hat{k}^2$ needs to be replaced by $\overline{k^2}$, throughout. We mention, in passing, that no saddle points exhibiting the spontaneous breaking of replica permutation symmetry have been found, to date, either for systems with S_{n+1} or S_n symmetry.

IV. FREE ENERGY IN TERMS OF THE DISTRIBUTION OF STATIC DENSITY FLUCTUATIONS

The aim of this section is to construct an expression for the disorder-averaged Landau free energy for the amorphous solidification transition, $\mathcal{F}$, in terms of the distribution of static density fluctuations. We present this approach as an alternative to the strategy of constructing a replica free energy $\mathcal{F}_n$ in terms of the replica order parameter Ω. In the familiar way, the equilibrium state will be determined via a variational principle: $\delta\mathcal{F}=0$ and $\delta^2\mathcal{F}>0$. What may be less familiar, however, is that in the present setting the *independent* variables for the variation (i.e., the distribution of static density fluctuations) themselves constitute a functional.

Our aim, then, is to work not with the replica order parameter $\Omega_{\hat{k}}$, but instead with the disorder-averaged probability density functional for the random static density fluctuations,[16,4] $\mathcal{N}(\{\rho_{\mathbf{k}}\})$, which is defined via

$$\mathcal{N}(\{\rho_{\mathbf{k}}\}) \equiv \left[\frac{1}{N} \sum_{i=1}^{N} \prod_{\mathbf{k}} \delta_c(\rho_{\mathbf{k}} - \langle \exp(i\mathbf{k}\cdot\mathbf{c}_i)\rangle_{\chi}) \right]. \tag{19}$$

Here, $\Pi_{\mathbf{k}}$ denotes the product over all d vectors $\mathbf{k}$, and the Dirac δ function of complex argument $\delta_c(z)$ is defined by $\delta_c(z) \equiv \delta(\mathrm{Re}\, z)\delta(\mathrm{Im}\, z)$, where $\mathrm{Re}\, z$ and $\mathrm{Im}\, z$, respectively, denote the real and imaginary parts of the complex number z. From the definition of $\mathcal{N}(\{\rho_{\mathbf{k}}\})$, we see that $\rho_{-\mathbf{k}} = \rho_{\mathbf{k}}^*$ and $\rho_0 = 1$. Thus we can take as independent variables $\rho_{\mathbf{k}}$ for all d vectors $\mathbf{k}$ in the half-space given by the condition $\mathbf{k}\cdot\mathbf{n}>0$ for a suitable unit d vector $\mathbf{n}$. In addition, $\mathcal{N}(\{\rho_{\mathbf{k}}\})$ obeys the normalization condition

$$\int \mathcal{D}\rho\, \mathcal{N}(\{\rho_{\mathbf{k}}\}) = 1. \tag{20}$$

It is straightforward to check that, for any particular positive integer g, the replica order parameter $\Omega_{\hat{k}}$ is a gth moment of $\mathcal{N}(\{\rho_{\mathbf{k}}\})$:

$$\int \mathcal{D}\rho\, \mathcal{N}(\{\rho_{\mathbf{k}}\}) \rho_{\mathbf{k}^1}\rho_{\mathbf{k}^2}\cdots\rho_{\mathbf{k}^g} = \Omega_{\mathbf{k}^1,\mathbf{k}^2,\dots,\mathbf{k}^g}. \tag{21}$$

We use $\mathcal{D}\rho$ to denote the measure $\Pi_{\mathbf{k}}\, d\mathrm{Re}\rho_{\mathbf{k}}\, d\mathrm{Im}\, \rho_{\mathbf{k}}$.

The merit of the distribution functional $\mathcal{N}(\{\rho_{\mathbf{k}}\})$ over the replica order parameter $\Omega_{\hat{k}}$ is that, as we shall soon see, it allows us to formulate a Landau free energy for the amorphous solidification transition, depending on $\mathcal{N}(\{\rho_{\mathbf{k}}\})$, in which replicated quantities do not appear, while maintaining the physical content of the theory. At the present time, this approach is not truly independent of the replica approach, in the following sense: we employ the replica approach to derive the free energy, Eq. (8), and only then do we transform from the language of order parameters to the language of the distribution of static density fluctuations. We are not yet in possession of either an analytical scheme or a set of physical arguments that would allow us to construct the Landau free energy directly. Nevertheless, we are able, by this indirect method, to propose a (replica-free) free energy, and also to hypothesize (and verify the correctness of) a stationary value of $\mathcal{N}(\{\rho_{\mathbf{k}}\})$. It would, however, be very attractive to find a scheme that would allow us to eschew the replica approach and work with the distribution of static density fluctuations from the outset.

To proceed, we take the replica Landau free energy $\mathcal{F}_n$, Eq. (8), in terms of the replica order parameter $\Omega_{\hat{k}}$, and replace $\Omega_{\hat{k}}$ by its expression in terms of the $(n+1)$th moment of $\mathcal{N}(\{\rho_{\mathbf{k}}\})$. Thus, we arrive at the replica Landau free energy:

$$nd\mathcal{F}_n=\epsilon-2+(3-\epsilon)\int \mathcal{D}\rho_1\mathcal{N}(\{\rho_{1,\mathbf{k}}\})\mathcal{D}\rho_2\mathcal{N}(\{\rho_{2,\mathbf{k}}\})\left(\sum_{\mathbf{k}}\rho_{1,\mathbf{k}}\rho_{2,-\mathbf{k}}\right)^{n+1}+\frac{1}{2}(n+1)\int \mathcal{D}\rho_1\mathcal{N}(\{\rho_{1,\mathbf{k}}\})\mathcal{D}\rho_2\mathcal{N}(\{\rho_{2,\mathbf{k}}\})$$
$$\times\left(\sum_{\mathbf{k}}k^2\rho_{1,\mathbf{k}}\rho_{2,-\mathbf{k}}\right)\left(\sum_{\mathbf{k}}\rho_{1,\mathbf{k}}\rho_{2,-\mathbf{k}}\right)^{n}-\int \mathcal{D}\rho_1\mathcal{N}(\{\rho_{1,\mathbf{k}}\})\mathcal{D}\rho_2\mathcal{N}(\{\rho_{2,\mathbf{k}}\})\mathcal{D}\rho_3\mathcal{N}(\{\rho_{3,\mathbf{k}}\})$$
$$\times\left(\sum_{\mathbf{k}_1,\mathbf{k}_2}\rho_{1,\mathbf{k}_1}\rho_{2,\mathbf{k}_2}\rho_{3,-\mathbf{k}_1-\mathbf{k}_2}\right)^{n+1}. \tag{22}$$

In order to obtain the desired (replica-independent) free energy, we take the limit $n\rightarrow0$ of Eq. (22):

$$d\mathcal{F}=d\lim_{n\rightarrow0}\mathcal{F}_n=(3-\epsilon)\int \mathcal{D}\rho_1\mathcal{N}(\{\rho_{1,\mathbf{k}}\})\mathcal{D}\rho_2\mathcal{N}(\{\rho_{2,\mathbf{k}}\})\left(\sum_{\mathbf{k}}\rho_{1,\mathbf{k}}\rho_{2,-\mathbf{k}}\right)\ln\left(\sum_{\mathbf{k}}\rho_{1,\mathbf{k}}\rho_{2,-\mathbf{k}}\right)+\frac{1}{2}\int \mathcal{D}\rho_1\mathcal{N}(\{\rho_{1,\mathbf{k}}\})\mathcal{D}\rho_2\mathcal{N}(\{\rho_{2,\mathbf{k}}\})$$
$$\times\left(\sum_{\mathbf{k}}k^2\rho_{1,\mathbf{k}}\rho_{2,-\mathbf{k}}\right)\ln\left(\sum_{\mathbf{k}}\rho_{1,\mathbf{k}}\rho_{2,-\mathbf{k}}\right)-\int \mathcal{D}\rho_1\mathcal{N}(\{\rho_{1,\mathbf{k}}\})\mathcal{D}\rho_2\mathcal{N}(\{\rho_{2,\mathbf{k}}\})\mathcal{D}\rho_3\mathcal{N}(\{\rho_{3,\mathbf{k}}\})$$
$$\times\left(\sum_{\mathbf{k}_1,\mathbf{k}_2}\rho_{1,\mathbf{k}_1}\rho_{2,\mathbf{k}_2}\rho_{3,-\mathbf{k}_1-\mathbf{k}_2}\right)\ln\left(\sum_{\mathbf{k}_1,\mathbf{k}_2}\rho_{1,\mathbf{k}_1}\rho_{2,\mathbf{k}_2}\rho_{3,-\mathbf{k}_1-\mathbf{k}_2}\right). \tag{23}$$

In deriving the above free energy we have employed the physical fact that the average particle density is uniform. In other words, the replica order parameter is zero if all but one of the replicated wave vectors is nonzero which, translated in the language of the distribution of static density fluctuations, means that the first moment of the static density distribution equals $\delta_{\mathbf{k},\mathbf{0}}$. It is worth noting that, within this formalism, the replica limit can already be taken at the level of the free energy, prior to the hypothesizing of an explicit form for the stationary value of the order parameter. On the one hand, this is attractive, as it leads to a Landau theory in which replicas play no role. On the other hand, the approach is, at present, restricted to replica-symmetric states.

We now construct the self-consistency condition that follows from the stationarity of the replica-independent free energy. We then proceed to solve the resulting functional equation exactly, by hypothesizing a solution having precisely the same physical content as the exact solution of the replica self-consistency condition discussed in Sec. III.

To construct the self-consistency condition for $\mathcal{N}(\{\rho_{\mathbf{k}}\})$ it is useful to make two observations. First, $\mathcal{N}(\{\rho_{\mathbf{k}}\})$ obeys the normalization condition (20). This introduces a constraint on the variations of $\mathcal{N}(\{\rho_{\mathbf{k}}\})$ which is readily accounted for via Lagrange's method of undetermined multipliers. Second, as mentioned above, not all the variables $\{\rho_{\mathbf{k}}\}$ are independent: we have the relations $\rho_{\mathbf{0}}=1$ and $\rho_{-\mathbf{k}}=\rho_{\mathbf{k}}^*$. In principle, one could proceed by defining a new distribution that only depends on the independent elements of $\{\rho_{\mathbf{k}}\}$, and re-express the free energy in terms of this new distribution. However, for convenience we will retain $\mathcal{N}(\{\rho_{\mathbf{k}}\})$ as the basic quantity to be varied, and bear in mind the fact that $\rho_{\mathbf{0}}=1$ and $\rho_{-\mathbf{k}}=\rho_{\mathbf{k}}^*$. By performing the constrained variation of $d\mathcal{F}$ with respect to the functional, $\mathcal{N}(\{\rho_{\mathbf{k}}\})$

$$0=\frac{\delta}{\delta\mathcal{N}(\{\rho_{\mathbf{k}}\})}\left(\mathcal{F}+\lambda\int \mathcal{D}\rho_1\mathcal{N}(\{\rho_{1,\mathbf{k}}\})\right), \tag{24}$$

where λ is the undetermined multiplier, we obtain the self-consistency condition obeyed by $\mathcal{N}(\{\rho_{\mathbf{k}}\})$:

$$0=\lambda d+2(3-\epsilon)\int \mathcal{D}\rho_1\mathcal{N}(\{\rho_{1,\mathbf{k}}\})$$
$$\times\left(\sum_{\mathbf{k}}\rho_{\mathbf{k}}\rho_{1,-\mathbf{k}}\right)\ln\left(\sum_{\mathbf{k}}\rho_{\mathbf{k}}\rho_{1,-\mathbf{k}}\right)+\int \mathcal{D}\rho_1\mathcal{N}(\{\rho_{1,\mathbf{k}}\})$$
$$\times\left(\sum_{\mathbf{k}}k^2\rho_{\mathbf{k}}\rho_{1,-\mathbf{k}}\right)\ln\left(\sum_{\mathbf{k}}\rho_{\mathbf{k}}\rho_{1,-\mathbf{k}}\right)$$
$$-3\int \mathcal{D}\rho_1\mathcal{N}(\{\rho_{1,\mathbf{k}}\})\mathcal{D}\rho_2\mathcal{N}(\{\rho_{2,\mathbf{k}}\})$$
$$\times\left(\sum_{\mathbf{k},\mathbf{k}'}\rho_{\mathbf{k}}\rho_{1,\mathbf{k}'}\rho_{2,-\mathbf{k}-\mathbf{k}'}\right)\ln\left(\sum_{\mathbf{k},\mathbf{k}'}\rho_{\mathbf{k}}\rho_{1,\mathbf{k}'}\rho_{2,-\mathbf{k}-\mathbf{k}'}\right). \tag{25}$$

To solve this self-consistency condition for $\mathcal{N}(\{\rho_{\mathbf{k}}\})$ we import our experience with the replica approach, thereby constructing the normalized hypothesis

$$\mathcal{N}(\{\rho_{\mathbf{k}}\})=(1-q)\delta_c(\rho_{\mathbf{0}}-1)\prod_{\mathbf{k}\neq\mathbf{0}}\delta_c(\rho_{\mathbf{k}})$$
$$+q\int\frac{dc}{V}\int_0^\infty d\tau p(\tau)\prod_{\mathbf{k}}\delta_c(\rho_{\mathbf{k}}-e^{ic\cdot\mathbf{k}-k^2/2\tau}), \tag{26}$$

in which q (which satisfies $0\leqslant q\leqslant1$) is the localized fraction and $p(\tau)$ (which is regular and normalized to unity) is the distribution of localization lengths of localized particles. It is straightforward to show that by taking the $(n+1)$th moment of $\mathcal{N}(\{\rho_{\mathbf{k}}\})$ we recover the self-consistent form of the replica order parameter, Eq. (10).

By inserting the hypothesis (26) into Eq. (25), making the replacement $\rho_{\mathbf{0}}\rightarrow1$, and performing some algebra, the self-consistency condition takes the form

$$0=\int \frac{dc}{V} \int_0^\infty d\tau\left(1+\sum_{\mathbf{k}\neq\mathbf{0}} \rho_{\mathbf{k}} e^{-ic\cdot\mathbf{k}-k^2/2\tau}\right)\ln\left(1+\sum_{\mathbf{k}\neq\mathbf{0}} \rho_{\mathbf{k}} e^{-ic\cdot\mathbf{k}-k^2/2\tau}\right)\left\{2q(-\epsilon+3q)p(\tau)-q\frac{d}{d\tau}[2\tau^2 p(\tau)]\right.$$
$$\left.-3q^2\int_0^\tau d\tau_1 p(\tau_1)p(\tau-\tau_1)\right\}-\frac{3}{2}dq^2\int_0^\infty d\tau_1 p(\tau_1)d\tau_2 p(\tau_2)\ln\{V^{2/d}\tau_1\tau_2/2\pi e(\tau_1+\tau_2)\}+\lambda d, \tag{27}$$

in terms of the undetermined multiplier λ. To determine λ we insert the choice $\rho_{\mathbf{k}}=\delta_{\mathbf{k},\mathbf{0}}$, which yields

$$\lambda=\frac{3}{2}q^2\int_0^\infty d\tau_1 p(\tau_1)d\tau_2 p(\tau_2)\ln\{V^{2/d}\tau_1\tau_2/2\pi e(\tau_1+\tau_2)\}. \tag{28}$$

By using this result to eliminate λ from the self-consistency condition, and observing that this condition must be satisfied for arbitrary $\{\rho_{\mathbf{k}}\}$, we arrive at a condition on q and $p(\tau)$, viz.,

$$0=2q(-\epsilon+3q)p(\tau)-q\frac{d}{d\tau}[2\tau^2 p(\tau)]$$
$$-3q^2\int_0^\tau d\tau_1 p(\tau_1)p(\tau-\tau_1). \tag{29}$$

We integrate this equation over all values of τ and use the normalization condition on $p(\tau)$ to arrive at the same equation relating q and ϵ as was found in Eq. (12) of the previous section, the appropriate solution of which is given by $q=2\epsilon/3$, i.e., Eq. (13). Finally, we use this result for q to eliminate it from Eq. (29), thus arriving at the same self-consistency condition on $p(\tau)$ as was found in Eq. (14) of the previous section. Thus, we see that these conditions, one for q and one for $p(\tau)$, are precisely the same as those arrived at by the replica method discussed in Sec. III.

V. COMPARISON WITH NUMERICAL SIMULATIONS: UNIVERSALITY EXHIBITED

The purpose of the present section is to examine the conclusions of the Landau theory, especially those concerning universality and scaling, in the light of the extensive molecular-dynamics simulations, performed by Barsky and Plischke.[10,11] These simulations address the amorphous solidification transition in the context of randomly cross-linked macromolecular systems, by using an off-lattice model of macromolecules interacting via a Lennard-Jones potential.

It should be emphasized that there are substantial differences between ingredients and calculational schemes used in the analytical and simulational approaches. In particular, the analytical approach: (i) invokes the replica technique; (ii) retains interparticle interactions only to the extent that macroscopically inhomogeneous states are disfavored (i.e., the one-replica sector remains stable at the transition); (iii) neglects order-parameter fluctuations, its conclusions therefore being independent of the space dimension; and (iv) is solved via an ansatz, which is not guaranteed to capture the optimal solution.

Nevertheless, and rather strikingly, the simulations yield an essentially identical picture for the transition to and properties of the amorphous solid state, inasmuch as they indicate that (i) there exists a (cross-link–density-controlled) continuous phase transition from a liquid state to an amorphous solid state; (ii) the critical cross-link density is very close to one cross link per macromolecule; (iii) q varies linearly with the density of cross links, at least in the vicinity of this transition (see Fig. 1); (iv) when scaled appropriately (i.e., by the mean localization length), the simulation data for the distribution of localization lengths exhibit very good collapse on to a universal scaling curve for the several (near-critical) cross-link densities and macromolecule lengths considered (see Figs. 2 and 3); and (v) the form of this universal scaling curve appears to be in remarkably good agreement with the precise form of the analytical prediction for this distribution.

It should not be surprising that by focusing on universal quantities, one finds agreement between the analytical and computational approaches. This indicates that the proposed Landau theory does indeed contain the essential ingredients necessary to describe the amorphous solidification transition.

Let us now look more critically at the comparison between the results of the simulation and the mean-field theory. With respect to the localized fraction, the Landau theory is only capable of showing the linearity of the dependence, near the transition, on the excess cross-link density, leaving undetermined the proportionality factor. The simulation results are consistent with this linear dependence, giving, in addition, the amplitude. There are two facets to the universality of the distribution of localization lengths, as mentioned in Sec. III. First, that the distributions can, for different systems and different cross-link densities, be collapsed on to a universal scaling curve, is verified by the simulations, as pointed out above. Second, the question of how the scaling parameter depends on the excess cross-link density is difficult to address in current simulations, because the dynamic range for the mean localization length accessible in them is limited, so that its predicted divergence at the transition is difficult to verify.

VI. SUMMARY AND CONCLUDING REMARKS

To summarize, we have proposed a replica Landau free energy for the amorphous solidification transition. The theory is applicable to systems of extended objects undergoing thermal density fluctuations and subject to quenched random translationally invariant constraints. The statistics of the quenched randomness are determined by the equilibrium density fluctuations of the unconstrained system. We have shown that there is, generically, a continuous phase transition between a liquid and an amorphous solid state, as a function of the density of random constraints. Both states are described by exact stationary points of this replica free energy, and are replica symmetric and macroscopically trans-

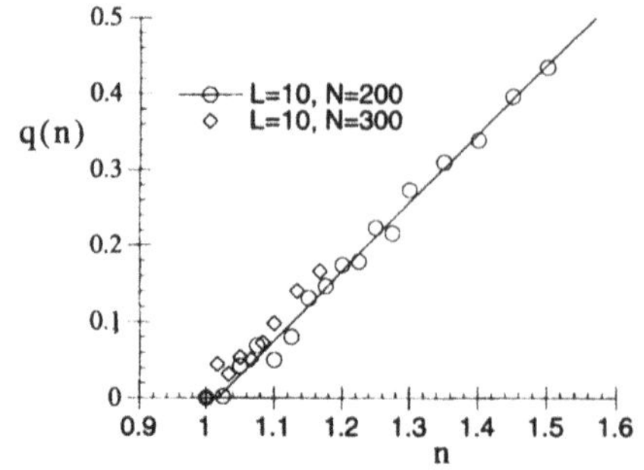

FIG. 1. Localized fraction q as a function of the number of cross links per macromolecule n, as computed in molecular-dynamics simulations by Barsky and Plischke (Ref. 11). L is the number of monomers in each macromolecule; N is the number of macromolecules in the system. The straight line is a linear fit to the $N=200$ data. Note the apparent existence of a continuous phase transition near $n=1$, as well as the apparent linear variation of q with n, both features being consistent with the mean-field description.

lationally invariant. They differ, however, in that the liquid is translationally invariant at the microscopic level, whereas the amorphous solid breaks this symmetry.

We have also shown how all these results may be recovered using an alternative formulation in which we focus less on the replica order parameter and more on the distribution of random static density fluctuations. In particular, we construct a representation of the free energy in terms of this distribution, and solve the resulting stationarity condition.

Lastly, we have examined our results in the light of the extensive molecular-dynamics simulations of randomly cross-linked macromolecular systems, due to Barsky and Plischke. Not only do these simulations support the general theoretical scenario of the vulcanization transition, but also they confirm the detailed analytical results for universal quantities, including the localized fraction exponent and the distribution of scaled localization lengths.

The ultimate origin of universality is not hard to understand, despite the apparent intricacy of the order parameter associated with the amorphous solidification transition. As we saw in Secs. II and III, there are two small emergent

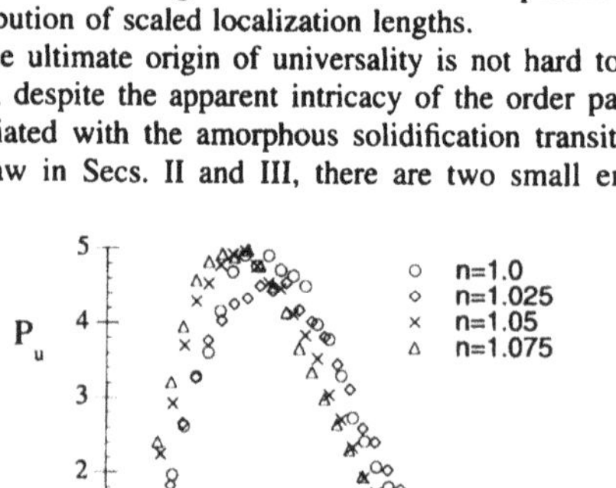

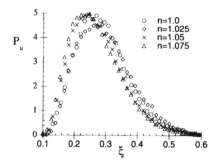

FIG. 2. Unscaled probability distribution P_u of localization lengths ξ (in units of the linear system size), as computed in molecular-dynamics simulations by Barsky and Plischke (Ref. 11). In the simulations the number of segments per macromolecule is 10; and the number of macromolecules is 200.

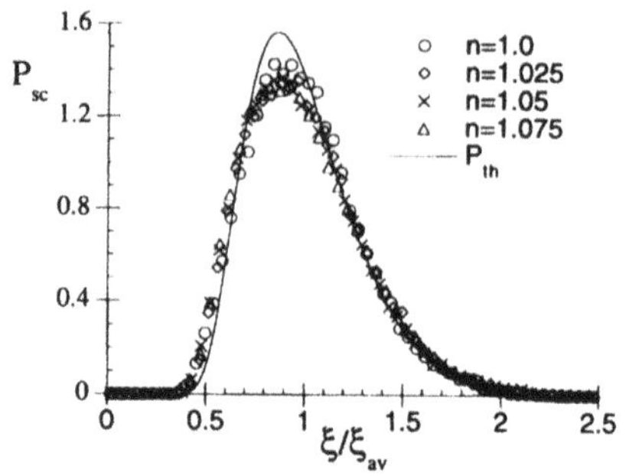

FIG. 3. Probability distribution (symbols) P_s of localization lengths ξ, scaled with the sample average of the localization lengths ξ_{av}, as computed in molecular-dynamics simulations by Barsky and Plischke (Ref. 11). Note the apparent collapse of the data on to a single universal scaling distribution, as well as the good quantitative agreement with the mean-field prediction for this distribution (solid line). In the simulation the number of segments per macromolecule is 10; and the number of macromolecules is 200. The mean-field prediction for $P_{sc}(\xi/\xi_{av})$ is obtained from the universal scaling function $\pi(\theta)$ by $P_{sc}(y)=(2s/y^3)\pi(s/y^2)$, where the constant $s \simeq 1.224$ is fixed by demanding that $\int_0^\infty dy\,y\,P_{sc}(y)=1$.

physical quantities, the fraction of localized particles and the characteristic inverse square localization length of localized particles. The smallness of the localized fraction validates the truncation of the expansion of the free energy in powers of the order parameter. The smallness of the characteristic inverse square localization length leads to a very simple dependence, $\Sigma_{\alpha=0}^{n}|\mathbf{k}^{\alpha}|^2$, on the $n+1$ independent wave vectors of the replica theory, well beyond the permutation invariance demanded by symmetry considerations alone. As a result, near the transition, the amorphous solid state is characterizable in terms of a single universal function of a single variable, along with the localized fraction.

Although throughout this paper we have borne in mind the example of randomly cross-linked macromolecular systems, the circle of ideas is by no means restricted to such systems. To encompass other systems possessing externally-induced quenched random constraints, such as networks formed by the permanent random covalent bonding of atoms or small molecules (e.g., silica gels), requires essentially no further conceptual ingredients (and only modest further technical ones).[17]

One may also envisage applications to the glass transition. Although it is generally presumed that externally-induced quenched random variables are not relevant for describing the glass transition, it is tempting to view the freezing-out of some degrees of freedom as the crucial consequence of the temperature-quench, with a form of quenched disorder thereby being developed spontaneously. The approach presented in the present paper becomes of relevance to the glass transition if one accepts this view of the temperature quench, and thus models the nonequilibrium state of the quenched liquid by the equilibrium state of a system in which some fraction of covalent bonds has been rendered permanent (the deeper the quench the larger the fraction).[18] This strategy, viz., the approximating of pure systems by ones with "self-induced" quenched disorder, has also been invoked in very interesting work on the Bernasconi model for binary sequences of low autocorrelation.[19] Interesting connections are also apparent with recent effective-potential approaches to glassy magnetic systems, in which one retains in the partition function only those configurations that lie close to the equilibrium state reached at the glass transition temperature.[20]

ACKNOWLEDGMENTS

We thank Nigel Goldenfeld and David Hertzog for useful discussions. We gratefully acknowledge support from the U.S. National Science Foundation through Grant Nos. DMR94-24511 (W.P., P.G.) and DMR91-57018 (P.G.), from the University of Illinois at Urbana-Champaign (H.C.), from NATO through CRG 94090 (P.G., A.Z.), and from the DFG through SFB 345 (A.Z.). Michael Plischke has generously provided us with unpublished results from his extensive computational studies of vulcanized macromolecular networks, performed in collaboration with Sandra J. Barsky, and has allowed us to report some of this work here. We are most grateful to him for this, and for his continued enthusiasm for the subject matter.

[1] P. M. Goldbart and N. Goldenfeld, Phys. Rev. Lett. **58**, 2676 (1987); Macromolecules **22**, 948 (1989); Phys. Rev. A **39**, 1402 (1989); **39**, 1412 (1989).

[2] P. M. Goldbart and A. Zippelius, Phys. Rev. Lett. **71**, 2256 (1993).

[3] H. E. Castillo, P. M. Goldbart, and A. Zippelius, Europhys. Lett. **28**, 519 (1994).

[4] P. M. Goldbart, H. E. Castillo, and A. Zippelius, Adv. Phys. **45**, 393 (1996).

[5] R. T. Deam and S. F. Edwards, Philos. Trans. R. Soc. London, Ser. A **280**, 317 (1976).

[6] See, e.g., M. Mézard, G. Parisi, and M. A. Virasoro, *Spin Glass Theory and Beyond* (World Scientific, Singapore, 1987); K. Binder and A. P. Young, Rev. Mod. Phys. **58**, 801 (1986).

[7] A. Zippelius and P. M. Goldbart, in *Spin Glasses and Random Fields*, edited by A. P. Young (World Scientific, Singapore, in press).

[8] M. Huthmann, M. Rehkopf, A. Zippelius, and P. M. Goldbart, Phys. Rev. E **54**, 3943 (1996).

[9] C. Roos, A. Zippelius, and P. M. Goldbart, J. Phys. A **30**, 1967 (1997).

[10] S. J. Barsky and M. Plischke, Phys. Rev. E **53**, 871 (1996).

[11] S. J. Barsky, Ph.D. thesis, Simon Fraser University, Canada, 1996; S. J. Barsky and M. Plischke (unpublished).

[12] As is conventional in statistical mechanics, we use the same notation, Ω_k, for the equilibrium value of the order parameter and for the fluctuating variable (of which it is the expectation value and which features in the Landau free energy). Which meaning is implied can be readily ascertained from the context.

[13] W. H. Stockmayer, J. Chem. Phys. **11**, 45 (1943). For further discussion, see Ref. 12.

[14] See, e.g., S.-K. Ma, *Modern Theory of Critical Phenomena* (Benjamin, Reading, MA, 1971).

[15] P. G. de Gennes, J. Phys. (France) Lett. **38**, L355 (1977); P. G. de Gennes, *Scaling Concepts in Polymer Physics* (Cornell University Press, Ithaca, 1979).

[16] A. Zippelius, P. M. Goldbart, and N. Goldenfeld, Europhys. Lett. **23**, 451 (1993).

[17] P. M. Goldbart and A. Zippelius, Europhys. Lett. **27**, 599 (1994); O. Thiessen, A. Zippelius, and P. M. Goldbart, Int. J. Mod. Phys. B **11**, 1945 (1996).

[18] This strategy has been discussed in greater depth in Ref. 7; it has been implemented in Ref. 17.

[19] E. Marinari, G. Parisi, and F. Ritort, J. Phys. A **27**, 7615 (1994); **27**, 7647 (1994); J.-P. Bouchaud and M. Mézard, J. Phys. I **4**, 1109 (1994).

[20] S. Franz and G. Parisi, J. Phys. I **5**, 1401 (1995).

Anomalous Viscous Loss in Emulsions

Andrea J. Liu,[1] Sriram Ramaswamy,[1,2] T. G. Mason,[3,4] Hu Gang,[3] and D. A. Weitz[3,5]
[1]*Department of Chemistry and Biochemistry, University of California, Los Angeles, California 90095*
[2]*Department of Physics, Indian Institute of Science, Bangalore 560 012 India*
[3]*Exxon Research & Engineering Company, Annandale, New Jersey 08801*
[4]*Centre de Recherche Paul Pascal, Ave. A. Schweitzer, F-33600 Pessac, France*
[5]*Department of Physics and Astronomy, University of Pennsylvania, 209 S. 33rd Street, Philadelphia, Pennsylvania 19104**
(Received 5 September 1995)

We propose a model for concentrated emulsions based on the speculation that a macroscopic shear strain does not produce an affine deformation in the randomly close-packed droplet structure. The model yields an anomalous contribution to the complex dynamic shear modulus that varies as the square root of frequency. We test this prediction using a novel light scattering technique to measure the dynamic shear modulus, and directly observe the predicted behavior over six decades of frequency and a wide range of volume fractions.

PACS numbers: 82.70.Kj, 61.43.–j, 62.20.Dc, 83.50.Fc

An emulsion is a dispersion of liquid droplets suspended in a second, immiscible liquid that contains a surfactant to stabilize the interfaces. Because they are deformable, droplets can be packed to very high volume fractions while still retaining integrity. These concentrated emulsions are remarkable because they are highly elastic, even though they are made up entirely of liquids [1,2]. They also possess viscous properties; this makes them viscoelastic. Such materials are often characterized by their complex dynamic shear modulus, $G^*(\omega) = G'(\omega) + iG''(\omega)$, where $G'(\omega)$ is the elastic or storage modulus and $G''(\omega)$ is the viscous or loss modulus. These moduli reflect the in-phase and out-of-phase responses to a small oscillatory shear at frequency ω. For concentrated emulsions, $G'(\omega)$ is large compared to $G''(\omega)$ at low frequencies, reflecting their elastic nature. However, $G'(\omega)$ is anomalously low compared to any reasonable expectation based on the droplet deformations [2]. By contrast, $G''(\omega)$ is anomalously large compared to any reasonable expectation based on the fluid viscosities [3]. While most work to date has focused on the storage modulus [1,2,4–7], the behavior of the loss modulus is equally important and interesting, and many dissipative mechanisms have recently been considered [3].

Here we suggest an alternate mechanism that may describe the origin of the puzzling behavior of the loss modulus. We speculate that a macroscopically applied shear strain does not produce an affine deformation because the droplets are randomly packed. Instead, some regions slip instead of deforming. This increases the viscous dissipation and decreases the elasticity. In this Letter, we focus on those regions that slip, and present a simple model to account for the anomalous behavior of the loss modulus. We show that, if the directions of the local planes of easy slip are randomly oriented, there is a contribution that varies as $\omega^{1/2}$. This contribution can dominate the usual viscous contribution proportional to ω (which controls the very high frequency behavior) as well as the constant elastic contribution (which controls the very low frequency behavior). We test this prediction by exploiting a new light scattering technique [8] that enables us to measure the moduli of concentrated emulsions over the required extended frequency range. We directly observe the predicted $\omega^{1/2}$ contribution over several decades of frequency for a wide range of volume fractions.

The central assumption of our model is that slip can occur in a local region, but only parallel to a certain plane. One physical realization of this weak plane is two well-aligned layers of droplets that can slip relative to each other. However, more complex realizations are also possible because the droplets can shift their positions. The essential requirement is that there is no linear restoring force, and therefore no local static shear modulus, for strains of the entire region where the displacement gradient is normal to the local weak plane. Because the droplet packing is disordered, the orientations of these weak planes vary randomly throughout the sample. Thus the material has randomly anisotropic shear moduli. This randomness leads to a broad range of stress relaxation rates, which yields the predicted power-law behavior of the macroscopic dynamic shear modulus.

To calculate the dynamic shear modulus, we construct the elastic free energy of a local region with a weak plane. We assume that the system is incompressible. We define a local Cartesian coordinate system $\{abc\}$ and take the local weak plane to be normal to the c axis. We assume that the region is isotropic in the a-b plane and has $c \to -c$ reflection symmetry; however, this degree of symmetry is not crucial. The elastic free energy of a region centered at $\vec{r}$ can then be written as [9]

$$F = \tfrac{1}{2}\lambda_1(\vec{r})u_{cc}^2 + \lambda_2(\vec{r})[(u_{aa} - u_{bb})^2 + 4u_{ab}^2] - \lambda_3(\vec{r})u_{aa}u_{bb} + \lambda_4(\vec{r})(u_{ac}^2 + u_{bc}^2), \quad (1)$$

where u_{ij} is the symmetric strain tensor [9]. The first term represents the energy cost of uniaxial compressions and must always be positive. The second and third terms represent costs of deformations within the a-b plane. Since the a-b plane is the weak plane, there is no energy cost for a shear gradient in the c direction, and the coefficient $\lambda_4(\vec{r})$ must vanish. To calculate the macroscopic mechanical response, we must average over the orientational variation in space of the $\{abc\}$ coordinate system with respect to the Cartesian coordinates in the lab frame, $\{xyz\}$. This is analogous to the problem of randomly oriented smectics, for which the free energy is given by Eq. (1) with $\lambda_2(\vec{r}) = \lambda_3(\vec{r}) = \lambda_4(\vec{r}) = 0$, and with $\hat{c}$ varying randomly in space [10,11]. Model (1) with $\lambda_4(\vec{r}) = 0$ has also been used to describe anisotropic glasses [12] and "decoupled" lamellar phases of tethered membranes [13]. Note that it is ill behaved when $\lambda_4(\vec{r}) = 0$; in the lamellar case, there are higher order gradient terms in the elastic free energy, namely, bending elastic terms, that stabilize the free energy. In the case of emulsions, however, it is unclear whether the appropriate stabilizing term is of the form of a bending energy. Fortunately, the bending modulus does not affect the dynamic shear modulus at frequencies above a low frequency cutoff, $\omega_{\min}$ [10]. We may therefore neglect the bending modulus for frequencies above $\omega_{\min}$, which is found experimentally to be about 10 Hz for emulsions.

To obtain the macroscopic dynamic shear modulus, we must average the stress-stress correlation function $S(\vec{r},t) = \langle \sigma_{xy}(\vec{r},t)\sigma_{xy}(0,0)\rangle$ over all $\vec{r}$. Thus $G^*(\omega)$ depends on the spatial Fourier transform of the stress-stress correlation function at zero wave vector,

$$\tilde{S}(k = 0,t) \equiv \langle \tilde{\sigma}_{xy}(k = 0,t)\tilde{\sigma}_{xy}(k = 0,0)\rangle . \quad (2)$$

Here, tildes refer to spatial Fourier transforms, and $\sigma_{xy}(\vec{r},t)$ is the xy component of the stress at position $\vec{r}$ and time t. The brackets correspond to both thermal and disorder averages [10]. We use the fluctuation-dissipation theorem to obtain the dynamic shear modulus,

$$G^*(\omega) = i\omega\eta(\omega) = \frac{i\omega}{k_BT}\int_0^\infty dt\, e^{-i\omega t}\tilde{S}(k = 0,t) . \quad (2)$$

The stress is the derivative of the free energy with respect to the strain, calculated in the laboratory coordinates $\{xyz\}$:

$$\begin{aligned}\sigma_{xy}(\vec{r},t) = {} & a_x(\vec{r})a_y(\vec{r})\frac{\delta F}{\delta u_{aa}(\vec{r},t)} + b_x(\vec{r})b_y(\vec{r})\frac{\delta F}{\delta u_{bb}(\vec{r},t)} \\ & + c_x(\vec{r})c_y(\vec{r})\frac{\delta F}{\delta u_{cc}(\vec{r},t)} \\ & + \frac{1}{2}[a_x(\vec{r})b_y(\vec{r}) + a_y(\vec{r})b_x(\vec{r})]\frac{\delta F}{\delta u_{ab}(\vec{r},t)} ,\end{aligned} \quad (4)$$

where $a_i(\vec{r})$, $b_i(\vec{r})$, and $c_i(\vec{r})$ are the components of the local Cartesian coordinates $\{abc\}$. Using Eq. (4) in Eq. (2), we can express the stress-stress correlation function $S(\vec{r},t)$ in terms of a sum over products of disorder averages of the form $\langle a_x(\vec{r})a_y(\vec{r})a_x(0)a_y(0)\rangle$ and strain-strain correlation functions of the form $\langle u_{aa}(\vec{r},t)u_{aa}(0,0)\rangle$ [10]. The essential approximation in this step is the breaking of averages over the coordinate axes $\{abc\}$ and strain variables u_{ij}. The strain-strain correlation functions are difficult to compute because $\hat{a}$, $\hat{b}$, and $\hat{c}$ also vary with r. We avoid this difficulty by assuming that the dominant contribution comes from regions over which the coordinate system $\{abc\}$ is highly correlated [10]. Therefore the response of a given correlated region to shear is calculated as if the region were infinite in extent, and individual regions are assumed to be uncorrelated from each other. This approximation is clarified by an alternate approach which considers the zero-temperature response of the system to an externally applied strain [11]. The desired quantity $\tilde{S}(k = 0,t)$ can therefore be expressed as a sum of integrals over wave vector $\vec{q}$ of products of disorder averages of the form $\langle \tilde{a}_x(\vec{q})\tilde{a}_y(\vec{q})\tilde{a}_x(-\vec{q})\tilde{a}_y(-\vec{q})\rangle$ and strain-strain correlation functions of the form $\langle \tilde{u}_{aa}(\vec{q},t)\tilde{u}_{aa}(-\vec{q},0)\rangle$ [10]. To evaluate the disorder averages, we assume that the coordinate axes $\{abc\}$ are correlated only over a characteristic distance ξ, the size of the local region of slip:

$$\langle \tilde{a}_x(\vec{q})\tilde{a}_y(\vec{q})\tilde{a}_x(-\vec{q})\tilde{a}_y(-\vec{q})\rangle = f(q^2\xi^2) . \quad (5)$$

Since the local axes are uncorrelated with each other, correlation functions of the form $\langle \tilde{a}_x(\vec{q})\tilde{a}_y(\vec{q})\tilde{b}_x(-\vec{q})\tilde{b}_y(-\vec{q})\rangle$ vanish. Finally, the strain is a derivative of the displacement, so the strain-strain correlation functions can be written in terms of displacement correlation functions: $\langle \tilde{u}_{aa}(\vec{q},t)\,\tilde{u}_{aa}(-\vec{q},0)\rangle = q_a^2\langle \tilde{u}_a(\vec{q},t)\,\tilde{u}_a(-\vec{q},0)\rangle$, where $q_a = \vec{q}\cdot\hat{a}$. The displacement correlations are calculated from the Langevin equation for $\tilde{u}(\vec{q},t)$, which includes contributions from both elastic and viscous forces [10]. The Langevin equation yields overdamped modes for frequencies low compared to $\omega_0 = \eta/\rho(2\pi/\xi)^2 \approx 10^6$ Hz, where $\rho \approx 1$ g/cm^3 is the mass density, $\eta \approx 1$ cp is the film viscosity, and $\xi \approx 3\ \mu$m, several droplet diameters, is the estimated size of a local slip region. There are only two modes due to incompressibility. Thus

$$\tilde{S}(k=0,t) = \int d\varphi\, d\cos(\theta) \times [f_1(\theta,\varphi)e^{-\Gamma_1(\theta,\varphi)t} + f_2(\theta,\varphi)e^{-\Gamma_2(\theta,\varphi)t}] , \quad (6)$$

where $\Gamma_i(\theta,\varphi)$ are the relaxation rates for the two modes with wave vector q oriented at Eulerian angles θ and φ relative to the local coordinates $\{abc\}$. The functions $f_i(\theta,\varphi)$ depend on integrals over the magnitude $q = |\vec{q}|$ of $f(q^2\xi^2)$ defined in Eq. (5). Since the low frequency behavior of $G^*(\omega)$ is controlled by the long time behavior of $\tilde{S}(k = 0,t)$, we need only consider the long time limit of Eq. (6). By low frequencies, we mean $\omega \ll \omega_c \equiv \sigma/\eta a$, where the high frequency cutoff $\omega_c \approx 10^7$ Hz is a characteristic relaxation frequency, given by the ratio of an elastic constant to the viscosity. For an emulsion, the elastic constants are of order σ/a, where σ is the surface tension and a is the droplet radius. Because there is a weak plane, one of the relaxation rates, Γ_1, vanishes for $\theta = \pi/2$, while the other always remains of order ω_c. Therefore values of θ near $\pi/2$ dominate the integral in Eq. (6) at long times, and we may neglect the second term in the integrand. Physically, the regions with weak planes nearly parallel to the direction $\hat{q} = \vec{q}/q$ of the shear gradient exhibit the most solidlike response and therefore have the slowest stress relaxation rates Γ_1. We find that Γ_1 depends on the projection $\hat{q}_c = \hat{q} \cdot \hat{c} \equiv \cos\theta \approx \pi/2 - \theta$. By symmetry, Γ_1 cannot depend on the sign of $\hat{q}_c$. Thus we find that at long times the integral is of the form $\tilde{S}(k = 0, t) \sim \int dx\, e^{-x^2 t}$, where $x = \pi/2 - \theta$. This Gaussian integral yields $\tilde{S}(k = 0, t) \sim t^{-1/2}$, and hence our main result $G^*(\omega) \sim (i\omega/\omega_c)^{1/2}$.

To test this prediction, we must measure the dynamic shear modulus over a wide frequency range, extending to frequencies higher than those accessible to traditional mechanical rheometers. We use a recently developed light scattering technique [8] that relies on the fluctuation-dissipation theorem, much in the spirit of our calculation. We measure the motion of a particle due to thermal fluctuations, as parametrized by the mean square displacement, $\langle \Delta r^2(t) \rangle$. In a viscoelastic material, this motion reflects both viscous loss and energy storage in the medium. The modulus is obtained by generalizing the Stokes-Einstein relation to finite frequencies,

$$\overline{G}(s) = s\overline{\eta}(s) = \frac{k_B T}{\pi a s \langle \Delta \overline{r}^2(s) \rangle}. \qquad (7)$$

Here, it is most convenient to work with the Laplace frequency, s, with bars representing Laplace-transformed quantities. The modulus $\overline{G}(s)$ is related to $G^*(\omega)$ through the analytic continuation $s = i\omega$ [8]. This relation provides physical intuition about the meaning of $\overline{G}(s)$: Nearly frequency-independent behavior reflects a large elastic contribution since $G'(\omega)$ dominates; linear behavior in s reflects a large viscous contribution since $G''(\omega)$ dominates. An intermediate frequency dependence reflects contributions to both components.

We use monodisperse emulsions [14] so that the droplets themselves can be used as the probes [8]. Our emulsion is comprised of 1 μm diameter silicone oil droplets in water, stabilized by sodium dodecylsulphate at a concentration of 10mM. The emulsion is concentrated by centrifugation, and the volume fraction is determined by weighing, before and after evaporating, the continuous water phase. We measure $\langle \Delta r^2(t) \rangle$ with diffusing-wave spectroscopy (DWS) in the transmission geometry [15,16]. The interpretation assumes that the scattered intensity can be determined from the product of a form factor and structure factor. Independent scattering measurements [17] of monodisperse emulsions, whose continuous phase has been adjusted by adding glycerol to index-match the oil phase, show that this factorization is valid even at the high volume fractions studied here.

To determine the modulus of the emulsion, we numerically calculate the Laplace transform of $\langle \Delta r^2(t) \rangle$ from our DWS measurements, and use Eq. (7). Typical results are shown by the open symbols in Fig. 1(a) for an emulsion with $\phi = 0.67$. The data extend over seven decades to very high frequencies, illustrating the utility of this technique. They also exhibit the expected behavior; at low frequencies, they are nearly independent of s, reflecting the dominant elastic behavior of a compressed emulsion; at high frequencies, they approach a linear dependence on s, reflecting the dominant viscous behavior of a system comprised solely of fluids. To determine the frequency dependence in the intermediate regime, we subtract from the data a constant, reflecting the low frequency elasticity, and a term proportional to s, reflecting the high frequency viscous loss. The sum of these terms, adjusted to match the asymptotic limits of the data, is shown by the dashed line in Fig. 1(a). The resulting difference, $\Delta\overline{G}(s)$, is shown as the solid symbols in Fig. 1(a). Its frequency dependence is $s^{1/2}$, as shown by the solid line through the data. This reflects a contribution to the modulus beyond the purely elastic and purely viscous components. Similar $s^{1/2}$ contributions are observed for higher volume fractions [see Fig. 1(b)]. The $s^{1/2}$ contribution results in an $(i\omega)^{1/2}$ contribution to $G^*(\omega)$, providing direct experimental support for our proposed model.

To quantify the dependence of the $s^{1/2}$ contribution on volume fraction, we fit the data by

$$\overline{G}(s) = G_p + A(\phi)s^{1/2} + \eta_\infty s, \qquad (8)$$

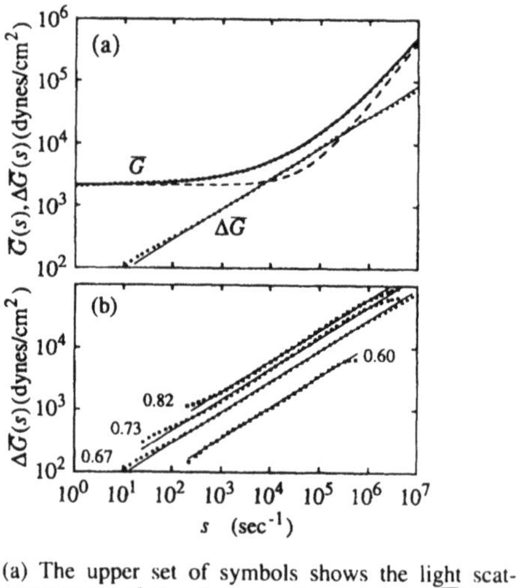

FIG. 1. (a) The upper set of symbols shows the light scattering measurements of the dynamic shear modulus $\overline{G}(s)$ as a function of the Laplace frequency s for an emulsion with $a = 0.5$ μm and $\phi \approx 0.67$. The dashed line represents the asymptotic contributions of an elastic component at low frequencies and a viscous component at high frequencies. The difference, $\Delta\overline{G}(s)$, shown by the lower set of solid points, represents the additional contribution to the modulus, and exhibits an $s^{1/2}$ frequency dependence, as shown by the solid line. (b) The $s^{1/2}$ contribution, $\Delta\overline{G}(s)$, for the same emulsion at several volume fractions.

where η_∞ is the high frequency viscosity and $A(\phi)$ the magnitude of the $s^{1/2}$ term. As shown in Fig. 2, $A(\phi)$ increases by a factor of 4 to 5 as ϕ increases above 0.64, the close-packing density. This trend can be understood in terms of the ϕ dependence of the characteristic stress relaxation frequency $\omega_c = \sigma/\eta a$. Comparing Eq. (8) to our prediction following Eq. (6), we find $A(\phi) \sim 1/\sqrt{\omega_c}$, or $A(\phi) \sim \sqrt{\eta(\phi)a/\sigma}$, where $\eta(\phi)$ is an effective viscosity arising from dissipation in the liquid films. It should depend inversely on the film thickness [3], which decreases by a factor of 4 over this range of ϕ [2]. Thus our model predicts a factor of 2 increase in $A(\phi)$ over the measured range, in reasonable agreement with the experimental results. In addition, our model predicts a high frequency cutoff to the $s^{1/2}$ behavior of $\omega_c = \sigma/\eta a \approx 10^7$ Hz. This is in excellent agreement with the cutoff observed experimentally, as shown in Fig. 1 (the vertical scale in Fig. 1 is logarithmic). Thus the experimental results are in accord with the predicted power law, the high-frequency cutoff on the power-law behavior, and the volume-fraction dependence of the amplitude of the power law. This offers strong support for the model.

A similar $s^{1/2}$ term in $\overline{G}(s)$ is also observed for emulsions with $\phi < 0.64$, where packing constraints do not prevent the droplets from diffusing locally, as well as for colloidal suspensions below close packing [18]. In those cases, it is believed to result from the high-frequency contribution of Brownian motion [19]. We emphasize, however, that this physical mechanism cannot give rise to the $s^{1/2}$ term described here for concentrated emulsions with $\phi > 0.64$, as the droplets cannot freely diffuse, even locally.

The experimental observation of the predicted behavior of the dynamic modulus strongly supports our speculation that slip regions exist. Since such regions do not contribute to the elastic modulus in the low-frequency limit, the existence of these regions may also reduce the *storage* modulus $G'(\omega)$. Recent experiments [2] show that $G'(\omega)$ is much smaller than one would expect based on periodic packings of droplets, especially at volume fractions just above the onset of elastic behavior. Thus our picture may also provide the key to understanding the anomalous storage modulus of emulsions.

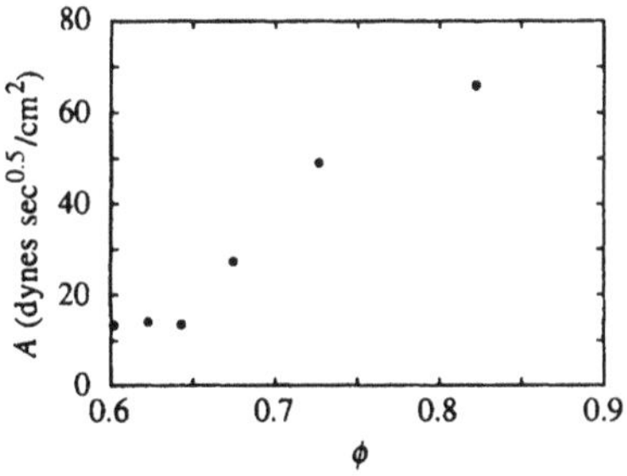

FIG. 2. The magnitude $A(\phi)$ of the $s^{1/2}$ contribution to the modulus, as a function of the volume fraction ϕ.

While our experiments focus on the high-frequency response, the dynamic modulus measured by our light scattering technique matches that measured by mechanical techniques at lower frequencies [2]. Since the light scattering measurements rely on thermal fluctuations, the measured moduli are definitely in the linear regime, even though motion along the weak planes can lead to changes in the droplet configurations. This is a direct consequence of disorder in the droplet packing. The success of the model in predicting an $\omega^{1/2}$ contribution to the dynamic shear modulus is especially remarkable since the loss modulus is sensitive to many different microscopic mechanisms of dissipation, including flow in the liquid films and flow in the surfactant films [3], which we have neglected. This suggests that our results may be applicable to other randomly close-packed materials, such as foams, dense colloidal suspensions, and non-Brownian particle suspensions.

We thank M. E. Cates, D. J. Durian, G. H. Fredrickson, S. T. Milner, and J. Toner for instructive discussions, and give special thanks to Durian, Fredrickson, and Milner for communicating unpublished results. We gratefully acknowledge support from NASA, the Petroleum Research Fund, and NSF Grant PHY 94-07194.

*New address.

[1] H. M. Princen and A. D. Kiss, J. Colloid Interface Sci. **112**, 427 (1986).
[2] T. G. Mason, J. Bibette, and D. A. Weitz, Phys. Rev. Lett. **75**, 2051 (1995).
[3] D. M. A. Buzza, C.-Y. D. Lu, and M. E. Cates, J. Phys. II (France) **5**, 37 (1995).
[4] A. M. Kraynik, Annu. Rev. Fluid. Mech. **20**, 325 (1988).
[5] S. A. Khan, C. A. Schnepper, and R. C. Armstrong, J. Rheol. **32**, 69 (1989).
[6] D. Weaire, F. Bolton, T. Herdtle, and H. Aref, Philos. Mag. Lett. **66**, 293 (1992); F. Bolton and D. Weaire, Phys. Rev. Lett. **65**, 3449 (1990).
[7] D. J. Durian, Phys. Rev. Lett. **75**, 4780 (1995).
[8] T. G. Mason and D. A. Weitz, Phys. Rev. Lett. **74**, 1250 (1995).
[9] L. D. Landau and E. M. Lifschitz, *Theory of Elasticity* (Pergamon Press, New York, 1986), Chap. 1.
[10] K. Kawasaki and A. Onuki, Phys. Rev. A **42**, 3664 (1990).
[11] S. T. Milner and G. H. Fredrickson (to be published).
[12] L. Golubović and T. C. Lubensky, Phys. Rev. Lett. **63**, 1082 (1989).
[13] L. Golubović and T. C. Lubensky, Phys. Rev. A **43**, 6793 (1991).
[14] J. Bibette, J. Colloid Interface Sci. **147**, 474 (1991).
[15] D. J. Pine, D. A. Weitz, E. Herbolzheimer, and P. M. Chaikin, Phys. Rev. Lett. **60**, 1134 (1988).
[16] D. J. Pine, D. A. Weitz, J. X. Zhu, and E. Herbolzheimer, J. Phys. (Paris) **51**, 2101 (1990).
[17] A. H. Krall, Hu Gang, and D. A. Weitz (unpublished).
[18] T. G. Mason and D. A. Weitz, Phys. Rev. Lett. **75**, 2770 (1995).
[19] R. A. Lionberger and W. B. Russel, J. Rheol. **38**, 1885 (1994).

Yielding and Rearrangements in Disordered Emulsions

P. Hébraud, F. Lequeux, and J. P. Munch
Laboratoire de Dynamique des Fluides Complexes, 3 rue de l'Université, 67070 Strasbourg, France

D. J. Pine
Departments of Chemical Engineering and Materials, University of California, Santa Barbara, California 93106
(Received 25 March 1997)

We use diffusing-wave spectroscopy to measure the motion of droplets in concentrated emulsions subjected to a periodic shear strain. The strain gives rise to periodic echoes in the correlation function which decay with increasing strain amplitude. For a given strain amplitude, the decay of the echoes implies that a finite fraction of the emulsion droplets never rearranges under periodic strain while the remaining fraction of droplets repeatedly rearranges. Yielding occurs when about 4–5% of the droplets rearrange. [S0031-9007(97)03423-6]

PACS numbers: 82.70.Kj

There are a number of soft materials that yield when subjected to a moderate shear stress. To a first approximation, the response of the material is elastic below some yield stress and plastic above. Some of these materials have a crystalline structure, and the yielding originates in the motion of dislocations [1–3]. But many other materials, such as concentrated emulsions and colloidal suspensions, foams, and clays, have an amorphous structure and also exhibit a yield stress [4]. The origin of the yield stress in these glassy materials is not well understood.

The structure of glassy materials typically evolves extremely slowly at rest because the particles explore restricted domains of space under the action of thermal fluctuations [5]. Flow dramatically changes the structure, forcing the particles to follow trajectories very different from the ones they explore under thermal fluctuations. This leads to very strong nonlinearities in the mechanical properties, including yielding and cracking. However, the particle trajectories associated with yielding in these disordered systems have never been properly determined.

In this Letter, we use diffusing-wave spectroscopy [6] (DWS) to probe the irreversible motion of droplets which accompanies yielding of concentrated emulsions subjected to an oscillating shear flow. In the absence of shear, DWS provides accurate information about the Brownian motion of particles which multiply scatter light. In concentrated glassy emulsions, however, Brownian motion is suppressed because of the crowding of droplets, and very little droplet motion is observed. The application of shear flow causes emulsion droplets to move, however, and this motion can be detected using DWS [7]. For oscillating shear flow, one expects the motion of the emulsion droplets to be elastic and largely reversible for sufficiently small amplitude strains. For strains large enough to cause the emulsion to yield, however, one expects there to be irreversible droplet movements. The key is to distinguish the irreversible droplet movements from the large background of reversible motion in an oscillating shear flow. We accomplish this by detecting "echoes" in correlation functions measured using DWS. Our method is described below.

In DWS, one measures the temporal decay of the intensity autocorrelation function of the multiply scattered light [6]. The correlation function is sensitive to very small relative motions of the scatters and completely decays when the scattering particles move a small fraction of the wavelength of light. Thus, in an oscillatory shear flow, the correlation function decays very rapidly upon the application of a straining motion. If the straining motion is reversed, however, the correlation function will fully recover to its initial value and an "echo" will be detected, provided each scatterer returns precisely to its initial position after one period. In fact, for a perfectly elastic system, a train of echoes would be observed at integral multiples of the period. By contrast, if each application of the strain causes some random fraction of the particles to rearrange and undergo irreversible motion, then the height of each subsequent echo will be diminished. The technique is reminiscent of NMR spin echo measurements where one probes how spins return to their initial orientation under the action of a symmetric magnetic field pulse sequence. Here, we detect the positions of the scatterers, and not the spins, and the macroscopic strain plays the role of the magnetic field.

In these experiments, we study yielding in simple concentrated emulsions consisting of hexadecane droplets in water, stabilized by sodium dodecyl sulfate at a concentration of 20 mM. These systems have been studied intensively both by rheometry and by DWS for samples at rest and are known to exhibit a yield stress [4]. We prepared emulsions with a mean droplet radius of $a = 0.9\ \mu$m and a polydispersity of about 10%, as measured by dynamic light scattering from a dilute suspension. Samples with volume fractions of oil ranging from $\phi_{\mathrm{oil}} = 0.67$ to 0.92 were obtained by centrifugation.

We built a shear cell consisting of two parallel glass plates with a variable gap. For most of our measurements, the gap was 2.0 mm. Oscillatory straining motion was realized by moving the upper plate, mounted on a precision translation stage, with a piezoelectric device. The motion of the piezoelectric device could be amplified by a lever which enabled the device to achieve amplitudes from 1 μm to 1 mm at a frequency of approximately 58 Hz. The glass plates were roughened by sandblasting to prevent slippage between the emulsion and the plates. We confirmed that the emulsions did not slip by visual observation and by checking that we obtained identical results independent of thickness. The roughening did not affect the light-scattering measurements, since the sample was already highly multiply scattering.

For the DWS measurements, we used an Ar-ion laser and collected multiply scattered light in both the backscattering and transmission geometries [6]. A key feature of our technique was the use of a Brookhaven Instruments BI-9000 correlator which allows one to specify the distribution of delay channels with great flexibility. For this experiment, this meant we could concentrate a high density of linearly spaced delay channels in the few narrow regions of delay times that we were interested in, specifically, at those delay times where the echoes would appear (at integral multiples of the oscillation period).

Figure 1(a) shows a typical correlation function obtained for a sinusoidal strain of constant amplitude. The correlation function decays and then exhibits very narrow echoes centered around multiples of the strain period. In Fig. 1(b), we show the shapes of the initial decays and echoes for three different strain amplitudes. The initial decay is caused by the affine motion of the particles, and the decay time scales with the mean inverse shear rate. Similarly, the widths of the echoes also scale with the mean inverse shear rate.

The height of the echoes is determined by the fraction of the sample which undergoes reversible motion. If all the scatterers were to exhibit a perfectly reversible periodic motion under the periodic strain, the correlation function would return to its initial value after each multiple of the oscillation period. To verify that this is indeed the case, we performed measurements on latex particles included in a polymer gel. In this case, the echoes recovered the amplitude of the initial decay for all strain amplitudes. By contrast, with emulsions, the height of the echoes decreases as the amplitude of the strain is increased. Thus, the strain induces irreversible rearrangements of the emulsion droplets.

The most surprising feature of the correlation functions we measure is that, after the decay of the first echo, *the amplitudes of the echoes remain constant* from the first to the sixteenth echo. This is observed for both backscattering and transmission measurements. This observation has strong implications. First, it means that there is a finite fraction of particles inside the sample which follows reversible trajectories over many periods. Second, it means that there is another finite fraction of the sample which follows irreversible chaotic trajectories. Third, it means that these two fractions of particles are disjoint sets. That is, the particles that rearrange between the initial decay and the first echo are the only particles that ever undergo rearrangements over the time scale of our measurements. If this were not so, and *different particles* underwent irreversible motions during the time between the first and subsequent echoes, then the subsequent echo heights would be lower than the first echoes. This is not what we observe. Thus, we conclude that some droplets always exhibit chaotic motion, while the others always exhibit purely periodic motion during oscillating shear flow of a fixed amplitude. Such a state might be realized, for example, if there were domains of chaotic droplets acting as boundary grains in polycrystals.

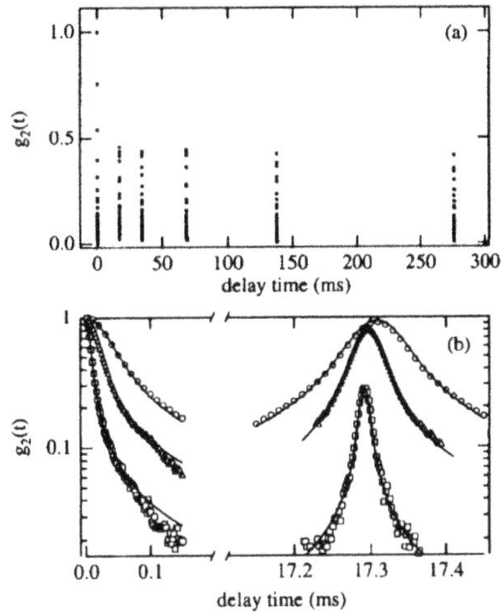

FIG. 1. Normalized intensity correlation functions obtained from an 85% oil-in-water emulsion subjected to oscillatory shear strain. (a) Correlation function displayed using an axis with a linear delay time spacing. To obtain the maximum delay, the correlator channels were set to measure only the initial decay and the first, second, fourth, eighth, and sixteenth echoes. Other echoes were omitted in order to devote a sufficient number of channels to each echo so that the echo heights could be accurately determined. Limitations in the correlator design prohibit accumulating echoes at delay times greater than 32 000 times the fundamental sample time. The strain amplitude is $\gamma_0 = 0.05$. (b) Expanded view of the echoes for different strain amplitudes: $\gamma_0 = 0.01$, top (circles); $\gamma_0 = 0.02$, middle (triangles); $\gamma_0 = 0.06$, bottom (squares). The solid lines through the data are computed from Eq. (3) for the different strain amplitudes without any adjustable parameters except for the echo heights.

In Fig. 2(a), we show the measured echo height as a function of strain amplitude. These data were obtained using the backscattering geometry [6] for a sample thickness of 3.0 mm. For a given oil concentration, the echo height decreases monotonically with increasing strain amplitude. Thus, an increasing fraction of the oil droplets undergoes irreversible rearrangements as the strain amplitude is increased.

At some critical value of the strain, we expect that the emulsion yields. Thus, it is instructive to compare

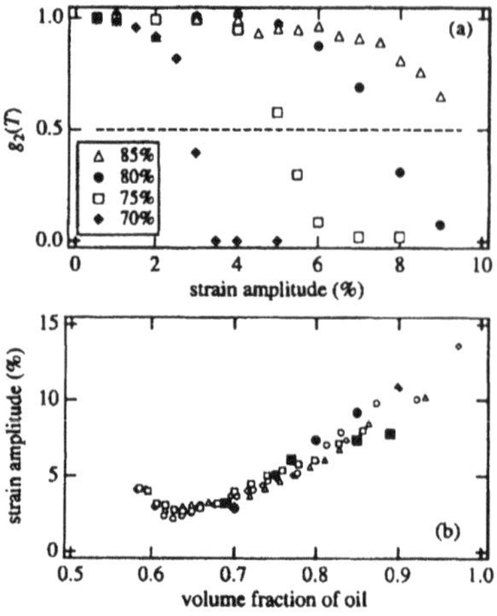

FIG. 2. (a) Echo height (normalized to $g_2(T)$ with no shear) obtained from backscattering measurements as a function of strain amplitude for samples with four different volume fractions of oil. The dashed line at $g_2(T) = 0.5$ is used to estimate the critical strains displayed in the lower plot. (b) Comparison of the critical strain obtained from the decay of echo heights (solid symbols) with the yield strains obtained by rheological measurements in Ref. [4] (smaller open symbols). The solid symbols are obtained from estimates of where the decay of the echo heights falls to 0.5. The solid circles are obtained from the data displayed above in (a); the solid squares in (b) are from data not displayed in (a) (for clarity).

mechanical measurements of the yield strain to the decrease in echo height with increasing strain amplitude. For this purpose, we plot in Fig. 2(b) the mechanical yield strain measured by Mason *et al.* [4] from the crossover between the linear elastic response and nonlinear behavior in oscillatory strain measurements. In the same figure, we also plot the strains at which the correlation functions displayed in Fig. 2(a) have decayed to half their initial value. From the data in Fig. 2(b), we see that the decrease in echo height with increasing strain scales approximately with the yield strain. While the criterion we use to estimate the yield strain is somewhat arbitrary, it lends support to our expectation that these measurements probe particle rearrangements that are associated with the yielding of the emulsion.

A more detailed analysis of our data allows us to gain insight and to determine quantitatively the volume fraction of the chaotic domains. To this end, we write down the general expression for the electric field correlation function for multiply scattered light [6]:

$$g_1(t) = \int_0^\infty P(s) e^{-x(t)s/l^*} ds, \tag{1}$$

where l^* is the transport mean free path of the scattered light, $P(s)$ gives the optical path length distribution through the sample, and $x(t)$ is a function that depends on the motion of the scatterers. The integration of Eq. (1) has been performed for both backscattering and transmission geometries [6]. For shear strain, $x(t) = [k_0 l^* \gamma(t)]^2/15$, where k_0 is the wave vector of the light and $\gamma(t)$ the time-dependent macroscopic strain [7]. For *oscillatory* shear strain, where $\gamma(t) = \gamma_0 \sin(\omega t)$, the strain is not stationary, and we must integrate the intensity correlation function over all the initial values of the strain. Thus, we substitute

$$x(t, t_0) = \frac{1}{15} [k_0 l^* |\gamma(t + t_0) - \gamma(t_0)|]^2 \tag{2}$$

into Eq. (1) and integrate t_0 over one oscillation period to obtain the intensity autocorrelation function:

$$g_2(t) = \frac{\omega}{2\pi} \int_0^{2\pi/\omega} |g_1[x(t, t_0)]| \, dt_0 . \tag{3}$$

Performing this integration numerically, we obtain the shapes of the decay of the correlation function and of the echoes. We have determined l^* (typically 100 μm) by measuring the transmitted intensity [6]. We have also determined l^* independently from our DWS measurements, as discussed below, and obtain consistent results.

For the transmission geometry, we find that the initial decay of $g_2(t)$ is completely insensitive to l^*, and depends only on the strain amplitude γ_0, as expected for shear flow [7]. This allows us to compare the expression for $g_2(t)$ calculated from Eq. (3) to the initial decay of our data without any adjustable parameters. The agreement between the data and Eq. (3) is excellent, as shown in Fig. 1(b). The excellent fit confirms our earlier assertion that there is no slip between the glass slides and the emulsions. It also confirms that on the time scale of the decay of $g_2(t)$, the flow is approximately affine. In backscattering, the initial decay of the correlation function depends on the value of l^* as well as γ_0. In this case, we obtain excellent fits to all the data when we use the single, strain-rate-independent value of l^* determined from the measurements of the transmitted intensity.

We can describe the full shape of the correlation functions, including the decrease in echo height, by taking into account three contributions: the oscillatory shear strain discussed above, the spontaneous decay in the absence of shear strain, and shear-induced rearrangements. The spontaneous decay in the absence of shear strain arises from thermal motion of the emulsion droplets and is significant only for $\phi_{\text{oil}} \lesssim 0.7$ for our systems. In this case, the contribution to $x(t)$ is $k_0^2 \langle r^2(t) \rangle /3$ [6] where $\langle r^2(t) \rangle$ is mean square displacement of a droplet arising from Brownian motion and can be determined from the correlation function in the absence of shear. The contribution to the decay of the correlation function from shear-induced rearrangements leads to an absorption-like term. In this case, the probability for an optical path to scatter from a region that has suffered a shear-induced rearrangement is $\exp[-n\Phi(t)]$, where $\Phi(t)$ is the volume fraction of the sample that has rearranged after a time t and n the number of times the photon scatters in traversing the sample.

Since $n = s/l^*$, we see from Eq. (1) that rearrangements contribute to $x(t)$ by the addition of $\Phi(t)$. This result is exactly equivalent to the result derived previously by Durian *et al.* [8] in order to describe spontaneous rearrangements in foams driven by coarsening. The difference is that in this case, the rearrangements are driven by shear strain rather than occurring spontaneously. Thus, including all three contributions, the full expression for $x(t)$ is

$$x(t) = \frac{1}{3} k_0^2 \langle r^2(t) \rangle + \frac{1}{15} [k_0 l^* |\gamma(t + t_0) - \gamma(t_0)|]^2 + \Phi(t) .$$

Substituting this expression into Eq. (3), and using the fact that $\Phi(t)$ is constant over the width of an echo, we obtain excellent fits to both the initial decay and to all the echoes, as shown in Fig. 1(b). For $\phi_{\mathrm{oil}} < 0.7$, we must take Brownian motion into account by including the effective diffusion coefficient D determined from equilibrium measurements. The only adjustable parameter, then, is $\Phi(t)$, which is varied to obtain a good fit to the echo height. The fact that the echo heights do not change after the first echo means that $\Phi(t)$ does not change after one strain period, confirming our earlier argument that the droplets that rearrange between the initial decay and the first echo are the only droplets that undergo rearrangements over the time scale of our measurements.

In Fig. 3, we plot the volume fraction Φ of the emulsion which rearranges versus strain amplitude γ_0. Data obtained in transmission and backscattering are plotted for oil volume fractions from 0.67 to 0.92. As expected, the softer emulsions, i.e., those with the smallest ϕ_{oil}, exhibit the most rearrangements with increasing strain amplitude. A more surprising feature of the data is that the volume fraction Φ that undergoes irreversible rearrangements seems to increase *linearly* with γ_0 for small γ_0 rather than quadratically. A quadratic dependence might be expected from the simple argument that the number of rearrangements should not depend on the sign of γ_0. It may be that a Φ does depend quadratically on γ_0, but only at strain amplitudes too small for us to resolve (i.e., $\gamma_0 \lesssim 0.01$). Alternatively, $\Phi(\gamma_0)$ may be nonanalytic at small γ_0. We also notice that for $\gamma_0 \lesssim 0.05$, the data obtained from backscattering measurements are higher than the data obtained from transmission measurements. Visual observations under a microscope rule out slippage at the wall. Thus, these data suggest that a greater fraction of the sample may be undergoing irreversible rearrangements *near* the wall than in the bulk.

By comparing the yield strain as a function of γ_0 by measured Mason *et al.* [4] to our determination of Φ shown in Fig. 3, we estimate that yielding occurs when

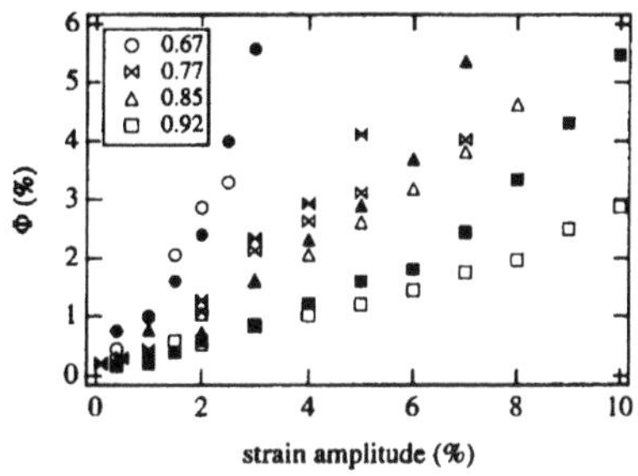

FIG. 3. Volume fraction Φ of emulsion which undergoes irreversible rearrangements as a function of strain amplitude for samples containing different volume fractions of oil droplets. The shear strain causes droplets to rearrange more readily as the volume fraction of droplets decreases. The open symbols were obtained from transmission data; the closed symbols were obtained from backscattering data.

$\Phi \sim 4\%$, i.e., when approximately 4% of the emulsion droplets have undergone irreversible rearrangements.

We note that rearrangements are observed down to $\gamma_0 \sim 0.01$, the smallest strain amplitudes for which we could measure an appreciable decrease in the echo heights in $g_2(t)$. Since there is no measurable decay of $g_2(t)$ in our experiments for $\phi_{\mathrm{oil}} \gtrsim 0.7$, this means that the number of rearrangements caused by shearing is greater than those resulting from thermal fluctuations when $\gamma_0 \gtrsim 0.01$. This is consistent with a simple estimate of the critical strain γ_c, which equates the mechanical energy for such a strain to the thermal energy, i.e., $G\gamma_c^2 a^3 \simeq k_B T$, where G is the elastic shear modulus and a is the radius of a droplet. Realistic values of G [9] give $\gamma_c \sim 10^{-3}$–10^{-4}.

We thank Dave Weitz for supplying the yield strain data reported by Mason *et al.* in Ref. [4].

[1] D. A. Weitz, W. D. Dozier, and P. M. Chaikin, J. Phys. Colloq. (France) **46**, 257 (1985).
[2] L. B. Chen *et al.*, Phys. Rev. Lett. **69**, 688 (1992).
[3] L. B. Chen, B. J. Ackerson, and C. F. Zukoski, J. Rheol. **38**, 193 (1994).
[4] T. G. Mason, J. Bibette, and D. A. Weitz, J. Colloid Interface Sci. **179**, 439 (1996).
[5] W. van Megen and P. N. Pusey, Phys. Rev. A **43**, 5429 (1991).
[6] D. A. Weitz and D. J. Pine, in *Dynamic Light Scattering: The Method and Some Applications* (Oxford University Press, Oxford, 1993), pp. 652–720.
[7] X.-l. Wu *et al.*, J. Opt. Soc. B **7**, 15 (1990).
[8] D. J. Durian, D. A. Weitz, and D. J. Pine, Science **252**, 686 (1991).
[9] T. G. Mason, J. Bibette, and D. A. Weitz, Phys. Rev. Lett. **75**, 2051 (1995).

Dynamics of viscoplastic deformation in amorphous solids

M. L. Falk and J. S. Langer
Department of Physics, University of California, Santa Barbara, Santa Barbara, California 93106
(Received 11 December 1997)

We propose a dynamical theory of low-temperature shear deformation in amorphous solids. Our analysis is based on molecular-dynamics simulations of a two-dimensional, two-component noncrystalline system. These numerical simulations reveal behavior typical of metallic glasses and other viscoplastic materials, specifically, reversible elastic deformation at small applied stresses, irreversible plastic deformation at larger stresses, a stress threshold above which unbounded plastic flow occurs, and a strong dependence of the state of the system on the history of past deformations. Microscopic observations suggest that a dynamically complete description of the macroscopic state of this deforming body requires specifying, in addition to stress and strain, certain average features of a population of two-state shear transformation zones. Our introduction of these state variables into the constitutive equations for this system is an extension of earlier models of creep in metallic glasses. In the treatment presented here, we specialize to temperatures far below the glass transition and postulate that irreversible motions are governed by local entropic fluctuations in the volumes of the transformation zones. In most respects, our theory is in good quantitative agreement with the rich variety of phenomena seen in the simulations. [S1063-651X(98)01306-3]

PACS number(s): 83.50.Nj, 62.20.Fe, 61.43.−j, 81.05.Kf

I. INTRODUCTION

This paper is a preliminary report on a molecular-dynamics investigation of viscoplastic deformation in a noncrystalline solid. It is preliminary in the sense that we have completed only the initial stages of our planned simulation project. The results, however, have led us to a theoretical interpretation that we believe is potentially useful as a guide for further investigations along these lines. In what follows, we describe both the simulations and the theory.

Our original motivation for this project was an interest in the physics of deformations near the tips of rapidly advancing cracks, where materials are subject to very large stresses and experience very high strain rates. Understanding the dissipative dynamics that occur in the vicinity of the crack tip is necessary to construct a satisfactory theory of dynamic fracture [1]. Indeed, we believe that the problem of dynamic fracture cannot be separated from the problem of understanding the conditions under which a solid behaves in a brittle or ductile manner [2–6]. To undertake such a project we eventually shall need sharper definitions of the terms "brittle" and "ductile" than are presently available; but we leave such questions to future investigations while we focus on the specifics of deformation in the absence of a crack.

We have chosen to study amorphous materials because the best experiments on dynamic instabilities in fracture have been carried out in silica glasses and polymers [7,8]. We know that amorphous materials exhibit both brittle and ductile behavior, often in ways that, on a macroscopic level, look very similar to deformation in crystals [9]. More generally, we are looking for fundamental principles that might point us toward theories of deformation and failure in broad classes of macroscopically isotropic solids where thinking of deformation in terms of the dynamics of individual dislocations [2,3] is either suspect, due to the absence of underlying crystalline order, or simply intractable, due to the extreme complexity of such an undertaking. In this way we hope that the ideas presented here will be generalizable perhaps to some polycrystalline materials or even single crystals with large numbers of randomly distributed dislocations.

We describe our numerical experiments in Sec. II. Our working material is a two-dimensional, two-component, noncrystalline solid in which the molecules interact via Lennard-Jones forces. We purposely maintain our system at a temperature very far below the glass transition. In the experiments, we subject this material to various sequences of pure shear stresses, during which we measure the mechanical response. The simulations reveal a rich variety of behaviors typical of metallic glasses [10–13] and other viscoplastic solids [14], specifically, reversible elastic deformation at small applied stresses, irreversible plastic deformation at somewhat larger stresses, a stress threshold above which unbounded plastic flow occurs, and a strong dependence of the state of the system on the history of past deformations. In addition, the molecular-dynamics method permits us to see what each molecule is doing at all times; thus we can identify the places where irreversible molecular rearrangements are occurring.

Our microscopic observations suggest that a dynamically complete description of the macroscopic state of this deforming body requires specifying, in addition to stress and strain, certain average features of a population of what we shall call "shear transformation zones." These zones are small regions, perhaps consisting of only five or ten molecules, in special configurations that are particularly susceptible to inelastic rearrangements in response to shear stresses. We argue that the constitutive relations for a system of this kind must include equations of motion for the density and internal states of these zones; that is, we must add new time-dependent state variables to the dynamical description of this system [15,16]. Our picture of shear transformation zones is based on earlier versions of the same idea due to Argon, Spaepen, and others who described creep in metallic alloys in terms of activated transitions in intrinsically heterogeneous materials [17–22]. These theories, in turn, drew on previous free-volume formulations of the glass transition by

Turnbull, Cohen, and others in relating the transition rates to local free-volume fluctuations [20,23–25]. None of those theories, however, were meant to describe the low-temperature behavior seen here, especially the different kinds of irreversible deformations that occur below and above a stress threshold, and the history dependence of the response of the system to applied loads.

We present the theory of the dynamics of shear transformation zones in Sec. III. This theory contains four crucial features that are not, so far as we know, in any previous analysis. First, once a zone has transformed and relieved a certain amount of shear stress, it cannot transform again in the same direction. Thus the system saturates and, in the language of granular materials, it becomes "jammed." Second, zones can be created and destroyed at rates proportional to the rate of irreversible plastic work being done on the system. This is the ingredient that produces a threshold for plastic flow; the system can become "unjammed" when new zones are being created as fast as existing zones are being transformed. Third, the attempt frequency is tied to the noise in the system, which is driven by the strain rate. The stochastic nature of these fluctuations is assumed to arise from random motions associated with the disorder in the system. Fourth, the transition rates are strongly sensitive to the applied stress. It is this sensitivity that produces memory effects.

The resulting theory accounts for many of the features of the deformation dynamics seen in our simulations. However, it is a mean-field theory that fails to take into account any spatial correlations induced by interactions between zones and therefore it cannot explain all aspects of the behavior that we observe. In particular, the mean-field nature of our theory precludes, at least for the moment, any analysis of strain localization or shear banding.

II. MOLECULAR-DYNAMICS EXPERIMENTS

A. Algorithm

Our numerical simulations have been performed in the spirit of previous investigations of deformation in amorphous solids [26–29]. We have examined the response to an applied shear of a noncrystalline, two-dimensional, two-component solid composed of either 10 000 or 20 000 molecules interacting via Lennard-Jones forces. Our molecular-dynamics algorithm is derived from a standard NPT (number, pressure, temperature) dynamics scheme [30], i.e., a pressure-temperature ensemble, with a Nose-Hoover thermostat [31–33] and a Parinello-Rahman barostat [34,35] modified to allow imposition of an arbitrary two-dimensional stress tensor. The system obeys periodic boundary conditions and both the thermostat and barostat act uniformly throughout the sample.

Our equations of motion are

$$\dot{\mathbf{r}}_n=\frac{\mathbf{p}_n}{m_n}+[\dot{\varepsilon}]\cdot(\mathbf{r}_n-\mathbf{R}_0), \tag{2.1}$$

$$\dot{\mathbf{p}}_n=\mathbf{F}_n-([\dot{\varepsilon}]+\xi[I])\mathbf{p}_n, \tag{2.2}$$

$$\dot{\xi}=\frac{1}{\tau_T^2}\left(\frac{T_{kin}}{T}-1\right), \tag{2.3}$$

$$[\ddot{\varepsilon}]=-\frac{1}{\tau_P^2}\frac{V}{Nk_BT}([\sigma_{av}]-[\sigma]), \tag{2.4}$$

$$\dot{\mathbf{L}}=[\dot{\varepsilon}]\cdot\mathbf{L}. \tag{2.5}$$

Here $\mathbf{r}_n$ and $\mathbf{p}_n$ are the position and momentum of the nth molecule and $\mathbf{F}_n$ is the force exerted on that molecule by its neighbors via the Lennard-Jones interactions. The quantities in square brackets, e.g., $[\dot{\varepsilon}]$ or $[\sigma]$, are two-dimensional tensors. T is the temperature of the thermal reservoir, V is the volume of the system (in this case, the area), and N is the number of molecules. T_{kin} is the average kinetic energy per molecule divided by Boltzmann's constant k_B. $[\sigma]$ is the externally applied stress and $[\sigma_{av}]$ is the average stress throughout the system computed to be

$$[\sigma_{av}]_{ij}=\frac{1}{4V}\sum_n\sum_m F^i_{nm}r^j_{nm}, \tag{2.6}$$

where F^i_{nm} is the ith component of the force between particles n and m, r^j_{nm} is the jth component of the vector displacement between those particles, and V is the volume of the system. $\mathbf{L}$ is the locus of points that describe the boundary of the simulation cell. While Eq. (2.5) is not directly relevant to the dynamics of the particles, keeping track of the boundary is necessary in order to properly calculate intermolecular distances in the periodic cell.

The additional dynamical degrees of freedom in Eqs. (2.1)–(2.5) are a viscosity ξ, which couples the system to the thermal reservoir, and a strain rate $[\dot{\varepsilon}]$, via which the externally applied stress is transmitted to the system. Note that $[\dot{\varepsilon}]$ induces an affine transformation about a reference point $\mathbf{R}_0$, which, without loss of generality, we choose to be the origin of our coordinate system. In a conventional formulation, $[\sigma]$ would be equal to $-P\,[I]$, where P is the pressure and $[I]$ is the unit tensor. In that case, these equations of motion are known to produce the same time-averaged equations of state as an equilibrium NPT ensemble [30]. By instead controlling the tensor $[\sigma]$, including its off-diagonal terms, it is possible to apply a shear stress to the system without creating any preferred surfaces that might enhance system-size effects and interfere with observations of bulk properties. The applied stress and the strain-rate tensor are constrained to be symmetric in order to avoid physically uninteresting rotations of the cell. Except where otherwise noted, all of our numerical experiments are carried out at constant temperature, with $P=0$, and with the sample loaded in uniform, pure shear.

We have chosen the artificial time constants τ_T and τ_P to represent physical aspects of the system. As suggested by Nose [31], τ_T is the time for a sound wave to travel an interatomic distance and, as suggested by Anderson [36], τ_P is the time for sound to travel the size of the system.

B. Model solid

The special two-component system that we have chosen to study here has been the subject of other investigations [37–39] primarily because it has a quasicrystalline ground state. The important point for our purposes, however, is that this system can be quenched easily into an apparently stable glassy state. Whether this is actually a thermodynamically stable glass phase is of no special interest here. We care only that the noncrystalline state has a lifetime that is very much longer than the duration of our experiments.

Our system consists of molecules of two different sizes, which we call "small" (S) and "large" (L). The interactions between these molecules are standard 6-12 Lennard-Jones potentials

$$U_{\alpha\beta}(r)=4e_{\alpha\beta}\left[\left(\frac{a_{\alpha\beta}}{r}\right)^{12}-\left(\frac{a_{\alpha\beta}}{r}\right)^{6}\right], \tag{2.7}$$

where the subscripts α,β denote S or L. We choose the zero-energy interatomic distances $a_{\alpha\beta}$ to be

$$a_{SS}=2\sin\left(\frac{\pi}{10}\right),\quad a_{LL}=2\sin\left(\frac{\pi}{5}\right),\quad a_{SL}=1, \tag{2.8}$$

with bond strengths

$$e_{SL}=1,\quad e_{SS}=e_{LL}=\frac{1}{2}. \tag{2.9}$$

For computational efficiency, we impose a finite-range cutoff on the potentials in Eq. (2.7) by setting them equal to zero for separation distances r greater than $2.5a_{SL}$. The masses are all taken to be equal. The ratio of the number of large molecules to the number of small molecules is half the golden mean

$$\frac{N_L}{N_S}=\frac{1+\sqrt{5}}{4}. \tag{2.10}$$

In the resulting system, it is energetically favorable for ten small molecules to surround one large molecule or for five large molecules to surround one small molecule. The highly frustrated nature of this system avoids problems of local crystallization that often occur in two dimensions where the nucleation of single-component crystalline regions is difficult to avoid. As shown by Lançon *et al.* [37], this system goes through something like a glass transition upon cooling from its liquid state. The glass transition temperature is $0.3T_0$, where $k_BT_0=e_{SL}$. All the simulations reported here have been carried out at a temperature $T=0.001T_0$, that is, at 0.3% of the glass transition temperature. Thus all of the phenomena to be discussed here take place at a temperature very much lower than the energies associated with the molecular interactions.

In order to start with a densely packed material, we have created our experimental systems by equilibrating a random distribution of particles under high pressure at the low temperature mentioned above. After allowing the system to relax at high pressure, we have reduced the pressure to zero and again allowed the sample to relax. Our molecular-dynamics procedure permits us to relax the system only for times on the order of nanoseconds, which are not long enough for the material to experience any significant amount of annealing, especially at such a low temperature.

We have performed numerical experiments on two different samples, containing 10 000 and 20 000 molecules, respectively. All of the simulation results shown are from the larger of the two samples; the smaller sample has been used primarily to check the reliablility of our procedures. We have created each of these samples only once; thus each experiment using either of them starts with precisely the same set of molecules in precisely the same positions. As will become clear, there are both advantages and uncertainties associated with this procedure. On the one hand, we have a very carefully controlled starting point for each experiment. On the other hand, we do not know how sensitive the mechanical properties of our system might be to details of the preparation process, nor do we know whether to expect significant sample-to-sample variations in the molecular configurations. To illustrate these uncertainties, we show the elastic constants of the samples in Table I. The moduli are expressed there in units of e_{SL}/a_{SL}^2. [Note that the Poisson ratio for a two-dimensional (2D) system has an upper bound of 1 rather than 0.5 as in the three-dimensional case.] The appreciable differences between the moduli of supposedly identical materials tell us that we must be very careful in drawing detailed conclusions from these preliminary results.

TABLE I. Sample sizes and elastic constants.

Sample	Molecules	Shear modulus	Bulk modulus	2D Poisson ratio	Young's modulus
1	10 000	9.9	31	0.51	30
2	20 000	16	58	0.57	50

C. Simulation results

1. Macroscopic observations

In all of our numerical experiments, we have tried simply to mimic conventional laboratory measurements of viscoplastic properties of real materials. The first of these is a measurement of stress at constant strain rate. As we shall see, this supposedly simplest of the experiments is especially interesting and problematic for us because it necessarily probes time-dependent behavior near the plastic yield stress.

Our results for two different strain rates are shown in Fig. 1. The strain rates are expressed in units proportional to the frequency of oscillation about the minimum in the Lennard-Jones potential, specifically, in units of $\omega_0\equiv(e_{SL}/ma_{SL}^2)^{1/2}$, where m is the particle mass. [The actual frequency for the SL potential, in cycles per second, is $(3\times2^{1/3}/\pi)\omega_0\cong1.2\omega_0$.] As usual, the sample has been kept at constant temperature and at pressure $P=0$. At low strain, the material behaves in a linearly elastic manner. As the strain increases, the response becomes nonlinear and the material begins to deform plastically. Plastic yielding, that is, the onset of plastic flow, occurs when the strain reaches approximately 0.7%. Note that the stress does not rise smoothly and monotonically in these experiments. We presume that most of this irregularity would average out in larger systems. As we shall see, however, there may also be more interesting dynamical effects at work here.

In all of the other experiments to be reported here, we have controlled the stress on the sample and measured the strain. In the first of these, shown in Fig. 2, we have increased the stress to various different values and then held it constant.

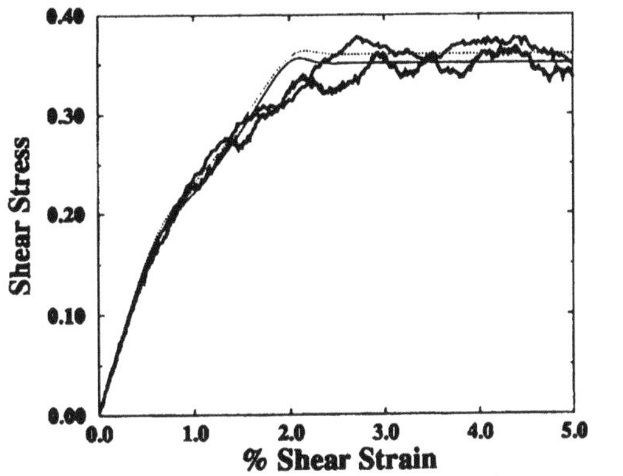

FIG. 1. Shear stress vs strain for strain rates of 10^{-4} (solid lines) and 2×10^{-4} (dotted lines). The thicker lines that denote the simulation results exhibit both linear elastic behavior at low strain and nonlinear response leading to yield at approximately $\sigma_s=0.35$. The thinner curves are predictions of the theory for the two strain rates. Strain rate is measured in units of $(e_{SL}/m\ a_{SL}^2)^{1/2}$; stress is measured in units of e_{SL}/a_{SL}^2.

In each of these experimental runs, the stress starts at zero and increases at the same constant rate until the desired final stress is reached. The graphs show both this applied stress (solid symbols) and the resulting strain (open symbols), as functions of time, for three different cases. Time is measured in the same molecular-vibration units used in the previous experiments, i.e., in units of $(ma_{SL}^2/e_{SL})^{1/2}$. The stresses and strain axes are related by twice the shear modulus so that, if the response is linearly elastic, the two curves lie on top of one another. In the case labeled by triangles, the final stress is small and the response is nearly elastic. For the cases labeled by circles and squares, the sample deforms plastically until it reaches some final strain, at which it ceases to undergo further deformation on observable time scales. (We cannot rule out the possibility of slow creep at much longer times.) In the case labeled by diamonds, for which the final stress is the largest of the three cases shown, the sample continues to deform plastically at constant stress throughout the duration of the experiment. We conclude from these and a number of similar experimental runs that there exists a well-defined critical stress for this material, below which it reaches a limit of plastic deformation, that is, it "jams," and above which it flows plastically. Because the stress is ramped up quickly, we can see in curves with squares and diamonds of Fig. 2 that there is a separation of time scales between the elastic and plastic responses. The elastic response is instantaneous, while the plastic response develops over a few hundred molecular vibrational periods. To see the distinction between these behaviors more clearly, we have performed experiments in which we load the system to a fixed, subcritical stress, hold it there, and then unload it by ramping the stress back down to zero. In Fig. 3, we show this stress and the resulting total shear strain, as functions of time, for one of those experiments. If we define the elastic strain to be the stress divided by twice the previously measured, as-quenched, shear modulus, then we can compute the inelastic strain by subtracting the elastic from the total. The result is shown in Fig. 4. Note that most, but not quite all, of the inelastic strain consists of nonrecoverable plastic deformation that persists after unloading to zero stress. Note also, as shown in Fig. 3, that the system undergoes a small dilation during this process and that this dilation appears to have both elastic and inelastic components.

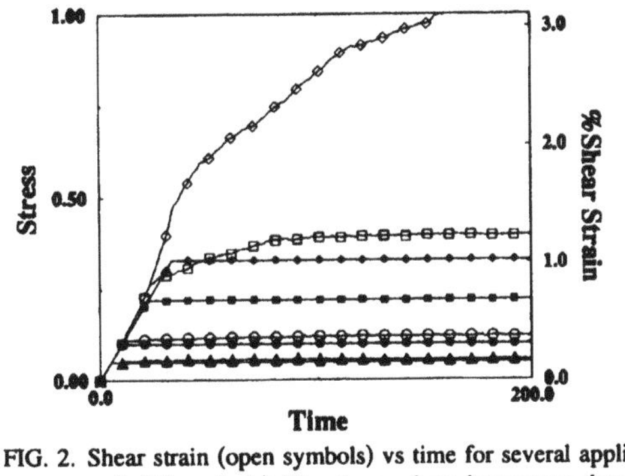

FIG. 2. Shear strain (open symbols) vs time for several applied shear stresses (solid symbols). The stresses have been ramped up at a constant rate until reaching a maximum value and then have been held constant. The strain and stress axes are related by twice the shear modulus so that, for linear elastic response, the open and closed symbols would be coincident. For low stresses the sample responds in an almost entirely elastic manner. For intermediate stresses the sample undergoes some plastic deformation prior to jamming. In the case where the stress is brought above the yield stress, the sample deforms indefinitely. Time is measured in units of $(ma_{SL}^2/e_{SL})^{1/2}$; stress is measured in units of e_{SL}/a_{SL}^2.

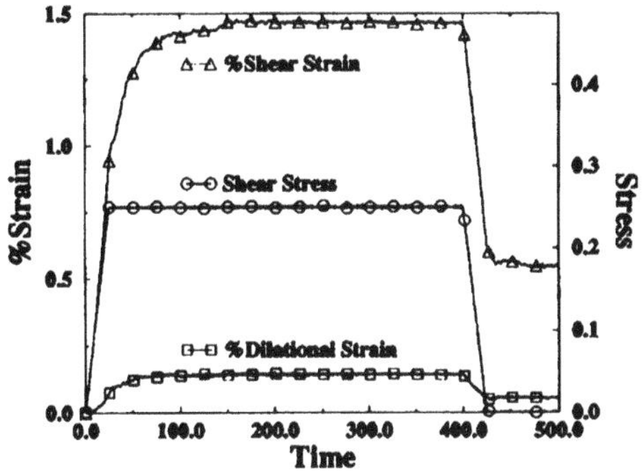

FIG. 3. Stress and strain vs time for one particular loading where the stress has been ramped up to $\sigma_s=0.25$, held for a time, and then released. Note that, in addition to the shear response, the material undergoes a small amount of dilation. Time is measured in units of $(ma_{SL}^2/e_{SL})^{1/2}$; stress is measured in units of e_{SL}/a_{SL}^2.

Using the simple prescription outlined above, we have measured the final inelastic shear strain as a function of shear stress. That is, we have measured the shear strain once the system has ceased to deform as in the subcritical cases in Fig. 2, and then subtracted the elastic part. The results are shown in Fig. 5. As expected, we see only very small amounts of inelastic strain at low stress. As the stress approaches the yield stress, the inelastic strain appears to diverge approximately logarithmically.

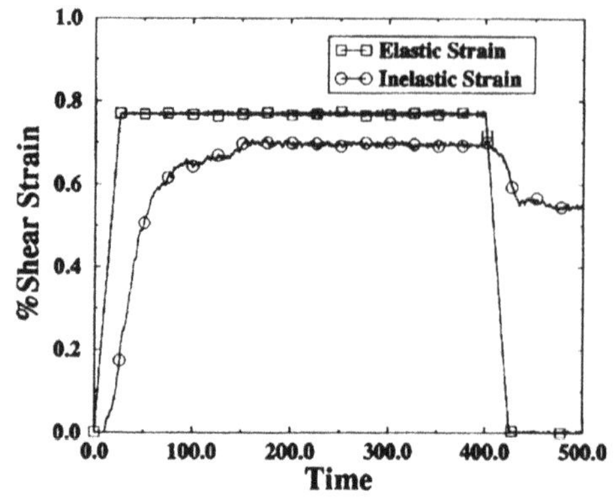

FIG. 4. Elastic and inelastic strain vs time for the same simulation as that shown in Fig. 3. The inelastic strain is found by subtracting the linearly elastic strain from the total strain. Note the partial recovery of the inelastic portion of the strain that occurs during and after unloading. Time is measured in units of $(ma_{SL}^2/e_{SL})^{1/2}$.

The final test that we have performed is to cycle the system through loading, reloading, and reverse loading. As shown in Fig. 6, the sample is first loaded on the curve from a to b. The initial response is linearly elastic, but, eventually, deviation from linearity occurs as the material begins to deform inelastically. From b to c, the stress is constant and the sample continues to deform inelastically until reaching a final strain at c. Upon unloading, from c to d, the system does not behave in a purely elastic manner but rather recovers some portion of the strain inelastically. While held at zero stress, the sample continues to undergo inelastic strain recovery from d to e.

When the sample is then reloaded from e to f, it undergoes much less inelastic deformation than during the initial loading. From f to g the sample again deforms inelastically, but by an amount only slightly more than the previously recovered strain, returning approximately to point c. Upon unloading again from g to h to i, less strain is recovered than

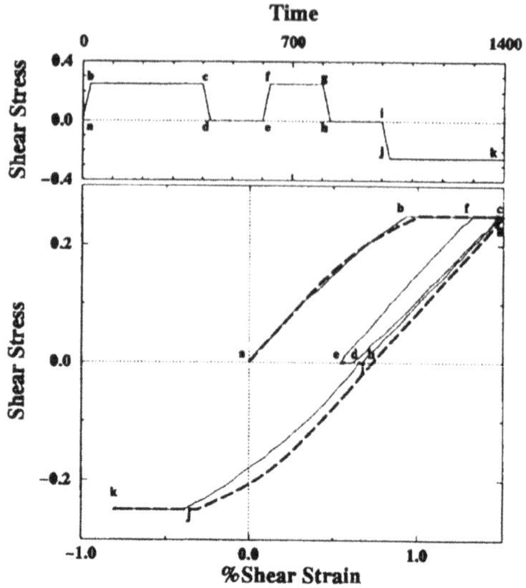

FIG. 6. Stress-strain trajectory for a molecular-dynamics experiment in which the sample has been loaded, unloaded, reloaded, unloaded again, and then reverse loaded, all at stresses below the yield stress. The smaller graph above shows the history of applied shear stress with letters indicating identical times in the two graphs. The dashed line in the main graph is the theoretical prediction for the same sequence of stresses. Note that a small amount of inelastic strain recovery occurs after the first unloading in the simulation, but that no such behavior occurs in the theory. Thus the theoretical curve from c through h unloads, reloads, and unloads again all along the same line. Time is measured in units of $(ma_{SL}^2/e_{SL})^{1/2}$; stress is measured in units of e_{SL}/a_{SL}^2.

in the previous unloading from c through e.

It is during reverse loading from i to k that it becomes apparent that the deformation history has rendered the amorphous sample highly anisotropic in its response to further applied shear. The inelastic strain from i to k is much greater than that from e to g, demonstrating a very significant Bauschinger effect. The plastic deformation in the initial direction apparently has biased the sample in such a way as to inhibit further inelastic yield in the same direction, but there is no such inhibition in the reverse direction. The material, therefore, must in some way have microstructurally encoded, i.e., partially "memorized," its loading history.

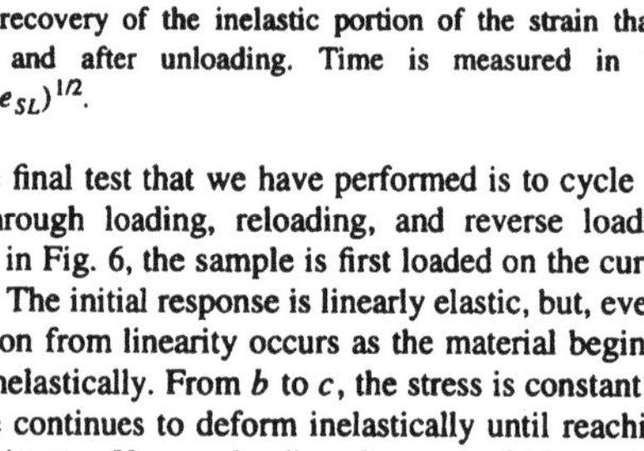

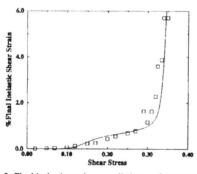

FIG. 5. Final inelastic strain vs applied stress for stresses below yield. The simulation data (squares) have been obtained by running the simulations until all deformation apparently had stopped. The comparison to the theory (line) was obtained by numerically integrating the equations of motion for a period of 800 time units, the duration of the longest simulation runs. Stress is measured in units of e_{SL}/a_{SL}^2.

2. *Microscopic observations*

Our numerical methods allow us to examine what is happening at the molecular level during these deformations. To do this systematically, we need to identify where irreversible plastic rearrangements are occurring. More precisely, we must identify places where the molecular displacements are nonaffine, that is, where they deviate substantially from displacements that can be described by a linear strain field.

We start with a set of molecular positions and subsequent displacements and compute the closest possible approximation to a local strain tensor in the neighborhood of any particular molecule. To define that neighborhood, we define a sampling radius, which we choose to be the interaction range $2.5a_{SL}$. The local strain is then determined by minimizing the mean-square difference between the actual displacements of the neighboring molecules relative to the central one and the relative displacements that they would have if they were in a region of uniform strain ε_{ij}. That is, we define

$$D^2(t,\Delta t)=\sum_n \sum_i \left(r_n^i(t)-r_0^i(t)-\sum_j (\delta_{ij}+\varepsilon_{ij}) \times[r_n^j(t-\Delta t)-r_0^j(t-\Delta t)]\right)^2, \qquad (2.11)$$

where the indices i and j denote spatial coordinates and the index n runs over the molecules within the interaction range of the reference molecule, $n=0$ being the reference molecule. $r_n^i(t)$ is the ith component of the position of the nth molecule at time t. We then find the ε_{ij} that minimizes D^2 by calculating

$$X_{ij}=\sum_n [r_n^i(t)-r_0^i(t)]\times[r_n^j(t-\Delta t)-r_0^j(t-\Delta t)], \qquad (2.12)$$

$$Y_{ij}=\sum_n [r_n^i(t-\Delta t)-r_0^i(t-\Delta t)] \times[r_n^j(t-\Delta t)-r_0^j(t-\Delta t)], \qquad (2.13)$$

$$\varepsilon_{ij}=\sum_k X_{ik}Y_{jk}^{-1}-\delta_{ij}. \qquad (2.14)$$

The minimum value of $D^2(t,\Delta t)$ is then the local deviation from affine deformation during the time interval $[t-\Delta t,\ t]$. We shall refer to this quantity as D^2_{min}.

We have found that D^2_{min} is an excellent diagnostic for identifying local irreversible shear transformations. Figure 7 contains four different intensity plots of D^2_{min} for a particular system as it is undergoing plastic deformation. The stress has been ramped up to $|\sigma_s|=0.12$ in the time interval [0,12] and then held constant in an experiment analogous to that shown in Fig. 2. Figure 7(a) shows D^2_{min} for $t=10$, $\Delta t=10$. It demonstrates that the nonaffine deformations occur as isolated small events. In Fig. 7(b) we observe the same simulation, but for $t=30$, $\Delta t=30$; that is, we are looking at a later time, but again we consider rearrangements relative to the inital configuration. Now it appears that the regions of rearrangement have a larger scale structure. The pattern seen here looks like an incipient shear band. However, in Fig. 7(c), where $t=30$, $\Delta t=1$, we again consider this later time but look only at rearrangements that have occurred in the preceding short time interval. The events shown in this figure are small, demonstrating that the pattern shown in Fig. 7(b) is, in fact, an aggregation of many local events. Finally, in Fig. 7(d), we show an experiment similar in all respects to Fig. 7(a) except that the sign of the stress has been reversed. As in Fig. 7(a), $t=10$, $\Delta t=10$, and again we observe small

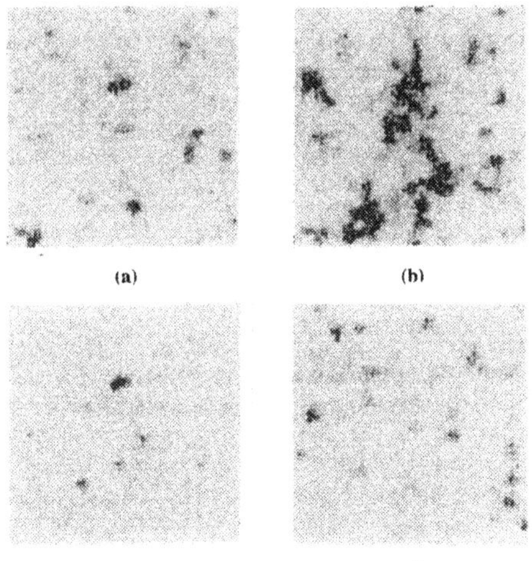

FIG. 7. Intensity plots of D^2_{min}, the deviation from affine deformation, for various intervals during two simulations. (a)–(c) show deformation during one simulation in which the stress has been ramped up quickly to a value less than the yield stress and then held constant. (a) shows deformations over the first 10 time units and (b) over the first 30 time units. (c) shows the same state as in (b), but with D^2_{min} computed only for deformations that took place during the preceding 1 time unit. In (d), the initial system and the time interval (10 units) are the same as in (a), but the stress has been applied in the opposite direction. The gray scale in these figures has been selected so that the darkest spots identify molecules for which $|D_{min}|\approx 0.5a_{SL}$.

isolated events. However, these events occur in different locations, implying a direction dependence of the local transformation mechanism.

Next we look at these processes in yet more detail. Figure 8 is a closeup of the molecular configurations in the lower left-hand part of the largest dark cluster seen in Fig. 7(c), shown just before and just after a shear transformation. During this event, the cluster of one large and three small molecules has compressed along the top-left to bottom-right axis and extended along the bottom-left to top-right axis. This deformation is consistent with the orientation of the applied shear, which is in the direction shown by the arrows on the

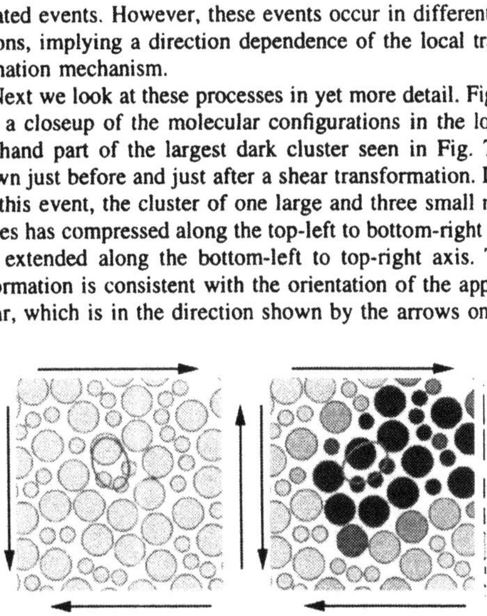

FIG. 8. Closeup picture of a shear transformation zone before and after undergoing transformation. Molecules after transformation are shaded according to their values of D^2_{min} using the same gray scale as in Fig. 7. The direction of the externally applied shear stress is shown by the arrows. The ovals are included solely as guides for the eye.

outside of the figure. Note that this rearrangement takes place without significantly affecting the relative positions of molecules in the immediate environment of the transforming region. This is the type of rearrangement that Spaepen identifies as a "flow defect" [20]. As mentioned in the Introduction, we shall call these regions shear transformation zones.

III. THEORETICAL INTERPRETATION OF THE MOLECULAR-DYNAMICS EXPERIMENTS

A. Basic hypotheses

We turn now to our attempts to develop a theoretical interpretation of the phenomena seen in the simulations. We shall not insist that our theory reproduce every detail of these results. In fact, the simulations are not yet complete enough to tell us whether some of our observations are truly general properties of the model or are artifacts of the ways in which we have prepared the system and carried out the numerical experiments. Our strategy will be first to specify what we believe to be the basic framework of a theory and then to determine which specific assumptions within this framework are consistent with the numerical experiments.

There are several features of our numerical experiments that we shall assume are fundamentally correct and which, therefore, must be outcomes of our theory. These are the following. (i) At a sufficiently small, fixed load, i.e., under a constant shear stress less than some value that we identify as a yield stress, the system undergoes a finite plastic deformation. The amount of this deformation diverges as the loading stress approaches the yield stress. (ii) At loading stresses above the yield stress, the system flows viscoplastically. (iii) The response of the system to loading is history dependent. If it is loaded, unloaded, and then reloaded to the same stress, it behaves almost elastically during the reloading, i.e., it does not undergo additional plastic deformation. On the other hand, if it is loaded, unloaded, and then reloaded with a stress of the opposite sign, it deforms substantially in the opposite direction.

Our theory consists of a set of rate equations describing plastic deformation. These include an equation for the inelastic strain rate as a function of the stress plus other variables that describe the internal state of the system. We also postulate equations of motion for these state variables. Deformation theories of this type are in the spirit of investigations by Hart [15], who, to the best of our knowledge, was the first to argue in a mathematically systematic way that any satisfactory theory of plasticity must include dynamical state variables, beyond just stress and strain. A similar point of view has been stressed by Rice [16]. Our analysis is also influenced by the use of state variables in theories of friction proposed recently by Ruina, Dieterich, Carlson, and others [40–45].

Our picture of what is happening at the molecular level in these systems is an extension of the ideas of Turnbull, Cohen, Argon, Spaepen, and others [17–21,23–25]. These authors postulated that deformation in amorphous materials occurs at special sites where the molecules are able to rearrange themselves in response to applied stresses. As described in Sec. II, we do see such sites in our simulations and shall use these shear transformation zones as the basis for our analysis. However, we must be careful to state as precisely as possible our definition of these zones because we shall use them in ways that were not considered by the previous authors.

One of the most fundamental differences between previous work and ours is the fact that our system is effectively at zero temperature. When it is in mechanical equilibrium, no changes occur in its internal state because there is no thermal noise to drive such changes. Thus the shear transformation zones can undergo transitions only when the system is in motion. Because the system is strongly disordered, the forces induced by large-scale motions at the position of any individual molecule may be noisy. These fluctuating forces may even look as if they have a thermal component [46]. The thermodynamic analogy (thermal activation of shear transformations with temperature being some function of the shear rate) may be an alternative to (or an equivalent of) the theory to be discussed here. However, it is beyond the scope of the present investigation.

Our next hypothesis is that shear transformation zones are geometrically identifiable regions in an amorphous solid. That is, we assume that we could, at least in principle, look at a picture of any one of the computer-generated states of our system and identify small regions that are particularly susceptible to inelastic rearrangement. As suggested by Fig. 8, these zones might consist of groups of four or more relatively loosely bound molecules surrounded by more rigid "cages," but that specific picture is not necessary. The main idea is that some such irregularities are locked in on time scales that are very much longer than molecular collision times. That is not to say that these zones are permanent features of the system on experimental time scales. On the contrary, the tendency of these zones to appear and disappear during plastic deformation will be an essential ingredient of our theory.

We suppose further that these shear transformation zones are two-state systems. That is, in the absence of any deformation of the cage of molecules that surrounds them, they are equally stable in either of two configurations. Very roughly speaking, the molecular arrangements in these two configurations are elongated along one or the other of two perpendicular directions, which, shortly, we shall take to be coincident with the principal axes of the applied shear stress. The transition between one such state and the other constitutes an elementary increment of shear strain. Note that bistability is the natural assumption here. More than two states of comparable stability might be possible but would have relatively low probability. A crucial feature of these bistable systems is that they can transform back and forth between their two states but cannot make repeated transformations in one direction. Thus there is a natural limit to how much shear can take place at one of these zones so long as the zone remains intact.

We now consider an ensemble of shear transformation zones and estimate the probability that any one of them will undergo a transition at an applied shear stress σ_s. Because the temperatures at which we are working are so low that ordinary thermal activation is irrelevant, we focus our attention on entropic variations of the local free volume. Our basic assumption is that the transition probability is proportional to the probability that the molecules in a zone have a sufficiently large excess free volume, say, ΔV^*, in which to rearrange themselves. This critical free volume must depend on the magnitude and orientation of the elastic deformation of the zone that is caused by the externally applied stress σ_s.

At this point, our analysis borrows in its general approach, but not in its specifics, from recent developments in the theory of granular materials [47] where the only extensive state variable is the volume Ω. What follows is a very simple approximation, which, at great loss of generality, leads us quickly to the result that we need. The free volume, i.e., the volume in excess of close packing that the particles have available for motion, is roughly

$$\Omega - N v_0 \equiv N v_f, \tag{3.1}$$

where N is the total number of particles, v_f is the average free volume per particle, and v_0 is the volume per particle in an ideal state of random dense packing. In the dense solids of interest to us here $v_f \ll v_0$ and therefore v_0 is approximately the average volume per particle even when the system is slightly dilated. The number of states available to this system is roughly proportional to $(v_f/h)^N$, where h is an arbitrary constant with dimensions of volume (the analog of Planck's constant in classical statistical mechanics) that plays no role other than to provide dimensional consistency. Thus the entropy, defined here to be a dimensionless quantity, is

$$S(\Omega,N) \cong N \ln\left(\frac{v_f}{h}\right) \cong N \ln\left(\frac{\Omega - N v_0}{N h}\right). \tag{3.2}$$

The intensive variable analogous to temperature is χ:

$$\frac{1}{\chi} \equiv \frac{\partial S}{\partial \Omega} \cong \frac{1}{v_f}. \tag{3.3}$$

Our activation factor, analogous to the Boltzmann factor for thermally activated processes, is therefore

$$e^{-(\Delta V^*/\chi)} \cong e^{-(\Delta V^*/v_f)}. \tag{3.4}$$

A formula like Eq. (3.4) appears in various places in the earlier literature [17,23–25]. There is an important difference between its earlier use and the way in which we are using it here. In earlier interpretations, Eq. (3.4) is an estimate of the probability that any given molecule has a large enough free volume near it to be the site at which a thermally activated irreversible transition might occur. In our interpretation, Eq. (3.4) plays more nearly the role of the thermal activation factor itself. It tells us something about the configurational probability for a zone, not just for a single molecule. When multiplied by the density of zones and a rate factor, about which we shall have more to say shortly, it becomes the transformation rate per unit volume.

Note what is happening here. Our system is extremely nonergodic and, even when it is undergoing appreciable strain, does not explore more than a very small part of its configuration space. Apart from the molecular rearrangements that take place during plastic deformation, the only chance that the system has for coming close to any state of equilibrium occurs during the quench by which it is formed initially. Because we control only the temperature and pressure during that quench, we must use entropic considerations to compute the relative probabilities of various molecular configurations that result from it.

The transitions occurring within shear transformation zones are strains and therefore they must, in principle, be described by tensors. For present purposes, however, we can make some simplifying assumptions. As described in Sec. II, our molecular-dynamics model is subject only to a uniform, pure shear stress of magnitude σ_s and a hydrostatic pressure P (usually zero). Therefore, in the principal-axis system of coordinates, the stress tensor is

$$[\sigma] = \begin{bmatrix} -P & \sigma_s \\ \sigma_s & -P \end{bmatrix}. \tag{3.5}$$

Our assumption is that the shear transformation zones are all oriented along the same pair of principal axes and therefore that the strain tensor has the form

$$[\varepsilon] = \begin{bmatrix} \varepsilon_d & \varepsilon_s \\ \varepsilon_s & \varepsilon_d \end{bmatrix}, \tag{3.6}$$

where ε_s and ε_d are the shear and dilational strains, respectively. The total shear strain is the sum of elastic and inelastic components

$$\varepsilon_s = \varepsilon_s^{el} + \varepsilon_s^{in}. \tag{3.7}$$

By definition, the elastic component is the linear response to the stress

$$\varepsilon_s^{el} = \frac{\sigma_s}{2\mu}, \tag{3.8}$$

where μ is the shear modulus.

In a more general formulation, we shall have to consider a distribution of orientations of the shear transformation zones. That distribution will not necessarily be isotropic when plastic deformations are occurring and very likely the distribution itself will be a dynamical entity with its own equations of motion. Our present analysis, however, is too crude to justify any such level of sophistication.

The last of our main hypotheses is an equation of motion for the densities of the shear transformation zones. Denote the two states of the shear transformation zones by the symbols + and − and let $n_\pm$ be the number densities of zones in those states. We then write

$$\dot{n}_\pm = R_\mp n_\mp - R_\pm n_\pm - C_1(\sigma_s \dot{\varepsilon}_s^{in}) n_\pm + C_2(\sigma_s \dot{\varepsilon}_s^{in}). \tag{3.9}$$

Here the $R_\pm$ are the rates at which $\pm$ states transform to $\mp$ states. These must be consistent with the transition probabilities described in the previous paragraphs.

The last two terms in Eq. (3.9) describe the way in which the population of shear transformation zones changes as the system undergoes plastic deformation. The zones can be annihilated and created, as shown by the terms with coefficients C_1 and C_2, respectively, at rates proportional to the rate $\sigma_s \dot{\varepsilon}_s^{in}$ at which irreversible work is being done on the system. This last assumption is simple and plausible, but it is not strictly dictated by the physics in any way that we can see. As a caveat we mention that in certain circumstances, when the sample does work on its environment, $\sigma_s \dot{\varepsilon}_s^{in}$ could be negative, in which case the annihilation and creation terms in Eq. (3.9) could produce results that would not be physically plausible. We believe that such states in our theory are dynamically accessible only from unphysical starting configurations. In related theories, however, that may not be the case.

It is important to recognize that the annihilation and creation terms in Eq. (3.9) are interaction terms and that they have been introduced here in a mean-field approximation. That is, we implicitly assume that the rates at which shear transformation zones are annihilated and created depend only on the rate at which irreversible work is being done on the system as a whole and that there is no correlation between the position at which the work is being done and the place where the annihilation or creation is occurring. This is, in fact, not the case as shown by Fig. 7(b) and is possibly the weakest aspect of our theory. With the preceding definitions, the time rate of change of the inelastic shear strain ε_s^{in} has the form

$$\dot{\varepsilon}_s^{in}=V_z\Delta\varepsilon[R_+n_+-R_-n_-], \tag{3.10}$$

where V_z is the typical volume of a zone and $\Delta\varepsilon$ is the increment of local shear strain.

B. Specific assumptions

We turn now to the more detailed assumptions and analyses that we need in order to develop our general hypotheses into a testable theory. According to our hypothesis about the probabilities of volume fluctuations, we should write the transition rates in Eq. (3.9) in the form

$$R_\pm=R_0\exp\left[-\frac{\Delta V^*(\pm\sigma_s)}{v_f}\right]. \tag{3.11}$$

The prefactor R_0 is an as-yet unspecified attempt frequency for these transformations. In writing Eq. (3.11) we have used the assumed symmetry of the system to note that if $\Delta V^*(\sigma_s)$ is the required excess free volume for a $+\to-$ transition, then the appropriate free volume for the reverse transition must be $\Delta V^*(-\sigma_s)$. We adopt the convention that a positive shear stress deforms a zone in such a way that it enhances the probability of a $+\to-$ transition and decreases the probability of a $-\to+$ transition. Then $\Delta V^*(\sigma_s)$ is a decreasing function of σ_s.

Before going any further in specifying the ingredients of R_0, ΔV^*, etc., it is useful to recast the equations of motion in the following form. Define

$$n_{tot}\equiv n_++n_-,\quad n_\Delta\equiv n_--n_+, \tag{3.12}$$

and

$$\mathcal{C}(\sigma_s)\equiv\frac{1}{2}\left[\exp\left(-\frac{\Delta V^*(\sigma_s)}{v_f}\right)+\exp\left(-\frac{\Delta V^*(-\sigma_s)}{v_f}\right)\right],$$

$$\mathcal{S}(\sigma_s)\equiv\frac{1}{2}\left[\exp\left(-\frac{\Delta V^*(\sigma_s)}{v_f}\right)-\exp\left(-\frac{\Delta V^*(-\sigma_s)}{v_f}\right)\right]. \tag{3.13}$$

[For convenience, and in order to be consistent with later assumptions, we have suppressed other possible arguments of the functions $\mathcal{C}(\sigma_s)$ and $\mathcal{S}(\sigma_s)$.] Then Eq. (3.11) becomes

$$\dot{\varepsilon}_s^{in}=R_0V_z\Delta\varepsilon[n_{tot}\mathcal{S}(\sigma_s)-n_\Delta\mathcal{C}(\sigma_s)]. \tag{3.14}$$

The equations of motion for n_Δ and n_{tot} are

$$\dot{n}_\Delta=\frac{2\dot{\varepsilon}_s^{in}}{V_z\Delta\varepsilon}\left(1-\frac{\sigma_s n_\Delta}{\bar{\sigma}n_\infty}\right) \tag{3.15}$$

and

$$\dot{n}_{tot}=\frac{2\sigma_s\dot{\varepsilon}_s^{in}}{V_z\Delta\varepsilon\bar{\sigma}}\left(1-\frac{n_{tot}}{n_\infty}\right), \tag{3.16}$$

where $\bar{\sigma}$ and n_∞ are defined by

$$C_1\equiv\frac{2}{V_z\Delta\varepsilon n_\infty\bar{\sigma}},\quad C_2\equiv\frac{1}{V_z\Delta\varepsilon\bar{\sigma}}. \tag{3.17}$$

From Eq. (3.16) we see that n_∞ is the stable equilibrium value of n_{tot} so long as $\sigma_s\dot{\varepsilon}_s^{in}$ remains positive. $\bar{\sigma}$ is a characteristic stress that, in certain cases, turns out to be the plastic yield stress. As we shall see, we need only the above form of the equations of motion to deduce the existence of the plastic yield stress and to compute some elementary properties of the system.

The interesting time-dependent behavior of the system, however, depends sensitively on the as-yet unspecified ingredients of these equations. Consider first the rate factor R_0. Our zero-temperature hypothesis implies that R_0 should be zero whenever the inelastic shear rate $\dot{\varepsilon}_s^{in}$ and the elastic shear rate $\dot{\varepsilon}_s^{el}=\dot{\sigma}_s/2\mu$ both vanish. Accordingly, we assume that

$$R_0\cong\nu^{1/2}[(\dot{\varepsilon}_s^{el})^2+(\dot{\varepsilon}_s^{in})^2]^{1/4}, \tag{3.18}$$

where ν is a constant that we must determine from the numerical data. Note that ν contains both an attempt frequency and a statistical factor associated with the multiplicity of trajectories leading from one state to the other in an active zone [48].

We can offer only a speculative justification for the right-hand side of Eq. (3.18). The rearrangements that occur during irreversible shear transformations are those in which molecules deviate from the trajectories that they would follow if the system were a continuous medium undergoing affine strain. If we assume that these deviations are diffusive and that the affine deformation over some time interval scales like the strain rate, then the nonaffine transformation rate must scale like the square root of the affine rate. (Diffusive deviations from smooth trajectories have been observed directly in numerical simulations of sheared foams [46], but only in the equivalent of our plastic flow regime.) In Eq. (3.18) we further assume that the elastic and inelastic strain rates are incoherent and thus write the sum of squares within the square brackets. In what follows, we shall not be able to test the validity of Eq. (3.18) with any precision. Most prob-

TABLE II. Values of parameters for comparison to simulation data.

Parameter	Value
$\bar{\sigma}$	0.32
$V_z\Delta\varepsilon n_\infty$	5.7%
ν	50.0
V_0^*/v_f	14.0
$\bar{\mu}$	0.25
$n_{tot}(0)/n_\infty$	2.0

ably, the only properties of importance to us for the present purposes are the magnitude of R_0 and the fact that it vanishes when the shear rates vanish.

Finally, we need to specify the ingredients of ΔV^* and v_f. For ΔV^* we choose the simple form

$$\Delta V^*(\sigma_s)=V_0^*\exp(-\sigma_s/\bar{\mu}), \tag{3.19}$$

where V_0^* is a volume, perhaps of order the average molecular volume v_0, and $\bar{\mu}$ has the dimensions of a shear modulus. The right-hand side of Eq. (3.19) simply reflects the fact that the free volume needed for an activated transition will decrease if the zone in question is loaded with a stress that coincides with the direction of the resulting strain. We choose the exponential rather than a linear dependence because it makes no sense for the incremental free volume V_0^* to be negative, even for very large values of the applied stresses.

Irreversibility enters the theory via a simple switching behavior that occurs when the σ_s dependence of ΔV^* in Eq. (3.19) is so strong that it converts a negligibly small rate at $\sigma_s=0$ to a large rate at relevant, nonzero values of σ_s. If this happens, then zones that have switched in one direction under the influence of the stress will remain in that state when the stress is removed.

In the formulation presented here, we consider v_f to be constant. This is certainly an approximation; in fact, as seen in Fig. 3, the system dilates during shear deformation. We have experimented with versions of this theory in which the dilation plays a controlling role in the dynamics via variations in v_f. We shall not discuss these versions further because they behaved in ways that were qualitatively different from what we observed in our simulations. The differences arise from feedback between inelastic dilation and flow that occur in these dilational models and apparently not in the simulations. A simple comparison of the quantities involved demonstrates that the assumption that v_f is approximately constant is consistent with our other assumptions. If we assume that the increment in free volume at zero stress must be of order the volume of a small particle $V_0^*\approx v_0\approx 0.3$ and then look ahead and use our best-fit value for the ratio $V_0^*/v_f\approx 14.0$ (see Sec. III D, Table II), we find $v_f\approx 0.02$. Since the change in free volume due to a dilational strain ε_d is $\Delta v_f=\varepsilon_d/\rho$, where ρ is the number density and ε_d $<0.2\%$ for all shear stresses except those very close to yield, it appears that, generally, $\Delta v_f\approx\varepsilon_d v_0\ll v_f$. Even when $\varepsilon_d=1\%$, the value observed in our simulations at yield, the dilational free volume is only about the same as the initial free volume estimated by this analysis.

As a final step in examining the underlying structure of these equations of motion, we make the scaling transformations

$$\frac{2\mu\varepsilon_s^{in}}{\bar{\sigma}}\equiv\mathcal{E},\quad \frac{n_\Delta}{n_\infty}\equiv\Delta,\quad \frac{n_{tot}}{n_\infty}\equiv\Lambda,\quad \frac{\sigma_s}{\bar{\sigma}}\equiv\Sigma. \tag{3.20}$$

Then we find

$$\dot{\mathcal{E}}=\bar{\mathcal{E}}\mathcal{F}(\Sigma,\Lambda,\Delta), \tag{3.21}$$

$$\dot{\Delta}=2\mathcal{F}(\Sigma,\Lambda,\Delta)(1-\Sigma\Delta), \tag{3.22}$$

$$\dot{\Lambda}=2\mathcal{F}(\Sigma,\Lambda,\Delta)\Sigma(1-\Lambda), \tag{3.23}$$

where

$$\mathcal{F}(\Sigma,\Lambda,\Delta)=R_0[\Lambda\mathcal{S}(\Sigma)-\Delta\mathcal{C}(\Sigma)] \tag{3.24}$$

and

$$\mathcal{C}(\Sigma)=\frac{1}{2}\left[\exp\left(-\frac{V_0^*}{v_f}e^{-A\Sigma}\right)+\exp\left(-\frac{V_0^*}{v_f}e^{A\Sigma}\right)\right],$$
$$\mathcal{S}(\Sigma)=\frac{1}{2}\left[\exp\left(-\frac{V_0^*}{v_f}e^{-A\Sigma}\right)-\exp\left(-\frac{V_0^*}{v_f}e^{A\Sigma}\right)\right]. \tag{3.25}$$

Here,

$$A\equiv\frac{\bar{\sigma}}{\bar{\mu}},\quad \bar{\mathcal{E}}\equiv\frac{2\mu V_z\Delta\varepsilon n_\infty}{\bar{\sigma}}. \tag{3.26}$$

The rate factor in Eq. (3.18) can be rewritten

$$R_0=\bar{\nu}^{1/2}(\Sigma^2+\dot{\mathcal{E}}^2)^{1/4}, \tag{3.27}$$

where

$$\bar{\nu}\equiv\frac{\bar{\sigma}}{2\mu}\nu. \tag{3.28}$$

C. Special steady-state solutions

Although, in general, we must use numerical methods to solve the fully time-dependent equations of motion, we can solve them analytically for special cases in which the stress Σ is held constant. Note that none of the results presented in this subsection, apart from Eq. (3.35), depend on our specific choice of the rate factor R_0.

There are two specially important steady-state solutions at constant Σ. The first of these is a jammed solution in which $\dot{\mathcal{E}}=0$, that is, $\mathcal{F}(\Sigma,\Lambda,\Delta)$ vanishes and therefore

$$\Delta=\Lambda\frac{\mathcal{S}(\Sigma)}{\mathcal{C}(\Sigma)}=\Lambda\mathcal{T}(\Sigma), \tag{3.29}$$

where

$$\mathcal{T}(\Sigma)\equiv 1-2\left[1+\exp\left(2\frac{V_0^*}{v_f}\sinh(A\Sigma)\right)\right]^{-1}. \tag{3.30}$$

Now suppose that instead of increasing the stress at a finite rate, as we have done in our numerical experiments, we let it jump discontinuously, from zero, perhaps, to its value Σ at time $t=0$. While Σ is constant, Eqs. (3.22) and (3.23) can be solved to yield

$$\frac{1-\Lambda(t)}{1-\Lambda(0)}=\frac{1-\Sigma\Delta(t)}{1-\Sigma\Delta(0)}, \tag{3.31}$$

where $\Lambda(0)$ and $\Delta(0)$ denote the initial values of $\Lambda(t)$ and $\Delta(t)$, respectively. Similarly, we can solve Eqs. (3.21) and (3.22) for $\mathcal{E}(t)$ in terms of $\Delta(t)$ and obtain a relationship between the bias in the population of defects and the change in strain,

$$\mathcal{E}(t)=\mathcal{E}(0)+\frac{\bar{\mathcal{E}}}{2\Sigma}\ln\left(\frac{1-\Sigma\Delta(0)}{1-\Sigma\Delta(t)}\right). \tag{3.32}$$

Combining Eqs. (3.29), (3.31), and (3.32), we can determine the change in strain prior to jamming. That is, for Σ sufficiently small that the following limit exists, we can compute a final inelastic strain $\mathcal{E}_f$:

$$\mathcal{E}_f\equiv\lim_{t\to\infty}\mathcal{E}(t)=\mathcal{E}(0)+\frac{\bar{\mathcal{E}}}{2\Sigma}\ln\left(1+\Sigma\,\frac{\mathcal{T}(\Sigma)\Lambda(0)-\Delta(0)}{1-\Sigma\mathcal{T}(\Sigma)}\right). \tag{3.33}$$

The right-hand side of Eq. (3.33), for $\mathcal{E}(0)=\Delta(0)=0$, should be at least a rough approximation for the inelastic strain as a function of stress as shown in Fig. 5.

The preceding analysis is our mathematical description of how the system jams due to the two-state nature of the shear transformation zones. Each increment of plastic deformation corresponds to the transformation of zones aligned favorably with the applied shear stress. As the zones transform, the bias in their population, i.e., Δ, grows. Eventually, all of the favorably aligned zones that can transform at the given magnitude and direction of the stress have undergone their one allowed transformation, Δ has become large enough to cause $\mathcal{F}$ in Eq. (3.24) to vanish, and plastic deformation comes to a halt.

The second steady state is a plastically flowing solution in which $\dot{\mathcal{E}}\neq 0$ but $\dot{\Delta}=\dot{\Lambda}=0$. From Eq. (3.22) and (3.23) we see that this condition requires

$$\Delta=\frac{1}{\Sigma},\quad \Lambda=1. \tag{3.34}$$

This leads us directly to an equation for the strain rate at constant applied stress,

$$\dot{\mathcal{E}}=\bar{\nu}\bar{\mathcal{E}}^2\left[\mathcal{S}(\Sigma)-\frac{1}{\Sigma}\mathcal{C}(\Sigma)\right]^2. \tag{3.35}$$

This flowing solution arises from the nonlinear annihilation and creation terms in Eq. (3.9). In the flowing state, stresses are high enough that shear transformation zones are continuously created. A balance between the rate of zone creation and the rate of transformation determines the rate of deformation.

Examination of Eqs. (3.22) and (3.23) reveals that the jammed solution (3.29) is stable for low stresses, while the flowing solution (3.34) is stable for high stresses. The crossover between the two solutions occurs when both Eqs. (3.29) and (3.34) are satisfied. This crossover defines the yield stress Σ_y, which satisfies the condition

$$\frac{1}{\Sigma_y}=\mathcal{T}(\Sigma_y). \tag{3.36}$$

Note that the argument of the logarithm in Eq. (3.33) diverges at $\Sigma=\Sigma_y$. Note also that, so long as $(2V_0^*/v_f)\sinh(A\Sigma_y)\gg 1$, Eq. (3.36) implies that $\Sigma_y\cong 1$. This inequality is easily satisfied for the parameters discussed in the following subsection. Thus the dimensional yield stress σ_y is approximated accurately by $\bar{\sigma}$ in our original units defined in Eq. (3.20).

D. Parameters of the theory

There are five adjustable system parameters in our theory: $\bar{\sigma}$, $V_z\Delta\varepsilon n_\infty$, ν, V_0^*/v_f, and $\bar{\mu}$. In addition, we must specify initial conditions for $\mathcal{E}$, Δ, and Λ. For all cases of interest here, $\mathcal{E}(0)=\Delta(0)=0$. However, $\Lambda(0)=n_{tot}(0)/n_\infty$ is an important parameter that characterizes the as-quenched initial state of the system and remains to be determined.

To test the validity of this theory, we now must find out whether there exists a set of physically reasonable values of these parameters for which the theory accounts for all (or almost all) of the wide variety of time-dependent phenomena seen in the molecular-dynamics experiments. Our strategy has been to start with rough guesses based on our understanding of what these parameters mean and then to adjust these values by trial and error to fit what we believe to be the crucial features of the experiments. We then have used those values of the parameters in the equations of motion to check agreement with other numerical experiments. In adjusting parameters, we have looked for accurate agreement between theory and experiment in low-stress situations where we expect the concentration of active shear transformation zones to be low and we have allowed larger discrepancies near and above the yield stress where we suspect that interactions between the zones may invalidate our mean-field approximation. Our best-fit parameters are shown in Table II.

The easiest parameter to fit should be $\bar{\sigma}$ because it should be very nearly equal to the yield stress. That is, it should be somewhere in the range 0.30–0.35 according to the data shown in Fig. 5. Note that we cannot use Eq. (3.33) to fit the experimental data near the yield point because both the numerical simulations and the theory tell us that the system approaches its stationary state infinitely slowly there. Moreover, we expect interaction effects to be important here. The solid curve in Fig. 5 is the theoretically predicted strain found by integrating the equations of motion for 800 time units, the duration of the longest of the simulation runs. The downward adjustment of $\bar{\sigma}$, from its apparent value of about 0.35 to its best-fit value of 0.32, has been made on the basis of the latter time-dependent calculations plus evidence about the effect of this parameter in other parts of the theory.

Next we consider $V_z\Delta\varepsilon n_\infty$, a dimensionless parameter that corresponds to the amount of strain that would occur if the density of zones were equal to the equilibrium concentration ($n_{tot}=n_\infty$) and if all the zones transformed in the same direction in unison. Alternatively, if the local strain increment $\Delta\varepsilon$ is about unity, then this parameter is the fraction of the volume of the system that is occupied by shear transformation zones. In either way of looking at this quantity, our best-fit value of 5.7% seems sensible.

The parameter ν is a rate that is roughly the product of an attempt frequency and a statistical factor. The only system-dependent quantity with the dimensions of inverse time is the molecular vibrational frequency, which we have seen is of order unity. Our best-fit value of 50 seems to imply that the statistical factor is moderately large, which, in turn, implies that the shear transformation zones are fairly complex, multimolecule structures. Lacking any first-principles theory of this rate factor, however, we cannot be confident about this observation.

Our first rough guess for a value of V_0^*/v_f comes from the assumption that ΔV^* must be about one molecular volume in the absence of an external stress and that v_f is likely to be about a tenth of this. Thus our best-fit value of 14.0 is reassuringly close to what we expected.

The parameter $\bar{\mu}$, a modulus that characterizes the sensitivity of ΔV^* to the applied stress, is especially interesting. Our best-fit value of 0.25 is almost two orders of magnitude smaller than a typical shear modulus for these systems. This means that the shear transformations are induced by relatively small stresses or, equivalently, the internal elastic modes within the zones are very soft. This conclusion is supported quite robustly by our fitting procedure. Alternative assumptions, such as control by variations in the average free volume v_f discussed earlier, produce qualitatively wrong pictures of the time-dependent onset of plastic deformation.

Finally, we consider $\Lambda(0)=n_{tot}(0)/n_\infty$, the ratio of the inital zone density to the equilibrium zone density. This parameter characterizes the transient behavior associated with the initial quench; that is, it determines the as-quenched system's first response to an applied stress. We can learn something about this parameter by looking at later behavior, i.e., the next few segments of a hysteresis loop such as that shown in Fig. 6. If, as is observed there, the loop narrows after the first leg, then we know that there was an excess of shear transformation zones in the as-quenched system and that this excess was reduced in the initial deformation. An initial excess means $\Lambda(0)>1$, consistent with our best-fit value of 2.0.

E. Comparisons between theory and simulations

We now illustrate the degree to which this theory can and cannot account for the phenomena observed in the numerical experiments.

Figure 9 summarizes one of the principal successes of the theory, specifically, its ability to predict the time-dependent onset of plastic deformation over a range of applied stresses below the yield stress. The solid lines in the figure show the shear strains in three different simulations as functions of time. In each simulation the stress is ramped up at the same controlled rate, held constant for a period of time, and then ramped down, again at the same rate. In the lowest curve the stress reaches a maximum of 0.1 in our dimensionless stress units (e_{SL}/a_{SL}^2), in the middle curve 0.2, and in the highest 0.3. The dashed lines show the predictions of the theory. The excellent agreement during and after the ramp up is our most direct evidence for the small value of $\bar{\mu}$ quoted above. The detailed shapes of these curves at the tops of the ramps, where $\dot{\sigma}_s$ drops abruptly to zero, provide some qualitative support for our choice of the rate dependence of R_0 in Eq. (3.18). As shown in Fig. 5 and discussed in Sec. III D, the final inelastic strains in these ramp-up experiments are also predicted adequately by the theory.

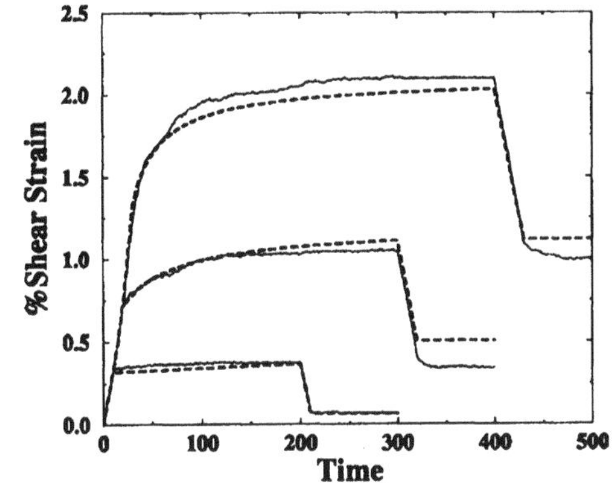

FIG. 9. Strain vs time for simulations in which the stress has been ramped up at a controlled rate to stresses of 0.1, 0.2, and 0.3, held constant, and then ramped down to zero (solid lines). The dashed lines are the corresponding theoretical predictions. Time is measured in units of $(ma_{SL}^2/e_{SL})^{1/2}$.

The situation is different for the unloading phases of these experiments, that is, during and after the periods when the stresses are ramped back down to zero. The theoretical strain rates shown in Fig. 9 vanish abruptly at the bottoms of the ramps because our transformation rates become negligibly small at zero stress. In the two experimental curves for the higher stresses, however, the strain continues to decrease for a short while after the stresses have stopped changing. Our theory seems to rule out any such recovery of inelastic strain at zero stress; thus we cannot account for this phenomenon except to remark that it must have something to do with the initial state of the as-quenched system. As seen in Fig. 6, no such recovery occurs when the system is loaded and unloaded a second time.

In Fig. 6, we compare the stress-strain hysteresis loop in the simulation (solid line) with that predicted by the theory (dashed line). Apart from the inelastic strain recovery after the first unloading in the simulation, the theory and the experiment agree well with one another at least through the reverse loading to point k. The agreement becomes less good in subsequent cycles of the hysteresis loop, possibly because shear bands are forming during repeated plastic deformations.

In the last of the tests of theory to be reported here, we have added in Fig. 1 two theoretical curves for stresses as functions of strain at the two different constant strain rates used in the simulations. The agreement between theory and experiment is better than we probably should expect for situations in which the stresses necessarily rise to values at or above the yield stress. Moreover, the validity of the comparison is obscured by the large fluctuations in the data, which we believe to be due primarily to small sample size.

Among the interesting features of the theoretical results in Fig. 1 are the peaks in the stresses that occur just prior to the establishment of steady states at constant stresses. These peaks occur because the internal degrees of freedom of the system, specifically $\Delta(t)$ and $\Lambda(t)$, cannot initially equilibrate fast enough to accommodate the rapidly increasing inelastic strain. Thus there is a transient stiffening of the material and a momentary increase in the stress needed to maintain the constant strain rate. This kind of effect may in part be the explanation for some of the oscillations in the stress seen in the experiments. In a more speculative vein, we note that this is our first direct hint of the kind of dynamic plastic stiffening that is needed in order to transmit high stresses to crack tips in brittle fracture. The strain rates near the tips of brittle cracks are at least of the same order of magnitude as the strain rates imposed here and may in fact be substantially higher.

IV. CONCLUDING REMARKS

The most striking and robust conclusion to emerge from this investigation, in our opinion, is that a wide range of realistic, irreversible, viscoplastic phenomena occur in an extremely simple molecular-dynamics model a two-dimensional, two-component, Lennard-Jones amorphous solid at essentially zero temperature. An almost equally striking conclusion is that a theory based on the dynamics of two-state shear transformation zones is in substantial agreement with the observed behavior of this model. This theory has survived several quantitative tests of its applicability.

We stated in our Introduction that this is a preliminary report. Both the numerical simulations and the theoretical analysis require careful evaluation and improvements. Most importantly, the work so far raises many important questions that need to be addressed in future investigations.

The first kind of question pertains to our molecular-dynamics simulations: Are they accurate and repeatable? We believe that they are good enough for present purposes, but we recognize that there are potentially important difficulties. The most obvious of these is that our simulations have been performed with very small systems; thus size effects may be important. For example, the fact that only a few shear transforming regions are active at any time may account for abrupt jumps and other irregularities sometimes seen in the simulations, e.g., in Fig. 1. We have performed the simulations in a periodic cell to eliminate edge effects. We also have tried to compare results from two systems of different sizes, although only the results from the larger system are presented here. Unfortunately, comparisons between any two different initial configurations are difficult because of our inability, as yet, to create reproducible glassy starting configurations (a problem that we shall discuss next). However, we have seen qualitatively the same behavior in both systems and assume that phenomena that are common to both systems can be used as a guide for theoretical investigations.

As noted in Sec. II B and in Table I, our two systems had quite different elastic moduli. (Remarkably, their yield stresses were nearly identical. It would be interesting to learn whether this is a repeatable and/or physically important phenomenon.) The discrepancy between the elastic properties of the two systems leads us to believe that, in future work, we shall have to learn how to control the initial configurations more carefully, perhaps by annealing the systems after the initial quenches. Unfortunately, straightforward annealing at temperatures well below the glass transition is not yet possible with standard molecular-dynamics algorithms, which can simulate times only up to about 1 μs for systems of this size even with today's fastest computers. Monte Carlo techniques or accelerated molecular-dynamics algorithms may eventually be useful in this effort [49–51]. An alternative strategy may be simply to look at larger numbers of simulations.

By far the most difficult and interesting questions, however, pertain to our theoretical analysis. Although Figs. 7 and 8 provide strong evidence that irreversible shear transformations are localized events, we have no sharp definition of a "shear transformation zone." So far, we have identified these zones only after the fact, that is, only by observing where the transformations are taking place. Is it possible, at least in principle, to identify zones before they become active?

One ingredient of a better definition of shear transformation zones will be a generalization to isotropic amorphous systems in both two and three dimensions. As we noted in Sec. III, our functions $n_{\pm}(t)$ should be tensor quantities that describe distributions over the ways in which the individual zones are aligned with respect to the orientation of the applied shear stress. We believe that this is a relatively easy generalization; one of us (M.L.F.) expects to report on work along these lines in the future.

Our more urgent reason for needing a better understanding of shear transformation zones is that, without such an understanding, we shall not be able to find first-principles derivations of several, as-yet purely phenomenological, ingredients of our theory. It might be useful, for example, to be able to start from the molecular force constants and calculate the parameters V_0^* and $\bar{\mu}$ that occur in the activation factor (3.19). These parameters, however, seem to have clear physical interpretations; thus we might be satisfied to deduce them from experiment. In contrast, the conceptually most challenging and important terms are the rate factor in Eq. (3.18) and the annihilation and creation terms in Eq. (3.9), where we do not even know what the functional forms ought to be.

Calculating the rate factor in Eq. (3.18), or a correct version of that equation, is clearly a very fundamental problem in nonequilibrium statistical physics. So far as we know, there are no studies in the literature that might help us compute the force fluctuations induced at some site by externally driven deformations of an amorphous material. Nor do we know how to compute a statistical prefactor analogous, perhaps, to the entropic factor that converts an activation energy to an activation free energy. [48] We do know, however, that that entropic factor will depend strongly on the size and structure of the zone that is undergoing the transformation.

As emphasized in Sec. III, the annihilation and creation terms in Eq. (3.9) describe interaction effects. Even within the framework of our mean-field approximation, we do not know with any certainty what these terms should be. Our assumption that they are proportional to the rate of irreversible work is by no means unique. (Indeed, we have tried other possibilities in related investigations and have arrived at qualitatively similar conclusions.) Without knowing more about the nature of the shear transformation zones, it will be difficult to derive such interaction terms from first principles.

A better understanding of these interaction terms is especially important because these are the terms that will have to be modified when we go beyond the mean-field theory to account for correlations between regions undergoing plastic deformations. We know from our simulations that the active zones cluster even at stresses far below the plastic yield stress and we know that plastic yield in real amorphous materials is dominated by shear banding. Thus, generalizing the present mean-field theory to one that takes into account spatial variations in the densities of shear transformation zones must be a high priority in this research program.

Finally, we return briefly to the question that motivated this investigation: How might the dynamical effects described here, which must occur in the vicinity of a crack tip, control crack stability and brittle or ductile behavior? As we have seen, our theoretical picture of viscoplasticity does allow large stresses to be transmitted, at least for short times, through plastically deforming materials. It should be interesting to see what happens if we incorporate this picture into theories of dynamic fracture.

ACKNOWLEDGMENTS

This research was supported by DOE Grant No. DE-FG03-84ER45108, by the DOE Computational Sciences Graduate Fellowship Program and, in part, by the MRSEC Program of the National Science Foundation under Grant No. DMR96-32716. We wish particularly to thank Alexander Lobkovsky for his attention to this project and for numerous useful suggestions. We also thank A. Argon and F. Lange for guidance in the early stages of this work, J. Cahn for directing us to the papers of E. Hart, and S. Langer and A. Liu for showing us their closely related results on the dynamics of sheared foams.

[1] J. Langer and A. Lobkovsky, J. Mech. Phys. Solids (to be published).
[2] J. Rice and R. Thomson, Philos. Mag. **29**, 73 (1974).
[3] J. Rice, J. Mech. Phys. Solids **40**, 239 (1992).
[4] M. Khantha, D. Pope, and V. Vitek, Phys. Rev. Lett. **73**, 684 (1994).
[5] L. Freund and Y. Lee, Int. J. Fract. **42**, 261 (1990).
[6] P. Steif, J. Mech. Phys. Solids **31**, 359 (1983).
[7] J. Fineberg, S. Gross, M. Marder, and H. Swinney, Phys. Rev. Lett. **67**, 457 (1991).
[8] J. Fineberg, S. Gross, M. Marder, and H. Swinney, Phys. Rev. B **45**, 5146 (1992).
[9] T.-W. Wu and F. Spaepen, Philos. Mag. B **61**, 739 (1990).
[10] P. Chaudhari, A. Levi, and P. Steinhardt, in *Glassy Metals II*, edited by H. Beck and H.-J. Guntherodt (Springer-Verlag, Berlin, 1983), Vol. 53, p. 127.
[11] F. Spaepen and A. Taub, in *Amorphous Metallic Alloys*, edited by F. Luborsky (Butterworths, London, 1983), p. 231.
[12] A. Taub, Acta Metall. **30**, 2117 (1982).
[13] H. Kimura and T. Matsumoto, in *Amorphous Metallic Alloys* (Ref. [11]), p. 187.
[14] E. Oleinikraux, O. Salamatina, S. Rudnev, and S. Shenogin, Polym. Sci. **35**, 1532 (1993).
[15] E. Hart, Acta Metall. **18**, 599 (1970).
[16] J. Rice, in *Constitutive Equations in Plasticity*, edited by A. Argon (MIT Press, Cambridge, MA, 1975), p. 23.
[17] F. Spaepen, Acta Metall. **25**, 407 (1977).
[18] A. Argon, Acta Metall. **27**, 47 (1979).
[19] A. Argon and H. Kuo, Mater. Sci. Eng. **39**, 101 (1979).
[20] F. Spaepen and A. Taub, in *Physics of Defects*, edited by J.-P. P. R. Balian and M. Kleman, 1981 Les Houches Lectures, Session XXXV (North-Holland, Amsterdam, 1981), p. 133.
[21] A. Argon and L. Shi, Acta Metall. **31**, 499 (1983).
[22] V. Khonik and A. Kosilov, J. Non-Cryst. Solids **170**, 270 (1994).
[23] M. Cohen and D. Turnbull, J. Chem. Phys. **31**, 1164 (1959).
[24] D. Turnbull and M. Cohen, J. Chem. Phys. **34**, 120 (1961).
[25] D. Turnbull and M. Cohen, J. Chem. Phys. **52**, 3038 (1970).
[26] D. Deng, A. Argon, and S. Yip, Philos. Trans. R. Soc. London, Ser. A **329**, 549 (1989).
[27] S. Kobayashi, K. Maeda, and S. Takeuchi, Acta Metall. **28**, 1641 (1980).
[28] K. Maeda and S. Takeuchi, Philos. Mag. A **44**, 643 (1981).
[29] D. Srolovitz, V. Vitek, and T. Egami, Acta Metall. **31**, 335 (1983).
[30] S. Melchionna, G. Ciccotti, and B. Holian, Mol. Phys. **78**, 533 (1993).
[31] S. Nose, J. Chem. Phys. **81**, 511 (1984).
[32] S. Nose, Mol. Phys. **52**, 255 (1984).
[33] S. Nose, Mol. Phys. **57**, 187 (1986).
[34] M. Parrinello and A. Rahman, J. Appl. Phys. **52**, 7182 (1981).
[35] M. Parrinello and A. Rahman, J. Chem. Phys. **76**, 2662 (1982).
[36] H. Anderson, J. Chem. Phys. **72**, 2384 (1980).
[37] F. Lançon, L. Billard, and P. Chaudhari, Europhys. Lett. **2**, 625 (1986).
[38] F. Lançon and L. Billard, J. Phys. (France) **49**, 249 (1988).
[39] R. Mikulla, J. Roth, and H.-R. Trebin, Philos. Mag. B **71**, 981 (1995).
[40] J. Dieterich, Pure Appl. Geophys. **116**, 790 (1978).
[41] J. Dieterich, J. Geophys. Res. **84**, 2161 (1979).
[42] J. Rice and A. Ruina, J. Appl. Mech. **105**, 343 (1983).
[43] A. Ruina, J. Geophys. Res. **88**, 10 359 (1983).
[44] J. Dieterich, PAGEOPH **143**, 283 (1994).
[45] J. Carlson and A. Batista, Phys. Rev. E **53**, 4153 (1996).
[46] S. Langer and A. Liu (unpublished).
[47] A. Mehta and S. F. Edwards, Physica A **157**, 1091 (1990).
[48] J. Langer, Ann. Phys. (N.Y.) **54**, 258 (1969).
[49] G. Barkema and N. Mousseau, Phys. Rev. Lett. **77**, 4358 (1996).
[50] A. Voter, J. Chem. Phys. **106**, 4665 (1997).
[51] A. Voter, Phys. Rev. Lett. **78**, 3908 (1997).

Part 4

EXPERIMENTAL MANIFESTATIONS OF MODE COUPLING

TRANSPORT THEORY AND STATISTICAL PHYSICS, 24(6-8), 801-853 (1995)

THE MODE COUPLING THEORY OF STRUCTURAL RELAXATIONS

W. Götze † and L. Sjögren ‡

† Physik-Department, Technische Universität München, D-85747 Garching; and Max-Planck-Institut für Physik, Werner-Heisenberg-Institut, P.O. Box 401212, D-80805 München, Germany

‡ Institutionen för teoretisk fysik, Göteborgs Universitet, S-41296 Göteborg, Sweden

ABSTRACT

The basic ideas and approximations underlying the derivation of the mode coupling theory of structural relaxation in simple liquids are summarized. We explain why in disordered many particle systems the infinite set of density fluctuation products constitutes slow modes whose interactions lead to bifurcation singularities. The singularities are connected with the appearance of spontaneous arrest of particle distributions in ideal glass states. These constitute an almost frozen potential landscape for the phonon assisted transport processes, which restore ergodicity in strongly supercooled or supercompressed liquids.

Some concepts needed for a description of structural relaxation are explained: glass transition singularities, non-ergodicity parameters, critical decay laws, separation parameters, critical temperature T_c or density n_c and α- and β-relaxation. It is emphasized that there are universality classes for the dynamics where the correlation functions within certain windows can be specified by a few number of parameters. Thereby it is shown that the theory can provide results for a quantitative description of experiments also for very complicated systems like polymers.

1 INTRODUCTION

Already Maxwell assumed that there are slow, strongly temperature dependent relaxation processes in liquids. Upon supercooling the time scale τ_α of these processes can become of macroscopic size. This means that τ_α becomes comparable to the time scale t_{exp}, characterizing the experiment which is designed to probe the system's dynamics. The origin of the mentioned processes is anticipated to be the cooperative motion of large clusters of molecules, and these processes are therefore referred to as structural relaxation. The slowest of these is called the $\alpha-$ relaxation process.

Within his visco–elastic theory Maxwell analyzed the implication of the $\alpha-$processes for the shear response. If $t_{\text{exp}} \gg \tau_\alpha$, the system exhibits liquid–like approach towards equilibrium; in particular shear deformations relax to zero. If however $t_{\text{exp}} \ll \tau_\alpha$, the system responds like a disordered solid, i.e. it appears as a glass, which in particular can sustain shear stresses within the probing time t_{exp}. Thereby his theory explains the liquid to glass transition as a transformation, which occurs if the temperature T reaches a value T_g, the calorimetric glass transition temperature, where $\tau_\alpha(T = T_g) = t_{\text{exp}}$. Many experiments have confirmed the generality of Maxwell's picture. Obviously, an understanding of the α–process is a prerequisit for an understanding of T_g–phenomena.

For more than hundred years many attempts to achieve a microscopic understanding of structural relaxation have failed. The question how this phenomenon evolves upon lowering T from above the melting temperature T_m to below it was not seriously studied. A major problem for a theory of structural relaxation is the stretching phenomenon. Let us therefore recall some aspects of this unique feature of the dynamics of disordered condensed matter. The simplest relaxation law is the exponential:

$$\phi_D(t) = f \exp[-t/\tau_\alpha] \; . \tag{1a}$$

This is equivalent to a Lorentzian susceptibility spectrum

$$\chi''_D(\omega) = f \; \omega\tau_\alpha/[1 + (\omega\tau_\alpha)^2] \; . \tag{1b}$$

These formulae were used by Maxwell; but in the context of supercooled liquid dynamics they are usually referred to as Debye's laws. The dynamical window of a Debye process spans about one decade. More precisely: the decay of ϕ_D from 90% to 10% of its initial value requires an increase of the time t by about a factor 22, i.e. 1.34 decades. Equivalently, the width of the χ''_D versus frequency ω curve at half height is about 1.14 decades. However, experiments show that 2 or more decades of time increase are needed to cover the mentioned 80% of decrease for ϕ. Similarly, experimental spectra are stretched over several orders of magnitude of frequency variation. Structural relaxation curves can be viewed properly only on logarithmic time or frequency abscissas.

Various more or less ad hoc formulae, which often describe structural relaxation data reasonably well, have been proposed in the past. The oldest is the Kohlrausch law, also known as the Kohlrausch–Williams–Watts function or stretched exponential. It modifies the exponential law (1a) by introduction of a stretching exponent $\beta < 1$:

$$\phi_K(t) = f \exp[-(t/\tau_\alpha)^\beta] \, . \tag{2}$$

Another example is the Cole–Davidson expression, which modifies the Lorentzian (1b) by introduction of a stretching exponent $b < 1$:

$$\chi_{CD}(\omega) = f/[1 - i\omega\tau_\alpha]^b \, . \tag{3}$$

This susceptibility implies a high frequency power law decay for the spectrum

$$\chi''(\omega\tau_\alpha \gg 1) \propto 1/(\omega\tau_\alpha)^b \, , \tag{4a}$$

which is equivalent to a corresponding power law variation for $\phi(t)$:

$$\phi(t \ll \tau_\alpha) - f \propto -(t/\tau_\alpha)^b \, . \tag{4b}$$

The formulae (4) were proposed by von Schweidler for the description of the high frequency wing of the α–peak of the susceptibility spectra or for the short time part of the α–decay function. There is a subtle connection between (2) and (4b) to Lévys generalization of the central limit theorem. If one considers the superposition of infinitely many infinitely small complexes, which relax independently for short times according to the von Schweidler law (4b), then the sum of all contributions is given by the Kohlrausch law (2) with $\beta = b$.

Occasionally another stretched peak in the susceptibility spectrum is found at some frequency $1/\tau_\beta$ above the α–peak position $1/\tau_\alpha$. This peak is referred to as the β–peak, and it can often be described by the Cole–Cole formula:

$$\chi_{cc}(\omega) = f/[1 + (-i\omega\tau_\beta)^a] \, . \tag{5}$$

These and other formulae introduce power law variations in certain limits, specified by anomalous exponents like a, b or β. Such power laws, or time fractals, are known from the mathematical theory of generalized Brownian motion, and they reflect some underlying singularity. It is a great challenge to understand the hidden singularity, responsible for the measured power laws and to derive the observed time fractals from the regular equations of motion which describe the dynamics of the molecules. Correlation functions behave regularly for short times, e.g. $\phi(t \to 0) = \phi(t = 0) + O(t^2)$. Thus (2) cannot be correct if the time becomes very short, say $t \sim t_0$ with t_0 specifying the scale for the microscopic dynamics. It is also unplausible that for extremely long times the decay should be different from the simple decay law (1a). An understanding of structural relaxation should therefore imply an understanding of the windows, where formulae like (2) are valid.

During the past ten years the so-called mode coupling theory (MCT) for structural relaxation has been developed. The MCT can be considered as a well defined mathematical ad hoc model for the dynamics of a disordered system. However, this theory grew out from preceding microscopic theories of classical liquids [1, 2, 3]. The MCT equations were obtained from a series of plausible approximations for the correlation functions which deal with the dynamics of simple one component fluids. The crucial result of the theory is the prediction of a singularity for the dynamics, called glass transition singularity (GTS). This is a bifurcation singularity connected with a novel dynamical scenario. The new results are caused by the subtle interplay between non-linearities and diverging retardation effects. Within the MCT the GTS is the origin for the evolution of structural relaxation. Time fractals, characterized by non-universal anomalous exponents, are obtained as exact limit results. The equations (2,4,5) have been derived with well specified ranges of validity. The MCT predicted a series of relations between measurable quantities valid for all systems. Thereby it stimulated new experiments and challenged tests of its applicability for a description of supercooled liquids.

Extensive light scattering work done for the standard glass forming mixed salt 0.4 $Ca(NO_3)_2$ 0.6 KNO_3 (CKN) [4] and for colloidal suspensions of hard spheres [5] provided essentially complete and affirmative tests of the general MCT results. Thus one can assume that MCT describes properly some essential features of structural relaxation for some glass forming systems. Reports on the present status of the discussion of the relevance of MCT can be inferred from recent conference proceedings [6, 7, 8].

The equations of motion of the MCT are complicated and we are far from understanding all important qualitative features of their solutions, leave aside quantitative details. But some mathematical properties of the equations have been worked out. The technique used is here – as in other singularity theories – the establishment of asymptotic expansions. The results of this work have been reviewed earlier

[9, 10, 11]. So far only the leading order expansions near the GTS have been compared with data. These results merely concern the general mathematical structure of the theory; microscopic information on the liquid is hidden in certain constants which are treated as fit parameters. Only for the hard sphere systems microscopic details have to a large extent been understood, and compared quantitatively with data. To study how a particular class of liquids emphasizes some general feature, like e.g. the von Schweidler law region, one will have to analyze the microscopic versions of the MCT, dealing with those complicated liquids like CKN, which are studied in experiments.

In this paper we review the basis of the MCT. We will explain in detail the ideas underlying the derivation of the theory with a minimum of formalism. We hope to convince the reader, that the MCT of structural relaxation is an approach towards a fascinating and subtle problem, which is well embedded in the body of knowledge on classical liquid dynamics [12, 13].

2 RELAXATION KERNELS

2.1 General Concepts

The structure of a many particle system is defined by the distribution of the positions $\vec{r}_1, \vec{r}_2, \ldots$ of the atoms. It can be characterized by the density fluctuations of wave vector $\vec{q}, \rho_{\vec{q}} = \sum_\ell \exp i\vec{q}\vec{r}_\ell$, by the density fluctuation pairs $\rho_{\vec{q}-\vec{k}}\rho_{\vec{k}}$, and by higher order products of the $\rho_{\vec{k}}$. The statistical description of the dynamics is conventionally provided by correlation functions or correlators. They are defined as canonically averaged products of some variable $A(t)$ at time t and the same variable at $t = 0$: $\langle A(t)^* A \rangle$. The time evolution is given by the microscopic equations of motion; i.e. by Hamilton's equations in the most general case. The simplest quantities dealing with structural dynamics are thus the density correlators:

$$\phi_q(t) = \langle \rho_{\vec{q}}(t)^* \rho_{\vec{q}} \rangle / S_q \ . \tag{6}$$

These are the basic quantities of the MCT for structural relaxation. Let us mention that $\phi_q(t)$ as well as the density spectrum $\phi_q''(\omega)$ can be measured directly in neutron and light scattering experiments. $S_q = \langle \rho_{\vec{q}}^* \rho_{\vec{q}} \rangle$ denotes the structure factor, the spatial Fourier transform of the averaged distribution of particle pairs. More complicated correlators like the ones for the pairs $\langle \rho_{\vec{q}-\vec{k}}(t)^* \rho_{\vec{k}}(t)^* \rho_{\vec{q}-\vec{p}} \rho_{\vec{p}} \rangle$ appear during the development of the theory e.g. as force correlators. Certain special combinations of those functions can also be measured e.g. as elastic moduli.

MCT is concerned with such stable or metastable states for $T > T_g$, where all equilibrium averages like S_q are smooth functions of temperature T and density n. Only such systems are considered, where there are no singularities appearing in S_q due to e.g. phase transitions. In such situations one can rely on established theories [13], which allow the evaluation of S_q as function of T, n and the pair potential. For large wave vectors the structure factor approaches the one of a gas of uncorrelated particles $S_{q\to\infty} = 1$. The strong short ranged correlations of dense liquids have two important implications. First, the structure factor for small wave vectors is strongly decreased below 1, since it is very difficult to produce homogeneous compressions. Second, the structure factor is enhanced compared to 1 for q close to some wave vector q_0, which is near the inverse of the interparticle distance. Spontaneous density fluctuations with wave vector q near q_0 are much more probable than those for other q. Structure fluctuations occur primarily for wave vectors of microscopic size $q \sim q_0 \sim 2\mathring{A}^{-1}$. They reflect rearrangements of atoms which are nearest and next nearest neighbours.

The equations of motion relate $\rho_{\vec{q}}$ to current fluctuations $\vec{j}_{\vec{q}}$, those to force fluctuations etc. The force can be written as some Hook restoring force $\Omega_q^2\rho_{\vec{q}}$ and a part $F'_{\vec{q}}$ perpendicular to the density fluctuations, $\langle F_{\vec{q}}^{*\prime}\rho_{\vec{q}}\rangle = 0$. Here $\Omega_q^2 = q^2v^2/S_q$, with $v = \sqrt{k_BT/m}$ denoting the thermal velocity of the particles of mass m. Frequency Ω_q provides a scale for the short time dynamics, since

$$\phi_q(t) = 1 - \frac{1}{2}(\Omega_q t)^2 + O(t^4) . \tag{7}$$

Frequency Ω_q plays the role of some "bare" phonon dispersion. For small wave vectors one gets $\Omega_q = c_0 q + 0(q^2)$ with c_0 denoting the isothermal sound velocity. For $q \sim q_0$ the dispersion exhibits a minimum indicating the trend to the building of soft vibrations.

The Zwanzig-Mori theory [12, 13] provides an algorithm to eliminate the explicit appearance of F'_q in the equations of motion. It yields for the density correlator the generalized oscillator equation:

$$\partial_t^2\phi_q(t) + \Omega_q^2\phi_q(t) + \Omega_q^2\int_0^t M_q^L(t-t')\partial_{t'}\phi_q(t')dt' = 0 . \tag{8a}$$

The coupling of $\rho_{\vec{q}}$ to the fluctuating forces $\vec{F}'_{\vec{q}}$ is here hidden in the relaxation or memory kernel M^L. This kernel is the constant of proportionality between a force at time t and a velocity at a preceding time t'. The appearance and proper handling of retardation effects, as described by the integral in (8a), is one essential ingredient of the MCT. If one introduces Fourier–Laplace transforms $\phi_q(\omega) = \phi_q'(\omega) + i\phi_q''(\omega)$, $M^L(\omega) = M^{L\prime}(\omega) + iM^{L\prime\prime}(\omega)$ for the correlator and kernel respectively, one can solve (8a) and express the former in terms of the latter. Let us note the equivalent result for the the susceptibility $\chi_q(\omega) = [1 + \omega\phi_q(\omega)]S_q$, which has the meaning of a dynamical compressibility for wave vector q. One gets

$$\chi_q(\omega) = -q^2v^2/[\omega^2 - \Omega_q^2 + \omega\Omega_q^2 M_q^L(\omega)] . \tag{8b}$$

For vanishing kernel one would get the susceptibility of a harmonic oscillator, i.e. the response for harmonic phonons. The imaginary part of the kernel, i.e. the relaxation spectrum $M^{L\prime\prime}(\omega)$, yields damping of the phonons and the real part $M^{L\prime}(\omega)$ leads to shifts of the phonon resonances. Thus $M_q^L(\omega)$ is a generalization of a Newtonian friction constant to a frequency and wave vector dependent function. However in the present context we will apply the equations (8a,b) in such parameter regimes, where the solutions are quite different from what one expects for oscillator motion.

The most obvious difference between a liquid and a solid appears for the shear response. While shear deformations disappear in a liquid via diffusion, they propagate as transversal sound waves in a solid. In the discussion of glassy dynamics one pays therefore special attention to shear dynamics described by the correlator for transversal current fluctuations $j_{\vec{q}}^T : \phi_q^T(t) = \langle j_{\vec{q}}^T(t)^* j_{\vec{q}}^T\rangle/(v^2 n)$. The Zwanzig–Mori formalism allows us to express this function in terms of some kernel M^T:

$$\phi_q^T(\omega) = -1/[\omega + q^2v^2 M_q^T(\omega)] . \tag{9}$$

The general theory provides closed expressions for the kernels in terms of fluctuating force correlators. There is a 3 by 3 matrix of such force correlation functions, but because of rotational symmetry this matrix is given by only two functions: the longitudinal relaxation kernel M^L and the transversal relaxation kernel M^T.

Let us mention that the fluctuating force for zero time and zero wave vector is a linear combination of density fluctuation pairs:

$$\vec{F}_0' = \sum_{\vec{k}} \vec{U}(\vec{k})\rho_{\vec{k}}\rho_{-\vec{k}} . \tag{10}$$

The coefficients $\vec{U}$ are given by the Fourier transforms of the interaction potential derivatives.

There is some arbitrariness in the introduction of kernels like M^L or M^T. One could have expressed $M_q^T(\omega)$ also as some fraction like (9). Thereby one would have obtained an equation for $\phi_q^T(t)$ with a leading second time derivative in analogy to (8a). Similarly, one could have expressed $M_q^L(\omega)$ as fraction, thereby replacing (8a) by one with a leading third order time derivative. The equations (8,9) as well as the indicated modifications are merely exact reformulations of the microscopic equations of motion. It will be explained below in section 3.2, why the MCT is based on (8,9) rather than on other versions of the Zwanzig–Mori theory.

2.2 Hydrodynamic Equations and Maxwell's Theory

Exact equations like (8,9) become fruitful starting points for the discussion, if they are complemented by additional input information. An important concept in this context is the separation of scales. Let us recall how this concept is applied for the case considered originally by Zwanzig and Mori themselves. The length scale ℓ_0 for variations of interparticle correlations in simple liquids is determined by the inverse of the width of the structure factor peak: $2\pi/\ell_0 \sim 0.2\mathring{A}^{-1}$. For a discussion of the dynamics on length scales longer than ℓ_0, one can ignore e.g. the variation of kernels with wave vector q and approximate $M_q^{L,T}(\omega) \approx M_0^{L,T}(\omega)$. Similarly one can replace the structure factor by its $q = 0$ limit $S_0 = \kappa n k_B T$, where κ denotes the isothermal compressibility. Hence one can simplify (8b,9) to

$$\chi_q(\omega) = -q^2 v^2/[\omega^2 - q^2 M_L^*(\omega)/(nm)] \ , \tag{11a}$$

$$\phi_q^T(\omega) = -\omega/[\omega^2 - q^2 M_T^*(\omega)/(nm)] \ , \quad q\ell_0 \ll 1 \ . \tag{11b}$$

Here the dynamical elastic moduli $M_{L,T}^*$ have been introduced:

$$M_L^*(\omega) = [1 - \omega M_0^L(\omega)]/\kappa \ , \quad M_T^* = -\omega M_0^T(\omega) v^2 nm \ . \tag{11c}$$

The Fourier back transform of (11) yields density and current correlations which are coarse grained over lengths of order ℓ_0. So they describe properly spatial variations on scales well separated from the microscopic length ℓ_0. The simplifications (11) are of interest, since a variety of experiments, like light scattering measurements or acoustic spectroscopy, probe the dynamics in the mentioned regime $q\ell_0 \ll 1$. Let us however emphasize, that within the MCT it is not possible to derive closed equations for the correlators coarse grained with respect to space variations. Merely the final results can be specialized to (11).

The time scale t_0 for excitations in normal liquids is the same as in crystalline solids: microscopic dynamics occurs in the THz–window. The characteristic motions of the molecules occur within some picoseconds or faster. Spectra within the THz band reflect the details which distinguish different glass formers, the salt mixture

CKN from the van der Waals liquid Salol say. The MCT like other microscopic theories of liquids yields results for the dynamics on scale t_0. However, in its present state of development attention is not focused on explaining dynamics on this scale. Rather one simplifies the discussion by coarse graining correlators on scale t_0, whenever possible; i.e. one simplifies formulae by the restriction $\omega t_0 \ll 1$ in appropriate places. There may be one or several scales specifying the dynamics within the regime well separated from the microscopic one. Let τ_α denote the longest of those scales. If we consider the dynamics coarse grained on scale τ_α, i.e. spectra for $\omega\tau_\alpha \ll 1$, we can ignore the frequency variation of the kernels and replace their spectra by their value for $\omega = 0$:

$$M_q^{L''}(\omega)/\kappa = \eta_L \ , \quad M_q^{T''}(\omega)v^2 nm = \eta \ , \quad \omega\tau_\alpha \ll 1 \ , q\ell_0 \ll 1 \ . \tag{12}$$

These formulae are equivalent to replacing the moduli by their leading Taylor expansion $M_L^*(\omega) = (1/\kappa) - i\eta_L\omega$, $M_T^*(\omega) = -i\omega\eta$. The regime specified by the inequalities $\omega\tau_\alpha \ll 1$, $q\ell_0 \ll 1$ is the hydrodynamic one. It deals with motion coarse grained on lengths of order ℓ_0 and times of order τ_α. Within the hydrodynamic regime the motion is specified by (11,12). These equations are equivalent to the Navier–Stokes equations of liquid dynamics, which are thus obtained within the Zwanzig–Mori theory with a well specified regime of validity. The relaxation kernels enter in form of two numbers η_L, η, the longitudinal and the shear viscosity respectively. These transport coefficients are proportional to zero frequency spectra of the relaxation kernels.

In undercooled liquids $1/\tau_\alpha$ becomes so small that the range of validity of the hydrodynamic description is too narrow for a complete discussion of experiments. To proceed beyond the Navier–Stokes equations the kernels M shall be split into one part m, describing the non-trivial features of the kernel spectra M'' on frequency scales below $1/t_0$, and a remainder: $M = m + m^0$. The latter can be simplified by the mentioned coarse graining, so that

$$M_q(\omega) = i\gamma_q + m_q(\omega) \ , \qquad \omega t_0 \ll 1 \ . \tag{13}$$

Kernel m can be written as a fraction: $m_q(\omega) = -A_q/[\omega + \mu_q(\omega)]$; and the Zwanzig–Mori theory provides an exact representation of $\mu_q(\omega)$ as some generalized fluctuating force correlator. Maxwell's theory is based on the hypothesis, that a single mode is responsible for the variations of m on scale τ_α, and that this mode is projected out of the dynamics of μ. Under these conditions one can coarse grain the μ–dynamics and gets:

$$m_q(\omega) = -A_q/[\omega + i/\tau_q] \ . \tag{14a}$$

Carrying out also coarse graining with respect to space variations one obtains rational functions for the moduli in (13), which are specified by six constants. There appear the zero wave vector limits of $\gamma_q, \gamma^{L,T}$, which specify a white noise background spectrum for the two fluctuating force correlators. There are the strength parameter $A^{L,T}$ of the structural relaxation resonance in the moduli. There are Maxwell's relaxation times $\tau^{L,T}$, which characterize the width of the relaxation peak in the force spectra $M_0^{L,T\prime\prime}(\omega)$. Because of (12) a combination of three constants yields for the longitudinal viscosity $\eta_L = [\gamma^L + A^L \tau^L]/\kappa$. A similar formula holds for the shear viscosity η. If the relaxation times become large, the structural relaxation peaks in the moduli become narrow and the viscosities become high. The formulated equations describe the cross over from hydrodynamic sound propagation with speed $c_0 = \sqrt{1/\kappa nm}$ to high frequency sound with the larger speed $c_\infty = c_0\sqrt{1 + A^L}$. Similarly, it describes the cross over from shear diffusion within the regime $\omega\tau^T \ll 1$ to shear wave propagation with a transversal sound speed $\sqrt{A^T/nm}$ for $\omega\tau^T \gg 1$.

Mountain [14] noticed an implication of Maxwell's theory, which is of utmost interest for an understanding of structural relaxation. In addition to the mentioned sound resonances, the density fluctuation spectrum exhibits a quasielastic peak of some strength f_q and some width Γ_q. From (8,13,14a) one gets for Mountain's resonance:

$$\phi_q(\omega) = f_q/[\omega + i\Gamma_q] \; ; \qquad \omega, 1/\tau_q^L \ll \Omega_q \; . \tag{14b}$$

The strength f_q of this relaxation resonance follows from

$$f_q/[1 - f_q] = A_q^L \; , \tag{14c}$$

and the width parameter Γ_q is given by $\Gamma_q = 1/[\tau_q^L(1 + A_q^L)]$. For small wave vectors the decay rate depends insensitively on q:

$$\Gamma_q = \Gamma > 0 \; , \qquad q\ell_0 \ll 1 \; . \tag{14d}$$

The preceding results describe relaxation properly for those cases, where it is caused by some single low lying mode, say by some soft intramolecular vibration. The formulae are not suited to describe structural relaxation in general, since they miss the stretching phenomenon.

It is important to understand that some of Mountain's findings are strict implications of the exact equations (8) which do not depend on formula (14a). To identify these results we give first a precise mathematical meaning to " $\ll$ " in (14b). To proceed we write $m_q(t) = \tilde{m}_q(\tilde{t})$ with $\tilde{t} = ts$. This is equivalent to $m_q(\omega) = \tilde{m}_q(\tilde{\omega})/s$ with $\tilde{\omega} = \omega/s$ and $\tilde{m}_q(\tilde{\omega})$ denoting the Fourier–Laplace transform of $\tilde{m}_q(\tilde{t})$. Now we consider the limit s tending to zero for fixed rescaled time $\tilde{t}$, rescaled frequency $\tilde{\omega}$ and shape function $\tilde{m}_q$. This limit procedure achieves the desired separation of the α-process from the other processes. From (8a) one derives $\phi_q(\omega)s \to \tilde{\phi}_q(\tilde{\omega})$ for $s \to 0$, where $\tilde{\phi}_q(\tilde{\omega}) = -1/[\tilde{\omega} - 1/\tilde{m}_q(\tilde{\omega})]$. This result is equivalent to $\phi_q(t) \to \tilde{\phi}_q(\tilde{t})$, where $\tilde{\phi}_q(\tilde{t})$ is the Fourier-Laplace back transform of $\tilde{\phi}_q(\omega)$, and:

$$\tilde{\phi}_q(\tilde{t}) = \tilde{m}_q(\tilde{t}) - (d/d\tilde{t}) \int_0^{\tilde{t}} \tilde{m}_q(\tilde{t} - \tilde{t}')\tilde{\phi}_q(\tilde{t}')d\tilde{t}' \ . \tag{15}$$

If one uses Maxwell's model (14a) for $\tilde{m}_q$, equation (15) is solved by Mountain's results (14b,c) for $\tilde{\phi}_q$. But independently of any specific model one concludes from (15): a quasi-elastic peak for the force spectrum of shape $\tilde{m}_q''(\tilde{\omega})$ causes a corresponding peak of shape $\tilde{\phi}_q''(\tilde{\omega})$ for the density spectrum and vice versa. The areas A_q and f_q of the α-peaks of force spectrum $\tilde{m}_q''(\tilde{\omega})$ and density spectrum $\tilde{\phi}_q''(\tilde{\omega})$ respectively are given by $A_q = \tilde{m}_q(\tilde{t} \to 0)$ and $f_q = \tilde{\phi}_q(\tilde{t} \to 0)$. The limit $\tilde{t} \to 0$ in (15) yields Mountain's result (14c). There are different ways to define a relaxation time τ_q for the force correlator or a relaxation rate Γ_q for the density correlator. In any case $\tau_q \propto 1/s$ and $\Gamma_q \propto s$, where the constants of proportionality depend on the shape functions $\tilde{m}_q$ and $\tilde{\phi}_q$ respectively. Equation (14d) holds in all cases.

Let us examine the ansatz for the small $\tilde{t}$ dynamics:

$$\tilde{m}_q(\tilde{t}) = A_q - H_q\tilde{t}^b + K_q\tilde{t}^{2b} + \dots \ , \quad \tilde{\phi}_q(\tilde{t}) = f_q - h_q\tilde{t}^b + k_q\tilde{t}^{2b} + \dots \ . \tag{16a}$$

In generalization of (14c) one derives from (15), that the coefficients of the $\tilde{m}_q$-expansion determine recursively those for the $\tilde{\phi}_q$-expansion and vice versa. In addition to (14c) one gets for example:

$$h_q = H_q/(1 + A_q)^2 \ , \ k_q = [K_q(1 + A_q) - \lambda H_q^2]/(1 + A_q)^3 \ . \tag{16b}$$

Here $\lambda = \Gamma(1 + b)^2/\Gamma(1 + 2b)$; Γ is the gamma function. For $b < 1$ one gets $\lambda > 1/2$. If $\tilde{m}_q(\tilde{t})$ or $\tilde{\phi}_q(\tilde{t})$ obey (2) one obtains the identity $2K_q = H_q^2/A_q$ or $2k_q = h_q^2/f_q$ respectively. The simultaneous validity of both identities is compatible with (14c, 16b) only if $\lambda = 1/2$. One concludes the following. If the α-process for the force relaxation follows the Kohlrausch formula (2) with exponent $b = \beta < 1$, the α-relaxation for the density correlator is not a Kohlrausch process and vice versa. The microscopic equations of motion establish exact relations between certain structural relaxation processes as is exemplified by (15). These relations exclude the possibility that the Kohlrausch function is a general law describing all α-processes.

The preceding discussion is somewhat oversimplified. The energy conservation law leads to low lying excitations in kernel $m_q(\omega)$ which cause a singular small q-variation. A complete treatment requires therefore the expansion of the formalism by considering also the entropy correlator, coupled to $\phi_q(t)$. As a result, the density spectrum gets an additional quasielastic contribution: the Rayleigh peak due to heat diffusion. A similar contribution appears in mixtures due to concentration fluctuations. These central spectral peaks appear in addition to the structural relaxation peaks. However, their width Γ'_q is proportional to q^2 for small wave vectors. Therefore these diffusion peaks can easily be separated from the structural relaxation peak, which is not of diffusive type.

3 THE MODE COUPLING FUNCTIONAL $\mathcal{F}$

3.1 The Factorization Approximation

Mountain's results and their generalizations show, that there is no separation of the time scale τ_q, which describes via (14a) the structural relaxation of the fluctuating forces, from the time scale $1/\Gamma_q$, which specifies via (14b) the structural relaxational part of the density fluctuations. In simple liquids there are no single distinguished low lying intramolecular modes, which could provide a single dangerous relaxation rate for kernel $m_q(\omega)$. Rather one can assume complexes of two, three etc. molecules as the origin of low lying excitations. One complex differs from the other and in particular one complex has a different neighbourhood than the other. One suspects therefore a large number of different but nearly degenerate modes to be responsible for the structural relaxation phenomena.

If $\rho_{\vec{q}}$ relaxes with rate Γ_q, the product modes $\rho_{\vec{k}_1}\rho_{\vec{k}_2}\cdots\rho_{\vec{k}_\ell}$ will relax in lowest order approximation with a rate $\Gamma_{k_1}+\Gamma_{k_2}+\cdots+\Gamma_{k_\ell}$. If ℓ is not too large all these rates are of the same order of magnitude, and therefore all density fluctuation products relax within the same dynamical window. The fluctuating force does not only couple to a single dangerous mode, but to an infinite set of them, as is obvious from (10). Because of the microscopic expression (10) for the fluctuating force, one cannot build a theory on the assumption, that a small number of modes leads to glassy relaxation in supercooled liquids. Rather one has to treat at least the infinite set of all density fluctuation pairs as dangerous modes connected with structural relaxation. Written in ordinary space the structural part of the fluctuating force takes the form

$$\vec{F}'(\vec{r},t)=\int d\vec{r}'\vec{\nabla}V(\vec{r}-\vec{r}')\rho(\vec{r}t)\rho(\vec{r}'t),$$

with $V(r)$ being the two particle interaction potential. It depends on the simultaneous density fluctuations at $\vec{r}$ and the surrounding positions $\vec{r}'$ within the interaction range. Hence, the force depends on the dynamics of a cluster of particles. Obviously, such local fluctuation clusters cannot be properly represented by only a few wave vectors.

The problem to calculate kernels, whose dynamics is caused by products of slow modes like $\rho_1\rho_2$ occurs in various areas of statistical physics. It was considered in detail first by Kawasaki [15], and the MCT is built on his work. The essence of his idea is the approximation of averages of products by products of averages: $\langle\rho_1(t)\rho_2(t)\rho_3\rho_4\rangle \sim \langle\rho_1(t)\rho_3\rangle\cdot\langle\rho_2(t)\rho_4\rangle + (3 \leftrightarrow 4)$. It is important to handle properly the projection of the forces on the pair modes, $\langle F'\rho_1\rho_2\rangle$, in order to account for the strong short ranged correlations in dense liquids. As a result one obtains $m_q(t)$ in (13) as a polynomial $\mathcal{F}_q$ of the correlators $\phi_k(t)$: $m_q(t) = \mathcal{F}_q(\phi_k(t))$. This polynomial is called mode coupling functional, since it describes the coupling of the fluctuating force correlators to the density correlators. If one considers only density fluctuation pairs the functional reads:

$$\mathcal{F}_q(\phi_k(t)) = \sum_{\vec{k}+\vec{p}=\vec{q}} V(\vec{q},\vec{k}\vec{p})\phi_k(t)\phi_p(t)\ . \tag{17a}$$

It represents in an approximate way the relevant cluster dynamics, where the two ϕ:s describe the motion of one particle together with the surrounding medium.

The mode coupling vertex V, i.e. the coefficient connecting in (17a) the density modes k and p with the force, is given in terms of the structure factor S_q. The interparticle potential, which enters the fluctuating forces (10) explicitly, can be eliminated by using exact identities for equilibrium averages. Introducing the direct correlation function c_q, given in terms of S_q via the Ornstein–Zernike equation $S_q = 1/[1 - nc_q]$, one finds [16]

$$V(\vec{q},\vec{k}\vec{p}) = nS_qS_kS_p\{\vec{q}[\vec{k}c_k + \vec{p}c_p]\}^2/(2q^3)\ . \tag{17b}$$

A similar expression is obtained for the transversal kernel

$$m_q^T(t) = \sum_{\vec{k}+\vec{p}=\vec{q}} V^T(\vec{q},\vec{k}\vec{p})\phi_k(t)\phi_p(t)\ , \tag{17c}$$

where the vertices V^T are also given in terms of S_q by a formula similar to (17b) [17]. Because of (10) there are no direct couplings of the fluctuating forces F_0' for wave vector $q = 0$ to density fluctuation products with $\ell \geq 3$. For other wave vectors such couplings occur and they would lead in (17a,c) to products of $\ell \geq 3$ density correlators in addition to the noted quadratic terms; but so far the corresponding vertices have not been evaluated.

The Zwanzig–Mori theory establishes the exact equations (8a,b) expressing the density correlator in terms of the relaxation kernel M^L or in terms of m via (13). The Kawasaki factorization approximation provids equations (17), expressing the kernels in terms of the density correlators. The Zwanzig–Mori equations (8b,9) have an elementary form, if written for the Fourier–Laplace transforms $\phi_q(\omega)$ and $m_q(\omega)$. The expressions (17) for the kernels have an elementary form if written for the time dependent functions $\phi_q(t)$ and $m_q(t)$. Equations (17) are the essence of the MCT.

Maxwell's theory is extended in several respects. It is acknowledged, that there is an infinite set of variables responsible for structural relaxation and these variables are explicitly identified. The subset of density fluctuation pairs $\rho_{\vec{q}-\vec{k}}\rho_{\vec{k}}$, the simplest and hence most important one, is kept and approximately treated within the microscopic theory of many particle systems. No ad hoc models for relaxation functions like equation (14a) are introduced. The kernels are expressed in terms of the correlators thereby avoiding assumptions on the separation of scales. The price to be paid for these extensions is very high, however. Maxwell's theory yields explicit elementary expressions for the correlators in terms of parameters like A_q, τ_q, γ_q. The factorization approximation yields only new equations, which are non-elementary. One has to develop a theory for the solution of these equations before one can judge the possible relevance of the approach for an understanding of liquid dynamics.

3.2 The Simplified MCT

Let us summarize the preceding results in order to have a concise basis for the following discussions. Substitution of (13) into (8a) gives the equation of motion for the density correlators:

$$\partial_t^2\phi_q(t) + \nu_q\partial_t\phi_q(t) + \Omega_q^2\phi_q(t) + \Omega_q^2\int_0^t m_q(t-t')\partial_{t'}\phi_q(t')dt' = 0 \;. \tag{18a}$$

This generalized oscillator equation has to be solved with the initial conditions $\phi_q(t=0) = 1$ and $\partial_t\phi_q(t=0) = 0$. The rate $\nu_q = \Omega_q^2\gamma_q$ represents a Newtonian friction term. Let us recall, that γ_q resulted from (13), since some correlators have been coarse grained on the time scale t_0. This implies, that the spectra for typical microscopic excitations around 1 THz are not treated properly by (18a). Sound excitations for small wave vectors can still be handled since their frequencies c_0q are small compared to $1/t_0$. Equation (18a) describes oscillatory excitations also for large wave vectors, say $q = q_0 \sim 2\mathring{A}^{-1}$. But it is not clear whether this is done in an oversimplified manner only.

Our main interest is the discussion of very low frequency phenomena. One could therefore restrict the frequency regime of interest even further by dropping the inertia term in (18a) so that one gets:

$$\gamma_q \partial_t \phi_q(t) + \phi_q(t) + \int_0^t m_q(t-t')\partial'_t \phi_q(t')dt' = 0 \; . \tag{18b}$$

This generalized relaxator equation has to be solved with the initial condition $\phi_q(t = 0) = 1$.

The equations (18) have to be complemented by the formulae expressing the kernel as polynomial of the correlators. Let us imagine the wave vector moduli to be taken discrete: $q = 1, 2 \ldots, M$. Let us also note the mode coupling functional $\mathcal{F}_q$ in a general form:

$$m_q(t) = \mathcal{F}_q(\phi_k(t)) \; , \tag{19a}$$

$$\mathcal{F}_q(f_k) = \sum_{r=1}^{r_0} \sum_{k_1 \cdots k_r} V_{q,k_1 \cdots k_r} f_{k_1} \cdots f_{k_r} \; . \tag{19b}$$

The mode coupling vertices $V_{q,k_1 \cdots k_r}$ enter the discussion as coupling constants. It is occasionally helpful to label them somehow as $V_1, V_2 \ldots, V_N$ and combine these N numbers to a control parameter vector $\vec{V} = (V_1, \ldots, V_N)$, which can be viewed as a point in some abstract control parameter space $\mathbf{R}$ of dimensionality N. Only non-negative values $V_i \geq 0$ enter in (19b). The vertices depend on temperature and density; and as these are varied we move along specified paths in the parameter space. In the work on simple liquids so far only quadratic terms, i.e. the $r = 2$ contributions, have been taken into account. In this case the vector $\vec{V}$ can be specified in terms of S_q from (17b).

The equations (18, 19) are closed. It was shown that they define for every set of $(2M + N)$ constants $\Omega_q, \nu_q, \vec{V}$, or $(M + N)$ constants $\gamma_q, \vec{V}$ respectively, a unique solution $\phi_1(t), \ldots, \phi_M(t)$. The solutions $\phi_q(t)$ have all the general properties of correlation functions, in particular the spectra $\phi''_q(\omega)$ are non negative. The solution of (18a) behaves for short times like $\phi_q(t) = 1 - \frac{1}{2}(\Omega_q t)^2 + \frac{1}{6}(\Omega_q t)^3(\nu_q/\Omega_q) + O(t^4)$. This result generalizes (7); the mode coupling effects influence only the terms of higher than cubic order in t for the short time expansion. The solution of (18b) starts like $\phi_q(t) = 1 - \gamma_q t + O(t^2)$. The dynamics on the microscopic time scale t_0 is modified in this case. This modification is analogous to what happens when replacing the Newtonian dynamics by that of simple Brownian motion. In the specified sense one can say that the parameters $\Omega_q, \nu_q, \gamma_q$ determine the transient motion in leading order. These parameters set the scales; otherwise they are of no particular mathematical interest for the discussion. From a mathematical point of view it is not important to introduce a friction term ν_q in (18a). The following discussion holds also with $\nu_q = 0$, since damping effects are introduced via kernel $m_q(t)$ anyway. Similarly, it is not necessary to introduce (18b); this equation is contained as special limit in (18a): $\Omega_q \to \infty$ with $\nu_q/\Omega_q^2 = \gamma_q$ fixed. Equation (18b) is helpful as a mathematically well defined tool to define a low frequency regime of (18a).

Equations (18, 19) define the basic version of the MCT for structural relaxation. In order to distinguish this theory from extended versions, developed in order to treat liquid dynamics in a less rudimentary form, it is called simplified MCT.

Our procedure to derive the basic equations of the MCT can be summarized as follows. Within the Zwanzig–Mori formalism correlation functions are expressed in the simplest manner in terms of those kernels for fluctuating forces, which couple directly to density fluctuation products. This requires a two step procedure for the density correlator leading to (8a). A one step reduction or a three step procedure would introduce generalized forces, which do not couple to density fluctuations because of time inversion symmetry. A four step procedure would yield the possibility for couplings, but would be unnecessarily complicated. Similarly one understands, that a one step reduction is the simplest way to couple the generalized forces for the transversal currents to density fluctuations and this leads to (9). This part of the theory is definitive and it reflects the understanding of the microscopic origin for structural relaxation. For correlation functions, which are not so directly related to the coordinates of the molecules like densities and currents, this part of the theory may be much more involved. This can be inferred from the papers dealing with heat fluctuations [18, 19] or with light scattering cross sections in cooled dense liquids [20].

The second part of the theory consists of expressing the kernels in terms of the density correlators. All structural relaxation phenomena are thus reduced to those for the density fluctuations. Shear dynamics does not play a distinguished role within the simplified MCT. Shear anomalies, in particular the structural relaxation anomalies of the shear viscosity, are obtained as straight–forward corollaries of those of $\phi_q(t)$, as is obvious from (12, 13, 17c). The density correlators are therefore the crucial objects. However the procedure outlined above does not give explicit results, but only relates $m_q(t)$ and henceforth ϕ_q to itself. To determine $\phi_q(t)$ one still has to solve the non–trivial equations (18, 19).

Let us stress that the motivation for expressing $\phi_q(t)$ in terms of $m_q(t)$ is almost opposite to the one appealed to originally by Zwanzig and Mori. These authors assumed that a separation of time scales exist. In our case we expect instead that the scales for the relaxation of density fluctuations are the same as those for the generalized force fluctuations. We do not actually assume absence of scale separation; but we construct the approximations so, that absence of scale separation does not render the equations inadequate.

3.3 General Features of the Simplified MCT

The equations (18, 19) are regular. No assumptions on singularities, transitions, stretching or fractals are made. It has been proven that the solutions $\phi_q(t)$ for every finite fixed time interval $0 \leq t \leq t_{max}$ depend smoothly on all parameters $\Omega_q, \nu_q, V_1, \ldots, V_N$ entering (18,19). There are no phenomenological parameters like the relaxation time τ_L in Maxwell's theory, which are assumed to vary with temperature changes over many orders of magnitude. Temperature T enters MCT essentially only indirectly via the temperature dependence of the structure factor S_q. This function depends smoothly on T; S_q varies by about 20% if T is lowered from T_m to T_g, as shown by experiment and explained by previous equilibrium structure theories.

The essential part of the MCT is the mode coupling functional $\mathcal{F}_q$. This functional is an equilibrium quantity. According to (17b), the polynomial $\mathcal{F}_q$ is given by S_q. It is determined by the Boltzmann factor of the many particle system in configuration space, i.e. by the so-called potential landscape. Notice, that $\mathcal{F}_q$ is the same for particles obeying Newtonian dynamics as for particles in colloidal suspensions. In the latter case the particle motion is ruled by a diffusion equation. Conventionally one starts the discussion of colloids with a Fokker–Planck or a Smoluchowski equation. To describe structural relaxation effects the detour via these equations can be avoided. A more refined work is necessary only if one wants to evaluate the time γ_q in (18b) from the properties of the solvent. All results depending only on $\mathcal{F}_q$ are therefore the same for a hard sphere liquid and for a colloidal suspension of hard spheres. Similarly, these results do not depend on the mass of the particles.

It has been shown [21], that the solutions of (18b, 19) are completely monotone functions. This means, that the correlators exhibit a positive or negative n'th derivative depending on whether n is even or odd: $(-\partial/\partial t)^n \phi_q(t) > 0$, $n = 0, 1, \ldots$. The same holds for the kernels $m_q(t)$. Such functions can be written as superposition of Debye functions, distributed with some weight function $\rho_q(\gamma) \geq 0$:

$$\phi_q(t) = \int_0^\infty \exp(-\gamma t)\rho_q(\gamma)d\gamma \; . \tag{20}$$

The integral is understood as one in Stieltjes' sense. It can be approximated for any finite time interval $0 \leq t \leq t_{max}$ to any desired accuracy by a finite number of Debye functions. Notice, that (20) implies a severe restriction within the class of correlation functions. This representation excludes oscillatory motion as expressed e.g. by equation (7) for the transient dynamics. The spectrum $\phi_q''(\omega)$ of a completely monotone function decreases with increasing ω; it cannot exhibit peaks for $\omega \neq 0$. On the other hand, the class of completely monotone functions is rather large. It contains for instance all the fit functions (1-5) quoted in the introduction.

We consider density fluctuations not because they enter a conservation law, but because they specify the structure of the system. Conservation laws like the ones for particle number or momentum lead to low lying hydrodynamic excitations. These have to be handled carefully if one wants to discuss for example sound propagation or shear diffusion, but they are not relevant for the explanation of structural relaxation. Structural relaxation processes in general and the glass transition in particular are not hydrodynamic phenomena. Systems with the same hydrodynamics may in principle exhibit quite different structural relaxation features. We rather have the opposite situation, where density fluctuations for wave vectors $\vec{q}$ from a shell $q \sim q_0 \sim 2\mathring{A}^{-1}$ are the crucial ones for the creation of structural relaxation. The phenomena then propagate down to small q via non-linear mode coupling and thereby cause e.g. the structural relaxation anomalies for the acoustic moduli in (11, 12), and in particular for the viscosities. The space correlations for interparticle distances $2\pi/q_0$ reflect the structure differences of e.g. CKN and Salol. Thereby MCT predicts structural relaxation phenomena to reflect microscopic details. The non-trivial result is, that there are also some features of general validity, which do not depend on the detailed form of the structure. This point will be discussed in more detail in section 6.

Contrary to what holds for the theory of phase transitions [15], the MCT vertices do not exhibit critical variations with wave vector q. There is no divergent length causing small q-singularities of S_q. Nevertheless one cannot neglect the q-dependence of the correlators in for example (17a), since this would lead to results where e.g. the fractal exponents are erroneously determined by large q cut off effects.

By deriving (18, 19) and the explicit expression (17b) for the vertices, a microscopic theory for structural relaxation has been established. This claim will be corroborated in section 5 with a discussion of a systematically derived extension of the theory. But all this work is based on Kawasaki's factorization approximation. The error of this approximation, and the propagation of this error to the final results, are not yet fully understood. The objection, that MCT is not rigorously derived from the microscopic equations of motion cannot be defeated at the present state of development. However, the indicated objection deals only with one facet of the MCT.

In conclusion of this subsection we wish to emphasize, that a more restricted view can be offered. One can consider MCT as a mathematical model for a dynamical system, defined by (18, 19). The model describes dynamics by M correlators $\phi_q(t)$. The state of the system is specified by the control parameter vector $\vec{V}$. One is interested to study the change of the solutions upon changes of $\vec{V}$. All results are to be derived from (18, 19) and no further hypothesis are to be invoked. It is a model for a non-linear dynamics. Non-linearities appear, because m_q in (18) depends on ϕ_q and also because $\mathcal{F}_q$ in (19) is a non-linear functional. If kernel $m_q(t-t')$ decreases to zero effectively for $t-t'$ approaching some time $t_{\rm ret}$, one has a special version of some non-linear dynamics with some given retardation time $t_{\rm ret}$. Such equations have been studied in the preceding literature. However, in our theory $t_{\rm ret}$ is not given but has to be calculated; and it turns out that $t_{\rm ret}$ can diverge for certain critical control parameter points $\vec{V}^c$. The interplay of non-linearities with divergent retardation scales $t_{\rm ret}$ yields new dynamical scenarios, which are unique features of the formulated theory. The existence of such novel features and the known correlation of these with experimental facts justifies the discussion of the specified mathematical model.

If one is interested in a quantitative liquid theory based on (17b), one has to introduce a sufficiently fine wave vector grid. To handle the hard sphere system for example, one has to deal with $M = 200$ to 300. Any quantitative discussion for realistic situations is therefore connected with serious numerical work. But some results of the theory are so robust, that they can be demonstrated for schematic models dealing with $M = 1,2$ or 3 correlators only. The simplest are the one component models based on the specialization of (18) to equations for one correlator $\phi(t)$ and read

$$\partial_t^2\phi(t) + \nu\partial_t\phi(t) + \Omega^2\phi(t) + \Omega^2\int_0^t m(t-t')\partial_{t'}\phi(t')dt' = 0\ , \tag{21a}$$

$$\gamma\partial_t\phi(t) + \phi(t) + \int_0^t m(t-t')\partial_t'\phi(t') = 0\ . \tag{21b}$$

respectively. The kernel m is a polynomial, e.g.

$$m(t) = v_1\phi(t) + v_2\phi(t)^2\ , \tag{21c}$$

or

$$m(t) = v_1\phi(t) + v_3\phi(t)^3\ , \tag{21d}$$

and the like. In these cases there are only two coupling constants; the control parameter space **R** is the plane, spanned by $\vec{V} = (v_1, v_2)$ or $\vec{V} = (v_1, v_3)$ respectively. The equations (21) make explicit that no input information is used, which is directly related to structural relaxation features. The equations (21) yield α–peaks. If $v_1 = 0$, these peaks describe Debye relaxation. In this particular case the assumptions of Maxwell's theory are reproduced by the solutions. But if $v_1 \neq 0$, the α–process exhibits stretching. For the model (21c) the α–peak is well described by a Kohlrausch law (2), where the stretching exponent β varies between 1 for $v_1 = 0$ and 0 for $v_1 = 1$. These results are quoted in order to demonstrate, that the simplified MCT is subtle and relevant. Some solutions of (21) are exhibited below in figs. 1–6.

3.4 The Cage Effect and the Back Flow Phenomenon

The equation (8b) and the results (17a,b) for the memory kernel or modifications thereof were discussed in the liquid literature before their relevance for the explanation of structure relaxation was appreciated. In order to understand the physical phenomena described by the mode coupling functional, it is helpful to recall two of the earlier results. Let us consider a tagged particle in its environment of the other liquid molecules. Let us also consider for a moment the surrounding particles as fixed in a frozen array. This picture on the dynamics of a liquid is called the Lorentz model. It focuses on the percolation problem of a single molecule and ignores the dynamics of the other constituents of the system. The latter produce a static random potential for the former. If the density n is that of the triple point or larger, the tagged particle could not move over distances seriously larger than the averaged interparticle spacing $n^{-1/3}$. The diffusion constant would vanish and the mean squared displacement of the tagged particle would approach a finite value for large times. The dense packing produces cages, formed typically by 10 to 14 neighbours, which keep the tagged particle trapped while performing some rattling motion. The tagged particle could not explore the whole phase space; it would exhibit non–ergodic dynamics. The Lorentz model predicts an ideal glass state for parameters (n, T), which are characteristic for a normal simple liquid.

The known motion of a tagged particle in a normal simple liquid, which manifests itself in a non vanishing diffusion constant D and a mean squared displacement increasing linearly with increasing time t, is possible only, because the surrounding particles are not fixed. The tagged particle can move because its neighbours move out of the way; and this is possible because their neighbours move out of the way etc. The tagged particle moves, because it can carry its cage along. The motion is connected with a complicated streaming pattern, or backflow, around the particle. The cage effect and the backflow are cooperative phenomena which exist because of the correlated action of large complexes of molecules.

Backflow was discussed by Feynman and Cohen [22] within their microscopic theory of liquid helium. In a first approximation they described backflow by a dipole streaming pattern analogous to the potential flow of an ideal incompressible fluid around a moving sphere. They used this picture as basis of a variational ansatz for the problem, thereby incorporating the non-zero compressibility of the liquid. As a result they obtained a theory for the phonon-roton excitation spectrum, which manifests itself as resonances for the density fluctuation spectrum $\phi''_q(\omega)$. The known roton minimum is caused by the structure factor peak for wavevectors near $q_0 \sim 2\text{Å}^{-1}$. The problem to evaluate $\omega\phi''_q(\omega) = \chi''_q(\omega)$ of liquid helium was later treated on the basis of the Zwanzig-Mori equation (8b) and the quantum-mechanical generalization of the mode coupling approximation (13, 17a) for the relaxation kernel [23]. The Feynman-Cohen result was reproduced as first approximation of the MCT equations of motion: $m_q^{(1)}(t) = \mathcal{F}_q(\phi_k^{(0)}(t))$; here the zeroth approximation $\phi^{(0)}$ was obtained by neglecting the relaxation kernel in (8b). The full solution of the MCT equations was shown to improve the original Feynman-Cohen results relative to the experimental findings.

The cage effect for the tagged particle motion has been studied repeatedly. The used attacks are all based on (8) and some approximation equivalent to $m_q(t) \sim \mathcal{F}_q(\phi_q^{(0)}(t))$, where $\phi_q^{(0)}$ is some reasonable approximation for the density correlator. It was shown for example [24], that the theory implies a suppression of the diffusivity D below its lowest order approximation $D^{(0)}$. This suppression of $D/D^{(0)}$ below unity increases with the density. The suppression comes about, because during the life time of the cages, which extends over several collision times, no diffusion is possible. If one would extrapolate the mentioned results, one would find some critical density n_c^*, where the diffusivity vanishes. This extrapolation indicates the existence of some singularity for the dynamics for $n = n_c \sim n_c^*$. However, for n near n_c^* the cited perturbation approach becomes unvalid. The cage is formed by the same particles, which are found to be trapped. Using $\phi_q^0(t)$ leads to qualitatively wrong results for $n \sim n_c^*$. One cannot analyze a singularity for density fluctuations with ad hoc models ϕ_q^0 for ϕ_q. Rather one must deal with the tagged particle on the same footing as with the particles forming the cages, and this is the purpose of the MCT equations of motion.

The formulated basic version of the MCT is the simplest theory of the cage effect, which is free of mathematical inconsistencies. The theory reproduces previous work for sufficiently small densities or high temperatures. The interesting outcome is that the solutions for high enough densities or low enough temperatures are quite different from the ones characterizing the normal liquid state. There appears indeed some singularity. To construct an improved MCT for structural relaxation it is necessary to understand some aspects of this singularity. An improved MCT will be discussed below in section 5; it has been developed in order to treat also other relaxation mechanisms which are ignored in the simplified MCT.

4 GLASS TRANSITION SINGULARITIES

4.1 The Non-Ergodicity Parameters

Let us consider the long time asymptote of the correlators $\phi_q(t \to \infty) = f_q$. Conventional liquid theories are constructed so that $f_q = 0$. This means that fluctuations disappear for long times, the system approaches its equilibrium state. If the long time limits of all correlators disappear, the system is ergodic. The simplified MCT yields also such solutions, albeit only for sufficiently small couplings $\vec{V}$, i.e. for sufficiently small densities n and for large temperatures T. The novel results of the theory are connected with the observation, that there are also solutions with $f_q > 0$; in this case f_q is called non-ergodicity parameter.

The long time limits f_q of the MCT solutions solve the following set of M coupled algebraic equations:

$$f_q/(1-f_q) = \mathcal{F}_q(\vec{V}, f_k)\,,\, 0 \leq f_q < 1\,, \qquad q = 1,\ldots,M\,. \tag{22}$$

To emphasize the dependence of the solution on the control parameters, the vertex vector $\vec{V}$ is indicated in the mode coupling functional $\mathcal{F}_q$. Equation (22) is always solved by $f_q = 0$, $q = 1,\ldots,M$; and for sufficiently small $\vec{V}$ this trivial solution is the only one. However for sufficiently large $\vec{V}$, i.e. for sufficiently large densities or low temperatures, there are also non-trivial solutions.

Usually, there are for given $\vec{V}$ several solutions of (22), say $f_q, f'_q, \ldots$. Upon change of $\vec{V}$ one can hit critical points $\vec{V}^c$, called bifurcation points, where $\ell \geq 2$ solutions appear out of complex solutions or where ℓ solutions coalesce and transform into complex ones. Near such control parameter points $\vec{V}^c$ some of the solutions $f_q, f'_q, \ldots$ exhibit a singularity if considered as function of $\vec{V}$. The branch of mathematics, dealing with the characterization of the singularities obtained from equations like (22), is called singularity theory. We wish to touch on some concepts of that theory, since we wish to relate MCT bifurcations to bifurcations studied in more familiar dynamical theories. Besides many other possibilities there is the class of so-called cuspoid singularities, whose members are denoted by $A_\ell, \ell = 2, 3, \ldots$ [25]. These are characterized by the Jacobian of the equations (22) to have a non-degenerate eigenvalue zero for $\vec{V} = \vec{V}^c$. For $\vec{V} \to \vec{V}^c$, ℓ solutions coalesce. The case A_2, the Whitney fold, and A_3, the Whitney cusp, are well known examples from the elementary theory of quadratic and cubic equations, respectively. The $M = 1$ schematic models (21) exhibit lines of A_2-bifurcations and points of A_3 bifurcations in the 2-dimensional plane $\mathbf{R}$ of control parameters, as one can find out easily.

Of all possible singularities of the solutions of (22) we are only interested in those critical points $\vec{V}^c$ which correspond to singularities in the long time limit f_q. For these cases the equations of motion (18, 19) imply the following remarkable results. The non-ergodicity parameter f_q is the maximum solution of (22) in the sense, that all other solutions f_q' obey the inequality: $f_q' \le f_q, q = 1, \ldots, M$. Furthermore, the maximum solution is obtained as limit $m \to \infty$ for the discrete evolution equation $f_q^{(m)}/(1 - f_q^{(m)}) = \mathcal{F}_q(f_k^{(m-1)}), m = 1, 2, \ldots, f_q^{(0)} = 1$. Finally, generically the singularities $\vec{V}^c$ are of type A_ℓ with some $\ell \ge 2$. One concludes therefore, that there are critical values n_c for the density or T_c for the temperature so, that $f_q = 0$ for $n < n_c$ or $T > T_c$ but $f_q > 0$ for $n \ge n_c$ or $T \le T_c$.

Edwards and Anderson [26] pointed out in connection with a discussion of spin glasses that the existence of a non-vanishing long time asymptote of some correlator is the definition of an ideal glass state. Therefore limits like f_q are also called Edwards-Anderson parameters. The simplified MCT predicts the existence of ideal glass states and sharp transitions from a liquid to an ideal glass state. Therefore this theory is also referred to as the ideal MCT. The bifurcation singularities $\vec{V}_c$, exhibited by f_q, are called glass transition singularities (GTS). This terminology is used independent of whether there is some transition at $\vec{V}_c$ from liquid to a glass or whether there is a transition from some glass state to another glass state. Let us recall, that the mode coupling functional $\mathcal{F}_q$ is an equilibrium quantity. Consequently f_q and the singularities $\vec{V}_c$ are equilibrium quantities as well. Contrary to the calorimetric glass transition temperature T_g, the critical temperature T_c of the MCT is an equilibrium concept. For a Lennard-Jones system the $n_c - T_c$ phase diagram is independent of the masses of the particles. The critical density n_c of a hard sphere liquid model is the same as for a colloidal suspension, if both densities are measured relative to the particle volume.

There are two properties of the MCT solutions for $n > n_c$ or $T < T_c$, which motivate the application of the word "glass". If $f_q \neq 0$, the Fourier-Laplace transform of the correlator exhibits a pole at zero frequency, called non-ergodicity pole: $\phi_q(\omega) = [-f_q/\omega] + o_q(\omega^{-1})$. This means, that the density fluctuation spectrum exhibits some elastic contribution

$$\phi_q''(\omega) = \pi f_q \delta(\omega) + \quad \text{(integrable function of } \omega) \,. \tag{23}$$

Like for solids but opposite to liquids, there is recoil-free scattering of probing particles. The non-ergodicity parameter f_q is the Debye–Waller factor of the $T \le T_c$ state of the system. The long ranged order in crystalline solids implies, that $f_q \neq 0$ only for wave vectors $\vec{q}$ on the discrete set of the reciprocal lattice points. The $T \le T_c$ state of the ideal MCT is characterized by a Debye–Waller factor f_q, which varies smoothly with q and is non-zero for all wave vectors. This means, that the found states are amorphous solids, and this is the conventional definition of a glass. From (19) one concludes that the kernel m_q for the longitudinal fluctuating forces also exhibits some non-vanishing long time limit $m_q(t \to \infty) = A_q$. One gets therefore

$$m_q''(\omega) = \pi A_q \delta(\omega) + \quad \text{(integrable function of } \omega) \,, \tag{24a}$$

$$A_q = \mathcal{F}_q(f_q) \,. \tag{24b}$$

The second property concerns the transversal current correlations. As above one derives from (17c) for the transversal force correlator the existence of a non-ergodicity pole

$$m_q^T(\omega) = [-A_q^T/\omega] + o_q(\omega^{-1}) \,, \tag{25a}$$

$$A_q^T = \sum_{\vec{k}+\vec{p}=\vec{q}} V^T(\vec{q}, \vec{k}\vec{p}) f_k f_p \,. \tag{25b}$$

The Edwards–Anderson parameter A_q^T is positive for all q, in particular for $q = 0$. Substitution of this finding into (11b, 13) yields

$$\phi_q^T(\omega) \propto \omega / [\omega^2 - c_T^2 q^2 + \omega q^2 o(\omega^{-1})] \,. \tag{25c}$$

The transversal current correlator has the form as found in an isotropic elastic medium. Low frequency long wave length excitations do not propagate via shear diffusion as in the liquid state. Rather they propagate as transversal sound waves with a speed c_T, given by $c_T^2 = A_0^T v^2$. For $T \le T_c$ the ideal glass state can sustain shear stress, a property which has always been anticipated in the past as an outstanding feature of the glass state.

The appearance of ideal glass states implies, that the established picture of liquid dynamics, as it was partly explained in section 3.4, is incomplete. Back flow accompanied motion is possible only for moderate densities $n < n_c$ or moderate cooling $T > T_c$. The simplified treatment by the ideal MCT brings out a different state for strong compression $n > n_c$ or strong cooling $T < T_c$. The cage effect leads then to the following result: a particle cannot move because its neighbours cannot move out of the way, because their neighbours cannot move etc. A back flow pattern cannot build up. The cages have a random distribution of sizes and shapes. The spectra are obtained by averaging the ones for localized motion in the cages. This leads to continuous spectra for $\omega \neq 0$ in addition to the mentioned elastic peaks. Such situation was indeed suggested above in section 3.4 in connection with the Lorentz model discussion. But in our case static cages are formed as result of cooperativity. One cannot consider the static random potential as given and then use percolation theory in order to discuss particle localization. The formation of the static cages and the trapping within these cages has to be handled simultaneously, as is done by (22). Static cages exist only for $n > n_c$ and they imply solidification. There are no static cages for $n < n_c$, but only slowly fluctuating ones.

The elastic contribution of strength A_q in (24a) is the analogue of Maxwell's structural relaxation resonance for the modulus in (14a). Similarly, the elastic contribution of strength f_q in (23) is the analogue of Mountain's peak intensity in (14b). In both cases the resonances are idealized to ones of width zero. Equations (22, 24b) show, that Mountain's relation (14c) between A_q and f_q remains valid. However, while Maxwell's theory introduces A_q as given model parameter, MCT derives the non-linear equation (22) which allows the evaluation of A_q from the liquid structure.

The ideal peaks in the spectra reflect, that the correlators do not relax to zero but to some plateau of height A_q or f_q, respectively. Let us imagine that $T > T_c$ or $n < n_c$. The quoted continuity of the MCT solutions for finite time intervals guarantee, that there is some time τ_α with the following significance. For $t_0 \ll t \ll \tau_\alpha$ the correlators will be close to their plateau values provided $| T - T_c |$ or $| n - n_c |$ are sufficiently small. For $T - T_c \to 0$ or $n - n_c \to 0$ the plateau length has to expand arbitrarily, i.e. τ_α will diverge. Within the extended MCT additional relaxation mechanism are incorporated and they yield a relaxation of the correlators to zero, no matter how large the cage effect couplings are. But if these ergodicity restoring processes, to be discussed in the following section 5, are sufficiently small, the scale τ_α for the decay from the plateau will be large. The decay from f_q to zero is the α-process for structural relaxation as described by MCT. It is the cage effect, which is the origin of the slow α-relaxation in liquids. Equation (22) treats already one feature of this process, viz. the intensity.

The limit $\tau_\alpha \to \infty$ is equivalent to considering the idealization, that the α-process is arbitrarily well separated from all other dynamical phenomena. This limit is the same as the limit $s \to 0$, considered above in order to derive (15). Equation (15) is therefore the starting equation for the MCT of the α-process. This equation has to be complemented by $\tilde{m}_q(t) = \mathcal{F}_q(\vec{V}_c, \tilde{\phi}_k(t))$. Thereby a basis is formulated for the evaluation of the α- -process shape functions $\tilde{\phi}_q, \tilde{m}_q$ from the equilibrium structure. We will not discuss this aspect of MCT any further in this paper. But let us mention, that the von Schweidler expansions (16) are an implication of our theory.

All traditional theories on liquid dynamics anticipate in some place or the other the separation of time scales as discussed above in section 2.2. This assumption becomes unvalid if parameters approach the GTS. Within the idealized MCT the time scale for the fluctuating force correlations is equal to the scale for the density fluctuations if $T \leq T_c$: both are infinite.

4.2 The Critical Dynamics

In the present paper we will not discuss the solution of the MCT equations of motion in detail. However, for the construction of the extended MCT it is necessary to understand one result for the dynamics, viz. the critical decay law at the GTS: $\vec{V} = \vec{V}^c$. It describes the decay of the density correlators from the initial value $\phi_q(t = 0) = 1$ to the non-ergodicity parameter at the singularity, which is called critical non-ergodicity parameter f_q^c. Besides f_q^c there appears a second quantity, called critical amplitude h_q, which specifies the wave vector dependence of the correlators. Both f_q^c and h_q are smooth positive functions of q. The critical dynamics is the germ of the structural relaxation; this holds independent of whether one studies the simplified or the extended version of the theory. The details of the critical dynamics depend on the nature of the singularity.

The simplest and thus most important singularity is the Whitney fold A_2. The crucial parameter specifying the time dependence in this case is the so-called exponent parameter λ, which obeys the inequality $0 \leq \lambda < 1$. It determines the anomalous exponents for the solutions, in particular the critical exponent a. The latter is the solution of the transcendental equation $\Gamma(1-a)^2/\Gamma(1-2a) = \lambda, 0 < a \leq 1/2$; where Γ denotes the gamma function. One gets for $\vec{V} = \vec{V}_c$:

$$\phi_q(t) - f_q^c = h_q(t_0/t)^a + O_q[(t_0/t)^{2a}]\,, \qquad A_2\,. \tag{26a}$$

This power law for the critical decay is equivalent to a sublinear variation of the susceptibility spectrum

$$\chi''_q(\omega) = h_q \Gamma(1-a)\sin(\pi a/2)(\omega t_0)^a + O_q[(\omega t_0)^{2a}] \ . \tag{26b}$$

The transient dynamics enters the preceding formulae only via the matching scale t_0. The regular properties of the equations for finite time intervals ensure, that t_0 depends smoothly on all the parameters like Ω_q, ν_q and $\vec{V}$, which specify the model.

For the schematic models (21) there are lines of A_2-singularities in the coupling parameter planes. Points $\lambda < 1$ specify inner points of these lines. The limit $\lambda \to 1$ specifies endpoints, where the lines terminate in points $\vec{V}^c$, which are A_3 singularities in those particular cases. Corresponding results hold for the general case of some N-dimensional space $\mathbf{R}$ of coupling vertices $(V_1, \ldots, V_N) = \vec{V}$. Parameters with $\lambda < 1$ specify inner points $\vec{V}^c$ of a bifurcation hypersurface; these are smooth $(N-1)$-dimensional sets in $\mathbf{R}$. The limit $\lambda \to 1$, i.e. $a \to 0$, signalizes that an edge of the hypersurface is approached, where there are singularities A_ℓ with $\ell > 2$. The dynamics at some higher order singularity A_ℓ is also specified by some number $\mu > 0$, called cusp parameter. This is the analogue of $1 - \lambda$, so that $\mu \to 0$ signalizes the approach towards a singularity $A_{\ell'}$, with $\ell' > \ell$. For A_ℓ with $\ell > 2$ one obtains logarithmic decay for $\vec{V} = \vec{V}_c$:

$$\phi_q(t) - f_q^c = h_q[\mu/(\ln t/t_0)^{\frac{2}{\ell-2}}][1 + O_q[\ln\ln(t/t_0)/\ln(t/t_0)]] \ . \tag{27}$$

The matching time t_0 has the same properties as mentioned above.

The parameters h_q, λ and μ can be evaluated from the mode coupling functional $\mathcal{F}_q$ in a straightforward but involved procedure. For schematic models like (21) the evaluation of h_q, λ and μ is trivial. Since $\mathcal{F}_q$ is an equilibrium quantity, the same holds for f_q^c, h_q, λ and μ. In particular the critical exponent a is a number solely determined by the equilibrium structure of the system. The phonon frequencies Ω_q or friction constants ν_q in (18a) or the time scales γ_q in (18b) enter the results merely via the single number t_0. Formulae (26) hold with the same numbers for f_q^c, h_q, a for the usual hard sphere model of a liquid and also for hard sphere colloids. The effect of hydrodynamic interactions and the viscosity of the solvent in colloids merely enter the quoted formulae via the scale t_0.

Power law decay like (26a) or critical spectra like (26b) are familiar from the dynamics near second order phase transition points or near percolation thresholds. In those cases the critical dynamics is a result of the critical clusters, which exhibit a fractal structure in space. Some response function for wave vector q can probe only the part of the critical cluster with diameter of order $1/q$. Therefore the critical decay laws in the cited familiar examples exhibit a critical dependence on wave vector q for $q \to 0$. The MCT deals with quite a different scenario. There are no critical clusters which introduce small q singularities in equilibrium susceptibilities like S_q. As a result, the critical decay depends only smoothly on q for all wave vectors. The anomalous exponent a, quantifying the time fractal (26), is not the result of some anomaly in configuration space, caused by some divergent length scale. The formulae (26, 27) hold also for schematic models like (21), where there is no wave vector or space variable at all. The familiar phase transition examples exhibit universality of the exponents like a in the sense, that these depend only on rather general symmetry properties of the system. This universality is intimately related to the mentioned $q \to 0$ singularities. For small-q-results most details of the microscopic properties of the system become irrelevant. Such conclusion does not hold for the MCT. The MCT vertex and thus the structure determine the value for λ and hence the exponents. A hard sphere system has an exponent $a = 0.30\ldots$, which differs e.g. slightly from that for a Lennard-Jones system. For a binary mixture the exponent depends somewhat on the relative size of the molecules and the like.

There are two reasons, why the exact asymptotic laws (26a, 27) cannot be tested directly for $t \to \infty$ in an experiment. Firstly, the coupling coefficients $\vec{V}$ in the mode coupling functional can differ from the critical values, e.g. if $T > T_c$. Secondly, there are other relaxation mechanisms besides those treated in the simplified MCT, which restore ergodicity and hinder the correlator to arrest at f_q^c. However, the continuity of the MCT results for fixed finite time intervals ensures, that there is a characteristic time, say τ_β, so that (26a, 27) hold for the window

$$t_0 \ll t \ll \tau_\beta \, . \tag{28}$$

The window becomes very large if $\vec{V} - \vec{V}_c$ and the mentioned ergodicity restoring processes are small. In particular, τ_β will diverge if $T - T_c$ or $n - n_c$ tend to zero and simultaneously the other relaxation channels are switched off. Let us recall, that a similar reasoning was used in the preceding section in order to understand the relevance of a time scale τ_α. Obviously $\tau_\beta < \tau_\alpha$. The existence of a control parameter sensitive large time scale τ_α results from the spontaneous arrest of density fluctuations with $f_q^c > 0$. The existence of a further control parameter sensitive large time scale τ_β results from the appearance of the critical decay.

Thus MCT predicts for systems near a GTS a dynamical window (28) where the correlators follow the critical laws (26a, 27). Within this window there is no sensitive dependence of the dynamics on the control parameters. The correlators may vary with temperature since for example the time scale t_0 can depend on T. But this dependence would be smooth, in particular it would not exhibit a singularity for

$T - T_c \to 0$. The sensitive dependence on control parameters enters via the sensitive variation of the scale τ_β, restricting the window for the critical decay.

The existence of the described critical dynamics was not noticed during more than a century of structural relaxation research. Apparently, it was always anticipated, that the α–process starts somehow after the transient time t_0 of some ps is passed. Equivalently, it seemed plausible, that the α–peak of the relaxation spectrum was superimposed on some white noise background spectrum: $\phi''_{\text{white}}(\omega) = C$. This means that the α–peakd of the susceptibility spectrum should be located on top of a linearly varying background spectrum $\chi''_{\text{white}}(\omega) = C\omega$. From (26b) and (28) one gets however a power law density fluctuation spectrum $\phi''_q(\omega)$ which increases with decreasing frequency:

$$\phi''_q(\omega) = h_q \Gamma(1-a) t_0 \sin(\pi a/2)/(\omega t_0)^{1-a} \; ; \qquad 1/\tau_\beta \ll \omega \ll 1/t_0 \; . \tag{29}$$

This spectrum, as opposed to the one for the α–process should not exhibit a sensitive dependence on density or temperature. Formula (29) describes an enhancement of the spectrum above the expected white noise level if ω falls below the band of microscopic excitations, say $1/t_0 \sim 1$ THz. The window, where (29) is valid is largest for T near T_c. The enhancement can be larger than an order of magnitude if ω decreases to, say 1 GHz, since $1 - a \geq 0.5$. The α–spectrum is superimposed for $\omega \ll 1/\tau_\beta$ on top of the fractal spectrum (29). It is simpler to distinguish the α–spectrum from the critical one if one considers the susceptibility. The high frequency wing of the α–peak decreases with increasing ω as described e.g. by (4a). The critical susceptibility spectrum (26b) increases with increasing frequency, leading to a minimum in $\chi''(\omega)$. The α–process is terminated by this minimum of the χ'' versus ω graph, located at some frequency ω_{min}. This minimum position is not characterized by the microscopic scale $1/t_0$, but it is much smaller. Obviously, the spectral intensity $\chi_{\text{min}} = \chi''(\omega_{\text{min}})$ is smaller than the one for the α–peak maximum: $\chi_{\text{min}} < \chi''(1/\tau_\alpha)$. The important implication of (29) is, however, that the intensity in the minimum is much enhanced compared to the white noise level:

$$\chi_{\text{min}} \gg C\omega_{\text{min}} \; . \tag{30}$$

We consider the detection of the critical dynamics in several recent experiments, e.g. the measurements for CKN by neutron scattering [27] and by light scattering [28] as very relevant arguments for an assessment of the MCT.

A fractal process like the critical decay has no characteristic time scale. Power law functions exhibit self-similarity: a change of the frequency by a factor x say, is equivalent to a change of the susceptibility scale h_q in (26b) by a factor x^a. The only scales relevant for a discussion of a time fractal are the ones specifying the dynamical window for the process: t_0 and τ_β for the case considered above.

It might be helpful to demonstrate the preceding results by some numerical solution of the MCT equations of motion. Figures 1 and 2 show correlators $\phi(t)$ and susceptibility spectra $\chi''(\omega)$ calculated for the schematic model (21b) and with (21c) for the memory function. Units are chosen such that $\gamma = 1$. The coupling constants in the $v_1 - v_2$ parameter plane are shifted on a straight line through the critical point $v_1^c = (2\lambda - 1)/\lambda^2$, $v_2^c = 1/\lambda^2$, $\lambda = 0.7$ such that $v_{1,2} = v_{1,2}^c(1+\epsilon)$, $\epsilon = \pm 1/4^n$ and $n = 0, 1, \ldots$. The relaxation pattern is characteristic for an A_2 singularity. For $\epsilon < 0$ the $\phi(t)$ curves decay to zero for $t \to \infty$ while for $\epsilon > 0$ one observes arrest $\phi(t \to \infty) = f > 0$. At the critical point $\epsilon = 0$ the solution is shown by the dashed curve, and for $t > 1$ this agrees very well with the predicted asymptotic result in (26a), shown as dashed-dotted curve. The correlators for $\epsilon < 0$, $n > 3$ exhibit a decay process from the plateau $f^c = 1 - \lambda$ to zero, whose time scale τ_α increases dramatically with decreasing ϵ. This decay produces the low frequency susceptibility peaks in fig. 2. A Debye peak (1b) is added as dotted line in fig. 2 with τ_α chosen so, that the peak maximum for $n = 7$ is matched. The corresponding Debye decay $f_c \exp -(t/\tau_\alpha)$ is also shown as dotted curve in fig. 1. The slow decay process can be fitted very well by the Kohlrausch law (2) with $f = f^c = 0.3, \beta = 0.58$. Thus the schematic model in (21b) exhibits the main feature of glassy relaxation: a slow control parameter sensitive α–process, which is stretched. For $\epsilon < 0$ the scales $\tau_{\alpha,\beta}$ can be defined by the position $\omega_{max}, \omega_{min}$ of maximum and minimum respectively of the susceptibility spectra: $\tau_\alpha = 2\pi/\omega_{max}$, $\tau_\beta = 2\pi/\omega_{min}$. Figure 3 shows the spectra ϕ'' corresponding to the data in fig. 1. For $\epsilon < 0$ and $\omega > 1$ they follow the critical decay given by the dashed curve, until the von Schweidler increase sets in for ω near ω_{min}. For low ω the curves levels off to a white noise spectrum.

In figures 4–6 we also show the relaxation pattern near a cusp singularity A_3. Here the schematic model in (21b) with (21d) for the memory function has been used with units chosen such that $\gamma = 1$. The coupling constants in the $v_1 - v_3$ parameter plane are now chosen on a straight line parallell to the v_1-axis and passing through the critical point at $v_1^c = \frac{9}{8}, v_3^c = \frac{27}{8}$ so that $v_1 = v_1^c(1+\epsilon)$, $\epsilon = \pm 1/4^n$ and $v_3 = v_3^c$. The decay curves near a cusp singularity are clearly qualitatively different from those near a fold singularity. The dominant feature exhibited in fig. 4 is a decay for $\epsilon < 0$ proportional to the logarithm of the time: $\phi(t) \propto -\ln(t/t_1)$. This logarithmic decay extends over several decades in time for the curves closest to the cusp point. The dashed curve shows the solution at the cusp point with $\epsilon = 0$, and the corresponding critical decay law in (27) with $\phi(t) = 1/3 + 2.2 \ln 10/[\log t + 3.5]^2$,

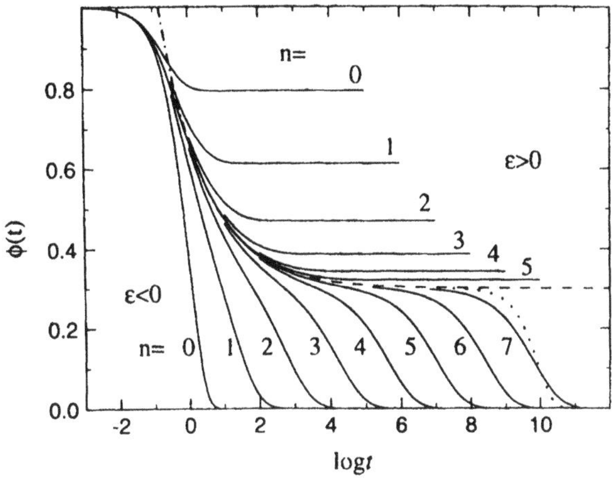

Figure 1: $\phi(t)$ versus $\log t$ from (21b) and with $m(t) = \mathcal{F}(\phi(t))$ where $\mathcal{F}(f) = v_1 f + v_2 f^2$ as in (21c). The filled curves correspond to parameter points $v_{1,2} = v_{1,2}^c(1+\epsilon)$ where $v_1^c = (2\lambda - 1)/\lambda^2, v_2^c = 1/\lambda^2$, $\lambda = 0.7$, $\epsilon = \pm 1/4^n$ and n-values as noted. The dashed curve is the solution at the critical point $\epsilon = 0$. The dotted curve shows the Debye decay $\phi_D(t) = 0.3\exp(-t/\tau_2)$ with $\tau_2 = 8.9 \times 10^9$. The chain curve shows the asymptotic critical decay from (26a): $\phi(t) = f^c + (t_0/t)^a$, $a = 0.327$, $t_0 = 0.05$.

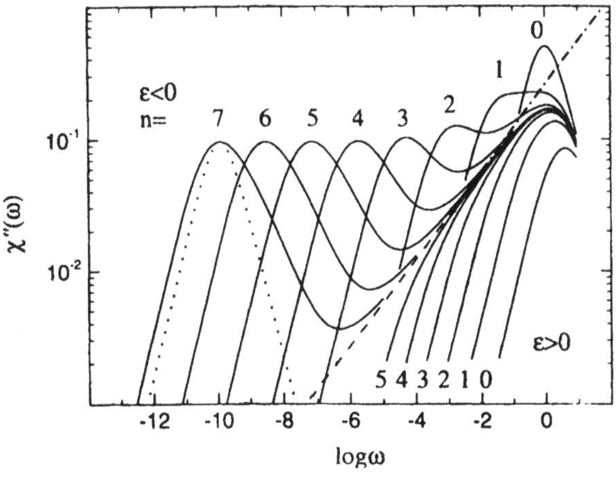

Figure 2: The susceptibility spectra $\chi''(\omega) = \omega\phi''(\omega)$ versus $\log\omega$ for the same data as in Fig. 1. Here the dotted Debye curve has been normalized to the same height as the full curve representing $n = 7$.

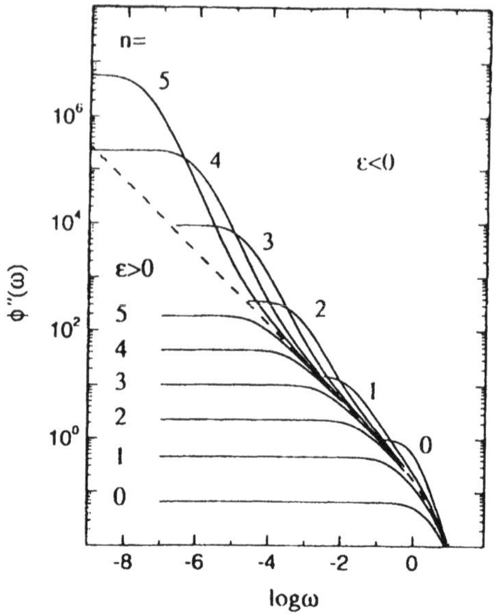

Figure 3: The spectra $\phi''(\omega)$ versus $\log\omega$ for the data in fig. 1. The small ω white noise parts of the curves for $\epsilon < 0$ are not shown for $n = 0-3$ in order not to overload the figure.

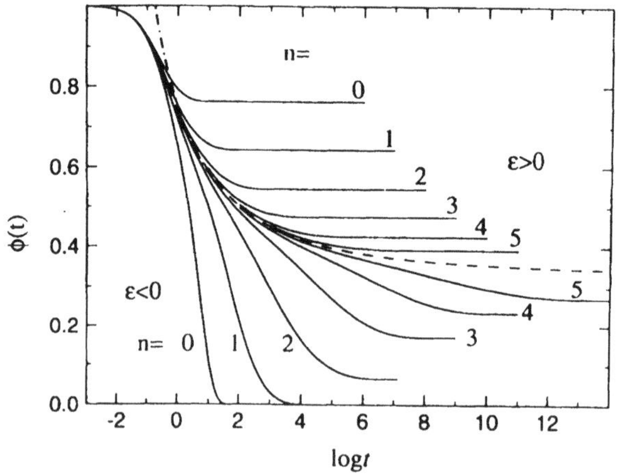

Figure 4: $\phi(t)$ versus $\log t$ from (21b) and with $m(t) = \mathcal{F}(\phi(t))$ where $\mathcal{F}(f) = v_1 f + v_3 f^3$ as in (21d). The filled curves correspond to parameter points $v_1 = v_1^c(1+\epsilon), v_3 = v_3^c$. Here $v_{1,3}^c$ denotes the cusp point at $(v_1^c, v_3^c) = (\frac{9}{8}, \frac{27}{8})$ and $\epsilon = \pm 1/4^n$, with n-values as noted. The dashed curve is the solution for $v_{1,3} = v_{1,3}^c$, and the chain curve shows the critical decay from (27): $\phi(t) = \frac{1}{3} + 2.2 \ln 10/(\log t + 3.5)^2$.

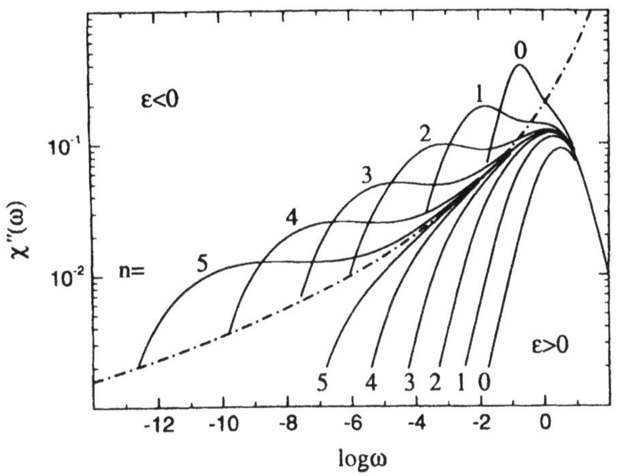

Figure 5: The susceptibility spectra $\chi''(\omega)$ versus $\log\omega$ for the same data as in fig. 4. The chain curve shows the critical decay $\chi''(\omega) = \pi 2.7/(-\log\omega + 3.5)^3$.

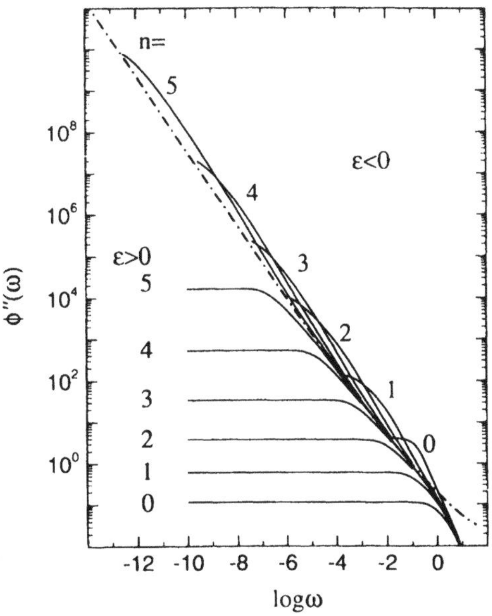

Figure 6: The spectra $\phi''(\omega)$ versus $\log\omega$ for the data in fig. 4. The constant plateau observed for $\epsilon < 0$ in the susceptibility curves in fig. 5 shows here up as a $1/f$-noise spectrum.

is shown as the dashed-dotted curve. The critical decay starts to extend the short time transient for $t \sim 1$. It represents a slow mesoscopic process, which preceds the mentioned logarithmic decay. The critical susceptibility spectrum is shown as the dashed-dotted curve in fig. 5 and corresponds to $\chi'' = \pi 2.7/[3.5 - \log\omega]^3$. The slight discrepancy in the coefficients (2.7 versus 2.2) reflects the uncertainty in the fitting procedure, but also the influence of corrections to the leading order terms. The logarithmic decay leads to a susceptibility spectrum, which is nearly frequency independent. The susceptibility minima of the fold scenario, shown in fig. 2, are nearly absent for the cusp scenario. The high frequency wing of the α-peaks of fig. 2 are increased for a cusp singularity so that the peak appears as the low frequency end of the susceptibility plateaus in fig. 5. The stretching of structural relaxation spectra is much more pronounced for the A_3 results than for the results of the A_2 glass transition singularity. The logarithmic decay in fig. 4 and the constant plateau in fig. 5 show up as $1/f$-noise spectra in fig. 6. The range of this $1/f$-region extends when approaching the cusp point.

5 THE EXTENDED MODE COUPLING THEORY

5.1 The Generalized Kinetic Equation

The derivation of the simplified MCT in sections 2, 3 was streamlined so, that the essential part of the theory, i.e. the cage effect induced GTS, was obtained as easily as possible. However, this approach is too narrow to be satisfactory. There is a series of well understood features of liquid dynamics, which are completely ignored in the simplified theory. A particular obvious - though not so important - example is the fact, that for certain parameter regions liquids exhibit the dynamics of free particles. The density correlator $\phi_q(\omega)$ of a free particle system is a non-elementary function [12]. The non- elementary features are all hidden in kernel $M_q^L(\omega)$ in (6), which for free particles is a singular function of q and ω. After the coarse graining and factorization approximations are carried out, the mentioned free gas singularities are lost. The derivation of sections (2, 3) focused on solid state concepts like phonon frequencies and elastic moduli. Concepts from the theory of gases like particle orbits and binary collision events have not been properly considered. The cage effect was looked at as a problem of localization of density fluctuations and suppression of diffusivity. It is of course equally appropriate to describe this phenomenon as a result of repeated correlated collision events, which force the particle orbits to coils.

In this section we wish to incorporate the simplified MCT into the more familiar transport theories. In addition we shall generalize the approximations, which lead to (18, 19).

A complete statistical description of single particle motion is provided by the phase space densities $\sum_\ell \delta(\vec{r} - \vec{r}_\ell(t))\,\delta(\vec{k} - \vec{p}_\ell(t))$. Here $\vec{r}_\ell$ and $\vec{p}_\ell$ denote positions and momenta of the particles. It is more convenient to consider phase space density fluctuations for wave vector $\vec{q}$, obtained by Fourier transformation with respect to $\vec{r}$: $f_{\vec{k}}(\vec{q}) = \sum_\ell \exp(i\vec{q}\vec{r}_\ell)\,\delta(\vec{k} - \vec{p}_\ell(t))$. The dynamics of single particles can then be described statistically by the phase space correlators for $\vec{q} \neq \vec{0}$:

$$C_{\vec{k}\vec{p}}(\vec{q}, t) = \langle f_{\vec{k}}(\vec{q}t)^* f_{\vec{p}}(\vec{q}) \rangle \ . \tag{31}$$

Instead of a single function (6) one considers now an infinite matrix of functions $C_{\vec{k}\vec{p}}$, where $\vec{k}$ and $\vec{p}$ serve as matrix indices. The initial condition for that matrix $C_{\vec{k}\vec{p}}(\vec{q}) = C_{\vec{k}\vec{p}}(\vec{q}, t = 0)$ can be expressed in terms of the structure factor S_q. The density correlator can be evaluated from C by a summation: $\phi_q(t) = \sum_{\vec{k}\vec{p}} C_{\vec{k}\vec{p}}(\vec{q}, t)$. The current correlators can also be obtained by averaging with proper weighting of $\vec{k}$ and $\vec{p}$ and so can be obtained more complicated functions like the correlators for temperature or pressure fluctuations.

The equation of motion for $f_{\vec{k}}(\vec{q})$ reads $i\partial_t f_{\vec{k}}(\vec{q}) = (\vec{k}\vec{q}/m) f_{\vec{k}}(\vec{q}) + F_{\vec{k}}(\vec{q})$. The first term describes free flight and the second one, $F_{\vec{k}}(\vec{q})$, is a force term, i.e. a combination of Fourier transforms of two particle potential derivatives and density fluctuation pairs $\rho_{\vec{\ell}-\vec{q}}\rho_{\vec{\ell}}$. The Zwanzig–Mori formalism provides a routine to eliminate the forces so that a generalized kinetic equation for the correlator is obtained [12]:

$$\partial_t C_{\vec{k}\vec{p}}(\vec{q}t) - \sum_{\vec{\ell}} \Omega_{\vec{k}\vec{\ell}}(\vec{q}) C_{\vec{\ell}\vec{p}}(\vec{q}t) + \sum_{\vec{\ell}} \int_0^t M_{\vec{k}\vec{\ell}}(\vec{q}, t - t') C_{\vec{\ell}\vec{p}}(\vec{q}t')dt' = 0 \ . \tag{32a}$$

This is equivalent to the equation for the functions, which are Fourier–Laplace transformed with respect to time:

$$\sum_{\vec{\ell}} [\omega\delta_{\vec{k}\vec{\ell}} - \Omega_{\vec{k}\vec{\ell}}(\vec{q}) + M_{\vec{k}\vec{\ell}}(\vec{q}, \omega)] C_{\vec{\ell}\vec{p}}(\vec{q}, \omega) = -C_{\vec{k}\vec{p}}(\vec{q}) \ . \tag{32b}$$

Matrix Ω can be expressed in terms of the structure factor S_q. All non-trivial features of the interaction of the many particle system are hidden in the memory kernel M. For $M = 0$ and $S_q = 1$ the result for the free particle gas is reproduced by (32b): $C^{(0)}_{\vec{k}\vec{p}}(\vec{q}, \omega) = -\varphi(k)\delta_{\vec{k}\vec{p}}/[\omega - \vec{q}\vec{k}/m]$, φ being Maxwell's velocity distribution. The exact equations (32) generalize the exact equations (8, 9). The second order differential equation (8a) is hidden in (32a) as a set of two coupled first order equations. While (8a) can be explicitly solved as (8b), the situation is more complicated for the generalized kinetic equation (32b). This is still an integral equation for the function of $\vec{k}$, $C_{\vec{k}\vec{p}}(\vec{q},\omega)$, where $\vec{p},\vec{q},\omega$ are considered as parameters. The solution of such linear equation is not the main problem of liquid dynamics. In practice however, it is a hard piece of work to get from (32b) the results for the interesting functions like $\phi_q(\omega)$. One can express the kernels M^L and M^T in (8, 9) in terms of the matrices Ω and M from (32); but the formulae involve the inverses of infinite matrices.

As a first step towards a specific theory we imagine, that a dangerous part $\Gamma_{\vec{k}\vec{p}}(\vec{q},\omega)$ can be split off from the relaxation kernel; Γ describes low frequency anomalies. The remainder shall be simplified by a coarse graining procedure as discussed in connection with (13), so that kernel Γ generalizes function m. Restricting the discussion to $\omega t_0 \ll 1$, the remainder can be evaluated for $\omega = 0$. It generalizes the constant γ in (13) to $K + iJ$, where K and J are equivalent to hermitian matrices. Kernel K renormalizes Ω in (32) to a matrix, which shall be called I. As a result one gets a reformulation of (32b) to

$$\sum_{\ell}[\omega\delta_{\vec{k}\vec{\ell}} - I_{\vec{k}\vec{\ell}}(\vec{q}) + iJ_{\vec{k}\vec{\ell}}(\vec{q})]C_{\vec{\ell}\vec{p}}(\vec{q},\omega) + \sum_{\ell}\Gamma_{\vec{k}\vec{\ell}}(\vec{q},\omega)C_{\vec{\ell}\vec{p}}(\vec{q},\omega) = -C_{\vec{k}\vec{p}}(\vec{q}) \,. \quad (33)$$

This equation makes sense only, after Γ is specified; it generalizes the first equation (18) of the simplified MCT to an equation for the matrix correlator C. If one drops kernel Γ, one gets a kinetic equation in the proper sense of this word: an integral equation where all dynamics is traced back to kernels I and J, which do not exhibit anymore frequency dependencies. Often one carries out also a coarse graining with respect to space variations. But this simplification is not possible here.

The most famous kinetic equation is Boltzmann's. It deals with dilute gases and is obtained by working out the integral kernels in leading order in the density n. This implies $I = \Omega$, $S_q = 1$ and reduces J to a q-independent kernel proportional to n. Kernel J can then be expressed by the two particle collision cross section. Enskog extended this theory by incorporating the fact, that colliding particles do not meet at the same place, but cannot approach each other more closely than the particle diameter. This leads to a q-dependence of J. From this work one understands that J describes effectively binary collision phenomena. A kinetic equation, which ignores J and replaces I by Ω is a modification of the Vlasov equation. From this observation one understands, that I describes particle interaction in a molecular field approximation: the paths are those of single molecules in an averaged field produced by the neighbours. For a first principle theory of liquids it is important, to handle properly the specified phenomena; and doing so one can indeed describe reasonably well the normal state liquid correlators, in particular the transport coefficients like the viscosity. But structural relaxations cannot be handled this way; these effects are hidden in the retardation phenomena, produced by kernel Γ.

5.2 The Equations of Motion of the Extended MCT

The fluctuating forces, whose correlations determine kernel M in (32), are combinations of products of two phase space densities $f_{\vec{p}}(\vec{q}-\vec{k})f_{\vec{t}}(\vec{k})$. The essential step to arrive at closed equations of motions consists again of Kawasaki's factorization approximation. The part of the average of products which is given by products of averages $C_{\vec{k}_1\vec{p}_1}(\vec{q}_1 t)C_{\vec{k}_2\vec{p}_2}(\vec{q}_2 t)$ is identified with kernel Γ in (33). The remainder contributes to I and J. The thus obtained equations are still too complicated to be directly applicable for the derivation of results, however.

For the final approximation steps, we transform from momentum distributions to irreducible tensors of the latter as is done in kinetic gas theory $f_{\vec{k}}(\vec{q}) \to A^{\alpha}_{\vec{q}}$. The index $\alpha = 0,1,2,\ldots$ labels the basis so that $A^0_{\vec{q}} = \rho_{\vec{q}}$ is the density fluctuation, $(A^1_{\vec{q}}, A^2_{\vec{q}}, A^3_{\vec{q}})$ are the current fluctuation coordinates etc. The equation (33) is then equivalent to an generalized kinetic equation for the infinite matrix of correlators $C^{\alpha\beta}(\vec{q},t) = \langle A^{\alpha}_{\vec{q}}(t)^* A^{\beta}_{\vec{q}}\rangle$. Now α,β serve as matrix indices and $\Gamma^{\alpha\beta}(\vec{q},t)$ is a combination of products $C^{\alpha_1\beta_1}(\vec{q}_1 t)C^{\alpha_2\beta_2}(\vec{q}_2 t)$. The coefficients $V^{\alpha\beta\alpha_1\beta_1\alpha_2\beta_2}_{(\vec{q},\vec{q}_1\vec{q}_2)}$ in those combinations can be viewed as coupling constants of the theory. If one replaces all coupling constants by zero except those for $\alpha_1 = \beta_1 = \alpha_2 = \beta_2 = 0$, kernel Γ in (33) is expressed as quadratic polynomial of density correlators $\phi_q(t) = C^{00}(\vec{q}t)/S_q$. The generalized kinetic equation can then be formally solved, and the result is the simplified MCT. Consequently, the findings of section 4 apply also to the present extended approach. In this sense one understands that there are ideal glass states and glass transition singularities for certain choices of the coupling constants.

At the GTS one finds critical decay towards the critical non-ergodicity parameter according to (26a): $C^{00}(\vec{q}t) - f^c_q S_q = O(1/t^a)$. All other correlators relax to zero; they do so algebraically with exponents larger than a : $C^{\alpha\beta}(\vec{q}t) = O(1/t^{a+x})$ with $x > 0$ unless $\alpha = \beta = 0$. For the mixed correlators between densities and longitudinal currents one gets $x = 1$. These enter the kernels quadratically, so that the corresponding contributions to Γ decay like $1/t^{a+2}$. For the current correlators one finds: $C^{\alpha\beta}(\vec{q}t) = O(1/t^{a+2})$, $\alpha,\beta = 1,2,3$. They contribute to Γ - after multiplication with $C^{00}(\vec{q}t)$-terms proportional to $1/t^{a+2}$. All other contributions to Γ decay like $1/t^{a+x}$ with $x \geq 3$. Now we recall the reasoning in section 4 and conclude: if all coupling constants are chosen close to the ones specifying the GTS, there will be a mesoscopic dynamical window $t_0 \ll t \ll \tau_\beta$, where all correlators and kernels decay algebraically with exponents as specified. To get a leading order description one neglects all kernels but those, which decay towards a plateau like $1/t^a$. Thereby the relaxation kernel $\Gamma(t)$ in (33) is expressed as mode coupling functional of $\phi_q(t)$. This treatment gives a leading approximation of structural relaxation within our theory. The resulting closed equations of motion for $\phi_q(t)$ are (17, 18), which specify the simplified MCT.

The simplest systematic extension of the leading order treatment of the mesoscopic relaxation incorporates all contributions to the kernels, which decay like $1/t^a$ and $1/t^{a+2}$. All other contributions, which decay like $1/t^{a+3}$ or faster, are neglected. Thereby the kernels Γ in (33) are combinations of the mentioned quadratic terms in the density fluctuations and terms like $C^{00}(\vec{q}t)C^{\alpha\beta}(\vec{k}t), C^{0\alpha}(\vec{q}t)C^{0\beta}(\vec{q}t)$; $\alpha, \beta = 1, 2, 3$. By use of the continuity equation one can express $C^{0\alpha}(\vec{q}t)$ in terms of the time derivative $\dot{\phi}_q(t)$ of the density correlator. The current correlators $C^{\alpha\beta}(\vec{q}, t)$ are given in terms of two functions: the transversal current correlator $\phi_q^T(t)$ and the longitudinal one $\phi_q^L(t)$. The latter can be expressed in terms of the second time derivative $\ddot{\phi}_q(t)$. From (33) one can thus obtain a closed set of equations for the two functions $\phi_q(t), \phi_q^T(t)$. As mentioned in section 2.1, these functions can be represented in terms of the two kernels $M_q^{L,T}(\omega)$ for longitudinal and transversal force fluctuations respectively. The equations (8, 9) remain valid also for the present discussion. For the kernels one gets the formulae:

$$M_q^{L,T}(\omega) = [i\gamma_q^{L,T} + m_q^{L,T}(\omega)]/\{1 - \delta_q^{L,T}(\omega)[i\gamma_q^{L,T} + m_q^{L,T}(\omega)]\} . \tag{34}$$

Here $m_q^{L,T}(t)$ are given by the mode coupling functionals (17); in particular the vertices are the same equilibrium quantities as considered before. For the ω–independent quantities $\gamma_q^{L,T}$ one gets involved expressions in terms of the integral operators I and J from (33). The new kernels $\delta_q(t)$ describe mode coupling to current excitations:

$$\begin{aligned} \delta_q^{L,T}(t) &= \sum_{kp}\{V_1^{\prime L,T}(q,kp)\phi_k(t)\partial_t^2\phi_p(t) \\ &\quad V_2^{\prime L,T}(q,kp)\phi_k(t)\phi_p^T(t) + V_3^{\prime L,T}(q,kp)\partial_t\phi_k(t)\partial_t\phi_p(t)\} . \end{aligned} \tag{35}$$

The vertices $V_i^{\prime L,T}(q,kp)$ in this formula depend on the solution of (33) with Γ neglected. These vertices are not equilibrium quantities. Differences between a hard sphere liquid obeying Newtonian dynamics on the one hand and a hard sphere colloidal suspension on the other hand lead to different results for the vertices V' in (35) for these two systems.

Equations (8, 9, 34, 35) define the mode coupling theory for structural relaxation in simple liquids. In order to distinguish it from the simplified MCT, which is reproduced by the additional approximation $\delta^{L,T} = 0$ in (34), it is referred to as extended MCT [3, 16]. These two versions of mode coupling theories appear as a first and second order treatment of the more general approach outlined above. The extended MCT alters some results of the simplified theory qualitatively. It would be interesting to work out also a next order extension within our scheme. So far this has not been attempted, however.

Within the ideal MCT structural relaxation evolves because the control parameters $\vec{V}$ approach some GTS. The key of the phenomenon was the decay of density fluctuations towards some plateau value. The plateau signalizes spontaneous arrest of structure fluctuation for mesoscopic times $t \sim \tau_\beta$. The α-process is the decay of these almost arrested fluctuations to the equilibrium. The same phenomena are described by the extended theory, provided the coupling constants V' in (35) are sufficiently small. Within the simplified MCT there are ideal glass states. The spontaneous arrest can be perfect in this idealization. This situation is reproduced by the extended theory albeit only in the trivial mathematical sense, that the coupling constants V' in (35) are put equal to zero. But for every temperature there are some non-vanishing coupling constants in (35). From this observation one derives $\delta_q(\omega = 0) \neq 0$ for some q and this excludes the appearance of those non-ergodicity poles in M_q^L or ϕ_q, discussed in connection with (23, 24). Thus the mode coupling to currents restores ergodicity; there is α-relaxation with a finite decay time τ_α in all cases. There cannot be an ideal glass state nor a singularity for those coupling parameters V and V', which follow from the microscopic equations of motion.

The construction of our theory was based on a classification of terms according to their relevance for the formation of an almost arrested structure for mesoscopic times. The classification was not done by looking for long time tails of the correlators or low frequency singularities of the spectra in a general sense. Indeed, it is well known that liquids exhibit algebraic long time decay processes. These are also produced by mode coupling to pair excitations, but now to terms like $\phi_k^L(t)\phi_p^L(t)$, $\phi_k^T(t)\phi_p^T(t)$ as well as couplings between current and temperature fluctuations [29]. In this known case the low lying hydrodynamic modes cause the singular decay processes. In dense gases and low density liquids these mode coupling effects dominate the low frequency spectra and they cause algebraic tails in the decay functions. These well known phenomena are ignored in our theory. Structural relaxation effects suppress the mentioned hydrodynamic long time tails. These tails are absent in the solid phase. Therefore it is plausible, that these hydrodynamic singularities are not of direct relevance for an understanding of structural arrest. However, it would be of interest to study the interplay of the hydrodynamic long time singularities with the anomalies caused by the GTS; but so far such study has not been done for supercooled liquids.

5.3 Hopping Effects

In certain limits the preceding equations produce an Arrhenius formula for the α-relaxation rate $1/\tau_\alpha \propto \exp -(E/k_BT)$. Such result is usually taken as evidence for transport via hopping events: a tagged particle jumps over saddle points in a potential landscape. In microscopic theories the hopping events appear because phonons kick the particles over barriers. Polaron-effects have to be taken into account in order to get the Arrhenius law for the rate [30]. In the present theory longitudinal and transverse phonons are described by the current correlators. It is not clear to us, whether all polaron effects are incorporated properly in the extended MCT equations of motion; but the back flow phenomenon, which was discussed in section 3.4, is at least an important part of this effect.

Phonon assisted hopping is the dominant transport mechanism in strongly disordered semiconductors. In those systems impurities produce a static random potential. The dynamics is usually discussed within Lorentz models, where one considers only tagged particle motion. A MCT treatment for this problem brought out [31], that coupling to phonons as described by (35) reproduces the results for the hopping transport phenomena, which had been achieved earlier by more conventional approximation techniques.

One concludes, that the kernels $\delta_q^{L,T}(t)$ describe approximately phonon assisted hopping, and they are therefore referred to as hopping kernels. The extended MCT deals with the interplay of two effects. Non-linear interactions of density fluctuations, as described by kernel m, lead to the cage effect with a trend to produce an ideal glass state and thereby a static random potential. Interactions between density fluctuations and currents, as described by kernel δ, lead to relaxation via phonon assisted hopping, i. e. via jumps over almost static potential barriers.

Tagged particle diffusion occurs also in crystalline solids. Hopping occurs e.g. from one place to some vacancy. Therefore the density fluctuation spectrum for wave vector q of a tagged particle in a crystal does not exhibit a strictly elastic peak, but only a quasielastic resonance. The width of the resonance is Dq^2 where D is the tagged particle diffusion constant. In crystals there is long ranged spatial order. The activation barriers for rearrangements of the frozen order parameter are therefore infinite. As a result, the density spectrum $\phi_q''(\omega)$ exhibits an ideal elastic peak and shear can propagate as wave. However, in liquids and glasses there is no long ranged order and therefore activation barriers for structural rearrangements are finite. As a result there cannot be an ideal glass state. This fact is obtained from the equations of motion of the extended MCT, as explained above.

Hopping effects depend crucially on structural details on length scales, corresponding to the interparticle spacing. The appearance of kernel δ is not a hydrodynamic phenomenon. The starting equation for the MCT of the α-process is (15), complemented by (17a) or (34,35) for $\vec{V} = \vec{V}_c$. The resulting α-peak reflects microscopic structure details, which enter via V, V'. The α-peak for the density fluctuation spectrum $\phi_q''(\omega)$ is not of diffusive type; for small wave vectors its width is q-independent as mentioned above in connection with (14d).

5.4 The Cross Over Temperature T_c

Molecular dynamics experiments, neutron scattering studies and theoretical work done during the past 25 years have achieved a detailed understanding of the dynamics of simple liquids in their normal state [12], [13]. It is known that binary collision events, excluded volume and molecular field effects as they are derived from extensions of Boltzmann's equation govern the dynamics in dense gases and high temperature liquids. Qualitatively new features appear due to non-analytic density dependencies of transport coefficients, due to hydrodynamic long time tails and due to the cage effect. These subtleties of normal state liquid dynamics can all be handled by one or the other version of mode coupling theories. The discussion of the diffusivity of the hard sphere liquid by Curkier and Mehaffey [24] is an illustrating example of the state of the art. One understands within the specified pictures of many particle physics the variations of the viscosity or of the diffusivity over one order of magnitude for temperatures T around the melting temperature T_m.

These achievements unvalidate the traditional way of fitting transport coefficients of simple liquids in the normal state by Arrhenius laws and of interpreting the data as results of hopping effects. There is no need to add such hopping relaxation mechanism to the cited ones and there is no justification for it either. The picture of phonon assisted hopping anticipates that there exists some potential landscape for the particles to move through. There must be a separation of time scales: the life time for the potential fluctuations must be much longer than the duration of the hops. Such separation of time scales does not exist for simple liquids at $T \sim T_m$.

In colloidal suspensions the dynamics is also understood for densities increasing up to close to the freezing point [32]. In this regime the diffusivity decreases with increasing density also by about a factor 10.

The quoted work on normal liquid dynamics fails for temperatures or densities outside the regime indicated above. The reason for this failure is the evolution of structural relaxation. This is a qualitatively new dynamical feature, which is outside the scope of the traditional theories. Our MCT explains these features as precursors of the formation of ideal glass states. Upon lowering T, the coupling parameters approach a glass transition singularity. As a result a decay towards some almost arrested array for the particle distribution occurs. The decay from this state to equilibrium constitutes the α-process. If one imagines hopping effects to be absent there would occur an abrupt transition to some ideal glass state for $T = T_c$. This temperature T_c is the ultimate limit for the normal liquid state behaviour.

For a Lennard-Jones system one can evaluate the temperature T_c and it is located below T_m. Maxwell's relaxation time τ for $T \approx T_c$, which can be inferred from the viscosity, is about 10^2 to 10^4 times larger there than in the normal state. The MCT explains this increase of τ by several orders of magnitude and implies that the evolution of structural relaxation in simple liquids occurs within the GHz-window for $T > T_c$. Conventional cooling experiments on glass forming liquids like CKN probe the dynamics with time scales t_{exp} of $10^2 - -10^4$ seconds; t_{exp} is 14 to 16 orders of magnitude larger than the scale t_0 for microscopic motions. Therefore the calorimetric glass transition temperature T_g is below T_c in those cases. For colloidal suspensions hopping effects are so small, that they cannot be measured. The increase of τ_α over 5 orders of magnitude can be observed for $n < n_c$ and the value n_g for the calorimetric glass transition cannot be distinguished from n_c. In computer experiments quenching rates are so large, that t_{exp} differs from t_0 by a few orders of magnitude only; therefore T_g can be larger than T_c in those situations.

The structural arrest is explained as result of non-linear interaction effects between density fluctuations. The arrest requires that the coupling constants for the interactions exceed certain critical values. Therefore the picture of a static potential landscape makes sense only for sufficiently low temperatures. Upon heating potential fluctuations smear out; the Debye-Waller factor f_q decreases as known for crystalline solids. However, there appears the critical temperature T_c, where the specified picture looses its justification. If one ignores the influence of hopping events on the dynamics, the frozen structure breaks down abruptly at T_c. As precursors of the melting of the ideal glass state, there appear dynamical anomalies like the critical decay process. Temperature T_c is the highest temperature for which the concept of a static potential landscape can be used.

The temperature T_c marks the cross over of transport mechanisms in liquids in agreement with an earlier conjecture by Goldstein [33]. In the normal and moderately supercooled state for $T > T_c$, relaxation is dominated by interaction effects among density fluctuations. In the strongly supercooled state for $T < T_c$, hopping events are the elementary steps leading to complex relaxation. Here is a separation of two time scales: the life time of the cages τ_α is much longer than the time for a jump over the barrier. This happens because there are two kernels ruling the dynamics; and the former time is set by m while the latter is related to δ. Near T_c all functions cross over smoothly from the ones reflecting the $T \gg T_c$ physics to ones reflecting $T \ll T_c$ physics. The cross over can be very rapid for some quantities like the viscosity.

The cross over near T_c is caused by a GTS, by the simplest one possible within the MCT. At $T = T_c$ the coupling constants, which control the dynamics, are closest to the critical values, characterizing the singularity. If one idealizes the picture by ignoring hopping effects, the system would be at the GTS for $T = T_c$. In such a simplified picture the spectra would exhibit true singularities for $\omega \to 0$. Acknowledging the possibility for hopping, the system cannot be driven at the GTS by changing the temperature; but it can be close to it. Therefore the spectra follow the singular ones only for certain frequency windows which exclude the $\omega = 0$ value. The GTS is the origin for the evolution of the structural relaxation phenomena, in particular for the α-process. Structural relaxation processes appear already for $T > T_c$. In this case the parameters are off the singularity and hopping effects are not essential there. The evolution can be understood already within the simplified MCT.

There are always some features of the dynamics, which are not strongly influenced by hopping effects, for example the mesoscopic critical spectrum. For T near and below T_c there are some features, which are dominated by hopping and which cannot even qualitatively be understood within the simplified MCT; an example is the viscosity η or the half width Γ_q of the Mountain peak. The frequency interval for the critical susceptibility spectrum (26b) is largest at the cross over temperature. This spectrum results from the cage effect, described by kernel m. The low frequency cut off causes the minimum of the χ'' versus ω curve; and for $T \leq T_c$ it is a result of the hopping effects, described by kernel δ. The dynamics within a frequency window around the minimum for $T = T_c$ is governed by a detailed balance between cage effects and hopping effects. This minimum for temperatures near the cross over T_c cannot be described within the ideal MCT, opposed to the minimum within the moderately supercooled state.

The aim of the MCT is the detailed description of the low frequency ($\omega t_0 \ll 1$) or long time ($t \gg t_0$) dynamics for temperatures or densities near the cross over parameters T_c or n_c, respectively.

6 UNIVERSALITY FEATURES

It was not yet possible to identify the range of parameter values where the MCT is applicable for a description of structural relaxation. At present it seems therefore more rewarding to concentrate on comparisons of MCT results with experimental data. A prerequisite for such comparisons is an understanding of the solutions of the MCT equations of motion.

The relevance of the MCT for a description of structural relaxation can easily be recognized by looking at some of the numerical solutions. The $\phi_q(t)$ versus $\log t$ graphs e.g. in fig. 1 exhibit α-relaxation stretching similar to many photon correlation data. The α-peaks for the calculated susceptibility spectra e.g. in fig. 2 exhibit stretching in the same clear manner as for example the absorptive parts measured for acoustic moduli. Logarithmic decay as shown e.g. in fig. 4 is often observed in photon correlation spectroscopy data for polymers; and flat susceptibility spectra as shown e.g. in fig. 5 are also quite common for those systems. But producing a numerical solution and noticing agreement with some experiments does not necessarily provide an understanding. There are also the following questions. Why should a MCT solution for a Lennard-Jones liquid be of relevance for the interpretation of polymer spectra? Why can a solution evaluated for some schematic model for few correlators be compared with light scattering data obtained for a system as complicated as the molten salt CKN? What are the dynamical features shared between a hard sphere colloidal suspension and a molecular liquid like Salol? Our discussion of the basis of the MCT for structural relaxation shall be concluded by some comments on the solution of the formulated equations of motion. Thereby it can be indicated why the formulated questions have answers which justify, for example, the quantitative analysis of CKN scattering data with MCT formulae.

It was explained above, that structural relaxation appears because the cage effect enforces partial arrest of the density fluctuations $\phi_q(t)$ near the plateau f_q^c. To exploit this information one can introduce $\delta\phi_q(t) = \phi_q(t) - f_q^c$ and reformulate the equations of motion to ones for $\delta\phi_q(t)$. From sections 4.2 and 5.2 one infers, that there is a dynamical window $t_0 \ll t \ll \tau_\alpha$, where $\delta\phi_q(t)$ is small. This holds, provided the various coupling constants are close to the critical ones, specifying a glass transition singularity. The dynamics within the indicated window is called

β-dynamics. One can perform an asymptotic analysis for the dynamics within the β-relaxation window. Such a mathematically well defined asymptotic solution for $\delta\phi_q(t) \to 0$ is the starting step of the analytical work on the implications of the MCT equations of motion. One finds that $\delta\phi_q(t)$ factorizes in two terms. The first factor is the critical amplitude h_q introduced in section 4.2. It is time independent and varies smoothly with wave vector q. The second factor, called β-correlator G, is independent of q. It depends on time t and varies sensitively with control parameters.

$$\phi_q(t) - f_q^c = h_q G(t) + O_q(G^2) \; . \tag{36a}$$

As a result, the problem of solving the coupled equations for the many functions $\phi_q(t)$, $\phi_q^T(t)$, $q = 1, 2, \ldots, M$ within the β-regime can be reduced to deriving and solving a single equation for one function G.

A digression on generalizations of (36a) seems useful. Let us consider some variable A which couples to density fluctuation products $\rho_1\rho_2\cdots$ in the same manner as was discussed in section 3.1 for the fluctuating forces. By coarse graining one can split off a white noise background term γ_A from the A-correlator: $\phi_A(\omega) = i\gamma_A + \varphi_A(\omega)$. This formula is the analogue of (13). The part φ_A deals with structural relaxation. By means of Kawasaki's factorization approximation it can be expressed as some mode coupling functional: $\varphi_A(t) = \mathcal{F}_A(\phi_k(t))$. This formula generalizes the expressions (17). Substituting (36a) in $\mathcal{F}_A$ one arrives at the expansion

$$\phi_A(t) - f_A^c = h_A G(t) + O_A(G^2) \; . \tag{36b}$$

Here $f_A^c = \mathcal{F}_A(f_k^c)$ is the plateau for the structural relaxation of variable A and the corresponding critical amplitude, $h_A = \sum_p [\partial\mathcal{F}_A(f_k)/\partial f_p] h_p$, is given by the Taylor coefficients of $\mathcal{F}_A$. Therefore the β-correlator G determines also the β-dynamics of variable A. The result (36b) is used, for example, for the interpretation of light scattering data [4]. In this case the measured depolarized photon scattering is caused by dipole induced dipole interaction. The cross section is a spectrum $\phi_A''(\omega)$, where in lowest approximation A is a combination of density fluctuation pairs $\rho_{\vec{q}-\vec{k}}\rho_{\vec{k}}$. Let us emphasize that usually a more sophisticated reasoning has to be applied in order to derive formulae like (36b), as can be inferred from a more detailed derivation of light scatttering cross sections within the MCT [20]. It is not clear, for example, under which circumstances structural relaxation can be probed by dielectric loss spectroscopy in the same direct manner as it can by neutron scattering or by dipole induced dipole scattering of light quanta.

The β-process starts with the critical decay (26, 27) as soon as the transient motion has disappeared. This part of the decay process is insensitive to variations of control parameters. The window for the critical decay extends up to some time τ_β. This scale τ_β and the correlator G for $t \geq \tau_\beta$ depend sensitively on control parameters like temperature T and density n. The dynamics for $t > \tau_\beta$ eventually leads to a decrease of $\phi_q(t)$ below the critical non-ergodicity parameter f_A^c. Hence the final stage of the β-process is the initial part of the α-decay. The short time transient dynamics does not belong to the β-regime, since $\delta\phi_q(t)$ is not small there: $\delta\phi_q(t \to 0) = 1 - f_q^c$. Similarly, the α-process for $t \gg \tau_\alpha$ is not included in a description by (36a), since $\delta\phi_q(t)$ is not small there either: $\delta\phi_q(t \to \infty) = -f_q^c$.

The space dependence of density fluctuations within the β-window is uncorrelated with the time dependence of the fluctuations, since there is the product representation (36a). This means, that β-dynamics deals with localized motion, a very clear manifestation of the cage effect. Such result does not hold for the dynamics within the microscopic window $t \leq t_0$, where phonon propagation or diffusion is observed. Let us add, that neither the full α-process, described by the solutions of (15), deals with localized dynamics. The initial part of the α-process for $\tau_\beta \ll t \ll \tau_\alpha$, is described by (36a) and reflects the localized motion of the particles in cages. But for $t \gg \tau_\alpha$ the particles can escape from the traps; and this can lead, for example, to the propagation of diffusion modes. Light scattering, caused by dipole induced dipole interaction, probes the same spectrum $h_A G''(\omega)$ as neutron scattering, which probes $S_q h_q G''(\omega)$. This holds up to factors h_A and $S_q h_q$ respectively. These factors quantify the projection of the probing variable on the localized relaxation modes. The comments of this paragraph exemplify, that MCT solutions can contribute to a physical interpretation of structural relaxation and that results with predictive power can be obtained.

The equation for the β-correlator is different for different glass transition singularities. Here only the canonical A_2-singularity shall be considered, which is relevant for the $T_c - n_c$ cross over phenomena. One gets

$$- \delta \cdot t + \sigma + \lambda G(t)^2 = (d/dt) \int_0^t G(t-t')G(t')dt' \;, \tag{37}$$

together with the initial condition $G(t)(t/t_0)^a \to 1$ for $t \to 0$. Here t_0 is the short time matching scale, λ is the exponent parameter, and a is the critical exponent as introduced in section 4.2. There are straightforward but lengthy formulae for the evaluation of σ and δ from the mode coupling functionals. The symbol σ denotes a smooth function $\sigma(\vec{V})$ of the cage effect control parameters; like f_q^c, h_q and λ also σ is an equilibrium quantity. The symbol δ denotes a non-negative function depending smoothly on all parameters of the model. The quantity δ vanishes if and only if all coupling coefficients V' in (35) are set to zero, which is equivalent to ignoring hopping effects. Thus the β-relaxation theory of the ideal MCT is obtained by specializing (37) to $\delta = 0$.

The solution G of (37) depends sensitively in a very subtle but well understood manner on three dimensionless arguments $t/t_0, \sigma$ and $\delta \cdot t_0$. The solution depends smoothly on λ. The solutions produce the critical fractal decay (26a) and also the von Schweidler fractal decay (4a). Computer programs have been developed to evaluate $G(t)$ versus $\log t$ and $G''(\omega), G'(\omega)$ versus $\log \omega$ tables. Understanding the solutions of (37) and of the solutions for the β-correlator G for the other singularities A_ℓ, $\ell \geq 3$, is the basis for deriving all so far identified general results of the MCT.

Let us stress the great simplifications exhibited by (37) in comparison to the original equations of motion (8, 9, 34, 35). The short time motion or the spectra within the microscopic excitation band are determined by many parameters like Ω_q, γ_q; but all these numbers enter the β-dynamics explicitly only via the matching scale t_0. Except for this single number t_0, the model (18a), which exhibits oscillatory excitations, and the model (18b), which deals with completely monotone functions, exhibit the same β-dynamics.

The cage effect comes about because an infinite number of pair modes $\rho_{\vec{q}-\vec{k}}\rho_{\vec{k}}$ couples to the fluctuating force $F_{\vec{q}}$; and the many constants $V(\vec{q}, \vec{k}\vec{p})$ quantify in (17) this coupling. Yet these coupling constants enter the β-dynamics explicitly only in form of a single combination σ. Within the ideal MCT the states $\sigma < 0$ and $\sigma > 0$ specify liquid and ideal glass states respectively, and the transition point is given by $\sigma = 0$. If the system is driven by changing the temperature, the coupling constants $\vec{V}$ change smoothly with T. Thus one can write $\sigma = -C\epsilon + O(\epsilon^2)$ with $\epsilon = (T - T_c)/T_c$ and $C > 0$. If the transition is driven by density changes one gets $\sigma = \hat{C}\hat{\epsilon} + O(\hat{\epsilon}^2)$ with $\hat{\epsilon} = (n - n_c)/n_c$ and $\hat{C} > 0$. The details of the microscopic structure determine the numerical values of $T_c, n_c, C, \hat{C}$; but otherwise the details are not relevant.

Hopping effects depend on phonon frequencies, friction constants and on the fluctuating free energy barriers. Yet all these many microscopic details condense to a single number δ. It is only this frequency δ which specifies the influence of phonon assisted hopping on the β-decay. Consequently, the complete sensitive dependence of the β-dynamics on control parameters is governed by the two parameters σ and δ, which are therefore called relevant control parameters. The parameter σ quantifies the distance from the transition and it is therefore called separation parameter. The number $\delta \cdot t_0$ is called hopping parameter.

Equations (36b, 37) formulate a law of corresponding states. Spectra agree if they are measured in units of h_A and if frequencies are measured in units of $(1/t_0)$, provided they agree in the pair of relevant control parameters $(\sigma, \delta t_0)$. Microscopic information enters via the scales only, which depend smoothly on, for example, the pair potential. The results are robust in the sense that changes of model properties or changes of densities merely induce smooth changes of constants like C, T_c and δ. The β-dynamics is determined quantitatively by the set of numbers $f_A^c, h_A, t_0, \lambda, \sigma$ and δ and none of these numbers changes very much if one drives the system from one thermodynamic state to the other, even if one changes from one system to some other. However, the spectra $\phi''_A(\omega)$ change dramatically; in particular the time scales τ_α, τ_β specifying various features of the dynamics vary over several orders of magnitude, as is exemplified by figs. 1–3. These strong variations and changes are described by the solution of (37). MCT provides an explanation of structural relaxation in the sense that complicated spectra or decay functions, which vary with T or n drastically, are reduced to trivial and small $T-n$-variations of microscopically well defined quantities like σ and δ.

It is the aim of singularity theory to identify robust results of the kind explained above and to identify the number of parameters quantifying the robust features. With $G(t)$ a universality class is identified, parametrized by λ. The other singularities A_ℓ, $\ell \geq 3$, define other classes. A class describes a certain dynamical scenario. If one understands all scenarios one understands the underlying theory, here the MCT.

The analytical work is not completed by deriving (36,37) and explaining function G, however. One wants, for example, to understand also the leading contribution to the correction terms $O_A(G^2)$. Thereby one can get an explanation, why for example the window for the β-process is small for one observable A while it is large for some other. It can happen, that one singularity is close to some other, as is exemplified by the elementary models (21). Then the disturbance of a leading order result of one singularity by the nearby other singularity can be large and the range of validity of asymptotic formulae like (36) can be very small. A useful analysis requires then the simultaneous handling of two singularities.

The bifurcation hypersurface can have self-intersecting pieces, as is the case already for the transition lines of the elementary model (21d). Near such crossing points the decay of $\phi_q(t)$ has the form of a cascade, which is governed by several plateaus $f_q, f'_q, \ldots$. As a result one gets a susceptibility spectrum, which exhibits several structural relaxation peaks.

A particular intricate phenomenon is the following. Let us denote by $X = (X_1, X_2, \ldots)$ all the mode coupling coefficients entering the equations of motion. Let us imagine, that these can be split in two subsets $X = (U, W)$. Let us furthermore restrict our view by considering the solutions as function of the K-dimensional control parameter vector $U = (U_1, \ldots, U_K)$ by specializing the theory to $X = (U, 0)$. There may appear some bifurcation singularity for $U = U^c$, which is absent whenever $W \neq 0$. Unless there is some symmetry which enforces the vanishing of the L coupling constants $W = (W_1, \ldots, W_L)$, one would be inclined to ignore the specified singularity $X^c = (U^c, 0)$ as non-generic. On a first glance one would see no reason to worry about the accidental vanishing of L coupling constants. It would be a mathematical idealization, which defines some simplified model only, if one restricts the theory to $W = 0$; for all normal choices the system would never be at the singularity X^c. However, it is a generic situation, that W is small. In such case the solutions will exhibit rapid variations of the dynamics as function of changes of control parameters, if the latter shift U through U^c. There appear new features of the dynamics, which can be understood as result of a new universality class, caused by the singularity X^c. An asymptotic solution for the limit $U \to U^c, W \to 0$ can characterize the dynamics by a set of well defined parameters, thereby establishing an understanding. A useful first step towards a discussion of the singularity X^c would be the study of a simplified model, defined by putting $W = 0$. An example illustrating the described phenomenon was discussed recently within the ideal MCT [34]. It deals with the problem of a liquid to glass transition for a subsystem occurring in the ideal glass state of a matrix. Indeed, our non-linear equations of motion contain an irritating variety of scenarios for complex relaxation; but so do experiments describing the spectra of glass forming systems.

Also the extended MCT exemplifies the scenario discussed in the preceding paragraph. One can use the coupling constants for the cage effect as the first parameter group U; in sections 3, 4 these were denoted by $\vec{V}$. The second group W consists of the vertices V' entering the mode coupling functionals (35) for the hopping effects. By changing the temperature T one probes generic results of the theory, and the decay functions $\phi_q(t)$ or spectra $\chi''_q(\omega)$ exhibit no singularity. Transport coefficients like the viscosity η vary smoothly with T, in particular they are finite for $T = T_c$. An Arrhenius plot shows a smooth cross over from some activation barrier for $T > T_c$ to some larger one for $T < T_c$. A few parameter fit formula can be invented to interpolate the $\log \eta$ versus $1/T$ graph. There appears some α-peak, which can be fitted reasonably by formula (2). There appears also some mesoscopic spectrum, which cannot be understood as part of microscopic excitation processes and does not fit to the familiar results for the α-process.

An understanding of the scenario above can be obtained in the conventional sense, that a large number of phenomena is reduced to a simple but subtle origin, namely a universality class caused by some GTS. This singularity cannot be identified compellingly from any figure, say a $\phi_q(t)$ versus $\log t$ or a $\chi_q''(\omega)$ versus $\log \omega$ graph. One cannot use a set of experimental data as proof or disproof of the existence of a singularity. A singularity is a theoretical concept, which is used here as in any other field of physics to summarize precisely a certain number of statements, in particular statements on relations between measurable quantities.

The MCT results for the hard sphere system can be compared in detail with the experimental data for hard sphere colloids [5]. Such comparison leads to an assessment of the theoretical results on f_q, h_q, λ, n_c and on the complete form of $G(t)$. However, all other experiments on structural relaxation are done for more complicated liquids. The choice of complicated system is necessary in order to bypass crystallization. It is therefore desirable to generalize the theory.

It is easy to generalize MCT to r–component mixtures of spherical molecules. Such systems are of direct interest for the interpretation of molecular dynamics experiments. Essentially, the theoretical generalizations amount to replacing functions like correlators ϕ, kernels M or frequencies Ω by $r \times r$ matrices. Mode coupling theories for non–spherical particles have been studied for molecular crystals and one can anticipate, that those results can be modified so, that van der Waals liquids like Salol can be handled. Beginning with the work of Hess [35], mode coupling theories have been derived also for polymers. All the approximate equations of motion derived in the indicated work contain the cage effect mode coupling functional, which we have identified as origin of glass transition singularities. Thus these more complicated theories can lead to structural relaxation for certain choice of parameters.

It is however much more difficult to discuss the solutions of MCT equations of motion for complex liquids than for the simple liquids considered up to now. For example, there is no general procedure known, to identify the Edwards–Anderson parameter among the many solutions of the matrix generalization of the bifurcation equation (22). So far one could not prove, that all bifurcations are cuspoids A_ℓ. Therefore one cannot exclude that the MCT for complex systems yields scenarios which do not lead to the factorization theorem (36), etc. But all the singularities identified for the simple models occur also for the complicated ones. One can examine, for example by studying (22) numerically, whether the system can be driven towards some A_2–bifurcation. If this is the case, the known theory can be used to derive (36), where only the non–ergodicity parameters f_q^c and amplitudes h_q have to be generalized. Equation (37) holds for the general models as it does for the simple ones. Such discussion was carried out in all numerical detail for a binary mixture [36]. The work for other systems remains to be done.

The MCT provides scenarios for the dynamics of disordered matter. The scenarios characterize structural relaxation for parameters near some GTS. Every type of singularity yields a universality class for the dynamics i.e. a scenario for the evolution of decay curves and spectra due to changes of temperature, density and other control parameters. One understands why colloids, polymers and CKN can exhibit the same dynamical pattern as schematic models. The specific details which discriminate a polymer from a hard sphere system do not enter the results formulated by (36, 37); they merely enter the smoothly varying scales like h_A or σ. One can look whether CKN light scattering data exhibit the T_c–cross over scenario described by the solution of (37). In case one can test quantitatively, whether the various predicted relations between measurable features are valid. One alternative would be, that the data violate for example the predicted relation between critical exponent a and von Schweidler exponent b. This would eliminate the MCT as a theory for structural relaxation in CKN.

A second alternative would be, that data can be understood qualitatively within a MCT scenario and that quantitative results are compatible with the predicted ones. Then one could continue and ask why a system like CKN exhibits $T_c \sim 370K$ and not $260K$ like Salol. One would like to understand, why the Debye–Waller factor f_q of the hard sphere system exhibits a strong maximum near the peak position q_0 of the structure factor while CKN exhibits only a weak one near $q_0/2$. One wonders why the measured exponent parameter λ for CKN is so much bigger than the one measured for the hard sphere colloids. One would like to explain why the CKN spectra can be interpreted better within the simplified set of formulae ignoring δ, than what is possible for Salol. One can formulate and appreciate these questions on the basis of the work done so far. But answering these questions will require a much better understanding of the microscopic structure details of complex liquids than is available at present.

ACKNOWLEDGEMENTS

We thank Alf Sjölander for numerous stimulating discussions and many helpful comments on the manuscript.

References

[1] E. Leutheusser, Phys.Rev. **A 29**, 2765 (1984).

[2] U. Bengtzelius, W. Götze, and A. Sjölander, J. Phys. C: Solid State Phys. **17**, 5915 (1984).

[3] W. Götze and L. Sjögren, Z. Phys. **B 65**, 415 (1987).

[4] G. Li, W.M. Du, X.K. Chen, H.Z. Cummins, and N.J. Tao, Phys.Rev. **A 45**, 3867 (1992).

[5] W. van Megen and S.M. Underwood, Phys.Rev.Lett. **70**, 2766 (1993) - Phys.Rev. **E49**, 4209 (1994).

[6] *Slow Dynamics in Condensed Matter*, APS conference proceeding 256, K. Kawasaki et al. eds., American Institute of Physics (1992).

[7] *Proceedings of the International Workshop on Dynamics of Disordered Materials II*, A.J. Dianoux et al. eds., Physica **201** (1993).

[8] *Proceedings of the International Discussion Meeting on Relaxation in Complex Systems*, K.L. Ngai et al. eds., J. Non-Cryst. Solids **172–174**, (1994).

[9] W. Götze and L. Sjögren, Rep. Prog. Phys. **55**, 241 (1992).

[10] L. Sjögren and W. Götze, J. Non-Cryst. Solids **172–174**, 7 (1994).

[11] W. Götze and L. Sjögren, J. Non-Cryst. Solids **172–174**, 16 (1994).

[12] J.P. Boon and S. Yip, *Molecular Hydrodynamics*, (McGraw-Hill, New York, 1980).

[13] J.P. Hansen and I.R. McDonald, *Theory of Simple Liquids*, 2nd edn. (Academic Press, London, 1986).

[14] R.D. Mountain, J. Res. Nat. Bur. Standards **70A**, 207 (1966).

[15] K. Kawasaki, Phys.Rev. **150**, 291 (1966).

[16] L. Sjögren, Phys.Rev. **A 22**, 2866 (1980).

[17] T. Munakata and A. Igarashi, Prog. Theor. Phys. **60**, 45 (1978).

[18] U. Bengtzelius and L. Sjögren, J. Chem. Phys. **84**, 1744 (1986).

[19] W. Götze and A. Latz, J. Phys.: Condensed Matter **1**, 4169 (1989).

[20] M. Fuchs and A. Latz, J. Chem. Phys. **95**, 7074 (1991).

[21] W. Götze and L. Sjögren, preprint (1993).

[22] R.P. Feynman and M. Cohen, Phys.Rev. **102**, 1189 (1956) - Phys. Rev. **107**, 13 (1957).

[23] W. Götze and M. Lücke, Phys.Rev. **B 13**, 3825 (1976).

[24] R.I. Curkier and J.R. Mehaffey, Phys.Rev. **A 18**, 1202 (1978).

[25] V.A. Arnold, *Catastroph Theory*, 2nd edn. (Springer, Berlin, 1986).

[26] S.F. Ewards and P.W. Anderson, J. Phys. F: Met. Phys. **5**, 965 (1975).

[27] W. Knaak, F. Mezei, and B. Farago, Europhys. Lett. **7**, 529 (1988).

[28] N.J. Tao, G. Li, and H.Z. Cummins, Phys.Rev. Lett. **66**, 1334 (1991).

[29] Y. Pomeau and P. Resibois, Phys.Rep. **19C**, 63 (1975).

[30] G.D. Mahan, *Many Particle Physics*, (Plenum Press, New York, 1981).

[31] D. Belitz and W. Schirmacher, J. Phys. **C 16**, 913 (1983) - J. Non-Cryst. Solids **61**, 1073 (1984).

[32] P.N. Pusey, *Liquids, Freezing and the Glass Transition*, J.P. Hansen et al. eds. (North-Holland Publ. Comp., Amsterdam, 1991) 763.

[33] M. Goldstein, J. Chem. Phys. **51**, 3728 (1969).

[34] T. Franosch and W. Götze, J. Phys.: Condensed Matter **6**, 4807 (1994).

[35] W. Hess, Macromolecules **19**, 1395 (1986) - **20**, 2587 (1987) - **21**, 2620 (1988).

[36] M. Fuchs and A. Latz, Physica **A201**, 1 (1993).

Received: October 25, 1994
Accepted: January 16, 1995

Light Scattering Spectroscopy of Orthoterphenyl

—— *Idealized and Extended Mode Coupling Analysis* ——

H. Z. CUMMINS, Gen LI,*) Weimin DU,**) Y. H. HWANG and G. Q. SHEN

Department of Physics, City College of the City University of New York New York 10031, U.S.A.

Depolarized light scattering spectra of orthoterphenyl were analyzed with both the idealized and extended versions of mode coupling theory. The idealized MCT analysis gave $\lambda = 0.72$ ($a = 0.318$) and $T_C \sim 290$ K, in agreement with previous neutron and light scattering results. The extended MCT analysis gave $\lambda = 0.76$ ($a = 0.30$) and $T_C \sim 276$ K, 14 K lower than the idealized result. For the ten spectra at $T \geq 320$ K where α-peaks were observed, direct KWW fits gave $\beta_K \sim 0.78$ with a slight trend of decreasing β_K with increasing temperature. Fits to $\tau_\alpha(T)$ gave $T_C \sim 274$ K, close to the extended MCT result. Finally, in an effort to resolve the ambiguity in α-peak fits due to overlap of the α and β spectral regions, we performed a combined $\alpha - \beta$ fit to six spectra for $T \geq 320$ K. The results suggest a temperature-independent $\beta_K = 0.79$.

§1. Introduction

Orthoterphenyl (OTP, $T_m = 329$ K, $T_g = 244$ K, M.W. = 230.31) is a fragile molecular glassforming material that has been studied extensively with a wide range of experimental techniques. Neutron scattering studies of OTP by Petry, Sillescu, and coworkers [1)-6)] revealed many of the dynamical features predicted by mode coupling theory (MCT), including two-step relaxation with critical exponents $a = 0.3$ and $b = 0.54$ (exponent parameter $\lambda = 0.76$), a crossover temperature $T_C = 293$ K, and time-temperature superposition of the α-relaxation dynamics at $T > 290$ K for both single-particle motion and for collective motion at the first maximum in $S(Q)$.

The dynamics of OTP have also been investigated with depolarized light scattering spectroscopy by Steffen and Patkowski and their coworkers. [7)-12)] In Ref. 9), $\theta = 90°$ depolarized spectra were converted to susceptibility spectra $\chi''(\omega)$ and fit to the MCT interpolation equation

$$\chi''(\omega) = \frac{\chi''_{\min}}{a+b}\left[b\left(\frac{\omega}{\omega_{\min}}\right)^{a} + a\left(\frac{\omega}{\omega_{\min}}\right)^{-b}\right]. \tag{1·1}$$

The fits gave $T_C = 290$ K, $a = 0.33$ and $b = 0.65$ ($\lambda = 0.7$), but the scaling prediction of the idealized MCT (I-MCT) were only partially satisfied. In Ref. 12), a detailed I-MCT analysis was presented which again gave $T_C = 290$ K, in good agreement with neutron scattering results.

In Ref. 7), a different approach was followed in which the intensity spectra were fitted with two Lorentzians representing a strongly temperature-dependent slow (α) process and a nearly temperature-independent fast process. In Ref. 8), the narrow

*) Present address: Ernst and Young LLP, 125 Chubb Ave., Lyndhurst, NJ 07071, U.S.A.

**) Present address: Beijing Dong Dan, Tai Ji Chang, Tao Tiao #2, Beijing, China 100005.

component was attributed to molecular rotations, while the broader component was attributed to dipole-induced-dipole interactions. This interpretation is related to a similar separation of the orientational and collision-induced components reported for the light-scattering spectrum of CS_2.[13)]

In previous light scattering studies of the glassformers CKN,[14)] salol[15)] and PC,[16)] it was found that $\chi''(\omega)$ data could be interpreted reasonably well with the I-MCT theory, but the fits were improved significantly by using the extended version of mode coupling theory (E-MCT) which includes activated hopping processes represented by the temperature-dependent hopping parameter $\delta(T)$.[17),18),16)] [For these materials $\delta(T)$ was found to approximately follow an Arrhenius temperature dependence.] Also, it was found that the stretching coefficient β_K did not increase with T, remaining approximately constant for all $T > T_C$.

In this paper we present an analysis of OTP depolarized light scattering spectra based on both the idealized and extended versions of MCT. While the spectra are understood to arise physically from a combination of orientational fluctuations and DID scattering, we will not attempt to analyze the data in terms of superpositions of different functions but will present fits to the asymptotic formulas provided by MCT in order to test the ability of MCT to explain our data.

The experiments are described briefly in §2; the idealized MCT analysis of the β-relaxation region is presented in §3. In §4 we describe the extended MCT analysis of the β-relaxation region. In §5 we first analyze the α-peaks directly with Kohlrausch (KWW) functions, and then describe a combined $\alpha - \beta$ analysis. Our summary and conclusions are given in §6.

§2. Depolarized light scattering spectroscopy

Depolarized near-backscattering spectra ($\theta = 173°$) of OTP were collected with a six-pass Sandercock tandem Fabry-Perot interferometer and with a Spex 1401 grating spectrometer. The apparatus and the experimental procedures are described in Refs. 14) and 15). Two samples were studied. One sample was prepared in Mainz by W. Steffen as described in Ref. 7). The second was prepared in New York from 99% OTP purchased from Aldrich, loaded into a 15 mm diameter glass sample tube without further purification, and placed under vacuum overnight before flame sealing. The spectra obtained from the two samples were indistinguishable.

Composite susceptibility spectra $\chi''(\omega) = I(\omega)/[n(\omega) + 1]$ found from the superimposed intensity spectra of the Aldrich sample at temperatures from 225 K to 435 K as described in Ref. 19) are shown in Fig. 1(a). At low temperatures, LA peaks near 15 GHz due to leakage of the intense Brillouin components through the polarizers become increasingly dominant. The height of the α-peak remains essentially constant while its frequency changes rapidly with temperature.

Depolarized OTP spectra were obtained by Steffen et al. in right angle scattering ($\theta = 90°$) with samples prepared by recrystallization and vacuum distillation as described in Ref. 12). The resulting $\chi''(\omega)$ spectra are shown in Fig. 1(b). Despite the visible differences in the spectra of Figs. 1(a) and 1(b), comparing individual spectra for the same temperature for the two sets indicates quite good agreement,

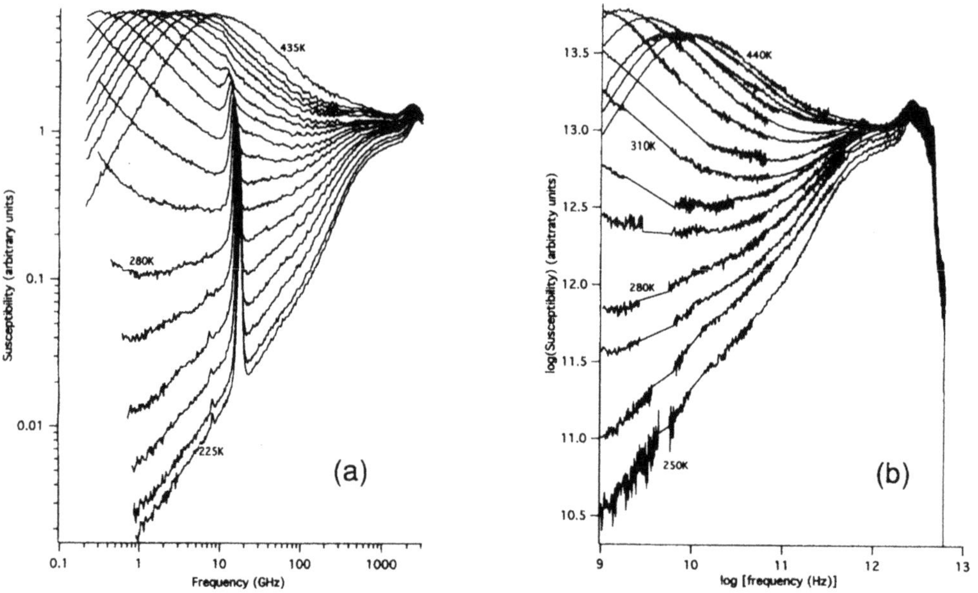

Fig. 1. (a) Composite OTP $\theta = 173°$ VH $\chi''(\omega)$ spectra (NY) at $T = 225$, 240, 250, 260, 270, 280, 290, 300, 310, 320, 330, 340, 350, 360, 370, 380, 395 and 435 K. (b) Composite OTP $\theta = 90°$ VH $\chi''(\omega)$ spectra (Mainz) at $T = 250$, 260, 270, 280, 290, 300, 310, 320, 340, 360, 380, 400, 420 and 440 K. [The Brillouin components in (b) have been removed.]

apart from the different positions of the Brillouin peaks due to the different scattering angles and an overall scale factor.

In this paper we will present an MCT analysis of the data in Fig. 1(a). A comparison of MCT analyses of the two data sets will appear in a future publication.

§3. Idealized MCT analysis of the β-relaxation region

The susceptibility spectra shown in Fig. 1(a) for $T \geq 300$ K were fit to the interpolation equation (1·1) with a and b constrained by the MCT gamma-function relation [20]

$$\lambda = \Gamma^2(1-a)/\Gamma(1-2a) = \Gamma^2(1+b)/\Gamma(1+2b) , \qquad (3\cdot1)$$

where $0 \leq a \leq 0.4$, $0 < b \leq 1$, and λ is the exponent parameter. Fits to individual spectra gave values of a ranging from 0.31 to 0.33. A single global fit to the five spectra for $T = 300$, 310, 320, 330 and 340 K gave $a = 0.318$ ($\lambda = 0.72$). This global fit is shown in Fig. 2. The fit values of $\chi''_{\min}$ and $\omega_{\min}$ plotted as $(\chi''_{\min})^2$ vs T and $(\omega_{\min})^{2a}$ vs T are shown in Fig. 3. Linear regression fits gave $T_C = 291.2$ K and 288.2 K respectively, in good agreement with previous estimates of $T_C \sim 290$ K. As found in previous studies of molecular glassformers, the fits of Eq. (1·1) to the susceptibility spectra do not extend very far above the minimum, presumably due to interference from the vibrational dynamics (boson peak). Since the asymptotic MCT results deal only with structural relaxation and do not include the microscopic

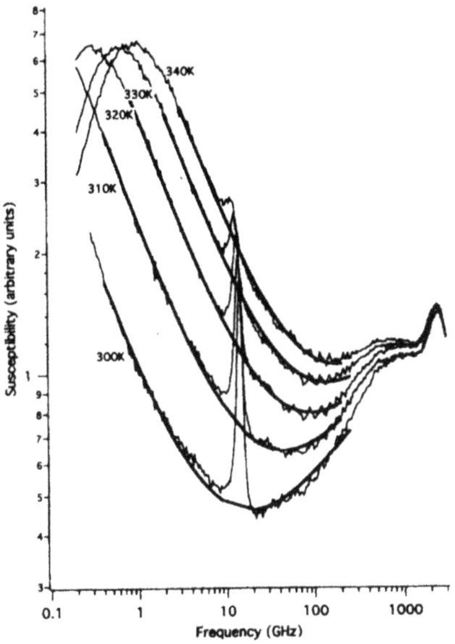

Fig. 2. Interpolation fits [Eq. (1·1)] to the $\chi''(\omega)$ spectra of Fig. 1(a) for $T = 300$, 310, 320, 330 and 340 K (fits are indicated by the heavy lines). The exponents a and b were constrained by Eq. (3·1) and adjusted globally. The fit gave $a = 0.318$, $b = 0.608$, $\lambda = 0.72$.

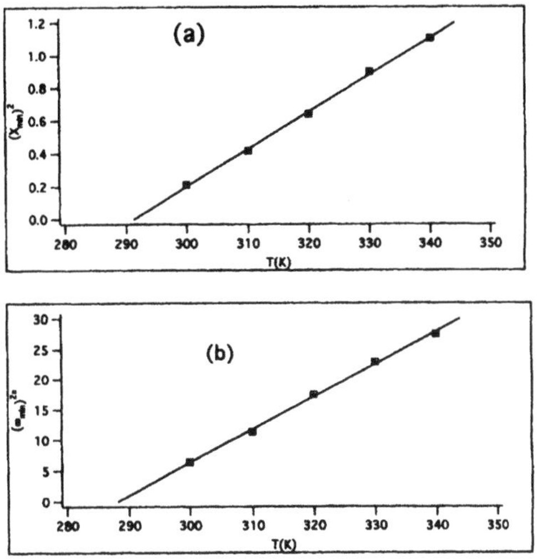

Fig. 3. (a) $(\chi''_{\min})^2$ vs T, and (b) $(\omega_{\min})^{2a}$ vs T with $a = 0.318$ from the global interpolation fits in Fig. 2. Linear regression fits gave $T_C = 291.2$ K and 288.2 K respectively.

dynamics, asymptotic MCT fits should not be extended into the boson peak region above $\sim$ 200 GHz.

In the I-MCT, the susceptibility spectrum $\chi''(\omega)$ obeys a one-parameter scaling relation

$$\chi''(\omega) = h_q|\sigma|^{1/2}\hat{\chi}''_{\pm}(\omega/\omega_\sigma)\,, \tag{3·2}$$

where

$$\omega_\sigma = \omega_0|\sigma|^{1/2a}\,. \tag{3·3}$$

(The microscopic frequency ω_0 is not a fitting parameter. In the computer programs used to solve the equation of motion for the I-MCT beta correlator $G(t)$, ω_0 is fixed at $2\pi \times 10^{12}$ sec^{-1}, so that $\nu_0 = 1{,}000$ GHz.)

The scaling property of Eq. (3·2) implies that $\chi''(\omega)$ spectra at different temperatures, when plotted on log-log plots, can be superimposed by appropriate horizontal and vertical displacements. Such a scaling plot of the $\chi''(\omega)$ spectra for $T = 280$ K through 340 K, together with a $\hat{\chi}''_{-}(\omega/\omega_\sigma)$ master curve for $a = 0.318$ shows that the scaling region expands as T decreases towards T_C as predicted by MCT. This scaling procedure can also be used to obtain estimates of T_C (see Refs. 14), 15) and 12)). However, these estimates are generally less reliable than the global fitting procedure shown in Figs. 2 and 3.

In Fig. 4(a), we show the I-MCT master function $\hat{\chi}''_{-}(\omega/\omega_\sigma)$ (top) and $\hat{\chi}''_{+}(\omega/\omega_\sigma)$ (bottom) for $a = 0.30$. Comparing these curves with the experimental data in Fig. 1, two problems are evident. First, as $T \to T_C$ the frequency of the $\chi''(\omega)$ minimum should theoretically approach zero, and the minimum should disappear at T_C along with the α peak since for $T < T_C$ the theoretical master function $\hat{\chi}''_+$ does not include a minimum. However, the minimum is still present in the experimental data at $T = 280$ K and 270 K, temperatures below the estimated $T_C = 290$ K. Second, the "knee" in the master function $\hat{\chi}''_+$, where there is a crossover from ω^1 to ω^a, is not seen in the $T < T_C$ experimental data. These disagreements, also found for some other materials studied previously, signal the importance of activated hopping processes which are not included in the I-MCT.

§4. Extended MCT analysis of the β-relaxation region

In the idealized MCT, the equations of motion for the normalized density fluctuation correlators $\phi_q(t) = \langle \rho_q(0)\rho_q(t)\rangle / \langle |\rho_q|^2 \rangle$ include nonlinear coupling between each ϕ_q and pairs of modes $\phi_{q_1}\phi_{q_2}$. These nonlinear interactions, which become stronger with increasing density (decreasing temperature), describe the cage effect which causes the drastic slowing down of the α-relaxation as $T \to T_C$. At $T = T_C$, a dynamic ergodic-nonergodic singularity occurs, the α-relaxation "freezes", and the $\hat{\chi}''(\omega)$ master spectra switch abruptly from the $\hat{\chi}''_-$ form to the $\hat{\chi}''_+$ form shown in Fig. 4(a).

In real glass-forming materials (with some exceptions such as colloidal glasses) this glass transition singularity is avoided, and the α-relaxation continues to move smoothly to lower frequencies with decreasing T. This avoided dynamical singularity occurs in the extended MCT which includes, in addition to the dominant nonlinearity of the I-MCT, another nonlinear interaction of coupling to currents $\dot{\phi}_q(t)$ which is formally equivalent to an activated hopping process. The resulting equation of motion for the β-correlator $G(t)$, which describes $\phi_q(t)$ in the plateau region between short-time microscopic dynamics and the α-relaxation process, is

$$\phi_q(t) = f_q^c + h_q G(t) \ , \tag{4·1}$$

where the β-correlator $G(t)$ obeys the equation of motion [20]

$$\sigma + \lambda G^2(t) - \delta t = \frac{d}{dt}\int_0^t G(t-t')G(t')dt' \ . \tag{4·2}$$

Note that the temperature-dependent hopping parameter $\delta(T)$ appears in Eq. (4·2) multiplied by t so that, even though δ may be very small, it will always become dominant at sufficiently long times. [In the I-MCT, $G(t)$ is given by Eq. (4·2) with $\delta = 0$.]

Numerical solutions to Eq. (4·2) with $\lambda = 0.76$ ($a = 0.30$), converted to $\chi''(\omega)$ by Fourier transformation and multiplication by ω, are shown in Figs. 4(b) and 4(c) for several values of δ. For $T > T_C$ {$\chi''_-(\omega)$ [Fig. 4(b)]}, the effect of $\delta \neq 0$ is a steepening on the low-frequency side of the minimum, while for $T < T_C$ [Fig. 4(c)], $\chi''_+(\omega)$ is changed dramatically by the inclusion of $\delta \neq 0$ which restores the minimum along with the α-relaxation process.

In order to fit the $\chi''_{\rm LS}(\omega)$ spectra of Fig. 1(a) using the E-MCT, we followed a two-stage procedure. Equation (4·2) obeys a two-parameter scaling law. If it is solved for $\hat{\sigma} = \pm 1$, then the resulting master function $\hat{\chi}''_{\pm}(\hat{\omega}, \pm 1, \hat{\delta t_0})$ is related to $\chi''(\omega, \sigma, \delta t_0)$ by

$$\chi''(\omega, \sigma, \delta t_0) = |\sigma|^{1/2} \hat{\chi}_{\pm}(\hat{\omega}, \pm 1, \hat{\delta t_0}) \,, \tag{4·3}$$

while χ'' is assumed to be proportional to the light scattering result:

$$\chi''_{\rm LS}(\omega, \sigma, \delta t_0) = h \chi''(\omega, \sigma, \delta t_0) \,. \tag{4·4}$$

In Eq. (4·3) [as in Eq. (3·2)] the scaled frequency $\hat{\omega} = \omega/\omega_\sigma$ (or ν/ν_σ) where $\omega_\sigma = \omega_0 |\sigma|^{1/2a}$, while the scaled hopping parameter $\hat{\delta t_0} = \delta t_0 / |\sigma|^{\frac{1+2a}{2a}}$. (Note that δt_0 is dimensionless since δ has dimensions of $\sec^{-1}$.) In the computer programs used to solve Eq. (4·2), the microscopic time t_0 was arbitrarily set to $t_0 = (1/2\pi) \times 10^{-12}$ sec so that $\nu_0 = 10^{12}$ Hz.

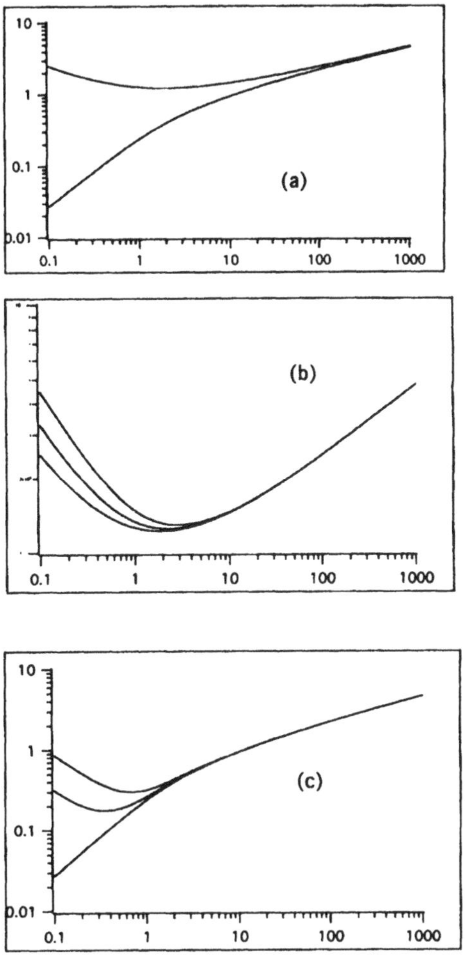

Fig. 4. MCT χ'' master functions for $a = 0.30$. (a) $\hat{\chi}''_{-}(\omega/\omega_\sigma)$ (top) and $\hat{\chi}''_{+}(\omega/\omega_\sigma)$ (bottom) for I-MCT ($\delta = 0$); (b) $\hat{\chi}''_{-}(\omega/\omega_\sigma)$ of the E-MCT with $\delta t_0 = 0$, 0.1 and 0.3 (bottom, middle, top); (c) $\hat{\chi}''_{+}(\omega/\omega_\sigma)$ of the E-MCT with $\delta = 0$, 0.03 and 0.1 (bottom, middle, top).

In the first step, we used the following procedure.[21] For a given value of λ (which fixes a and b), and with $\hat{\sigma} = -1$ (liquid) or $+1$ (glass), the program is run for a series of $\hat{\delta t_0}$ values to generate a set of master functions ($\hat{\chi}''$ vs $\hat{\omega}$) such as those for $\lambda = 0.76$ ($a = 0.30$) shown in Figs. 4(b) and 4(c). The $\chi''_{\rm LS}(\omega)$ spectra shown in Fig. 1(a) are then overlaid with sets of such master functions and displaced along both axes until a reasonable fit is found. With the overlap optimized, $\nu = 1$ GHz on the $\chi''_{\rm LS}$ spectrum corresponds to $\hat{\omega} = C_1$ on the master function, and $\chi''_{\rm LS} = 1$ corresponds to $\hat{\chi}'' = C_2$. (Because $\delta \neq 0$ tends to raise $\hat{\chi}''_{-}(\omega)$ on the low-frequency side, this procedure favors values of a somewhat smaller than that which optimizes the I-MCT fits, as previously found for CKN and salol in Ref. 17).) Then

$$|\sigma| = (1{,}000 C_1)^{-2a} \,. \tag{4·5}$$

For our OTP data optimum fits were found with $a = 0.30$. (The sign of σ is determined by the choice of $\hat{\sigma} = +1$ or -1.) Once σ has been determined, $\delta t_0 = \hat{\delta t_0}|\sigma|^{\frac{1+2a}{2a}}$ and $h = (C_2|\sigma|^{1/2})^{-1}$. This scaling procedure produces approximate values for $\sigma(T)$, $h(T)$ and $\delta t_0(T)$. The results, with $\lambda = 0.76$ ($a = 0.30$) are given on the left side in Table I. The fits and temperature-dependent parameters are shown in Fig. 5(a) and 5(b).

Table I. OTP extended MCT analysis: $a = 0.30$ ($\lambda = 0.76$).

	Master Function Overlay (1)			Nonlinear Least Squares Fits (2)		
T	σ	δt_0	h	σ	δt_0	h
225	0.063	6.3×10^{-7}	0.22	0.0775	7.6×10^{-7}	0.252
240	0.048	8.8×10^{-7}	0.21	0.0695	1.016×10^{-6}	0.290
250	0.033	1.1×10^{-6}	0.28	0.0500	1.44×10^{-7}	0.379
260	0.024	1.42×10^{-6}	0.43	0.0314	1.97×10^{-6}	0.506
270	0.069	1.7×10^{-6}	0.69	0.00887	2.426×10^{-6}	0.673
280	−0.0082	8.1×10^{-7}	0.92	-5.15×10^{-3}	2.376×10^{-6}	0.975
290	−0.030	8.4×10^{-6}	1.22	−0.035	4.39×10^{-7}	1.24
300	−0.067	2.2×10^{-5}	1.48	−0.071	4.8136×10^{-7}	1.445
310	−0.11	8.1×10^{-5}	1.59	−0.115	7.61×10^{-5}	1.55
320	−0.166	2.5×10^{-4}	1.64	−0.156	3.941×10^{-4}	1.64
330	−0.20	3.9×10^{-4}	1.73	−0.171	1.08×10^{-3}	1.84
340	−0.23	5.6×10^{-4}	1.83	−0.213	1.43×10^{-3}	1.86
350	−0.31	1.33×10^{-3}	1.63	−0.289	2.14×10^{-3}	1.70

The second stage of the fitting procedure employs the same computer program used to generate the master functions discussed above, linked to a nonlinear least-squares fitting routine to optimize the fitting parameters λ, σ, δt_0 and h. The exponent parameter λ is treated as a single global parameter, while σ, δt_0 and h are optimized independently for each temperature.

In carrying out this procedure, the fitting range for each spectrum must be specified separately. For the fits shown in Fig. 6, the fitting range was varied in order to test the effect on the resulting parameters. The fits for the lowest five temperatures (225–270 K), carried out globally, gave $\lambda = 0.76$. As shown in Fig. 7, the parameters σ, h and δt_0 for these five fits vary smoothly with temperature. For $T > 270$ K, global fits were unsuccessful, so individual fits were carried out with λ fixed at 0.76. The fitting ranges were chosen arbitrarily. The resulting values for σ and h, shown in Fig. 7, are nevertheless smooth functions of T. However, δt_0 is very sensitive to change in the fitting range and is therefore not reliably determined for fits at $T > 270$ K. This result is not surprising since for temperatures above T_C relaxation dynamics are dominated by the cage effect and the effects of hopping become progressively less important.

From a linear fit to the $\sigma(T)$ values in Fig. 7, we estimate that $\sigma = 0$ at $T_C \sim 276$ K, compared to 290 K obtained from the I-MCT fits. [$\sigma(T)$ found from procedure (1), shown in Fig. 5(b), also indicates $T_C \sim 277$ K.]

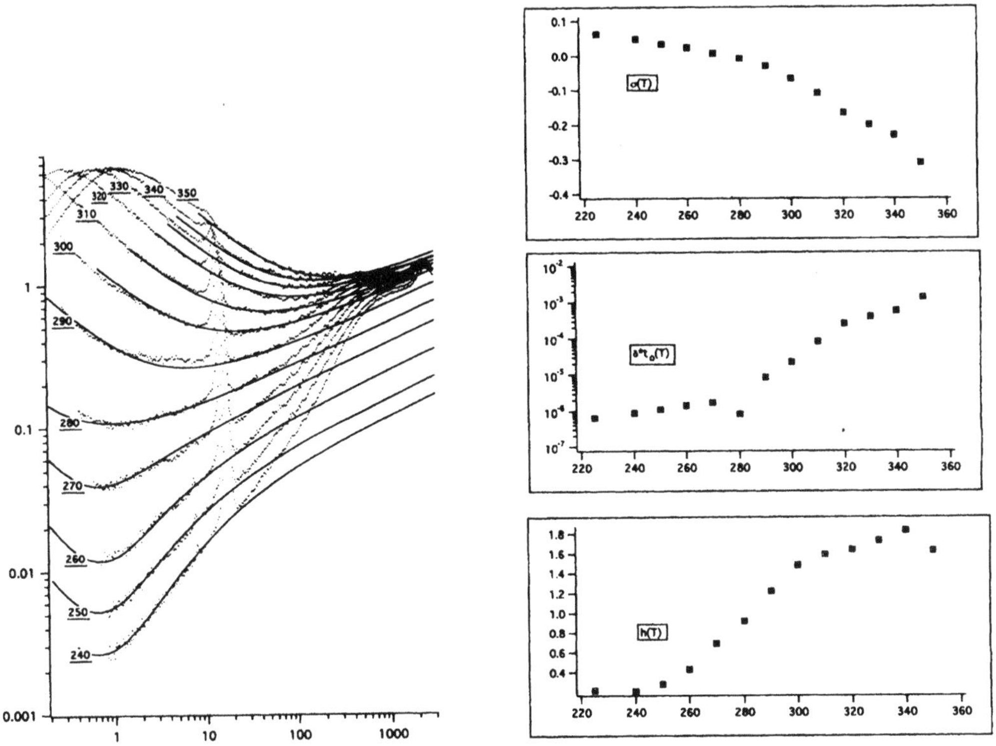

Fig. 5. (a) OTP $\chi''_{\mathrm{LS}}(\omega)$ data fit with E-MCT master functions found with the scaling procedure described in the text. (b) MCT parameters $\sigma(T)$, $h(T)$ and $\delta t_0(T)$ from the OTP master function scaling fits in (a).

§5. The α-relaxation region

5.1. *Kohlrausch fits*

The $\chi''(\omega)$ spectra for $T \geq 320$ K in Fig. 1(a) exhibit a strongly T-dependent α-relaxation peak. (For $T < 320$ K, the α peak has moved out of our currently available experimental window.) The α-peaks for the spectra at temperatures between 320 and 435 K were fit using Fourier-transformed KWW functions

$$\phi(t) = f e^{-(t/\tau_\alpha)^{\beta_K}} \tag{5·1}$$

as shown in Fig. 8. The stretching parameter β_K found from these ten fits had an average value of $\langle \beta_K \rangle = 0.78$ with a slight tendency to *decrease* with increasing T as shown in Fig. 9.

The spectra were also replotted as χ'' vs $\omega/\omega_{(\mathrm{MAX})}$ using the $\omega_{(\mathrm{MAX})}$ values found from the KWW fits. The scaled spectra are shown in Fig. 10 and are seen to obey time-temperature superposition (α scaling) for at least one decade on either side of the maximum. We also include in Fig. 10 the KWW prediction for $\beta = 0.78$ as well as a Debye curve ($\beta = 1.0$) for comparison. Evidently there is no tendency for the spectra to approach the Debye form with increasing T.

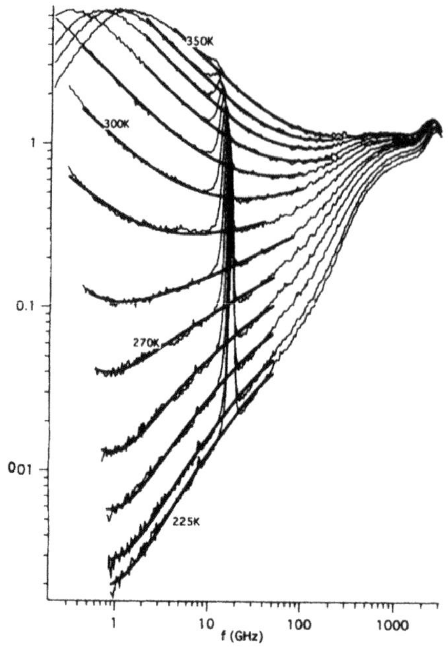

Fig. 6. Nonlinear least squares E-MCT fits to OTP $\chi''(\omega)$ spectra at $T = 225$, 240, 250, 260, 270, 280, 290, 300, 310, 320, 330, 340 and 350 K. The fit results are indicated by heavy solid lines which extend over the frequency range included in the fit for each spectrum.

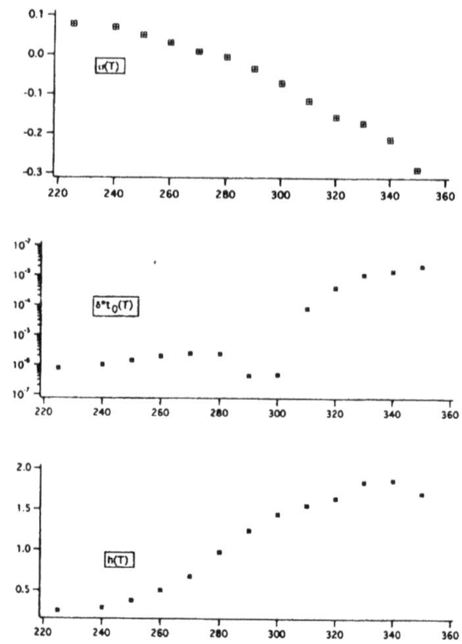

Fig. 7. Parameter values determined from the E-MCT fits shown in Fig. 6: $\sigma(T)$, $h(T)$ and $\delta t_0(T)$. From the $\sigma(T)$ values,we estimate $T_C \sim 277$ K.

I-MCT predicts that

$$\tau_\alpha \propto |\sigma|^{-\gamma} \tag{5·2}$$

so that, assuming that $|\sigma| \propto (T - T_C)$, a plot of $\tau_\alpha^{-1/\gamma}$ vs T should produce a straight line extrapolating to zero at $T = T_C$. [In Eq. (5·2), $\gamma = \frac{1}{2a} + \frac{1}{2b}$.]

The τ_α values (in $n$sec) found from the KWW fits are plotted as $\tau_\alpha^{-1/\gamma}$ vs T in Fig. 11. The two sets of points correspond to the values of a found from the I-MCT fits ($a = 0.318$, $b = 0.608$, $\lambda = 0.72$, $\gamma = 2.79$) and from the E-MCT fits ($a = 0.30$, $b = 0.542$, $\lambda = 0.76$, $\gamma = 2.59$.) Both give reasonably linear behavior, extrapolating to zero at $T_C^\alpha = 277$ K (for $a = 0.3$) and $T_C^\alpha = 270$ (for $a = 0.318$).

5.2. *Combined $\alpha - \beta$ fits*

In Ref. 12), Steffen et al. found that direct fits of the α-peaks to KWW functions produced values of β_K that tend to decrease with increasing temperature as we have again found here. However, they noted that there is ambiguity in performing these fits since the high-frequency wing of the α-peak overlaps with the β-relaxation region. In order to avoid this ambiguity, some criterion is required to fix the upper limit to the frequency region for the KWW fits.

We have attempted a combined $\alpha - \beta$ fit using the following procedure as illus-

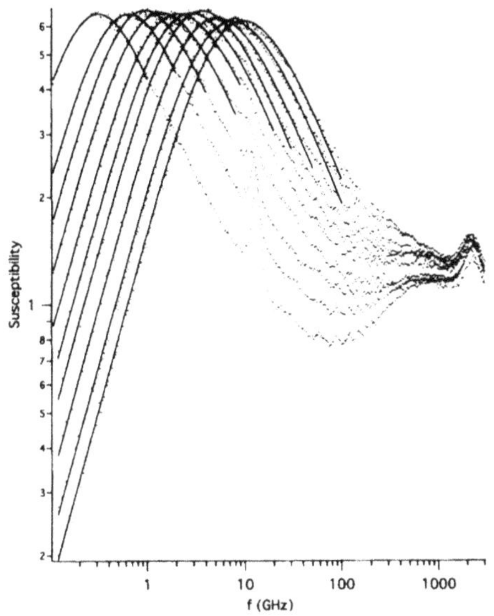

Fig. 8. OTP $\chi''(\omega)$ spectra for $T = 320$, 330, 340, 350, 360, 370, 380, 395, 415 and 435 K. The solid lines are fits using Fourier-transformed KWW functions.

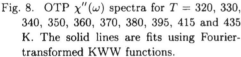

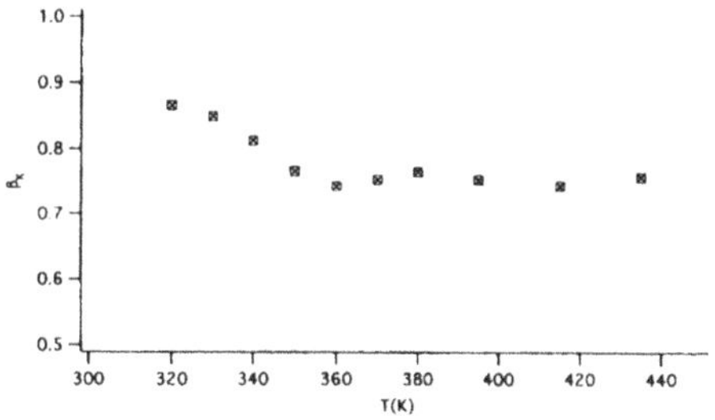

Fig. 9. Stretching exponent β_K from the KWW fits shown in Fig. 8.

trated for the 320 K $\chi''(\omega)$ spectrum in Fig. 12. First, the E-MCT fit found in the fitting procedure of Fig. 6 is extended into the α-peak range (a). Second, KWW fits are performed using several different fitting ranges centered at the α-peak (b). The α and β fits are then superimposed, and the α fit is selected that joins smoothly onto the β fit with the smallest change in slope. The E-MCT and KWW functions are then joined at their crossing point to produce a single theoretical $\alpha - \beta$ fit curve (c). Since β_K tends to decrease as the fitting range increases, this procedure uses the requirement that the two asymptotic fits join smoothly to resolve the ambiguity in choosing the upper frequency limit for fitting the α-peak.

In Fig. 13 we show the combined $\alpha - \beta$ fit results obtained with this procedure for $T = 320$, 340, 360, 380 and 415 K. The β_K values found from the fits are shown in the inset. While there is still some indication of a gradual decrease in β_K with increasing T, the data are also consistent with a temperature-independent $\beta_K = 0.79$.

We emphasize that this procedure is *not* a superposition of different processes. It is an approach to representing the two-step relaxation process of MCT by constructing a function with the correct asymptotic behavior in two regions: the E-MCT β-correlator in the intermediate region and the Kohlrausch function at long times. It is thus an approximate representation of the full solution to the MCT equations.

§6. Summary and conclusions

While there is still uncertainty concerning the microscopic origin of the depolarized light scattering spectra of anisotropic molecular liquids like OTP, mode coupling

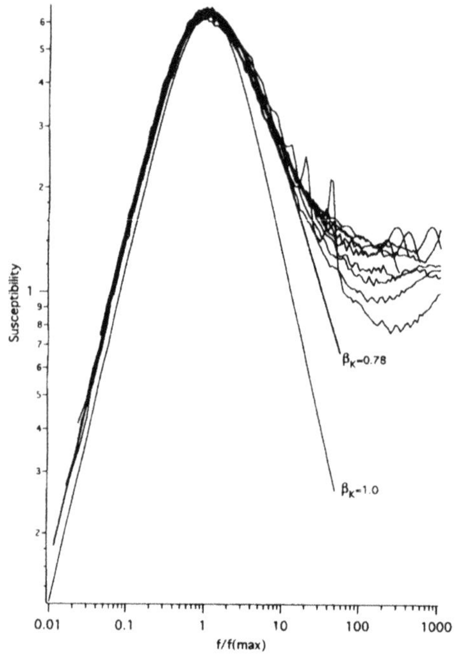

Fig. 10. $\chi''(\omega)$ spectra frequency scaled by f_{MAX} from the fits in Fig. 8. KWW master functions for $\beta_K = 0.78$ and $\beta_K = 1.0$ (Debye) are also shown.

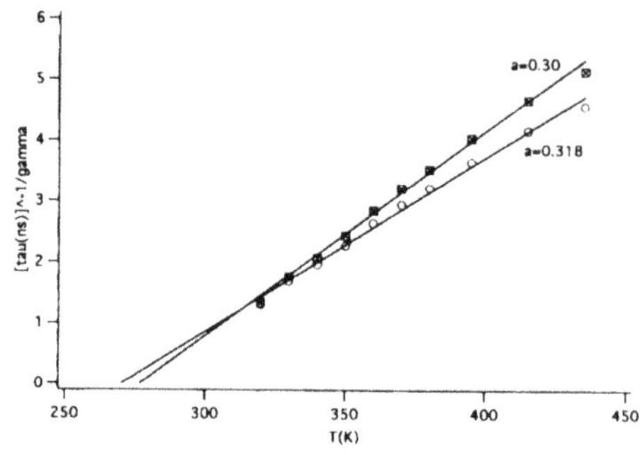

Fig. 11. $[\tau_\alpha(ns)]^{-1/\gamma}$ vs T with τ_α from the fits in Fig. 8. $\gamma = 2.79$ corresponds to $a = 0.318$ found in the I-MCT fits; $\gamma = 2.59$ corresponds to $a = 0.30$ found in the E-MCT fits. The straight lines are linear regression fits giving $T_C = 277$ (for $a = 0.30$) and 270 (for $a = 0.318$).

theory is able to give excellent fits to the experimental OTP light-scattering data with the assumption that $\chi''_{LS}(\omega)$ is proportional to χ'' of MCT [Eq. (4·4)]. The quality of the fits for the β-relaxation region are significantly improved by including $\delta \neq 0$ of the extended theory.

We note that the E-MCT fits shown in Figs. 5 and 6 only describe the $\chi''(\omega)$ data in the limited intermediate frequency range where Eq. (4·2) is valid, missing both the α-peaks and the microscopic structure (boson peak). This limitation, inherent in the use of asymptotic approximations to the MCT equations, could in principle be overcome by fitting the complete spectrum to solutions to the full MCT equations. Although the necessary coupling constants needed for this procedure are not yet available, they may be accessible using the model potential for OTP developed by Lewis and Wahnstrom.[22] We intend to pursue this approach in a future publication.

We have presented a preliminary attempt to resolve the ambiguity in fitting the α-relaxation peaks arising from overlap of the α and β regions. By requiring that the asymptotic E-MCT $\chi''(\omega)$ curves join smoothly to the KWW α spectra, we

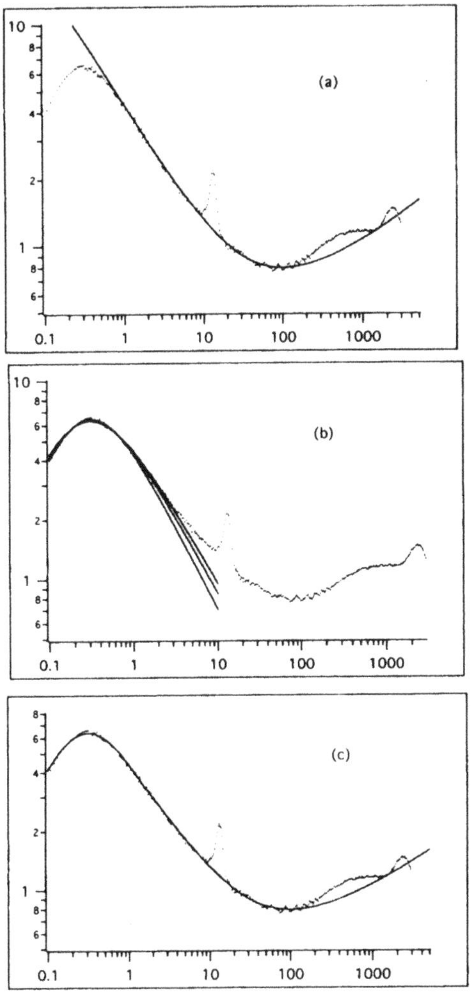

Fig. 12. Combined $\alpha - \beta$ fit for $T = 320$ K $\chi''(\omega)$ spectrum. (a) E-MCT fit to the β region. (b) KWW fits to the α peak for three different fitting ranges: top: 0.1 – 1.0 GHz, $\beta_K = 0.86$; center: 0.05 – 2.0 GHz, $\beta_K = 0.81$; bottom: 0.02 – 4.0 GHz, $\beta = 0.78$. (c) Combined E-MCT fit and KWW fit joined at $f = 0.832$.

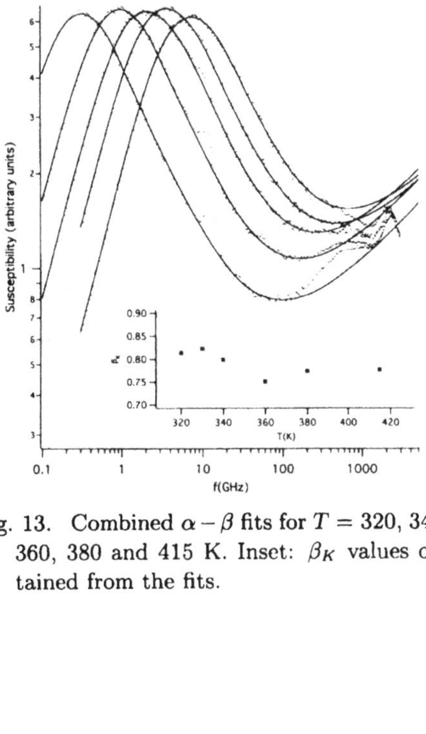

Fig. 13. Combined $\alpha - \beta$ fits for $T = 320$, 340, 360, 380 and 415 K. Inset: β_K values obtained from the fits.

obtained combined $\alpha - \beta$ fits for spectra for frequencies from below the α-peak to above the susceptibility minimum. The stretching coefficient β_K obtained with this procedure is essentially temperature independent, and shows no trend of increasing with increasing temperature.

The fits described in this paper, like MCT analyses of other light scattering data, are least satisfactory on the high-frequency side of the minimum where they tend to generally fall below the experimental data [see Fig. 5(a)]. Recently, Franosch et al.[23] have shown that this problem arises in general from the limited range of validity of the asymptotic formulas of MCT. They showed that a particular two-correlator schematic form of MCT can accurately describe light scattering spectra of glycerol, while the asymptotic formulae produce fits that fall systematically below both the experimental spectra and the theoretical curves at frequencies above the minimum. Therefore, to progress beyond the level of MCT analysis presented here based on the asymptotic approximations, the best strategy would appear to lie in comparing experimental data with numerical solutions to the MCT equations, perhaps in some appropriate simplified form.

Acknowledgements

We thank W. Götze and M. Fuchs for many helpful discussions and for providing the computer programs used to generate the E-MCT $\hat{\chi}''(\omega)$ spectra. We thank W. Steffen and A. Patkowski for providing an OTP sample and the data shown in Fig. 1(b), and for several discussions of experimental and theoretical aspects of this work. Research at CCNY was supported by the NSF under grant DMR-9315526.

References

1) W. Petry, E. Bartsch, F. Fujara, M. Kiebel, H. Sillescu and B. Farago, Z. Phys. **B83** (1991), 175.
2) O. Debus, H. Zimmerman, E. Bartsch, F. Fujara, M. Kiebel, W. Petry and H. Sillescu, Chem. Phys. Lett. **180** (1991), 271.
3) M. Kiebel, E. Bartsch, O. Debus, F. Fujara, W. Petry and H. Sillescu, Phys. Rev. **B45** (1992), 10301.
4) J. Wuttke, M. Kiebel, E. Bartsch, F. Fujara, W. Petry and H. Sillescu, Z. Phys. **B91** (1993), 357.
5) E. Bartsch, F. Fujara, B. Geil, M. Kiebel, W. Petry, W. Schnauss, H. Sillescu and J. Wuttke, Physica **A201** (1993), 223.
6) E. Bartsch, F. Fujara, J. F. Legrand, W. Petry, H. Sillescu and J. Wuttke, Phys. Rev. **E52** (1995), 738.
7) W. Steffen, A. Patkowski, G. Meier and E. W. Fischer, J. Chem. Phys. **96** (1992), 4171.
8) D. Kivelson, W. Steffen, G. Meier and A. Patkowski, J. Chem. Phys. **95** (1991), 1943.
9) W. Steffen, G. Meier, A. Patkowski and E. W. Fischer, Physica **A201** (1993), 300.
10) A. Patkowski, W. Steffen, G. Meier and E. W. Fischer, J. Non-Cryst. Solids **172-174** (1994), 52.
11) W. Steffen, B. Zimmer, A. Patkowski, G. Meier and E. W. Fischer, J. Non-Cryst. Solids **172-174** (1994), 37.
12) W. Steffen, A. Patkowski, H. Glaser, G. Meier and E. W. Fischer, Phys. Rev. **E49** (1994), 2992.
13) P. A. Madden and D. J. Tildesley, Mol. Phys. **55** (1985), 969.
14) G. Li, W. M. Du, X. K. Chen, H. Z. Cummins and N. J. Tao, Phys. Rev. **A45** (1992), 3867.
15) G. Li, W. M. Du, A. Sakai and H. Z. Cummins, Phys. Rev. **A46** (1992), 3343.
16) W. M. Du, G. Li, H. Z. Cummins, M. Fuchs, J. Toulouse and L. A. Knauss, Phys. Rev. **E49** (1994), 2192.
17) H. Z. Cummins, W. M. Du, M. Fuchs, W. Götze, S. Hildebrand, A. Latz, G. Li and N. J. Tao, Phys. Rev. **E47** (1993), 4223.
18) H. Z. Cummins, W. M. Du, M. Fuchs, G. Götze, A. Latz, G. Li and N. J. Tao, Physica **A201** (1993), 207.
19) H. Z. Cummins, G. Li, W. M. Du, J. Hernandez and N. J. Tao, Transport Theory and

Statistical Physics **24** (1995), 981.
20) W. Götze and L. Sjogren, Rep. Prog. Phys. **55** (1992), 241.
21) This fitting procedure, suggested by M. Fuchs and A. Latz, was used in Ref. 17).
22) L. J. Lewis and G. Wahnstrom, Phys. Rev. **E50** (1994), 3865.
23) T. Franosch, W. Götze, M. Mayr and A. P. Singh, Phys. Rev. **E55** (1997), 3183.

Glass Transition in Colloidal Hard Spheres: Mode-Coupling Theory Analysis

W. van Megen and S. M. Underwood
Department of Applied Physics, Royal Melbourne Institute of Technology, Melbourne, Victoria 3000, Australia
(Received 1 December 1992)

Coherent intermediate scattering functions are measured by dynamic light scattering for several wave vectors around the structure factor peak on metastable colloidal fluids and glasses of hard spherical particles. The results are quantitatively described by a combination of the α and β processes. The scaling laws and factorization property predicted by mode-coupling theory are verified.

PACS numbers: 64.70.Pf, 61.20.Ne, 82.70.Dd

Much of the recent work on the liquid-glass transition (GT) has been stimulated by the predictions of mode-coupling theory (MCT) [1]. This theory predicts a dynamical singularity during supercooling of a liquid where its structure is arrested [2,3]. Suspensions of identical colloidal spheres stabilized by thin macromolecular surface coatings offer several advantages over other materials for both fundamental studies of the GT and assessment of the detailed dynamics predicted by MCT. First, they appear to be the simplest experimental system to show a GT since the interparticle forces are like those of hard spheres [4-6]. Second, the motion of the particles is heavily damped by the suspending liquid so that phonon-activated hopping motions, which may restore ergodicity in molecular glasses, should be strongly reduced and, consequently, the GT in suspensions should be close to the ideal GT prediction by the basic version of MCT. Third, the dynamics of suspensions can be studied by dynamic light scattering (DLS). This gives access to correlations in time over a range of more than nine decades and, through varying the wave vector q, a range of correlations in space.

MCT predicts that beyond the time scale t_0 of microscopic motions the relaxation of the intermediate scattering function (ISF or number density autocorrelator), $f(q,\tau)$, proceeds in two stages. These are connected by two time scales, τ_α and τ_β, which diverge as the separation parameter $\sigma = c_0(\phi-\phi_c)/\phi_c$ approaches zero; $\tau_\alpha = t_0|\sigma|^{-\gamma}$ and $\tau_\beta = t_0|\sigma|^{-\delta}$, with $\gamma=(1/2a)+(1/2b)$ and $\delta=(1/2a)$ [2,3]. Here ϕ is the particle volume fraction and ϕ_c its critical value for the GT. The dynamics in the regime $t_0 \ll \tau \ll \tau_\alpha$ is governed by the β process where

$$f(q,\tau) = f_c(q) + |\sigma|^{1/2} h(q) g_\pm(\tau/\tau_\beta). \quad (1)$$

Neither the universal master functions $g_\pm(\tau/\tau_\beta)$ (the subscript "$\pm$" denoting the sign of σ), the nonergodicity parameter $f_c(q)$, nor the critical amplitude $h(q)$ depend on concentration. The above factorization of spatial and temporal variables suggests that localized dynamics of caged particle clusters promote relaxation of density fluctuations to $f_c(q)$. On the fluid side of the transition ($\sigma<0$) τ_β marks the crossover to the second relaxation stage, the α process, describing cage breakdown and allowing large-scale diffusion, for which a second scaling law holds,

$$f(q,\tau) = f_c(q) G(q,\tau/\tau_\alpha). \quad (2)$$

In the glass ($\sigma>0$) the α process is arrested but the β process persists and saturates at long times to the value $g_+(\tau\to\infty) = (1-\lambda)^{-1/2}$. The quantities a and b are specified by the exponent parameter λ, in turn determined by the static structure factor $S(q)$. In comparing the dynamics of molecular glass formers with MCT, λ is generally treated as a free parameter [3]. For the hard sphere system approximations for $S(q)$ are available and the exponents and the functions appearing in Eqs. (1) and (2) have been evaluated (in particular, $a=0.301$, $b=0.545$, $\lambda=0.758$, and $c_0 \simeq 1.2$) [7,8].

As in previous work [4-6], we use suspensions of polymer particles (radius $R=205$ nm) made nearly transparent, and therefore suitable for light scattering studies, by matching the refractive index of the suspending liquid to that of the particles. Thin steric barriers chemically grafted to the particle surfaces provide steeply repulsive interparticle forces. Identification of freezing and melting volume fractions, $\phi_f=0.494$ and $\phi_m=0.543$ (± 0.002), in accord with computer results for the perfect hard spheres, confirms the viability of these suspensions as model hard sphere systems [4,6,9].

Where previous MCT analyses of $f(q,\tau)$ measured on concentrated metastable colloidal fluids could be explained by the β process only [5,7], the more extensive measurements reported here show that *both* α and β processes are necessary for the description of the slow relaxation of the ISF's in a simple hard sphere system on the fluid side of the GT. In addition to verification of the scaling properties of these processes, the analysis provides the first comprehensive confirmation of the factorization property of the β relaxation.

Measurements of duration $T=1000$ s were made of the intensity autocorrelation function $g^{(2)}(q,\tau) = \langle I(q,0) \times I(q,\tau)\rangle_T / \langle I(q)\rangle_T^2$ on eight metastable fluid samples in the concentration range 0.494–0.587. In the ergodic case all spatial configurations are accessible in the course of the measurement, so that the time average $\langle\cdots\rangle_T$ constitutes an ensemble average and the ISF is calculated in the usual way from $sf^2(q,\tau) = g^{(2)}(q,\tau)-1$ [10] ($s \simeq 0.2$ is the spatial coherence factor). The transition to nonergodicity (fluid to glass) is indicated by a strong variation in $\langle I \rangle_T$ for different scattering volumes in the sample as well as a significant reduction in the mean square amplitude, $g^{(2)}(q,0)$, of intensity fluctuations. In the nonergodic glass phase, calculation of the (ensemble-averaged) ISF from $g^{(2)}(q,\tau)$ must then take into account both the fluctuating and nonfluctuating components of the scattered radiation associated with fluctuating and arrested structure, respectively. The derivation of a procedure for achieving this and its experimental verification have been described previously [6,10,11].

Figure 1 shows a representative set of ISF's, made at wave vectors below, near, and above the position q_m of the main peak in $S(q)$. Increasing the concentration from freezing ($\phi_f = 0.494$) to $\phi = 0.574$ lengthens the overall decay time of $f(q,\tau)$ by nearly four decades, from about 10^5 to 10^9 μs, and significantly three relaxation stages become increasingly apparent; the crossover times between these stages are indicated by points of inflection which, for $\phi = 0.574$, occur at around 10^4 and 10^7 μs. The fastest of these processes [indicated by the dashed curve in Fig. 1(b)] is associated with the microscopic diffusive motions of particles within their instantaneous neighbor cages and accounts for the initial few percent of the decay of $f(q,\tau)$. This is followed by two slower processes whose time scales lengthen and separate with increasing concentration. When the concentration is increased further by only 1%, from 0.574 to 0.581, $f(q,\tau)$ saturates to an almost constant value indicating the presence of concentration fluctuations whose duration significantly exceeds 1000 s. Thus, the occurrence of the kinetic GT is indicated at a concentration ϕ_c ($0.574 < \phi_c < 0.581$) by the cessation of large-scale diffusion and arrest of the fluid structure on the experimental time scale.

A significant feature of these colloidal systems is that the kinetic GT coincides with the suppression of homogeneously nucleated crystallization [5,6]. Below ϕ_c roughly isometric crystals are nucleated homogeneously throughout the sample. However, just above ϕ_c much larger crystals grow slowly on highly asymmetric nuclei. We have suggested elsewhere [12] that these asymmetric nuclei are shear-induced structures, resulting from the tumbling process used to randomize the particle positions prior to DLS measurements, which remain frozen in the structurally arrested glass. We speculate that infrequent small-scale collective particle rearrangements are responsible not only for the slow growth of these asymmetric crystals but also for the small remnant decay in $f(q,\tau)$ at long times when $\phi > \phi_c$ (see Fig. 1).

On the fluid side of the GT ($\sigma < 0$) MCT has been fitted to the data, shown in Fig. 1, by combining the α and β processes of Eqs. (1) and (2) using the master functions $g_{\pm}(\tau/\tau_\beta)$ and $G(q,\tau/\tau_\alpha)$ calculated for the hard sphere system [7,8]. In this procedure the amplitudes $h(q)$ and $f_c(q)$, the separation parameter σ, and the time scale τ_α are treated as adjustable parameters,

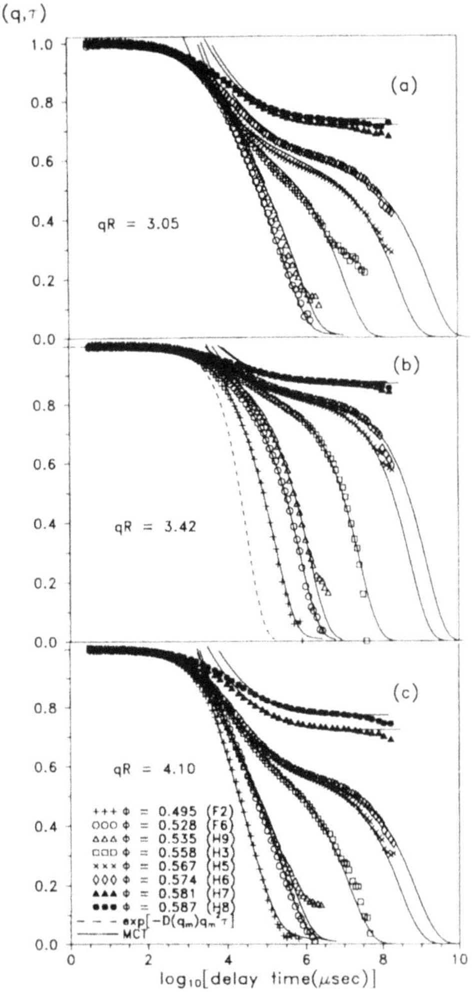

FIG. 1. Intermediate scattering functions for indicated values qR, the product of the scattering vector and the particle radius; the main maximum in $S(q)$ for the hard sphere fluid at freezing is located at $q_m R = 3.46$. The symbols refer to the experimental data for volume fractions indicated. The solid curves are the MCT fits. The dashed curve in (b) is the quantity $\exp[-q^2 D(q)\tau]$ representative of the microscopic dynamics, where $D(q)$ is the short-time q-dependent collective diffusion coefficient.

subject to the constraints of MCT that $h(q)$ and $f_c(q)$ are independent of concentration and that σ is independent of the scattering vector. Thus the only global parameter is the scaling time τ_α and it therefore absorbs most of the random and systematic errors in the data.

We reiterate that the MCT predictions apply only to the slow structural relaxation processes that emerge at high concentration and whose time scales lie beyond that

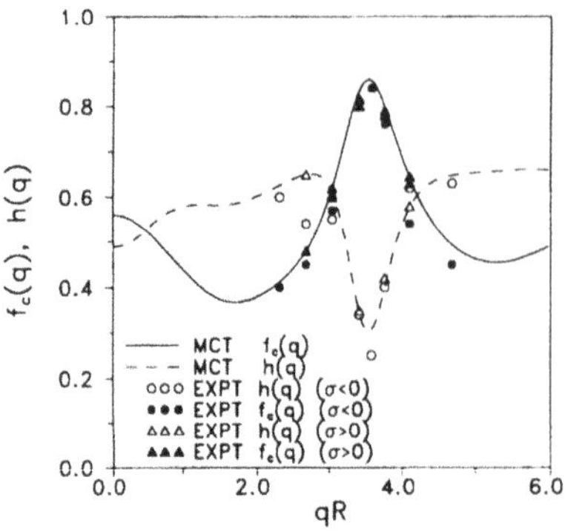

FIG. 2. Nonergodicity parameters $f_c(q)$ and critical amplitudes $h(q)$.

of the microscopic motions. It is evident from Fig. 1 that beyond the influence ($\simeq 10^4$ μs) of the microscopic diffusive motions MCT fits are possible over a matrix of scattering vectors and suspension concentrations. At $q = 3.05$ small systematic differences between MCT and experiment are evident. Here the amplitude $S(q)$ of the (un-normalized) coherent ISF, $S(q)f(q,\tau)$, is roughly one-tenth its peak value at $q = q_m$ and, as suggested elsewhere [11], both incoherent scattering associated with the small ($\simeq 5\%$) spread in particle size and multiple scattering may influence the experimental results.

A detailed MCT analysis of colloidal glasses is contained in a previous publication [6] but, for completeness, we include some results here. We assume, for $\phi > \phi_c$, that the α process is arrested and we accordingly analyze the data in terms of the β process only [Eq. (1)]. Because of the statistical errors in the experimental data, small variations in $f_c(q)$ and $h(q)$ from those found for $\sigma < 0$ have to be tolerated to obtain the theoretical results

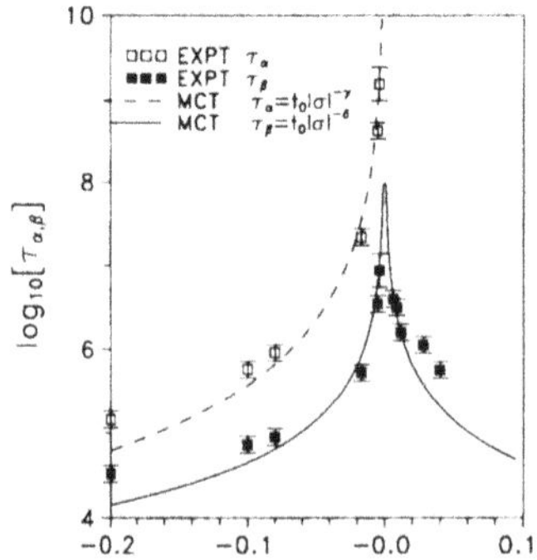

FIG. 3. Scaling times τ_α and τ_β vs the separation parameter σ. The MCT results are given by $\tau_\alpha = t_0|\sigma|^{-\gamma}$ and $\tau_\beta = t_0|\sigma|^{-\delta}$ and t_0 was calculated from τ_α and σ.

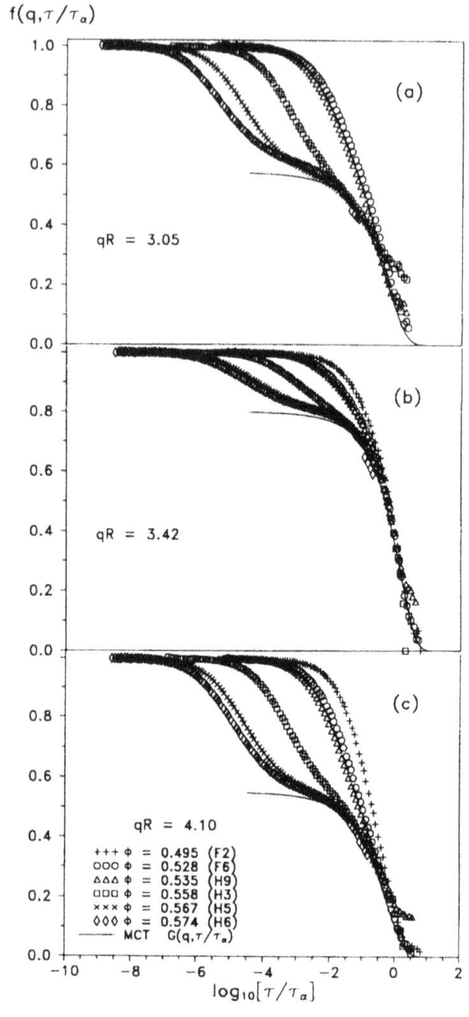

FIG. 4. Intermediate scattering functions expressed in terms of the scaled time τ/τ_α. The solid curves are the q-dependent master functions, $f_c(q)G(q,\tau/\tau_\alpha)$, of the α process.

in Fig. 1.

Figures 2 and 3 show the amplitudes and scaling times required for the MCT fits to the data; both are in good agreement with predictions. The predicted universality of the scaling times is confirmed by our finding that, apart from a few percent random variation, τ_α and by implication τ_β are independent of scattering vectors. τ_α and τ_β also show the predicted divergence and separation; for $\sigma = -0.1$, $\tau_\alpha/\tau_\beta \simeq 8$ while for $\sigma = -0.0035$, the smallest negative separation from the GT, $\tau_\alpha/\tau_\beta \simeq 180$. The characteristic times of the β process are consistent with the predicted symmetry about the GT. However, as one

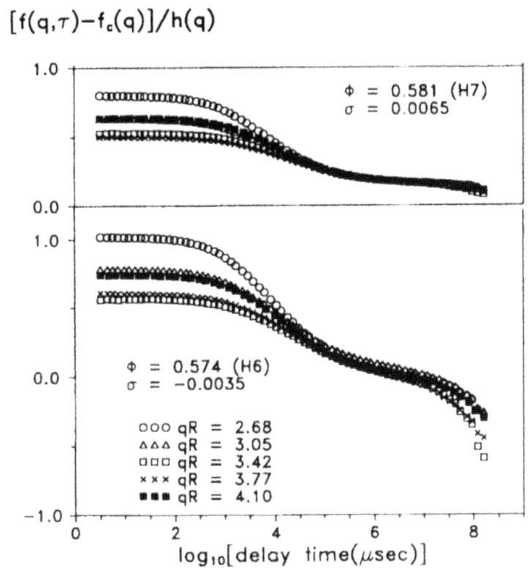

FIG. 5. Intermediate scattering functions scaled according to Eq. (1). The top and bottom figures represent results for samples just above ($\sigma=0.0065$) and below ($\sigma=-0.0035$) the GT.

goes deeper into the hard sphere glass ($\sigma > 0.03$) deviations from the theory possibly due to the failure of the asymptotic results of MCT are apparent. Note that no parameters are involved in the consistency checks illustrated in Figs. 2 and 3.

A conventional test of the superposition principle for the α process is to plot the ISF's in terms of the rescaled time τ/τ_α, as shown in Fig. 4. The sections of the ISF's which obey the principle also follow the predicted master functions. However, there is a significant fraction of the ISF's, amounting to about 15% at the structure factor peak and 40% at the other wave vectors, which does not scale. This departure from scaling corresponds to the β process and spans a time window from about 10^4 μs (the time scale of the microscopic motion) to the crossover time τ_β which, for $\phi=0.574$ ($\sigma=-0.0035$), is about 10^7 μs. As shown in Fig. 5 (lower figure), this is precisely the temporal range over which the factorization property of the β process is satisfied, i.e., where the quantity $[f(q,\tau)-f_c(q)]/h(q)$ is independent of q. The small but systematic deviations from this factorization seen at long times on the fluid side of the GT can be attributed to the α process. These deviations are not apparent in the glass phase (shown for $\phi=0.581$, $\sigma=0.0065$ in the top part of Fig. 5), where the α process is arrested.

We conclude that the GT observed in these suspensions approximates the ideal GT predicted by the basic version of MCT. On the fluid side of the GT, Figs. 4 and 5, respectively, show that neither the α process alone nor the β process alone can account for the slow relaxation of concentration fluctuations. At a concentration ϕ_c the α process is arrested but the β process persists.

We thank W. Götze, M. Fuchs, and R. O'Sullivan for helpful discussions and P. Francis for his technical assistance. This work was supported by the Australian Research Council.

[1] *Slow Dynamics in Condensed Matter,* edited by K. Kawasaki, M. Tokuyama, and T. Kawakatsu, AIP Conf. Proc. No. 256 (AIP, New York, 1992).

[2] W. Götze, in *Liquids, Freezing and the Glass Transition,* edited by J. P. Hansen, D. Lesvesque, and J. Zinn-Justin (North-Holland, Amsterdam, 1991), p. 287.

[3] W. Götze and L. Sjögren, Rep. Prog. Phys. **55**, 241 (1992).

[4] P. N. Pusey and W. van Megen, Nature (London) **320**, 340 (1986).

[5] W. van Megen and P. N. Pusey, Phys. Rev. A **43**, 5429 (1991).

[6] W. van Megen and S. M. Underwood, Phys. Rev. E **47**, 248 (1993).

[7] W. Götze and L. Sjögren, Phys. Rev. A **43**, 5442 (1991).

[8] M. Fuchs, I. Hofacker, and A. Latz, Phys. Rev. A **45**, 898 (1992).

[9] S. E. Paulin and B. J. Ackerson, Phys. Rev. Lett. **64**, 2663 (1990).

[10] P. N. Pusey and W. van Megen, Physica (Amsterdam) **157A**, 705 (1989).

[11] W. van Megen, S. M. Underwood, and P. N. Pusey, Phys. Rev. Lett. **67**, 1586 (1991).

[12] W. van Megen and S. M. Underwood (to be published).

Emulsion glasses: A dynamic light-scattering study

Hu Gang,[1] A. H. Krall,[2] H. Z. Cummins,[3] and D. A. Weitz[2]
[1]Exxon Research and Engineering Corporation, Route 22E, Annandale, New Jersey 08801
[2]Department of Physics and Astronomy, University of Pennsylvania, 209 S 33rd Street, Philadelphia, Pennsylvania 19104
[3]Department of Physics, City College of CUNY, New York, New York 10031
(Received 11 June 1998)

A liquid-glass transition was observed experimentally in a new system, an oil-in-water emulsion. Dynamic light scattering was employed to obtain the intermediate scattering function $f(q,t)$ for a range of volume fractions ϕ and scattering vectors q. The results are compared with predictions of the mode coupling theory. While the usual idealized version of the theory provides accurate fits to the data on the liquid side of the transition, fits for volume fractions near the transition and in the glass phase were found to require the extended version, presumably due to an additional decay mechanism related to the deformability of the oil droplets. [S1063-651X(99)04901-6]

PACS number(s): 64.70.Pf, 82.70.Kj, 83.10.Pp

INTRODUCTION

The liquid-glass transition occurs in a wide range of materials including the silicates, molecular liquids, molten salts, and polymers. At temperatures well below the glass transition temperature T_g, glasses are amorphous solids with nonzero shear modulus on any measurable time scale, while their structures exhibit no evidence of long-range order. It is generally (though not universally) believed that the glass transition is a purely kinetic transition, and that no thermodynamic phase transition is involved.

The hallmarks of the glass transition include extremely rapid increases in the shear viscosity and structural relaxation time with decreasing temperature, and the quenching in of disorder as the material undergoes an ergodic to nonergodic transition. While many experimental studies of these properties with a variety of experimental techniques have been reported, there is still no universally accepted theoretical explanation for the glass transition; experimental data continue to be analyzed with a variety of different theoretical models [1,2]. Thus further model systems that undergo a glass transition, and whose properties can be probed in detail, would be of great value to further elucidate this important transition.

Mode coupling theory (MCT) [3,4] has been very successful in explaining the evolving dynamics of the relaxation process in liquids approaching T_g, but comparisons of experiment and theory have generally been hampered by the complex structure of most real glass-forming materials. An important exception is provided by colloidal glasses for which the solid colloidal particles interact primarily via simple hard-sphere interaction potentials. In this case, particle volume fraction, ϕ, plays the role of temperature as the thermodynamic variable which controls the onset of the glass transition. As ϕ increases toward the glass transition, the viscosity of a suspension of colloidal particles diverges, and the frequency-dependent viscoelastic moduli can be well described within the framework of MCT [5]. Moreover, because colloidal particles typically have sizes of several thousand angstroms, light-scattering spectroscopy can directly probe the dynamics of these systems on the length scale $qa \sim 1$, where q is the scattering vector and a is the particle radius. Detailed dynamic light-scattering studies of colloidal glass formers have been reported and compared with the predictions of MCT, particularly by van Megen and his co-workers [6–8] and by Bartsch and co-workers [9,10]. Since MCT calculations can be carried out with the important coupling constants evaluated explicitly for the hard-sphere system, the dynamics of colloidal glasses as probed by dynamic light scattering provide a possibility for direct comparison of experiment and theory with no adjustable parameters. There are, however, hydrodynamic interactions present in colloidal dispersions that are not included in MCT. To further explore the relevance of MCT to the liquid-glass transition, it is clearly desirable to investigate other systems for which the interaction potentials can be treated explicitly.

In this paper we describe a new system that, somewhat unexpectedly, exhibits the hallmarks of the glass transition. It is an emulsion consisting of two liquids—oil and water—with a small concentration of stabilizing surfactant. The oil is dispersed as surfactant-covered droplets in the continuous (water) phase, forming an emulsion. The emulsion droplets are further purified to make them monodisperse in size [11,12], making possible more detailed study of their behavior and more exact comparison to theoretical predictions of packing and dynamics. We show that this emulsion clearly undergoes a glass transition as the volume fraction of droplets, ϕ, is increased. We study the glass transition with dynamic light scattering, and show that its behavior is well described within the formalism of mode coupling theory [3]. However, in marked contrast to the behavior of hard-sphere colloids, the extended version of MCT must be used for the emulsion, presumably reflecting the consequences of the deformability of the liquid droplets.

Besides providing a new system that can be used to test the validity of MCT, these data also provide important insight into the elastic properties of emulsions. As the droplet volume fraction increases, these emulsions undergo a pronounced transition from viscous fluids to highly elastic solids [12]. At the highest volume fractions, the shear modulus of the emulsions is controlled by the energy of deformation of the droplet shapes, or the surface tension [12]. The transition to this behavior is governed by the deformation of the droplets, which first occurs at $\phi_c \approx 0.64$, or random close packing, the highest volume fraction at which undeformed spheres can be randomly packed. By contrast, the results presented here show that, in fact, these emulsions first become a solid at a significantly lower volume fraction, $\phi_g \sim 0.58$, determined by the colloidal glass transition, where the spatial packing of the droplets becomes so large that they are no longer able to freely move over all the space.

EXPERIMENT

Our emulsions were comprised of silicone oil droplets in water, stabilized by sodium dodecylsulfate. The method of crystallization fractionation [11] was used to obtain monodisperse droplets with a radius of $a \approx 0.25\ \mu m$. By replacing about half the continuous phase water with glycerol, we matched the indices of refraction of the droplets and continuous phase, thereby eliminating multiple scattering. The structure and dynamics of the emulsions could then be probed with light scattering, using the 514.5 nm line of an Ar^+ laser. A desktop centrifuge was used to concentrate the emulsion to $\phi \approx 0.7$; the excess solvent was removed, and then used to dilute the sample to the desired volume fraction which was set by careful control of the mass of the sample components, and was determined by drying and weighing the constituents of a portion of the concentrated sample. The emulsion clearly exhibited a sharp transition in its behavior with increasing ϕ, going from a freely flowing fluid, to a very viscous fluid to a solid as ϕ varied from about 0.5 to 0.7. Although the surface of the droplets is charged by the surfactant, the concentration of surfactant in the continuous phase is sufficient to reduce the screening length to a sufficiently small value that the droplet packing behaves very nearly as hard spheres. Nevertheless there is a very small difference between the phase volume fraction, which determines the phase behavior of the droplet packing, and the absolute volume fraction, determined by the weight of the constituents. For the droplets used in these experiments, this difference was less than 1% [12]; here, we quote the phase volume fraction.

The structure of the emulsions can be characterized by means of their structure factor $S(q)$, which can be measured quantitatively with static light scattering, because the droplets are monodisperse, and because they can be index matched to the solvent. The form factor of the individual droplets is first determined from the scattering intensity of a low volume fraction; this is used to normalize the measured scattering intensity from higher volume fractions, allowing $S(q)$ to be determined. A typical example of $S(q)$ for the samples used in these studies is shown in Fig. 1. It is obtained from a sample with $a \approx 0.25\ \mu m$ and $\phi \approx 0.54$, a volume fraction that is near the colloidal glass transition. The structure factor exhibits the typical behavior of a concentrated hard-sphere system. There is a pronounced peak at $q_0 \sim 13\ \mu m^{-1}$, corresponding to $qa \sim \pi$, and a weaker peak at $q_0 \sim 2q_0$. The height of the strong first peak is $S(qa \sim \pi) \sim 2.8$, close to the value expected for a colloidal glass. Interestingly, the structure factor remains well determined even at volume fractions well above the glass transition, although the height of the first peak becomes significantly greater [13]. This reflects the fact that the disordered structure of the emulsion droplets is quenched in as the volume fraction is increased.

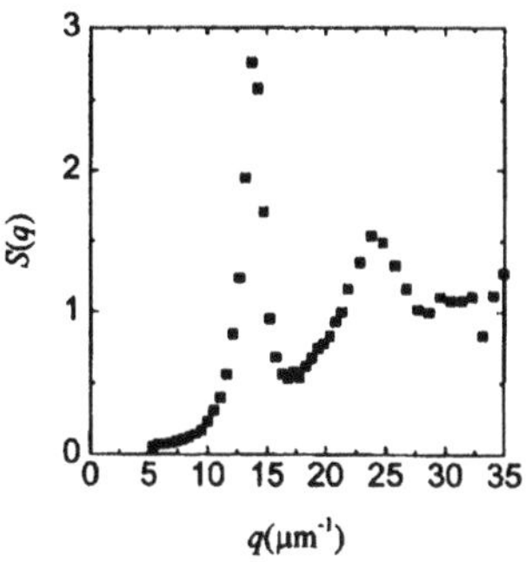

FIG. 1. (a) Static structure factor of a monodisperse emulsion with a droplet radius of $a \approx 0.25\ \mu m$ and a volume fraction of $\phi \approx 0.54$.

To probe the dynamics of the emulsion as it goes through the glass transition, we use dynamic light scattering (DLS), and collect data at values of qa below, at, and above the first peak in the structure factor, for samples whose volume fraction ranges from well below to well above the colloidal glass transition. We thereby obtain a comprehensive dynamical measure of the system as its dynamics change with increasing volume fraction ϕ, and as it goes through the glass transition. We measure the intermediate scattering function (ISF) $f(q,t)=S(q,t)/S(q)$. Because the dynamics of the emulsion become exceedingly slow with increasing volume fractions, time averaging of the data no longer corresponds to the true ensemble average required for comparing to theoretical predictions, reflecting the nonergodicity of the system on the time scale of the measurement. Thus, during the course of a measurement, some fraction of the scattered light will fluctuate completely, while a second fraction will remain unchanged. The exact correlation function measured will depend on the relative fraction of these two contributions for the light collected by the detector. The relative contributions of the static and dynamic contributions must be properly weighted to obtain a true ensemble-averaged ISF.

Different techniques can be used to properly average the data. Perhaps the simplest is to slowly rotate the sample while the correlation function is being measured [14]. This ensures that an average over all relative contributions of static and dynamic scattering is collected. However, the motion of the sample also results in a decay of the correlation function as the speckles move across the detector, limiting the time scale over which a true decay of the sample is observed. A second method to obtain properly ensemble-averaged data entails the collection of a correlation function at a single point, and then rotating the sample more rapidly to measure the average scattered intensity; this is then used to correct the measured correlation function to obtain the ensemble-averaged ISF [15]. While this method does not directly average the data, it also does not result in any additional decay due to sample motion.

We tried both methods in these experiments; however, contrary to claims that the two techniques are equivalent [14], we found that we were able to collect considerably better data using the latter method. The data for the samples studied here decay very slowly, requiring long collection times to observe the full extent of the decay. As a result, it was essential to measure the longest decay times possible. Using the rotation method, the limitation to measuring long decay times is set by the combination of the total duration of the experiment and by the requirement that a large number of independent speckles must be measured to obtain the average scattered intensity with sufficient accuracy to normalize the data. It is this combination of requirements that ultimately determines the slowest speed at which the sample must be rotated. Thus, for example, 10^4 independent speckles must be averaged to determine the average intensity to an accuracy of about 1%. Since it is quite important to accurately determine the average intensity, at least 10^4 speckles must be measured. For a given experiment duration, this sets the speed at which the sample must be rotated, and therefore the longest decay time that can be probed before sample motion obscures further decay. This decay time must always be significantly less than 10^{-4} of the total duration of the experiment, severely limiting the decay times that can be probed in any experiment. By contrast, in the second technique, the measurement of the average intensity is separate from the measurement of the correlation function; hence the average intensity can be measured with a very high degree of accuracy by rotating the sample quite rapidly, to collect the data over a very large number of independent speckles. Then the correlation function can be measured to much longer times, without the introduction of any spurious decay due to sample motion. Although there will be increasing uncertainty in the data at these longer times, the data are nevertheless of sufficient quality to provide valuable information even at the longest delay times. Thus we found experimentally that the latter method was far better, and we used it to collect all the data presented here.

To collect our data, we typically collected static scattering for about 1 h, while the sample was slowly rotated. This provided a very accurate determination of the average static scattering. We then collected correlation functions from several different points, corrected each one individually to obtain a measure of the ensemble-averaged ISF, and finally averaged these together. We ensured that the acceptance angle of the detector was very small, so that the coherence factor, measured with the same optics as the experiments, was well above 0.9, as determined by the intercept of correlation functions from purely ergodic samples. Of course the apparent value of the intercept varied considerably for the emulsion sample, depending on the relative magnitude of the static and dynamic contributions. However, after the correction, the ISF's varied only very slightly, indicating that the averaging that we performed was sufficient.

Data collected from a series of samples, with different volume fractions ranging from $\phi=0.54$ to 0.62 are shown in Fig. 2. For each volume fraction, we plot data obtained at three values of qa, below, at, and above the first peak of the structure factor; the qa values are indicated by arrows in Fig. 1. The $f(q,t)$ data clearly exhibit two distinct decays on well separated time scales. Moreover, as ϕ increases, the separation of these time scales also increases significantly, while the plateau in the data between the decays becomes more extended. In addition, there is a pronounced change in the data between $\phi=0.57$ and 0.58; while the volume fraction changes by only 0.01, the separation of the time scales of the two decay processes of the ISF increases by several decades, particularly at the highest qa. Nevertheless, at all values of both ϕ and qa, the data always decay at the longest time scales measured. Finally, there is also a marked difference in the behavior as qa is varied; the amount of the initial decay is clearly the least for qa at the peak of the structure factor, increasing both below and above.

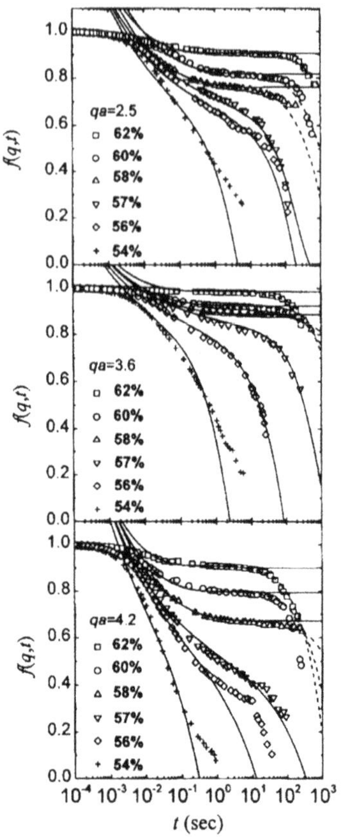

FIG. 2. Dynamic light-scattering data $f(q,t)$ (symbols) from monodisperse emulsions collected at $qa\approx2.5$ (top panel), 3.6 (middle panel), and 4.2 (lower panel), for a series of volume fractions. These values of qa correspond to values of $q\approx10$, 14.4, and 16.8 μm^{-1} in the structure factor shown in Fig. 1. The lowest value of qa is below the first peak, the intermediate value is approximately at the first peak, while the highest value is above the first peak in structure factor. Fits to the data using the idealized MCT are shown by the solid lines, while the improved fits for volume fractions above the glass transition obtained with the extended MCT are shown by the dashed lines.

While the shapes of the ISF's shown in Fig. 2 are quite complex, they possess all the hallmarks expected for a system undergoing a colloidal glass transition [6–8,16]. The physical picture that accounts for the shape is that of cages formed around each particle by its neighbors. As ϕ increases, any given particle becomes trapped within its cage for increasing periods of time, and any relaxation mechanism for the particle to escape from its cage must become increasingly more cooperative, involving the collective motion of particles over larger distances. This cooperativity results in the pronounced slowing down of the relaxation of density fluctuations evident in the data. Qualitatively, the initial decay of the ISF corresponds to local motion of the particle within its cage, the slowly decaying plateau region corresponds to relaxation of the cage, called the β relaxation, while the final decay corresponds to the breakup of the cage and escape of the particle, designated as the α relaxation. To quantitatively describe the data, we use mode coupling theory, which has been successfully applied to describe light-scattering data from colloidal suspensions near the glass transition [6–8,16].

Mode coupling theory describes the dynamics of a liquid approaching the liquid-glass transition through equations of motion for $\Phi_q(t)$, the normalized autocorrelation functions of the density fluctuations $\rho_q(t)$:

$$\ddot{\Phi}_q(t)+\Omega_q^2\Phi_q(t)+\int_0^t M_q(t-t')\dot{\Phi}_q(t')dt'=0. \quad (1)$$

These equations, one for each q, are formally exact, with the important physics hidden in the memory functions $M_q(t) = \gamma_q\delta(t)+m_q(t)$ where γ_q is the "regular" damping constant, and $m_q(t)$ represents the slowly varying part of $M_q(t)$. MCT then utilizes the Mori-Zwanzig projection operator formalism and Kawasaki's factorization approximation to express $m_q(t)$ in terms of products of two or more other modes. In the original "idealized" version of MCT, only the largest such term is retained:

$$m_q(t)=\sum_{q_1,q_2} V(q,q_1,q_2)\Phi_{q_1}(t)\Phi_{q_2}(t), \quad (2)$$

where $q_1+q_2=q$, and the coupling constants $V(q,q_1,q_2)$ are given in terms of the static structure factors $S(q)$, $S(q_1)$, $S(q_2)$ which are, in turn, determined by the intermolecular potentials. For hard spheres, the coupling constants can be evaluated analytically and the coupled set of Eqs. (1) and (2) can then be solved numerically. Such solutions were obtained for both hard-sphere [17] and Lennard-Jones liquids [18]. These complete solutions to the MCT equations can, in principle, be compared directly to experimental data, providing a critical test of the theory. However, preliminary attempts to carry out such a comparison for both colloidal dispersions and emulsions have been only moderately successful; while they capture the behavior at the time scales comparable to the β decay, they do not properly account for the behavior at short times, and thus do not capture the full behavior of the data.

The origin of this discrepancy is that the systems modeled by these MCT expressions do not correspond exactly to the experimental colloidal or emulsion systems. Since the particles or droplets are immersed in a fluid, the diffusive motion of the particles at short time scales must be incorporated into the MCT. Recently, Mayr and co-workers have reanalyzed the hard-sphere liquid using more powerful algorithms in an attempt to do so [19]. However, in addition to the diffusive dynamics, hydrodynamic interactions between the particles also play a crucial role in determining the dynamics at short-time scales. These interactions are not yet included in the MCT calculations, although attempts to do so are currently underway [20]. Therefore, as in all other previous comparisons of experimental dynamical data with MCT predictions, we employ the asymptotic MCT results rather than the full solutions to the MCT equations. However, in contrast to the case of hard-sphere colloids, the deformability of the emulsion droplets also affects the dynamics. To account for this, we will use the extended version of the MCT.

Near ϕ_c, $\Phi_q(t)$, which is assumed to be equal to $f(q,t)$, exhibits a two step decay, corresponding to the α and β relaxation processes characterized by two time scales, τ_α and τ_β, respectively, which are both greater than the microscopic time scale τ_0. In the idealized version of MCT, the two decay times are well separated, and $f(q,t)$ decays fully to zero at long times for $\phi<\phi_c$. In contrast, for $\phi>\phi_c$, the α process is frozen out, leaving only the β process, and $f(q,t)$ saturates to a finite value, $f(q,\infty)$, at long times. The transition at ϕ_c signifies an ergodic to nonergodic transition that typifies a glass transition.

As $\phi\to\phi_c$, the divergence of τ_β inherent in the MCT equations permits asymptotic expansions to be carried out that provide the MCT asymptotic results usually exploited for fitting data [7,8,16]. For $\tau_0\ll t\ll\tau_\alpha$, $f(q,t)$ can be factorized into time- and q-dependent functions,

$$f(q,t)=f_c(q)+h(q)G(t), \quad (3)$$

where $f_c(q)$ is the nonergodicity parameter which represents the amplitude of the arrested structure at ϕ_c, and $h(q)$ is the critical amplitude that describes the contribution of the β correlator,

$$G(t)=|\sigma|^{1/2}g_\pm(t/\tau_\beta), \quad (4)$$

where the master scaling function $g_\pm(t/\tau_\beta)$ is independent of both q and concentration, and where $\sigma=c_0(\phi-\phi_c)/\phi_c$ is the separation parameter which describes the approach to the critical volume fraction ϕ_c of the glass transition. Here, c_0 is a material-dependent constant. Both $f_c(q)$ and $h(q)$ can, in principle, be calculated from $S(q)$ at ϕ_c, and are only very weakly dependent on ϕ. Volume fraction dependence enters the dynamics only through σ, and the two scaling times, which diverge at ϕ_c,

$$\tau_\alpha=t_0|\sigma|^{-\gamma}, \quad \gamma=1/2a+1/2\beta, \quad (5)$$

$$\tau_\alpha=t_0|\sigma|^{-\delta}, \quad \delta=1/2\alpha, \quad (6)$$

where the critical exponents a ($0<a<0.5$) and b ($0<b<1$) are related to the exponent parameter λ by

$$\lambda=\frac{\Gamma^2(1-a)}{\Gamma^2(1-2a)}=\frac{\Gamma^2(1+b)}{\Gamma^2(1+2b)}, \quad (7)$$

where Γ is the gamma function. The final, long-time relaxation of $f(q,t)$, reflecting the α process, is described by a second scaling law with the α correlator, and can be well approximated by a stretched exponential,

$$f(q,t)=f_c(q)\exp\{-(t/\tau_\alpha)^{\beta_q}\}, \quad (8)$$

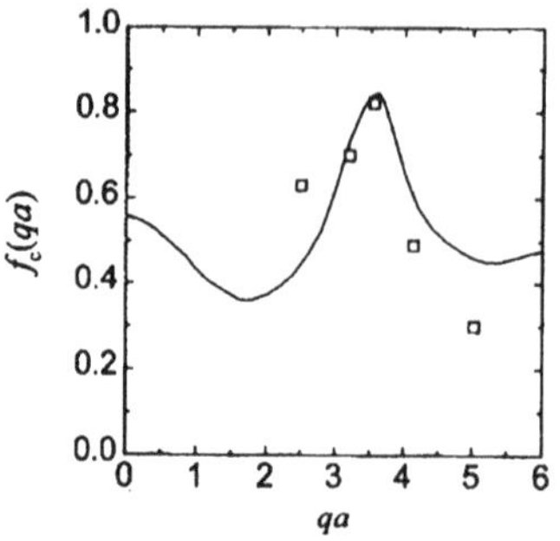

FIG. 3. Fitted values of the nonergodicity parameter $f_c(q)$ compared with the prediction for hard spheres, shown by the solid line.

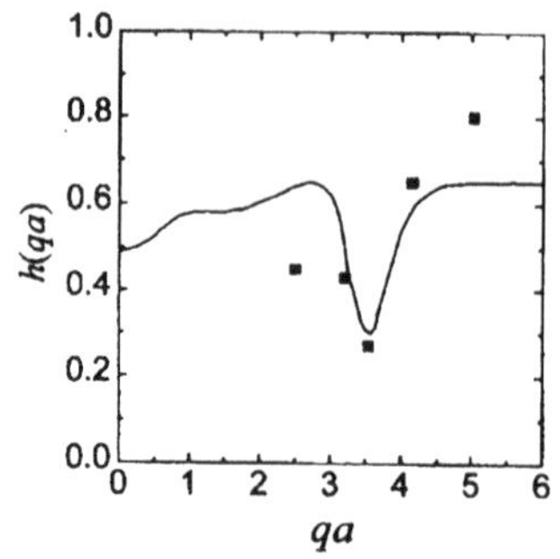

FIG. 4. Fitted values of the critical amplitude $h(q)$ compared with the prediction for hard spheres, shown by the solid line.

where β_q is the stretching exponent. Within the ideal MCT, α relaxation exists only on the liquid side, where for times greater than τ_β, the relaxation of the density fluctuations is shared by both the β and the α processes, with τ_β marking the crossover between the initial part of the β relaxation (the critical decay $\sim t^{-a}$) and the von Schweidler decay ($\sim -t^b$) which joins smoothly to the long-time α decay [Eq. (5)]. By contrast, on the glass side, the α process is arrested and the β process saturates at a definite value at long times.

The complete structural arrest at ϕ_c predicted by the idealized MCT does not occur in simple structural glasses, where the α-relaxation process moves to longer times with decreasing temperature but does not disappear. Retention of the next significant term in the memory function, beyond the leading term of Eq. (1), leads to the extended MCT [21,22]. The additional term, representing coupling to currents, is usually included through a temperature-dependent "hopping parameter" $\delta(T)$. With this parameter included, much better fits to experimental data can be obtained in the transition region and in the glass phase [23,24]. For our emulsion system, the droplet deformability is physically similar to the activated hopping process underlying $\delta(T)$, so we will also employ the extended MCT to analyze data at concentrations above ϕ_c.

DATA ANALYSIS

To quantitatively analyze our data, we begin by calculating the β correlator $G(\sigma,t)$ for hard spheres, for a trial exponent parameter λ, as a function of ϕ. Then, for the data for each q, we approximate $f_c(q)$ by the experimental value of the plateau of the ISF between 0.57 and 0.58, where it becomes flat. We then close a trial value of $h(q)$ to fit the initial decay of the data using Eq. (2). Below ϕ=0.58, we combine the α process using Eq. (5) with τ_α as a fitting parameter; since in the idealized MCT the α process is frozen out above ϕ_g, we do not include it for ϕ>0.57. This fitting process is iterated for each data set to obtain the best fit to all the data. In doing this fit, we hold both $f_c(q)$ and $h(q)$ fixed for all ϕ for each value of q, since MCT predicts that they are only very weakly dependent on ϕ; this ensures that the fit satisfies the constraints of MCT for the factorization of $f(q,t)$. The results of the fit are shown by the solid lines in Fig. 2. The fit to the data is very good for $\phi<\phi_g$; however, at higher ϕ, in the glass region, the data clearly continue to exhibit a decay at longer times, contrary to the expectations of the idealized MCT. To account for this decay, we use the extended MCT [21,22]. Within extended MCT, the final decay is also well described by a stretched exponential above ϕ_g [although the simple scaling of Eq. (5) is not expected to apply], and we use this form here, choosing the value of τ_α to ensure a continuous curve through the data. With this addition, all the data are fit very well; the new fits at long times for $\phi>\phi_g$ are shown by the dashed lines in Fig. 2. These fits are indistinguishable from the simple MCT fits at shorter times, and correctly account for the decay observed at longer-time scales.

The values obtained for $f_c(qa)$ and $h(qa)$ are shown in Figs. 3 and 4, respectively. The data for $f_c(qa)$ are compared to the values predicted for hard spheres, shown by the solid line in Fig. 3. The fitted values exhibit the same trend as the prediction, with $f_c(qa)$ exhibiting a sharp peak near the first peak in the structure factor. Similarly, $h(qa)$ exhibits a pronounced dip near the first peak in the structure factor. In Fig. 5 we plot the ϕ dependences of the fitted time scales, τ_α (circles) and τ_β (squares), and compare these with the predictions of MCT for hard spheres shown by the solid and dashed lines, respectively. The agreement is again quite good. There is a pronounced divergence for 0.57<ϕ<0.58,

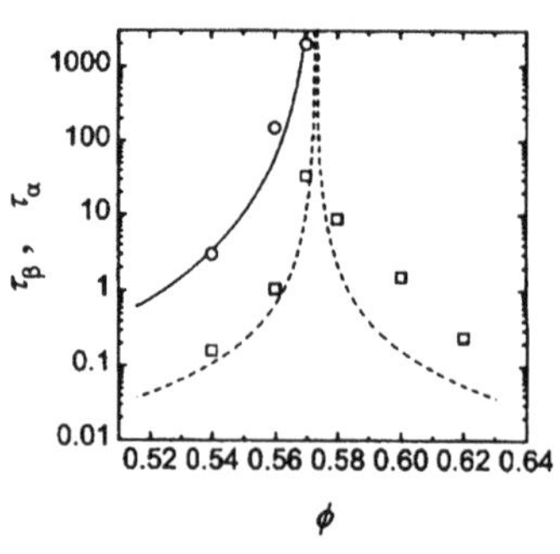

FIG. 5. The ϕ dependence of the two time scales obtained from the MCT fits, τ_α (circles) and τ_β (squares), compared to the predictions for hard spheres shown by the solid and dashed lines, respectively. Data for the α process are shown only for $\phi<\phi_g$. The divergence between ϕ=0.57 and 0.58 is strong evidence of the existence of the colloidal glass transition at this volume fraction.

providing a clear measure of ϕ_g. We do not show the values of τ_α obtained from the extended mode coupling fits above ϕ_g. These reflect the time scales for long-time relaxation for the emulsions in the glassy state, and are in marked contrast to the behavior observed for other colloidal systems. Hard spheres exhibit no long-time decay above ϕ_g [7], while soft gel spheres do exhibit a decay [9], but it cannot be described within extended MCT [10]. Thus the emulsion data are the first colloidal system that can be described by the extended MCT. This additional decay presumably arises because of the deformability of the liquid droplets [25,26]; which may enable thermal effects to cause the ultimate decay of $f(q,t)$, consistent with the use of a phonon-assisted hopping term. In addition, we believe that there are stresses built up within the emulsion as it is loaded into the light-scattering cell, as the observed time scale increases very slowly over the course of several weeks, indicating slow aging of the sample. Consistent with this, we observe a slow increase in the characteristic time of the final decay of the correlation function. These effects may be interesting to study as an example of the consequences of sample aging.

The asymptotic MCT equations [in contrast to the full solutions to Eq. (1)] do not account for the decay of the ISF at the shortest times where the particles are diffusing within their local cages. Furthermore, the data shown in Fig. 2 do not provide an accurate measure of this initial decay. However, some insight into this behavior can be obtained by using the same emulsion, without index matching, for diffusing wave spectroscopy which probes the very short-time motion of the particles [27]. Interestingly, for all the volume fractions shown in Fig. 2, the mean square displacement is subdiffusive, increasing with time more slowly than linearly, even at time scales as short as 1 μsec, and length scales of order 1 Å [25,26]. Thus the effects of the neighboring particles are felt through hydrodynamic interactions at the very shortest of time scales. This emphasizes the importance of including these interactions in any complete treatment of colloids and emulsions by the MCT.

CONCLUSIONS

These results demonstrate the utility of these monodisperse emulsions as a new system in which to study the glass transition. Because of the flexibility of the droplets, it is a simple matter to load samples at all volume fractions, allowing the study of their properties both below and above ϕ_g; by contrast, the very long relaxation times of hard-sphere suspensions make it very difficult to work with glassy samples. Furthermore, by suitable choice of oils, emulsion droplets can be made with different indices of refraction; thus it is a simple matter to mix small concentrations of particles with the same size, but with different scattering intensities, allowing tracer measurements to be made to extend the range of these studies. Furthermore, these samples are also amenable to mechanical measurements to determine their rheological properties; these properties also exhibit all the hallmarks of a colloidal glass transition [5]. Thus this system is ideal for a detailed test of current theoretical descriptions of the glass transition. Finally, these results highlight the importance of the droplet deformability in controlling the final relaxation of the droplets. Similar long-time relaxation is expected in other systems which show glassy relaxations, such as gels or foams. These long relaxations also have a strong influence on the rheological behavior of these systems, leading to a loss modulus that is surprisingly independent of frequency [12]. Recent work has suggested that these long-time scale relaxations are directly related to the frequency independence of the loss modulus [28]. The data presented here provide a direct measure of the dynamics and the relaxations which are speculated to be essential for this unusual rheological behavior.

ACKNOWLEDGMENTS

We thank W. Götze for helpful discussions and M. Mayr for providing numerical data from recent MCT hard-sphere computations. This work was partially supported by the NSF (Grant No. DMR96-31279) and NASA (Grant No. NAG3-2058).

[1] Dynamics of Glass Transitions and Related Topics, edited by T Odagaki, Y. Hiwatari, and J. Matsui [Prog. Theor. Phys. Suppl. **126**, R1 (1997)].

[2] Second International Discussion Meeting on Relaxations in Complex Systems, edited by K. Ngai [J. Non-Cryst. Solids **172-174** (1994)].

[3] W. Götze and L. Sjogren, Rep. Prog. Phys. **55**, 241 (1992).

[4] *Transport Theory and Statistical Physics: Special Issue Devoted to Relaxation Kinetics in Supercooled Liquids-Mode Coupling Theory and its Experimental Tests*, edited by S. Yip (Dekker, New York, 1995).

[5] T. G. Mason and D. A. Weitz, Phys. Rev. Lett. **75**, 2770 (1995).

[6] W. van Megen and P. N. Pusey, Phys. Rev. A **43**, 5429 (1991).

[7] W. van Megen and S. M. Underwood, Phys. Rev. Lett. **70**, 2766 (1993).

[8] W. van Megen and S. M. Underwood, Phys. Rev. E **49**, 4206 (1994).

[9] E. Bartsch, M. Antonietti, W. Schupp, and H. Sillescu, J. Phys. Chem. **97**, 3050 (1992).

[10] E. Bartsch, V. Frenz, J. Baschnagel, W. Schartl, and H. Sillescu, J. Phys. Chem. **106**, 3743 (1997).

[11] J. Bibette, J. Colloid Interface Sci. **147**, 474 (1991).

[12] T. G. Mason, J. Bibette, and D. A. Weitz, Phys. Rev. Lett. **75**, 2051 (1995).

[13] T. G. Mason, A. H. Krall, H. Gang, J. Bibette, and D. A. Weitz, in *Encyclopedia of Emulsion Technology*, edited by P. Becher (Dekker, New York, 1996), Vol. 4, p. 299.

[14] J.-Z. Xue, D. J. Pine, S. T. Milner, X.-L. Wu, and P. M. Chaikin, Phys. Rev. A **46**, 6550 (1992).

[15] P. N. Pusey and W. van Megen, Physica A **157**, 705 (1989).

[16] W. Götze and L. Sjogren, Phys. Rev. A **43**, 5442 (1991).

[17] U. Bengtzelius, W. Götze, and A. Sjolander, J. Phys. C **17**, 5915 (1984).

[18] U. Bengtzelius, Phys. Rev. A **34**, 5059 (1986).

[19] T. Franosch, M. Fuchs, W. Götze, M. R. Mayr, and A. P.

Singh, Phys. Rev. E **55**, 7153 (1997).
[20] W. Götze (private communication).
[21] W. Götze and L. Sjogren, Z. Phys. B **65**, 415 (1987).
[22] M. Fuchs, W. Götze, S. Hildebrand, and A. Latz, J. Phys.: Condens. Matter **4**, 7709 (1992).
[23] H. Z. Cummins *et al.*, Phys. Rev. E **47**, 4223 (1993).
[24] H. Z. Cummins, G. Li, W. Du, Y. H. Hwang, and G. Q. Shen, Prog. Theor. Phys. **S126**, 21 (1997).
[25] H. Gang, A. H. Krall, and D. A. Weitz, Phys. Rev. Lett. **73**, 3435 (1994).
[26] H. Gang, A. H. Krall, and D. A. Weitz, Phys. Rev. E **25**, 6289 (1995).
[27] D. J. Pine, D. A. Weitz, P. M. Chaikin, and E. Herbolzheimer, Phys. Rev. Lett. **60**, 1134 (1988).
[28] P. Sollich, F. Lecqueux, P. Hebraud, and M. E. Cates, Phys. Rev. Lett. **78**, 2020 (1997).

Part 5

GLASS TRANSITION

Signatures and Models

Scaling concepts for the dynamics of viscous liquids near an ideal glassy state

T. R. Kirkpatrick and D. Thirumalai
Institute for Physical Science and Technology, University of Maryland, College Park, Maryland 20742

P. G. Wolynes
Noyes Laboratory, University of Illinois at Urbana–Champaign, Urbana, Illinois 61801
(Received 23 December 1988; revised manuscript received 22 February 1989)

Motivated by recent mean-field theories of the structural glass transition and of the Potts glass model we formulate a scaling and droplet picture of an assumed ideal structural glass transition. The phase transition is a random first-order phase transition where the supercooled-liquid phase is composed of glassy clusters separated by interfaces or domain walls. Because of entropic driving forces the glassy clusters are continually being created and destroyed. As the ideal transition temperature is approached the entropic driving force vanishes and the size of the glassy clusters diverges with an exponent of $\nu=2/d$. All long-time dynamical processes are activated and the Vogel-Fulcher law is obtained for the liquid-state relaxation time.

I. INTRODUCTION

It is exceedingly tempting to try to relate the dramatic changes in various properties of a liquid undergoing a glass transition in the laboratory at finite cooling rates to an underlying ideal structural glass (STG) transition which would occur (at least in good glass formers) at a finite temperature upon infinitely slow cooling.[1] Many scenarios of this kind have been constructed over the years.[1–7] Recently, we have developed a picture of an ideal STG transition based on mean-field theories of the STG transition[8–10] and on the mean-field theory of the random Potts glass[11,12] (PG) and related *p*-spin models.[13] Having given in to the temptation of assuming an ideal STG transition, one is forced to consider what the scaling arguments used for equilibrium phase transitions might say about such an ideal glass transition and, *perhaps*, about laboratory glass transitions. In this paper we will further explore the concept of an ideal STG transition using scaling notions together with concepts arising from our earlier investigations of mean-field theories of structural glasses and of mean-field spin-glass models without reflection symmetry.

Due to the heuristic nature of our arguments, their range of validity and to what glass forming systems they apply is not clear. Our arguments always assume the liquid state is close to (metastable) equilibrium and consequently our ideas most naturally apply to generic-glass-forming materials and not to, say, metallic glasses which are formed by very rapid quenching from the melt and hence are systems far from equilibrium.[14] In addition, some of our arguments are based on large-scale droplet (or glassy clusters) ideas. Near the laboratory glass-transition temperature, T_g (which depends on the cooling rate), this is problematic because the size of these droplets (if they exist) is not very large. One of the main problems in constructing a theory for the glass transition is that if the laboratory glass transition is controlled by an ideal glass transition then one must come to grips with the fact that T_g is never asymptotically close to the ideal transition temperature which we call T_K. The main point is that because dynamics is controlled by activated processes even small correlation lengths lead to relaxation times that exceed typical experimental times. Even in the most favorable cases (say, *o*-terphenyl) the distance from an assumed ideal glass transition can be estimated to be $t=(T_g-T_K)/T_K>0.1$.[15] Nevertheless, we take the point of view here that the observed glassy phenomenology is controlled by an ideal transition, with, possibly, crossover effects playing a major role. We also point out that with minor modifications, our scenario is also valid if it turns out that the ideal glass transition is rounded by, for example, frustration effects. This is discussed further below.

The organization of this paper is as follows. In Sec. II we present the phenomenology relevant for our purposes. To motivate the scaling theory we also review our previous mean-field results. These considerations lead us to argue that an ideal glass transition would be a random first-order phase transition. In Sec. III we discuss the nature of the driving force for activated transport in viscous liquids. In Sec. IV we discuss the (finite-size) scaling exponents for a random first-order phase transition. In Sec. V we combine the ideas presented in Secs. II–IV with scaling ideas and derive the Vogel-Fulcher (VF) law for transport in a viscous liquid. In Sec. VI we discuss some of the other experimental consequences of our picture and conclude with some additional remarks.

II. REVIEW OF RELEVANT PHENOMENOLOGY AND MEAN-FIELD THEORY

The most obvious mystery of glassy behavior is the strongly non-Arrhenius slowing down of transport properties as the temperature is lowered. In the usual liquid regime above the melting point the isochoric activation energies are very small but the apparent activation energies grow dramatically in the supercooled regime. The viscosity for at least fragile glasses is often fit by the Vogel-Fulcher equation:[1,16]

$$\eta=\eta_0\exp[A/(T-T_0)] . \tag{2.1}$$

At the lowest temperatures a weaker divergence is often (but not always) reported.[17] This may be due in some cases to nonequilibrium effects but a true equilibrium crossover may sometimes be valid.

In general, at T_g there is no observed latent heat or change in volume. On the other hand the heat capacity and other susceptibilities change very rapidly in a nearly discontinuous manner.[1,16] The laboratory transition is a dynamic phenomenona dependent on the time scale of measurement and on the cooling rates. If the VF equation were to remain valid at low temperature the motions responsible for transport must freeze out at T_0 and the glass transition would be independent of the experimental time scale. Naively, we might expect this ideal glass transition to have the limiting characteristics of the laboratory transition—no latent heat but discontinuous susceptibilities. Generally the larger the discontinuities in heat capacity, etc., the more dramatic is the deviation from Arrhenius behavior.

Another mystery of the glass transition crucial to our picture is the behavior of the configurational entropy of the supercooled liquid.[3] It is important to note that, in general, the configurational entropy of a liquid is not a well-defined theoretical concept that different researchers mean different things by it (cf. Secs. III and VI). As pointed out by Simon[18] and Kauzmann[19] the extrapolation of heat-capacity data for supercooled liquids suggests that if a glass transition did not intervene the entropy of a supercooled liquid would be less than that of the corresponding crystal, at a temperature T_K less than the laboratory T_g. Since the vibrational entropies should be comparable for a crystal and a glass at these temperatures, this suggests a vanishing of a configurational contribution to the entropy at T_K. This temperature T_K is generally believed to be close to the T_0 in the Vogel-Fulcher equation and, as already remarked, we denote the ideal glass-transition temperature in this paper by T_K.

In order to better appreciate our arguments leading to Eq. (2.1) we give a qualitative description of the crucial features of the free-energy surface in a viscous liquid. Formally, the concept of a free-energy surface is not well defined because it involves unjustified analytic continuations. The surface we discuss can be viewed as the energy surface of a coarse-grained Hamiltonian in terms of order parameters. Because such a Hamiltonian contains entropic contributions we refer to the energy surface as a free-energy surface. Our picture is motivated by recent mean-field theories of the STG transition and spin-glass models without reflection symmetry.[9–13]

As a liquid is cooled (quenched) to low temperatures, say, below its equilibrium crystallization temperature, $T_{\rm cry}$, one intuitively expects a rough free-energy landscape in a very high-dimensional order-parameter space. The characterization of this high-dimensional free-energy surface is difficult. We begin by describing what some of the valleys on this surface are. By a state, s, we mean that set of configurations for the system as a whole which connects to s. Note that implicit in this definition of a state is a time scale argument. For infinite times, there is probably only the crystal state. We believe that this global thermodynamically stable phase is irrelevant to the glass-transition problem. It has been shown that mean-field theory leads to a multivalley structure in which an infinite number of aperiodic crystal phases can coexist, giving a finite "configuration" entropy above the ideal glass-transition temperature. For models with finite-ranged interactions this entropy will lead to an instability of the global mean-field states.[11] A state will break up into droplets. The resulting set of mosaic structures describes the liquid.

The main problem in glass formation is to understand how and why the relaxation time, τ, in the supercooled-liquid state grows so rapidly. As already mentioned, experiments indicate that for certain glass-forming materials τ appears to diverge at a temperature T_0 $(T_0<T_g)$ and the divergence is exponential. Exponentially large relaxation times are most naturally obtained by considering activated transitions between different mosaic states. Just as in ordinary nucleation theory, glass dynamics can be understood by using restricted statistical mechanics for metastable states.

This can be done by introducing restricted phase-space average: In statistical mechanical averages replace the trace Tr by Tr→Tr′ where Tr′ means including only those configurations belonging to the metastable glassy state.[9–13] The global metastable glassy states are essentially frozen liquid states with nonzero Debye-Waller factors which indicates they have elastic properties. To describe them we introduce two related key ideas.[10] First we imagine an order-parameter description in terms of frozen-density fluctuations, $\delta n=n-n_l$. Here n is the number density and n_l is the liquid-state number density. Other order parameters can be used but frozen-density fluctuations are the simplest and most directly related to the most trivial characterization of a solid: a nonzero Debye-Waller factor. Since the glassy state is amorphous or aperiodic the frozen-density order parameter is most naturally specified by a (functional) probability measure $DP[\delta n]$. Within a mean-field approach the glassy states are characterized by the first and second moments of this measure,[10]

$$\overline{\delta n(\mathbf{x})}=\int DP[\delta n]\delta n(\mathbf{x})=\frac{1}{V}\int d\mathbf{x}\,\delta n(\mathbf{x})=0 , \tag{2.2a}$$

$$q\equiv\overline{[\delta n(\mathbf{x})]^2}=\int DP[\delta n][\delta n(\mathbf{x})]^2 =\frac{1}{V}\int d\mathbf{x}[\delta n(\mathbf{x})]^2 . \tag{2.2b}$$

The final equalities in Eqs. (2.2) assumes self-averaging and from Eq. (2.2b) it follows that the zero in Eq. (2.2a) is really a term of $O(V^{-1/2})$ with V the volume and the bulk limit is always taken. The Edwards-Anderson order parameter[20] q in Eq. (2.2b) is zero in the metastable liquid phase and is nonzero in metastable glassy states. Some features of this order parameter will be discussed below. The second key notion we introduce is that, in general, one expects a large number of distinct glassy metastable states. We denote a particular glassy state by the label s, with the frozen-density field in that state given by $n_s = n_l + \delta n_s$ and the free energy equal to F_s [the density fields in Eq. (2.1) should also be labeled by the index s]. Calculations indicate[10,11] that below a temperature we will call T_A, there are an extensive number (the number of states scales like $\exp[\alpha N]$ for an N-particle system) of global statistically similar incongruent metastable glassy states. Statistically similar states have the same spatially averaged correlation functions and incongruent states have zero overlap:[10,21]

$$q_{ss'} = \delta_{ss'} q = \frac{1}{V} \int d\mathbf{x}\, \delta n_s(\mathbf{x}) \delta n_{s'}(\mathbf{x}) \; . \tag{2.3}$$

Note that because these states are statistically similar, one cannot use an external field to pick out a particular state. Consequently, the partition for the global metastable states is given by a sum over all s:[21,22]

$$Z = \mathrm{Tr} \exp(-\beta H) = \sum_s \exp(-\beta F_s) \; . \tag{2.4}$$

Technically Eq. (2.4) is correct (in the restricted ensemble Tr′) if the barriers between states diverge in the bulk limit. Physically it is a reasonable equation for a restricted time interval if the barriers are large but finite. We argue below that even in this doubly restricted ensemble, entropic driving forces always lead to nucleation processes which in turn lead to mosaic states rather than global metastable glassy states. Nevertheless, the phase-space decomposition given by Eq. (2.4) is a useful intermediate step.

With Eq. (2.4) one can define a canonical free energy,[21,22]

$$F_c = -\frac{1}{\beta} \ln Z \tag{2.5a}$$

and a component averaged free energy for the global metastable states,

$$\bar{F} = \sum_s P_s F_s \tag{2.5b}$$

with P_s the probability of being in the s component,

$$P_s = \frac{1}{Z} \exp(-\beta F_s) \; . \tag{2.5c}$$

F_c'' and $\bar{F}$ are related by[21]

$$F_c = \bar{F} + k_B T \sum_s P_s \ln P_s \equiv \bar{F} - TI \; . \tag{2.6}$$

Here I is usually called the complexity, but we will argue it can also be interpreted as a state entropy which is bounded from above by a configurational entropy in non-mean-field models. In general I is related to the solution degeneracy and it is extensive (and $F_c \neq \bar{F}$) if there are an exponentially large number of states. Note that the physical free energy, if the barriers are infinite, is $\bar{F}$ because the term, $-TI$ in Eq. (2.6) is a entropy term which is a measure of parts of state space not probed in a finite amount of time. Since a physical entropy should only be associated with accessible configurations it follows that F_c is not the physically meaningful free energy of the global glassy metastable states.

If we ignore the nonperturbative droplet (cf. below) fluctuations that lead to the mosaic states then a mean-field-like calculation gives the following.[10] (1) For $T < T_A$ there exist an extensive number of statistically similar incongruent global glassy metastable states. For $T = T_A^+$, q drops discontinuously to zero and the restricted ensemble Z' is no longer physical meaningful for higher temperatures. The only global state is the liquid state. (2) For $T < T_A$, F_c'' is identically equal to the liquid-state free energy, F_L, and $\bar{F} - F_L = TI \sim O(V)$, for all temperatures such that (defining T_K) $T_K < T < T_A$. This indicates that in this temperature range the glassy states are metastable with respect to the liquid phase. (3) At T_K, $\bar{F} = F_L$, and the complexity vanishes (or becomes nonextensive). The glassy states are then thermodynamically preferred to the liquid state. We identify T_K with an ideal STG transition temperature. It will be interpreted further below. (4) A mean-field dynamical theory leads to a glassy freezing for all $T < T_A$. Physically this is because the extensive solution degeneracy implies that with probability one the initial configuration of the system will be in a global glassy metastable state. Because nucleation out of a metastable state is a nonperturbative fluctuation effect that is ignored in a mean-field dynamical theory the system will stay in a particular metastable forever. Note that the equality of F_c'' and F_L is consistent with this result. The canonical free energy of all the glassy states is equal to the liquid-state free energy because of the entropy or complexity term in Eq. (2.6). Conceptually the existence of a temperature (region) T_A is important because it indicates (cf. Sec. III) that for all $T < T_A$, long-time dynamical processes are activated.

III. ENTROPIC DRIVING FORCES FOR ACTIVATED TRANSPORT

In this section we give a preliminary discussion of the expected nonperturbative droplet or domain-wall dynamics for $T_K < T < T_A$. The scenario given here will set the stage for the more refined arguments given in the subsequent sections of this paper. It should be pointed out that because the equation for the saddle point for the field theory we have proposed for the structural glass transition,[10] has multiple solutions, fluctuation effects cannot be accounted for by self-consistent variational (or perturbative) schemes. What is needed is to compute the instanton contributions and physically this corresponds to the large-scale droplets considered here. Crucial to our arguments is the nature of the entropic driving force for activated transport. We first discuss this driving force and then we will use it in a naive way to obtain some preliminary results.

The relevant entropy[11] for our scenario is associated with the multiplicity of disjoint ergodic states above T_K. In several exactly soluble models[11,13,23] this entropy does vanish at a temperature T_K, where a random first-order phase transition occurs. We assume a similar behavior for structural glasses. We shall refer to this entropy as the state entropy S_s since it is a measure of the number of disjoint (at least in mean-field theory) ergoidc states. A precise discussion about the meaning of S_s can be given by considering the model problem of the Potts glass in the mean-field limit.[11,12] For this model both the canonical free energy, F_c, and the component-averaged free energy, $\bar{F}$, can be exactly computed and written as

$$F_c = -\frac{1}{\beta}\ln\left[\sum_s \exp(-\beta F_s)\right] = -\frac{1}{\beta}\ln N\int df\, e^{N[\alpha(f)-\beta f]} \tag{3.1a}$$

and

$$\bar{F} = \sum_s P_s F_s = \frac{N}{Z''}\int df\, f e^{N[\alpha(f)-\beta f]} . \tag{3.1b}$$

Here $f = F/N$ is an intensive free-energy variable and $g(f) = \exp[N\alpha(f)]$ is the density of free-energy states. In the thermodynamic limit, the integrals in Eq. (3.1) can be evaluated by saddle-point methods and one gets

$$F_c = \bar{F} - \frac{N}{\beta}\alpha(\bar{f}) , \tag{3.2a}$$

with $\alpha(\bar{f}) \neq 0$ for $T_K < T < T_A$. Here $\bar{f}$ satisfies the relation $\beta = \partial\alpha(\bar{f})/\partial\bar{f}$. Notice that the existence of $\bar{F}$ in non-mean-field theories is hard to justify in the region $T_K < T < T_A$ because the states being considered are metastable. We have assumed that the relaxation times are extremely long compared to the typical experimental time scales and that one can calculate F_s by considering those configurations belonging to s. From Eq. (2.6) and the discussion below it, the state entropy is

$$S_s = k_B N\alpha(\bar{f}) , \tag{3.2b}$$

i.e., S_s is related to the number of free-energy states. In the droplet arguments given below we shall assume that S_s is the important driving force for large-scale domain dynamics and when $S_s \neq 0$, there is liquid-like relaxation. Note that S_s is not precisely the configurational entropy, S_c, considered in the Adams-Gibbs[3] picture of the glass transition. The distinction between S_s and S_c can be seen in several ways. First, as already mentioned in Sec. II, S_s (there called the complexity) is not a physically meaningful entropy in mean-field theory. However, even in this case, in a given ergodic component one would imagine that both a vibrational and a configurational entropy exist. The configurational entropy would be associated with different defect arrangements in a particular ergodic state. In non-mean-field models S_s would become a physical entropy[11] because domains of a distinct glassy state inside an original glassy state are allowed if droplet fluctuations are taken into account. Within each droplet there will again be vibrational and configurational entropy. These arguments indicate that the total configurational entropy as is usually discussed is greater than or equal to the state entropy, S_s. Physically, we imagine that the distinction between the configurational entropy associated with defects and the "configurational" entropy associated with states is related to length scales. The configurational entropy associated with defects should lead to motion involving only a few particles because that is all it does in the crystal phase and consequently it does not lead to liquid-like relaxation. We argue below, on the other hand, that S_s leads to large-scale liquid-like transport involving many particles.

Assuming that S_s is an entropic driving force for activated transport we next give a preliminary argument on the expected behavior of the liquid-state relaxation time as S_s decreases.[11] Consider a region of size L^d in a single glassy state and estimate the probability of a different glassy state forming inside this region. Because the different glassy states have roughly the same free energy the driving force for droplet formation is entropic: Nucleation of a new state occurs because there are so many states to escape to. The effective free-energy driving force for this process is of magnitude $Ts_s L^d$. Here s_s is the state entropy per unit volume. Opposing the droplet formation is a surface free energy cost, F_{surface}. For large L, F_{surface} can scale at most as σL^{d-1}, with σ the surface tension between two different glassy states. Comparing these two forces one sees that at large enough L a droplet will always form. Repeating this argument for every fluid region leads to the mosaic state already discussed. With this argument the typical size of a glassy cluster will be of order $L^* \sim \sigma T/s_s$ and the free-energy barriers for activated transport are of order $\Delta F^* \sim (\sigma T/s_s)^{d-1}$. Note that this picture predicts a divergent length scale, L^*, and a divergent relaxation time, $\tau \sim \exp(\beta\Delta F^*)$, as the ideal STG transition at T_K $[s_s(T=T_K)=0]$ is approached. Also note that if we consider transitions into the liquid state then the picture is unchanged because the free-energy driving force for this process is also $\bar{F} - F_L = \bar{F} - F_c'' = TI = TS_s$. In fact, because the formation and destruction of the glassy clusters will continuously occur, the mosaic state for $T_K < T < T_A$ is also a liquid state.

These considerations lead to the following picture of a viscous liquid at low temperatures (well below T_A). First the liquid consists of glassy clusters or amorphons separated by interfaces or domain walls. As the temperature is decreased the size of the glassy clusters, $L^* = \xi$, increases and the coherence length of the clusters, ξ, diverges st an ideal STG transition. Secondly, the long-time dynamics is controlled by activated processes where the glassy clusters are destroyed and created. The time scale for these processes diverges exponentially as the ideal STG transition is approached. Motivated by the appearance of the divergent coherence length, ξ, we use scaling ideas in the following sections to refine these ideas.

We stress the dynamical picture given above is very different than the nucleation and growth picture near a usual first-order phase transition.[24] The glassy clusters we consider remain small (on a scale given below) because once a droplet nucleates, another droplet can nucleate inside the previous one (because of the entropic driving force) and consequently the system will consist of many percolating domains. Thus the kinetics is both controlled and limited by fluctuation effects. In usual nucleation theory the growth is exponentially fast immediately subsequent to nucleation and it is not controlled by fluctuations. These arguments also suggest that unlike usual first-order phase transitions, fluctuations are very important in the droplet picture of the glass transition and the physics should be controlled by a single divergent length scale $\xi \sim t^{-\nu}$, with ν a correlation or coherence length exponent. Because a domain within domain picture makes sense only on scales greater than or equal to ξ, we expect ξ is the typical size of a glassy cluster. In general we can also argue that because the driving forces for activated transport scale as

$$F_{\text{driving}} \sim -Ts_s L^d \sim -tL^d \sim -t\xi^d \sim -t^{1-\nu d} ,$$

the activation barriers for critical transport near T_K must scale as

$$\Delta F^* \sim t^{1-\nu d} . \tag{3.3}$$

Here $t=(T-T_K)/T_K$ and we assume s_s vanishes linearly at T_K. If one prefers throughout this paper one could write s_s where $t=(T-T_K)/T_K$ occurs. This may be preferable if one looks at variations with other parameters other than the temperature such as pressure or if the transition really eventually becomes rounded. If the transition is rounded, our physical picture still makes sense, as long as large (but finite in this case) amorphous clusters determine transport in the glass-transition region and these clusters increase in size as temperature is decreased.

IV. SCALING EXPONENTS NEAR RANDOM FIRST-ORDER PHASE TRANSITIONS

As already mentioned we imagine an ideal STG transition that is in part characterized by glassy clusters whose typical size ξ is assumed to diverge as the ideal transition temperature is approached. The following arguments, as well as earlier ones by others, give estimates of the correlated volumes as containing 50–500 particles at the laboratory glass transition. In general, this indicates that T_g is a considerable distance from a possible ideal STG transition. In fact, $(T_g-T_K)/T_K$ is typically estimated to be ≥ 0.1. All of this implies that droplet arguments based on large length scales are somewhat problematic near the experimental T_g. With this caveat in mind, we now proceed to investigate the supposed ideal STG transition.

In this section we provide two different arguments for the value of the exponent ν which characterizes the divergence of ξ near T_K. The first argument is heuristic and is based on an approximate use of the fluctuation formula. The argument was originally hinted at by Donth.[25] We assume that the correlated volumes are large enough that they can be treated as independent thermodynamic systems. Because of our belief that an ideal glass transition exists, the correlation length becomes anamolously large near T_K and hence for $t \ll 1$, such an assumption is valid. The temperature fluctuation in a correlated volume is given by[26]

$$\delta T^2 = k_B T^2/NC , \tag{4.1a}$$

where N is the number of particles in a correlated volume of dimension ξ, C is the specific heat per particle, and the difference between constant volume and constant pressure density is ignored. Let the mean temperature of a correlated volume be $T=T_K(1+t)$ with $t \ll 1$. We now assume that mean-temperature fluctuation δT is less than the deviation of T from T_K, i.e., $\delta T \leq tT_K$. The rationale for this is that the temperature fluctuation is related to entropy fluctuation and this implies that if δT is large then there is sufficient entropic driving force at $T=T_K(1+t)$ for relaxation. Since, $t \ll 1$ this is not very probable. This assumption allows us to write Eq. (4.1a) as

$$k_B/NC \leq t^2 . \tag{4.1b}$$

The number of particles N scales like $\xi^d \sim t^{\nu d}$ and using this in Eq. (3.1b) we get the inequality,

$$\nu d + \alpha \geq 2 , \tag{4.2a}$$

or

$$\nu \geq 2/d . \tag{4.2b}$$

In obtaining Eq. (4.2b) we have assumed that the specific-heat exponent $\alpha=0$. The specific heat at $T=T_g$ for most glass-forming materials is discontinuous at $T=T_g$ and presumably it is also discontinuous at $T=T_K$. This allows one to estimate $\alpha=0$. Note that the equality form of Eq. (4.2) is just a hyperscaling law.[27] This connection justifies our assumption that $\delta T \leq tT_K$ for T near T_K. Notice that Donth applied the fluctuation formula at $T=T_g$. His expression for the mean size of the domain ξ involves the experimental decrease in specific heat at constant pressure and T_g.

The second argument that leads to $\nu=2/d$ is more formal. Mean-field calculations based on a model Hamiltonian written in terms of density fields, indicate that at the ideal STG transition temperature, T_K, the Edwards-Anderson order parameter [Eq. (2.2)] is discontinuous. Because q is a second moment of a probability measure we call such a transition a random first-order phase transition. We also expect that, in general, there is a discontinuity in q at T_K. As a liquid is cooled above T_K (and above T_g) it is well documented that at finite time or frequency scales there exist solidlike behavior, e.g., shear waves. This effectively implies the presence of finite-frequency elastic coefficients, or a finite-frequency Debye-Waller factor. It would seem that any reasonable dynamical scaling law which contains solid behavior at *short times* and liquid behavior at *long times* will imply a discontinuous Debye-Waller factor as the relaxation time diverges. This, in turn, implies a discontinuous q at T_K.

Scaling exponents for regular first-order phase transitions and their interpretation have been discussed before by many authors. Here we generalize the finite-size-scaling arguments of Fisher and Berker[28] to random first-order phase transitions. We first give the argument and then interpret the results.

In a finite system of size $L^d = V$ there can be no sharp phase transition. To describe the growth of a sharp random first-order phase transition as $L \to \infty$, we assume a finite-size-scaling hypothesis. We again use a density order parameter. For the conjugate field we use a chemical potential $h=\mu(T,p)-\mu(T_K,p_K)$. Here we allow for the fact that, in general, an ideal STG transition will depend on both the temperature T and the pressure p. We postulate that the scaling part of the free-energy density behaves as

$$f(h,L)\sim L^{-\zeta}Z(L/\tilde{\xi})\equiv L^{-\zeta}Y(hL^{\theta}) , \tag{4.3}$$

where $\tilde{\xi}$ (the finite-size correlation or coherence length) is the length scale where finite-size rounding takes place. The last equality in Eq. (3.1) follows from the definition $\tilde{\xi}\sim|h|^{-\tilde{\nu}}$ (where $\tilde{\nu}$ is the finite-size-coherence-length exponent) and $\theta\equiv 1/\tilde{\nu}$. The exponents $\tilde{\nu}$ and ζ can be determined from the scaling ansatz given by Eq. (3.3) and Eqs. (2.2) with q discontinuous at T_K. We first consider

$$\left[\frac{\partial f}{\partial h}\right]^2=\frac{1}{L^{2d}}\left[\int d\mathbf{x}\,\delta n(\mathbf{x})\right]^2 \sim L^{-d}q\sim L^{-2\zeta+2\theta}[Y'(hL^{\theta})]^2 . \tag{4.4a}$$

Using the discontinuity of q at T_K^- this implies the equality

$$\theta-\zeta=-\frac{d}{2} . \tag{4.4b}$$

Another equality can be obtained by considering the susceptibility squared. Following Fisher and Berker[28] we obtain the equality

$$2\theta-\zeta=0 . \tag{4.4c}$$

Combining these equations yields

$$\tilde{\nu}=1/\theta=\frac{2}{d},\ \ \zeta=d . \tag{4.5}$$

The corresponding result for a regular first-order phase transition is $\tilde{\nu}=1/d$. The difference arises because the conjugate field for the Edwards-Anderson order parameter is effectively $\sim h^2$. If we identify $\tilde{\nu}=2/d$ with the correlation-length exponent ν and use hyperscaling $\alpha=2-\nu d$, then we conclude that at T_K there is no latent heat and the specific heat is discontinuous (or perhaps logarithmically divergent). We remark the first-order-like phase transitions with these properties are somewhat unusual. However, there are several exactly soluble random-spin models with precisely these properties: the random-energy model,[23] mean-field p-spin-interaction spin-glass models,[13] and mean-field Potts glass models.[11]

There is another way of looking at this scaling physically.[11] Above T_K, the state entropy per particle is finite and the number of thermally accessible states grows linearly with the volume of the region. Below T_K the state entropy per particle vanishes yet the number of thermally accessible state is not *one* for a macroscopic sample. Rather it is a large but finite number diverging at T_K from below. There will be a characteristic size $V^*\sim\xi^d$ where we will be able to distinguish these two behaviors, that is, below T_K the number of states will increase exponentially with size until V^* is reached and saturation observed. Using the scaling ansatz above and below the transition then gives $\xi^d\sim(T-T_K)^{-2}$ and hence leads to the exponent $\nu=2/d$. In obtaining this result we have assumed that for $T>T_K$, $S_s\sim Vt$, and that for $T<T_K$, $S_s\sim V^0|t|^{-1}$. The exponent for the divergence of S_s below T_K has been taken from a mean-field calculation.[11] Above the transition, V^* will be the size after which exponential growth of the number of states clearly applies.

We also note that these exponents are consistent with the observed phenomenology at the laboratory glass transition where a latent heat is not found but there is fairly sharp (at least in fragile glass formers) discontinuity in the specific heat.[1] In fact the scaling exponent $\alpha=0$ (and by hyperscaling $\nu=2/d$) follows phenomenologically if we naively assume that the ideal STG transition has the limiting characteristics of the laboratory glass transition. This assumption is actually more reasonable than it probably appears at first glance. In both the ideal and laboratory transitions the crucial physical picture is the (effective) vanishing of the configurational entropy: The system gets trapped in a region of phase space. In the ideal transition there is a real phase transition because the number of liquid states becomes nonextensive and because the global glassy state is free energetically preferred to the liquid state. In the laboratory transition the effective number of liquid states become nonextensive because other states are not accessible on the experimental time scale. The similarity of these transitions is also indicated by the fact that glass transitions observed in computer simulations are similar to laboratory glass transitions.

Finally, we note that if the starting STG Hamiltonian has a random part in it and if the STG transition was due to this randomness then the rigorous inequality $\nu\geq 2/d$ must be obeyed.[29(a)] The equality $\nu=2/d$ is most naturally associated with a random first-order phase transition.[29(b)] It is not clear, however, whether these arguments are relevant for the STG transition where the randomness is self-generated. Within our picture, however, the equality, $\nu=2/d$, is however necessary to give relaxation times in accord with the Vogel-Fulcher law. Thus, insofar as the experimental fits to viscosity data obeys the VF equation, it seems that the STG transition mimics the behavior seen in systems where the transition is random first order.

V. SCALING AND ACTIVATED TRANSPORT

In this section we combine the physical picture set up in Sec. II with the scaling ideas given in Sec. III, to obtain an expression for the relaxation time in the liquid state. The fundamental assumption is that there are glassy domains that are metastable, and consequently they continuously undergo transitions with a characteristic relaxation time which determines transport in the viscous liquid. For large scale domains, we first argue that this notion automatically implies that the free energy opposing the transition must scale as

$$F_{\text{opposing}}\simeq\gamma L^{\theta} \tag{5.1a}$$

with

$$\theta \le \frac{1}{\nu}(\nu d - 1) , \tag{5.1b}$$

or $\theta \le d/2$ if $\nu = 2/d$. In Eq. (5.1a), γ is a positive constant. The inequality given by Eq. (5.1b) can be rationalized by the following argument. In general, for $T_K < T \ll T_A$, we expect the driving force of activated transport in a volume of size L^d to be entropic and of order

$$F_{\text{driving}} \simeq -Ts_s L^d \simeq -tAL^d . \tag{5.2a}$$

Here $t = (T - T_K)/T_K$ and we have assumed that s_s vanishes linearly at T_K and A is a positive constant. If the typical size of a glassy cluster scales like $L \sim \xi \sim t^{-\nu}$ then Eq. (5.2a) becomes

$$F_{\text{driving}} \sim -At^{-(\nu d - 1)} . \tag{5.2b}$$

Similarly, Eq. (5.1a) scales as $F_{\text{opposing}} \sim \gamma t^{-\nu\theta}$. Therefore, in order for the droplets to be unstable at large enough scales, the inequality given by Eq. (5.1b) must be satisfied.

We next give two arguments for the expected behavior of the liquid-state relaxation time near T_K. The most naive argument is as follows. If the typical size of a glassy cluster behaves like $L \sim \xi \sim t^{-\nu} \sim t^{-2/d}$ then Eq. (3.3) implies that the free-energy barriers for activated transport must behave like

$$\Delta F^* \sim t^{-(\nu d - 1)} \sim t^{-1} . \tag{5.3a}$$

This immediately leads to a relaxation time of order,

$$\tau \simeq \tau_0 \exp[DT/(T - T_K)] . \tag{5.3b}$$

Here τ_0 is a microscopic relaxation time, D is a positive constant and we note that Eq. (5.3) is just the Vogel-Fulcher law. Also note that this is a consistent result only if the free-energy force opposing activated transport scales as

$$F_{\text{opposing}} \simeq \gamma L^{d/2} , \tag{5.4}$$

with γ a positive constant. This should be contrasted with the ansatz $F_{\text{surface}} = F_{\text{opposing}} \sim L^{d-1}$ used in Sec. III.

We next give a more careful discussion of the free-energy forces which oppose activated transport. We generalize an argument due to Villian[30] for the random-field Ising model (RFIM) to the structural glass problem. Our main conclusion will be that the surface forces which oppose activated transport do scale as Eq. (5.4) for the length scales of interest. Equations (5.1a) and (5.4) then independently of Sec. IV lead to Eq. (5.3) and all of our arguments are consistent with each other. Our arguments also lead to other features which are of experimental consequence which we discuss in Sec. VI and elsewhere.

In general in disordered phases one does not expect a finite macroscopic surface energy. This implies that the force opposing activated transport should be written $F_{\text{opposing}} \sim \sigma(L)L^{d-1}$ with $\sigma(L)$ meaningless beyond a length scale related to ξ. To understand the vanishing surface tension we start with a small droplet with a finite surface tension and then use renormalization-group-like ideas to determine the larger-scale behavior of $\sigma(L)$. The basic idea is to note that a smooth domain wall can decrease its free energy, $F_{\text{DW}} = F_{\text{surface}}$, by forming bumps of a third glassy state (say 3) if the original smooth domain wall separates two other glassy states (say, 1 and 2). Because there are an extensive number of different glassy states they have a free-energy distribution that is Gaussian and the typical free-energy difference between any two states in a volume of size L^d is of order $\pm L^{d/2}$. Thus the typical free-energy gain for a bump of radius r and height ζ is

$$\delta F_1 \sim -Hr^{d/2} \left[\frac{\zeta}{r} \right]^{1/2} \tag{5.5a}$$

with H a positive constant. The free-energy loss due to the additional surface tension σ is (here ζ is assumed to be less than r and that the bump is not very rough)

$$\delta F_2 \sim \sigma r^{d-1} \left[\frac{\zeta}{r} \right]^2 . \tag{5.5b}$$

Finally, as for the bulk processes, there is also an entropic driving force

$$\delta F_3 \sim -tAr^{d-1}\zeta . \tag{5.5c}$$

We avoid explicitly including δF_3 by considering the surface free energy at T_K where δF_3 vanishes. If we assume scaling, then these results will be shown to lead to Eq. (5.3a). Minimizing $\delta F_{\text{DW}} = \delta F_1 + \delta F_2$ with respect to ζ yields

$$\zeta \sim \left[\frac{H}{\sigma} \right]^{2/3} r^{(5-d)/3} , \tag{5.6a}$$

and the resulting free-energy gain per unit area is

$$\frac{\delta F_{\text{DW}}}{r^{d-1}} \sim -H \left[\frac{H}{\sigma} \right]^{1/3} r^{-2(d-2)/3} . \tag{5.6b}$$

Equation (5.6b) can be interpreted as a modification of σ due to the elimination of degrees of freedom of wavelength of order r. In a renormalization-group spirit we write Eq. (5.6b) in differential form as

$$d\sigma(r) = -K^{4/3} H \left[\frac{H}{\sigma} \right]^{1/3} r^{-2(d-2)/3} \frac{1}{r} dr , \tag{5.7}$$

with K a constant. Integrating Eq. (5.7) yields

$$\sigma(r) \sim KH/r^{(d-2)/2} . \tag{5.8}$$

In obtaining Eq. (5.8) we have imposed the boundary condition $\sigma(r \to \infty) \to 0$. At $T = T_K$ this is required to obtain a scale-invariant surface tension. With Eq. (5.8) and using that r scales with L, the surface-free-energy opposing activated transport at T_K is

$$F_{\text{opposing}} \sim \sigma(L)L^{d-1} \sim KHL^{d/2} , \tag{5.9}$$

which is just Eq. (5.4). Note that Eq. (5.8) in Eq. (5.6a) yields $\zeta \sim r$.

Using Eq. (5.9) and assuming scaling we write the surface term opposing activated transport as $F_{\text{opposing}} = L^{d/2}h(L/\xi)$, where $h(x)$ is a scaling function with $h(0)$ a constant. Minimizing $F_{\text{opposing}} + F_{\text{driving}}$ then self-consistently gives $L^* \sim \xi \sim t^{-2/d}$. We conclude that the Vogel-Fulcher form, Eq. (5.3), is probably the asymptotic relaxation time near T_K and that the typical size of the cooperatively rearranging regions (CRR's) diverges as $\xi \sim t^{-2/d}$ as T_K is approached from above. Note that the typical cluster size, $\xi \sim t^{-2/d}$, is less than that indicated by the naive arguments of Sec. III for all $d > 2$.

Finally, it is easy to verify that if Eq. (5.5c) is retained then this picture is not changed. As already mentioned Eq. (5.8) in Eq. (5.6a) yields $\zeta \sim r$. As a consequence of this δF_3 can be interpreted as a renormalization of F_{driving}. We stress, that we have generalized only the simplest of Villian's arguments for the RFIM problem.[30] More sophisticated arguments could lead to a more complicated scaling behavior characterized by additional exponents. In particular, the bulk driving force could also be renormalized in a nontrivial way. Physically, the main conclusion we draw is that there are mechanisms that decrease the surface free energy from its naive behavior $\sim L^{d-1}$.

VI. DISCUSSION

The arguments we have put forward in this paper are speculative and in some respects similar to previous qualitative pictures of the liquid-glass transition. Our picture is, however, grounded in the study of some exactly solved, albeit, mean-field models of random-phase transitions which have transition characteristics bearing a strong resemblance to observations on the glass transition. In this section we would like to highlight the contrasts between our picture and previous ones and discuss experimental tests that might distinguish alternative hypotheses. There are several independent points that arise in this consideration and they are enumerated below.

(1) One of the hallmarks of our picture is the presence of a length scale diverging near T_K and that rearrangements of regions of this size have barriers dependent on that size. Such a divergent length connected with an impending entropy crisis also plays a role in the Adams-Gibbs-DiMarzio theory of the glass transition.[3] Divergent or nearly divergent lengths unconnected with an entropy crisis also enter into other theories of glass transitions such as those for metallic glasses[4] and, of course, continue to be invoked in entirely qualitative accounts of the glass transition. The divergence according to our analysis, $\xi \sim (T-T_K)^{-2/d}$ is stronger than the Adams-Gibbs-DiMarzio divergence $\xi \sim (T-T_K)^{-1/d}$. The $1/d$ exponent would ordinarily be associated with a traditional first-order transition rather than a random one. Thus, the Adams-Gibbs-DiMarzio exponent would give extremeley small connected clusters. On the other hand, the $2/d$ exponent would, in favorable cases, give lengths of several nanometers. Thus, experiments on glass formation in confined geometries may give some insight. Experiments along these lines are progress,[31] although it is clear that many complicating features can and do enter, such as surface effects and the difficulty of achieving a homogeneous distribution for confined systems. Other direct probes of such length scales are also conceivable. One important possibility is the use of the scanning tunnel microscope or various tribiological experiments[32] on friction in ultrathin films where nonlinear effects may be connected with phenomena at the correlation length scale.

(2) An issue we have not addressed in our scenario is the distribution of relaxation times. The problem of distributed relaxation times has attracted a great deal of attention lately in the study of many systems. The droplet scenario presented here puts great store in the random parts of the free energies of different droplets. It therefore suggests a rather wide distribution of activation energies and, therefore, relaxation times. Although there is a distribution of activation energies in glasses, others have emphasized the relative narrowness of this distribution in their speculations on the glass transition. A calculation of the width in our picture requires assessing the relative size of the renormalized surface energy and the randomness energies themselves. They are dimensionally the same but need not be numerically comparable. There is some evidence that the distribution gets broader in fragile glasses with decreasing temperature and this would be consistent with our picture.[33] Furthermore we have not dealt with droplet interaction effects and/or the question of the pinning of the walls in our so-called mosaic states. An *ad hoc* explanation of a narrow relaxation-time distribution would be a motional narrowing of the randomness due to fluctuations of other droplets because there is no quenched randomness in the glass Hamiltonian. Such a phenomenon would be however outside the scope of our single-droplet calculations and would require a complete dynamical analysis.

(3) It is worth emphasizing the distinction between configurational entropy S_c and the entropic driving force we have considered here. As stated before these differences become transparent by considering the mean-field spin-glass models without reflection symmetry. In these models, there is an Edwards-Anderson order parameter for $T < T_K$ even though the thermodynamics of this model is not frozen entirely below this transition temperature. Hence, there are still relaxing degrees of freedom, which can, for example, give a β relaxation. These can be further frozen out by the time the system reaches zero temperature. Any estimation of configurational entropy in real glasses is also complicated by the existence of additional relaxing degrees of freedom below the laboratory glass-transition temperature. We would agree with the recent observation[7] that as usually defined, the configurational entropy will not vanish in any realistic model with finite-ranged forces. This is because point defects are always present and they lead to a finite contribution to what is usually called configurational entropy.[34] Still, this in itself does not make a random first-order phase transition impossible. Even in the case of a crystalline solid, where we known there is a first-order phase transition from liquid to solid, there is a finite "configurational" entropy in the crystal phase associated with point defects.

(4) The remarks about the ambiguity of configurational entropy (see Sec. III) must translate into some ambiguity about the value of T_K or even its existence as a sharp point. Suffice it to say that our arguments correlate the impending vanishing of an entropy at T_K with the Vogel-Fulcher behavior of the activation barriers. Thus, even if some crossover to another type of phase transition or if rounding of the transition were to occur below the laboratory transition, we believe our arguments are relevant to the temperature range usually studied. We also point out, that for the exactly soluble spin models which we considered previously,[11–13] T_K is a temperature where both S_s vanishes and where the free energy of the high-temperature phase is equal to the free energy of the low-temperature glassy phase. Thus, within our picture, at T_K the liquid-state free energy is equal to the free energy of a global glassy state. This is another independent and less ambiguous condition for an ideal glass transition to occur.

In this regard, the connection of the glassy state and the random-field Ising model alluded to in Sec. V (as well as by others[6]) may conceal a deeper analogy. The renormalization-group analysis of nonrandom Potts systems proceeds more readily with the introduction of new Ising-like variables in addition to the original Potts variables.[35] If such an analysis is applied to a random Potts system with short-range interactions these Ising variables attain both random and regular fields upon renormalization.[36] Thus, on longer length scales there may be a natural connection of the Potts glass and the random-field Ising magnet. The presence of regular, as well as random fields in this analogy should remind us of another possibility. A random Ising model with an average field will probably only show a critical point when that field vanishes. In the glass analogy this is equivalent to having a vanishing configurational entropy not along a line in the temperature-pressure phase diagram but only at a point, where the critical behavior will occur. We point out that such a possibility allows one to accommodate in a simple way deviations of the Prigogine-deFay ratio from unity.[1] Such deviations, of course, can also be explained in a variety of dynamical fashions at the laboratory, as opposed to the ideal glass transition.

(5) The role of crystallization in any scenario needs to be addressed. First, macroscopic nucleation theory is applicable if the critical embryo for the periodic crystalline phase is larger than the correlation length ξ. In this case the growth of the critical nuclei is severely impeded by the large value of viscosity.[37] If the critical embryo is smaller than ξ then new considerations apply. In this case nucleation is not strongly impeded by slow transport but the thermodynamics of the transition is modified. The situation would be like solid-in-solid nucleation. The surface energies (barriers) for this situation are very large and consequently these processes are not very likely.

Finally we remark that the basic ideas of the droplet pictures require the existence of some sort of coherent structure on a moderate length scale. Once simulations on good glass formers are done on long length scales and rather long times this may become more apparent.

ACKNOWLEDGMENTS

This work was supported by National Science Foundation Grant Nos. DMR-86-07605, CHE-86-57356, and CHE-84-18619.

[1]For a recent review of structural glasses see, J. Jackle, Rep. Prog. Phys. **49**, 171 (1986).

[2](a) D. Turnbull and M. H. Cohen, J. Chem. Phys. **29**, 1049 (1958); (b) G. S. Grest and M. H. Cohen, Adv. Chem. Phys. **48**, 454 (1981).

[3](a) J. H. Gibbs and E. A. DiMarzio, J. Chem. Phys. **28**, 373 (1958); (b) G. Adam and J. H. Gibbs, J. Chem. Phys. **43**, 139 (1965).

[4](a) D. R. Nelson, Phys. Rev. B **28**, 5515 (1983); (b) J. F. Sadoc, J. Non-Cryst. Solids **75**, 103 (1985), and references therein.

[5]R. G. Palmer and D. Stein, Phys. Rev. B **38**, 12 035 (1988).

[6]J. Sethna, Europhys. Lett. **6**, 529 (1988).

[7]F. Stillinger, J. Chem. Phys. **89**, 6461 (1988); **88**, 7818 (1988).

[8]Y. Singh, J. P. Stoessel, and P. G. Wolynes, Phys. Rev. Lett. **54**, 1059 (1985).

[9]T. R. Kirkpartrick and P. G. Wolynes, Phys. Rev. A **35**, 3072 (1987).

[10]T. R. Kirkpartrick and D. Thirumalai, J. Phys. A **22**, L149 (1989).

[11]T. R. Kirkpartick and P. G. Wolynes, Phys. Rev. B **36**, 8552 (1987).

[12](a) T. R. Kirkpatrick and D. Thirumalai, Phys. Rev. A **37**, 4439 (1988); (b) Phys. Rev. B **37**, 5342 (1988); (c) D. Thirumalai and T. R. Kirkpartrick, *ibid.* **38**, 4881 (1988).

[13]T. R. Kirkpartrick and D. Thirumalai Phys. Rev. Lett. **58**, 2091 (1987); Phys. Rev. B **36**, 5388 (1987).

[14]See, for example, the papers in *Proceedings of the Third International Conference on Rapidly Quenched Metals, Brighton 1978*, edited by B. Cantor (The Metals Society, London, 1978).

[15]S. S. Chang, J. A. Horman, and B. B. Bestul, J. Res. Natl. Bur. Stand. Sect. A **71**, 293 (1967).

[16]C. A. Angell and W. Sichina, Ann. N.Y. Acad. Sci. **279**, 53 (1976).

[17]W. T. Laughlin and D. R. Uhlmann, J. Phys. Chem. **76**, 2317 (1972).

[18]F. Simon and Z. Anorg, Algemine. Chem. **203**, 217 (1931).

[19]W. Kauzmann, Chem. Rev. **43**, 219 (1948).

[20]This terminology is taken from the spin-glass literature. See K. Binder and A. P. Young, Rev. Mod. Phys. **58**, 801 (1986).

[21]R. G. Palmer, Adv. Phys. **31**, 669 (1982).

[22]C. DeDominicis and A. P. Young, J. Phys. A **16**, 2063 (1983).

[23]B. Derrida, Phys. Rev. B **24**, 2613 (1981).

[24]J. D. Gunton, M. San Miguel, and P. S. Sahni, in *Phase Transitions and Critical Phenomena*, edited by C. Domb and J. L. Lebowitz (Academic, New York, 1983), Vol. 8, p. 269.

[25]E. Donth, J. Non-Cryst. Solids **53**, 325 (1982).

[26]L. D. Landau and E. M. Lifshitz, *Statistical Physics* (Per-

gamon, London, 1980).
[27]S. K. Ma, *Modern Theory of Critical Phenomena* (Benjamin, Reading, 1976).
[28]M. E. Fisher and A. N. Berker, Phys. Rev. B **26**, 2507 (1982).
[29](a) J. T. Chayes, L. Chayes, D. S. Fisher, and T. Spencer, Phys. Rev. Lett. **57**, 2999 (1986); (b) D. Huse, Ref. 7 in Ref. 29(a).
[30]J. Villian, J. Phys. **46**, 1843 (1985).
[31]J. Jonas (unpublished).
[32]J. N. Israelachvilli, P. M. McGuiggon, and A. M. Hourola, Science **240**, 189 (1988).
[33]P. K. Dixon and S. Nagel (unpublished).
[34]C. P. Flynn, *Point Defects and Diffusion* (Clarendon Press, Oxford, 1972).
[35]B. Nienhuis, E. K. Riedel, and M. Schick, Phys. Rev. B **23**, 6055 (1981).
[36]P. G. Wolynes (unpublished).
[37]C. A. Angell, D. R. MacFarlane, and M. Oguim, Ann. N.Y. Acad. Sci. **484**, 241 (1986).

A Topographic View of Supercooled Liquids and Glass Formation

Frank H. Stillinger

Various static and dynamic phenomena displayed by glass-forming liquids, particularly those near the so-called "fragile" limit, emerge as manifestations of the multidimensional complex topography of the collective potential energy function. These include non-Arrhenius viscosity and relaxation times, bifurcation between the α- and β-relaxation processes, and a breakdown of the Stokes-Einstein relation for self-diffusion. This multidimensional viewpoint also produces an extension of the venerable Lindemann melting criterion and provides a critical evaluation of the popular "ideal glass state" concept.

Methods for preparing amorphous solids include a wide range of techniques. One of the most prominent, both historically and in current practice, involves cooling a viscous liquid below its thermodynamic freezing point, through a metastable supercooled regime, and finally below a "glass transition" temperature T_g. A qualitative understanding, at the molecular level, has long been available for materials produced by this latter preparative sequence; however, many key aspects of a detailed quantitative description are still missing. Fortunately, focused and complimentary efforts in experiment, numerical simulation, and analytical theory currently are filling the gaps.

The author is at AT&T Bell Laboratories, Murray Hill, NJ 07974, USA.

The present article sets forth a descriptive viewpoint that is particularly useful for discussing liquids and the glasses that can be formed from them, although in principle it applies to all condensed phases. Conceptual precursors to this viewpoint can be found throughout the scientific literature (1), but most notably in the work of Goldstein (2). The objective here is to classify and unify at least some of the many static and kinetic phenomena associated with the glass transition.

Interaction Potentials

Condensed phases, whether liquid, glassy, or crystalline, owe their existence and measurable properties to the interactions between the constituent particles: atoms, ions, or molecules. These interactions are comprised in a potential energy function $\Phi(\mathbf{r}_1 \cdots \mathbf{r}_N)$ that depends on the spatial location $\mathbf{r}_i$ for each of those particles. The potential energy includes (as circumstances require) contributions from electrostatic multipoles and polarization effects, covalency and hydrogen bonding, short-range electron-cloud-overlap repulsions and longer range dispersion attractions, and intramolecular force fields. Obviously, the chemical characteristics of any substance of interest would substantially influence the details of Φ. Time evolution of the multiparticle system is controlled by the interactions, and for most applications of concern here the classical Newtonian equations of motion (incorporating forces specified by Φ) provide an adequate description of the particle dynamics.

In order to understand basic phenomena related to supercooling and glass formation, it is useful to adopt a "topographic" view of Φ. By analogy to topographic maps of the Earth's features, we can imagine a multidimensional topographic map showing the "elevation" Φ at any "location" $\mathbf{R} \equiv (\mathbf{r}_1 \cdots \mathbf{r}_N)$ in the configuration space of the N particle system. This simple change in perspective from conventional three-dimensional space to a space of much higher dimension [$3N$ for structureless particles, and even more for particles that are asymmetric or nonrigid or both (*3*)] intrinsically creates no new information, but it facilitates the description and understanding of collective phenomena that operate in condensed phases, particularly in liquids and glasses.

An obvious set of topographic questions to ask concerns the extrema of the Φ surface, such as maxima ("mountain tops"), minima ("valley bottoms"), and saddle points ("mountain passes"). The minima correspond to mechanically stable arrangements of the N particles in space, with vanishing force and torque on every particle; any small displacement from such an arrangement gives rise to restoring forces to the undisplaced arrangement. The lowest lying minima are those whose neighborhoods would be selected for occupation by the system if it were cooled to absolute zero slowly enough to maintain thermal equilibrium; for a pure substance, this would correspond to a virtually perfect crystal. Higher lying minima correspond to amorphous particle packings and are sampled by the stable liquid phase above the melting temperature T_m (*4*).

Figure 1 shows a highly schematic illustration of the multidimensional "Φ-scape." Such a simplified representation can be misleading, but it serves to stress several key points. First, the minima have a substantial variation in depth and are arranged in a complex pattern throughout configuration space. Second, each minimum is enclosed in its own "basin of attraction," most simply defined as the set of all configurations in its "valley," that is, all locations that strict downhill motion would connect to that minimum. Third, contiguous basins share a boundary containing at least one saddle point, or transition state. Fourth, equivalent minima can be attained by permutations of identical particles. This last point implies, for a pure substance, that every minimum belongs to a group of $N!$ equivalent minima distributed regularly throughout the multidimensional configuration space.

An important issue concerning the Φ-scape topography that remains largely unresolved concerns the number of minima $\Omega(N)$, and in particular how fast it rises with N. Rather general arguments (bolstered by exact calculations for some special theoretical models) yield a simple generic form for the large-N limit in a single-component system (*4*, *5*):

$$\Omega(N) \sim N! \exp(\alpha N) \qquad (1)$$

where $\alpha > 0$ depends significantly on the chemical nature of the substance considered. The permutational factor $N!$ has already been explained; the challenge is to predict α reliably from known molecular structures and interactions.

Melting and Freezing Criteria

The topographic view of the Φ-scape advocated above leads to a clean separation between the inherent structural aspects of the substance under consideration (that is, the classification of potential energy local minima), and the "vibrational" aspects that concern motions within and among the basins defined by those inherent structures. Such a separation leads naturally to a fresh examination of an old but very useful idea, namely, the Lindemann melting criterion for crystalline solids, first formulated in 1910 (*6*).

The Lindemann criterion concerns the dimensionless ratio, $\ell(T)$, of the root-mean-square (rms) vibrational displacement of particles from their nominal lattice positions, to the nearest-neighbor spacing a. It asserts that when temperature rise causes $\ell(T)$ to reach a characteristic instability value, melting occurs. X-ray and neutron diffraction experiments provide measurements of $\ell(T)$ (through the Debye-Waller factor) for a wide variety of real substances, and computer simulations can be used to calculate $\ell(T)$ for model systems. These results show that $\ell(T_m)$ varies a bit with crystal structure: It is approximately 0.13 for face-centered-cubic crystals and 0.18 for body-centered-cubic crystals (*7*). In any event, it is substantially constant across substances in a given crystal class and provides a good account of the pressure dependence of T_m for a given substance.

Vibrational motions contributing to $\ell(T_m)$ have a significant anharmonic character. But aside from a very small concentration of thermally created point defects in the crystal, these vibrations are confined to the basins surrounding the $N!$ absolute minima. At any temperature T, then, the Lindemann ratio for the crystal can be expressed:

$$\ell(T) = \langle(\Delta \mathbf{R})^2\rangle^{1/2}/(Na) \qquad (2)$$

where $\Delta\mathbf{R}$ is the intrabasin vibrational displacement from the absolute minimum in the multidimensional configuration space, and the brackets denote a thermal average confined to that basin at temperature T.

Although it is not obvious in the usual way of presenting the Lindemann melting criterion, Eq. 2 has a straightforward extension to the liquid phase. One simply recognizes that the thermal average and the displacements refer to inherent structures of the amorphous packing and their associated Φ-scape basins that predominate after melting. The mean nearest-neighbor distance a for the liquid phase (obtained from the measured radial distribution function) typically is close to that of the unmelted crystal.

No laboratory experiment has yet been devised to measure $\ell(T)$ for liquids. Nevertheless, computer simulations for models of real substances can be designed to supply the needed information. Numerically, these simulations are required to generate a representative collection of system configurations for the temperature of interest and to evaluate the rms particle displacements that return each configuration to its corresponding inherent structure. Although only a small number of simulations of this kind have thus far been carried out (*8*), the main features of this extension are clear, and are summarized qualitatively in Fig. 2. The crystal and liquid $\ell(T)$ curves are distinct; both monotonically increase with T, with the curve for the liquid located well above that for the crystal. At the melting-freezing point, ℓ_{liq} is approximately three times that for the crystal; equivalently, the rms particle displacement is approximately one-half that of the nearest-neighbor spacing.

In its conventional form, the Lindemann criterion advances an asymmetric, one-

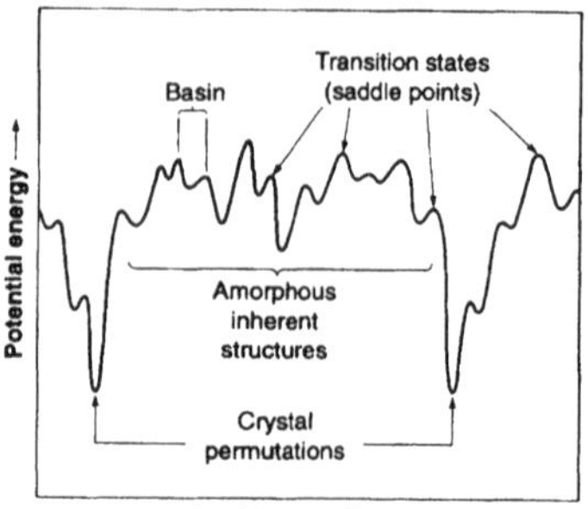

Fig. 1. Schematic diagram of the potential energy hypersurface in the multidimensional configuration space for a many-particle system.

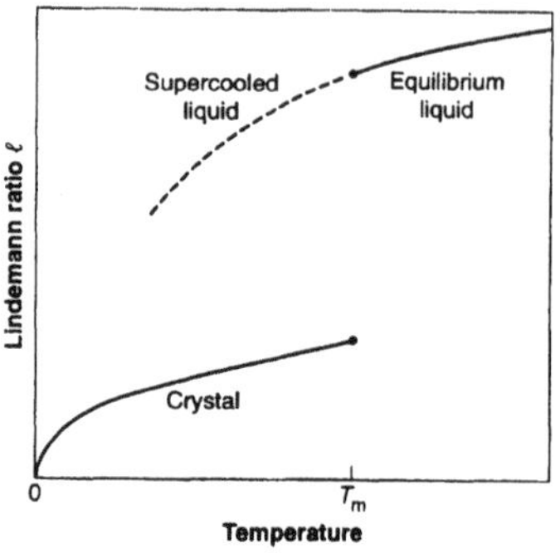

Fig. 2. Root-mean-square particle displacement divided by mean neighbor separation, versus temperature, for crystal and liquid phases. The value of this ratio for the crystal at the melting point, $\ell(T_m)$, is specified by the Lindemann melting criterion.

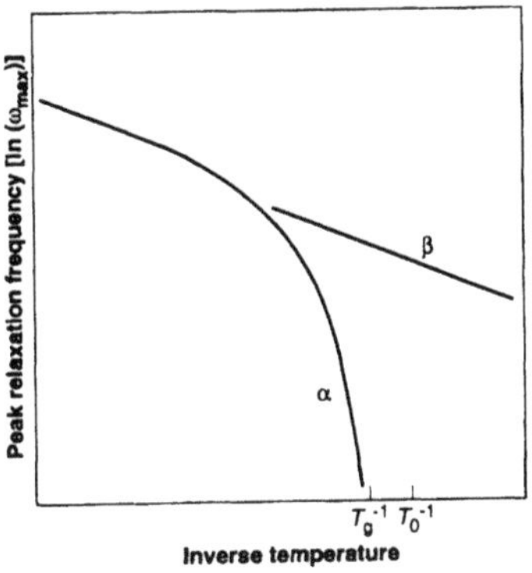

Fig. 3. Temperature dependence of peak relaxation (or absorption) frequencies for glass-forming liquids.

phase view of the first-order crystal-liquid transition. This model contrasts markedly with the thermodynamic description that calls for two-phase equality of pressure and chemical potential at the transition. However, the extension just described effectively restores two-phase symmetry. It supplements the melting criterion with an exactly analogous Lindemann-like freezing criterion for the liquid, specifically, that thermodynamic instability with respect to freezing occurs when cooling causes $\ell_{liq}(T)$ to decline to the cited transition value.

Although it is relatively difficult to superheat crystals above their T_m, supercooling of the melt is commonplace, particularly with viscous liquids. Figure 2 shows the extension of $\ell_{liq}(T)$ into the $T < T_m$ regime of supercooling, under the assumption that crystal nucleation has been avoided. In this extension, the system's configuration point $\mathbf{R}(t)$ continues to wander as time t progresses among the basins for amorphous inherent structures, without discovering entrance channels to any of the absolute-minimum basins of the crystalline state. Upon cooling to absolute zero, the system becomes trapped almost at random in one of these inherent-structure basins of the amorphous state, ℓ_{liq} becomes small (vanishing for classical statistics), and the system becomes a rigid glass.

Arrested Kinetics

The entire collection of $\Omega(N)$ basins can be classified by their depths. Let $\phi \equiv \Phi/N$ denote the inherent-structure potential energy for any given basin on a per-particle basis. In analogy to Eq. 1 above, it can be shown that the distribution of basins by depth has the form $N!\exp[\sigma(\phi)N]$ in the large-N limit (5). A mean vibrational free energy per particle, f_v, can then be defined for basins of depth ϕ. The equilibrium state of the system at temperature T corresponds to preferential occupation of basins with depth $\phi^*(T)$, which is the ϕ value that maximizes the simple combination (5, 9):

$$\sigma(\phi) - (k_BT)^{-1}[\phi + f_v(\phi,T)] \quad (3)$$

where k_B is Boltzmann's constant. When T is near T_m, this combination has two local maxima with respect to ϕ; the first-order melting transition corresponds to a discontinuous change in ϕ^* as the role of the absolute maximum switches from one to the other at T_m.

Supercooling the liquid phase below T_m kinetically avoids switching back to deep crystalline basins, but rather the liquid remains in those that correspond to the other local maximum of Eq. 3, given by $\phi^*_{liq}(T)$, that continues to refer to higher lying amorphous inherent structures. So long as the system configuration point $\mathbf{R}(T)$ can move more or less freely among the higher lying amorphous inherent structures and attain a representative sampling while avoiding crystal nucleation, the system remains in a reproducible quasi-equilibrium state of liquid supercooling (9).

The individual transitions that carry the system between contiguous (boundary-sharing) basins apparently almost always involve localized particle rearrangements. In other words, only order O(1) out of N of the particles undergo substantial location shifts. This is true whether the basins are those for substantially crystalline inherent structures (and the transition involves creation or destruction of point defects), or whether they refer to a pair of amorphous packings. Consequently, the activation barrier that must be surmounted, and the final change in basin depth will also only be O(1). Because the total kinetic energy is much larger (roughly Nk_BT), enough thermal energy is virtually always available in principle to surmount the intervening barrier. The bottleneck is that this kinetic energy is distributed throughout the many-particle system, and it may take a very long time for a proper fluctuation to concentrate enough kinetic energy at the required location to effect the transition between basins.

Relaxation response functions in the time domain, $g_\nu(t)$, to any of a variety of weak external perturbations ν (mechanical, electrical, thermal, optical, and so forth) provide an indication of the actual restructuring kinetics resulting from interbasin transitions. If the assumption that the initial-time normalization $g_\nu(0) = 1$ is imposed, the areas under the $g_\nu(t)$ curve along the positive time axis define mean relaxation times $\tau_\nu(T)$. These depend somewhat on property ν, but in supercooled liquids they tend to display essentially a common rapid rise with declining temperature and can often be fitted to a Vogel-Tammann-Fulcher (VTF) form (10):

$$\tau_\nu(T) \cong \tau_0 \exp[A/(T - T_0)] \quad (4)$$

where τ_0 is in the picosecond range and A and T_0 are positive constants.

So long as all τ_ν are substantially shorter than the time available for experimental measurement, the supercooled liquid remains in a state of quasi-equilibrium, and in particular inhabits and moves among basins whose depths are closely clustered around $\phi^*_{liq}(T)$. But as T declines, the mean relaxation times represented by the VTF form increase strongly and all cross the time scale of experimental measurement at essentially a common glass transition temperature $T_g > T_0$. Further reduction in T fails to lower the depth of the inhabited basins below $\phi^*_{liq}(T_g)$. The supercooled liquid then has fallen out of quasi-equilibrium. The ratio T_g/T_m for most good glass-forming liquids falls in the range from 0.60 to 0.75.

Careful examination of the relaxation functions $g_\nu(t)$ above T_g reveals the presence of distinct processes. At very short times (comparable to vibrational periods), intrabasin relaxation predominates. This domain is followed by a much more extended time regime during which interbasin structural relaxation processes occur, and in the long-time limit of this regime the relaxation inevitably seems to display a Kohlrausch-Williams-Watts (KWW) "stretched exponential" decay (11):

$$g_\nu(t) \sim \Gamma_\nu \exp[-(t/t_\nu)^\theta] \quad (5)$$

in which $0 < \theta \leq 1$ and Γ_ν is a constant; the characteristic time t_ν is comparable to mean relaxation time τ_ν when T is near T_g.

The $\theta = 1$ limit in Eq. 5 corresponds to simple Debye relaxation, with t_ν serving as the single structural relaxation time. However, smaller θ values lead to a broad distribution of relaxation times, and by transforming g_ν to the frequency domain this breadth becomes explicit (12). Peaks in the frequency-dependent absorption then correspond approximately to the dominant relaxation times.

As Fig. 3 illustrates, the temperature dependence of peak relaxation frequency for liquids often exhibits a bifurcation (13). In the equilibrium liquid range and into the moderately supercooled regime, there is a single absorption maximum frequency. Upon approaching T_g, this peak splits into a pair of maxima, the slow α ("primary") and faster β ("secondary") relaxations. The former are non-Arrhenius and kinetically frozen out at T_g, and the latter are more nearly Arrhenius and remain operative at T_g.

The α, β relaxation bifurcation has a straightforward interpretation in terms of the Φ topography. As T declines, the configuration point $\mathbf{R}(t)$ is forced into regions of increasingly rugged and heterogeneous topography in order to seek out the ever deeper basins that are identified by $\phi^*_{liq}(T)$. The lower the temperature, the rarer and more widely separated these basins must be. However, the elementary transition processes that connect contiguous basins continue to require only local rearrangements of small numbers of particles. Evidently, the basins are geometrically organized to create a two-scale length and potential energy pattern; Fig. 4 illustrates this feature.

The β processes are identified with the elementary relaxations between neighboring basins, whereas the α processes entail escape from one deep basin within a large-scale "crater" or "metabasin" and eventually into another (*14*). Because the latter requires a lengthy directed sequence of elementary transitions, it will acquire a net elevation change (activation energy) many times that of the former. Also, the vast intervening stretch of higher lying basins produces a large activation entropy.

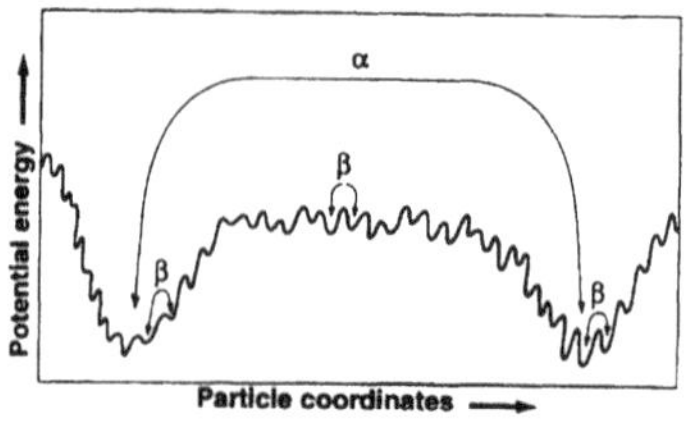

Fig. 4. Two-scale potential energy topography characteristic of regions of configuration space explored by fragile glass formers near T_g. The elementary interbasin transitions are associated with β relaxations, and large distance intercrater transitions are associated with α relaxations.

Viscosity and Self-Diffusion

The relaxational response of a fluid to shear stress is often described by a specific instance of Eq. 4, the Maxwell relaxation time $\tau_M(T)$, defined by the ratio of the shear viscosity $\eta(T)$ to the elastic modulus G, which is nearly temperature independent. Consequently, the VTF form in Eq. 4 is also a useful representation of shear viscosity [indeed, this was its original application (*10*)]:

$$\eta(T) \cong \eta_0 \exp[B/(T - T_0)] \qquad (6)$$

with virtually the same divergence temperature T_0 that obtains for other types of relaxation. Experimentally, T_g's often correspond to η in the range from 10^{11} to 10^{13} poise for nonpolymeric liquids.

The VTF representation for viscosity generates a useful classification of glass-forming liquids between "strong" and "fragile" extremes (*15*), depending on the value of the ratio T_0/T_m (or alternatively T_0/T_g). The strong limit has $T_0/T_m \cong 0$, and $\eta(T)$ displays Arrhenius temperature behavior; real material examples that appear to be close to this limit are the oxide glasses SiO_2 and GeO_2. The fragile limit displays dramatic non-Arrhenius $\eta(T)$, and is illustrated by orthoterphenyl, and the ionic material $K_{0.6}Ca_{0.4}(NO_3)_{1.4}$.

The variation in behavior between the strong and fragile extremes can be traced back to topographic differences in the Φ-scapes for the respective materials. The extreme limit of strong glass formers presents a uniformly rough (single energy scale) topography, in which only the β transitions of Fig. 4 have relevance. Little or no coherent organization of the individual basins into large and deep craters, associated with the low-temperature α transitions, appears to be present. It is no surprise then that the α,β bifurcation is weak or absent in the strong cases. In contrast, the most fragile glass formers indeed exhibit significant Φ-scape cratering and distinctive α,β bifurcation. The relatively large effective singularity temperature T_0 for fragile materials reflects the larger and wider net barriers that they must surmount, as T declines toward T_0, for the system configuration to pass from the interior of one inhabited crater to another of comparable or greater depth.

The self-diffusion constant D measures the rate of increase with time, because of Brownian motion, of the mean square displacement of a tagged particle in the medium. The Stokes-Einstein relation connects D to η and to an effective hydrodynamic radius b for the particle:

$$D(T) = k_B T/[6\pi b \eta(T)] \qquad (7)$$

(the constant 6π assumes a "sticking" boundary condition at the particle surface). This relation has been remarkably successful at correlating independent measurements of $D(T)$ and $\eta(T)$ for many liquids as both vary over many orders of magnitude in the stable and supercooled liquid regimes. However, some recent experiments on fragile glass formers near T_g show a striking breakdown of Eq. 7; in these cases $D(T)$ becomes two orders of magnitude or more larger than the measured $\eta(T)$ would indicate (*16*). Paradoxically, this dramatic effect occurs while the corresponding rotational diffusion rate is still linked to $\eta(T)$ or only weakly decoupled (*17*).

Once again, the Φ-scape cratering characteristic for fragile glass formers offers an explanation. When the temperature is low and the system configuration point emerges from one deep crater to search for another, it has been emphasized that a long sequence of elementary interbasin transitions will be involved. In three dimensions, this sequence would appear as a local structural excitation running around in a microscopic domain that encompasses many particles, temporarily fluidizing that domain compared to the surroundings. Because such domains are expected to be large on the molecular scale near T_g and to have long lifetimes, translational diffusion receives a disproportionate enhancement compared to rotational diffusion, and a detailed analysis (*18*) shows that the latter continues to adhere more closely than the former to the temperature dependence of $\eta(T)$.

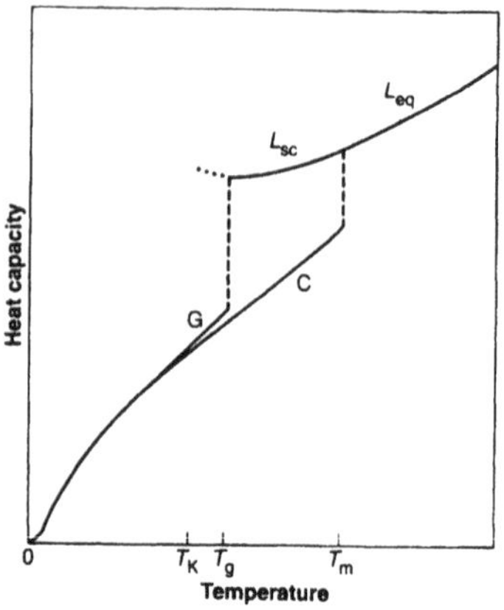

Fig. 5. Typical heat capacity curves for fragile glass formers in crystal (C), supercooled liquid (L_{sc}), equilibrium liquid (L_{eq}), and glass (G) phases; T_K is the Kauzmann temperature.

Ideal Glass State?

Figure 5 illustrates schematically variation with temperature of the heat capacity (C_p, constant pressure) typically observed for fragile glass formers. Notable features are: (i) C_p rises discontinuously when the crystal melts at T_m; (ii) supercooling the liquid down toward T_g increases the discrepancy between C_p (liquid) and C_p (crystal); and (iii) further cooling produces a nearly discontinuous drop in C_p (liquid) so that C_p (glass) is very close to that of C_p (crystal) at the same low temperature. In view of the basin-trapping interpretation of the glass transition, it can be concluded that the intrabasin vibrational properties are largely the same for all basins, whether corresponding to crystalline or to amorphous inherent structures. Furthermore, most of the enhancement of C_p (liquid) over C_p (crystal) above T_g stems from the temperature variation of ϕ^*_{liq}, the depth of the basins predominately inhabited by the supercooled liquid.

The latent heat of melting causes the liquid at T_m to possess a substantially higher entropy than the crystal. However, the C_p discrepancy illustrated in Fig. 5 means that the liquid loses entropy faster than the crystal as both are cooled below T_m. The strongly supercooled liquid at T_g still has the higher entropy, although the difference has been substantially reduced. Smooth extrapolation of C_p (liquid) below T_g and of the corresponding entropy indicate the existence of a positive "Kauzmann temperature" T_K at which the crystal and the extrapolated liquid attain equal entropies (*19*). Considering that vibrational entropies are nearly the same for the two phases, and that the inherent structural entropy of the ordered crystal vanishes, the fully relaxed glass at T_K (the extrapolated liquid) must also have vanishing inherent structural entropy. This realization, coupled with the empirical observation that $T_K \cong T_0$, the mean relaxation-time divergence temperature, has generated the concept of an "ideal glass state" that could be experimentally attained if only sufficiently slow cooling rates were available (*20*).

If indeed it exists, the ideal glass state must correspond to the inherent structure with the lowest potential energy (deepest "crater") that is devoid of substantial regions with local crystalline order. Unfortunately, the details of this noncrystallinity constraint are unclear but may be crucial: Qualifying inherent structures may depend on the maximum size and degree of perfection permitted crystalline inclusions in otherwise amorphous structures. This ambiquity, or non-uniqueness of choice criterion, would seem to undermine the concept of a substantially unique ideal glass state.

It has also been argued (*9*) that the seemingly innocuous extrapolations that identify a positive Kauzmann temperature T_K (and by implication T_0) are flawed. Localized particle rearrangements (associated with β relaxations) are always possible, even in a hypothetical ideal glass structure, and raise the potential energy only by O(1). These structural excitations in the strict sense prevent attaining the ideal glass state at positive temperature, in conflict with the usual view (*20*).

In spite of these formal reservations, the ideal glass state concept remains valuable. Careful and systematic experiments on the most fragile glass formers should help to remove some of the obscuring uncertainties.

Conclusions

By focusing attention on the topographic characteristics of Φ in the multidimensional configuration space, a comprehensive description becomes available for static and kinetic phenomena exhibited by supercooled liquids and their glass transitions. In particular, this viewpoint rationalizes the characteristic properties of fragile glass formers, including non-Arrhenius viscosity, primary-secondary relaxation bifurcation, and the enhancement of self-diffusion rates over the Stokes-Einstein prediction. This multidimensional topographic representation has the further benefit of uncovering and promoting several basic research topics that need sustained experimental and theoretical-simulational attention. Examples of the latter are the material-specific enumeration of inherent structures (Φ minima), the exploitation of the Lindemann-like freezing criterion for liquids and its relation to the glass transition, and the critical evaluation of the ideal-glass-state concept.

REFERENCES AND NOTES

1. J. D. Bernal, *Nature* **183**, 141 (1959); H. Eyring *et al.*, *Proc. Natl. Acad. Sci. U.S.A.* **44**, 683 (1958); M. R. Hoare, *Adv. Chem. Phys.* **40**, 49 (1979).
2. M. Goldstein, *J. Chem. Phys.* **51**, 3728 (1969); *ibid.* **67**, 2246 (1977).
3. For constant-pressure circumstances, system volume becomes an additional coordinate and Φ includes a pressure-volume contribution *PV*.
4. F. H. Stillinger and T. A. Weber, *Science* **225**, 983 (1984).
5. ———, *Phys. Rev. A* **25**, 978 (1982).
6. F. A. Lindemann, *Z. Phys.* **11**, 609 (1910).
7. H. Löwen, *Phys. Rep.* **237**, 249 (1994).
8. R. A. LaViolette and F. H. Stillinger, *J. Chem. Phys.* **83**, 4079 (1985).
9. F. H. Stillinger, *ibid.* **88**, 7818 (1988).
10. G. W. Scherer, *J. Am. Ceram. Soc.* **75**, 1060 (1992).
11. K. L. Ngai, Comments *Solid State Phys.* **9**, 127 and 141 (1979); R. Böhmer, K. L. Ngai, C. A. Angell, D. J. Plazek, *J. Chem. Phys.* **99**, 4201 (1993).
12. E. W. Montroll and J. T. Bendler, *J. Stat. Phys.* **34**, 129 (1984).
13. G. P. Johari and M. Goldstein, *J. Chem. Phys.* **53**, 2372 (1970).
14. F. H. Stillinger, *Phys. Rev. B* **41**, 2409 (1990).
15. C. A. Angell, in *Relaxations in Complex Systems*, K. Ngai and G. B. Wright, Eds. (National Technical Information Service, U.S. Department of Commerce, Springfield, VA, 1985), p. 1.
16. F. Fujara, B. Geil, H. Sillescu, G. Fleischer, *Z. Phys. B* **88**, 195 (1992); M. T. Cicerone and M. D. Ediger, *J. Phys. Chem.* **97**, 10489 (1993); R. Kind *et al.*, *Phys. Rev. B* **45**, 7697 (1992).
17. M. T. Cicerone, F. R. Blackburn, M. D. Ediger, *J. Chem. Phys.* **102**, 471 (1995).
18. F. H. Stillinger and J. A. Hodgdon, *Phys. Rev. E* **50**, 2064 (1994).
19. W. Kauzmann, *Chem. Rev.* **43**, 219 (1948).
20. J. Jäckle, *Rep. Prog. Phys.* **49**, 171 (1986).

ELSEVIER

Physica A 219 (1995) 27–38

PHYSICA A

A thermodynamic theory of supercooled liquids

Daniel Kivelson[a,]*, Steven A. Kivelson[b], Xiaolin Zhao[a], Zohar Nussinov[b], Gilles Tarjus[c]

[a]*Department of Chemistry and Biochemistry, University of California, Los Angeles, CA 90024, USA*
[b]*Department of Physics, University of California, Los Angeles, CA 90024, USA*
[c]*Laboratoire de Physique Theorique des Liquides, Université Pierre et Marie Curie, 4 Place Jussieu, 75252 Paris Cedex 05, France*

Received 3 May 1995

Abstract

A novel thermodynamic theory of the properties of supercooled liquids as they get glassy is presented. It is based on the postulated existence of a narrowly avoided thermodynamic phase transition at a temperature $T^* \geqslant T_m$, where T_m is the melting point, and the "avoidance" is due to geometric frustration. We show that as a consequence two large emergent length scales develop at temperatures less than T^*, and we also show that this picture is consistent with appropriate statistical mechanical models. A theoretical expression is obtained which permits collapse of the viscosity versus temperature of all known glass-formers onto a single master-curve.

1. Introduction

The salient feature defining a supercooled liquid is the extraordinary increase of its characteristic structural (α) relaxation times as the liquid is cooled below its melting point T_m. For instance, the coefficient of shear viscosity (η) increases smoothly by as much as 15 orders of magnitude over a temperature (T) range, often less than 100 K [1]. As a consequence, the range of temperatures over which locally equilibrated supercooled liquids can be studied is bounded below by the glass transition, T_g, which is that temperature below which the local equilibration time becomes longer than the experimentally observable time. (If the liquid is successfully supercooled, the more stable crystalline phase is dynamically inaccessible.) Despite the dramatic change in the dynamics, thermodynamic evidence of a growing length scale or a diverging susceptibility has been weak [2].

Not only do the α-relaxation times increase spectacularly with decreasing temperature, but the relaxation also seems to have a number of universal properties,

* Corresponding author.

SSDI 0378-4371(95)00140-9

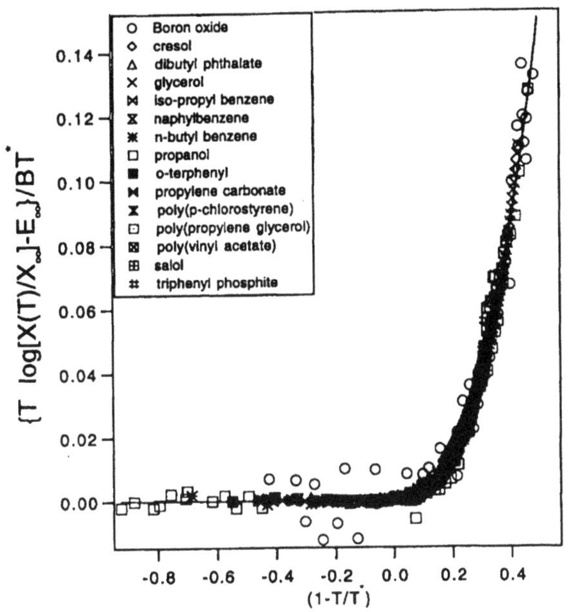

Fig. 1. Reduced collective activation energy, $\{T \log[X(T)/X_\infty] - E_\infty\}/BT^*$, versus $(T^* - T)/T^*$, where $X = \eta$ or τ, for 14 liquids (see Ref. [3] for the specification of the symbols). Parameters E_∞, X_∞, B and T^* are evaluated from fits to Eqs. (4) and (5) and the dashed line represents the theoretical curve.

i.e., to be describable in terms of a very few adjustable parameters. So, for example, for many glass-forming liquids (including molecular fragile [1], network strong [1], polymeric, and H-bonded) the strongly T-dependent activation energy associated with these relaxation processes, when expressed as a function of T, can be scaled to a common curve [3, 4]. See Fig. 1. Additionally, Dixon et al. have proposed a scaling prescription which permits data collapse of the frequency-dependent susceptibilities of many supercooled liquids [5]. The apparent universality of the dynamics of microscopically very different molecular liquids leads us to look for a scaling theory of supercooled liquids with characteristic length-scales larger than the typical molecular length, a_0.

There have been two distinct approaches to the study of supercooled liquids:

(1) Thermodynamic approaches, which are typically based on the assumed existence of an ideal glass transition temperature, $T_0 < T_g$, which is never reached because of dynamical constraints, but which nonetheless produces critical fluctuations of increasing importance as the temperature approaches T_0 from above T_g [6]. The existence of such a low-temperature critical point is most strongly supported by the success in fitting the viscosity at temperatures above T_g to the Vogel–Fulcher–Tammann (VFT) formula, $\eta(T) = \eta_0 \exp[DT_0/(T - T_0)]$, with its implied divergence were the system to remain equilibrated to low enough temperature [1]. This dynamical argument for a low-temperature phase transition is further supported by a thermodynamic one in which it is noted that the entropy of melting decreases markedly as the liquid is supercooled, with an extrapolated value of zero at a temperature in the vicinity of the VFT T_0 [1, 7, 8].

(2) Mode-coupling theories which envisage a purely dynamical *avoided* transition at an intermediate critical temperature, T_c, where $T_g < T_c < T_m$, and for which all thermodynamic properties are smooth analytic functions of temperature [9]. Whether either of these approaches provides understanding of the salient features of supercooled liquids is still an open question.

2. A new physical "picture"

Here we propose a radically different approach, based on the postulate that there exists an isolated narrowly "avoided" critical point (described below) at a temperature T^* which is usually above the melting point T_m, and which controls the physics over the relevant temperature range. Then, on the basis of straightforward thermodynamic analysis augmented by a simple dynamical scaling assumption, we can account for the fundamental properties of supercooled liquids. Lastly we introduce a uniformly frustrated spin model which embodies the essential features of our physical picture and which we argue has an avoided critical point of the sort envisaged.

Although the validity of our theory is not dependent on a specific physical model, we motivate the above postulate by imagining that in the liquid there is a *locally preferred structure* which differs greatly from the local structure in the actual crystalline phase. (For example, it is known that Lennard-Jones liquids have an icosahedral locally preferred structure, but this is not their crystal structure.) Were it possible, the system would prefer to crystallize into a "reference" crystal with the locally preferred structure; it is prevented from doing so by the fact that the local structure will not tile ordinary three-dimensional space, i.e. the system is geometrically *frustrated*. Were we to force a large volume of the system into this ideal crystalline structure, a superextensive strain would build up; in the thermodynamic limit the system must thus break up into domains. This picture can be implemented by conceiving of an *unfrustrated* reference Hamiltonian which allows continuous freezing [10] (as can occur in two dimensions) into the locally preferred structure, plus a small perturbation which accounts for the frustration. In some cases the frustration, i.e., the difference between the reference and the physical Hamiltonian, can be given a simple geometric interpretation as representing the curvature of a reference space which can be tiled by the locally preferred structure [11]; at least in these cases we expect the strain energy density in flat space to grow with system size like L^2 for L small compared to the radius of curvature, reflecting the fact that any curved space is locally flat [3].

3. The postulate of an avoided transition

We consider a frustration-temperature phase diagram with an avoided transition. (As usual in theories of supercooled liquids, we exclude that part of phase space associated with the real crystal.) See Fig. 2. The temperature T is the physical control

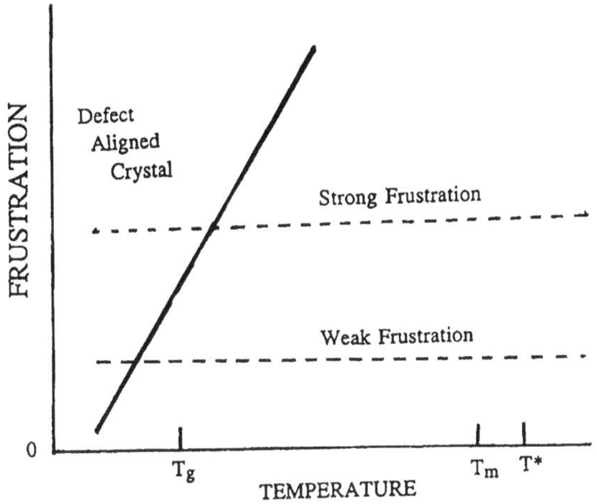

Fig. 2. Frustration (K)–temperature phase diagram (schematic). Note the possibility of a defect-aligned phase-transition indicated by the solid line.

parameter, and the frustration K is taken as having a given, small value for each liquid. The reference state is that for which $K = 0$; this state exhibits a finite-temperature critical phase transition at T^* such that for $T > T^*$ the system is disordered (liquid), while for $T < T^*$ the system is ordered. At finite K, it is argued that *no* relevant continuous phase transition occurs for T's ranging from well above T^* down to T_g [12]; this is what we denote as an "avoided transition". However, since K is small, the fact that the system passes near the critical point ($T = T^*$, $K = 0$) has dramatic consequences. We show that for temperatures somewhat below T^*, there are two emergent length scales that are long compared to a_0. The first of these is the critical correlation length ξ_0 which governs the fluctuations in the absence of frustration; $\xi_0/a_0 \propto \varepsilon^{-\nu}$, where $\varepsilon = (T^* - T)/T^*$. The second length, R_D, characterizes the scale at which the otherwise ordered structure is broken up due to frustration, i.e. the *frustration-limited domain* size. Within a domain the order looks somewhat like that of the unfrustrated system; as $K \to 0$, the length scale R_D on which the frustration makes itself felt must diverge. Thus, for any $\varepsilon > 0$ and K sufficiently small, $R_D \gg \xi_0$, which means that the frustration has no effect on length scales as short as ξ_0. In particular, if $\boldsymbol{O}(\boldsymbol{r})$ is the local order parameter for the *reference* crystal, this means that

$$\langle \boldsymbol{O}(\boldsymbol{r}) \cdot \boldsymbol{O}(\boldsymbol{r}') \rangle \approx \begin{cases} A_1 a_0 |\boldsymbol{r} - \boldsymbol{r}'|^{-1} \exp[-|\boldsymbol{r} - \boldsymbol{r}'/\xi_0] + m^2 & \text{for } \xi_0 \ll |\boldsymbol{r} - \boldsymbol{r}'| \ll R_D, \\ A_2 R_D m^2 |\boldsymbol{r} - \boldsymbol{r}'|^{-1} \exp[-|\boldsymbol{r} - \boldsymbol{r}'|/R_D] & \text{for } R_D \ll |\boldsymbol{r} - \boldsymbol{r}'|, \end{cases} \tag{1}$$

where m, which vanishes as $\varepsilon \to 0$, is the expectation value of the order parameter in the absence of frustration, A_j are numbers of order one, and we have assumed that the connected correlations always decay asymptotically with an Ornstein–Zernicke form [13].

4. Dynamics

Once it is recognized that the avoided critical point produces a system with two long length scales, it is rather straightforward to understand how this can give rise to a stupendous slowing of the dynamics as T is lowered *below* T^*. We recall from finite size studies of ordinary systems *below* their critical temperatures, that there is a time-scale τ, associated with relaxation of the order parameter, which proceeds via nucleation and motion of a domain wall, and that the divergence of τ in the thermodynamic limit is the signal of a broken symmetry state; for system size $L \gg \xi_0$, one expects $\ln[\tau]$ to be a linearly increasing function of $\sigma L^2/T$, where the domain-wall surface tension is $\sigma \propto T^*(\xi_0)^{-2}$. In the case of actual interest to us here, an extended system spontaneously breaks up into domains of size R_D, but we assume that the restructuring dynamics for a domain of size L is the same as that for a system of finite size L, and consequently the long-time relaxation dynamics have a characteristic time scale

$$T \ln[\tau/\tau_0] \propto (R_D/\xi_0)^2 T^* , \qquad (2)$$

where τ_0 is a molecular time scale. In other words, *the existence of two large length scales provides a mechanism for generating exponentially long and strongly temperature-dependent time scales for a system with rather modest spatial correlations* [14]. We assume that this same time is characteristic of all α-relaxations, in particular, rotational and shear relaxation.

We shall show that at least for the class of models of frustrated systems introduced below,

$$R_D \propto \xi_0^{-1} K^{-1/2} , \qquad (3)$$

or, consequently, that $\log[\tau/\tau_0] \propto |\varepsilon|^{4\nu}$. The precise value of ν depends on the universality class of the transition to the reference crystal, but we note that for ordinary three-dimensional critical phenomena in the absence of quenched disorder, $\nu \approx 2/3$. Thus we find

$$T \log[\tau/\tau_0] \approx T \log[\eta/\eta_0] \approx BT^*\varepsilon^{4\nu} , \qquad T < T^* , \qquad (4)$$

where $4\nu \approx 8/3$, and both $B \propto 1/K$ and T^* are species-specific constants. We take the molecular quantity η_0 to be

$$\eta_0 \approx \eta_\infty \exp[E_\infty/k_B T] , \qquad (5)$$

where η_∞ and E_∞ are species-specific molecular constants characteristic of the liquid at $T > T^*$. An analogous expression holds for τ_0 with τ_∞ replacing η_∞. Collapse of viscosity and relaxation data by means of these relations for all the glass-forming liquids (both fragile and strong) [1] for which we could obtain data is shown in Fig. 1 [4].

In assessing the expressions in Eqs. (4) and (5), we note that whenever data exist over a wide enough range, η_∞ (or τ_∞) and E_∞ are evaluated by means of Eq. (5) with data only in the range with T well above T^*, and then with these two constants,

Eqs. (4) and (5) are used to fit the rest of the data and to determine B and T^*. Thus ideally there are two adjustable parameters in each of two independent experiments [14]. We believe [4] this to be *the most parameter-lean of existing expressions that fit all the data.* By contrast, the VFT expression has three parameters which must be evaluated simultaneously, and since for many liquids it fits only the low T data, one must also include a cut-off or cross-over temperature; to extend the description over the entire temperature-range, additional parameters are needed. The "universal" exponent given by our theory is $4\nu \approx 8/3$; if the exponent is treated as an adjustable universal parameter, the best fit to the data is, in fact, close to 8/3, but reasonable fits [4] can be obtained with exponents between about 7/3 and 3. (Eq. (4) involves a simple interpolation in the immediate neighborhood of T^*, where the inequality $R_D \gg \xi_0$ is violated, and since Eq. (4) retains only leading order behavior in powers of ε, it is expected to become decreasingly accurate far below T^*.)

In addition to the slow processes involving domain-relaxation on length scales of order R_D (which we have associated with α-relaxation), there may exist in our model relaxations associated with material not in domains and others that specifically occur on length scales of order ξ_0. The latter are faster and largely unaffected by the frustration, and should thus be well described by conventional dynamical scaling, even below T_g, so long as $\xi_0 \gg a_0$; we tentatively associate these with the well-known β-relaxations which are observed to be faster than the α-relaxations above T_g and to persist below T_g [1].

5. Model calculations

Our goal, only partially achieved, is to develop and study a theoretical model which yields a frustration-avoided critical point below which the two length scales, as well as thermodynamic properties, can be analyzed quantitatively.

Consider a model frustrated spin system, which for purposes of illustration we take to be an Ising model, or a more general clock model, in which the spin variable $\boldsymbol{S}_i$ on the ith site takes on a set of discrete values $\{\boldsymbol{S}\}$, which corresponds to the possible different "orientations" of the local reference crystalline order parameter. We define our model along lines originally suggested by L. Chayes [16] to be

$$H = -J \sum_{\langle ij \rangle} \boldsymbol{S}_i \cdot \boldsymbol{S}_j + \tfrac{1}{2} K \sum_{i \neq j} \frac{\boldsymbol{S}_i \cdot \boldsymbol{S}_j}{|\boldsymbol{R}_i - \boldsymbol{R}_j|^x}, \tag{6}$$

where the sum over $\langle ij \rangle$ is taken over nearest neighbor sites, J and K are both > 0, and $0 < x \leqslant 3$. The first (short-range, ferromagnetic) term favors long-range order of the locally preferred structure; we consider only models in which for $K = 0$ there is a single second-order phase transition which occurs at a critical temperature T^* of order J. The second (long-range, antiferromagnetic) term represents the extent to which this ideal ordered state is frustrated; because of the antiferromagnetic nature of the long-ranged interactions, the model has a well-behaved thermodynamic limit in the same way as does a neutral plasma. Elsewhere the strong evidence that this model has an avoided critical point is reviewed [17, 18]. Since we are interested in a narrowly avoided critical point (weak frustration), we will always consider $K \ll J$.

Even in the absence of a complete solution, we can gain insight by examining the perturbation theory (cumulant expansion) in powers of K for the free energy, F, of a large finite system. (To eliminate surface effects, we consider the model with periodic boundary conditions.)

Thus

$$F = F_0 + KF_1 + K^2F_2 + \dots , \tag{7}$$

where F_0 is the free energy of the unfrustrated system,

$$F_1 = \sum_{i>j} \frac{\langle \boldsymbol{S}_i \cdot \boldsymbol{S}_j \rangle_0}{|\boldsymbol{R}_i - \boldsymbol{R}_j|^x} \tag{8}$$

$$F_2 = -\frac{1}{2} \sum_{i>j} \sum_{p>q} \frac{\langle (\boldsymbol{S}_i \cdot \boldsymbol{S}_j)(\boldsymbol{S}_p \cdot \boldsymbol{S}_q) \rangle_0 - \langle \boldsymbol{S}_i \cdot \boldsymbol{S}_j \rangle_0 \langle \boldsymbol{S}_p \cdot \boldsymbol{S}_q \rangle_0}{|\boldsymbol{R}_i - \boldsymbol{R}_j|^x |\boldsymbol{R}_p - \boldsymbol{R}_q|^x}, \tag{9}$$

and $\langle\rangle_0$ signifies the thermal average with respect to the unfrustrated Hamiltonian. For $x < 3$ (long-range interactions) and $T < T^*$, we obtain the leading behavior of the above terms in the limit of large system size $L \gg \xi_0$ by approximating the unperturbed two-spin correlator by an Ornstein–Zernicke form over the relevant range of positions and by factoring the long-distance pieces of the higher spin correlators:

$$F_1 \sim m^2 L^{6-x} + O(\xi_0^{2-x} L^3), \tag{10}$$

$$F_2 \sim m^2 (\xi_0)^2 L^{9-2x} + O(\xi_0)^4 L^{6-2x}). \tag{11}$$

(Corrections to the unperturbed correlator at distances shorter than ξ_0 might introduce small shifts in the power of ξ_0.)

From the above analysis we conclude that the strain energy to produce a region with the reference crystal structure grows superextensively, like L^{2d-x}. We have argued above that geometric frustration should result in a strain energy which grows like L^{d+2}, where d is the dimensionality of the physical space, so on the basis of this observation, we identify $x = d - 2$ as the physically relevant value. With this choice, the frustration has the same form as the Coulombic interaction, which is, of course, very natural; induced forces of precisely this form occur in a variety of circumstances in which the forces arise from the long-range piece of a strain field. We also conclude that the characteristic length scale $L = R_D$ is that at which the first- and second-order perturbative terms are comparable, i.e., the size beyond which adiabatic behavior breaks down. The result is that given in Eq. (3). This is the system size at which the unperturbed structure gets significantly broken up. Similar estimates are obtained by comparing consecutive higher order terms.

6. What have we accomplished?

We have presented a thermodynamic scaling theory which is built about a weakly frustration-avoided critical point. Explicit calculations have been carried out for specific models, in particular, one in which a ferromagnetic critical system is frustrated by a weak, long-range antiferromagnetic perturbation; these calculations need to be extended and perfected. The applicability of the model to supercooled liquids is supported by the considerable success of Eq. (4) in fitting data.

The concept of domains or some form of correlated structural or dynamical regions that give rise to heterogeneity is not new, and, in fact, it underscores many (even most) of the models used to explain the observed slow, non-exponential relaxation in supercooled liquids [6, 19]. However, we believe that our theory is unique in that the heterogeneity need not be imposed, as for example in spin glasses and random field spin models [20], but arises as a direct consequence of an avoided critical point. A novel (and essential) feature of our model at temperatures below T^* is the emergence, as a consequence of the narrowly avoided critical point, of two non-molecular scale lengths, ξ_0 and R_D; the latter is proportional to ξ_0^{-1} and can be associated with self-forming domains. Therefore the properties (or at least sizes) of the domains are constrained and cannot be further modified to fit the phenomenology. Whereas in many models the physical arguments motivating the imposition of heterogeneity or quenched disorder are ambiguous, the physical reasons leading to avoidance of the critical point, i.e., structural frustration, are rather clear. Although ours is a thermodynamic theory, it is the relaxation behavior that we seek to explain; we have assumed that α-relaxation is controlled by domain relaxation or restructuring, which in turn occurs by means of wall mechanisms.

Although the concept of an inaccessible critical point that governs the dominant fluctuations is not new, the placement of this avoided critical point as a cross-over between molecular and collective behavior is, we believe, new. And although the concept of frustration as relevant to the behavior of supercooled liquids has been explored, its application as a perturbation leading to the avoidance of a critical point is, we believe, novel. As a consequence of the arguments above, we believe the present analysis yields the first physically-based model in which the nature and behavior of the domains is determined by the theory.

In constrasting our model to others, we note that $T^* \geqslant T_m > T_c > T_g > T_0$, where the "ideal glass" T_0 and the mode-coupling T_c represent low-temperature limits beyond which ultimate collective congestion or order sets in, while in our theory "reference crystallization" at T^* represents a high-temperature cross-over below which collective behavior initiates. Whereas in the mode-coupling theory of glasses the structural and thermodynamic properties show no anomalous or non-analytic behavior, the ideal glass theories exhibit divergent correlation lengths and activation energies as T is lowered towards T_0, in our model, the domain size (R_D) and activation energy ($E_\infty + T^* B\varepsilon^{8/3}$) both increase with decreasing T below T^*, but in a mild, non-divergent manner.

7. What remains to be done?

Our model makes use of scaling about an avoided critical point T^*. In order that our scaling assumptions be valid, it is clear that $(T^* - T)/T^*$ must be sufficiently small. On the other hand, our analysis of the dynamics rests on the assumption, $\xi_0 < R_D$, which is valid only sufficiently far from the critical point, i.e., for $(T^* - T) > (T^* - T_1)$, where T_1 is the temperature at which $\xi_0 = R_D$. These two conditions are, in principle, compatible for sufficiently weak frustration since $(T^* - T_1)/T^* \to 0$ as the frustration $K \to 0$. What is more to the point is the question of how well the scaling works in real physical situations where the frustration is probably only moderately weak and we are asking the theory to work, at least semi-quantitatively, over the rather broad range of temperatures from T_m to T_g, at which point $(T^* - T_g)/T^* \approx 1/3$. It is known in more conventional critical phenomena that the range of reduced temperatures over which scaling can be observed varies greatly according to circumstance, in some cases the range being rather large. In the present case, we take as a working hypothesis that the scaling regime is sufficiently wide to include the entire relevant range of temperatures, and as supporting evidence for this point of view we take the success of Eq. (4) in fitting the measured viscosities [4].

Because the critical point in our model is avoided, the expected critical divergences near T^* in ξ_0, in the order parameter susceptibility, and in the thermodynamic properties should be rounded, i.e., limited. Furthermore, we have not as yet identified the order parameter, nor do we know which structural or thermodynamic variables are strongly coupled to it, all of which suggests that if there are anomalies around T^* in the measured structural and thermodynamic properties they will be very muted unless the frustration happens to be exceedingly weak. See Fig. 3. At present both the theoretical and experimental status of thermodynamic singatures of an avoided critical point near T^* is uncertain; if such a signature is confirmed, it would provide dramatic support for the theory. However, for the frustration actually present in any real liquid, it is not clear whether a detectable signature should be expected, even within the theory. Grimsditch and Rivier [21] have reported a clear critical-like anomaly in C_v/C_p at T's above T_m, but their analysis has been questioned because it excludes viscoelastic effects. Others have noticed [22] that slightly above the melting point there seems to be a change in slope in the C_p versus T curves of glass-forming liquids, but there is not enough combined viscosity and heat capacity data over the range from T_g to well above T_m to determine whether this change is relevant to the model.

Our model suggests that the entropy of melting should decrease rather strongly with increase in domain size and number of domains; indeed this is observed, but the model attaches no significance to the Kauzmann limit, the temperature at which the *extrapolated* entropy of melting vanishes [1, 7]. One hopes that more detailed calculations based on models such as those in Eq. (6) will yield understanding of all the thermodynamic properties; these properties are less dependent on the avoided critical scaling than are the dynamic properties.

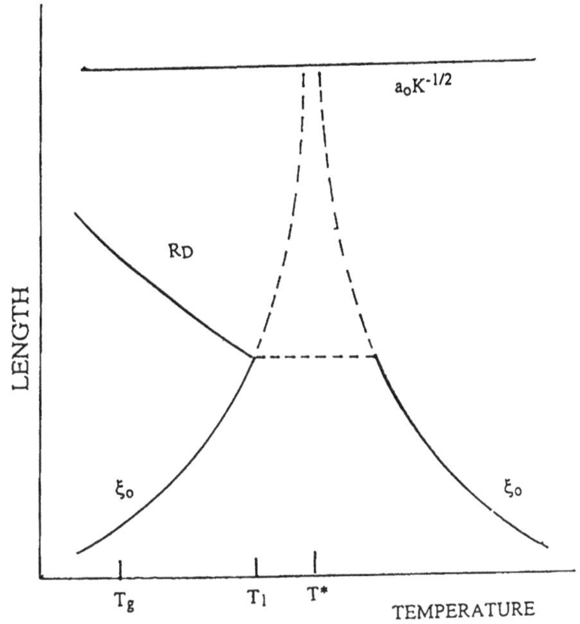

Fig. 3. Lengths versus T (schematic). Note that around T^* the critical behavior is rounded, and note that only below T_1, where $R_D > \xi_0$, are the dynamic assumptions valid.

Domains or clusters play a role in our model, as they do in many others [6, 19]. One would like to estimate their size, in our theory the length R_D. Based on various questionable self-consistent estimates, we believe that R_D at T_g should be not much greater than $10a_0$, i.e., the largest domains should contain of the order of 10^3 molecules. But despite the fact that most models incorporate such regions, and despite the fact that considerable indirect evidence of such domains exist [19], there is little extent diffraction data to support this concept [23]. Of course, near T^* the domains should be much smaller.

There are many relevant dynamic and structural properties that have not been addressed here. Primarily, there is the frequency scaling of the susceptibility developed by Dixon et al. [5]. There is also the "high-frequency" susceptibility minimum explored by Cummins and coworkers [24]. There is the break-down of the Stokes–Einstein relation for translational, but not for rotational diffusion [19, 25]; we address this problem elsewhere [26].

Acknowledgements

We would like to thank James Sethna, Joe Rudnick, Sidney Nagel, Anders Karlhede, Lincoln Chayes, Charles Knobler, and Thomas Fischer for their many valuable comments. We are grateful to Christopher Sorensen for showing us some relevant unpublished work. This work was supported, in part, by the National

Science Foundation under grants Nos. CHE 91-17192 and DMR 93-120606 at UCLA, and LPTL is C.N.R.S. URA No. 765.

References

[1] A. Angell, J. Non-Cryst. Solids 131–133 (1991) 13. This is a good summary with a reasonably complete bibliography.
[2] R.M. Ernst, S.R. Nagel, and G.S. Grest, Phys. Rev. B 43 (1991) 8070.
[3] S.A. Kivelson, D. Kivelson, X.-L. Zhao, T. Fischer and C.M. Knobler, J. Chem. Phys. 101 (1994) 2391.
[4] D. Kivelson, G. Tarjus, X.-L. Zhao and S.A. Kivelson, Phys. Rev. E (submitted). A detailed analysis of fits is given here.
[5] P.K. Dixon, L. Wu, S.R. Nagel, B.D. Williams and J.P. Carini, Phys. Rev. Lett. 65 (1990) 1108;
L. Wu, P.K. Dixon, S.R. Nagel, B.D. Williams and J.P. Carini, J. Non-Cryst. Solids 131–133 (1991) 32;
S.R. Nagel, in: Lectures for the NATO ASI on Phase Transitions and Relaxations in Systems with Competing Energy Scales, Geilo, Norway, 1993, Ed. T. Riste and D. Sherrington;
D.D. Leslie-Pelecky and N.O. Birge, Phys. Rev. Lett. 72 (1994) 1232.
[6] See, for example, G. Adam and J.H. Gibbs, J. Chem. Phys. 43 (1965) 139;
M. Hoare, Ann. NY Acad. Sci. 279 (1976) 186;
P.W. Anderson, in: Ill Condensed Matter, Les Houches Lectures, 1978;
M.H. Cohen and G.S. Grest, Phys. Rev. B 20 (1979) 1077;
T.A. Weber and F.H. Stillinger, Phys. Rev. B 36 (1987) 7043;
J.P. Sethna, Europhys. Lett. 6 (1988) 529;
P.G. Wolynes, in: Proceedings of the International Symposium on Frontiers in Science, AIP Conference proceedings No. 180, S.S. Chan and P.G. Debrunner, Eds. AIP, 1988) p. 39;
T.R. Kirkpatrick, D. Thirumulai and P.G. Wolynes, Phys. Rev. A 40 (1989) 1045;
J.P. Sethna, J.D. Shore and M. Huang, Rev. B 44 (1991) 4943;
R.D. Mountain and D. Thirumulai, Phys. Rev. A 45 (1992) R3380;
R.V. Chamberlin, Phys. Rev. B 48 (1993) 15638
[7] W. Kauzmann, Chem. Rev. 43 (1948) 219
[8] See N. Menon and S.R. Nagel, Phys. Rev. Lett. 74 (1995) 1230 for an argument that the dielectric constant diverges near T_0;
and see D. Kivelson, W. Steffen, G. Meier and A. Patkowski, J. Chem. Phys. 95 (1991) 1943 for an argument that the dipole–induced-dipole correlations vanish near T_0.
[9] W. Götze, in: Liquids, Freezing and Glass Transition, J.P. Hansen, D. Levesque and Zinn-Justin, Eds. (Elsevier, 1991) p. 287;
W. Götze and L. Sjögren, Rep. Prog. Phys. 55 (1992) 241.
[10] Because most actual transitions are first order (but some weakly so), we comment on our statements concerning continuous crystallization. If the ordered phase is characterized by an order parameter which admits cubic invariants, the transition would almost certainly be first order; there is no reason to expect this to be the case in the present situation. If the structural rearrangement associated with a transition involves great changes in local symmetry or volume, the transition is generally strongly first order; this is explicitly not the case in the present situation. A weakly first order transition can be interpreted in terms of a nearby continuous transition (which could be the transition to the "reference state" in our model) thwarted because of a weak "first-order-inducing perturbation" (which should not be confused with the "frustration" in our model). Alternatively, if one were to perturb the critical system with "frustration," the critical point would be "avoided," and no nearby transition would occur, continuous or first order; the same would be true in the presence of a weak "first-order-inducing perturbation" provided the frustration were sufficient to prevent the first order effect. With all this said, it should be pointed out, that "by conceiving of an unfrustrated reference Hamiltonian which allows continuous freezing," we actually require only that a Hamiltonian of the form of Eq. (6), which includes frustration, describe the physical phenomena, not that continuous freezing be physically realizable.

[11] M. Kleman and J.F. Sadoc, J. Phys. (Paris) Lett. 40 (1979) L569;
D.R. Nelson, Phys. Rev. Lett. 50 (1983) 982; Phys. Rev. B 28 (1983) 5515;
J.P. Sethna, Phys. Rev. Lett. 51 (1983) 2198; Phys. Rev. B 31 (1985) 6278.
[12] Under appropriate circumstances, weak frustration gives rise to a system of dilute line or planar defects which can ultimately order, as they apparently do in Frank–Kasper bimetallic or cholesteric liquid crystal blue phases [11]. However, in contrast with a commonly made assumption (see for example, S. Sachdev, in: Bond Orientational Order in Condensed Matter Systems, K.J. Standburg, Ed. (Springer New York., 1992) we do not believe that this putative defect ordering temperature, T_D, typically approaches T^* as $K \to 0$; rather, the generic behavior should be like that conjectured (D. Nelson, J. Stat. Phys. 57 (1989) 511) for a highly anisotropic type II superconductor with the Meisner-phase suppressed, where $T_D < T^*$ in the $K \to 0$ limit.
[13] Note that an Ising ferromagnet in a uniform field has an avoided critical point, but does not behave as in Eq. (1) since the correlation function approaches a finite value at large distances at all temperatures.
[14] The slow time τ is that associated with restructuring of domains, i.e., of changes in the local order parameter of a domain; the actual *lifetime* of a domain should be at least as long and may be longer.
[15] For those liquids for which limited data are available, the four parameters must be evaluated simultaneously by fitting Eqs. (4) and (5).
[16] L. Chayes, private communication.
[17] Related frustrated spin models were also shown to form meso-scale structures at low enough temperature; see e.g. U. Low, V.J. Emery, K. Fabricus and S.A. Kivelson, Phys. Rev. Lett. 72 (1994) 1918 and D. Wu, D. Chandler and B. Smit, J. Phys. Chem. 96 (1992) 4077.
[18] S.A. Kivelson and L. Chayes, unpublished
[19] M.T. Cicerone, F.R. Blackburn and M.D. Ediger, J. Chem. Phys. 102 (1995) 471.
[20] J. Villain, J. Physique 46 (1985) 1843;
D.S. Fischer, Phys. Rev. Lett. 56 (1986) 416;
R.G. Caflisch, H. Levine and J.R. Banavar, Phys. Rev. Lett. 57 (1986) 2679;
K. Binder and A.P. Young, Rev. Mod. Phys. 58 (1986) 801;
B. Bässler, Phys. Rev. Lett. 58 (1987) 767.
[21] M. Grimsditch and N. Rivier, Appl. Phys. Lett. 58 (1991) 2345.
[22] F.H. Stillinger, J. Chem. Phys. 89 (1988) 6461;
D.L. Sidebottom and C.M. Sorensen (unpublished);
X. Zhao and D. Kivelson, J. Phys. Chem. 99 (1995) 6721.
[23] A. Uhlherr and S.R. Elliot, J. Phys. Condensed Matter 6 (1994) L99;
N. Menon and S.R. Nagel (unpublished).
[24] H.Z. Cummins, W.M. Du, M. Fuchs, W. Götze, S. Hilderbrand, A. Latz, G. Li and N.J. Tao, Phys. Rev. E 47 (1993) 4223.
[25] I. Chang, F. Fujara, G. Heuberger, T. Mangel and H. Sillescu, J. Non-Cryst. Solids 172 174, (1994) 248.
[26] G. Tarjus and D. Kivelson, J. Chem. Phys., in press.

Long-Lived Structures in Fragile Glass-Forming Liquids

Andrew I. Mel'cuk,[1] Raphael A. Ramos,[2,*] Harvey Gould,[1] W. Klein,[2] and Raymond D. Mountain[3]
[1]*Department of Physics, Clark University, Worcester, Massachusetts 01610*
[2]*Department of Physics, Boston University, Boston, Massachusetts 02215*
[3]*Thermophysics Division, National Institute of Standards and Technology, Gaithersburg, Maryland 20899*
(Received 22 September 1994)

We present molecular dynamics results for the existence of long-lived clusters near the glass transition in a two component, two-dimensional Lennard-Jones supercooled liquid. Several properties of this system are similar to a mean-field glass-forming liquid near the spinodal. This similarity suggests that the glass "transition" in the supercooled liquid is associated with an incipient thermodynamic instability. Our results also suggest that single particle properties are not relevant for characterizing the instability, but are relevant to the kinetic transition that occurs at a lower temperature than the glass transition.

PACS numbers: 64.70.Pf

The characterization of a supercooled liquid near the glass transition is an active area of research [1]. Outstanding unsolved problems include the possible existence of an underlying thermodynamic transition [2], the history dependence of the glass, and the mechanisms responsible for the large increase in relaxation times.

The primary focus of this Letter is on our molecular dynamics (MD) simulations of a two component, two-dimensional ($d = 2$) Lennard-Jones (LJ) supercooled liquid. To help the reader understand our interpretation of the MD data and the questions we pose, we first review the behavior of a mean-field (MF) model of a structural glass transition.

We have shown [3] that the MF model has a well-defined thermodynamic glass transition associated with a spinodal. The static properties of the glass phase are dominated by localized structures (clumps). A new result is that the dynamical properties that depend directly on the clumps, e.g., the diffusion coefficient associated with their center of mass motion, go to zero at the thermodynamic glass transition. However, the transition does not affect dynamical quantities that depend on single particle motion.

In the MF model, particles interact via a repulsive, two-body potential of the form [4] $V(r) = \gamma^d \phi(\gamma r)$, where $\phi(x) = 1$ if $x \leq 1$, $\phi(x) = 0$ if $x > 1$, and r is the distance between particles. The range of the interaction is $R = \gamma^{-1}$. In the limit $R \to \infty$, the static properties of the uniform fluid are described exactly by MF theory [5]. For fixed density ρ, the system has a spinodal singularity [6] at a temperature T_s, which is defined by the condition $1 + \beta\rho\hat{\phi}(k_0) = 0$, where $\beta^{-1} = k_B T$, $\hat{\phi}(k)$ is the Fourier transform of $\phi(x)$, and k_0 is the location of the minimum of $\hat{\phi}(k)$.

The singularity is well defined only in the MF limit ($R \to \infty$). For $d = 3$ and $\rho = 1.95$, the spinodal is at $T_s = 0.705$ ($k_B = 1$). Our Monte Carlo (MC) simulations at $\rho = 1.95$ with $R = 3$ show that the measured static structure function $S(k)$ has a maximum at $k \neq 0$ that increases rapidly as T is decreased until $T \approx 0.75$, below which the peak ceases to increase as T is lowered. This behavior is characteristic of a pseudospinodal. As R increases, the pseudospinodal better approximates a true singularity and has measurable effects if R is sufficiently long.

If the system is equilibrated at $T > T_s$ and quenched to $T < T_s$ where the uniform phase is unstable [6], the particle immediately form clumps with order ρR^3 particles in each clump. The arrangement of the clumps is noncrystalline, and their number depends on the quench history [3,6]. The free energy has been calculated numerically in the MF limit and has many minima corresponding to different numbers of clumps [3,7]. These properties suggest that the MF model has a metastable glass phase for $T < T_s$. A new result is that the glassy dynamics of the MF model is associated with the *clumps*. In contrast, its single particle dynamical properties do not show the usual signature of the approach to a glass. A simple argument [8], based on the fact that all potential barriers in the MF model are finite, implies that the self-diffusion coefficient $D > 0$ for all $T > 0$. That is, the *particles* are not localized in the metastable glass phase. Hence, if the observation time is sufficiently long, the mean square displacement of the particles increases without bound.

A similar argument implies that the MF model is ergodic for all $T > 0$ if single particle properties are probed. The ergodic behavior can be characterized by several fluctuation metrics [9]. The single particle energy fluctuation metric $\Omega_e(t)$ is given by [9]

$$\Omega_e(t) = \frac{1}{N}\sum_{i=1}^{N}[\epsilon_i(t) - \overline{\epsilon(t)}]^2, \qquad (1)$$

where $\overline{\epsilon(t)} = (1/N)\sum_{i=1}^{N}\epsilon_i(t)$, and $\epsilon_i(t)$ is the mean energy of particle i over the time interval t. If the system is ergodic, $\Omega_e(t) \sim 1/t$ for t sufficiently large [9]. We find that $\Omega_e(t)$ exhibits ergodic behavior at $T = 0.4$, a value of $T < T_s$. For $T \leq 0.15$, $\Omega_e(t)$ exhibits nonergodic behavior during our longest runs, and the measured D is indistinguishable from zero at $T = 0.15$. Given our theoretical prediction that $D \neq 0$ for all $T > 0$,

 0031-9007/95/75(13)/2522(4)$06.00

the observed nonergodic behavior implies only that the time for a particle to leave a clump is much longer than our observation time. Neither $\Omega_e(t)$ nor D show evidence of the glass-spinodal transition, and we interpret the change from ergodic to nonergodic behavior as an apparent kinetic transition.

How can we reconcile the T dependence of D and $\Omega_e(t)$ with our identification of T_s as the spinodal-glass transition? The answer lies in the dynamical properties of the *clumps*. For example, the diffusion coefficient of the center of mass of the clumps D_c is zero for $T < T_s$ in the MF limit. To understand this behavior, note that n, the mean number of particles in a clump, diverges as R^d as $R \to \infty$. In this limit, the number of clumps does not change with time, and the mean numbers of particles exiting and entering a clump are equal. From the central limit theorem, the relative fluctuations of these quantities go to zero as R and $n \to \infty$. We conclude that $D_c \leq D/\sqrt{n}$, and hence $D_c = 0$ in the MF limit. Our simulations of D_c for finite R are consistent with this prediction, and also indicate that the clumps do not see the same local environment, i.e., they have different numbers of nearest-neighbor clumps. This difference in the local environment persists as $R \to \infty$ because the clumps do not diffuse and cannot sample different local environments. Hence, the system is nonergodic on a clump (mass $\sim R^d$) scale for $T < T_s$ in the MF limit.

In summary, the clumps are long lived and localized for $T < T_s$, even though particles move from clump to clump. The time scale for the motion of the clumps diverge in the MF limit, and the spinodal-glass transition is seen dynamically only on a clump scale; single particle dynamical properties show no evidence of the underlying spinodal–metastable glass transition. However, we observe an *apparent* kinetic transition that is associated with the slow diffusion of the particles and the finite duration of our runs. The temperature of this apparent transition is less than T_s and depends on the observation time.

The well-characterized behavior of the MF model motivates us to ask if similar behavior occurs with more realistic interactions, and we consider a two component, $d = 2$ system of LJ particles of mass m. We take the LJ length parameter of the minority component to be 1.5 times larger than the length parameter σ of the majority component; the values of the energy parameter ϵ are the same for both. The role of the minority component is to inhibit nucleation [10]. We choose units such that lengths are measured in terms of σ, energies in terms of ϵ, and the time in terms of $\tau = (m\sigma^2/\epsilon)^{1/2}$. We cut off the LJ potential at $r = 3\sigma$. The MD simulations [11] are for $N = 500$ particles with 80% of the total being the majority component. The simulations are at fixed volume with the central cell size at each T chosen so that the pressure $P \approx 70$ ($\rho \sim 1$). The time step $\Delta t = 0.005\tau$, and averages are over a duration ranging from 2000τ at $T = 5.5$ to $20\,000\tau$ at $T = 2.15$. The following results are for the majority particles only.

The single particle energy and velocity fluctuation metrics show ergodic behavior for $2.15 \leq T \leq 5.5$. $D(T)$ is consistent with the Vogel-Fulcher form, $D \sim e^{-B/(T-T_0)}$ with $T_0 \approx 1.5$ (see Fig. 1). This form implies that the system loses ergodicity at $T \sim T_0$ for our runs. This behavior is similar to that of a two component, $d = 3$ LJ system [9].

If the LJ system exhibits pseudospinodal effects, we should find behavior analogous to that observed in the MF glass model [3] and in Ising models with long, but finite range interactions [12]. In these systems, $S(k)$ appears to diverge if its behavior is extrapolated from high T or small magnetic field, respectively, but the extrapolated singularity is not observed if measurements are made too close to the apparent singularity. For the LJ system we find a diffraction peak in $S(k)$ at $k = k_0 \approx 7.1$. The height of the peak $\chi(k_0, T)$ increases by a factor of ≈ 1.8 and the width $w(k_0, T)$ decreases by a factor of ≈ 1.5 as T is lowered from 5.5 to 2.15 (see Fig. 1). Because this range of T is limited, we can fit $\chi(k_0, T)$ and $w(k_0, T)$ by a variety of functional forms. If we consider the data only for $T \geq 3.1$, the most consistent fit is $\chi(k_0, T) \sim (T - 2.5)^{-0.2}$ and $w(k_0, T) \sim (T - 2.5)^{0.25}$. These fits suggest that the increase of the first peak of $S(k)$ is influenced by a weak singularity at $T = T_s \approx 2.5$. Given the limited range of T, this fit is justified only in the context of our rigorous results for $S(k)$ in the MF model [3]. If we fit the data for $2.15 \leq T \leq 5.5$, we find $\chi(k_0, T) \sim T^{-0.4}$ and $w(k_0, T) \sim T^{0.4}$, and we see no evidence of a spinodal-like singularity. This behavior is consistent with the pseudospinodal interpretation; i.e., if measurements are

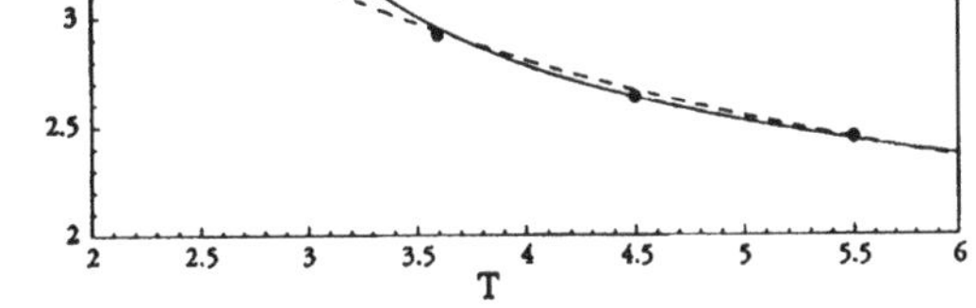

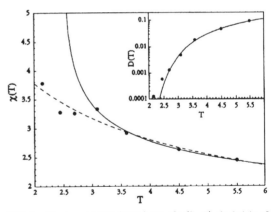

FIG. 1. The temperature dependence of $\chi(k_0, T)$, the height of the diffraction peak in the static structure function $S(k)$ at $k = k_0 \approx 7.1$. Note that $\chi(k_0, T)$ increases by approximately 1.8. The solid line represents the best fit in the range $3.1 \leq T \leq 5.5$ and has the form $(T - 2.5)^{-0.2}$; the dotted line represents the best fit in the range $2.15 \leq T \leq 5.5$ and has the form $T^{-0.4}$. The inset shows the T dependence of the self-diffusion coefficient D. The solid line represents the fit to the Vogel-Fulcher form $e^{-B/(T-T_0)}$. $T_0 = 1.5$ if we omit the two lowest values of T for which the data are limited; and $T_0 = 1.0$ if all data points are included.

made too close to the apparent singularity, its effects vanish.

For T near or below the pseudospinodal the system should show signs of an instability. We look for long-lived structures whose constituent particles remain in close proximity to each other over extended times at sufficiently low T. Because the LJ potential diverges at small interparticle separations, these structures are not identical to the clumps found in the MF model. A visual examination of the configurations shows evidence of a partial phase separation in which a significant fraction of the majority particles form clusters of hexagonal-like structures, which become better defined at T is decreased. To characterize these clusters, we determine the Voronoi structure. For each particle with six Voronoi neighbors we measure [13] $\Delta_i = (1/\langle \ell_i \rangle^2)[\langle \ell_i^2 \rangle - \ell_i \rangle^2]$, where $\langle \ell_i \rangle$ is the mean edge length of the Voronoi hexagon of particle i. If i is in an ideal crystalline environment, $\Delta_i = 0$. Such a particle belongs to a cluster if its combination of Δ and kinetic energy is sufficiently low. (The criterion assumes a linear relation, with a larger Δ implying a more stringent requirement for the kinetic energy.) Including the kinetic energy in the cluster criterion reduces the effect of thermal fluctuations. The qualitative properties of the clusters are *independent* of the cutoffs over a wide range of values.

The mean cluster lifetime is measured by dividing the system into boxes and computing the number of particles that belongs to any cluster in each box. The idea is to find if the mean number of cluster particles in each box becomes approximately the same as $t \gg 1$. If n_α, the number of cluster particles in box α, is greater than a threshold value, box α is said to be occupied; the corresponding nonzero p_α is used to compute a time-displaced cluster correlation function $G_c(t)$ and a cluster fluctuation metric $\Omega_c(t)$. (The latter is defined analogously to the energy fluctuation metric.) At $T = 5.5$, the decay of $G_c(t)$ to its equilibrium value can be fitted to an exponential function with a relaxation time $\tau_c \approx 50\tau$; similarly, Ω_c exhibits ergodic behavior. At $T = 3.1$, $G_c(t)$ does not reach its equilibrium value for $t \leq 5000\tau$, and $\Omega_c(t)$ exhibits nonergodic behavior; i.e., τ_c is much longer than our runs. At $T = 3.1$, $D > 0$ and the system appears ergodic if single particle properties are probed. This qualitatively different behavior of the clusters and the particles is analogous to the behavior of the clumps and the particles in the MF model.

The clusters also exhibit interesting static properties. Because the width of the peak of $S(k)$ decreases as $T \rightarrow 0$, we expect that the mean size of the clusters grows until they become "frustrated" by the minority component and the different cluster orientations. Our results for n_s, the number of clusters of size s, can be fit by the form (see Fig. 2)

$$n_s \sim s^{-3/2} e^{-s/m(T)}, \qquad (2)$$

where $m(T)$ is a parameter that increases as T is lowered until $T \approx 3.1$; below $T \approx 3.1$ $m(T)$ does not increase. The form (2) is robust and independent of the cutoffs used to determine the clusters. Similar scaling behavior has been found near the freezing transition in a $d = 2$ LJ system using a different cluster criterion [14]. The scaling behavior of the clusters, in particular the exponent of 3/2, is similar to the MF behavior found in other systems [15].

As noted, τ_c, the cluster lifetime, is longer than our observation time for $T \lesssim 3.1$. We assume $\tau_c \sim e^{-b/(T-T_{cl})}$, and find that $2.5 \leq T_{cl} \leq 2.9$; T_{cl} is the extrapolated temperature at which τ_c becomes infinite. Although our estimates for τ_c are only qualitative and the value of T_{cl} is the least accurate of our parameters, this value of T_{cl} is consistent with the value of $T_s \approx 2.5$ obtained from the extrapolation of the T dependence of the peak of $S(k)$.

To summarize our MD results, we find that there is a range of T for which $D > 0$ and τ_c is too long to estimate. (Our longest runs are for $20\,000\tau$, runs that are relatively long in comparison to most simulations of glasses.) This qualitatively different dynamical behavior of the single particle and cluster properties is analogous to the behavior of our MF model. From our extrapolations of the T dependence of the peak of $S(k)$ and τ_c, we see evidence of an incipient "transition" at $T_s \approx 2.5$. These two results provide indirect evidence for our identification of T_s with an apparent ergodic to nonergodic transition associated with the dynamics of the clusters and the presence of a pseudospinodal [16]. That is, we find evidence of an incipient thermodynamic glass "transition" at $T_s \approx 2.5$, a value of $T > T_0$ at which the single particle dynamical properties indicate an ergodic to nonergodic transition. Our extrapolation of $D(T)$ to 0 at $T_0 \approx 1.5$ suggests that T_0 can be interpreted as the kinetic transition and is distinguishable from the incipient thermodynamic transition.

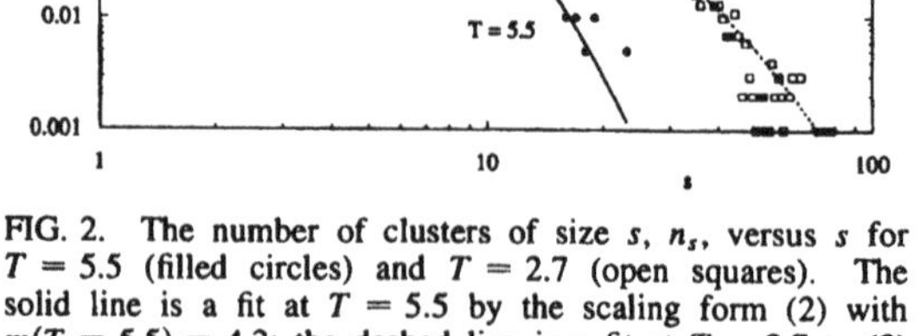

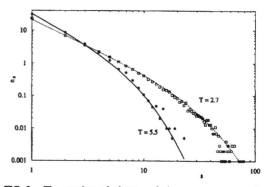

FIG. 2. The number of clusters of size s, n_s, versus s for $T = 5.5$ (filled circles) and $T = 2.7$ (open squares). The solid line is a fit at $T = 5.5$ by the scaling form (2) with $m(T = 5.5) = 4.2$; the dashed line is a fit at $T = 2.7$ to (2) with $m(T = 2.7) = 21$. These values for $m(T)$ are consistent with the mean size (mass) of the clusters that is observed directly.

The power-law scaling of n_s might be important for understanding the relaxation processes in glasses on experimental time scales. Nagel and collaborators [17] have fitted their measured dielectric susceptibility to a single scaling curve over 13 decades of frequency for a wide range of T and for many glass formers. Their results together with our simulation and theoretical results suggest that there might be a hierarchy of time and length scales and hence a hierarchy of clusters, i.e., a geometrical basis for the observed scaling.

We stress that the effects of the pseudospinodal and the incipient thermodynamic glass transition will be more or less apparent depending on the interaction range, the details of the interaction, and d [13]. We do not expect to find spinodal-like effects in all supercooled liquids. In particular, there are MF models, e.g., the MF limit of the Gaussian potential [3], which have no spinodal and hence no thermodynamic glass transition. These considerations suggest that there is a class of materials for which the observed glass transition is associated with a pseudospinodal and an incipient thermodynamic transition, and other materials for which the observed glass transition is not associated with such effects. We do not expect the behavior of the clusters we have found to be observed in all glass formers.

Based on our MC and theoretical studies of a MF model and our MD results for a $d = 2$ LJ system, we suggest that the latter is in the class of systems whose behavior can be attributed to an incipient thermodynamic instability (the pseudospinodal). We emphasize that a true thermodynamic glass transition does not exist in the LJ system, even though the pseudospinodal has measurable effects including increasing length and time scales as the pseudospinodal is approached. In addition to this glass-pseudospinodal transition, there is a temperature (for fixed density) that can be interpreted as a kinetic transition below which the diffusion coefficient is not measurable during our observation time.

Research at Boston University was supported in part by the NSF DMR. Acknowledgment is made to the donors of the Petroleum Research Foundation, administrated by the American Chemical Society, for partial support of this research at Clark University. We thank Louis Colonna-Romano for useful conversations.

*Present address: Florida State University, Supercomputer Computations Research Institute, Tallahassee, FL 32306-4052.

[1] Reviews of the properties of structural glasses can be found, for example, in *Liquids, Freezing, and the Glass Transition,* edited by J.P. Hansen, D. Levesque, and J. Zinn-Justin (North-Holland, Amsterdam, 1991); W. Götze and L. Sjögren, Rep. Prog. Phys. **55**, 241 (1992); R.D. Mountain, Int. J. Mod. Phys. C **5**, 247 (1994).

[2] J.P. Sethna, J.D. Shore, and M. Huang, Phys. Rev. B **44**, 4943 (1991).

[3] W. Klein *et al.*, Physica (Amsterdam) **205A**, 738 (1994).

[4] M. Kac, G.E. Uhlenbeck, and P.C. Hemmer, J. Math Phys. (N.Y.) **4**, 216 (1961).

[5] N. Grewe and W. Klein, J. Math. Phys. **18**, 1729 (1977); **18**, 1735 (1977).

[6] R.A. Ramos, Ph.D. thesis, Boston University, 1994.

[7] I. Clejan, Ph.D. thesis, Boston University, 1994.

[8] Because the clumps are order R apart, the interaction of particles in a given clump with particles in neighboring clumps is minimized. Inside a clump, a particle undergoes a restricted random walk. A particle that tries to leave a clump experiences a potential barrier due to the close proximity of other clumps. Hence, such a particle must interact with particles in the same clump and with particles in at least one other clump at some time during its possible escape. The upper bound of the potential barrier is $c\gamma^d R^d$, where the constant c depends on d. (R^d is proportional to the number of particles in a clump, and γ^d is the strength of the interaction $\gamma^d R^d = 1$.) The probability of leaving a clump in a MC simulation is bounded from below by $\frac{1}{2}e^{-c/k_BT}$ for all R and hence $D > 0$. Similar arguments hold for a MD simulation of the same system.

[9] D. Thirumalai and R.D. Mountain, Phys. Rev. E **47**, 479 (1993); R.D. Mountain and D. Thirumalai, Physica (Amsterdam) **210A**, 453 (1994). In this context a system is "ergodic" if time averages and ensemble averages are equal.

[10] Y.J. Wong and G.W. Chester, Phys. Rev. B **35**, 3506 (1987).

[11] A.I. Mel'cuk, Ph.D. thesis, Clark University, 1994.

[12] D.W. Heermann, W. Klein, and D. Stauffer, Phys. Rev. Lett. **50**, 1062 (1983).

[13] N.N. Medvedev, A. Geiger, and W. Brostow, J. Chem. Phys. **93**, 8337 (1990).

[14] M.A. Glaser and N.A. Clark, Adv. Chem. Phys. **83**, 543 (1993), use sixfold bond orientational order to define the clusters in a one component, $d = 2$ LJ system.

[15] T.S. Ray and W. Klein, J. Stat. Phys. **168**, 891 (1990). See also D. Stauffer and A. Aharony, *Introduction to Percolation Theory* (Taylor & Francis, London, 1992).

[16] Pseudospinodal effects have been observed in a $d = 3$ deeply supercooled LJ liquid undergoing nucleation [J. Yang *et al.*, J. Chem. Phys. **93**, 711 (1990)] and in the fractal structures that develop in the early stages of spinodal decomposition in a $d = 2$ LJ fluid [R.C. Desai and A.R. Denton, in *Growth and Form,* edited by H.E. Stanley and N. Ostrowsky (Martinus Nijhoff Press, Dordrecht, 1986); W. Klein, Phys. Rev. Lett. **65**, 1462 (1990)].

[17] P.K. Dixon *et al.*, Phys. Rev. Lett. **65**, 1108 (1990).

Authors' note

See also G. Johnson, A. Mel'cuk, H. Gould, W. Klein and R. Mountain, "Molecular Dynamics Study of Long-Lived Structures in a Fragile Glass Forming Liquid," Phys. Rev. E 57, 5707 (1998).

Classification
Physics Abstracts
05.40+j — 64.70-p — 61.20-p

Short Communication

On a Dynamical Model of Glasses

Jean-Philippe Bouchaud(1), Alain Comtet(2,3) and Cécile Monthus(2,3)

(1)Service de Physique de l'Etat Condensé, CEA-Saclay, Orme des Merisiers, 91191 Gif s/ Yvette Cedex, France
(2)Division de Physique Théorique, IPN, Batiment 100, Université de Paris-Sud, 91406 Orsay Cedex, France
(3)LPTPE, Université P. et M. Curie, 4 Place Jussieu, 75231 Paris Cedex 05, France

(Received 13 June 1995, accepted in final form 28 September 1995)

Abstract. — We analyze a simple dynamical model of glasses, based on the idea that each particle is trapped in a local potential well, which itself evolves due to hopping of neighbouring particles. The glass transition is signalled by the fact that the equilibrium distribution ceases to be normalisable and dynamics becomes non-stationary. We generically find stretching of the correlation function at low temperatures and a Vogel-Fulcher like behaviour of the terminal time.

Glasses have a number of fascinatingly universal properties which are still not satisfactorily accounted for theoretically [1, 2]. A common experimental feature is the 'shouldering' of the relaxation laws. More precisely, the relaxation of — say — the density fluctuations evolves from a simple Debye exponential at high temperature (liquid) to a two-step process at lower temperature, where the correlation function decays fast to a plateau value, from which it subsequently decays on a much longer time scale. These two regimes are called, respectively, the β and α relaxations; the α decay is often described in terms of a 'stretched exponential' with a characteristic time scale τ diverging faster than exponentially with the temperature, and controlling the transport properties such as the viscosity. The most popular description of this divergence is the Vogel-Fulcher law: $\tau \sim \tau_0 e^{\frac{\Delta}{T-T_0}}$, where τ_0 is a microscopic time scale [3]. One has to note that despite its tremendous phenomenological success, this law predicts such an abrupt divergence when T is lowered that it cannot be tested near T_0. Correspondingly, other functional forms, such as $\tau \sim \tau_0 e^{(\Delta/T)^2}$, give reasonable fits of the data [4, 5]. Furthermore, the Vogel-Fulcher law has only been justified on rather heuristic grounds [6].

Up to now, the most comprehensive theory of dynamical processes in glasses is the so-called mode-coupling theory, developed by Gotze and others [5]. It is based on a family of schematic equations coupling the density fluctuations in a non-linear and retarded way. Generically, these equations have a singularity which is associated to an 'ideal glass' transition temperature T_c, below which the correlation function does not decay to zero ('broken ergodicity'). This theory describes satisfactorily the overall shape of the relaxation function, at least for $T > T_c$ — in particular the existence of the two regimes β and α mentionned above, and a power-law divergence of the 'terminal' time scale τ as $(T-T_c)^{-\gamma}$. However, comparison with experiments [7] shows that the transition temperature T_c, if it exists, is much higher than the Vogel-Fulcher temperature, leaving a whole temperature interval $[T_0, T_c]$ where mode-coupling predicts a 'fluctuation arrest' (i.e. the α regime disappears) while the experimental relaxation time is still finite (and behaves à la Vogel-Fulcher). A way out of this contradiction is to argue that the mode coupling theory leaves out 'activated processes' which are responsible for the long time relaxation, and act to blur out the power-law divergence of τ near T_c. Although not unconceivable, this possibility requires the introduction of at least one extra free parameter to fit the data, which would be zero in the ideal transition scenario and is not found to be particularly small in the experiments. In other words, one major aspect of the mode-coupling theory is the existence of a singular temperature which however does not manifest itself very directly experimentally — in particular, the terminal time τ does not reveal any accident around T_c [7].

Another — rather more subtle — difficulty associated with the mode-coupling theory is that the dynamical equations are formally identical [8, 9] to those describing *exactly* some mean field models of spin-glasses [8, 10], where the presence of *quenched* disorder is assumed from the start. In glasses, however, this quenched disorder must be in some sense 'self-induced'. Although some progress has recently been made to substantiate such a scenario [11, 12], it is not yet clear whether the glassiness found in mode-coupling theories is or not an artefact of the very approximation.

In this paper, we propose and solve a simple model of glasses. Although still rather abstract, we believe that it captures at least part of the physics involved in the glass transition. The shape of the correlation function evolves, as the temperature is decreased, precisely as in experimental glasses — in particular, the terminal time diverges according to the Vogel-Fulcher law. The glass transition is signalled by the fact that the equilibrium distribution ceases to be normalisable; correlatively, as argued in [9, 13, 14], aging effects are present in the glass phase.

The progressive freezing of a liquid can be thought as follows: each particle is in a 'cage', i.e. a potential well of depth ϵ created by its neighbours, from which it can escape through thermal activation [15]. However, since *a priori* all particles can move, the (random) potential well trapping any one of them is in fact time dependent, further enhancing the probability of moving. In order to understand the glass transition, one must describe how, in a self-consistent way, all motion ceases. We thus introduce a density of local potential depth $\rho(\epsilon)$, describing the fact that the efficiency of the 'traps' depend on the environment [15]. Now, the basic object on which we shall focus is the probability $P(\epsilon, t)$ that a given particle is in a trap of depth ϵ at time t. This probability evolves because a given particle, with rate $\Gamma_0 \exp -\epsilon/T$, leaves its trap and chooses a new one with weight $\rho(\epsilon)$. Doing so, all the neighbouring 'traps' are affected by the motion which has taken place. In a mean field description, the resulting evolution of $P(\epsilon, t)$ is described by the following equation:

$$\frac{\partial P(\epsilon,t)}{\partial t} = -\Gamma_0 \exp -(\frac{\epsilon}{T})\, P(\epsilon,t) + \Gamma(t)\rho(\epsilon) + \Gamma(t) D \frac{\partial}{\partial \epsilon}\left[\rho(\epsilon)\frac{\partial P(\epsilon,t)}{\partial \epsilon} - P(\epsilon,t)\frac{\partial \rho(\epsilon)}{\partial \epsilon}\right] \quad (1)$$

where $\Gamma(t) \equiv \Gamma_0 \langle \exp -\epsilon/T \rangle$ is the average hopping rate ($\langle ... \rangle$ means an average over $P(\epsilon, t)$ itself). The two first terms describe the direct effect of leaving a trap, while the third one expresses the fact that every 'hop' induces a small change in all the neighbouring ϵ's. Assuming that the transition rate is proportional to the density of final states, the balance equation reads: $\Gamma(t) \int d\epsilon'\; T(\epsilon - \epsilon')\{P(\epsilon', t)\rho(\epsilon) - P(\epsilon, t)\rho(\epsilon')\}$. The fact that the change is small, justified in a mean-field limit where the number of neighbours is large, allows one to write this term in a diffusion like fashion, with an effective diffusion constant D proportional to the width of $\mathcal{T}$. More general forms for this diffusion term can also be considered, and will be discussed in [18].

With no incidence on the following results, we shall restrict ϵ to be positive — in line with our trap picture. Equation (1) is then supplemented by the boundary condition:

$$\left(\rho(\epsilon)\frac{\partial P(\epsilon,t)}{\partial \epsilon} - P(\epsilon,t)\frac{\partial \rho(\epsilon)}{\partial \epsilon}\right) = 0 \qquad \text{for} \quad \epsilon = 0 \tag{2}$$

which means that $\epsilon = 0$ is a 'reflecting' point. In fact, equations (1) and (2) can be taken as a definition of our dynamical model for glasses, which must be supplemented by an initial condition $P(\epsilon, t = 0)$.

Immediate properties of equations (1) and (2) are that:

- $\int_0^{+\infty} \mathrm{d}\epsilon P(\epsilon,t)$ is a conserved quantity, as it should.
- When $T \to \infty$, $\exp -\epsilon/T = 1$ and the equilibrium distribution is simply given by $P_{\mathrm{eq}}(\epsilon) \equiv \rho(\epsilon)$, as expected since there is no Boltzmann factor biasing the *a priori* weights.

Let us now study the equilibrium distribution at finite T. Setting $\frac{\partial P_{\mathrm{eq}}(\epsilon)}{\partial t} = 0$, one obtains an inhomogeneous Schrodinger equation for $P_{\mathrm{eq}}(\epsilon)$:

$$-D\frac{\partial^2 P_{\mathrm{eq}}(\epsilon)}{\partial \epsilon^2} + \mathcal{V}(\epsilon)P_{\mathrm{eq}}(\epsilon) = 1 \qquad \mathcal{V}(\epsilon) \equiv \frac{\Gamma_0 \exp -(\frac{\epsilon}{T}) + \Gamma D\frac{\partial^2 \rho(\epsilon)}{\partial \epsilon^2}}{\Gamma \rho(\epsilon)} \tag{3}$$

with the boundary condition equation (2). As long as this equation admits a 'bound state', a normalisable $P_{\mathrm{eq}}(\epsilon)$ exists and we shall call the resulting state 'liquid'. However, as the temperature decreases, the effective potential $\mathcal{V}$ tends to push $P_{\mathrm{eq}}(\epsilon)$ towards larger ϵ, and depending on the shape of $\rho(\epsilon)$, an 'extended', non normalisable state may appear — corresponding to a glass phase. It is easy to show that if $\rho(\epsilon)$ decreases slower than exponentially for large ϵ, the bound state ceases to exist as soon as $T < \infty$, while if $\rho(\epsilon)$ decays faster than exponentially, the bound state remains down to $T = 0$ (we shall come back to this case below). We shall thus focus on the case where $\rho(\epsilon)$ is a simple exponential: $\rho(\epsilon) \equiv \frac{1}{T_0}\exp -\epsilon/T_0$, where T_0 turns out to be the glass transition temperature. The reason for this is quite simple: it is the temperature at which the Boltzmann weighting factor exactly compensates the fact that deep potential wells are *a priori* extremely rare. Such a scenario is reminiscent of Derrida's random energy model [16], where the transition temperature is also defined by balancing the locally exponential density of states with the Boltzmann factor.

Solving equation (3) for $P_{\mathrm{eq}}(\epsilon)$ in this case leads to a well defined equilibrium state for $T > T_0$, which reads [17]:

$$P_{\mathrm{eq}}(\epsilon) = \mathcal{N}\left[K_\nu(x)\frac{I_{\nu-1}(x_0)}{K_{\nu-1}(x_0)}\mathcal{K}_\nu(x_0) + K_\nu(x)(\mathcal{I}_\nu(x_0) - \mathcal{I}_\nu(x)) + I_\nu(x)\mathcal{K}_\nu(x)\right] \tag{4}$$

where $\nu = \frac{2T}{T-T_0}$, $x \equiv x_0 \exp\frac{\epsilon}{\nu T_0}$, and $x_0 = \frac{\nu T_0^{3/2}}{\sqrt{D}}\frac{\sqrt{\Gamma_0}}{\sqrt{\Gamma}}$. I_ν and K_ν are the Bessel functions of order ν, and $\mathcal{K}_\nu(x) \equiv \int_x^\infty \frac{\mathrm{d}u}{u}K_\nu(u)$, $\mathcal{I}_\nu(x) \equiv \int_0^x \frac{\mathrm{d}u}{u}I_\nu(u)$. $\mathcal{N}$ and Γ are fixed by the normalisation of $P_{\mathrm{eq}}(\epsilon)$ and the boundary condition (2), which lead to the following equation:

$$\frac{D}{\nu^3 T_0^3} = \frac{I_{\nu-1}(x_0)}{K_{\nu-1}(x_0)}[\mathcal{K}_\nu(x_0)]^2 + 2\int_{x_0}^\infty \frac{\mathrm{d}u}{u}I_\nu(u)\mathcal{K}_\nu(u) \tag{5}$$

One can check, using the properties of Bessel functions, that this last equation is identically satisfied when $T \to \infty$, where $\nu = 2$ and $\Gamma = \Gamma_0$. For $T \to T_0$, on the other hand, we find that the average hopping rate Γ vanishes linearly, as $\Gamma_0 \frac{T-T_0}{T}$, which is numerically found to be a very good approximation for all temperatures. More interestingly, however, one finds that for $T \to T_0$, $P_{\text{eq}}(\epsilon)$ decays as $P_{\text{eq}}(\epsilon) \simeq \frac{T-T_0}{TT_0} \exp - \frac{\epsilon(T-T_0)}{TT_0}$ [19], which means that the characteristic energy scale is $\epsilon^* = \frac{2TT_0}{T-T_0}$.

In order to make contact with experimental observables one must further define a (two-time) correlation function. The simplest one to consider, corresponding to large wavevectors q, is such that only particles which have not moved at all contribute to the correlation, i.e.

$$C(t,t') = \int_0^\infty \mathrm{d}\epsilon P(\epsilon,t') \exp - [\Gamma_0 \exp - (\frac{\epsilon}{T})(t-t')] \tag{6}$$

but other choices, corresponding e.g. to particles hopping on a d-dimensional lattice and finite q, are possible [18, 20]. Equation (6) assumes in particular that the 'width' of the potential wells is zero. In order to be more realistic and take into account the fast vibration of the particles in their 'cages', a simple modification is to multiply $C(t,t')$ defined in equation (6) by a 'Debye-Waller' factor $C_\beta(t,t') = \exp - \frac{q^2 r^2(t-t')}{2}$, where q is the probing wave vector. $r(t)$ describes the diffusive motion in an harmonic potential well: $r^2(t) = \xi_0^2[1 - \exp(-\frac{t}{\tau_0})]$; ξ_0 can be thought as the 'size' of the cage, and ξ_0^2/τ_0 of the order of the high temperature (liquid) diffusion constant [21]. One should however emphasize that this way of introducing the β peak, although physically motivated, is rather *ad-hoc*. An important success of the Mode-Coupling Theory is that the β regime appears naturally, and is furthermore predicted to be intimately connected to the α relaxation [5].

For $T > T_0$, i.e. when $P_{\text{eq}}(\epsilon)$ exists, the correlation function only depends on the difference $t - t'$. One finds that $C(t)$ (defined by Eq. (6)) behaves as

$$C(t) = \begin{cases} 1 - \Gamma t & \text{for } t \ll \Gamma^{-1} \\ (\Gamma_0 t)^{\frac{T_0 - T}{T_0}} & \text{for } t \gg \tau(T) \end{cases} \tag{7}$$

where $\tau(T) = \Gamma_0^{-1} \exp(\frac{T_0}{T-T_0})$ is the Vogel-Fulcher time, which very naturally appears within the present model (although no diverging length scale is involved). From equation (7), and Figure 1, one sees that $C(t)C_\beta(t)$ has precisely the shape observed in most experimental situations, provided one takes into account the Debye-Waller factor C_β defined above. Note the presence of two characteristic time scales, a short (microscopic) one τ_0, corresponding to the cage vibrations (β peak), and a long one $\tau(T)$, corresponding to the α peak; these two time scales separate extremely fast as the temperature is reduced.

When $T < T_0$, on the other hand, no normalisable $P_{\text{eq}}(\epsilon)$ can be found, which corresponds to the weak ergodicity breaking situation described in [10, 13, 14]. In this situation, $P(\epsilon, t)$ never reaches a stationary limit, but continuously drifts towards larger and larger energies. Time translational invariance is spontaneously broken as $C(t,t')$ never becomes a function of $t - t'$ alone, a situation now referred to as 'aging'. In fact, one can show [18] that $C(t_{\text{w}} + t, t_{\text{w}}) \equiv \mathcal{C}(t/t_{\text{w}})$, with $1 - \mathcal{C}(u) \propto u^{1-\frac{T}{T_0}}$ for $u \to 0$ and $\mathcal{C} \propto u^{-\frac{T}{T_0}}$ for $u \to \infty$, precisely as in the 'trap' model studied in [13]. This suggests that the difference between the 'quenched' model considered in [13] (corresponding to $D \equiv 0$ in Eq. (1)) and the 'annealed' model considered here is, to some extent, irrelevant.

All the above results are still expected to hold if the density of states $\rho(\epsilon)$ is approximatively exponential below a certain cut-off ϵ_c, provided that $T - T_0 > 2TT_0/\epsilon_c$. If $\rho(\epsilon)$ decays faster than exponentially, for example as $\rho(\epsilon) \propto \exp - (\epsilon/\epsilon_c)^2$, then strictly speaking the glass temperature is pushed down to zero. This choice for $\rho(\epsilon)$ is interesting since it really corresponds

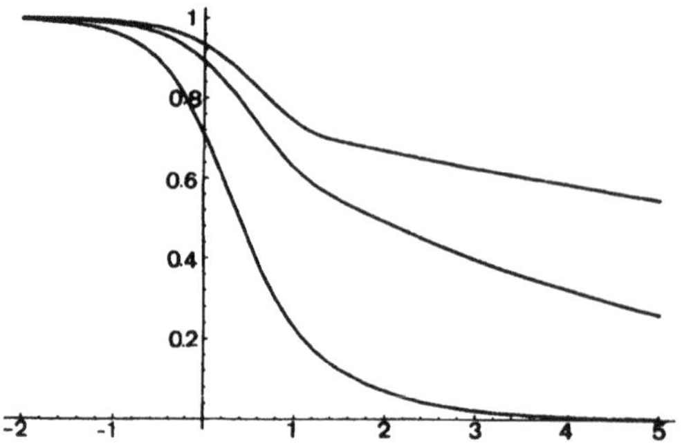

Fig. 1.

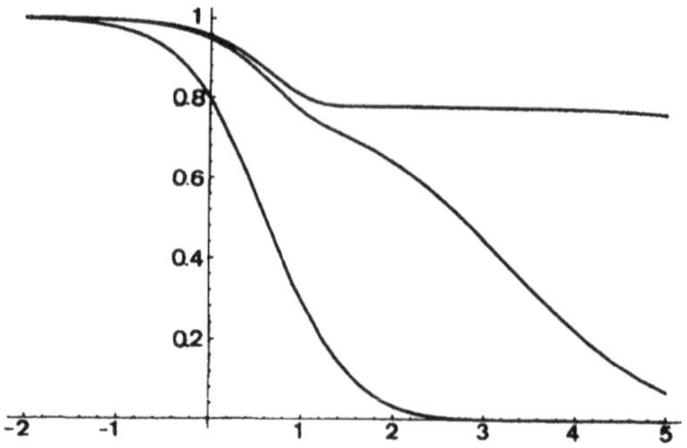

Fig. 2.

Fig. 1. — Exponential density of states. Plot of $C(t)C_\beta(t)$ *versus* $\log_{10}(\Gamma_0 t)$ for $q\xi_0 = 0.5$, $\tau_0 = 5\Gamma_0^{-1}$, and $\frac{T}{T_0} = 2., 1.1, 1.03$.

Fig. 2. — Gaussian density of states. Plot of $C(t)C_\beta(t)$ *versus* $\log_{10}(\Gamma_0 t)$ for $q\xi_0 = 0.5$, $\tau_0 = 5\Gamma_0^{-1}$, and $\frac{T}{\epsilon_c} = 0.5, 0.25, 0.15$. Note that the plateau observed for the lowest temperature eventually decays to zero.

to a mean-field limit where the local trap strength is obtained as a sum of contributions from the (large) number of neighbours. The equilibrium distribution $P_{\rm eq}(\epsilon)$ is then, for large ϵ, given by $\exp\left[\frac{\epsilon}{T} - (\frac{\epsilon}{\epsilon_c})^2\right]$, and the corresponding $C(t)$ exhibits a considerable amount of stretching at low enough temperatures. $C(t)$ is indeed very well fitted by a stretched exponential at intermediate times [22]. The long time fall off of $C(t)$ is in this case given by $C(t) \propto (\frac{\tau(T)}{t})^\mu$ with $\mu = (\frac{T}{\epsilon_c})^2 \log \Gamma_0 t$. The terminal time $\tau(T)$ diverges as $\exp\frac{\epsilon_c^2}{T^2}$, i.e. much faster than an activated law, and, as mentioned in the introduction, also compatible with the experimental data [4, 5]. The shape of $C(t)C_\beta(t)$ is plotted in Figure 2 for different T/ϵ_c. It is, again, very similar to the experimental data. In particular, the α relaxation for different temperatures can be approximatively [23] rescaled onto a unique master curve when plotted as a function of $\frac{t}{\tau(T)}$, as observed in experiments and numerical simulations [24], and predicted by the mode-coupling theory [5].

In conclusion, we have proposed a very idealized model for the glass transition, which we have solved in the high temperature phase. The results are found to reproduce two important observations: the shouldering and stretching of the correlation and the Vogel-Fulcher like divergence of the terminal time scale. The former feature is usually accounted for by the mode-coupling theory on the basis of a 'phantom' singularity which is removed by exogeneous processes, in turn responsible for the finiteness of the terminal time at low temperatures. In line with previous analysis [4, 15], the present 'trap' model suggests that the hypothesis of a intermediate temperature transition might not be necessary, although a more detailed comparison with experimental data is obviously desirable, in particular the relation between the (aging) α and β regime deep in the glass phase [9].

Acknowledgments

JPB wants to thank M. Adam, C. Alba-Simionescu, A. Barrat, L. Cugliandolo, D. Dean, J. Kurchan, D. Lairez, J. M. Luck, E. Vincent and especially M. Mézard for many inspiring discussions on these subjects.

References

[1] For a review, see the interesting series of papers in *Science*, **267** (1995) 1924.

[2] 'Liquids, freezing and glass transition', Les Houches 1989, J.P. Hansen, D. Levesque, J. Zinn-Justin, Eds. (North Holland, 1989) in particular reference [5].

[3] Vogel H., *Z. Phys.* **22** (1921) 645; Fulcher G.S., *J. Am. Ceram. Soc.* **6** (1925) 339.

[4] Ferry J.D., Grandine L.D. and Fitzgerald E.R., *J. Appl. Phys.* **24** (1953) 911; Bassler H., *Phys Rev. Lett* **58** (1987) 767 and references therein.

[5] Gotze W., in reference [2], p. 403 et sq.

[6] See Kirkpatrick T.R., Thirumalai D. and Wolynes P.G. in reference [8]; Vilgis T., *J. Phys. Cond Matter* **2** (1990) 3667; Parisi G., Slow dynamics in glasses, preprint cond-mat 941115 and 9412034

[7] For a enlightening introduction to the experimental controversy, see the series of Comments ir *Phys. Rev. E*: Zeng X.C., Kivelson D. and Tarjus G., *Phys. Rev. E* **50** (1994) 1711; Dixon P.K. Menon N. and Nagel S.R., *Phys. Rev. E* **50** (1994) 1717; Cummins H.Z., Li G., *Phys. Rev. E* **5**[cut off] (1994) 1720 and references therein, in particular Cummins H.Z., Du W.M., Fuchs M., Gotze W. Hildebrand S., Latz A., Li G. and Tao N.J., *Phys. Rev. E* **47** (1993) 4223.

[8] Kirkpatrick T.R., Thirumalai D. and Wolynes P.G., *Phys. Rev. A* **40** (1989) 1045, and references therein. See also Nieuwenhuizen Th., preprint (1995).

[9] Bouchaud J.P., Cugliandolo L., Kurchan J. and Mézard M., in preparation.

[10] Cugliandolo L. and Kurchan J., *Phys. Rev. Lett.* **71** (1993) 173; Cugliandolo L. and Le Doussa[cut off] P., Large time off-equilibrium dynamics of a particle diffusing in a random potential, preprint cond-mat 9505112; Cugliandolo L., Le Doussal P. and Kurchan J., preprint cond-mat 9509008 Franz S. and Mézard M., *Europhys. Lett.* **26** (1994) 209; *Physica A* **209** (1994) 1.

[11] Kirkpatrick T.R. and Thirumalai D., *J. Phys. A* **22** (1989) L149.

[12] Bouchaud J.P. and Mézard M., *J. Phys. I France* **4** (1994) 1109; Marinari E., Parisi G. and Ritort F., *J. Phys. A* **27** (1994) 7615; *J. Phys. A* **27** (1994) 7647; Cugliandolo L.F., Kurchan J., Paris[cut off] G. and Ritort F., *Phys. Rev. Lett.* **74** (1995) 1012; Franz S. and Hertz J., *Phys. Rev. Lett.* **7**[cut off] (1995) 2114. See also: Ritort F., Glassiness in a Model without energy barriers, preprint cond-ma[cut off] 9504081.

[13] Bouchaud J.P., *J. Phys. I France* 2 (1992) 1705; Bouchaud J.P. and Dean D.S., *J. Phys. I Franc*[cut off] **5** (1995) 265; Barrat A. and Mézard M., *J. Phys. France* **5** (1995) 941.

[14] Bardou F., Bouchaud J.P., Emile O., Cohen-Tannouji C. and Aspect A., *Phys. Rev. Lett.* **7**[cut off] (1994) 203.

[15] Note that this one-particle density of state can *a priori* depend on the density of the liquid an[cut off] the temperature itself.

[16] Derrida B., *Phys. Rev. B* **24** (1981) 2613.

[17] The problem of the convergence towards this equilibrium state is interesting and will be discusse[cut off] in [18]

[18] Monthus C., Bouchaud J.P. and Comtet A., in preparation.

[19] This simple exponential form is actually a good numerical approximation for all values of ϵ.

[20] See e.g. Bouchaud J.P. and Georges A., *Phys. Rep.* **195** (1990) p. 142-143.

[21] This model emerged from discussions with M. Lairez and M. Adam: see Lairez D., Raspaud E. Adam M., Carton J.P. and Bouchaud J.P., to appear in 'Die Makromol. Chem. Symp.' (1995). ξ_0 and τ_0 are expected to depend somewhat on temperature, but this dependence is much weake[cut off] than that of the terminal time $\tau(T)$.

[22] On this point, see, e.g. Castaing B. and Souletie J., *J. Phys. I France* **1** (1991) 403.

[23] Note that $\mu(t = \tau(T))$ is indeed independent of temperature.

[24] Kob W. and Andersen H.C., *Phys. Rev. Lett.* **73** (1994) 1376.

EUROPHYSICS LETTERS 1 August 1997
Europhys. Lett., **39** (4), pp. 447-452 (1997)

High-frequency asymptotic shape of the primary relaxation in supercooled liquids

R. L. LEHENY and S. R. NAGEL
The James Franck Institute and The Department of Physics
The University of Chicago - Chicago, IL 60637 USA

(received 2 October 1996; accepted in final form 27 June 1997)

PACS. 77.22Gm – Dielectric loss and relaxation.
PACS. 64.60–i – General studies of phase transitions.
PACS. 64.70Pf – Glass transitions.

Abstract. – We characterize the high-frequency asymptotic shape of the primary dielectric relaxation for four supercooled liquids: glycerol, propylene carbonate, D-propylene glycol and tricresyl phosphate. Two distinct power laws, with an exact relationship between their exponents, exist above the peak in the imaginary part of the permittivity, $\varepsilon''(\nu)$. The second power law is a robust feature of the spectra, extending at least 13 decades above the peak frequency, for all the liquids. The change with temperature of $\varepsilon''(\nu)$ in this high-frequency region provides direct evidence for a proposed divergence in the static susceptibility in supercooled liquids and suggests that in the simplest analysis the approach to this divergence obeys Curie-Weiss behavior.

As a liquid supercools, its characteristic relaxation times increase dramatically. No consensus has formed for a theoretical picture of this phenomenon, and controversy surrounds interpretations of experimental results [1]. Due to its sensitivity and dynamic range, dielectric susceptibility has been a popular spectroscopic technique for studying these slowing dynamics. However, despite numerous studies of dielectric relaxation of supercooled liquids, a characterization of the spectral shape remains incomplete, particularly at frequencies much larger than the mean response frequency. Popular fitting forms for the primary relaxation, such as Havriliak-Negami and KWW (the Fourier transform of a stretched exponential), contain a single asymptotic power law above the peak frequency, ν_p, in the imaginary part of the permittivity, $\varepsilon''(\nu)$. While systematic deviations from a single power law at high frequencies have been known for a long time [2], few studies have made a careful evaluation of the precise response in this regime [3].

However, a recent argument has suggested that this high-frequency region is intimately related to the slowing dynamics [4]. A study of the primary relaxation in a large number of liquids led Dixon *et al.* [5] to develop a three-parameter scaling procedure which successfully collapses $\varepsilon''(\nu)$ for all the liquids at all temperatures onto a single curve [6]. Analyzing the shape of $\varepsilon''(\nu)$ implied by this scaling curve as a function of temperature, Menon and

Nagel [4] argued that the static dielectric susceptibility in supercooled liquids diverges at the temperature at which extrapolations suggest that relaxation times diverge. These divergences imply a phase transition in the liquid at low temperature.

The central feature of this argument is the identification of the deviations from a single power law above ν_p with a second power law region. The shape of the scaling plot implies an exact relationship between the exponents of the two power laws above ν_p:

$$\frac{\sigma+1}{\beta+1}=\gamma, \tag{1}$$

where $-\beta$ and $-\sigma$ are the exponents of the lower- and higher-frequency power laws, respectively; and where $\gamma = 0.72 \pm 0.02$, independent of temperature and liquid. One consequence of eq. (1) is that, as β decreases to 0.38 (for $\gamma = 0.72$), σ goes to zero [7]. Noting the trend in supercooled liquids of β decreasing with decreasing temperature, Menon and Nagel argued that σ becomes zero close to the temperature, T_0, at which ν_p goes to zero. At T_0, they concluded, a constant contribution to $\varepsilon''(\nu)$ extends from zero frequency up to an infrared cut-off. Therefore, the integrated spectral strength of this second power law region becomes infinite, leading to a divergence in the static susceptibility, $\Delta\varepsilon$. The evolution of the dynamic magnetic susceptibility of a paramagnet [8] cooled through the spin-glass transition reveals striking similarities with this scenario proposed for liquids. Two assumptions enter this argument for a diverging $\Delta\varepsilon$ in liquids: 1) this second power law is a universal feature of dielectric spectra and represents the asymptotic behavior of the primary response (up to the approximately temperature-independent infrared cut-off); and 2) trends seen in β at high temperature continue as the liquid approaches T_0.

We present here a detailed study of the high-frequency asymptotic shape of the primary dielectric relaxation in glycerol, propylene carbonate, D-propylene glycol [9], and tricresyl phosphate. These four liquids are well suited for this study due to the absence of secondary relaxations which overlap the high-frequency tail of the primary response [10]. (Propylene carbonate lies near the fragile extreme for simple liquids, while the other liquids are more intermediate glass formers.) We demonstrate the presence of two distinct power law regions above ν_p and show that the second power law is a robust feature, extending over at least 8 decades for all of the liquids. Its exponent, σ, follows the relationship predicted in eq. (1). In addition, from measurements of the decreasing value of σ with temperature, we find evidence for a divergence of $\Delta\varepsilon$ directly in the spectral response.

Figures 1 (a) to (d) show $\varepsilon''(\nu)$ for the four liquids at several temperatures. Our measurements span from 10^{-3} Hz to 10^7 Hz. Rather than extend our range to higher frequency to determine the shape $\varepsilon''(\nu)$ above ν_p, we have measured the response at the lowest possible temperatures, thus positioning the peak at very low frequency. This strategy has three advantages: 1) it allows us to characterize the relaxation as close as possible to the proposed low-temperature transition; 2) it avoids overlap in the spectrum with any microscopic peaks or other features at high frequency [3], [11]; and 3) it enables us to measure $\varepsilon''(\nu)$ through the entire frequency range with one capacitor geometry, thus avoiding the introduction of scaling factors to connect different regions of the spectra, as is typically necessary with higher-frequency measurements.

The principle limitation to measuring at arbitrarily low temperature is that, upon cooling, the sample falls from equilibrium below T_g, the calorimetric glass transition temperature, and only after a long period of equilibration, with the liquid held at constant temperature, is the system in a proper thermodynamic state. During equilibration the dielectric response shows time dependence characteristic of physical aging [12] in quenched glass formers. At our lowest measurement temperatures this equilibration required us to wait with the liquid

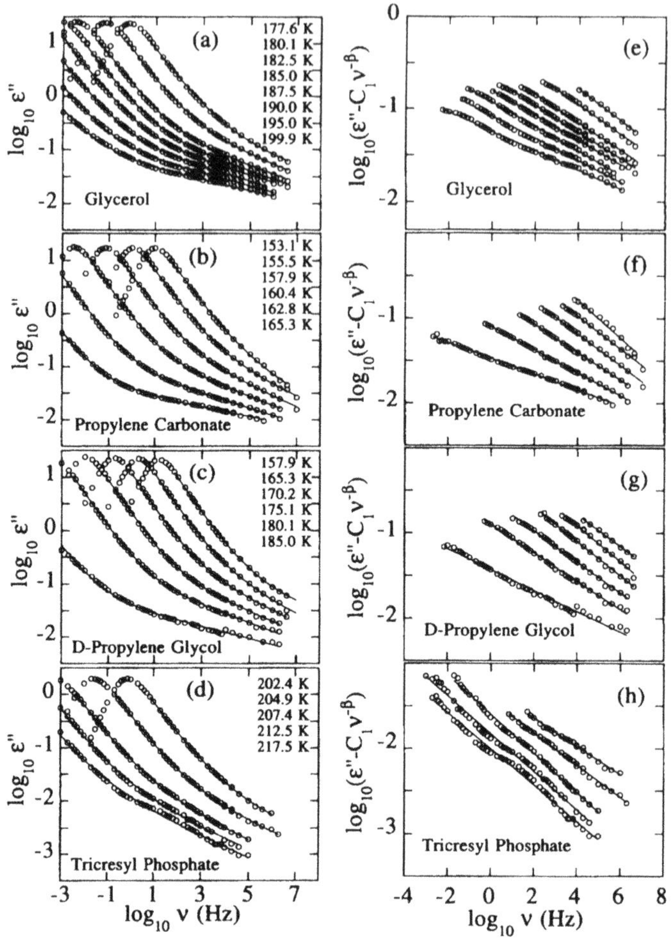

Fig. 1. – The imaginary part of the dielectric permittivity, $\varepsilon''(\nu)$, for (a) glycerol, (b) propylene carbonate, (c) D-propylene glycol and (d) tricresyl phosphate. The solid lines in (a) to (d) are fits to two power laws above the peak (eq. (2)). (e) to (h) show the $\varepsilon''(\nu)$ for the four liquids with the first power law region subtracted to highlight the high-frequency power law. T_g is 190 K for glycerol, 158 K for propylene carbonate, 167 K for propylene glycol, and 205 K for tricresyl phosphate.

at fixed temperature for more than 40 days before we collected data. Efforts to extend the measurements to significantly lower temperature are clearly impractical. However, at these temperatures ν_p is on the order of 10^{-7} Hz, more than 13 decades below our highest measured frequency.

Throughout the temperatures at which ν_p falls in our frequency window, a Vogel-Fulcher fit, $\nu_p = \nu_\infty \exp[-A/(T - T_0)]$, provides a good description of the peak position for all four liquids. $\nu_\infty \approx 10^{12}$ Hz is a characteristic phonon frequency. (This form implies that the relaxation times, characterized by the inverse of ν_p, diverge at a finite temperature, T_0.)

The solid lines in fig. 1 (a) to (d) are fits to two power laws:

$$\varepsilon''(\nu) = C_1 \nu^{-\beta} + C_2 \nu^{-\sigma} . \qquad (2)$$

For all the liquids at all temperatures this expression provides a good description of the

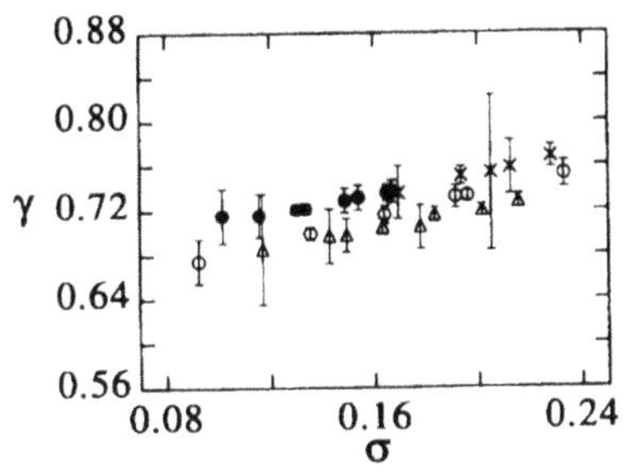

Fig. 2

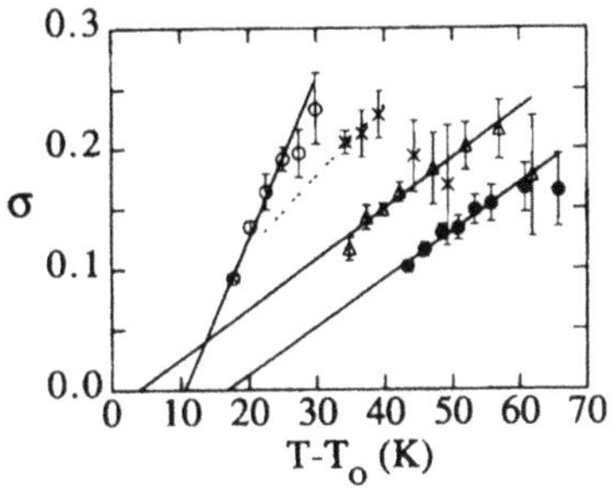

Fig. 3

Fig. 2. – The parameter, γ, from eq. (1) which relates β and σ, exponents of the low- and high-frequency power laws, respectively, for (•) glycerol, (○) propylene carbonate, (△) D-propylene glycol and (×) tricresyl phosphate as a function of σ.

Fig. 3. – The exponent of the high-frequency power law for (•) glycerol, (○) propylene carbonate, (△) D-propylene glycol and (×) tricresyl phosphate as a function of the difference in temperature from the Vogel-Fulcher temperature, T_0. The T_0 values are 134.1 K for glycerol, 135.1 K for propylene carbonate, 123.1 K for D-propylene glycol, and 168.2 K for tricresyl phosphate. The solid lines are linear fits to the exponents. The dashed line for tricresyl phosphate is drawn through the lowest three points to suggest the low-temperature behavior.

spectrum above ν_p. We estimate $-\beta$ by the minimum in $\mathrm{d}\log(\varepsilon'')/\mathrm{d}\log(\nu)$; σ is treated as a free parameter in all the fits. Figure 2 shows the resulting values for γ. While the results suggest γ may decrease weakly with σ, they are consistent with a constant value. The mean for γ is 0.72 ± 0.02, confirming the scaling prediction. (Indeed, constraining the exponents so that $\gamma = 0.72$ leads to fits virtually indistinguishable from those in fig. 1.) In figs. 1 (e)-(h), we replot the spectra with the first power law subtracted to highlight the high-frequency response. For all the liquids, this response is robust, persisting to the high-frequency limit of our data (more than 13 decades above ν_p for the lowest temperatures). These results support the first assumption of Menon and Nagel that the second power law is asymptotic.

Figure 3 displays σ as a function of the difference in temperature from T_0, the temperature at which ν_p goes to zero. The solid lines are linear fits to σ for glycerol, propylene carbonate, and D-propylene glycol; they extrapolate to zero slightly above T_0. These direct measurements of σ, along with the validity of eq. (1), extend to lower temperature the trend of β decreasing on cooling for these liquids [13]. Thus, our results support the second assumption of the divergence argument that trends seen at high temperature continue as liquids approach T_0. Although no theoretical justification exists for these linear extrapolations, the behavior of the data suggests that this ansatz is reasonable. (In $\varepsilon''(\nu)$ for tricresyl phosphate, a broad feature resembling a Johari-Goldstein beta peak centered near 10^2 Hz appears just above the limit of our experimental resolution and interferes with the determination of σ at the lowest temperatures. Also, the spectral shape for tricresyl phosphate shows little change throughout the temperatures at which we have data (202 K to 253 K), contrasting with the general trend among supercooled liquids [5] of β decreasing with decreasing temperature. We speculate that tricresyl phosphate begins to conform to this trend at low temperature, as suggested by the dashed line in fig. 3.) The argument for the divergence in $\Delta\varepsilon$ requires that $\sigma = 0$ when $\nu_p = 0$. The extrapolation of σ to zero slightly above T_0 demonstrates a trend strongly consistent with this contention.

From the Kramers-Kronig relation we determine the contribution to $\Delta\varepsilon$ from the high-

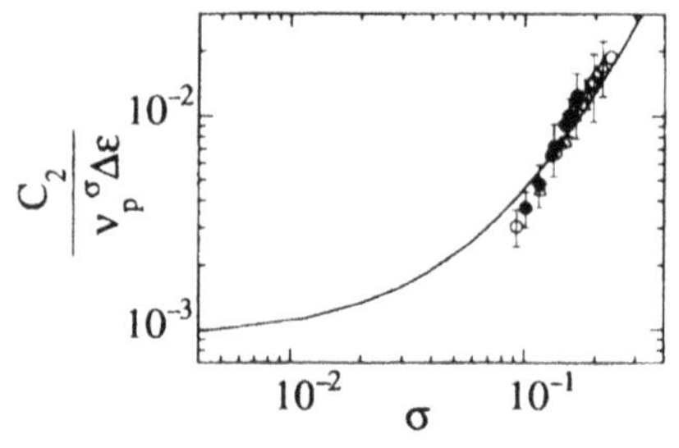

Fig. 4. – The normalized strength of the high-frequency power law, C_2, as a function of its exponent, σ, for (•) glycerol, (○) propylene carbonate and (△) D-propylene glycol. (For temperatures at which $\nu_\mathrm{p} < 10^{-3}$ Hz, we obtain ν_p by extrapolating Vogel-Fulcher fits and $\Delta\varepsilon$ by extrapolating Curie laws.) The solid line, given by $\log\left(C_2/\nu_\mathrm{p}^\sigma \Delta\varepsilon\right) = -0.977/(\sigma + 0.32)$, is the strength predicted by the shape of the scaling curve for $\varepsilon''(\nu)$ [5].

frequency power law alone, $\delta\Delta\varepsilon$,

$$\delta\Delta\varepsilon = \frac{2}{\pi}\int_{\nu_T}^{\nu_0} C_2 \nu^{-\sigma}\frac{\mathrm{d}\nu}{\nu}\,, \tag{3}$$

where ν_0 is the high-frequency cut-off for the power law and ν_T is the frequency at which the dominant contribution to $\varepsilon''(\nu)$ switches from the first power law, $\nu^{-\beta}$, to the second, $\nu^{-\sigma}$. The temperature dependence of ν_T, defined by $C_1\nu_T^{-\beta} = C_2\nu_T^{-\sigma}$, follows the same form as does ν_p, $\nu_T \approx \nu_0 \exp[-A/(T - T_0)]$. Substituting $\sigma = D(T - T_0)$ into eq. (3) gives $\delta\Delta\varepsilon \sim C_2/(T - T_0)$. Thus, a divergence in $\Delta\varepsilon$ depends on the temperature dependence of C_2. If C_2 remains greater than zero as σ goes to zero, as the shape of the scaling curve introduced by Dixon *et al.* [5] predicts, then $\delta\Delta\varepsilon \sim 1/(T - T_0)$; that is, the susceptibility follows a Curie-Weiss divergence. If C_2 goes to zero along with σ, $C_2 \sim (T - T_0)^\alpha$, then $\delta\Delta\varepsilon$ still diverges for $\alpha < 1$.

Figure 4 shows C_2 as a function of σ for the three liquids in which σ changes appreciably with temperature. We normalize C_2 by $\nu_\mathrm{p}^\sigma \Delta\varepsilon$ to account both for the change in position of the peak with temperature and for the differing strengths of the relaxation in the different liquids. As fig. 4 indicates, C_2 decreases with σ. If this rapid decrease were to continue to $\sigma = 0$, then the purported divergence in $\delta\Delta\varepsilon$ would disappear. However, the values of C_2 implied by the shape of the scaling curve, shown by the solid line in fig. 4 (which we calculate from the scaling plot in ref. [5] with no free parameters), both closely agree with our direct measurements and remain finite for $\sigma = 0$. To resolve whether C_2 continues to follow this behavior predicted by the scaling curve as σ goes to zero, measurements on liquids with smaller σ must be performed. (Preliminary results, based on reanalysis of the broad peak in $\varepsilon''(\nu)$ for salol [5], indicate values of C_2 for σ near 0.02 consistent with the scaling curve prediction in fig. 4.) The agreement between the scaling curve and our direct measurements (along with the extrapolated behavior of the scaling curve) is consistent with the conclusion that the spectral weight grows without bound as σ goes to zero. We note, however, that this conclusion relies on the validity of the scaling curve at frequencies above the peak.

In conclusion, we have evaluated precisely the high-frequency shape of the primary relaxation in $\varepsilon''(\nu)$ for several supercooled liquids confirming that the spectrum above the peak contains two power law regions, with exponents that are uniquely related to each other, independent of liquid or temperature. Thus, we have characterized a region of the spectrum whose shape has been poorly understood. Our results are consistent with the argument for a diverging static susceptibility at low temperature in supercooled liquids and demonstrate that evidence for this divergence can be seen directly in the spectra.

***

We thank R. Deegan and N. Menon for many helpful discussions. Funding for this work was provided by NSF DMR 94-10478.

REFERENCES

[1] For a review, see: Ediger M. D., Angell C. A. and Nagel S. R., *J. Phys. Chem.*, **100** (1996) 13200.

[2] Davidson D. W. and Cole R. H., *J. Chem. Phys.*, **19** (1951) 1484.

[3] Lunkenheimer P., Pimenov A., Schiener B., Böhmer R. and Loidl A., *Europhys. Lett.*, **33** (1996) 611. This paper includes broad frequency measurements of $\varepsilon''(\nu)$ in glycerol, including a study of the second power law region which we discuss. For the temperatures at which they overlap, our results are consistent with those measurements.

[4] Menon N. and Nagel S. R., *Phys. Rev. Lett.*, **74** (1995) 1230.

[5] Dixon P. K., Wu L., Nagel S. R., Williams B. D. and Carini J. P., *Phys. Rev. Lett.*, **65** (1990) 1108.

[6] The general applicability (see Schonhals A., Kremer F. and Schlosser E., *Phys. Rev. Lett.*, **67** (1991) 999; Menon N. and Nagel S. R., *Phys. Rev. Lett.*, **71** (1993) 4095) and strict validity (see Kudlik A., Benkhof S., Lenk R. and Rössler E., *Europhys. Lett.*, **32** (1995) 511; Leheny R. L., Menon N. and Nagel S. R., *Europhys. Lett.*, **36** (1996) 473) of the scaling procedure have been the subject of controversy. However, these debates do not take issue with the high-frequency shape of the scaling curve from which the arguments we present have been developed.

[7] The original divergence argument [4] was constructed from an equation similar to eq. (1) relating σ to the scaling parameter, w. $1/w$ is closely related, but not identical, to β.

[8] Bitko D., Menon N., Nagel S. R., Rosenbaum T. F. and Aeppli G., *Europhys. Lett.*, **33** (1996) 489.

[9] The data we present is for fully deuterated propylene glycol. We have measured $\varepsilon''(\nu)$ for the undeuterated liquid and find the shape and position of the spectra identical at all temperatures to that for the deuterated liquid.

[10] See, *e.g.*, Deegan R. D. and Nagel S. R., *Phys. Rev. B*, **52** (1995) 5653; Wu L., *Phys. Rev. B*, **43** (1991) 9906.

[11] Lunkenheimer P., Pimenov A., Dressel M., Goncharov Y. G., Böhmer R. and Loidl A., *Phys. Rev. Lett.*, **77** (1996) 318.

[12] Struik L. C. E., *Physical Aging in Amorphous Polymers and Other Materials* (Elsevier, Amsterdam) 1978.

[13] Schönhals A., Kremer F., Hofmann A., Fischer E. W. and Schlosser E., *Phys. Rev. Lett.*, **70** (1993) 3459.

PHYSICAL REVIEW E VOLUME 60, NUMBER 1 JULY 1999

Ergodicity-breaking transition and high-frequency response in a simple free-energy landscape

M. Ignatiev and Bulbul Chakraborty
The Martin Fisher School of Physics, Brandeis University, Waltham, Massachusetts 02254
(Received 29 January 1999)

We present a simple dynamical model described by a Langevin equation in a piecewise parabolic free-energy landscape, modulated by a temperature-dependent overall curvature. The zero-curvature point marks a transition to a phase with broken ergodicity. The frequency-dependent response near this transition is reminiscent of observations near the glass transition. [S1063-651X(99)51007-6]

PACS number(s): 64.70.Pf, 64.60.My, 82.20.Mj

Supercooled liquids can undergo an ergodicity-breaking transition at which the time required to adequately sample the allowed phase space becomes longer than observation times [1,2] and the liquid freezes into a glassy state [1]. Recent experiments indicate that the approach to this glass transition has some universal features [3] when viewed in terms of the frequency-dependent response of the system. In both supercooled liquids [3] and spin glasses [4], the approach to the glass transition is characterized by two generic features, a frequency-independent behavior at high frequencies and the presence of three distinct regimes in the relaxation spectrum. These features are different from those observed near a critical point where the high-frequency response is thought to be uninteresting [5]. Existing theories of the glass transition [1,6] offer no satisfactory explanation of the unusual frequency-dependent response. In this Rapid Communication, we present a simple, dynamical model which is able to describe the frequency-dependent response observed near the glass transition.

Time-dependent fluctuations in a system approaching a critical point are described, within the spirit of Landau theory, by a Langevin equation for the order parameter [5]. Adopting a similar framework for describing the fluctuations near a glass transition, we study the Langevin dynamics of a collective variable, ϕ, relaxing in a multivalleyed free-energy landscape. This collective variable is envisioned to be one of the slow variables in a viscoelastic liquid, such as a density fluctuation mode [6] or a component of the average strain field [7]. The multivalleyed free-energy surface is modulated by a temperature-dependent overall curvature. The introduction of this curvature was inspired by simulations of a frustrated spin system in which the role of ϕ is played by an elastic strain field, and the curvature arises from a coupling between the frustrated spin variables and the strain field [8]. The vanishing of the overall curvature is identified in our model with the glass transition.

A schematic picture of the free energy is shown in Fig. 1. All valleys (including the megavalley) in the free-energy surface are assumed to be parabolic. Each valley is parametrized by its curvature r_n, width Δ_n, position of the center ϕ_n^0 and position of the minimum C_n. Consequently, each valley is characterized not only by the time it takes to escape from it but also by its internal relaxation time. The set of $\{C_n\}$ is fixed by the requirement that the free energy, $F(\phi)$, is a continuous function. To simplify the picture even further, we set all $\Delta_n=\Delta$, which then automatically fixes $\phi_n^0=n\Delta$. The curvatures of the valleys are taken to be independent random variables picked from a distribution $P(r,n)$. This defines a free-energy function

$$F(\phi)=\frac{1}{2}R\phi^2+\frac{1}{2}\sum_{n=-\infty}^{\infty}\mu_n\{r_n(\phi-\phi_n^0)^2+C_n\}. \quad (1)$$

Here $\{\mu_n\}$ is the set of functions specifying the range of each subwell, i.e., $\mu_n=1$ if $\phi_n^0-\Delta/2\leqslant\phi\leqslant\phi_n^0+\Delta/2$ and zero otherwise. The curvature of the megavalley is denoted by R.

The dynamics is modeled by relaxation in this free-energy surface and is defined by the Langevin equation

$$\frac{\partial\phi}{\partial t}=-R\phi-\sum\mu_n r_n(\phi-\phi_n^0)+\eta(t), \quad (2)$$

where η is a Gaussian noise with zero average and variance $\langle\eta(t)\eta(t')\rangle=\Gamma\delta(t-t')$. The temperature scale is set by $\beta=\Delta^2/\Gamma$. In the absence of any subvalley structure, (all $r_n=0$), Eq. (2) results in a Debye relaxation spectrum with a

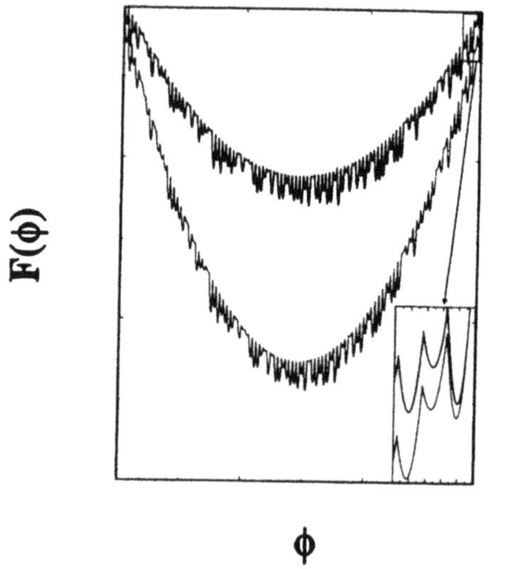

FIG. 1. Free-energy landscape for two different values of R with a fixed distribution of $\{r_n\}$. The inset shows how the overall curvature modifies the heights of the barriers between the valleys.

relaxation time of $1/R$. If R is taken to be of the form assumed in Landau theory, such that it vanishes linearly at the critical temperature, then Eq. (2) provides a mean-field description of critical slowing down [5]. The effect of the subvalley structure on the relaxation spectrum, and the nature and existence of phase transitions in the two-dimensional space spanned by R and β are the subjects of this paper.

The dynamics of systems approaching the glass transition has been modeled previously by random walks in the environment of traps [9]. What distinguishes our model is the presence of an overall curvature modulating the landscape and the description of the dynamics *within* the valleys. As shown below, both these features have nontrivial effects on the relaxation spectrum.

Specific features of the distribution $P(r,n)$ affect the detailed nature of the response. The assumption that the curvature of each valley is uncorrelated with its position, $P(r,n) \sim P(r)$ is the easiest to implement and is the scenario that we examine in detail. A natural candidate for $P(r)$ is an exponential distribution $P(r)=e^{-\beta_0 r}/\beta_0$, observed in many spin-glass models [9]. The frustrated spin model that motivated the introduction of the overall curvature also indicated an exponential distribution of barrier heights [8].

The occurrence of an ergodicity-breaking transition in our model can be demonstrated explicitly [11]. The equilibrium probability distribution, predicted by Eq. (2), is $\exp[-\beta F(\phi)]$ as long as the integral of this function over ϕ remains finite [10]. A simple calculation [11] shows that this integral diverges as $1/(\beta_0-\beta)\sqrt{R}$. As $R\rightarrow 0$, an equilibrium distribution can no longer be defined [10] and correlation functions become power law in nature with no characteristic time scale. The point $R=0$, $\beta=\beta_0$ is special in that the power-law correlations no longer decay to zero and the system falls out of equilibrium. The approach to this special point, where ergodicity is broken, can be studied by analyzing the equilibrium correlation function, $C(t)=\langle\phi(0)\phi(t)\rangle$, and the response function associated with it through the fluctuation dissipation relation [5]. In the absence of the overall curvature, the $1/\sqrt{R}$ factor gets replaced by the system size, the probability *density* diverges as $\beta\rightarrow\beta_0$, and the correlation functions are always characterized by power laws [9]. The line with $R=0$ is, therefore, special and in the present work, we are mainly interested in studying the $R=0$, $\beta=\beta_0$ point as it is approached along a generic line in the (R,β) space.

The dynamical processes contributing to the correlation function can be roughly subdivided into the internal relaxation within each subvalley and the activated motion between subvalleys, modulated by the presence of the overall curvature R. The correlation functions along the $R=0$ line can be calculated by using the well known mapping of the Langevin equation to a quantum mechanical model [10] and leads to [11],

$$C(t)=\sum_n e^{\beta r_n}\left(\frac{e^{-r_n t}}{r_n}+e^{-(t/\beta)e^{-\beta r_n}}\right). \qquad (3)$$

Physically, the first term within the parentheses represents the superposition of independent relaxations within each valley, while the second term represents activated motion.

Generalization of these results to nonzero R involves evaluating the eigenvalue spectrum of the quantum Hamiltonian as modified by the overall curvature. Taking the curvature into account perturbatively leads to the following generalization of $C(t)$:

$$C^{trap}(t)=\sum_n \theta(\Delta F_n)e^{-\beta F_n^{min}}\left(\frac{e^{-R_n t}}{R_n}+e^{-(t/\beta)e^{-\beta\Delta F_n}}\right). \qquad (4)$$

Here $R_n=r_n+R$, $F_n^{min}=n^2R-r_n$, and the θ-function excludes the valleys for which the effective free-energy barrier, $\Delta F_n=(R_n/2)(1/2-nR/R_n)^2$, becomes zero due to the presence of the overall curvature. The contribution of these zero-barrier valleys cannot be calculated perturbatively. The time evolution of ϕ involves hopping over barriers, relaxation within the subvalleys, and, in the "free" regions, relaxation in response to only the overall curvature. Since the free region is expected to be small in the vicinity of $R=0$, we have simplified this complex relaxation process by including the free relaxation within a mean-field type approximation that neglects the positional relationship between the free valleys and the valleys with barriers. The total correlation function then reduces to a sum of C^{trap} and C^{free} with

$$C^{free}(t)=\sum_n \theta(-\Delta F_n)\frac{e^{-\beta n^2R-Rt}}{R}. \qquad (5)$$

One of the most interesting aspects of our model is the high-frequency response. The origin of this can be understood from an analysis of the results along the $R=0$ line [cf. Eq.(3)]. The hopping term in $C(t)$ is then identical to the one analyzed in [9], and in frequency domain, leads to an imaginary part of the susceptibility, $\chi''(\omega)=\omega\tilde{C}(\omega)$, of the form $\omega^{(\beta_0-\beta)/\beta}$ at low frequencies and decaying as $1/\omega$ at high frequencies. There is, however, a new feature that arises from the internal dynamics and drastically changes the high-frequency behavior. The internal relaxation part of $\tilde{C}(\omega)$ is given by

$$\tilde{C}^{int}(\omega)=\int_{r^*}^{\infty}dr\frac{e^{-(\beta_0-\beta)r}}{\omega^2+r^2}. \qquad (6)$$

The contribution of this term to $\chi''(\omega)$ behaves as $[\pi/2-\arctan(r^*/\omega)]$ [12] for $\omega\leqslant 1/(\beta-\beta_0)$; a function that decays extremely slowly with frequency as $\beta\rightarrow\beta_0$. The total frequency-dependent response, therefore, behaves as $\omega^{(\beta_0-\beta)/\beta}$ for $\omega\rightarrow 0$ and is a slowly decaying function at high frequencies.

The overall curvature changes the effective barriers heights and the effective distribution of r. The parameter space of our model is spanned by R and β. In order to simplify the analysis, we study the response along the family of lines defined by $a(\beta_0-\beta)=\beta_0\beta R$, with $a=1$ for most of the calculations. The relaxation spectrum, obtained from $C^{trap}(t)$, is given by

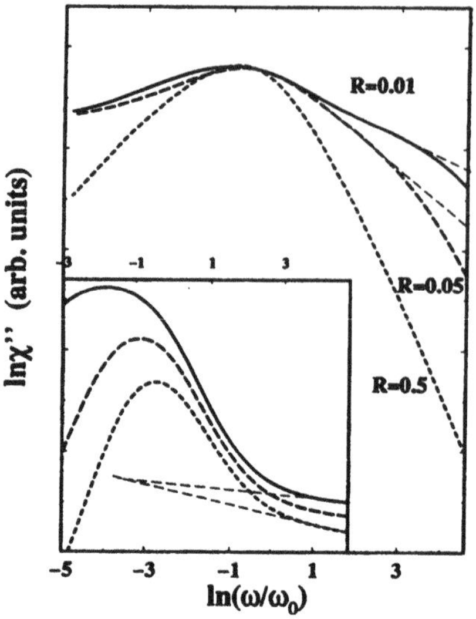

FIG. 2. Imaginary part of the susceptibility, $\chi''(\omega) = \omega\tilde{C}^{trap}(\omega)$ [cf. Eq. (7)] for different values of R. The peak intensities have been matched. Frequencies have been normalized to ω_0, a microscopic frequency scale determined by the parameters of the model. The inset shows the full relaxation spectra, *including* the contribution from the "free" valleys for $R=0.1, 0.05$ and 0.01. The long-dashed lines show the high-frequency power law extending up to $\omega \simeq a/R$. The parameter $a=1$ in the main figure but $a=100$ in the inset.

$$\tilde{C}^{trap}(\omega)=\sum_n e^{-\beta n^2 R}\int_{nR}^{\infty} dr e^{-(\beta_0-\beta)r} \times\left(\frac{1}{\omega^2+r^2}+\frac{e^{-\beta\Delta F_n/\beta}}{\omega^2+e^{-2\beta\Delta F_n/\beta^2}}\right). \quad (7)$$

This expression in not analytically tractable and the complete frequency-dependent response can be obtained only from a numerical calculation. The basic features can, however, be understood from a simplified analysis. For small R, the primary effect of the overall curvature is to introduce an upper and a lower cutoff to the distribution $P(r)$. The $e^{-\beta n^2 R}$ term of Eq. (7) makes the system finite with an "effective" number of wells $\simeq 1/\sqrt{R}$, which in turn leads to an upper cutoff of $r_{max}=-\log R$ [11]. The lower cutoff arises from the elimination of the zero-barrier valleys that is the source of the lower limit on the integral in Eq. (7). The primary effect of the upper cutoff is to alter the exponent of the low-frequency power law from $(\beta_0-\beta)/\beta$ to one which goes to zero more rapidly with R. The high-frequency response is affected by the overall width of the distribution of r which increases as $R\rightarrow 0$ and leads to an essentially frequency-independent response for $\sqrt{R}\ll\omega\ll 1/R$. A less significant effect of R is to alter the exponent of the power law in this frequency range.

The results from a complete numerical analysis of Eq. (7) are shown in Fig. 2. It is clear from these results that there is a slowly decaying high-frequency response. The exponent characterizing this high-frequency power law is a function of R and approaches zero as $R\rightarrow 0$. It is also evident that the response is characterized by a low-frequency power law with the exponent approaching zero as $R\rightarrow 0$.

The free part of the relaxation spectrum is given by

$$\tilde{C}^{free}(\omega)=\frac{1}{R^2+\omega^2}\sum_n e^{-\beta n^2 R}\int_0^{nR} dr e^{-\beta_0 r}, \quad (8)$$

and leads to a Debye spectrum with peak at $\omega=R$. There is a tradeoff between the hopping dynamics and free relaxation, with the hopping contribution decreasing as R increases, and more valleys become effectively "free." The inset of Fig. 2 demonstrates the effect of the free part with a shift in peak frequency as well as the appearance of an intermediate regime.

Three distinct regimes of the response emerge: a low-frequency power law, associated with hopping between different valleys; a Debye-like peak coming from barrierless relaxation; and a high-frequency power-law decay resulting from the superposition of many single-relaxation-time processes. The high-frequency power law is intimately related to the fact that the system explores more valleys as the curvature decreases. The low-frequency power law is controlled by the effective distribution of barriers, which also depends on R. The curvature, therefore, controls both the high- and low-frequency behavior of the dynamical response. This picture of the dynamical response is very similar to the experimentally observed response in spin glasses [4], with the response flattening out at both high- and low-frequency ends and the peak moving to zero, but slower than exponentially.

In supercooled liquids, only the high-frequency response flattens out, and the peak shifts towards zero according to a Vogel-Fulcher law [1,3]. In our model, this difference could be ascribed to a difference in the nature of correlations in the distribution $P(r,n)$. In a structural glass, the crystalline state is the absolute global free-energy minimum. The valleys of our model correspond to metastable states and the megavalley represents the states accessible in the supercooled phase, with the crystalline minimum lying outside this region. This suggests that the depth of a valley and its position are correlated, with the deeper valleys situated further away from the minimum of the megavalley. A correlation of this form in $P(r,n)$ alters the R dependence of the upper cutoff in the distribution of barrier heights and leads to a maximum escape time $t_{esc}\sim\exp(1/R^{\alpha})$. Here $\alpha>0$ is the exponent of a power law describing the correlation between the depth and the position of valleys. For times longer that t_{esc}, there is only barrier-free motion in our model. At frequencies lower than $\omega_c=(t_{esc})^{-1}$, we, therefore, predict that $\chi''(\omega)\sim\omega$, with no flattening out of the low-frequency response. The upper cutoff does not have a large influence on the response arising from the internal dynamics of the valleys. The high-frequency cutoff is still $\approx 1/R$, as seen from Eq. (6).

The original motivation for constructing the dynamical model came from observations in a nonrandomly frustrated spin system whose phenomenology is remarkably similar to structural glasses [8]. Simulations of this system indicated a free-energy surface with an overall curvature, and the vanishing of this curvature was accompanied by the appearance of broken ergodicity and "aging" [8]. The dynamics was effectively one dimensional, with the shear strain playing the role of ϕ. This frustrated spin system, can, therefore, be viewed as a microscopic realization of the dynamical model presented here, and could provide the connection between our simple toy model and the dynamics of real glasses.

In conclusion, we have demonstrated that the basic features of the frequency-dependent response near a glass transition can be understood on the basis of a multivalleyed free-energy surface with an overall curvature, which goes to zero at the glass transition. This is reminiscent of models where the glass transition is associated with an instability [13]. In our model, the spectrum crosses over from being pure Debye at large curvatures to one with three distinct regimes. The asymptotic, high-frequency power law is characterized by an exponent approaching zero as the curvature approaches zero. Our analysis also suggests that the relaxation spectra of spin glasses and structural glasses can be described by the same underlying model with different correlations in the distribution of valleys.

The authors would like to acknowledge the hospitality of ITP, Santa Barbara, where a major portion of this work was performed. This work has been partially supported by NSF Grant No. NSF-DMR-9520923.

[1] M. D. Ediger, C. A. Angell, and Sidney R. Nagel, J. Phys. Chem. **100**, 13 200 (1996).

[2] D. Thirumalai, R. D. Mountain, and T. R. Kirkpatrick, Phys. Rev. B **39**, 3563 (1989); R. D. Mountain and D. Thirumalai, Physica A **192**, 543 (1993), and references therein.

[3] P. K. Dixon *et al.*, Phys. Rev. Lett. **65**, 1108 (1990); N. Menon and S. R. Nagel, *ibid.* **74**, 1230 (1995).

[4] D. Bitko *et al.*, Europhys. Lett. **33**, 489 (1996).

[5] Nigel Goldenfeld, *Lectures on Phase Transitions and the Renormalization Group* (Addison-Wesley, New York, 1992).

[6] U. Bengtzelius, W. Gotze, and A. Sjolander, J. Phys. C **17**, 5915 (1984); W. Kob, e-print cond-mat/9702073, and references therein.

[7] S. Dattagupta and L. Turski, Phys. Rev. E **47**, 1222 (1993).

[8] Lei Gu and Bulbul Chakraborty, Mater. Res. Soc. Symp. Proc. **455**, 229 (1997), and (unpublished).

[9] C. Monthus and J. P. Bouchaud, J. Phys. A **29**, 3847 (1996); J. P. Bouchaud, J. Phys. I **2**, 1705 (1992).

[10] J. Zinn-Justin, *Quantum Field Theory and Critical Phenomenon* (Oxford, England, 1989).

[11] M. Ignatiev, Ph.D. thesis, Brandeis University, 1998 (unpublished).

[12] The lower cutoff r^* is introduced to ensure that one can define an internal relaxation process. This arbitrary cutoff is necessary only for $R=0$.

[13] A. I. Mel'cuk *et al.*, Phys. Rev. Lett. **75**, 2522 (1995).

Part 6

INHOMOGENEITIES NEAR JAMMING

E OPH SICS LETTE S 1 October 1996

Europhys. Lett., **36** (1), pp. 55-60 (1996)

Dynamic heterogeneity in supercooled ortho-terphenyl studied by multidimensional deuteron NMR

R. BÖHMER[1], G. HINZE[1], G. DIEZEMANN[1], B. GEIL[2] and H. SILLESCU[1]

[1] *Institut für Physikalische Chemie, Johannes Gutenberg-Universität 099 Main , Germany*

[2] *Fachbereich Physik der Universität Dortmund - 221 Dortmund, Germany*

(received 9 May 1996; accepted in final form 19 August 1996)

PACS. 64.70Pf – Glass transitions.
PACS. 76.60Lz – Spin echoes.
PACS. 05.40+j – Fluctuation phenomena, random processes, and Brownian motion.

Abstract. – Using deuteron NMR, we have studied molecular reorientation rates and rate exchange processes in supercooled ortho-terphenyl. We monitor the re-equilibration of differently selected subensembles through four-time stimulated echo experiments. A comparison of the two-time with the four-time echoes suggests that the characteristic time scales for reorientation and dynamical exchange are relatively similar. The four-time correlation functions were described using various multi-state rate exchange models.

One of the most salient features accompanying the glass transition of supercooled liquids and amorphous polymers is their pronounced non-exponential α-relaxation dynamics. While the phenomenon itself is abundantly documented in the literature, its origin has been discussed controversially for a long time [1]-[3]. Among different concepts some start from a heterogeneous structure with different correlation times which are thought to be associated with spatial regions differing in density or local molecular arrangements and packing efficiencies. Others start from identical subunits which relax in an intrinsically non-exponential fashion. The question as to which of these limiting scenarios yields a more appropriate description of the α-relaxation remained unresolved for a long time, since most experimental techniques only measure properties unselectively averaged over the entire ensemble. Rather what is required to address the problem properly is a study of the dynamical behavior of specifically selected subensembles. Although locally selective schemes can be devised [4] the selection is usually performed spectrally, *i.e.* with respect to orientational correlation times.

The first successful experiment based on this idea was carried out using reduced 4-dimensional ^{13}C-NMR by Schmidt-Rohr and Spiess who investigated the amorphous polymer poly(vinylacetate) (PVAc) above the glass transition ($T = T + 20\,\mathrm{K}$)[5]. They were able to select slow subensembles and thus demonstrated the heterogeneity of the α-relaxation [6]. Subsequent work on PVAc was interpreted to indicate that the selected subensemble returns to represent the full ensemble after the polymer segments typically have reoriented twice, only [7]. In terms of the so-called rate memory parameter Q introduced by Heuer *et al.* this can be expressed as $Q \approx 2$ [8].

More recently Cicerone and Ediger [9] developed an optical deep bleaching technique for selecting a slow subensemble of fluorescent dye tracers and monitoring its return to equilibrium. They concluded that this "structural" relaxation occurred on a time scale of about 10^3 times the average rotational correlation time in the supercooled liquid ortho-terphenyl (OTP) at T . No indications for such long-lived heterogeneity were found using non-resonant dielectric spectral hole burning experiments carried out for various other supercooled liquids, including glycerol and propylene carbonate, near T [10].

In the present article we report on a ^{2}H-NMR investigation of deuterated OTP performed at $T\ + 10\,\mathrm{K}$. In deuteron NMR the molecular orientation Ω is directly probed via the angular dependence of the quadrupole frequency $\omega(\Omega) = (\delta/2)(3\cos^2\theta - 1)$. Here θ is the angle enclosed by the direction of a C-D bond and the external magnetic field and $\delta = (1.16\,\mu\mathrm{s})^{-1}$ is the quadrupole coupling. Previous stimulated echo experiments of OTP have shown that the reorientational process proceeds via rotation steps characterized by an average angle of about 10° [11]. The basic three-pulse sequence used in these studies contains an evolution period t_p and a mixing time $t_{\mathrm{m}1}$ and is schematically depicted in fig. 1 *a*), *b*). By using appropriately chosen pulse phases the functions $F_2^{\mathrm{cc}}(t_{\mathrm{m}1}) = \langle\cos[\omega(0)t_\mathrm{p}]\cos[\omega(t_{\mathrm{m}1})t_\mathrm{p}]\rangle / \langle\cos^2[\omega(0)t_\mathrm{p}]\rangle$ and $F_2^{\mathrm{ss}}(t_{\mathrm{m}1}) = \langle\sin[\omega(0)t_\mathrm{p}]\sin[\omega(t_{\mathrm{m}1})t_\mathrm{p}]\rangle / \langle\sin^2[\omega(0)t_\mathrm{p}]\rangle$ can be measured. They correlate the orientation-dependent Larmor frequencies before and after the mixing time and for stationary Markov processes F_2^{cc} (and analogously F_2^{ss}) can be formulated as [12]

$$F_2^{\mathrm{cc}} \propto \langle\cos(\omega_1 t_\mathrm{p})\cos(\omega_2 t_\mathrm{p})\rangle, \tag{1a}$$

$$F_2^{\mathrm{cc}} \propto \sum_{,} \int \mathrm{d}\Omega_1 \int \mathrm{d}\Omega\ \cos[\omega(\Omega_1)t_\mathrm{p}]\cos[\omega(\Omega_2)t_\mathrm{p}]P_{1\ 1}(\Omega_2, x\ , t_{\mathrm{m}1}|\Omega_1, x\ ), \tag{1b}$$

with, *e.g.*, $\omega_1 = \omega(\Omega_1) = \omega[\Omega(t = 0)]$. $P_{1\ 1}$ is a probability conditional upon molecular orientations Ω *and* dynamical states x . The dynamical states are characterized by different rotation rates and by their lifetimes. The latter are limited by exchange processes like, *e.g.*, local environmental fluctuations [13].

By adding the results of the two experiments, one obtains $F_2(t_{\mathrm{m}1}) \equiv (F_2^{\mathrm{cc}} + F_2^{\mathrm{ss}})/2 = \langle\cos[(\omega_1 - \omega_2)t_\mathrm{p}]\rangle$. This function presents itself as a filter for the orientational motion and it is important to note that the action of this filter is twofold since it involves geometric and dynamic selectivity: Using the evolution time t_p the sensitivity on jump angles is adjusted. If t_p is chosen sufficiently large ($\delta t_\mathrm{p} \gg 1$), then any rotational jump with a non-enhanced return probability, like, *e.g.*, a random jump, will alter the NMR frequency so much that the jumped molecule will not contribute to F_2. The *dynamic* filter effect arises because the magnetization only of those molecules with typical jump rates smaller than $1/t_{\mathrm{m}1}$ will not have decayed in a time $t_{\mathrm{m}1}$.

The basic idea of the 4D experiment is to apply the same filter twice with an interval of length $t_{\mathrm{m}2}$ inserted (cf. fig. 1 *c*)) and to determine the four-time (4*t*) stimulated echo $E_4(t_{\mathrm{m}2}) \doteq \langle\cos[(\omega_1 - \omega_2)t_\mathrm{p}]\cos[(\omega_3 - \omega_4)t_\mathrm{p}]\rangle$. If $t_{\mathrm{m}2}$ is sufficiently short so that no dynamic exchange of rotation rates can take place, then all selected slow molecules will remain slow and contribute to $E_4(t_{\mathrm{m}2})$. If, however, some molecules change their reorientation rate at longer $t_{\mathrm{m}2}$, an attenuation of E_4 will result. Hence the normalized echo height $F_4(t_{\mathrm{m}2}) = E_4(t_{\mathrm{m}2})/E_4(0)$ may be viewed as the correlation function for dynamical exchange and can be shown to be independent of molecular reorientation at least for the case of random rotational jump motion [12]. It has to be noted that F_4 is not measured directly but obtained by adding the results of four subexperiments, F_4^{cccc}, F_4^{ccss}, F_4^{sscc}, and F_4^{ssss} with, *e.g.*, F_4^{ccss}

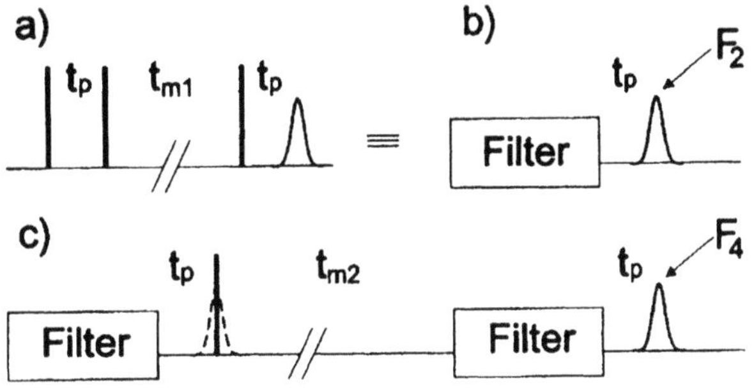

Fig. 1. –) The 2 -stimulated echo as produced by three $\pi/2$ pulses may be viewed as a filter for the magnetization from those C-D bonds which are slow on the time scale set by $_{\mathrm{m1}}$. The evolution time $_{\mathrm{p}}$ in our experiments was chosen to be 25 μs. *b*) Schematic representation of the filter for which the input, as in), is the magnetization from all deuterons.) The output from the first filter is stored using another $\pi/2$ pulse and is then subject to dynamic rate exchange in the mixing time $_{\mathrm{m2}}$. Only those C-D bonds that have remained slow all the time will pass also the second filter and thus contribute to the 4 echo $_4$.

being proportional to $\langle\cos(\omega_1 t_\mathrm{p})\cos(\omega_2 t_\mathrm{p})\sin(\omega_3 t_\mathrm{p})\sin(\omega_4 t_\mathrm{p})\rangle$. In the past only the correlation function $F_4 \equiv (F_4^{\mathrm{cccc}} + F_4^{\mathrm{sscc}})/2 = \langle\cos[(\omega_1 - \omega_2)t_\mathrm{p}]\cos(\omega_3 t_\mathrm{p})\cos(\omega_4 t_\mathrm{p})\rangle$ was measured [7] and has received particular attention in theoretical analyses [12]. Therefore, a critical comparison of the various possible combinations of $4t$ echoes seems to be warranted. As another new feature in this letter the dependence of the $4t$ echoes on the filter time t_{m1} will be explored in detail.

To obtain well-equilibrated OTP samples, they were annealed at about 360 K for 12 h prior to the NMR investigations. These were carried out at a Larmor frequency of 40.24 MHz. The length of 90° pulses was 2.5 μs and the evolution time was set to $t_\mathrm{p} = 25$ μs. Temperature stability could be maintained to within ±0.1 K during the entire set of experiments. All measurements reported in this article were taken at $T = 253.7$ K, *i.e.* about 10 K above T . The spin-lattice relaxation turned out to be exponential with $T_1 = 1.33$ s and was used to normalize the $4t$ echoes ([1]).

In fig. 2 *a*) the function F_2 is plotted. It can be characterized by a Kohlrausch function $F_2(t_{\mathrm{m1}}) = \exp[-t_{\mathrm{m1}}/\tau]^\beta$ with $\tau = 17$ ms and $\beta = 0.42$, leading to an average apparent rotational time scale $\langle\tau\ \rangle = \tau\Gamma(1/\beta)/\beta = 49.7$ ms. F_2^{cc} is characterized by the same β but a slightly smaller $\langle\tau\rangle = 44.2$ ms ([2]). Since for ideal random rotational jump motion F_2^{cc} and F_2 are expected to be identical [12], the similarity of the two functions confirms that with the chosen evolution time this idealization is closely approached. Nevertheless, the remaining small difference indicates that a minor fraction of the molecules either exhibit jump angles significantly smaller than 10° (ref. [11]) or that some of them possess an enhanced return probability.

In fig. 2 *b*) various $4t$ echoes with $t_{\mathrm{m1}} = 15$ ms ($\approx \tau$) are shown. They are seen to decay slower and in a more exponential fashion than F_2. This difference in time scale reflects the ability to select slow subensembles and hence confirms the heterogeneous nature of the α-relaxation in

([1]) Longitudinal magnetization arising from non-selected components during long $_{\mathrm{m2}}$ can give rise to additional contributions to the 4 echoes, and therefore was eliminated in our experiments using appropriate phase cycling.

([2]) It should be noted that the average rotational correlation time corresponding to the second Legendre polynomial, as measured via $_2$ in the limit $_\mathrm{p} \to 0$, is about 10 τ_R , see ref. [11].

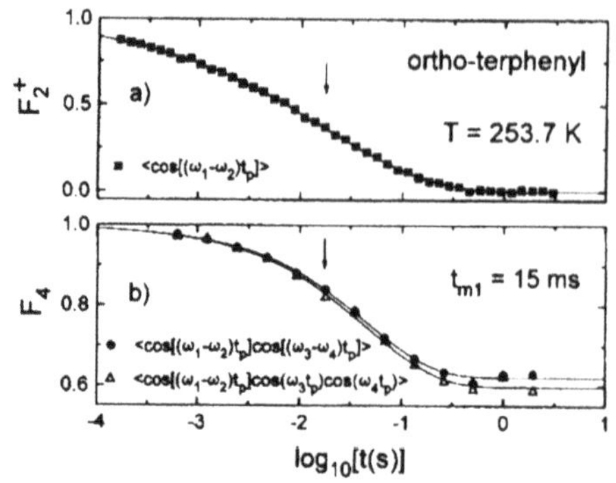

Fig. 2.

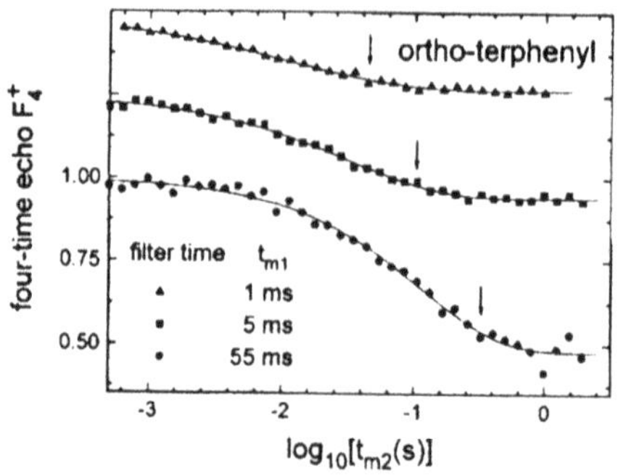

Fig. 3.

Fig. 2. –) Filter function $_2($ $_{m1})$ with a fit by a ohlrausch function shown as a solid line. *b*) 4 echoes $_4($ $_{m2})$ (•) and $_4($ $_{m2})$ ( ) with fits using eq. (2). The arrows indicate where $_2$ has decayed to $1/e$ of its initial value. It is evident that the 4 echoes decay slower than $_2$.

Fig. 3. – 4 -echoes $_4$ measured for ortho-terphenyl at $T = 253.7$ for several filter times $_{m1}$. The results for $_{m1} = 5\,\mathrm{ms}$ and $1\,\mathrm{ms}$ have been shifted upwards for clarity by 0.25 and 0.5, respectively. The solid lines are fits using eq. (2) and the parameters shown in fig. 4. The arrows indicate the characteristic time scales set by κ^{-1}.

supercooled OTP. In accord with the case of PVAc [7] the time scales for rotation and exchange are relatively similar. Furthermore the non-exponential decay of F_4 indicates that OTP has to be characterized by a distribution of exchange rates. The different $4t$ echoes are almost identical with small but significant differences in the final values. Thus one finds that the simpler function $F_4 \equiv (F_4^{\mathrm{cccc}} + F_4^{\mathrm{sscc}})/2$ provides a good approximation to the full expression $F_4 = (F_4^{\mathrm{cccc}} + F_4^{\mathrm{sscc}} + F_4^{\mathrm{ccss}} + F_4^{\mathrm{ssss}})/4$.

Figure 3 shows the $4t$-echoes F_4 as taken for several filter times $1\,\mathrm{ms} \le t_{\mathrm{m1}} \le 55\,\mathrm{ms}$. It is seen that F_4 depends on t_{m1} in a characteristic manner. In order to parameterize the data we have used the Kohlrausch function

$$F_4(t_{\mathrm{m2}}) = F_4 + (1 - F_4)\exp[-2(\kappa t_{\mathrm{m2}})^{\beta_4}]\,. \tag{2}$$

The results of the fits are shown as solid lines in fig. 3. Figure 4 summarizes the three fitting parameters *vi* . the characteristic exchange time κ^{-1}, the stretching β_4, and the final value F_4. It is seen that the exchange time κ^{-1} is proportional to the filter time t_{m1} and it is of the same order of magnitude as the corresponding apparent reorientation time from the F_2 decay. This obviously means that the F_2 filter not only selects molecules that rotate slowly but also ones that exchange slowly. Hence, if by increasing the filter time t_{m1} one selects a progressively smaller part of the distribution of exchange rates, then it can be expected that the *effective* distribution which determines the shape of F_4 gets narrower. This expectation is nicely reflected by the stretching exponents β_4 that are seen to approach unity for large t_{m1}.

At first glance the results presented in fig. 4 seem to suggest that those (selected) molecules with small rotation rates also exhibit small exchange rates. As we will show in the following this apparent correspondence of rates, while being consistent with our data, is not really necessary to interpret them. A quantitative analysis of our experimental results should be possible in terms of exchange models which can be defined by specifying the conditional probability appearing in eq. (1*b*) and allow the calculation of $2t$ echoes (cf. eq. (1*a*)) and corresponding expressions for the $4t$ echoes [12]. The conditional probability expanded in terms of Legendre

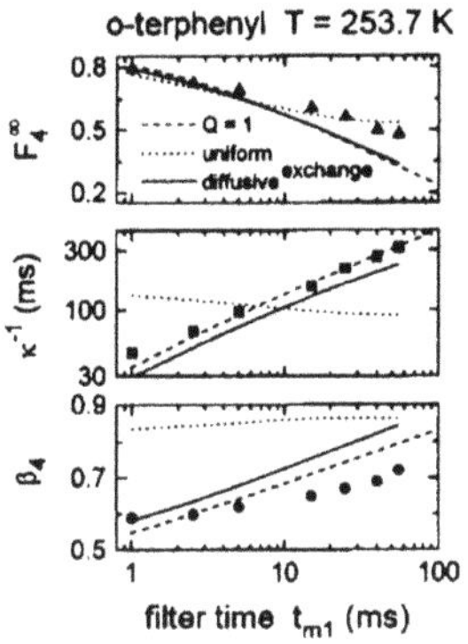

Fig. 4. – Final values $_4$, characteristic exchange time scales κ^{-1}, and the stretching β_4 from the analysis of the 4 echoes $_4$ using eq. (2). All parameters show a pronounced dependence on the filter time $_{m1}$. The lines are model calculations using eq. (3) based on diffusive exchange according to models i) with $\kappa_0 = 0.25\,\mathrm{ms}^{-1}$ (dotted lines) and ii) with $= 25$ (solid lines), or they have been derived using the concept of rate memory [8] ($= 1$, dashed lines).

polynomials of order L can generally be written as $P_{1\ 1} = \sum D\ (\Omega) C\ (t)$, where $D\ (\Omega)$ and $C\ (t)$ are terms specifying molecular orientations and dynamical states (involving rotation rates κ and exchange rates κ), respectively (for details see ref. [12]). Using the random rotational jump approximation for the angular-dependent part the problem is reduced to finding a solution of the master equation that has to be satisfied by $P_{1\ 1}$, which here reads

$$\dot{C}\ (t) = \sum_{1} R\ C\ (t)\,. \tag{3}$$

For the exchange rates, R , between different rotational rates, κ , one obtains $R = \kappa$ for $j = k$ and otherwise one finds $R = (\delta_0 - 1)\kappa - \Sigma\kappa$, where the sum runs over $j = k$. In ref. [12] the symmetric two-state problem ($N = 2$) with only two rotational rates and a single exchange rate was discussed. A more realistic description can be expected from numerical solutions of eq. (3) for *large* N. Different N-state models are now obtained by defining various scenarios for the exchange matrix κ and the distribution of rotational rates κ . The rotational rates κ were chosen according to a Kohlrausch distribution [14]. Among the exchange models that we have considered are the following: diffusive exchange i) with a uniform exchange rate $\kappa_{,\,+1} = \kappa_0/\Delta^2$ and ii) with exchange rates depending on the rotational rates $\kappa_{,\,+1} = K(\kappa + \kappa_{+1})/(2\Delta^2)$, and iii) random exchange where κ as well as κ are chosen at random from logarithmic Gaussian distributions and an appropriate average is performed subsequently ([3]). The parameters entering into this latter model are the centers and the widths of these distributions. For the various scenarios eq. (3) was diagonalized numerically for $N = 50\ldots100$. The calculated 4t echoes were parameterized using eq. (2). A simultaneous optimization of the calculated 2t and 4t correlation functions leads to the results represented by lines in fig. 4. While the parameters derived from the diffusive model ii) (solid line) describe our experimental data well, the assumption of a uniform exchange rate (dotted

([3]) A more rigorous formulation of the models and computational details will be published elsewhere. As one technical detail we only note here that the factor $\Delta = \log(\kappa_i^{\mathrm{rot}}/\kappa_{i\ 1}^{\mathrm{rot}})$ is due to the discretization of the distribution of reorientation rates.

line) is evidently inadequate. The random exchange scenario iii) yielded results similar to those obtained from the diffusive model ii) which, therefore, were not plotted in fig. 4.

In a related effort Heuer has considered generalized two-state models in order to justify the introduction of the parameter Q [8]. Without invoking explicit exchange mechanisms, he has formulated general expressions for the $4t$ echo (eqs. (25), (26) in ref. [8]). We have evaluated the cited expressions for the case of minimal rate memory ($Q = 1$) for various $t_{\rm m1}$, fitted the results using eq. (2), and plotted the fit parameter as dashed lines in fig. 4. It is seen that also this model reproduces the trends seen in the experimental data.

A very remarkable result of our simulations is that scenarios with and without direct correspondences of rotation and exchange rates describe the experimental data equally well. This is probably a consequence of the fact that the slow exchange rates are not much smaller than the slow *apparent* rotation rates selected by the F_2 filter. In other words, immobilized molecules can only rotate after exchange-mediated release from constraints that have previously locked their orientation. If this "constraint release" is the dominating feature of structural relaxation, it also governs the behavior of F_4 which then might be no longer sensitive to other details of the dynamics.

To summarize, we have measured and critically compared different two-time and four-time correlation functions in OTP at about 10 K above T . By varying the time constant of the F_2 filter over almost two decades we were able to alter the effective decay rates of the $4t$ echoes significantly.

We thank A. HEUER and S. KUEBLER for stimulating discussions. This project was supported by the Deutsche Forschungsgemeinschaft (SFB 262/project D9).

REFERENCES

[1]) WILLIAMS G., COOK M. and HAINS P. J., *hem oc araday rans* , **68** (1972) 1045; *b*) NGAI K. L. and RENDELL R. W., in *elaxation in omplex ystems and elated opics* , edited by I. A. CAMPBELL and C. GIOVANELLA (Plenum Press, New ork, N. .) 1990, p. 309;) BUTTLER S. and HARROWELL P., *hem hys* , **95** (1991) 4466; DONATI C. and JÄCKLE J., *hys* , **8** (1996) 2733.

[2] CHAMBERLIN R. V., *urophys ett* , **33** (1996) 545.

[3] PHILLIPS J. C., *ep rog hys* , **59** (1996) 1.

[4] BÖHMER R., SANCHEZ E. and ANGELL C. A., *hys hem* , **96** (1992) 9089.

[5] SCHMIDT-ROHR K. and SPIESS H. W., *hys ev ett* , **66** (1991) 3020.

[6] The heterogeneity of the β-process was demonstrated by SCHNAUSS W., FUJARA F., HARTMANN K. and SILLESCU H., *hem hys ett* , **166** (1990) 381.

[7] HEUER A., WILHELM M., ZIMMERMANN H. and SPIESS H. W., *hys ev ett* , **75** (1995) 2851.

[8] HEUER A., *hys ev* , preprint (1996).

[9] CICERONE M. T. and EDIGER M. D., *hem hys* , **103** (1995) 5684.

[10] SCHIENER B., BÖHMER R., LOIDL A. and CHAMBERLIN R. V., to be published.

[11] CHANG I., FUJARA F., GEIL B., HEUBERGER G., MANGEL T. and SILLESCU H., *on ryst olids* , **172-174** (1994) 248; GEIL B. *et al* , unpublished.

[12] SILLESCU H., *hem hys* , **104** (1996) 4877.

[13] ANDERSON J. E. and ULLMAN R., *hem hys* , **47** (1967) 2178.

[14] LINDSEY C. P. and PATTERSON G. D., *hem hys* , **73** (1980) 3348.

How Long Do Regions of Different Dynamics Persist in Supercooled *o*-Terphenyl?

Chia-Ying Wang and M. D. Ediger*

Department of Chemistry, University of Wisconsin-Madison, 1101 University Avenue, Madison, Wisconsin 53706

Received: October 23, 1998; In Final Form: December 31, 1998

In supercooled *o*-terphenyl (OTP), subsets of probe molecules in more mobile environments can be selectively photobleached. The time required for the remaining slower-than-average probes to be redistributed into an equilibrium set of environments has been measured. At $T_g + 4$ K ($T_g = 243$ K), this exchange time is 6 times greater than the average probe rotational correlation time. These results are compared to previous optical measurements at $T_g + 1$ K (Cicerone, M. T.; Ediger, M. D. *J. Chem. Phys.* **1995**, *103*, 5684), which showed that the exchange time is more than 100 times the rotational correlation time, and to multidimensional NMR experiments on deuterated OTP at $T_g + 10$ K (Bohmer, R.; et al. *Europhys. Lett.* **1996**, *36*, 55), which showed that the exchange time is nearly equal to the correlation time. These results in aggregate suggest that a new relaxation process in equilibrium supercooled liquids emerges only at temperatures very near T_g. A model for selective photobleaching is proposed. The model reproduces the experimental data reasonably well and indicates that the photobleaching efficiency of a given probe is only weakly correlated with its rotational correlation time.

I. Introduction

The fundamental origin of the characteristic nonexponential relaxation processes in supercooled liquids and amorphous polymers has been in dispute for decades. Interest in this issue was revived recently as new experimental techniques have been developed to provide more explicit investigations.[1–7] Nonexponential relaxation can be interpreted in two fundamentally different ways. One possibility is that a heterogeneous set of environments exists in supercooled liquids; relaxation within a single environment is essentially exponential, but the relaxation time varies significantly among environments. Alternatively, environments are homogeneous in supercooled liquids and each molecule relaxes nearly identically in an intrinsically nonexponential manner. Most recent experiments indicate that the heterogeneous explanation is more nearly correct,[2–7] although this view is not unanimous.[8] The size of heterogeneous dynamic regions is estimated to be 2–3 nm,[9–11] and the width of the distribution of relaxation times is proposed to be 2–3 orders of magnitude.[9,12] However, this heterogeneity cannot last forever; i.e., exchange must occur between domains of different dynamics since supercooled liquids are ergodic. Is this exchange time much longer than or comparable to the ensemble-averaged relaxation time? We will focus on the issue of the lifetime of the heterogeneous dynamic regions in this report.

Spiess and co-workers[13,14] first measured the lifetime of the heterogeneous dynamic domains in polymeric glass formers using a reduced four-dimensional NMR experiment which selectively observed a slow subensemble of C–H vectors. In studies on poly(vinyl acetate), they observed that the slow subensemble gradually became an average subensemble on a time scale which we can identify as the exchange time. At $T_g + 20$ K, the exchange time was twice the average relaxation time of a C–H vector. These observations demonstrated that poly(vinyl acetate) exhibits a heterogeneous distribution of relaxation times above T_g and that this heterogeneity possesses a lifetime comparable to the ensemble-averaged relaxation time. Recently, similar observations for another polymeric glass former, polystyrene, were reported by the same group.[15] At $T_g + 10$ K, the exchange time is comparable to the average relaxation time of polymer segments.

For low molecular weight glass formers, Cicerone and Ediger first reported direct observation of spatially heterogeneous dynamics. They used a photobleaching technique to measure probe reorientation times in supercooled *o*-terphenyl (OTP).[4] Cicerone and Ediger were able to selectively photobleach probe molecules in more mobile environments and observe the dynamics of the remaining probes. With time, the environments of these remaining probe molecules evolve into an equilibrium set of environments. At $T_g + 1$ K, an exceedingly long time is required to realize this reequilibration; some initially "slow" environments remain slow even after the molecules in these environments have reoriented more than 100 times. Since the probe, tetracene, and the OTP host molecules have similar rotational behavior,[16] it is expected that similarly long exchange times exist for OTP molecules at $T_g + 1$ K.

Soon thereafter, Bohmer et al. reported multidimensional NMR experiments on deuterated OTP at $T_g + 10$ K.[2] They found that the characteristic time scales for molecular reorientation and dynamic exchange are similar in supercooled OTP. This conclusion seems inconsistent with the results of the photobleaching experiments described in the previous paragraph. However, there is a temperature difference of nearly 10 K between the optical measurements and the NMR experiments. Could the exchange time exhibit such a strong temperature dependence?

Here we report experiments which explore the effect of temperature on the lifetimes of heterogeneous dynamic domains in supercooled liquids. We performed photobleaching experiments analogous to those reported in ref 4 on tetracene in OTP at $T_g + 4$ K. Strikingly, the exchange time is determined to be 6.5 times the average probe correlation time at $T_g + 4$ K whereas the exchange time is 540 times slower than the probe reorienta-

10.1021/jp984149x CCC: $18.00
Published on Web 02/19/1999

tion at $T_g + 1$ K. These results allow a consistent interpretation with NMR measurements at $T_g + 10$ K and suggest that a new relaxation process may emerge in supercooled liquids only at temperatures very near T_g.

To develop a more quantitative understanding of the photobleaching selectivity, we introduce a model for the photobleaching experiments. The model reasonably reproduces the experimental data for tetracene in OTP at $T_g + 4$ K and suggests a weak photobleaching selectivity among probes in different dynamic environments.

II. Experimental Section

A complete description of the photobleaching technique and method of sample preparation can be found in reference 16. Here we briefly review this technique since some details are pertinent to the interpretation and discussion of results presented in this paper.

A. Standard Photobleaching. Many different probe molecules may be photobleached with intense visible light to yield photoproducts which do not absorb visible wavelengths. When a probe molecule with a strongly polarized electronic transition is photobleached with a linearly polarized writing beam, an orientational anisotropy in the probe population is created. A weak reading beam whose polarization is either parallel or perpendicular to the polarization of the writing beam can probe this orientational anisotropy. With time, the unbleached probe molecules reorient and reach an isotropic distribution. As a result, the difference between the parallel and perpendicular fluorescence intensities decreases and eventually becomes zero. This fluorescence modulation decay defines the anisotropy $r(t)$ of the probe orientational distribution. *Only the reorientation of unbleached probe molecules is observed in the photobleaching experiment*; bleached probes do not absorb the reading beam and thus do not contribute to the fluorescence signal. The bleach depth, the fraction of probe molecules which have been bleached, is ≤ 0.1 in the standard photobleaching experiment. When the bleach depth is this small, the acquired anisotropy function describes essentially the entire ensemble of probes. Figure 1a schematically illustrates the write/read sequences used in the standard-photobleaching experiment.

The probe rotational correlation function, CF(t), is defined in terms of the anisotropy:

$$\mathrm{CF}(t) \equiv r(t)/r(0) \qquad (1)$$

where $r(0)$ is the extrapolation of $r(t)$ to time zero. Typically, the rotational correlation function decays nonexponentially and is usually fit with the Kohlrausch–Williams–Watts (KWW) function:

$$\mathrm{CF}(t) \approx e^{-(t/\tau_{\mathrm{KWW}})^{\beta}} \qquad (2)$$

A model-independent rotational correlation time, τ_c, is defined as the integral of the correlation function and can be calculated from the KWW fitting result:

$$\tau_c \equiv \int_0^{\infty} \mathrm{CF}(t)\, dt = \frac{\tau_{\mathrm{KWW}}}{\beta}\Gamma\left(\frac{1}{\beta}\right) \qquad (3)$$

B. Deep Photobleaching: Varying Bleach Depth. With an increase in the bleaching time, a larger fraction of the probe molecules are photobleached. In the deep-photobleaching experiment, an intense linearly polarized writing beam is used to produce bleach depths >0.1 (see Figure 1b). The rotational correlation time of the remaining probe molecules, τ_{obs}, is then determined via the measurements of the fluorescence intensities from a weak modulated reading beam, following the same procedure as the standard-photobleaching experiment (eqs 1–eq 3).

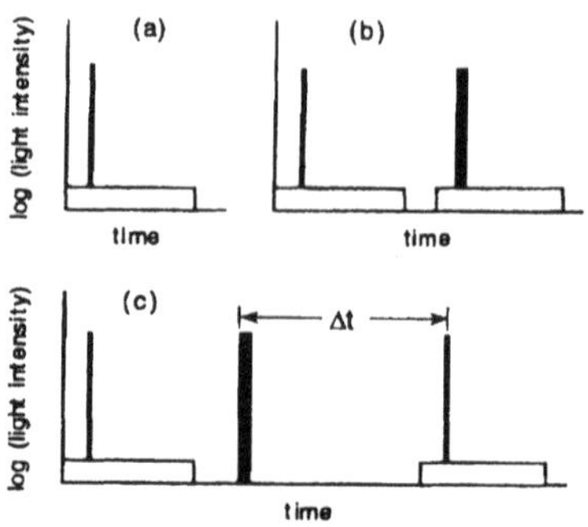

Figure 1. Schematic representation of write/read sequences for photobleaching experiments. In each part of the figure, the high-intensity light represents the bleaching beams and the low-intensity light represents the reading beams. (a) Standard photobleaching. The sample is bleached 10%, and τ_c is determined. (b) Deep photobleaching with varying bleach depth. First, a standard-photobleaching experiment is performed to obtain τ_c. Then, on the same spot, the sample is bleached >10% with longer bleaching time to obtain τ_{obs}. (c) Deep photobleaching with varying time. After the equilibrium correlation time, τ_c is determined from a standard-photobleaching experiment, the sample is bleached 65%, creating a nonequilibrium distribution of unbleached probe mobilities. After a waiting time Δt, the sample is bleached again at 10% to obtain $\tau_{obs}(\Delta t)$. The entire procedure is performed on a single spot in the sample.

Previously Cicerone and Ediger reported that the rotational correlation time τ_{obs} observed after a deep bleach is longer than the correlation time τ_c obtained from the standard-photobleaching experiment for tetracene in OTP at $T_g + 1$ K.[4] They suggested that a distribution of probe mobilities exists and that the observation of longer rotation times after a deep bleach can be explained by the selective destruction of probes in more mobile environments.

How is this selective photobleaching to be understood? The photobleaching reaction of tetracene with oxygen likely changes the planar tetracene molecule into a nonplanar peroxide.[17] More mobile environments may result from lower-than-average local matrix densities, and probe molecules in these environments may be more likely to undergo the conformational change required for photoreaction. Figure 2 illustrates this possibility schematically with small circles representing matrix molecules, rectangles representing unbleached probe molecules, and the black distorted shapes representing photobleached probes. In region a, lower density results in higher photobleaching efficiency whereas in region b, higher density results in lower photobleaching efficiency. Since it is likely that the mobility is higher in regions of low density, this scenario predicts a tendency toward selective bleaching of more mobile probes. The above description is one of several possible explanations for selective photobleaching. Another possibility is that the diffusion of oxygen molecules (needed for photobleaching) is enhanced in the more mobile environments.

C. Deep Photobleaching: Varying Time. After a deep bleach, the remaining slower-than-average probes will reestablish the equilibrium distribution through exchange between regions of different dynamics. That is, the observed correlation time τ_{obs} immediately after a deep bleach is longer than the equilibrium correlation time τ_c and gradually decays toward τ_c as the remaining probes are redistributed into an equilibrium set of environments.

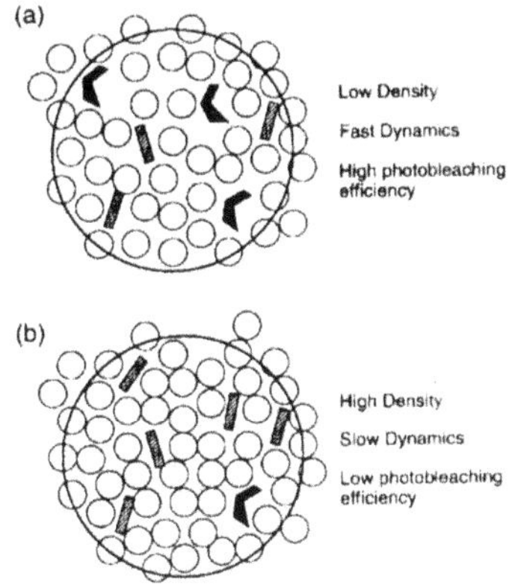

Figure 2. Schematic illustration of different photobleaching efficiencies in regions of different local densities. Small circles represent matrix molecules, and the rectangles are unbleached probe molecules. The black distorted shapes represent photobleached probes. In region a, lower density results in higher photobleaching efficiency since the probe is more likely to be able to make the change in conformation required by photobleaching. In region b, higher density results in lower photobleaching efficiency.

The time scale for the above process is the exchange time τ_{ex} and can be determined via the following procedure (see Figure 1c). First, we measure the equilibrium correlation time τ_c using the standard-photobleaching technique (reading after a shallow bleach with bleach depth ≤ 0.1). Second, a deep bleach (bleach depth ≈ 0.65) at the same spot of the sample is performed with circularly polarized light. Circularly polarized light is used to avoid any orientational anisotropy induced by this deep bleach. Third, after a waiting time Δt, another shallow bleach with linear polarization is performed at the same spot to create the orientational anisotropy. The fluorescence modulation is then measured in order to determine the rotational correlation time τ_{obs} at time Δt. By varying Δt, the time evolution of τ_{obs} is obtained and the exchange time between regions of different dynamics can be determined.

All of these experiments were performed on well-equilibrated samples of OTP. Typically, samples were held at constant temperature for 1 day before beginning measurements. Standard-photobleaching experiments yielded the same rotational correlation times whether the samples had been held isothermally for 1 day or 2 weeks. $T_g = 243$ K is the onset of the glass transition, determined by differential scanning calorimetry (10 K/min).

Samples with optical densities ranging from 0.4 to 0.7 were held in a 5 mm cuvette, and 100 mW/mm^2 of 476-nm light from a continuous-wave Ar^+ laser was used to photobleach tetracene, with the duration of the bleaching pulse being 0.1−0.2 s for the 10% linearly polarized shallow bleach. For tetracene in supercooled OTP at $T_g + 4$ K, the equilibrium correlation time τ_c was measured to be 13 s and the bleaching time for the 65% deep bleach ranged from 1 to 1.5 s. At $T_g + 1$ K, τ_c was 125 s and the bleaching time for the 65% deep bleach ranged from 2 to 6 s. At both temperatures, the initial anisotropy $r(0)$ obtained for the shallow-bleaching experiments was about 0.2 irrespective of whether the shallow bleach preceded or followed the deep bleach. For the deep-bleaching experiments, typically seven experiments were performed at each Δt to improve precision.

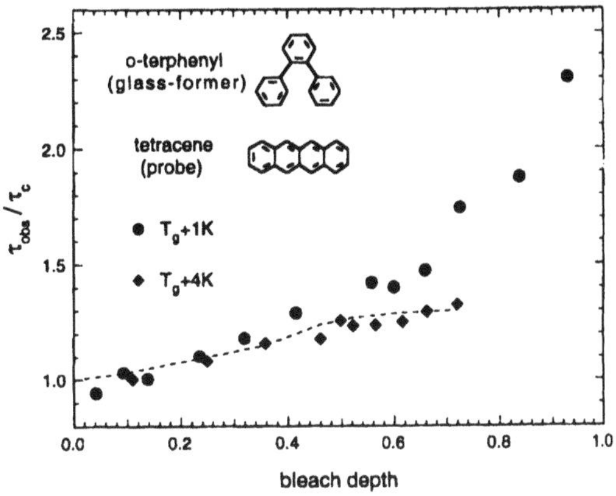

Figure 3. Values of observed correlation times (τ_{obs}) measured immediately following a bleach as a function of bleach depth. These values have been divided by those obtained in the limit of a shallow bleach (τ_c). Solid circle = 244 K (T_g + 1 K)[4] and solid diamond = 247 K (T_g + 4 K) for tetracene in OTP. The dashed line is the model result for T_g + 4 K discussed in section IV.

TABLE 1: KWW β Parameters for Tetracene in OTP at 247 K (T_g + 4 K) and 244 K (T_g + 1 K)

exptl method	temp, (K)	β(exptl fit)	β(model)[a]
shallow bleach	247	0.62 ± 0.02	0.59
	244	0.62 ± 0.02	
65% deep bleach	247	0.67 ± 0.03	0.66
	244	0.69 ± 0.03	
shallow bleach following a 65% deep bleach	247	0.60 ± 0.03	0.64
	244	0.59 ± 0.02	

[a] These values of β are calculated from the model (section IV). The input β for the model is 0.6.

III. Results and Discussion

This section presents photobleaching results for tetracene in OTP. We reiterate that we expect the dynamics of OTP molecules to be very similar to what is observed for tetracene, on the basis of the similarity between their rotation behavior.[16]

A. τ_{obs} vs Bleach Depth. Figure 3 displays the ratio τ_{obs}/τ_c as a function of bleach depth for tetracene in OTP at T_g + 1 K[4] and T_g + 4 K. Values of τ_{obs} represent the average rotation time of the probes which remain after the indicated fraction of the probe molecules have been destroyed by photobleaching (experimental method described in section IIB). For both temperatures, probe reorientation becomes significantly slower as bleach depth is increased. As discussed above in connection with Figure 2, we are able to selectively bleach probe molecules in more mobile environments and observe the dynamics of those probes which remain in slower subensembles. However, the selectivity of photobleaching is decreased with increasing temperature. As shown in Figure 3, the bleach depth dependence of τ_{obs}/τ_c at T_g + 4 K is not as strong as that at T_g + 1 K.

Table 1 lists the KWW β parameters obtained from fitting the experimental data and from calculations using the model discussed in section IV. For those values of β obtained after a shallow bleaching following the 65% deep bleach, no trend was observed for different waiting times Δt. Consequently, only the average of the fitting results over all waiting times is reported.

B. Exchange Time. The observation that the average probe rotation time τ_{obs} increases as a function of bleach depth demonstrates that a nonequilibrium distribution of reorientation

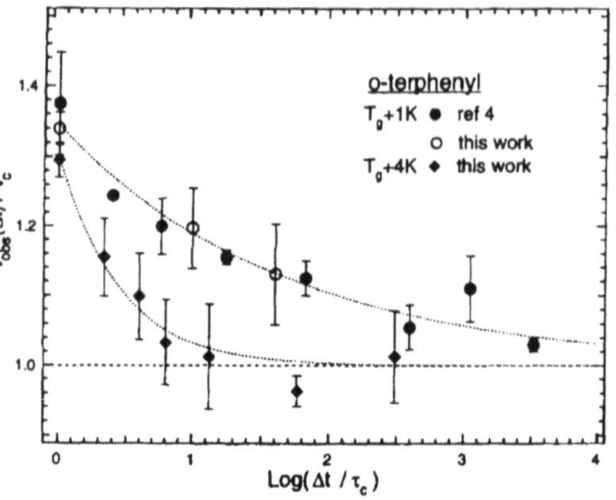

Figure 4. $\tau_{obs}(\Delta t)$ as a function of the delay time Δt after a 65% deep bleach for tetracene in OTP at $T_g + 1$ K (solid circle)[4] and $T_g + 4$ K (solid diamond). The dotted lines are fits used for calculating the exchange time, i.e., the time required for a slow subset of probes to become an average subset. Open circles at $T_g + 1$ K represent new measurements and confirm the reliability of the original measurements.

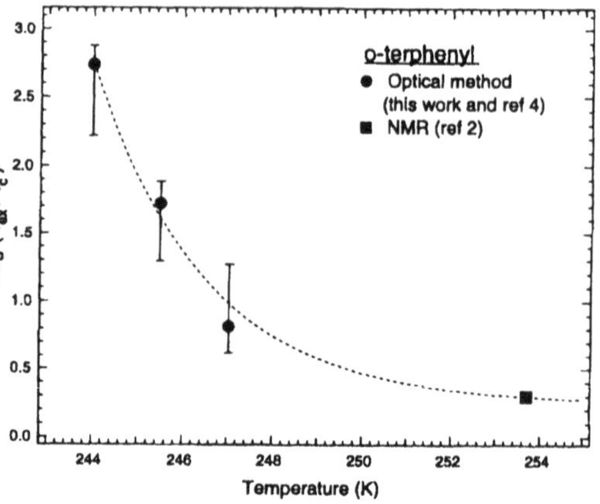

Figure 5. Temperature dependence of the exchange time (τ_{ex}) divided by the rotational correlation time (τ_c). Even after normalization to τ_c, the time required for a molecule to change dynamic environments is a strong function of temperature near T_g.

times within the probe molecule population is created by a deep bleach. Eventually, the remaining probes will be redistributed into a random set of environments because of exchange between regions of different dynamics. Thus the values of τ_{obs} should tend toward their equilibrium value τ_c as time increases following the initial deep bleach. This experiment does not establish the mechanism by which exchange occurs. Exchange could be due to a fluctuation in the dynamic character of a region, the translational diffusion of a probe out of a region with particular dynamic characteristics, or a process which couples these two.

After a deep bleach (bleach depth ≈ 0.65), $\tau_{obs}(\Delta t)$ was measured with various waiting times Δt for tetracene in OTP at $T_g + 4$ K by following the method described in section IIC. The results are displayed in Figure 4 with a log time scale. The time evolution of $\tau_{obs}(\Delta t)$ results from the exchange between regions of different dynamics. Hence, the exchange time is defined and can be calculated by integration of the area under $[\tau_{obs}(\Delta t)/\tau_c - 1]/[\tau_{obs}(\Delta t = 0)/\tau_c - 1]$. As shown in Figure 4, an exponential function (in log time) fits the $\tau_{obs}(\Delta t)$ data reasonably well. The exchange time is then calculated by integrating the fit function with the maximum waiting time as a long time cutoff. An exchange time of $6.5\tau_c$ is obtained for tetracene in OTP at $T_g + 4$ K. Data at $T_g + 1$ K obtained by Cicerone and Ediger[4] is also plotted in Figure 4, and the exchange time is calculated as $540\tau_c$ by employing the same approach. Some measurements were repeated at $T_g + 1$ K (shown as open circles in Figure 4), and these new results are consistent with the results reported by Cicerone and Ediger[4] even though different excitation wavelengths were used (476 nm for the new measurements vs 488 nm previously).

At $T_g + 1$ K, the dynamics of the slower environments in which the probes initially reside have not completely randomized even after these probes and the surrounding OTP molecules have reoriented more than 100 times. Strikingly, only 3 K higher, the environments reach equilibrium after probe and host molecules reorient roughly 6 times!

Since tetracene molecules and OTP molecules exhibit similar rotational characteristics in OTP over a wide temperature range,[16] we assume that the probe rotational dynamics observed in these deep-bleaching experiments reasonably represents the matrix dynamics and can be compared to measurements on OTP. Figure 5 displays the temperature dependence of the exchange time obtained from optical experiments at $T_g + 1$ K,[4] $T_g + 2.5$ K, and $T_g + 4$ K. The exchange time at $T_g + 2.5$ K was estimated from measurements of $\tau_{obs}(\Delta t)$ for tetracene in OTP at few waiting times employing the method described in section IIIB. Also shown in Figure 5 are NMR results at $T_g + 10$ K from Sillescu and co-workers.[2] To interpret the multidimensional NMR experiments, a dimensionless rate memory parameter Q[18] was introduced as the ratio of slow relaxation rates to the exchange rate between slow and fast regions. For deuterated OTP, Sillescu and co-workers fit the experimental data at $T_g + 10$ K with the minimal rate memory ($Q = 1$). We interpret this as indicating that the exchange time τ_{ex} and the correlation time τ_c are very nearly equal. While experimental results are not available for OTP for temperatures greater than $T_g + 10$ K, it is likely that $\log(\tau_{ex}/\tau_c)$ continues to be ≈ 0 up to much higher temperatures.

Figure 5 illustrates that the photobleaching results very near T_g can be fully reconciled with the NMR results at $T_g + 10$ K. The results indicate that an extremely slow relaxation process emerges in supercooled OTP as the temperature is lowered to very near T_g. This striking result was obtained from experiments performed on samples in the ***equilibrium*** supercooled state. Thus, the emerging slow relaxation process cannot be associated in any way with the kinetic aspects of the laboratory glass transition. This new time scale might imply that a thermodynamic singularity is nearby. Many theories of the glass transition are based on a thermodynamic singularity somewhat below the laboratory T_g,[19–24] but the experimental evidence (i.e., the emergence of a new time scale or length scale as the temperature is lowered) for such a singularity has been quite indirect.

As a test of the generality of this phenomenon among glass-forming materials, similar photobleaching experiments are being performed on a polymeric glass former, polystyrene. Preliminary results indicate that the exchange time (as probed by tetracene reorientation) is substantially longer than the α-relaxation time. Thus these preliminary results on polystyrene are consistent with the results reported here for OTP.

C. Comparison with Work on Other Glass-Formers. Is the new relaxation process implied by Figure 5 consistent with experiments on other materials? Recently, MacPhail and co-workers also observed an ultraslow relaxation process in their studies of nonequilibrium dynamics in supercooled glycerol by stimulated Brillouin gain spectroscopy.[25] These authors argued that this ultraslow process could result either from the heterogeneous nature of dynamics in supercooled liquids or from the relaxation of mechanical strain built up in the sample during the measurements.

Schiener et al. successfully performed dielectric hole-burning experiments of propylene carbonate and glycerol near their glass transitions.[3] This observation provides evidence for the existence of a distribution of relaxation times in these supercooled liquids. The observed time scale for hole refilling is very similar to that characterizing the α-relaxation at temperatures close to T_g. Although this result has sometimes been interpreted as indicating that the exchange time is equal to the α-relaxation time, we support this view expressed in ref 3: "The present finding of a reequilibration time scale of the order of the α-relaxation time does therefore not necessarily imply that the exchange among different parts of the relaxation time spectrum also occur relatively fast." If the exchange time were equal to the α-relaxation time, hole broadening would be observed as the hole fills. Since this is not observed, we interpret these dielectric results as indicating that the exchange time is longer than the α-relaxation time and could be much longer.

Very recently, Russell et al. investigated thermal fluctuations in the dielectric properties of a poly(vinyl acetate) film near a scanning probe tip.[6] Anomalous variations in the noise spectrum were interpreted as arising from fluctuations in the relaxation time spectrum of a 50 nm volume of the film at the surface. These authors modeled their results with a correlation size of 10 nm and heterogeneity lifetimes roughly comparable to the α-relaxation time. Since these experiments were performed near and below the usually reported T_g for poly(vinyl acetate), they address the same temperature range as the measurements reported here. The apparent difference in the conclusions regarding the heterogeneity lifetimes certainly invites further investigation and might be attributed to the different glass-formers used. Experiments on OTP with the nanoscale dielectric fluctuation technique would be particularly interesting.

A number of simulation studies recently showed evidence for spatially heterogeneous dynamics in supercooled liquids.[26-28] In cases where the heterogeneity lifetimes were investigated, it was found that exchange times are roughly comparable to the α-relaxation times. Since these simulations represent systems far above the laboratory glass transition, these results should be viewed as consistent with the results presented here. On the other hand, recent calculations by Spiess and co-workers using a lattice model glass suggest the exchange time between regions of heterogeneous dynamics has a weak temperature dependence.[29] This result is difficult to reconcile with the current observations.

D. Possible Artifacts. Since the results shown in Figure 5 are so striking, we have spent considerable effort investigating whether these experiments are reproducible and whether any artifact might be influencing our interpretation. We are confident that the results are reproducible and show evidence in Figure 4 that different experimentalists using different samples produce the same results within the reported error bars. It is impossible to be completely confident that no artifact is influencing the interpretation of these (or any other) experiments. Cicerone and Ediger considered some possible artifacts and argued that the observed heterogeneous domains are not induced by the probes and are not the result of local heating.[4] Here we consider three additional possible artifacts. Our conclusion is that none of these potential artifacts provide an alternate interpretation for the data in Figure 4.

Do tetracene dimers play a role in the deep photobleaching experiment? Angell and co-workers reported that the equilibration times for a probe monomer–dimer equilibrium in sorbitol at temperatures near T_g were 10^4–10^5 times longer than the matrix shear relaxation times.[30] One might wonder if the extraordinarily long decay of $\tau_{obs}(\Delta t)$ at $T_g + 1$ K is due to the tetracene monomer–dimer reequilibration. For this to be the case, a mixture of monomers and dimers would need to exist at equilibrium and the monomers would need to be selectively photobleached. In this scenario, the longer τ_{obs} would be attributed to the remaining slower dimer molecules. Two investigations were performed to test this possibility, both based on the observation that tetracene dimers have an absorption spectrum distinctly different from that of tetracene monomers in the range of wavelengths used to perform the deep-photobleaching experiment.[31] First, the absorption spectrum of tetracene in OTP under the conditions of these experiments did not show any evidence of dimer formation. Second, since the dimer spectrum is quite different, using different excitation wavelengths would change the ratio of dimer/monomer photobleached and thus different values of the average correlation time τ_{obs} would be obtained. Nevertheless, as shown in Figure 4, measurements at $T_g + 1$ K using two different excitation wavelengths yielded consistent results. Consequently, it is implausible to presume that the evolution of $\tau_{obs}(\Delta t)$ results from probe monomer–dimer reequilibration.

Does volume relaxation of the matrix influence the measurements of the probe correlation times? During a deep bleach, most probes are converted to a different chemical species, the photoproduct. We then examined the rotational behavior of the unbleached probes at various times. On average, each unbleached probe is about 25 nm from the nearest bleached probe. Could a molecular volume change associated with photobleaching induce an internal pressure in the sample which would modify the behavior of the unbleached species? A simple calculation of the effect indicates that the resulting ΔP is less than 0.15 bar and that the associated change in relaxation time, $\Delta\tau/\tau$, is less than 0.003. This calculation indicates that this effect is 2 orders of magnitude too small to account for the experimental observation. As a check, we measured the volume relaxation at $T_g + 1$ K by monitoring the equilibration of the probe dynamics after a temperature jump from 3 K higher. The observed probe correlation time reached its equilibrium value τ_c by a waiting time of $50\tau_c$ after the temperature jump.[32] In contrast, in the deep-photobleaching experiments shown in Figure 4, $\tau_{obs}(\Delta t)$ is still significantly slower than τ_c at the $50\tau_c$ waiting time at the same temperature $T_g + 1$ K. Thus we conclude that the photobleaching of some probe molecules is unlikely to influence the dynamics of the remaining probes.

Could oxygen diffusion play a role in the deep photobleaching experiment? Conceivably oxygen might be depleted in some regions of the sample as a result of the deep bleach. Assuming an oxygen diffusion coefficient of 10^{-10} cm^2/s, reequilibration on the length scale of 3 nm would occur in 10^{-4} s. Clearly, oxygen diffusion is too fast to be responsible for the very slow exchange times observed here.

None of these considerations to our knowledge contribute to the observed relaxation of the probe correlation time $\tau_{obs}(\Delta t)$ after a deep bleach or influence our interpretation of Figure 4.

IV. Model of Selective Bleaching

To develop a more quantitative understanding of the selectivity of photobleaching, we introduce a model to simulate the bleach-depth dependence of tetracene rotational correlation times in OTP at $T_g + 4$ K (shown in Figure 3). The results reasonably reproduce the experimental data and suggest that the photobleaching efficiency of a given probe is only weakly correlated with its rotational correlation time.

A. Model Assumptions. The following assumptions are made to construct the model: (i) there is a distribution of regions with different dynamics; (ii) relaxation is exponential within a region; (iii) dynamic heterogeneity is static on the time scale of reorientation (i.e., no exchange occurs among regions of heterogeneous dynamics); (iv) the dynamics and photophysical behavior of one set of probes are not influenced by bleaching another set; (v) all regions are equally illuminated during the bleaching period; and (vi) photobleaching efficiency only depends on the mobility of the environments where probe molecules reside, i.e. the probe reorientation time in a given region.

On the basis of the above assumptions, the bleach depth (bd_i) for regions with probe rotational correlation time τ can be expressed as a function of $\Phi(\tau)$, where $\Phi(\tau)$ is the photobleaching efficiency function. For convenience, the photobleaching efficiency function is chosen to be a power law of the probe correlation time τ:

$$\Phi(\tau) \propto 1/\tau^{\alpha} \tag{4}$$

If $\alpha = 0$, bleaching is not selective. If $\alpha > 0$, probe molecules in more mobile environments are being bleached more effectively.

The equilibrium distribution of probe rotation times is taken to be given by a KWW distribution function $g(\tau)$ with the KWW parameter β equal to 0.6.[33,34] The bleach depth of the whole ensemble is then calculated using the weighted summation of the bleach depths from regions of different dynamics:

$$\text{bleach depth} = \frac{\sum_i g_i(bd_i)}{\sum_i g_i} = \frac{\sum_\tau g(\tau) f(\Phi(\tau))}{\sum_\tau g(\tau)} \tag{5}$$

The Appendix presents details for the calculation of the bleach depth and the rotational correlation time from the remaining distribution of unbleached probes.

Probe reorientation during the bleaching period is considered in order to model more closely the actual photobleaching experiments. Subsets of probe molecules with rotational correlation times much shorter than the bleaching time reorient and approach an isotropic distribution of orientations during the bleaching period and thus make no contribution to the anisotropy measurement. Furthermore, the photobleaching efficiency is enhanced in those regions where probe molecules partially or completely reorient during the bleaching duration. These two effects are included in an approximate manner in the model, which uses the experimental bleaching time as an input (see Appendix).

B. Model Solution. We start with an initial guess of the power α in the photobleaching efficiency function $\Phi(\tau)$. τ_{obs} is calculated as a function of bleach depth. The equilibrium rotational correlation time τ_c is obtained for a bleach depth near zero, e.g., 0.01. The value of α is iterated until the calculated results for bleach depth dependence of τ_{obs}/τ_c match the experimental data (Figure 3) as well as possible.

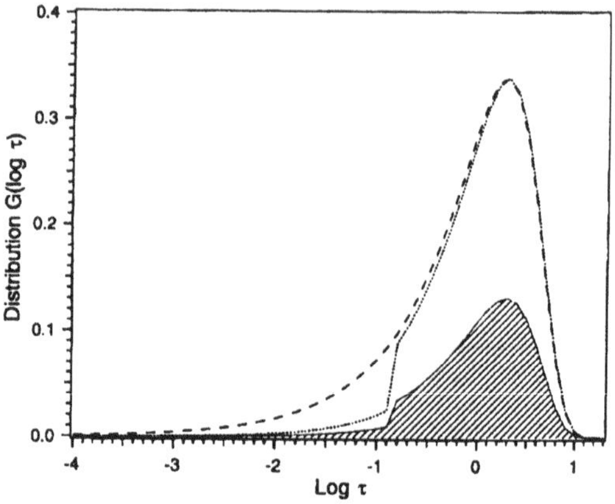

Figure 6. Distributions of probe relaxation times. The dashed curve represents the equilibrium KWW distribution with $\beta = 0.6$. The shadowed area represents the distribution remaining after a 65% deep bleach with the photobleaching efficiency function $\Phi(\tau) \propto 1/\tau^{0.08}$. The dotted curve shows this postbleach distribution normalized to the peak of the original distribution. The postbleach distribution differs from the original distribution both because of the photobleaching efficiency function and because of the fact that molecules which can reorient during bleaching are more efficiently bleached.

C. Model Results. *(a) Bleaching Selectivity.* The dashed line in Figure 3 displays results with $\alpha = 0.08$. This choice reproduces the observed bleach-depth dependence of the average probe rotational correlation time τ_{obs}. The photobleaching efficiency function $\Phi(\tau) \propto 1/\tau^{0.08}$ indicates a weak photobleach selectivity among probes in heterogeneous dynamic domains. In addition, for those probes with correlation times shorter than the bleaching time, bleaching is even more efficient, as discussed above. Figure 6 illustrates the distribution remaining after a 65% deep bleach. This postbleach distribution shows that the photobleaching selectivity is very subtle; e.g., for regions with correlation times differing by 1.2 orders of magnitude, the photobleaching efficiency in the slower regions is only 25% less than that in the faster regions.

Once α is determined, the initial anisotropy $r(0)$ of the whole ensemble can be calculated for various bleach depths (eq A7). The calculated functions exhibit the same trend as those observed in the experiment (see Figure 7).

(b) Relationship between True and Apparent Distributions. For the standard-photobleaching experiments, we have assumed that the measured correlation time τ_c can characterize the dynamics of the entire ensemble as long as the bleach depth is small enough, e.g., ≤ 0.1. To test this assumption, we calculated the "true" average correlation time, τ_{true}, directly from the distribution of rotational correlation times. The calculated average correlation times τ_{obs} at bleach depths 0.01 (defined as τ_c in the simulation) and 0.1 both differ from τ_{true} by less than 10%. As a result, within our experimental error of 10%, the correlation time τ_c obtained after a shallow bleach (bleach depth ≤ 0.1) reasonably represents the true average correlation time of the entire ensemble even with a biased photobleaching efficiency.

Furthermore, the KWW β parameter of the distribution following a shallow bleach (bleach depth 0.1) is calculated to be 0.59, consistent with the value of 0.6 for the initial equilibrium distribution. Hence, this result verifies the assump-

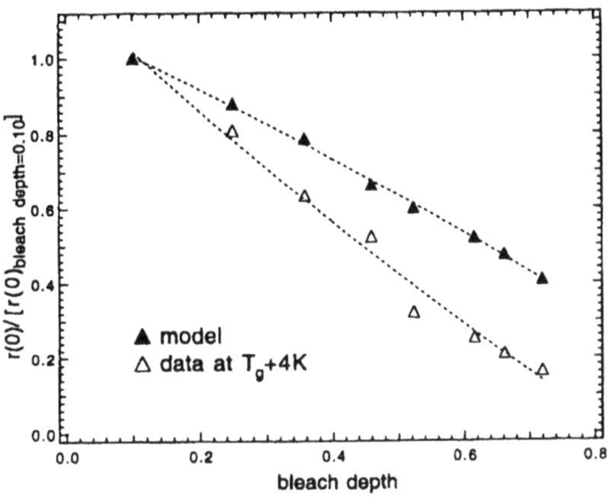

Figure 7. Initial anisotropy $r(0)$ obtained from the experiments (open triangle) and the simulation (solid triangle) as a function of bleach depth for tetracene in OTP at $T_g + 4$ K. These values are normalized to those observed for a bleach depth of 0.10.

tion that, for very small bleach depths (≤0.1), the observed postbleach distribution is not significantly different from the true distribution of relaxation times. The model also makes predictions for the dependence of β upon bleach depth and for the β value observed in a shallow-bleaching experiment after a deep bleach. These predictions are compared to experiment in Table 1. The model agrees with experiment in that β values near 0.6 are consistently observed.

(c) Interpretation of Deep-Bleaching Experiments. For the measurements of the time evolution of $\tau_{obs}(\Delta t)$, it is impossible to perform a shallow bleach immediately after a 65% deep bleach and acquire the average correlation time at short waiting times ($\tau_{obs}(\Delta t \approx 0)$). Instead, we used the average correlation time obtained from the anisotropy decay after a linearly polarized 65% deep bleach (τ_{obs}) as the first data point in Figure 4, $\tau_{obs}(\Delta t = 0)$. However, we were able to calculate $\tau_{obs}(\Delta t = 0)$ in the simulation. The correlation time $\tau_{obs}(\Delta t = 0)$ was calculated with a 0.1 bleach depth for a normalized distribution of relaxation times, which represents the remaining population following a 65% deep bleach. The correlation time τ_{obs} for a 0.65 bleach depth was calculated from the original equilibrium distribution. The results show that these two numbers are only 7% different. It is notable that a 7% error in the $\Delta t = 0$ point in Figure 4 does not affect the exponential fit enough to significantly change the integrated exchange time.

V. Concluding Remarks

Using photobleaching techniques, we were able to selectively photobleach probe molecules in more mobile environments and observe a longer rotational correlation time for tetracene in supercooled o-terphenyl (OTP) near T_g. This observation supports the idea that spatially heterogeneous dynamics play an important role near T_g. In addition, after selectively photobleaching those probes with faster dynamics, we can measure the time required for the remaining slower-than-average probes to be redistributed into an equilibrium set of environments. At $T_g + 4$ K ($T_g = 243$ K), this exchange time is 6.5 times greater than the average probe rotational correlation time, τ_c, whereas Cicerone and Ediger reported an exchange time of $540\tau_c$ at $T_g + 1$ K in previous work.[4] Considering the experimental temperature range, these results are consistent with multidimensional NMR measurements for deuterated OTP[2] and suggest that there is a new relaxation process in supercooled liquids emerging only at temperatures very near T_g. This new relaxation process is suggestive of an underlying thermodynamic singularity such as the ones assumed in many theories of the glass transition.

Our new results are consistent with much but not all of the existing literature on exchange times in supercooled liquids near T_g. It may be that the emergence of a new relaxation process is not a universal phenomenon or that it happens in a somewhat different temperature range relative to T_g in different materials. We are currently performing similar photobleaching experiments on polystyrene.

Acknowledgment. This work was supported by the National Science Foundation (Grant CHE-9618824). We wish to thank Maurice Leutenegger for the investigation of the tetracene absorption spectrum in o-terphenyl at temperatures near T_g.

Appendix

In this section, we show how to calculate the bleach depth on the basis of the model described in section IVA, which assumes a distribution of relaxation times, $g(\tau)$, and a photobleaching efficiency function, $\Phi(\tau) \propto 1/\tau^{\alpha}$. Also shown is the calculation of the average correlation time τ_{obs} and the initial anisotropy $r(0)$ after a bleach. Calculations are first performed for a homogeneous environment and then averaged to produce the results for a heterogeneous system.

A. Relaxation within a Homogeneous Environment. It is assumed in this model that the relaxation is exponential in each region. That is, probe molecules within one region have the same rotational correlation time τ.

For a first-order photoreaction during the bleaching period, the change in concentration of probes with an absorption dipole moment at angle θ relative to the polarization of the bleaching beam is:

$$dc(\theta)/dt = -a(\cos^2\theta)\, c(\theta) \qquad (A1)$$

The variable a depends on the absorption strength, the light intensity, and the photobleaching yield, which is a function of probe relaxation time τ; i.e., $a = k\Phi(\tau)$. The photobleaching efficiency function $\Phi(\tau)$ is chosen as a power law, $1/\tau^{\alpha}$. The value of α was iterated to fit the experimental results and determines the photobleaching selectivity (see sections IVA and IVB). For a given α, the parameter k was varied to obtain different bleach depths.

The difference between the fluorescence intensities immediately before and after a bleach can then be written as

$$\Delta I_{\parallel} = 2\pi b\left[\frac{2}{3} + \frac{e^{-a}}{a} - \frac{1}{2a}\sqrt{\frac{\pi}{a}}\,\mathrm{erf}(\sqrt{a})\right] \qquad (A2)$$

$$\Delta I_{\perp} = 2\pi b\left[\frac{2}{3} + \frac{e^{-a}}{2a} - \left(\frac{1}{4a} - \frac{1}{2}\right)\sqrt{\frac{\pi}{a}}\,\mathrm{erf}(\sqrt{a})\right] \qquad (A3)$$

Parallel and perpendicular designations refer to the polarization of the reading beam relative to the bleaching beam. The constant b is a function of the initial probe concentration, the light intensity of the reading beam, and the probe fluorescence quantum yield; b includes factors not related to the probe relaxation time. The bleach depth of molecules with rotation time τ (bd_i) is calculated as follows:

$$\mathrm{bd}_i \equiv \frac{(2\Delta I_\perp + \Delta I_\parallel)/3}{I_0} = 1 - \frac{1}{2}\sqrt{\frac{\pi}{a}}\,\mathrm{erf}(\sqrt{a}) \quad \text{(A4)}$$

I_0 is the fluorescence intensity before the bleaching. The initial anisotropy $r(0)_i$ can be calculated on the basis of the definition

$$r(0)_i \equiv \frac{\Delta I_\parallel - \Delta I_\perp}{\Delta I_\parallel + 2\Delta I_\perp} \quad \text{(A5)}$$

B. Heterogeneous Environment. The entire ensemble consists of a distribution of homogeneous regions whose relaxation times are significantly different. As a result, the bleach depth for the whole ensemble is the weighted sum of the bleach depths from homogeneous regions with different correlation times:

$$\text{bleach depth} = \frac{\sum_i g_i(\mathrm{bd}_i)}{\sum_i g_i} \quad \text{(A6)}$$

where g_i is the probability of finding probe molecules in regions with a correlation time τ and is obtained from the KWW distribution of relaxation times. The initial anisotropy of the entire ensemble is also calculated from the $r(0)_i$ from regions of different dynamics:

$$r(0) = \frac{\sum_i g_i(r(0)_i)}{\sum_i g_i} \quad \text{(A7)}$$

Finally, the ensemble-averaged correlation time τ_{obs} can be calculated as follows:

$$\tau_{obs} = \frac{1}{r(0)}\left[\frac{\sum_i g_i\,\mathrm{bd}_i\, r(0)_i\,\tau_i}{\sum_i g_i\,\mathrm{bd}_i}\right] \quad \text{(A8)}$$

C. Consideration of Probe Reorientation during the Photobleaching Period. As discussed in section IVA, for subsets of probes with relaxation times faster than the bleaching time, it is not appropriate to assume that these probes maintain their orientation during the bleaching. Therefore, two refinements are added to the above calculation. First, $r(0)_i$ is assumed to be zero for these regions. Second, bleach depth in these regions is calculated using the following equation instead of eq A4:

$$\mathrm{bd}_i = 1 - e^{-a/3} \quad \text{(A9)}$$

Equation A9 is obtained by starting with the following replacement for eq A1:

$$\mathrm{d}c/\mathrm{d}t = -a\langle\cos^2\theta\rangle c \quad \text{(A10)}$$

This treatment is approximate since we consider probe molecules to be either rapidly rotating or rigidly fixed during the bleaching pulse. Although in reality there are probes with intermediate dynamics, this population can be ignored if the distribution of relaxation times is sufficiently broad.

References and Notes

(1) Ediger, M. D.; Angell, C. A.; Nagel, S. R. *J. Phys. Chem.* **1996**, *100*, 13200.
(2) Bohmer, R.; Hinze, G.; Diezemann, G.; Geil, B.; Sillescu, H. *Europhys. Lett.* **1996**, *36*, 55.
(3) Schiener, B.; Chamberlin, R. V.; Diezemann, G.; Bohmer, R. *J. Chem. Phys.* **1997**, *107*, 7746.
(4) Cicerone, M. T.; Ediger, M. D. *J. Chem. Phys.* **1995**, *103*, 5684.
(5) Richert, R. *J. Phys. Chem. B* **1997**, *101*, 6323.
(6) Russell, E. V.; Israeloff, N. E.; Walther, L. E.; Gomariz, H. A. *Phys. Rev. Lett.* **1998**, *81*, 1461.
(7) Vogel, M.; Rossler, E. *J. Phys. Chem. A* **1998**, *102*, 2102.
(8) Arbe, A.; Colmenero, J.; Monkenbusch, M.; Richter, D. *Phys. Rev. Lett.* **1998**, *81*, 590.
(9) Moynihan, C. T.; Schroeder, J. *J. Non-Cryst. Solids* **1993**, *160*, 52.
(10) Cicerone, M. T.; Blackburn F. R.; Ediger, M. D. *J. Chem. Phys.* **1995**, *102*, 471.
(11) Tracht, U.; Wilhelm, M.; Heuer, A.; Feng, H.; Schmidt-Rohr, K.; Spiess, H. W. *Phys. Rev. Lett.* **1998**, *81*, 2727.
(12) Cicerone, M. T.; Blackburn, F. R.; Ediger, M. D. *Macromolecules* **1995**, *28*, 8224.
(13) Schmidt-Rohr, K.; Spiess, H. W. *Phys. Rev. Lett.* **1991**, *66*, 3020.
(14) Heuer, A.; Wilhelm, M.; Zimmermann, H.; Spiess, H. W. *Phys. Rev. Lett.* **1995**, *75*, 2851.
(15) Kuebler, S. C.; Heuer, A.; Spiess, H. W. *Phys. Rev. E* **1997**, *56*, 741.
(16) Cicerone, M. T.; Ediger, M. D. *J. Phys. Chem.* **1993**, *97*, 10489.
(17) Stevens, B.; Algar, B. E. *J. Phys. Chem.* **1968**, *72*, 3794.
(18) Heuer, A. *Phys. Rev. E* **1997**, *56*, 730.
(19) Gibbs, J. H.; DiMarzio, E. A. *J. Chem. Phys.* **1958**, *28*, 373.
(20) Adam, G.; Gibbs, J. H. *J. Chem. Phys.* **1965**, *43*, 139.
(21) Edwards, S. F.; Vilgis, Th. *Phys. Scr.* **1986**, *T13*, 7.
(22) Binder, K.; Young, A. P. *Rev. Mod. Phys.* **1986**, *58*, 801.
(23) Shlesinger, M. F.; Montroll, E. W. *Proc. Natl. Acad. Sci. U.S.A.* **1984**, *81*, 1280. Bendler, J. T.; Shlesinger, M. F. *J. Stat. Phys.* **1988**, *53*, 531.
(24) Kirkpatrick, T. R.; Thirumalai, D.; Wolynes, P. G. *Phys. Rev. A* **1989**, *40*, 1045. Kirkpatrick, T. R.; Thirumalai, D. *Transp. Theory Stat. Phys.* **1995**, *24*, 927.
(25) Miller, R. S.; MacPhail, R. A. *J. Chem. Phys.* **1997**, *106*, 3393.
(26) Perera, D.; Harrowell, P. *J. Non-Cryst. Solids*, in press.
(27) Kob, W.; Donati, C.; Plimpton, S. J.; Glotzer, S. C.; Poole, P. H. *Phys. Rev. Lett.* **1997**, *79*, 2827. Donati, C.; Douglas, J. F.; Kob, W.; Plimpton, S. J.; Poole, P. H.; Glotzer, S. C. *Phys. Rev. Lett.*, in press.
(28) Doliwa, B.; Heuer, A. *Phys. Rev. Lett.* **1998**, *80*, 4915.
(29) Heuer, A.; Tracht U.; Spiess, H. W. *J. Chem. Phys.* **1997**, *107*, 3813.
(30) Barkatt, A.; Angell, C. A. *J. Chem. Phys.* **1979**, *70*, 901.
(31) Fournie, G.; Dupuy, F.; Martinaud, M.; Nouchi G.; Turlet, J. M. *Chem. Phys. Lett.* **1972**, *16*, 332.
(32) As discussed in ref 16, these aging times are influenced by the sample cell and do not represent a time scale intrinsic to the material.
(33) The value of β was obtained from the KWW fit for tetracene reorientation in OTP near T_g.
(34) The KWW distribution function is calculated using Mathamatica by following the work of: Lindsey, C. P.; Patterson, G. D. *J. Chem. Phys.* **1980**, *73*, 3348.

Origin of the Difference in the Temperature Dependences of Diffusion and Structural Relaxation in a Supercooled Liquid

Donna N. Perera* and Peter Harrowell†
School of Chemistry, University of Sydney, NSW 2006, Australia
(Received 15 January 1998)

A supercooled binary mixture of soft disks in 2D, studied using molecular dynamics simulations, exhibits a breakdown of scaling between the self-diffusion constant and a structural relaxation time below a crossover temperature. Evidence is presented that this breakdown arises from two causes. First, the small displacements involved in structural relaxation correspond to the transition between ballistic and diffusive motion. Second, the different physical quantities averaged to obtain transport and relaxation time scales sample different parts of the increasingly heterogeneous distribution of local relaxation times and, hence, exhibit different temperature variation. [S0031-9007(98)06504-1]

PACS numbers: 61.20.Lc, 61.20.Ja, 61.43.Fs, 64.70.Pf

In this paper we address the relationship between various time scales of a liquid as it approaches the glassy state. The comparison of the self-diffusion constant D_T and shear viscosity η is probably the most studied of these relationships [1]. Well above any glass transition, the scaling $D_T \eta / T$ = const (the Stokes-Einstein relation) holds for a wide range of liquids, supporting the assumption that particle friction at these temperatures arises from the viscous dissipation of the associated velocity field in the surrounding liquid. Recently, a number of studies [2,3] on a range of organic glass formers have shown that this relation breaks down as the supercooling is increased, with D_T proving to be larger than expected. The rotational diffusion constant D_R, however, continues to scale with η well below the temperature at which the scaling breaks down for D_T.

The most popular explanation of this difference in the temperature dependences of these two time scales (either D_T^{-1} and η, or D_T and D_R) invokes the existence of a spatially heterogeneous and transient distribution of local relaxation times [1,2,4–6]. The mean-squared particle displacement (MSD) and the relaxation of the stress correlation function corresponds to different averages over this distribution. The former average is dominated by the more mobile particles, while the stress correlation function is dominated by the slower regions. As the distribution broadens (and becomes more asymmetric) on cooling, so does the difference between the different averages increase. The proposal makes sense, particularly as the existence of the dynamic heterogeneities have been established experimentally (as long-lived kinetic subpopulations [7]) and from computer simulations (as spatially resolved domains [8,9]). This proposal, however, has yet to be directly tested.

We present evidence, obtained from molecular dynamics simulations of a two dimensional (2D) binary mixture of soft disks, of markedly different temperature dependences of a structural relaxation time, and the self-diffusion constant as the temperature drops below a crossover temperature. As we can also explicitly resolve the heterogeneous distribution of local relaxation times, the 2D mixture provides a valuable opportunity to examine the origin of the difference between the time scales and the role of heterogeneities.

Simulations have been carried out in 2D on an equimolar ($x_1 = N_1/N = 0.5$) binary mixture of particles with diameters $\sigma_2 = 1.4$ and $\sigma_1 = 1.0$, and with the same mass m. The pairwise additive interactions are given by the repulsive soft-core potential,

$$u_{ab}(r) = \epsilon \left[\frac{\sigma_{ab}}{r} \right]^{12}, \qquad a, b = 1, 2, \tag{1}$$

where $\sigma_{aa} = \sigma_a$ and $\sigma_{ab} = (\sigma_a + \sigma_b)/2$. The cutoff radii of the interactions are $4.5\sigma_{ab}$. The units of mass, length, and time are m, σ_1, and $\tau = \sigma_1 \sqrt{m/\epsilon}$, respectively. A total of $N = 1024$ particles was enclosed in a square box with periodic boundary conditions. The simulations were carried out at constant N, pressure $P^* = P\sigma_1^2/\epsilon$, and temperature $T^* = k_B T/\epsilon$, where k_B is Boltzmann's constant [10]. A time step of 0.0025τ was used for $T^* > 1.0$ and 0.005τ for $T^* \leq 1.0$. The pressure was fixed at $P^* = 13.5$. At this pressure, the single component systems of small and large particles freeze at $T^*_{f,1} = 0.95$ and $T^*_{f,2} = 1.70$, respectively [9,11]. The mean-squared displacements and scattering functions reported in this Letter have been averaged over 200 time origins, each separated by regular time intervals of 5000 time steps.

It has been established [8,9] that the supercooled mixture is a stationary metastable state. Structural relaxation is characterized by the intermediate scattering function $F^a(q_a, t)$ for the two species ($a = 1, 2$) evaluated at a wave vector q_a corresponding to the first peak in the appropriate partial structure factor. We have found little significant qualitative difference between the total and self-intermediate scattering functions [9] and shall refer in this paper to the self-intermediate scattering function

 0031-9007/98/81(1)/120(4)$15.00

$F_s^1(q_1, t)$ in measuring structural relaxation, where

$$F_s^1(q_1,t) = \frac{1}{N_1}\left\langle \sum_{j=1}^{N_1} \exp\{i\mathbf{k}\cdot[\mathbf{r}_j(t) - \mathbf{r}_j(0)]\}\right\rangle. \quad (2)$$

The same conclusions are arrived at in this paper if the time correlation functions of the large particles are used instead. A structural relaxation time τ_e is defined as the time at which $F_s^1(q_1, t) = 1/e$. (This definition results in a τ_e which scales as the relaxation time of the slow or α process.) For $T^* \leq 0.5$ we observe the development of a step or plateau in the relaxation function at intermediate times. For convenience we shall refer to the temperature at which this feature of slow dynamics is first observed as a crossover temperature. We emphasize that there is no evidence for any singular behavior at this temperature. It is introduced simply as a convenient reference point.

What is the relationship between the structural relaxation characterized by the incoherent scattering function and self-diffusion? In Fig. 1, the long-time self-diffusion constant of the small particles $D_1 = \frac{1}{4}\lim_{t\to\infty}\frac{dR_1^2(t)}{dt}$, where $R_1^2(t) = \langle |r_{(1)}(t) - r_{(1)}(0)|^2\rangle$, is plotted against τ_e. In the 1970s it was shown that diffusion coefficients diverged in low density 2D liquids due to the coupling of a particle's motion to circulating hydrodynamic flows [12]. At high densities, however, we find [13] the hydrodynamic modes to be strongly damped and a well defined limiting slope of the appropriate MSD persists over distances of $> 10\sigma_1$. It can be clearly seen in Fig. 1 that D_1 deviates positively from the high temperature extrapolation below the crossover temperature. The plot shows a striking qualitative similarity to the relationship between D_T and η in Ref. [2]. The same data in Fig. 1 are plotted against inverse temperature in Fig. 2. Below $T^* = 0.5$, the diffusion constant exhibits only a weak deviation from Arrhenius temperature dependence while the structural relaxation time shows a strong non-Arrhenius variation.

The different temperature dependences of the two time scales at low temperatures suggest a growing complexity of the supercooled liquid and it poses something of a puzzle. After all, both quantities, τ_e and D_1^{-1}, characterize the self-motion of particles, the only differences being the characteristic length scales over which the particles move and the quantity being averaged over the single particle distribution of relaxation times. To separate the contributions of these two features we look for two new times, corresponding to D_1^{-1} and τ_e, but defined in terms of the *same* property, the MSD. In Fig. 2 we have also plotted $\tau_{4.0}$ and $\tau_{0.145}$, the time required for the MSD of the small particles, $R_1^2(t)$, to equal 4.0 and 0.145, respectively. $\tau_{4.0} \approx D_1^{-1}$ for all temperatures by definition, while $\tau_{0.145} \approx \tau_e$ for temperatures above the crossover temperature. This latter result underscores the observation that structural relaxation is dominated by very small particle motions. This is also true of the time correlation of the shear stress (the integral of which is proportional to the shear viscosity) which we find [9] to have relaxed to less than 20% of its initial value in times of the order of $\tau_{0.145}$.

$\tau_{4.0}$ and $\tau_{0.145}$ represent an important step in our analysis. Differing only in the associated length scale, these new times are not expected to vary significantly in the way they sample any heterogeneous distribution of kinetics. The longer time, $\tau_{4.0}$, will certainly involve individual particles sampling more of any distribution of kinetic environments than is possible for the shorter time. On top of this, the averaging over all particles involves an average over the ensemble of kinetic environments. We have previously shown [14] that diffusion through a medium of fluctuating local mobilities results in the same average diffusion as that obtained via an ensemble

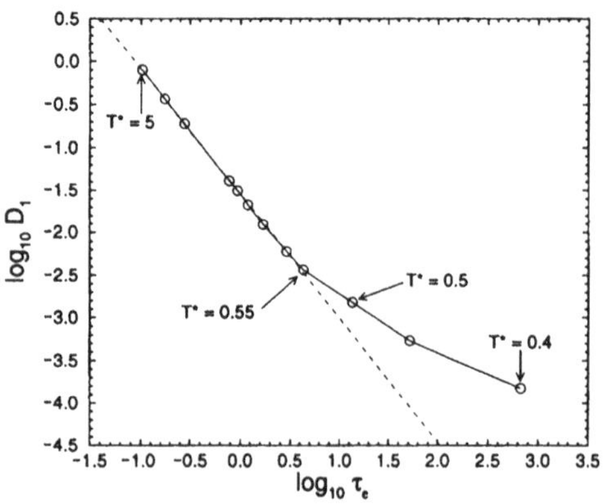

FIG. 1. A plot of $\log_{10} D_1$ against $\log_{10}\tau_e$. The dashed line represents the extrapolation of the high temperature relationship. Note the positive deviation of D_1 from this extrapolation for $T^* \leq 0.5$.

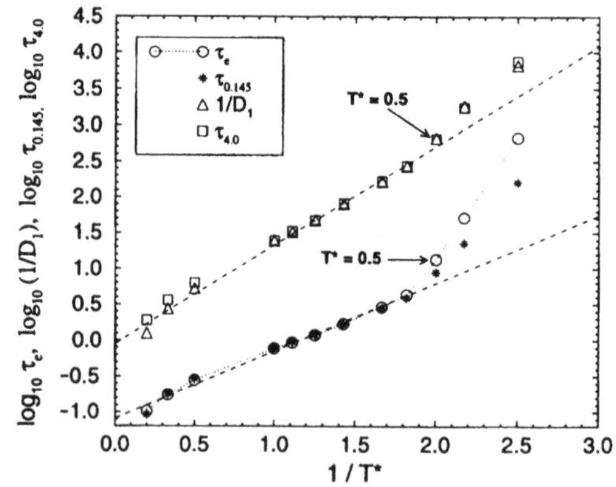

FIG. 2. Arrhenius plots of D_1^{-1} and τ_e. The relaxation time τ_e exhibits a striking deviation from Arrhenius behavior for $T^* \leq 0.5$. Also shown are the relaxation times $\tau_{4.0}$ and $\tau_{0.145}$ as defined in the text.

average over a distribution of mobilities. Hence, we argue that the different length scales of $\tau_{4.0}$ and $\tau_{0.145}$ will not result in different averages over the heterogeneities. From Fig. 2, however, $\tau_{0.145}$ still clearly exhibits a greater deviation from Arrhenius behavior than $\tau_{4.0}$. Having rejected dynamic heterogeneity as an explanation, we must now consider the significance of the crossover between ballistic and diffusive motion on the dynamics over the different length scales.

Consider a simplified model of particle dynamics where, following the initial ballistic motion, diffusion abruptly appears after a time τ_0 and a MSD $= \Delta_0$. If τ_Δ is the time required for the MSD $= \Delta$, then τ_Δ can be written as

$$\tau_\Delta = \tau_0 + (\Delta - \Delta_0)/4D, \qquad (3)$$

with D, the diffusion coefficient. When $\Delta \gg \Delta_0$, we have

$$\tau_\Delta \approx \Delta/4D \qquad (4)$$

and τ_Δ scales as D^{-1} as is the case for $\Delta = 4.0$. For smaller Δ's, however, we must use Eq. (3) and the temperature dependence of the associated relaxation time will reflect the temperature dependence of both τ_0 and D. As we find for $\tau_{0.145}$, it is only the short distance dynamics which will reflect this more complex temperature dependence. This argument can also be made graphically. Figure 3 shows that at $R_1^2(t) = 4$ the slope has settled to one and the only temperature dependence lies in the diffusion constant itself. For $R_1^2(t) = 0.145$, however, we are in the transition between ballistic and diffusive motion and the variation of the slope here provides an *additional* temperature dependence. As the temperature is lowered and D approaches zero, there must come a point, even for $\Delta = 0.145$, when the time scale τ_0 for ballistic motion is insignificant compared to the slow diffusive motion. Below this temperature (presumably a temperature lower than we can follow in our simulations) $\tau_{0.145}$ and $\tau_{4.0}$ are expected to show similar temperature dependences.

While this explanation accounts, at least qualitatively, for the difference between $\tau_{4.0}$ and $\tau_{0.145}$, we still must test our proposition that the growing difference between τ_e and $\tau_{0.145}$ observed below the crossover temperature in Fig. 2 is due to dynamic heterogeneities. This increasing gap between the relaxation rate and the time scale of particle displacement below the crossover is highlighted in Fig. 4 where $F_s^1(q_1, t)$ has been plotted against $R_1^2(t)$ for a range of temperatures. Above $T^* = 0.5$, we find the relaxation functions collapsing onto a single non-Gaussian master curve. At and below the crossover temperature, the scaling in Fig. 4 breaks down with the structural relaxation lagging increasingly behind the MSD as the temperature is lowered. A similar behavior has been observed in a recent simulation of a polymer melt [15]. This feature can be accounted for by the presence of dynamic heterogeneities [9]. It is proposed here that, irrespective of the type of system, the plot of the incoherent scattering function against the MSD in Fig. 4 provides a useful diagnostic for dynamic heterogeneity without the need for explicit filtering of kinetic subpopulations.

To test the proposal that the difference in the temperature dependence of τ_e and $\tau_{0.145}$ is a consequence of the broad distribution of local mobilities, we can examine the relation between analogous times in kinetically resolved subsets of the particles. By construction, such subsets will have narrower distributions of relaxation times. A relaxation time per particle is defined as the time taken to

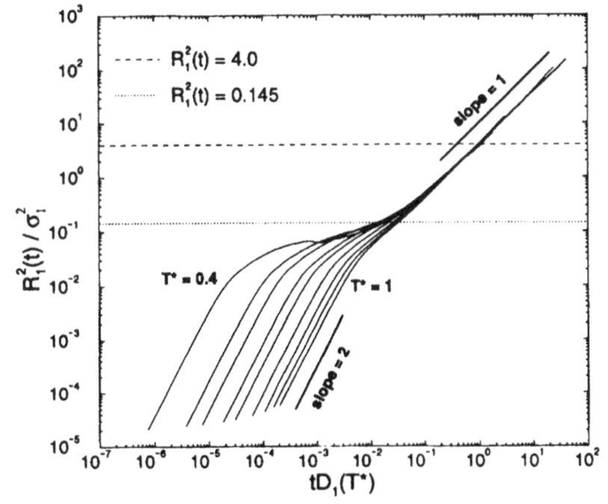

FIG. 3. A log-log plot of $R_1^2(t)$, against time scaled by $D_1(T^*)$. A dashed line at $R_1^2(t) = 4.0$ and a dotted line at $R_1^2(t) = 0.145$ are shown to assist in visualizing the different temperature dependences of $\tau_{4.0}$ and $\tau_{0.145}$. The temperature of the curves from left to right is $T^* = 0.4$, 0.46, 0.5, 0.55, 0.6, 0.7, 0.8, 0.9, and 1.0, respectively.

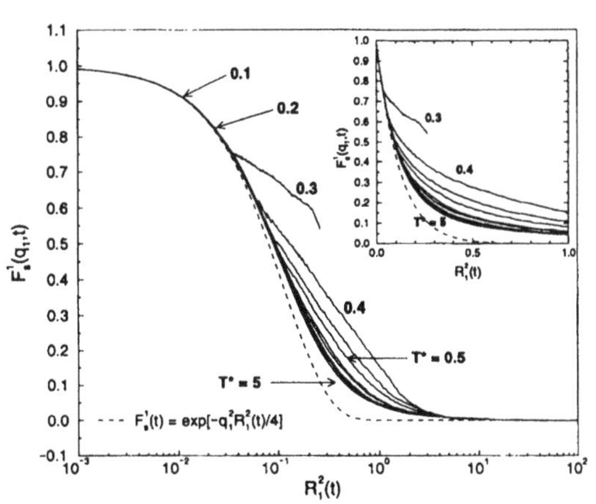

FIG. 4. $F_s^1(q_1, t)$, where $q_1 = 7.17\sigma_1^{-1}$, is plotted against $R_1^2(t)$ in a log-linear plot for $0.1 \le T^* \le 5.0$. Observe the scaling at high temperatures. The inset shows the same data on a linear scale for $R_1^2(t)$. The Gaussian approximation for $F_s^1(q_1, t)$ is presented as a dashed line.

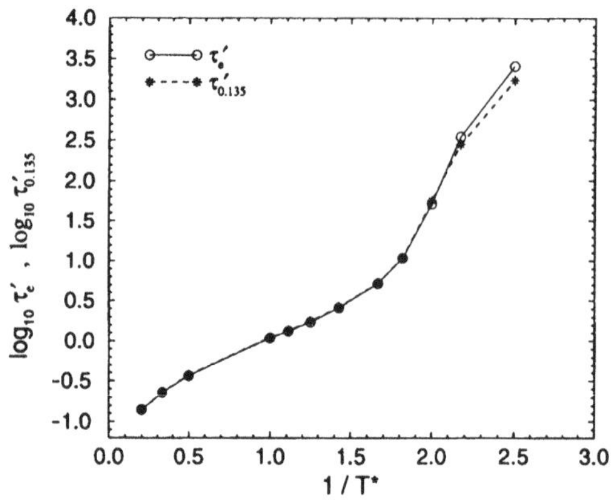

FIG. 5. Arrhenius plots of τ_e' and $\tau_{0.135}'$, both averaged over the slow subpopulation of small particles as described in the text. Note the agreement between the two times below $T^* = 0.5$. See text for discussion.

first achieve a displacement of σ_1. (A detailed discussion concerning this definition is presented in Refs. [9,16].) Consider the subset of particles corresponding to the slowest 40% of particles (irrespective of particle type) in any given run. A relaxation time τ_e' is defined as the time taken by the incoherent scattering function of the small particles in this slow subset to decay to $1/e$. Above the crossover temperature, the structural relaxation of this subset of small particles corresponds to a MSD, averaged over this slow subpopulation, of 0.135. In Fig. 5, τ_e' and $\tau_{0.135}'$ are plotted against inverse temperature, where $\tau_{0.135}'$ is the time the MSD for the slow small particles equals 0.135. We find that the two time scales of the slow subset coincide for T^* well below the crossover. This result provides explicit support for the proposal that the growing dynamic heterogeneity is responsible, in part, for the different temperature dependences of structural relaxation and the short distance mobility at large supercoolings.

In conclusion, we have established that the structural relaxation time in the simulated mixture exhibits a striking positive deviation from Arrhenius behavior at low temperatures which is not shared by the self-diffusion constant. The difference in behavior between the structural relaxation and the diffusion constant is traced to two sources: the small displacements associated with structural relaxation and the increasing heterogeneity of the local kinetic environment. As particles, on average, experience increasing transient localization ("caging") as the system is cooled, the transition from ballistic to diffusive dynamics involves larger changes in the temperature dependence of relaxation times on short length scales within this regime. Structural relaxation, at least in simple liquids, will always be governed by displacements of the same order as those associated with this transition in dynamics. This transition length, on the other hand, is insignificant when compared to the characteristic mean-squared displacements over which the self-diffusion constant is defined. Dynamic heterogeneities also contribute to the difference between structural relaxation and short length scale diffusion as the two processes become dominated by the opposite ends of the broadening spectrum of relaxation times. As the diffusion constant drops below the value accessible by simulations, it is expected that the short time associated with the crossover ceases to represent a significant contribution to the overall structural relaxation time. Below such temperatures, dynamic heterogeneities provide the only remaining explanation of any difference in the properties of the different average time scales.

*Electronic address: d.perera@chem.usyd.edu.au
†Electronic address: peter@chem.usyd.edu.au
Author to whom correspondence should be addressed.

[1] C. Z.-W. Liu and I. Oppenheim, Phys. Rev. E **53**, 799 (1996); J.-L. Barrat, J.-N Roux, and J.-P Hansen, Chem. Phys. **149**, 197 (1990).
[2] F. Fujara, B. Geil, H. Sillescu, and G. Fleischer, Z. Phys. B **88**, 195 (1992); I. Chang, F. Fujara, B. Geil, G. Heuberger, T. Mangel, and H. Sillescu, J. Non-Cryst. Solids **172–174**, 248 (1994).
[3] M. T. Cicerone and M. D. Ediger, J. Chem. Phys. **104**, 7210 (1996); F. R. Blackburn, C. Y. Wang, and M. D. Ediger, J. Phys. Chem. **100**, 18 249 (1996); M. T. Cicerone, F. R. Blackburn, and M. D. Ediger, Macromolecules **28**, 8224 (1995).
[4] F. H. Stillinger and J. A. Hodgdon, Phys. Rev. E **50**, 2064 (1994).
[5] D. N. Perera and P. Harrowell, J. Chem. Phys. **104**, 2369 (1996).
[6] M. D. Ediger, J. Non-Cryst. Solids (to be published).
[7] K. Schmidt-Rohr and H. W. Spiess, Phys. Rev. Lett. **66**, 3020 (1991); M. T. Cicerone and M. D. Ediger, J. Chem. Phys. **103**, 5684 (1995).
[8] D. N. Perera and P. Harrowell, J. Non-Cryst. Solids (to be published); R. Yamamoto and A. Onuki, Europhys. Lett. **40**, 61 (1997); J. Phys. Soc. Jpn. **66**, 2545 (1997); W. Kob *et al.*, Phys. Rev. Lett. **79**, 2827 (1997).
[9] D. N. Perera and P. Harrowell (to be published).
[10] D. J. Evans and G. P. Morriss, Comput. Phys. Rep. **1**, 297 (1984).
[11] J. Q. Broughton, G. H. Gilmer, and J. D. Weeks, Phys. Rev. B **25**, 4651 (1982).
[12] B. J. Alder and T. E. Wainwright, Phys. Rev. A **1**, 18 (1970).
[13] D. N. Perera and P. Harrowell, Phys. Rev. Lett. **80**, 4446 (1998).
[14] M. M. Hurley and P. Harrowell, J. Chem. Phys. **105**, 10521 (1996).
[15] K. Okun *et al.*, Macromolecules **30**, 3075 (1997).
[16] M. M. Hurley and P. Harrowell, Phys. Rev. E **52**, 1694 (1995).

Dynamics of highly supercooled liquids: Heterogeneity, rheology, and diffusion

Ryoichi Yamamoto and Akira Onuki
Department of Physics, Kyoto University, Kyoto 606-8502, Japan
(Received 20 March 1998)

Highly supercooled liquids with soft-core potentials are studied via molecular-dynamics simulations in two and three dimensions in quiescent and sheared conditions. We may define bonds between neighboring particle pairs unambiguously owing to the sharpness of the first peak of the pair correlation functions. Upon structural rearrangements, they break collectively in the form of clusters whose sizes grow with lowering the temperature T. The bond lifetime τ_b, which depends on T and the shear rate $\dot{\gamma}$, is on the order of the usual structural or α relaxation time τ_α in weak shear $\dot{\gamma}\tau_\alpha \ll 1$, while it decreases as $1/\dot{\gamma}$ in strong shear $\dot{\gamma}\tau_\alpha \gg 1$ due to shear-induced cage breakage. Accumulated broken bonds in a time interval ($\sim 0.05\tau_b$) closely resemble the critical fluctuations of Ising spin systems. For example, their structure factor is well fitted to the Ornstein-Zernike form, which yields the correlation length ξ representing the maximum size of the clusters composed of broken bonds. We also find a dynamical scaling relation, $\tau_b \sim \xi^z$, valid for any T and $\dot{\gamma}$ with $z=4$ in two dimensions and $z=2$ in three dimensions. The viscosity is of order τ_b for any T and $\dot{\gamma}$, so marked shear-thinning behavior emerges. The shear stress is close to a limiting stress in a wide shear region. We also examine motion of tagged particles in shear in three dimensions. The diffusion constant is found to be of order $\tau_b^{-\nu}$ with $\nu = 0.75 \sim 0.8$ for any T and $\dot{\gamma}$, so it is much enhanced in strong shear compared with its value at zero shear. This indicates a breakdown of the Einstein-Stokes relation in accord with experiments. Some possible experiments are also proposed. [S1063-651X(98)16409-7]

PACS number(s): 64.70.Pf, 83.50.Gd, 61.43.Fs

I. INTRODUCTION

Particle motions in supercooled liquids are severely restricted or jammed, thus giving rise to slow structural relaxations and highly viscoelastic behavior [1,2]. Recently much attention has been paid to the mode-coupling theory [3,4], which is an analytic scheme describing the onset of slow structural relaxations considerably above T_g. There, the density fluctuations with wave numbers around the first peak position of the structure factor are of the most importance and no long-range correlations are predicted. For a long time, however, it has been expected [5–8] that rearrangements of particle configurations in glassy materials should be cooperative, involving many molecules, owing to configuration restrictions. In other words, such events occur only in the form of *clusters* whose sizes increase at low temperatures. In normal liquid states, on the contrary, they are frequent and uncorrelated among one another in space and time. Such an idea was first put forth by Adam and Gibbs [5], who invented a frequently used jargon, *cooperatively rearranging regions* (CRR). However, it is difficult to judge whether or not such phenomenological models are successful in describing real physics and in making quantitative predictions.

Molecular-dynamics (MD) simulations can be powerful tools to gain insights into relevant physical processes in highly supercooled liquids. Such processes are often masked in averaged quantities such as the density time correlation functions. As a marked example, we mention kinetic heterogeneities observed in recent simulations [9–18]. Using a simple two-dimensional fluid, Muranaka and Hiwatari [9] visualized significant large-scale heterogeneities in particle displacements in a relatively short time interval, which was supposed to correspond to the β relaxation time regime. In liquid states with higher temperatures, Hurley and Harrowell [11] observed similar kinetic heterogeneities but the correlation length was still on the order of a few particle diameters. The characterization of these patterns has not been made in these papers. Recently our simulations on model fluid mixtures in two and three dimensions [13–15] have identified *weakly bonded* or *relatively active* regions from breakage of appropriately defined bonds. Spatial distributions of such regions resemble the critical fluctuations in Ising spin systems, so the correlation length ξ can be determined. It grows up to the system size as T is lowered, but no divergence seems to exist at nonzero temperatures [13,19–21]. Donati *et al.* have observed *stringlike* clusters whose lengths increase at low temperatures in a three-dimensional (3D) binary mixture [17]. In addition, Monte Carlo simulations of a dense polymer by Ray and Binder showed a significant system size dependence of the monomer diffusion constant, which indicates heterogeneities over the system size [18].

Most previous papers so far have been concerned with near-equilibrium properties, such as relaxation of the density time correlation functions or dielectric response. From our point of view, these quantities are too restricted or indirect, and there remains a rich group of unexplored problems in far-from-equilibrium states. For example, nonlinear glassy response against electric field, strain, etc. constitutes a future problem [22]. In this paper we apply a simple shear flow $v_x = \dot{\gamma}y$ in the x direction and realize steady states [23]. The velocity gradient $\dot{\gamma}$ in the y direction is called the shear rate or simply shear. We shall see that it is a relevant perturbation drastically changing the glassy dynamics when $\dot{\gamma}$ exceeds the inverse of the structural or α relaxation time τ_α. As is well known, τ_α increases dramatically from microscopic to macroscopic times in a rather narrow temperature range [1,2]. Generally, in near-critical fluids and various complex fluids, nonlinear shear regimes are known to emerge when $\dot{\gamma}$ exceeds some underlying relaxation rate [23]. However, in supercooled liquids, it is unique that even very small shear can greatly accelerate the *microscopic* rearrangement processes. Similar effects are usually expected in systems composed of very large elements such as colloidal suspensions.

Though rheological experiments on glass-forming fluids have not been abundant, Simmons *et al.* found that the viscosity $\eta(\dot\gamma)=\sigma_{xy}/\dot\gamma$ exhibits strong shear-thinning behavior,

$$\eta(\dot\gamma)\cong\eta(0)/(1+\dot\gamma\tau_\eta), \tag{1.1}$$

in soda-lime-silica glasses in steady states under shear [24–26], τ_η being a long rheological time. After application of shear, they also observed overshoots of the shear stress before approach to steady states. Our previous reports [14,15] have treated nonlinear rheology in supercooled liquids, in agreement with these experiments. Interestingly, similar *jamming dynamics* has begun to be recognized also in rheology of foams [27–29] and granular materials [30] composed of large elements. Shear-thinning behavior and heterogeneities in configuration rearrangements are commonly observed also in these macroscopic systems.

As a closely related problem, understanding of mechanical properties of amorphous metals such as $Cu_{57}Zr_{43}$ has been of great technological importance [31–35]. They are usually ductile in spite of their high strength. At low temperatures $T\lesssim 0.6\sim 0.7T_g$, localized bands ($\lesssim 1$ μm), where zonal slip occurs, have been observed above a yield stress. At relatively high temperatures $T\gtrsim 0.6\sim 0.7T_g$, on the other hand, shear deformations are induced *homogeneously* (on macroscopic scales) throughout samples, giving rise to viscous flow with strong shear thinning behavior. In particular, in their 3D simulations, Takeuchi *et al.* [34,35] followed atomic motions after application of a small shear strain to observe heterogeneities among *poorly and closely packed regions*, which are essentially the same entities we have discussed. Our simulations under shear in this paper correspond to the *homogeneous* regime at relatively high temperatures in amorphous metals.

Another interesting issue is as follows. Several experiments have revealed that the translational diffusion constant D of a tagged particle in a fragile glassy matrix becomes increasingly larger than the Einstein-Stokes value $D_{ES}=k_BT/2\pi\eta a$ with lowering T, where η is the (zero-shear) viscosity and a is the diameter of the particle [2,36–38]. In particular, the power-law behavior $D\propto\eta^{-\nu}$ with $\nu\cong 0.75$ was observed at sufficiently low temperatures [36]. Thus D/D_{ES} increases from of order 1 up to order 10^2–10^3 in supercooling experiments. Furthermore, smaller probe particles exhibit a more pronounced increase of the ratio $D\eta/T\propto D/D_{ES}$ with lowering T [37]. It is generally believed that η is proportional to the α relaxation time τ_α or the rotational relaxation time $1/D_{rot}$ for anisotropic molecules (D_{rot} being the rotational diffusion constant) [36,37,39]. Therefore, individual particles are much more mobile at long times $t\gtrsim\tau_\alpha$ than expected from the Stokes-Einstein formula. In a MD simulation on a 3D binary mixture with $N=500$ in 3D [40], the same tendency was apparently seen despite their small system size. Very recently, in a MD simulation in a 2D binary mixture with $N=1024$, Perera and Harrowell have observed clear deviation from the linear relation $D\propto\tau_\alpha$, where τ_α is obtained from the decay of the time correlation function as in our case in Sec. VI [41]. We will examine this problem in a much larger 3D system with $N=N_1+N_2=10^4$ generally in the presence of shear, where the viscosity and the diffusion constant both vary tremendously in strong shear ($\dot\gamma\gtrsim 1/\tau_\alpha$).

The organization of this paper is as follows. In Sec. II, our model binary mixtures and our simulation method will be explained. In Sec. III, *bonds* among particle pairs will be introduced at distances close to the first peak position of the pair correlation functions. Breakage of such bonds will then be followed numerically, which exhibits heterogeneities enhanced at low temperatures. Their analysis will yield the correlation length in Sec. IV. Rheology of supercooled liquids will be studied in Sec. V. These effects were briefly reported in our previous reports [13–15]. In Sec. VI, results on the motion of tagged particles will be presented.

II. MODEL AND SIMULATION METHOD

We performed MD simulations in two dimensions (2D) and three dimensions (3D) on binary mixtures composed of two different atomic species, 1 and 2, with $N_1=N_2=5000$ particles with the system volume V being fixed. Parameters chosen are mostly common in 2D and 3D. They interact via the soft-core potential [9–15,41–43],

$$v_{\alpha\beta}(r)=\epsilon(\sigma_{\alpha\beta}/r)^{12},\quad \sigma_{\alpha\beta}=\frac{1}{2}(\sigma_\alpha+\sigma_\beta), \tag{2.1}$$

where r is the distance between two particles and $\alpha,\beta=1,2$. The interaction is truncated at $r=4.5\sigma_1$ in 2D and $r=3\sigma_1$ in 3D. The leapfrog algorithm is used to integrate the differential equations with a time step of $0.005\tau_0$, where

$$\tau_0=(m_1\sigma_1^2/\epsilon)^{1/2}. \tag{2.2}$$

The space and time are measured in units of σ_1 and τ_0. The mass ratio is $m_2/m_1=2$, while the size ratio is

$$\sigma_2/\sigma_1=1.4\quad (d=2),\qquad \sigma_2/\sigma_1=1.2\quad (d=3), \tag{2.3}$$

where d is the space dimensionality. This size difference prevents crystallization and produces amorphous states in our systems at low temperatures.

We fixed the particle density at

$$n=0.8/\sigma_1^d, \tag{2.4}$$

where $n=n_1+n_2$ is the total number density. The system linear dimension is $L=118$ in 2D and $L=23.2$ in 3D. Then our systems are highly compressed. In fact, the volume fraction of the particles may be estimated as $\pi(\sigma_1^2n_1+\sigma_2^2n_2)=0.93$ in 2D and as $\frac{4}{3}\pi(\sigma_1^3n_1+\sigma_2^3n_2)=0.57$ in 3D, where overlapped regions are doubly counted. In such cases, according to the Henderson and Leonard theory [43–45], our binary mixtures may be fairly mapped onto one-component fluids with the soft-core potential with an effective radius defined by

$$\sigma_{eff}^d=\sum_{\alpha,\beta=1,2}x_\alpha x_\beta\sigma_{\alpha\beta}^d, \tag{2.5}$$

TABLE I. Simulations in 2D.

$\Gamma_{\rm eff}$	1.0	1.1	1.2	1.3	1.4
k_BT/ϵ	2.54	1.43	0.85	0.526	0.337
p/nk_BT-1	15.1	22.6	33.5	50.2	75.1

where $x_1=n_1/n$ and $x_2=n_2/n=1-x_1$ are the compositions of the two components and are 1/2 in our case. As in the one-component case, the thermodynamic state is characterized by a single parameter (effective density),

$$\Gamma_{\rm eff}=n(\epsilon/k_BT)^{d/12}\sigma_{\rm eff}^d. \tag{2.6}$$

For example, Bernu *et al.* [43] confirmed that the equilibrium pressure p may be well fitted to the scaling form, $p/nk_BT-1\cong 6+6.848(\Gamma_{\rm eff})^4$, at all x_1 in 3D. In Tables I and II we list $\Gamma_{\rm eff}$ chosen in our simulations together with the corresponding scaled temperatures and pressures in 2D and 3D. Our pressure data agree very well with the above scaling form for 3D.

We introduce here the pair correlation functions $g_{\alpha\beta}(r)$ by

$$\langle\hat{n}_\alpha(\mathbf{r})\hat{n}_\beta(\mathbf{0})\rangle=n_\alpha n_\beta g_{\alpha\beta}(r)+n_\alpha\delta_{\alpha\beta}\delta(\mathbf{r}), \tag{2.7}$$

where

$$\hat{n}_\alpha(\mathbf{r})=\sum_j \delta(\mathbf{r}-\mathbf{r}_{\alpha j}) \quad (\alpha=1,2) \tag{2.8}$$

are the number densities in terms of the particle positions $\mathbf{r}_{\alpha j}$ ($\alpha=1,2,j=1,\ldots,N/2$). The time dependence is suppressed for simplicity. In a highly compressed state the interparticle distances between the α and β particles are characterized by

$$l_{\alpha\beta}=\sigma_{\alpha\beta}(\epsilon/k_BT)^{1/12}. \tag{2.9}$$

The last factor $(\epsilon/k_BT)^{1/12}$ represents the degree of expulsion or penetration from or into the soft-core regions ($r<\sigma_{\alpha\beta}$) on particle encounters, though it is not far from 1 in our case. The one-fluid approximation may be justified if the pair correlation functions satisfy

$$g_{\alpha\beta}(r)=G(r/l_{\alpha\beta},\Gamma_{\rm eff}). \tag{2.10}$$

Namely, $g_{\alpha\beta}(r)$ are independent of α, β, and x_1 once the distance is scaled by $l_{\alpha\beta}$. The pressure is then expressed as [45]

TABLE II. Simulations in 3D.

$\Gamma_{\rm eff}$	1.15	1.3	1.4	1.45	1.5	1.55
k_BT/ϵ	0.772	0.473	0.352	0.306	0.267	0.234
p/nk_BT-1	18.9	26.7	33.4	37.2	41.4	46.3

$$\frac{p}{nk_BT}-1=-\frac{n}{2dk_BT}\sum_{\alpha,\beta}\int d\mathbf{r}\,x_\alpha x_\beta v'_{\alpha\beta}(r)rg_{\alpha\beta}(r)$$
$$=6V_d\Gamma_{\rm eff}\int_0^\infty ds\frac{1}{s^{13-d}}G(s,\Gamma_{\rm eff}), \tag{2.11}$$

where $v'_{\alpha\beta}(r)=dv_{\alpha\beta}(r)/dr$ and V_d is the volume of a unit sphere, so it is $4\pi/3$ in 3D and π in 2D. We shall see that Eq. (2.10) holds very well around the first peak of the pair correlation functions in our simulations. This fairly supports the one-fluid approximation, because the soft-core potential and the pair correlation functions decrease very abruptly for $r\gtrsim l_{\alpha\beta}$ and for $r\lesssim l_{\alpha\beta}$, respectively, and the dominant contribution arises from $r\sim l_{\alpha\beta}\sim\sigma_{\alpha\beta}$.

In our systems the structural relaxation time becomes very long at low temperatures. Therefore, the annealing time was taken to be at least 10^5 in 2D and 10^4 in 3D. No appreciable aging effect was detected in the course of taking data in various quantities such as the pressure or the density time correlation function except for the lowest temperature cases, $\Gamma_{\rm eff}=1.4$ in 2D and $\Gamma_{\rm eff}=1.55$ in 3D. A small aging effect remained in the density time correlation function in these exceptional cases, however.

Our simulations were performed in the absence and presence of shear flow [46,47]. In the unsheared case ($\dot{\gamma}=0$) we performed simulations under the microcanonical (constant energy) condition. However, in the sheared case ($\dot{\gamma}>0$), we kept the temperature at a constant using the Gaussian constraint thermostat to eliminate the viscous heating effect. No difference was detected between the profile-based and profile-unbased thermostats [47], so results with the profile-based thermostat will be presented in this paper. Our method of applying shear is as follows: The system was at rest for $t<0$ for a very long equilibration time and was then sheared for $t>0$. Here we added the average velocity $\dot{\gamma}y$ to the velocities of all the particles in the x direction at $t=0$ and afterwards maintained the shear flow by using the Lee-

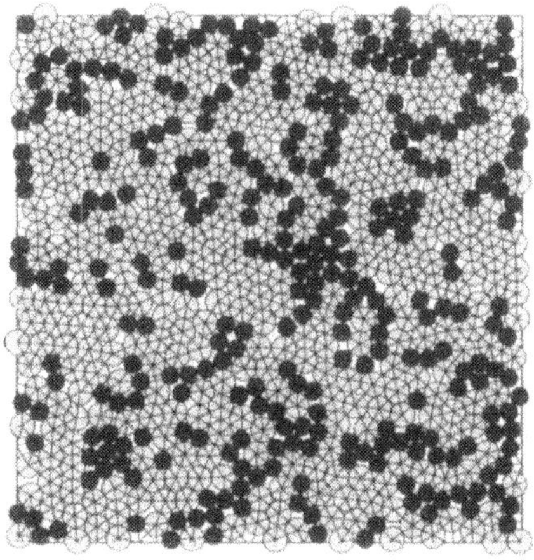

FIG. 1. A typical particle configuration and the bonds defined at a given time at $\Gamma_{\rm eff}=1.4$ in 2D. The diameters of the circles here are equal to σ_α. The areal fraction of the soft-core regions is 93%. A 1/16 of the total system is shown.

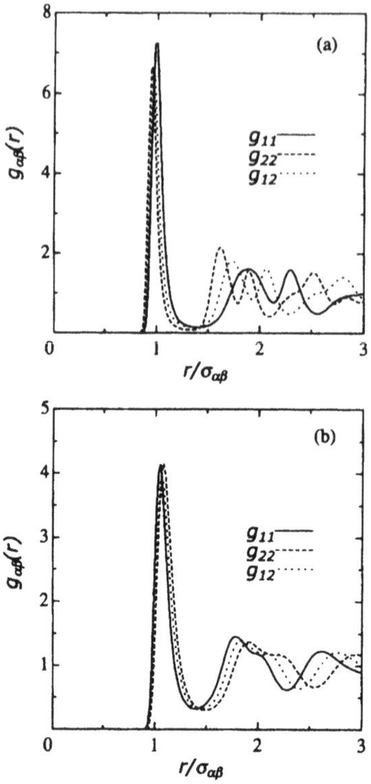

FIG. 2. The pair correlation functions $g_{\alpha\beta}(r)$ in quiescent states as functions of $r/\sigma_{\alpha\beta}$ at $\Gamma_{\rm eff}=1.4$ in 2D (a) and at $\Gamma_{\rm eff}=1.55$ in 3D (b).

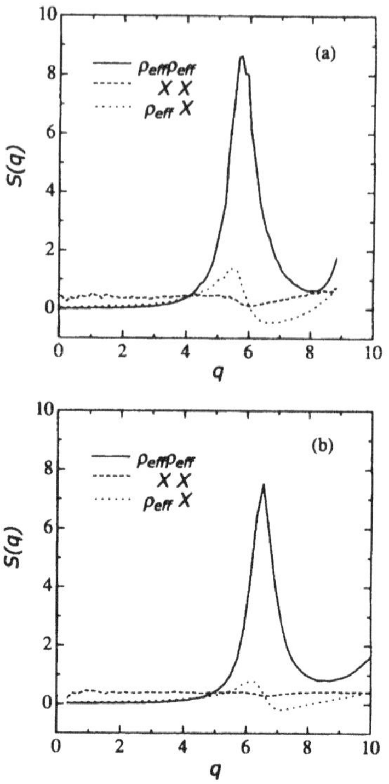

FIG. 3. The structure factors $S(q)$ defined in Eqs. (3.3)–(3.5) in quiescent states at $\Gamma_{\rm eff}=1.4$ in 2D (a) and at $\Gamma_{\rm eff}=1.55$ in 3D (b). The dimensionless wave number q is measured in units of σ_1^{-1}. The solid, dashed, and dotted lines correspond to $\rho_{\rm eff}$-$\rho_{\rm eff}$, X-X, and $\rho_{\rm eff}$-X correlations, respectively.

Edwards boundary condition [46,47]. Then steady states were realized after a transient time. In our case shear flow serves to destroy glassy structures and produces no long-range structure.

III. PAIR CORRELATIONS AND BOND BREAKAGE

A. Pair correlations

Because of the convenience of visualization in 2D, we first present a snapshot of particles at $\Gamma_{\rm eff}=1.4$ in 2D in Fig. 1, which gives an intuitive picture of the particle configurations. We can see that each particle is touching mostly six particles and infrequently five particles at distances close to $\sigma_{\alpha\beta}(=1.095^{-1}l_{\alpha\beta})$. Similar jammed particle configurations can also be found in 3D, where the coordination number of other particles around each particle is about 12. Then it is natural that the pair correlation functions $g_{\alpha\beta}(r)$ $(\alpha,\beta=1,2)$ have a very sharp peak at $r\cong\sigma_{\alpha\beta}$, as displayed in Fig. 2 for $\Gamma_{\rm eff}=1.4$ in 2D and $\Gamma_{\rm eff}=1.55$ in 3D. Furthermore, the heights of these peaks are all close to 7 in 2D and 4 in 3D. This confirms the scaling form (2.10) around the first peak.

We introduce a density variable representing the degree of particle packing by

$$\hat{\rho}_{\rm eff}(\boldsymbol{r})=\sigma_1^d\hat{n}_1(\boldsymbol{r})+\sigma_2^d\hat{n}_2(\boldsymbol{r}), \tag{3.1}$$

in terms of which the local volume fraction of the soft-core regions is $\pi\hat{\rho}_{\rm eff}(\boldsymbol{r})$ in 2D and by $(4\pi/3)\hat{\rho}_{\rm eff}(\boldsymbol{r})$ in 3D. We also consider the local composition fluctuation,

$$\delta\hat{X}(\boldsymbol{r})=\frac{1}{n}[x_2\hat{n}_1(\boldsymbol{r})-x_1\hat{n}_2(\boldsymbol{r})], \tag{3.2}$$

where $x_1=x_2=1/2$ in our case. In Fig. 3 we show the corresponding, dimensionless structure factors,

$$S_{\rho\rho}(q)=\sigma_1^{-d}\int d\boldsymbol{r}e^{i\boldsymbol{q}\cdot\boldsymbol{r}}\langle\delta\hat{\rho}_{\rm eff}(\boldsymbol{r})\delta\hat{\rho}_{\rm eff}(\boldsymbol{0})\rangle, \tag{3.3}$$

$$S_{\rho X}(q)=\sigma_1^{-d}\int d\boldsymbol{r}e^{i\boldsymbol{q}\cdot\boldsymbol{r}}\langle\delta\hat{\rho}_{\rm eff}(\boldsymbol{r})\delta\hat{X}(\boldsymbol{0})\rangle, \tag{3.4}$$

$$S_{XX}(q)=\sigma_1^{-d}\int d\boldsymbol{r}e^{i\boldsymbol{q}\cdot\boldsymbol{r}}\langle\delta\hat{X}(\boldsymbol{r})\delta\hat{X}(\boldsymbol{0})\rangle, \tag{3.5}$$

where $\delta\hat{\rho}_{\rm eff}=\hat{\rho}_{\rm eff}-\langle\hat{\rho}_{\rm eff}\rangle$. They are linear combinations of the usual structure factors,

$$S_{\alpha\beta}(q)=n_\alpha n_\beta\int d\boldsymbol{r}e^{i\boldsymbol{q}\cdot\boldsymbol{r}}[g_{\alpha\beta}(r)-1], \tag{3.6}$$

from the definitions (3.1) and (3.2). The temperatures in Fig. 3 are common to those in Fig. 2. Note that the dimensionless wave number q is measured in units of σ_1^{-1}. The $S_{\rho\rho}(q)$ has a pronounced peak at $q\sim 6$ and becomes very small (~ 0.01) at smaller q both in 2D and 3D. In this sense our systems are highly incompressible at long wavelengths. On the other hand, $S_{XX}(q)$ has no peak and is roughly a constant over a very wide q region, suggesting no enhancement of the composition fluctuations and no tendency of phase separation at least in our simulation times.

From Fig. 3 we may estimate the magnitude of the isothermal compressibility $K_{TX}=(\partial n/\partial p)_{TX}/n$. In equilibrium it is expressed in terms of the fluctuation variances as

$$k_B T K_{TX} = n^{-4} \lim_{q\to 0} [S_{11}(q)S_{22}(q)-S_{12}(q)^2]/S_{XX}(q)$$

$$= \frac{\sigma_1^d}{(\sigma_1^d n_1+\sigma_2^d n_2)^2} \lim_{q\to 0} [S_{\rho\rho}(q)-S_{\rho X}(q)^2/S_{XX}(q)]. \tag{3.7}$$

The first line was the expression in Ref. [48], and the second line follows if use is made of Eqs. (3.1) and (3.2). The dimensionless combination nk_BTK_{TX} is equal to 0.0028 in 2D and 0.0067 in 3D. If we assume that the adiabatic compressibility $K_{sX}=(\partial n/\partial p)_{sX}/n$ is of the same order as K_{TX}, the sound speed c turns out to be of order 10 in units of σ_1/τ_0.

Our structure factors were obtained by time averaging ovor very long times, which are 10^5 for 2D and 10^4 for 3D. However, irregular shapes of $S_{XX}(q)$ persisted at long wavelengths $q\lesssim 1$. Such large-scale composition fluctuations have very long lifetimes ($\gg\tau_\alpha$) and are virtually frozen throughout the simulation. Therefore, we admit the possibility that our supercooled states at low temperatures might phase-separate to form crystalline regions on much longer time scales. On the contrary, the long-wavelength fluctuations of $\hat{\rho}_{\rm eff}$ have much shorter time scales; probably they vary on acoustic time scales $\sim 1/cq$.

As is well known, the temperature dependence of the static pair correlation functions is much milder than that of the dynamical quantities. Similarly, their shear dependence is also mild even for $\dot{\gamma}\tau_\alpha \gg 1$ as long as $\dot{\gamma}\ll 1$. In particular, their spatially anisotropic part is at most a few percent of their isotropic part around the first peak positions $r\cong\sigma_{\alpha\beta}$ in our case. This is consistent with the fact that the attained shear stress in our simulations is at most a few percent of the particularly high pressure p of our systems. Note that the average shear stress σ_{xy} in sheared steady states may be related to the steady-state pair correlation functions $g_{\alpha\beta}(\mathbf{r})$ as [45]

$$\sigma_{xy} = -\frac{1}{2}\sum_{\alpha,\beta} n_\alpha n_\beta \int d\mathbf{r} v'_{\alpha\beta}(r)\frac{r_x r_y}{r} g_{\alpha\beta}(\mathbf{r}), \tag{3.8}$$

where r_x and r_y are the x and y components of the vector $\mathbf{r}$ connecting particle pairs. The dominant contribution here arises from the anisotropy at $r\cong\sigma_{\alpha\beta}$.

B. Bond breakage

Because of the sharpness of the first peak of $g_{\alpha\beta}(\mathbf{r})$ in our systems, we can unambiguously define *bonds* between particle pairs at distances close to $\sigma_{\alpha\beta}$ in the absence and presence of shear. Such bonds will be broken on the structural (α) relaxation time, because the bond breakage takes place on local configurational rearrangements. We define the bonds as follows. For each atomic configuration given at time t_0, a pair of particles i and j is considered to be bonded if

$$r_{ij}(t_0)=|\mathbf{r}_i(t_0)-\mathbf{r}_j(t_0)|\le A_1\sigma_{\alpha\beta}, \tag{3.9}$$

where i and j belong to the species α and β, respectively. We have set $A_1=1.1$ for 2D and 1.5 for 3D. The resultant bond numbers between α and β pairs, $N_{b\alpha\beta}$, are related to the first peak structure of $g_{\alpha\beta}(r)$ as follows. We consider the coordination number $\nu_{\alpha\beta}$ of β particles around an α particle within the distance $A_1\sigma_{\alpha\beta}$ [43],

$$\nu_{\alpha\beta}=n_\beta\int_{r<A_1\sigma_{\alpha\beta}} d\mathbf{r}\, g_{\alpha\beta}(r)\sim C n_\beta\sigma_{\alpha\beta}^d, \tag{3.10}$$

where C is about 5 in 2D and 12 in 3D. Then we simply have

$$N_{b\alpha\alpha}=\frac{1}{2}N_\alpha\nu_{\alpha\alpha}\quad(\alpha=1,2),\quad N_{b12}=\frac{1}{2}N_1\nu_{12}+\frac{1}{2}N_2\nu_{21}. \tag{3.11}$$

In 2D at $\Gamma_{\rm eff}=1.4$, we find $\nu_{11}=2.19$, $\nu_{12}=\nu_{21}=2.54$, and $\nu_{22}=3.41$, which are consistent with the bond numbers, $N_{b11}=5514$, $N_{b11}=13\,135$, and $N_{b22}=8436$, counted in a simulation. In 3D at $\Gamma_{\rm eff}=1.55$, these numbers are $\nu_{11}=5.57$, $\nu_{12}=\nu_{21}=6.90$, and $\nu_{22}=8.30$, which are again consistent with $N_{b11}=13\,925$, $N_{b11}=34\,476$, and $N_{b22}=20\,744$ in a simulation. We stress that our bond definition is insensitive to A_1, owing to the sharpness of the first peak, as long as it is somewhat larger than 1 and smaller than the second peak distances divided by $\sigma_{\alpha\beta}$.

After a lapse of time Δt, a pair is regarded to have been broken if

$$r_{ij}(t_0+\Delta t)>A_2\sigma_{\alpha\beta} \tag{3.12}$$

with $A_2=1.6$ for 2D and 1.5 for 3D. This definition of bond breakage is also insensitive to A_2 as long as $A_2\gtrsim A_1$ and $A_2\sigma_{\alpha\beta}$ is shorter than the second peak position of $g_{\alpha\beta}(r)$. We have followed the relaxation of the total surviving (unbroken) bonds $N_{\rm bond}(\Delta t)$ from the initial number

$$N_{\rm bond}(0)=N_{b11}+N_{b12}+N_{b22} \tag{3.13}$$

to zero with increasing Δt. No significant difference has been found between the bond breakage processes of the three kinds of bonds, 1-1, 1-2, and 2-2, so we consider their sum only. We define the bond breakage time τ_b by

$$N_{\rm bond}(\tau_b)=N_{\rm bond}(0)/e. \tag{3.14}$$

The relaxation is not simply exponential at low temperatures, apparently because of large-scale heterogeneities composed of relatively weakly and strongly bonded regions. If we fit $N_{\rm bond}(\Delta t)$ to the stretched exponential form, $N_{\rm bond}(\Delta t)\sim\exp[-(\Delta t/\tau_b)^{a'}]$, the exponent a' is close to 1 at relatively high temperatures but is considerably smaller than 1 at the lowest temperatures (for example, $a'\sim 0.6$ at $\Gamma_{\rm eff}=1.55$ in 3D).

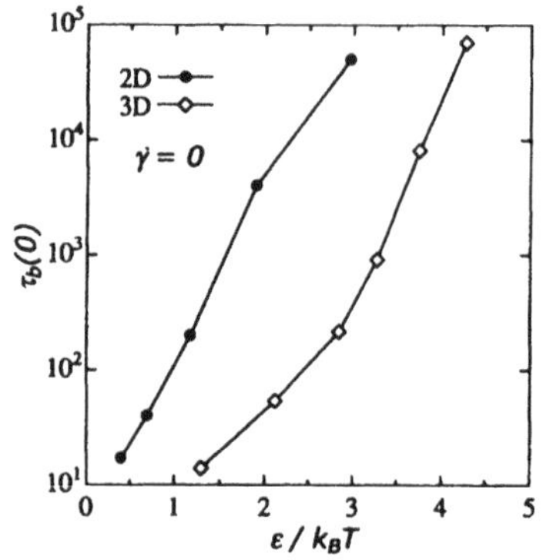

FIG. 4. Temperature dependence of the bond breakage time $\tau_b(0)$ at zero shear (●) in 2D and (◇) in 3D. The ϵ is the potential parameter in the soft-core potentials (2.1). The time is measured in units of τ_0 in Eq. (2.2), so $\tau_b(0)$ is dimensionless.

In Fig. 4 we show the bond breakage time $\tau_b = \tau_b(T)$ in the absence of shear as a function of the temperature. It grows strongly with decreasing the temperature. As will be shown in Eq. (6.8) in Sec. III, the bond breakage time τ_b is proportional to the α relaxation time τ_α obtained from the decay of the self-part of the time correlation function $F_s(q,t)$ at $q = 2\pi$. The shear dependence of the bond breakage time $\tau_b = \tau_b(\dot{\gamma})$ is also of great interest. As shown in Fig. 5, the bond breakage rate $1/\tau_b(\dot{\gamma})$ consists of the thermal breakage rate $1/\tau_b(0)$ strongly dependent on T and a shear-induced breakage rate proportional to $\dot{\gamma}$. It is expressed in the simplest conceivable form,

$$1/\tau_b(\dot{\gamma}) \cong 1/\tau_b(0) + A_b \dot{\gamma}, \tag{3.15}$$

where $A_b = 0.57$ in 2D and 0.80 in 3D. In the strong shear condition $\dot{\gamma}\tau_b(0) > 1$, jump motions are induced by shear on the time scale of $1/\dot{\gamma}$. We shall see that the bond breakage occurs more homogeneously with increasing shear. Therefore, it is natural that, when the strain $\gamma = \dot{\gamma}\Delta t$ reaches 1, a large fraction of bonds have been broken by shear.

IV. HETEROGENEITY IN BOND BREAKAGE

Following the bond breakage process, we can visualize the kinetic heterogeneity without ambiguity and quantitatively characterize the heterogeneous patterns. In Fig. 6 we show spatial distributions of broken bonds in a time interval of $[t_0, t_0 + 0.05\tau_b]$ in 2D, where about 5% of the initial bonds defined at $t = t_0$ have been broken. The dots are the center positions $\mathbf{R}_{ij} = \frac{1}{2}[\mathbf{r}_i(t_0) + \mathbf{r}_j(t_0)]$ of the broken pairs at the initial time t_0. The broken bonds are seen to form *clusters* with various sizes. The heterogeneity is marked in the glassy case (b) with $\Gamma_{\rm eff} = 1.4$ and $\dot{\gamma} = 0$, whereas it is much weaker for the liquid case (a) with $\Gamma_{\rm eff} = 1$ and $\dot{\gamma} = 0$. The bond breakage time τ_b is 17 in (a) and 5×10^4 in (b). In (c) we set $\dot{\gamma} = 0.25 \times 10^{-2}$ and $\Gamma_{\rm eff} = 1.4$ with $\tau_b = 32 \sim 1/\dot{\gamma}$. The heterogeneity is known to become much suppressed by shear, while its spatial anisotropy remains small. Notice that even in normal liquids, bond breakage events frequently occur in the form of strings involving a few to several particles, obviously because of the high density of our system. In glassy states, such strings become longer and aggregate, forming large-scale clusters. In 3D we also observe stringlike jump motions in accord with Ref. [17] and aggregation of such strings at low temperatures.

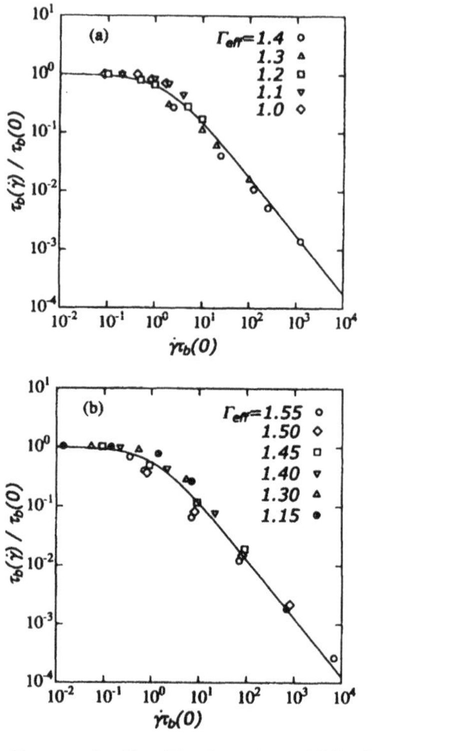

FIG. 5. The normalized bond breakage time $\tau_b(\dot{\gamma})/\tau_b(0)$ versus $\dot{\gamma}\tau_b(0)$ for various $\Gamma_{\rm eff}$ in 2D (a) and 3D (b). All the data collapse on the curve $1/(1 + A_b x)$ with $x = \dot{\gamma}\tau_b(0)$.

In Fig. 7 we write the broken bonds in two consecutive time intervals, $[t_0, t_0 + 0.05\tau_b]$ and $[t_0 + 0.05\tau_b, t_0 + 0.1\tau_b]$ at $\Gamma_{\rm eff} = 1.4$ and $\dot{\gamma} = 0$. The clusters of broken bonds in the two time intervals mostly overlap or are adjacent to one another. This demonstrates that *weakly bonded regions* or *collectively rearranging regions* (CRR) follow complex space-time evolution on the scales of ξ and τ_b. We do not know its evolution laws but will encounter a dynamical scaling law between ξ and τ_b in Eq. (4.4) below.

We define the structure factor $S_b(q)$ of the broken bonds as

$$S_b(q) = \frac{1}{N_b} \left\langle \left| \sum_{\langle i,j \rangle} \exp(i\mathbf{q} \cdot \mathbf{R}_{ij}) \right|^2 \right\rangle, \tag{4.1}$$

where the summation is over the broken pairs, N_b is the total number of the broken bonds in a time interval $[t_0, t_0 + \Delta t]$, and the angular average over the direction of the wave vector has been taken. Furthermore, we have averaged over 5–50 $S_b(q)$ data calculated from sequential configurations of broken bonds. Figure 8 displays the resultant $S_b(q)$ after these averaging procedures on logarithmic scales at several $\Gamma_{\rm eff}$

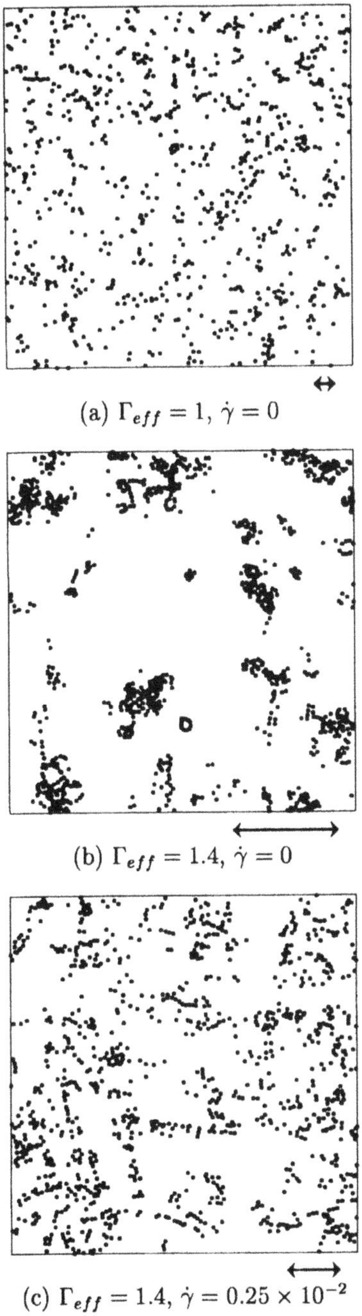

FIG. 6. Snapshots of the broken bonds in 2D without and with shear. The system length is $118\sigma_1$. Here $\Gamma_{eff}=1$ with weak heterogeneity (a), and $\Gamma_{eff}=1.4$ with enhanced heterogeneity (b). For $\dot{\gamma}=2.5\times 10^{-2}$ (c), the heterogeneity is much suppressed. The flow is in the upward direction and the velocity gradient is in the horizontal direction from left to right. The arrows indicate the correlation length ξ obtained from Eq. (4.2).

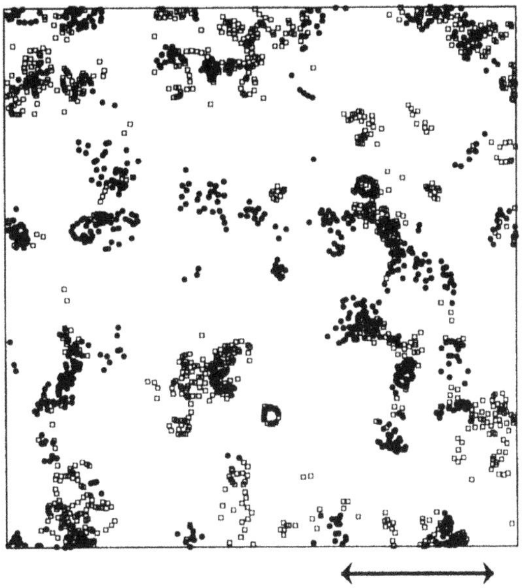

FIG. 7. Broken bond distributions in two consecutive time intervals, $[t_0, t_0+0.05\tau_b]$ (□) and $[t_0+0.05\tau_b, t_0+0.1\tau_b]$ (●), at $\Gamma_{eff}=1.4$ in 2D. The arrow indicates ξ.

without shear. The enhancement of $S_b(q)$ at small q arises from large-scale kinetic heterogeneities growing with increasing Γ_{eff} both in 2D and 3D. From a plot of $1/S_b(q)$ versus q^2 in our previous reports [13], we already found that $S_b(q)$ can be nicely fitted to the Ornstein-Zernike (OZ) form:

$$S_b(q)=S_b(0)/(1+\xi^2q^2). \tag{4.2}$$

The correlation length ξ is determined from this expression. It grows up to the system length at the lowest temperatures and is insensitive to the width of the time interval Δt as long as it is considerably shorter than the bond breakage time τ_b [13]. The agreement of our $S_b(q)$ with the OZ form becomes more evident in the plots of $S_b(q)/S_b(0)$ versus $q\xi$ in Fig. 9, in which all the data collapse onto a single OZ master curve both in 2D and 3D. In particular, in 3D the deviations are very small, although $\xi\sim L$ for low T and small $\dot{\gamma}$ in our case.

We also notice that $S_b(q)$ is insensitive to the temperature at large q, so from the OZ form (4.2) we find

$$S_b(0)\sim\xi^2. \tag{4.3}$$

The clusters of the broken bonds are thus very analogous to the critical fluctuations in Ising spin systems. In fact, small-scale heterogeneities with sizes l in the region $1\ll l\ll\xi$ are insensitive to the temperature. The relation (4.3) is analogous to the relation $\chi\propto\xi^{2-\eta}$ in Ising spin systems between the magnetic susceptibility $\chi=\lim_{q\to 0}S(q)$ and the correlation length ξ near the critical point. Here $S(q)$ is the spin structure factor and η is the Fisher critical exponent ($\ll 1$ in 3D).

Obviously, ξ represents the order of the maximum length of the clusters. However, Adam and Gibbs [5] intuitively expected that the *minimum* size of CRR increases as $\exp(\text{const}/(T-T_0))$ on lowering T towards T_0. It has also

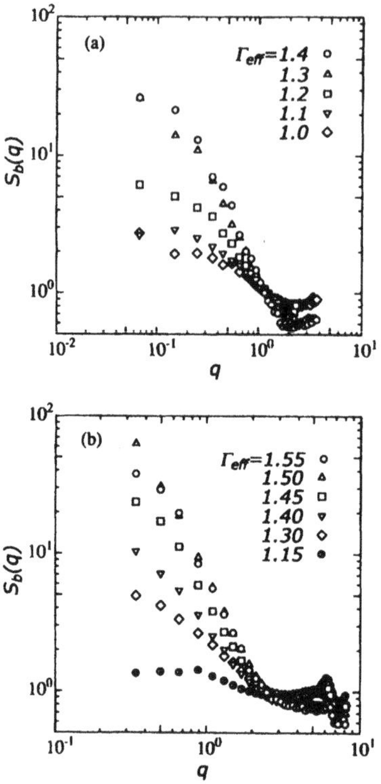

FIG. 8. $S_b(q)$ versus q on logarithmic scales for various $\Gamma_{\rm eff}$ at $\dot{\gamma}=0$ in 2D (a) and 3D (b). Its long-wavelength limit is of order ξ^2 as Eq. (4.3).

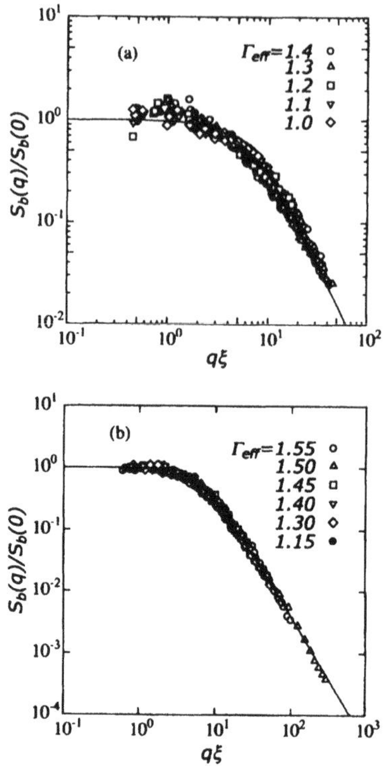

FIG. 9. $S_b(q)/S_b(0)$ on logarithmic scales for various $\Gamma_{\rm eff}$ and $\dot{\gamma}$ in 2D (a) and 3D (b). The solid line is the Ornstein-Zernike form $1/(1+x^2)$ with $x=q\xi$.

been discussed as to whether or not there is an underlying thermodynamic phase transition at a nonzero temperature T_0 in highly supercooled liquids [19–21]. From our data we cannot detect any divergence of ξ at a nonzero temperature, although this is not conclusive due to the finite-size effect arising from $\xi \sim L$.

Furthermore, as in critical dynamics, we have confirmed a dynamical scaling relation between the bond breakage time τ_b and the correlation length ξ,

$$\tau_b \cong A\xi^z, \tag{4.4}$$

where $z=4$ in 2D [41] and $z=2$ in 3D. The coefficient A is independent of $\Gamma_{\rm eff}$ and $\dot{\gamma}$ chosen in our simulations, as shown in Figs. 10(a) and 10(b). Notice that the data points at the largest ξ in Fig. 10 are those at zero shear for each $\Gamma_{\rm eff}$. At present we cannot explain the origin of these simple numbers for z. We may only argue that z should be larger in 2D than in 3D because of stronger configurational restrictions in 2D. It is surprising that Eq. (4.4) holds even in strong shear $\dot{\gamma}\tau_b(0) \gg 1$, where the correlation length is independent of T and is determined by shear as

$$\xi \sim \dot{\gamma}^{-1/z}. \tag{4.5}$$

In Fig. 10(b) for 3D, however, we notice $\xi > L$ at $\Gamma_{\rm eff} = 1.50$ and 1.55 for weak shear. At present we cannot assess the influence of this finite-size effect.

In a zeroth-order approximation, therefore, the kinetic heterogeneities are characterized by a single parameter, ξ or τ_b, owing to the small space anisotropy induced by shear in our systems. The shear rate $\dot{\gamma}$ is apparently playing a role similar to a magnetic field h in Ising spin systems. Thus, $\dot{\gamma}$ and T are two relevant external parameters in supercooled liquids, while h and the reduced temperature $(T-T_c)/T_c$ are two relevant scaling fields in Ising systems.

V. SUPERCOOLED LIQUID RHEOLOGY

We next examine nonlinear rheology in our fluid mixtures in supercooled amorphous states. We first display in Fig. 11 the shear-dependent viscosity $\eta(\dot{\gamma})$ (in units of $\epsilon\tau_0/\sigma_1^d$) versus $\dot{\gamma}$ in steady states at various $\Gamma_{\rm eff}$ in 2D and 3D. This rheological behavior is similar to those in the experiments [24–26]. The viscosity is much enhanced at large $\Gamma_{\rm eff}$ (low T) and at low shear, but it tends to be independent of T at very high shear. Remarkably, glassy states exhibit large non-Newtonian behavior even when $\dot{\gamma}$ is much smaller than the microscopic frequency $1/\tau_0=1$, whereas such large effects are expected to appear only for $\dot{\gamma} \sim 1/\tau_0$ in normal liquids far from the critical point [47,49].

In Fig. 12 we demonstrate that the viscosity $\eta(\dot{\gamma})$

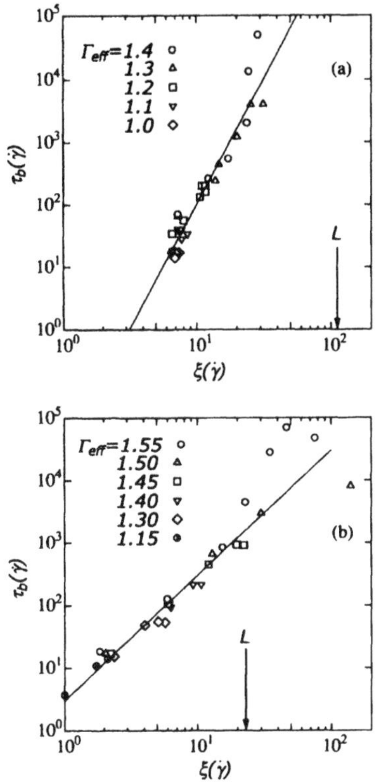

FIG. 10. Universal relation between the correlation length $\xi(\dot\gamma)$ in units of σ_1 and the bond breakage time $\tau_b(\dot\gamma)$ in units of τ_0 in Eq. (2.2). In 2D (a), the line of the slope 4 is a viewing guide and L is the system length. The corresponding 3D plot is shown in (b) with the slope being 2.

$=\sigma_{\alpha\beta}/\dot\gamma$ is determined solely by the bond breakage time $\tau_b(\dot\gamma)$ in Eq. (3.15) as

$$\eta(\dot\gamma)\cong A_\eta\tau_b(\dot\gamma)+\eta_B, \tag{5.1}$$

where A_η and η_B are 0.34 and 6.25 in 2D, and 0.24 and 2.2 in 3D, respectively. Because the linearity $\eta\propto\tau_b$ is systematically violated at small τ_b, the presence of the background viscosity η_B independent of $\Gamma_{\rm eff}$ and $\dot\gamma$ may be concluded. Note that the effective exponent $(\dot\gamma/\eta)(d\eta/d\dot\gamma)$ remains about -0.8 in Fig. 11. As well as the kinetic heterogeneities, steady state rheology is determined only by a single parameter, τ_b or ξ. This suggests that a sheared steady state can be fairly mapped onto a quiescent state with a higher temperature but with the same ξ.

Substitution of Eq. (3.15) then yields

$$\eta(\dot\gamma)\cong A_\eta/[\tau_b(0)^{-1}+A_b\dot\gamma]+\eta_B. \tag{5.2}$$

This form coincides with the empirical law (1.1) by Simons *et al.* [24,25]. Figure 13 shows that the ratio $[\eta(\dot\gamma)-\eta_B]/[\eta(0)-\eta_B]$ can be fitted to the universal curve $1/(1+A_bx)$ with $x=\dot\gamma\tau_b(0)$ independently of $\Gamma_{\rm eff}$ both in 2D and 3D. In strong shear $\dot\gamma\tau_b(0)\gg1$, we have the temperature-

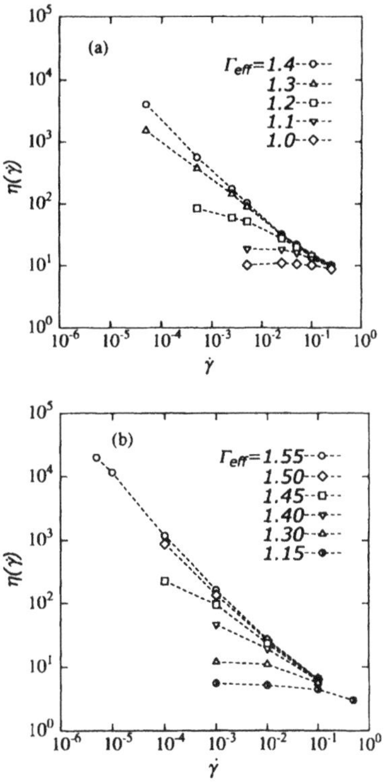

FIG. 11. The viscosity $\eta(\dot\gamma)$ in units of $\epsilon\tau_0/\sigma_1^d$ versus the shear rate $\dot\gamma$ in units of $1/\tau_0$ at various $\Gamma_{\rm eff}$ in 2D (a) and 3D (b). The data tend to become independent of $\Gamma_{\rm eff}$ at high shear.

independent behavior $\eta(\dot\gamma)\cong(A_\eta/A_b)/\dot\gamma+\eta_B$, which is evidently seen in Fig. 11. If the background viscosity is negligible, a constant limiting stress follows as

$$\sigma_{xy}\cong\sigma_{\rm lim}=A_\eta/A_b, \tag{5.3}$$

which holds for

$$1/\tau_b(0)\ll\dot\gamma\ll\sigma_{\rm min}/\eta_B\sim0.1/\tau_0. \tag{5.4}$$

Here, $\sigma_{\rm lim}$ is 0.59 in 2D and 0.30 in 3D in units of ϵ/σ_1^d and is typically a few percent of the pressure in our systems. The upper bound in Eq. (5.4) is very large in the usual glass-forming liquids but should be attainable in colloidal systems, while the lower bound can be very small with lowering T.

We will argue to derive the above behavior intuitively. Supercooled liquids behave as solids against infinitesimal strain on time scales shorter than $\tau_b(0)$ even if the temperature is considerably above the so-called glass transition temperature. Fluidlike behavior is realized only after the bond breakage processes. It is natural that the viscosity is of order $\tau_b(0)$ in the linear regime. This is usually justified from the time correlation function expression for the viscosity in terms of the xy component of the stress tensor [45]. In strong shear, on the other hand, the bond breakage occurs on the time scale of $1/\dot\gamma$. Upon each bond breakage induced by shear, the particles involved release a potential energy ϵ_r

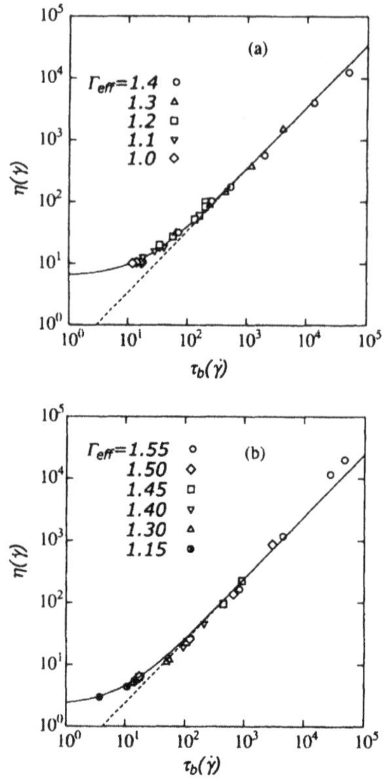

FIG. 12. $\eta(\dot\gamma)$ versus $\tau_b(\dot\gamma)$ for various $\Gamma_{\rm eff}$ in 2D (a) and 3D (b). The $\eta(\dot\gamma)$ is determined by $\tau_b(\dot\gamma)$ only irrespective of $\Gamma_{\rm eff}$.

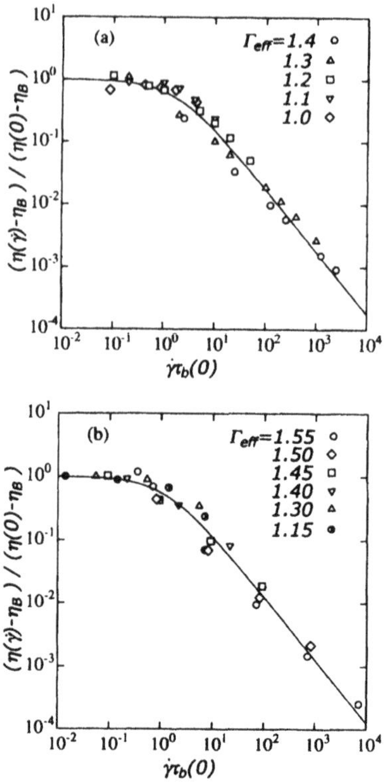

FIG. 13. $[\eta(\dot\gamma)-\eta_B]/[\eta(0)-\eta_B]$ vs $\dot\gamma\tau_b(0)$ in 2D (a) and 3D (b). The solid curve is $1/(1+A_b x)$ with $x=\dot\gamma\tau_b(0)$.

whose maximum is ϵ. There should be a distribution of ϵ_r, but let us assume $\epsilon_r \sim \epsilon$ for simplicity. It is then instantaneously changed into energies of random motions (and probably sounds) supported by the surrounding particles. The heat transport is rapid in this dissipative process. Because of this and also because of the background thermal motions superposed, we have not detected clear temperature inhomogeneities such as *hot spots* around broken bonds in our simulations. The heat production rate is estimated as

$$Q \sim n\epsilon/\tau_b(\dot\gamma) \sim n\epsilon\dot\gamma, \tag{5.5}$$

where n is the number density. Because Q is related to the viscosity by $Q=\sigma_{xy}\dot\gamma=\eta(\dot\gamma)\dot\gamma^2$, we obtain

$$\sigma_{xy}=\eta(\dot\gamma)\dot\gamma \sim n\epsilon \tag{5.6}$$

in high shear, so $\sigma_{\rm lim} \sim n\epsilon$. Due to disordered particle configurations, however, it is natural to consider a distribution of the released energy ϵ_r, which will explain the viscosity behavior at lower shear. Such a distribution was calculated for a model foam system in shear flow by Durian [28].

VI. MOTION OF TAGGED PARTICLES

In this section we will follow the motion of tagged particles in a glassy matrix both in the absence and presence of shear in 3D. We will present results only in three dimensions. We first plot in Fig. 14 the self-part of the density time correlation function for various $\Gamma_{\rm eff}$ in the usual zero shear condition,

$$F_s(q,t)=\frac{1}{N_1}\left\langle \sum_{j=1}^{N_1} \exp[i\boldsymbol{q}\cdot\Delta\boldsymbol{r}_j(t)] \right\rangle, \tag{6.1}$$

where $q=2\pi$, $\Delta\boldsymbol{r}_j(t)=\boldsymbol{r}_j(t)-\boldsymbol{r}_j(0)$, and the summation is taken over all the particles of the species 1. This function is

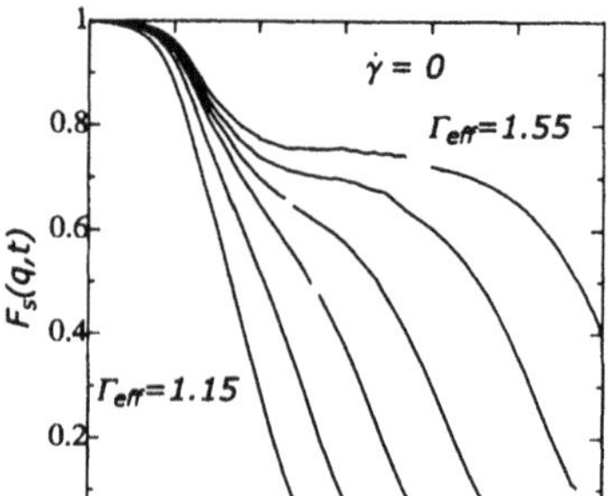

FIG. 14. The self-part of the intermediate scattering function $F_s(q,t)$ at $q=2\pi$ and $\dot\gamma=0$ in 3D. $\Gamma_{\rm eff}$ increases from left.

proportional to the (incoherent) scattering amplitude from labeled particles. As is well known, this function has a plateau at low temperatures ($\Gamma_{eff} \gtrsim 1.45$ in our case), during which the particle is trapped in a cage. After a long time the cage eventually breaks, resulting in diffusion with a very small diffusion constant D. In this paper we define the α relaxation time τ_α such that $F_s(q,\tau_\alpha)=e^{-1}$ at $q=2\pi$.

We generalize the time correlation function (6.1) in the presence of shear flow by introducing a new displacement vector of the jth particle as

$$\Delta \boldsymbol{r}_j(t)=\boldsymbol{r}_j(t)-\dot{\gamma}\int_0^t dt' y_j(t')\boldsymbol{e}_x-\boldsymbol{r}_j(0), \qquad (6.2)$$

where $\boldsymbol{e}_x$ is the unit vector in the x (flow) direction. In this displacement, the contribution from convective transport by the average flow has been subtracted, which can be known from the time derivative,

$$\frac{\partial}{\partial t}\Delta \boldsymbol{r}_j(t)=\boldsymbol{v}_j(t)-\dot{\gamma}y_j(t)\boldsymbol{e}_x. \qquad (6.3)$$

To get a clear understanding of the meaning of this subtraction, let us consider a Brownian particle placed in shear flow as a simple example. On time scales longer than the relaxation time of its velocity, its position $\boldsymbol{r}(t)$ obeys

$$\frac{\partial}{\partial t}\boldsymbol{r}(t)=\dot{\gamma}y(t)\boldsymbol{e}_x+\boldsymbol{f}(t), \qquad (6.4)$$

where $\boldsymbol{f}(t)$ is the Gaussian random force characterized by $\langle f_\mu(t)f_\nu(t')\rangle=2D\delta_{\mu\nu}\delta(t-t')$ $(\mu,\nu=x,y,z)$. Then the modified displacement vector reads

$$\Delta \boldsymbol{r}(t)\equiv\boldsymbol{r}(t)-\dot{\gamma}\int_0^t dt' y(t')\boldsymbol{e}_x-\boldsymbol{r}(0)=\int_0^t dt'\boldsymbol{f}(t'). \qquad (6.5)$$

Here the convective effect does not appear explicitly and the diffusion behavior follows as

$$\langle \Delta \boldsymbol{r}(t)^2\rangle=6Dt. \qquad (6.6)$$

On the other hand, in the incoherent scattering amplitude, $\Delta \boldsymbol{r}_j(t)$ in Eq. (6.1) should be taken as the net displacement $\boldsymbol{r}_j(t)-\boldsymbol{r}_j(0)$ even in shear flow. If $q_x\neq 0$, it strongly depends on the thickness of the scattering region in the y (velocity gradient) direction due to a position-dependent Doppler effect [23,50]. Only for $q_x=0$, it is proportional to $F_s(q,t)$ in the above definition.

Figure 15 shows $F_s(q,t)$ at $q=2\pi$ for various $\dot{\gamma}$ with a fixed temperature, $\Gamma_{eff}=1.5$ or $k_BT/\epsilon=0.267$ in 3D. Comparison of this figure with Fig. 14 suggests that applying shear is equivalent to raising the temperature. Recall that we have made the same statement in analyzing the bond structure factor $S_b(q)$ and the nonlinear rheology. Also, we may define the shear-dependent α relaxation time $\tau_\alpha=\tau_\alpha(\dot{\gamma})$ by

$$F_s(q,\tau_\alpha)=e^{-1}. \qquad (6.7)$$

In Fig. 16 we recognize that τ_α is proportional to the bond lifetime τ_b as

$$\tau_\alpha\cong 0.1\tau_b. \qquad (6.8)$$

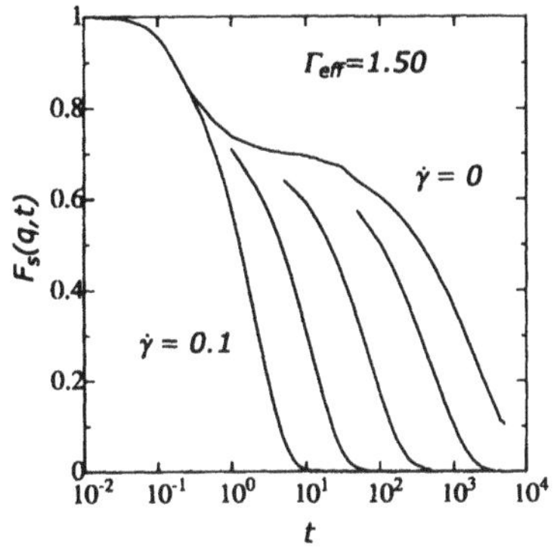

FIG. 15. The time correlation function $F_s(q,t)$ at $q=2\pi$ defined by Eqs. (6.1) and (6.2) in shear flow, where $\dot{\gamma}=0$, 10^{-4}, 10^{-3}, 10^{-2}, and 10^{-1} from the right. The temperature is fixed at $k_BT/\epsilon=0.267$ ($\Gamma_{eff}=1.5$). Increasing $\dot{\gamma}$ is equivalent to raising T.

This relation holds for any Γ_{eff} and $\dot{\gamma}$ in our 3D simulations. The decay of $F_s(q,t)$ is not exponential for large τ_α. If it is fitted to the stretched exponential form $\exp[-(t/\tau_\alpha)^a]$ around $t\sim\tau_\alpha$, the exponent a is increased from values about 0.8 to 1 with increasing $\dot{\gamma}$ as well as with raising T. Furthermore, the time correlation function (6.1) has turned out to be almost independent of the direction of the wave vector $\boldsymbol{q}$.

Next, it is convenient to analyze the mean square displacement of tagged particles of the species 1,

$$\langle[\Delta \boldsymbol{r}(t)]^2\rangle=\frac{1}{N_1}\sum_{j=1}^{N_1}\langle[\Delta \boldsymbol{r}_j(t)]^2\rangle. \qquad (6.9)$$

Figure 17 shows the transition from the ballistic behavior $\langle[\Delta \boldsymbol{r}(t)]^2\rangle\cong 3(k_BT/m_1)t^2$ to the diffusion behavior $\langle[\Delta \boldsymbol{r}(t)]^2\rangle\cong 6Dt$ in shear flow at $\Gamma_{eff}=1.5$. The arrows in the figure indicate the α relaxation time $\tau_\alpha(\dot{\gamma})$. The diffusion behavior is almost attained at $t\sim\tau_\alpha$. Figure 18 demonstrates the surprising isotropy of the statistical distribution of $\Delta r_i(t)$, where the mean square displacements of the x, y, and z components of the vector $\Delta \boldsymbol{r}_j(t)$ are separately displayed. We can thus determine D from the mean square displace-

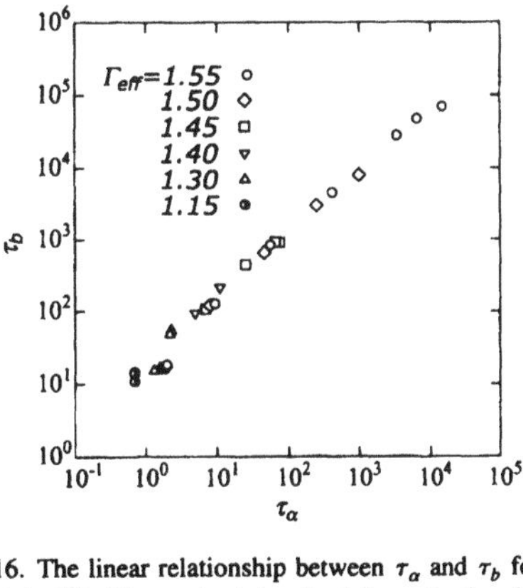

FIG. 16. The linear relationship between τ_α and τ_b for various Γ_{eff} and $\dot{\gamma}$ in 3D. The τ_b is determined from the bond breakage (3.14), and τ_α from the decay of the time correlation function (6.7).

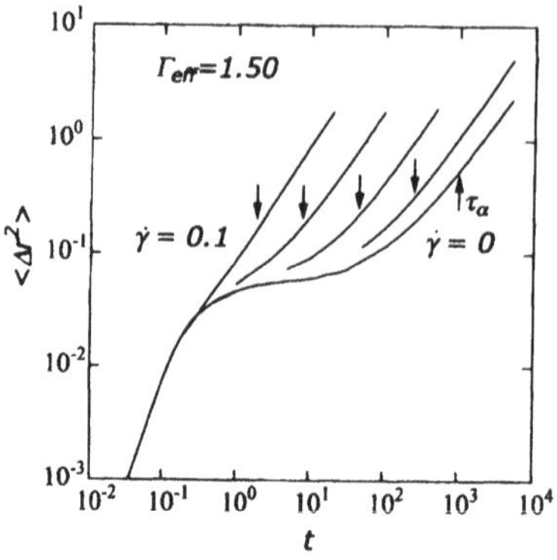

FIG. 17. The mean square displacement in sheared states at $\Gamma_{\rm eff}=1.5$. The shear rate $\dot\gamma$ is 0, 10^{-4}, 10^{-3}, 10^{-2}, and 10^{-1} from the right. Increasing $\dot\gamma$ is equivalent to raising T. The arrows indicate τ_α for each $\dot\gamma$. The diffusion (linear) behavior is attained at $t \sim \tau_\alpha$.

ment in addition to τ_α in shear flow. Note that the x component in Fig. 18 is not the usual mean square displacement due to the second term in Eq. (6.2). In the Appendix we will consider the variances of the net displacement vector $\boldsymbol{r}_j(t) - \boldsymbol{r}_j(0)$.

Figure 19 shows the shear rate dependences of the viscosity η ($\sim \tau_\alpha$) and the inverse diffusion constant D^{-1} from the linear ($\dot\gamma \lesssim 10^{-5}$) to the non-Newtonian regime at $\Gamma_{\rm eff} = 1.55$ in 3D, where D is measured in units of σ_1^2/τ_0 and η in units of $\epsilon\tau_0/\sigma_1^d$. We deduce the relation $D^{-1} \sim \dot\gamma^{-\nu}$ with $\nu = 0.75$–0.80 in agreement of the experiment [36], which is appreciably milder than the viscosity decrease $\eta \sim \tau_\alpha \sim \dot\gamma^{-1}$. In Fig. 20, we plot D versus η/k_BT (in units of τ_0/σ_1^d) obtained for various $\Gamma_{\rm eff}$ and $\dot\gamma$. The Einstein-Stokes formula, which holds excellently in normal liquids, appears to be violated in supercooled liquids as the other simulations have suggested [40,41]. It is widely believed that this breakdown is a natural consequence of the dynamic heterogeneity in glassy states [36–38]. Detailed numerical analysis will appear in a forthcoming paper.

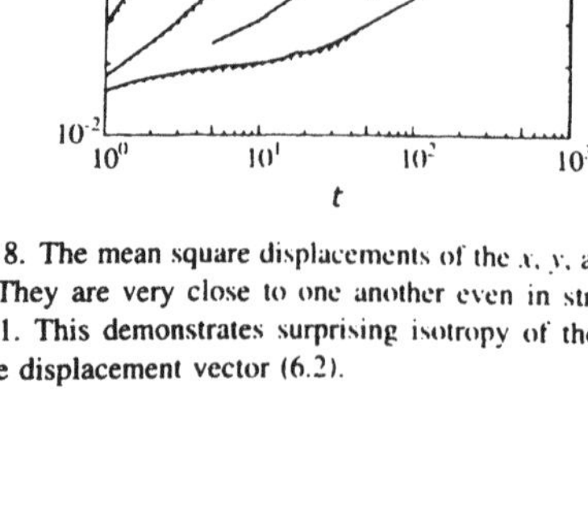

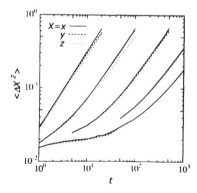

FIG. 18. The mean square displacements of the x, y, and z components. They are very close to one another even in strong shear $\dot\gamma\tau_b(0) \gg 1$. This demonstrates surprising isotropy of the distribution of the displacement vector (6.2).

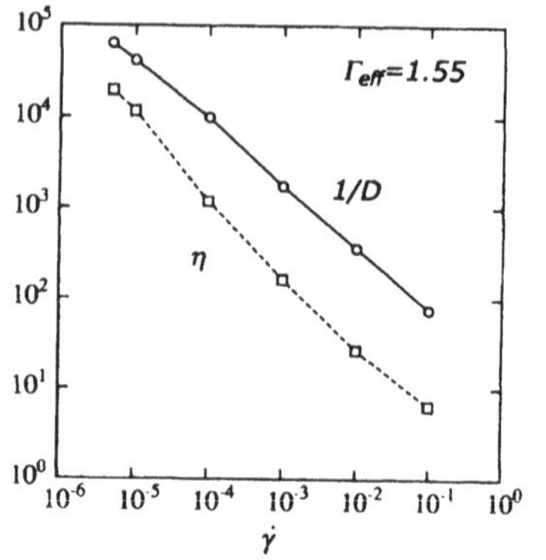

FIG. 19. Shear rate dependences of the inverse diffusion constant D and the viscosity η at the lowest temperature, $\Gamma_{\rm eff}=1.55$. The slope of D^{-1} is noticeably smaller than that of η.

In our case η/k_BT changes over 4 decades until ξ reaches the system dimension L, whereas it has been changed over 12 decades in the experiments [36,37]. Though the same tendency indicating the breakdown of the Einstein-Stokes relation has been obtained in our simulation, we should admit that our system size in 3D is not yet sufficiently large and our data at $\Gamma_{\rm eff}=1.5$ and 1.55 might be somewhat affected by the system size effect. It is worth noting that the Monte Carlo simulation of a dense polymer by Ray and Binder [18] shows that the monomer diffusion constant decreases with increasing system size.

VII. SUMMARY AND DISCUSSIONS

Most of our findings in this work have been obtained from numerical analysis only without first-principles derivations. Nevertheless, we believe that they pose new problems and suggest new experiments. We make some discussions mentioning possible experiments below.

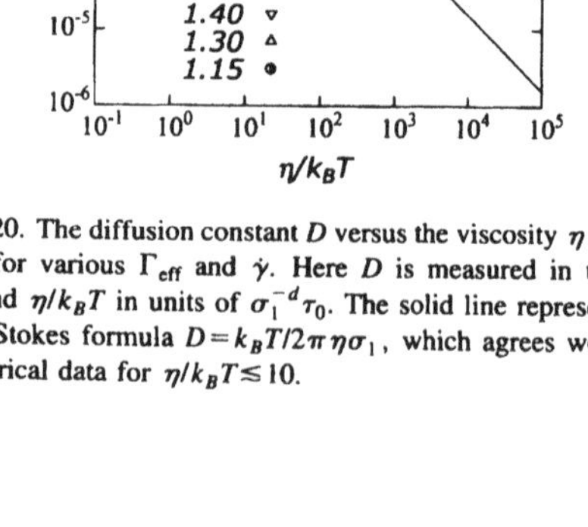

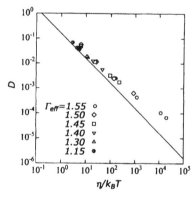

FIG. 20. The diffusion constant D versus the viscosity η divided by k_BT for various $\Gamma_{\rm eff}$ and $\dot\gamma$. Here D is measured in units of $\sigma_1^2\tau_0^{-1}$ and η/k_BT in units of $\sigma_1^{-d}\tau_0$. The solid line represents the Einstein-Stokes formula $D = k_BT/2\pi\eta\sigma_1$, which agrees well with the numerical data for $\eta/k_BT \lesssim 10$.

(i) Introducing the concept of bond breakage, we have succeeded to quantitatively analyze the kinetic heterogeneities in simple model systems, which have been witnessed by a number of the authors. As shown in Fig. 6, strings composed of broken bonds are very frequent and they aggregate at low temperatures to form clusters. The bond breakage time τ_b is related to the correlation length ξ as Eq. (4.4). In future work we should clarify the relationship of our patterns in the α relaxation and those by Muranaka and Hiwatari [9] on a much shorter time scale.

(ii) The weakly bonded regions identified by the bond breakage are purely dynamical objects. Large-scale heterogeneities have not been clearly detected in snapshots of the usual physical quantities such as the densities, the stress tensor, the kinetic energy (=temperature), etc. On the other hand, in granular matters in shear flow [30], stress heterogeneities have been observed optically by using birefringent materials. We admit the possibility that such stress heterogeneities also exist in supercooled liquids but are masked by the thermal fluctuations. We will check this point in the future.

(iii) It is of great interest how the kinetic heterogeneities, which satisfy the dynamic scaling (4.4), evolve in space and time and why they look so similar to the critical fluctuations in Ising systems in the mean field level. In our steady-state problem T and $\dot{\gamma}$ are two relevant scaling fields, the *critical point* being located at $T=\dot{\gamma}=0$. No divergence has been detected at a nonzero temperature in our simulations.

(iv) In his experiments, Fischer [51] has reported large excess light scattering with a correlation length ξ (20–200 nm) that increases on approaching the glass transition from a liquid state. This indicates the presence of very large-scale *density* heterogeneities in supercooled liquids, which is often called Fischer's clusters. Motivated by this effect, Weber *et al.* [52] performed Monte Carlo simulations on a dense polymer and found that short-range nematic orientational order can give rise to enhancement of long-range density fluctuations. They expected that such anisotropic interactions could be the origin of Fischer's clusters. This suggests that Fischer's clusters do not exist in liquids composed of structureless particles.

(v) We have examined nonlinear rheology in glassy states. The rheological relations obtained are simplest among those consistent with the experiments [24–26]. The mechanism of the non-Newtonian behavior in supercooled liquids is conceptually new and should be further examined in experiments such as in colloidal systems in glassy states. In particular, polymers should exhibit pronounced non-Newtonian behavior, as the glass transition is approached, even without entanglement. Rheology of chain systems remains totally unexplored near the glass transition.

(vi) In our systems, small anisotropic changes of the pair correlation functions $g_{\alpha\beta}(r)$ near the first peak ($\sim\sigma_{\alpha\beta}$) can give rise to the limiting shear stress σ_{lim}, which is 3–5 % of the pressure in our case. Note that our systems are highly compressed with high pressure. However, the pressure need not be very high in supercooled liquids in the presence of an attractive part of the potential. Even in such cases, we expect that σ_{lim} is a few percent of the shear modulus. This is suggested by the previous work on amorphous alloys [31–35], where the yield stress σ_y in the inhomogeneous case (in which shear bands appear) is known to be 2–3 % of the shear modulus.

(vii) Stillinger expected that in fragile glass-forming liquids shear flow occurs by *tear and repair* of slipping walls separating strongly bonded regions [7]. We have not observed such localization of slips or jump at least in our temperature range. But there might be a tendency that broken bonds form surfaces at low temperatures in 3D, though not conspicuous, which should be checked in the future.

(viii) There is no tendency of phase separation for the parameters used. However, there are many cases in which the composition fluctuations are enhanced towards the glass transition temperature. It is of great interest how the two transitions influence each other [53,54]. It is also known that shear flow can induce composition fluctuation enhancement in asymmetric viscoelastic mixtures, when emergence of less viscous regions can reduce the effective viscosity [23]. We expect that this effect can come into play also in supercooled liquids, for example, for large enough size ratios or in the presence of small attraction between the two components. Experiments to detect this effect seem to be promising in colloidal systems.

(ix) We have introduced the time correlation function $F_s(q,t)$ in shear and found its simple relaxation behavior in Fig. 15. It coincides with the usual time correlation function for $q_x=0$ or when the scattering vector is perpendicular to the flow direction. Dynamic scattering experiments in shear flow would be very informative to detect the shear-induced diffusion [23]. A direct diffusion measurement in sheared supercooled fluids, which will be analyzed in the Appendix, is also very interesting. Though our system size is still too small, we have detected a tendency of the breakdown of the Einstein-Stokes relation in 3D to obtain $D\sim\eta^{-\nu}$ with $\nu=0.75$–0.8.

(x) In strong shear the structural relaxation is characterized by $\tau_\alpha\sim 0.1\tau_b\sim 0.1/\dot{\gamma}$ as Eq. (6.8). This nonlinear effect could be measured as a drastic reduction of the rotational relaxation time by dielectric response or by more sophisticated techniques [36,37] from sheared supercooled liquids. The same effect is expected for periodic shear flow.

(xi) Understanding transient mechanical response in terms of the kinetic heterogeneities is of great importance. For example, we have found a stress overshoot after application of shear strain in accord with the experiments [24,25]. We should also understand glassy behavior of the complex shear modulus against small periodic shear [55]. We will report on these topics shortly.

(xii) In our systems, we have not yet found essential differences between 2D and 3D except for the difference in the value of the dynamic exponent z in Eq. (4.4). We believe that a large part of the essential ingredients of glassy dynamics can be understood even in two dimensions.

(xiii) In a forthcoming paper, we will focus our attention on jump motions of particles over distances longer than σ_1. They will be shown to occur heterogeneously in space and determine the diffusion constant. These heterogeneity structures are essentially the same as those in the bond breakage processes studied in this paper.

ACKNOWLEDGMENTS

We thank Dr. T. Muranaka, Professor M. Y. Hiwatari, Professor K. Kawasaki, Professor P. Harrowell, Dr. D. Perera, and Professor A.J. Liu for helpful discussions. Thanks are also due to Professor S. Takeuchi, who kindly sent us his work on amorphous alloys. This work is supported by a Grant-in-Aid for Scientific Research from the Ministry of Education, Science and Culture. Calculations have been carried out at the Supercomputer Laboratory, Institute for Chemical Research, Kyoto University and the Computer Center of the Institute for Molecular Science, Okazaki, Japan.

APPENDIX

Let us calculate the variances among the x, y, and z components, $x_j(t)-x_j(0)$, $y_j(t)-y_j(0)$, and $z_j(t)-z_j(0)$, of the net displacement vector $\boldsymbol{r}_j(t)-\boldsymbol{r}_j(0)$ of the jth particle in shear flow. We fix its initial position $\boldsymbol{r}_j(0)$ at $\boldsymbol{r}_0=(x_0,y_0,z_0)$. The average displacement arises from the convection as

$$\langle \boldsymbol{r}_j(t)-\boldsymbol{r}_j(0)\rangle = \dot{\gamma} t y_0 \boldsymbol{e}_x. \tag{A1}$$

Assuming the isotropy of the subtracted displacement (6.2), which is suggested by Fig. 18, we may write the variances of the y and z components as

$$G(t)=\langle [y_j(t)-y_j(0)]^2\rangle = \langle [z_j(t)-z_j(0)]^2\rangle. \tag{A2}$$

The variance of the x component then becomes

$$\langle [x_j(t)-x_j(0)-\dot{\gamma} t y_0]^2\rangle = G(t)+2\dot{\gamma}^2\int_0^t dt_1(t-t_1)G(t_1). \tag{A3}$$

The cross correlation exists between the x and y components as

$$\langle [x_j(t)-x_j(0)][y_j(t)-y_j(0)]\rangle = \dot{\gamma}\int_0^t dt_1 G(t_1). \tag{A4}$$

In the diffusion time regime $t \gtrsim \tau_\alpha$ we may set $G(t)=2Dt$ to obtain

$$\langle [x_j(t)-x_j(0)-\dot{\gamma} t y_0]^2\rangle \cong 2Dt(1+\tfrac{1}{3}\dot{\gamma}^2 t^2), \tag{A5}$$

$$\langle [x_j(t)-x_j(0)][y_j(t)-y_j(0)]\rangle \cong D\dot{\gamma} t^2. \tag{A6}$$

Note that D is strongly dependent on $\dot{\gamma}$ in strong shear as shown in Fig. 19. Measurements of the above variances are very informative.

[1] J. Jäckle, Rep. Prog. Phys. **49**, 171 (1986).
[2] M. D. Ediger, C. A. Angell, and S. R. Nagel, J. Phys. Chem. **100**, 13 200 (1996).
[3] U. Bengtzelius, W. Götze, and A. Sjölander, J. Phys. C **17**, 5915 (1984).
[4] E. Leutheusser, Phys. Rev. A **29**, 2765 (1984).
[5] G. Adam and J. H. Gibbs, J. Chem. Phys. **43**, 139 (1965).
[6] M. H. Cohen and G. S. Grest, Phys. Rev. B **20**, 1077 (1979).
[7] F. H. Stillinger, J. Chem. Phys. **89**, 6461 (1988).
[8] K. L. Ngai, R. W. Rendell, and D. J. Plazek, J. Chem. Phys. **94**, 3018 (1991).
[9] T. Muranaka and Y. Hiwatari, Phys. Rev. E **51**, R2735 (1995).
[10] T. Muranaka and Y. Hiwatari, Prog. Theor. Phys. Suppl. **126**, 403 (1997).
[11] M. M. Hurley and P. Harrowell, Phys. Rev. E **52**, 1694 (1995).
[12] D. N. Perera and P. Harrowell, Phys. Rev. E **54**, 1652 (1996); Proceedings for the Vigo Meeting, Vigo, 1997 [J. Non-Cryst. Solids (to be published)].
[13] R. Yamamoto and A. Onuki, J. Phys. Soc. Jpn. **66**, 2545 (1997).
[14] R. Yamamoto and A. Onuki, Europhys. Lett. **40**, 61 (1997).
[15] A. Onuki and R. Yamamoto, Proceedings for the Vigo Meeting, Vigo, 1997 (Ref. [12]); *Proceedings of the 2nd Tohwa University International Meeting "Statistical Physics,"* edited by M. Tokuyama and I Oppenheim (World Scientific, Singapore, 1998).
[16] W. Kob, C. Donati, S. J. Plimton, P. H. Poole, and S. C. Glotzer, Phys. Rev. Lett. **79**, 2827 (1997).
[17] C. Donati, J. F. Douglas, W. Kob, S. J. Plimton, P. H. Poole, and S. C. Glotzer, Phys. Rev. Lett. **80**, 2338 (1998).
[18] P. Ray and K. Binder, Europhys. Lett. **27**, 53 (1994).
[19] C. Dasgupta, A. V. Indrani, S. Ramaswamy, and M. K. Phani, Europhys. Lett. **15**, 307 (1991).
[20] R. M. Ernst, S. R. Nagel, and G. S. Grest, Phys. Rev. B **43**, 8070 (1991).
[21] S. S. Ghosh and C. Dasgupta, Phys. Rev. Lett. **77**, 1310 (1996).
[22] D. J. Salvino, S. Rogge, B. Tigner, and D. D. Osheroff, Phys. Rev. Lett. **73**, 268 (1994); S. Rogge, D. Natelson, and D. D. Osheroff, J. Low Temp. Phys. **106**, 767 (1997).
[23] A. Onuki, J. Phys. C **9**, 6119 (1997).
[24] J. H. Simmons, R. K. Mohr, and C. J. Montrose, J. Appl. Phys. **53**, 4075 (1982).
[25] J. H. Simmons, R. Ochoa, K. D. Simmons, and J. J. Mills, J. Non-Cryst. Solids **105**, 313 (1988).
[26] Y. Yue and R. Brückner, J. Non-Cryst. Solids **180**, 66 (1994).
[27] T. Okuzono and K. Kawasaki, Phys. Rev. E **51**, 1246 (1995).
[28] D. J. Durian, Phys. Rev. E **55**, 1739 (1997).
[29] S. A. Langer and A. J. Liu, J. Phys. Chem. B **101**, 8667 (1997).
[30] B. Miller, C. O'Hern, and R. P. Behringer, Phys. Rev. Lett. **77**, 3110 (1996); R. Khosropour, J. Zirinsky, H. K. Pak, and R. P. Behringer, Phys. Rev. E **56**, 4467 (1997).
[31] H. S. Chen and M. Goldstein, J. Appl. Phys. **43**, 1642 (1971).
[32] F. Spaepen, Acta Metall. **25**, 407 (1977).
[33] A. S. Aragon, Acta Metall. **27**, 47 (1979).
[34] K. Maeda and S. Takeuchi, Phys. Status Solidi A **49**, 685 (1978); S. Kobayashi, K. K. Maeda, and S. Takeuchi, Acta Metall. **28**, 1641 (1980); K. Maeda and S. Takeuchi, Philos. Mag. A **44**, 643 (1981).
[35] S. Takeuchi and K. Maeda, in *Metallic and Semiconducting Glasses*, edited by A. Bhatnagar (Trans. Tech Pub., Aedermannsdorf, 1987), and references quoted therein.
[36] F. Fujara, B. Geil, H. Sillescu, and G. Fleischer, Z. Phys. B **88**, 195 (1992); I. Chang, F. Fujara, B. Geil, G. Heuberger, T. Mangel, and H. Sillescu, J. Non-Cryst. Solids **172-174**, 248 (1994).

[37] M. T. Cicerone, F. R. Blackburn, and M. D. Ediger, Macromolecules **28**, 8224 (1995); M. T. Cicerone and M. D. Ediger, J. Chem. Phys. **104**, 7210 (1996).
[38] F. H. Stillinger and A. Hodgdon, Phys. Rev. E **50**, 2064 (1994).
[39] D. Richter, R. Frick, and B. Farago, Phys. Rev. Lett. **61**, 2465 (1988).
[40] D. Thirumalai and R. D. Mountain, Phys. Rev. E **47**, 479 (1993).
[41] D. Perera and P. Harrowell, Phys. Rev. Lett. **81**, 120 (1998). They have also reproduced the dynamical scaling exponent $z \cong 4$ in a 2D simulation using a different definition of ξ.
[42] J. Matsui, T. Odagaki, and Y. Hiwatari, Phys. Rev. Lett. **73**, 2452 (1994).
[43] B. Bernu, Y. Hiwatari, and J. P. Hansen, J. Phys. C **18**, L371 (1985); B. Bernu, J. P. Hansen, Y. Hiwatari, and G. Pastore, Phys. Rev. A **36**, 4891 (1987).
[44] D. Henderson and P. J. Lenard, *Physical Chemistry, an Advanced Treatise*, edited by D. Henderson (Academic, London, 1971), p. 414.
[45] J. P. Hansen and I. R. McDonald, *Theory of Simple Liquids* (Academic, London, 1986).
[46] M. P. Allen and D. J. Tildesley, *Computer Simulation of Liquids* (Clarendon, Oxford, 1987).
[47] D. J. Evans and G. P. Morriss, *Statistical Mechanics of Nonequilibrium Liquids* (Academic, London, 1990).
[48] J. G. Kirkwood and F. P. Buff, J. Chem. Phys. **19**, 774 (1951).
[49] H. J. M. Hanley, D. J. Evans, and S. Hess, J. Chem. Phys. **78**, 1440 (1983).
[50] A. Onuki and K. Kawasaki, Phys. Lett. **72A**, 233 (1979); B. J. Ackerson and N. A. Clark, J. Phys. (France) **42**, 929 (1981).
[51] E. W. Fischer, Physica A **201**, 183 (1993).
[52] H. Weber, W. Paul, W. Kob, and K. Binder, Phys. Rev. Lett. **78**, 2136 (1997).
[53] A. Sappelt and J. Jäckle, Physica A **240**, 453 (1997).
[54] G. Meier, D. Vlassopoulos, and G. Fytas, Europhys. Lett. **30**, 325 (1995).
[55] N. Menon, S. R. Nagel, and D. C. Venerus, Phys. Rev. Lett. **73**, 963 (1994).

Part 7

UNSTEADY RHEOLOGY

Liquid to solidlike transitions of molecularly thin films under shear

Michelle L. Gee, Patricia M. McGuiggan, and Jacob N. Israelachvili
Department of Chemical and Nuclear Engineering, and Materials Department, University of California, Santa Barbara, California 93106

Andrew M. Homola
IBM Almaden Research Center, San Jose, California 95120

(Received 27 November 1989; accepted 26 April 1990)

We have measured the shear forces between two molecularly smooth solid surfaces separated by thin films of various organic liquids. The aim was to investigate the nature of the transitions from continuum to molecular behavior in very thin films. For films whose thickness exceeds ten molecular diameters both their static and dynamic behavior can usually be described in terms of their bulk properties, but for thinner films their behavior becomes progressively more solidlike and can no longer be described, even qualitatively, in terms of bulk/continuum properties such as viscosity. The solidlike state is characterized by the ordering of the liquid molecules into discrete layers. The molecular ordering is further modified by shear, which imposes a preferred orientation. All solidlike films exhibit a yield point or critical shear stress, beyond which they behave like liquid crystals or ductile solids undergoing plastic deformation. Our results on five liquids of different molecular geometry reveal some very complex thin-film properties, such as the quantization of various static and dynamic properties, discontinuous or continuous solid–liquid transitions, smooth or stick–slip friction, and two-dimensional nucleation. Quantitatively, the "effective" viscosity in molecularly thin films can be 10^5 times the bulk value, and molecular relaxation times can be 10^{10} times slower. These properties depend not only on the nature of the liquid, but also on the atomic structure of the surfaces, the normal pressure, and the direction and velocity of sliding. We also conclude that many thin-film properties depend on there being two surfaces close together and that they cannot be understood from a consideration of a single solid–liquid interface. The results provide new fundamental insights into the states of thin films, and have a bearing on understanding boundary friction, thin-film lubrication and the stress–strain properties of solids at the molecular level.

INTRODUCTION

Interactions of surfaces separated by very thin liquid films

The properties of liquids confined within very small spaces, such as narrow pores or thin films, are generally quite different from their bulk properties whenever the pore dimensions or film thickness falls below a few molecular dimensions.[1-9] Both experimentally and theoretically it is found that molecules become progressively more ordered[4,8] and that their mobility sharply decreases[3,6,7] in films less than about ten molecular diameters.

The liquid density across molecularly thin films is not uniform but has an oscillatory profile. The periodicity of the oscillations is close to the diameter of the liquid molecules[3-9] and reflects the forced ordering of the liquid molecules into quasidiscrete layers between the two surfaces. The closer two surfaces approach each other the sharper these density oscillations become, and when only one layer remains between the two surfaces it can become difficult to distinguish its properties from that of a solid layer. This appears to be consistent with recent simulations which indicate that abrupt liquid–solid transitions may occur in thin liquid films at certain thicknesses,[10-12] but the question as to whether these may be considered as true liquid or solid phases or whether the transitions are truly first-order (i.e., discontinuous) has not been explored either experimentally or theoretically.

Closely associated with the layering phenomenon is the oscillatory "structural" or "solvation" force between two surfaces in liquids.[4,5,8,9] These forces have been studied both experimentally[4,5,13] and theoretically,[8,9] and are now well-understood, at least for simple liquids. In particular, both theory and experiment show that solvation forces generally dominate over continuum or mean-field forces (e.g., the monotonically attractive van der Waals force) at surface separations below 5–10 molecular diameters.[4,5,9,13,14]

Shear properties of thin films: experimental background

Turning now to dynamic properties, it has likewise been found that when two surfaces are farther apart than about ten molecular diameters a simple liquid in the gap retains its bulk Newtonian behavior and that the shear plane remains coincident with the physical solid–liquid interface (to within about one molecular diameter) even at shear rates as high as 10^5 s^{-1}.[7,15-17] However, for thinner films the "effective" viscosity rises dramatically. In experiments with simple liquids such as cyclohexane[6] it was found that when two molecularly smooth mica surfaces slide past each other with only one or two layers of liquid molecules between them

(Couette flow) the "effective" shear viscosity is 5–7 orders of magnitude higher than the bulk value. More importantly, it was found that the whole concept of a Newtonian viscosity breaks down for such thin films: for example, they exhibit a yield point and the shear stress no longer depends on the shear rate. In addition, the sliding now occurs with the surfaces separated by an integral number of liquid layers at surface separations coinciding roughly with the energy minima in the oscillatory force curves, and the shear stress and other tribological properties are "quantized" with the number of layers.[6,18-21] Clearly, we are once again faced with a situation where the properties of "liquid" films only one or two layers thick appear to be more solidlike than liquidlike.[6,10-12,21,22]

Shear properties of thin films: theoretical background (computer simulations)

Monte Carlo and molecular dynamics simulation techniques have recently been used to investigate shear in molecularly thin liquid films. Davis and co-workers[3] and Schoen and co-workers[10-12,21] simulated two shearing surfaces separated by Lennard-Jones liquids and found that for films thinner than 6–10 molecular diameters the molecules are expected to order into discrete layers between the surfaces and that the diffusion rate decreases and the viscosity increases. Schoen *et al.*[10,11] using atomically structured (i.e., corrugated) surfaces predicted the existence of abrupt liquid to solid transitions in films thinner than 6 molecular diameters, with the molecules becoming ordered both perpendicular and parallel to the surfaces. However, the horizontal order disappeared for unstructured, i.e., mathematically smooth, surfaces. Strong, quantized yield points were also found,[21] as observed experimentally.[6,18] These, too, disappeared for unstructured surfaces. These simulations clearly indicate the profound importance of the atomic-scale structure or "granularity" of real surfaces when considering the properties of films that are themselves of similar dimensions.

Phenomenological equations of friction

The study of friction is as old as time. The basic law of friction, Amonton's Law, states that

$$\text{Frictional force: } F = \mu L, \qquad (1)$$

where F is the frictional force, L the load, and μ the coefficient of friction. Equation (1) applies to most real situations where sliding occurs between nonadhering damaged surfaces. In the present study we are more concerned with the sliding of undamaged, molecularly smooth surfaces, commonly referred to as "boundary" or "interfacial" friction, for which Eq. (1) does not apply.

A friction model was originally proposed by Tabor[23] to explain the boundary friction of two solid hydrocarbon surfaces sliding past each other. In this model the frictional force was calculated from the energy needed to overcome the attractive intermolecular forces between the surface methyl groups when one surface is dragging over the other. The resulting equation is of the form

$$F = S_c A, \qquad (2)$$

where the friction is now proportional to the area of contact, A, and where S_c is the critical shear stress which depends on the adhesion force between the two surfaces.

We have recently proposed an extension of this model[18,20] to two sliding surfaces separated by a layer of liquid molecules. In this model (which we have termed the "Cobblestone" model) the case of two surfaces sliding past each other while separated by a thin liquid film is considered analogous to the motion of a cart on a cobblestone road. When the two surfaces are close together the liquid molecules between them order to fit within the spaces between the atoms of the two surfaces, in an analogous manner to the self-positioning of the cartwheels in the grooves of the cobblestones, so as to minimize the energy of the system. A tangential force applied to one surface will not immediately result in the sliding of that surface since the attractive van der Waals forces between the two surfaces must first be overcome by the surfaces separating by a small amount (cf. dilatancy). It is a simple matter to show that for this type of sliding Eq. (2) applies, and that the critical shear stress needed to initiate motion with one layer of liquid molecules between the surfaces is[20]

$$S_c = \text{frictional force/area} = F/A \approx \epsilon(2\gamma/d), \qquad (3)$$

where d is some lattice dimension ($d \approx 0.5$ nm), γ a typical surface energy ($\gamma \approx 50$ mN/m), and where $\epsilon < 1$ ($\epsilon \approx 0.1$ for $n = 1$, decreasing with n, where n is the number of molecular layers). Putting these values into the above we obtain $S_c \approx 2 \times 10^7$ N/m^2. This value compares well with typical experimental values obtained for simple spherical molecules (see Table I later). It is intuitively clear that lower frictional values will be obtained the larger the size of the liquid molecules (at constant n), and as n increases, as observed experimentally.

In addition to the contribution arising from the (internal) adhesive forces across the film there is also a contribution from the externally applied load L. This contribution to F turns out to be proportional to L, as in Eq. (1), so that we finally obtain for the case of interfacial sliding:[20]

$$F = S_c A + CL \qquad (4)$$

which, in terms of the total shear stress, becomes[23]

$$S = F/A = S_c + CL/A = S_c + CP, \qquad (5)$$

where S_c and C are constants, and $P = L/A$ is the pressure. The friction coefficient C is related to the atomic granularity of the surfaces and on the size, shape, and configuration of the liquid molecules in the gap. In general, the smoother the surfaces the smaller should be the value of C. Note that the second term in the above two equations has the same form as Amonton's Law: $F = \mu L$. However, for damaged surfaces μ has a completely different origin and magnitude from C.[20]

Equations (4) and (5) provide a simple starting point for analyzing the results of interfacial sliding experiments, though they do not include such effects as time-dependent friction or stick–slip sliding as occur in real systems.

Aim of present experiments

The aim was to measure shear stress, relaxation phenomena, and other tribological properties under various conditions with the expectation that the results would provide new insights into the conformations of differently shaped molecules between shearing surfaces. Such information should enhance our understanding of friction, lubrication, rheology, and other tribological mechanisms at the molecular level—areas which are currently of particular concern for producing low adhesion, low friction surfaces, designing better lubricants, reducing head-disc interactions in computers, etc.

More fundamental issues are also involved that bring us to the heart of the definitions of and distinctions between a solid and a liquid. There are two common definitions: one pertaining to the long-range positional order of molecules in solids but not in liquids, the other to the finite shear modulus of solids which vanishes for liquids. The first definition[24] implies the absence of a critical point for the solid-to-liquid transition due to the breaking of the order–disorder symmetry during that transition. Solid–liquid transitions are therefore of first order and one cannot go continuously from one state to the other. The short-range oscillatory forces characteristic of interactions across thin liquid films indicate that such a simple picture (which, of course, strictly applies to an infinite bulk phase) cannot be unambiguously applied to molecularly thin films that can become continuously more ordered and solidlike with decreasing film thickness. Indeed, it is well known[25] that in two dimensions it is possible to have a continuous solid-to-liquid transition, as well as other types of phases that have no three-dimensional counterpart. (We are not referring to *glasses*, which are simply liquids whose relaxation processes are extremely slow, i.e., they are liquids not at equilibrium; nor to *amorphous solids*, which are crystalline solids not at equilibrium.)

The second definition is that solids resist shear but liquids do not. Here again we find that this simple definition cannot be applied unambiguously to molecularly thin films which can resist shear but can also relax and flow. It is from our studies of the shear properties of thin fluid films, reported here, that we have obtained the greatest insights into this complex phenomenon, where we find that molecularly thin films can be in a state that is neither liquid nor solid and that the properties can change continuously from liquidlike to solidlike as a function of film thickness and shear rate.

Liquids used in this study

Four hydrocarbon liquids and one silicone liquid were chosen for this study: cyclohexane, octamethylcyclotetrasiloxane (OMCTS), *n*-octane, *n*-tetradecane, and the branched isoparaffin 2-methyloctadecane. These liquids were chosen to investigate the effects of molecular geometry. Both cyclohexane and OMCTS are approximately spherical molecules (of diameters 0.55 and 0.85 nm, respectively) and are representative of simple isotropic liquids. Octane and tetradecane are symmetrical chain molecules. 2-methyloctadecane is a branched chain and does not have an axis of symmetry. It is of the type commonly used in certain lubricant oils. A comparison of the shear properties of these four liquids is therefore of both fundamental and practical interest.

Another advantage of using these liquids is that the equilibrium force laws between mica surfaces across them have previously been measured. The results, summarized in Fig. 1, show that the spherical and chain molecules form quasidiscrete layers when confined between two mica surfaces, while the branched 2-methyloctadecane molecules do not form layers even under high compression. Instead, they appear to remain in some form of disordered glassy/amorphous state near each surface. A study of the shear properties of these liquids therefore allows an examination of how their equilibrium interactions and conformations are correlated with their dynamic interactions.

EXPERIMENT

Sliding attachment

A recently developed lateral sliding mechanism for use with the surface forces apparatus is described in Ref. 6 and is shown schematically in Fig. 2. This attachment allows for two flattened molecularly smooth surfaces to be slid past each other while simultaneously controlling and measuring the surface separation D (to ± 0.1 nm), sliding speed v, normal load L, molecular contact area A, and the lateral shear (or frictional) force F. In addition, any elastic deformations of the moving surfaces can be continually monitored in real time by recording the multiple-beam interference fringes used for measuring the surface separation.

The basic geometry and principle of the method is shown in the lower part of Fig. 2. One surface slides at velocity v and the shear or frictional force F experienced by the upper surface is recorded on a chart recorder from the displacement, x, of a spring of stiffness K, so that $F = Kx$.

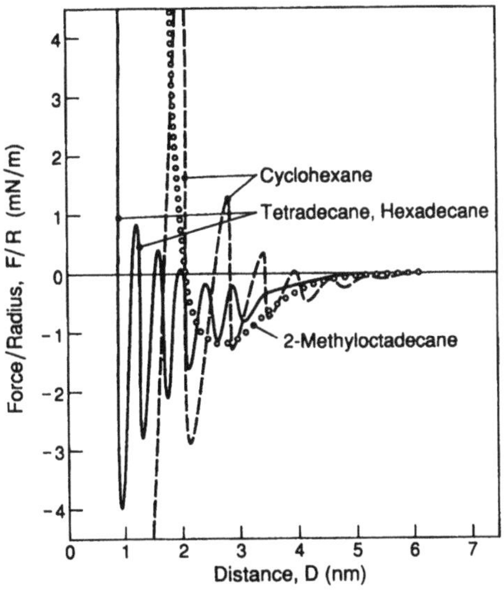

FIG. 1. Previously measured forces between mica surfaces immersed in cyclohexane of mean molecular diameter 0.55 nm (solid curve, Ref 35), *n*-alkanes of molecular width 0.4 nm (dashed curve, Ref. 36) and 2-methyloctadecane (dotted curve, Ref. 28). OMCTS, Refs. 35, 37 (data not shown) also exhibits an oscillatory force of periodicity 0.85 nm. The last one or two force barriers were usually difficult to surmount on compression only, but could be surmounted during compression with shear (see Results).

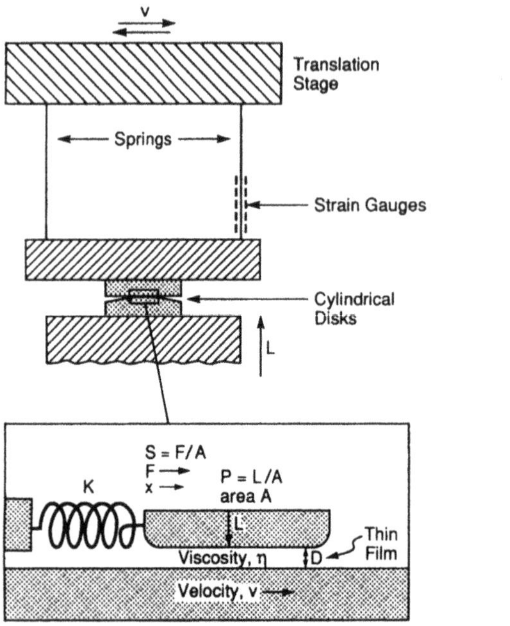

FIG. 2. Schematic drawing of sliding attachment (full details given in Ref. 6). During experiments the experimental parameters varied as follows: normal loads L were typically $+100$ g (compression) to -100 g (tension), areas of molecular contact A were 10^{-6}–10^{-3} cm^2, frictional forces F were 4×10^{-3}–10 g, force constants of springs were $K = 10^2$–10^3 N/m, sliding velocities v were 0.5–50 μm/s (in both directions), surface separations D ranged from 0 to above 3 nm, and shear rates τ_s from 10^2 to 10^5 s^{-1}.

Purification of Liquids

The cyclohexane (Aldrich, HPLC grade), *n*-octane and *n*-tetradecane (Sigma, >99% pure), OMCTS (Aldrich, >98% pure) and the 2-methyloctadecane (Wiley Organics, >98% pure) were each distilled under dry, pure nitrogen before use and dried over P_2O_5.

Experiments were carried out with a macroscopic droplet of liquid between the two surfaces inside the apparatus chamber whose humidity and temperature could be controlled. Most experiments were done at 22 °C, and between three and six independent experiments were done with each liquid.

RESULTS AND DISCUSSION

Theoretical aspects of data analysis

In order to put the results into perspective we shall first show what would be expected if the intervening material between the two surfaces in Fig. 2 were (i) a Newtonian liquid and (ii) a solid.

(i) Liquid Film. For the case of a liquid of shear viscosity η, if the spring is displaced by a distance x during sliding then at any time t

$$F = Kx = A\,[v - dx/dt\,]\eta/D. \tag{6}$$

This is the usual expression for Couette flow. The solution to Eq. (6) is

$$x = Av\eta[1 - e^{-KDt/A\eta}]/KD, \tag{7}$$

where $t = 0$ corresponds to the start of sliding. In the steady state, when $dx/dt = 0$, the shear (or frictional) force is given by

$$F = Kx = Av\eta/D \tag{8}$$

which, in terms of Eq. (5), gives the following shear stress:

$$S = F/A = v\eta/D. \tag{9}$$

Note that in the steady state S is proportional to v, and the shear rate is

$$\tau_s = v/D\ s^{-1}. \tag{10}$$

If the sliding is suddenly stopped, the shear force will relax exponentially according to

$$F = Kx = \frac{Av\eta}{KD}\,e^{-KDt/A\eta}, \tag{11}$$

where $t = 0$ now corresponds to the time when sliding stops.

Figure 3 shows how the shear stress of a purely liquid film rises with time after the commencement of sliding and then falls back to zero after sliding stops, as would be measured by the chart recorder.

(ii) Solid Film. Figure 3 also shows a typical stress curve for a solidlike film, such as a ductile metal under shear, though the exact form of this curve beyond the yield point depends on the material and sliding velocity,[26] and so is less general than the liquid curve.

Two important differences between the shapes of the two curves in Fig. 3 are (i) the uniform initial slope of the solid curve, and (ii) the decay to zero of the liquid curve on stopping. Another important difference concerns the magnitude of the shear stress, which is much higher for solids than for liquids. For example, with a liquid of bulk viscosity 0.1 poise, in a film of thickness $D = 0.5$ nm and sliding velocity of $v = 5\mu$m/s, we expect from Eq. (9): $S = 10^2$ N/m^2 if we assume that the viscosity of the liquid in the film is as in the

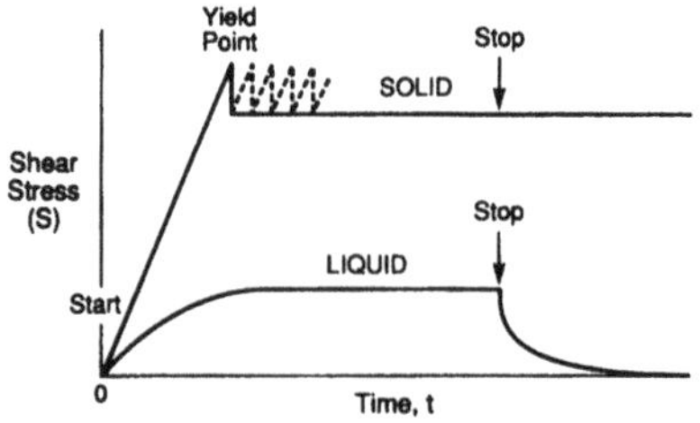

FIG. 3. Shear stress (or frictional force) for liquid film and solid films. The liquid is considered ideal, i.e., describable in terms of bulk/continuum properties. The curves are only schematic; the actual values of S can differ by many orders of magnitude, with S_{solid} usually much larger than S_{liquid}. Note that the "start" at $t = 0$ is when the translation stage of Fig. 2 starts moving with velocity v; the surfaces themselves start sliding relative to each other later, for example, at the yield point for solids or solidlike films. Beyond the yield point ductile solids such as metals or alloys show either a smooth "yield elongation zone," or a single saw-toothed spike having "upper" and "lower" yield points, or a series of successive spikes known as "serrations." (Ref. 26). The serrations are very sensitive to the experimental conditions such as strain rate, temperature, etc.

bulk. This is about seven orders of magnitude less than typical values for solids. As we shall see, for films of simple liquids their shear stresses can rise to values approaching those of solids when only one layer of "liquid" molecules remains between the surfaces.

The two characteristic curves of Fig. 3—one for a liquid, the other for a solid—serve as suitable reference curves for comparing and discussing the experimental results.

Summary of previous results with these liquids

We have already mentioned that the equilibrium forces between mica surfaces across the five liquids have previously been measured (Fig. 1). The shear viscosities of these liquids have also been measured but only in films thicker than ten molecular diameters.[15,16,17,27] The results showed that the liquids exhibit their bulk viscosity and that the shear plane is within about one molecular diameter of the solid–liquid interface except for 2-methyloctadecane, where it is located at $D = 3$ nm from each surface.[27] This again indicates that the branched hydrocarbon molecules are in a quasiimmobilized/glassy configuration near the surfaces—most likely because of the entanglement of their branched chains.[28]

General features of results: quantization of properties

Between three and six separate experiments were performed with each liquid. Two types of data were always recorded simultaneously. First, the interference fringes give the contact diameter and the distance between the surfaces as well as the changing profiles of the elastically deformed surfaces during sliding. Second, the chart-recorder traces give the frictional force at any instant. In a typical experiment, contact areas A were measured to $\pm 5\%$, loads L to $\pm 3\%$, film thicknesses D to ± 0.1 nm, frictional forces F to $\pm (5–15)\%$, and sliding velocities v to $\pm 3\%$. The data obtained from one experiment to another for a particular liquid varied by as much as 50%. We believe this to be due to the different angles between the surface lattices of the two mica sheets[29] and the different sliding directions relative to these lattices in the different experiments.[30] This is something over which we had no control and plan to investigate in the future. These data are illustrated schematically in Figs. 4 and 5 which show the more common types of surface deformations and frictional forces that were observed.

Figure 4 shows six fringe patterns typically observed during sliding and the surface profiles which correspond to these patterns. Here is a typical scenario for the case of spherical molecules: in Fig. 4(a) the initially curved elastic sheets have become flattened due to the external load, but are still separated by two layers of liquid. Sliding begins in this configuration at $t = 0$ with a steady frictional force. At some finite time t, as shown in Fig. 4(b), part of the contact zone is moving in to one layer as sliding proceeds at constant L and v. During the transition, the frictional force rises steadily but there is no significant change in the contact area. The transition from $n = 2$ layers to $n = 1$ layer is complete by Fig. 4(c). It should be noted that this observed transition takes less time than it takes for the surfaces to slide across their area of contact. At $t > 50$ s the confined molecules have

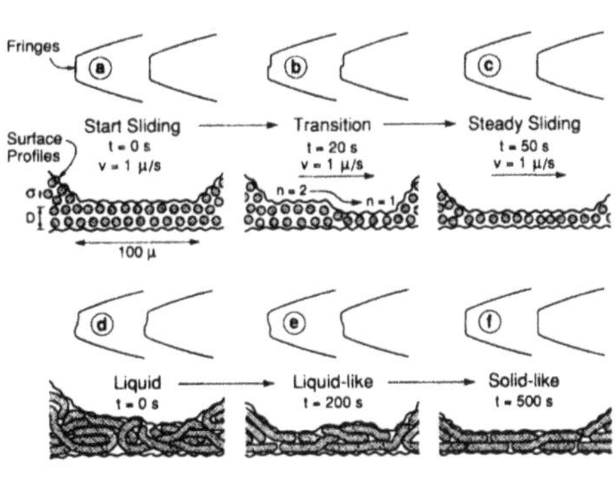

FIG. 4. Typical fringe patterns and corresponding configurations of surfaces and molecules during "interfacial" sliding–i.e., reversible sliding with no surface damage. During interlayer transitions from n layers to $n \pm 1$ layers, as in (b), it was noted that the interlayer boundary line (a sort of edge dislocation) usually moves backwards faster than v. (Layer thicknesses measured to ± 0.1 nm clearly identify n.)

reached a new steady state configuration and the frictional force is now constant (smooth sliding) or intermittent (stick–slip sliding).

With the linear chained and branched molecules the transitions between different layers are usually not as well defined as with spherical molecules, and two surfaces often slide with a number of local regions separated by different numbers of layers. This is shown in Figs. 4(d)–4(f). In addition, longer sliding times are needed with such liquids before the steady state is attained, i.e., before the molecules become ordered enough for sliding to proceed with a fixed number of layers between the surfaces. In some cases, different regions of the surfaces continue to sample different layers during sliding, even in the steady state.

Figure 5 shows the shear forces, as measured by the

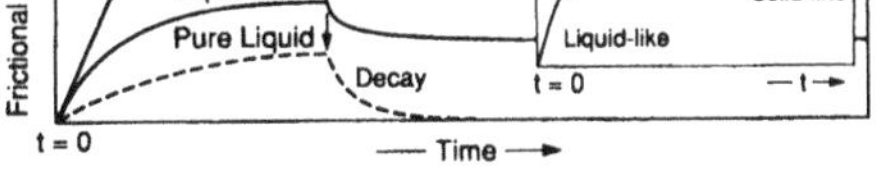

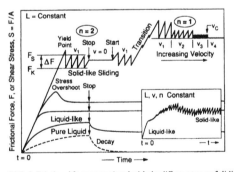

FIG. 5. Frictional forces associated with the different types of sliding modes of Fig. 5. "Pure liquid" sliding occurs with surfaces farther apart than 5–10 σ. "Liquidlike" sliding occurs with configurations as in Figs. 5(d) and 5(e), while "solidlike" sliding is associated with Figs. 5(a), 5(c), and 5(f). With certain liquids the sliding starts by being liquidlike and becomes progressively more solidlike during sliding; this is generally accompanied by a decrease in the film thickness and a "stress overshoot." An example of this given in the inset which shows measured data during an experiment with tetradecane. Note that a single stick–slip occurs over many microns and should not be confused with atomic scale stick–slip (Ref. 38) which may also be occurring but is beyond our resolution.

chart recorder, typically associated with the different sliding configurations of Fig. 4. The frictional forces go from being liquidlike to solidlike as we go from the lower to the upper curves, corresponding, for example, to the molecular configurations of Fig. 4 from (d) to (f). Often, the liquid-to-solid transition occurs during sliding, as shown in the inset of Fig. 5. The upper curve also shows a typical transition from n layers to $n-1$ layers during sliding in the purely solidlike regime.

Both the magnitude of the force F_K and the superimposed stick–slip component $(\Delta F = F_{\text{static}} - F_{\text{kinetic}} = F_s - F_k)$ generally increases as the number of layers falls. In addition, with increasing sliding velocity, while F_K does not change much, the frequency of the stick–slip always increases and the amplitude ΔF eventually falls, until at some critical velocity v_c the stick–slip disappears altogether and the frictional force is steady at $F = F_K$. (Note that the velocity v, the stick–slip amplitude ΔF and frequency ν are related by $v = \Delta F \cdot \nu$.) As we shall see, the stick–slip phenomenon provides valuable information on the relaxation mechanisms and relaxation times of the liquid molecules.

A particularly important phenomenon is the "quantization" of many properties associated with solidlike sliding. We have found that steady state sliding generally occurs with an integral number of molecular layers between the surfaces ("quantized spacing"). Likewise, the shear stress is also often quantized with the number of layers, i.e., $\dot{S}_c$ = constant (at constant n). This is especially pronounced for the small spherical molecules, but for the more complex liquids and water this is not always the case even though the molecules may still slide in layers.

We now proceed with more detailed descriptions and analyses of the results obtained with the five liquids.

Transition from liquid-like to solid-like properties during shear

All the five liquids studied exhibited liquid or liquidlike behavior for gap thicknesses greater than 5 layers, and solidlike behavior for less than 5 layers. For small spherical molecules, such as cyclohexane and OMCTS, solidlike sliding occurred immediately after the surfaces were brought together (Fig. 6): the film thickness was quantized by a multiple (n) of the molecular diameter σ, and the frictional force F was proportional to A, indicative of sliding in an adhesive well. However, with tetradecane, and even more so with the branched alkane 2-methyloctadecane, the initial sliding after the surfaces were brought together was liquidlike, and only became solidlike after the surfaces had been sliding for some time over the same area (Fig. 5, inset). The branched alkane showed the greatest tendency to remain in the liquidlike state, followed by the unbranched tetradecane and octane, while the two spherical molecules studied did not exhibit liquidlike behavior in such molecularly thin films.

For the branched alkane, when sliding in the liquidlike regime the gap thickness D was not uniform or quantized but varied across the contact area, as shown in Figs. 4(d) and 4(e). Under such conditions F was always proportional to L rather than A, i.e., the second term in Eq. (4) dominates (Table I). This is indicative of a weak or shallow adhesive minimum, consistent with the static force measurements (Fig. 1). After some sliding, it was often possible to shear order the molecules into a more aligned, layered configuration, as illustrated in Figs. 5(e) and 5(f). This was inferred from the constancy of D across the film and the development of an irregular stick–slip component to the friction which was always taken as indicative of a solidlike film. The changes in F and D reflecting the transition from liquidlike to solidlike sliding are shown in Figs. 7 and 8.

With the chain molecules we have noted that, once the film is in the solidlike state, if the two surfaces are separated, kept apart for more than 1–5 minutes, and brought together again, the film returns to its original liquidlike state. This shows that with the chain molecules their solidlike conformation near an isolated surface can persist for a short time but is eventually lost. More importantly, it shows that two confining surfaces are needed for the liquid to become solidlike, i.e., an isolated surface cannot order the liquid molecules adjacent to it to the same degree. Similar memory effects have previously been observed in the equilibrium force laws after the close approach and separation of two surfaces in OMCTS and water.[31]

The results of Figs. 5–8 indicate that spherical molecules in molecularly thin films are ordered into layers and exhibit solidlike behavior right from the start, but that asymmetric chainlike molecules are initially much less ordered, with no preferred orientation and become oriented only when sheared (shear ordering).

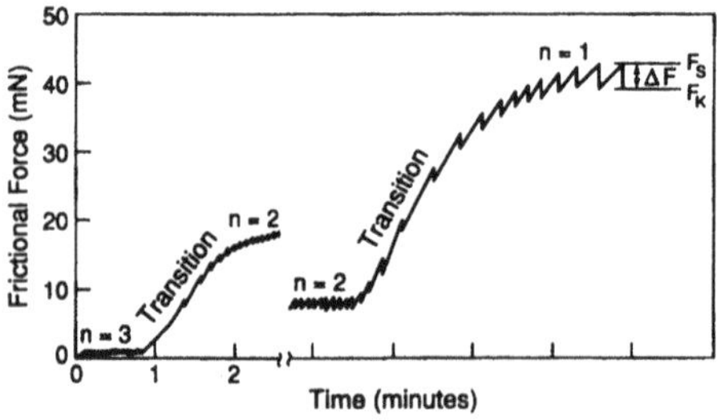

FIG. 6. Measured shear forces during interlayer transitions of OMCTS. Note that while the shear force F may be different at fixed n, this is due to the changing contact area (A) with load and position (time). The shear stress S, given by $S = F/A$, remained constant so long as n was constant (see Table I). Qualitatively similar recordings were obtained with cyclohexane.

TABLE I. Frictional force coefficients.[a]

Liquid	$S_c(n)(10^6\ \text{N/m}^2)$ $n=1$	$n=2$	$n=3$	C
Cyclohexane	23	1	0.5	
OMCTS	6	4	1	
n-octane	~10	~6		1.5
n-tetradecane				0.8
2-methyloctadecane				0.3

[a] The frictional force is given by Eq. (4) as $F_K = S_c(n)A + CL$. Note: As we go from cyclohexane to 2-methyloctadecane the friction becomes less dependent on n (i.e., less quantized), and increasingly dominated by the second term in the above equation. Data shown is from one experiment (Fig. 9).

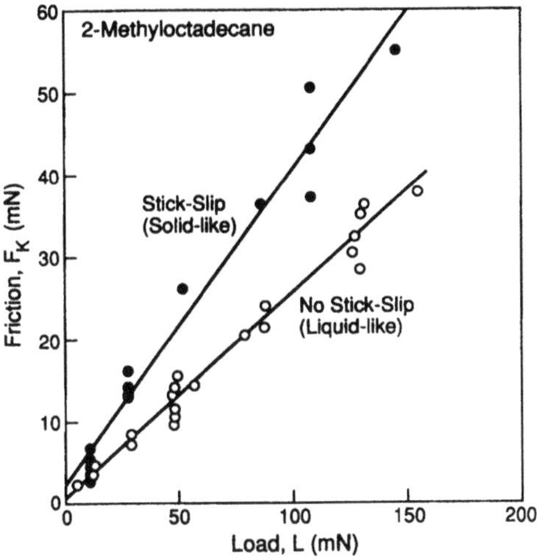

FIG. 7. Frictional force for 2-methyloctadecane. Initially, the film is in the liquidlike state, as in Fig. 5(d), and the sliding is smooth. The film thickness is not uniform but averages $D \approx 2.4$ nm (i.e., $n \approx 5$) and the friction coefficient is $C \approx 0.25$ (open circles). After some sliding the film is massaged into the solidlike state, as in Fig. 5(f), and there is stick–slip. The film thickness has fallen to $D = 0.8$ nm ($n = 2$) and the friction coefficient has risen to $C \approx 0.35$ (black circles). Values for the other liquids are given in Fig. 9 and Table I.

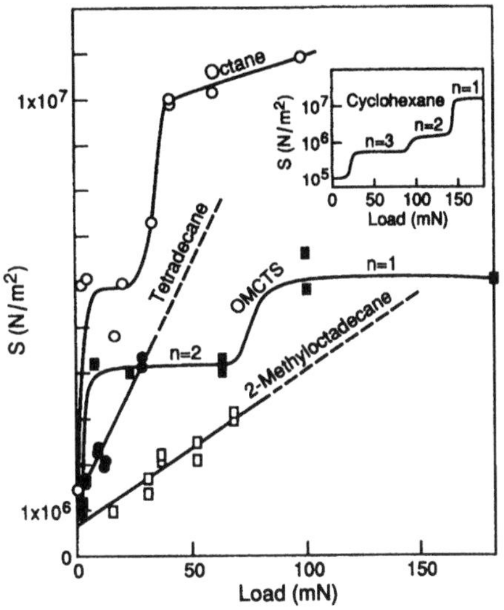

FIG. 9. Variation of shear stress, $S = F_K/A$, with load L in the steady state for films in the solidlike state. Each curve was obtained from a single experiment with a different liquid. There were quantitative variations in the data from experiment to experiment, probably due to the different orientations (twist angles) between the two surface lattices and the different sliding directions in different experiments (Errors in F: $\pm$ 10%.)

Properties of solidlike films

For all the five liquids studied, solidlike behavior was characterized by a total pinning of the surfaces until the yield point, after which there was stick–slip motion at low velocities and smooth sliding at high velocities (Fig. 5, upper curve). The transition from stick–slip to smooth sliding is shown in Fig. 10 for tetradecane and in Fig. 11 for OMCTS. If the sliding was stopped, the two surfaces remained pinned indefinitely, i.e., there is no stress relaxation of the surfaces, even though the molecules in the gap do relax (discussed later). Another feature of solidlike sliding was that the film thickness was quantized with an integral number of molecular layers, viz., multiples of the molecular diameter with spherical molecules or the molecular width with chain molecules.

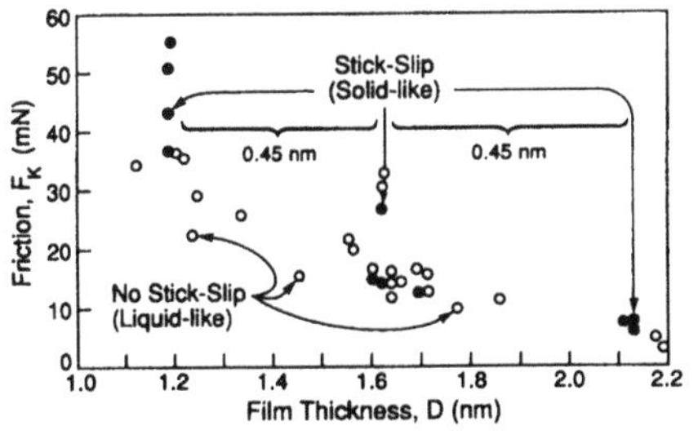

FIG. 8. Measured film thickness for 2-methyloctadecane in the liquidlike and solidlike states (see Fig. 7 for corresponding friction data). Note the quantization of spacing in the solidlike state but not in the liquidlike state.

Smooth friction (no stick–slip)

We first compare the different solidlike properties of the five liquids in the absence of stick–slip, when $v > v_c$ and the frictional force F_K is largely independent of v (at constant n). From Fig. 9 and Table I, we see that for the smaller, more spherical, molecules the friction is dominated by S_c (i.e., $F \propto A$), while for the longer chain and branched chain molecules it is dominated by C (i.e., $F \propto L$). These trends are a reflection of the relative contributions made by the adhesive forces,[20] and liquid and surface structure[36,20] in determining the stresses of the liquid films.

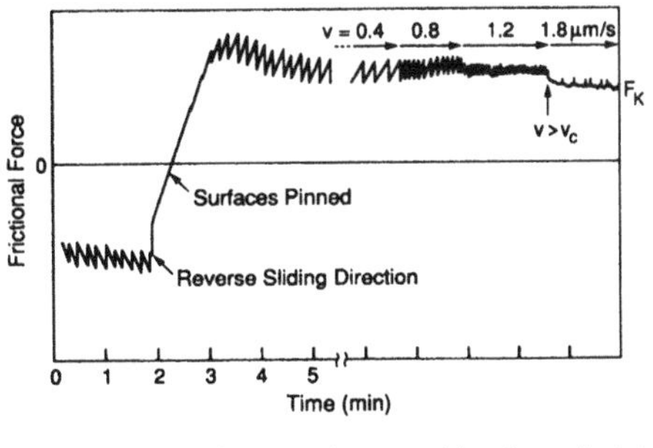

FIG. 10. Measured chart recorder traces giving the steady state frictional force between two mica surfaces separated by two layers of tetradecane sliding under a constant load. Also shown is the effect of reversing the sliding direction ($v = +0.4 \to -0.4$ μm/s). With increasing sliding velocity v, the stick–slip frequency increases while the stick–slip amplitude ΔF does not change. But as v approaches v_c the amplitude begins to fall rapidly, reaching zero at $v = v_c$. A similar trend was observed with the other liquids.

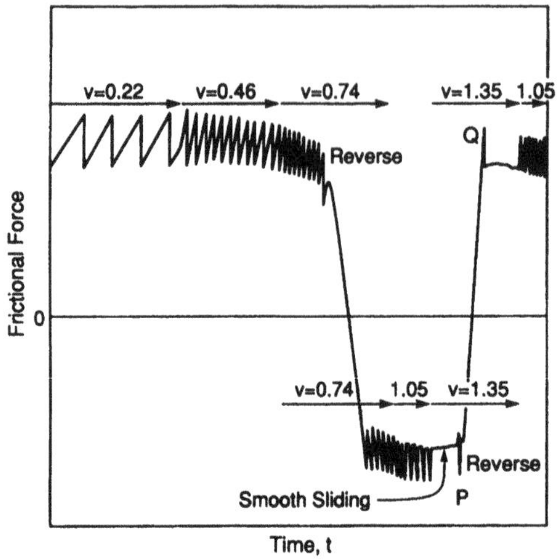

FIG. 11. Effect of increasing the sliding velocity on the friction of a two-layer film of OMCTS. On increasing v from 0.22 to 1.05 μm/s the stick–slip frequency v increases while both the kinetic friction F_K and the stick–slip amplitude ΔF fall, but only slightly (note that $v = \Delta F \cdot v$). However, once v exceeds the critical velocity v_c (between 1.05 and 1.35 μm/s) the stick–slip disappears abruptly, and returns equally abruptly when the velocity falls below v_c again. Note the "stopping spike" at P and the "starting spike" at Q.

Stick–slip and molecular relaxations

We now turn our attention to the stick–slip component of the frictional force. This has already been shown in Figs. 5, 6, 10, and 11. The stick–slip was very sensitive to sliding velocity and the immediate previous history of sliding, and provided the greatest insights into the molecular ordering and relaxation mechanisms within the films. At slow sliding speeds the stick–slip was generally quite regular, viz., a periodic saw-tooth function. Only for the branched alkane was the stick–slip aperiodic or 'irregular," i.e., smooth sliding punctuated by isolated saw-tooth "spikes" at irregular intervals. As already mentioned, for each liquid, above some critical velocity v_c the stick–slip totally disappears and the sliding becomes smooth (Figs. 10 and 11). With the spherical molecules, such as cyclohexane, the stick–slip persisted to quite high velocities, but with the larger and chain molecules the stick–slip vanished at much lower velocities. This suggests that some shear-induced molecular ordering effects are very sensitive to the shear rate. This matter was investigated in detail for tetradecane and 2-methyloctadecane.

With either tetradecane or 2-methyloctadecane, if the sliding is suddenly stopped for a short time and then restarted, a remarkable phenomenon is observed: if the stopping time is less than a certain time τ there is no change in the friction on restarting, i.e., sliding proceeds *as if there had been no interruption*. But if the resting time exceeds τ a single fully developed stick–slip spike occurs. This indicates that well after the surfaces have stopped moving relative to each other, the molecules in the gap are still relaxing and that some dramatic change in their configuration occurs at time τ after stopping. This is shown in the upper two curves of Fig. 12, where we refer to this time τ as the "latency time."

Figure 12 also shows the characteristic shape of the spikes, which always start with a straight line of constant slope (the stick) followed by a decay curve (the slip). From the slope of the straight line we can deduce that the two surfaces remain locked together or "pinned" in the "sticking" regime. The "slip" regime starts by a sudden rapid slipping of the two surfaces which gradually decays over a time τ_0 when the surfaces once again move at the steady velocity v relative to each other.

All these phenomena clearly show that the conformation of the molecules in the gap must be changing as soon as they stop sliding, even though the surfaces do not continue to move relative to each other, and that conformational changes are also occurring during the "slip" part of the stick–slip. Indeed, we may identify four distinct regimes associated with a full cycle as follows (see also Fig. 12):

Sliding regime. The surfaces slide smoothly (if $v > v_c$), and the confined molecules are probably rolling. Chain molecules have most likely become shear aligned normal to the direction of motion and have orientational order perhaps akin to a two-dimensional nematic, but no long-range positional order. However, at any instant during sliding, both spherical molecules and individual segments of a chain molecule may have some short-range positional order, determined by the instantaneous coordinates of the two surface lattices, the load, velocity, etc. We may refer to this as the "sliding" configurations (or "kinetic" configuration, since it is associated with F_K).

Resting regime. On stopping, the surfaces remain pinned to each other, but the configuration of the molecules within the film relaxes to a more solidlike state where there is long-range order. Thus, there is a transition from the kinetic configuration to a static ("frozen" or

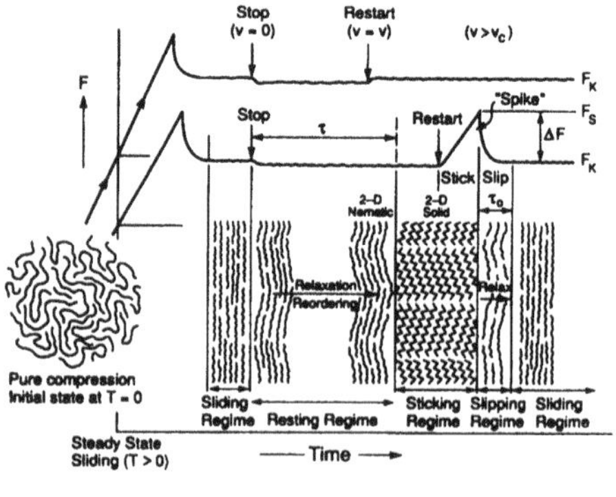

FIG. 12. Effect of stopping and starting (strictly, changing the velocity of the translational stage of Fig. 2) on F as a function of time for the chain molecules tetradecane and 2-methyloctadecane. The magnitude of the spikes were only weakly dependent on the stopping time in excess of τ. Lower part of figure: proposed scenario of changing conformations of chain molecules in the SLIDING, RESTING, STICKING and SLIPPING regimes. The drawings are schematic and are for illustrative purposes only. The conformations in each regime would depend critically on the shape and size of the liquid molecules, the surface structure, the twist angle between the two surface lattices, and the direction and velocity of sliding.

"locked") configuration where $F = F_S$, the latter being determined by the lattice coordinates of the two surfaces. The enhanced solidlike ordering leads to an increased adhesion and subsequent frictional force by ΔF (from F_{kinetic} to F_{static}, though F_{static} may not be the same as F_{equil} unless the stopping time is very long). On restarting, fully developed spikes appear only if the stopping time exceeds τ, but no spikes appear for shorter stopping times (see Fig. 13). This suggests that the molecular relaxations occurring in the resting regime may be more in the nature of a nucleation phenomenon (first order transition) occurring at time $t = \tau$ after stopping. This would be consistent with the proposed disorder–order nature of the transition.

Sticking regime. On starting to strain, the surfaces remain pinned until the yield point. The frozen solidlike configurations of the molecules remain unchanged (simply elastically stressed) up to the yield point when melting followed by slip occurs.

Slipping regime. The yield point is analogous to the onset of plastic flow of ductile solids.[26] Sliding begins by a slip, during which the molecules reorder back to their sliding mode configuration. This takes a finite time τ_0 which is different from τ (Fig. 12).

The above model of four distinct configuration regimes is further supported by a number of observations, some qualitative, some quantitative. First, for the same number of layers, the latency times τ were longest for 2-methyloctadecane (many seconds), shorter for tetradecane ($t \sim 1$ s), and too short for us to measure for cyclohexane and OMCTS ($t \ll 0.1$ s). This is expected since any relaxation associated with the trapped molecules would be expected to occur faster as we go from branched chain to unbranched chain to spherical molecules. Second, for any particular liquid τ increased with the externally applied load or pressure (Fig. 13). This, too, is consistent with the above model, since the relaxation time τ is also expected to increase the more the molecules are compressed.

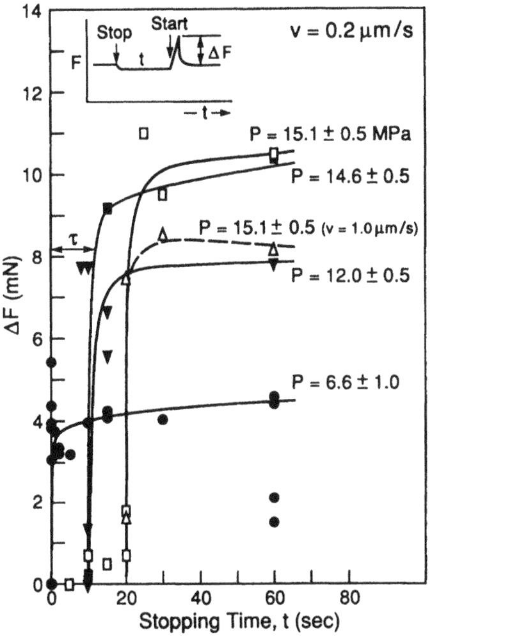

FIG. 13. Spike height ΔF against stopping time for two mica surfaces sliding across two layers of 2-methyloctadecane under different pressures P. The sliding velocity was the same in each case ($v = 0.2\,\mu$m/s), except where indicated by the dashed curve. Note how τ increases with P. Similar results were obtained with n-tetradecane, where τ varied from 0.3 to 5 s over the same range of loads.

Another interesting correlation was observed between the spike height and sliding velocity. This is shown in Fig. 13 for 2-methyloctadecane (see the two curves for $P = 15$ mPa). Generally, the higher the velocity the lower the subsequent spike height (for the same resting time). This is consistent with the notion that the greater the sliding velocity, the less favorable is the "kinetic configuration" for adhesion and sticking, so that ΔF decreases.

We have also noted a slight increase in ΔF as the stopping time increases above τ (Fig. 13), which indicates that even when at rest the static configuration is not totally frozen, but continues to relax to even more favorable adhesive configurations (as F_{static} presumably approaches F_{equil}).

There is another characteristic relaxation time associated with the stick–slip phenomenon; this is the "slip time," τ_0 spent in the slipping regime (see Fig. 12), and is the time it takes the molecules to go from the sticking configuration back to the sliding configuration. One might expect τ_0 to be correlated with τ, and, in general, it was indeed found that molecules having longer latency times in the resting regime also had longer slip times in the slipping regime. For example, τ and τ_0 were longest for the branched molecules and shortest for the small spherical molecules. But some trends were not the same for τ and τ_0. For example, τ_0 does not increase with load whereas τ does, and τ_0 decreases with increasing sliding velocity whereas τ increases with v. We may note that unlike τ, which depends only on the (intrinsic) properties of the films, τ_0 probably also depends on the spring constant K (Fig. 2) and thus on the overall mechanical, or extrinsic, properties of the system.

Transition between stick–slip and smooth sliding with increasing sliding velocity

On increasing the sliding velocity from some low value the stick–slip frequency increases while the amplitude ΔF does not initially change (Figs. 10 and 11). However, above some velocity close to v_c the amplitude falls rapidly with increasing v, reaching zero at $v = v_c$. The constancy of ΔF at $v < v_c$ is consistent with the proposed model, viz., during the slip the molecules have enough time τ to nucleate into their frozen configuration so that ΔF is fully developed. But at $v > v_c$ there is not enough time to nucleate and there is no stick–slip. At some intermediate range of velocities, just below v_c, there is just enough time to nucleate a spike, but not a fully developed one, and in this regime the amplitude falls sharply over a narrow velocity range.

We have found that stick–slip also disappears when the lubricating film is contaminated, or contains another component. This is described later.

properties: the films become structured and are able to sustain a shear stress.

The properties of molecularly thin films can be very complex. They depend on (i) the structure of the liquid, (ii) the structure and commensurability of the surfaces, (iii) the surface–liquid interaction potential, (iv) the pressure between the surfaces, (v) the direction of shear, (vi) the shear rate, and—particularly for large asymmetric molecules—(vii) their history. Films in the solidlike state behave more like two-dimensional nematics or ductile solids undergoing plastic deformation. The transition from liquid to solid can be a continuous one, and so is unlike a normal liquid–solid transition, which is discontinuous. Also, unlike macroscopic states of matter, the positional and orientational ordering of the molecules in such films are imposed not only by the molecules themselves but also by the close proximity of the two surfaces. Thus, it is unlikely that either their equilibrium or dynamic properties can be properly described in terms of conventional concepts applicable to bulk phases.

The enhanced molecular order in very thin films is further affected by shear, which now imposes a preferred direction. Nonspherical molecules, such as chain molecules, can become more aligned during shear and therefore structurally even more solidlike (shear ordering). However, the dynamic properties of such films may become more liquidlike in the sense that their resistance to shear is lowered, resulting in shear thinning and stress overshoot. One must distinguish between films shearing at constant thickness, where shear thinning may occur, and films shearing under a constant external pressure, where it may or may not occur. In our experiments, as in most tribological experiments, the latter holds, while in other types of experiments and in many computer simulations the former constraint holds.

The static and dynamic properties of molecules in molecularly thin films are very different from those in the bulk liquids. The effective shear stress can be more than five orders of magnitude higher, and the relaxation times of short chain molecules can be many seconds, i.e., more than ten orders of magnitude slower than in the bulk liquids. This means that in many rheological situations the liquid solvent molecules may take longer to relax than the time of the actual shearing event, e.g., the collision time of two particles in a colloidal dispersion under shear. The dynamic interactions in such systems would be far from equilibrium.

Our results lead us to identify four distinct regimes and two characteristic relaxation times for the configurations of the film molecules during a shearing cycle: a slipping regime where the molecules become shear ordered over a time τ_0, a steady state sliding regime, a resting regime after sliding stops during which the molecules in the film first relax for time τ, and then "freeze" into an adhesive configuration (the sticking regime). Our descriptive model (Fig. 12) is consistent with some recent computer simulations of liquids under shear. Thus Heyes *et al.*,[33] in a molecular dynamics study of bulk liquids at high shearing rates, found that the liquid molecules shear order into layers to facilitate flow. They also found that this more solidlike appearance is accompanied by a decreased shear viscosity, which is normally associated with enhanced fluidity. They suggested this as a possible mechanism for "stress overshoot" (see Fig. 5). Schoen and co-workers[10–12] studied the equilibrium and dynamic behavior of Lennard-Jones (L-J) liquids between two structured solid L-J surfaces. They concluded that a liquid film can progressively freeze and melt (i.e., order to disorder) as the film thins from n layers to $n-1$ layers, but that this depends critically on the surface structure and the registry (commensurability) of the two surface lattices. The liquid–solid transitions and the long-range lateral order disappears for smooth, unstructured surfaces, even though the layering remains. Recent molecular dynamics simulations by Stevens *et al.*[34] of the lateral sliding of two solid surfaces separated by four liquid layers, resulted in a similar order–disorder cycle. These findings indicate the important effect of the surface structure on the properties of thin films confined between them.

Schoen *et al.*[21] also found that films thinner than five molecular diameters exhibited a critical shear stress when one structured surface is sheared past the other. The critical shear stress for one molecular layer was 2×10^8 N/m^2, which is about ten times the value we obtain for cyclohexane. The simulated value is, however, expected to be higher than any experimental value since in the simulation the two lattices were in perfect registry, the solid and liquid molecules were of the same size, and the gap distance remained constant during sliding (instead of the load or pressure remaining constant as occurs during experiments). Clearly, the simulated constraints were much more severe than the experimental.

Our results also indicate that two confining surfaces are needed for a liquid to become solidlike, i.e., that an isolated surface will not order or immobilize the liquid molecules adjacent to it to the same extent. This is consistent with earlier viscosity results on films thicker than ten molecular diameters[17] where the first layer of molecules adjacent to each surface was found to have the same viscosity as the bulk. But as soon as the second surface came closer than 10σ, the viscosity of the whole film increases dramatically. Thus, any experiment or theory on the state of a liquid adjacent to an isolated surface may not throw much light on the properties of the liquid when it is confined between two such surfaces.

Finally, concerning the more practical world of tribology, materials science and colloid science, our findings show that a liquid film of finite thickness can withstand both a normal hydrostatic pressure and a shear force, and that below a certain thickness one cannot apply continuum theories even with modified parameters. A modified or "effective" viscosity will not account for the rheological properties of molecularly thin liquid films when these behave more like solids than liquids. Solidlike behavior is also manifested by the occurrence of single stick–slip events or "spikes" during which the friction increases to a high value then suddenly relaxes. We have found that such spikes can occur not only when sliding starts but also when it stops. Friction accompanied by intermittent sticking is believed to be a prime cause of unwanted surface damage in a variety of different systems. Preliminary results with mixtures and impurities show that a second component, even in small amounts, can have a drastic effect on interfacial friction. This may have important implications for understanding the effects of humidity and atmospheric vapors on the action of lubricants, and on the friction at grain boundaries in solids. Our measured stress–strain curves for solidlike films appear to be very similar to the stress–strain curves associated with the plastic deformations of ductile solids.[26] The correlations between the behavior of molecularly thin liquid films with tribological and rheological phenomena, and material properties, will be pursued in a future paper.

ACKNOWLEDGMENTS

We are thankful to the Department of Energy for financial support to carry out this research project under DOE grant number DE-FG03-87ER45331, though this support does not constitute an endorsement by DOE of the views expressed in this article. We also thank Robin Bruinsma and Mark Robbins for their constructive comments.

[1] R. Evans and U. Marini Bettolo Marconi, J. Chem. Phys. **86**, 7138 (1987).
[2] B. K. Peterson, J. P. R. B. Walton, and K.E. J. Gubbins, J. Chem. Soc., Faraday Trans 2. **82**, 1789 (1986).
[3] I. Bitsanis, J. J. Magda, M. Tirrell, and H. T. Davis, J. Chem. Phys. **87**, 733 (1987); T. K. Vanderlick and H. T. J. Davis, *ibid.* **87**, 1791 (1988).
[4] J. N. Israelachvili, *Intermolecular and Surface Forces* (Academic, London, 1985).
[5] J. N. Israelachvili, Accounts Chem. Res. **20**, 415 (1987).
[6] J. N. Israelachvili, P. M. McGuiggan, and A. M. Homola, Science **240**, 189 (1988).
[7] D. Y. C. Chan and R. G. Horn, J. Chem. Phys. **83**, 5311 (1985).
[8] W. van Megen and I. K. Snook, J. Chem. Soc., Faraday Trans II. **75**, 1095 (1979); J. Chem. Phys. **74**, 1409 (1981).
[9] D. Henderson and M. Lozada-Cassou, J. Colloid Interface Sci. **114**, 180 (1986).
[10] C. L. Rhykerd, Jr., M. Schoen, D. J. Diestler, and J. H. Cushman, Nature **330**, 461 (1987).
[11] M. Schoen, D. J. Diestler, and J. H. Cushman, J. Chem. Phys. **87**, 5464 (1987).
[12] M. Schoen, J. H. Cushman, D. J. Diestler, and C. L. Rhykerd, Jr., J. Chem. Phys. **88**, 1394 (1988).
[13] H. K. Christenson and R. G. Horn, Chemica Scripta. **25**, 37 (1985).
[14] H. K. Christenson, J. Phys. Chem. **90**, 4 (1986).
[15] J. N. Israelachvili, J. Colloid Interface Sci. **110**, 263 (1986).
[16] J. N. Israelachvili, Colloid Polymer Sci. **264**, 1060 (1986).
[17] J. N. Israelachvili and S. J. Kott, J. Colloid Interface Sci. **129**, 461 (1989).
[18] P. M. McGuiggan, J. N. Israelachvili, M. L. Gee, and A. M. Homola, Mat. Res. Soc. Symp. Proc. **140**, 79 (1989).
[19] A. M. Homola, J. N. Israelachvili, M. L. Gee, and P. M. McGuiggan, J. Tribology. **111**, 675 (1989).
[20] A. M. Homola, J. N. Israelachvili, P. M. McGuiggan, and M. L. Gee, Wear. **136**, 65 (1990).
[21] M. Schoen, C. L. Rhykerd, Jr., D. J. Diestler, and J. H. Cushman, Science **245**, 1223 (1989).
[22] J. van Alsten and S. Granick, Phys. Rev. Lett. **61**, 2570 (1988).
[23] D. Tabor, in *Microscopic Aspects of Adhesion and Lubrication, Tribology Series 7*, edited by J. M. Georges (Elsevier, New York, 1982), pp. 651–682; B. Briscoe and D. Tabor, J. Adhesion. **9**, 145 (1978); B. J. Briscoe and D. C. B. Evans, Proc. R. Soc. London. A. **380**, 389 (1982).
[24] L. D. Landau and E. M. Lifshitz, *Statistical Physics*, 3rd Ed. (Pergamon, New York, 1980); Vol. 5, Part 1, Chap. 8.
[25] R. Bruinsma, Nature **341**, 486 (1989).
[26] R. W. K. Honeycombe, *The Plastic Deformation of Metals* (Edward Arnold, London, 1984); Chap. 6.
[27] M. L. Gee and J. N. Israelachvili, Faraday Trans. (submitted).
[28] J. N. Israelachvili, S. J. Kott, M. L. Gee, and T. A. Witten, Langmuir. **5**, 1111 (1989); J. N. Israelachvili, S. J. Kott, M. L. Gee, and T. A. Witten, Macromolecules **22**, 4247 (1989).
[29] P. M. McGuiggan and J. N. Israelachvili, Mat. Res. Soc. Symp. Proc. **138**, 349 (1988); J. Materials Research (submitted).
[30] D. H. Buckley and K. Miyoshi, in *Structural Ceramics Vol. 29 Tribological Properties of Structural Ceramics*, edited by J. B. Wachtman, Jr., (Academic, San Diego, 1989), p. 300.
[31] R. G. Horn and J. N. Israelachvili, J. Chem. Phys. **75**, 1400 (1981).
[32] H. K. Christenson, J. Fang, and J. N. Israelachvili, Phys. Rev. B **39**, 11750 (1989).
[33] D. M. Heyes, J. J. Kim, C. J. Montrose, and T. A. Litovitz, J. Chem. Phys. **73**, 3987 (1980).
[34] M. J. Stevens and M. O. Robbins (private communication).
[35] H. K. Christenson, J. Chem. Phys. **78**, 6906 (1983).
[36] H. K. Christenson, D. W. R. Gruen, R. G. Horn, and J. N. Israelachvili, J. Chem. Phys. **87**, 1834 (1987).
[37] H. K.Christenson and C. E. Blom, J. Chem. Phys. **86**, 419 (1987).
[38] C. M. Mate, G. M. McClelland, R. Erlandsson, and S. Chiang, Phys. Rev. Letts. **59**, 1942 (1987).

Structure and Shear Response in Nanometer-Thick Films

PETER A. THOMPSON,[a] MARK O. ROBBINS,[b,*] AND GARY S. GREST[c]
[a]Department of Mechanical Engineering and Materials Science, and Center for Nonlinear Dynamics and Complex Systems, Duke University, Durham, North Carolina 27708-0300, USA
[b]Department of Physics and Astronomy, The Johns Hopkins University, Baltimore, Maryland 21218, USA
[c]Corporate Research Science Laboratories, Exxon Research and Engineering Co., Annandale, New Jersey 08801, USA

(Received 29 January 1995; accepted in revised form 23 February 1995)

Abstract. Simulations of the structure and dynamics of fluid films confined to a thickness of a few molecular diameters are described. Confining walls introduce layering and in-plane order in the adjacent fluid. The latter is essential to transfer of shear stress. As the film thickness is decreased, by increasing pressure or decreasing the number of molecular layers, the entire film may undergo a phase transition. Spherical molecules tend to crystallize, while short-chain molecules enter a glassy state with strong local orientational and translational order. These phase transitions lead to dramatic changes in the response of the film to imposed shear velocities υ. Spherical molecules show an abrupt transition from Newtonian response to a yield stress as they crystallize. Chain molecules exhibit a continuously growing regime of non-Newtonian behavior where the shear viscosity drops as $\upsilon^{-2/3}$ at constant normal load. The same power law is found for a wide range of parameters, and extends to lower and lower velocities as a glass transition is approached. Once in the glassy state, chain molecules exhibit a finite yield stress. Shear may occur either within the film or at the film/wall interface. Interfacial shear dominates when films become glassy and when the film viscosity is increased by increasing the chain length.

I. INTRODUCTION

The rheological behavior of fluids in highly confined geometries is a subject of immense technological importance. Knowledge of the behavior of confined fluids is crucial to understanding flows and phase transitions in porous media, the stability and dynamics of coatings, and friction and wear in boundary lubrication and nanomachines. Our knowledge of such thin films has increased greatly in recent years due to the development of faster computers and new experimental techniques. Examples of the latter include atomic force microscopy,[1,2] the quartz crystal microbalance,[3] and the surface forces apparatus (SFA).[4–10]

In this paper we use molecular dynamics simulations to study the structure and dynamics of fluids in a geometry that models the SFA.[4–10] In the SFA, fluids are confined between mica plates that are atomically flat over contact diameters of order 100 μm. The separation h between the plates can be varied with angstrom resolution from contact to many nanometers, and is monitored using optical interferometry. Normal and shear forces on the film are measured as the plate separation is decreased, or as the plates are sheared past each other at velocity υ.

SFA experiments have focused primarily on films of simple hydrocarbon liquids, such as cyclohexane, dodecane, tetradecane, hexadecane, and the silicone-liquid octamethlycyclotetrasiloxane (OMCTS).[4–12] Cyclohexane and OMCTS are roughly spherical molecules, while dodecane, tetradecane, and hexadecane are short, symmetric chains. Other fluids that have been studied include the branched molecules squalane and isoparaffin 2-methyloctadecane, and polymers, such as polydimethylsiloxane (PDMS) and perfluoropolyether (PFPE).[13,14] The latter is commonly used in the computer industry as a lubricant for thin-film magnetic disks where operating conditions typically require film thicknesses of order 1–5 nm.

These experiments have shown that the static and dynamic properties of fluids in highly confined geometries can be remarkably different from bulk properties. In most cases, bulk behavior persists until separations of order ten molecular diameters. At smaller

*Author to whom correspondence should be addressed.

separations there are pronounced oscillations in the normal load as a function of film thickness.[15,16] These are evidence of strong layering within the film.[17,18] Changes in the dynamic response are even more dramatic. The effective viscosities μ of thin films rise more than five orders of magnitude above bulk values. Relaxation times are fractions of a second rather than nanoseconds. When films are sheared rapidly, a wide variety of molecules exhibits the same power-law decrease of viscosity with velocity,[9,10,12] $\mu \propto \upsilon^{-2/3}$. In some cases, the response of thin films becomes solid-like: shear stresses do not relax to zero and there is a substantial yield stress.[4,6–8] Even those long-chain and branched molecules which do not exhibit layering still develop a yield stress after long equilibration or shearing. When the yield stress is exceeded, "solid" films often exhibit oscillatory stick-slip dynamics.[4,6,8]

Computer simulations have proved to be an effective tool in understanding the origin of some of these experimental observations. In addition to revealing the layering that was inferred from experiment and analytic work,[15,17–21] simulations show that the periodic potential of the wall lattice induces in-plane order in the adjacent fluid.[22–26] The degree of in-plane order was found to correlate with the viscosity at the solid/fluid interface and thus to determine the boundary condition for fluid flow.[23,27] As the film thickness decreases, confinement can induce a phase transition to a crystalline or glassy phase.[25,28–34] Near the glass transition, films exhibit a non-Newtonian response with the same power-law viscosity seen in experiments.[9,10,29–32,35] Studies of crystalline films of spherical molecules showed that stick-slip motion resulted from periodic phase transitions between sliding fluid and static solid phases.[28,31] This suggested that the pervasive phenomenon of stick-slip motion is generally associated with phase transitions between distinct sliding and static states.

In this paper we take a deeper look at the effect of molecular geometry on the structure and dynamics of thin films. We present results for both spherical molecules and freely-jointed, linear-chain molecules of varying length n. We first describe the changes in equilibrium structure and diffusion that occur as films enter a glassy or crystalline state. The transition pressure and temperature are shifted from bulk values by confinement. In some cases the nature of the transition is also different than in the bulk; i.e., the system enters a glassy rather than crystalline state. We next consider the dynamic response to shear. Changing the wall/fluid interaction parameters leads to very different types of flow profile: shear at the wall/fluid interface, uniform shear within the film, and shear between layers that are strongly adsorbed onto each wall. However, in each case we find the same power-law drop in viscosity with velocity at constant normal load, $\mu \propto \upsilon^{-2/3}$. Simulations at constant film thickness produce a less rapid fall in viscosity. Data for the diffusion and relaxation times are fit to the free volume theory for glass transitions.[36] Results for slower velocities and higher pressures than studied previously[29–32] reveal that the glassy phase exhibits a yield stress (implying $\mu \propto \upsilon^{-1}$). This has also been observed in recent experiments.[37]

The organization of the rest of the paper is as follows. The next section describes the simulation techniques used in our study. We then present results that illustrate the effect of confinement on the equilibrium properties of spherical and chain molecules. Section IV examines the corresponding changes in the dynamic response of films to an imposed shear velocity. Section V contains a brief summary and discussion.

II. SIMULATION GEOMETRY AND METHOD

In order to mimic the SFA, our simulations were performed in a planar, Couette geometry (Fig. 1). A thin film of spherical or short-chain molecules was confined between two solid plates parallel to the xy plane. Edge effects were minimized by applying periodic boundary conditions within this plane. Each wall consisted of $2N_w$ atoms forming two [111] planes of an fcc lattice with nearest-neighbor spacing d. Since mica is much more rigid than the confined films, we simplified the simulations described below by fixing wall atoms to

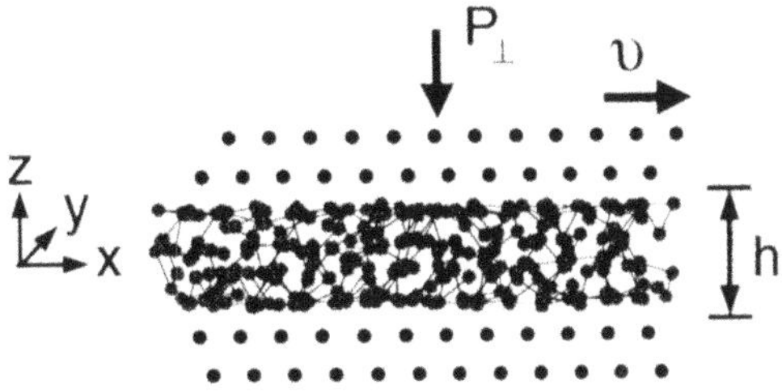

Fig. 1. Projection of particle positions onto the xz plane illustrating the simulation geometry. Periodic boundary conditions are imposed in the x and y directions. Gray circles connected by lines show the short-chain molecules that make up the film of thickness h. Solid circles indicate wall atoms. They sit on the sites of fcc [111] layers in the xy plane. A constant load $P_\perp$ is applied in the direction normal to the top wall, and the bottom wall is held fixed. Shear may be applied by moving the top wall at fixed velocity υ in the x direction. In most simulations, the wall was allowed to optimize its registry in the y direction as sliding occurred (eq 2.4).

their lattice sites. Previous work[23] shows that allowing the wall atoms to move does not produce qualitative changes in the behavior of the film. The main effect is to produce a small decrease in the ability of the wall to order the adjacent fluid at a given temperature T and pressure P, and a corresponding decrease in interfacial viscosity. The definition of the film thickness h becomes ambiguous at the atomic level. As shown in the figure and discussed below, h was defined to exclude the volume occupied by wall atoms.

The chain molecules comprising the fluid were simulated using a well-studied bead-spring model.[29,38] Monomers separated by distance r interacted through a truncated Lennard-Jones (LJ) potential

$$V^{LJ}(r) = \begin{cases} 4\varepsilon[(\sigma/r)^{12} - (\sigma/r)^6], & r < r_c \\ 0, & r > r_c \end{cases} \quad (2.1)$$

characterized by energy and length scales ε and σ, respectively. In some simulations the interaction was made purely repulsive by setting the cutoff radius $r_c/\sigma = 2^{1/6}$. This minimized the computation time and the temperature dependence. In other simulations $r_c/\sigma = 2.2$. Increasing r_c decreases the pressure needed to reach a given density, and increases the viscosity at that density. These changes do not alter the nature of the transitions in structure and dynamics that we describe below. However, they do shift the location of transitions in the (P,T) phase diagram.

The number n of monomers in a chain was varied from 1 (spherical molecules) to 24. Adjacent monomers along each chain were coupled through an additional strongly attractive, anharmonic potential

$$V^{CH}(r) = \begin{cases} -\frac{1}{2}kR_0^2\ln[1 - (r/R_0)^2], & r < R_0 \\ \infty, & r \geq R_0 \end{cases} \quad (2.2)$$

with $R_0 = 1.5\sigma$ and $k = 30\varepsilon\sigma^{-2}$. These values of R_0 and k have been used in previous studies[29,38] of polymer melts at comparable density and temperature. They were shown to eliminate unphysical bond crossings and bond-breaking. Another important aspect of this choice of parameters was that k was not so strong that there was a separation of timescales between motions induced by V^{LJ} and V^{CH}.

Wall and fluid atoms also interacted with a LJ potential, but with different energy and length scales: ε_w, σ_w, and r_{cw}. Increasing the ratio $\varepsilon_w/\varepsilon$ increases the viscosity of the wall/fluid interface relative to that within the fluid. The effective viscosity at the interface is also strongly affected by the relative size of wall and fluid atoms.[23] When the sizes are equal, it is easy for fluid atoms to lock into epitaxial registry with the substrate, and the viscous coupling is maximized. When the sizes are mismatched, the coupling is weaker. Chain molecules have two characteristic sizes. The intermolecular potential V^{LJ} has a minimum at $2^{1/6}\sigma$, while V^{CH} reduces the intramolecular separation to about 0.96σ in our simulations. The presence of an extra length scale frustrates epitaxial order, as discussed below.

The simulations were performed in an ensemble where the number of monomers N_f, the system temperature $T = 1.1\varepsilon/k_B$, and the normal pressure or load $P_\perp$ exerted on the top wall were all constant. The latter constraint allowed the plate separation h to vary, as in experiments. To maintain constant load in the z direction, we added the following equation of motion for the top wall

$$\ddot{Z} = (P_{zz} - P_\perp)A/M \quad (2.3)$$

where Z is the average z coordinate of atoms in the top wall, P_{zz} is the zz-component of the instantaneous, microscopic pressure-stress tensor[39] at the wall, and A is the area of the wall. The mass M is an adjustable parameter that controls the oscillation frequency of the wall. A small M leads to unphysically rapid oscillations, while a large M leads to a slow exploration of the parameter space associated with h. The times associated with the actual mass of mica sheets in an SFA are prohibitively long. We found that our results were relatively insensitive to M as long as the period of height oscillations was longer than that of the longest phonon period in our finite films. This condition was satisfied in the simulations described below by setting M equal to twice the product of the monomer mass and the number of wall atoms.

Shear was imposed by pulling the top plate at a fixed velocity υ. The shear stress was obtained from two independent and consistent methods: direct calculation of the forces exerted on the walls by fluid molecules, or from the xz component of the the microscopic pressure-stress tensor.[39] In the SFA, the transverse alignment of walls in the y direction is not rigidly fixed. The top wall can displace in this direction as it slides in response to stress from the film. We incorporated this motion in most of the simulations described below by coupling the y degree of freedom of the top plate to a spring of force constant κ_y. The extra equation of motion is:

$$\ddot{Y} = (P_{yz}A - \kappa_y Y)/M \quad (2.4)$$

where Y is the average y coordinate of atoms in the top wall. The transverse stress, P_{yz}, can generally be minimized by displacements of order d. We used a value of $\kappa_y/M = 0.05\varepsilon m^{-1}\sigma^{-2}$ that was small enough to allow such displacements, and large enough to prohibit displacements that were comparable to the system size.

Simulations with variable Y gave slightly smaller shear stresses than those with fixed Y, but did not change the scaling of shear stress with shear rate.

Constant temperature was maintained by coupling the y component of the velocity to a thermal reservoir.[40] Langevin noise and frictional terms were added to the equation of motion in the y direction

$$m\ddot{y} = -\frac{\partial}{\partial y}(V^{LJ} + V^{CH}) - m\Gamma\dot{y} + W \qquad (2.5)$$

where Γ is the friction constant that controls the rate of heat exchange with the reservoir, and W is the Gaussian-distributed random force from the heat bath acting on each monomer. Note that the variance of W is related to Γ via the fluctuation-dissipation theorem

$$< W(t)W(t') > = 2mk_BT\Gamma\delta(t-t'). \qquad (2.6)$$

If Γ is too small, the energy dissipated in shearing the system will cause the temperature to rise. If Γ is too large, the random force is not integrated accurately. We found the system was well-behaved with $\Gamma = 2\tau^{-1}$, where $\tau = (m\sigma^2/\varepsilon)^{1/2}$ is the characteristic LJ time. The equations of motion were integrated using a fifth-order, Gear predictor-corrector algorithm with a time step $\Delta t = 0.005\tau$.

The major difference between our theoretical ensemble and experiment is the constraint of constant N_f. In experiments, fluid can drain from the compressed region between the mica plates to equilibrate with the external chemical potential. However, the equilibration time may be extremely long due to the small ratio between h (~1 nm) and the contact diameter (~100 μm) and the immense viscosities of confined films.[41] Equilibration is more likely in steady sliding experiments where the contact area changes completely,[4,6,11] than in the experiments with small (~10 nm) sinusoidal motions that we compare to here.[7,10,37]

To determine N_f we used the fact that the film orders into well-defined layers of fairly constant density ρ_L. We set $N_f = m_l A \rho_L$, where m_l was the desired number of layers. This guaranteed that the nominal density of films was independent of the number of layers. Equilibrium configurations were obtained through a sequence of runs. The load was decreased to a small value for $10^4\Delta t$, so that chains could move rapidly. The load was then increased sequentially, allowing at least $10^5\Delta t$ for equilibration. To check that equilibrium was reached, we compared runs with different equilibration times and starting configurations. We also compared results with increasing and decreasing load, and runs equilibrated at fixed load and high temperature. The film thickness h was defined in terms of the distance h_0 between the innermost wall layers as $h = h_0 m_l/(m_l + 1)$. This is equivalent to subtracting a distance of half the layer spacing from h_0 for each wall to account for the volume taken up by wall atoms. Other methods of correcting for the size of wall atoms give essentially the same results. At most, the effective viscosity (eq 4) is changed by a constant multiplicative factor of order unity.

III. EQUILIBRIUM STRUCTURE

Two types of structure are induced in the fluid adjacent to a flat solid surface: Layering normal to the surface and epitaxial order in the plane of the surface. When a fluid film is confined between two solid surfaces, both types of order may span the entire film. Experimental consequences are oscillations in the normal force with film thickness,[15] and phase transitions to solid states that resist shear forces.[4,6-8]

Layering has been studied extensively for a variety of molecular geometries.[17-24,34,42-45] It is induced by the monomer pair correlation function $g(r)$, and the sharp cutoff in fluid density at the wall. Figure 2 shows plots of the density as a function of the distance between walls for monomers and 6-mers. Note the well-defined density peaks near the walls, and the decay of oscillations into the center of the film.

In general, the sharpness and height of the first density peak are determined mainly by the wall/fluid interaction, and increase with ε_w or $P_\perp$. The rate at which the density oscillations decay is determined by the decay of correlations in $g(r)$. Except near critical points, the decay length is a few molecular diameters. For fluids of simple spherical molecules, pronounced density

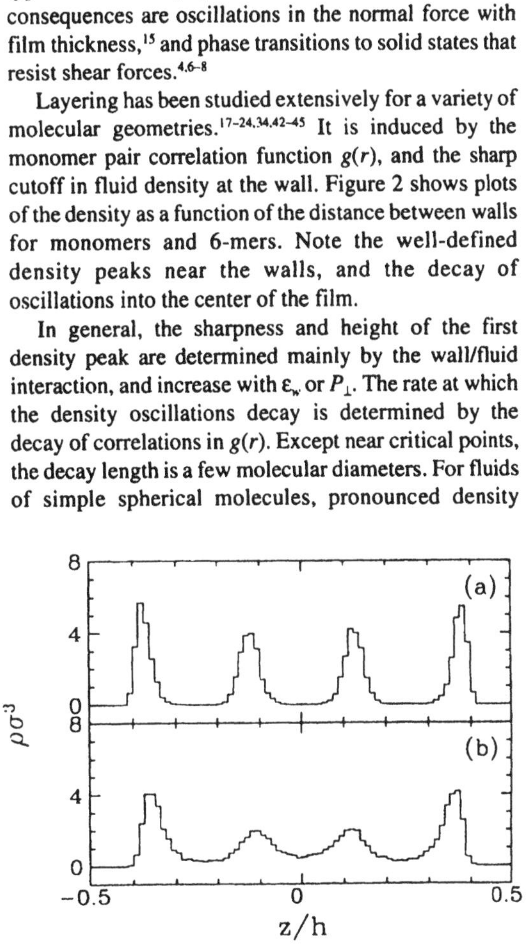

Fig. 2. Profiles of the monomer number density ρ as a function of z at $\upsilon = 0$ for films with (a) $n = 1$, $h = 3.41\sigma$ and (b) $n = 6$, $h = 3.22\sigma$. Other parameters were $\varepsilon_w/\varepsilon = 1$, $r_{cw} = r_c = 2.2\sigma$, $m_l = 4$, $P_\perp = 8\varepsilon\sigma^{-3}$, $k_BT = 1.1\varepsilon$, $N_f = 288$, $N_w = 72$, and $d = 1.2\sigma$. The number of monomers in bins of width $h/80$ along the z direction was averaged over 150τ. Note that z is normalized by the average plate separation h.

oscillations can extend up to ~5σ from an isolated solid surface. The competition between intra- and intermolecular spacings leads to less pronounced layering with chain molecules (Fig. 2). The degree of layering is fairly independent of chain length once n exceeds 6. Simulations with realistic potentials for alkanes show more pronounced layering near the wall due to a transition to an extended state of the chains in the first layer.[44,45]

In-plane order is induced by the in-plane variation, or corrugation, in the potential from wall atoms. As for layering, the rate of decay is determined by $g(r)$. It has been studied extensively for spherical molecules[17,18,22,23,25,26] but only recently for chain molecules.[35,46,47] In part, this is because most studies of chain molecules near surfaces have neglected the discrete lattice structure of the solid surface. In-plane order plays an essential role in determining the dynamics of fluids near solid interfaces. For example, we have shown that flow boundary conditions near solids are well correlated with the amount of in-plane order, and not with the degree of layering.[23,27]

The amount of in-plane order depends upon the relative size of wall and fluid atoms, as well as the strength of their interaction.[23] If the spacing between fluid atoms is equal to that of wall atoms, the atoms can lock in to all the local minima in the wall potential. In an earlier study of simple spherical molecules[23] we found crystallization of one or two fluid layers adjacent to a solid interface. If the wall and fluid atoms have different sizes, epitaxial order is frustrated.[23] As we now show, the presence of two different length scales in our chain molecules also frustrates in-plane order.

Two measures were used to quantify the degree of in-plane ordering. The first was the spatial probability distribution $\rho_l(x,y)$ of monomers in the lth layer relative to the unit cell of the solid lattice. The second was the two-dimensional static structure factor $S(k_x, k_y)$ evaluated in the lth layer according to

$$S_l(\vec{k}) = 1/N_l \left| \sum_{j=1}^{N_l} e^{i\vec{k}\cdot\vec{r}_j} \right|^2 \tag{3.1}$$

where N_l was the number of monomers i within the layer.

Figures 3 and 4 show $\rho(\vec{r})$ and $S(\vec{k})$ at two loads for two-layer films with $n = 1$ and 6, respectively. Values of S and ρ for each layer were averaged for 250τ. Then results for layers related by symmetry about the middle of the film were averaged to improve statistics. At the lowest applied load, $P_\perp = 4\varepsilon\sigma^{-3}$, there is little evidence of any in-plane order in ρ for either $n = 1$ or 6. The structure factor S shows weak rings that are characteristic of liquid order, and small peaks at the reciprocal lattice vectors $\vec{G}_i$ of the solid surface. These peaks represent a linear response to the surface potential. Their strength is best expressed[23] in terms of the Debye–Waller factor $e^{-2W} \equiv S(\vec{G}_0)/N_l$ at the smallest reciprocal lattice vector $\vec{G}_0$. The Debye–Waller factor is unity in an ideal crystal at zero temperature. Thermal fluctuations decrease e^{-2W} to about 0.6 at the melting point of a bulk crystal.[48] The value of e^{-2W} at $P_\perp = 4\varepsilon\sigma^{-3}$ is only 0.06 for both $n = 1$ and 6. Such small values are clear evidence of liquid structure.

As the load increases, the walls induce more in-plane order. At $P_\perp = 16\varepsilon\sigma^{-3}$, there are well-defined modulations in $\rho(\vec{r})$. The peaks in ρ are located above gaps in the adjacent solid layer. For $n = 1$, the monomers are nearly always over these gaps. Less order is induced in films of chain molecules because of the different intra- and intermolecular spacings. Note the pronounced ridges that connect the peaks in $\rho(\vec{r})$. These follow the lowest energy path between the gaps in wall atoms.

The monomer results for $S(\vec{k})$ and $\rho(\vec{r})$ clearly indicate crystalline order at $P_\perp = 16\varepsilon\sigma^{-3}$. For example, the mean-squared displacement of a fluid atom from the peaks in $\rho(\vec{r})$ is ~0.017 d^2, which is below the Lindemann criterion for melting.[48,49] Furthermore, $S(\vec{k})$ shows many sharp Bragg peaks, including several higher-order harmonics. The Debye–Waller factor, $e^{-2W} = 0.78$, is above the bulk melting criterion[48] and is consistent with the mean-squared displacement. In contrast, the chain molecules never crystallize within our simulation times. The degree of order increases with $P_\perp$ and then saturates. The Debye–Waller factor is only 0.43 at $P_\perp = 16\varepsilon\sigma^{-3}$ in Fig. 4.

Figure 5 shows the load dependence of film thickness, Debye–Waller factor, and in-plane diffusion constant D in 2-layer films.[29] Note the sharp transitions in the monomer results at $P_\perp = 7\varepsilon\sigma^{-3}$. At lower loads the structure is liquid-like and diffusion is rapid. At higher loads the film has crystallized and diffusion is suppressed. The lowest value of e^{-2W} within the crystalline phase (0.62) is very close to that in bulk crystals at their melting point.[48] There is a similar phase transition in bulk films, but it occurs at $P_\perp = 12\varepsilon\sigma^{-3}$. Using typical values for $\varepsilon = 200\ K$ and $\sigma = 3$ Å, this corresponds to a confinement-induced shift of 0.5 GPa. Note that the absolute loads are sensitive to the potential cutoffs r_c and r_{cw}, but the shift in the transition point is less so.

Results for chain molecules show more gradual changes.[29] The Debye–Waller factor saturates at about 0.4, and remains nearly unchanged if the load is increased by an additional order of magnitude. The only evidence of a transition to a new phase comes from measurements of dynamic quantities, such as the

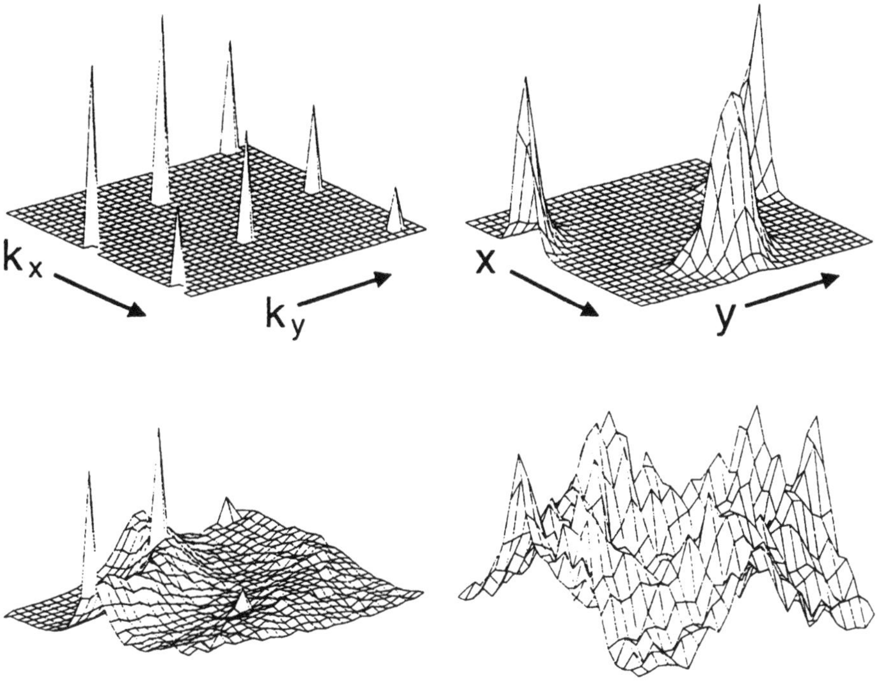

Fig. 3. Values of $S(\vec{k})$ (left) and $\rho(\vec{r})$ (right) for a two-layer monomer film at $P_{\perp} = 4$ (bottom) and $16\varepsilon\sigma^{-3}$ (top). The scale for each quantity is adjusted to fill the figure. All data were averaged over 250τ with $\upsilon = 0$, $\varepsilon_w/\varepsilon = 1$, $r_{cw} = r_c = 2^{1/6}\sigma$, $N_w = 144$, $N_f = 288$, $k_BT = 1.1\varepsilon$, and $d = 1.2\sigma$. Wall atoms lie on the sites of a triangular lattice and are at the corners and center of the unit cell shown for ρ. Sharp Bragg peaks are found in $S(\vec{k})$ at the reciprocal lattice vectors of the wall surface. At $P_{\perp} = 4\varepsilon\sigma^{-3}$, the peaks are barely above the circular ridge that is characteristic of fluid structure.

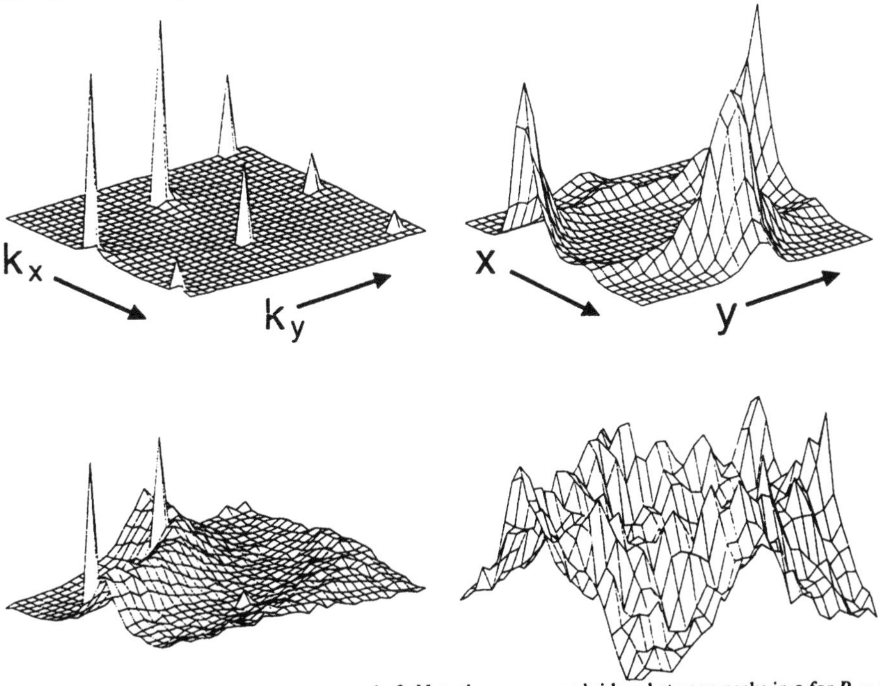

Fig. 4. Results for 6-mers with the same parameters as in 3. Note the pronounced ridges between peaks in ρ for $P_{\perp} = 16\varepsilon\sigma^{-3}$.

diffusion constant. The value of D drops rapidly below that in a bulk fluid at the same pressure. Studies of the viscous response, described in the next section, also show a dramatic slowing of molecular motions. These findings are evidence that the film undergoes a glass transition. Extrapolation to $D = 0$ using free volume theory indicates a transition near $P_\perp^G = 16\epsilon\sigma^{-3}$ (see Fig. 10). At loads above this value we find that films exhibit solid behavior over the timescales of our simulations: there is no relaxation of applied shear forces, and diffusion is undetectably small. Note that diffusion in bulk films extrapolates to zero at a much higher pressure, $P_{bulk}^G \approx 24\epsilon\sigma^{-3}$. As for spherical molecules, confinement produces large shifts (~0.8 GPa) in the bulk transition pressure. These shifts are comparable to the glass transition pressures (~0.5 GPa) of bulk lubricants at room temperature,[50] and one may expect that many lubricants vitrify in thin films.

Although chain molecules do not crystallize, there are pronounced changes in their configuration and orientation as they approach the glass transition. Figure 6 shows the distribution of end-to-end distances R_{1n} for $n = 6$ at three different pressures. At the lowest load, $P_\perp = 1\epsilon\sigma^{-3}$, the film thickness $h = 2.94\sigma$ is larger than the radius of gyration of a free three-dimensional chain,[51] and the density is nearly independent of z. As a result, the distribution of end-to-end distances is close to that for an ideal Gaussian chain.[52] The major difference is a cutoff in the distribution at small separations due to the hard-core repulsion between monomers. As $P_\perp$ increases, the distribution becomes more structured, indicating that monomers are being locked into a discrete set of locations. One also sees that highly stretched configurations become less likely. Previous studies of polymers confined to two dimensions show that they maximize their entropy when each polymer

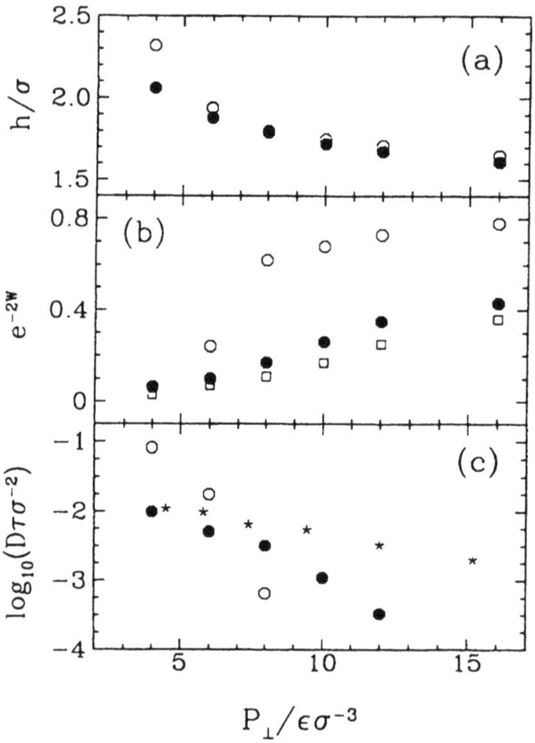

Fig. 5. Effect of varying load on (a) plate separation, (b) Debye–Waller factor, and (c) diffusion constant for motion parallel to the walls. All data are for $\upsilon = 0$ with $\epsilon_w/\epsilon = 1$, $r_{cw} = r_c = 2^{1/6}\sigma$, $d = 1.2\sigma$, $k_BT = 1.1\epsilon$, $N_w = 144$, $N_f = 288$, and $m_l = 2$. The open and filled circles denote films with $n = 1$ and 6, respectively. Stars indicate the bulk diffusion constant for $n = 6$. Results for $n = 20$ (squares) overlap $n = 6$ results in (a).

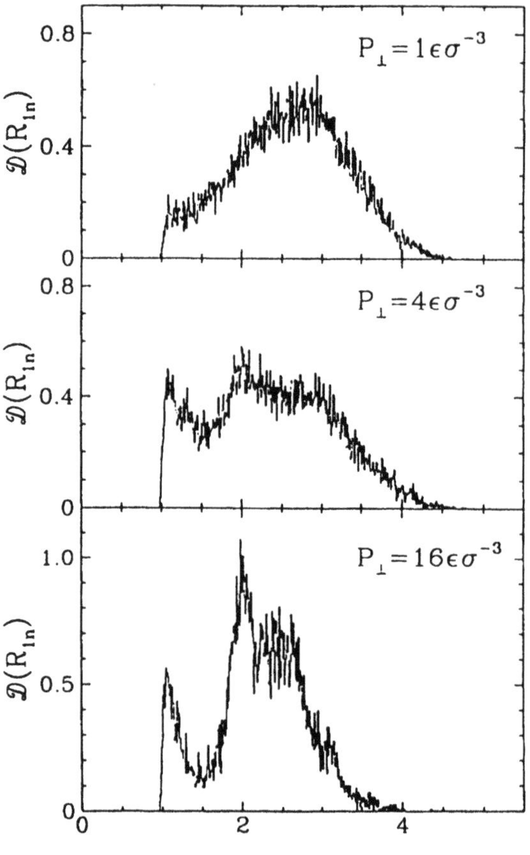

Fig. 6. Effect of varying $P_\perp$ on the distribution $\mathcal{D}$ of end-to-end distances, R_{1n}, of $n = 6$ chains in systems with $\upsilon = 0$. All parameters as in Fig. 5.

segregates into its own region of the plane.[53] The result is a collapse into a compact configuration of the chain. As $P_\perp$ increases from 1 to $16\varepsilon\sigma^{-3}$, h drops from a value greater than the radius of gyration to a smaller value, and the behavior crosses over from three- to two-dimensional.[51]

Figure 7 shows how the orientation of intramolecular bonds varies with load. We use the conventional coordinate system where θ is the polar angle relative to the z axis and ϕ is the azimuthal angle within the xy plane and relative to the x axis. Since the sequence along the chain is arbitrary, the sign of the vector connecting nearest neighbors is not defined. Thus we calculated the distribution of $\cos^2\theta$ for all intramolecular bonds. The distribution for completely random orientations, $1/(2\cos\theta)$, is shown by dashed lines in the plots. At $P_\perp = 4\varepsilon\sigma^{-3}$, the distribution is nearly random. For $P_\perp = 16\varepsilon\sigma^{-3}$, one finds sharp peaks at preferred orientations. The suppression of other orientations limits the ability of molecules to realign and results in a decreased diffusion rate. At the preferred values of $\cos^2\theta = 0$ and 0.6, bonds connect monomers in the same ($\theta = 90°$) or adjacent ($\theta \approx 40°$) layers. Similar peaks appear in the distribution of $\cos^2\phi$ near 0 and 0.75, or $\phi = 90$ and 30°. These are the directions of bonds between monomers that have locked epitaxially into minima in the wall potential or lie along the density ridges in Fig. 4.

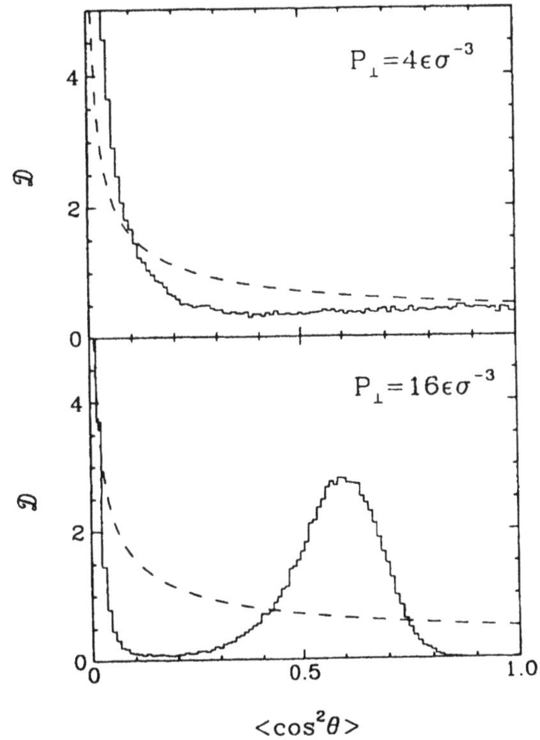

Fig. 7. Effect of $P_\perp$ on chain orientation. Here θ is the angle between the z-axis and a bond between adjacent monomers on a chain. Results for the distribution function $\mathcal{D}$ were averaged over 250τ for the system of Fig. 5.

The structural changes in 2-layer films of chains with $n > 6$ are nearly the same as the 6-mer results just described. In general, the structure becomes insensitive to chain length when the film thickness is much smaller than the radius of gyration in a bulk melt. As we now show, the shear response of the system may also become insensitive to n in this limit.

IV. RESPONSE TO STEADY SHEAR

SFA experiments probe the dynamic response of films by measuring the shear force on the walls as a function of velocity.[6,10,54] This measurement necessarily combines information about the shear response of the film and the interface. There is no way to determine whether shear occurs primarily within the film or is localized at the interface. It has also been impossible to examine structural changes in shearing films.

Experimentalists generally assume a no-slip boundary condition at the wall/fluid interface, i.e., that all shear occurs within the film. The shear rate $\dot{\gamma} \equiv \delta \upsilon_x/\delta z$ is then given by υ/h. An effective viscosity of the film μ can be determined using the macroscopic relation between force and shear rate in a lubricant film

$$f = \mu\dot{\gamma} = \mu\upsilon/h \tag{4.1}$$

where f is the force per unit area or shear stress. In a simple Newtonian fluid, μ is independent of shear rate

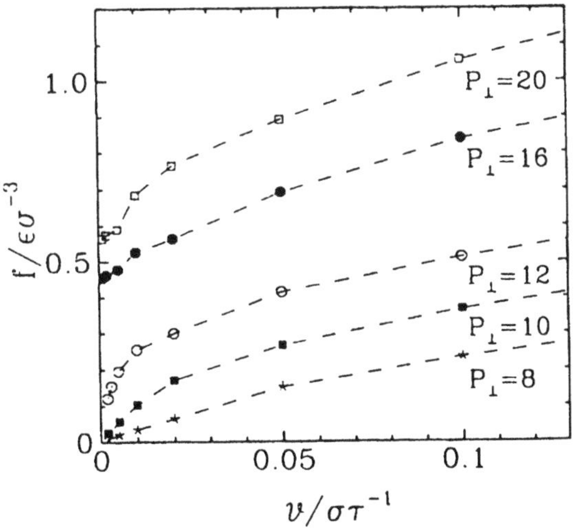

Fig. 8. Force per unit wall area, f, as a function of υ at the indicated loads. Other parameters as in Fig. 5. The force saturates at large velocities. In the limit $\upsilon \to 0$, f goes to 0 when $P_\perp < P_\perp^G$ and to a constant when $P_\perp > P_\perp^G$.

and f rises linearly with υ. In most non-Newtonian fluids, μ decreases with υ, and f rises sublinearly.[55,56]

Typical force–velocity curves for chain molecules are shown in Fig. 8, where $n = 6$, $\varepsilon_w = \varepsilon$, and $m_l = 2$. The top wall was sheared at constant velocity υ in the x direction, and allowed to adjust its registry in the y direction in response to shear stresses, as described above. Results for $P_\perp = 8$ and $12\varepsilon\sigma^{-3}$ rise rapidly from zero and then saturate as υ increases. The film is highly non-Newtonian and, as shown below, f scales roughly as $\upsilon^{1/3}$. At $P_\perp = 16$ and $20\varepsilon\sigma^{-3}$, the force approaches a constant value or yield stress as $\upsilon \rightarrow 0$. This behavior is characteristic of solid-on-solid friction,[57] and is further evidence that the film is in a glassy state at these loads.

Granick and coworkers[9,10] have plotted their shear response curves in terms of the effective viscosity and shear rate defined in eq 4.1. Figure 9a shows the data of Fig. 8 replotted in this manner. A Newtonian regime with constant viscosity μ_0 is seen at the lowest loads and shear rates. As $\dot{\gamma}$ increases, the system can no longer respond rapidly enough to keep up with the sliding walls. When $P_\perp$ is below the glass transition load $P_\perp^G$, there is a well-defined crossover shear rate $\dot{\gamma}_c$ at which the viscosity begins to drop. The shear-thinning appears to follow a universal power law $\mu \propto \upsilon^{-2/3}$ (implying $f \propto \upsilon^{-1/3}$) that is indicated by a dashed line on the figure.[9,10,29]

The inverse of $\dot{\gamma}_c$ corresponds to the longest structural relaxation time of the film.[55,56] As $P_\perp$ rises toward $P_\perp^G$, $\dot{\gamma}_c$ drops to zero and the relaxation time diverges. This slowing of dynamics in the film is directly related to the decrease in diffusion constant with $P_\perp$ seen in Fig. 5, and the rapid rise in μ_0 with $P_\perp$ seen in Fig. 9.

One common description of glass transitions is the free volume model.[36] It states that $\dot{\gamma}_c$ and D should vanish as $\exp(-h_0/(h-h_c))$ where h_c is the film thickness at the glass transition.[31] Figure 10 shows fits of both quantities to this form, with h_c corresponding to a glass transition near $P_\perp^G = 16\varepsilon\sigma^{-3}$. There is considerable debate about the proper scaling of dynamics near a glass transition, and even about the existence of a sharp transition. The success of the fit in Fig. 10 should be taken as evidence that the transition we see is like other glass transitions, rather than evidence for the free volume theory.

When $P_\perp$ exceeds $P_\perp^G$, there is a qualitative change in the response at small $\dot{\gamma}$. As shown in Fig. 9, μ falls more steeply with υ at $P_\perp = 16$ and $20\varepsilon\sigma^{-3}$. The slope on this log–log plot approaches -1 as $\dot{\gamma}$ decreases. This implies a constant shear force or yield stress, as already seen for these loads in Fig. 8. A constant shear force is typical of solid-on-solid sliding.[57] We show below that the film is nearly rigid at these loads, and that shear is largely confined to the wall/film interface. This constant shear force regime was not evident in previous simulations[29-32]

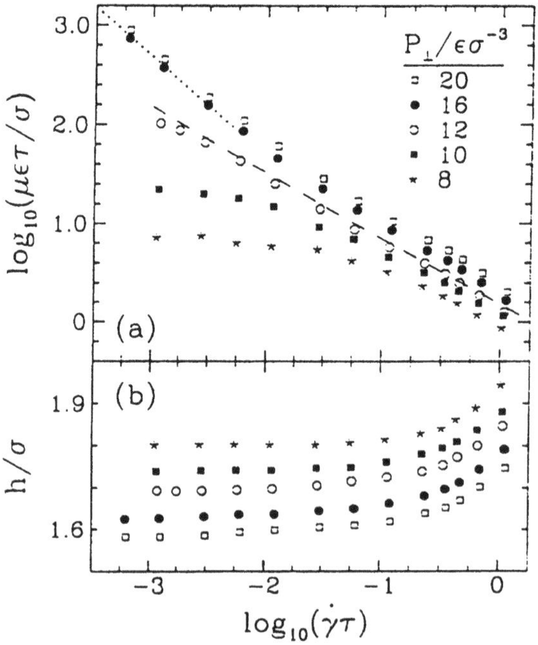

Fig. 9. Data of Fig. 8 replotted in (a) as $\log_{10}\mu$ vs. $\log_{10}\dot{\gamma}$. The corresponding film thickness at each load is shown in (b). At low $P_\perp$ and $\dot{\gamma}$ the value of μ is independent of $\dot{\gamma}_c$. Above $\dot{\gamma}_c$, μ begins to drop as $\dot{\gamma}^{-2/3}$, and h starts to rise. For $P_\perp > P_\perp^G$, μ drops as γ^{-1} at low $\dot{\gamma}$. The dashed and dotted lines indicate slopes of -2/3 and -1, respectively.

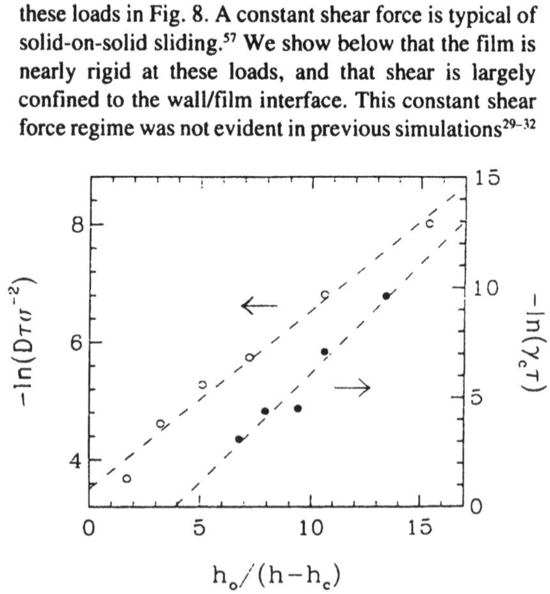

Fig. 10. Fit of variation in dynamic quantities $\dot{\gamma}_c$, and D to free volume theory. The logarithm of the inverse of both quantities should scale like $h_0/(h_c-h)$ where h_c is the thickness at $P_\perp^G$. In the fit, $h_0 = 1.58\sigma$ and $h_c = 1.57\sigma$, implying $P_\perp^G = 16\varepsilon\sigma^{-3}$.

because they did not extend to high enough loads and low enough velocities.

Granick and coworkers have found a very similar series of changes with load.[9,10,37] Indeed, they are even more dramatic, because a larger range of timescales is accessible to experiments. Increasing the load produced a decrease in $\dot{\gamma}_c$ by at least 10 orders of magnitude, and a rise in the Newtonian viscosity μ_0 by more than 5 orders of magnitude. At loads where $\dot{\gamma}_c$ was still nonzero,[9] there was a pronounced non-Newtonian regime where μ dropped as $\upsilon^{-2/3}$. At higher loads there is a crossover to constant shear force.[37]

A decrease in viscosity with increasing shear rate is normally accompanied by structural changes that reflect the fact that the system no longer has time to relax back to the equilibrium state.[55,56] The main structural change in our constant-load simulations is an increase in the film thickness with shear rate. Shear produces an additional normal force that separates the walls. This facilitates sliding and lowers μ. Note in Fig. 9b that h is constant in the Newtonian regime for a given load and begins to rise only when $\dot{\gamma}$ exceeds the value of $\dot{\gamma}_c$. Once the film is in the glassy phase, changes in h extend to the lowest practical shear rates. The changes in h may be coupled to structural changes within the film, but these are difficult to detect until relatively high shear rates ($\dot{\gamma} > 0.1\tau^{-1}$). In this regime, the number of intramolecular bonds which connect layers begins to decrease to facilitate flow.[35,46,47]

The above results imply that μ should drop less rapidly with $\dot{\gamma}$ if h is fixed. This is verified by the constant h data shown in Fig. 11. We find that μ decreases with a power law near –0.5 in this ensemble.[29,35] Previous experiments have been done at fixed $P_\perp$, and it would be interesting to test our simulation results against experiments at fixed h. Studies of structural changes in our fixed h simulations show that the adjacent layers of monomers move away from the walls to facilitate interfacial shear, and the layering at the wall becomes slightly more defined. As above, there is little structural change within the films.

Figure 12 shows how f depends on chain length in 2-layer films at $P_\perp = 16\epsilon\sigma^{-3}$. Note that the results become independent of chain length for large n. Only the results for $n = 1$ and 2 are noticeably different. This length independence may seem surprising since the bulk viscosity rises rapidly with n. However, at these high loads, films of long chains have frozen into a glassy state, and all the shear occurs at the interface. As noted in the previous section, the structure of the interface and the resulting shear coupling are insensitive to n when chains are long enough to span the system.

In Fig. 9, μ and $\dot{\gamma}$ were obtained from f and υ by assuming a no-slip flow boundary condition (eq 4.1). This assumption is invalid when shear localizes at the interface. Mechanical equilibrium requires that the shear *stress* be uniform throughout the system, however the shear *rate* may vary with position. In Fig. 13 we show the variation of υ_x with z for a number of systems. The actual shear rate within the film, $\dot{\gamma}_{film} = \delta\upsilon_x/\delta z$, is given by the slope of the flow profiles. For the monomer results shown in Fig. 13a, there is relatively little shear at the walls and $\dot{\gamma}_{film} \approx \dot{\gamma}$. As n increases to 6 (Fig. 13a), shear becomes predominantly confined to the wall/film interface. Figure 14 shows the variation of $\dot{\gamma}_{film}/\dot{\gamma}$ with n at $P_\perp = 16\epsilon\sigma^{-3}$. This ratio would be unity if

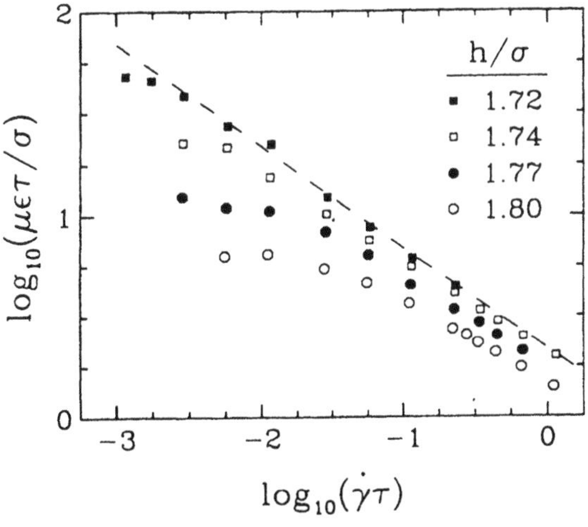

Fig. 11. Data for the system of Fig. 8, but with fixed h instead of fixed load. Note that the data now indicate that μ falls roughly as $\dot{\gamma}^{-1/2}$ (dashed line) above $\dot{\gamma}_c$.

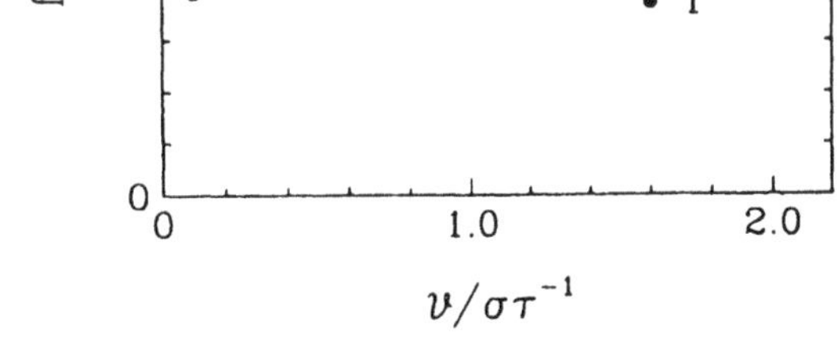

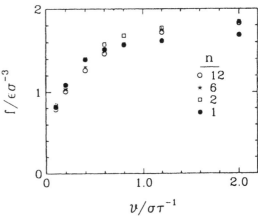

Fig. 12. Frictional force per unit area f as a function of υ and n at $P_\perp = 16\epsilon\sigma^{-3}$. Other parameters as in Fig. 5.

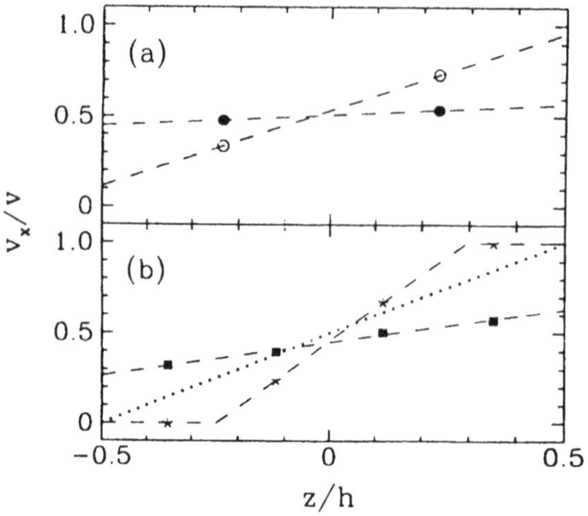

Fig. 13. Flow profiles for different interaction potentials. Panel (a) shows results for the system of Fig. 5 at $\upsilon = 0.05\sigma\tau^{-1}$ with $n = 1$ (open circles) and $n = 6$ (closed circles). Stars in (b) are for $n = 6$, $m_l = 4$, $P_\perp = 6\epsilon\sigma^{-3}$, $\epsilon_w/\epsilon = 3$, $r_c = 2^{1/6}\sigma$, $r_{cw} = 2.2\sigma$, and $\upsilon = 0.02\sigma\tau^{-1}$ (see also Fig. 15). Squares are for $n = 6$, $m_l = 4$, $P_\perp = 8\epsilon\sigma^{-3}$, $\epsilon_w/\epsilon = 1$, $r_c = 2.2\sigma$, $r_{cw} = 2.2\sigma$, and $\upsilon = 0.08\ \sigma\tau^{-1}$ (see also Fig. 16). Particle velocities in the x direction, υ_x, were averaged within layers and over at least 250τ. These averages were then normalized by the wall velocity υ and plotted against the center of each layer z divided by h. The flow profile for a no-slip boundary condition is indicated by the dotted line in (b).

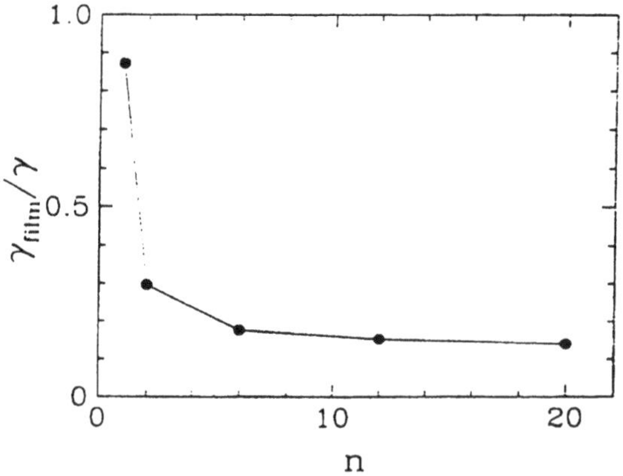

Fig. 14. Ratio of the actual shear rate within the film to the value inferred from a no-slip boundary condition as a function of n. All other parameters as in Fig. 5. For $n > 3$ only about 15% of the shear is within the film. The rest is at the two interfaces.

a no-slip boundary condition applied. While the no-slip approximation is very accurate for monomers, it fails rapidly as n increases to 2 and beyond. For long chains, only 15% of the shear occurs within the film at $P_\perp = 16\epsilon\sigma^{-3}$. This type of flow profile is frequently called plug-like flow.

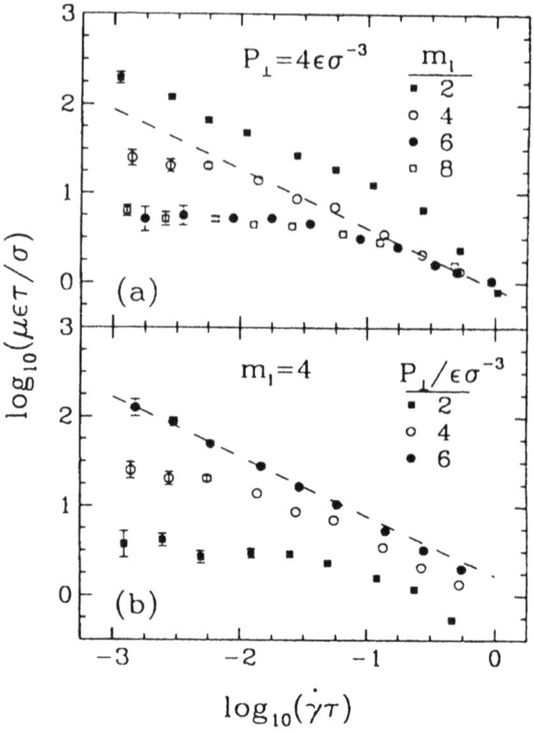

Fig. 15. Plots of μ vs. $\dot{\gamma}$ at (a) fixed $P_\perp = 4\epsilon\sigma^{-3}$ and varying m_l, and (b) fixed $m_l = 4$ and varying $P_\perp$. Dashed lines have slope −2/3. Both panels show the same behavior seen in Fig. 9a as the glass transition is approached. Other parameters were $\epsilon_w/\epsilon = 3$, $r_{cw} = 2.2\sigma$, $k_BT = 1.1\epsilon$, $r_c = 2^{1/6}\sigma$, $n = 6$, and $d = \sigma$.

The power-law shear-thinning shown for 2-layer films in Fig. 9 is quite universal. Figure 15 shows results for a system where ϵ_w was increased to 3ϵ in order to decrease the amount of shear at the interface.[29] Note that the glass transition can be approached in two different ways because the same load can be obtained with different numbers of layers.[15] Panel (a) shows results at fixed load with a decreasing number of layers, and panel (b) shows the effect of increasing load at $m_l = 4$. In either case, there is an increase in the Newtonian viscosity and a decrease in $\dot{\gamma}_c$ as the fluid becomes more confined. The shear-thinning region shows the same power-law dependence seen in Fig. 9 and experiment.[9,10] If the load is increased further into the glassy regime, shear will localize at the interface. In this limit one finds a regime where the shear force is velocity independent.

A typical flow profile for the system just discussed is shown in Fig. 13b. Note that while all the shear occurs within the film, it is not spread uniformly throughout the film. Over the entire range of power-law shear-thinning, shear is localized at a plane near the center of the film. Closer examination of particle motions reveals that each molecule adsorbs tightly to one of the two walls. These adsorbed films then slide past each other. Hence, the power-law shear-thinning is still associated with interfacial sliding. The only difference from the system of Fig. 5 is that the interface is within the film rather than at the wall/film interface.

To determine whether uniform shear could produce the same power-law response, we examined several other sets of parameters. Figure 13b also shows a flow profile for $\varepsilon_w/\varepsilon = 1$, $P_\perp = 8\varepsilon\sigma^{-3}$, and $r_{cw} = r_c = 2.2\sigma$. Note that the shear rate is constant within the film (υ_x is linear). Roughly half of the shear occurs in the film and the other half at the two wall/film interfaces. Figure 16 shows the variation of μ with $\dot{\gamma}$. As for other systems, there is a substantial range where $\mu \propto \dot{\gamma}^{-2/3}$. Thus we conclude that even films with uniform shear exhibit the same power law near the glass transition. Note that if the interface and film had different power laws the flow profile would change with $\dot{\gamma}$. We find the same division of shear between interface and film over the entire range of shear rates.

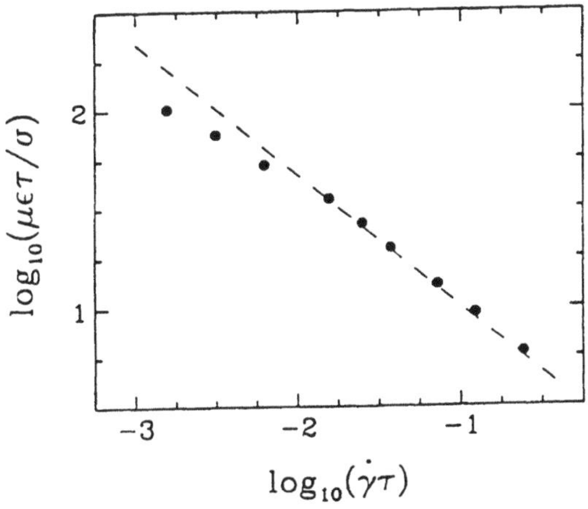

Fig. 16. Plot of μ vs. $\dot{\gamma}$ at fixed $P_\perp = 8\varepsilon\sigma^{-3}$ for $m_l = 4$, $\varepsilon_w/\varepsilon = 1$, $n = 6$, $r_{cw} = r_c = 2.2\sigma$, $d = 1.2\sigma$, and $k_BT = 1.1\varepsilon$. As before, the dashed line has slope −2/3.

DISCUSSION AND SUMMARY

The simulations reported here, and other related work, show that confinement can produce a rich variety of phenomena in nanometer-scale films. Structural changes including layering and in-plane order are now well documented.[18,20,21,23,42] We have shown here that both types of ordering are smaller in chain molecules due to the presence of more than one length scale. This effect should be even more pronounced in films of branched or other more complicated molecules.

Confinement may eventually lead to a phase transition within the film (Fig. 5).[25,29,31,32] In some cases, the transition is just a bulk phase transition that is shifted to a lower pressure or higher temperature. This is the case for crystallization of spherical molecules by walls whose atoms have nearly the same size. In other cases, the nature of the transition is changed. Chain molecules tend to be trapped in a glassy state. Spherical molecules can form a glassy phase or a new crystal structure when the walls are amorphous or when epitaxy is frustrated due to size differences.[23,28] Glassy phases also occur when the crystalline axes of the confining walls are not aligned, because the walls induce contradictory ordering tendencies on the film.[58] The mica walls in the SFA are generally not aligned and glassy phases may be the rule in experimental studies. This would explain why the same power-law non-Newtonian behavior is observed in films of chain molecules, such as alkanes, and of more spherical molecules, such as OMCTS,[9,10] An interesting observation from our simulations is that thicker films (3 or 4 layers) are more able to accommodate the conflicting influences of the walls and may have a higher yield stress than films with only 1 or 2 layers. Such films typically melt when they begin to slide and will be the topic of a later paper.

Three distinct types of dynamic shear response were found. Newtonian behavior occurs at large thicknesses and temperatures and at low pressures. In the opposite limits there is a constant shear stress or yield stress as $\upsilon \to 0$. The transition between the two types of behavior occurs abruptly if the film crystallizes, and gradually if it vitrifies. The third type of response occurs near the glass transition where there is an extended range of power-law shear-thinning, $\mu \propto \dot{\gamma}^{-2/3}$ ($f \propto \upsilon^{1/3}$). The power law holds from a crossover shear rate $\dot{\gamma}_c$ up to the frequencies of typical phonons. As the glass transition is approached, $\dot{\gamma}_c$ drops rapidly to zero. There is a corresponding drop in the diffusion constant and a rise in the value of μ at $\dot{\gamma} < \dot{\gamma}_c$.

These types of response describe the behavior of the entire system, including film and wall/film interfaces. Although the shear stress is independent of position, the shear rates within the film and at the interface may be very different. The shear rate at the interface decreases as the strength of the wall/fluid coupling (ε_w and/or r_{cw}) increases, and when d is such that monomers can easily lock into epitaxy with the wall. The shear rate within the film depends strongly on n, m_l, and their relative sizes. When chains are strongly adsorbed, shear may become localized at the midpoint of the film.

All flow profiles seem to be associated with the same power-law shear-thinning. A different power law was obtained only when the ensemble was changed from constant load to constant thickness. For constant film thickness, μ decreases as $\dot{\gamma}^{-1/2}$. Recent simulation studies of the bulk shear viscosity for the same molecular potential find a very similar shear-thinning exponent at constant volume.[59] It would be interesting to compare these results to bulk simulations at fixed pressure.

Several models have been developed to explain the power-law shear-thinning observed in experiment and simulations.[60–62] All the models find shear-thinning with an exponent of 2/3 in certain limits. However, they start from very different sets of assumptions, and it is not clear if any of these correspond to our systems. Two of the models work at constant film thickness[60,61] where we find an exponent of 1/2. One of these[61] also finds that the exponent depends on the velocity profile, while our results do not. The final model[62] is based on scaling results for the stretching of polymers under shear. While it may be appropriate for thick films, it can not describe the behavior of films which exhibit plug-like flow. It remains to be seen if the 2/3 exponent has a single explanation or arises from different mechanisms in different limits.

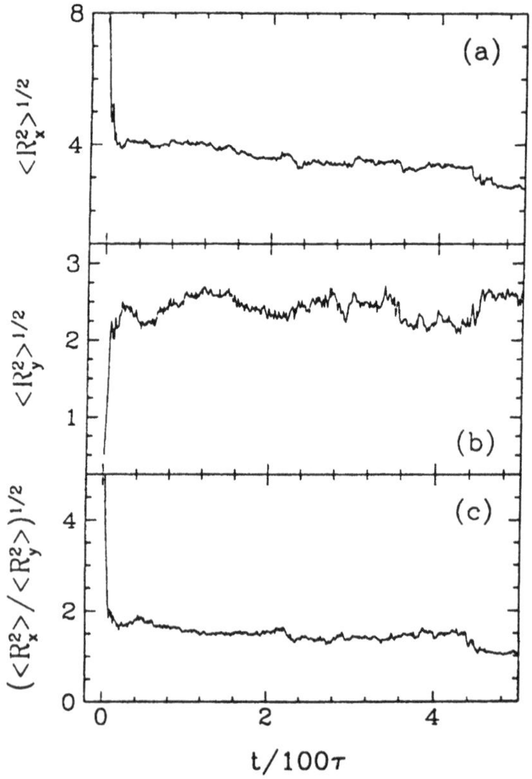

Fig. 17. Time dependence of mean-squared projections on to the x and y axes of the vectors connecting ends of chains. The initial state was fully aligned along the x-direction. The ratio of the projections decays to unity by the end of the figure, showing that the final state is isotropic. Longer runs and runs from other starting states show the same final isotropy.

Shear is known to stretch and align bulk polymers. One might expect similar changes in thin films, but confinement limits the ability of polymers to rotate in the xz plane. It has been suggested that chains might prefer to align normal to the xz plane in this case.[6] We found no tendency for shear-induced alignment in any direction when the film was much thinner than the bulk radius of gyration, and the shear rate was less than $0.1\tau^{-1}$. Figure 17 shows the decay of order in a film where all molecules were originally aligned in the x direction. The projections of the vector between the ends of the chains on to the x and y axes are plotted against time. There is a rapid decrease in anisotropy over ~10τ, and the ratio of the projections (Fig. 17c) reaches unity after about 450τ. There does appear to be memory of the direction of sliding for even longer times, but this must be reflected in more subtle details of the structure, such as the relative positions of monomers in the xy plane. Studies of the development of orientational order as the degree of confinement is reduced are underway.[63]

Acknowledgments. We thank S. Granick, J.N. Israelachvili, P.M. McGuiggan, and J. Klein for useful discussions. Support from National Science Foundation Grant DMR-9110004 and the Pittsburgh Supercomputing Center is gratefully acknowledged.

REFERENCES AND NOTES

(1) Mate, C.M.; McClelland, G.M.; Erlandsson, R.; Chiang, S. *Phys. Rev. Lett.* 1987, **59**: 1942.
(2) Meyer, E.; Overney, R.; Brodbeck, D.; Howald, L.; Lüthi, R.; Frommer, J.; Güntherodt, J.-J. *Phys. Rev. Lett.* 1992, **69**: 1777.
(3) Watts, E.; Krim, J.; Widom, A. *Phys. Rev. B* 1990, **41**: 3466. Krim, J.; Solina, D.H.; Chiarello, R. *Phys. Rev. Lett.* 1991, **66**: 181.
(4) Israelachvili, J.N.; McGuiggan, P.M.; Homola, A.M. *Science* 1988, **240**: 189.
(5) McGuiggan, P.M.; Israelachvili, J.N.; Gee, M.L.; Homola, A.M. *Mater. Res. Soc. Symp. Proc.* 1989, **140**: 79.
(6) Gee, M.L.; McGuiggan, P.M.; Israelachvili, J.N. *J. Chem. Phys.* 1990, **93**: 1895.
(7) Van Alsten, J.; Granick, S. *Phys. Rev. Lett.* 1988, **61**: 2570.
(8) Van Alsten, J.; Granick, S. *Langmuir* 1990, **6**: 877.
(9) Hu, H.-W.; Carson, G.A.; Granick, S. *Phys. Rev. Lett.* 1991, **66**: 2758.
(10) Granick, S. *Science* 1992, **253**: 1374.
(11) Israelachvili, J.N.; McGuiggan, P.M.; Gee, M.;

Homola, A.; Robbins, M.O.; Thompson, P.A. *J. Phys.: Condens. Matter* 1990, **2:** SA89.
(12) Carson, G.A.; Hu, H.-W.; Granick, S. *Tribol. Trans.* 1992, **35**: 405.
(13) Van Alsten, J.; Granick, S. *Macromolecules* 1990, **23**: 4856.
(14) Homola, A.M.; Nguyen, H.V.; Hadziioannou, G. *J. Chem. Phys.* 1991, **94**: 2346. Homola, A.M.; Street, G.B.; Mate, M. *MRS Bull.* 1990, **15**(3): 45.
(15) Horn, R.G.; Israelachvili, J.N. *J. Chem. Phys.* 1981, **75**: 1400.
(16) Israelachvili, J.N. *Intermolecular and Surface Forces.* 2nd ed.; Academic Press: London, 1991.
(17) Abraham, F.F. *J. Chem. Phys.* 1978, **68**: 3713.
(18) Toxvaerd, S. *J. Chem. Phys.* 1981, **74**: 1998.
(19) Plischke, M.; Henderson, D. *J. Chem. Phys.* 1986, **84**: 2846.
(20) Bitsanis, I.; Somers, S.A.; Davis, H.T.; Tirrell, M. *J. Chem. Phys.* 1990, **93**: 3427.
(21) Bitsanis, I.; Hadziioannou, G. *J. Chem. Phys.* 1990, **92**: 3827.
(22) Heinbuch, U.; Fischer, J. *Phys. Rev. A* 1989, **40**: 1144.
(23) Thompson, P.A.; Robbins, M.O. *Phys. Rev. A* 1990, **41**: 6830.
(24) Schöen, M.; Cushman, J.; Diestler, D.; Rhykerd, C. *J. Chem. Phys.* 1988, **88**: 1394.
(25) Schöen, M.; Rhykerd, C.L.; Diestler, D.; Cushman, J.H. *Science* 1989, **245**: 1223.
(26) Landman, U.; Luedtke, W.D.; Ribarsky, M.W. *J. Vac. Sci. Technol. A* 1989, **7**: 2829.
(27) Cieplak, M.; Smith, E.D.; Robbins, M.O. *Science* 1994, **265**: 1209.
(28) Thompson, P.A.; Robbins, M.O. *Science* 1990, **250**: 792. Robbins, M.O.; Thompson, P.A. *Science* 1991, **253**: 916.
(29) Thompson, P.A.; Grest, G.S.; Robbins, M.O. *Phys. Rev. Lett.* 1992, **68**: 3448.
(30) Thompson, P.A.; Robbins, M.O.; Grest, G.S. In *Computations for the Nano-Scale*; Blöch, P.; Joachim, C.; Fisher, A.J., Eds.; Kluwer: Dordrecht, 1993, p. 127.
(31) Thompson, P.A.; Robbins, M.O.; Grest, G.S. In *Thin Films in Tribology;* Dowson, D.; Taylor, C.M.; Godet, M., Eds.; Elsevier: Amsterdam, 1993, p. 347.
(32) Robbins, M.O.; Thompson, P.A.; Grest, G.S., Eds. *MRS Bull.* 1993, **18**(10): 45.
(33) Lupowski, M.; van Swol, F. *J. Chem. Phys.* 1991, **95**: 1995.
(34) Bitsanis, I.; Pan, C. *J. Chem. Phys.* 1993, **99**: 5520.
(35) Manias, E.; Hadziioannou, G.; ten Brinke, G. To be published.
(36) Grest, G.S.; Cohen, M.H. In *Advances in Chemical Physics*; Prigogine, E.; Rice, S.A., Eds.; Wiley: New York; 1981, Vol. 48, p. 455.
(37) Reiter, G.; Demirel, A.L.; Peanasky, J.; Cai, L.; Granick, S. *J. Chem. Phys.* 1994, **101**: 2606. Reiter, G.; Demirel, A.L.; Granick, S. *Science* 1994, **26**: 1741.
(38) Kremer, K.; Grest, G.S. *J. Chem. Phys.* 1990, **92**: 5057.
(39) Allen, M.; Tildesley, D. *Computer Simulation of Liquids*; Clarendon Press: Oxford, 1987.
(40) Grest, G.S.; Kremer, K. *Phys. Rev. A* 1986, **33**: 3628.
(41) Horn, R.G.; Hirz, S.J.; Hadziioannou, G.; Frank, C.W.; Catala, J.M. *J. Chem. Phys.* 1989, **90**: 6767.
(42) Magda, J.; Tirrell, M.; Davis, H.T. *J. Chem. Phys.* 1985, **83**: 1888.
(43) Kumar, A.K.; Vacatello, M.; Yoon, D.Y. *J. Chem. Phys.* 1988, **89**: 5206.
(44) Ribarsky, M.W.; Landman, U. *J. Chem. Phys.* 1992, **97**: 1937.
(45) Xia, T.K.; Ouyang, J.; Ribarsky, M.W.; Landman, U. *Phys. Rev. Lett.* 1992, **69**: 1967.
(46) Manias, E.; Hadziioannou, G.; Bitsanis, I.; ten Brinke, G. *Europhys. Lett.* 1993, **24**: 99.
(47) Manias, E.; Hadziioannou, G.; ten Brinke, G. *J. Chem. Phys.* 1994, **101**: 1721.
(48) Stevens, M.J.; Robbins, M.O. *J. Chem. Phys.* 1993, **98**: 2319.
(49) See, for example, Hansen, J.P.; McDonald, I.R. *Theory of Simple Liquids.* 2nd ed.; Academic Press: New York, 1986.
(50) Alsaad, M.A.; Winer, W.O.; Medina, F.D.; O'Shea, D.C. *J. Lubr. Technol.* 1978, **100**: 419.
(51) The results for Fig. 6 are for a number of monomers that is fixed by $m_1 = 2$, but at $P_\perp = 1\epsilon\sigma^{-3}$ the film thickness h is much larger than the thickness of two layers.
(52) Flory, P.J. *Statistical Mechanics of Chain Molecules;* Interscience: New York, 1969.
(53) Carmesin, I.; Kremer, K. *J. Phys. (Fr.)* 1990, **51**: 915.
(54) Georges, J.M.; Millot, S.; Loubet, J.L.; Tonck, A.; Mazuyer, D. In *Thin Films in Tribology;* Dowson, D.; Taylor, C.M.; Godet, M., Eds.; Elsevier: Amsterdam, 1993, p. 443.
(55) Graessley, W.W. *Adv. Polym. Sci.* 1974, **16**: 1.
(56) Ferry, J.D. *Viscoelastic Properties of Polymers.* 3rd ed.; Wiley: New York, 1980.
(57) Bowden, F.P.; Tabor, D. *Friction and Lubrication;* Oxford University Press: Oxford, 1958.
(58) Thompson, P.A.; Robbins, M.O.; Grest, G.S. To be published.
(59) Kröger, M.; Loose, W.; Hess, S. *J. Rheology* 1993, **37**: 1057.
(60) Rabin, Y.; Hersht, I. *Physica A* 1993, **200**: 708.
(61) Urbakh, M.; Daikhin, L.; Klafter, J. *Phys. Rev. E* 1995, **51**: 2137.
(62) deGennes, P.G. Private communication.
(63) Baljon, A.; Robbins, M.O.; Thompson, P.A. To be published.

Glasslike Transition of a Confined Simple Fluid

A. Levent Demirel and Steve Granick
Materials Research Laboratory, University of Illinois, Urbana, Illinois 61801
(Received 15 May 1996)

A simple globular-shaped liquid (octamethylcyclotetrasiloxane, OMCTS) was placed between two solid plates at variable spacings comparable to the size of this molecule and the linear shear viscoelasticity of the confined interfacial film was measured. Strong monotonic increase of the shear relaxation time, elastic modulus, and effective viscosity were observed at spacings less than about 10 molecular dimensions. Frequency dependence showed good superposition at different film thickness. The observed smooth transition to solidity is inconsistent with a first-order transition from bulk fluid to solidity. [S0031-9007(96)01142-8]

PACS numbers: 68.45.–v, 47.27.Lx, 62.20.Dc, 82.65.Dp

The literature is conflicting on the question whether the observed reduction of molecular mobility, as a fluid is confined in one or more directions between solid surfaces to be molecularly thin, stems from a phase transition [1–16]. For fluid particles of regular shape, some interpretations show the possibility of surface-induced crystallization [3–7], others of a glass transition [9–11], both at temperatures well above the respective bulk transition temperature. This problem, with ramifications in tribology [14], geology [17], and biology [18], is important to decide in order to resolve the senses in which confinement-induced solidity has a thermodynamic or kinetic origin. It is unsatisfactory that the same basic experiment [1,13–16] has been interpreted in both terms.

The ambiguity stems from the problem that previous measurements and simulations have both been made at fairly high rates or frequencies—leaving unclear the response at sufficiently long times (or low frequencies) to actually measure the relaxation time, the characteristic time of Brownian motion. Such a measurement should involve linear response in order to rule out the possibility of shear-induced changes.

Here we provide the needed data by considering the linear-response shear viscoelastic spectrum of the liquid that has been considered to be the prototypical small-molecule fluid for experimental studies. This is octamethylcyclotetrasiloxane (OMCTS), considered to be a model liquid for comparison with "Lennard-Jones" liquids because of the globular shape of this ring-shaped nonpolar molecule [19,20]. The data presented below definitively rule out the possibility of the abrupt first-order transition whose possibility has been much discussed. Instead, the solidification is continuous with diminishing film thickness and the viscoelastic spectra at different film thickness appear to scale with reduced variables.

The frequency dependence of the shear modulus was measured using a modified surface force apparatus described in detail previously [21–23]. Briefly, a droplet of sample fluid was confined between two atomically smooth crystals of muscovite mica, glued onto crossed cylindrical lenses whose separation was controlled using a surface forces apparatus. The surface separation was measured by interferometry. The radius of curvature of the mica sheets was large, ≈2 cm, giving a slitlike geometry when the surface separation was molecularly thin. To apply shear, sinusoidal shear forces were applied to one piezoelectric bimorph and the resulting sinusoidal displacement was monitored by a second piezoelectric bimorph; details of the piezoelectric circuitry are described elsewhere [21]. A lock-in amplifier (Stanford Research Instruments 850) was used to decompose the output into one component in phase with the drive and a second component out of phase with it, giving the elastic and dissipative shear forces as a function of the excitation frequency. The apparatus compliance was calibrated separately and taken into account in analyzing the data [22,23]. However, films thicker than three molecular dimensions were unaffected by this correction because the apparatus was then so much stiffer than the sample. Linear response, obtained for shear displacements less than ≈20% of the sample thickness, was verified. It is well known that a linear viscoelastic response characterizes the Brownian relaxation rate of a sample at rest [24].

Experiments were performed at 27 °C with P_2O_5 (a highly hygroscopic chemical) inside the sealed sample chamber. The sample of OMCTS (Fluka, purim grade) was used as received after control experiments showed similar behavior following further purification [25].

As smooth solid surfaces separated by OMCTS were pushed together, fluid drained smoothly until oscillatory forces of alternative attraction and repulsion were first detected at thickness ≈7 molecular dimensions. As shown by others in earlier reports [19,20], the period of oscillatory forces was an integral multiple of the molecular dimension, ≈9 Å. These oscillatory forces extended to larger thickness than in our earlier preliminary study [15]; the present measurements agree with literature values [19,20], as shown in the inset of Fig. 2. These oscillatory forces arise from the tendency of fluid to form layers parallel to the surface. Application of pressure caused the fluid to drain in discrete steps corresponding to squeezing out of successive molecular layers.

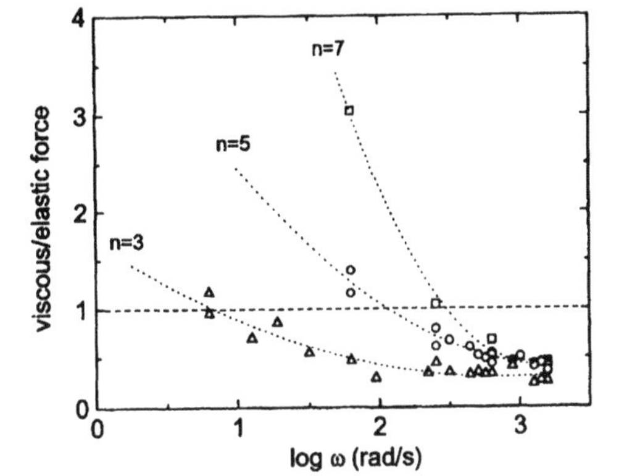

FIG. 1. Ratio of dissipative to elastic shear force plotted against logarithmic frequency for OMCTS confined between mica sheets at film thickness $n = D/\sigma = 3$, 5, and 7 molecular dimensions of the OMCTS molecule (indicated by numerals next to lines in the figure). Temperature was 27 °C. The relaxation time was quantified as $\tau_1 = 1/\omega_c$, with ω_c the frequency at which the ratio of dissipative and elastic force was unity. The associated rigidity at ω_c (1, 3, and 60 μN, respectively, for 2 Å deflection) is analyzed in Fig. 3.

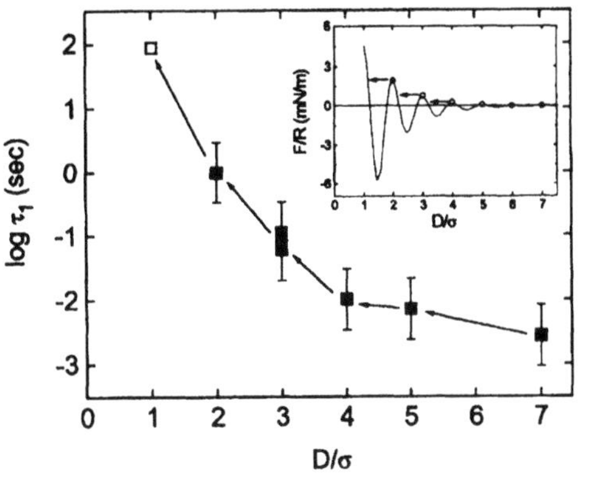

FIG. 2. Longest relaxation time τ_1 plotted logarithmically against film thickness expressed in multiples of the molecular dimension of OMCTS. Data at different thickness were acquired by squeezing the surfaces together progressively as shown in the inset (data for $D/\sigma = 6$ are missing owing to thermal drifts). Inset shows the force F (normalized by the radius of curvature of the mica sheets, R) required to squeeze the films to the given thickness. As the normal force was increased beyond the values indicated by circles, film thickness decreased abruptly by the thickness of a single molecule, as indicated schematically by the arrows. Control experiments show that for a given D/σ but $F/R < 0$, τ_1 was slightly less than plotted in the figure, but still within the indicated error bars. Open square: extrapolated τ_1 used for frequency-thickness superposition shown in Fig. 4. Filled squares: measured values.

Figure 1 shows shear measurements at three illustrative levels of film thickness: 3, 5, and 7 molecular dimensions. In order to avoid the need for normalization (problematical aspects of normalizing shear forces between curved surfaces were discussed previously [22]), the dimensionless ratio of dissipative to elastic force is plotted against logarithmic frequency.

The viscoelastic frequency spectra in Fig. 1 are typical of liquids in a frequency range near $1/\tau_1$, where τ_1 is the inverse longest relaxation time. The films relaxed shear stress predominantly in the manner of viscous liquids when the frequency was less than $1/\tau_1$, but predominantly in the manner of solids when the frequency was higher; at high frequencies the perturbation was not relaxed during the period of oscillation and energy was stored. From common experience, the analogous viscoelastic phenomenon is known for silly putty: It bounces when dropped on a floor (high-frequency response) but flows when it sits on a table.

These raw data show definitively that the change of viscoelastic response was continuous rather than abrupt with decreasing film thickness. To quantify τ_1, a useful measure was the inverse frequency at which the ratio of dissipative and elastic force was unity. In Fig. 2, τ_1 is plotted semilogarithmically against film thickness expressed as multiples of the molecular diameter of OMCTS. The inset of Fig. 2 shows the normal forces required to produce this change of thickness. The sudden slowing down of τ_1 by 3 orders of magnitude, as the film was squeezed from a thickness of 7 to 2 molecular dimensions, is striking. Though the limited range of thickness admits various functional fits, the increase of τ_1 appears to be more rapid than exponential. Well-known viscoelastic relations show that the effective viscosity of a viscoelastic body is proportional to τ_1 [24], indicating a similarly sudden increase of the viscosity. Control experiments confirmed this result also for noncrystalline surfaces [26].

Why are the error bars so large? The large error bars in Fig. 2 were intrinsic to these experiments: In repeated experiments, variations of the compression rate resulted in somewhat different τ_1 at a given film thickness. Others also have noticed dependence on the compression rate for experiments involving chain molecules [27]. This dependence on sample history is consistent with the hypothesis of a glassy state. If liquid exchange between the confined volume and the reservoir outside were suppressed by the sluggish mobility upon confinement, molecules might become trapped within the gap when the solid surfaces were squeezed together rapidly.

We now discuss elastic rigidity of these confined films: the in-phase component of the viscoelastic response. The elastic shear forces were normalized by the displacement amplitude (0.1–0.5 nm) to give the elastic shear constant $g'(\omega)$, which is roughly proportional to the shear storage modulus [24]. The dissipative shear forces were normalized in the same fashion to give the viscous shear constant $g''(\omega)$, which is roughly proportional to the shear loss modulus [24]. Figure 3 shows that the longer the relaxation time, the stiffer the confined fluid at $\omega = 1/\tau_1$.

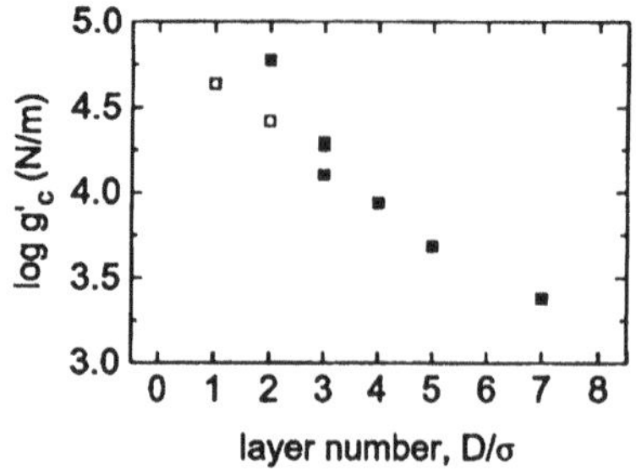

FIG. 3. Elastic spring constant of confined OMCTS at the frequency $(1/\tau_1)$ where elastic and dissipative shear forces are equal, plotted logarithmically against film thickness expressed in multiples of the molecular dimension $n = D/\sigma$. The elastic spring constant is the measured elastic shear force normalized by shear displacement (0.1–0.5 Å). Open squares: extrapolated values used for frequency-thickness superposition shown in Fig. 4. Filled squares: measured values.

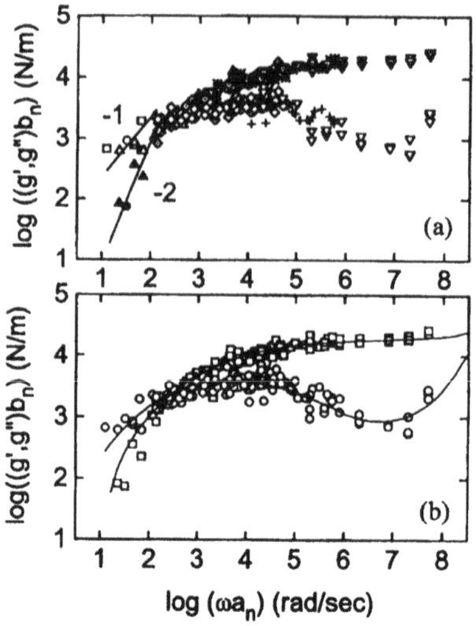

FIG. 4. Master curve describing frequency-thickness superposition of the viscoelastic response. The g' and g'' are the elastic and viscous shear forces, respectively, normalized by shear displacement (approximately 2 Å). The reference state is $n \equiv D/\sigma = 7$ molecular layers. Data at other film thickness were shifted vertically (shifts denoted b_n) and horizontally (shifts denoted a_n) as described in text, to make the elastic and dissipative spring constants coincide at ω_c. (a) The following symbols indicate real and imaginary components, respectively. Filled and open squares $(D/\sigma = 7)$. Filled and open circles $(D/\sigma = 5)$. Filled and open triangles to right $(D/\sigma = 4)$. Diamonds with and without center dot $(D/\sigma = 3)$. Stars and plus sign $(D/\sigma = 2)$. Triangles with and without center dot $(D/\sigma = 1)$. The logarithmic slopes of +1 and +2 are the classical response of a fluid dictated by the Kramers-Kronig relations. (b) Imaginary component (dotted curve) calculated from the real component (solid curve) using the Kramers-Kronig relation. Data taken at different D/σ are identified in (a).

This elastic rigidity g'_c was rather soft. It can be normalized roughly as follows. The effective contact area between curved surfaces in this experiment is on the order of 10 μm on a side; this gives an estimate of the frequency-dependent elastic shear modulus, with units of energy per unit volume. From this reasoning, it follows that the volume attributable to a k_BT of elastic energy ranged from ≈300 molecules (thickness of 7 molecules) to ≈60 molecules (thickness of 2 molecules). Clearly, an explanation should not be sought in single molecule-molecule interactions, but rather in collective interactions at larger length scales than molecular—again consistent with the hypothesis of glassy response.

To describe the viscoelastic spectrum at *all* ω (not just at the crossover frequency $1/\tau_1$), it was tempting to search for a method of reduced variables. To describe bulk liquids in the vicinity of a glass transition, reduced variables that relate time, temperature, and pressure are fundamental [24,28]. Though initial attempts to superpose the data solely by shifts along the frequency axis were not successful, superposition was possible with an additional vertical shift, as we now discuss.

Figure 4 shows a master curve consisting of data taken at six different film thicknesses but shifted on both the horizontal and vertical scales to coincide at $\omega = \omega_c$. The elastic and viscous spring constants are plotted against frequency on log-log scales. The reference state is the film of thickness $n \equiv D/\sigma = 7$ molecular dimensions. For films of lesser n, the data are shifted on the frequency scale by $a_n \equiv \tau_1(n)/\tau_1(7)$. On the vertical scale, they are shifted by $b_n \equiv g'(1/\tau_1(n))/g'(1/\tau_1(7))$. A viscoelastic property measured at frequency ω and a given thickness was equivalent to the same viscoelastic property measured at a different frequency ωa_n and different thickness n.

This superposition was self-consistent as shown by the following test. The raw data [Fig. 4(a)] were consistent with the classical slopes of +1 (viscous component g'') and +2 (elastic component g') required for a fluid by the Kramers-Kronig relation (the necessary relation between the real and imaginary parts of the response to a sinusoidal excitation) [24]. Furthermore, from the Kramers-Kronig relation, one spring constant could be calculated from the other; Fig. 4(b) shows g'' calculated from g'. The quantitative overlap of measured and calculated quantities shows consistency of the frequency-thickness superposition over the reduced frequency scale of 7 logarithmic decades. The shifts were not arbitrary; the measured viscoelastic response at $1/\tau_1$ determined the shift for superposition at every other ω.

We now discuss the main features of the phenomenological master curve and ignore aspects of fundamental interpretation. The key point is wide separation between two families of viscoelastic relaxation. Figure 4 shows a broad maximum in $g''(\omega a_n)$ at low reduced frequency, indicating one cluster of relaxation times. At much higher reduced frequency the $g''(\omega a_n)$ rises sharply, indicating the onset of additional relaxation processes. It is curious that these two clusters of relaxation times seem to be separated by a constant distance on the reduced frequency scale. This constitutes an important difference from the well-known α and β relaxations of bulk glasses, whose separations certainly increase with increasing pressure or diminishing temperature [24,28]. A point for future work will be to expand the experimental frequency window to better probe the relaxation processes at high ωa_n, to test more strictly whether the observed frequency-thickness superposition will hold as well at high frequency as at low. This, however, is not yet feasible with our present experimental setup.

In summary, the linear viscoelastic responses described here, for a confined molecule selected because of its ideally simple globular shape, are inconsistent with a first-order freezing transition from bulk fluidity to crystallinity. The nature of solidification is instead a smooth transition towards slower relaxation and increasing rigidity with decreasing film thickness. There is obvious phenomenological similarity to a bulk glass transition [24,28]; the usual roles of temperature or pressure are played by the film thickness, which in turn depends on the pressure that squeezes the films together. The frequency spectra of viscoelastic relaxation, measured at different confined film thickness, appear to scale with reduced variables.

We are indebted to Lenore Cai, Ali Dhinojwala, Jack F. Douglas, Jacob Klein, Mark O. Robbins, and Kenneth S. Schweizer for discussions. This work was supported by grants from the Exxon Research & Eng. Corp., the U.S. National Science Foundation (Tribology Program), and the U.S. Air Force (AFOSR-URI-F49620-93-1-02-41).

[1] For a survey of recent investigations, see *Dynamics in Small Confining Systems,* edited by J. M. Drake, J. Klafter, S.M. Troian, and R. Kopelman (Materials Research Society, Pittsburgh, 1995).

[2] Yu.B. Mel'nichenko, J. Schüller, R. Richert, B. Ewen, and C.-K. Loong, J. Chem. Phys. **103**, 2016 (1995).

[3] M. Schoen, C.L. Rhykerd, D.J. Diestler, and J.H. Cushman, Science **245**, 1223 (1989).

[4] P.A. Thompson and M.O. Robbins, Science **250**, 792 (1990).

[5] M.W. Ribarsky and U. Landman, J. Chem. Phys. **97**, 1937 (1992).

[6] Y. Rabin and I. Hersht, Physica (Amsterdam) **200A**, 708 (1993).

[7] M. Urbakh, L. Daikhin, and J. Klafter, Phys. Rev. E **51**, 2137 (1995).

[8] I.A. Bitsanis and C. Pan, J. Chem. Phys. **99**, 5520 (1993).

[9] P.A. Thompson, G.S. Grest, and M.O. Robbins, Phys. Rev. Lett. **68**, 3448 (1992).

[10] P.A. Thompson, M.O. Robbins, and G.S. Grest, Isr. J. Chem. **35**, 93 (1995).

[11] M. Schoen, S. Hess, and D.J. Diestler, Phys. Rev. E **52**, 2587 (1995).

[12] J.P. Gao, W.D. Luedtke, and U. Landman, Science **270**, 605 (1995).

[13] S. Granick, Science **253**, 1374 (1991).

[14] B. Bhushan, J.N. Israelachvili, and U. Landman, Nature (London) **374**, 607 (1995).

[15] S. Granick, A.L. Demirel, L. Cai, and J. Peanasky, Isr. J. Chem. **35**, 75 (1995).

[16] J. Klein and E. Kumacheva, Science **269**, 816 (1995).

[17] J. Byerlee, Geophys. Res. Lett. **17**, 2109–2112 (1990).

[18] M.S.P. Sansom, I.D. Kerr, J. Breed, and R. Sankararamakrishnan, Biophys. J. **70**, 693 (1996).

[19] H.K. Christenson, D.W.R. Gruen, R.G. Horn, and J.N. Israelachvili, J. Chem. Phys. **87**, 1834 (1987).

[20] R.G. Horn and J. Israelachvili, J. Chem. Phys. **75**, 1400 (1981).

[21] J. Peachey, J. Van Alsten, and S. Granick, Rev. Sci. Instrum. **62**, 463 (1991).

[22] S. Granick and H.-W. Hu, Langmuir **10**, 3857 (1994).

[23] A. Dhinojwala and S. Granick, J. Chem. Soc. Faraday Trans. **4**, 619 (1996).

[24] J.D. Ferry, *Viscoelastic Properties of Polymers* (Wiley, New York, 1980), 3rd ed.

[25] The critical element of sample purification was to avoid moisture. In control experiments with alkane liquids, the sample was also distilled, to remove potential nonpolar contaminants, or passed through a chromatography column consisting of silica beads, to remove potential polar contaminants. These additional procedures did not noticeably change the results.

[26] A similar glasslike transition was observed when the mica was coated with a self-assembled methyl-terminated monolayers of octadecyl hydrocarbon chains (OTE), rendering the surface noncrystalline and with low energy surface energy. Therefore a surface crystalline structure played no essential role in producing the observed smooth transition from fluidity to solidity [L. Cai and S. Granick (to be published)].

[27] S. Hirz, A. Subbotin, C. Frank, and G. Hadziioannou, Macromolecules (to be published).

[28] W. Götze and L. Sjögren, Rep. Prog. Phys. **55**, 241 (1992).

Nonlinear Bubble Dynamics in a Slowly Driven Foam

A. D. Gopal and D. J. Durian
Department of Physics and Astronomy, University of California, Los Angeles, California 90095-1547
(Received 12 June 1995)

Sudden topological rearrangement of neighboring bubbles in a foam occur during coarsening, and can also be induced by applied forces. Diffusing-wave spectroscopy measurements are presented of such dynamics before, during, and after an imposed shear strain. The rate of rearrangements is proportional to the strain rate, and the shape of the correlation functions shows that they are spatially and temporally uncorrelated. Macroscopic deformation is thus accomplished by a nonlinear microscopic process reminiscent of dynamics in the propagation of earthquake faults or the flow of granular media.

PACS numbers: 82.70.Rr, 05.40.+j, 83.70.Hq

Aqueous foams consist of a dense random packing of gas bubbles stabilized by surface active macromolecules [1,2]. The bubble shapes can vary from nearly spherical to nearly polyhedral, forming a complex geometrical structure insensitive to details of the liquid composition or the average bubble size. As a form of matter, foams exhibit remarkable mechanical properties that arise from this structure in ways that are not well understood. Namely, foams can support static shear stress, like a solid, but can also flow and deform arbitrarily, like a liquid, if the applied stress is sufficiently large [3,4]. The solidlike properties are due to surface tension and the shape distortion of bubbles in linear response to a small applied strain. The liquidlike properties, however, cannot be similarly understood by linear response since large deformations, though macroscopically homogeneous, are accomplished by microscopically inhomogeneous neighbor-switching rearrangements of bubbles from one tightly packed configuration to another. Intermittent structural rearrangements also occur in quiescent foams due to the alteration of packing conditions from the diffusion of gas from smaller to larger bubbles [5–8]. No matter what the driving force, all such dynamics are highly nonlinear and complex, involving abrupt topology changes and large local motions that depend on structure at the bubble scale. For example, in the Princen-Prud'homme model of foam as a two-dimensional periodic array of hexagonal bubbles, topological rearrangements happen instantaneously and simultaneously throughout the entire sample [9,10]. In a more realistic dense random packing of bubbles, however, the rearrangement events can be *localized,* occurring with variable size and duration in different regions at different times. The influence of randomness on the link between microscopic structure and macroscopic deformation has been studied by computer simulation [5,11–14]. Recently, Okuzono and Kawasaki [15] predicted that rearrangements in a slowly driven foam have a broad, power-law, distribution of event rate vs energy release, and thus exhibit self-organized criticality.

Experimentally, rearrangement phenomena are difficult to study because the opaque nature of foams restricts direct visualization to bubbles near a surface. Here, we report multiple light scattering measurements that take advantage of this property to probe the microscopic, bubble-scale, response of a bulk foam to a macroscopically imposed shear strain. The average bubble size is monitored from the average transmitted intensity, while the nature and rate of rearrangements are monitored from the intensity fluctuations via diffusing-wave spectroscopy (DWS) [16]. In addition to studying the link between macroscopically homogeneous shear strain and microscopically inhomogeneous bubble rearrangements, we also investigate the interrelationships between shear-induced dynamics and stability. Since foams are nonequilibrium systems, it is crucial to understand how foam rheology and bubble dynamics are affected by evolution, and, conversely, how foam stability is affected by flow.

Our measurements are performed on a commercial shaving foam which has been previously characterized by DWS [6,17]. It consists of polydisperse gas bubbles, 92% by volume, and coarsens by the diffusion of gas from smaller to larger bubbles; drainage and film rupture are not significant. The foam structure and its evolution are monitored from the probability T for incident light to be transmitted through an opaque slab of thickness L using the diffusion theory prediction, $T = (1 + z_e)/(L/l^* + 2z_e)$ where $z_e = 0.88$ and l^* is the photon transport mean free path [18]. For our material, l^* is 3.5 times the average bubble diameter, and transmission measurements show that it grows as nearly the $\frac{1}{2}$ power of time after 20 min. Such growth indicates a highly reproducible, self-similar bubble size distribution that is independent of initial conditions. Most measurements reported here were therefore taken after the foam had aged about 100 min, well into the scaling regime, where the average bubble diameter is 60 μm.

In order to shear the foam while simultaneously probing its bubble-scale response with DWS, we confine samples between parallel glass plates, 13×46 cm^2, arranged such that the narrow ends are open and that one plate can be slid at constant velocity over the other. To ensure that the strain is homogeneous and that the light scattering measurements and analyses are sound, we vary the plate spacing to be $L = 6$, 8, or 10 mm, and apply shear by

 0031-9007/95/75(13)/2610(4)$06.00

motion of the plate on either the incident or transmission sides. To minimize wall slip, both plates are rendered hydrophobic by treatment with dichlorodimethylsilane, and different combinations of smooth and roughened (via sandblasting) surfaces are employed. A previous study of strain-induced dynamics used flow through a constriction, where the strain is inhomogeneous and even involves plug glow [19,20].

The bubble rearrangement dynamics induced by either coarsening or application of shear were probed by fluctuations in the intensity of a speckle of transmitted light at $\lambda = 488$ nm. Results are expressed by the normalized electric field autocorrelation function, $g_1(\tau)$, vs the delay time τ. The inset of Fig. 1 shows typical data for 2 min collection durations before, during, and after shear was applied at rate 0.5 s^{-1}. In all three cases, the shape of $g_1(\tau)$ is nearly a single exponential as described by the thick-sample prediction of the theory of DWS for uncorrelated dynamics, $\sqrt{6\Gamma_1\tau}/\sinh\sqrt{6\Gamma_1\tau}$, where the first cumulant, Γ_1, or initial decay rate, is adjusted to fit the data. The quality of such fits, while always satisfactory, is best for quiescent samples and small strain rates. Introducing a second parameter for correlated motion due to a velocity gradient, as in Refs. [21,22], improves the fits but does not affect the result for Γ_1, showing that it provides a robust characterization of $g_1(\tau)$. Its time evolution is shown in the main plot of Fig. 1, and is contrasted with results for a quiescent, unsheared, foam. Evidently Γ_1 is larger during shear and smaller afterwards, as seen in previous work [19], but recovers to the quiescent behavior within about 10 min, demonstrating that any change caused by shear heals away in a coarsening time.

Results from monitoring the foam structure via the transmission probability, T, simultaneously with $g_1(\tau)$, show that application of shear affects the dynamics but not the average bubble size. The evolution of Γ_1 and l^* for a foam sheared at a steady rate of 0.5 s^{-1} until 130 min of age, by which time the average bubble diameter has grown by a factor of 3, is contrasted in Fig. 2 with results for a quiescent foam. The results for Γ_1 are larger during shear, and recover to the quiescent behavior soon after the shear is stopped. However, the results for l^*, and therefore the average bubble size, are identical for sheared and quiescent samples. Thus, in contrast with a recent prediction [13], *shear has minimal affect on foam structure and none at all on the coarsening processes.*

The effect of shear is primarily to rearrange bubbles without causing change in average size or rate of coarsening. Nevertheless, Γ_1 is temporarily suppressed following the cessation of shear. The magnitude and duration of the effect can be understood as follows. Since coarsening-induced rearrangements occur at random, there must be a random spatial distribution of regions under different local strains. Some regions have recently undergone rearrangement and are therefore under low strain, while in others the bubbles are distorted away from their relaxed shapes and are on the threshold of rearrangement. To the extent that flow homogenizes the strain field, the rate of coarsening-induced rearrangements should be suppressed and should remain so until inhomogeneities are reestablished during coarsening. This would require on the order of 10 min for 100 min old foam samples, consistent with all our recovery data independent of the size of the suppression. Furthermore, the number of rearrangements induced by sudden shear, and the corresponding suppression in dynamics afterwards, should depend on the total applied strain. This is tested in Fig. 3, where the fractional suppression in the cumulant following shear is plotted as a function of applied strain. These data were obtained with the roughened glass plates, and the strain rate was always larger than 0.5 s^{-1}. Data for the two sample thicknesses are indistinguishable, and show the expected behavior. First, there is no effect below a strain of about 5%, which may therefore be identified as the yield strain below which no rearrangements

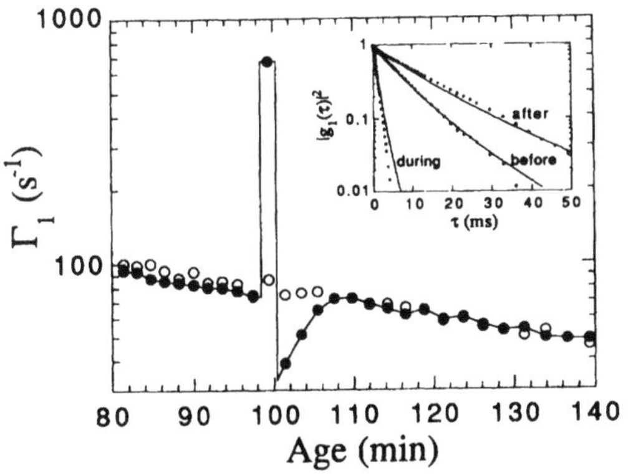

FIG. 1. First cumulant vs foam age for a quiescent sample (open circles) and one that is sheared from 98 to 100 min (solid circles connected by lines); the plate velocity for the latter is $V = 4$ mm/s, and the plate separation is $L = 8$ mm. The inset shows raw autocorrelation data (symbols) and fits (curves) before, during, and after application of shear.

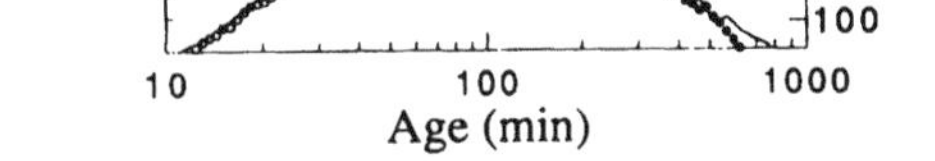

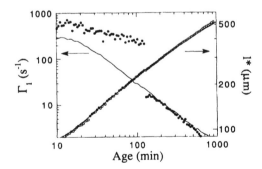

FIG. 2. First cumulant (closed circles, left axis) and transport mean free path (open circles, right axis) vs age for foam sheared at rate 0.5 s^{-1} until 130 min; results for an unsheared sample are shown for comparison by solid curves. Note that flow affects Γ_1 but not l^*.

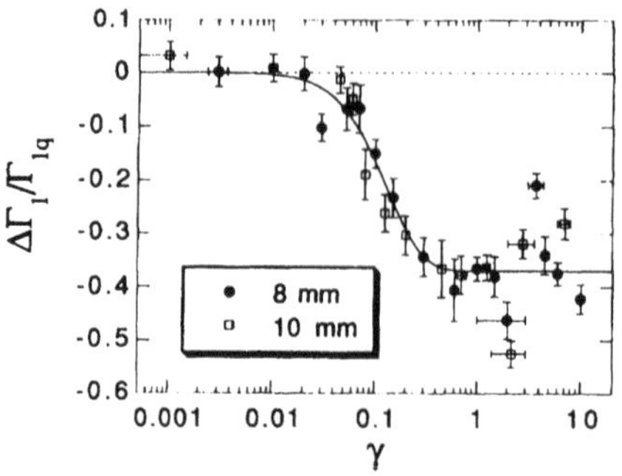

FIG. 3. Fractional change in the cumulant immediately following a shear step strain of size γ; the plate separations are as labeled, and the solid curve is a guide to the eye. Results are for roughened plates, and uncertainty in strain due to both uncertainty in total plate motion and wall slip is denoted by horizontal error bars.

are induced. Second, the observed suppression saturates above a strain of about 1, which may therefore be identified as the strain required to induce rearrangements everywhere and thus destroy correlations caused by coarsening. It is not understood why the degree of suppression is limited to only about 45%, but may reflect the difficulty of achieving an optimal random close-packed configuration with minimal strain inhomogeneity. Nevertheless, it is clear that the influence of shear on our foam is limited to inducing rearrangements and thereby destroying subtle correlations in the strain field that arise from the evolution process.

The nature and the rate of the rearrangement dynamics caused by shear can now be studied quantitatively through $g_1(\tau)$. If the motion of scattering sites in a diffuse photon path are random and uncorrelated, the theory of DWS predicts $g_1(\tau)$ to have the observed nearly exponential form with $\Gamma_1 \cong (L/l^*)^2/\tau_0$, where τ_0 is the average time for a single scattering site to move by a distance $k^{-1} = \lambda/2\pi$ [16]. This has been previously observed for both Brownian motion, where $\tau_0 = 1/Dk^2$ is set by the particle diffusion coefficient, and the sudden rearrangement of neighboring bubbles induced by coarsening, where τ_0 is proportional to the average time between rearrangements at a single scattering site [6]. For a quiescent sample at 100 min, the value of this characteristic time scale is $\tau_{0q} = 20$ s [17], much shorter than both the 2 min run durations and the 10 min needed for noticeable coarsening. The essentially exponential decay observed in Fig. 1 for the same foam under shear shows that strain-induced rearrangements are also uncorrelated. By contrast, if the shear deformation were homogeneous down to the bubble scale, then $g_1(\tau)$ would decay exponentially with τ^2 [21,22]. Therefore, macroscopically homogenous shear deformation is accomplished by a series of discrete microscopic bubble rearrangements that occur intermittently throughout the foam.

The average time τ_0 between random, uncorrelated bubble rearrangements can be gauged using the relation $\Gamma_1 \cong (L/l^*)^2/\tau_0$. Since flow has no effect on l^*, the entire change in Γ_1 is due to a change in the rate $1/\tau_0$. Therefore, the ratio of this rate in a sheared foam to that of a quiescent, unsheared, sample with the same thickness is given by $\tau_{0q}/\tau_0 = \Gamma_1/\Gamma_{1q}$, independent of L, and can be deduced for a single sample by measurements of the type in Fig. 1. Results are shown in Fig. 4 as a function of plate velocity for three slab thicknesses and smooth vs rough surface preparations. The two relevant time scales are V/L, the strain rate in the absence of wall slip, and $1/\tau_{0q}$, set by the rate of rearrangements in a quiescent sample; therefore, data are plotted vs the Deborah number $(V/L)\tau_{0q}$. The good collapse of data for different thicknesses, but not for different surface separations, implies that the strain rate is macroscopically uniform across the sample, though it may be less than or equal to V/L. In all cases, Γ_1/Γ_{1q} approaches 1 for small $(V/L)\tau_{0q}$, and increases monotonically for larger values of $(V/L)\tau_{0q}$.

The behavior seen in Fig. 4 can be explained as follows. The rate of *coarsening-induced* rearrangements is always proportional to $1/\tau_{0q}$, while the rate of *shear-induced* rearrangements, on the other hand, is a yield strain divided by the time needed to accumulate that strain via homogenous deformation, and is therefore equal to the applied strain rate. The latter has been observed in computer simulation [14] and suggested in analysis of flow experiments [20,23]. If these two processes are independent, the total rate of rearrangements is the sum of their individual rates, giving $\Gamma_1/\Gamma_{1q} = 1 + A(V/L)\tau_{0q}$ where A is a constant that increases with decreasing wall slip and with increasing rearrangement event size. By adjusting A, an excellent fit is obtained to the data in Fig. 4 for the case of roughened surfaces; the systematic deviation for the case of smooth surfaces could be due, e.g., to a degree of wall slip that increases with

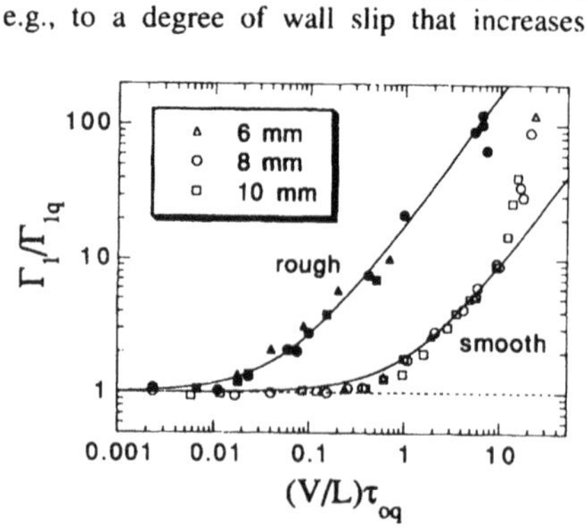

FIG. 4. Ratio of the first cumulant for a foam under shear to that of a quiescent sample of the same thickness, L, vs Deborah number; V is the velocity difference imposed between top and bottom plates and $\tau_{0q} = 20$ s. Open (solid) symbols are for smooth (roughened) plates. Solid curves are fits, indicating that coarsening-induced rearrangements dominate at small $(V/L)\tau_{0q}$ while shear-induced rearrangements increase proportional to the strain rate.

V/L. This supports our picture and also shows that flow at small strain rates is accomplished through time evolution according to the rate of coarsening-induced rearrangements rather than the rate of change of the bubble size distribution. Furthermore, the fitting result $A = 17$ for the case of minimal wall slip is sensible. The rearrangement event size as measured by DWS is the entire region where bubble displacement exceeds λ; this includes a core region where bubbles undergo topological rearrangement as well as a shell of surrounding bubbles that respond elastically. For the case of no wall slip, A must equal the ratio of event volume to core volume since only the latter contributes to the flow. If the event volume is 10 bubbles across, as estimated for coarsening [6], then the inner core undergoing rearrangement is $10/\sqrt[3]{A} \cong 4$.

While the observed dynamics is a nonlinear stick-slip process reminiscent of avalanches in models of granular media and earthquakes, our DWS measurements show that there is a single characteristic time scale, τ_0, between shear-induced events. The simplest explanation is that events have a characteristic volume ν, as discussed above, and are initiated at rate per unit volume R such that $\tau_0 = 1/R\nu$. However, this would contradict the self-organized criticality prediction of Ref. [15], since a broad distribution of event sizes and rates would generally give rise to a broad distribution of times between events and a correspondingly stretched exponential decay of $g_1(\tau)$. These simulations are for a two-dimensional foam with zero liquid content, ignoring coarsening and conservation of film material, but are otherwise realistic. Nevertheless, some such omission may probe unwarranted because it is difficult to reconcile the predictions with our observations. For example, large events may be killed by the finite system size or strain rate, and the cutoff would correspond to the characteristic time observed here. The finite system size is not a possibility, however, since our results are independent of sample thickness. To determine whether the strain rate is finite, we compare with the time scale for completion of rearrangements, which for our material is 0.1 s [23]. The strain rate is therefore infinitesimal for $(V/L)\tau_{0q} \ll 200$, which is well satisfied here. Another possibility is that there could be a spectrum of event sizes such that $R\nu \propto V/L$; however, an intrinsic cutoff would still be required since we see no dependence of the dynamics on L. These scenarios could be explored by further simulation of the mean-squared change in position of the scattering sites, as well as the event rate vs spatial size. Alternative models relaxing the assumptions of Ref. [15] would also be useful.

The study of sudden rearrangements of bubbles in a slowly driven foam offers a unique opportunity for general insight into stick-slip dynamics. By contrast with systems such as granular media or earthquake faults, where microscopic details can only be caricaturized, the microscopic physics of soap films is well understood and the structural randomness of the bubble packing is a natural and highly reproducible feature. Furthermore, the nonlinear rearrangement dynamics in foams can be directly probed by DWS as demonstrated here. Reconciling the contradictions between our observations and computer simulations should lead to a deeper understanding of not only stick-slip dynamics in general, but of the interrelationships between the structure, stability, and rheology of an important class of disordered materials as well.

We thank NASA for support of this work.

[1] J. H. Aubert, A. M. Kraynik, and P. B. Rand, Sci. Am. **254**, 74 (1989).

[2] D. J. Durian and D. A. Weitz, in *Kirk-Othmer Encyclopedia of Chemical Technology*, edited by J. I. Kroschwitz (Wiley, New York, 1994), Vol. 11, p. 783.

[3] J. P. Heller and M. S. Kuntamukkula, Ind. Eng. Chem. Res. **26**, 318 (1987).

[4] A. M. Kraynik, Annu. Rev. Fluid Mech. **20**, 325 (1988).

[5] D. Weaire and N. Rivier, Contemp. Phys. **25**, 55 (1984).

[6] D. J. Durian, D. A. Weitz, and D. J. Pine, Science **252**, 686 (1991).

[7] J. Lucassen, S. Akamatsu, and F. Rondelez, J. Colloid Interface Sci. **144**, 434 (1991).

[8] J. Stavans, Rep. Prog. Phys. **56**, 733 (1993).

[9] R. K. Prud'homme, Ann. Meet. Soc. Rheol., Louisville, KY (1981).

[10] H. M. Princen, J. Colloid Interface Sci. **91**, 160 (1983).

[11] D. Weaire and J. P. Kermode, Philos. Mag. B **50**, 379 (1984).

[12] K. Kawasaki, T. Okuzono, and T. Nagai, J. Mech. Behavior Mat. **4**, 51 (1992).

[13] D. Weaire, F. Bolton, T. Herdtle, and H. Aref, Philos. Mag. Lett. **66**, 293 (1992).

[14] T. Okuzono and K. Kawasaki, J. Rheol. **37**, 571 (1993).

[15] T. Okuzono and K. Kawasaki, Phys. Rev. E **51**, 1246 (1995).

[16] D. A. Weitz and D. J. Pine, in *Dynamic Light Scattering: The Method and Some Applications*, edited by W. Brown (Clarendon Press, Oxford, 1993), p. 652.

[17] D. J. Durian, D. A. Weitz, and D. J. Pine, Phys. Rev. A **44**, R7902 (1991).

[18] D. J. Durian, Phys. Rev. E **50**, 857 (1994).

[19] J. C. Earnshaw and A. H. Jafaar, Phys. Rev. E **49**, 5408 (1994).

[20] J. C. Earnshaw and M. Wilson, J. Phys. Condens. Matter **7**, L49 (1995).

[21] X.-l. Wu, D. J. Pine, P. M. Chaikin, J. S. Huang, and D. A. Weitz, J. Opt. Soc. Am. B **7**, 15 (1990).

[22] D. Bicout and G. Maret, Physica (Amsterdam) **210A**, 87 (1994).

[23] S. S. Park and D. J. Durian, Phys. Rev. Lett. **72**, 3347 (1994).

Kinetic theory of jamming in hard-sphere startup flows

R. S. Farr,[1,*] J. R. Melrose,[2,†] and R. C. Ball[1,‡]
[1]*Theory of Condensed Matter Group, Cavendish Laboratory, Madingley Road, Cambridge CB3 0HE, United Kingdom*
[2]*Polymer and Colloid Group, Cavendish Laboratory, Madingley Road, Cambridge CB3 0HE, United Kingdom*
(Received 25 July 1996; revised manuscript received 4 February 1997)

We consider the problem of hard spheres shearing from rest with hydrodynamic lubrication, but no Brownian forces. A theoretical model is presented, in terms of the aggregation of elongated clusters of particles, and predicts a jamming transition, where stress and average cluster size tend to infinity after a finite amount of strain. The model is compared with simulation data [Europhys. Lett. **32**, 535 (1995)], and predicts a critical volume fraction above which jamming will occur in macroscopic systems. [S1063-651X(97)01806-0]

PACS number(s): 47.50.+d, 83.50.−v

I. INTRODUCTION

A canonical problem in the rheology of colloids is the shear of a suspension of monodisperse hard spheres interacting hydrodynamically through a Newtonian solvent of viscosity η_0. We imagine the bulk material to be driven in simple shear by distant rheometer plates or, more elegantly, by Lees-Edwards boundary conditions [1] applied to an arbitrarily large periodic cell. The key issue in this paper is that for high enough volume fractions, a steady shear rate cannot be achieved. Shearing from rest, a logjam occurs at a finite strain angle.

For our system, the deformation rate is sufficiently small that inertial effects are negligible, that is, a Reynolds number defined on the particle diameter a is effectively zero. Stick boundary conditions hold at the particle surfaces. Hydrodynamics provides the only interactions; there are no Brownian, buoyancy, or other conservative forces, and indeed even the hard-sphere repulsion never actually affects the particles. This follows from the fact that the mobility for a pair of lubricating spheres falls to zero at contact so that the viscous solvent alone is sufficient to prevent particle overlap. The Péclet number Pe, defined by

$$\mathrm{Pe}=\frac{6\pi\dot{\gamma}\eta_0 a^3}{k_B T},$$

where $\dot{\gamma}$ is the rate of strain, is therefore formally infinite.

The problem of steady shear rate behavior has a venerable history starting with Einstein [2], who solved the case of infinite dilution, finding that the suspension viscosity η was increased over that of the solvent by a factor of $1+(5/2)\phi_v$ by the presence of hard spheres at a volume fraction $\phi_v \ll 1$. Following many phenomenological expressions at intermediate volume fractions, Frankel and Acrivos [3], assuming a cage model and lubrication interactions, proposed an expression for the viscosity close to the maximum packing limit ϕ_m, suggesting that it diverges close to this point as

$$\eta \propto \eta_0 (1-\phi_v/\phi_m)^{-1}.$$

This result was supported theoretically by Nunan and Keller [4], who derived numerical results for periodic arrays of spheres, but was challenged as a result for continuously sheared random dispersions by Marrucci and Denn [5]. They argued that hydrodynamics could only provide a much weaker divergence, logarithmic in $(1-\phi_v/\phi_m)$, which is unable to account for the large viscosities observed experimentally in hard-sphere colloids. They further pointed out [5] that if hydrodynamics were to generate such large viscosities it must be via the formation of extended structures in the flow.

The arguments of [3] and [5] are mentioned here as they are both incorporated into the model presented in the next section; a context in which they are no longer irreconcilable.

Turning to computer simulations, Bossis and Brady [6], who approximated the hydrodynamics by a low order moment expansion and lubrication terms, indeed observed clustering among the particles in two dimensions. Ball and Melrose [7,8], who modeled the hydrodynamics by just retaining the lubrication terms and so were able to simulate much larger three-dimensional (3D) systems, found that the clusters (defined by a criterion on the gaps) consisted of irregular chains of particles forming along the compression axis, and growing until they hit their own periodic images. Since the code [8] rigorously imposed the no overlap constraint on the spheres, these percolating clusters locked up the system at a finite strain at which the stress should tend to infinity and the gaps in the cluster to zero. In practice the gaps collapsed catastrophically to below machine accuracy and so the simulations had to be stopped before the stress had grown by more than an order of magnitude.

This "hydrodynamic logjam" is an intrinsically many body effect in which gaps collapse more quickly than can be accounted for by any pair theory. The structures involved are, however, tied to the size of the simulation cell, which with current techniques is limited to of order 10^3 particles, so that in even the largest cells the clusters are still small, comprising no more than of order 10^1 particles. This leaves open the effect that clustering may have on a macroscopic system, but is suggestive that at Pe$=\infty$ hard spheres will not flow; the response to an applied strain being transient and leading to a logjam.

*Electronic address: rsf10@phy.cam.ac.uk
†Electronic address: jrm23@phy.cam.ac.uk
‡Electronic address: rcb1@phy.cam.ac.uk

In this paper we present a model of the growth and aggregation of clusters upon shearing from rest, in an attempt to clarify the concept of the jam for an infinite system. The model provides predictions for a range of characteristics of the system, which may be tested against computer simulations and agree semiquantitatively for the early stages of the flow. We predict that above a lower critical volume fraction $\phi_l=0.515\pm0.02$, the logjam is not confined to small systems, but occurs in macroscopic flows, being characterized by the formation of an infinite cluster before a strain of 1. Below this volume fraction we presume, but do not show, that steady flow may be achieved.

From the computer simulations it appears that achieving steady flow may be facilitated by repulsive conservative interactions between the particles [7]. These may be provided by polymer coats, particle deformability, or Brownian forces which from the second order Langevin equation lead to a repulsive interaction [7]. At present, the best model for the flow of hard spheres with hydrodynamic and Brownian forces is that due to Brady [9,10], who derives a pair theory from a truncation of the hierarchy of integral equations for the pair distribution function. He finds a viscosity which diverges close to maximum packing as

$$\eta \propto \eta_0(1-\phi_v/\phi_m)^{-2}, \tag{1}$$

in accordance with the standard phenomenological expression as found, for example, in [11], where $\phi_m\approx0.63$ at low Pe and $\phi_m\approx0.71$ for an ordered system at high Pe. This leading order divergence is due solely to the Brownian forces, one factor of $1/(1-\phi_v/\phi_m)$ coming from the vanishing of the short time self-diffusivity, and the other from the divergence of the pair distribution function $g_2(r)$ at contact.

If this is true as claimed, for all Pe no matter how large, then it appears that the limit $\text{Pe}\to\infty$ is qualitatively different to the case of $\text{Pe}=\infty$. This receives backing from recent work of Brady [12] also in the context of a pair theory, who finds that at high Pe, boundary layers form in $g_2(r)$ in the compression directions when $r\approx a$. However, the pair theory is necessarily blind to any many body instabilities such as are considered here, and there is some evidence that a hydrodynamic jam of this type may be relevant to real systems at finite Pe and volume fractions substantially below the random close packed volume fraction $\phi_c\approx0.63$ or the ϕ_m of Eq. (1). For example, D'Haene [13] observes a sudden *discontinuous* shear thickening (jump in stress) at high Pe in controlled strain rate experiments on hard spheres above a certain volume fraction, while Frith *et al.* [14] observe the same phenomenon in controlled stress experiments. Computer simulations [7] with conservative interparticle repulsion at finite shear rate scaled on this force are also prone to locking up. We suggest that our model of a hydrodynamic jam represents the physical mechanism underlying these observations. Once the jam gets under way, the stress is formally divergent and swamps the stress [Eq. (1)] due to continuous shear, although in the experimental systems, other effects such as particle deformability must intervene to curb this growth.

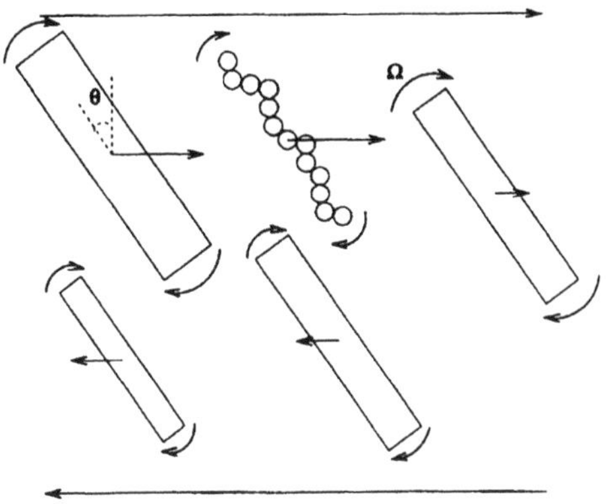

FIG. 1. Schematic picture of structures in the flow.

II. KINETIC MODEL

The principal structures in evidence in computer simulations of lubricating spheres [7] are irregular chains of nearly touching particles, which first form nearly parallel to the "compression axis" lying at 45° to the "flow" direction in the "flow"-"gradient" plane.

Figure 1 shows a schematic picture of the flow in which particles not belonging to the clusters are omitted. These clusters or rods of particles, forming at the start of the transient motion, are roughly parallel to one another and for simplicity we assume that all the rods which subsequently form and grow by aggregation constitute an approximately parallel population. It will turn out that this is a fairly consistent approximation as the jam occurs sufficiently quickly that any scatter in angles generated by the rotational component of the flow is minimal.

The rods contain extremely small interparticle gaps and are defined by this separation of length scales. We therefore take the rod lengths to be incompressible, and for a rod of j particles, each of diameter a, the length will be $L\approx ja$.

The bulk average deformation of the sample is simple shear; that is, a velocity field given by

$$\mathbf{v}=(0,0,\dot{\gamma}x),$$

where the x, y, and z axes are the "gradient," "flow," and "vorticity" directions, respectively. We make a mean field approximation and imagine each rod to be embedded in this average flow, composed of the other rods, particles, and solvent. We further imagine that the rod translates and rotates as would a rigid line of zero thickness or a streak of material composing the mean flow. Thus its center of mass moves with the mean velocity at that point, and the rod rotates at angular velocity

$$\Omega=-\dot{\gamma}\cos^2\theta, \tag{2}$$

where $\dot{\gamma}$ is the shear rate, and θ is the angle the rods make with the gradient direction, in the flow-gradient plane.

Viewed in a frame comoving and corotating with a rod, the surrounding mean flow is seen to be extension and compression along nonorthogonal axes (Fig. 2). This brings other rods towards it, tending to produce head to head collisions

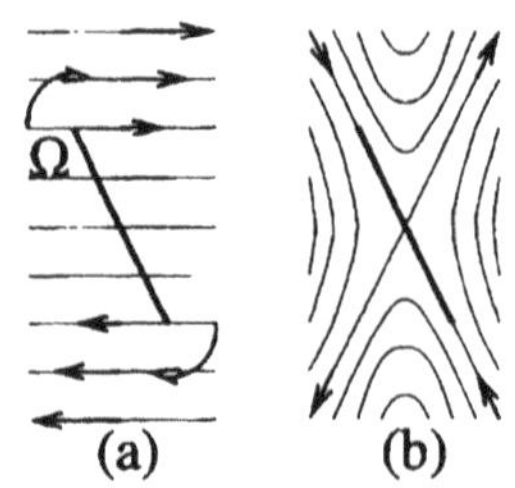

FIG. 2. (a) Mean flow field about a one dimensional rod immersed in a simple shear flow. (b) Same flow viewed in a frame corotating with the rod at angular velocity Ω [Eq. (2)]; as the rod does not fully corotate with the fluid, the apparent axes of compression and extension are not orthogonal.

and a new, longer rod. This growth by aggregation allows the large rods to rapidly increase in size and since the length of the rods determines their effect on the system, this in turn leads to a rapid increase in the bulk stress.

If we consider two rods about to collide, of lengths i and j particle diameters, respectively, we find that in this mean flow approximation, they approach one another with a relative velocity given by

$$V_{\rm rel}=\frac{\dot{\gamma}}{2}(ai+aj)\cos\theta\ \sin\theta.$$

In the collision we expect there to be some lateral offset. On a monomeric scale therefore, the transverse displacements of the particles from the long axis of the rod must form a two dimensional random walk. This fractal arrangement will necessarily be generated after repeated collision events.

The transverse span of the walk for j particles will be

$$W \propto a j^{1/2},$$

and so one might expect a collision cross section for two rods with spans W_1 and W_2 to be $\pi(W_1+W_2)^2$. This is problematical, since in the context of the aggregation equation to be presented below, it leads to an "instantaneously gelling kernel" [15]. It is clear, however, that this expression is an overestimate. The span of a walk is its maximum width at any point along its length; in fact, during a collision resulting in a longer rod, only a length of the order of the shorter of the two rods can become entangled. Thus a more reasonable estimate of the collision cross section Σ is

$$\Sigma=\omega_1[a\sqrt{\min(i,j)}]^2,$$

where ω_1 is a dimensionless constant, controlling the width of the walks.

If we now take the rod lengths to be approximately additive on collision, we will obtain a form of Smoluchowski's aggregation equation [16]:

$$\frac{dn_k}{dt}=\sum_{i,j}^{\infty}\Sigma V_{\rm rel}C(\phi_v)n_in_j(\delta_{i+j,k}-\delta_{i,k}-\delta_{j,k}). \quad (3)$$

Here n_k is the number of rods per unit volume with length ka, and $C(\phi_v)$ is a "crowding factor" representing the increase in the collision frequency due to the volume excluded by the particles. Near the random close packed volume fraction $\phi_c \approx 0.63$ at which the dispersion is jammed at the start and cannot flow, one would expect the crowding factor to diverge as $1/[1-(\phi_v/\phi_c)^{1/3}]$. Note in Eq. (3) that the combinatorial factor of 1/2 normally present is canceled by the fact that rods may collide at both ends.

We may nondimensionalize in terms of $X_k=n_k/n_0$ where n_0 is the initial concentration of particles, and the scaled time

$$dT=dt\left(\frac{6\phi_v}{2\pi}\right)\dot{\gamma}\ \cos\theta\ \sin\theta C(\phi_v)\omega_1, \quad (4)$$

where $\phi_v=\pi n_0 a^3/6$, to obtain the standard form of Smoluchowski's equation [15]:

$$\frac{dX_k}{dT}=\frac{1}{2}\sum_{i,j}^{\infty}K_{ij}X_iX_j(\delta_{i+j,k}-\delta_{i,k}-\delta_{j,k}). \quad (5)$$

This has the homogeneous kernel

$$K_{ij}=2(i+j)\min(i,j)=2ij+2[\min(i,j)]^2. \quad (6)$$

In the notation of Van Dongen [15] this kernel has scaling exponents $\lambda=2$ and $\mu=\nu=1$, which places it in the class that undergoes a gel transition where the average cluster size tends to infinity after a finite amount of reduced time. Van Dongen also derives scaling solutions applying close to this gel point, and in particular, we have the divergence in the third moment of the cluster size,

$$M_3=\sum_{j=1}^{\infty}j^3X_j \sim (1-T/T_{\rm gel})^{-3}, \quad (7)$$

which we will show later, gives the asymptotic form of the stress close to the hydrodynamically jammed state. It is also important that the buildup of this divergence results from aggregation among the larger prevailing rods [17]; this justifies our earlier neglect, in discussing Eq. (3), of events where a long rod captures shorter ones without increasing in length, since the short rod population is relatively unimportant.

For our purposes, we will also need an explicit solution to Eq. (5), which will require the initial rod orientation. This is taken to be $\theta_0=\pi/4$, being a "typical" angle of first collision, and the orientation for which rod growth in this model is fastest (so that if rod growth starts in various directions, this one is singled out by the kinetics).

III. SOLUTION OF SMOLUCHOWSKI'S EQUATION

We approach the set of Eqs. (5) by obtaining the Taylor series solution about the origin, for monomeric initial conditions $[X_1(T=0)=1]$.

Van Dongen [15] provides the first term in the expansion for each cluster size:

$$X_n=N_nT^{n-1}+O(T^n),$$

where the numbers N_n are given by the recurrence relation:

$$(n-1)N_n=\frac{1}{2}\sum_{i+j=n} K_{ij}N_iN_j .$$

This is readily generalized, to provide the full Taylor expansion, up to a given order, for

$$X_n(T)=\sum_k x_n^{[k+1]}T^k.$$

With $x_1^{[1]}=1$, and $x_a^{[b]}=0\ \forall a>b$, then

$$(k+r-1)x_k^{[k+r]}=\frac{1}{2}\sum_{a=1}^{k-1}\left[K_{a,k-a}\sum_{c=0}^{r} x_{k-a}^{[k+r-a-c]}\right]-\sum_{c=0}^{r-1}\left[x_k^{[k+c]}\sum_{m=1}^{r-c} K_{km}x_m^{[r-c]}\right]. \quad (8)$$

This is used by first setting $r=0$, and applying the formula, for $k=1,2,\ldots$, then proceeding to $r=1$, etc. filling in the matrix of coefficients in a diagonal manner.

The Taylor expansion of any desired moment is then obtained trivially.

For reasons to follow, we are interested in the third moment of the cluster size; a Domb-Sykes plot shows this to have an unphysical nonpolar singularity, at $T\approx-0.093$, so to obtain a solution out to the gel point, a continued fraction expansion was derived from the Taylor series. This has the advantage of being able to reproduce subtle analytic properties of the solution [18]; building branch cuts out of alternating poles and zeroes radiating from the origin of the complex T plane, and placing polar singularities, in roughly the right position, as the order increases.

For the kernel in Eq. (6), pole zero plots were drawn, for the continued fraction expansion of M_3, using up to 50 Taylor terms. A cluster of three poles was observed, which converged towards $T\approx 0.33$ as the order was increased, in accordance with Van Dongen's scaling results. This value is somewhat less than the gel time of 0.5 for the product kernel $K_{ij}=2ij$, as would be expected from the form of Eq. (6).

IV. JAMMING AND THE BEHAVIOR OF THE STRESS

To calculate the stress, we imagine the rods to be immersed in the mean effective fluid composed of the other rods, particles, and solvent. From Fig. 2, we see that in the corotating frame of a thin rod, the velocity of the background flow relative to the rod and close to it is parallel to this rod and increases linearly with distance from the center of mass. Let x be this distance; then the relative speed is readily shown to be

$$v=\dot{\gamma}(\sin\theta)(\cos\theta)x. \quad (9)$$

In the mean field approximation this passing fluid exerts a frictional force per unit length dF/dx given by

$$\frac{dF}{dx}=E(\phi_v)\dot{\gamma}(\sin\theta)(\cos\theta)x, \quad (10)$$

where $E(\phi_v)$ is a parameter for each volume fraction, with the dimensions of viscosity.

From this ansatz and Eq. (8), we readily calculate the viscous power dissipated around a rod of j particles, as

$$\int_{-ja/2}^{ja/2} E(\phi_v)[\dot{\gamma}(\sin\theta)(\cos\theta)x]^2dx$$

and therefore the stress σ_{rod} in the fluid due to the rods (which contain at least two particles) is

$$\sigma_{\text{rod}}=\frac{E(\phi_v)}{12}\dot{\gamma}(\cos^2\theta)(\sin^2\theta)\left(\frac{6\phi_v}{\pi}\right)\sum_{j=2}^{\infty} j^3X_j . \quad (11)$$

Since from Eq. (7) this stress will dominate at late times, the model predicts that colloids with purely hydrodynamic interactions can jam up, after a finite amount γ_{jam} of strain, with the stress diverging with the third moment of the cluster size, that is, as $(\gamma-\gamma_{\text{jam}})^{-3}$ close to this point.

In this model, the condition for jamming is that the reduced time reaches a value of $T_{\text{gel}}=0.33$, before the rods "tumble" by reaching $\theta=0$ and are no longer subject to a compressional flow field (Fig. 2).

We expect our discussion above to apply quantitatively close to the jamming point; unfortunately computer simulations are restricted to the initial phase of the flow, where we can no longer assume that the stress is dominated by large rods. These rods, once they form, will have disrupted the local affine flow, carving out "channels" in the effective fluid, in which their constituent particles may have large velocity components in the "gradient" direction. In contrast the monomers, whose contribution to the stress is omitted from Eq. (11), will by and large partake of the average affine flow.

The monomers will dissipate a power P given by

$$P=\sigma_{\text{mon}}\dot{\gamma}=n_b\langle P_b\rangle, \quad (12)$$

where n_b is the concentration of actively deforming nearest neighbor bonds, and $\langle P_b\rangle$ is the average power dissipated by such a bond. Initially, when the stress in the dispersion is $\sigma(0)$, we will have $n_b=(1/2)\nu_{\text{NN}}n_0$ where ν_{NN} is the number of nearest neighbors per particle, and is approximately 10 near to random close packing.

At later times, some of the monomers have been incorporated into rods, and so one might expect

$$\sigma_{\text{mon}}\approx X_1\sigma(0) \quad (13)$$

so that the final expression for the stress is

$$\sigma=\sigma_{\text{mon}}+\sigma_{\text{rod}}, \quad (14)$$

which involves the three parameters ω_1, $E(\phi_v)$, and $\sigma(0)$, theoretical estimates of which will be given below.

It must be pointed out that the form of Eq. (13) is rather arguable, depending upon how many gaps around and within a rod one considers to be either no longer active in dissipating power, or to have been counted by Eq. (10). Most of the results that follow, and in particular the "lower critical volume fraction" ϕ_l, are somewhat sensitive to the details of this equation; for example, the fitted values of $E(\phi_v)$ are most sensitive, and may vary by as much as a factor of 2 upon making different assumptions.

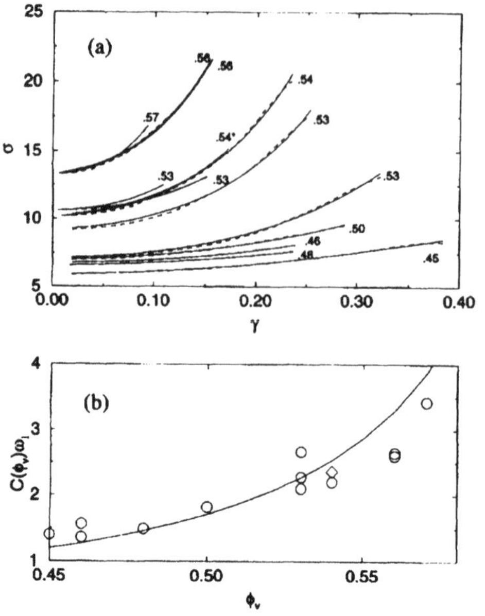

FIG. 3. (a) Simulation (dashed), and fitted theoretical (solid) graphs of σ against γ, at the 13 volume fractions indicated next to the curves. For this and the following figures units are chosen so that $a=\eta_0=1=\dot{\gamma}$: Thus the physical stress σ is in units of $\eta_0\dot{\gamma}$. (b) Fitted values of the product $C(\phi_v)\omega_1$, where the diamond is for 1400 particles. The solid line is a fit to Eq. (15) in the text.

To test Eq. (14), computer simulations involving pure hydrodynamic interactions were run, for several volume fractions and 700 particles, in an initially cubic simulation cell, using the code of Ref. [8]. The resulting stress vs strain curves, were fitted to Eq. (14), by choosing $E(\phi_v)$, and the product $C(\phi_v)\omega_1$ for each volume fraction as shown in Fig. 3(a). The simulation results at each volume fraction are shown dashed, and the fitted theoretical results, which in each case lie essentially on top of the simulation data, are shown as solid curves. Figure 3(b) shows the best values of the product $C(\phi_v)\omega_1$ as a function of ϕ_v, and a fit to the form

$$\frac{\omega_1}{1-(\phi_v/\phi_c)^{1/3}}, \qquad (15)$$

where ω_1 is constant for all ϕ_v ($\omega_1=0.13\pm0.02$ fits best). This value of ω_1 is *a priori* reasonable, since the step size of the random walk in particle diameters is then of order $\omega_1^{1/2}\approx 0.35$.

Since we are dealing with large scale structures in the flow, finite size effects may well be important. To test their significance, a 1400 particle simulation was run, in an initially cuboidal simulation cell of aspect ratio 1:1:2—the longer side being in the flow direction, and this geometry chosen to reduce the risk of a growing cluster hitting one of its periodic images. The results are shown alongside the 700 particle runs in Figs. 3 and 5, and the model is able to fit them, with essentially the same values for the parameters.

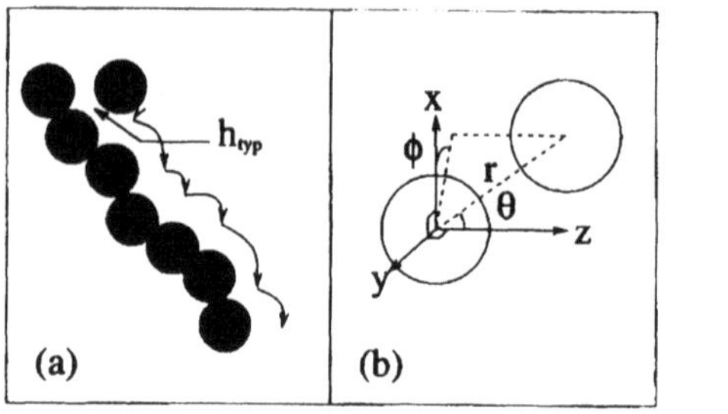

FIG. 4. (a) Schematic picture of a particle sliding along a cluster in the flow, generating the friction parameter $E(\phi_v)$. (b) Two particles in general position, with centers separated by a distance $r\approx a+h_{\text{typ}}$.

V. ESTIMATES OF THE MODEL PARAMETERS

So far, the values of $E(\phi)$ and $\sigma(0)$ have simply been extracted from the simulation data; however, we do have some theoretical handle upon them. Consider first a typical gap between nearest neighbors; this falls to zero at random close packing, and considering the low concentration limit, one readily sees that a reasonable estimate is

$$h_{\text{typ}}\approx a\left(\frac{\pi}{6}\right)^{1/3}\left[\left(\frac{1}{\phi_v}\right)^{1/3}-\left(\frac{1}{\phi_c}\right)^{1/3}\right], \qquad (16)$$

which vanishes as $(1-\phi_v/\phi_c)^{-1}$ near ϕ_c.

To estimate the value of $E(\phi_v)$ we follow an argument similar to Marrucci and Denn [5], and imagine a particle in the effective fluid sliding along the cluster [Fig. 4(a)] with a relative speed v [Eq. (9)] and maintaining a distance h_{typ} from it. When it separates from one of the particles in the cluster to move to the next, the gap will change roughly as $h=h_{\text{typ}}+vt$, and so the impulse along the separation direction, from the instantaneous "squeeze" force F, is

$$\mathcal{I}=\int F\,dt=\int_{t=0}^{a/v}\frac{3\pi\eta_0 v a^2}{8h}dt.$$

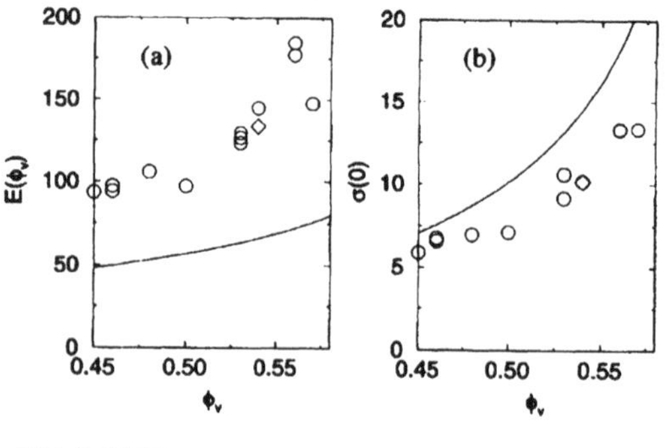

FIG. 5. (a) Plot of $E(\phi_v)\eta_0$ against ϕ_v obtained from the fit to the stress-strain data in Fig. 3. The circles for 700 particles and the diamond for 1400 particles show good agreement. The solid line for comparison is a theoretical prediction [Eq. (17) in the text]. (b) Same for $\sigma(0)\dot{\gamma}\eta_0$; the theoretical prediction is from Eq. (18).

The rate of encounter of particles in the cluster is a/v and if we assume each particle of the rod has eight particles sliding past, and that $a \gg h_{\rm typ}$, then the sliding force on a length a of the rod due to squeeze lubrication interactions is

$$F_{\rm sq} = 8\mathcal{I}\frac{v}{a} \approx 3\pi\eta_0 av \,\ln\left(\frac{a}{h_{\rm typ}}\right).$$

This force is logarithmic in the gap, and so one needs to include the logarithmic "shear" lubrication forces of the sliding particles, which contribute

$$F_{\rm sh} = 8\frac{\pi\eta_0 av}{2}\ln\left(\frac{a}{2h_{\rm typ}}\right)$$

for a length a of the rod. When compared with Eq. (10) this gives, ignoring factors of order unity inside the logarithm,

$$E(\phi_v) \approx 7\pi\eta_0 \ln\left(\frac{a}{h_{\rm typ}}\right). \tag{17}$$

Figure 5(a) shows the fitted values of $E(\phi_v)$ from the simulation plotted with this theoretical prediction, and shows that Eq. (17) underestimates the fitted values, by a factor of about 2. This is quite acceptable for an argument based purely on dimensional grounds, and indeed, one might expect an underestimate on the part of Eq. (17), since we have assumed that all gaps surrounding the rods are of size $h_{\rm typ}$, while in practice, the rod will have to fight through many gaps that are significantly smaller than this.

Next we turn to $\sigma(0)$, which is the stress at the moment the simulation is started. Consider two particles, as in Fig. 4(b), moving with the affine flow, and separated by a position vector given in polar coordinates by

$$\mathbf{r} = \begin{pmatrix} r\ \sin\theta\ \cos\phi \\ r\ \sin\theta\ \sin\phi \\ r\ \cos\theta \end{pmatrix},$$

Where the gradient, vorticity, and flow directions are the x, y, and z axes, respectively, and $r \approx a + h_{\rm typ}$. The relative velocity of the particles is therefore

$$\mathbf{v}_{\rm rel} = \dot{\gamma} r(\sin\theta)(\cos\phi)\hat{\mathbf{z}}.$$

The "squeeze" and "shear" components of this are

$$|v_{\rm sq}| = \dot{\gamma} r\ \sin\theta\ \cos\theta\ \cos\phi,$$

$$|v_{\rm sh}| = \dot{\gamma} r\ \sin^2\theta\ \cos\phi,$$

and so the power dissipated in the bond will be

$$\langle P_b \rangle = \frac{3\pi\eta_0 a^2}{8h_{\rm typ}}\langle v_{\rm sq}^2\rangle + \frac{\pi\eta_0 a}{2}\langle v_{\rm sh}^2\rangle,$$

where the angle brackets denote an average over all orientations with equal weight.

This leads from Eq. (12) to

$$\sigma(0) \approx \frac{3\phi_v \nu_{\rm NN}\dot{\gamma}^2}{a^3\pi}(a+h_{\rm typ})^2\left[\frac{\pi\eta_0 a^2}{40h_{\rm typ}} + \frac{2\pi\eta_0 a}{15}\ln\left(\frac{a}{2h_{\rm typ}}\right)\right], \tag{18}$$

which, as discussed shortly, diverges near random close packing as $(1-\phi_v/\phi_c)^{-1}$.

This prediction is plotted with the actual values from the simulation data in Fig. 5(b) and shows that the agreement is reasonable, given the crudeness of the approximations.

The derivation of Eq. (18) is essentially that due to Frankel and Acrivos [3] who intended to predict the steady state rheology of a concentrated dispersion. As pointed out by Marrucci and Denn [5], the argument makes the implicit assumption that

$$\left\langle \frac{v^2}{h} \right\rangle \approx \frac{\langle v^2 \rangle}{\langle h \rangle}, \tag{19}$$

where v is the relative velocity of a pair of particles and the average is performed over all nearest neighbor pairs in the system. This is unjustified at late times, when the velocities and gaps will be highly correlated [5]. In our case we apply the argument only at the start of the flow, when the gaps are prescribed by the initial configuration and Eq. (19) is a fair approximation.

Equations (17) and (18) together with the reasonable guess that $\omega_1^{1/2} \approx 0.5$, or a little less, therefore provide a semiquantitative description of hydrodynamic jamming with no free parameters.

Although these parameters may be predicted theoretically, the lower critical volume fraction ϕ_l derived in Sec. VII below is obtained from a fit to simulation data, and is not dependent on Eqs. (17) and (18).

VI. CLOSURE OF THE SMALLEST GAPS

Given the $E(\phi_v)$'s and products $C(\phi_v)\omega_1$ fitted to Eq. (14), the dynamics of the infinite system are determined and it should be possible to calculate the distribution of gap sizes as a function of strain γ. In practice, however, this is a difficult problem as it involves taking an average over all the possible histories of gaps lying in rods (a possibility space of very large dimension) while the solution of Sec. III only gives us access to the populations of different rod lengths as a function of strain. Instead, we are forced to make an estimate based on a scaling argument.

We start by noting that in the model, there is a typical longest rod length, which will scale as

$$L_{\rm max} = \lambda a(1-T/T_{\rm gel})^{-2} \tag{20}$$

for some factor λ, of order unity (λ is likely to be system-size dependent). We then assert that the smallest gaps have always been near the center of a rod of this typical maximum size. Such a gap, h, will be squeezed by the compressive force F_c in the rod, so that

$$\dot{h} = -\frac{8hF_c}{3\pi\eta_0 a^2}, \tag{21}$$

and so, using Eq. (10) we obtain an expression for the minimum gap:

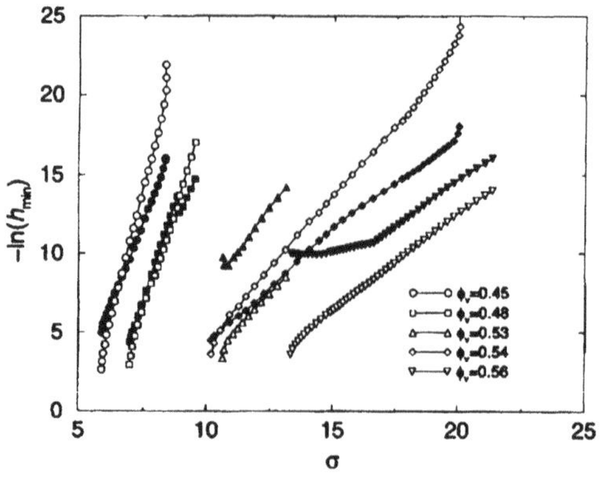

FIG. 6. Simulation, and theoretical curves of $-\ln(h_{min}a)$ against $\sigma\eta_0\dot{\gamma}$, for various ϕ_v with $\lambda=2.1$. The open symbols show fitted theoretical results [Eq. (22)], and the closed symbols of the same shape are the simulation data at the same volume fraction.

$$-\ln\left[\frac{h_{min}}{h_{min}(0)}\right]=\frac{\lambda^2E(\phi_v)}{9C(\phi_v)\ \omega_1\eta_0\phi_v}[(1-T/T_{gel})^{-3}-1]\frac{T_{gel}}{3}. \tag{22}$$

We may also (over) estimate the size of the initial minimum gap $h_{min}(0)$ (which will be highly sample dependent) from

$$h_{min}(0)\approx h_{typ}.$$

The model thus predicts that the smallest gap will collapse catastrophically, falling as $\exp[(\gamma-\gamma_{jam})^{-3}]$ near the gel point. It is this singular dependence which makes computer simulations of this regime extremely difficult. However, it also has a bearing on the self-consistency of the model, for the rods are only defined by a separation of length scales, between gaps and particle diameters. A rapid collapse of the gaps is therefore essential to this approximation.

Figure 6 shows $-\ln h_{min}$ plotted against the stress σ, for simulations of 700 particles and the predicted curves superposed at various volume fractions. A value of 2.1 was used for λ.

VII. LOWER CRITICAL VOLUME FRACTION

Our model predicts a lower volume fraction ϕ_l below which the rods will tumble (reaching $\theta=0$ and $\gamma=1$) before the gel transition occurs (at $T=T_{gel}\approx0.33$). Since the reduced time T is related to the angle θ the rods make with the gradient direction, by [Eqs. (2) and (4)]

$$T=\frac{6\phi_v}{2\pi}\dot{\gamma}C(\phi_v)\omega_1\ln\left(\frac{\cos\theta}{\cos\theta_0}\right),$$

with $\theta_0=\pi/4$, it follows from Eq. (15) and the fitted value of ω_1 that $\phi_l=0.515\pm0.015$. It should be noted, however, that the error on ϕ_l is only in the context of the precise model presented here; its value is quite sensitive to the detailed assumptions made about, e.g., the form of the stress [Eq. (13)].

We suggest that this concept of a lower critical volume fraction ϕ_l may be of relevance to sudden shear thickening in real colloids, with finite values of Péclet number. D'Haene [13], for example, finds that in controlled strain rate experiments on hard-sphere systems, there is a discontinuous jump in stress as a function of Pe, but only above a volume fraction of 53–54 %. In the context of our model, this would correspond to the onset of a jammed state, with the formation of system spanning clusters leading to the irregular stress fluctuations he observes after the discontinuous thickening. Since the stress he observes does not grow arbitrarily, the consequences of this putative logjam must be mitigated in his experiments by some other effect such as particle deformation or fracture.

VIII. CLUSTERS IN THE SIMULATION

The computer simulations contain all the information about the flow, and so can be used to visualize, and perform statistics on any structures that may be present. In order to define a cluster, we choose a criterion based upon interparticle gaps, which is that two particles belong to the same cluster if they are separated by a gap h less than some critical gap h_c, for which we take a "theory-driven" value. In the context of this theory, we choose for h_c a value dependent upon the expected gap h_d in a dimer forming at the start of the flow. This is predicted to collapse like

$$h_d=h_{typ}\exp\left[\frac{2E(\phi_v)}{3\pi\eta_0}\ln\left(\frac{\cos(\pi/4)}{\cos\theta}\right)\right], \tag{23}$$

in which the time dependence enters through θ, and we take

$$h_c=\frac{h_d}{\beta}, \tag{24}$$

where β is some parameter of order unity.

A simulation was run at $\phi_v=0.53$, using 700 particles and $E(\phi_v)=125$, and the configurations partitioned into clusters on the basis of Eq. (23). From these partitions we may define moments of the cluster sizes: Let $\mathcal{P}_j$ be the fraction of particles belonging to clusters of size j particles, then we define the qth moment M_q by

$$M_q=\sum_{j=1}^{\infty}\mathcal{P}_j\ j^q.$$

Figure 7 shows plots of M_2 and M_3, averaged over various strain intervals, as a function of strain, γ for five values of β. The error bars show the predicted values, using $\omega_1=0.13\pm0.02$.

Lastly, in order to obtain some pictorial impression of these clusters, six configurations equally spaced in strain were chosen from this simulation at $\phi_v=0.53$ using $\beta=3$. The clusters in the flow were then analyzed by finding for each the "radius of gyration" tensor $\mathcal{R}$. This is defined in the following manner: Let k label each of the N particles belonging to the cluster in turn. Let $\mathbf{r}_k$ be the position vector of the kth particle and

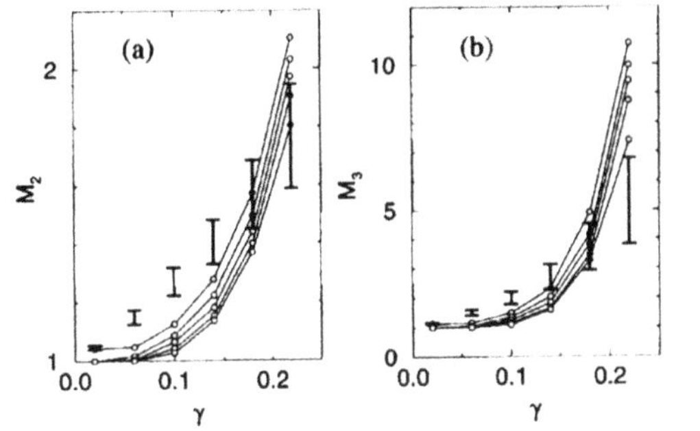

FIG. 7. (a) Plot of M_2 against strain γ for the five values β = 2.6, 3.0, 3.5, 4.0, 4.5. Each point in strain is actually an average over four equally spaced configurations in one of the six strain ranges, $\gamma=0-0.04$, $0.04-0.08$, $0.08-0.12$, $0.12-0.16$, $0.16-0.20$, and $0.20-0.24$. The six error bars show the predicted values of the moment M_2, for each strain range, using $\omega_1=0.13\pm0.02$. (b) Same for M_3.

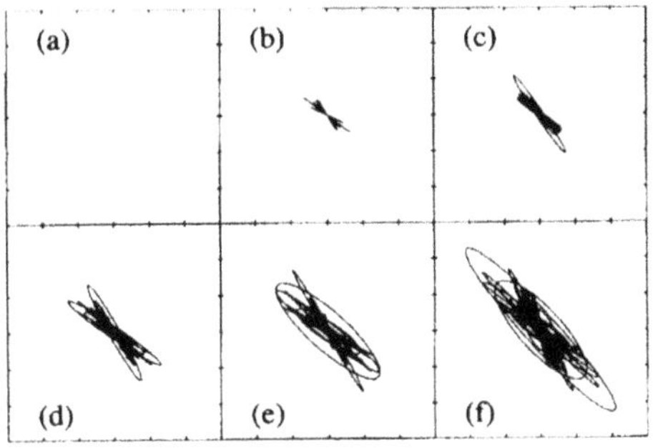

FIG. 8. Plots of projections of the ellipsoids from Eq. (25) for all the clusters in a configuration. Tick marks on the axes are at a spacing of one particle diameter. Configurations are at (a)–(f) $\gamma=0.04$, 0.08, 0.12, 0.16, 0.20, and 0.24.

$$\mathbf{r}_{\text{c.m.}}=\sum_k \mathbf{r}_k/N$$

the center of mass of the cluster. Then

$$\mathcal{R}=\sum_k \frac{(\mathbf{r}_k-\mathbf{r}_{\text{c.m.}})(\mathbf{r}_k-\mathbf{r}_{\text{c.m.}})}{N}.$$

This tensor defines an ellipsoid approximating the cluster's true distribution in space. To project this ellipsoid onto the flow-gradient plane, let the x, y, and z axes be the flow, gradient, and vorticity directions, then define a new 2 by 2 tensor $\mathcal{R}_P$, by the upper left hand corner of $\mathcal{R}$. The projection onto the plane $z=0$ is then the set of points $\{\mathbf{x}\}$ given by

$$\mathbf{x}=\mathcal{R}_P^{1/2}\hat{\mathbf{n}}, \qquad (25)$$

where $\hat{\mathbf{n}}$ is a vector lying on the unit circle in the plane $z=0$.

These projections are plotted for the chosen configurations in Fig. 8, and clearly show narrow "rods" near the "compression" axis ($\theta=\pi/4$). However, since Fig. 8 suppresses all information about the "vorticity" direction, one may wonder how good is the approximation that the rods are confined to the $z=0$ plane. This was tested, by calculating the average value, and standard deviation of the vector representing the principal axis of each ellipsoid. We find that this vector has a component in the vorticity direction with a mean of less than 0.02 particle diameters, and a standard deviation increasing from $0.10a$ at $\gamma=0.08$ to $0.22a$ at $\gamma=0.24$, entirely consistent with the clusters lying in the flow-gradient plane.

IX. CONCLUSION

In this paper we have constructed a model of the stress carrying "fabric" in a concentrated dispersion of hard spheres at Pe$=\infty$. The model predicts that above a lower volume fraction ϕ_l substantially less than 0.63, a hydrodynamic logjam occurs even in an infinite system. This is characterized by an elongated cluster or rod, first forming parallel to the compression axis, growing to infinite size before it tumbles by passing $\theta=0$ and entering a region of extensional flow. The model quantifies the hitherto qualitative notion of hydrodynamic clustering and jamming, which we believe to be a new physical phenomenon underlying discontinuous shear thickening.

At volume fractions below ϕ_l, the clusters in the model will tumble, passing $\theta=0$ and failing to jam at a "first attempt." The assumption that all clusters in the flow remain roughly parallel will then break down, as it is closely tied to the transience of the motion. Whether a steady state is then eventually achieved is not covered by our model and remains an open question. It is therefore unclear whether for a pure hydrodynamic model a divergence such as Eq. (1) can be studied below ϕ_l.

ACKNOWLEDGMENT

Thanks are due to Dr. C. Nex, for his help and advice on Padé approximants.

[1] A. W. Lees and S. F. Edwards, J. Phys. C **5**, 1921 (1972).
[2] A. Einstein, Ann. Phys. (Leipzig) **19**, 289 (1906).
[3] N. A. Frankel and A. Acrivos, Chem. Eng. Sci. **22**, 847 (1967).
[4] K. C. Nunan and J. B. Keller, J. Fluid. Mech. **142**, 269 (1984).
[5] G. Marrucci and M. M. Denn, Rheol. Acta **24**, 317 (1985).
[6] G. Bossis and J. F. Brady, J. Chem. Phys. **80**, 5141 (1984).
[7] R. C. Ball and J. R. Melrose, Adv. Colloid Interface Sci. **59**, 19 (1995).
[8] J. R. Melrose and R. C. Ball, Europhys. Lett. **32**, 535 (1995).
[9] J. F. Brady, J. Chem. Phys. **99**, 567 (1993).
[10] J. F. Brady, J. Fluid Mech. **272**, 109 (1994).
[11] W. B. Russel, D. A. Saville, and W. R. Schowalter, *Colloidal*

Dispersions (Cambridge University Press, Cambridge, England, 1989).

[12] J. F. Brady (private communication).

[13] P. D'Haene, Ph.D. thesis, Katholieke Universitiet Leuven, 1992.

[14] W. J. Frith, P. D'Haene, R. Buscall, and J. Mewis, J. Rheol. **40**, 531 (1996).

[15] P. G. J. Van Dongen, Ph.D. thesis, Rijksuniversiteit Utrecht, 1983.

[16] M. von Smoluchowski, Z. Phys. Chem. **92**, 129 (1917).

[17] M. H. Ernst, *Fractals in Physics* (North-Holland, Amsterdam, 1986), p. 289.

[18] E. J. Hinch, *Perturbation Methods* (Cambridge University Press, Cambridge, England, 1991).

Friction in Granular Layers: Hysteresis and Precursors

S. Nasuno,* A. Kudrolli, and J. P. Gollub†
Physics Department, Haverford College, Haverford, Pennsylvania 19041
and Department of Physics, University of Pennsylvania, Philadelphia, Pennsylvania 19104
(Received 9 April 1997)

Sensitive and fast force measurements are performed on sheared granular layers with simultaneous optical imaging. Discrete displacements that can be much smaller than the particle diameter occur for low imposed velocity gradients, with a transition to continuous motion at a higher gradient. The instantaneous frictional force measured within individual slip events is a multivalued function of the instantaneous velocity. Localized microscopic rearrangements precede (and follow) macroscopic slip events; the accumulation of localized rearrangements leads to macroscopic creep. [S0031-9007(97)03732-0]

PACS numbers: 83.70.Fn, 47.55.Kf, 62.40.+i, 81.05.Rm

The response of a granular medium to shear forces is important both as a fundamental physical property of these materials [1,2] and for understanding earthquake dynamics [3,4], where slipping events occur along faults that are often separated by a granular "gouge." Stick-slip motion often occurs at low shear or strain rates; its occurrence can be explained phenomenologically by a friction law that declines with velocity v, and a stiffness (see below) that is less than a critical value [5]. A molecular dynamics simulation of granular friction [6] suggests that repetitive shear-induced fluidization accompanies stick-slip motion. Though models exist for steady state forces [7], the actual time-dependent frictional force during individual slip events appears not to have been measured or computed for granular materials. Previous measurements [8–12] of frictional forces in sheared granular media have mostly emphasized large systems, high velocity flows, and large normal stresses, in contrast to those reported here, which also allow direct imaging of the particle dynamics.

In this Letter, we report sensitive measurements of frictional forces produced by sheared layers of polydisperse particles, with simultaneous optical imaging to reveal microscopic particle displacements. We are able to determine the force variations *within slip events lasting as little as 40 ms.* We find $F_f(v)$ to be multivalued, with the instantaneous force being less for decreasing than for increasing velocities. This hysteresis, which occurs in conjunction with a slight dilation of the material, is important in understanding the observed dynamics.

Very slow displacement (creep) is known to precede the rapid slip events [4,13–15], but its microscopic mechanism remains unclear, and may be quite different for granular materials than for solid-on-solid friction. In the present work, we find that significant microscopic rearrangement events occur during the sticking intervals. Their accumulation leads to creep, and their frequency rises dramatically as the slips are approached. Similar patterns are known in connection with earthquakes [16,17]. We also study the effects of various system parameters on the dynamics, and we show that, in some cases, the stick-slip dynamics can be nonperiodic and the discrete displacements surprisingly small.

Apparatus—A schematic diagram of our experimental setup is shown in Fig. 1. The granular layer typically consists of spherical glass beads whose size distribution includes particles 70–110 μm in diameter. Shear stress is applied by pushing a transparent glass cover plate with a leaf spring of adjustable stiffness and a stainless steel ball bearing coupler glued to the cover plate. The spring mount is attached to a translating stage whose micrometer is driven by a computer-controlled stepping motor. The translator moves at a pushing speed V, while the cover plate moves at speed $\dot{x}(t) \equiv v(t)$. The relative position $\delta x(t) = Vt - x(t)$ of the cover plate (relative to the spring mount) is determined to within 0.1 μm, using inductive displacement sensors similar to those used in Ref. [14]. This signal is equal to the spring deflection and is hence proportional to the spring force applied to the plate. We also monitor the vertical displacement of the cover plate with a separate inductive sensor.

The main part of the apparatus is mounted on a microscope, which allows the observation of microscopic motion of granular particles through the transparent top plate. The particle motion can be captured on video tape for

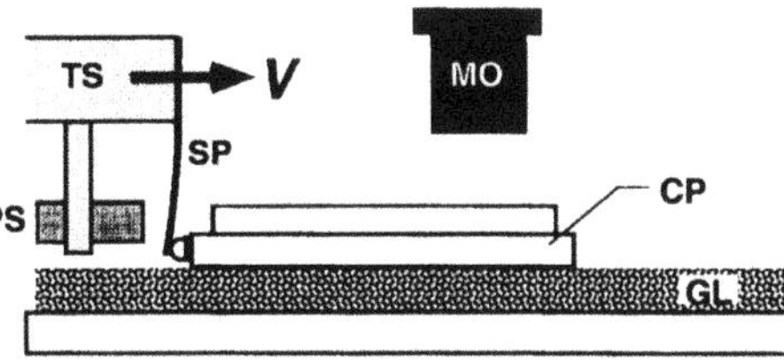

FIG. 1. Schematic diagram of the apparatus, showing the granular layer (GL) and a transparent cover plate (CP) pushed by a leaf spring (SP) connected to a translating stage (TS). An inductive sensor (PS) detects the deflection of the leaf spring. The microscope objective (MO) is also shown.

later quantitative analysis, or digitized directly. Synchronization allows events in the images to be correlated with structure in the force measurements. The apparatus and microscope are contained within a temperature controlled box and maintained somewhat above room temperature (37.5 ± 0.2 °C) and at a reduced humidity (20% ± 2%) to minimize absorbed water on the particles. We typically use a layer thickness of 2 mm, which is uniform to about 50 μm initially. For our samples, cohesive forces are negligible.

The horizontal dimensions of the upper plate are generally 75 × 50 mm; the lower one is much larger. The surfaces of the plates can be treated in several ways to transmit the shear force to the granular layer without interfacial slipping. We roughen the bottom plate and either (a) glue a layer of particles to a glass top plate, or (b) rule a Plexiglas top plate with parallel grooves, 2 mm apart, so that imaging measurements can be made in the interstices. Similar results are found in the two cases. Particles are used only for a few runs to avoid the effects of wear.

Detailed studies reported here utilized the following ranges of the major parameters: imposed translation or pushing velocity V of the spring mount (10^0–10^4 μm/s); layer thickness (2 mm); spring constant k (10^2–10^4 N/m); mass M of the upper plate ($1-5 \times 10^{-2}$ kg). Edge effects (due to the finite horizontal size of the layer) are believed to be negligible. Limited experiments were also performed on hollow ceramic beads, whose surfaces are rough, and on clean filtered sand.

Stick-slip phenomena.—For relatively low V, the top plate alternately sticks and slips (Fig. 2). During the static intervals, the top plate is at rest, and the spring deflection δx increases linearly in time [Fig. 2(a)] up to a maximum value corresponding to the maximum static friction force F_s. When this value is reached, the granular layer can no longer sustain the imposed shear stress, and the plate suddenly accelerates, while the friction force F_f decreases. When the imposed stress becomes

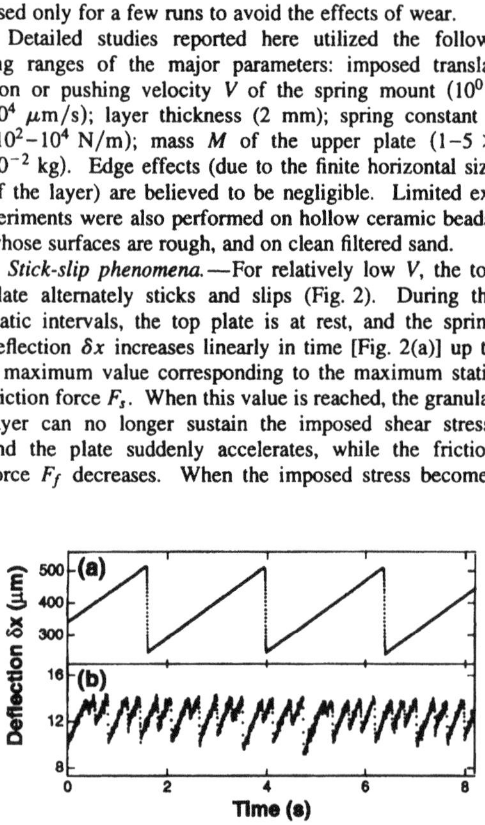

FIG. 2. (a) Spring deflection $\delta x(t)$ as a function of time for periodic stick-slip motion at pushing speed $V = 113.33$ μm/s, spring constant $k = 135$ N/m, and mass $M = 1.09 \times 10^{-2}$ kg. (b) Irregular stick-slip motion at $V = 11.33$ μm/s and $k = 3636$ N/m.

sufficiently small, sticking recurs. The slip duration τ_s is comparable to the characteristic inertial time $\tau_{\text{in}} = 2\pi\sqrt{M/k}$ and is substantially shorter than the time between slip events. The granular flow during slip seems to occur predominantly in the top few layers of particles. Visual observation through the top plate reveals that the onset of sliding involves fluidization of granular particles in the upper layers. Measurement of *vertical* displacement confirms that the layer thickness increases by roughly 15 μm during the slip events, which is less than the mean particle diameter.

The observed stick-slip motion is almost periodic for a wide range of parameter values. However, strongly nonperiodic motion occurs at very low V and large k, as shown in Fig. 2(b). The mean period is short, and the displacement per cycle (about 2 μm) is less than the particle diameter. Under these conditions, the maximum static force is almost the same for each cycle, but the degree to which the stress is unloaded (and hence the time of resticking) varies from cycle to cycle, probably because the degree of fluidization is low and many microscopic configurations are possible in a disordered medium.

As V increased, the mean period T decreases while the slip distance per event remains fairly constant, until an inertia-dominated regime is reached, where the oscillations of both the spring force and the frictional force are nearly (but not precisely) sinusoidal. Inertial effects in friction studies have been considered in Ref. [18]. A transition to continuous sliding (with fluctuations) occurs at a critical value V_c (6 mm/s for $k = 1077$ N/m and $M = 1.09 \times 10^{-2}$ kg). This threshold decreases with increasing k and also varies with M. The oscillation amplitude fluctuates strongly near V_c. A detailed study of the transition from inertial oscillation to continuous sliding will be published elsewhere [19].

We now consider the instantaneous velocity variations of the moving plate during individual slip events as shown in Fig. 3(a). It is remarkable that the behavior during slip for different V is nearly identical when k and M are fixed. The acceleration and deceleration portions of the pulse are different, and resticking is sudden. The maximum slip speed is of the order of $F_s/\sqrt{km}$ and decreases with increasing k.

Frictional force.—We have determined the instantaneous normalized frictional force $\mu(t) = F_f(t)/Mg$ during the slip events. This quantity is derived from the measured force data by first obtaining the position $x(t)$ and the acceleration $\ddot{x}(t)$ of the top plate from the measured displacement signal $\delta x(t) = Vt - x(t)$, and then using the force balance equation $M\ddot{x} = k\delta x - F_f$. In Fig. 3(b) we plot the instantaneous frictional force as a function of the instantaneous sliding velocity $v(t) = \dot{x}(t)$. During the sticking period, μ increases linearly in time from a to b. When μ reaches the static threshold $\mu_s = F_s/Mg$, the top plate starts to slide. During the acceleration phase b to c, μ decreases monotonically from μ_s as the slip speed

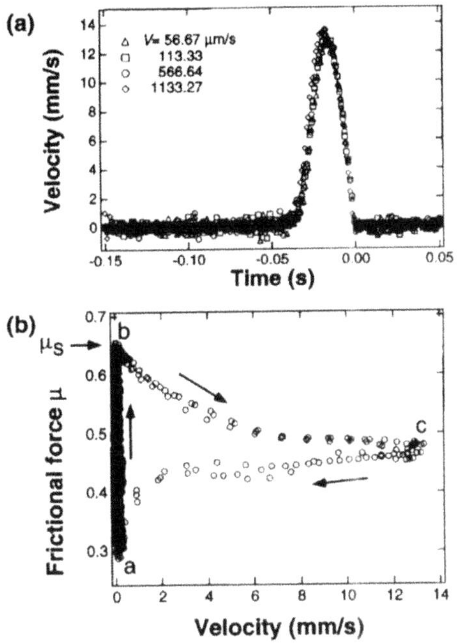

FIG. 3. (a) Instantaneous velocity $\nu(t)$ of the cover plate during slippage for various pushing speeds in the stick-slip regime ($k = 135\ \mathrm{Nm}^{-1}$; $M = 1.09 \times 10^{-2}$ kg). The time origins of the pulses are forced to agree at the end of each event. (b) The instantaneous normalized frictional force $\mu(t) = F_f(t)/M_g$ as a function of $\nu(t)$, for three slip events at $V = 113\ \mu$m/s.

increases. During the *deceleration* portion c to a, μ continues to decrease slowly at first and then rapidly as the slider comes to rest. This loop is almost identical for all events in a run.

The frictional force is not a single valued function of velocity; it is larger for increasing than for decreasing velocity within individual events. Delay or memory effects [20] also occur in solid-on-solid friction, but have a different origin. One source of delay here is the time (about 2 ms) required for the plate to fall a distance equal to the observed vertical dilation during the slip pulse (about 15 μm). The force may possibly be modeled by introducing state variables in addition to $\nu(t)$, as is done in studies of rock friction; see Refs. [4,5].

Precursors.—Microscopic observations permit us to detect local particle motion within the granular layer. To allow optical access, we use a ruled Plexiglas plate, as discussed earlier. We find that local rearrangements of granular particles occur even during the sticking interval. These small motions can be detected by taking differences between successive images. We coarse grain the image over a size comparable to the particle diameter. Then the number of spots for which the difference image is nonzero is a measure of the number of particles that have been displaced between the two images. An image of a portion of the layer (about 10 grains or 1 mm in width) is shown in Fig. 4, along with a sample difference image showing local displacements in 1 s. We find that the microscopic rearrangement rate, as indicated by the average number $\langle n(t)\rangle$ of microscopic displacement sites (in a 1 s interval and a 4 mm^2 area), increases rapidly near the time of a slip event, as shown in Fig. 5(a). The data has been averaged over 416 major slip events. Note that the event rate is asymmetric, with precursors being somewhat more likely than microscopic events at an equivalent time *after* a major slip occurs at $t/T = 0$.

The precursors are detectable macroscopically as well because they contribute to a small amount of creep before the major events. This creep process is shown in Fig. 5(b). The accumulated precursors produce a displacement about 1% as large as that occurring during a slip event. There is substantial variability from one event to the next; the probability distribution $P(N)$ of the *total* number N of macroscopic slip sites occurring at any time before or after a single major slip event is broad, approximately exponential, with a decay constant of about 40 events. The most likely explanation for the local rearrangement events is the breaking of chains of particles that sustain most of the stress [1,2,21].

In conclusion, within slip events, fluidization and dilation of the granular material cause the frictional force to decline with increasing instantaneous velocity, and the lower value is maintained for at least 10 ms even as $\nu(t)$ subsequently decreases. The measured dilation is small, much less than the particle size. We have also observed striking localized precursors and post-slip rearrangements, which are the microscopic origin of slow creep between slip events. It may be possible to determine by faster imaging whether some of the precursors grow spatially into macroscopic slip events. There is a need for improved phenomenological models that apply on the short time scales of individual slip events, and microscopic models that can explain the small slip

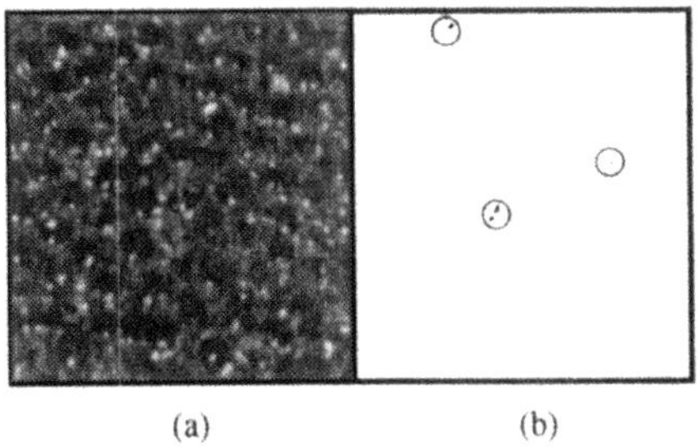

FIG. 4. (a) Image of a portion of a granular layer (about 1 mm across), and (b) difference image ($\Delta t = 1$ s) showing localized particle rearrangements (circled) between major slip events.

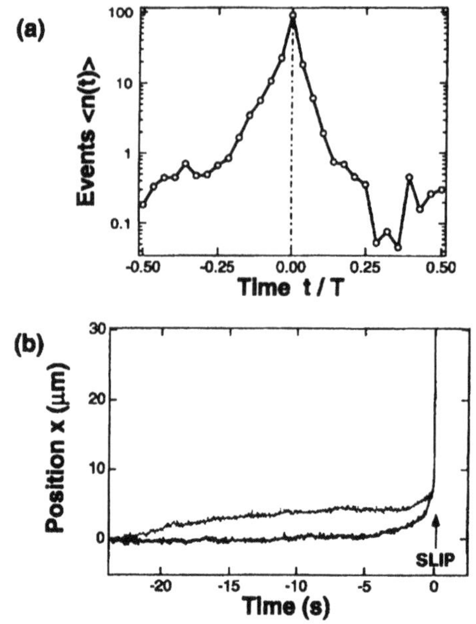

FIG. 5. (a) Average number of microscopic slip sites (per sec in a 4 mm^2 area) as a function of time before or after a major slip event, in units of the period T of the stick-slip cycle. (b) Displacement creep (of variable amount) before several major slip events.

displacements (much less than a particle diameter) that we noted for high stiffness. It is interesting to note that small irregular displacements also occur in thin lubricated films due to constrained dynamics [22].

We have varied many aspects of the experiments, including the translation velocity, spring constant, mass of the upper plate, and the mean size of the particles. The qualitative effects reported here are robust. A more detailed report, including studies of the inertial regime, the transition to continuous motion, and rough particles, will be published elsewhere [19]. Though these experiments are conducted at much lower normal stress than situations relevant to geophysics, they provide insight into the dynamics of sudden slip events, and about localized precursors.

This work was supported in part by the National Science Foundation Grant No. DMR 9319973. S. N. acknowledges the support of the Ministry of Education, Science, Sports, and Culture of Japan. We appreciate the assistance of Anthony Bak with some of the experiments, and we thank P. Molnar, G. S. Grest, and J. M. Carlson for helpful suggestions.

*Permanent address: Department of Electrical Engineering, Kyushu Institute of Technology, Tobata, Kitakyushu 804, Japan.
Electronic address: nasuno@ele.kyutech.ac.jp
†To whom correspondence should be addressed.
Electronic address: jgollub@haverford.edu

[1] H. Jaeger, S. R. Nagel, and R. P. Behringer, Rev. Mod. Phys. **68**, 1259 (1996).
[2] B. Miller, C. O'Hern, and R. P. Behringer, Phys. Rev. Lett. **77**, 3110 (1996).
[3] J. M. Carlson, J. S. Langer, and B. E. Shaw, Rev. Mod. Phys. **66**, 657 (1994).
[4] C. H. Scholz, *The Mechanics of Earthquakes and Faulting* (Cambridge University Press, Cambridge, England, 1990).
[5] A. Ruina, J. Geophys. Res. **88**, 10 359 (1983).
[6] P. A. Thompson and G. S. Grest, Phys. Rev. Lett. **67**, 1751 (1991).
[7] H. M. Jaeger, C. H. Liu, S. R. Nagel, and T. A. Witten, Europhys. Lett. **11**, 619 (1990).
[8] D. M. Hanes and D. Inman, J. Fluid Mech. **150**, 357 (1985).
[9] D. R. Scott, C. J. Marone, and C. G. Sammis, J. Geophys. Res. **99**, 7231 (1994).
[10] D. G. Wang and C. S. Campbell, J. Fluid Mech. **244**, 527 (1992).
[11] S. B. Savage and M. Sayed, J. Fluid Mech. **142**, 391 (1984).
[12] N. M. Beeler, T. E. Tullis, M. L. Blanpied, and J. D. Weeks, J. Geophys. Res. **101**, 8697 (1996).
[13] J. H. Dietrich, J. Geophys. Res. **83**, 3940 (1978).
[14] F. Heslot, T. Baumberger, B. Perrin, B. Caroli, and C. Caroli, Phys. Rev. E **49**, 4973 (1994).
[15] T. Baumberger, F. Heslot, and B. Perrin, Nature (London) **367**, 544 (1994).
[16] L. M. Jones and P. Molnar, J. Geophys. Res. **84**, 3596 (1979).
[17] B. E. Shaw, J. M. Carlson, and J. S. Langer, J. Geophys. Res. **97**, 479 (1992).
[18] J. R. Rice and S. T. Tse, J. Geophys. Res. **91**, 521 (1986).
[19] S. Nasuno, A. Bak, A. Kudrolli, and J. P. Gollub, *Time-Resolved Studies of Stick-Slip Friction in Sheared Granular Layers* (to be published).
[20] J. A. C. Martins, J. T. Oden, and F. M. F. Simões, Int. J. Eng. Sci. **28**, 29 (1990).
[21] C. G. Sammis and S. J. Steacy, Pure Appl. Geophys. **142**, 777 (1994).
[22] A. Demirel and S. Granick, Phys. Rev. Lett. **77**, 4330 (1996).

Kinematics of a two-dimensional granular Couette experiment at the transition to shearing

C. T. Veje*

PMMH, Ecole Supérieure de Physique et de Chimie Industrielle, 10 Rue Vauquelin, 75231 Paris Cedex 05, France

Daniel W. Howell and R. P. Behringer

Department of Physics and Center for Nonlinear and Complex Systems, Duke University, Durham, North Carolina 27708-0305

(Received 2 June 1998)

We describe experiments on a two-dimensional granular Couette system consisting of photoelastic disks undergoing slow shearing. The disks rest on a smooth surface and are confined between an inner wheel and an outer ring. Only shearing from the inner wheel is considered here. We obtain velocity, particle rotation rate (spin), and density distributions for the system by tracking positions and orientations of individual particles. At a characteristic packing fraction, $\gamma_c \simeq 0.77$, the wheel just engages the particles. In a narrow range of γ, $0.77 \leq \gamma \leq 0.80$ the system changes from just able to shear to densely packed. The transition at γ_c has a number of hallmarks of a critical transition, including critical slowing down, and an order parameter. For instance, the mean stress grows from 0 as γ increases above γ_c, and hence plays the role of an order parameter. Also, the mean particle velocity vanishes at the transition point, implying slowing down at γ_c. Above γ_c, the mean azimuthal velocity decreases roughly exponentially with distance from the inner shearing wheel, and the local packing fraction shows roughly comparable exponential decay from a highly dilated region next to the wheel to a denser but frozen packing further away. Approximate but not perfect shear rate invariance occurs; variations from perfect rate invariance appear to be related to small long-time rearrangements of the disks. The characteristic width of the induced "shear band" near the wheel varies most rapidly with distance from the wheel for $\gamma \simeq \gamma_c$, and is relatively insensitive to the packing fraction for the larger γ's studied here. The mean particle spin oscillates near the wheel, and falls rapidly to zero away from the shearing surface. The distributions for the tangential velocity and particle spins are wide and show a complex shape, particularly for the disk layer nearest to the shearing surface. The two-variable distribution function for tangential velocity and spin reveals a separation of the kinematics into a slipping state and a nonslipping state consisting of a combination of rolling and translation. [S1063-651X(99)06001-8]

PACS number(s): 81.05.Rm, 45.05.+x

I. INTRODUCTION

The pioneering work of Reynolds in 1885 [1] and the more elaborate investigations by Bagnold [2] were among the first experiments to closely address the problem of granular shearing. Recently, granular flows in which shearing plays a key role have regained much interest in the physics community. Examples of these flows include pipe and chute flow [4,5], avalanches [6,7], compression [8], and to some extent, convection [3]. In addition, there are interesting connections between granular flows and crack formation and earthquakes [9]. For a recent overview on the general physics of granular media see Jaeger, Nagel and Behringer [10], Behringer and Jenkins [11], and Luding and Herrmann [12].

Here, we study both the mean and statistical kinetic properties of slow shear flow in a two-dimensional (2D) Couette cell filled with photoelastic polymer disks. Using particle tracking techniques, we measure the spin, defined below, transport velocities, and the associated density variations during steady-state shearing. Because the particles are photoelastic, it is also possible to deduce information on the stress state of the system, a topic that we will consider elsewhere [13].

*Present address: Center for Chaos and Turbulence Studies, Niels Bohr Institute, Blegdamsvej 17, DK-2100 Copenhagen Ø, Denmark.

In the traditional picture of shearing for a dense granular material, grains are assumed to be relatively hard. If shear is applied to a sample subject by a normal load, the system responds elastically (if the individual grains are not infinitely stiff) up to the yield point. At the yield point, the system dilates enough to allow shear displacement, resulting in grains slipping over each other. This picture is referred to as Reynolds dilatancy. The dilation often occurs in narrow spatial regions known as shear bands. After the initial dilation, it is generally assumed that the system can attain a steady state under continued shearing without significant temporal variation [14].

In the present experiments, we consider the evolution of the system both during the initial dilation process and for extended times thereafter. The initial stage in this process is the dilation of the material near the shearing wheel. We find that after the formation of this dilated region, the system approaches an approximately statistically stationary state, but that fluctuations can be large. Also, there are indications that some changes in the system continue to occur even over very long times.

Previous experiments on granular shearing have primarily focused on the mean properties of the system. Typical examples include triaxial tests where localized shear bands are formed upon failure of the material [15,16], or shearing in annular cells where shear bands are formed by imposing a moving boundary for unlimited shearing [17–19]. (See also Savage [20] and references therein.) Other experiments have

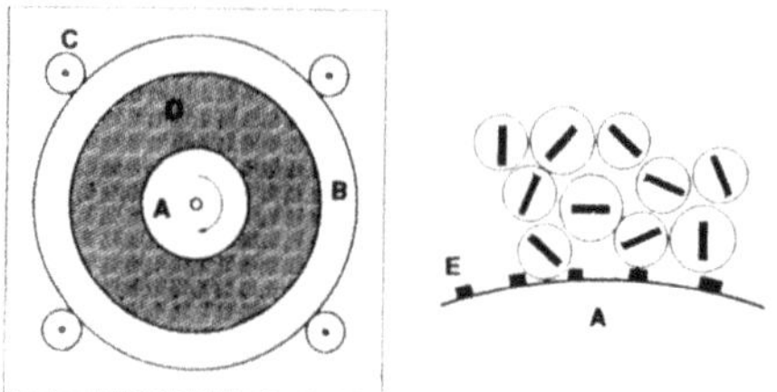

FIG. 1. Left: schematic top view of the experimental setup. Right: schematic drawing of the disks close to the inner shearing wheel.

explored the kinematics of shear zones for fast collisional flows; for example, flow in inclined or vertical chutes and pipes [4,5,7,21]. There have been a number of studies of 2D systems, and of these, the most closely related work to ours is that of Buggisch [22], who investigated mixing mechanisms due to shearing in a 2D annular cell roughly similar to the one described in this paper.

There are also numerical studies involving shearing; these typically center on the stress characteristics of the material [23,24]. However, Schöllmann, Luding and Herrmann [25] have recently performed molecular- dynamics (MD) simulations to investigate the kinematic and force properties on a model system that was structured to be as parallel as possible to the physical system discussed here. A comparison between this model and the present experiment will be presented elsewhere [26].

II. EXPERIMENTAL SETUP AND PROCEDURE

The apparatus, as sketched in Fig. 1, consists of an inner shearing wheel (A) of diameter $D=20.32$ cm and an outer shearing ring (B) of inner diameter 50.8 cm confined by planetary gears (C). Flat, 6-mm-thick disks (region D) are confined between the ring, the wheel, and two smooth horizontal Plexiglas sheets. The separation between the Plexiglas sheets is 2 mm greater than the thickness of the disks, so that there will be no vertical compression of the disks while still allowing little movement out of the plane. The side of the wheel and the inner surface of the ring are coated with plastic "teeth" spaced 0.7 cm apart and 0.2 cm deep (E). Both the wheel and the outer ring can be used to shear, although here we focus on shearing with the inner wheel. The surfaces of the Plexiglas sheets are covered with a fine powder (baking powder) that serves as a nonliquid lubricant. There is still some remaining friction between the disks and sheet, but the friction force is typically at least an order of magnitude smaller than the force in a stress chain [13]. The disks are made of a transparent photoelastic polymer, nominal Young's modulus, $Y=4.8$ MPa, and these disks are softer than the Plexiglas shearing wheel and ring (nominal Young's modulus $Y\sim 3$ GPa). Forces in the experiment are not sufficient to cause any appreciable deformation of the disks out of the plane. Each disk has a small dark bar on top which allows us to track the individual disks using conventional video. Frame-grabbed images are analyzed to find each bar and its orientation. Using computer analysis, we follow the trajectories of particles from frame to frame, determining kinematic quantities for each disk, specifically, the azimuthal velocity V_θ, the radial velocity V_r and the spin S. Here, S corresponds to the angular rotation of a disk in a frame that moves with the disk center of mass and has a fixed orientation relative to a lab coordinate system. The subscripts r and θ refer to a polar coordinate system with the origin at the center of the inner shearing wheel. Throughout this work, we subtract the wheel radius from r, so that $r=0$ corresponds to the location of the wheel edge. By V_θ, we mean the azimuthal angular velocity, $\dot{\theta}$ of the center of a grain. The most interesting quantities are V_θ and S as functions of r; it is these quantities that we present below. V_r is typically much smaller than V_θ because the grains are largely confined in the radial direction, and we do not consider V_r here.

In these experiments, a bimodal distribution of disks was used, with roughly 400 larger disks of diameter 0.9 cm, and roughly 2500 smaller disks of diameter 0.74 cm. An inhomogeneous distribution is useful, since it limits the formation of hexagonally ordered regions over large scales. One concern was that the disks would segregate by size. However, we did not observe any tendency for this to happen over the course of a typical experiment. We use the diameter d of the smaller disks as a characteristic length scale.

The packing fraction γ (fractional area occupied by disks versus total area) was varied over the range $0.77\leq\gamma\leq 0.80$. As we varied γ, we maintained the ratio of small to large disks roughly constant. We also varied the rotation period, τ of the inner wheel over the range $10^2\ \text{s}\leq\tau\leq 2\times 10^3$ s. This period defines a shear rate, $\Omega=2\pi/\tau$.

After placing the disks in the experiment, but before beginning to take data, we typically ran the inner wheel for ~ 60 min. This corresponded to about 20 rotations of the inner wheel and was necessary, because as a consequence of shearing the local packing fraction γ became nonuniform radially, i.e., a shear band formed, as shown in Fig. 2. Much of the evolution of the resulting shear band occurred in less than a wheel revolution and after ~ 5 rotations the mean density distributions were usually well established. However, some evolution of the density was observed on rare occasions after much longer times.

The transient time for the formation of the shear band also depended on γ. In fact, there are actually two slow times, one associated with the transition between shearing and nonshearing at a critical γ_c, discussed below, and the other associated with long-time rearrangements of a dense packing of grains, and therefore to the parking problem [27,28].

To obtain distributions for V_θ and S, we observed a region of the apparatus that covered roughly 1/4 to 1/5 of the azimuthal span, and all of the radial range. Typically, we observed for times long enough to accumulate $>10^5$ velocity and spin samples over the complete azimuthal and radial span. Typically, this corresponded to ~ 5 rotations of the shearing wheel. Spatial information on all quantities was determined by binning radially (with a bin width $\sim d$); all the data within a radial bin were placed together without regard to the azimuthal coordinate.

It is also important to determine the local packing fraction. We measured this quantity by using the fact that "black light," i.e., fluorescent light with a largely blue color, was strongly attenuated by the polymer disks. Specifically, the local transmission was determined by the open space. We

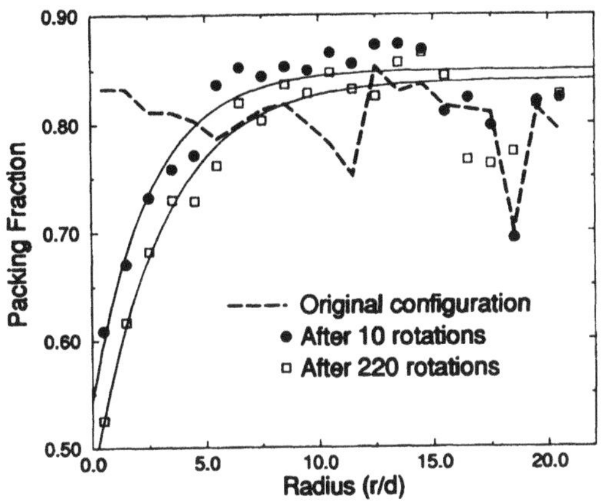

FIG. 2. Results for the local packing fraction after 10 and 220 rotations. The dashed line is the initial configuration. In this case, $\gamma=0.788$. Solid curves show least squares fits to the form $\rho=A-B\exp(-Cr/d)$. Since we are only looking at a quarter of the total cell the area under each curve need not be the same. Most of the relaxation has occurred after 10 rotations; the coefficient C is essentially the same for 10 and 220 rotations. Notably, however, some variation in the packing structure at large r (filling the gap in density near $r=18$) occurred even after very long times—between 20 and 40 rotations—corresponding to a displacement of the shearing wheel by ~ 2200 to ~ 4400 disk diameters.

calibrated this transmission by measurements through various packings of predetermined γ. We estimate that errors associated with these measurements are roughly 1%–2%. We resolve relative changes in packing fraction exactly, since we remove or add exactly known numbers of disks.

To the extent that the shearing rate Ω is small compared to relaxation rates for the system, we would expect rate invariance. In the results presented below, we use normalized units V_θ/Ω for velocity. Then, rate invariance is manifest if data for different Ω, but otherwise similar conditions, are indistinguishable. In the same spirit, we normalize S by $\Omega D/d$, which is the gear-ratio factor for rolling of the disks on the shear wheel.

III. RESULTS

A. Packing fraction profiles

We consider now the density or packing fraction profile, the kinematic distributions, and their means. We begin with the local packing fraction profiles and their time evolution. Figure 2 shows typical results for the local packing fraction as a function of the distance to the shearing wheel. The profiles clearly indicate a dilated region close to the inner wheel. To obtain a characteristic measure of the width of the dilated region we fit the data for $r/d<14$ to the form $\rho=A-B\exp(-Cr/d)$, where A, B, and C are fit coefficients. Results are shown for the initial state, after 10 rotations and after 220 rotations.

The process of dilation begins with the motion of disks adjacent to the shearing wheel, where yield/failure begins. As time evolves, the density in the vicinity of the failure region decreases, leading to azimuthal flow further from the shearing wheel, and the evolution of a shear band. The angle through which the inner wheel must rotate, or more usefully, the number of disk diameters traversed, is surprisingly large. Several wheel rotations are necessary to reasonably approach the steady-state profile, (with more than $10^2 d$ per wheel rotation). Thus, in many settings where shear bands evolve, the result may actually be a transient rather than the steady state. The nature and length of this transient is interesting, since it relates to failure in other granular flows and possibly to situations such as earthquake fault zones.

Clearly the effect of continued shearing for more than 10 rotations is small. For the most part, the structure far from the shearing wheel remains largely frozen. However, there can be changes in the packing at large r/d that impact other features of the flow. For instance, in Fig. 2, there is clearly an observable rearrangement at $r/d\simeq 17$, a region where the mean velocity is too small to measure easily. The impact on the rest of the profile is also observable, although simple quantitative characterizations, such as the widths, C^{-1}, are virtually indistinguishable within experimental error for 10 and 220 rotations ($3.2d$ and $2.7d$, respectivly). The possibility of changes like those near $r/d\simeq 17$ imply the existence of very long time scales in the system, similar to those that are observed in compaction [27,28,30]. These very long times appear to play a role in moderate but clearly observable variations in the velocity profiles over very long experiments.

We now sketch the qualitative properties of this system as we increase γ. Below γ_c, the sample is quiescent all the time. Thus at very low packing fractions, we expect that no motion will occur after initial transients (assuming, for example, a uniform initial packing). In particular, the disks are easily pushed away from the wheel, where after some rearranging, they remain without further contact. Just above a critical packing fraction γ_c the system is just strong enough to allow shearing indefinitely. The displacement of particles, and/or the formation of stress chains, occur at isolated and temporally fluctuating angular locations. For γ well above γ_c stresses are large, and chains appear in many places throughout the system. The displacement of particles occurs at many different angular positions simultaneously.

One of the key aspects of the present experiments is the exploration of packing fractions near γ_c. For $\gamma<\gamma_c$, disks are pushed away from the shearing wheel, and after transients, the systems remains at rest indefinitely. Just above γ_c, disks at a given position are sheared only intermittently, but once γ_c has been crossed, the disks will always be under shear somewhere in the system. For γ well above γ_c, the system is in the densely packed regime that is typical of usual gravitationally compacted 3D systems.

The range of γ studied was chosen so that at the lowest $\gamma\simeq\gamma_c$, the disks were only rarely in contact with the shearing wheel, and at high γ the system was densely packed, with the innermost disks constantly in contact with the shearing wheel. Notably, only a small variation in γ, less than 4%, was sufficient to traverse this regime.

The transition at γ_c is related to Reynolds dilatancy. However, in the usual picture of Reynolds dilatancy, the system is under normal load, and significant shear stresses are needed to deform the system. At the yield point, the system then tends to dilate at the mechanically weakest point. In the present experiments, the system as a whole is actually quite soft, easily sheared, and highly compressible. In order to distinguish the present case from the conventional picture, we will refer to the transition at γ_c as the softening (or strengthening) transition (ST).

In an ordinary 3D granular material, the corresponding γ_c is not generally accessible, because gravity compacts the material. This compaction occurs readily near γ_c, because the material is so highly compressible. Since there are no gravitational effects in this experiment, we can adjust the mean packing fraction, and hence use it as the control parameter, rather than the normal load.

There is one final aspect of these experiments that facilitates the observations of the ST, namely, the relative softness of the disks. To understand why this is important, we envision what would happen if we were to make the disks increasingly stiffer. We would expect to find that properties such as the mean stress would increase faster for γ increasing above γ_c, for increasingly stiffer disks. The rapid increase in mean stress above γ_c for stiff disks would make observations of the near-γ_c properties harder.

The precise experimental determination of γ_c is difficult because near γ_c, forces are weak, and changes occur intermittently over long times. The reason for this slow evolution is obvious on reflection. Although slow relaxation is observed for compaction [27,28], the phenomena here relate to the critical nature of the transition, rather than to the difficulty of rearranging the packing. Specifically, the packing in this case is highly compressible, as opposed to the long-time limit of the compaction case.

It is interesting, and perhaps not surprising, that γ_c is very close to the packing fraction for a square lattice, $\gamma_s = \pi/4 \simeq 0.785$. By contrast, a hexagonally ordered packing of disks occurs for $\gamma_h = \pi/(2\cdot 3^{1/2}) = 0.907$. Our system becomes stiff well below this packing fraction at a value comparable to those found for a random close packing of disks ($\gamma_{rcp} \approx 0.82$) [29]. There are analogous packing fractions for 3D systems. However, as noted above, in typical earth-bound experiments, gravity invariably leads to systems that have higher values of γ than the corresponding critical value, and it is very difficult to approach this transition.

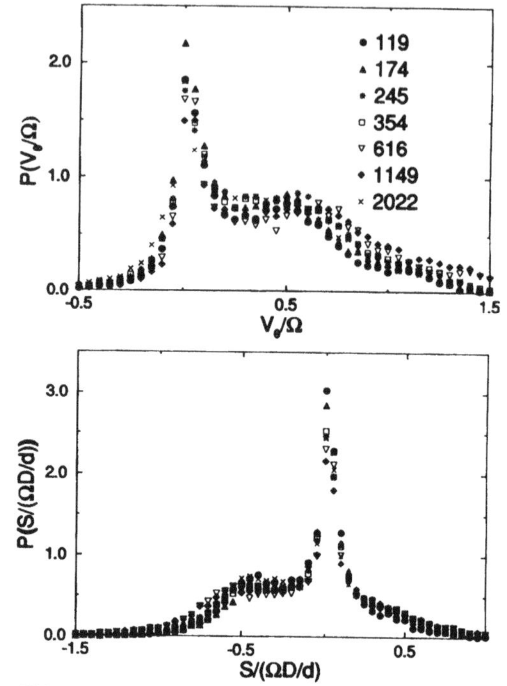

FIG. 3. Normalized azimuthal velocity and spin distributions close to the inner wheel for different rotation times, for $\gamma = 0.788$. Data are taken from radii $0 < r/d < 1$. The symbols refers to different values of the rotation time τ in seconds. If there were perfect rate invariance, these distributions would collapse perfectly.

B. Distributions for V_θ and S

We turn next to the velocity and spin distributions for $\gamma \geqslant \gamma_c$. The azimuthal velocities, V_θ/Ω, and particle spin, $S/(\Omega D/d)$, are particularly interesting. Distributions for V_θ/Ω and $S/(\Omega D/d)$ close to the shearing wheel are shown in Fig. 3 and their mean values versus r/d in Fig. 4 for different values of $\tau = 2\pi\Omega^{-1}$. All distributions have been normalized to unity, and the packing fraction, $\gamma = 0.788$, is in the midrange of the values studied here.

The mean of V_θ/Ω as a function of r/d, Fig. 4, shows a roughly exponential profile corresponding to a shear zone with a width of a few d. By $r/d = 7$, the data have reached the noise level. This is consistent with previous observations for quasistatic flow [7,22]. Although the profile is nearly exponential in distance from the wheel, there is also the suggestion of some downward curvature when V_θ/Ω is viewed semilogarithmically. The solid curve in Fig. 4 shows a fit of $\ln(V_\theta/\Omega)$ to $-A - B(r/d) - C(r/d)^2$ in the range $0 < r/d < 6$. The fit coefficients are $A = 0.503 \pm 0.148$, $B = 0.417 \pm 0.096$, and $C = 0.105 \pm 0.013$. The quoted errors reflect the statistical uncertainty. The second-order correction is relatively weak, and becomes comparable to the leading term only for $r/d = \simeq 6$. The characteristic decay length in this case, $B^{-1} = 2.4d$, is somewhat smaller than the corresponding decay length for the density profiles. Presumably, the reason for the more rapid decay for V_θ than for γ is the rapid drop in disk mobility with increasing γ. The approximately exponential form for V_θ/Ω is interesting, since it suggests the possibility that disks past the first few layers may eventually move, but only after exponentially long-time scales.

The mean profiles for $S/(\Omega D/d)$ are also interesting. The disks nearest to the inner wheel rotate backwards on average—i.e., in the direction opposite to the wheel rotation. However, the next layer rotates in same direction as the wheel on average. These oscillations in the mean spin damp very quickly with distance from the wheel. In order to determine how fast the spin decays to zero, we fit the spin profile to $S/(\Omega D/d) = A \exp[-B(r/d)]\cos(\pi r/d + \phi)$. Here, the key fit coefficients are $A = 0.760 \pm 0.116$, $B = 1.67 \pm 0.19$, and $\phi = 0.790 \pm 0.080$. Here, the errors represent the statistical uncertainty. Again the second-order correction is very small. The characteristic length in this case is just above $0.5d$, which is even faster than the velocity decay. This is likely related to the fact that the spin of the disks in a given layer is driven in one direction by neighboring disks in the next-closest layer to the wheel, while at the same time impeded by

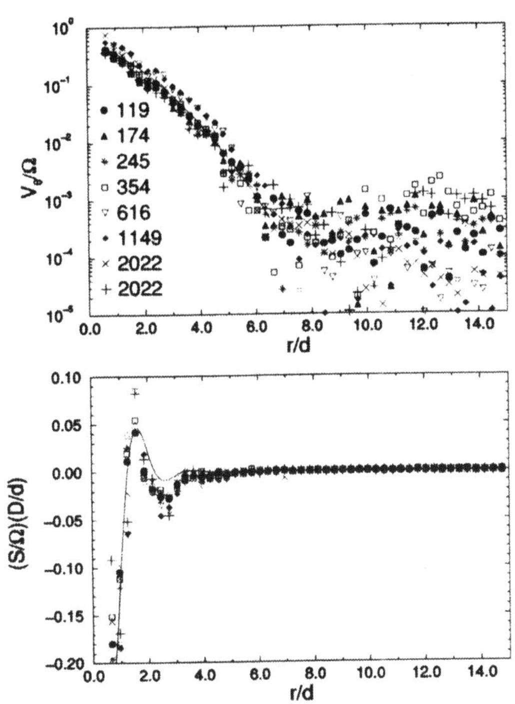

FIG. 4. Mean azimuthal velocity and mean spin for different rotation rates at $\gamma = 0.788$. The symbols refer to different values of the rotation time τ in seconds. Two sets of data at $\tau = 2022$ sec show that deviations between the different rates are comparable to reproducibility from run to run at the same rate. The solid lines correspond to the fits described in the text.

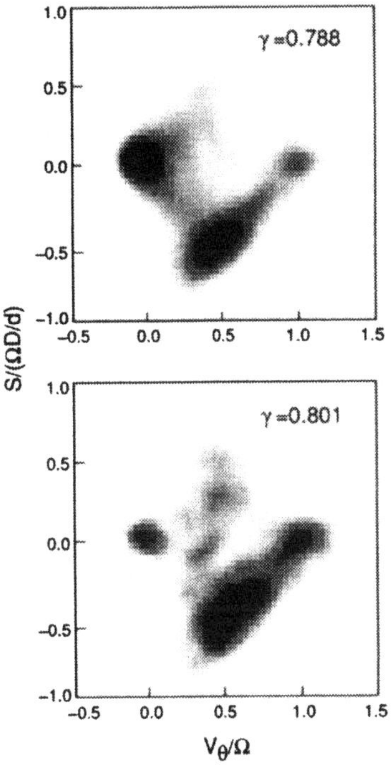

FIG. 5. Gray scale representation of the 2D probability density for V_θ/Ω and $S/(\Omega D/d)$ for $0 < r/d < 1$. Data are shown for two indicated values of γ.

the neighboring disks that lie one layer further from the wheel.

An inspection of Fig. 4 shows that the data for different τ for the means of both V_θ/Ω and $S/(\Omega D/d)$ do not fall exactly on top of each other. For a given Ω, the data are smooth in r, so that the variability in scaled velocity is not simply experimental uncertainty in determining V_θ.

This suggests a breakdown of the rate invariance and the quasistatic limit. However, the reasons for the noncollapse of the data are more subtle. Specifically, there is no clear trend in the data for either of these quantities with increasing Ω. Although the widths of the distributions, are large, the smooth variation of the mean velocity with r again suggests that this alone cannot account for the departures from scaling. At this point, we believe that departure from rate independence in the mean profiles is related to relatively small fluctuations in the packing profiles over long times—i.e., after the wheel has moved many disk diameters. Thus, we might expect that if data were obtained over observation times much longer than what we have used, rate independence might well be seen, although the widths of the distributions would be little changed. Put differently, our results suggest that there are long time scales over which small packing adjustments can occur, and that these adjustments affect the kinematic properties of the system. We expect to pursue even longer time scales in future experiments.

The distributions $P(V_\theta/\Omega)$ and $P(S/(\Omega D/d))$ for moderate γ reveal additional interesting features. $P(V_\theta/\Omega)$ for the innermost layer, Fig. 3, shows that there is some motion in the negative direction, i.e., for $V_\theta/\Omega < 0$ and in the positive direction for $V_\theta/\Omega > 1$. These types of motion occur because disks slip backwards or forwards as a chain fails. There is a large peak at $V_\theta/\Omega = 0$ that corresponds to disks that remain at rest. The next-largest feature is at $V_\theta/\Omega \simeq 0.5$, corresponding to disks that move at approximately 1/2 the wheel speed, and there also appears to be some structure at $V_\theta/\Omega \simeq 1.0$. These various features lead to a complex distribution for V_θ/Ω reflecting a combination of slip and nonslip events, as discussed below. However, for practical purposes, a fit of the distribution $P(V_\theta/\Omega)$ to three Gaussians centered at 0.0, 0.5, and 1.0 leads to a good quantitative characterization of the data shown in Fig. 3.

The distribution for $S/(\Omega D/d)$ near the shearing wheel shows qualitatively similar structure except that it occurs for negative S, corresponding to the backward spin seen in the mean profile. In addition, there are only two peaks for the spin distribution. However, a fit to two Gaussians for the spin is not quite as good as either a two- or three-Gaussian fit to the V_θ distribution.

A key understanding of the structure of the one-variable distributions for V_θ and S comes from Fig. 5, which shows two examples, for two different packing fractions, of the two-variable distribution $P(V_\theta/\Omega, S/(\Omega D/d))$ for grains within one particle diameter of the wheel. Here, $P(V_\theta/\Omega, S/(\Omega D/d))$ is coded in gray scale (dark is high probability density.) Figure 5 shows that there are two dis-

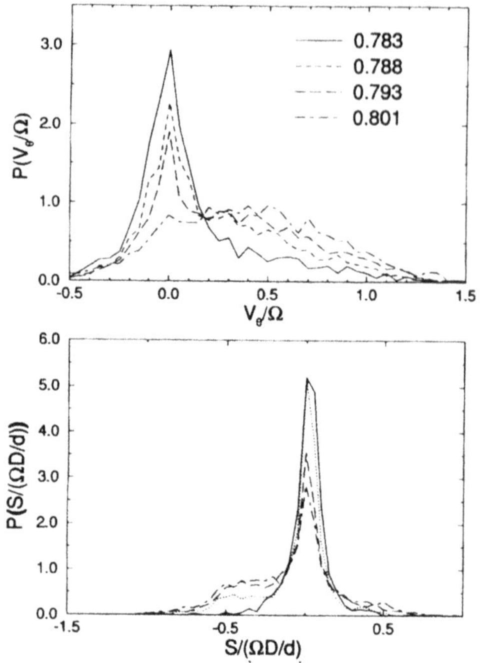

FIG. 6. Azimuthal velocity and spin distributions close to the inner wheel. Data are taken from radii $0<r/d<1$. Different lines are for different packing fractions γ, as noted.

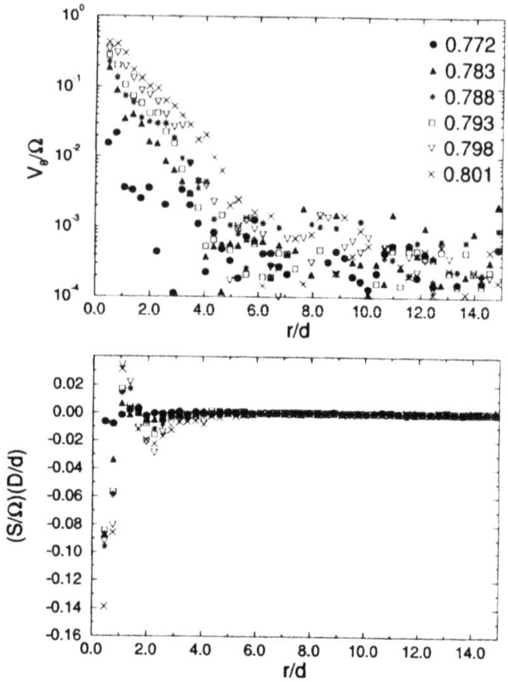

FIG. 7. Mean azimuthal velocity and mean spin. Different symbols are for different γ.

tinct features corresponding to two qualitatively different processes. Which of these processes is most important changes with increasing γ. One of these features, centered around (0,0), corresponds to a state where disks are essentially at rest, without either translation or spin. The other is clustered around a line $V_\theta/\Omega=1+S/(\Omega D/d)$ that corresponds to nonslip motion of grains relative to the wheel, with much of the weight near $(0.5,-0.5)$. By no slip, we mean that particles execute a combination of backwards rolling and translation such that the wheel surface and the disk surfaces remain in continued contact. The peak at $V_\theta/\Omega=0$ is strong for lower γ, but it has almost vanished at $\gamma=0.801$, and the data fall around the no slip line. Thus the transition at γ_c can also be thought of as one in which slip at the shearing surface is replaced by nonslip in a continuous way.

In Fig. 6 we show quantitative data for the distributions for various γ's. With increasing γ, the peaks in $P(V_\theta/\Omega)$ and $P(S/(\Omega D/d))$ near zero weaken, and the corresponding regions with negative spin and nonzero V_θ grow. A surprising feature is that the part of the distribution satisfying nonslip does not extend to velocities much lower than $V_\theta/\Omega\simeq0.5$. This may be related to specific properties of this system, such as frictional or geometric properties. We will pursue this point in future experiments.

The distributions narrow rapidly with distance away from the wheel, and approach an instrumentally sharp distribution centered around $V_\theta=0$. In the large-r regime, past the first seven layers, the distribution width reflects primarily experimental errors from finite pixel size, tape registry errors, etc. In this nearly stationary regime, rolling of the disks is impossible, and S is no longer an important variable. However, there may well be some fast-slipping events that are beyond the resolution of the imaging equipment used here.

From the above discussion, it is clear that changing the packing fraction must change not only the distributions, but also the mean properties. In Fig. 7 we show quantitative data for the profiles for several values of γ. We average data for different values of Ω to obtain statistics from almost 10^6 data points. The V_θ/Ω profiles show a roughly exponential decay with a comparable characteristic length for all γ. The amplitude of the exponential term grows steadily from zero as γ increases above γ_c. The profile for $S/(\Omega D/d)$ also evolves in qualitatively the same manner with γ. This evolution of the spin and velocity profiles with γ is, of course, tied intimately to the density profiles, Fig. 8.

IV. CONCLUSIONS

To conclude, this work is an exploration of the kinematic statistical properties for slow 2D granular shear flow. We have focused on the transition to shearing that first occurs for packing fractions greater than a critical value, $\gamma_c\simeq0.77$. This transition has some features reminiscent of a second-order thermodynamic phase transition, including the existence of an order parameter, critical slowing down, and large compressibility.

With increasing γ, the following qualitative behavior occurs. Below γ_c, the disks can remain indefinitely at rest. Just above γ_c, the disks move very slowly, so that necessarily, the system shows critical slowing down as $\gamma\to\gamma_c$ from above. This slow behavior is qualitatively different from the

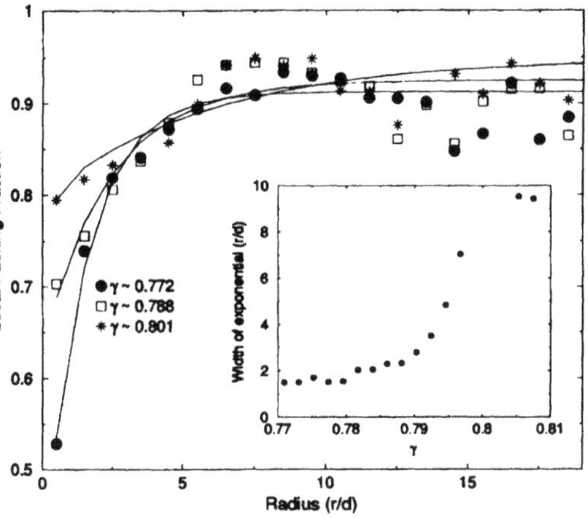

FIG. 8. Steady-state density profiles for different values of γ. Solid lines are exponential fits (see text). Inset: The exponential width of the steady-state profiles vs the overall packing fraction.

slow time that appears in densely packed systems where rearrangement of the grains becomes increasingly more difficult with time. Of course, rearrangements of this type also occur particularly at dense packings, and influence the statistical and mean properties of the flow on relatively long time scales.

The density, and azimuthal velocity obey roughly exponential profiles with distance from the wheel, and these profiles grow as γ rises above γ_c. The spin profiles show interesting oscillations and disks next to the wheel have a high probability of rotating backwards. This feature is particularly clear in the bimodal distributions for S and V_θ; these show that the transition near γ_c is characterized by a change from complete slipping to increasingly nonslip dynamics. The latter behavior is qualitatively similar to the usual nonslip condition for conventional fluids, except that here the rolling component must also be considered. These distributions also show forward and backward slipping events at velocities less than 0, and greater than the wheel velocity.

There are a number of other interesting aspects of this system that we will describe elsewhere [13]. These include the force distributions, and other measures associated with dynamical properties.

ACKNOWLEDGMENTS

C.T.V. and R.P.B. are indebted to the École Superieure de Physique et Chemie Industrielle, PMMH, where parts of this work were carried out. We particularly appreciate interactions with Professor Hans Herrmann, as well as useful comments by Joshua Socolar, David Schaeffer, Stéfane Roux, Eric Clément, Jean Rajchenbach, and P.-G. de Gennes. The work of D.W.H. and R.P.B. was supported by the National Science Foundation under Grant Nos. DMR-9321791 and DMS-9504577, and by NASA under Grant No. NAG3-1917. The work of C.T.V. was supported by the Danish Science Foundation (Statens Naturvidenskabelige Forskningsråd).

[1] O. Reynolds, Philos. Mag. 5, **50**, 496 (1885).
[2] R. A. Bagnold, Proc. R. Soc. London, Ser. A **225**, 49 (1954).
[3] J. B. Knight, Phys. Rev. E **55**, 6016 (1997).
[4] R. M. Nedderman and C. Laohakul, Powder Technol. **25**, 91 (1980).
[5] T. G. Drake, J. Geophys. Res. **95**, 8681 (1990).
[6] S. B. Savage and K. Hutter, J. Fluid Mech. **199**, 177 (1989).
[7] O. Pouliquen and R. Gutfraind, Phys. Rev. E **53**, 552 (1996).
[8] A. Ngadit and J. Rachenbach, Phys. Rev. Lett. **80**, 273 (1998).
[9] H. J. Herrmann, A. N. B. Poliakov, and S. Roux, Fractals **3**, 821 (1995).
[10] H. M. Jaeger, S. R. Nagel, and R. P. Behringer, Phys. Today **49**, 32 (1996); Rev. Mod. Phys. **68**, 1259 (1996).
[11] *Powders and Grains 97*, edited by R. P. Behringer and J. T. Jenkins (Balkena, Rotterdam, 1997).
[12] *Physics of Dry Granular Media*, Vol. 350 of *NATO Advanced Study Institute Series E: Applied Sciences*, edited by H. J. Herrmann, J. P. Hovi, and S. Luding (Kluwer, Amsterdam, 1998).
[13] D. W. Howell, C. T. Veje, and R. P. Behringer (unpublished).
[14] For a review see R. Jackson, in *Theory of Dispersed Multiphase Flow*, edited by R. Meyer (Academic, New York, 1983).
[15] K. H. Roscoe, Geotechnique **20**, 129 (1970).
[16] H.-B. Mühlhaus and I. Vardoulakis, Geotechnique **37**, 271 (1987).
[17] J. F. Carr and D. M. Walker, Powder Technol. **1**, 369 (1967).
[18] S. B. Savage and S. McKeown, J. Fluid Mech. **127**, 453 (1983).
[19] B. Miller, C. O'Hern, and R. P. Behringer, Phys. Rev. Lett. **77**, 3110 (1996).
[20] S. B. Savage, Adv. Appl. Mech. **24**, 289 (1984).
[21] E. Azana, F. Chevoir and P. Moucheront, in *Powders and Grains 97* (Ref. [11]).
[22] H. Buggisch and G. Löffelmann, Chemical Engineering and Processing **26**, 193 (1989).
[23] C. S. Campbell and C. E. Brennen, J. Fluid Mech. **151**, 167 (1985).
[24] H.-J. Tillemans and H. J. Herrmann, Physica A **217**, 261 (1995).
[25] S. Schöllmann, S. Luding, and H. J. Herrmann (unpublished).
[26] C. T. Veje, D. W. Howell, R. P. Behringer, S. Schöllmann, S. Luding, and H. J. Herrmann, in *Dry Granular Media* (Ref. [12]).
[27] J. B. Knight, C. G. Fandrich, C. N. Lau, H. M. Jaeger, and S. R. Nagel, Phys. Rev. E **51**, 3957 (1995).
[28] E. Ben-Naim, J. B. Knight, and E. R. Nowak, J. Chem. Phys. **100**, 6778 (1996).
[29] L. Rouillé, J-M. Missiaen, and G. Thomas, J. Phys.: Condens. Matter **2**, 3041 (1990).
[30] Edmund R. Nowak, J. B. Knight, E. Ben-Naim, H. M. Jaeger, and S. R. Nagel, Phys. Rev. E **57**, 1971 (1998).

Part 8

EFFECTIVE TEMPERATURE

Physica A 157 (1989) 1080–1090
North-Holland, Amsterdam

THEORY OF POWDERS

S.F. EDWARDS and R.B.S. OAKESHOTT
Cavendish Laboratory, Madingley Road, Cambridge CB3 0HE, UK

Received 20 February 1989

Starting from the observations that powders have a large number of particles, and reproducible properties, we show how statistical mechanics applies to powders. The volume of the powder plays the role of the energy in conventional statistical mechanics with the hypothesis that all states of a specified volume are equally probable. We introduce a variable X – the compactivity – which is analogous to the temperature in thermodynamics. Some simple models are considered which demonstrate how the problems involved can be tackled using the concept of compactivity.

1. Introduction

There is an increasing interest in applying the methods of statistical mechanics and of transport theory to systems which are neither atomistic, nor in equilibrium, but which still fulfil a remaining tenet of statistical physics which is that systems can be completely defined by a very small number of parameters and can be constructed in a reproducible way. Powders fall into this category. If a powder consists for example of uniform cubes of salt, and is poured into a container, falling at low density uniformly from a great height, one expects a salt powder of a certain density. Repeating the preparation reproduces the same density. A treatment such as shaking the powder by a definite routine produces a new density and the identical routine applied to another sample of the initial powder will result in the same final density. Clearly a Maxwell demon could arrange the little cubes of NaCl to make a material of different properties to that of our experiment, but if such demonics are ignored, and we restrict ourselves to extensive operations such as stirring, shaking, compressing – all actions which do not act on grains individually – then well defined states of the powder result.

In this paper we will set up a framework for describing the state of the powder, basing our development on anologies with statistical mechanics. Some attempts have been made to apply information theory ideas directly to powders

0378-4371/89/$03.50

[1–3], but we want to introduce a formulation closer to conventional statistical mechanics in order to use the powerful ideas developed there. Although much information is available on powders and a notable literature exists [4–6], one is struck by the rather advanced situations discussed. It seems to us that the most elementary problems have not been fully addressed, so that this paper will study some powder problems at a very elementary level of statistical mechanical technique. We aim at new physical ideas and offer no new mathematical ideas.

The number density of a powder lies between $10^0\,\mathrm{ml}^{-1}$ and $10^{16}\,\mathrm{ml}^{-1}$. Although much less than the 10^{22} of atoms, this range takes one into the region where Van der Waals forces and plain electric charges become dominant. Although we do not need to refer to a density, we have in mind the range 10^3 to 10^6, i.e. friction is important but attractive forces are not dominant. For bulk samples, the numbers are still very large so that statistical arguments are entirely appropriate.

We will argue that powders have an entropy, but it will be the volume which plays the role of the energy in normal statistical mechanics: the energy corresponding to the powder's thermal temperature 300 K is negligible. In statistical mechanics the most fundamental entry is via the microcanonical ensemble. This is a closed system, contained in a volume V, which is assumed to take up all configurations subject to the Hamiltonian function taking the value E with equal probability. Thus the distribution function is

$$e^{-S/k_B}\delta(E-H)\,,$$

where the entropy S is defined in terms of the total number of configurations

$$\Omega = \int \delta(E-H)\,, \qquad S = k_B \ln \Omega\,,$$

k_B being Boltzmann's constant which harmonizes units of measurement.

Although the microcanonical ensemble is the most fundamental, E is not easy to measure and the δ function is difficult to handle, so it is easier to use the canonical ensemble and define the free energy F by the distribution function

$$e^{(F-H)/k_B T}\,,$$

where

$$e^{-F/k_B T} = \int e^{-H/k_B T}$$

and

$$T = \frac{\partial E}{\partial S} \,.$$

The energy is now given by

$$E = F + TS$$
$$= F - T \frac{\partial F}{\partial T} \,.$$

The most important variable describing the state of the powder is its density, or equivalently, but more usefully its volume V. The volume here is the actual volume taken up by the powder – i.e. as would be measured by letting a piston rest on the top of the powder. The volume therefore depends on the configuration of the particles – unlike the conventional case where the volume is set externally, and only the energy depends on the configuration of the particles. In principle other variables may be necessary to describe the state of a powder, i.e. if a powder is sheared, then it may develop an anisotropic texture, but we will not consider this possibility here.

We now introduce a function W of the coordinates of the grains which specifies the volume of the system in terms of the positions and orientations of the grains. The form of W will depend on how the configuration of the grains is specified, and on the shape and size of the grains. We also introduce a function Q which picks out valid configurations of the grains – that is stable arrangements, where the particles can remain at rest under the influence of the confining forces, and with no overlap between particles. On the assumption that for a given volume all these configurations are equally probable, a table can be drawn up developing the analogy. (See table I.)

There are model systems with very simple Hamiltonians available, e.g. the perfect gas or the Ising model, but in general H is not simple, for example a true ferromagnet has complicated many ion potentials. For powders, W and Q are never simple, e.g. hard spheres already offer a major problem from this point of view (though they can be written simply as is well known and discussed below).

In table I we have introduced two new variables X and Y. X is the analogue of temperature and measures the "frothiness" or "fluffiness" of the powder. Since $X = 0$ corresponds to the most compact powder and $X = \infty$ to the least, we may name it the compactivity of the powder; it is the inverse of the compaction. A powder prepared with a larger volume than that corresponding to $X = \infty$, and so with $X < 0$, would be unstable: given any vibration the powder would tend to compact, increasing its entropy, and reducing its

Table I

Statistical mechanics	Powders
E	V
H	W
S	S
k_B (which gives S the dimensions of energy)	λ (which gives S the dimension of volume)
$\int \prod \mathrm{d}r_i \prod \mathrm{d}p_i$	$\int \prod \mathrm{d}r_i\, Q$
$\mathrm{e}^{-S/k_B}\delta(E-H)$	$\mathrm{e}^{-S/\lambda}\delta(V-W)$
$T=\dfrac{\partial E}{\partial S}$	$X=\dfrac{\partial V}{\partial S}$
$\mathrm{e}^{-F/k_BT}=\int \mathrm{e}^{-H/k_BT}$	$\mathrm{e}^{-Y/\lambda X}=\int \mathrm{e}^{-W/\lambda X}$
$F=E-TS$	$Y=V-XS$
$E=F-T\dfrac{\partial F}{\partial T}$	$V=Y-X\dfrac{\partial Y}{\partial X}$

potential energy. Our assumption that all configurations of a given volume are equally probable implies that the same X should characterize arrangements around particles in the powder with different sizes densities, coefficients of friction, etc. An obvious name for Y would be the free volume, but that phrase is already appropriated in glass and liquid theory, hence we propose the name "effective volume" for Y. Notice that our new functions have fewer dependencies than those of thermodynamics, e.g.

$$E=E(S,V,N)$$

whereas

$$V=V(S,N)\,.$$

Similarly

$$F=F(T,V,N)\,,\qquad Y=Y(X,N)\,,$$

as a consequence, there are no analogies of the Maxwell relations of classical thermodynamics.

For a thermodynamic system

$$\frac{\partial S}{\partial V}=\frac{\partial S}{\partial E}\frac{\partial E}{\partial V}=\frac{P}{T}\,,$$

but since $P=0=T$ in our powder, there is no value in an identification of X in these terms. The general problem of pressure in the resistance of a powder to compression is more advanced than anything at the level of this paper. This is because whereas the exertion of pressure will decrease volume irreversibly, there comes a point when the volume can no longer decrease by rearrangement, but requires deformation of the grains. This involves stress finding (multiply connected) "percolating" pathways through the solid, and this requires a different kind of analysis which is downstream from the present problem. Here we can omit a discussion of the pressure because we always assume that frictional thresholds are such that the powder can rest in the specified configuration. For simple frictional forces ($F \leqslant \mu N$), and hard particles, the absolute gravitational field, or confining pressure does not matter.

In a stable configuration the particles must be touching, so that these states are a subset of measure zero of the total phase space of a hard particle gas. Of itself this is no problem – it is only if states with different numbers of degrees of freedom exist that there is a problem because, since this is a purely classical problem, there is no natural scale analogous to Planck's constant, h. For frictionless particles stable states are local volume, minima, and so form a discrete set of states – the only remaining possibility is that the minima should have different weights.

Two simple criteria for stability are possible. The simplest local criterion for stability is that no single particle is free to move. A global criterion is obtained if we use the Maxwell condition: the number of constraints greater than or equal to the number of variables. For spheres this implies that the average number of contacts per particle is at least 6. If the particles are not frictionless, then the true condition will lie between the local and global criteria. In particular states with fewer contacts will be stable, and maximising the entropy will select those states which minimise the number of contacts, whilst still being stable.

To illustrate these ideas we study the packing of a simple one species powder. The crude model illustrates overall features, rather than detailed study of particular local environments.

2. The volume of a simple powder studied by the compactivity concept

We start with the highly artificial, but instructive model of a powder in one dimension. Now powders obviously cannot exist in one dimension as gravity will always fully compact them, but we can still learn useful lessons. If the grains are rods of length a, whose midpoints are x_n, with $x_n < x_{n-1}$, then clearly trivially

$$W = x_N - x_0 + a .$$

However we can also write W as a sum of local volumes,

$$W = \sum_{1}^{N} (x_n - x_{n-1}) + a \, .$$

The volume exclusion implies that

$$x_n - x_{n-1} \geqslant a \, .$$

If we put as the stability condition that each rod touches its neighbour, so that

$$Q = \prod \delta(x_n - x_{n-1} - a)\Theta(x_n - x_{n-1} - a) \, ,$$

where

$$\Theta(x) = \begin{cases} 1 \, , & x \geqslant 0 \, , \\ 0 \, , & x < 0 \, , \end{cases}$$

then the problem is trivial; there is one configuration, and $e^{S/\lambda} = \delta(V - Na)$.

We can produce a model for a real – two- or three-dimensional – powder, if we consider our one-dimensional system as a section of the actual powder. The grains need not be touching in the section, but can have a range of separations up to a maximum b. Q then becomes

$$Q = \Theta(a + b - (x_n - x_{n-1}))\Theta(x_n - x_{n-1} - a) \, ,$$

and the integral for S is

$$e^{S/\lambda} = \int \delta\left(V - \sum_{1}^{N} (x_n - x_{n-1}) - a\right) \prod \Theta(a + b - (x_n - x_{n-1})) \times \Theta(x_n - x_{n-1} - a) \prod \mathrm{d}x_n \, .$$

Using the canonical ensemble we have

$$e^{-Y/\lambda X} = \left(\int_{a}^{a+b} e^{-v/\lambda X} \, \mathrm{d}v \right)^{N} ,$$

so that

$$Y = N\left(\frac{a+b}{2}\right) - N\lambda X \ln(\lambda X(e^{b/2\lambda X} - e^{-b/2\lambda X})) \, ,$$

and

$$V = N\left(\frac{a+b}{2}\right) + N\lambda X - N\frac{b}{2}\coth\frac{b}{2\lambda X}\,.$$

In two or three dimensions, this form using the particle coordinates is intractable: we have taken the hard particle fluid form, and added some difficult conditions. To do anything we must choose a smaller set of variables, with the same dimensionality as the set of stable states, and so making the stability and compatibility conditions simpler. By doing this we also bring in a Jacobian.

The simplest problem in three dimensions is the compactivity of a simple powder of uniform grains of the same material in approximately spherical form. Each grain will have neighbours touching it with a certain coordination and angular direction. There will also be a certain number of near neighbours which do not touch. If we want to set up an analogy with the statistical mechanics of alloys or magnetism, we want to consider that each grain has a certain property, and this property interacts with its neighbours, e.g. in an AB alloy there are A and B type atoms with interactions v_{aa}, v_{ab}, v_{bb}. Suppose we take the coordination of a grain as such a property, i.e. suppose we assign a volume v_c to any grain with c neighbours. Clearly this is not a comprehensive description of the powder, but we can suppose that

$$W = \sum_c v_c n_c + \sum_{c,c'} v_{cc'} n_{cc'} \ldots ,$$

where n_c is the number of grains with c neighbours, and $v_{cc'}$ the refinement of the volume function when there are $n_{cc'}$ pairs of neighbouring grains with one with c and one with c' neighbours. A problem arises: there must be some labelling of which grain is which, unlike the situation where a lattice exists and can be referred to. This is not as serious as may at first sight appear, for given the overall density one can consider the site of each grain as the distortion of a lattice which has the correct mean coordination (which is still of course to be discovered). But even cruder one can label the grains according to say a simple cubic lattice with the right lattice spacing, i.e. one can say this particular grain is the lth in the mth row of the nth column. One only needs a label: the quantification of distance and volume comes from the formula for W. At the level of this paper, this labelling issue will not actually arise and we discuss it here only to assure the reader that there is no real problem. There are still compatibility conditions in this formulation: clearly one cannot have a grain with coordination 4 next to a grain with coordination 12. It is difficult to quantify these conditions, and we will not attempt to do so here, but hope to return to it in a later paper.

Suppose then we label our grains i meaning that their true positions can be deformed affinely onto lattice points r_i. Suppose that grain i has a coordination c_i so that

$$W = \sum_i v_{c_i} + \sum_{\substack{i,j \\ \text{neighbours}}} v_{c_i c_j} + \cdots .$$

The final form of the integral will then be

$$\sum_{c_i=0}^{1} \delta(W - V) J(\{c_i\}) ,$$

where $J(\{c(r_i)\})$ expresses the combined effect of the Jacobian from the change of variables, and the stability and no-overlap conditions. We will not attempt to evaluate J in this paper, but work through the very simplest examples: firstly that of one kind of grain, and with the volume depending only on the coordinations of the individual grains. Since we are illustrating a point rather than being realistic we can be even simpler and say that there are just two types of coordination c_0 and c_1 which leads to an Ising model,

$$\begin{aligned} e^{-Y/\lambda X} &= \sum_{c_i=0}^{1} e^{-\Sigma v_{c_i}/\lambda X} \\ &= (e^{-v_0/\lambda X} + e^{-v_1/\lambda X})^N , \end{aligned}$$

or

$$Y = N\left(\frac{v_0 + v_1}{2}\right) - N\lambda X \ln \cosh\left(\frac{v_0 - v_1}{\lambda X}\right),$$

$$V = N\left(\frac{v_0 + v_1}{2}\right) + N(v_0 - v_1) \tanh\left(\frac{v_0 - v_1}{\lambda X}\right).$$

The two limits of V are Nv_0 corresponding to $X = 0$ and having the maximum density and $N(v_0 + v_1)/2$ which corresponds to $X = \infty$ and is the lowest density. Although this is a very simplified model, it clearly will be related to the real problem of an array of c's and n's and compatibilities. There will be a maximum density possible which will correspond to the highest coordination but then the other extreme is that of all the (stable) coordinations being equally likely. This crude analysis does not address the subtleties of sphere packings in three dimensions which arise because almost thirteen spheres can touch a central sphere; so that for small clusters one can achieve higher densities than the 0.7405 of face centred cubic. (Perhaps the demon we have forbidden might be able to beat f.c.c. in the large but not in extensive operations.) In practice the reproducible maximum density obtained by packing for example ball-bearings is lower [7] (0.6366, "random close packing"). A large literature exists discussing the details of sphere packings (see Gray [8], or Cumberland and Crawford [9] for a review), although there is no satisfactory intrinsic – rather than operational – definition of the random close packed state. The general result will remain that the highest density concentrates on a particular type of coordination whereas the lowest has all coordinations.

3. Mixtures of particles

As another example, at the simplest level, we consider a mixture of powders of type A and type B. Let us focus on the fact that when an A is next to an A there is a contribution to the volume which is different from an A next to a B and a B next to a B. Thus although the nature of coordination number matters as in the example above, we just concern ourselves with the nearest neighbour quality. This one aspect is enough to throw interesting light on the separation and miscibility of powders. Suppose the number of A type at r_i is m^i_{A} (=0 or 1), and similarly for B, so that

$$W = \sum_{\substack{i,j \\ \text{neighbours}}} m^i_{\mathrm{A}} m^j_{\mathrm{A}} v_{\mathrm{AA}} + \cdots .$$

Then following Bragg and Williams we write

$$v_{\mathrm{AA}} + v_{\mathrm{BB}} = 2v_{\mathrm{AB}} = v$$

and since $m^i_{\mathrm{A}} + m^i_{\mathrm{B}} = 1$, then if we write

$$m_i = 2m^i_{\mathrm{A}} - 1 ,$$

$$-m_i = 2m^i_{\mathrm{B}} - 1 ,$$

we have

$$m_i = \pm 1 .$$

We now quote Bragg and Williams who give $\phi = \tanh v\phi/\lambda X$ where $\phi = \langle m_i \rangle$ so that for

$$\frac{v}{\lambda X} < 1 , \qquad \phi = 0 ,$$

$$\frac{v}{\lambda X} > 1 , \qquad \phi \text{ small} , \qquad \phi = \pm\sqrt{\frac{3\lambda^3 X^3}{v^3}\left(\frac{v}{\lambda X} - 1\right)} ,$$

until for

$$\frac{v}{\lambda X} \gg 1 , \qquad \phi = \pm 1 .$$

Thus the A and B grains are miscible for $v/\lambda X < 1$, but, for $v/\lambda X > 1$, the powder forms domains of unequal concentrations, until at $X = 0$ the material separates into domains of pure A and pure B.

If A and B are particles of different sizes, then we expect $v < 0$ and that the powders will be miscible for all X – at least as long as the pure phases are expected to be disordered. The well-known phenomenon where large particles rise to the top in a vibrated powder is a purely dynamical phenomenon [10]. The analysis here is relevant if we invert the powder at intervals, so that there is no extrinsic bias, or if we consider particles of the same size, but different shapes (e.g. cubes and spheres). An interesting computational study of this problem has been given by Barker [11].

4. Discussion

This paper argues that, although detail of local structure is needed for a complete theory of powders, there are large scale behaviours which fit into the structure of statistical mechanics in the sense that analogues exist of Gibbs type integrals and relationships. The compactivity provides a useful way to characterize theoretically states of a powder with different densities, although, unlike the case with energy and temperature, the volume is the easier experimental quantity.

We have illustrated the problems by some rather trivial examples, but have in hand some more detailed and significant examples for future publications.

Acknowledgements

R.B.S. Oakeshott would like to thank British Petroleum for the generous award of a Research Studentship.

S.F.E. thanks the Enrico Fermi International School of the Italian Physical Society for an invitation to a lecture on this problem, which much clarified his thinking [12].

References

[1] M. Shahinpoor, Powder Tech. 25 (1980) 163.
[2] M. Shahinpoor, Bulk Solids Handling 1 (1981) 1.
[3] C.B. Brown, Proc. U.S.-Japan Seminar on Continuum-Mechanical and Statistical Approaches to the Mechanics of Granular Materials (Gakujutsu Bunken, Fukyu-Kai, Tokyo, 1978), p. 98.
[4] R.L. Brown and J.C. Richards, Principles of Powder Mechanics (Pergamon, Oxford, 1970).
[5] Advances in the Flow of Granular Materials, vols I, II, M. Shahinpoor, ed. (Trans Tech Publications, 1983).
[6] B.J.B. Briscoe and M.J. Adams, in: Triboloty in Particulate Technology, M.J. Adams, ed. (Adam Hilger, Bristol, 1987).
[7] G.D. Scott and D.M. Kilgour, J. Phys D: Appl Phys. 2 (1969) 863.
[8] W.A. Gray, The Packing of Solid Particles (Chapman and Hall, London, 1968).
[9] D.J. Cumberland and R.J. Crawford, The Packing of Particles, Handbook of Powder Technology, vol. 6 (North-Holland, Amsterdam, 1987).
[10] A. Rosato, H.J. Strandburg, F. Prinz and R.H. Swendensen, Phys. Rev. Lett. 58 (1987) 1038.
[11] G.C. Barker and M.J. Grimson, Proc. of Royal Society of Chemistry Food Group meeting: Food Colloids (1988), to be published in a special volume.
[12] S.F. Edwards, in: Proc. Enrico Fermi School of Physics (to be published by the Italian Physical Society).

Density fluctuations in vibrated granular materials

Edmund R. Nowak, James B. Knight,* Eli Ben-Naim,† Heinrich M. Jaeger, and Sidney R. Nagel
The James Franck Institute and the Department of Physics, The University of Chicago, Chicago, Illinois 60637
(Received 22 May 1997; revised manuscript received 23 October 1997)

We report systematic measurements of the density of a vibrated granular material as a function of time. Monodisperse spherical beads were confined to a cylindrical container and shaken vertically. Under vibrations, the density of the pile slowly reaches a final steady-state value about which the density fluctuates. We have investigated the frequency dependence and amplitude of these fluctuations as a function of vibration intensity Γ. The spectrum of density fluctuations around the steady state value provides a probe of the internal relaxation dynamics of the system and a link to recent thermodynamic theories for the settling of granular material. In particular, we propose a method to evaluate the compactivity of a powder, first put forth by Edwards and co-workers, that is the analog to temperature for a quasistatic powder. We also propose a stochastic model based on free volume considerations that captures the essential mechanism underlying the slow relaxation. We compare our experimental results with simulations of a one-dimensional model for random adsorption and desorption. [S1063-651X(98)07602-8]

PACS number(s): 81.05.Rm, 05.40.+j, 46.10.+z, 81.20.Ev

I. INTRODUCTION

One of the salient features of noncohesive granular materials is that they can be packed over a range of densities and still retain their resistance to shear. For example, a stable conglomeration of monodisperse spheres can exist with a packing fraction ρ ranging from $\rho \approx 0.55$ (the random loose packed limit) to $\rho \approx 0.64$ (the random close packed limit) and even to $\rho \approx 0.74$ (the crystalline state). Because thermal energies, $k_B T$, are insignificant when compared to the energy it takes to rearrange a single particle, each metastable configuration will persist indefinitely until an external vibration comes along to knock it into another state. Thus, no thermal averaging takes place to equilibrate the system. The density of the material is determined both by its initial preparation and by the manner in which it was handled or processed, since such activities normally introduce some vibrations into the material. The phase space for the granular medium is explored not by fluctuations induced by ordinary temperature but by fluctuations induced by external noise sources, such as vibrations. It is the goal of this paper to provide an experimental foundation for the use of such fluctuations as a probe of the dynamics as well as the microstructure of granular media in the quasistatic, densely packed limit.

Granular compaction involves the evolution from an initial low-density packing state to one with higher final density and provides a model system for nonthermal relaxation in a disordered medium. In a previous study [1], we focused on the approach to a final steady-state density as vibrations were applied to the system. In particular, we studied the density of monodisperse spherical particles in a tall cylindrical tube as a series of external excitations, consisting of discrete, vertical shakes or "taps," were applied to the container. Such data indicate that the compaction process is exceedingly slow: the density approaches its final steady-state value approximately logarithmically in the number of taps. A typical example of such behavior, in Fig. 1, shows that in excess of 10^4 taps may be required before the density has relaxed to its steady-state value. However, if one vibrates for a long enough time

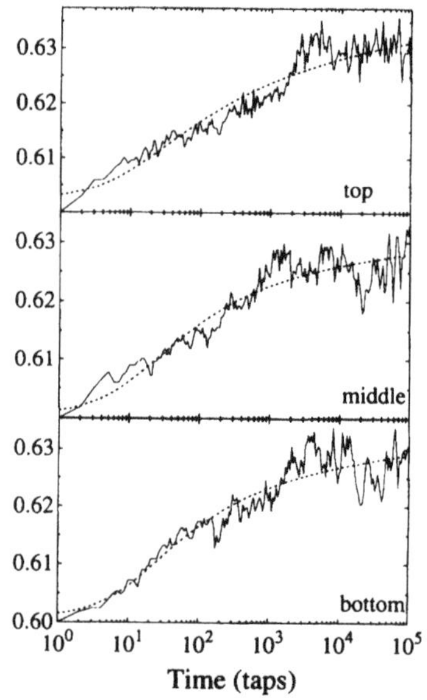

FIG. 1. The time evolution of the volume density ρ at three different depths near the top, middle, and bottom of the pile of beads. The curves represent a single run (no ensemble averaging) at a vibration intensity $\Gamma = 6.8$. The pile settles slowly from its initial low density configuration towards a higher steady-state density at long times, $t > 10^4$ taps. The dashed lines are fits to Eq. (1) with typical values of parameters: $0.637 \leqslant \rho_\infty \leqslant 0.647$, $0.036 \leqslant \Delta\rho_\infty \leqslant 0.044$, $0.20 \leqslant B \leqslant 0.40$, $10 \leqslant \tau \leqslant 18$.

*Present address: Department of Physics, Princeton University, Princeton, NJ 08540.

†Present address: Theoretical Division and Center for Nonlinear Studies, Los Alamos National Laboratory, Los Alamos, NM 87545.

a steady-state density, depending on the intensity of the taps, will be attained. Even after the density reaches the steady-state value, one can discern fluctuations in the density about that value: after each "tap," the density will be slightly higher or lower than it was before. These fluctuations are reminiscent of thermal fluctuations about an equilibrium state, yet such a connection so far has not been investigated experimentally.

In statistical mechanics the study of fluctuations is of great physical interest. The fluctuation-dissipation theorem relates the dissipative response of a system to an external perturbation with the microscopic dynamics of the system in a state of equilibrium. Energy fluctuations in thermal systems can be used to investigate the set of distinct, microscopic states that are accessible to a system maintained at a fixed temperature. Likewise, a study of density fluctuations in granular media may provide a framework for understanding the physical phenomenon of compaction, i.e., how a vibrated powder, that is not in a steady state, finally approaches a steady state.

In a granular system, density fluctuations from the steady state represent the different volume configurations accessible to particles subject to an external vibration. It is desirable to develop an analogy between the role that vibrations play in nonthermal systems, such as granular media, and the role of temperature in thermal systems. Theoretically, this issue was addressed by Edwards and co-workers [2–4] who introduced a statistical mechanics for powders. The idea is based on the assertion that an analogy can be drawn between the energy of a thermal system and the volume V occupied by a powder. The entropy S of a powder is defined in the usual sense, by the logarithm of the number of available configurations. Edwards and co-workers then put forth the concept of an effective temperature for a powder, called the compactivity X, which is defined as $X \equiv \partial V / \partial S$. The significance of this effective temperature is that it allows for the characterization of a *static* granular system. This is distinct from the case of rapid granular flows where a "granular temperature" given by the mean-square value of the fluctuating component of the particle velocities can be written down [5–7]. The compactivity is then a measure of "fluffiness" in the powder: when $X=0$, the powder is in its most compact configuration, whereas for $X=\infty$ the powder is the least dense.

Recently, another approach [8–10] that describes the static packing of powders has adapted a statistical model that contains geometric frustration as an essential ingredient. For granular materials, frustration arises in the form of hard-core repulsive constraints and the interlocking of grains of different shapes, which prevents local rearrangements. Both the static and dynamic (in the presence of vibration and gravity) properties of this model exhibit complex behavior with features that are common to granular packing, such as the logarithmic relaxation of density under tapping [1].

In this paper, we make contact with these ideas through a detailed study of the process of granular compaction. In particular, we propose a method for evaluating the compactivity of a vibrated powder through a definition of a "granular specific heat" and measurements of density fluctuations observed in the reversible regime of steady-state behavior. We also elaborate on a theoretical model [11,12], based on the idea that the rate of increase in volume density is exponentially reduced by the free volume, which captures many of the significant features of our experiments. A model addressing the compaction of binary mixtures consisting of grains with very different sizes was recently proposed by de Gennes [13]. That model is similar to ours in that it incorporates free volume constraints and also exhibits a similar inverse logarithmic dependence for the density relaxation.

In the next section we will describe the experimental details of the system, review how to obtain reproducible and reversible densities, and present our results for the density fluctuations. In Sec. III we discuss several models in relation to our experimental results and motivate the relevance of free volume constraints for granular compaction. In Sec. IV, we present the theoretical model and the results of related simulations of compaction. Finally, in the last section we discuss the central result of this paper, namely, how our data can be related to thermodynamic approaches for understanding granular media.

II. EXPERIMENTAL RESULTS

Experimental method

The details of the experimental apparatus and measurement technique were published elsewhere [1]. Monodisperse, spherical soda-lime glass beads (of 2 mm diameter) were confined to a 1.88 cm diameter Pyrex tube measuring 1 m in height. The tube was subjected to discrete vertical shakes (or "taps") each consisting of one complete cycle of a 30 Hz sine wave. The vibration intensity was parametrized by Γ, which is the ratio of the peak acceleration A that occurs during a single tap to the gravitational acceleration $g = 9.8\ \mathrm{m/s^2}$: $\Gamma = A/g$. The beads were baked prior to loading in the tube and precautions were taken to minimize complications resulting from electrostatic charging, convection, and external humidity fluctuations. The column of beads was prepared in a low density initial state by flowing high pressure, dry nitrogen gas from the bottom to the top of the tube. The top layer of the beads was free to move, i.e., there was no load or dead-weight surcharge applied to the column of beads. The density, or equivalently the packing fraction ρ, which is the percentage of volume occupied by the beads, was determined either by a measurement of the total height of the beads within the tube or using capacitors that were mounted on the outside wall of the tube. For the latter, the capacitance was found to vary linearly with packing fraction. Each capacitor averaged the density over sections containing approximately 6000 beads. Measurements of ρ were taken as a function of time, i.e., number of taps t and as a function of the intensity of the vibrations, Γ. Corrections for instrumental drift were made by using simultaneously acquired data from a second, stationary tube (identically prepared with the same type of beads and connected to the same vacuum system). Our instrumentation allowed shaking intensities up to $\Gamma \approx 7$ and provided a resolution $\Delta\rho = 0.0006$ in measured packing fraction changes.

The desired outcome of a shake cycle is to provide clearly defined periods of uniform dilation of the bead assembly. During these periods of dilation the beads have some freedom to rearrange their positions relative to their neighbors and thereby replace one stable close-packed configuration by another. Previously [1,14], we have shown that the overall

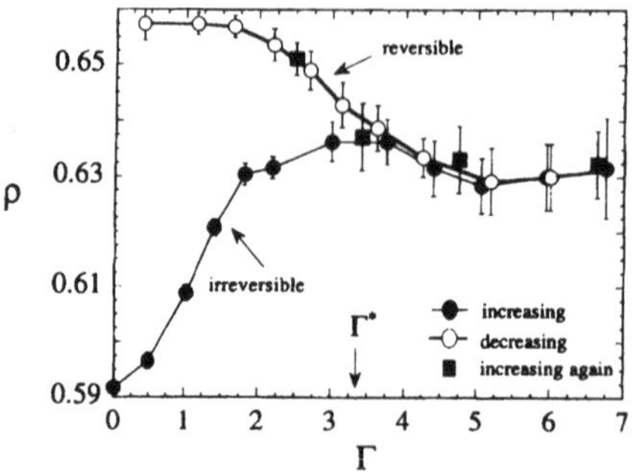

FIG. 2. The dependence of ρ on the vibration history. The beads were prepared in a low density initial configuration and then the acceleration amplitude Γ was slowly first increased (solid symbols) and then decreased (open symbols). At each value of Γ the system was tapped 10^5 times after which the density was recorded and Γ was subsequently incremented by $\Delta\Gamma \approx 0.5$. The upper branch that has the higher density is reversible to changes in Γ, see square symbols. Γ^* denotes the irreversibility point (see text).

behavior of the compaction process is qualitatively similar at different depths into the container (see also Fig. 1). Spurious effects from continuous vibrations, such as period doubling or surface waves [12], were avoided by spacing the taps sufficiently far apart in time to allow the system to come to complete rest between taps. Also, by using a tall container with smooth, low-friction interior walls shear-induced dilation and granular convection were suppressed [15]. Although friction between beads and with the tube walls can affect the mechanical stability of a bead configuration, we argue below that the motion of beads is limited primarily by geometric constraints imposed by the presence of other beads, particularly at the high densities investigated here.

The ratio of the container diameter to the bead diameter can also influence the compaction process. For small values of this ratio, ordering (crystallization) induced by the container walls [16] will increase the measured packing fraction over its bulk value, leading to densities that can exceed the random close-packed limit. This may be responsible for the high maximum packing fractions seen in Fig. 2. Previous studies [1,14] indicate that the qualitative behavior of the compaction process is similar for varying bead sizes. The container walls can also place constraints on the density fluctuations. Since it is our aim to investigate these density fluctuations, the choice of bead size was a compromise between maximizing the container-to-bead diameter ratio and not having the amplitude of the density fluctuations be obscured by statistical averaging over a large number of particles.

Reaching the steady state

At a high acceleration Γ the steady-state density, ρ_{ss} can be approached by simply applying a very large number of taps (often greater than 10^4-10^5). An example is shown in Fig. 1 for $\Gamma=6.8$. The three panels correspond to the capacitively measured density near the top, middle, and bottom sections of the pile of beads. (The tap number t is offset by $+1$ tap so that the initial density can be included on the logarithmic axis.) Note that these curves represent a *single* run, and separate runs starting from the same initial density differ in the details of the density fluctuations but show a similar overall behavior. The behavior of $\rho(t)$, obtained by averaging many runs of this kind, appears to be homogeneous throughout the pile at these high accelerations. As discussed in Ref. [1], the time evolution of this ensemble averaged density is well fitted by the expression

$$\rho(t)=\rho_\infty-\frac{\Delta\rho_\infty}{[1+B\ \ln(1+t/\tau)]}, \qquad (1)$$

where the parameters ρ_∞, $\Delta\rho_\infty$, B, and τ depend only on the acceleration Γ. Equation (1) was found to fit the ensemble averaged density over the whole range $0<\Gamma<7$ better than other functional forms that were tried (i.e., exponential, stretched exponential, or algebraic forms, see Ref. [1]). The dashed lines in Fig. 1 show a fit to Eq. (1). Here, the value of the final density, ρ_∞, is approximately equal to the observed steady-state density ρ_{ss}.

For small values of Γ, however, ρ_∞ corresponds to a metastable state and not the steady-state density. In particular, for values of the applied acceleration $\Gamma<3$, it is difficult, if not experimentally impossible, to reach the steady-state by merely applying a sufficiently large number of taps of identical intensity. In this case, the steady state can be reached by "annealing" [14] the system. The annealing is controlled by the ramp rate, $\Delta\Gamma/\Delta t$, at which the vibration intensity is varied over time. Experimentally, we slowly raise the value of Γ from 0 to a value beyond Γ^* in increments of $\Delta\Gamma \approx 0.5$. At each intermediate value of Γ we apply $\Delta t = 10^5$ taps. Γ^* defines an "irreversibility point" in the sense that, once it has been exceeded, subsequent increases as well as *decreases* in Γ at a sufficiently slow rate $\Delta\Gamma/\Delta t$ lead to reversible, steady-state behavior. We found that $\Gamma^* \approx 3$ for 1, 2, and 3 mm beads [14]. A typical run is shown in Fig. 2. Here we have used 2 mm beads, and started with an initial density of $\rho \approx 0.59$. The highest densities are achieved by annealing the system, i.e., decreasing Γ slowly from Γ^* back down to $\Gamma=0$. If Γ is rapidly reduced to 0 (large $\Delta\Gamma/\Delta t$) then the system falls out of "equilibrium." This leads to lower final densities and a curve for $\rho(\Gamma)$ that is not reversible. A crucial result emerging from data such as in Fig. 2 is that along the reversible branch, the density is monotonically related to the acceleration. We note that in 3D simulations of granular compaction by Mehta and Barker [17] a similar monotonic decrease in steady-state volume fraction as a function of shaking intensity was found. Thus, only once the steady-state has been reached is there a single-valued correspondence between the average density and the applied acceleration.

Density fluctuations about the steady state

After the granular material has been vibrated for a sufficiently long time, it reaches a steady-state density ρ_{ss}. Although there is a well-defined average density, Fig. 1 already hints that there are large fluctuations about this value. The magnitude of the fluctuations depends on the vibration intensity and depth within the container. Figure 3 shows in more detail an example of these fluctuations as a function of time, $\delta\rho(t)=\rho(t)-\rho_{ss}$. In Fig. 3(a) we plot $\delta\rho(t)$ for a fixed

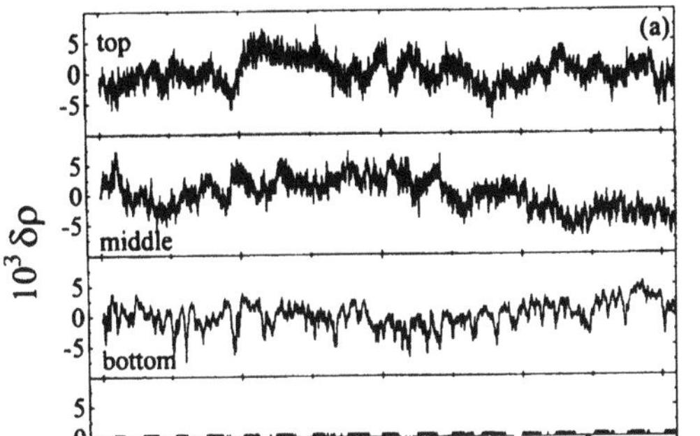

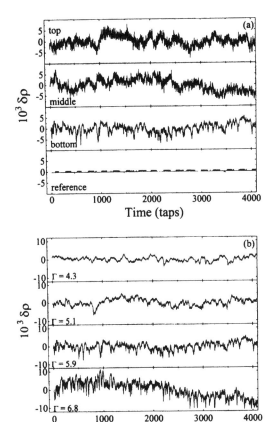

FIG. 3. Fluctuations in the volume density $\delta\rho(t)=\rho(t)-\rho_{ss}$ after the system has had sufficient time to relax to a steady-state density ρ_{ss}. In (a) the fluctuations at three different depths are shown for $\Gamma=5.9$. The reference capacitor is used to correct for any instrumental drift. The dependence of the fluctuations on Γ is shown in (b) for the beads near the bottom of the pile. Fluctuations over a broad range of time scales are evident.

value of acceleration, $\Gamma=5.9$, but measured at different depths in the container. Note that the rate at which the density varies in time decreases with depth into the pile. That is, the top of the pile has more high frequency noise than the bottom. The curve marked "reference" is the reference capacitor to which no vibrations are applied. This last curve is essential to compensate for drifts that could occur in the electronics over the very long period of our measurements. Each record shown here is 4096 taps long and up to 132 successive such records were assembled to produce one very long time sequence. Figure 3(b) shows the fluctuations in the density measured at the bottom capacitor as a function of acceleration Γ. As Γ is increased both the magnitude of the fluctuations and the amount of high-frequency noise increase.

From data as in Fig. 3 we can obtain the shape of the distribution function for the fluctuation amplitudes. We plot in Fig. 4 the logarithm of the relative probability of occurrence $D(\delta\rho)$ versus $\Psi^2=(\rho-\rho_{ss})^2\,\mathrm{sgn}(\rho-\rho_{ss})$ so that a Gaussian random process will have a triangular shape. In that figure we plot $D(\delta\rho)$ for the entire range of accelerations, $4<\Gamma<7$, for which fluctuations could be reliably measured with our equipment. All data records were corrected for instrumental drifts using the reference capacitor. As can be seen in Fig. 4, the majority of data shows Gaussian character. For a small fraction of runs (e.g., $\Gamma=5.9$), however, we find significant deviations from Gaussian behavior, particularly near the middle and bottom of the pile. When such deviations are present they tend to preferentially occur for positive values of Ψ^2, i.e., higher densities. The deviations could be due to a metastable state, away from the mean, in which the system gets trapped. Fluctuations about this metastable state may even be distributed in a Gaussian fashion. The reason why such metastable states favor the lower portion of the column and why they are prominent at certain values of Γ is unclear.

We can qualitatively check whether the distribution functions correspond to a stationary random process or whether they conceal a slow drift away from an originally well-defined mean density. (Strictly speaking, a stationary Gaussian process is one for which correlation functions of order higher than second are zero, see Ref. [18]). This is done by dividing each time record into two equal length halves and then determining the distribution functions for each half separately, as shown for selected values of Γ and depths by the open symbols in Fig. 4. We find that in practically all cases the fluctuations do appear to be stationary and, moreover, that in the very few nonstationary cases observed, the Gaussian character is recovered at *later* times (i.e., in the second half of the record).

By assembling 132 successive time traces of the type shown in Fig. 3, we can obtain continuous time records containing 540 672 data points. From such records we calculate both the density autocorrelation function and the power spectrum for the density fluctuations, $S_\rho(\omega)$, where the frequency ω is measured in units of inverse taps. In Fig. 5 we plot $S_\rho(\omega)$ versus ω for the three depths at various values of acceleration, $\Gamma=4.3$, 5.1, 5.9, and 6.8. We note several distinctive features to these power spectra. In particular, three characteristic regimes emerge: (i) a white noise regime, $S_\rho(\omega)\propto\omega^0$ below a low-frequency corner ω_L, (ii) an intermediate-frequency regime with nontrivial power-law behavior, and (iii) a simple roll-off $S_\rho(\omega)\propto\omega^{-2}$ above a high-frequency corner, ω_H. This classification appears to apply to all traces shown in Fig. 5. It is most pronounced for the spectrum in the lower right hand panel. For spectra where ω_L and ω_H are sufficiently separated in frequency, the data show that the spectral dependence between ω_L and ω_H cannot be approximated by just a simple superposition of two separate Lorentzians each having a frequency dependence $S\propto\tau/(1+\omega^2\tau^2)$ but different characteristic times τ. A comprehensive analysis of this region reveals that the most consistent description for all the data is obtained with a Lorentzian tail, $S_\rho(f)\propto\omega^{-2}$ just above ω_L, followed by a region with $S_\rho(\omega)\propto\omega^{-\delta}$ (with $\delta\approx0.9\pm0.2$) stretching up to ω_H, the high-frequency corner.

One result from the data in Fig. 5 is the dependence of both corner frequencies on the acceleration Γ. To determine these frequencies we used a combination of two methods,

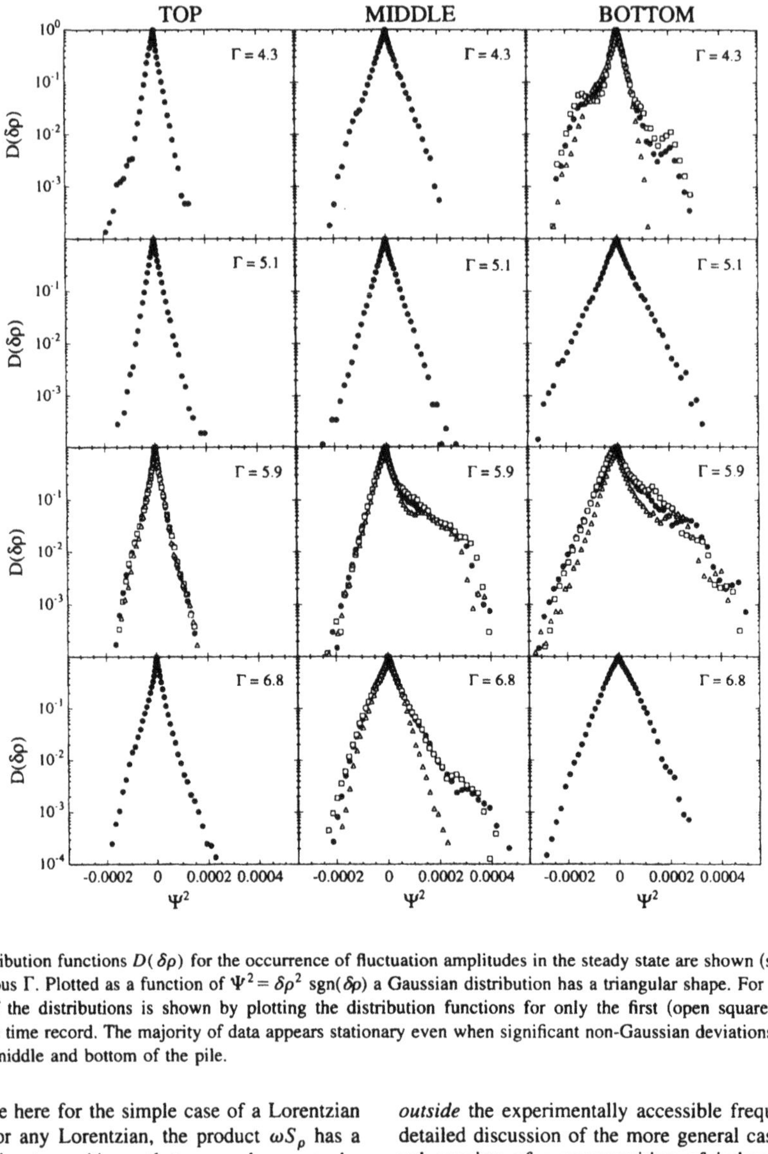

FIG. 4. The distribution functions $D(\delta\rho)$ for the occurrence of fluctuation amplitudes in the steady state are shown (solid circles) for the three depths at various Γ. Plotted as a function of $\Psi^2 = \delta\rho^2\ \mathrm{sgn}(\delta\rho)$ a Gaussian distribution has a triangular shape. For selected panels, the time dependence of the distributions is shown by plotting the distribution functions for only the first (open squares) or second (open triangles) half of the time record. The majority of data appears stationary even when significant non-Gaussian deviations are observed, e.g., at Γ = 5.9 near the middle and bottom of the pile.

which we illustrate here for the simple case of a Lorentzian spectrum. First, for any Lorentzian, the product ωS_ρ has a maximum precisely at $\omega = 1/\tau$ so that ω_L and ω_H can be associated with the frequencies at which ωS_ρ exhibits peaks. Second, even though we were using extremely long time records they are still of finite length. Figure 5 clearly indicates cases where ω_L is difficult to obtain because of the large statistical variance ($\sim 25\%$) in S_ρ throughout the lowest decade in frequencies. In these instances we employed the additional information contained in the one-sided sine transform of the density-density autocorrelation function. For example, for a single Lorentzian for which the autocorrelation function is simply $\propto e^{-t/\tau}$, the ratio of sine to cosine transform of the autocorrelation function is given by $\omega\tau$, which depends only on τ. A plot of this ratio versus ω then allows one to obtain $\omega_L = 1/\tau$ even if this frequency falls *outside* the experimentally accessible frequency window. A detailed discussion of the more general case, where the signal consists of a superposition of independent fluctuators with a distribution of relaxation times τ will be presented elsewhere [19].

Figure 6 plots the resulting corner frequencies as a function of applied acceleration. The trend is for both ω_L and ω_H to increase as a function of increasing Γ and with decreasing depth into the pile, see Fig. 6(a). We note that over the relatively small available range of Γ, the variation of ω_H is consistent with behavior reminiscent of thermal activation: $\omega_H = \omega_0 \exp(-\Gamma_0/\Gamma)$. In this context, Γ_0 would represent an energy barrier and ω_0 would be an attempt frequency. We find that a value of $\Gamma_0 \approx 15$ is consistent with all the data, and that the greatest variation is in the parameter ω_0, which varies from 2×10^{-3} to 7×10^{-2} for ω_L and 1 to 15 for ω_H.

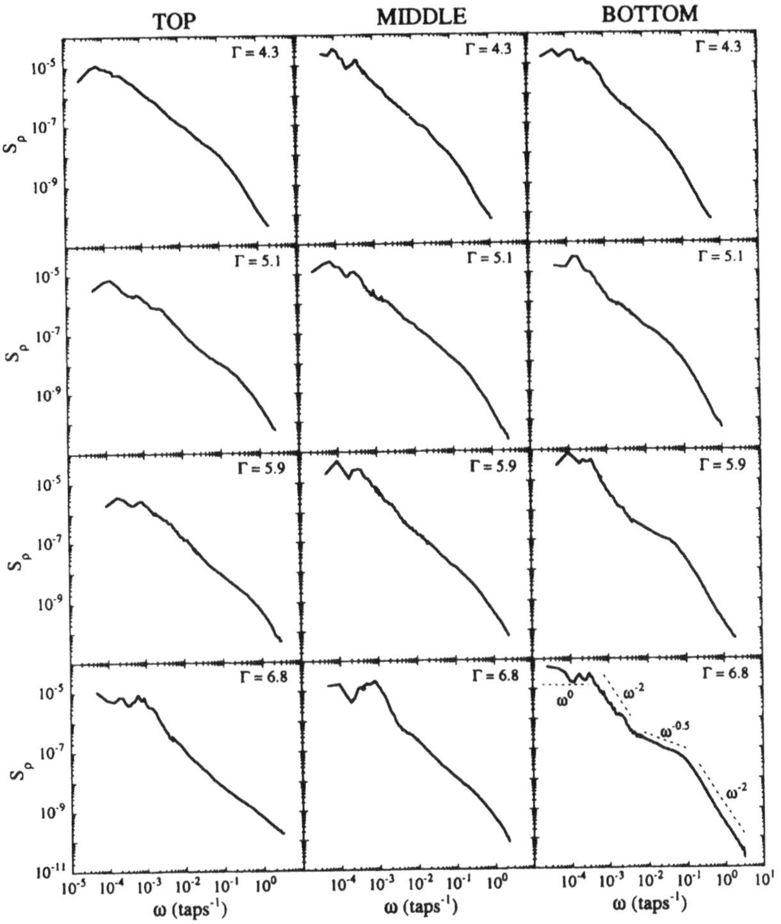

FIG. 5. The power spectral density S_ρ (taps) of the fluctuations as a function of frequency ω in the steady state is shown for the three depths at various Γ. For most spectra, two characteristic corner frequencies, ω_L and ω_H, are discernible which shift to higher frequencies for increasing Γ and decreasing depth. The characteristic regimes of behavior are denoted by the dashed lines in the lower right hand panel, which are guides to the eye.

III. DISCUSSION

Several mechanisms [17,20–22] have been proposed to explain the kinetics of compaction. Although the proposed mechanisms are compelling, their quantitative predictions fail to describe the time dependence observed experimentally [1]. In light of our experimental results, we pay special attention to models based on free volume considerations as it appears that they not only capture the experimentally observed slow relaxation towards the steady state, but may also provide a valid framework for understanding the fluctuation spectrum. Such models [8–10,12,13] include strong nearest-neighbor repulsive interactions between particles that effectively block the occupation of adjacent sites. On very general grounds it is reasonable to assume that for the case of granular compaction, the rate of increase in volume density is exponentially reduced by free volume [23–25]. One alternative approach to the compaction problem is due to Linz [26] who proposes a phenomenological decay law for successive inverse packing fractions. Moreover, recent models based on the dynamics of crystalline clusters in the material have been proposed by Gavrilov [27] and by Head and Rogers [28]. These approaches lead to a time evolution that is essentially equivalent to Eq. (1).

A simple heuristic argument [12,24] for the compaction process illustrates how the effects of free volume can lead to the observed inverse logarithmic behavior. If each grain has a volume V_g and we start with a number n of grains per unit volume, then the volume fraction is given by $\rho = nV_g$. In general there exists a maximum possible volume fraction, ρ_{max}, corresponding to the configuration of grains that occupies the least amount of volume. Then, at some volume fraction ρ, the average free volume available to each grain for

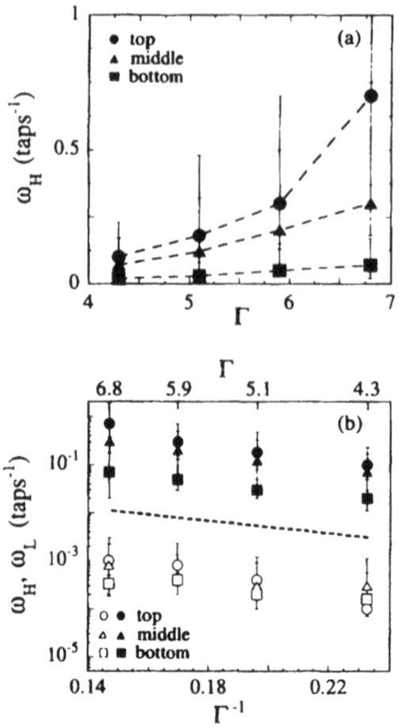

FIG. 6. The characteristic frequencies, ω_L (open symbols) and ω_H (solid symbols), in the power spectra plotted as a function of $1/\Gamma$. The general trend is for both ω_L and ω_H to increase with increasing Γ and decreasing depth into the pile. The dashed line in (b) is a guide to the eye, indicating that the trend is consistent with an activated-like behavior $\omega \propto \exp(-\Gamma_0/\Gamma)$, with $\Gamma_0 \approx 15$. For comparison, (a) shows the dependence of ω_H on Γ.

rearrangements is $V_f = V_g(1/\rho - 1/\rho_{max})$. During compaction, individual hard-core grains move, and when a void large enough to contain a grain is created, it is quickly filled by a new particle. When the volume fraction is large, voids the size of a particle are rare and a large number of voids must rearrange to accommodate an additional particle. We can estimate the rate of compaction by assuming that N grains must rearrange in such a way that they contribute their entire free volume to create a grain-sized void, $NV_f = V_g$. We find that this number increases as $N = \rho\rho_{max}/(\rho_{max} - \rho)$ during the compaction process. For independent, random grain motion during a tap, the probability for N grains to rearrange and open up a grain-size void is then e^{-N}. Consequently,

$$d\rho/dt \propto (1-\rho)e^{-N} = (1-\rho)e^{-\rho\rho_{max}/(\rho_{max}-\rho)}. \quad (2)$$

The rate at which the density increases is proportional to the void volume and the probability for such a rearrangement. The latter exponential factor reduces the rate and dominates for large ρ. In the limit $\rho \to \rho_{max}$ we have $N \approx (\rho_{max})^2/(\rho_{max} - \rho)$ and the solution of this equation is given asymptotically by $\rho(t) = \rho_{max} - A/[B + \ln(t)]$, where A and B are constants [10,13,24,25]. This result closely approximates our experimentally based fitting form, Eq. (1), for the ensemble-averaged $\rho(t)$ as it approaches the steady state.

The solution to Eq. (2) always approaches the maximum density ρ_{max} and does not allow for a lower steady-state density. The reason that this model leads to jamming is the absence of any void-creating mechanism that would be represented by a competing term on the right-hand side of Eq. (2). (The consequences of including a mechanism for the generation of voids are discussed in the next section for the "parking lot model" and in Appendix A.) The competition between void annihilation and creation during tapping naturally leads to density fluctuations. We can also examine our data for the dependence of the corner frequencies on the acceleration Γ [Fig. 6(b)], where we found that $\omega_H = \omega_0 \exp(-\Gamma_0/\Gamma)$. We use the fact that ρ is a monotonic function of Γ in the reversible steady-state regime (Fig. 2) and write to first order $\rho(\Gamma) \approx \rho_{max} - m\Gamma$, where m is a positive constant, locally approximating the slope $\Delta\rho/\Delta\Gamma$ (see Ref. [14] for data on other bead sizes). Substituting in for Γ, the expression for ω_H can then be rewritten as $\omega_H = \omega_0 \exp[-m\Gamma_0/(\rho_{max} - \rho)]$. This has the same form as the right-hand side of Eq. (2) (in the limit that $\rho \to \rho_{max}$) and indicates that the kinetics depend sensitively on the available free volume, $\propto (1/\rho_{max} - 1/\rho)$. As Γ is reduced and the density approaches the maximum density, the kinetics slow down rapidly. The manner in which the kinetics slow down is reminiscent of the Vogel-Fulcher form used to describe another class of disordered metastable materials, namely, glasses [29]. Similarities to glasses have recently been found in another approach to the compaction process [8–10].

IV. THEORETICAL MODEL AND SIMULATIONS OF COMPACTION

The parking lot model

In an attempt to explicitly work out some of the consequences of the free volume approach to granular compaction, we next discuss a simplified model. The model was previously studied in the context of chemisorption [23–25,30,31] and protein binding [32]. Despite its simple nature, it gives remarkably good qualitative agreement with the experimental data, both for the approach to the steady state and for the spectrum of fluctuations in the steady state. This model has the advantage that it readily lends itself to computer simulations; we restrict ourselves to the one-dimensional (1D) case, but extensions to higher dimensions are straightforward. Moreover, much is known about its low-density limit, for which mean-field equations exist that are amenable to analytic treatment (see the Appendix). In 1D, the model can be compared to parallel curbside parking where there are no marked parking spaces. For the person wishing to park a vehicle, the familiar situation is that there exist large, but not quite large enough, spaces between previously parked cars. The analogous question to the one we have been asking is "How many other cars have to be moved just a bit for the additional one to fit in?"

The model is defined as follows: identical particles of unit length adsorb uniformly from the bulk onto a substrate with rate k_+ and desorb with rate k_-. In other words, k_+ adsorption attempts are made per unit time per unit length, and similarly, the probability that an adsorbed particle desorbs in an infinitesimal time interval between t and $t+dt$ is $k_- dt$. While the desorption process is unrestricted, the adsorption process is subject to free volume constraints, i.e., particles cannot adsorb on top of previously adsorbed particles; see Fig. 7. This stochastic process is well-defined in arbitrary

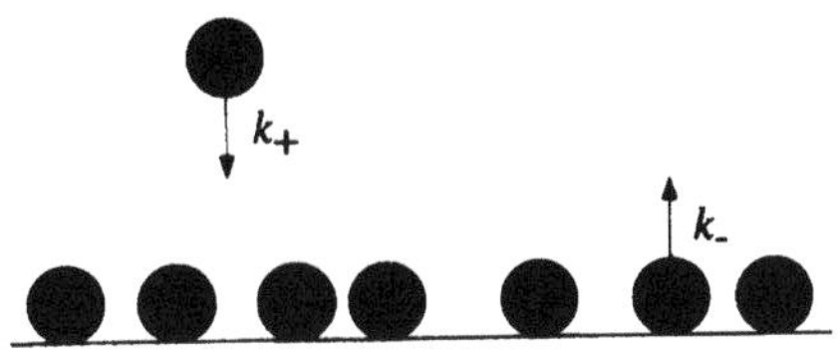

FIG. 7. The adsorption-desorption process. Adsorption is successful only in spaces large enough to accommodate a particle. Desorption of a particle, on the other hand, is unrestricted.

dimension and clearly satisfies detailed balance so that the system eventually reaches a steady-state density. In one dimension, $\rho_{max}=1$.

Mapping the model on to the experiment, we associate an adsorption event with the annihilation or filling of a void within the pile of beads, whereas a desorption event is associated with the creation of a void. The ratio of adsorption to desorption rates, $k=k_+/k_-$, determines the final steady-state density in the model (see also the Appendix). Thus one can associate k in the model with the magnitude of the acceleration Γ in our experiment.

Simulation of compaction based on the parking lot model

In this section we compare the experimental results with Monte Carlo simulations of the 1D parking process. The details of the simulation are described elsewhere [24]. Here we report our results for a system size of 100; similar results were found for a system size of 25.

The simulations were started from a zero density initial state and allowed to evolve to various steady-state densities by varying k_- at a fixed value of $k_+=1$. In Fig. 8 we show the time evolution of the density as it approaches a steady-state density $\rho_{ss}=0.84$. The steady-state densities obtained after equilibration coincide with those predicted by Eq. (A2a) in the Appendix. We find that the simulations reproduce the slow logarithmic relaxation towards the steady state in agreement with Eq. (1).

We now turn to the density fluctuations. In this case, the simulations ran long enough to ensure that a steady-state

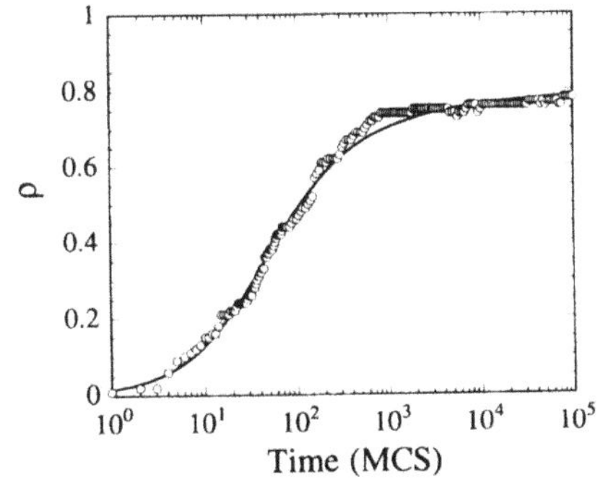

FIG. 8. The time evolution of the density in the simulation for $k=10^3$ ($\rho_{ss}=0.84$). Time is in units of Monte Carlo steps (MCS). The solid line represents a fit to Eq. (1).

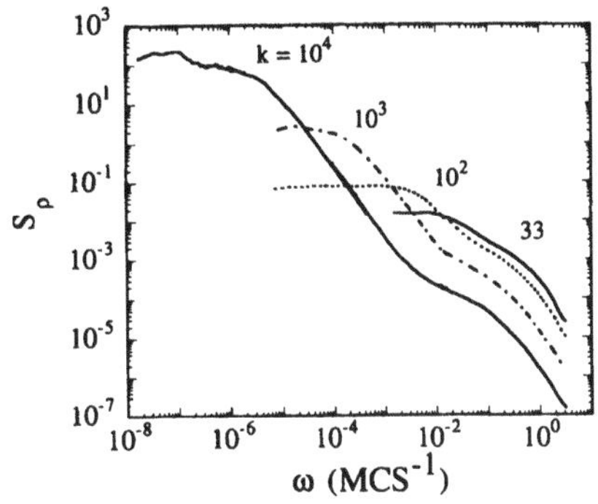

FIG. 9. Power spectra, S_ρ (MCS), of the density fluctuations in the simulation of the one-dimensional parking process. The evolution of the spectral dependence is shown for values of the ratio $k=33,10^2,10^3,10^4$, corresponding to final steady-state densities $\rho_{ss}=0.72$, 0.77, 0.84, and 0.88, respectively (see text). The strongest dependence on k is for ω_L, which decreases rapidly as the density increases. Such spectra are similar to those found in the experiment, see Fig. 5.

density was attained before density fluctuations were recorded. For low densities $\rho_{ss}<0.37$ we find that in the simulation the power spectra of the fluctuations are best described by a Lorentzian having a single characteristic time scale, as expected from the mean-field analysis [see Eq. (A4) in the Appendix]. However, at higher densities (higher k) local fluctuations dominate the dynamics and the power spectra show the emergence of two distinct corner frequencies, which become progressively more separated. This is shown in Fig. 9 for a wide range of ratios k, where k_+ was fixed at a value of 1. Most notable is the low-frequency corner, which shifts rapidly to lower frequencies for small increments in density. By comparison, the high-frequency corner decreases much more slowly.

Our simulations find power spectra, $S_\rho(\omega)$, strikingly similar in shape to those obtained experimentally when $\rho_{ss}>0.50$. Again we see three distinct regimes (Fig. 9). Below a corner frequency, ω_L, there is white noise $[S_\rho(\omega)\propto\omega^0]$. Above a high-frequency corner, ω_H, $S_\rho(\omega)\propto\omega^{-2}$. The simulations offer the advantage of allowing the separation between ω_L and ω_H to be tuned by increasing the value of k or, equivalently, increasing the density. This allows the systematic investigation of the spectral dependence in the intermediate regime between the two corner frequencies. As in the experimental data, we find that there is a Lorentzian tail, $S_\rho(\omega)\propto\omega^{-2}$ just above ω_L. At higher frequencies, stretching up to the high-frequency corner ω_H, we find a power-law regime $S_\rho(\omega)\propto\omega^{-\delta}$. The exponent δ appears to depend slightly on the separation between the two corner frequencies. For the largest separations that span nearly 5 decades in frequency we find $\delta\approx0.5$. This value is smaller than that found in the experimental data ($\delta\approx0.9$), but again is inconsistent with a simple superposition of two Lorentzians having characteristic time scales ω_L^{-1} and ω_H^{-1}.

V. ANALOGY WITH THERMAL FLUCTUATIONS

In ordinary statistical mechanics, the fluctuation-dissipation theorem allows the determination of the response of a system to a small perturbation from its thermal fluctuations about equilibrium. In this section, we will explore the possibility that we can derive similar information about the granular system from its fluctuations about its steady-state density. In the granular thermodynamics theory developed by Edwards and co-workers, an analogy is made between granular and thermal systems. The basic assumption is that the *volume* V of a powder is analogous to the *energy* of a statistical system (we note that V here refers to the total volume and not just to the free volume). Instead of a Hamiltonian, there is a function that specifies the volume of the system in terms of the positions of the individual grains. The "entropy" is thus the logarithm of the number of configurations: $S=\lambda \ln \int d$ (all configurations) where λ is the analog of Boltzmann's constant. Using this they defined a quantity analogous to a temperature in a thermal system, which they call the "compactivity" X: $X=\partial V/\partial S$. In contrast to the notion of "granular temperature," which depends on the random motion of the particles, the compactivity characterizes the static system after it has reached a steady-state density via some preparation algorithm. Such an algorithm would be one as we have described above, where we have vibrated the granular system until it has reached the reversible steady-state density. If this theory is valid, then we should be able to define an equilibrium such that two systems in equilibrium with a third system are also in equilibrium with each other. That is, no net volume will be transferred between the two systems when they are placed in contact with each other if they have the same value of X.

In a thermal system we can write the specific heat in two ways as follows:

$$C_V=dE_0/dT|_V=\langle(E-E_0)^2\rangle/k_BT^2, \quad (3)$$

where E_0 is the equilibrium average of the energy E of the system, k_B is Boltzmann's constant, T is the temperature, and $\langle\cdots\rangle$ represents the time average. In Edwards' theory for a powder the analogous quantity to the specific heat of a thermal system given in Eq. (3) becomes

$$C=dV_{ss}/dX=\langle(V-V_{ss})^2\rangle/\lambda X^2, \quad (4)$$

where V_{ss} is the steady-state volume. Since we have measured the density fluctuations in the steady state (Figs. 3–5), we are in a position to explicitly calculate the variance, $\langle(V-V_{ss})^2\rangle$, of volume fluctuations for a given steady-state volume V_{ss} defined here as $V_{ss}=1/\rho_{ss}$. We can then write

$$\int_{V_1}^{V_2} dV_{ss}/\langle(V-V_{ss})^2\rangle=\int_{X_1}^{X_2} dX/\lambda X^2=1/\lambda X_1-1/\lambda X_2. \quad (5)$$

Equation (5) allows us to measure the difference in compactivities for any two volumes as long as we know the fluctuations of the volumes (i.e., densities) as a function of the average volume. This is equivalent to obtaining the difference in temperatures for a thermal system between any two energies. Clearly, as V_{ss} increases X is expected to increase as well. Equation (5) allows the determination of an absolute value for the compactivity only once a suitable point of reference can be found.

In Fig. 10 we show the experimentally obtained values of

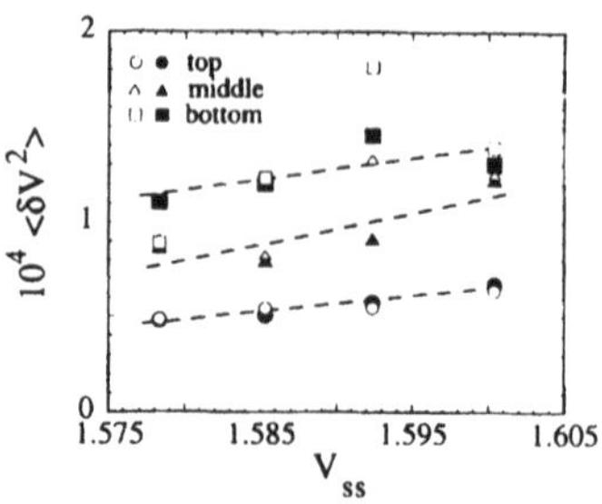

FIG. 10. The average variance of the experimental volume fluctuations (open symbols) as a function of the steady-state volume. The trend is for the variance to increase with increasing volume and *depth* into the pile. The solid symbols represent the variance as determined from the distribution of fluctuation amplitudes in Fig. 4 (see text). The dashed lines are linear fits to the solid symbols.

$\langle(V-V_{ss})^2\rangle$ for several steady-state volumes along the reversible branch of Fig. 2 over the range $4<\Gamma<7$ [33]. The solid curve through the data for the top of the pile represents a linear fit to the function:

$$\langle(V-V_{ss})^2\rangle=a+bV_{ss},$$

where $a=-7.2\times10^{-4}$ and $b=4.9\times10^{-4}$. This implies that the magnitude of the fluctuations goes to zero at $\rho\approx0.68$, that is, near the close-packed density. Using this form for the dependence of the fluctuations on V_{ss} in Eq. (5) we find that

$$1/\lambda X \propto \ln(a+bV_{ss}). \quad (6)$$

This functional dependence is valid only over the limited range of experimental data, and may not be an adequate description of the general behavior. Below, we discuss a similar analysis for the simulation data for which a broader range of volumes can be explored. Using Eq. (6), we can evaluate the difference in inverse compactivities between any two steady-state volumes. We find that $1/\lambda X_1-1/\lambda X_2=0.04$ where the subscripts 1 and 2 refer to the smallest and largest volumes for which we have data. This result explicitly demonstrates how the compactivity increases for larger volumes (smaller densities).

It is also interesting to consider the size of the fluctuating volumes that give rise to the observed variance. This can be estimated by assuming that the fractional fluctuations scale as $\langle\delta\rho^2\rangle/\rho_{ss}^2=\langle\delta V^2\rangle/V_{ss}^2=\kappa^2/N$, where $\delta V=V-V_{ss}$. This is the usual $N^{-1/2}$ classical self-averaging property of N independently fluctuating variables. The parameter κ accounts for the fact that there exists a maximum range of density changes for each grain that does not compromise the mechanical stability of the granular assembly. Figure 10 indicates that $\langle\delta V^2\rangle/V_{ss}^2\cong 1/40\,000$, and as an upper bound we let $\kappa=\rho_{xtal}-\rho_{RLP}=0.74-0.55\approx0.2$. We then find that $N\cong1600$. Since we know that each capacitor averages over a volume corresponding to 6000 beads, this suggests that there are roughly 1600 independently fluctuating clusters each consisting of ~4 beads (lower bound).

An important feature that can be seen in Fig. 10 is that the variance becomes systematically larger the deeper into the pile one goes [see also $D(\delta\rho)$ for $\Gamma=5.1$ in Fig. 4]. For the middle and bottom sections of the column, the variance appears nonmonotonic with a peak near $V_{ss}=1.592$, see open symbols in Fig. 10. Examination of the corresponding distributions of fluctuation amplitudes ($\Gamma=5.9$ in Fig. 4) indicates that the increased variance is due to a non-Gaussian tail in the distributions (see the description of Fig. 4, above). For comparison, we have also determined the variance from the slopes of the distribution functions in Fig. 4. We used the slopes corresponding to the low-density side of the distributions because these were most consistent with a Gaussian form over all accelerations and depths. In this way, the effect of the non-Gaussian tails in some of the distributions can be avoided. These results are shown as solid symbols in Fig. 10 and the dashed lines correspond to linear fits through the data. Here too, it is evident that the variance is larger for larger depths. A larger variance implies a correspondingly larger phase space. At first this seems counterintuitive because the time records [Fig. 3(a)] and power spectra (Fig. 5) indicate that density fluctuations are slower at the bottom of the pile. Although the kinetics near the bottom of the pile may be slower, there is a greater number of configurations with different volumes that are accessible to those beads.

A depth dependence to the variance also suggests the presence of a gradient in the compactivity. In Fig. 10, we used the average steady-state volumes V_{ss}, obtained from optical measurements of the total column height. One possibility is that this average volume density does not accurately represent the density in the different sections of the pile. If so, the larger compactivity near the bottom of the pile then implies that the bottom beads are actually in a less compact state than those at the top. However, from the trend in $\langle\delta V^2\rangle$ versus V_{ss} in Fig. 10 the difference in packing fraction between the top and bottom of the pile that would be necessary to have the variances be equal would be $\Delta\rho\approx0.035$. Since this difference is huge on the scale of $\rho(\Gamma)$ for the reversible branch in Fig. 2 we do not believe this to be a plausible explanation. Rather, it appears that there is another variable, such as pressure, in addition to the volume, that controls the depth dependence of the fluctuations. Indeed, supporting evidence to this effect can also be seen in Fig. 6, which shows that the high-frequency corner ω_H decreases with increasing depth into the pile. Nevertheless, we expect that the system is entirely jammed $\langle\delta V^2\rangle\to0$ at the same density (i.e., ρ_{max}) for all depths in the pile.

With the simulation described above, a broader range of densities can be explored than that which is experimentally accessible. Figure 11 shows the dependence of the variance in volume fluctuations as a function of steady-state volume for the 1D parking lot model. The rapid decrease in variance near $V_{ss}=1$ suggests that there may be a diverging length or time scale as the system approaches its most compact state. Indeed, plotting the normalized variance as a function of the free volume $(V_{ss}-1)$ does reveal power-law-like behavior $\langle\delta V^2\rangle/V_{ss}^2\propto(V_{ss}-1)^\beta$ with $\beta\approx1.4$. This is shown in the inset of Fig. 11. Proceeding with the compactivity analysis, the data in Fig. 11 was numerically integrated to yield the left-hand side of Eq. (5) to within a constant. An absolute value cannot be established with just our data. In Fig. 12 we plot the difference in inverse compactivity as a function of volume. This difference is with respect to the state $V_{ss}=1.1$.

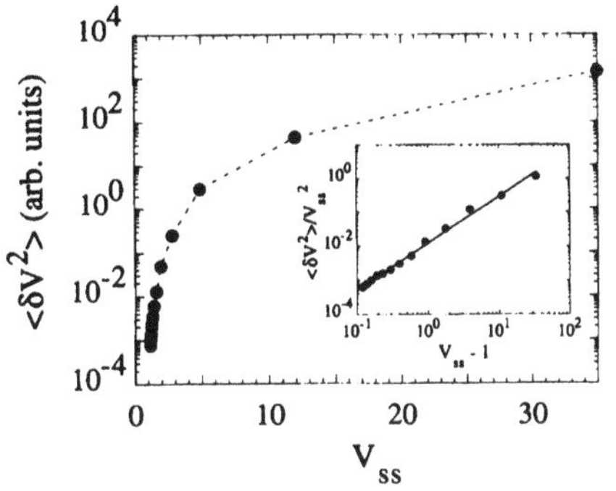

FIG. 11. The average variance of the volume fluctuations in the simulation as a function of the steady-state volume, V_{ss}. The fractional variance as a function of $V_{ss}-1$ is plotted on logarithmic axes in the inset. The solid line is a power-law fit given by $0.0124(V_{ss}-1)^{1.37}$.

Figure 12 indicates a nontrivial functional dependence to the increase in compactivity with system volume.

For comparison, the 3D experimental results correspond to relatively high densities in the 1D simulation because in 3D the available void volume is with respect to the random close packed limit (≈0.64) while the corresponding limit in 1D is 1. Taking this into account, the 30% increase in $\langle\delta V^2\rangle$ in the experimental results shown in Fig. 10 compares well with the simulation data in Fig. 11 over a similarly restricted range in V_{ss}.

VI. CONCLUSIONS

In this paper, we have examined the volume fluctuations about a steady-state density for a granular system. For these measurements to provide a useful analogy with a thermal system, it is essential that the fluctuations be measured in steady-state conditions. For this reason, we have explicitly taken data on the reversible density line as shown in Fig. 2. From these measurements, we have been able to determine experimentally the compactivity, which is the quantity analogous to the temperature in the theory of Edwards *et al.*

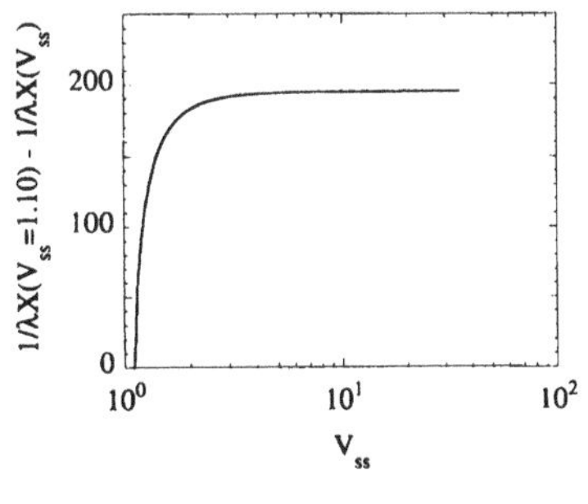

FIG. 12. The left hand side of Eq. (5) is numerically evaluated and plotted as a function of steady-state volume for the simulation data. As plotted, the difference in inverse compactivities between the highest density configuration ($V_{ss}=1.1$) and a low density configuration (higher V_{ss}) can be read off directly.

Theories based on free volume seem particularly well suited for describing the data. As the system approaches its final state, a growing number of particles have to be rearranged in order for the density to be increased locally. The rate of increase in the density is exponentially reduced by this number leading to a logarithmically slow approach to the steady-state density as observed experimentally. Monte Carlo simulations of a one-dimensional adsorption-desorption process based on these ideas show fluctuations about the steady state density that are strikingly similar to those observed experimentally. These results attest to the importance of volume exclusion for granular relaxation and steady-state dynamics under vibration.

Despite this model's simplicity and obvious shortcomings, it appears to capture an essential mechanism underlying the remarkably slow relaxation and the nature of the density fluctuations. This mechanism is associated with the reduction of free volume available for particle motion as the density increases. Although our simple model cannot predict the experimental values of the fitting parameters in Eq. (1), the inverse logarithmic density relaxation towards the steady state is the same one observed for granular compaction (see Fig. 1).

In the simulation model, our treatment was restricted to one-dimensional processes, but we expect that the results hold in higher dimensions as well. There are other important distinctions between the model and real granular media. One difference is that in the structure of a granular assembly the particles form contact networks. The creation of a void (a desorption event in the model) therefore requires the rearrangement of several particles and is thus restricted just as is the annihilation of voids. Another difference is the mechanical stability of a granular assembly. This property will place limits on the free volume available for large void creation. For instance, for spherical particles in three dimensions and in the presence of gravity the available free volume is determined by the restricted range of accessible volume fractions, namely, between the random close packed $\rho \approx 0.64$ and random loose packed $\rho \approx 0.55$ configurations.

It is interesting to speculate whether the reduction in free volume leads to a crossover from a simple independent particle picture for compaction to a more complex process at higher densities, presumably involving correlations over increasingly longer length and time scales. In this regard, we have demonstrated that density fluctuations are an important probe of the underlying microscopic dynamics. Indeed, the study of fluctuations may elucidate the physics of independent- and cooperative-particle motions, which lead to the macroscopic response of a powder subject to vertical vibrations. For instance, it is interesting to note that from both the experimental and simulation data there appear to be two characteristic time scales, related to the corner frequencies ω_L and ω_H in the power spectra, that characterize the steady-state dynamics. This behavior may be related to the results found in 3D simulations of vibrated powders by Mehta and Barker [17,20,34] and in simulations of a frustrated lattice gas [8–10]. Those results suggested the existence of two exponential relaxation mechanisms: the faster of the two involves the motions of independent particles while the slower involves collective particle motions, which were found to be diffusive. However, we emphasize again that both our experimental and simulation data are not consistent with a superposition of two independent exponentially decaying processes.

In this paper we have presented results for monodisperse spherical particles subject to vertical shaking. Realistic powders are far more complicated with properties that depend on factors like cohesive forces, polydispersity in size, and irregularity or anisotropy in shape. Nevertheless, our results can provide a valuable benchmark for evaluating the predictions of theoretical models and simulations. The applicability of concepts such as compactivity or "granular temperature" in the description of quasistatic granular media requires further exploration. In particular, it would be interesting to examine the properties of granular systems comprised of particles with shape anisotropies and subject to isotropic shaking. Furthermore, our experimental data suggest that the steady-state properties of a granular assembly cannot be fully described by a single state variable, i.e., the volume. Rather, another variable is required to account for the depth dependence of the volume fluctuations.

Note added in proof. The width of the density fluctuations in the parking lot model can be calculated in the mean-field approach described in the Appendix. For details see E. Ben-Naim *et al.*, Physica D (to be published). Such calculation predicts a power law as seen in the inset to Fig. 11 for the simulation data, but with an exponent $\beta=2$.

ACKNOWLEDGMENTS

We are grateful to M. L. Povinelli and S. Tseng for assistance on certain aspects of this work. It is a pleasure to acknowledge stimulating discussions with S. Coppersmith and T. Witten. This work was supported by the NSF through MRSEC Grant DMR-9400379 and through Grant No. CTS-9710991. We acknowledge additional support from the David and Lucile Packard Foundation, and from the Research Corporation.

APPENDIX

For the one-dimensional parking lot model an analytical mean-field description exists. On the continuum, an approximate rate equation for the density evolution was constructed from the exact steady-state void distribution. This equation yields an approach to the steady state that is essentially identical to that found in the experiment [i.e., Eq. (1)]. However, it is less successful in capturing the fluctuation behavior, particularly for the high densities relevant here. We summarize the salient analytic results for the model. Details can be found in Refs. [23–25]. A modified Langmuir equation can be written for the rate of change in density [24]:

$$\frac{d\rho}{dt} = k_+(1-\rho)e^{-\rho/(1-\rho)} - k_-\rho. \tag{A1}$$

The gain term is proportional to the fraction of unoccupied space, which is modified by an "excluded volume" constraint. It was previously shown that in steady state the probability $s(\rho)$ that an adsorption event is successful is given by $s(\rho)=e^{-\rho/(1-\rho)}$ [24]. This so-called "sticking coefficient" vanishes exponentially as $\rho \to 1$. This effectively reduces the sticking rate, $k_+ \to k_+(\rho)=k_+s(\rho)$. The desorption process, on the other hand, is unrestricted and so the loss term is proportional to the density itself.

The steady-state density ρ_{ss}, which is obtained by imposing $d\rho/dt=0$, can be determined as a function of the adsorption to desorption rate ratio, $k=k_+/k_-$, from the following transcendental equation:

$$\alpha e^{\alpha}=k, \quad \text{with} \quad \alpha=\rho_{ss}/(1-\rho_{ss}). \tag{A2a}$$

The following leading behavior in the two limiting cases is found

$$\rho_{ss}(k)\cong\begin{cases} k & \text{for } k\ll 1 \\ 1-(\ln k)^{-1} & \text{for } k\gg 1. \end{cases} \tag{A2b}$$

The effect of the volume exclusion constraint is striking, a huge adsorption to desorption rate ratio, $k\cong 10^9$, is necessary to achieve a 0.95 steady-state density.

We now focus on the relaxation properties of the system. The granular compaction process corresponds to the high density limit, and we thus consider the desorption-controlled case, $k\gg 1$. Hence, let us fix $k_+=1$ and consider the limit $k_-\to 0$ of Eq. (A1). For $t\gg 1/k_+$, it can be shown that the system approaches complete coverage, $\rho_\infty=1$, according to [24,25]

$$\rho(t)\cong\rho_\infty-1/(\ln k_+t). \tag{A3}$$

This is confirmed by numerical simulations in one dimension (see Ref. [24] and Sec. IV). We conclude that the excluded volume constraint gives rise to a slow relaxation.

Equation (A3) holds indefinitely only for the truly irreversible limit of the parking process, i.e., for $k=\infty$. For large but finite rate ratios, the final density is given by Eq. (A2b). By computing how a small perturbation from the steady state decays with time, an exponential relaxation towards the steady state is found $|\rho_{ss}-\rho(t)|\propto e^{-1/T}$ for $t\gg 1/k_-$. The relaxation time is

$$T=(1-\rho_{ss})^2/k_-. \tag{A4}$$

The above results can be simply understood: the early time behavior of the system follows the irreversible limit of $k_-=0$. Once the system is sufficiently close to the steady state, the density relaxes exponentially to its final value.

[1] J. B. Knight *et al.*, Phys. Rev. E **51**, 3957 (1995).
[2] S. F. Edwards and R. B. S. Oakeshott, Physica A **157**, 1080 (1989).
[3] A. Mehta and S. F. Edwards, Physica A **157**, 1091 (1989).
[4] S. F. Edwards and C. C. Mounfield, Physica A **210**, 290 (1994).
[5] S. Ogawa, in *Proceedings of the U.S.-Japan Symposium on Continuum Mechanics and Statistical Approaches in the Mechanics of Granular Materials*, edited by S. C. Cowin and M. Satake (Gakujutsu Bunken Fukyukai, Tokyo, 1979).
[6] C. S. Campbell, Annu. Rev. Fluid Mech. **22**, 57 (1990).
[7] I. Ippolito *et al.*, Phys. Rev. E **52**, 2072 (1995).
[8] A. Coniglio and H. J. Herrmann, Physica A **225**, 1 (1996).
[9] M. Nicodemi, A. Coniglio, and H. J. Herrmann, J. Phys. A **30**, L379 (1997).
[10] M. Nicodemi, A. Coniglio, and H. J. Herrmann, Phys. Rev. E **55**, 3962 (1997).
[11] H. M. Jaeger, S. R. Nagel, and R. P. Behringer, Phys. Today **49** (4), 32 (1996).
[12] H. M. Jaeger, S. R. Nagel, and R. P. Behringer, Rev. Mod. Phys. **68**, 1259 (1996).
[13] P. G. de Gennes (unpublished); T. Boutreux and P. G. de Gennes, Physica A **224**, 59 (1997).
[14] E. R. Nowak *et al.*, Powder Technol. **94**, 79 (1997).
[15] J. B. Knight, H. M. Jaeger, and S. R. Nagel, Phys. Rev. Lett. **70**, 3728 (1993).
[16] K. Shinohara, in *Handbook of Powder Science and Technology*, edited by M. E. Fayed and L. Otten (van Nostrand Reinhold Co., New York, 1984).
[17] A. Mehta and G. C. Barker, Phys. Rev. Lett. **67**, 394 (1991).
[18] M. Nelkin and A.-M. S. Tremblay, J. Stat. Phys. **25**, 253 (1981).
[19] E. R. Nowak, S. R. Nagel, and H. M. Jaeger (unpublished).
[20] G. C. Barker and A. Mehta, Phys. Rev. E **47**, 184 (1993).
[21] H. S. Caram and D. C. Hong, Phys. Rev. Lett. **67**, 828 (1991).
[22] D. C. Hong *et al.*, Phys. Rev. E **50**, 4123 (1994).
[23] G. Tarjus and P. Viot, Phys. Rev. Lett. **68**, 2354 (1992).
[24] P. L. Krapivsky and E. Ben-Naim, J. Chem. Phys. **100**, 6778 (1994).
[25] X. Jin, G. Tarjus, and J. Talbot, J. Phys. A **27**, L195 (1994).
[26] S. J. Linz, Phys. Rev. E **54**, 2925 (1996).
[27] K. Gavrilov (unpublished).
[28] D. A. Head and G. J. Rodgers (unpublished).
[29] M. D. Ediger, C. A. Angell, and S. R. Nagel, J. Phys. Chem. **100**, 13200 (1996).
[30] V. Privman and M. Barma, J. Chem. Phys. **97**, 6714 (1992).
[31] J. W. Evans, Rev. Mod. Phys. **65**, 1281 (1993).
[32] J. D. McGhee and P. H. von Hippel, J. Mol. Biol. **86**, 469 (1974).

[33] The variance is also equal to the integrated power spectrum over all frequencies. Because the power spectral dependence is weaker than ω^{-1} at lowest frequencies and tends to fall off as ω^{-2} at the highest frequencies we can have confidence that the variance is finite.

[34] G. C. Barker and A. Mehta, Phys. Rev. A **45**, 3435 (1992).

Diffusing-Wave Spectroscopy of Dynamics in a Three-Dimensional Granular Flow

Narayanan Menon and Douglas J. Durian

Diffusing-wave spectroscopy was used to measure the microscopic dynamics of grains in the interior of a three-dimensional flow of sand. The correlation functions show that minutely separated grains fly from collision to collision with large random velocities. On a time scale 10^3 to 10^4 times longer than the average time between collisions, the grains displayed slow, collective rearrangements, which, at the long-time limit, produced diffusive dynamics.

Sand dunes, grain silos, hourglasses, catalytic beds, filtration towers, river beds, ice fields, and many foods and building materials are granular systems (*1*). They consist of large numbers of randomly arranged, distinct, macroscopic grains that are too large to be moved by thermal energies but can be driven into flow by external forces. We do not have an understanding of the fluid state of a granular medium analogous to that for the macroscopic flow properties of a liquid. In a series of seminal papers, Bagnold (*2*) made the first efforts toward creating a phenomenological "fluid mechanics" for sand, identifying inertia of the grains and their collisions as significant elements in the dynamics. Since then there has been considerable theoretical effort (*3*) in formulating a continuum description of granular flows based on the kinetic theory of dense gases. However, in contrast to molecular fluids, kinetic energy in grain flows is irreversibly lost in the inelastic collisions of the grains. A further complication is that the scale of velocity fluctuations in the material (referred to as the "granular temperature") is nonthermal and has been difficult to measure, especially in three-dimensional (3D) flows (*4*). Also, recent computer simulations and theoretical work (*5*) show that 1D and 2D inelastic systems spontaneously form inhomogeneities that potentially restrict the applicability of hydrodynamic approaches to grain flow.

Department of Physics and Astronomy, University of California, Los Angeles, CA 90095-1547, USA.

We used diffusing-wave spectroscopy (DWS) (*6*) to probe the local, short-time dynamics of grains in a 3D gravity-driven flow and examine the physical basis of hydrodynamic models. DWS is a multiple-light-scattering technique that yields two-particle correlation functions at time intervals greater than 10^{-8} s and spatial separations greater than 1 Å. These capabilities are necessary because the collisional dynamics we studied are at time scales of 10^{-6} to 10^{-4} s and length scales of 0.01 to 1 μm. Because granular materials strongly scatter light, earlier experiments have chiefly studied quasi-2D flows (*7*) or highly diluted flows (where the short-time dynamics are collisional by construction). Experiments in dense flows (*8–10*) have been analyzed (*8*, *9*) with the assumptions that short-time dynamics are collisional and that the quantities of interest may be inferred from long-time, spatially averaged motions obtained by direct imaging of tracer beads.

The granular material we studied consisted of dry, cohesionless, monodisperse, smooth, spherical glass beads (*11*) 95 or 194 μm in diameter. The flow was gravity-driven (*12*) and confined to a vertical channel 30 cm high, 10 cm wide, and 0.3 to 1 cm thick (Fig. 1). Video and DWS measurements showed that spatial gradients in all three dimensions were small and that the flow field everywhere in the channel was characterized by a single average flow velocity V_f. We varied V_f from 0.03 to 3 cm/s by changing the mesh size at the bottom of the channel. The arrangement of the beads in flow showed no evidence of density inhomogeneities or crystalline packing.

For DWS measurements, we illuminated the sample with an Ar^+ laser of 488- or 514-nm wavelength and 3-mm beam waist. Incident photons were multiply scattered by the glass beads, performed random walks through the sample, and interfered, producing a speckled pattern. Grain motions caused this pattern to fluctuate, decorrelating the intensity measured at the detector. To infer the dependence of the dynamics of the beads on time τ from the autocorrelation function $g_1(\tau)$, we described photon transport as a random walk through the medium with a step length l^* and an absorption length l_A (which were determined by measuring the fraction of light transmitted through the sample as a function of its thickness). For example, the normalized electric-field autocorrelation function in transmission (Fig. 2A) is $g_1(\tau) \approx \exp[-(L/l^*)^2k^2\langle\Delta r^2(\tau)\rangle]$, where L is the sample thickness, $\langle\Delta r^2(\tau)\rangle$ is the mean-squared displacement of the scatterers, k is the wave vector of light in the medium, and the factor ($L/$

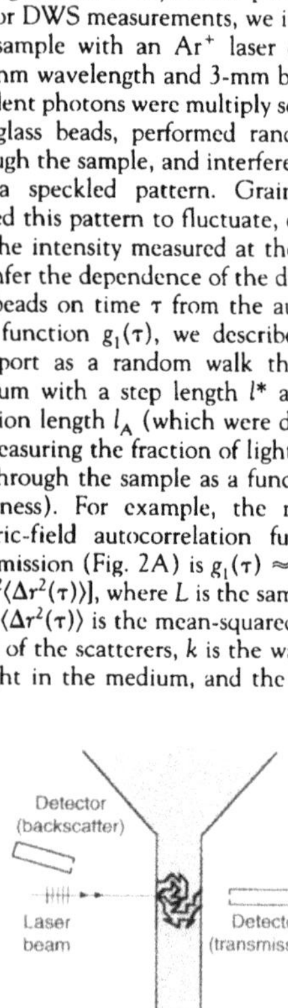

Fig. 1. Side view of the flow and light-scattering geometry.

$l^*)^2$ represents the number of random steps of length l^* in an average photon path. The complete solution of the photon diffusion equation with boundary conditions—specified by the geometry of the sample, the illuminating beam, and the detection optics—is weighted by paths of all lengths and includes absorption effects (6).

In such an inversion, the curves for $\langle \Delta r^2(\tau) \rangle$ from backscattering and transmission coincided (Fig. 2B) even though they were derived from markedly different $g_1(\tau)$, which shows that the analysis was reliable. Furthermore, this result implies that the dynamics of grains are uniform across the thickness of the channel, because $g_1(\tau)$ samples very different distributions of photon paths in these two scattering geometries: Photons transmitted through the sample were scattered by grains through the bulk of the sample, whereas photons backscattered from the sample were scattered mainly within a few l^* of the illuminated wall. The result in Fig. 2, B and C, demonstrates that the motion of sand grains at short times was ballistic, that is, $\langle \Delta r^2(\tau) \rangle = (\delta V)^2 \tau^2$, where δV is a randomly directed velocity. Each grain remained in free flight for a mean free time τ_c until a collision with a neighboring grain randomized the direction of the next ballistic flight. From δV and τ_c, we can determine the mean free path $s = \delta V \tau_c$, corresponding to the average distance between the surfaces of neighboring grains. Repeated collisions eventually changed the relative positions of grains. This movement was reflected in the slow increase in $\langle \Delta r^2(\tau) \rangle$ at times longer than τ_c. This interpretation of $\langle \Delta r^2(\tau) \rangle$ as a single-particle quantity is independent of spatial correlations between beads because in $g_1(\tau)$ the structure factor is weighted heavily to large scattering wave vectors (13). These features in the signal only reflect relative motions of beads; the effect of the average downward drift of beads was to continuously change the set of beads being sampled and contributed to the decay of $g_1(\tau)$ at a longer time scale (estimated by $D/V_f \approx 0.1$ to 10 s, where D is the beam waist).

The data in Fig. 2B show that the dynamics of grains are dominated by collisions rather than sliding contacts, even in dense, slow flows. The average velocity fluctuations were large ($\delta V \approx 0.31$ cm/s) and comparable to the overall flow velocity ($V_f \approx 0.32$ cm/s). The collisional frequency was high, and the interparticle separation was small: For the data in Fig. 2B, the average collision time and distance were $\tau_c \approx 9$ μs and $s \approx 0.028$ μm, respectively. Our data show that the dilation of sand in flow (14) may be tiny compared to the particle diameter d ($s/d \approx 10^{-4}$). Thus, although the bulk density of the sand may be almost unchanged by flow, the relative motions of the particles reflect a state of great activity. Our measurement establishes that there can be a "granular temperature" in the absence of a shear gradient. This result contradicts hydrodynamic models (15), which find $\delta V \rightarrow 0$ in this geometry, except in a shear layer at the boundary. Our experiment, however, does not establish that $(\delta V)^2$ has all the attributes of a temperature; in particular, we do not know if the velocity fluctuations are isotropic and Boltzmann-distributed. However, because the short-time limit of $\langle \Delta r^2(\tau) \rangle / \tau^2 = (\delta V)^2$ is well defined, the distribution at large velocities is stronger than a power law, which is consistent with a Boltzmann distribution.

We measured the dependencies of each of these microscopic quantities—δV, τ_c, and s—on V_f. The mean velocity fluctuation δV was the same order of magnitude as V_f for the range covered in this experiment (Fig. 3A). The data can be approximated by a power law: $\delta V \propto V_f^{2/3}$ over this range. The origin of this power law and its exponent are unknown. The mean collision time τ_c showed only a weak dependence on V_f (Fig. 3B). The collision frequency ranged from about 500 kHz (for $d = 95$ μm) to 10 kHz (for $d = 194$ μm). From τ_c and δV, we obtained s (Fig. 3C). The dilation ranged from 0.01 to 0.1% of the sphere diameter.

These data (Fig. 3) are consistent with energy and momentum balance for typical values of the coefficient of restitution e for glass spheres. For example, even the small dilation in Fig. 3C is sufficient for gravity to produce the large random velocities measured. However, energy and momentum budgets do not fully capture the dependence of the measured parameters on the driving velocity V_f. An energy balance between gravitational energy gained and kinetic energy lost because of inelasticity gives $(1 - e^2)[m(\delta V)^2/2]/\tau_c \sim mgV_f$, where m is the particle mass and g is acceleration due to gravity. Likewise, if we assume that the velocities of neighboring grains are uncorrelated and that collisions completely decorrelate velocity autocorrelations, we obtain an estimate $em\delta V \propto mg\tau_c$. Together, these give a scaling $\delta V \propto V_f$, which differs from the experimental scaling. A treatment of particle motions as uncorrelated thus seems

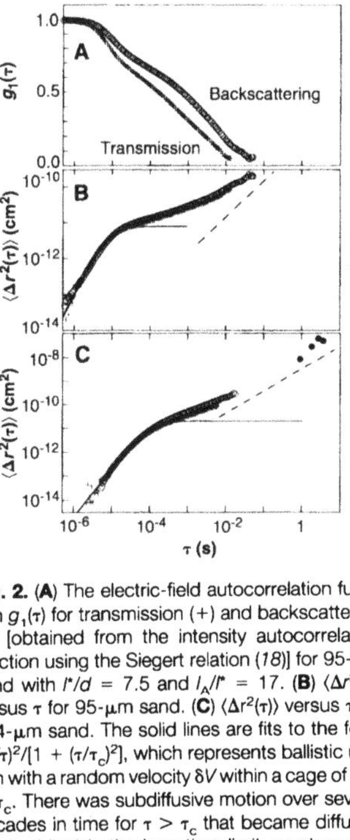

Fig. 2. (**A**) The electric-field autocorrelation function $g_1(\tau)$ for transmission (+) and backscattering (○) [obtained from the intensity autocorrelation function using the Siegert relation (18)] for 95-μm sand with $l^*/d = 7.5$ and $l_A/l^* = 17$. (**B**) $\langle \Delta r^2(\tau) \rangle$ versus τ for 95-μm sand. (**C**) $\langle \Delta r^2(\tau) \rangle$ versus τ for 194-μm sand. The solid lines are fits to the form $(\delta V \tau)^2/[1 + (\tau/\tau_c)^2]$, which represents ballistic motion with a random velocity δV within a cage of size $\delta V \tau_c$. There was subdiffusive motion over several decades in time for $\tau > \tau_c$ that became diffusive (dashed line) in the long-time limit, as shown by video measurements (●) of single-particle diffusion in one dimension perpendicular to the flow.

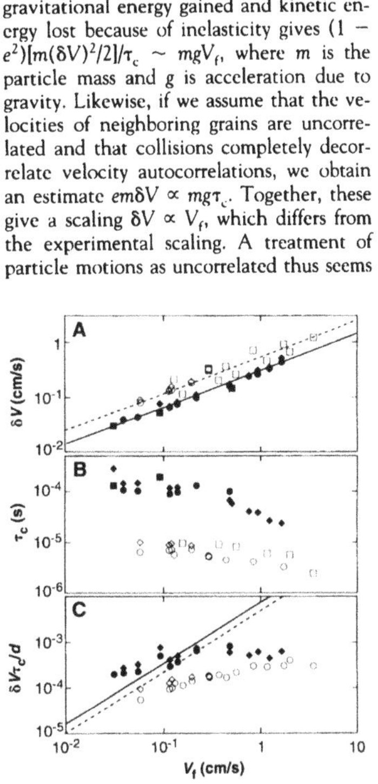

Fig. 3. Microscopic scales of motion versus macroscopic flow velocity V_f. (**A**) Mean velocity fluctuation δV versus V_f. The lines are power-law fits (with an exponent of ⅔) to the 95-μm (dashed line and open symbols) and 194-μm (solid line and filled symbols) data. (**B**) Mean collision time τ_c versus V_f. (**C**) Mean free path $s = \delta V \tau_c$, scaled to the particle diameter d. The lines show $(\delta V)^2/2gd$, the fraction of its own diameter a particle must fall under the influence of gravity to attain a speed δV. The conformity of measurements in various flow geometries [cells with $L = 0.32$ cm (open squares), 0.625 cm (circles), and 0.92 cm (filled squares)] and various scattering geometries [transmission (squares and circles) and backscattering (diamonds)] indicates that spatial gradients perpendicular to the walls of the channel were small enough that all measured quantities are functions of only V_f (19).

insufficient, suggesting that collective effects, such as velocity correlations between particles (*16*) or slowly decaying velocity autocorrelations, may be important.

All of the features in Fig. 3 refer to motions at times less than τ_c. At times longer than τ_c, the relative displacement of grains is characterized by subdiffusive motion, which presumably corresponds to gradual rearrangements of neighboring grains, over several orders of magnitude in time (Fig. 2, B and C). We did not obtain the diffusive limit $\langle \Delta r^2(\tau) \rangle \propto \tau$, even at the end of the range available to DWS. Measurements by long-range video microscopy (solid circles in Fig. 2C) show that relative motions of sand grains in the direction transverse to V_f ($\hat{y}$) and in the plane of the channel were diffusive (*17*). The time for rearrangement of nearest neighbors was 0.1 to 10 s, so that if relative motion is diffusive, it is a consequence of complex collective behavior. Hunt and others (*8*) reported experiments on video imaging of granular material in channel flow from which they extracted diffusion constants; similar experiments have been performed on vibrated granular systems (*9*). Our data indicate that the parameters of collisional dynamics cannot be deduced from such measurements using conventional prescriptions such as the Langevin equation as a bridge between short-time ballistic motion and long-time diffusion; for example, the diffusion coefficient of the grains in our experiment was much less than $s\delta V$.

The wide separation of time scales between collisional dynamics and the long-time diffusive limit suggest that complex collective dynamics occur even in the absence of long-wavelength clustering instabilities (*5*), reminiscent of dynamics in glassy systems such as viscous liquids and dense colloids. Our simple realization of a granular flow accentuates the contrast between a molecular fluid, where viscous loss occurs only in shear gradients, and sand, where a region of uniform flow dissipates energy because, as our results show, velocity fluctuations exist even in the absence of macroscopic gradients.

REFERENCES AND NOTES

1. H. M. Jaeger and S. R. Nagel, *Science* **255**, 1523 (1992); ______ and R. P. Behringer, *Phys. Today* **49**, 32 (April 1996); C. S. Campbell, *Annu. Rev. Fluid Mech.* **22**, 57 (1990).
2. R. A. Bagnold, *Proc. R. Soc. London Ser. A* **225**, 49 (1954); *ibid.* **295**, 219 (1966).
3. S. Ogawa, in *Proceedings of the U.S.–Japan Symposium on Continuum Mechanics and Statistical Approaches in the Mechanics of Granular Materials*, S. C. Cowin and M. Satake, Eds. (Gakujutsu Bunken Fukyukai, Tokyo, 1979), pp. 208–217; J. T. Jenkins and S. B. Savage, *J. Fluid Mech.* **130**, 187 (1983); P. K. Haff, *ibid.* **134**, 401 (1983); J. Schofield and I. Oppenheim, *Physica A* **196**, 209 (1993).
4. G. D. Cody, D. J. Goldfarb, G. V. Storch Jr., and A. N. Norris [*Powder Technol.* **87**, 211 (1996)] inferred velocity fluctuations in a fluidized bed from the acoustic response of the container to discrete collisions of particles.
5. M. A. Hopkins and M. Y. Louge, *Phys. Fluids A* **3**, 47 (1991); I. Goldhirsch and G. Zanetti, *Phys. Rev. Lett.* **70**, 1619 (1993); S. McNamara and W. Young, *Phys. Rev. E* **50**, R28 (1994); D. R. M. Williams and F. C. MacKintosh, *ibid.* **54**, R9 (1996).
6. G. Maret and P. E. Wolf, *Z. Phys. B* **65**, 409 (1987); D. J. Pine, D. A. Weitz, P. M. Chaikin, E. Herbolzheimer, *Phys. Rev. Lett.* **60**, 1134 (1988); D. J. Pine, D. A. Weitz, J. X. Zhu, E. Herbolzheimer, *J. Phys. (France)* **51**, 2101 (1990).
7. T. G. Drake, *J. Fluid Mech.* **225**, 121 (1991).
8. S. S. Hsiau and M. L. Hunt, *ibid.* **251**, 299 (1993); V. V. R. Natarajan, M. L. Hunt, E. D. Taylor, *ibid.* **304**, 1 (1995).
9. O. Zik and J. Stavans, *Europhys. Lett.* **16**, 255 (1991).
10. M. Nakagawa *et al.*, *Exp. Fluids* **16**, 54 (1993); E. E. Ehrichs *et al.*, *Science* **267**, 1632 (1995).
11. Jaygo (Union, NJ) and Cataphote (Jackson, MS).
12. The effect of the ambient air was small because the weight of the grains was at least 10^4 times the Stokes drag. Dissipation was primarily through inelastic collisions, which (for typical grain velocities and a coefficient of restitution of 0.9) was greater by a factor 10^4 than the energy lost to the viscous drag of the air.
13. S. Fraden and G. Maret, *Phys. Rev. Lett.* **65**, 512 (1990); X. Qiu *et al.*, *ibid.*, p. 516; J.-Z. Xue *et al.*, *ibid.* **69**, 1715 (1992).
14. O. Reynolds, *Philos. Mag.* **20**, 469 (1885).
15. K. Hui and P. K. Haff, *Int. J. Multiphase Flow* **12**, 189 (1986).
16. Y. H. Taguchi and H. Y. Takayasu, *Europhys. Lett.* **30**, 499 (1995).
17. These points are obtained from the width of $[\mathbf{r}(0) - \mathbf{r}(\tau)]\cdot\hat{\mathbf{y}}$ averaged over many particles and starting times. For comparison with the value of $\langle \Delta r^2(\tau) \rangle$ obtained from DWS, this quantity is multiplied by a factor of 3.
18. B. J. Berne and R. Pecora, *Dynamic Light Scattering: With Applications to Chemistry, Biology and Physics* (Wiley, New York, 1976).
19. We directly confirmed that the gradients in the other two directions were small by moving the beam across the channel.

5 August 1996; accepted 28 January 1997

Rheology of Soft Glassy Materials

Peter Sollich,[1,*] François Lequeux,[2] Pascal Hébraud,[2] and Michael E. Cates[1]
[1]Department of Physics and Astronomy, University of Edinburgh, Edinburgh EH9 3JZ, United Kingdom
[2]Laboratoire d'Ultrasons et de Dynamique des Fluides Complexes, 4 rue Blaise Pascal, 67070 Strasbourg Cedex, France
(Received 2 December 1996)

We attribute similarities in the rheology of many soft materials (foams, emulsions, slurries, etc.) to the shared features of structural disorder and metastability. A generic model for the mesoscopic dynamics of "soft glassy matter" is introduced, with interactions represented by a mean-field noise temperature x. We find power-law fluid behavior either with ($x < 1$) or without ($1 < x < 2$) a yield stress. For $1 < x < 2$, both storage and loss modulus vary with frequency as ω^{x-1}, becoming flat near a glass transition ($x = 1$). Values of $x \approx 1$ may result from marginal dynamics as seen in some spin glass models. [S0031-9007(97)02673-2]

PACS numbers: 83.20.–d, 05.40.+j, 83.50.–v

Many soft materials, such as foams, emulsions, pastes, and slurries, have intriguing rheological properties. Experimentally, there is a well-developed phenomenology for such systems: Their nonlinear flow behavior is often fit to the form $\sigma = A + B\dot{\gamma}^n$ where σ is shear stress and $\dot{\gamma}$ strain rate. This is the Herschel-Bulkeley equation [1,2]; or (for $A = 0$) the "power-law fluid" [1–3]. For the same materials, linear or quasilinear viscoelastic measurements often reveal storage and loss moduli $G'(\omega)$, $G''(\omega)$ in nearly constant ratio (G''/G' is usually about 0.1) with a frequency dependence that is either a weak power law (clay slurries, paints, microgels) or negligible (tomato paste, dense emulsions, dense multilayer vesicles, colloidal glasses) [4–10]. This behavior persists down to the lowest accessible frequencies (about 10^{-3}–1 Hz depending on the system), in apparent contradiction to linear response theory [11], which requires that $G''(\omega)$ should be an odd function of ω.

That similar anomalous rheology should be seen in such a wide range of soft materials suggests a common cause. Indeed, the frequency dependence indicated above points strongly to the generic presence of slow "glassy" dynamics persisting to arbitrarily small frequencies. This feature is found in several other contexts [12–14], such as elastic manifold dynamics in random media [15,16]. The latter is suggestive of rheology: Charge density waves, vortices, contact lines, etc., can "flow" in response to an imposed "stress." In this Letter we argue that glassy dynamics is a natural consequence of two properties shared by all the soft materials mentioned above: *structural disorder* and *metastability*. In such materials, thermal motion alone is not enough to achieve complete structural relaxation. The system has to cross energy barriers (for example, those associated with rearrangement of droplets in an emulsion) that are very large compared to typical thermal energies. Therefore the system adopts a disordered, metastable configuration even when (as in a monodisperse emulsion or foam) the state of least free energy would be ordered [17]. While the importance of disorder has been noted before for specific systems [7,11,18–21], we feel that its unifying role in rheological modeling has not been appreciated.

To test these ideas, we construct a minimal "generic model" for soft glassy matter. For simplicity, we ignore tensorial aspects, restricting our analysis to simple shear strains. Consider first the behavior of a foam or dense emulsion under shear. We focus on a *mesoscopic* region, large enough for a local strain variable l to be defined, but small enough for this to be approximately uniform within the region, whose size we choose as the unit of length. As the system is sheared, droplets in this region will first deform elastically from a local equilibrium configuration, giving rise to a stored elastic energy (due to surface tension, in this example [18]). This continues up to a yield point, characterized by a strain l_y, whereupon the droplets rearrange to new positions in which they are less deformed, thus relaxing stress. The mesoscopic strain l *measured from the nearest equilibrium position* (i.e., the one which can be reached by purely elastic deformation) therefore executes a saw-tooth motion as the macroscopic strain γ is increased [22]. Neglecting nonlinearities before yielding, the local shear stress is given by kl, with k an elastic constant; the yield point defines a maximal elastic energy $E = \frac{1}{2}kl_y^2$. A similar description obviously extends to many others of the soft materials discussed above.

We now ascribe to each mesoscopic region not only its own strain variable l, but also its own maximal yield elastic energy $E > 0$. We model the effects of structural disorder by assuming a *distribution* of such yield energies E, rather than a single value common to all regions. The state of a macroscopic sample is then characterized by a probability distribution $P(l, E; t)$. We propose the following dynamics for the time evolution of P:

$$\frac{\partial}{\partial t}P = -\dot{\gamma}\frac{\partial}{\partial l}P - \Gamma_0 e^{-(E-\frac{1}{2}kl^2)/x}P + \Gamma(t)\rho(E)\delta(l). \tag{1}$$

The first term on the right-hand side (r.h.s.) arises from the elastic deformation of the regions. This embodies a mean-field assumption that between successive local yield events, changes in local strain follow those of the macroscopic deformation: $\dot{l} = \dot{\gamma}$. Note, however, that due to stochastic yielding events the stress kl is spatially inhomogeneous (as is the local strain l). The macroscopic stress is defined as an average over regions

$$\sigma(t) = k\langle l\rangle \equiv k \int l\, P(l, E; t)\, dl\, dE\,. \tag{2}$$

The second term on the r.h.s. of (1) describes the yielding of our mesoscopic regions. We have written the yielding rate as the product of an "attempt frequency" Γ_0, and an exponential probability for activation over an energy barrier $E - \frac{1}{2}kl^2$ (the excess of the yield energy over that stored elastically). However, the resemblance to thermal activation is formal: We expect these "activated" yield processes to arise primarily by coupling to structural rearrangements elsewhere in the system. In a mean-field spirit, all such interactions between regions are subsumed into an effective "noise temperature" x. We first regard Γ_0, x as arbitrary constants, but later discuss their meaning and their possible dependences on other quantities.

Finally, the third term on the r.h.s. of (1) describes the relaxation of regions to new local equilibrium positions after yielding, which we treat as effectively instantaneous. The first factor in this term is simply the total yielding rate $\Gamma(t) = \Gamma_0\langle \exp[-(E - \frac{1}{2}kl^2)/x]\rangle_P$. The remaining two factors incorporate further mean-field assumptions as follows. First, the yield energy E for distortions about any equilibrium configuration is uncorrelated with the previous one for this region; it is drawn randomly from the prior distribution ("density of states") $\rho(E)$ which we assume to be time independent. Second, immediately after yielding, a region always finds itself in a completely unstressed state of local equilibrium with $l = 0$ [hence the Dirac delta function $\delta(l)$]. This latter simplification is not essential, as shown elsewhere [23].

In the absence of flow [$\gamma(t) = 0$], the model (1) describes activated hopping between "traps" of depth $E' = E - \frac{1}{2}kl^2$ with density $\rho(E')$. This corresponds to Bouchaud's model for glassy dynamics [12–14], whose predictions we briefly recall. For high (noise) temperatures x the system evolves towards the Boltzmann distribution $P_{\rm eq}(E') \sim \rho(E')\exp(E'/x)$. As x is lowered, this distribution may cease to be normalizable, leading to a glass transition at $x_g^{-1} = -\lim_{E\to\infty}(\partial/\partial E)\ln\rho(E)$. For $x < x_g$, no equilibrium state exists, and the system shows "weak ergodicity breaking" and various aging phenomena. A finite value of x_g implies an exponential tail in the density of states, $\rho \sim \exp(-E/x_g)$, which corresponds to a Gaussian distribution of yield strains $l_y = (2E/k)^{1/2}$.

A major attraction of the model defined by (1) and (2) is that an exact constitutive equation, relating the stress $\sigma(t)$ to the strain-rate history [$\dot{\gamma}(t' < t)$], can be obtained [23]. Since this is quite complicated, we restrict ourselves here to two standard rheological tests, for which the full form is not required. We use nondimensional units for time and energy by setting $\Gamma_0 = x_g = 1$; we also rescale our strain variables (l, γ) so that $k = 1$. In these units, $\rho(E) = \exp\{-E[1 + f(E)]\}$ with $f(E) \to 0$ for $E \to \infty$. Up to sub-power-law factors such as logarithms, all power laws reported below are valid for any $f(E)$; numerical examples use $f \equiv 0$. Analytical and numerical support for our results will be detailed elsewhere [23].

Consider first the complex dynamic shear modulus $G^*(\omega) = G' + iG''$, which describes the stress response to small shear strain perturbations around the equilibrium state. As such, it is well defined (i.e., time independent) only above the glass transition $x > 1$. Expanding (1) to first order in the amplitude γ of an oscillatory strain $\gamma(t) = \gamma\cos\omega t$, we find $G^*(\omega) = \langle i\omega\tau/(i\omega\tau + 1)\rangle_{\rm eq}$. This corresponds to a distribution of Maxwell modes whose spectrum of relaxation times $\tau = \exp(E/x)$ is given by the equilibrium distribution $P_{\rm eq}(E) \sim \exp(E/x)\rho(E)$. The relaxation time spectrum thus exhibits power-law behavior for large τ: $P(\tau) \sim \tau^{-x}$. This leads to power laws for G^* in the low frequency range (Fig. 1):

$$\begin{aligned} G'' &\sim \omega \quad \text{for } 2 < x, \qquad \sim \omega^{x-1} \quad \text{for } 1 < x < 2, \\ G' &\sim \omega^2 \quad \text{for } 3 < x, \qquad \sim \omega^{x-1} \quad \text{for } 1 < x < 3, \end{aligned} \tag{3}$$

For $x > 3$ the system is Maxwell-like at low frequencies, whereas for $2 < x < 3$ there is an anomalous power law in the elastic modulus. Most interesting is the regime $1 < x < 2$, where G' and G'' have constant ratio; both vary as ω^{x-1}. Behavior like this is observed in a number of soft materials [4–7,10]. Moreover, the frequency exponent approaches zero as $x \to 1$, resulting in essentially constant values of G'' and G', as reported in dense emulsions, foams, and onion phases [6–8]. Note, however, that the ratio $G''/G' \sim x - 1$ becomes small as the glass transition is approached. This increasing dominance of the elastic response G' prefigures the onset of a yield stress for

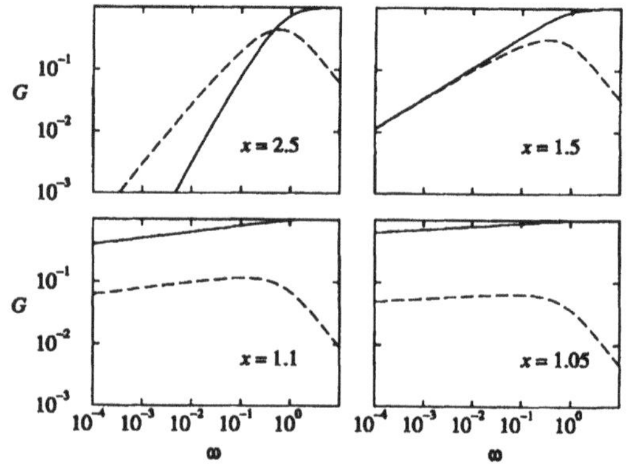

FIG. 1. Linear moduli G' (solid line) and G'' (dashed) vs frequency ω at various noise temperatures.

$x < 1$ (discussed below) [24]. If a high energy cutoff E_{max} is imposed on $\rho(E)$ (giving an upper limit on local yield strains), the above results remain valid down to $\omega_{min} = \exp(-E_{max}/x)$. Well-defined equilibrium values of the linear moduli then exist also for $x < 1$; one still finds $G'' \sim \omega^{x-1}$ for $\omega_{min} \ll \omega \ll 1$. For x just below $x_g = 1$, a log-log plot of $G''(\omega)$ therefore exhibits a small *negative* slope (whereas G' is constant). This may again be compatible with recent experimental data [7–10].

We now turn to the case of steady shear flow, $\dot{\gamma} =$ const, for which the steady-state distribution $P_{ss}(l, E)$ can be obtained analytically. After integrating over E, one finds $P_{ss}(l) \sim \Theta(l)g(z(l))$ with

$$z(l) = \frac{1}{\dot{\gamma}} \int_0^l e^{\gamma^2/2x}\, d\gamma,$$

$$g(z) = \int \rho(E) \exp(-ze^{-E/x})\, dE.$$

In the large z limit, $g(z) \sim z^{-x}$. Figure 2 shows that for large shear rates $\dot{\gamma} \gtrsim 1$, σ increases very slowly for all x $[\sigma \sim (x \ln \dot{\gamma})^{1/2}]$. More interesting is the small $\dot{\gamma}$ behavior, where we find three regimes: (i) For $x > 2$, the system is Newtonian, $\sigma = \eta\dot{\gamma}$. The viscosity is simply the average relaxation time $\eta = \langle \exp(E/x) \rangle_{eq} = \langle \tau \rangle_{eq}$ taken over the equilibrium distribution of energies, $P_{eq}(E) \sim \exp(E/x)\rho(E)$. Hence $\eta \sim \langle \exp(2E/x) \rangle_\rho$, which diverges at $x = 2$. (ii) For $1 < x < 2$ one finds power-law fluid behavior $\sigma \sim \dot{\gamma}^{x-1}$. (iii) For $x < 1$, the system shows a yield stress $\sigma(\dot{\gamma} \to 0) = \sigma_y > 0$. (This has a linear onset near the glass transition $\sigma_y \sim 1 - x$.) Beyond yield, the stress again increases as a power law of shear rate, $\sigma - \sigma_y \sim \dot{\gamma}^{1-x}$ (for $\dot{\gamma} \ll 1$). The behavior of our model in regimes (ii) and (iii) therefore matches, respectively, the power-law fluid [1–3] and Herschel-Bulkeley [1,2] scenarios as used to fit the nonlinear rheology of pastes, emulsions, slurries, etc.

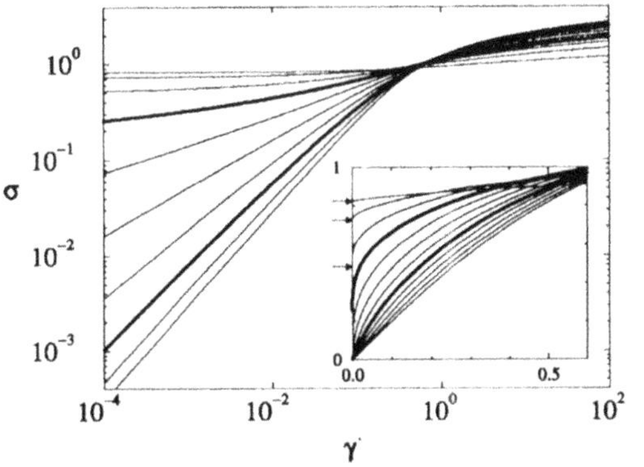

FIG. 2. Shear stress σ vs shear rate $\dot{\gamma}$, for $x = 0.25, 0.5, \ldots, 2.5$ (top to bottom on left); $x = 1, 2$ are shown in bold. Inset: small $\dot{\gamma}$ behavior, with yield stresses for $x < 1$ shown by arrows.

We now speculate on the origin and magnitude of the attempt frequency Γ_0 and the noise temperature x. First note that the parameter Γ_0 is the only source of a characteristic time scale (chosen as the time unit above). We have approximated it by a constant value: $\Gamma_0(\dot{\gamma}) = \Gamma_0(0)$. One possibility is that the intrinsic rate constant Γ_0 arises from *true* thermal processes. If so, it can be estimated as $\Gamma_{loc} k_B T P_{eq}(0) Q$ with Γ_{loc} a local diffusive attempt rate (for 1 μm emulsions this might be 0.01 s); $k_B T P_{eq}(0)$ is the (small) fraction of regions in which true thermal activation can surmount the yield barrier. The factor Q denotes the number of neighboring regions perturbed as a result of one such thermal event. A more detailed analysis (involving an extension to our model [23]) then shows that $k_B T P_{eq}(0) Q$ must be large enough (at least of order unity) to avoid depletion of the low energy part ($E \leq k_B T$) of the barrier distribution. This mechanism may arise in systems (such as foams) in which one local rearrangement can trigger a long sequence of others [20,21]. If so, the resulting intrinsic rate $\Gamma_0 \sim \Gamma_{loc}$ provides a plausible rheological time scale. (If Q is too small, Γ_0 will instead be of order $\Gamma_{loc} e^{-\overline{E}/k_B T}$, which for typical barrier energies $\overline{E} = \langle E \rangle_\rho$ is unfeasibly slow.)

We emphasize, however, that Γ_0 may be strongly system dependent, and any specific interpretation of it remains speculative. Nonetheless, we may view the activation factor in Eq. (1) as the probability that a perturbative "kick" to a given mesoscopic region (from events elsewhere) causes it to yield. We believe this activation factor should be primarily geometric in origin and hence depend on the disorder, but not on any intrinsic energy scale. Accordingly, (in our units) x values generically of order unity can be expected. We argue next that x values *close to* unity may be normal.

Consider first a steady shear experiment. For soft metastable materials, the rheological properties of a sample freshly loaded into a rheometer are usually not reproducible; they become so only after a period of shearing to eliminate memory of the loading procedure. In the process of loading one expects a large degree of disorder to be introduced; the initial dynamics under flow should therefore involve a high noise temperature $x \gg 1$. As the sample approaches the steady state, the flow will (in many cases) tend to eliminate much of this macroscopic disorder [25] so that x will decrease. But as this occurs, the noise-activated processes will slow down; as $x \to 1$, they become negligible. Assuming that, in their absence, the disorder cannot be reduced further, x is then "pinned" at a steady-state value at or close to the glass transition. This scenario, although extremely speculative, is strongly reminiscent of the "marginal dynamics" seen in some mean-field spin glass models [26].

There remain several ambiguities within this picture—for example, whether the steady-state value of x should depend on $\dot{\gamma}$; if it does so strongly, our results for steady flow curves will of course be changed. If a steady flow is stopped and a linear viscoelastic spectrum measured, the behavior observed should presumably pertain to the x characterizing the preceding steady flow (assuming that x reflects structure only). But unless the strain amplitude is extremely small the x value obtained in steady state could be affected by the oscillatory flow itself [27].

Also uncertain is to what extent a steady energy input is needed to sustain the nonlinear dynamics. Although not represented in the model, a small finite strain rate amplitude might be needed to balance the gradual dissipation of energy in yield events. In its absence, one might expect the sample to show aging [i.e., $P(l, E; t)$ nonstationary in time]. Within the model, aging in fact occurs only for $x < 1$ [14] (the regime for which we predict a yield stress). Conversely, we saw above that, even in this regime, for finite $\dot{\gamma}$ a well-defined steady-state distribution is recovered: *flow interrupts aging* [13]. This can be understood by considering the distribution of energies. Without flow, one obtains a Boltzmann distribution $P(E) \sim \rho(E)e^{E/x}$ up to (for $x < 1$) a cutoff which shifts to higher and higher energies as the system ages [12]. This cutoff, and hence the most long-lived traps visited (which have a lifetime comparable to the age of the system), dominate the aging behavior [14]. The presence of flow leads to a steady-state value of this cutoff of $E \sim x \ln(\dot{\gamma}^{-1}x^{1/2})$, while for higher energies one has $P_{ss}(E) \sim \rho(E)E^{1/2}$. Hence flow prevents regions from getting stuck in progressively deeper traps and the aging process is truncated after a finite time.

We are currently investigating more complicated nonlinear strain histories [23]. In future work, explicit spatial structure and interactions between regions must be added so as to understand better the mutual dynamical evolution of the attempt rate, the effective noise temperature, and the disorder. One issue concerns the relative importance of localized [19,28–33] versus avalanche-like [20,21] events in the relaxation of stress.

The authors are indebted to J.-P. Bouchaud for various seminal suggestions. They also thank him, M. O. Robbins, and D. Weaire for helpful discussions, and the Newton Institute, Cambridge, for hospitality. P. S. is a Royal Society Dorothy Hodgkin Research Fellow.

*Electronic address: P.Sollich@ed.ac.uk

[1] S. D. Holdsworth, Trans. Inst. Chem. Eng. **71**, 139 (1993).
[2] E. Dickinson, *An Introduction to Food Colloids* (Oxford University Press, Oxford, 1992).
[3] H. A. Barnes, J. F. Hutton, and K. Walters, *An Introduction to Rheology* (Elsevier, Amsterdam, 1989).
[4] M. R. Mackley, R. T. J. Marshall, J. B. A. F. Smeulders, and F. D. Zhao, Chem. Eng. Sci. **49**, 2551 (1994).
[5] R. J. Ketz, R. K. Prudhomme, and W. W. Graessley, Rheol. Acta **27**, 531 (1988).
[6] S. A. Khan, C. A. Schnepper, and R. C. Armstrong, J. Rheol. **32**, 69 (1988).
[7] T. G. Mason, J. Bibette, and D. A. Weitz, Phys. Rev. Lett. **75**, 2051 (1995).
[8] P. Panizza *et al.*, Langmuir **12**, 248 (1996).
[9] H. Hoffmann and A. Rauscher, Colloid Polym. Sci. **271**, 390 (1993).
[10] T. G. Mason, and D. A. Weitz, Phys. Rev. Lett. **75**, 2770 (1995).
[11] D. M. A. Buzza, C. Y. D. Lu, and M. E. Cates, J. Phys. II (France) **5**, 37 (1995).
[12] C. Monthus and J. P. Bouchaud, J. Phys. A **29**, 3847 (1996).
[13] J. P. Bouchaud and D. S. Dean, J. Phys. I (France) **5**, 265 (1995).
[14] J. P. Bouchaud, J. Phys. I (France) **2**, 1705 (1992).
[15] V. M. Vinokur, M. C. Marchetti, and L. W. Chen, Phys. Rev. Lett. **77**, 1845 (1996).
[16] P. LeDoussal and V. M. Vinokur, Physica (Amsterdam) **254C**, 63 (1995).
[17] Soft systems may also be intrinsically metastable in a more drastic sense (for example, with respect to coalescence in emulsions)—we ignore this here.
[18] D. Weaire and M. A. Fortes, Adv. Phys. **43**, 685 (1994).
[19] M. D. Lacasse *et al.*, Phys. Rev. Lett. **76**, 3448 (1996).
[20] T. Okuzono and K. Kawasaki, Phys. Rev. E **51**, 1246 (1995).
[21] D. J. Durian, Phys. Rev. Lett. **75**, 4780 (1995).
[22] Note that precisely this motion is predicted, on a global rather than mesoscopic scale, for perfectly ordered foams. See, e.g., Ref. [34].
[23] P. Sollich (unpublished).
[24] It does not mean, however, that G'' for fixed (small) ω always decreases with x; in fact, it first *increases* strongly as x is lowered and only starts decreasing very close to the glass transition (when $x - 1 \sim |\ln \omega|^{-1}$) [23].
[25] D. Weaire, F. Bolton, T. Herdtle, and H. Aref, Philos. Mag. Lett. **66**, 293 (1992).
[26] After a quench from $T = \infty$ to *any* temperature $0 < T_0 < T_g$, the spin glass is dynamically arrested in regions of phase space characteristic of T_g itself, rather than the true temperature T_0. See, e.g., L. F. Cugliandolo and J. Kurchan, Phys. Rev. Lett. **71**, 173 (1993).
[27] This might allow "flat" moduli $G^*(\omega)$ ($x \approx 1$) to be found alongside a nonzero yield stress with power-law flow exponent around $\frac{1}{2}$ ($x \approx \frac{1}{2}$) [7,35,36].
[28] P. Hébraud, J. P. Munch, F. Lequeux, and D. J. Pine (unpublished).
[29] A. J. Liu *et al.*, Phys. Rev. Lett. **76**, 3017 (1996).
[30] S. Hutzler, D. Weaire, and F. Bolton, Philos. Mag. B **71**, 277 (1995).
[31] A. D. Gopal and D. J. Durian, Phys. Rev. Lett. **75**, 2610 (1995).
[32] D. J. Durian, D. A. Weitz, and D. J. Pine, Science **252**, 686 (1991).
[33] J. C. Earnshaw and M. Wilson, J. Phys. II (France) **6**, 713 (1996).
[34] A. M. Kraynik, Annu. Rev. Fluid Mech. **20**, 325 (1988).
[35] T. G. Mason, J. Bibette, and D. A. Weitz, J. Colloid Interface Sci. **179**, 439 (1996).
[36] H. M. Princen and A. D. Kiss, J. Colloid Interface Sci. **128**, 176 (1989).

PHYSICAL REVIEW E VOLUME 55, NUMBER 4 APRIL 1997

Energy flow, partial equilibration, and effective temperatures in systems with slow dynamics

Leticia F. Cugliandolo*
Laboratoire de Physique Théorique des Liquides, 4 place Jussieu, F-75005 Paris, France

Jorge Kurchan†
École Normale Supérieure de Lyon, 46 Allée d'Italie, F-69364 Lyon Cedex 07, France

Luca Peliti‡
Groupe de Physico-Chimie Théorique, CNRS URA 1382, ESPCI, 10 rue Vauquelin, F-75231 Paris Cedex 05, France and Dipartimento di Scienze Fisiche, Unità INFM, Università "Federico II," Mostra d'Oltremare, Pad. 19, I-80125 Napoli, Italy

(Received 11 November 1996)

We show that, in nonequilibrium systems with small heat flows, there is a time-scale-dependent effective temperature that plays the same role as the thermodynamical temperature in that it controls the direction of heat flows and acts as a criterion for thermalization. We simultaneously treat the case of stationary systems with weak stirring and of glassy systems that age after cooling and show that they exhibit very similar behavior provided that time dependences are expressed in terms of the correlations of the system. We substantiate our claims with examples taken from solvable models with nontrivial low-temperature dynamics, but argue that they have a much wider range of validity. We suggest experimental checks of these ideas.
[S1063-651X(97)05903-5]

PACS number(s): 05.20.−y, 75.40.Gb, 75.10.Nr, 02.50.−r

I. INTRODUCTION

No physical system is ever in thermodynamical equilibrium. When we apply thermodynamics or statistical mechanics, we idealize the situation by assuming that "fast" processes have taken place, and "slow" ones will not: hence, we define an observation time scale which distinguishes these two kinds of processes [1]. It follows that the same system can be at equilibrium on one scale, and out of equilibrium on another, and, more strikingly, that it can be at equilibrium, but exhibiting different properties, on two scales at once.

Since the assumption of thermal equilibrium lies at the heart of statistical mechanics, it is usually hard to make these considerations without a strong appeal to one's intuition. We show in the following that they can be made, in fact, quite precise for a class of systems characterized by very slow energy flows. These systems are out of equilibrium, either because they are very gently "stirred," i.e., work is constantly done on them, or because they have undergone a quench from higher temperatures a long time ago.

The most typical example of such a system is a piece of glass that has been in a room at constant temperature for several months. Since the glass itself is not in equilibrium, we have, in principle, no right of talking about "the temperature of the glass," but only about the temperature of the room. However, we may legitimately ask what temperature would indicate a thermometer brought into contact with the glass and, again, we would be very surprised if it did not coincide with the room temperature. We would be even more surprised if, putting two points of the glass in contact with both ends of a copper wire, a heat flow were established through it.

In other words, although we know that equilibrium thermodynamics does not apply for the glass, we implicitly assume that some concepts that apply for equilibrium are still relevant for it. This is not because the glass is "near equilibrium" but rather because it has been relaxing for a long time, and therefore thermal flows are small.

Many attempts have been done to extend the concepts of thermodynamics to nonequilibrium systems—such as systems exhibiting spatiotemporal chaos or weak turbulence [2,3]. In this context, Hohenberg and Shraiman [3] have defined an effective "temperature" for stationary nonequilibrium systems through an expression involving the response, the correlation, and the temperature of the bath. A closely related expression appears naturally in the theory of nonstationary systems exhibiting aging [4,5], such as glasses.

We show here that this expression indeed deserves the name of temperature, because (i) the effective temperature associated with a time scale is the one measured on the system by a thermometer, in contact with the system, whose reaction time is equal to the time scale, (ii) it determines the direction of heat flows within a time scale, and (iii) it acts as a criterion for thermalization.

We shall here consider simultaneously two different conditions in which a regime with small flows of energy exists.

(1) Ordinary thermodynamical systems in contact with a heat bath at temperature T that are slowly driven ("stirred") mechanically. The driving force is proportional to a small number which we shall denote D. The observation time

---

*Also at Service de Physique de l'Etat Condensé, Saclay, CEA, Saclay, France. Permanent address: Laboratoire de Physique Théorique de l'École Normale Superieure de Paris, Paris, France. Electronic addresses: leticia@:lptl.jussieu.fr, leticia@spec.saclay.cea.fr
†Electronic address: Jorge. Kurchan@enslapp.ens-lyon.fr
‡Electronic address: luca@turner.pct.espci.fr, peliti@na.infn.it

and the manner of the stirring are such that, for long enough times, the system enters a *stationary*, time-translational invariant (TTI) regime [6]: one-time average quantities are independent of time, two-time quantities depend only upon time differences, etc. Stationarity is a weaker condition than thermodynamical equilibrium, since it implies loss of memory of the initial condition but not all other properties that are linked with the Gibbs-Boltzmann distribution.

(2) Purely relaxational systems that have been prepared through some cooling procedure ending at time $t=0$ and are kept in contact with a heat bath at a constant temperature T up to a (long) waiting time t_w (as in the example of the glass). t_w is also usually called "annealing time" in the glass literature. In this case physical quantities need not be TTI, and in interesting cases they will keep a dependence upon t_w (and also, in many cases, upon the cooling procedure) for all later times $t=\tau+t_w$.

We shall treat in parallel the "weak stirring" ($D\to 0$) and the "old age" ($t_w\to\infty$) limits: both taken *after the thermodynamical limit of infinite number of degrees of freedom*. We show that they lead to the same behavior, from the point of view of thermalization and effective temperatures, provided that one expresses time dependences in terms of the correlations of the system [7].

As a test of our ideas, we discuss thermalization in the context of the mean-field theory of disordered systems, or the low-temperature generalization of the mode coupling equations, but the nature of our results makes us confident that they have a much wider range of validity.

In Sec. II, we recall the generalization of the fluctuation-dissipation relation to the nonequilibrium case. In Sec. III, we consider the reading of a thermometer coupled to a system: when the system is in equilibrium we show that the thermometer measures the temperature of the heat bath, while when it is out of equilibrium it measures different effective temperatures depending on the observation time scale. These effective temperatures are equal or higher than the one of the bath and are closely related to the FDT violation factor [8–13,4,5] introduced to describe the out-of-equilibrium dynamics of glassy systems. In Sec. IV, we recall how time scales or correlation scales are defined in systems with slow dynamics. We then argue that, if the FDT violation factor is well defined within a time scale, a single degree of freedom thermalizes within that time scale to the corresponding effective temperature. In Sec. V, we extend this analysis to several degrees of freedom and show that the effective temperature determines the direction of heat flows, and can be used as a thermalization criterion. In Sec. VI, we discuss various phenomenological "fictive temperature" ideas that have been used for a long time in the theory of structural glasses. Our conclusions are summarized in Sec. VII, where some experimental implications of our work are suggested.

II. THE FLUCTUATION-DISSIPATION RELATION OUT OF EQUILIBRIUM

Let us consider a system with N degrees of freedom $(s_1,\ldots,s_N)$, whose dynamics is described by Langevin equations of the form

$$\dot{s}_i=b_i(s)+\eta_i(t), \tag{2.1}$$

where $\eta_i(t)$ is the Gaussian thermal noise. For *unstirred* systems, we consider purely relaxational dynamics, where the average velocity $b_i(s)$ is proportional to the gradient of the Hamiltonian $E(s)$:

$$b_i(s)=-\sum_j \Gamma_{ij}\frac{\partial E(s)}{\partial s_j}. \tag{2.2}$$

The symmetric matrix Γ is related to the correlation function of the noise η by the Einstein relations

$$\langle \eta_i(t)\eta_j(t')\rangle=2T\Gamma_{ij}\delta(t-t'), \tag{2.3}$$

where T is the temperature of the heat bath. Averages over the thermal history, i.e., averages over many realizations of the same experiment with different realizations of the heat bath, will be denoted by angular brackets. We assume of course $\langle \eta_i(t)\rangle=0$, $\forall i,t$. We have chosen the temperature units so that Boltzmann's constant is equal to 1. The equilibrium distribution is then proportional to the Boltzmann factor $\exp(-E/T)$.

For *stirred* systems, we add to b_i a perturbation proportional to D, that cannot be represented as the gradient of a function (i.e., is not purely relaxational), but is otherwise generic. We then have $W\equiv\langle\Sigma_i b_i(s)s_i\rangle>0$ at stationarity, meaning that work is being done on the system [14,15].

We denote the observables (energy, density, magnetization, etc.) by $O(s)$. Throughout this work we shall denote by t_w or t' the earliest time (to be related to the waiting time), t the latest time, and τ the relative time $t-t_w$. These times are measured, in the case of the unstirred systems which exhibit aging, from the end of the cooling procedure.

Given two observables O_1 and O_2, we define their correlations $C_{12}(t,t_w)\equiv\langle O_1(t)O_2(t_w)\rangle-\langle O_1(t)\rangle\langle O_2(t_w)\rangle$, and their mutual response

$$R_{12}(t,t_w)\equiv\frac{\delta O_1(t)}{\delta h_2(t_w)}, \tag{2.4}$$

where h_2 appears in a perturbation of the Hamiltonian of the form $E\to E-h_2(t)O_2$. Obviously, causality implies $R_{12}(t,t_w)=0$ for $t<t_w$. It is also useful to introduce the integrated response (susceptibility)

$$\chi_{12}(t,t_w)\equiv\int_{t_w}^{t}dt'R_{12}(t,t'). \tag{2.5}$$

Let us now make a parametric plot [5] of $\chi(t,t_w)$ vs $C(t,t_w)$ for several increasing values of t_w. We thus obtain a limit curve $\lim_{t_w\to\infty}(t,t_w)=\chi(C)$. In the case of a weakly driven system, we wait for stationarity and plot $\chi(t-t_w)$ vs $C(t-t_w)$ for several decreasing values of the driving D. We thus obtain the curve $\lim_{D\to 0}\chi(t-t_w,D)=\chi(C)$.

The fluctuation-dissipation theorem (FDT) relates the response and correlation function at equilibrium. One has

$$R_{12}(t-t_w)=\frac{1}{T}\frac{\partial C_{12}(t-t_w)}{\partial t_w}, \tag{2.6a}$$

$$\chi_{12}(t-t_w)=\frac{1}{T}[C_{12}(0)-C_{12}(t-t_w)]. \tag{2.6b}$$

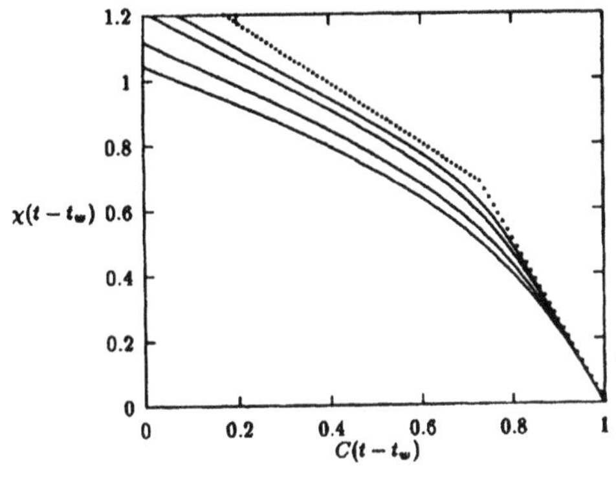

FIG. 1. The susceptibility $\chi(t-t_w)$ vs the autocorrelation function $C(t-t_w)$ for the model of Appendix A once stationarity is achieved. The parameter D is equal to 0.05, 0.375, 0.025, and 0.0125, respectively, from bottom to top. The dots represent the analytical solution for the limit $D\to 0$. One sees that, in this limit, the FDT violation factor $X(C)$ tends *continuously* to the dotted straight lines. The value of C at the breakpoint is C^{EA}, the Edwards-Anderson order parameter or the ergodicity breaking parameter in the language of the MCT.

If the equilibrium distribution is asymptotically reached for $t_w\to\infty$ (or $D\to 0$ in the case of stirred systems) the FDT implies that the limit curve $\chi(C)$ is a straight line of slope $-1/T$.

However, there is a family of systems for which the limiting curve $\chi(C)$ *does not* approach a straight line (see Figs. 1 and 2). For driven systems this means that the slightest stirring is sufficient to produce a large departure from equilibrium *even at stationarity* (Fig. 1), while in the case of a relaxational system it means that the system is unable to equilibrate within experimental times (Fig. 2).

Let us denote by $-X(C)/T$ the slope of the curve $\chi(C)$:

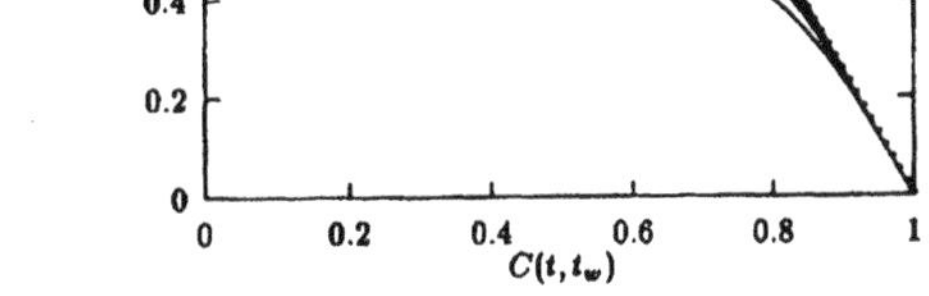

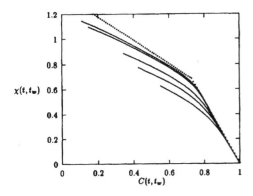

FIG. 2. The susceptibility $\chi(t,t_w)$ vs the autocorrelation function $C(t,t_w)$ for the model of Appendix A at $T<T_g$ (D is strictly zero). The full curves correspond to different total times t, equal, from bottom to top, to 12.5, 25, 37.5, 50, and 75, respectively ($t_w>t/4$ throughout). The dots represent the analytical solution when $t_w\to\infty$. Neither $\chi(t,t_w)$ nor $C(t,t_w)$ achieve stationarity (see Fig. 3).

$$\frac{d\chi(C)}{dC}\equiv-\frac{X(C)}{T}. \tag{2.7}$$

This corresponds to

$$R(t,t_w)=\frac{X(C)}{T}\frac{\partial C(t,t_w)}{\partial t_w}, \tag{2.8}$$

where the derivative is taken with respect to the earlier time. We have thus defined $X(C)$, the FDT violation factor [4,5], for nonequilibrium systems with slow dynamics.

When FDT holds, for $D\to 0$ (or $t\gg t_w\to\infty$, respectively) we can treat the system as being in equilibrium and X tends to 1 in the limit. When this does not happen, we may inquire about the physical meaning of $X(C)$. In order to answer this question, let us first recall the relationship there is between the FDT and the equipartition of energy.

III. FREQUENCY-DEPENDENT THERMOMETERS THAT MEASURE EFFECTIVE TEMPERATURES

We use a harmonic oscillator of frequency ω_0 to measure the "temperature" of a degree of freedom $O(s)$ [$O(s)$ may be the energy, or some spatial Fourier component of the magnetization]. At the waiting time t_w we weakly couple the oscillator to the system via $O(s)$, while we keep the system in contact with a heat bath at temperature T. We wait for a short time until the average energy of the oscillator has stabilized. If the system were in equilibrium, by the principle of equipartition of energy we would have $\langle E_{osc}\rangle=T$.

Assuming linear coupling, the Hamiltonian reads

$$E_{total}=E(s)+E_{osc}+E_{int}, \tag{3.1}$$

where

$$E_{osc}=\frac{1}{2}\dot{x}^2+\frac{1}{2}\omega_0^2x^2, \tag{3.2}$$

$$E_{int}=-aO(s)x. \tag{3.3}$$

The equation of motion of the oscillator reads

$$\ddot{x}=-\omega_0^2x+aO(t). \tag{3.4}$$

In the presence of the coupling, if $ax(t)$ is sufficiently small (an assumption we have to verify *a posteriori*), we can use linear response theory to calculate the action on O of the oscillator:

$$O(t)=O_b(t)=a\int_0^t dt' R_O(t,t')x(t'), \tag{3.5}$$

where $O_b(t)$ is the fluctuating term and where the response function R_O is defined by

$$R_O(t,t')=\frac{\delta\langle O\rangle(t)}{\delta ax(t')}. \tag{3.6}$$

We assume moreover that the average $\langle O_b(t)\rangle$ exists (we set it to zero by a suitable shift of x) and that the fluctuations of O_b (in the absence of coupling) are correlated as

$$\langle O_b(t)O_b(t_w)\rangle = C_O(t,t_w), \tag{3.7}$$

where $C_O(t,t_w)$ is a quantity of $O(N)$.

The equation for x then reads

$$\ddot{x} = -\omega_0^2 x + aO_b(t) + a^2\int_0^t dt' R_O(t,t')x(t'). \tag{3.8}$$

Thus the oscillator takes up energy from the fluctuations of O, and dissipates it through the response of the system. Equation (3.8) is linear and easy to solve in the limit of small a^2 by Fourier-Laplace transform. One thus obtains the following results, whose proof is sketched in Appendix B.

Consider first the case of stirred systems at stationarity, in which both the correlation and the response are TTI. The average potential energy of the oscillator reaches the limit

$$\frac{1}{2}\,\omega_0^2\langle x^2\rangle = \frac{1}{2}\,\langle E_{\rm osc}\rangle = \frac{\omega_0 C_O(\omega_0)}{2\chi''_O(\omega_0)}, \tag{3.9}$$

after a time $\sim t_c$ given by

$$t_c = \frac{2\omega_0}{a^2\chi''(\omega_0)}. \tag{3.10}$$

We have defined

$$\chi''(\omega) \equiv {\rm Im}\int_0^\infty dt\; R(t)e^{i\omega t},$$

$$\tilde{C}(\omega) \equiv {\rm Re}\int_0^\infty dt\; C(t)e^{i\omega t}. \tag{3.11}$$

If we now define the temperature $T_O(\omega)$ as

$$T_O(\omega) \equiv \langle E_{\rm osc}\rangle, \tag{3.12}$$

we obtain

$$T_O(\omega) = \frac{\omega_0\tilde{C}_O(\omega_0)}{\chi''_O(\omega_0)}. \tag{3.13}$$

This is precisely the temperature defined by Hohenberg and Shraiman [3] for the case of weak turbulence [the spatial dependence is encoded in $O(s)$].

Let us now turn to the case of relaxational dynamics where TTI is violated. Here we have to take into account the time t_w at which the measurement is performed, and consider $\omega_0^{-1} \ll t_c \ll t_w$. Then, a similar calculation yields for the energy of the oscillator

$$\frac{1}{2}\,\omega_0^2\langle x^2\rangle_{t_w} = \frac{1}{2}\,\langle E_{\rm osc}\rangle_{t_w} = \frac{\omega_0\tilde{C}_O(\omega_0,t_w)}{2\chi''_O(\omega_0,t_w)}, \tag{3.14}$$

where the average $\langle x^2\rangle_{t_w}$ is taken on a comparatively short time stretch after t_w. This definition of the frequency- and waiting-time-dependent correlation $C_O(\omega_0,t_w)$ and out-of-phase susceptibility $\chi''_O(\omega_0,t_w)$ closely follows the actual experimental procedure for their measure: one considers a time window around t_w consisting of a few cycles (so that phase and amplitude can be defined) and small enough respect to t_w so that the measure is "as local as possible in time." In fact, these two quantities are standard in the experimental investigation of aging phenomena in spin glasses [16,18].

The natural definition of the frequency- and time-dependent temperature of O is then

$$T_O(\omega_0,t_w) \equiv \frac{\omega_0\tilde{C}_O(\omega_0,t_w)}{\chi''_O(\omega_0,t_w)}. \tag{3.15}$$

If equilibrium is achieved, then the temperature is independent of t_w, of the frequency ω_0 and of the observable O, and coincides with that of the heat bath. The index O recalls that the effective temperature may depend on the observable. The frequency-dependent temperature defined by either Eq. (3.13) or (3.15) is compatible with the Fourier transformed expression of the FDT violation factor (2.8) provided that it does not vary too fast with ω_0. We show later that this is indeed the case for systems with slow dynamics.

For this definition we have chosen somehow arbitrarily an oscillator as our thermometer. However, we show in Appendix C that the same role can be played by any small but macroscopic thermometer, weakly coupled to the system. The role of the characteristic frequency ω_0 is then played by the inverse of the typical response time of the thermometer.

Now $T_O(\omega,t_w)$ only deserves to be called a "temperature" if it controls the direction of heat flow. A first way to check whether this is the case is to consider an experiment in which we connect the oscillator to an observable O_1, and let it equilibrate at the temperature $T_{O_1}(\omega,t_w)$. We then disconnect it and connect it to another observable O_2 and let it equilibrate at the temperature $T_{O_2}(\omega,t_w)$. This is not like a Maxwell demon, since the times of connection and disconnection are unrelated to the microscopic behavior of the system. The net result is that an amount of energy $T_{O_1}(\omega,t_w) - T_{O_2}(\omega,t_w)$ has been transferred from the degrees of freedom associated with O_1 to those associated with O_2: therefore the flow goes from high to low temperatures. This is an actual realization of the idea of "touching two points of the glass with a copper wire" described in the Introduction.

This observation also suggests a possible explanation of the fact that all FDT violation factors that we know of are smaller than one: if $T_O(\omega,t_w)$ were smaller than the temperature of the heat bath, it could be possible, in principle, to extract energy from the bath by connecting between it and our system a small Carnot engine. The argument can be made sharper by considering a stationary situation in a weakly stirred system: but to argue that this situation would lead to a violation of the second principle one needs to prove that the equilibration times of the Carnot engine are short enough to make the power it produces larger than the dissipated one.

IV. EQUILIBRATION WITHIN A TIME SCALE

Before discussing the effective temperature as an equilibration factor we need to introduce some general features of the time evolution of systems with slow dynamics. We first define the time correlation scales and we then argue that, if the FDT violation factor is well defined within a time scale, a single degree of freedom thermalizes within that time scale at the corresponding effective temperature.

A. Correlation and response scales

Systems having a long-time out-of-equilibrium dynamics tend to have different behaviors in different time scales. Let us start by describing them for long-time dynamics in the purely relaxational case. In Appendix D we give a formal definition of correlation scale, following [5].

Consider first, as an example, the dynamics of domain growth [19] for the Ising model at low but nonzero temperatures. The autocorrelation function $C(t,t_w) = (1/N)\Sigma_i s_i(t) s_i(t_w)$ exhibits two regimes.

At long times t,t_w, such that $t-t_w \ll t_w$, the correlation function shows a fast decay from 1 (at equal times) to m^2, where m is the magnetization. This regime describes the fast relaxation of the spins within the bulk of each domain.

At long and well-separated times $(t-t_w \sim t_w)$, the correlation behaves as a function of $L(t_w)/L(t)$, where $L(t)$ is some measure of the typical domain size at time t.

We refer to these two scales as "quasiequilibrium" and "coarsening" (or "aging"), respectively. After a given time, the correlation rapidly decays to a plateau value m^2, and the speed with which it falls below that value becomes smaller and smaller as t_w grows.

This example helps to stress the fact that different scales are defined as a function of both times [in this case $t-t_w$ finite, $L(t_w)/L(t)$ finite] and are well separated only in the limit where both times are large.

In Fourier space, this separation of scales is achieved by considering several frequencies ω and increasing times t_w. Then, if we consider $\omega \sim$const for increasing t_w, we probe the quasiequilibrium scale, while if we want to probe the aging or coarsening scale we have to consider smaller and smaller ω, keeping $\omega L(t_w)/L'(t_w) \sim$const.

If we now consider two frequencies ω_1 and ω_2 and keep

$$\frac{\omega_1}{\omega_2} \sim \text{const}, \tag{4.1}$$

we shall probe, as $t_w \to \infty$, the *same* scale: it will be the quasiequilibrium scale if they both remain finite, or the coarsening (aging) one if we keep $\omega_{1,2} L(t_w)/L'(t_w) \sim$const.

These considerations can be generalized to other systems with slow dynamics. In general there may be more than two relevant scales [9,4,5,20], for example ω=const, $\omega t_w^{1/2}$=const, ωt_w=const, etc. In any case, the condition for looking into the same scale via two successions of frequencies remains Eq. (4.1).

In Fig. 3 we show the numerical solution of our test model, defined in Appendix A. The autocorrelation function $C(t,t_w)$ is plotted vs the time difference $t-t_w$ for several waiting times in log-log scale. These plots (which are standard in the Monte Carlo simulations of spin glasses [21]), show (i) that the system is out of equilibrium, since we have an explicit dependence on t_w, and (ii) that there are at least two time correlation scales. For short time differences $t-t_w$ the decay is fast, the autocorrelation function is TTI and it falls from 1 to C^{EA}. It then decays further from C^{EA} to zero, more and more slowly as the waiting time increases. The quantity C^{EA} is known in the language of spin glasses as the

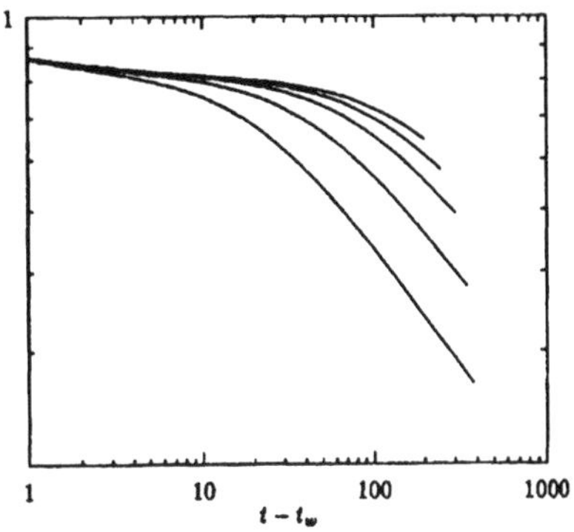

FIG. 3. The correlation function $C(t,t_w)$ vs $t-t_w$ for the same model as in Fig. 2. From bottom to top t_w=20, 50, 100, 150, and 200. The correlation decays rapidly from 1 to $C^{EA} \sim 0.73$ and then more slowly from C^{EA} to 0. This second decay becomes slower and slower as t_w increases.

Edwards-Anderson order parameter and in mode coupling theory (MCT) as the nonergodicity parameter: it measures the strength of the fast correlations ($C^{EA}=m^2$ in domain growth). Quasiequilibrium and aging regimes are experimentally observed in real spin glasses and polymer glasses [22,18].

Let us now turn to a similar analysis for driven systems in the limit of weak driving energy $D \to 0$. In that case, even if the system reaches a TTI regime provided we wait long enough (the amount of waiting increases when the stirring rate D decreases), it sometimes happens that some correlations and responses acquire a nontrivial low-frequency behavior in the limit $D \to 0$ [23]. For example, in Fig. 4 we show the same plot as in Fig. 3, i.e., $C(t-t_w)$ vs $t-t_w$, for different, but small, stirring rates. We see that there are at least two time scales: one, for short time differences, where the correlation decays rapidly from 1 to C^{EA} and one, for long time differences, in which it slowly decays from C^{EA} to

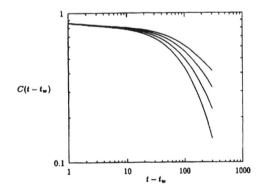

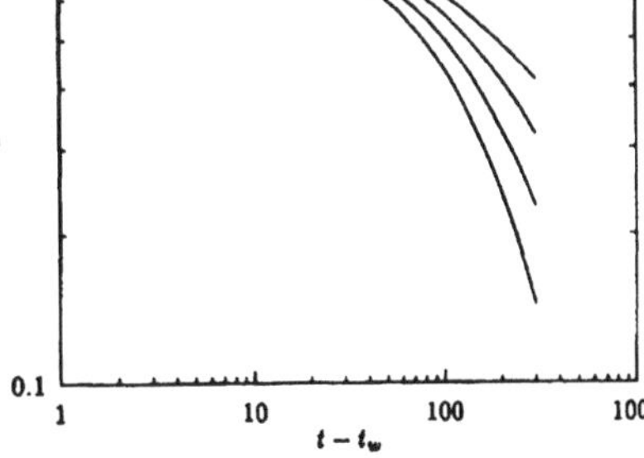

FIG. 4. The correlation function $C(t-t_w)$ vs $t-t_w$ for the same model as in Fig. 1. From bottom to top D=0.1, 0.075, 0.05, and 0.025. The correlation decays rapidly from 1 to $C^{EA} \sim 0.73$ and then more slowly from C^{EA} to 0. This second decay becomes slower and slower as D decreases.

zero. The smaller the stirring, the slower the decay of C from C^{EA} to zero.

It is useful to consider, in these cases, frequencies that go to zero as some function of D. The condition that two sequences of frequencies ω_1, ω_2 correspond to the same scale now reads

$$\lim_{D\to 0} \frac{\omega_1}{\omega_2} = \text{const.} \tag{4.2}$$

B. Thermalization criterion for a single degree of freedom within a scale

Let us now consider a system at a given waiting time t_w or stirring rate D. Heat flows tend to zero as $t_w \to \infty$ or $D \to 0$, respectively. It is reasonable to assume that, in these limits, the *effective temperatures* associated with any given observable O on a given time scale tend to equalize, provided that their limit is finite. We thus have, for example, in the case of relaxational dynamics, for ω_1/ω_2=const,

$$\lim_{\substack{t_w\to\infty \\ \omega_1\to 0}} T_O(\omega_1, t_w) = \lim_{\substack{t_w\to\infty \\ \omega_2\to 0}} T_O(\omega_2, t_w). \tag{4.3}$$

Similarly, we expect that each fixed frequency sooner or later thermalizes with the bath:

$$\lim_{\substack{t_w\to\infty \\ \omega=\text{const}}} T_O(\omega, t_w) = T. \tag{4.4}$$

In some important cases the effective temperatures defined by Eqs. (3.13) and (3.15) diverge in the limit $t_w \to \infty$ $(D \to 0)$. In these cases the effective temperature $T_O(\omega, t_w)$ should diverge for the whole time scale. We shall dwell on this problem in Sec. IV C.

Equations (4.3) and (4.4) are trivially true in a system that reaches thermal equilibrium, where all effective temperatures eventually reach the temperature of the reservoir. However, they also describe the situation in which smaller frequencies take longer to reach the temperature of the heat bath, in such a way that at any given (long) waiting time there are low enough frequencies that have a temperature substantially different (in all cases we know, higher) from that of the bath. In particular, Eq. (4.4) allows us to answer a question we asked at the beginning: if a piece of glass has been kept at room temperature for several months, a thermometer whose response time is of order of a few seconds would measure the room temperature, but it would read a higher temperature if its response time is of the order of weeks.

In fact, with the appropriate handling of time scales [5] (see Appendix D), Eqs. (4.3) and (4.4) make it possible to calculate the out-of-equilibrium relaxation of mean-field spin glasses, and also the low-temperature generalization of the mode-coupling equations for one single mode. We thus obtain a solvable example where Eqs. (4.3) and (4.4) hold.

For the case of stirred systems, the $D \to 0$ limit plays the same role as the $t_w \to \infty$ limit in relaxational systems provided that the time scales are suitably redefined as in Eq. (4.2). In particular, at each fixed value of ω one has $\lim_{D\to 0} T_O(\omega, D) = T$, and, for ω_1/ω_2=const one has

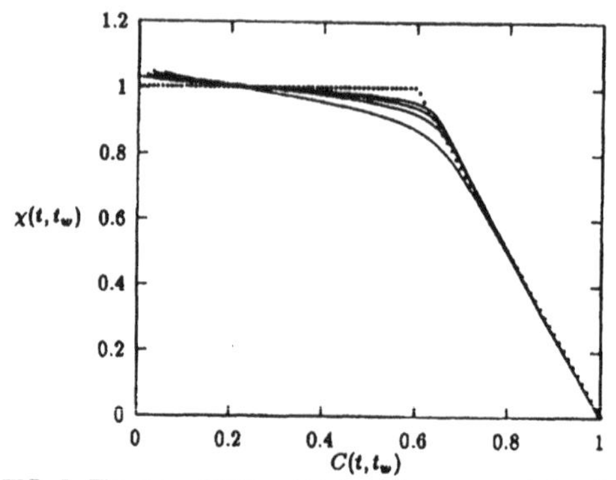

FIG. 5. The susceptibility $\chi(t,t_w)$ vs the autocorrelation function $C(t,t_w)$ for the purely relaxational $p=2$ spherical model [equivalent to the $O(N)$, $N \to \infty$ ferromagnetic coarsening in $d=3$]. The dots represent the analytical solution when $t_w \to \infty$. The total time t is equal to 20, 50, 100, and 200 ($t_w > t/4$ throughout).

$$\lim_{\substack{D\to 0 \\ \omega_1\to 0}} T_O(\omega_1, D) = \lim_{\substack{D\to 0 \\ \omega_2\to 0}} T_O(\omega_2, D), \tag{4.5}$$

provided that both limits are finite.

If one looks into a time scale within which the temperature is almost constant, one can indifferently use a temperature defined in terms of a frequency or a time. In the relaxational case one has

$$T_O(\omega, t_w) = T_O(t, t_w), \tag{4.6}$$

provided that $t - t_w \sim \omega^{-1}$. In the same situation one has

$$R(t, t_w) = \frac{1}{T_O(t, t_w)} \frac{\partial C(t, t_w)}{\partial T_w}. \tag{4.7}$$

Similar relations hold for the driven case. If, on the contrary, one considers values of frequency or time for which the temperature defined by Eq. (4.7) is not constant, one cannot directly relate it with the reading of a thermometer coupled to the system.

C. Coarsening and the case of infinite effective temperature

An important physical situation which deserves a special discussion is that of domain growth. In this case the effective temperature (associated, say, with the total magnetization) in the "coarsening scale," $\omega L(t_w)/L'(t_w)$=const, tends to infinity as $t_w \to \infty$. The curve χ (magnetic susceptibility) C (magnetization correlation) looks like in Fig. 5. This figure shows the results for a model that is equivalent to the $O(N)$ ferromagnetic coarsening in three dimensions. The integrated response χ becomes flat, as $t_w \to \infty$, in the aging regime. In other words, the long-term memory *tends to disappear*.

Because experimental measures of aging are in general related to the response, this kind of system is sometimes referred to as exhibiting aging "in the correlations" (correlations are not TTI) but not in the response. One should stress, however, that the divergence of the effective temperature in the aging time scale can be extremely slow: for example, in the Fisher-Huse model for spin glasses [24]

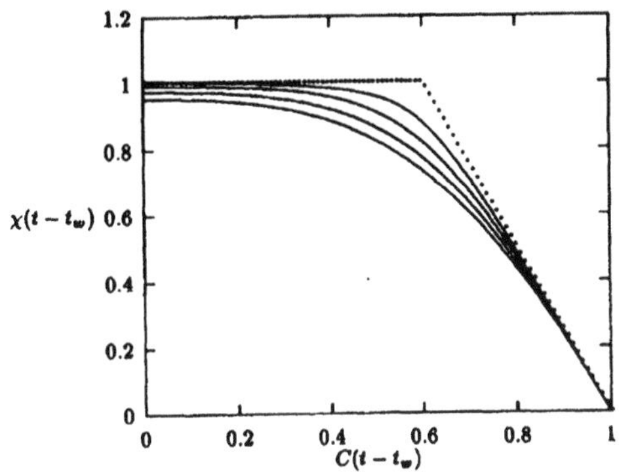

FIG. 6. The susceptibility $\chi(t-t_w)$ vs the autocorrelation function $C(t-t_w)$ for the $p=2$ spherical asymmetric model [''stirred'' $O(N)$, $N\to\infty$ ferromagnetic coarsening in $d=3$] for different levels of asymmetry (stirring): from bottom to top $D=0.8$, 0.6, 0.4, and 0.2. The dots represent the analytical solution in the limit of zero asymmetry.

$T_O(\omega,t_w)$ grows like a power of $\ln t_w$.

The fact that the effective temperature in the aging regime tends to infinity means that if we measure it at ω small and fixed, the temperature $T(\omega,t_w)$ grows with t_w (while we are still probing the aging regime) and then starts falling, finally reaching the temperature of the bath. If we repeat the experiment with a smaller ω, the overall behavior is the same, but the highest temperature reached will be higher; and so on without an upper limit. The situation is similar, but experimentally more subtle, for ''stirred'' systems for which the effective temperatures diverge as $D\to 0$. In such cases we measure a system that has a given (small) driving D. If the system is such that for $D\to 0$ the effective temperature of all but the fastest scale tend to infinity, we may never realize that this is the case in an experimental situation, as long as we are not able to repeat the experiment with smaller and smaller stirring.

We show one such case in Fig. 6, which represents the same coarsening problem as before, but made stationary by a ''stirring'' term in the equation of motion proportional to D. If one had performed an experiment on such a system for a fixed D, one would have observed high effective temperatures for separated times, and no evidence of thermalization. Only by letting D take smaller and smaller values one can notice that in fact these temperatures tend to infinity.

D. Systems with finite effective temperatures

In mean-field models one can close the Schwinger-Dyson equations into a set of dynamical equations involving only the correlation and response functions. This allows us to solve them analytically for large times, and also to obtain their full solution, via a numerical integration, starting from a pair of initial times $t_w=t=0$. We thus obtain the curve $\chi(C)$ by integrating the response function and plotting it vs C. Two cases in which there are only two time-correlation scales are the test model [4] considered here (cf. Appendix A) and the case of a particle moving in an infinite-dimensional random medium with short-range correlations [25]. Both the analytical and the numerical results exhibit thermalization within the aging regime.

In the Sherrington-Kirkpatrick spin-glass model there is a full hierarchy of time scales and effective temperatures, a fact also confirmed numerically [26] and by Monte Carlo simulations [5]. Another model with infinite many scales and with a full hierarchy of effective temperatures is that of a particle moving in an infinite-dimensional random medium with long-range correlations [20,25].

In Ref. [27], the Monte Carlo simulation of the ''realistic'' 3D Edwards-Anderson model for a spin glass was used to obtain the $\chi(C)$ curves, which seem to approach a nontrivial curve for increasing t_w. This suggests the existence of a hierarchy of time-correlation scales.

The ''trap model'' for spin-glass dynamics [28] violates Eq. (4.3) when an unusual choice of a parameter (α) is made. However, it is difficult to interpret the model, with this choice of parameter, as a phenomenological model stemming from a reasonable microscopic dynamics.

V. THERMALIZATION OF DIFFERENT DEGREES OF FREEDOM WITHIN A TIME SCALE

As we remarked in Sec. III, a *bona fide* temperature should control heat flow and thermalization. It is the aim of this section to show that this is indeed the case for the effective temperature we have defined. If this were not the case, it would be possible to use our small oscillator to transfer heat from some degrees of freedom to others: in other words, by decorating our copper wire with a suitable frequency filter, we would observe heat flowing through it when it touches the two ends of the glass.

We shall argue that if different degrees of freedom *effectively interact* on a given time scale, then they thermalize on that scale. A useful—though not universal—criterion for ''effective interaction'' is that their mutual response (the response of one of the degrees of freedom to an oscillating field of the given frequency conjugate to the other degree of freedom) is of the same order as their self-response on that time scale.

The argument is essentially the same both for relaxational systems that have evolved for a long time, or in stationary driven systems in the limit of small driving energy. We shall thus focus on the relaxational case only.

We emphasize again that the thermalization of different degrees of freedom is well defined only in the limit of vanishing heat flow, i.e., long waiting times or vanishing stirring rates, respectively. This is witnessed by the appearance of one (or more) well-separated plateaus in the decay of the two-time correlation functions.

Let us consider n modes (labeled by $a,b=1,\dots,n$). One can write, in general, some Schwinger-Dyson equations for their correlations $\mathsf{C}=(C_{ab})$ and responses $\mathsf{R}=(R_{ab})$:

$$\frac{\partial C_{ab}(t,t_w)}{\partial t}=-\sum_c \mu_{ac}(t)C_{cb}(t,t_w)+2TR_{ab}(t_w,t)+\sum_c\int_0^{t_w}dt''D_{ac}(t,t'')R_{cb}(t_w,t'')+\sum_c\int_0^{t}dt''\Sigma_{ac}(t,t'')C_{cb}(t'',t_w), \quad (5.1a)$$

$$\frac{\partial R_{ab}(t,t_w)}{\partial t} = -\sum_c \mu_{ac}(t) R_{cb}(t,t_w) + \delta(t-t_w)\delta_{ab} + \sum_c \int_{t_w}^{t} dt'' \Sigma_{ac}(t,t'') R_{cb}(t'',t_w). \tag{5.1b}$$

As they stand, Eqs. (5.1) are just a way to hide our difficulties under the Σ, D carpet. Several approximations extensively used in the literature amount to various approximations of the kernels Σ and D.

In equilibrium, the symmetries [29] of the original problem allow us to write

$$\Sigma_{ab}(t-t_w) = \frac{1}{T}\frac{\partial D_{ab}}{\partial t_w}(t-t_w), \tag{5.2a}$$

$$R_{ab}(t-t_w) = \frac{1}{T}\frac{\partial C_{ab}}{\partial t_w}(t-t_w). \tag{5.2b}$$

If we now make for Σ and D the approximation that are ordinary functions (instead of general functionals) of the correlations and the responses, we obtain the mode-coupling approximation (MCA)

$$D_{ab}(\mathrm{C}) = F_{ab}(\mathrm{C}), \quad \Sigma_{ab}(\mathrm{C}) = \sum_{c,d} F_{ab,cd}(\mathrm{C}) R_{cd}, \tag{5.3}$$

where there is a model-dependent function $F(\mathrm{q})$ such that

$$F_{ab}(\mathrm{q}) = \frac{\partial F}{\partial q_{ab}}, \quad F_{ab,cd}(\mathrm{q}) = \frac{\partial^2 F}{\partial q_{ab}\partial q_{cd}}. \tag{5.4}$$

If the system equilibrates, one recovers the usual form of the mode-coupling theory (MCT) that is applied for supercooled liquids [30]. It will be instructive to recall how equilibrium is reached within the framework of Eq. (5.4). One recalls the FDT conditions (5.2); time translational invariance: functions of one time are just constant and functions of two times depend upon time differences only; reciprocity: $C_{ab}(t-t_w) = C_{ba}(t-t_w)$. Putting this information in Eqs. (5.1) and (5.4) one obtains

$$\frac{\partial C_{ab}(t-t_w)}{\partial t} = -\sum_c \mu_{ac} C_{cb}(t-t_w) + \frac{1}{T}\sum_c [D_{ac}(0) C_{cb}(t_w-t_w) - D^{\infty}_{ac} C^{\infty}_{cb}] + \frac{1}{T}\sum_c \int_{t_w}^{t} dt'' D_{ac}(t-t'') \frac{\partial C_{cb}(t''-t_w)}{\partial t''}, \tag{5.5}$$

where $D^{\infty}_{ac}, C^{\infty}_{cb}$ stand for limits of widely separated times, i.e.,

$$C^{\infty}_{ac} \equiv \lim_{t\to\infty} C_{ac}(t), \quad D^{\infty}_{ac} \equiv \lim_{t\to\infty} D_{ac}(t). \tag{5.6}$$

This is the usual equilibrium MCT equation [30]. The equation for the response (5.1b) with the same equilibrium assumptions becomes the time derivative (divided by T) of Eq. (5.1a), as it should.

In a case in which the system is unable to reach thermal equilibrium, like the low-temperature phase of spin glasses, the solution of Eqs. (5.1) will exhibit several time scales. The question as to how many scales one has to consider in order to close the dynamical equations can be answered for each model unambiguously by the construction in Ref. [5], suitably generalized to several modes.

Here, for simplicity, we assume that there are only two relevant time scales (as in the coarsening example of the last section). We discuss later a model system that acts as an explicit example of this situation. We propose an ansatz for the long-time asymptotics, and then verify that it closes the equations [4]. We assume (and later verify) that all two-time functions can be separated in two regimes.

Finite time differences with respect to the long waiting time, i.e., $t-t_w$ finite and positive, and $t_w \to \infty$. In this regime TTI and FDT hold.

Aging regime, corresponding to long and widely separated times, i.e., $t \sim t_w \to \infty$. In this regime neither TTI nor FDT hold.

For finite time differences and a long waiting time t_w we have

$$C^{\mathrm{FDT}}_{ab}(t-t_w) \equiv \lim_{t_w\to\infty} C_{ab}(t,t_w), \tag{5.7a}$$

$$R^{\mathrm{FDT}}_{ab}(t-t_w) \equiv \lim_{t_w\to\infty} R_{ab}(t,t_w), \tag{5.7b}$$

with

$$R^{\mathrm{FDT}}_{ab}(t-t_w) = \frac{1}{T}\frac{\partial C^{\mathrm{FDT}}_{ab}}{\partial t_w}(t-t_w), \tag{5.8}$$

$$C^{\mathrm{EA}}_{ab} \equiv \lim_{t-t_w\to\infty} \lim_{t_w\to\infty} C^{\mathrm{FDT}}_{ab}(t-t_w). \tag{5.9}$$

The aging regime is defined as the time domain in which the correlations fall (more and more slowly) below C^{EA}_{ab}. For these times we denote

$$C_{ab}(t,t_w) = \tilde{C}_{ab}(t,t_w), \quad R_{ab}(t,t_w) = \tilde{R}_{ab}(t,t_w). \tag{5.10}$$

The separation (5.7) induces within the MCA a similar separation for Σ and D [cf. Eqs. (5.3) and (5.4)], namely, for close times,

$$D^{\mathrm{FDT}}_{ab}(t-t_w) = \lim_{t_w\to\infty} D_{ab}(t,t_w),$$

$$\Sigma^{\mathrm{FDT}}_{ab}(t-t_w) = \lim_{t_w\to\infty} \Sigma_{ab}(t,t_w), \tag{5.11}$$

where FDT holds:

$$\Sigma^{\mathrm{FDT}}_{ab}(t-t_w) = \frac{1}{T}\frac{\partial D^{\mathrm{FDT}}_{ab}}{\partial t_w}(t-t_w). \tag{5.12}$$

Again, we can define

$$\lim_{t-t_w\to\infty}\lim_{t_w\to\infty} D_{ab}^{\rm FDT}(t-t_w)\equiv D_{ab}^{\rm EA}, \tag{5.13}$$

and mark with the tilde the aging part of the kernels:

$$D_{ab}(t,t_w)=\tilde{D}_{ab}(t,t_w), \quad \Sigma_{ab}(t,t_w)=\tilde{\Sigma}_{ab}(t,t_w). \tag{5.14}$$

In order to close the dynamical equations, we make an ansatz for the aging parts $\tilde{C}_{ab}(t,t_w)$ and $\tilde{R}_{ab}(t,t_w)$. For a problem in which the correlations vary only within two time scales, the natural generalization of the solution in [4] is

$$\tilde{C}_{ab}(t,t_w)=\tilde{C}_{ab}[h_{ab}(t_w)/h_{ab}(t)], \tag{5.15a}$$

$$\tilde{R}_{ab}(t,t_w)=\frac{X_a}{T}\frac{\partial\tilde{C}_{ab}}{\partial t_w}(t,t_w), \tag{5.15b}$$

where the X_{ab} are constants and the $h_{ab}(t)$ are functions to be determined from the dynamical equations. Note that the derivative in Eq. (5.15) is taken with respect to the earliest time.

Remarkably, it turns out that one can close the equations with two different types of ansatz for the long-time aging behavior. In Appendix E we show how this is done. In terms of the effective temperatures the meaning of these two possibilities is the following.

(1) Thermalized aging regime. The effective temperatures associated with the observables O_1, O_2 are equal to each other for frequencies and waiting times in the aging regime: they are not necessarily equal to the temperature of the bath. At higher frequencies, they both coincide with the one of the bath:

$$\begin{aligned}&T_1(\omega,t_w)=T_2(\omega,t_w)=T, \quad t_w\to\infty, \quad C_{ab}>C_{ab}^{\rm EA}, \quad \text{quasiequilibrium},\\ &T_1(\omega,t_w)=T_2(\omega,t_w)\neq T, \quad \omega\to 0, \quad t_w\to\infty, \quad C_{ab}<C_{ab}^{\rm EA}, \quad \text{aging}.\end{aligned} \tag{5.16}$$

Not surprisingly, in this case we find that O_1 and O_2 are strongly coupled (*also in the aging regime*) in the sense that the mutual responses

$$\tilde{R}_{12}(t,t_w)=\frac{X}{T}\frac{\partial\tilde{C}_{12}}{\partial t_w}, \quad \tilde{R}_{21}(t,t_w)=\frac{X}{T}\frac{\partial\tilde{C}_{21}}{\partial t_w}, \tag{5.17}$$

where $X>0$, are of the same order of the self response functions.

(2) Unthermalized aging regime. The effective temperatures associated with the observables O_1, O_2 for combinations of frequencies and waiting times corresponding to the aging regime are neither equal to each other nor to that of the bath, while for higher frequencies they both coincide with the one of the bath.

$$\begin{aligned}&T_1(\omega,t_w)=T_2(\omega,t_w)=T, \quad t_w\to\infty, \quad C_{ab}>C_{ab}^{\rm EA}, \quad \text{quasieq.},\\ &T_1(\omega,t_w)\neq T_2(\omega,t_w)\neq T, \quad \omega\to 0, \quad t_w\to\infty, \quad C_{ab}<C_{ab}^{\rm EA}, \quad \text{aging}.\end{aligned} \tag{5.18}$$

In this case, O_1 and O_2 are effectively uncoupled (*in the aging regime*), in the sense that

$$\tilde{R}_{12}(t,t_w)=\frac{X_{12}}{T}\frac{\partial\tilde{C}_{12}}{\partial t_w}, \quad \tilde{R}_{21}(t,t_w)=\frac{X_{21}}{T}\frac{\partial\tilde{C}_{21}}{\partial t_w}, \tag{5.19}$$

where

$$X_{12}, X_{21}\to 0. \tag{5.20}$$

We have not found any other way of closing the equations [31,32]. We shall show below, in a particular case which can be numerically solved, that indeed either case (1) or case (2) take place. Let us remark that a solution with $X_{12}=X_{21}\neq 0$, $X_{11}=X_{22}=X\neq 0$, and $X\neq X_{12}$ is not compatible with the interpretation of X as an inverse temperature. We do not find such a solution in our test models and believe that it is not realizable in general.

The considerations of this section can also be made in the limit of small stirring.

A. An explicit model with thermalization

In order to test in a particular example that these asymptotic solutions are not only consistent, but are in fact reached, we solve numerically the mode coupling equations with two coupled modes, with F given by

$$F(\mathbf{q})=q_{11}^p+K^2 q_{22}^p. \tag{5.21}$$

We impose normalization at equal times of the autocorrelation of both modes:

$$C_{11}(t,t)=C_{22}(t,t)=1. \tag{5.22}$$

As usual with the mode-coupling equations [(33)–(36) and (29)], one can find a disordered mean field model whose dynamics is exactly given by these equations. This is the model defined in Appendix A.

Figures 7 and 8 show the numerical solution of the exact system of coupled integrodifferential equations whose large-time asymptotic can be analytically obtained as in Appendix E. In Figs. 7 and 8 we plot $\chi(C)$ for two cases. In Fig. 7 we consider two uncoupled systems that evolve from the initial

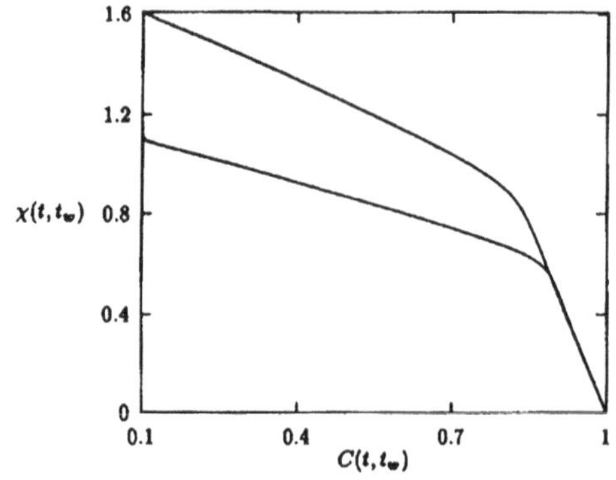

FIG. 7. The susceptibility $\chi_{11}(t,t_w)$ and $\chi_{22}(t,t_w)$ vs the corresponding autocorrelation functions $C_{11}(t,t_w)$ and $C_{22}(t,t_w)$ for the *uncoupled*, aging systems. The slopes of the curves, i.e., the FDT violation factors and hence the effective temperatures are different. This corresponds to the *unthermalized* case.

condition $C_{11}(0,0)=C_{22}(0,0)=1$ and $C_{12}(0,0)=C_{21}(0,0)=0$. We see that the systems evolve independently and $X_{11}\neq X_{22}$ while $X_{12}=X_{21}=0$ (unthermalized case). In Fig. 8 we consider the same global system but now including a weak coupling between the two individual systems. Clearly, after a short transient associated to short times, all curves $\chi_{ij}(C_{ij})$ get parallel. The systems aging to regime temperatures for the two subsystems have become equal (thermalized case).

It is interesting to note that, in this example, Onsager's reciprocity relations hold: $R_{12}=R_{21}$, $C_{12}=C_{21}$. However, one can imagine situations in which they do not hold separately, but where however $X_{12}=X_{21}$.

For comparison, we show in Fig. 9 the corresponding plot for two weakly coupled systems in the limit of small stirring.

B. Many scales

Let us briefly discuss what would happen in the presence of many scales, each one with its own temperature.

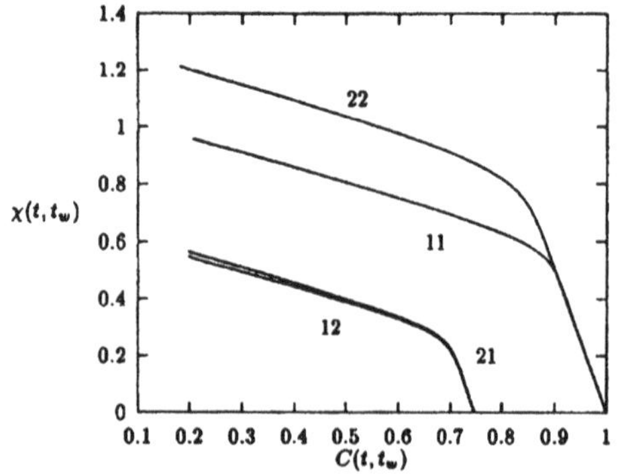

FIG. 8. The susceptibilities $\chi_{ij}(t,t_w)$ vs the correlation functions $C_{ij}(t,t_w)$ for the two aging systems of Fig. 2, this time *weakly* coupled. The FDT violation factors $X_{ij}(C_{ij})$ are almost parallel: thermalization is almost complete.

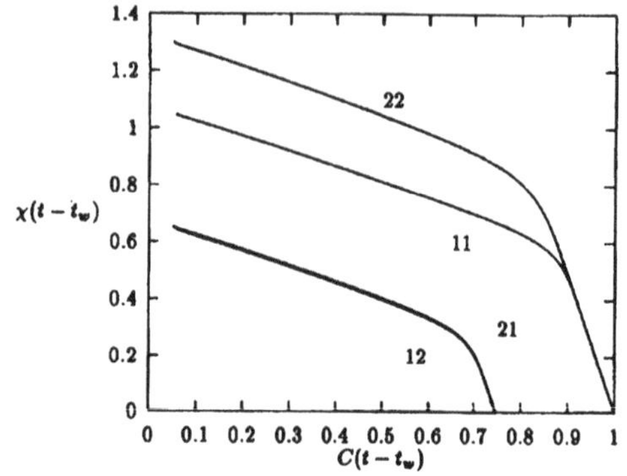

FIG. 9. The susceptibilities $\chi_{ij}(t-t_w)$ vs the correlation functions $C_{ij}(t-t_w)$ for two slightly driven systems ($D=0.1$), *weakly* coupled. The FDT violation factors $X_{ij}(C_{ij})$ given by the slopes of these curves are now the same: after a short transient corresponding to finite times all the curves become parallel. The aging regimes have *thermalized*.

Consider, for instance, two observables O_1 and O_2, their associated autocorrelation functions $C_{11}(t,t_w)$, $C_{22}(t,t_w)$, their integrated self-responses $\chi_{11}(t,t_w)$, $\chi_{22}(t,t_w)$ and the effective temperatures $T_1(C_{11})$, $T_2(C_{22})$. We can plot $T_1(C_{11})$ vs C_{11} and $T_2(C_{22})$ vs C_{22}. These two plots need not be the same, even if all scales are thermalized. Consider now a parametric plot of $C_{11}(t,t_w)$ vs $C_{22}(t,t_w)$, in the limit of very large t_w: it defines a function $C_{11}=\mathcal{H}(C_{22})$ that allows one to calculate C_{11} for large t,t_w, given $C_{22}(t,t_w)$.

The condition for thermalization in every time scale is then that the curve $T_2(C_{22})$ coincides with the curve $T_1(\mathcal{H}(C_{22}))$, both considered as functions of C_{22}:

$$T_1(\mathcal{H}(C_{22}))=T_2(C_{22}). \qquad (5.23)$$

The deviation of $T_1(\mathcal{H}(C_{22}))$ from $T_2(C_{22})$ is a measure of the degree in which the two observables are not thermalized.

Actually, Eq. (5.23) was obtained as an ansatz for the dynamics of a manifold in a random medium within the Hartree approximation [37], where the role of different observables is played by the displacements at different spatial wavelengths:

$$C_k(t,t')=\mathcal{H}_k(C_{k=0}),$$

$$T_k(\mathcal{H}_k(C_{k=0}))=T_{k=0}(C_{k=0}). \qquad (5.24)$$

In this case, the modes at different k thermalize at the same effective temperature, although their mutual responses vanish, as implied by translational symmetry in mean.

VI. COMPARISON WITH OTHER EFFECTIVE TEMPERATURES

The idea of "fictive temperatures" [38] T_f in glasses goes back to the 1940s and it has developed since [39–41]. Here we recall it briefly, for the sake of comparison with the effective temperature that we have discussed.

When cooling a liquid the time needed to establish equilibrium grows and, eventually, the structural change cannot keep pace with the rate of cooling: the system falls out of equilibrium and enters the glass transition region. It is then said that "the structure is frozen" at a temperature characterized by a fictive temperature T_f. The fictive temperature defined in terms of different quantities of interest, e.g., the enthalpy, the thermal expansion coefficient, etc., do not necessarily coincide. Furthermore, it has been experimentally observed that glasses with the same fictive temperatures arrived at through different preparation paths may have different molecular structures. The fictive temperature is hence a phenomenological convenience and should not be associated with a definite molecular structure [41].

The fictive temperature is a function of the temperature of the bath T. At high temperature, when the sample is in the liquid phase, $T_f = T$. When the liquid enters the transition range T_f departs from T and $T_f > T$; and, finally, deep below the transition range, where the relaxation is fully stopped, $T_f \to T_g$. The detailed bath-temperature dependence of the fictive temperature in the region of interest is usually expressed by [40,41]

$$\dot{T}_f = -\frac{T_f - T}{\tau(T_f, -T)}, \tag{6.1}$$

where the characteristic time $\tau(T_f, T)$ depends both upon $T_f(t)$ and upon the thermal history of the sample given by $T(t)$. At equilibrium $T_f = T$ and $\dot{T}_f = 0$. This nonlinear differential equation determines $T_f(t)$ once one *chooses* $\tau(T_f, T)$. A commonly used expression is the Narayanaswamy-Moynihan equation

$$\tau(T_f, T) = \tau_0 \exp\left[\frac{xA}{T} + \frac{(1-x)A}{T_f}\right], \tag{6.2}$$

where τ_0, A and $x \in [0,1]$ are some constants. All the information about the dynamics of the system enters into T_f through these constants.

In order to obtain the time relaxation of the quantity of interest, the picture is completed by proposing a given relaxation function, like the stretched exponential, the Davidson-Cole function, etc., and by introducing the fictive temperature through the characteristic time $\tau(T_f, T)$. See [41] for an extensive discussion about the applications of T_f to the description of experimental data.

One may wonder whether is a relation between this *fictive* temperature of glass phenomenology and the *effective* temperature we have been discussing in this paper.

First of all, one notes that the fictive temperatures are defined through the relaxation of the observables, unlike the one we consider here which are defined rather in terms of fluctuations and responses. Both temperatures may depend upon the observable and upon the thermal history of the sample.

The dependence of the effective temperature defined in Eq. (3) upon T depends on the model. In all cases FDT holds in the high-temperature phase and $T_O(\omega, t_w) = T$, the effective temperature is equal to the bath temperature. When entering the low-temperature phase, the temperature dependence of the effective temperature observed at fixed low frequency depends on the model. In certain simple mean-field (or low-temperature mode-coupling) models whose dynamics following a quench into the glass phase can be solved, one can compute $T_O(\omega, t_w)$ explicitly. Three different behaviors are found.

(1) In the simple model we have been using as a test example in the previous sections [4], $T_O(\omega, t_w)$ starts from $T_O(\omega, t_w) = T_g$ at $T = T_g$ and slightly *increases* when the bath temperature decreases below T_g. One would instead expect a fictive temperature to remain stuck to T_g in a (mean-field) model in which the glass transition is sharp.

(2) There are other mean-field models such as the Sherrington-Kirkpatrick spin-glass model in which there are infinitely many different effective temperatures. The lowest aging-regime temperature appears discontinuously $[T_O(\omega, t_w) > T_g]$ as one crosses T_g and can be shown [5] to *decrease* with decreasing temperature. Another example of this kind is the model of a particle moving in a random potential with long-range correlations [20,25].

(3) In all cases we know of domain growth models [19] one obtains $X(C) \to 0$ for $t_w \to \infty$, in the whole low-temperature phase. Hence $T_O(\omega, t_w) \to \infty$ in the aging (coarsening) regime for all heat bath temperatures below the ordering transition. This behavior also holds for certain extremely simple disordered systems such as the spherical Sherrington-Kirkpatrick spin glass [42], the toy domain growth model we used in this paper.

It is important to remark that these simple examples have the (sometimes unrealistic) feature that nothing depends permanently upon the cooling procedure. One expects, however, that more refined models that go beyond the mean-field approximation will capture a cooling rate dependence that will also become manifest in the effective temperature.

Another attempt to identify and relate a microscopic effective temperature to the fictive temperature of glasses was put forward by Baschnagel, Binder, and Wittmann [43] in the context of a lattice model for polymer melts. They have pointed out that in this model the usual FDT relation between the specific heat with the energy fluctuations is broken at low temperatures and have tried to identify the FDT violation factor with an expression they propose for an *internal temperature*. The internal temperature defined in [43] has, however, the unpleasant property of not reducing to the bath temperature in the high temperature phase.

In [44] Franz and Ritort have also discussed the possibility of relating the FDT ratio with an effective temperature, in the particular case of the Backgammonn model. They have compared the value of $X(t, t_w)$ for finite times t, t_w with the temperature arising from an adiabatic approximation they used to solve the model. The result is negative, in the sense that $T/X(t, t_w)$ does not coincide with the "adiabatic temperature."

These examples suggest that the phenomenological fictive temperatures act essentially as parameters for describing an out-of-equilibrium "equation of state" while the effective temperature we have discussed plays a role closer to the thermodynamical one.

VII. DISCUSSION, EXPERIMENTAL PERSPECTIVES, AND CONCLUSIONS

Although our discussion has been biased by the models we can solve at present, we feel that the concept of the ef-

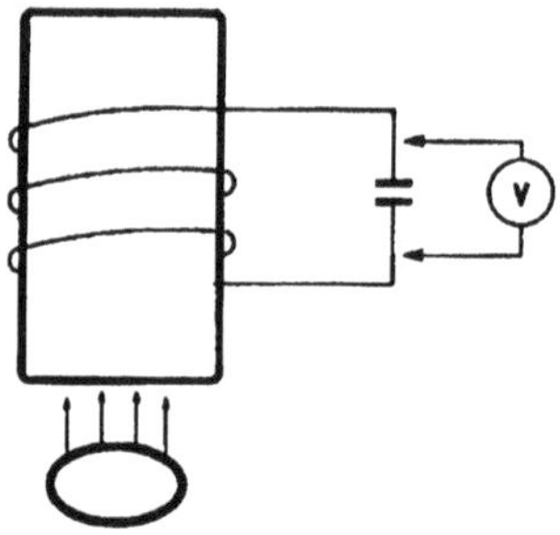

FIG. 10. An effective temperature measurement for a magnetic system. The coil is wound around the sample, which is in contact with the bath. The coil and capacitor have zero resistance.

fective temperature that we discussed should be relevant for many systems *with small energy flows*. Indeed, the key observation we make is that this temperature actually controls thermalization and heat flows within a time scale. Therefore, the effective temperature can be a starting point for the investigation of the thermology and, hopefully, the thermodynamics of out-of-equilibrium systems with small energy flows.

In order to use this idea as a guiding concept for the planning of experiments, one has to take some care: obviously, waiting times can be large but not infinite in experiments, and stirring rates can be small but not infinitesimal. Moreover, the models that we have explicitly discussed do not exhibit dependence on the cooling history of the sample. Now, one would expect in general the effective temperature to depend on t,t_w and, e.g., on the cooling rate. One can hope to catch this aspect in more refined models.

One should also pay attention to the equilibration times. The effective temperature measured by a small oscillator is related to the FDT violation factor provided that the equilibration time t_c [Eq. (3.10)] remains much smaller than t_w. It sometimes happens that an observable O which should thermalize at an effective temperature $T_O(\omega,t_w)$ exhibits so small fluctuations at this frequency that this time becomes unbearably long. This is the case, for instance, of high k modes in the aging time scale for a manifold in a random potential. Therefore, although all k modes nominally thermalize at the same effective temperature, only low enough k modes can be effectively used to measure it. How low k must be will depend on the time scale one is looking at.

Several numerical and real experiments in structural glasses can be envisaged. For example, one can compare density fluctuations and compressibility in different length scales in order to check if they are equilibrated within a time scale. Since the low-temperature extension of the MCT makes definite predictions on the value of the first nontrivial effective temperature appearing as one crosses the transition (see Sec. VI), this provides a concrete ground for an experimental testing of MCT.

We close by considering the following (slightly *Gedanken*) spin-glass experiment of Fig. 10: Currents are induced in the coil by the magnetization noise of the spin glass, which is in contact with a heat bath. Apart from the interaction with the sample, the L-C circuit of coil and capacitor is without losses. This is exactly a realization of the oscillator as a thermometer of Sec. II. From what we know from the time scales of real spin glasses [16,18], if the time after the quench is of the order of 10 min, and the period of the L-C circuit is of the order of the second, we are probing (at least partially) the aging regime: the temperature (defined as the average energy of the capacitor) should be different from the bath temperature. We believe that it would be interesting to return to the magnetization noise experiments [17] with the purpose of measuring the effective temperature: this would give us, for instance, useful insights into the nature of the spin-glass transition.

ACKNOWLEDGMENTS

We are especially indebted to Andrea Baldassarri and Gilles Tarjus since this work initiated from discussions with them. We also thank Gilles Tarjus for introducing us to the old ideas of "fictive temperatures" used in glass phenomenology. We acknowledge useful discussions with S. Ciliberto, D. Dean, P. Le Doussal, M. Mézard, R. Monasson, and P. Viot. L.P. fondly remembers discussions with the late S.-K. Ma, who impressed upon him the concept expressed by the first sentence of this paper, and acknowledges the support of a Chaire Joliot de l'ESPCI and of INFN. L.F.C. is thankful for the hospitality of the Service de Physique Théorique at Saclay, where this work was started, and the European Union for financial support through the Contract No. ERBCHRXCT920069. J. K. acknowledges the support of CNRS.

APPENDIX A: THE TEST MODEL

Throughout this paper we have used as a test model a spherical disordered model with p-spin interactions. The model has been introduced in its purely relaxational version by Crisanti and Sommers [45] as a simple spin-glass model with several advantages, in particular, that exact dynamical equations can be written for it in the thermodynamic limit. As shown by Franz and Hertz [36], these dynamical equations are also those obtained from the MCA to the Amit-Roginsky model [34]. When considered in full generality, the two-time dynamical equations correspond to the low temperature extension [36,29] of the simplest mode coupling theory for the supercooled liquid phase proposed by Leutheusser [46] and Bengtzelius, Gotze, and Sjölander [47,30]. A thorough discussion of the physical principles underlying mode-coupling theory is found in [48], Chap. 9. It has been also recently shown by Chandra *et al.* that this model is related to a mean-field approach to Josephson junction arrays [49].

We consider a system of N variables $\mathbf{s}=(s_1\ldots,s_N)$, subject to forces F_i^J given by

$$F_i^J(\mathbf{s})=\sum_{\{j_1,\ldots,j_{p-1}\}} J_i^{j_1\cdots j_{p-1}} s_{j_1}\cdots s_{j_{p-1}}, \tag{A1}$$

where the couplings are random Gaussian variables. For different sets of indices $\{i,j_1,\ldots,j_{p-1}\}$ the J's are uncorrelated, while for permutations of the same set of indices they are correlated so that

$$\overline{F_i^J(\mathbf{s}')F_j^J(\mathbf{s})}=\delta_{ij}f_1(q)+s_is_j'f_2(q)/N, \tag{A2}$$

where $q=(\mathbf{s}\cdot\mathbf{s}')/N$. In the purely relaxational case, one has $f_2(q)=f_1'(q)$. We take here $f_2(q)=(1-D)f_1'(q)$, where

$f_1(q)=pq^{p-1}/2$. The couplings $J_i^{j_1\cdots j_{p-1}}$ are symmetric under the permutation $i\leftrightarrow j_k$ in the purely relaxational case ($D=0$). On the other hand, if $J_i^{jj_2\cdots j_{p-1}}$ and $J_j^{ij_2\cdots j_{p-1}}$ are uncorrelated, one has $D=1$.

The dynamics is of the Langevin type:

$$\dot{s}_i=-F_i^J(\mathbf{s})-\mu(t)s_i+h_i(t)+\eta_i(t), \tag{A3}$$

where η is a white noise of variance $2T$, $\mu(t)$ is a Lagrange multiplier enforcing the spherical constraint $\Sigma_{i=1}^N s_i^2=1$, and $h_i(t)$ is an external field (usually set to zero), needed to define the response functions.

In Figs. 1 and 2 we plot the $\chi(C)$ curves for the asymmetrical and symmetrical $p=3$ model, respectively. In Figs. 3 and 4 we plot the autocorrelation decays for the symmetrical and asymmetrical versions, respectively. Figures 5 and 6 show the $\chi(C)$ curves for the $p=2$ version that is analogous to the $O(N)$ model in $D=3$ when $N\rightarrow\infty$.

In order to check thermalization in this particular example we consider the evolution of two such systems with spins s and σ, and uncorrelated realizations of disorder J and J' and thermal noise η and η', respectively. They are coupled via the term proportional to μ_{12} and μ_{21} in the Langevin equations

$$\frac{\partial s_i}{\partial t}=-F_i^J(s)-\mu_{11}(t)s_i-\mu_{12}\sigma_i+h_i+\eta_i, \tag{A4a}$$

$$\frac{\partial \sigma_i}{\partial t}=-KF_i^{J'}(\sigma)-\mu_{22}(t)\sigma_i-\mu_{21}s_i+h_i'+\eta_i'. \tag{A4b}$$

We set $\mu_{12}=\mu_{21}$, so that the coupling does not contribute to the stirring. The coefficients μ_{ii} are the Lagrange multipliers for each system. The factor K is introduced to break the symmetry between the subsystems. The correlations

$$C_{11}(t,t_w)=\frac{1}{N}\sum_{i=1}^N \langle s_i(t)s_i(t_w)\rangle,$$

$$C_{22}(t,t_w)=\frac{1}{N}\sum_{i=1}^N \langle \sigma_i(t)\sigma_i(t_w)\rangle,$$

$$C_{12}(t,t_w)=\frac{1}{N}\sum_{i=1}^N \langle s_i(t)\sigma_i(t_w)\rangle,$$

$$C_{21}(t,t_w)=\frac{1}{N}\sum_{i=1}^N \langle s_i(t)\sigma_i(t_w)\rangle \tag{A5}$$

and responses

$$R_{11}(t,t_w)=\frac{1}{N}\sum_{i=1}^N \delta\langle s_i(t)\rangle/\delta h_i^s(t_w),$$

$$R_{22}(t,t_w)=\frac{1}{N}\sum_{i=1}^N \delta\langle \sigma_i(t)\rangle/\delta h_i^\sigma(t_w),$$

$$R_{12}(t,t_w)=\frac{1}{N}\sum_{i=1}^N \delta\langle s_i(t)\rangle/\delta h_i^\sigma(t_w),$$

$$R_{21}(t,t_w)=\frac{1}{N}\sum_{i=1}^N \delta\langle s_i(t)\rangle/\delta h_i^s(t_w) \tag{A6}$$

precisely satisfy Eqs. (5.1a)–(5.4) with F given by Eq. (5.21).

The results for the effective temperatures obtained from the numerical integration of the exact evolution equations of these systems are shown in Figs. 7 and 8. See the main text for the discussion.

APPENDIX B: ENERGY OF THE OSCILLATOR

In this appendix we solve Eq. (3.8), we compute $\frac{1}{2}\omega_0^2\langle x^2(t)\rangle$, the average potential energy of the oscillator. We thus prove Eq. (3.14) and its form Eq. (3.13) valid for the stationary case.

Let us define

$$\chi(\omega,t)\exp(i\omega t)\equiv\int_0^t dt' R(t,t')\exp(i\omega t'), \tag{B1a}$$

$$C(\omega,t)\exp(i\omega t)\equiv\int_0^t dt' C(t,t')\exp(i\omega t'). \tag{B1b}$$

If $\omega^{-1}\ll t$, we can assume that $\chi(\omega,t)$ and $C(\omega,t)$ are functions that vary slowly with t, thus defining a Fourier component that is "local" in time t.

In general

$$\langle x^2(t)\rangle=a^2\int_0^t dt'\int_0^t dt'' G(t,t')G(t,t'')C(t',t''), \tag{B2}$$

where $G(t,t')$ is the Green's function for the oscillator plus the term representing the response of the system and $G(t,t')$ is the system's auto-correlation function. Using the definition in Eq. (B1a) one can show that the damped oscillator's Green's function reads

$$G(\omega,t)=\frac{1}{-\omega^2+\omega_0^2-a^2\chi''(\omega,t)}, \tag{B3}$$

$$G(t,t'')=\exp\left(-\frac{t-t'}{t_c(t)}\right)\sin[\omega_0(t-t')]\theta(t-t'). \tag{B4}$$

We have here replaced $\chi(t,\omega)$ by $\chi''(t,\omega)$ using the fact that $a^2N\ll 1$. The characteristic time $t_c(t)$ of the damped oscillator is given by

$$t_c(t)=\frac{2\omega_0}{a^2\chi''(\omega_0 t)}. \tag{B5}$$

We can now study Eq. (B2) by using the above expressions for $G(t,t')$. After a simple change of variables, using causality and the fact that $G(t,t')$ decays exponentially as a function of time differences, Eq. (B2) can be rewritten as

$$\langle x^2(t)\rangle = a^2 \int_{-\infty}^{\infty} \frac{d\omega}{2\pi} \int_{-\infty}^{\infty} d\tau \int_{-\infty}^{\infty} d\tau' G(t,t-\tau) G(t,t-\tau') C(\omega,t-\tau) \exp[i\omega(\tau-\tau')]. \tag{B6}$$

The fast exponential decay of the Green's function allows us to replace $C(\omega,t-\tau)$ by $C(\omega,t)$. Thus,

$$\langle x^2(t)\rangle = a^2 \int_{\infty}^{\infty} \frac{d\omega}{2\pi} G(\omega,t) G(-\omega,t) C(\omega,t)$$

$$= a^2 \int_{\infty}^{\infty} \frac{d\omega}{2\pi} \frac{C(\omega,t)}{\chi''(\omega,t)-\chi''(-\omega,t)} \left[\frac{1}{\omega_0^2-\omega^2-a^2N\chi''(-\omega,t)} - \frac{1}{\omega_0^2-\omega^2-a^2N\chi''(\omega,t)} \right]. \tag{B7}$$

This integral can be calculated by the method of residues. We can close the circuit on the upper complex half plane. Since $\chi''(\omega,t)$ is analytic in the upper half plane, the only singularities are the zeroes of the denominators. Assuming that $a^2\chi''$ is small, they lie in the vicinity of $\omega=\pm\omega_0$. In fact one can check that only two poles penetrate inside the circuit of integration. We obtain therefore

$$\langle x^2(t)\rangle = \frac{2\tilde{C}(\omega_0,t)}{\omega_0 \chi''(\omega_0,t)}. \tag{B8}$$

This is the general result for the temperature. The particular result (3.13) that holds for the stationary case is recovered from Eq. (B8) by letting $\tilde{C}(t,\omega)$ and $\chi''(t,\omega)$ be independent of t. Thus, Eq. (B8) reduces to Eq. (3.13).

APPENDIX C: SMALL BUT MACROSCOPIC THERMOMETERS

In this appendix we show that the thermometric considerations made in Sec. III do not crucially depend on the choice of an oscillator as a thermometer. We use here a small but *macroscopic* thermometer defined by the variables y_i, $i=1,\ldots,n$, and we couple it only to the observable $O(s)$ of the system through a degree of freedom $x(y)$.

Our measurement procedure is as follows: we first thermalize the thermometer with an auxiliary bath at temperature T^*. We then disconnect it from the bath and we connect it to the system through O. If there is no flow of energy between thermometer and system, then we conclude that the measured temperature is T^*.

The energy of the thermometer plus its coupling with the system is

$$H = E(y) - aO(s)x(y). \tag{C1}$$

The net power gain of the thermometer is then $\dot{Q}(t)$, given by

$$\dot{Q} = a\langle \dot{O}x\rangle = a\partial_{t'}\langle x(t)O(t')\rangle|_{t'\to t^-}. \tag{C2}$$

We look for the condition that ensures stationarity for the thermometer. The thermometer is characterized by a temperature-dependent correlation $C_x(t,t')=\langle x(t)x(t')\rangle$ and its associated response $R_x(t,t')$. Using linear response one has

$$O(t) = O_b(t) + a\int_0^t dt' R_O(t,t')x(t'), \tag{C3a}$$

$$x(t) = x_b(t) + a\int_0^t dt' R_x(t,t')O(t'). \tag{C3b}$$

To leading order in a, $\langle Ox\rangle$ is given by

$$\langle Ox\rangle = a\int_0^t dt''[R_x(t,t'')C_O(t'',t') + R_O(t',t'')C_x(t,t'')]. \tag{C4}$$

Assuming now that T^* is such that the thermometer can be considered to be almost in equilibrium (note that the coupling a is small and that we have chosen a thermometer that is not itself a glass), we obtain

$$\dot{Q} = a^2\int_0^t dt' R_x(t-t')\left(\frac{\partial C_O(t,t')}{\partial t'} - T^* R_O(t,t') \right). \tag{C5}$$

The condition for having no flow is then that the average of the parenthesis in the integral is zero. The weight function for this average is $R_x(t-t')$ which contains the characteristic time of the thermometer.

APPENDIX D: DEFINITION OF TIME CORRELATION SCALES

In this appendix we review briefly the definition of correlation "time scale" introduced in [5] for a correlation function that depends nontrivially upon two times.

Given a correlation function $C(t,t_w)$, which we assume normalizable in the large-time limit $C(t,t)\to C_\infty>0$, we consider three increasing times $t_1<t_2<t_3$, and the limit in which they all go to infinity, but in a way to keep $C(t_3,t_2)=a$ and $C(t_2,t_1)=b$ const. We thus define the limit

$$\lim_{\substack{t_1,t_2,t_3\to\infty \\ C(t_3,t_2)=a,\, C(t_2,t_1)=b}} C(t_3,t_1) \equiv f(a,b). \tag{D1}$$

The mere existence of the limit "triangle relation" f has extremely strong consequences: considering four times one can easily show that f is associative

$$f(a,f(b,c)) = f(f(a,b),c). \tag{D2}$$

The form of an associative function on the reals is very restricted, and a classification of all possible forms can be made [5].

It is sometimes convenient to work with the "inverse" $\bar{f}$ of f defined as

$$f(a,b)=c \Rightarrow \bar{f}(a,c)=b. \tag{D3}$$

We can now define a correlation scale in the following way: given two values of the correlation at large times $C(t,t_w)=C^1$ and $C(t,t_w')=C^2$, $t>t_w'>t_w$ and $C^1<C^2$, they are in a different correlation scale if

$$\bar{f}(C^2,C^1)=C^1, \tag{D4}$$

and are in the same scale otherwise. In other words, the time it takes the system to achieve C^2 is negligible with respect to the time it takes to achieve C^1.

APPENDIX E: SOLUTION OF THE TWO-COUPLED-MODE EQUATION

Using the separation (5.7)–(5.14) and the mode-coupling approximation (5.3) and (5.4) we obtain a similar separation for D_{ab} and Σ_{ab}:

$$D_{ab}^{\rm FDT}(t-t_w)=F_{ab}(\mathbf{C}^{\rm FDT}(t-t_w)), \tag{E1a}$$

$$\Sigma_{ab}^{\rm FDT}(t-t_w)=\sum_{c,d} F_{ab,cd}(\mathbf{C}^{\rm FDT}(t-t_w))R_{cd}^{\rm FDT}(t-t_w), \tag{E1b}$$

where $\mathbf{C}^{\rm FDT}$ stands for the set $C_{ab}^{\rm FDT}(t-t_w)$. One similarly obtains, in the aging regime,

$$\tilde{D}_{ab}(t,t_w)=F_{ab}(\tilde{\mathbf{C}}(t,t_w)), \tag{E2a}$$

$$\tilde{\Sigma}_{ab}(t,t_w)=\sum_{c,d} F_{ab,cd}(\tilde{\mathbf{C}}(t,t_w))\tilde{R}_{cd}(t,t_w). \tag{E2b}$$

We can now write two coupled sets of equations, valid in the quasiequilibrium regime and in the aging regime, respectively. For $t-t_w$ finite, and large times $t>t_w$, we have

$$\frac{\partial C_{ab}^{\rm FDT}(t-t_w)}{\partial t}=-\sum_c\left[\mu_{ac}^{\infty}+\frac{D_{ac}^{\rm FDT}(0)}{T}\right]C_{cb}^{\rm FDT}(t-t_w)-\frac{1}{T}\sum_c D_{ac}^{\rm EA}C_{cb}^{\rm EA}+M_{ab}^{\infty}+\frac{1}{T}\sum_c\int_{t_w}^{t}dt''D_{ac}^{\rm FDT}(t-t'')\frac{\partial C_{cb}^{\rm FDT}(t''-t_w)}{\partial t_w}, \tag{E3}$$

where $\mu_{ac}^{\infty}\equiv\lim_{t\to\infty}\mu_{ac}(t)$ and

$$M_{ab}^{\infty}\equiv\sum_c\lim_{t\to\infty}\int_0^t dt''[\tilde{D}_{ac}(t,t'')\tilde{R}_{cb}(t,t'')+\tilde{\Sigma}_{ac}(t,t'')\tilde{C}_{cb}(t,t'')]. \tag{E4}$$

In the aging regime, for $t>t_w$ we have

$$\frac{\partial\tilde{C}_{ab}(t,t_w)}{\partial t}=-\sum_c\left[\mu_{ac}(t)+\frac{D_{ac}^{\rm FDT}(0)-D_{ac}^{\rm EA}}{T}\right]\tilde{C}_{cb}(t,t_w)+\sum_c\tilde{D}_{ac}(t,t_w)\frac{C_{cb}^{\rm FDT}(0)-C_{cb}^{\rm EA}}{T}+\sum_c\int_0^{t_w}dt''\tilde{D}_{ac}(t,t'')\tilde{R}_{cb}(t_w,t'')+\sum_c\int_0^{t}dt''\tilde{\Sigma}_{ac}(t,t'')\tilde{C}_{cb}(t'',t_w), \tag{E5a}$$

$$\frac{\partial\tilde{R}_{ab}(t,t_w)}{\partial t}=-\sum_c\left[\mu_{ac}(t)+\frac{D_{ac}^{\rm FDT}(0)-D_{ac}^{\rm EA}}{T}\right]\tilde{R}_{cb}(t,t_w)+\sum_c\int_{t_w}^{t}dt''\tilde{\Sigma}_{ac}(t,t'')\tilde{R}_{cb}(t'',t_w)+\sum_c\tilde{\Sigma}_{ac}(t,t_w)\frac{C_{cb}^{\rm FDT}(0)-C_{cb}^{\rm EA}}{T}. \tag{E5b}$$

Equation (E4), for given M_{ab}^{∞}, is very similar to the high-temperature mode-coupling equations [30], and can be solved in the same way. An asymptotic solution for the aging regime can be obtained by using the generalization to more than one mode of the ansatz in [4], Eq. (5.15). The derivative terms in Eqs. (E5) can be then dropped provided that $X_{11}\neq 0$ and $X_{22}\neq 0$, a fact to be verified *a posteriori*. We shall find in this way two different solutions for Eqs. (E5).

In the unthermalized case X_{11} and X_{22} are different from zero, and possibly different from each other, while $X_{12}=X_{21}=0$. It is then easy to see that Eqs. (E5) become effectively uncoupled *in this regime*, and can be solved [4] as two separated one-mode equations, with the ansatz

$$\tilde{C}_{aa}(t,t_w)=\tilde{\mathcal{C}}_{aa}(h_{aa}(t_w)/h_{aa}(t)), \tag{E6a}$$

$$\tilde{R}_{aa}(t,t_w)=\frac{X_{aa}}{T}\frac{\partial\tilde{C}_{aa}}{\partial t_w}(t,t_w), \tag{E6b}$$

$$\tilde{C}_{12}(t,t_w)=\tilde{C}_{21}(t,t_w)=0. \tag{E6c}$$

In the thermalized case, one assumes

$$X_{11}=X_{22}=X_{12}=X_{21}=X\neq 0, \tag{E7a}$$

$$h_{11}=h_{22}=h_{12}=h_{21}=h. \tag{E7b}$$

Making the change of variables

$$\lambda\equiv\frac{h(t_w)}{h(t)},\quad \lambda'\equiv\frac{h(t'')}{h(t)}, \tag{E8}$$

one finds that only the dependence on λ survives in the equations, and that they reduce to a set of four (instead of eight) consistent equations for the correlations (see [4,25] for a systematic approach).

In this way, both for the thermalized and the unthermalized case one can obtain X_{ab}, M_{ab}, and $G^{\rm EA}_{ab}$. One then has to check that $X_{11}\neq 0$ and $X_{22}\neq 0$. If this is not the case the equations become identities, and one cannot anymore neglect the derivative term. One has therefore to use a more refined long-time limit. These values have to be substituted in Eq. (E4), in order to complete the solution in both regimes.

Let us remark here that the problem of selecting the functions h_{ab} remains open. This is an asymptotic matching problem in a non-local equation, and does not appear to be easily solvable.

We have thus found the long-time limit of the correlations and responses. If there is more than one asymptotic solution (even modulo h), we do not know for the time being which asymptotic form is selected by the unique solution of the evolution equations, without resorting to explicit numerical integration.

In short we have the following.

(1) Unthermalized aging regime:

$$X_{12}\rightarrow 0,\quad X_{21}\rightarrow 0,\quad X_{11}\neq X_{22}. \tag{E9}$$

We have, therefore,

$$\tilde{R}_{11}(t,t_w)=\frac{X_{11}}{T}\frac{\partial\tilde{C}_{11}}{\partial t_w},\quad \tilde{R}_{22}(t,t_w)=\frac{X_{22}}{T}\frac{\partial\tilde{C}_{22}}{\partial t_w}, \tag{E10}$$

and

$$\tilde{\Sigma}_{11}(t,t_w)=\frac{X_{11}}{T}\frac{\partial\tilde{D}_{11}}{\partial t_w},\quad \tilde{\Sigma}_{22}(t,t_w)=\frac{X_{22}}{T}\frac{\partial\tilde{D}_{22}}{\partial t_w}. \tag{E11}$$

(2) Thermalized aging regime:

$$X_{11}=X_{12}=X_{21}=X_{22}\equiv X, \tag{E12}$$

$$\tilde{\Sigma}_{ab}(t,t_w)=\frac{X}{T}\frac{\partial\tilde{D}_{ab}}{\partial t_w},\quad \forall a,b, \tag{E13}$$

where the aging time scales "lock in," i.e., there is the *same* function $h(t)$ for all a, b, such that

$$\tilde{C}_{ab}(t,t_w)=\tilde{C}_{ab}(h(t_w)/h(t)). \tag{E14}$$

This property can also be stated by saying that as $t,t_w\rightarrow\infty$ a plot of $\tilde{C}_{11}(t,t_w)$ vs $\tilde{C}_{22}(t,t_w)$ yields a single smooth curve: i.e., that there is a function $\mathcal{H}(C)$ such that

$$\tilde{C}_{11}(t,t_w)=\mathcal{H}(\tilde{C}_{22})(t,t_w). \tag{E15}$$

[1] S. K. Ma, *Statistical Mechanics* (World Scientific, Singapore, 1985), Chap. 1.

[2] R. H. Kraichnan and S. Chen, Physica D **37**, 160 (1989).

[3] P. C. Hohenberg and B. I. Shraiman, Physica D **37**, 109 (1989); M. C. Cross and P. C. Hohenberg, Rev. Mod. Phys. **65**, 851 (1993); M. S. Bourzutschky and M. C. Cross, CHAOS **2**, 173 (1992); M. Caponeri and S. Ciliberto, Physica D **58**, 365 (1992).

[4] L. F. Cugliandolo and J. Kurchan, Phys. Rev. Lett. **71**, 173 (1993); Philos. Mag. B **71**, 50 (1995).

[5] L. F. Cugliandolo and J. Kurchan, J. Phys. A **27**, 5749 (1994).

[6] In the case of stirred systems, it sometimes happens that the stationary regime sets in a time that grows with decreasing D. In these cases, we shall take the observation time large enough to guarantee stationarity.

[7] L. F. Cugliandolo, J. Kurchan, P. Le Doussal, and L. Peliti, Phys. Rev. Lett. **77**, 350 (1997).

[8] To the best of our knowledge, the first to propose that violations of FDT where relevant for glassy dynamics was Sompolinsky [9], albeit within a framework that is now believed not to represent purely relaxational systems. In fact, Sompolinsky's "barrier crossing" dynamics is not an equilibrium dynamics (it yields a wrong value for the equilibrium correlation functions) and it is not an out-of-equilibrium dynamics either (it assumes TTI, which is only valid in equilibrium). The dynamics for times that are not very separated [10], however, can be well justified, and FDT holds in that regime of times. See Refs. [11–13] for a discussion.

[9] H. Sompolinksy, Phys. Rev. Lett. **47**, 935 (1981).

[10] H. Sompolinsky and A. Zippelius, Phys. Rev. B **25**, 6860 (1982).

[11] A. Houghton, S. Jain, and A. P. Young, J. Phys. C **16**, L375 (1983); Phys. Rev. B **28**, 290 (1983).

[12] K. Binder and P. Young, Rev. Mod. Phys. **58**, 801 (1986).

[13] M. Mézard, G. Parisi, and M. A. Virasoro, *Spin-Glass Theory and Beyond* (World Scientific, Singapore, 1987).

[14] A closely related situation is when the parameters describing the Hamiltonian are made to be varied slowly. This approach has been advocated by Horner [15] as a tool to "regularize" (i.e., render stationary) the dynamics of spin glasses, thus avoiding having to deal with waiting-time dependences.

[15] H. Horner, Z. Phys. B **57**, 29 (1984); **57**, 39 (1984).

[16] L. Lundgren, P. Svedlindh, P. Nordblad, and O. Beckman, Phys. Rev. Lett. **51**, 911 (1983); E. Vincent, J. Hammann, and M. Ocio, in *Recent Progress in Random Magnets*, edited by D. H. Ryan (World Scientific, Singapore, 1992), p. 207.

[17] M. Alba, J. Hammann, M. Ocio, Ph. Refregier, and H. Bouchiat, J. Appl. Phys. **61**, 3683 (1987); H. Bouchiat and M. Ocio, Comments Condens. Matter Phys. **14**, 163 (1988).

[18] E. Vincent, J. Hammann, M. Ocio, J.-P. Bouchaud, and L. F. Cugliandolo, (unpublished)].

[19] For a review, see, e.g., A. J. Bray, Adv. Phys. **43**, 357 (1994).
[20] S. Franz and M. Mézard, Europhys. Lett. **26**, 209 (1994); Physica A **209**, 1 (1994).
[21] H. Rieger, J. Phys. A **26**, L615 (1993); J. Phys. (France) I **4**, 883 (1994); Annu. Rev. Comput. Phys. **2**, 295 (1995).
[22] L. C. E. Struik, *Physical Aging in Amorphous Polymers and Other Materials* (Elsevier, Houston, 1978).
[23] A. Crisanti and H. Sompolinsky, Phys. Rev. A **36**, 4922 (1987).
[24] D. S. Fisher and D. Huse, Phys. Rev. B **38**, 373 (1988).
[25] L. F. Cugliandolo and P. Le Doussal, Phys. Rev. E **53**, 1525 (1996).
[26] G. Ferraro, Ph.D. thesis, University of Roma I, 1995 (unpublished).
[27] S. Franz and H. Rieger, J. Stat. Phys. **79**, 749 (1995).
[28] J.-P. Bouchaud and D. S. Dean, J. Phys. (France) I **5**, 265 (1995). We thank S. Franz for this remark.
[29] J.-P. Bouchaud, L. F. Cugliandolo, J. Kurchan, and M. Mézard, Physica A **226**, 243 (1996).
[30] For a review on mode-coupling theory see W. Götze, in *Liquids, Freezing and Glass Transition*, 1989 Les Houches Lectures, edited by J.-P. Hansen, D. Levesque, J. Zinn-Justin (North-Holland, Amsterdam, 1991); W. Götze, L. Sjögren, Rep. Prog. Phys. **55**, 241 (1992).
[31] An algebraically similar phenomenon occurs in the static treatment of two coupled systems with replicas, when an ansatz is made as in Ref. [32]. In that context, however, the meaning of the ansatz is still uncertain. We still lack a probabilistic interpretation in terms of pure states, as in the ordinary Parisi ansatz.
[32] S. Franz, G. Parisi, and M. A. Virasoro, J. Phys. (France) I **2**, 1869 (1992); J. Kurchan, G. Parisi, and M. A. Virasoro, *ibid.* **3**, 1819 (1993).
[33] R. Kraichnan, J. Fluid. Mech. **7**, 124 (1961).
[34] D. J. Amit and D. V. Roginsky, J. Phys. A **12**, 689 (1979).
[35] H.-J. Sommers and K. H. Fischer, Z. Phys. B **58**, 125 (1985); T. R. Kirkpatrick and D. Thirumalai, Phys. Rev. B **36**, 5388 (1987).
[36] S. Franz and J. Hertz, Phys. Rev. Lett. **74**, 2114 (1995).
[37] L. F. Cugliandolo, J. Kurchan, and P. Le Doussal, Phys. Rev. Lett. **76**, 2390 (1996).
[38] A. Q. Tool, J. Am. Ceram. Soc. **29**, 240 (1946).
[39] R. Gardon and O. S. Narayanaswamy, J. Am. Ceram. Soc. **53**, 380 (1970); O. S. Narayanaswamy, *ibid.* **54**, 491 (1971); C. T. Moynihan *et al.*, *ibid.* **59**, 12 (1976); G. W. Scherer, *ibid.* **67**, 504 (1984); I. M. Hodge, Macromolecules **20**, 2897 (1987).
[40] J. Jäckle, Rep. Prog. Phys. **49**, 171 (1986); G. W. Scherer, J. Non-Cryst. Solids **123**, 75 (1990).
[41] I. Hodge, J. Non-Cryst. Solids **169**, 211 (1994); Science **267**, 1945 (1996).
[42] J. M. Kosterlitz, D. J. Thouless, and R. C. Jones, Phys. Rev. Lett. **36**, 1217 (1976); P. Shukla and S. Singh, J. Phys. C **14**, L81 (1981); S. Ciuchi and F. De Pasquale, Nucl. Phys. B **300**, 31 (1988); L. F. Cugliandolo and D. S. Dean, J. Phys. A **28**, 4213 (1995); **28**, L453 (1995).
[43] J. Baschnagel, K. Binder, and H.-P. Wittmann, J. Phys. C **5**, 1597 (1993).
[44] S. Franz and F. Ritort (unpublished).
[45] A. Crisanti and H.-J. Sommers, Z. Phys. B **87**, 341 (1992); A. Crisanti, H. Horner, and H.-J. Sommers, *ibid.* **92**, 257 (1993).
[46] E. Leutheusser, Phys. Rev. A **29**, 2765 (1984).
[47] U. Bengtzelius, W. Götze, and A. Sjölander, J. Phys. C **17**, 5915 (1984); W. Götze and L. Sjögren, *ibid.* **17**, 5759 (1984).
[48] J.-P. Hansen and I. R. McDonald, *The Theory of Simple Liquids*, 2nd ed. (Academic, New York, 1986).
[49] P. Chandra, M. V. Feigel'man, M. E. Gershenson, and L. B. Ioffe (unpublished).

Authors' note

What exactly happens when two systems with different effective temperatures are coupled (in particular if the coupling is very weak) is discussed further in: Leticia F. Cugliandolo, Jorge Kurchan, Physica A263, 242 (1999).

The proof of the fact that FDT-temperature corresponds to a true temperature is made more properly in:

'A scenario for the dynamics in the small entropy production limit' Leticia F. Cugliandolo, Jorge Kurchan cond-mat/9911086, to appear in "Frontiers in Magnetism", special issue of the Journal of the Physical Society of Japan.

There are two further works dealing with granular matter that are a direct offspring of the discussion in Santa Barbara:

"Emergence of macroscopic temperatures in systems that are not thermodynamical microscopically: towards a thermodynamical description of slow granular rheology" J. Kurchan cond-mat/9909306.

"Edwards measures for powders and glasses" A. Barrat, J. Kurchan, V. Loreto, M. Sellitto cond - mat/0006140.

Frequency-domain study of physical aging in a simple liquid

Robert L. Leheny* and Sidney R. Nagel
The James Franck Institute and The Department of Physics, The University of Chicago, Chicago, Illinois 60637
(Received 8 September 1997)

We characterize the time dependence of the frequency-dependent dielectric susceptibility of glycerol following quenches to low temperature. For quenches to temperatures not far below the glass-transition temperature, the spectral shape and position during equilibration closely approximate those of equilibrium spectra at higher temperatures. For deeper quenches the correspondence with equilibrium spectra breaks down, and we observe more subtle behavior. Most of the susceptibility's change with time occurs within one *equilibrium* relaxation time after the quench, and our results do not support the presence of a significant ultraslow component to the equilibration of the dielectric response. We find some qualitative agreement between the time dependence of the susceptibility and analogous measurements on spin glasses; however, closer inspection reveals that the liquid fails to reproduce many of the salient features of the spin-glass behavior. [S0163-1829(98)05309-0]

I. INTRODUCTION

A general feature of systems with long relaxation times, such as a supercooled liquid in its glass transition region, is the protracted approach to equilibrium following a temperature (or pressure) change. During this equilibration, which is often termed physical aging, the liquid's bulk properties, such as its volume and enthalpy, vary as the system approaches equilibrium. In addition, the dynamics of the liquid, such as its mechanical or dielectric response, similarly evolve in time.

In an effort to characterize aging in a simple liquid, we have measured the change with time of the frequency-dependent dielectric susceptibility in supercooled glycerol following quenches through its glass-transition temperature, T_g. We have also performed dielectric studies of aging in three other simple liquids: propylene glycol, propylene carbonate, and tricresyl phosphate. (We use the term simple liquid to describe glycerol and these other liquids to distinguish them from polymeric systems. Glycerol is a strongly hydrogen-bonding molecular liquid.) Because our most extensive studies are on glycerol, the results we present here are for this liquid. Our measurements on the other liquids are consistent with these results. Extensive research on aging in polymers[1] has characterized the time dependence of response functions in these systems. However, the vast majority of these studies have been time-domain measurements, and our frequency-domain study on simple liquids complements this approach. (Although we are aware of one short account of a frequency-domain, dielectric study of aging in a polymer,[2] we know of no comprehensive examination of aging in the frequency domain on polymers or simple liquids.)

Recently, aging behavior has played an increasingly important role in efforts to understand the microscopic physics of disordered systems. Aging in spin glasses has been analyzed in an effort to distinguish among the theoretical pictures proposed for these systems.[3] Our measurements of the dielectric response in liquids provide a direct analogy with the magnetic-susceptibility aging studies in spin glasses. We can thus examine if theories proposed for spin glasses have relevance for liquids.

In the case of deeply supercooled liquids, the equilibrium linear-response dynamics is typically nonexponential, and several experiments[4] have indicated that this broadening results from dynamic heterogeneities. Do these heterogeneous dynamics imply spatially extended, structurally heterogeneous regions in the liquids? Miller and Macphail[5] have suggested that the presence of such extended regions could have consequences in the aging of supercooled liquids. One goal of our study has been to search for evidence in the aging of the dielectric response that might support the existence of extended domains.

This paper is organized as follows. Section II reviews the properties of the dielectric susceptibility of equilibrated supercooled liquids. Section III describes the experimental procedure. Section IV summarizes our results. Specifically, after quenches to temperatures not far below T_g, the shape and position of $\varepsilon''(\nu)$ at different times during aging mimic the equilibrium response at higher temperatures. However, at lower temperatures this scenario breaks down. We compare our results with the aging behavior in spin glasses and find that the aging in simple liquids possesses qualities distinct from those in the spin systems. Finally, in Sec. V we conclude.

II. EQUILIBRIUM SPECTRAL SHAPE

We have measured the dielectric susceptibility $\varepsilon(t_a,\nu)=\varepsilon'(t_a,\nu)+i\varepsilon''(t_a,\nu)$ as a function of frequency ν and aging time t_a. As Fig. 1 illustrates, the imaginary part of the spectrum for glycerol in equilibrium $\varepsilon''_{\mathrm{eq}}(\nu)\equiv\varepsilon''(t_a\rightarrow\infty,\nu)$ contains a broad, asymmetric peak characteristic of supercooled liquids. The frequency of the peak ν_p for glycerol follows a Vogel-Fulcher law

$$\log_{10}(\nu_p)=\log_{10}(\nu_0)-A/(T-T_0), \qquad (1)$$

with $T_0\approx 134$ K, $\nu_0\approx 10^{13}$ Hz, and $A\approx 870$ K. At T_g, $\nu_p\approx 10^{-3}$ Hz, which lies below our lowest measurement frequency of 4×10^{-2} Hz. Therefore, in the aging experiments we probe the evolution with time of the susceptibility above the peak.

A detailed study[6] of the equilibrium response for glycerol and several other liquids near T_g has revealed three power-

0163-1829/98/57(9)/5154(9)/$15.00

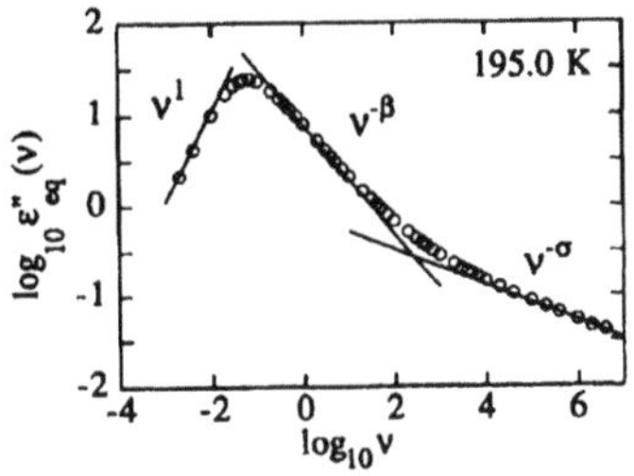

FIG. 1. The equilibrium dielectric spectrum for glycerol, $\varepsilon''_{eq}(\nu)\equiv\varepsilon''(t_a\to\infty,\nu)$, at 195.0 K. The spectrum contains two distinct power-law regions above ν_p.

law regions: one below ν_p and two above the peak, as indicated in Fig. 1. The exponent of the power law in the highest frequency region $-\sigma$ has a linear temperature dependence near T_g,

$$\sigma=B(T-T_\sigma), \tag{2}$$

where $T_\sigma\approx T_0$. For glycerol we find $B\approx0.0039$. The exponent of the power law just above the peak $-\beta$ also changes with temperature to maintain the relation

$$(\sigma+1)/(\beta+1)=0.72. \tag{3}$$

This temperature dependence of the spectral shape forms the basis of a recent argument that the static dielectric susceptibility in supercooled liquids diverges at T_0.[6,7] That argument assumes that σ approaches 0 at T_0, as predicted by Eq. (2). One interest in our aging studies has been to determine whether the aging-time dependence of the spectrum can provide information about the equilibrium behavior of σ in the temperature region where measurements on the fully equilibrated liquid are not feasible.

III. EXPERIMENTAL PROCEDURE

We obtained the glycerol (99.5%) from Aldrich Chemical Co. Because the liquid is hygroscopic, we handled it strictly in a dry nitrogen environment. To determine $\varepsilon(t_a,\nu)$, we measured the complex impedance of a capacitor filled with the liquid relative to its impedance when empty. The capacitor consisted of two gold covered parallel plates 2.5 cm in diameter and separated by a 0.025 cm-thick annular Teflon spacer, which held the sample between the plates. We placed the filled capacitor within a heating coil in a copper can, filled the can with a dry nitrogen gas, and submerged it in a liquid-nitrogen dewar. The temperature of the capacitor, measured with a platinum resistance temperature detector, was stable during experiments to better than ±0.02 K for up to 120 h (the hold time of the dewar). Temperature gradients across the sample were less than 0.05 K.

We measured $\varepsilon(t_a,\nu)$ from $\nu=4\times10^{-2}$ Hz to $\nu=10^6$ Hz. For measurements up to 10^4 Hz, we placed the reference output of a Stanford 850 digital lock-in amplifier across the capacitor in series with a Keithley 428 current amplifier, and obtained the complex impedance of the circuit by measuring the output of the current amplifier with the lock-in. Above 10^4 Hz, we measured the impedance using a Hewlett Packard 4275A LCR meter.

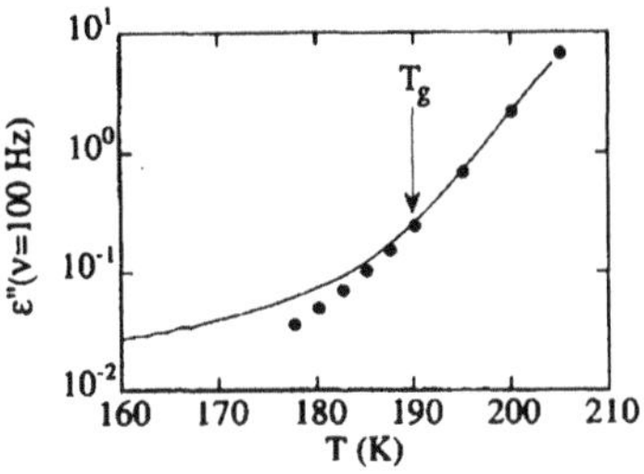

FIG. 2. ε''_{eq} ($\nu=100$ Hz) vs temperature (circles) compared with $\varepsilon''(t_a\approx0,\nu=100\text{ Hz})$ (solid line). The two measurements depart from one another below the glass-transition temperature, $T_g=190$ K.

We began each measurement by holding the liquid at 206.2 K for several hours to be sure it was well equilibrated there. (At this temperature, equilibrium relaxation times are less than 0.02 s.) We then rapidly cooled the liquid to a low temperature where we monitored the aging. The cooling rate during the quenches was 1.5 K/min. Figure 2 shows $\varepsilon''(t_a\approx0,\nu=100\text{ Hz})$ measured during cooling in comparison with the equilibrium values, $\varepsilon''(t_a\to\infty,\nu=100\text{ Hz})$. The value during cooling begins to diverge from the equilibrium value near 190 K. This divergence signifies that structural relaxation times in the liquid have grown larger than the experimental time scale set by the cooling rate and that the liquid has fallen from equilibrium. We equate this temperature with T_g.

During cooling the temperature typically undershot the set point by about 4 K and then steadily increased until the final temperature was reached. The total thermal equilibration took approximately 5000 s. In our analysis we include data taken only after the liquid reaches thermal equilibrium at the final temperature. Because the sample experienced this complicated thermal history before reaching thermal equilibrium, the correct origin for the aging time t_a has some uncertainty; we equate t_a with the time elapsed since the liquid passed through T_g. For most quenches, we monitored aging until $t_a\approx4\times10^5$ s (the hold time of the liquid nitrogen dewar). However, at some temperatures we carefully replenished the liquid nitrogen to track aging to $t_a\approx2.5\times10^6$ s.

IV. TIME DEPENDENCE OF SUSCEPTIBILITY FOLLOWING A QUENCH

We apply two complementary methods for analyzing $\varepsilon(t_a,\nu)$. (1) In Sec. IV A we compare the spectral shape and position at one aging time with those at later times. Essentially, this method attempts to scale the response at different aging times by shifting $\varepsilon(t_a,\nu)$ *horizontally* along the $\log_{10}\nu$ axis. The degree to which such scaling fails indicates how the spectral shape changes with time. The study of physical aging in polymers focuses on the aging-time dependence of response functions and takes primarily this point of view. Viewing the changing spectral shape with aging time also offers a suitable perspective for placing our results in the context of phenomenological models that have been developed to describe aging. (2) In Sec. IV B we consider the aging-time dependence of $\varepsilon(t_a,\nu)$ at a fixed frequency. (That is, we shift the data *vertically* along the $\log_{10}\varepsilon$ axis.)

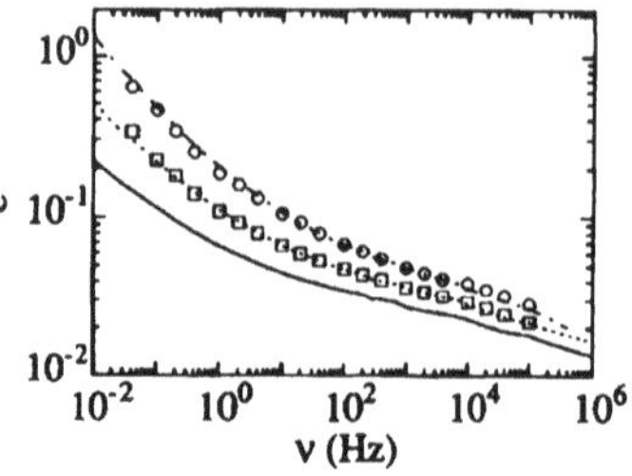

FIG. 3. $\varepsilon''(t_a,\nu)$ vs frequency following a quench from 206.2 to 177.6 K at two different aging times: $t_a=200$ s (circles) and $t_a=2\times10^4$ s (squares). The lines are equilibrium spectra at three temperatures: 177.6 K (solid), 180.1 K (dotted), and 182.5 K (dot-dashed). The spectrum after 200 s at 177.6 K closely approximates the equilibrium spectrum at 182.5 K. Likewise, the spectrum after the longer aging time of 2×10^4 s closely resembles the equilibrium spectrum at the lower temperature, 180.1 K.

This approach is the method commonly taken in aging studies on spin glasses, and in Sec. IV C we analyze our results in the context of spin-glass models. In Sec. IV D we describe an experiment that looks for "pre-aging" phenomena like those recently found in simulations.

A. Aging-time dependence of the spectral shape

We focus on the imaginary part of the susceptibility $\varepsilon''(t_a,\nu)$, since this term shows a much larger fractional change during aging than does the real part, $\varepsilon'(t_a,\nu)$. Figure 3 shows $\varepsilon''(t_a=200\text{ s},\nu)$ and $\varepsilon''(t_a=2\times10^4\text{ s},\nu)$ following a quench from 206.2 K to 177.6 K. [We construct the spectrum $t_a=200$ s by extrapolating $\varepsilon''(t_a,\nu)$ at each ν with Eq. (5) given below.] The lines in the figure are the equilibrium spectra, $\varepsilon''_{\text{eq}}(\nu)\equiv\varepsilon''(t_a\to\infty,\nu)$, measured at 177.6 K (solid), 180.1 K (dotted), and 182.5 K (dot-dashed). As the figure illustrates, the spectral shape and position at different times during aging at 177.6 K closely approximate those of equilibrium spectra at higher temperature. As mentioned above, a detailed study of $\varepsilon''_{\text{eq}}(\nu)$ has revealed that the exponents of the two power-law regions above the peak β and σ decrease with decreasing temperature. Therefore, the association of $\varepsilon''(t_a,\nu)$ with $\varepsilon''_{\text{eq}}(\nu)$ at different temperatures (with earlier t_a corresponding to higher temperature) implies that during aging the spectral shape is similarly changing with time.

We note that this changing spectral shape with aging time in glycerol contradicts the "time/aging-time superposition principle" often applied to aging in polymers. In his extensive study of aging in polymers, Struik[1] developed this principle, in which the (time-domain) response functions maintain a shape on a log-time scale independent of t_a and shift to longer time (or, equivalently, lower frequency) with aging as t/t_a^{μ}. Several studies have indicated that time/aging-time superposition is not universally held in polymers. In particular, two recent time-domain dielectric experiments have reported aging-time dependence in the shape of the dielectric response.[8,9] However, neither of these studies associated the shape at different aging times with higher temperature equilibrium shapes as we have done here for the aging in simple liquids.

To illustrate more concretely the aging-time dependence of the spectral shape in glycerol, we display in Fig. 4 the frequency $\nu'(t_a)$, at which $\varepsilon''(t_a,\nu')$ takes on a particular value after a quench to 177.6 K normalized by the frequency ν_∞, at which $\varepsilon''_{\text{eq}}(\nu_\infty)$ at 177.6 K has this same value. Specifically, we plot $\nu'(t_a)/\nu_\infty$ for which

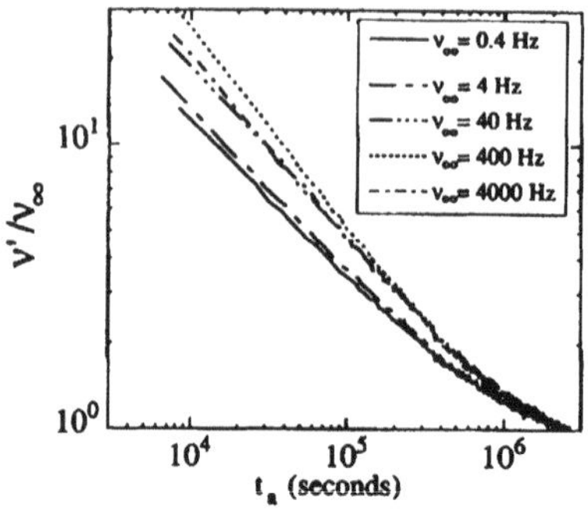

FIG. 4. The ratio of the frequencies $\nu'(t_a)$ and ν_∞ for which $\varepsilon''(t_a,\nu')=\varepsilon''_{\text{eq}}(\nu_\infty)$ vs aging time t_a, after a quench to 177.6 K. The different curves are for ν_∞ ranging from 0.4 to 4000 Hz. If the spectral shape were aging-time-independent, these curves would all lie on top of one another. The departure of these curves from one another indicates the degree to which the dielectric response violates time/aging-time superposition.

$$\varepsilon''(t_a,\nu')=\varepsilon''_{\text{eq}}(\nu_\infty) \tag{4}$$

for ν_∞ ranging from 0.4 to 4000 Hz. [In graphical terms, we are plotting the horizontal shift (on the logarithmic frequency axis) that is necessary to bring $\varepsilon''(t_a,\nu')$ for a given value of t_a on top of the equilibrium curve, $\varepsilon''_{\text{eq}}(\nu_\infty)$.] If the spectral shape were aging-time-independent, then the change in $\nu'(t_a)/\nu_\infty$ with t_a would be independent of $\varepsilon''_{\text{eq}}(\nu_\infty)$, and these curves would collapse onto a single line. The variation with ν_∞ displayed in Fig. 4 is consistent with the spectral shape mimicking higher temperature equilibrium shapes during aging. For example, the circles in Fig. 5 are $\nu'(t_a)/\nu_\infty$ at $t_a=2\times10^4$ s, and the solid line is ν'/ν_∞ obtained by substituting $\varepsilon''(t_a,\nu')$ in Eq. (4) with $\varepsilon''_{\text{eq}}(\nu')$ at 180.1 K. In each case the ratio follows the same trend, illustrating the close association between the spectral shape during aging and equilibrium shapes at higher temperatures.

This effective temperature that the evolving spectral shape assumes during aging bears a close resemblance to the fictive temperature from the phenomenological model Narayanaswamy[10] and Moynihan *et al.*[11] developed to describe the aging-time dependence of bulk properties in quenched liquids. In this model the fictive temperature, T_F, accounts for the effects of the nonequilibrium structure of the liquid on the equilibration rate of these properties. For example, following a quench through T_g, a liquid's density is smaller than its equilibrium density, and T_F is defined as the temperature at which this nonequilibrium density is the equilibrium value. Although T_F has proven a useful concept in analyzing aging data, a correspondence between T_F and the nonequilibrium state of the liquid is often questioned because, for example, the time dependence one obtains for T_F typically varies with the quantity under observation.[11] However, the behavior of $\varepsilon''(t_a,\nu)$ seems to give physical substance to the notion of a fictive temperature, since this tem-

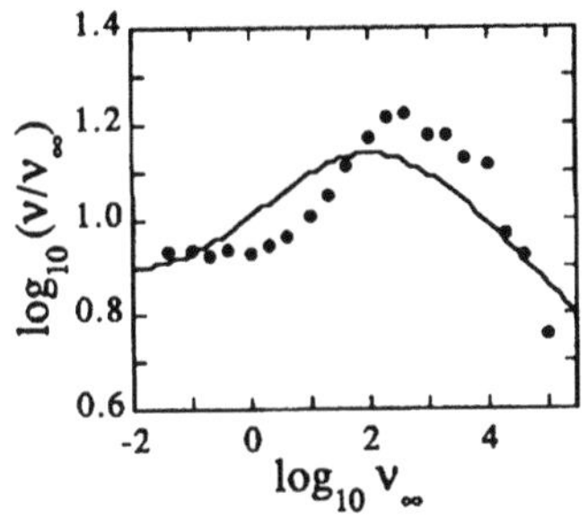

FIG. 5. The ratio of the frequencies $\nu'(t_a=2\times10^4$ s) and ν_∞ for which $\varepsilon''(t_a,\nu')=\varepsilon''_{eq}(\nu_\infty)$ after a quench to 177.6 K as a function of ν_∞. The solid line is the frequency ratio, $\nu_{\infty,T=180.1}$ and $\nu_{\infty,T=177.6}$, necessary to bring the higher temperature equilibrium spectrum at 180.1 K on to the equilibrium spectrum at 177.6 K. The two ratios have very similar frequency dependencies.

perature can be defined not only for a single quantity, which is always possible by definition, but to an entire response function.

While this association between $\varepsilon''(t_a,\nu)$ at different t_a and $\varepsilon''_{eq}(\nu)$ at different temperatures appears accurate for quenches to temperatures not far below T_g, deviations become increasingly apparent with decreasing temperature. Figure 6 shows $\varepsilon''(t_a=10^4\ \text{s},\nu)$ for a series of temperatures ranging from 140.9 K to 177.6 K. The spectra at low temperature in Fig. 6 are not consistent with an equilibrium spectrum at any temperature. For example, at 140.9 K $\varepsilon''(t_a=10^4\ \text{s},\nu)$ above 0.4 Hz appears well fit by a single asymptotic power law: $\varepsilon''(t_a=10^4\ \text{s},\nu)=D\nu^{-\sigma}$ with $\sigma=0.10$. This value for σ corresponds through Eq. (2) to an equilibrium temperature near 177 K. However, the small value of D suggests that the spectrum reaches a peak at a frequency orders of magnitude below the equilibrium value, $\nu_p\approx7\times10^{-8}$ Hz, at 177 K.

Thus, after deep quenches the phenomenology becomes more complicated. The spectral shape continues to resemble an equilibrium shape, but its position in frequency is much lower than the equilibrium position corresponding to that shape. Thus, two effective temperatures seem necessary to produce a full description of the response during aging, and with increasing quench depth these temperatures become increasingly separated. The more rapid decrease of the spectral position with quench depth indicates that the mean dielectric

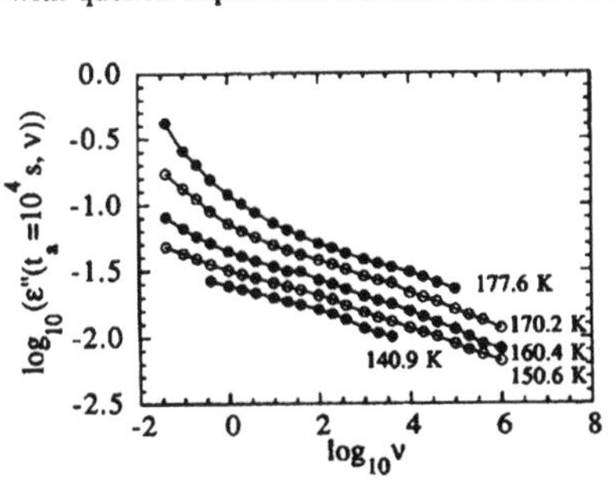

FIG. 6. $\varepsilon''(t_a=10^4\ \text{s},\nu)$ for a series of temperatures ranging from 140.9 to 177.6 K. The spectra at low temperature are not consistent with an equilibrium spectrum at any temperature.

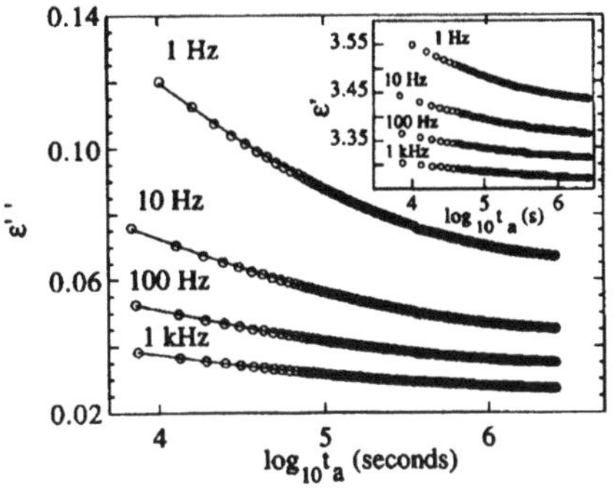

FIG. 7. The change with t_a of $\varepsilon''(t_a,\nu)|_\nu$ at several fixed frequencies for a quench to 177.6 K. The inset shows the real part of the susceptibility, $\varepsilon'(t_a,\nu)$, for the same frequencies. The shape of the decays is typical of the aging-time dependence seen at all temperatures and frequencies. Fits to a stretched exponential function, given by the solid lines, provide a good description of $\varepsilon''(t_a,\nu)$ at fixed ν over the entire range of t_a that we access.

response time is sensitive to the liquid's temperature. The shape of the response, on the other hand, continues to resemble equilibrium shapes for temperatures not far below T_g, and thus depends less strongly on temperature and more on the nonequilibrium structural state created in the liquid by quenching through T_g.

B. Aging-time dependence of the susceptibility at fixed frequency

In addition to tracking the changing spectral shape with aging time, we can also examine the change that the susceptibility at fixed frequency experiences after a quench. Figure 7 shows the change with t_a of $\varepsilon''(t_a,\nu)|_\nu$ at several fixed frequencies for a quench to 177.6 K. The inset to the figure shows the real part of the susceptibility $\varepsilon'(t_a,\nu)$ for the same frequencies. The shape of the decays in Fig. 7 is typical of the aging-time dependence seen at all temperatures and frequencies. In general, the curves are not well fit by a straight line, which would have implied logarithmic relaxation. We find that a stretched exponential function in the aging time provides an excellent description of $\varepsilon''(t_a,\nu)$ at fixed ν over the entire range of t_a that we access:

$$\varepsilon''(t_a,\nu)|_\nu=\Delta\varepsilon''_\nu\exp[-(t_a/\tau)^\beta]+\varepsilon''^{eq}_\nu. \quad (5)$$

The decay time, τ, follows an Arrhenius temperature dependence, as shown in Fig. 8(a) for $\nu=1$ Hz. Figure 8(b) shows the temperature dependence of β for $\nu=1$ Hz. Except at temperatures just below T_g, β maintains a temperature independent value near 0.25. (The slight upward trend in β with decreasing temperature below 179 K is not statistically significant. Below this temperature all of the decay curves at all frequencies are adequately fit assuming a constant value of β near 0.25.)

In a recent study of aging in glycerol by stimulated Brillouin scattering, Miller and MacPhail[5] observed that aging at a temperature continued for many times longer than the equilibrium structural relaxation times measured in the linear-response regime. They described the time dependence of the Brillouin peak shift with a stretched-exponential decay plus

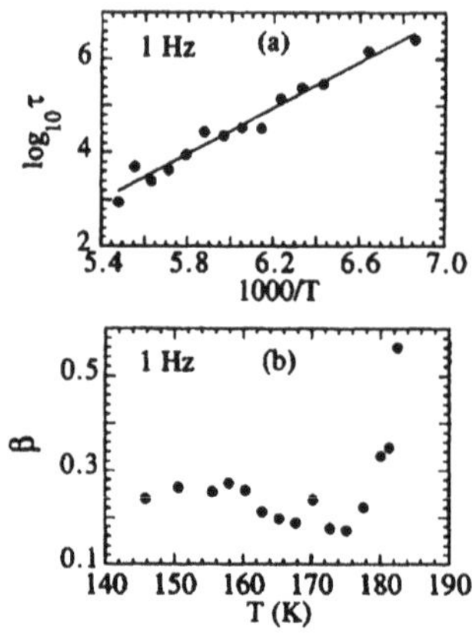

FIG. 8. (a) The decay time τ and (b) β from stretched exponential fits to $\varepsilon''(t_a, \nu = 1\text{ Hz})$ at various temperatures. The Arrhenius temperature dependence of τ is typical of the behavior at all frequencies. Below 179 K, β is temperature-independent.

an additional exponential decay to account for an ultraslow component in the aging. Introducing such an additional term to Eq. (5) gives

$$\varepsilon''(t_a,\nu)|_\nu = \Delta\varepsilon''_u \exp(-t_a/\tau_u) + \Delta\varepsilon''_s \exp[-(t_a/\tau_s)^\beta] + \varepsilon''^{\text{eq}}_\nu. \quad (6)$$

Miller and MacPhail noted that the existence of such a long-time component is consistent with models for the nonexponential relaxation in supercooled liquids that predict spatially extended dynamically heterogeneous regions in the liquids. We have searched for the presence of a long-time component in the decays of $\varepsilon''(t_a,\nu)|_\nu$ at temperatures for which we are able to monitor the aging to what we believe is equilibrium. Figure 9 shows $\varepsilon''(t_a, \nu = 200\text{ Hz})$ after quenches to 180.1 and 182.5 K with t_a scaled by τ_{eq} for each temperature. (We use $\tau_{\text{eq}} = 8.1\times10^3$ s at 182.5 K and $\tau_{\text{eq}} = 6.3\times10^4$ s at 180.1 K. These values for the equilibrium relaxation times result from setting $\tau_{\text{eq}} = 1/2\pi\nu_p$, where we obtain ν_p from extrapolations of higher temperature data through Eq. (1). In supercooled liquids the dielectric relaxation time, defined in this way, closely tracks with temperature the structural relaxation times extracted from frequency-dependent specific heat and shear viscosity measurements.[12]) As Fig. 9 illustrates, all of the measurable aging occurs at $t_a/\tau_{\text{eq}} < 10$, and beyond this point the susceptibility is independent of time. (We estimate the precision of the experiments would allow us to distinguish changes in ε'' greater than 1% on the time scale of these measurements.) The inset of Fig. 9 shows the region at large t_a on an expanded scale to emphasize this point. While this time scale for equilibration appears roughly consistent with that observed by Miller and MacPhail, our data do not support the addition of an ultraslow component in the description of the decay. Attempting to fit $\varepsilon''(t_a,\nu)|_\nu$ with Eq. (6), we find $\Delta\varepsilon''_u$, the strength of an ultraslow component, averages less than $0.04\Delta\varepsilon''_s$ at both 180.1 and 182.5 K. (To reduce the number of free parameters in these fits, we set

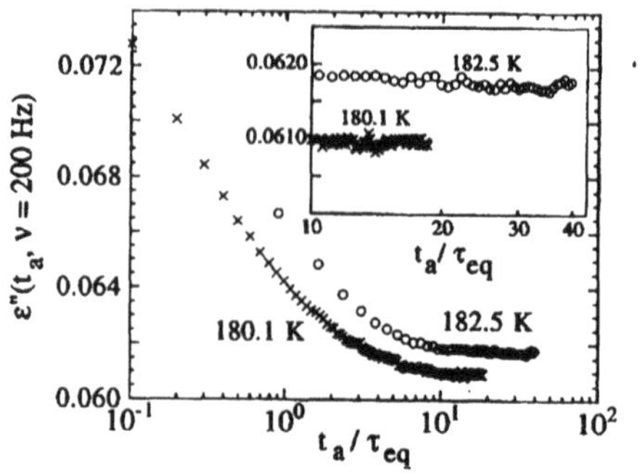

FIG. 9. $\varepsilon''(t_a, \nu = 200\text{ Hz})$ after quenches to 180.1 and 182.5 K vs t_a/τ_{eq} for each temperature. The data for 180.1 K have been offset by 0.017 for clarity. All of the measurable aging occurs at $t_a/\tau_{\text{eq}} < 10$, and beyond this point the susceptibility is independent of time. Inset shows the region at large t_a on an expanded scale.

$\tau_u = 35\tau_s$, the ratio Miller and MacPhail obtained.) Constraining $\Delta\varepsilon''_u$ to a larger value (for example, $0.5\Delta\varepsilon''_s$ as Miller and MacPhail obtained) gives markedly worse fits than Eq. (5), with the same number of free parameters. The fit results from Eq. (5) give averages for $\exp[-(\tau_{\text{eq}}/\tau)^\beta]$ of 0.25 at 182.5 K, 0.10 at 180.1 K, and 0.03 at 177.6 K. (Miller and Macphail, on the other hand, obtain a value for $\exp[-(\tau_{\text{eq}}/\tau)^\beta]$ near 0.35 at 179.8 K after quenching from above 192 K.) These numbers suggest that the significant portion of the aging, as measured by dielectric response, occurs within one τ_{eq} of the quench. Therefore, from this analysis we conclude that our results are not consistent with an ultraslow component to the aging. This observation, that the linear response behavior of glycerol reaches equilibrium following a quench faster than does the sound velocity (which depends directly on density), has been noted for an epoxy glass by Santore and co-workers[13] who simultaneously measured the mechanical response and density during aging.

Finally we note that the shapes of $\varepsilon''(t_a,\nu)|_\nu$ at a given temperature have a slight frequency dependence. This dependence appears as an increase in τ with ν when β in Eq. (5) is held fixed, such as in Fig. 10(a). Alternatively, fixing τ leads to a decrease in β with increasing ν, as in Fig. 10(b). (Allowing both parameters to vary simultaneously introduces scatter that obscures these trends.)

C. Comparison with aging in spin glasses

1. Scaling the susceptibility

Analysis of aging in spin glasses has provided insights into the microscopic physics of these systems, and one motivation for our study has been to compare the behavior we observe in simple liquids with aging in spin glasses. A number of studies on spin glasses have monitored the aging-time dependence of the frequency-dependent magnetic susceptibility, $\chi(t_a,\nu)$, following temperature quenches. The time dependence of $\varepsilon''(t_a,\nu)$ at fixed ν, such as shown in Fig. 7, appears, at least superficially, very similar to that observed for $\chi''(t_a,\nu)$ at fixed ν.[14,15] Indeed, over a limited range of time the behavior of $\varepsilon''(t_a,\nu)$ closely resembles both the

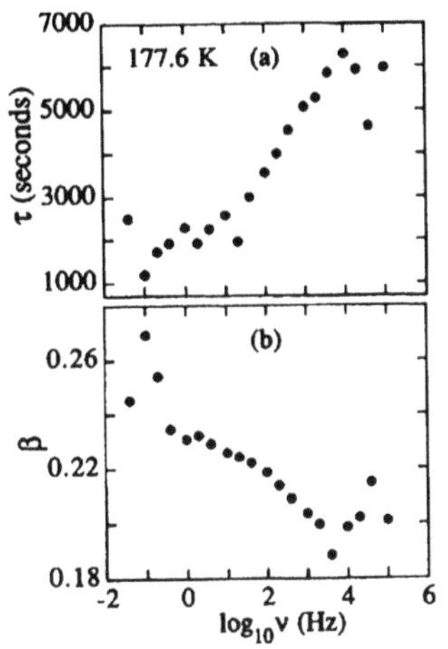

FIG. 10. (a) τ vs ν for the decays at 177.6 K (such as in Fig. 7) when β in the stretched-exponential form [Eq. (5)] is held fixed. (b) β vs ν for the same decays when τ is held fixed. These results indicate that the shape of the decay has a slight frequency dependence.

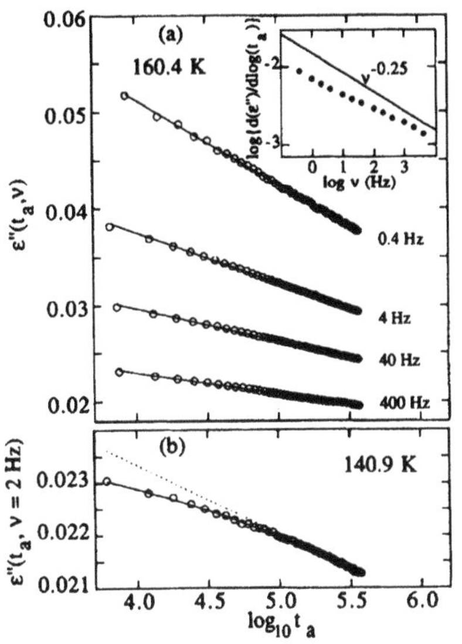

FIG. 11. (a) $\varepsilon''(t_a,\nu)$ for several frequencies at 160.4 K vs $\log_{10} t_a$. The inset shows the rate of decay, $d(\varepsilon'')/d(\log_{10} t_a)$, in the range $0.1\tau<t_a<10\tau$. The frequency dependence of the decay rate is similar to the $\nu^{-0.25}$ dependence reported for the decay of χ'' in a spin glass (Ref. 14). However, we observe deviations from this logarithmic decay both for $t_a \ll \tau$ and $t_a \gg \tau$. (b) $\varepsilon''(t_a,\nu=2\text{ Hz})$ at 140.9 K on a $\log_{10} t_a$ scale. (At this temperature $\tau>10^8$ s.) The dashed line in the figure is a guide to highlight the deviations from logarithmic behavior. The solid line through the data is a fit to Eq. (5).

$\log_{10} t_a$ decay[14] and power-law decay[15] reported for $\chi''(t_a,\nu)$ in spin glasses. [For $\beta\approx 0.25$, Eq. (5) is essentially indistinguishable from a $\log_{10} t_a$ decay for $0.1\tau<t_a<10\tau$.] However, by quenching to very low temperatures, so that we can access the aging at relatively early times, we observe deviations from log decay and power-law decay, consistent with Eq. (5).

As an illustration of the similarity between the liquid and spin-glass aging behaviors, Fig. 11(a) shows $\varepsilon''(t_a,\nu)$ for several frequencies at 160.4 K on a $\log_{10} t_a$ scale. The rate of decay, $d(\chi'')/d(\log_{10} t_a)$, for a spin glass was reported to vary as $\nu^{-0.25}$.[14] Fitting within the range $0.1\tau<t_a<10\tau$, we find a similar frequency dependence for the decay of ε'', as the inset to Fig. 11(a) illustrates. [In terms of Eq. (5), $d(\varepsilon'')/d(\log_{10} t_a)=-\Delta\varepsilon''_\nu\beta/(e\log_{10} e)$ at $t_a=\tau$. Thus, this frequency dependence simply reflects the frequency dependence of $\Delta\varepsilon''_\nu$ and β.] We stress again, however, that we observe deviations from logarithmic decay, consistent with Eq. (5), both for $t_a\ll\tau$ and $t_a\gg\tau$. The deviations at large t_a are apparent in Fig. 7. To illustrate the deviations at small t_a, we show in Fig. 11(b) $\varepsilon''(t_a,\nu=2\text{ Hz})$ at 140.9 K on a $\log_{10} t_a$ scale (At this temperature $\tau>10^8$ s.) The dashed line in the figure is a guide to highlight the deviations from logarithmic behavior. The solid line through the data is a fit to Eq. (5).

To illustrate the range of applicability of a power-law description to $\varepsilon''(t_a,\nu)|_\nu$, we show in Fig. 12 $\varepsilon''(t_a,\nu=2\text{ Hz})$ on a log-log scale for (a) $T=140.9$ K, (b) $T=172.7$ K, and (c) $T=180.1$ K. At these low, intermediate, and high temperatures we monitor the aging at relatively early, intermediate and late times, respectively. The solid lines in Figs 12(a) and 12(c), which are fits to a stretched exponential decay [Eq. (5)], again demonstrate its good agreement to the data throughout the aging times we access. The dotted lines in Figs. 12(b) and 12(c) are fits to a power-law decay:

$$\varepsilon''(t_a,\nu)|_\nu = A_\nu t_a^{-\alpha} + \varepsilon''^{\text{eq}}_\nu . \tag{7}$$

As Fig. 12(b) illustrates, the decay of $\varepsilon''(t_a,\nu)|_\nu$ is consistent with a power law over an extended period during aging. Equation (7) also provides a fair description of the final approach to equilibrium, as Fig. 12(c) illustrates. (Fits to the stretched-exponential form, which benefits from one additional free parameter, are superior in this region.) However, clear deviations from a power-law decay are apparent during the early period of aging accessed at low temperature, as in Fig. 12(a). At these aging times, the rate of decay of $\varepsilon''(t_a,\nu)|_\nu$ on a log-log scale increases with aging time, in contrast to the behavior predicted by Eq. (7).

We have attempted to scale $\varepsilon''(t_a,\nu)$ at different frequencies following a procedure suggested for $\chi''(t_a,\nu)$ of spin glasses[3] in which the difference in susceptibility from its equilibrium value is scaled against νt_a. (We note that in the spin glasses the equilibrium spectra are not measured, and consequently some freedom exists for choosing the subtracted values.) Figure 13(a) which displays an effort to scale the data at 180.1 K, demonstrates that the scaling does not work for glycerol. Figure 13(b) shows the νt_a scaling applied to $\varepsilon''(t_a,\nu)$ at a lower temperature, 162.8 K. Because we do not know $\varepsilon''_{\text{eq}}(\nu)$ at this temperature, we have chosen offsets, $C(\nu)$, for each frequency to optimize the scaling. However, even introducing this free parameter fails to pro-

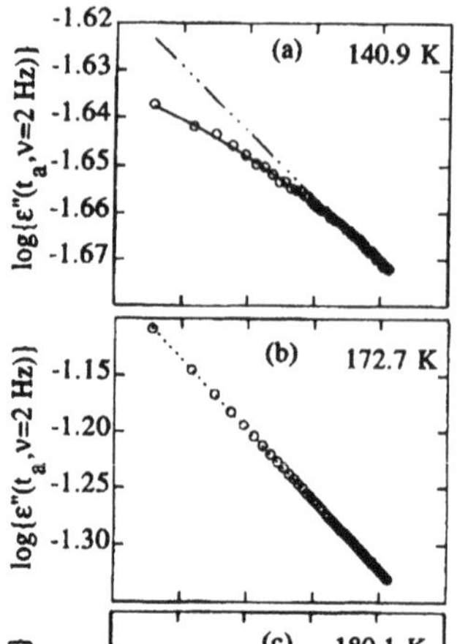

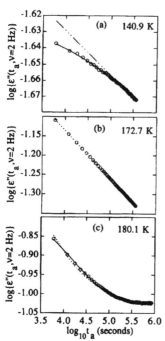

FIG. 12. $\varepsilon''(t_a,\nu=2\text{ Hz})$ on a log-log scale for quenches to (a) $T=140.9$ K, (b) $T=172.7$ K, and (c) $T=180.1$ K. At these low, intermediate, and high temperatures we monitor the aging at relatively early, intermediate and late times, respectively. The solid lines in (a) and (c) are fits to a stretched exponential decay. The dotted lines in (b) and (c) are fits to a power-law decay. The dash-dotted line in (a) is a straight line "guide to the eye."

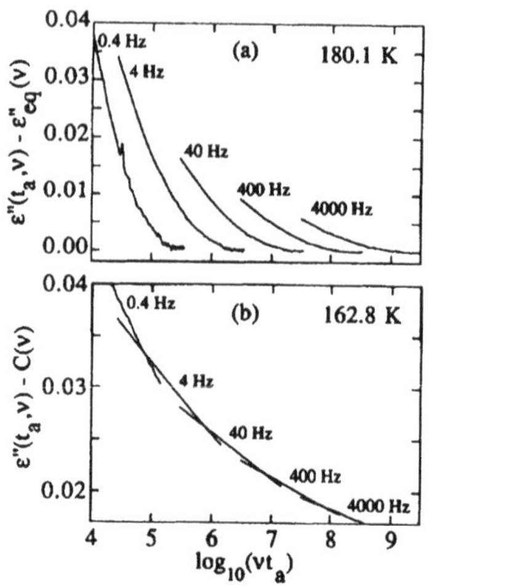

FIG. 13. The difference in susceptibility from its equilibrium value, $\varepsilon''(t_a,\nu)-\varepsilon''_{eq}(\nu)$, plotted vs νt_a for (a) 180.1 K and (b) 162.8 K. For 162.8 K $\varepsilon''_{eq}(\nu)$ is not known; therefore, we choose values, $C(\nu)$, to optimize the overlap between frequencies. This scaling procedure, proposed for spin glasses, does not work for glycerol.

vide good scaling. (At this temperature, a $\log_{10} t_a$ dependence approximates the aging well for the entire time we record the decay. Since the slope of $\varepsilon''(t_a,\nu)$ vs $\log_{10} t_a$ decreases with increasing frequency [see the inset to Fig. 11(a)], we should not expect the data at this temperature to obey νt_a scaling.) Successful νt_a scaling of the transient response implies that, at sufficiently high frequency, the susceptibility reaches its equilibrium value arbitrarily rapidly. This scenario appears valid for spin glasses for which $\chi''(t_a,\nu)$ reaches equilibrium quickly at high frequency.[3,14] However, in glycerol we observe changes with aging time in $\varepsilon''(t_a,\nu)$ up to $\nu=10^6$ Hz (our highest frequency) through out the measurement time ($t_a>4\times10^5$ s). From these observations we conclude that, in spite of the similarities highlighted by Figs. 11 and 12, important qualitative differences exist between aging in liquids and in spin glasses and, in particular, that νt_a scaling for the transient portion of the susceptibility is not appropriate for the aging spectra in liquids.

2. *Temperature jumps during aging*

Many of the insights in the study of spin glasses have come from monitoring the response of $\chi''(t_a,\nu)$ during aging to temperature changes.[3,16] We have attempted the same type of experiments on glycerol to compare its behavior with that of the spin glasses. Figure 14 shows the result of aging the liquid at 177.6 K for two days, decreasing the temperature to 170.2 K for two days ($\Delta T=-7.4$), and then returning the temperature to 177.6 K. (Figure 14 displays $\varepsilon''(t_a,\nu=2\text{ Hz})$. However, the features are independent of frequency.) The behavior of the liquid in these experiments shows important qualitative differences with that observed for the spin glasses. When the temperature of an aging spin glass is dropped, the susceptibility typically jumps suddenly to a large value and begins a rapid decay, as if the aging process were being restarted. This dramatic behavior constrasts with the drop in susceptibility and slow decay shown at 170.2 K in Fig. 14, and for no value of ΔT have we found such a "restart of aging" in the susceptibility of the liquid. (We have performed six such measurements all with an initial temperature of 177.6 K and with $\Delta T=-2.5$, -4.9, -7.4, -12.3, -17.2, and -27.0 K.)

Upon returning to the initial temperature, the liquid, like the spin glasses, has aged at the lower temperature less than it would have if the temperature had not been lowered, and one might attempt to define an effective time, t_{eff}, to characterize the degree of aging experienced at the lower temperature, as is done for the spin glasses.[16] However, in the liquid the determination of t_{eff} is complicated by the shape of $\varepsilon''(t_a,\nu)$ with aging time following the return to the initial temperature. For sufficiently small ΔT, $\varepsilon''(t_a,\nu)$ increases for a period immediately following the return to the initial temperature before resuming its decrease, as the inset of Fig. 14 illustrates. This increase appears to be an example of the "memory effect" that one observes for a quenched liquid's bulk properties in similar temperature jump experiments.[11] (A phenomenological explanation of this effect was a principal motivation for the development of the Narayanaswamy-Moynihan model.) Therefore, for these studies of temperature jumps, we conclude that the salient features revealing the underlying physics of spin glasses are not reproduced with the liquids.

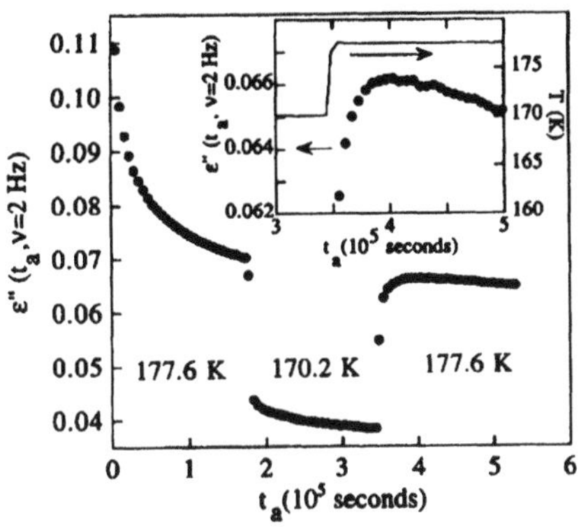

FIG. 14. $\varepsilon''(t_a, \nu=2\ \text{Hz})$ vs t_a during a measurement in which the temperature was held at 177.6 K for two days, decreased to 170.2 K for two days ($\Delta T=-7.4$), and then returned to 177.6 K. The behavior of the liquid in these experiments shows important qualitative differences with that observed for the spin glasses. The inset shows that, for sufficiently small ΔT, $\varepsilon''(t_a, \nu)$ demonstrates the "memory effect," increasing for a period immediately following its return to the initial temperature before resuming its decrease.

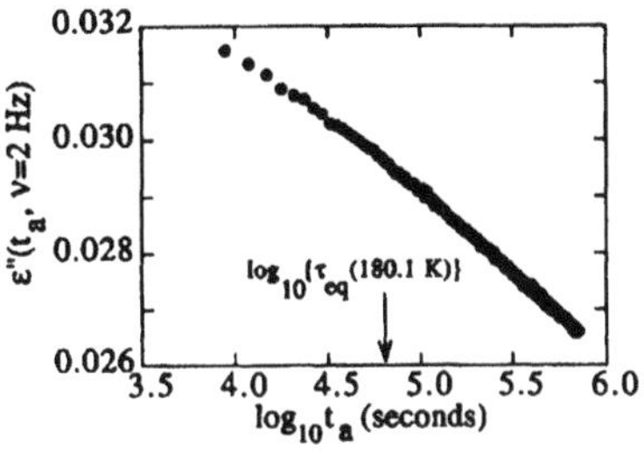

FIG. 15. $\varepsilon''(t_a, \nu=2\ \text{Hz})$ after a quench from 180.1 to 160.4 K. Aging has commenced before our earliest measurement times, which are approximately one tenth of the equilibrium relaxation time of the initial temperature. We observe no pre-aging effects [i.e., changes in the decay rate or other features as t_a passes through $\tau_{eq}(T=180.1\ \text{K})$].

D. Quenches from low initial temperatures

In a recent simulation study of aging in a Lennard-Jones liquid, Kob and Barrat[17] analyzed the dependence on the initial temperature of the aging behavior following a quench. Monitoring the changes with t_a of the self-intermediate scattering function, they observed a period at early t_a in which the scattering function showed no change. Aging commenced only as t_a approached the equilibrium relaxation time at the *initial* temperature. They explained this "pre-aging" period with the physically appealing idea that the liquid requires on the order of one relaxation time to realize that the quench has created a nonequilibrium condition and to respond structurally.

Following the quench procedure that we describe in Sec. III, we are not able to access this pre-aging regime. For these quenches, T_g, the temperature at which the sample falls from equilibrium, is effectively the initial temperature, and equilibrium relaxation times at T_g are approximately 100 s, much shorter than the 5000 s the experiment requires for thermal equilibration. Therefore, to search for this pre-aging regime, we have performed a quench to 160.4 K after equilibrating the sample at 180.1 K, where the equilibrium relaxation time is 6.3×10^4 s. (We held the sample at 180.1 K for 14 days before the quench.) Figure 15 shows the resulting decay, $\varepsilon''(t_a, \nu=2\ \text{Hz})$. In contrast to the simulation, we observe aging at our earliest measurement times, which are approximately one tenth of the equilibrium relaxation time of the initial temperature. No change in the decay rate or other feature appears as t_a passes through τ_{eq} (180.1 K).

The behavior we observe is qualitatively identical to that seen with the other quench procedure (i.e., quenching from above T_g), and Eq. (5) accurately describes $\varepsilon''(t_a, \nu)$ at fixed ν. However, the magnitude and time scale of the decays of $\varepsilon''(t_a, \nu)$ at fixed ν are different. Fits to Eq. (5) give $\tau\approx7\times10^5$ s for this quench from 180.1 K (averaging over 0.4 to 4000 Hz), while its value following a quench from 206.2 K through T_g to this temperature is $\tau\approx1\times10^5$ s. Also, $\Delta\varepsilon''_\nu$ is approximately a factor of two smaller, independent of frequency (between 0.4 and 4000 Hz), after the quench from 180.1 K than it is after a quench from the higher temperature.

V. CONCLUSIONS

In this frequency domain study of aging in a simple liquid we have stressed comparisons with other techniques which monitor equilibration in liquids as well as with aging behavior in spin glasses. These comparisons have not only provided information about the nonequilibrium dynamics but have also touched on the equilibrium properties of deeply supercooled simple liquids. In particular, the absence in the dielectric response of a significant ultraslow component to the aging, such as that seen with stimulated Brillouin scattering, places constraints on the source of this component since it does not appear to couple strongly to the dielectric susceptibility. Also, our effort to make detailed comparisons between aging in liquids and that in spin glasses has highlighted both the similarities between their behaviors as well as important differences. The aging behavior near T_g seems to fit well within the phenomenological framework of the Narayanaswamy-Moynihan model. Indeed, we see direct physical manifestation of an effective temperature, like the fictive temperature in this model, in the aging-time dependence of the spectral shape. However, the failure of this analysis when extended to results from deeper quenches may indicate its limitations. At these temperatures two effective temperatures seem necessary to characterize the out-of-equilibrium response, one describing the spectral shape and a lower one describing its position. An understanding of this contrasting dependence on quench depth, in particular the relative insensitivity of the initial spectral shape to quench depth, should provide insight not only to the mechanisms that drive aging, but also to the nature of the spectral broadening in equilibrated supercooled liquids.

ACKNOWLEDGMENTS

We gratefully acknowledge L. Cugliandolo, R. Deegan, W. Kob, J. Kurchan, and R. MacPhail for valuable discussions. Funding was provided by NSF-DMR 94-10478.

*Present address: Dept. of Physics, Room 13-2154, Massachusetts Institute of Technology, Cambridge, MA 02139.

[1] For an extensive study of aging in polymers, see L. C. E. Struik, *Physical Aging in Amorphous Polymers and Other Materials* (Elsevier, Amsterdam, 1978); For reviews of more recent progress, see S. Matsuoka, *Relaxation Phenomena in Polymers* (Hanser, New York, 1992); and G. B. McKenna, J. Res. Natl. Inst. Stand. Technol. **99**, 169 (1994).

[2] V. S. Kästner and M. Dittmer, Kolloid Z. Z. Polym. **204**, 74 (1965).

[3] E. Vincent, J. Hammann, M. Ocio, J-P. Bouchaud, and L. F. Cugliandolo, in *Complex Behavior of Glassy Systems*, edited by M. Rubí and C. Perez-Vicente (Springer-Verlag, New York, 1997).

[4] Evidence for heterogeneous dynamics come from flourescent probe studies: M. T. Cicerone and M. D. Ediger, J. Chem. Phys. **103**, 5684 (1995); and NMR studies: R. Böhmer, G. Hinze, G. Diezemann, B. Geil, and H. Sillescu, Europhys. Lett. **36**, 55 (1996).

[5] R. S. Miller and R. A. MacPhail, J. Chem. Phys. **106**, 3393 (1997).

[6] R. L. Leheny and S. R. Nagel, Europhys. Lett. **39**, 447 (1997).

[7] N. Menon and S. R. Nagel, Phys. Rev. Lett. **74**, 1230 (1995).

[8] E. Schlosser and A. Schönhals, Polymer **32**, 2138 (1991).

[9] A. Alegría, L. Goitiandia, I. Tellería, and J. Colmenero, J. Non-Cryst. Solids **131-133**, 457 (1991).

[10] O. S. Narayanaswamy, J. Am. Ceram. Soc. **54**, 491 (1971).

[11] C. T. Moynihan, P. B. Macedo, C. J. Montrose, P. K. Gupta, M. A. DeBolt, J. F. Dill, B. E. Dom, P. W. Drake, A. J. Easteal, P. B. Elterman, R. P. Moeller, H. Sasabe, and J. A. Wilder, Ann. (N.Y.) Acad. Sci. **279**, 15 (1976).

[12] P. K. Dixon, Phys. Rev. B **42**, 8179 (1990); N. Menon, J. Chem. Phys. **105**, 5246 (1996); L. Wu, P. K. Dixon, S. R. Nagel, B. D. Williams, and J. P. Carini, J. Non-Cryst. Solids **131**, 32 (1991).

[13] M. M. Santore, R. S. Duran, and G. B. McKenna, Polymer **32**, 2377 (1991).

[14] P. Svedlindh, K. Gunnarsson, J-O. Andersson, H. A. Katori, and A. Ito, Phys. Rev. B **46**, 13 867 (1992).

[15] J-P. Bouchaud and D. S. Dean, J. Phys. I **5**, 265 (1995).

[16] E. Vincent, J-P. Bouchaud, J. Hammann, and F. Lefloch, Philos. Mag. B **71**, 489 (1995).

[17] W. Kob and J-L. Barrat, Phys. Rev. Lett. **78**, 4581 (1997).

EUROPHYSICS LETTERS 1 January 2000

Europhys. Lett., **49** (1), pp. 68–74 (2000)

Sheared foam as a supercooled liquid?

S. A. LANGER[1] and A. J. LIU[2]

[1] *Information Technology Laboratory, NIST - Gaithersburg, MD 20899, USA*
[2] *Department of Chemistry and Biochemistry, University of California Los Angeles, CA 90095, USA*

(received 5 May 1999; accepted in final form 1 November 1999)

PACS. 64.70.Pf – Glass transitions.
PACS. 83.50.Ax – Steady shear flows.
PACS. 83.70.Hq – Heterogeneous liquids: suspensions, dispersions, emulsions, pastes, slurries, foams, block copolymers, etc.

Abstract. – We conduct numerical simulations on a simple model of a two-dimensional steady-state sheared foam, and define a quantity Γ that measures stress fluctuations in the constant-area system. This quantity reduces to the temperature in an equilibrium system. We find that the relation between the viscosity and Γ is the same as that between viscosity and temperature in a very different system, namely a supercooled liquid. This is the first evidence of a common phenomenon linking these two systems.

A liquid foam consists of gas or liquid bubbles suspended in an immiscible liquid at a packing fraction that exceeds random close-packing. A quiescent foam has a nonzero static shear modulus, because the thermal energy is negligible compared to the energy barrier required to change the relative positions of bubbles. When a foam is sheared, however, it can be characterized by a well-defined viscosity. As the shear rate is lowered towards zero, the viscosity diverges. This behavior is reminiscent of a supercooled liquid, where the viscosity increases rapidly as the temperature is lowered towards the glass transition [1]. These two systems are completely different: a sheared foam is a driven, athermal system at steady state, while a supercooled liquid is a quiescent thermal system. However, in both cases, the systems are jamming (*i.e.* the systems are spontaneously restricting themselves to a small part of phase space). Other driven, athermal systems such as granular materials [2] also jam. We therefore ask: does the common phenomenon of jamming lead to any common behavior in driven, athermal systems and in quiescent, thermal systems?

Foam is a particularly simple athermal system that jams, because it can be driven homogeneously by steady shear flow. As foam is sheared, there are fluctuations caused by rearrangement events where bubbles change their relative positions [3–5]. In an *equilibrium* system, temperature is a measure of the size of fluctuations relative to how easy it is to create a fluctuation (a response function). This property of temperature is embodied in the linear response relation that connects fluctuations in the stress on the boundary normal to the y-direction, σ_{yy}, to the yy compression modulus [6, 7]

$$h\frac{\partial\langle\sigma_{yy}\rangle}{\partial h} = \frac{1}{\rho k_{\mathrm{B}}T}\left\langle(\sigma_{yy} - \langle\sigma_{yy}\rangle)^2\right\rangle, \tag{1}$$

where h is the fixed height of the system in the y-direction and ρ is its density. One way to compare the non-equilibrium sheared foam to a supercooled liquid would be to compare the functional form of the divergence of viscosity *vs.* temperature in the supercooled liquid to the form of viscosity *vs.* shear strain rate in the foam. However, strain rate and temperature are very different variables and cannot be compared directly.

We conduct simulations on a simple model of foam [8], and measure a quantity Γ that characterizes stress fluctuations due to rearrangement events in a two-dimensional, constant-area system that is steadily sheared in the x-direction with the shear gradient in the y-direction. In our non-equilibrium, steady-state system, we define Γ by analogy to eq. (1), so that it corresponds to the temperature in a non-driven, equilibrium system [9]:

$$\Gamma = \frac{\langle(\sigma_{yy} - \langle\sigma_{yy}\rangle)^2\rangle_h}{\rho h \partial\langle\sigma_{yy}\rangle/\partial h}. \tag{2}$$

Note that Γ depends on shear rate and should not be interpreted as an effective temperature since there is no reason to believe that the equilibrium relation eq. (1) holds for our system. However, Γ does provide a measure of the size of fluctuations relative to how easy it is to create a fluctuation in our system, just as temperature does for an equilibrium system. In this sense, Γ provides a reasonable measure of the ability of the system to overcome energy barriers.

We measure Γ and the viscosity η as a function of shear rate. Our central result is that we find that the relation between viscosity and Γ is the *same* as the relation between viscosity and temperature in a supercooled liquid. This remarkable finding is the first quantitative evidence that a common mechanism may underlie jamming in a driven, athermal system and supercooled liquids.

Our simulations are carried out on a model introduced by Durian [8]. In two dimensions this model treats bubbles as circles interacting via two types of interactions. The first interaction is purely repulsive and originates physically in the energy cost to distort bubbles. If the distance between the centers of two bubbles is less than the sum of their radii, they will distort to avoid overlap. This gives rise to a harmonic repulsive force [10, 11]. The model therefore assumes that bubbles always remain circles, but includes the elastic energy of a compressed spring between two overlapping bubbles. The second interaction is a frictional force proportional to the velocity difference between neighboring bubbles. By modeling the dissipation in the thin liquid films and Plateau borders as a dynamic friction, we neglect lubrication effects that could be important in determining the stress/strain-rate relations for real foams [12]. However, this simple model already gives rise to remarkably complex behavior. Moreover, the regime in which the viscosity is high (the low shear rate regime) corresponds to the regime where elastic effects dominate viscous ones, so the specific dissipation mechanism may be unimportant there.

In the discussion that follows, we will scale all energies (such as Γ) by kr^2, where k is a characteristic spring constant and r is the average bubble radius. The strain rate $\dot{\gamma}$ is scaled by a characteristic time $\tau = b/k$, where b is the friction coefficient. Thus, $\dot{\gamma}$ is the Deborah number, or equivalently for this system, the capillary number.

Given the two types of forces on bubbles, we can solve the equations of motion [8] numerically and propagate the bubble positions forward in time, as in a molecular-dynamics simulation. Our simulations are conducted on polydisperse bubbles whose radii are drawn from a flat distribution of a specified width, and whose spring constants vary inversely with radius. We use square samples, with periodic boundary conditions in the x-direction and confining walls (separated by the fixed distance h) in the y-direction. The bubbles along the

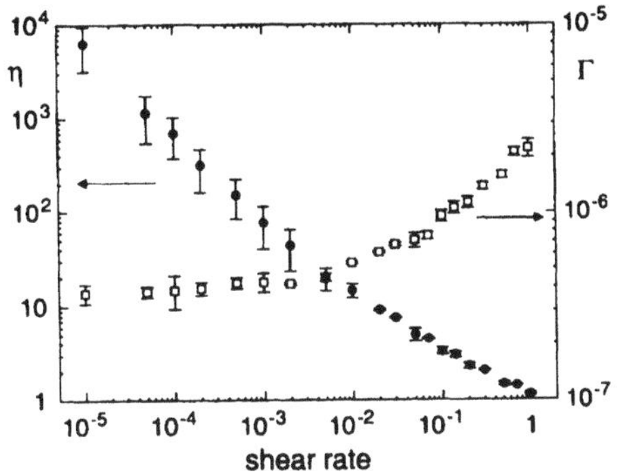

Fig. 1 – The quantity Γ (open squares, right axis) and the viscosity η (solid circles, left axis) as functions of the dimensionless shear rate $\dot{\gamma}$ (Deborah number). Neither one obeys simple power law behavior. Γ approaches a constant and η diverges in the limit $\dot{\gamma} \to 0$. All quantities are dimensionless as explained in the text.

confining walls are glued to them, and the system is sheared by moving one of the two walls at fixed velocity in the x-direction. We then measure the force per length in the y-direction on the wall (this is $-\sigma_{yy}$) and the force per length in the x-direction required to keep the wall moving at fixed velocity (this is σ_{xy}, the xy-component of the stress). These fluctuate in time as the system is sheared. The viscosity is defined by $\eta = \langle\sigma_{xy}\rangle/\dot{\gamma}$, where $\dot{\gamma}$ is the strain rate. In order to measure $\partial\langle\sigma_{yy}\rangle/\partial h$, we perturb the y-separation, h, of the confining walls and measure the average value of σ_{yy} during shear. We study fairly small systems, ranging in size from 63 to 621 bubbles. To obtain adequate statistics, averages (indicated by angle brackets) were taken over configurations as well as over time. Thus, each data point shown in figs. 1-3 corresponds to an average over 10000 time steps covering a total strain of at least 10, and at least 9 different initial configurations. We find that at sufficiently long times, the time average does approach the configurational average, but it proves more efficient to use a combination of configurational and time averaging.

In fig. 1, we show the viscosity, η, and Γ as functions of the strain rate, $\dot{\gamma}$. We have measured these quantities over 5 decades of strain rate for a 238-bubble system at a packing fraction of $\phi = 0.95$ (the close-packing fraction is roughly 0.84). Note that η, which diverges as $\dot{\gamma} \to 0$, is not a pure power law as a function of $\dot{\gamma}$; it requires two different power laws and an additive constant to fit the data over the range shown. In the infinite shear rate limit, η approaches a constant, η_∞, which can be calculated analytically by assuming perfectly laminar flow and integrating the viscous forces over the time that bubbles overlap.

Figure 1 also shows that Γ appears to level off at low strain rates. Although we cannot deduce the limiting behavior of Γ directly from the data, we know that it should approach a nonzero constant, Γ_0, in the limit $\dot{\gamma} \to 0$, by the following argument. As $\dot{\gamma}$ approaches zero, the system reduces to a simple network of springs, so the elastic constant $\partial\langle\sigma_{yy}\rangle/\partial h$ must approach a finite, nonzero constant. The stress fluctuations $\langle(\sigma_{yy} - \langle\sigma_{yy}\rangle)^2\rangle_h$ also approach a finite, nonzero constant, because as $\dot{\gamma} \to 0$ (*i.e.* the quasistatic limit), the system still explores its configuration space and the stress fluctuates, however slowly. (In the quasistatic limit the system is sheared infinitesimally and allowed to equilibrate before being sheared again.) *At* $\dot{\gamma} = 0$, of course, there are no stress fluctuations. In this sense, $\dot{\gamma} \to 0$ is a singular limit. Since both $\partial\langle\sigma_{yy}\rangle/\partial h$ and $\langle(\sigma_{yy} - \langle\sigma_{yy}\rangle)^2\rangle_h$ are nonzero and finite in the quasistatic limit, it follows that the ratio Γ defined in eq. (2) *must* approach a finite, nonzero constant, Γ_0. We

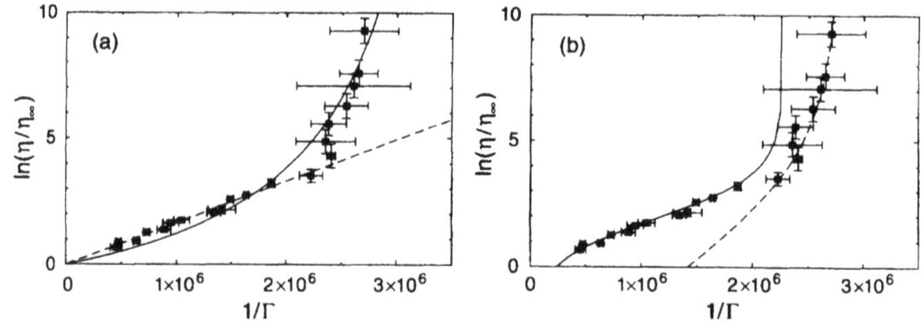

Fig. 2 – (a) An Arrhenius plot of $\ln \eta$ *vs.* $1/\Gamma$ from our simulations (solid squares) at $\phi = 0.95$. The dashed line is a fit to the Arrhenius form, $\eta/\eta_\infty = \exp[A_1/\Gamma]$, at high values of Γ. The solid line is a fit to the Vogel-Fulcher form, which is often used to fit data on supercooled liquids. The deviation from the dashed line indicates super-Arrhenius behavior. (b) Power law fits to the high-Γ (solid) and low-Γ (dashed) data. The solid curve corresponds to an exponent of 0.9 and the dashed curve to an exponent of 2.5. The latter exponent is consistent with mode-coupling theory, which is unsurprising given our limited dynamic range.

find that Γ_0 is comparable to the average elastic energy per bubble. However, our dynamic range is limited compared to that of experiments, so we cannot estimate the value of Γ_0 with much reliability.

Our aim is to determine whether jamming leads to common behavior in sheared foams and supercooled liquids. The most striking feature of jamming in a supercooled liquid is the stupendous rise in the viscosity with decreasing temperature. We have measured a quantity Γ that characterizes stress fluctuations in our driven system in the same way that temperature measures stress fluctuations in an equilibrium system. We therefore study the behavior of η as a function of Γ. This is shown in the solid symbols of fig. 2(a), where we have plotted $\ln \eta$ *vs.* $1/\Gamma$ [13]. A straight line on this plot would correspond to Arrhenius behavior. At high Γ, the behavior is indeed Arrhenius, as shown by the fit to the dashed line, which has the form $\eta/\eta_\infty = \exp[A_1/\Gamma]$, with $A_1 = 1.6 \times 10^{-6}$. The Arrhenius form implies that A_1 should characterize the height of energy barriers to bubble rearrangements in the foam. In our case, we can measure the height of these barriers by measuring the elastic energy per bubble as a function of strain as the system is sheared very slowly. As bubbles overlap (distort) the elastic energy rises, and when they rearrange the elastic energy drops. Thus, the distribution of energy rises measures the heights of barriers that the bubbles cross as the system is strained. The resulting distribution of barrier heights per bubble is shown in fig. 3. At low shear rates, the elastic energy rise distribution approaches a well-defined quasistatic limit [14]. The distribution is quite broad, with a power law region that is cut off at the high-barrier end. The average barrier height is $\langle \delta E \rangle = 5.3 \times 10^{-6}$, which is quite close to our measured value of A_1. Note that A_1 represents some unknown moment of the distribution and that the distribution is fairly broad, so it is not surprising that the two numbers are not exactly the same.

Clearly, the observed behavior is not Arrhenius over the entire range of Γ. Figure 2(a) shows that the behavior is super-Arrhenius at low Γ, in that the viscosity increases more rapidly than predicted by the Arrhenius law. The solid line is a fit to the Vogel-Fulcher form, $\eta/\eta_\infty = \exp[A/(\Gamma - \Gamma_0)]$, with $A = 9 \times 10^{-7}$ and $\Gamma_0 = 2.5 \times 10^{-7}$. *This is the same form that is often used to fit glass transition data* [15]. The ratio $A/\Gamma_0 \approx 3.6$ is a measure of fragility, and corresponds to an extremely fragile glass-forming system [13]. Note that the Vogel-Fulcher form does not fit particularly well at the high-Γ end; this is also typical of

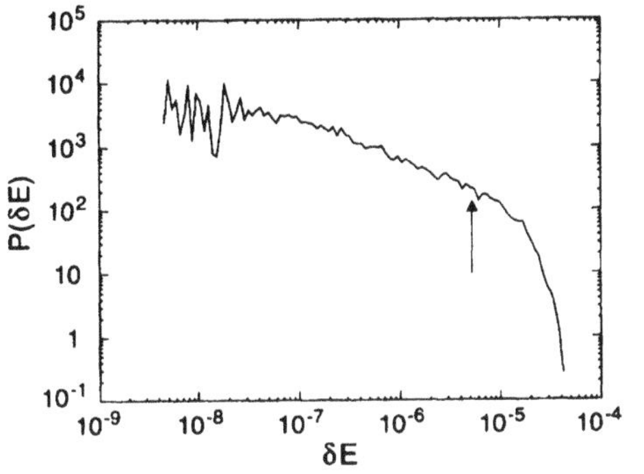

Fig. 3

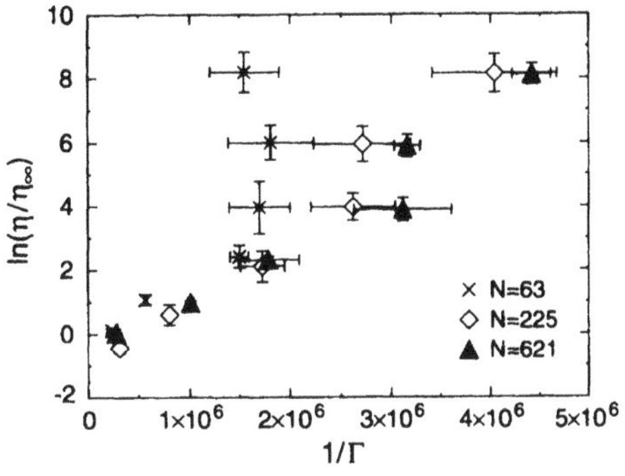

Fig. 4

Fig. 3 – The distribution of energy barriers (measured on a per-bubble basis) crossed by the system as it is strained in the low-shear-rate limit. The average of this distribution (marked by an arrow) provides an estimate of the characteristic energy barrier height that is independent of the Arrhenius parameter A_1.

Fig. 4 – Arrhenius plot of our data for different system sizes, where N is the total number of bubbles. Here, $\phi = 0.9$. The two larger systems yield nearly the same results, but there is a discernible trend towards more Arrhenius behavior with increasing size. Even in the infinite system size limit, however, the behavior must be super-Arrhenius, as discussed in the text.

supercooled liquids, where the Arrhenius form is often used to fit the high-temperature data, and the Vogel-Fulcher form is used to fit the lower-temperature data. Many other functional forms have also been used successfully for supercooled liquids [17]; these forms work equally well for our system.

Skeptics have asked whether our data can be fit by a power law, $\eta \propto 1/(\Gamma - \Gamma_c)^{\alpha}$. We find that a single power law does not fit the data well over the entire range. However, the high-Γ and low-Γ data can be fit reasonably well by two different power laws, as shown in fig. 2b. The high-Γ fit corresponds to an exponent of $\alpha = 0.9$ and diverges at $1/\Gamma_c = 2.3 \times 10^6$; this evidently underestimates the value of $1/\Gamma_c$. The low-Γ fit, which covers a viscosity range of nearly 3 decades, corresponds to an exponent of $\alpha = 2.5$ and diverges at $1/\Gamma_c = 2.8 \times 10^6$. The latter exponent is in good agreement with the prediction of mode-coupling theory [18]. Viscosity measurements near the colloidal glass transition, which cover a similar dynamic range, are consistent with a similar power law [19]. Data for supercooled liquids also can be fit by a comparable exponent over the same dynamic range [18], but this power law fit predicts a divergence well above the glass transition temperature so it is known to fail badly at higher viscosities.

It is not possible to provide a definite answer to whether our data should be interpreted in terms of a power law or super-Arrhenius form. The same difficulty plagues studies of the colloidal glass transition [19], which cover a similar dynamic range, as well as studies of supercooled liquids, which cover 10 more decades of dynamic range. However, given that a single power law does not fit our data well over the whole range, and that the Arrhenius form at high Γ agrees with an independent measurement of the barrier height, we believe that it is more natural to interpret our data as super-Arrhenius rather than power law.

Since our data are taken for a small system (238 bubbles), it is important to study the system size dependence of the observed trend. This is shown in fig. 4 for three different system sizes at a packing fraction of $\phi = 0.9$. The 63-bubble system is clearly too small, and shows a very different dependence of η on Γ. There is also a difference between the results for 224 and 621 bubbles. It is small compared to the error bars but there is a systematic trend towards more Arrhenius behavior with increasing system size. Thus one might wonder if the behavior is Arrhenius for infinite systems. We argue that it must still be super-Arrhenius in that limit, because the viscosity diverges as $\dot{\gamma} \to 0$, but Γ approaches Γ_0, a nonzero constant. We note, however, that even if the behavior were Arrhenius, it would still be nontrivial and remarkable.

We find that the super-Arrhenius behavior depends only weakly on packing fraction. We have studied 3 different packing fractions: $\phi = 0.85$, just above close-packing, $\phi = 0.90$ and $\phi = 0.95$. At the lowest packing fraction, the behavior is nearly Arrhenius. Above $\phi = 0.85$, the behavior is fairly insensitive to ϕ, but Γ_0 increases somewhat with ϕ. This trend, as well as the much weaker dependence of viscosity on ϕ than on Γ, is consistent with experiments on supercooled liquids at constant density [20].

Finally, we have studied the effect of frustration on the super-Arrhenius behavior. A monodisperse system will crystallize under shear. We can therefore increase frustration by increasing the polydispersity. The results presented so far are for a flat distribution of width $w = 0.8$, so that the bubbles range from $1 - w = 0.2$ to $1 + w = 1.8$ times the average radius. We have also studied the case $w = 0.2$. We find that the curvature on an Arrhenius plot (with Γ on the abscissa instead of Γ/Γ_0) is greater for the less-frustrated system ($w = 0.2$); this reflects the increase of Γ_0 with decreasing w. This trend cannot be compared to experimental results for supercooled liquids, since there is no way to measure frustration for those systems. However, we note that a recent theoretical picture of the glass transition does predict the same trend [21].

In this paper, we have suggested a measure of fluctuations in an athermal system under steady-state shear, Γ defined in eq. (2), which reduces to $k_B T$ in an equilibrium system. We have shown that the viscosity of a model foam obeys the same relation with Γ that the viscosity of a supercooled liquid obeys with temperature. While there is no *a priori* reason that fluctuations in a sheared foam should be described by an effective temperature, we note that this assumption appears implicitly in a recent approach to the rheology of soft glassy materials [22], which introduces a temperature variable into systems that are definitely athermal. The concept of a "granular temperature" has also been used extensively to describe driven granular flows [23]. Finally, there is numerical evidence that applying shear may be similar to raising the temperature in supercooled liquids. Molecular-dynamics simulations on a model supercooled liquid, namely a bidisperse repulsive Lennard-Jones system, study kinetic heterogeneities in the steady-state sheared supercooled liquid [24]. These simulations show that the kinetic heterogeneities in the sheared system are the same as those in an unsheared system at a higher temperature.

One reason that the concept of an effective temperature may be of some use to driven athermal systems is that the systems explore many different configurations. A different type of flow could explore a different set of bubble-packing configurations, but perhaps a limited amount of ergodicity is enough for statistical mechanics to be useful, within limits that must be determined. We note also that a steady-state driven system obeys a type of fluctuation-dissipation relation in the sense that the average flow gives rise to fluctuations, but the average flow and fluctuations must be determined self-consistently subject to the constraint that the total energy dissipated must be the same as the energy fed into the system. Clearly, a systematic study of the applicability of the concept of a temperature to driven systems is needed. We are currently investigating whether other definitions of an effective temperature yield results for our model that are consistent with the one proposed here. If there is indeed a common phenomenon of jamming underlying driven, athermal systems and supercooled liquids, however, it may not be necessary for the concept of temperature to be valid; it is possible that *any* reasonable measure of fluctuations in the driven system will lead to qualitatively similar behavior.

* * *

We thank D. J. DURIAN, D. KIVELSON, J. S. LANGER, D. LEVINE, S. R. NAGEL, J. P. SETHNA and G. TARJUS for stimulating discussions. The support of the National Science Foundation under Grant. Nos. PHY94-07194 (SAL and AJL) and CHE-9624090 (AJL) is gratefully acknowledged.

REFERENCES

[1] EDIGER M. D., ANGELL C. A. and NAGEL S. R., *J. Phys. Chem.*, **100** (1996) 13200.
[2] JAEGER H. M., NAGEL S. R. and BEHRINGER R. P., *Rev. Mod. Phys.*, **68** (1996) 1259.
[3] WEAIRE D. and RIVIER N., *Cont. Phys.*, **25** (1984) 59.
[4] GOPAL A. D. and DURIAN D. J., *Phys. Rev. Lett.*, **75** (1995) 2610.
[5] DENNIN M. and KNOBLER C. M., *Phys. Rev. Lett.*, **78** (1997) 2485.
[6] LANDAU L. D. and LIFSCHITZ E. M., *Statistical Physics* (Pergamon Press, New York) 1980.
[7] See eq. (67) in SCHOFIELD P., *Proc. Phys. Soc. (London)*, **88** (1966) 149.
[8] DURIAN D. J., *Phys. Rev. Lett.*, **75** (1995) 4780; *Phys. Rev. E*, **55** (1997) 1739.
[9] It might appear more straightforward to define Γ using the velocity fluctuation distribution. However, the velocity distribution in an equilibrium system defines the ratio of the temperature to a mass, and in our system, the bubbles are massless. This is why we have chosen a definition of Γ that reduces simply to the temperature in an equilibrium system.
[10] MORSE D. C. and WITTEN T. A., *Europhys. Lett.*, **22** (1993) 549.
[11] In three dimensions, there is a logarithmic correction [10]. For the effects of this correction on static properties, see LACASSE M. D., GREST G. S., LEVINE D., MASON T. G. and WEITZ D. A., *Phys. Rev. Lett.*, **76** (1996) 3448; LACASSE M. D., GREST G. S. and LEVINE D., *Phys. Rev. E*, **54** (1996) 5436.
[12] LI X. F. and POZRIKIDIS C., *J. Fluid. Mech.*, **286** (1995) 379.
[13] ANGELL C. A., *J. Non-Cryst. Solids*, **131** (1991) 13.
[14] TEWARI S., SCHIEMANN D., DURIAN D. J., KNOBLER C. M., LANGER S. A. and LIU A. J., preprint (1999).
[15] We have also compared our results directly to experiments on supercooled di-n-butylphthalate [16]. We adjust our scale for Γ so that our ratio Γ/Γ_0 corresponds to their T/T_g. We then find that our data collapses with theirs. However, their dynamic range is 10 orders of magnitude higher!
[16] MENON N., NAGEL S. R. and VENERUS D. C., *Phys. Rev. Lett.*, **73** (1994) 963.
[17] KIVELSON D., TARJUS G., ZHAO X. and KIVELSON S. A., *Phys. Rev. E*, **53** (1996) 751.
[18] GÖTZE W. and SJÖGREN L., *Rep. Prog. Phys.*, **55** (1992) 241.
[19] CHENG Z.-D., Ph. D. dissertation (Princeton University) 1998.
[20] FERRER M. L., LAWRENCE C., DEMIRJIAN B. G., KIVELSON D., ALBA-SIMIONESCO C. and TARJUS G., *J. Chem. Phys.*, **109** (1998) 8010.
[21] KIVELSON D., KIVELSON S. A., ZHAO X., NUSSINOV Z. and TARJUS G., *Physica A*, **219** (1995) 27.
[22] SOLLICH P., LEQUEUX F., HÉBRAUD P. and CATES M. E., *Phys. Rev. Lett.*, **78** (1997) 2020; SOLLICH P., *Phys. Rev. E*, **58** (1998) 738.
[23] SAVAGE S. B. and JEFFREY D. J., *J. Fluid Mech.*, **110** (1981) 255; for a recent reference, see SELA N. and GOLDHIRSCH I., *J. Fluid Mech.*, **361** (1998) 41; for an experimental measurement, see MENON N. and DURIAN D. J., *Science*, **275** (1997) 1920.
[24] YAMAMOTO R. and ONUKI A., *Europhys. Lett.*, **40** (1997) 61; *Phys. Rev. E*, **58** (998) 3515.

Part 9

TRAFFIC

Journal of Statistical Physics, Vol. 69, Nos. 3/4, 1992

An Exact Solution of a One-Dimensional Asymmetric Exclusion Model with Open Boundaries

B. Derrida,[1] E. Domany,[2] and D. Mukamel[3]

Received February 27, 1992; final May 21, 1992

A simple asymmetric exclusion model with open boundaries is solved exactly in one dimension. The exact solution is obtained by deriving a recursion relation for the steady state: if the steady state is known for all system sizes less than N, then our equation (8) gives the steady state for size N. Using this recursion, we obtain closed expressions (48) for the average occupations of all sites. The results are compared to the predictions of a mean field theory. In particular, for infinitely large systems, the effect of the boundary decays as the distance to the power $-1/2$ instead of the inverse of the distance, as predicted by the mean field theory.

KEY WORDS: Asymmetric exclusion process; steady state; phase diagram.

1. INTRODUCTION

Systems of particles with stochastic dynamics and exclusion interactions have been studied for a long time in statistical mechanics.[1] Despite their simplicity, relatively few exactly soluble cases are known,[2-7] especially when the invariant measure does not factorize.

An important class of problems deals with asymmetric exclusion processes with periodic boundary conditions. In these cases the system reaches a stationary state of constant density, and one is interested in density fluctuations and their correlations.[8-10] Recent interest in these problems is at least partly due to their close relationship to growth models,[9] whose continuum version, the KPZ equation,[10] is related in turn to the exactly soluble (in one dimension!) Burgers equation. A direct connection between

[1] Service de Physique Théorique, CE Saclay F-91191 Gif sur Yvette, France.
[2] Department of Electronics, Weizmann Institute of Science, 76100 Rehovot, Israel.
[3] Department of Physics, Weizmann Institute of Science, 76100 Rehovot, Israel.

0022-4715/92/1100-0667$06.50/0

driven lattice gas models and the noisy Bugers equation was also established.[20] On the other hand, exclusion processes in discrete space-time are related to vertex models.[6]

The purpose of this paper is to present a case which can be solved exactly: the fully asymmetric exclusion process on a segment of finite length with *open boundaries* with particles injected at one end with probability α and removed at the other end with probability β. Here, unlike the case of periodic boundary conditions, the density is *not* uniform.[11] This type of problem is related to growth models with a defect or inhomogeneity.[12-14] These models are interesting for a number of reasons. They provide examples of systems far from thermal equilibrium with long-range spatial and temporal correlations. In some cases they exhibit phase transitions in one dimension, a phenomenon which does not usually occur in systems at thermal equilibrium.

The model studied here is defined as follows. Consider a one-dimensional system of N sites. Each site i, $1 \leqslant i \leqslant N$, is either occupied by a particle ($\tau_i = 1$) or is empty ($\tau_i = 0$). This system evolves in time according to the following rule: At each time step $t \to t+1$, one chooses at random an integer $0 \leqslant i \leqslant N$ with probability $1/(N+1)$. If the integer i is between 1 and $N-1$, then the particle on site i (if there is one) jumps to site $i+1$ (if this site is empty), i.e.,

$$\begin{aligned} \tau_i(t+1) &= \tau_i(t)\,\tau_{i+1}(t) \\ \tau_{i+1}(t+1) &= \tau_{i+1}(t) + [1-\tau_{i+1}(t)]\,\tau_i(t) \end{aligned} \tag{1}$$

If the integer chosen is $i=0$, then site 1 remains occupied at time $t+1$ if it was occupied at time t, and it gets occupied with probability α if it was empty at time t. Therefore

$$\begin{aligned} \tau_1(t+1) &= 1 \qquad \text{with probability} \quad \tau_1(t) + \alpha[1-\tau_1(t)] \\ &= 0 \qquad \text{with probability} \quad (1-\alpha)[1-\tau_1(t)] \end{aligned} \tag{2}$$

Similarly, if the integer chosen is $i=N$, then site N remains empty at time $t+1$ if it was empty at time t, and it gets empty with probability β if it was occupied. So

$$\begin{aligned} \tau_N(t+1) &= 1 \qquad \text{with probability} \quad (1-\beta)\,\tau_N(t) \\ &= 0 \qquad \text{with probability} \quad 1-(1-\beta)\,\tau_N(t) \end{aligned} \tag{3}$$

In the present paper, the exact steady state is obtained for arbitrary α and β (Section 2 and the Appendix). The result is expressed as a recursion relation which relates the steady state of a system of length $N-1$ to that of a system of length N. This relation allows one to calculate the exact expression of the average occupation $\langle \tau_i \rangle_N$ of any site i of a chain of N sites. The expression of the $\langle \tau_i \rangle_N$ is derived in the present paper only for the case $\alpha = \beta = 1$ (Section 3), but there is in principle no obstacle (other than a long calculation) to extend the results to arbitrary α and β. In Section 4, the results are discussed in several asymptotic limits (large N, large i). Finally, these exact results are compared with the predictions of a mean field theory in Section 5.

2. CONSTRUCTION OF THE STEADY STATE

From the stochastic dynamics defined in the introduction [Eqs. (1)–(3)], it is easy to write the equation satisfied by the steady state. If one denotes by $P_N(\tau_1,..., \tau_N)$ the steady-state probability of finding the system in configuration $\{\tau_1,..., \tau_N\}$, then the probability P_N satisfies

$$\begin{aligned}
P_N(\tau_1, \tau_2,..., \tau_N) &= \frac{1-\alpha}{N+1} P_N(\tau_1, \tau_2,..., \tau_N) \\
&+ \frac{\alpha}{N+1} \tau_1 [P_N(0, \tau_2,..., \tau_N) + P_N(1, \tau_2,..., \tau_N)] \\
&+ \frac{1}{N+1} [P_N(\tau_1, \tau_2,..., \tau_N) + (\tau_2 - \tau_1) P_N(1, 0, \tau_3,..., \tau_N)] \\
&+ \cdots \\
&+ \frac{1}{N+1} [P_N(\tau_1, \tau_2,..., \tau_N) + (\tau_N - \tau_{N-1}) P_N(\tau_1, \tau_2,..., \tau_{N-2}, 1, 0)] \\
&+ \frac{1-\beta}{N+1} P_N(\tau_1, \tau_2,..., \tau_N) \\
&+ \frac{\beta}{N+1} (1-\tau_N)[P_N(\tau_1,..., \tau_{N-1}, 0) + P_N(\tau_1,..., \tau_{N-1}, 1)]
\end{aligned} \tag{4}$$

which can be written as

$$\begin{aligned}
&\alpha(2\tau_1 - 1) P_N(0, \tau_2,..., \tau_N) \\
&\quad + (\tau_2 - \tau_1) P_N(1, 0, \tau_3,..., \tau_N) \\
&\quad + \cdots \\
&\quad + (\tau_N - \tau_{N-1}) P_N(\tau_1, \tau_2,..., \tau_{N-2}, 1, 0) \\
&\quad + \beta(1 - 2\tau_N) P_N(\tau_1, \tau_2,..., \tau_{N-1}, 1) = 0
\end{aligned} \tag{5}$$

The problem of finding the steady state amounts to solving the 2^N coupled equations (5). As explained in the Appendix, it is possible to write a recursion relation which gives $P_{N+1}(\tau_1, \tau_2,..., \tau_{N+1})$ if one knows the $P_N(\tau_1, \tau_2,..., \tau_N)$. In what follows, it will be more convenient to write recursions for weights $f_N(\tau_1, \tau_2,..., \tau_N)$ which are equal to the $P_N(\tau_1, \tau_2,..., \tau_N)$, up to a multiplicative constant. Clearly, since Eqs. (4) and (5) are linear, they are also satisfied by the f_N and the probabilities P_N will then be given by

$$P_N(\tau_1,..., \tau_N) = f_N(\tau_1,..., \tau_N)/Z_N \tag{6}$$

where

$$Z_N = \sum_{\tau_1=0,1} \cdots \sum_{\tau_N=0,1} f_N(\tau_1,..., \tau_N) \tag{7}$$

We prove in the Appendix that all the $f_N(\tau_1, \tau_2,..., \tau_N)$ can be constructed by the following recursion rule:

$$\begin{aligned} f_N(\tau_1, \tau_2,..., \tau_N) &\\ = \alpha\tau_N f_{N-1}(\tau_1, \tau_2,..., \tau_{N-1}) &\\ + \alpha\beta(1-\tau_N)\,\tau_{N-1}[f_{N-1}(\tau_1, \tau_2,..., \tau_{N-2}, 1) &\\ + f_{N-1}(\tau_1, \tau_2,..., \tau_{N-2}, 0)] &\\ + \cdots &\\ + \alpha\beta(1-\tau_N)(1-\tau_{N-1})\cdots(1-\tau_2)\,\tau_1 &\\ \times [f_{N-1}(1, \tau_2,..., \tau_{N-1}) + f_{N-1}(0, \tau_2,..., \tau_{N-1})] &\\ + \beta(1-\tau_N)(1-\tau_{N-1})\cdots(1-\tau_1)\, f_{N-1}(\tau_1, \tau_2,..., \tau_{N-1}) & \end{aligned} \tag{8}$$

The problem is, of course, easy to solve directly for small N: for example, for $N = 1$, one can show that

$$f_1(0) = \beta \qquad \text{and} \qquad f_1(1) = \alpha \tag{9a}$$

and for $N = 2$ the choice

$$f_2(0, 0) = \beta^2; \quad f_2(1, 0) = \alpha\beta(\alpha+\beta); \quad f_2(0, 1) = \alpha\beta; \quad f_2(1, 1) = \alpha^2 \tag{9b}$$

solves the steady state equation (4) or (5). So (8) together with (9) enables one to obtain recursively the f_N for all N.

Because the recursion (8) is rather complicated, the calculation of expectations in the steady state is not immediate. In Section 3 we shall see how the average occupations $\langle \tau_i \rangle_N$ can be obtained from (8) in the case $\alpha = \beta = 1$. It is, however, interesting to note that since the recursion (8) is valid for any choice of α and β, there is no difficulty, other than doing a long calculation, to extend the results of Section 3 to the more general case of arbitrary α and β.

Before discussing how the calculation of average occupations can be done from (8), it is worth noting that in the case

$$\alpha + \beta = 1 \tag{10}$$

there is a simple closed expression of the f_N

$$f_N(\tau_1, \tau_2, ..., \tau_N) = \alpha^T (1-\alpha)^{N-T}, \qquad T = \sum_i \tau_i \tag{11}$$

which solves (4) and (5).

This means that when (10) holds, the steady state factorizes and leads to very simple expressions of all the correlation functions,

$$\langle \tau_i \rangle = \alpha \qquad \text{and} \qquad \langle \tau_i \tau_j \cdots \tau_k \rangle = \langle \tau_i \rangle \langle \tau_j \rangle \cdots \langle \tau_k \rangle \tag{12}$$

3. THE AVERAGE OCCUPATIONS $\langle \tau_i \rangle_N$

In this section we will obtain closed expressions for the average occupation $\langle \tau_i \rangle_N$ of site i for a system of length N. From the recursion (8) on f_N, it is clear that the only quantities needed to obtain the $\langle \tau_i \rangle_N$ are the following sums:

$$Z_N = \sum_{\tau_1} \cdots \sum_{\tau_N} f_N(\tau_1, ..., \tau_N) \tag{13}$$

and

$$T_{N,i} = \sum_{\tau_1} \cdots \sum_{\tau_N} \tau_i f_N(\tau_1, ..., \tau_N) \tag{14}$$

Once these sums are known, the expression for $\langle \tau_i \rangle_N$ is simply given by

$$\langle \tau_i \rangle = T_{N,i}/Z_N \tag{15}$$

The difficulty in computing the Z_N and the $T_{N,i}$ from the recursion (8) is that one cannot get closed recursions for these quantities. It is necessary to compute at the same time other quantities: the $Y_N(K)$ and the $X_N(K, p)$ defined by

$$Y_N(K) = \sum_{\tau_1} \cdots \sum_{\tau_N} (1-\tau_N)(1-\tau_{N-1}) \cdots (1-\tau_K) f_N(\tau_1, ..., \tau_N) \tag{16}$$

and

$$X_N(K, p) = \sum_{\tau_1} \cdots \sum_{\tau_N} (1-\tau_N) \cdots (1-\tau_K)\, \tau_p f_N(\tau_1, \ldots, \tau_N) \tag{17}$$

By extension of the definition of the $Y_N(K)$ and of the $X_N(K, p)$ we shall define for convenience

$$Y_N(N+1) = Z_N \tag{18}$$

and

$$X_N(N+1, p) = T_{N,p} \tag{19}$$

Having defined the $Y_N(K)$ for $1 \leqslant K \leqslant N+1$ and the $X_N(K, p)$ for $p+1 \leqslant K \leqslant N+1$, one can obtain from (8) closed recursions for these quantities. For the Y_N one gets

$$\begin{aligned} Y_N(1) &= \beta Y_{N-1}(1) \\ Y_N(K) &= Y_N(K-1) + \alpha\beta Y_{N-1}(K) \qquad \text{for} \quad 2 \leqslant K \leqslant N \\ Y_N(N+1) &= Y_N(N) + \alpha Y_{N-1}(N) \end{aligned} \tag{20}$$

This recursion, together with the initial conditions

$$Y_1(1) = \beta \qquad \text{and} \qquad Y_1(2) = \alpha + \beta \tag{21}$$

determines all the $Y_N(K)$.

Clearly, the length of the calculation to determine all the $Y_N(K)$ increases like N^2 (instead of exponentially with N, as we would have needed had we had to calculate all the f_N).

Once the Y_N are determined, one can determine recursively all the X_N from (8):

$$\begin{aligned} X_N(p+1, p) &= \alpha\beta Y_{N-1}(p+1) && \text{for} \quad 1 \leqslant p \leqslant N-1 \\ X_N(K, p) &= X_N(K-1, p) + \alpha\beta X_{N-1}(K, p) && \text{for} \quad p+2 \leqslant K \leqslant N \\ X_N(N+1, p) &= X_N(N, p) + \alpha X_{N-1}(N, p) && \text{for} \quad 1 \leqslant p \leqslant N-1 \\ X_N(N+1, N) &= \alpha Y_{N-1}(N) \end{aligned} \tag{22}$$

This recursion, together with the initial condition

$$X_1(2, 1) = \alpha \tag{23}$$

and the knowledge of the $Y_N(K)$ obtained from (20) and (21), determine all the $X_N(K, p)$.

At this stage we see that one can obtain the exact values of all the $Y_N(K)$ and the $X_N(K, p)$ for arbitrary N, for any choice of α and β, by iterating the recursions (20) and (22) on a computer.

In order to obtain closed analytic expressions, we will limit our calculations to the case

$$\alpha = \beta = 1 \tag{24}$$

The reason for this choice is to avoid manipulation of too complicated expressions in the calculations below. However, since the recursions (20) and (22) are valid for any α and β, there is nothing to prevent extension of the results to arbitrary α and β.

A possible way to proceed with the recursions (20) and (22) is to introduce generating functions. If one defines for $p \geqslant -1$

$$L_p(\lambda) = \sum_{N=p+1}^{\infty} \lambda^N Y_N(N-p) \tag{25}$$

with the convention that $Y_0(1) = 1$, the recursion (20) on $Y_N(K)$ becomes

$$L_p(\lambda) - \lambda^{p+1} Y_{p+1}(1) = L_{p+1}(\lambda) + \lambda[L_{p-1}(\lambda) - \lambda^p Y_p(1)] \qquad \text{for} \quad p \geqslant 0 \tag{26}$$

and

$$L_{-1}(\lambda) = \lambda L_{-1}(\lambda) + 1 + L_0(\lambda) \tag{27}$$

Because the $Y_p(1)$ are easy to calculate [$Y_p(1) = 1$ for all p], one immediately concludes from (26) that the general solution of (26) is

$$L_p = A_1(\lambda)\left(\frac{1-(1-4\lambda)^{1/2}}{2}\right)^{p+1} + A_2(\lambda)\left(\frac{1+(1-4\lambda)^{1/2}}{2}\right)^{p+1} \tag{28}$$

The problem now is to determine the constants $A_1(\lambda)$ and $A_2(\lambda)$. From the definition (25), it is clear that for any p the first term in the sum $L_p(\lambda)$ is proportional to λ^{p+1}. This is compatible with the result (28) only if

$$A_2(\lambda) = 0 \tag{29}$$

The expression of $A_1(\lambda)$ is then easy to find from the boundary condition (27) and one ends up with

$$L_p(\lambda) = \left(\frac{1-(1-4\lambda)^{1/2}}{2}\right)^{p+3} \frac{1}{\lambda^2} \tag{30}$$

From the definitions (18) and (25), it is clear that this allows one to obtain all the Z_N since

$$\sum_{N=0}^{\infty} Z_N \lambda^N = L_{-1}(\lambda) = \left(\frac{1-(1-4\lambda)^{1/2}}{2\lambda}\right)^2 = \frac{1-2\lambda-(1-4\lambda)^{1/2}}{2\lambda^2} \tag{31}$$

This leads to the following expression for the Z_N:

$$Z_N = \frac{(2N+2)!}{(N+1)!\,(N+2)!} \tag{32}$$

One can also treat the recursions on the $X_N(K, p)$ using generating functions. If one defines

$$M_{q,r}(\lambda) = \sum_{N=r+1}^{\infty} \lambda^N X_N(N-q, N-r) \tag{33}$$

then it follows from the recursions (22) that

$$\begin{aligned}
M_{r-1,r}(\lambda) &= \lambda L_{r-2}(\lambda) - \lambda^r && \text{for } r \geqslant 1 \\
M_{q,r}(\lambda) &= M_{q+1,r}(\lambda) + \lambda M_{q-1,r-1} && \text{for } 0 \leqslant q \leqslant r-2 \\
M_{-1,r}(\lambda) &= M_{0,r}(\lambda) + \lambda M_{-1,r-1}(\lambda) && \text{for } r \geqslant 1 \\
M_{-1,0}(\lambda) &= \lambda L_{-1}(\lambda)
\end{aligned} \tag{34}$$

To proceed, one can introduce new generating functions $H_q(\mu, \lambda)$ defined by

$$H_q(\mu, \lambda) = \sum_{r=q+1}^{\infty} \mu^r M_{q,r}(\lambda) \tag{35}$$

Then the recursions (34) become

$$H_{q+1} - H_q + \lambda\mu H_{q-1} = \lambda\mu^{q+1}(\lambda L_{q-2} - L_{q-1}) \qquad \text{for } q \geqslant 1 \tag{36}$$

$$H_1 - H_0 + \lambda\mu H_{-1} = \lambda\mu(\lambda - 1)\, L_{-1} + \lambda\mu \tag{37}$$

$$H_{-1} = H_0 + \lambda\mu H_{-1} + \lambda L_{-1} \tag{38}$$

The general solution of (36) with the boundary conditions (37) and (38) is

$$\begin{aligned}
H_q(\mu, \lambda) = {} & \frac{1}{\lambda(1-\mu)}\left(\frac{1-(1-4\lambda)^{1/2}}{2}\right)^{q+2} \mu^{q+1} \\
& + B_1(\mu, \lambda)\left(\frac{1-(1-4\lambda\mu)^{1/2}}{2}\right)^{q+1} \\
& + B_2(\mu, \lambda)\left(\frac{1+(1-4\lambda\mu)^{1/2}}{2}\right)^{q+1}
\end{aligned} \tag{39}$$

As before, one can argue that

$$B_2(\mu, \lambda) = 0 \tag{40}$$

because this is the only way that the leading order in H_q is of the order μ^{q+1}. Then from the boundary conditions (37) and (38), one determines the constant $B_1(\mu, \lambda)$ and one ends up with the following final expression:

$$\begin{aligned} H_q(\mu, \lambda) = & \frac{1}{\lambda(1-\mu)} \left(\frac{1-(1-4\lambda)^{1/2}}{2} \right)^{q+2} \mu^{q+1} \\ & + \frac{1}{(1-\mu)\,\lambda^2 \mu^2} \left(\frac{1-(1-4\lambda\mu)^{1/2}}{2} \right)^{q+3} \\ & \times \left[\mu \frac{1-(1-4\lambda)^{1/2}}{2} - 1 \right] \end{aligned} \tag{41}$$

To obtain the average occupations, it turns out that only $H_{-1}(\mu, \lambda)$ is needed [see (19), (33) and (35)] and one finally gets from (41)

$$H_{-1}(\mu, \lambda) = \frac{1}{1-\mu} \frac{1-(1-4\lambda\mu)^{1/2}}{2\lambda\mu} \left[\frac{1-(1-4\lambda)^{1/2}}{2\lambda} - \frac{1-(1-4\lambda\mu)^{1/2}}{2\lambda\mu} \right] \tag{42}$$

This expression is the generating function of all the $T_{N,i}$ [see (14), (19), (33), and (35)] since

$$H_{-1}(\mu, \lambda) = \sum_{r=0}^{\infty} \mu^r \sum_{N=r+1}^{\infty} \lambda^N T_{N,N-r} \tag{43}$$

Using expression (42) and the identity

$$\frac{1-(1-4x)^{1/2}}{2x} = \sum_{n=0}^{\infty} \frac{(2n)!}{n!\,(n+1)!} x^n \tag{44}$$

one can rewrite H_{-1} as

$$\begin{aligned} H_{-1}(\mu, \lambda) &= \frac{1}{1-\mu} \sum_{q=0}^{\infty} \frac{(2q)!}{q!\,(q+1)!} (\lambda\mu)^q \sum_{n=1}^{\infty} \frac{(2n)!}{n!\,(n+1)!} [\lambda^n - (\lambda\mu)^n] \\ &= \sum_{q=0}^{\infty} \sum_{n=1}^{\infty} \frac{(2q)!}{q!\,(q+1)!} \frac{(2n)!}{n!\,(n+1)!} \lambda^{q+n} \mu^q \sum_{m=0}^{n-1} \mu^m \\ &= \sum_{q=0}^{\infty} \sum_{m=0}^{\infty} \sum_{n=m+1}^{\infty} \frac{(2q)!}{q!\,(q+1)!} \frac{(2n)!}{n!\,(n+1)!} \lambda^{q+n} \mu^{q+m} \end{aligned} \tag{45}$$

By making the change of variables $q+m=r$ and $N=n+q$, one ends up with

$$H_{-1}(\mu, \lambda)=\sum_{r=0}^{\infty} \sum_{N=1}^{\infty} \lambda^N \mu^r \sum_{q=0}^{r} \frac{(2q)!}{q!\,(q+1)!} \frac{(2N-2q)!}{(N-q)!\,(N-q+1)!} \qquad (46)$$

By comparing with (43), we see that this gives closed expressions for the $T_{N,i}$ and therefore from (15) and (32), one obtains

$$\langle \tau_{N-r} \rangle_N = \sum_{q=0}^{r} \frac{(2q)!}{q!\,(q+1)!} \frac{(2N-2q)!}{(N-q+1)!\,(N-q)!} \frac{(N+2)!\,(N+1)!}{(2N+2)!} \qquad (47)$$

It is easy to prove by recursion over r that (47) can be simplified and rewritten as

$$\langle \tau_K \rangle_N = \frac{1}{2} + \frac{1}{4} \frac{(2K)!}{(K!)^2} \frac{(N!)^2}{(2N+1)!} \frac{(2N-2K+2)!}{[(N-K+1)!]^2} (N-2K+1) \qquad (48)$$

This expression is the main result derived in the present paper. The initial method used to obtain it had nothing to do with the above derivation. It was first obtained by solving exactly the problem on a computer for small sizes ($N \leqslant 10$). By looking at the numerical results, it appeared that some $\langle \tau_i \rangle_N$ were simple rational numbers and (48) was first obtained by trying to find an expression which would agree with these numerical results. It is only after this formula had been guessed as a conjecture that the above method was developed to demonstrate its validity.

It is worth noting that in the whole problem (when $\alpha=\beta=1$), there is a particle–hole symmetry which implies that

$$\langle \tau_K \rangle_N = 1 - \langle \tau_{N+1-K} \rangle_N \qquad (49)$$

This relation, which could have been guessed from the very beginning, is of course satisfied by (48).

4. SOME LIMITING CASES

From the exact expression (48), it is easy to obtain the expression for $\langle \tau_K \rangle_N$ in some asymptotic limits corresponding to large systems.[15,16]

First let us consider the limit $N \to \infty$ with fixed K. One gets

$$\langle \tau_K \rangle_\infty = \frac{1}{2} + \frac{(2K)!}{(K!)^2} \frac{1}{2^{2K+1}} \qquad (50)$$

Then if K becomes large, keeping $N \to \infty$ first, one obtains

$$\langle \tau_K \rangle_\infty - \frac{1}{2} \simeq \frac{1}{2\sqrt{\pi}} K^{-1/2} \tag{51}$$

This $K^{-1/2}$ convergence is consistent with a conjecture that was verified by simulations[11] and with what was found recently in a model of growth with an inhomogeneity.[14]

Another limit of interest is when N and K are large, with $|N-K| \gg 1$ and $K \gg 1$. Then, using Stirling's formula, one gets

$$\langle \tau_K \rangle_N - \frac{1}{2} \simeq \frac{1}{\sqrt{\pi}} \frac{1}{\sqrt{N}} \frac{1+N-2K}{[4K(N-K)]^{1/2}} \tag{52}$$

which becomes, in the scaling limit $K = Nx$,

$$\langle \tau_K \rangle_N - \frac{1}{2} \simeq \frac{1}{\sqrt{\pi}} \frac{1}{\sqrt{N}} \frac{1-2x}{2[x(1-x)]^{1/2}} \tag{53}$$

5. COMPARISON TO THE MEAN FIELD PREDICTIONS

In this section we see how a mean field theory can be developed for the model defined in the introduction. To do so, it is useful to first obtain some exact relations which relate correlation functions in the steady state.

The easiest relation one can obtain is by requiring that in the steady state the average occupation $\langle \tau_i(t) \rangle$ remains unchanged.[17] Consider a site i $(2 \leqslant i \leqslant N-2)$ at time t with an occupation number $\tau_i(t)$. Then the occupation $\tau_i(t+1)$ at time $t+1$ is given by

$$\begin{aligned} \tau_i(t+1) &= \tau_i(t) && \text{with probability } 1 - \frac{2}{N+1} \\ &= \tau_i(t) + [1-\tau_i(t)]\, \tau_{i-1}(t) && \text{with probability } \frac{1}{N+1} \\ &= \tau_i(t)\, \tau_{i+1}(t) && \text{with probability } \frac{1}{N+1} \end{aligned} \tag{54}$$

These three possibilities correspond respectively to the cases where the site updated is different from $i-1$ and i, equal to $i-1$, or equal to i.

Averaging (54), one gets

$$\begin{aligned} \langle \tau_i(t+1) \rangle = \langle \tau_i(t) \rangle + \frac{1}{N+1} [&\langle \tau_i(t)\, \tau_{i+1}(t) \rangle - \langle \tau_i(t)\, \tau_{i-1}(t) \rangle \\ &+ \langle \tau_{i-1}(t) \rangle - \langle \tau_i(t) \rangle] \end{aligned} \tag{55}$$

In the steady state the expectations are time independent and one obtains

$$\langle \tau_i \rangle - \langle \tau_{i+i} \tau_i \rangle = \langle \tau_{i-1} \rangle - \langle \tau_i \tau_{i-1} \rangle \tag{56}$$

This could have been guessed, since it expresses the conservation of the flux of particles.

By similar reasoning one can treat the special cases of sites $i = 1$ and $i = N$, and one gets

$$\begin{aligned} \langle \tau_1 \rangle - \langle \tau_1 \tau_2 \rangle &= \alpha(1 - \langle \tau_1 \rangle) \\ \beta \langle \tau_N \rangle &= \langle \tau_{N-1} \rangle - \langle \tau_N \tau_{N-1} \rangle \end{aligned} \tag{57}$$

The steady-state equations (56) and (57) are exact. They are, however, not very useful to calculate the $\langle \tau_i \rangle$ because the $\langle \tau_i \rangle$ are related to higher correlations ($\langle \tau_i \tau_{i+1} \rangle$) which themselves are related to other correlations.

An approximation scheme very often used in statistical mechanics is the mean field theory, in which the effect of correlations is neglected, i.e., correlations like $\langle \tau_i \tau_j \rangle$ are replaced by $\langle \tau_i \rangle \langle \tau_j \rangle$. If we denote by t_i the value of $\langle \tau_i \rangle$ in the mean field theory, one gets from (56) and (57) the following equations for the t_i:

$$t_i - t_i t_{i+1} = t_{i-1} - t_{i-1} t_i \tag{58a}$$

$$t_1 - t_1 t_2 = \alpha(1 - t_1) \tag{58b}$$

$$\beta t_N = t_{N-1} - t_N t_{N-1} \tag{58c}$$

The solution of these N equations (with N unknowns) determines for any finite N the average occupations t_i.

These equations are nonlinear and therefore are at first sight difficult to solve. There is, however, a simple way of looking at Eqs. (58). We can rewrite (58a) as a recursion

$$t_{i+1} = 1 - \frac{C}{t_i} \tag{59}$$

C is a constant (to be determined later). In fact C is the current of particles in the chain. This recursion is shown in Figs. 1a–1c. For $C < 1/4$ there are two fixed points

$$t_{\pm} = \tfrac{1}{2}[1 \pm (1 - 4C)^{1/2}] \tag{60}$$

t_+ is stable and t_- is unstable.

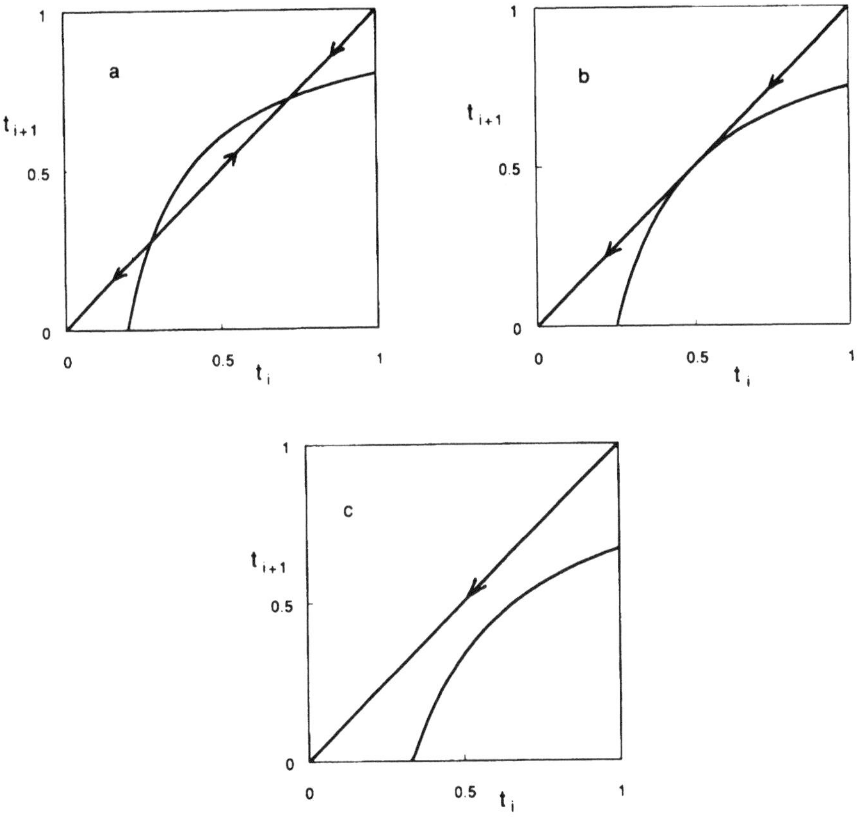

Fig. 1. Graphical representation of the mean-field recursion relation (59) for (a) $C < 1/4$, (b) $C = 1/4$, and (c) $C > 1/4$.

When $C = 1/4$ there is only one marginal fixed point, and there are no real-valued fixed points when $C > 1/4$. The recursion (59) can be solved (using the fact that t_{i+1} is a homographic function of t_i),

$$t_i = \frac{-t_+ t_-(t_+^{i-1} - t_-^{i-1}) + (t_+^i - t_-^i)\, t_1}{-t_+ t_-(t_+^{i-2} - t_-^{i-2}) + (t_+^{i-1} - t_-^{i-1})\, t_1} \tag{61}$$

Clearly t_i depends on C and on t_1; for $i = N$ this relation can be cast as

$$t_N = f(C, t_1) \tag{62}$$

This solution holds for any α, β; these parameters enter through the boundary conditions (58b), (58c), which, together with (62), determine,

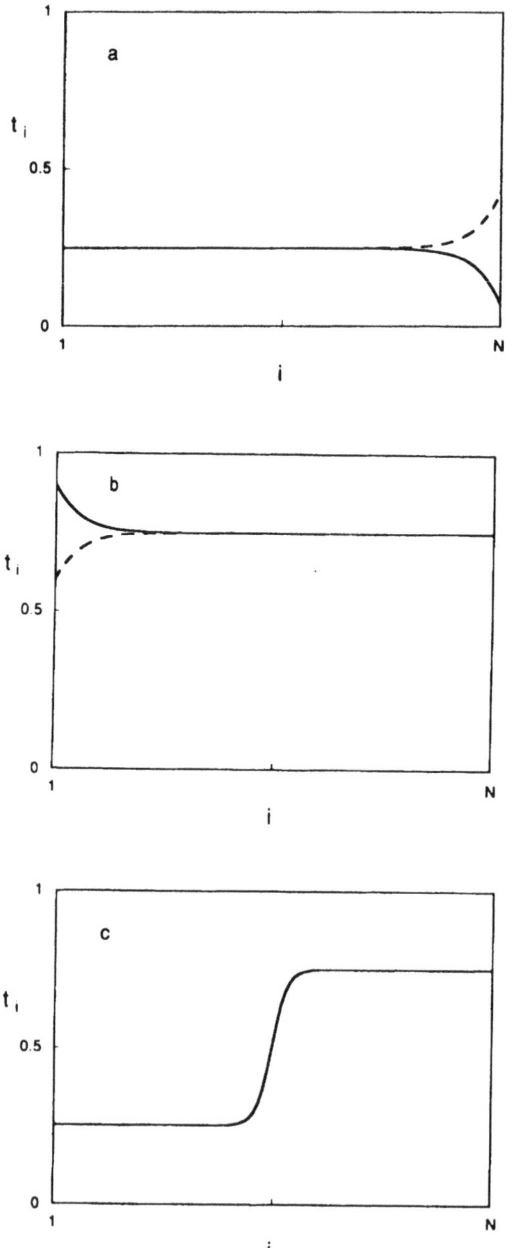

Fig. 2. Schematic mean-field density profiles in (a) the low-density phase, (b) the high-density phase, (c) on the coexistence line, and (d) in the maximal current phase.

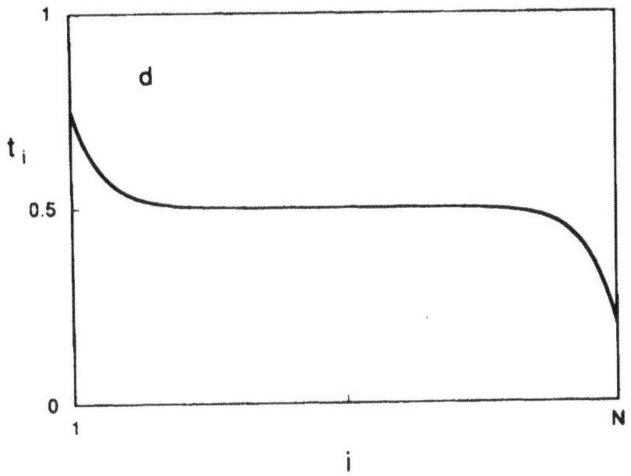

Fig. 2. *(Continued)*

t_1, t_N, and C. We now proceed to show graphically the different kinds of solutions that can be obtained. These correspond to different phases of the system.

Low-Density Phase. $t_1 = t_- + 0^{\pm}$; $t_N < t_+$. In this phase t_1 is set infinitesimally close to the unstable fixed point t_-. For many iterations t_i stays there, and deviates from t_- only when $i \simeq N$, either up or down, depending on t_1. The resulting density profile is shown on Fig. 2a.

The conditions $t_1 = t_-$ and $t_N < t_+$ are consistent with Eqs. (58b), (58c), and (62) provided

$$\alpha \leqslant 1/2, \qquad \beta > \alpha \tag{63}$$

In this regime we have the solution

$$t_1 = \alpha, \qquad t_N = \frac{\alpha(1-\alpha)}{\beta}$$
$$C = \alpha(1-\alpha) \tag{64}$$

High-Density Phase. $t_N = t_+ + 0^{\pm}$, $t_1 > t_-$. In this phase t_1 is set in the domain of attraction of the stable fixed point. Hence t_i relaxes toward t_+, which is (almost) reached by the time we get to t_N. The corresponding density profile is shown on Fig. 2b. This phase occurs when

$$\beta \leqslant 1/2, \qquad \beta < \alpha \tag{65}$$

and the solution is characterized by

$$t_N = 1-\beta, \qquad t_1 = 1 - \frac{\beta(1-\beta)}{\alpha}$$
$$C = \beta(1-\beta) \tag{66}$$

Coexistence Line. $t_1 = t_- + 0^+$, $t_N = t_+ + 0^-$. The high- and low-density phases described above coexist when the recursion starts at t_1 infinitesimally above t_- and iterates to the stable fixed point t_+. The solution, shown in Fig. 2c, contains a front or "domain wall." Such a solution occurs when

$$\alpha = \beta < 1/2 \tag{67}$$

As usual in coexistence regions, the position of this wall depends on the manner in which the limits $\alpha \to \beta$ and $N \to \infty$ are taken. On the coexistence line we have

$$\begin{gathered} t_1 = \alpha, \qquad t_N = 1 - \alpha \\ C = \alpha(1 - \alpha) \end{gathered} \tag{68}$$

Maximal Current Phase. $t_1 \geqslant 1/2$, $t_N \leqslant 1/2$. In this phase the system attains the maximal current it can support,

$$C = 1/4 \tag{69}$$

throughout the phase. The recursion is described by Fig. 1b; actually for any finite N one has the situation of Fig. 1c, with that of Fig. 1b approached in the $N \to \infty$ limit. The exact nature of this limit will be discussed below. The resulting density profile is shown in Fig. 2d.

Except near $i = 1$ and $i = N$, the density is near 1/2. This kind of solution exists when

$$\alpha \geqslant 1/2 \qquad \text{and} \qquad \beta \geqslant 1/2 \tag{70}$$

and is characterized by

$$t_1 = 1 - \frac{1}{4\alpha}, \qquad t_N = \frac{1}{4\beta} \tag{71}$$

The transition to this phase, along the $\beta = 1$ line, was recently studied by Krug.[11]

Phase Diagram. Collecting the results of Eqs. (63), (65), (67), and (70), we obtain the phase diagram shown in Fig. 3. The low- and high-density phases are separated by a first-order transition line. Each of these phases undergoes continuous transitions to the phase at maximal current.

The Density Profile in the $N \to \infty$ Limit. We return now to the maximal current phase, and obtain expressions for the density profile in the

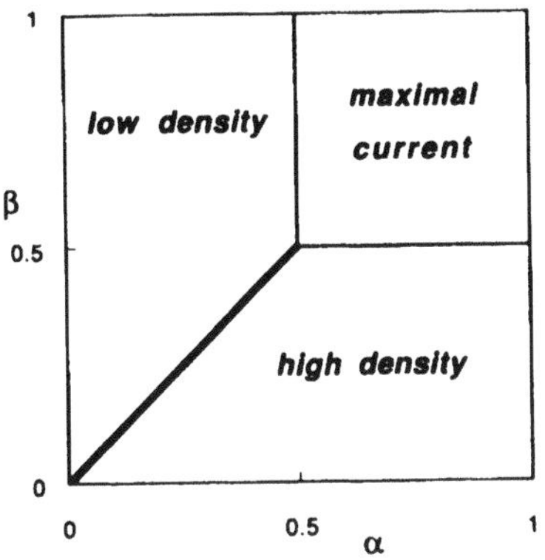

Fig. 3. Mean-field phase diagram. The high- and low-density phases coexist on the heavy line.

$N \to \infty$ limit. These (mean field) expressions can be compared with their exact analogs given in Section 4. For simplicity we restrict the discussion to the case

$$\alpha = \beta = 1 \tag{72}$$

In this case the boundary conditions imply

$$t_1 = 1 - C, \qquad t_N = C, \qquad C = \tfrac{1}{4} + O^+ \tag{73}$$

The density profile, as given by (61) in this case, takes the form

$$t_i = \frac{1}{2} + \frac{1}{2}(t_+ - t_-)\frac{t_+^i + t_-^i}{t_+^{i+1} - t_-^{i+1}} \tag{74}$$

Note that for $C > 1/4$ the fixed points $t_\pm$ are complex. One can now compare the result (74) with those of Section 4.

First, if i is fixed and $N \to \infty$, one finds that $C \to 1/4$ and that t_i is given by

$$t_i = \frac{i+2}{2(i+1)} \tag{75}$$

This implies that the mean field prediction[11] is

$$t_K - \frac{1}{2} = \langle \tau_K \rangle_\infty - \frac{1}{2} \simeq \frac{1}{2K} \tag{76}$$

instead of the $K^{-1/2}$ decay found in the exact solution (51).

Another prediction[11] of the mean field theory which follows from (74) is that as $N \to \infty$,

$$C - \frac{1}{4} = \frac{\pi^2}{4N^2} \tag{77}$$

and

$$\langle \tau_K \rangle - \frac{1}{2} \simeq \frac{\pi}{2N} \frac{\cos(\pi x)}{\sin(\pi x)} \tag{78}$$

for $K = Nx$, to be compared with (53). We see that the mean field prediction is a $1/N$ correction instead of the $1/\sqrt{N}$ correction obtained in the exact solution (53).

6. CONCLUSION

The main results of the present paper are the recursion relation (8), which gives the steady state for systems of arbitrary size N, and the exact expression (48) of the average occupations in the steady state.

Since the recursion relation (8) is valid for arbitrary α and β, it is certainly possible to generalize some of the results of this paper to this more general case. From the knowledge of the steady state one should be able to calculate also the correlation functions, in addition to the average densities.

The phase diagram shown in Fig. 3 was obtained by a mean field calculation. Since the steady state is known exactly for arbitrary α and β, it is certainly possible to obtain from (8) the true phase diagram to check that it remains identical to the mean field phase diagram as conjectured by Krug.[21]

By exploiting the results in terms of the corresponding interface growth model,[12-14] one should be able to obtain explicit expressions describing the fluctuating properties of a growing interface.

Looking at the above results, one can wonder whether exact results could be obtained for the partially asymmetric case (where particles have a nonzero probability of jumping to their left), or to other geometries. A case of interest would be the ring geometry with one special bond having a different hopping rate.[18] At the moment, a direct generalization of our results to these other cases seems difficult. One can try, however, to follow an approach similar to ours, by solving exactly the steady-state equations on a computer for small $N \leqslant 10$ (easy part), guessing the general result from these finite-N results (difficult part), and then, it is hoped, finding a proof (tedious part).

Another possibility would be to try Bethe ansatz techniques,[8,19] which were used recently to calculate the gap for the ring geometry.

APPENDIX

The goal of this Appendix is to prove that recursion (8) together with the initial condition (9) gives the steady state for all N. To prove the recursion (8), we assume that it gives the solution of the steady-state equations for all systems of size up to $N-1$ and show that f_N built from (8) is also a solution of (5). To do so we need to distinguish three cases:

Case 1. $\tau_N = 1$. In this case, according to (8), f_N is given by

$$f_N(\tau_1, \tau_2,..., \tau_N) = \alpha f_{N-1}(\tau_1, \tau_2,..., \tau_{N-1}) \tag{A1}$$

By using the expression (A1) in (5) and the fact that f_{N-1} is a solution of (5), one finds that the condition for f_N to solve (5) is

$$\begin{aligned}(1-\tau_{N-1})\, & f_N(\tau_1,..., \tau_{N-2}, 1, 0) \\ & - \beta f_N(\tau_1, \tau_2,..., \tau_{N-1}, 1) \\ & = \alpha\beta(1-2\tau_{N-1})\, f_{N-1}(\tau_1, \tau_2,..., \tau_{N-2}, 1)\end{aligned} \tag{A2}$$

which is always satisfied, as f_N is given by recursion (8).

Case 2. $\tau_N = \tau_{N-1} = \cdots \tau_{K+1} = 0$ *and* $\tau_K = 1$ *for* $1 \leqslant K \leqslant N-1$. Using Eq. (8), one finds

$$\begin{aligned} & f_N(\tau_1, \tau_2,..., \tau_{K-1}, 1, 0,..., 0) \\ & \quad = \alpha\beta[f_{N-1}(\tau_1, \tau_2,..., \tau_{K-1}, 1, 0,..., 0) \\ & \qquad + f_{N-1}(\tau_1, \tau_2,..., \tau_{K-1}, 0, 0,..., 0)]\end{aligned} \tag{A3}$$

For f_N to satisfy (5), one needs to prove that

$$\begin{aligned}(1-\tau_{K-1})\, & f_N(\tau_1, \tau_2,..., \tau_{K-2}, 1, 0,..., 0) \\ & - f_N(\tau_1, \tau_2,..., \tau_{K-2}, \tau_{K-1}, 1, 0,..., 0) \\ & + \beta f_N(\tau_1, \tau_2,..., \tau_{K-1}, 1, 0,..., 0, 1) \\ & = \alpha\beta[(1-\tau_{K-1})\, f_{N-1}(\tau_1, \tau_2,..., \tau_{K-2}, 1, 0,..., 0) \\ & \quad - f_{N-1}(\tau_1,..., \tau_{K-2}, \tau_{K-1}, 1, 0,..., 0) \\ & \quad + \beta f_{N-1}(\tau_1,..., \tau_{K-2}, \tau_{K-1}, 1, 0,..., 0, 1) \\ & \quad - \tau_{K-1} f_{N-1}(\tau_1,..., \tau_{K-2}, 1, 0,..., 0) \\ & \quad + \beta f_{N-1}(\tau_1, \tau_2,..., \tau_{K-2}, \tau_{K-1}, 0,..., 0, 1)]\end{aligned} \tag{A4}$$

By replacing f_N by its expression (8) and (A4), one gets that (A4) is satisfied if

$$\begin{aligned}(1-\tau_{K-1})\,&f_{N-1}(\tau_1,..., \tau_{K-2}, 0, 0,..., 0)\\ &-f_{N-1}(\tau_1, \tau_2,..., \tau_{K-1}, 0,..., 0)\\ &= -f_{N-1}(\tau_1,..., \tau_{K-2}, \tau_{K-1}, 1, 0,..., 0)\\ &-\tau_{K-1} f_{N-1}(\tau_1,..., \tau_{K-2}, 1, 0,..., 0)\\ &+\beta f_{N-1}(\tau_1,..., \tau_{K-2}, \tau_{K-1}, 1, 0,..., 0, 1)\\ &+\beta f_{N-1}(\tau_1, \tau_2,..., \tau_{K-1}, 0,..., 0, 1)\end{aligned} \tag{A5}$$

Using once more the recursion (8) to replace f_{N-1} by its expression in terms of f_{N-2}, one completes the proof of (A5).

Case 3. $\tau_N=\tau_{N-1}=\cdots=\tau_1=0$. In this case,

$$f_N(0, 0,..., 0)=\beta f_{N-1}(0,..., 0) \tag{A6}$$

For f_N to satisfy (5) in this case, one needs that

$$-\alpha f_N(0,..., 0)+\beta f_N(0, 0,..., 0, 1)=0 \tag{A7}$$

which is a clear consequence of (A6) and of the fact that

$$f_N(0,..., 0, 1)=\alpha f_{N-1}(0, 0,..., 0) \tag{A8}$$

This completes the proof that the recursion (8) gives all the steady states for arbitrary N.

ACKNOWLEDGMENTS

B.D. would like to acknowledge the kind hospitality of the Weizmann Institute, where this work was started, and the Einstein Center of Theoretical Physics for support. We also thank C. Kipnis, J. L. Lebowitz, and H. Spohn for their helpful and encouraging comments.

This research was partially supported by the United States–Israel Binational Science Foundation.

REFERENCES

1. T. M. Liggett, *Interacting Particle Systems* (Springer-Verlag, New York, 1985).
2. R. Kutner, *Phys. Lett. A* **81**:239 (1981).
3. H. van Beijeren, K. W. Kehr, and R. Kutner, *Phys. Rev. B* **28**:5711 (1983).

4. P. A. Ferrari, *Ann. Prob.* **14**:1277 (1986).
5. A. De Masi and P. Ferrari, *J. Stat. Phys.* **36**:81 (1984).
6. D. Kandel and E. Domany, *J. Stat. Phys.* **58**:685 (1990); D. Kandel, E. Domany, and B. Nienhuis, *J. Phys. A* **23**:L755 (1990).
7. J. P. Marchand and P. A. Martin, *J. Stat. Phys.* **44**:491 (1986).
8. D. Dhar, *Phase Transitions* **9**:51 (1987).
9. J. Krug and H. Spohn, in *Solids far from Equilibrium: Growth, Morphology and Defects*, C. Godreche, ed. (Cambridge University Press, Cambridge, 1991).
10. M. Kardar, G. Parisi, and Y. Zhang, *Phys. Rev. Lett.* **56**:889 (1986).
11. J. Krug, *Phys. Rev. Lett.* **67**:1882 (1991).
12. D. E. Wolf and L. H. Tang, *Phys. Rev. Lett.* **65**:1591 (1990).
13. D. Kandel and D. Mukamel, *Europhys. Lett.* (1992), in press.
14. J. Cook and D. E. Wolf, *J. Phys. A* **24**:L351 (1991).
15. H. Rost, *Z. Wahrsch. Verw. Geb.* **58**:41 (1981).
16. A. Galves, C. Kipnis, C. Macchioro, and E. Presutti, *Commun. Math. Phys.* **81**:127 (1981).
17. B. Derrida, J. L. Lebowitz, E. R. Speer, and H. Spohn, *Phys. Rev. Lett.* **67**:165 (1991); *J. Phys. A* **24**:4805 (1991).
18. S. A. Janowsky and J. L. Lebowitz, *Phys. Rev. A* **45**:618 (1992).
19. L. H. Gwa and H. Spohn, *Phys. Rev. Lett.* **68**:725 (1992).
20. H. van Beijeren, R. Kutner, and H. Spohn, *Phys. Rev. Lett.* **54**:2026 (1985).
21. J. Krug, Private communication.

PHYSICAL REVIEW LETTERS

VOLUME 70 22 FEBRUARY 1993 NUMBER 8

Jamming and Kinetics of Deposition-Evaporation Systems and Associated Quantum Spin Models

Mustansir Barma,[a] M. D. Grynberg, and R. B. Stinchcombe
Department of Theoretical Physics, University of Oxford, 1 Keble Road, Oxford OX1 3NP, United Kingdom
(Received 8 October 1992)

A class of models of deposition and evaporation of dimers, trimers, ..., k-mers is studied analytically and by simulation. Correlation functions decay as power laws in time, related to broken symmetries in associated spin Hamiltonians. For $k \geq 3$, the number of jammed and evolving steady states increases exponentially with size. Finite size scaling studies support a phenomenological diffusive picture for dynamics and indicate universality over k in many subspaces.

PACS numbers: 05.50.+q, 02.50.−r, 75.10.Jm, 82.20.Mj

Stochastic models of lattice gas dynamics can provide valuable insight into nonequilibrium behavior and complex dynamics. In this Letter, we introduce a class of such models motivated by, and capturing some basic aspects of, simple deposition and evaporation processes involving k particles at a time. Despite their simplicity, the models exhibit unusually strong nonergodic behavior, a rich variety of partially and fully jammed steady states, and power law decays of dynamical correlation functions.

The models are studied analytically and numerically. They are equivalent to an interesting new class of quantum spin systems [1], whose simplest nontrivial member is the Heisenberg ferromagnet. These equivalences elucidate conservation laws and symmetries, and in some cases relate power laws in correlation functions to Goldstone modes. A phenomenological picture of the dynamics based on random walks of unjammed regions through jammed backgrounds is developed, and suggests that correlation functions have a diffusive tail. This is supported quantitatively by a Monte Carlo study of finite size scaling of the dynamical correlation function; a universal scaling function is shown to describe dynamics in several steady states for various k.

The basic process is the deposition and evaporation of k-mers on a d-dimensional lattice, where $k = 1, 2, 3, \ldots$ represents monomers, dimers, trimers, etc. Deposition of k-mers at rate ε and evaporation at rate ε' are attempted at random locations; a deposition attempt is successful if k successive sites are vacant, while evaporation requires k successive occupied sites. The rule for evaporation allows for reconstitution of k-mers. Our model includes as a special case (ε'=0) random sequential adsorption of k-mers on a lattice [2], and is related to lattice models of chemical reactions [3]. It differs from coordination models [4] and adsorption-desorption models considered earlier [5, 6] in that both deposition and evaporation involve k particles in our model—a crucial feature.

The case d=1 (linear k-mers on a chain) is typical and already exhibits very rich behavior. Consequently, we consider only this case in detail. The operator which describes the stochastic evolution of the system is $\exp(-Ht)$ where the "Hamiltonian" H is, for general k,

$$H = \sum_n (R_n - Q_n),$$
$$R_n \equiv \varepsilon \prod [\tfrac{1}{2}(1 - \sigma_l^z)] + \varepsilon' \prod [\tfrac{1}{2}(1 + \sigma_l^z)], \quad (1)$$
$$Q_n \equiv \varepsilon \prod \sigma_l^+ + \varepsilon' \prod \sigma_l^-,$$

where $\prod \equiv \prod_{l=n}^{n+k-1}$ and $\sigma_l^+ (\sigma_l^-)$ is a spin-$\frac{1}{2}$ raising (lowering) operator at site l. This form of Hamiltonian arises from having represented a particle (or vacancy) at site l by a pseudospin operator $\sigma_l^z = +1$ (or -1). The deposition (or evaporation) of a k-mer at k adjacent empty (or full) sites is equivalent to the flip of k

adjacent spins from down to up (or up to down), and this is given by the operator Q_n. Since this process only occurs with probability ε (or ε') ≤ 1, conservation of probability requires the appearance of a second operator R_n which does not change the up or down state of any spin. Since $\exp(-Ht)$ is a stochastic matrix [7], the eigenvalues E of H have a non-negative real part. The steady states are states with E=0 while positive energy eigenstates decay with lifetime $1/E$. States without k adjacent spins up (particles) or down (vacancies) are unchanged by Q_n and therefore also by R_n. These are the fully jammed states. Dynamics in a jammed environment is important for the long-time kinetics, and thus will be discussed later.

A family of conservation laws holds, if, as will be assumed throughout, the number of sites L is a multiple of k. Divide the chain into k sublattices $\alpha = 1, 2, \ldots, k$ such that site $l \in \alpha$ if $l = k\gamma + \alpha$ where γ is an integer. An important symmetry property can be inferred from the observation that deposition or evaporation of k-mers changes the occupation of all k sublattices by the same amount. The quantities $M_{\alpha\beta} \equiv M_\alpha - M_\beta$, where $M_\alpha \equiv \sum_{l\in\alpha} \sigma_l^z$ is a measure of occupation of sublattice α, are therefore constants of the motion. Thus they commute with H. The quantities $M_{\alpha\beta}$ are the infinitesimal generators of rotations around the z axis of all spins on each of the k different sublattices by angles θ_α ($\alpha = 1, 2, \ldots, k$) provided that

$$\sum_{\alpha=1}^{k} \theta_\alpha = 0 . \tag{2}$$

The vanishing commutators $[M_{\alpha\beta}, H]$ imply that H is invariant under such rotation. This can be easily checked by noting that $\sigma_l^\pm \to \exp(\pm i\theta_l)\sigma_l^\pm$. This symmetry plays a crucial role in the following. For the random monomer case $k = 1$, there are no cooperative effects [6]. There is a unique steady state with average coverage (particle concentration) $\varepsilon/(\varepsilon + \varepsilon')$. Dynamical correlation functions decay exponentially.

For $k \geq 2$, cooperative effects strongly affect the number and nature of steady states and concomitant dynamics. Figure 1 shows Monte Carlo results for the particle number autocorrelation function $C(t) = \sum_l [\langle \sigma_l^z(0)\sigma_l^z(t)\rangle - \langle\sigma_l^z\rangle^2]/4L$ where the expectation value is taken in steady state. $C(t)$ and higher order space- and time-dependent correlation functions describe completely the full adsorption-desorption kinetics. Evidently $C(t)$ decays as a diffusive power law for the deposition-evaporation process for $k \geq 2$. This is at first sight surprising, since the process contains no explicit particle diffusion terms. Also shown in Fig. 1 is the effect of adding particle diffusion: the decay changes from power law to exponential. The spin Hamiltonian helps to explain these initially puzzling behaviors.

For the case k=2 (dimers), H involves nearest-neighbor spin-pair interactions. For the symmetric case $\varepsilon = \varepsilon'$, the mapping $\sigma \to \hat{\sigma} \equiv (\hat{\sigma}^x, \hat{\sigma}^y, \hat{\sigma}^z) = (\sigma^x, -\sigma^y, -\sigma^z)$ on one sublattice takes H into the Heisenberg Hamiltonian

$$\hat{H} = \sum_l \frac{\varepsilon}{2}(1 - \hat{\sigma}_l \cdot \hat{\sigma}_{l+1}) . \tag{3}$$

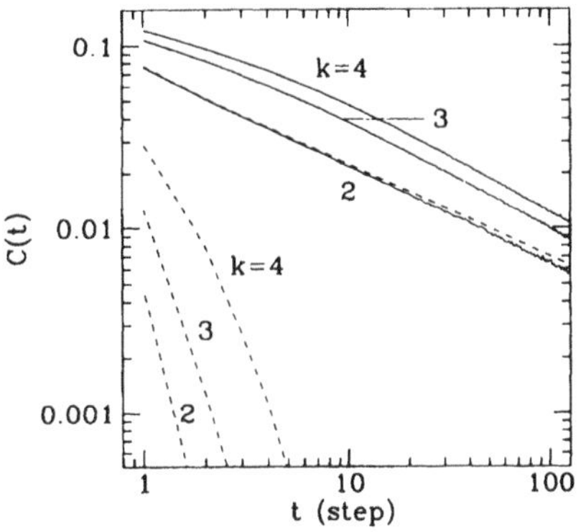

FIG. 1. The autocorrelation function for the deposition-evaporation process for $\varepsilon = \varepsilon'$, $L = 1.2 \times 10^5$, averaged over 100 histories (solid curves) shows power law decay ($\sim t^{-\frac{1}{2}}$) for $k = 2, 3, 4$. The dashed curve is an analytic determination for $k = 2$. Single particle diffusion changes the decay to exponential (dashed curves at the bottom left).

Each of the $L + 1$ ground states of $\hat{H}$ corresponds to a steady state, and is labeled by distinct values of $\hat{M} = \sum_l \hat{\sigma}_l^z$. The calculation of the associated autocorrelation function $C(t)$ is straightforward but lengthy. It is carried out using selection rules based on the conservation of total spin, and is related to the dynamics of local active patches on an otherwise jammed background. The calculation shows that the asymptotic long-time adsorption-desorption kinetics is completely contained in this pair correlation function. Higher order (for example, dimer-dimer) correlation functions can be treated by generalizations of the method. The result [8] is $C(t) = (\frac{1}{4} - \hat{m}^2)\exp(-2\varepsilon t) I_0(2\varepsilon t)$, where $\hat{m} = \hat{M}/L$ and I_0 is the Bessel function of imaginary argument. Figure 1 shows that the Monte Carlo results agree well with this formula. In terms of the evaporation-deposition model, the full rotational symmetry of $\hat{H}$ arises from the conservation of probability, which implies equal coefficients of the transverse terms (describing transitions) and the longitudinal terms (corresponding to no change). Since the steady state breaks the rotational symmetry of $\hat{H}$, the Goldstone theorem implies the existence of low-lying bosons (spin waves) which are responsible for the asymptotic $t^{-\frac{1}{2}}$ decay of $C(t)$.

Explicit particle diffusion adds an exchange anisotropy term to $\hat{H}$ and destroys the rotation symmetry and associated conservation of $\hat{M}$. The resulting gap in the spin-wave spectrum leads to the exponential decay observed (Fig. 1) for this case.

For $\varepsilon \neq \varepsilon'$, the sublattice-transformed Hamiltonian $\hat{H}$ is non-Hermitian and not fully rotationally symmetric. Nevertheless, all $L+1$ steady states can be found explicitly. Further, $C(t)$ can be found in the sectors $\hat{M} = \pm(L/2 - 1)$ by solving a single-excitation problem. For arbitrary $\hat{M}$ the long-time behavior of $C(t)$ can be found [8] by studying the linear response of steady states to a driving field. The diffusion constant is $D = 2[\varepsilon\varepsilon'(1-m_A^2)(1-m_B^2)]^{\frac{1}{2}}[(1-m_A^2)+(1-m_B^2)]$ with m_A, m_B satisfying $m_A + m_B = 2\hat{M}/L$ and $(1+m_A)(1-m_B)\varepsilon = (1-m_A)(1+m_B)\varepsilon'$. The associated power law decay is consistent with the breaking of the continuous symmetry described above Eq. (2).

The possibility of an exact solution is hinted by the properties of the operator $R_n - Q_n$ in Eq. (1): for k=2, the operator satisfies two of the three requirements of a Temperley-Lieb algebra [9], while the third condition takes a generalized form, reducing to the standard one for $\varepsilon = \varepsilon'$.

The situation for $k \geq 3$ is more complex and interesting. The full phase space of 2^L microscopic configurations splits into a very large number $I(k,L)$ of invariant subspaces which are not connected to each other by the dynamics. $I(k,L)$ grows exponentially with L if $k \geq 3$, in contrast to $I(2,L) = L+1$. The exponential growth can be established as follows. Write $I(k,L) = I_1(k,L) + I^*(k,L)$, where $I_1(k,L)$ is the number of subspaces of size 1 (each corresponding to a completely jammed configuration) and $I^*(k,L)$ is the number of larger subspaces. $I_1(k,L)$ may be calculated using a recurrence relation, as each completely jammed configuration has no more than $k-1$ succesive parallel spins. With open boundary conditions, the result is $I_1(k,L) = 2F_k(L)$ where $F_k(L)$ are generalized Fibonacci numbers defined by $F_k(L) = \sum_{j=1}^{k-1} F_k(L-j)$ with $F_k(0) = 1$, $F_k(L) = 0$ for $-(k-2) \leq L < 0$. Asymptotically, $I_1(k,L) \sim \lambda^L$ where λ is the largest eigenvalue of $\lambda^k = 2\lambda^{k-1} - 1$.

Further, the number I^* of nontrivial invariant subspaces also grows exponentially with L. Evidence for this comes from studies of the form of H in the Ising (site-occupation) basis $|\{\sigma_l^z\}\rangle$, on finite rings with lengths in the range $3 \leq L \leq 18$. We find $I^* \sim \mu^L$ with $\mu > 1.4$ for $k = 3$ and $\mu > 1.6$ for $k = 4$. This exponential proliferation of subspaces with nontrivial evolution, indicating strongly broken ergodicity, is quite unusual in systems without quenched disorder. Each subspace Λ has a unique [7] steady state $|\Lambda, 0\rangle$, which, in the symmetric case ($\varepsilon = \varepsilon'$), is an equal-weight linear combination of all configurations $|\{\sigma_l^z\}\rangle \in \Lambda$.

It is also possible to form steady states involving linear combinations of the form $|\psi\rangle = \sum_\Lambda c_\Lambda |\Lambda, 0\rangle$. For instance, several exact steady states follow from rewriting H as

$$H = \sum_n R_n \left(1 - \prod \xi_l\right), \tag{4}$$

where $\xi_l \equiv \alpha\,\sigma_l^+ + (1/\alpha)\sigma_l^-$ with $\alpha = (\varepsilon/\varepsilon')^{1/k}$. Product eigenstates of ξ_l with eigenvalues $m_l = \pm 1$ such that $m_l = m_\alpha$ for l on sublattice α, with $\prod_{\alpha=1}^{k} m_\alpha = 1$, are steady states which involve all the states $|\Lambda, 0\rangle$ in linear combinations. From these new steady states, many more can be generated by applying the rotations specified by (2) since those operations do not change H. All these states do not share the rotation invariance of the Hamiltonian. This broken symmetry requires the existence of Goldstone modes, which are responsible for the asymptotic slow kinetics of the deposition-evaporation system. Indeed, distorted versions of those rotations [rotations by θ_l of the form $A_\alpha \exp(iql)$ for $l \in \alpha$] generate the Goldstone modes and also provide their energies.

In the Ising basis, steady states $|\Lambda, 0\rangle$ corresponding to different subspaces Λ differ from each other in several respects. The majority of subspaces have nonzero values of the conserved quantities $M_\alpha - M_\beta$, etc. implying broken translational invariance in the steady state, i.e., $\rho_l \equiv \langle \Lambda, 0 | \frac{1}{2}(1+\sigma_l^z) | \Lambda, 0\rangle$ depends on site l.

Turning to the dynamics, it is useful to introduce the notion of local jamming—the inability to deposit or evaporate owing to the absence of k successive parallel spins. Steady state dynamics entails a succession of stochastic transitions between states $|\{\sigma_l^z\}\rangle$ in Λ. The mean rate of such transitions $J(\Lambda) = \langle \Lambda, 0 | \sum_n R_n | \Lambda, 0\rangle / L$ gives a quantitative measure of the lack of jamming in that steady state. For nonevolving states, we have $J(\Lambda) = 0$ (maximal jamming). A study of finite systems shows that for $k \geq 3$ even the least jammed steady state (that reached from an initially empty lattice) has fairly low J ($\simeq 0.36\varepsilon$ for $k = 3$, $\varepsilon = \varepsilon'$) indicating quite a large degree of jamming.

We studied dynamics in a jammed environment by writing and solving evolution equations for a localized deviation (e.g., a patch of $k+1$ parallel spins) in an otherwise completely jammed background. Unlike mixed states such as those described under Eq. (4), the completely jammed states do not break the continuous symmetry of H expressed in Eq. (2) so in their case no power law decay can be inferred from the Goldstone theorem. Nevertheless, the deposition-evaporation kinetics induces a random walk of descendants of the parallel-spin patch on the lattice. The precise stepping rule and the associated diffusion constant depend on the details of the jammed background. A power law decay $\propto (Dt)^{-\frac{1}{2}}$ of the spin autocorrelation function $C_n(t)$ follows.

When there is a finite low density of unjammed patches in a jammed background, there are collisions between patches (Fig. 2). A plausible hypothesis is that at long times the only effect of collisions is to modify the diffusion constant. This would imply a diffusive tail ($\sim t^{-\frac{1}{2}}$) in $C(t)$, even away from the single-walker limit. This hypothesis has been tested by studying the autocorrelation function $C(L,t)$ in finite systems of size L. On general grounds we expect $C(L,t)$ to conform with finite size scaling in the limit $t \to \infty$, $L \to \infty$, t/L^z constant,

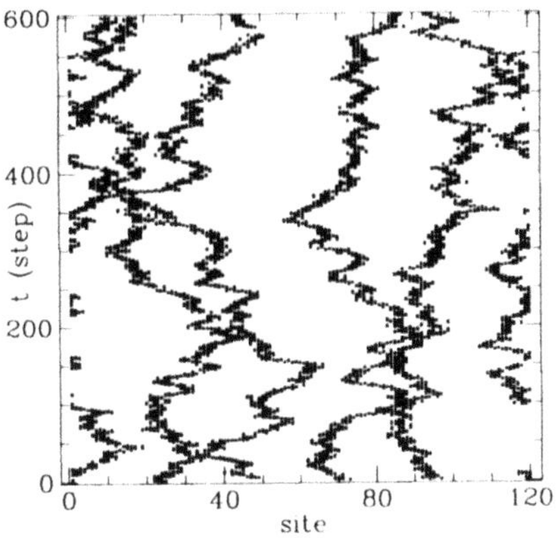

FIG. 2. A particular history of unjammed regions through a jammed background (here taken to be antiferromagnetic), showing characteristic random walk behavior; only updated spins are shown.

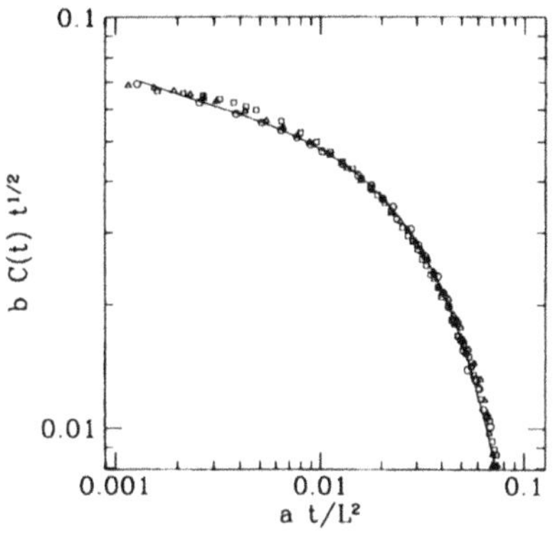

FIG. 3. Evidence for universality of k-mer kinetics from finite size scaling, for steady states reached from an initially empty lattice. Data (averaged over 10^6 histories) for k=2, L=28 (circles), k=3, L=54 (triangles), k=4, L=52 (squares) are shown; a and b are metric factors referred to in the text. The solid curve is the theoretical prediction for the scaling function.

i.e.,

$$bC(L,t) \simeq t^{-\theta} Y(at/L^z) , \quad (5)$$

where θ and z are critical exponents and Y is a universal scaling function. a and b are system-dependent (nonuniversal) metric factors. If the "diffusion hypothesis" is valid, (5) should hold irrespective of the value of k. Figure 3 shows the collapse of numerical data for k=2, 3, and 4 in the least jammed subspace, on setting $\theta = \frac{1}{2}$ and $z = 2$ as implied by the diffusion hypothesis. The solid curve is an analytical determination of the scaling function (possible because the problem is exactly solvable for k=2). Explicitly, we have $Y(y) = \frac{1}{4} y^{\frac{1}{2}} \sum_{n \neq 0} \exp(-4\pi n^2 y)$. The good agreement between k=2, 3, and 4 lends support to the diffusive picture, even in the subspace with the least jamming, where collisions are most frequent. Similar numerical studies reveal that while the diffusive picture continues to hold in other subspaces (including translationally noninvariant ones), it may fail in yet others, where $z < 2$ is suggested.

Generalizations of the deposition-evaporation models to include particle diffusion (already mentioned), k-mer diffusion, and nearest-neighbor cooperative effects are clearly of interest. Other interesting extensions to mixed k-mer, k'-mer cases and to higher dimensions are under consideration. For example, dimers with $\varepsilon = \varepsilon'$ on the square lattice share many of the properties detailed above for the case k=2, $\varepsilon = \varepsilon'$, $d = 1$, because sublattice mapping to the Heisenberg model is again possible. While the general k-mer problem in higher dimensions remains quite open, it will be shown elsewhere that the quantum spin analogy is very useful in such special cases as: linear k-mers on the square lattice, mixed k-mer k'-mer cases with k, k' both integer multiples of a common integer, or addition of particle diffusion or particle deposition or evaporation to the dimer case.

It is a pleasure to acknowledge fruitful discussions with Professor V. Privman. We also thank Professor J. W. Evans for sending us a copy of his review before publication. The support of the SERC under Grants No. GR/G02741 and No. GR/G02727 is acknowledged.

[a] On sabbatical leave from Tata Institute of Fundamental Research, Homi Bhabha Road, Bombay 400005, India.

[1] For references to known results in this area see, e.g., M. Fowler, in *Nonlinearity in Condensed Matter*, edited by A.R. Bishop *et al.*, Springer Series in Solid State Sciences Vol. 69 (Springer, Berlin, 1987); I. Affleck, J. Phys. Condens. Matter **1**, 3047 (1989).

[2] P.J. Flory, J. Am. Chem. Soc. **61**, 1518 (1939); for recent reviews, see, e.g., M.C. Bartelt and V. Privman, Int. J. Mod. Phys. B **5**, 2883 (1991); and J.W. Evans (unpublished).

[3] R.M. Ziff, E. Gulari, and Y. Barshad, Phys. Rev. Lett. **56**, 2553 (1988); D. Ben-Avraham and J. Kohler, J. Stat. Phys. **65**, 839 (1991).

[4] J. Toner and G.Y. Onoda, Phys. Rev. Lett. **69**, 1481 (1992).

[5] A.H. Bretag, B.R. Davis, and D.I.B. Kerr, J. Membr. Biol. **16**, 363 (1974); A.H. Bretag, C.A. Hurst, and D.I.B. Kerr, J. Chem. Biol. **73**, 367 (1978); R. Dickman and R. Burschka, Phys. Lett. A **127**, 132 (1988).

[6] J.W. Evans and C.A. Hurst, Phys. Rev. A **40**, 3461 (1989).

[7] N.G. van Kampen, *Stochastic Processes in Physics and Chemistry* (North-Holland, Amsterdam, 1981).

[8] R.B. Stinchcombe, M.D. Grynberg, and M. Barma (to be published).

[9] H.N.V. Temperley and E.H. Lieb, Proc. R. Soc. London A **322**, 251 (1971); R.J. Baxter, *Exactly Solved Models in Statistical Mechanics* (Academic, London, 1982).

Slow Relaxation in a Model with Many Conservation Laws: Deposition and Evaporation of Trimers on a Line

Mustansir Barma and Deepak Dhar
Tata Institute of Fundamental Research, Homi Bhabha Road, Bombay 400 005, India
(Received 12 April 1994)

We study the slow decay of the steady-state autocorrelation function $C(t)$ in a stochastic model of deposition and evaporation of trimers on a line with infinitely many conservation laws and sectors. Simulations show that $C(t)$ decays as different powers of t, or as $\exp(-t^{1/2})$, depending on the sector. We explain this diversity by relating the problem to diffusion of hard core particles with conserved spin labels. The model embodies a matrix generalization of the Kardar-Parisi-Zhang model of interface roughening. In the sector which includes the empty line, the dynamical exponent z is 2.55 ± 0.15.

PACS numbers: 82.20.Mj, 02.50.−r, 05.40.+j, 68.45.Da

It is well known that conservation laws strongly affect the decay in time of fluctuations in equilibrium. For example, in magnetic systems, the rate of decay of the spin autocorrelation function is quite different depending on whether or not the spin dynamics conserves magnetization [1]. What happens with more than one—indeed an infinity of—conservation laws? We address this question in this paper by studying the decay of autocorrelation functions in the steady state of a simple stochastic model of deposition and evaporation of trimers on a line, which has been introduced earlier [2,3].

This model, which is defined more precisely later, has been shown to possess an infinite number of independent constants of motion [4]. Here we show that it exhibits a very rich variety of slow relaxations: Depending on the initial conditions, we can get a very large number of power-law decays, or even stretched exponential decay. We present numerical evidence from Monte Carlo simulations for these different kinds of decay, and also provide an analytical understanding of this remarkable diversity by relating the dynamics to that of the diffusive motion of hard core particles with spin. In this reformulation, the infinity of conservation laws become the much simpler conservation law of spin labels of the diffusing particles.

Our model can be viewed as a generalization of the well-known Kardar-Parisi-Zhang (KPZ) problem [5] of roughening of a one-dimensional interface, where the scalar-height variables in the KPZ model are replaced by matrix-valued variables (say 2×2 complex matrices). In the steady state reached after starting from an empty lattice, we find that the density-density autocorrelation function decays as a power law in time, and the dynamical exponent $z \simeq 2.5$, very different from the KPZ value $z = 3/2$. The model thus includes a new universality class for nonlinear one-dimensional evolution models.

In the deposition-evaporation (DE) model under consideration, there is a site variable n_i, taking values 0 or 1, at each site i of a line, $1 \leq i \leq L$; it may be thought of as an occupation number variable of a lattice gas. We define pseudospin variables $S_i \equiv 2n_i - 1$. The time evolution is Markovian, specified by the following rates: In a small time dt, a triplet of spins at adjacent sites i, $i+1$, $i+2$ can flip simultaneously only if $S_i = S_{i+1} = S_{i+2}$. The rate is ϵ if the spins were originally -1, and ϵ' if they were originally $+1$. These rates satisfy the detailed balance condition corresponding to a noninteracting lattice gas Hamiltonian with chemical potential $\frac{1}{3}\ln(\epsilon/\epsilon')$. However, the long-time steady states in the present model are nontrivial, as the dynamics is strongly nonergodic. We have shown [4] that the total phase space of 2^L configurations breaks up into a large number N_L of mutually disconnected sectors, and N_L increases exponentially with L as $[(\sqrt{5}+1)/2]^L$ for large L.

This decomposition of phase space is a consequence of the existence of an infinity of conservation laws. A compact representation of these is provided by the "irreducible string" (IS) defined as follows: Regard the configuration as a string of up and down spins. Scan the string left to right, and stop at the first triplet of parallel spins encountered. If no such triplet is found, the string is the IS. If a triplet is encountered, simply delete it, reducing the length of string by 3, and repeat the process. The IS which finally results is a constant of motion, and different sectors of phase space correspond to different IS's. For details, see [4].

Under time evolution, an initial configuration evolves into the steady state of the corresponding sector. For instance, if $\epsilon = \epsilon'$, in the steady state all configurations having the same IS as that of the initial configuration occur with equal probability. It is then interesting to ask how a dynamical quantity such as the steady-state autocorrelation function

$$C_i(t) = \langle n_i(t + t_0) n_i(t_0) \rangle - \langle n_i \rangle^2 \quad (1)$$

varies from one sector to another. Figure 1 shows the behavior of $C(t)$ obtained from Monte Carlo simulation studies in a number of different representative sectors. The data depicted in the figure pertain to four sectors in which the IS constitutes a finite fraction of all sites, and one sector in which the fraction is zero. These sectors

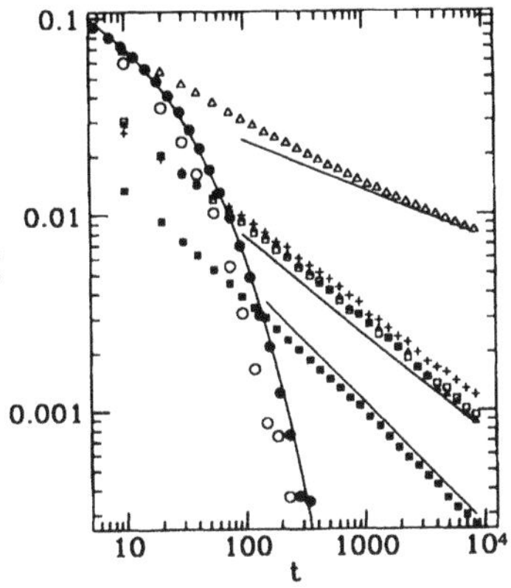

FIG. 1. Variation of the decay of the autocorrelation function $C_i(t)$ in sectors (1)-(5) (defined in text). Sector (1) (open triangles); sector (2), sublattice A (+), sublattices B and C (filled triangles); sector (3) (filled circles); sector (4), sublattices A and C (open squares), sublattice B (open circles); sector (5) (filled squares). The straight lines show power-law decays $t^{-\theta}$ with $\theta = 0.25$ (top), 0.50 (middle), 0.62 (bottom); the curve shows a stretched exponential $[\sim \exp(-t^{1/2})]$ decay.

correspond to (1) a random initial configuration, (2) an IS formed by repeating [↑↓↓] $L/6$ times, (3) an IS formed by repeating [↑↓] $L/4$ times, (4) an IS formed by repeating [↑↑↓↑↓↓] $L/12$ times, and (5) the null sector (vanishing IS). Lattice sizes $L = 120\,000$ were used and averages over 30–60 different histories were taken. From the figure, we note the following points: (i) If the lattice is divided into 3 sublattices A, B, C, the autocorrelation function $C_i(t)$ depends, in general [e.g., in sectors (2) and (4)], on the sublattice to which site i belongs. (ii) In several sectors and sublattices $C_i(t)$ decays as a power law $\sim t^{-\theta}$; the power θ is sector dependent. In sector (1), $\theta \simeq 0.25$; in sector (2) and on two sublattices of sector (4), $\theta \simeq 0.5$; in sector (5), $\theta \simeq 0.6$. (iii) In some sectors [e.g., sector (3) and one sublattice in sector (4)], the decay is faster than a power law, and fits well with a stretched exponential form. Evidently, there is a very wide range of possible behaviors. Understanding the source and nature of this dynamical diversity is one of the main points of this Letter.

We now relate deposition-evaporation dynamics to that of diffusing interacting particles with conserved spin labels. Consider time evolution in a sector labeled by an IS with elements $\{\alpha_n\} \equiv \{\alpha_1, \alpha_2, \ldots, \alpha_\ell\}$, where the length ℓ is a nonzero fraction $\rho = \ell/L$ of the total length L of the lattice. Let $C(t)$ be the full string corresponding to the configuration at time t. If we apply the deletion algorithm to $C(t)$, some characters are eventually deleted, others not. If $X_j(t)$ is the location of the jth site (counting from the left) that remains undeleted, we may look upon the set $\{X_j(t)\}$ as the positions of ℓ interacting random walkers on a line; DE dynamics induces the time evolution of $\{X_j(t)\}$. The walkers cannot cross each other $[X_{j+1}(t) > X_j(t)$, for all j, all $t]$ and they carry a conserved spin label $S_{X_j(t)} = \alpha_j$ for all t. Under time evolution $X_j(t)$ changes in jumps by a multiple of 3 spaces at a time.

In the steady state, the joint probability distribution $\mathrm{Prob}(\{X_j\})$ is proportional to the number of different configurations of the lattice, consistent with the positions of walkers being $\{X_j\}$. The substring between sites X_{j+1} and X_j should be a string reducible to a null string, and all such configurations are easily enumerated. The result is

$$\mathrm{Prob}(\{X_j\}) = \mathcal{N} \prod_{j=0}^{\ell} g(X_{j+1} - X_j), \qquad (2)$$

where $\mathcal{N}$ is a normalization constant, and $X_0 = 0$, $X_{\ell+1} = L + 1$. The separation probability $g(r)$ can be computed using the generating function method of [4]. The result is $g(r) \sim r^{-3/2} \exp(-\kappa r)$ for large r, where κ is the reduced pressure, which depends on the density ℓ/L, and tends to zero as $(\ell/L)^2$ for small ℓ/L. We see that $\{X_j\}$ constitute the slow modes of the system as they evolve diffusively, and are linked to conserved quantities. If the typical relaxation time of a string of length λ reducible to the null string varies as λ^z, then for times $t \gg (L/\ell)^z$ we can assume that all the degrees of freedom other than $\{X_j\}$ are in instantaneous local equilibrium. Thus $g(r) \sim \exp[-V(r)]$, where the effective interaction $V(r)$ between walkers is attractive, and increases linearly with separation if $\kappa \neq 0$, and as $\ln r$ if $\kappa = 0$.

While the logarithmic part of the interaction is crucial for the understanding of sectors in which the walker density $\ell/L \to 0$, it appears not to be very important if ℓ/L, and hence κ, is finite. In the latter case, we are led to consider a simpler system: ℓ random walkers on a line of length L, with each walker carrying a spin α_j which is unchanged under dynamics. Each walker jumps left or right by 3 steps only if no other walkers intervene. The point is that this simpler system of hard core random walkers with spin (HCRWS) has exactly the same conservation laws as the original deposition-evaporation model. Thus we are led to conjecture that the long-time behavior of $C_i(t)$ in a particular sector of the DE model is in the same universality class as the spin-spin autocorrelation function $D_i(t)$ in the HCRWS problem with the corresponding spin sequence.

It is then easy to understand the sublattice dependence of $C_i(t)$ observed in simulations. As each element α_n of the IS moves by multiples of 3 lattice spacings, it stays on the same sublattice, say A. The long-time behavior of $C_i(t)$ for $i \in A$ is governed only by the subset $\{\alpha_{n'}\}_A$ on sublattice A, independent of $\{\alpha_{n'}\}_B$ and $\{\alpha_{n'}\}_C$, despite the fact that the elementary step of deposition or evaporation couples all sublattices strongly.

In the steady state of the HCRWS model, the mean squared difference between the number of particles up to point r, at times t_0 and $t_0 + t$, grows as $t^{1/2}$ [6]. To test for a similar behavior in the DE system, we monitored $\sigma^2(t) \equiv \langle [N(r, t_0 + t) - N(r, t_0)]^2 \rangle$, where $N(r,t)$ is the length of irreducible string between points 0 and r at time t. This quantity is roughly equal to number of walkers to the left of r, but remains well defined even in the sector where there are no walkers $\ell = 0$. Figure 2 shows Monte Carlo data for $\sigma^2(t)$ in a number of sectors. For all sectors with ℓ/L finite, $\sigma^2(t)$ is seen to grow as $t^{1/2}$, lending strong support to the conjecture of equivalence of the DE and HCRWS models. In the sector where the IS vanishes, on the other hand, σ^2 is found to grow as $t^{2\beta}$ with $\beta \simeq 0.19$. If the irreducible string is very short, then κ is very small, and fluctuations in the separation $(X_{i+1} - X_i)$ between near neighbors in the diffusing, interacting lattice gas are large. The low value of the exponent β shows that these fluctuations are slowly decaying.

In the HCRWS model, the spin-spin autocorrelation function in the steady state is defined as

$$D_i(t) = \langle S_i(t + t_0) S_i(t_0) \rangle - \langle S_i \rangle^2, \tag{3}$$

where $S_i(t)$ is the spin at site i, and an unoccupied site is considered to have zero spin. A nonzero contribution to D_i results from events in which site i is occupied at both times t_0 and $t + t_0$. Let $G(m \mid t)$ be the conditional probability that in steady state site i is occupied at time $t + t_0$ by the $(j + 3m)$th particle [i.e., $X_{j+3m}(t + t_0) = i$], given that at time t it was occupied by the jth particle [i.e., $X_j(t_0) = i$] [7]. Evidently, $G(m \mid t)$ is independent of the spin configuration $\{\alpha_n\}$ of the IS, and, for a system with periodic boundary conditions, is also independent of j. Then $D_i(t)$ can be written in the form

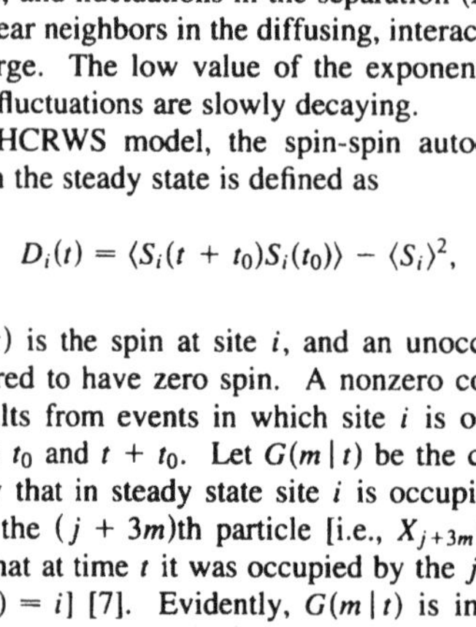

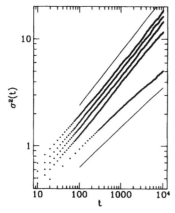

FIG. 2. Time dependence of the mean squared fluctuations in the length of irreducible string $N(r,t)$ up to point r. The four sets of data (from top) pertain to sectors (1)–(4) (shifted vertically for clarity) while the fifth pertains to sector (5). The straight lines have slopes 0.50 and 0.38.

$$D_i(t) = \sum_{m=-\infty}^{+\infty} G(m \mid t)\, \gamma_i(m), \tag{4a}$$

where $\gamma_i(m)$ is a time-independent, sector-dependent function of m which is given in terms of the IS of the sector by

$$\gamma_i(m) = \overline{\alpha_{i+3m+3m'} \alpha_{i+3m'}}, \qquad i = 1 \text{ to } 3, \tag{4b}$$

where the overbar represents an average over m'. The different dependences of $D_i(t)$ on time in different sectors comes entirely from the different dependences of $\gamma_i(m)$ on m.

Now $G(m \mid t)$ may be evaluated for large t using the equivalence of HCRW dynamics to a stochastic harmonic model [8,9]. The result is

$$G(m \mid t) \approx \frac{1}{\sqrt{2\pi\delta^2}} e^{-m^2/2\delta^2}, \tag{5}$$

where δ^2 has been evaluated in closed form for large t in [9]. The only feature of the result that we need is that because of caging effects δ^2 grows asymptotically as $\sqrt{t}$. Using Eqs. (4) and (5) we can determine the behavior of $D_i(t)$ in the HCRWS model in different sectors.

Sector (1) (random initial condition): In this case $\gamma_i(m)$ is significant only for very small values of m. For m fixed, $G \sim t^{-1/4}$ for large t, whence $D_i(t)$ also varies as $t^{-1/4}$. Note that this result is also true for other sectors in which the correlations in the IS decay fast with distance.

Sector (2) [↑↓↓]: In this case $\alpha_k = \alpha_{k+3m}$ for all m. Hence $D_i(t)$ reduces to a density-density correlation which decreases as $t^{-1/2}$ for large t. The same behavior is expected in all periodic IS where the magnetization in each sublattice is nonzero.

Sector (3) [↑↓]: In this case $\gamma_i(m) = (-1)^m$, and this implies that $D_i(t)$ has the stretched exponential form $\exp[-(t/\tau)^{1/2}]$ at large time at all sites. The same behavior would occur whenever the IS is periodic with zero net magnetization on each sublattice.

Sector (4) [↑↑↓↑↓↓]: On sublattices A and C we have $\alpha_k = \alpha_{k+3m}$, and thus, as in sector (2), we have $D_i(t) \sim t^{-1/2}$. On sublattice B, $\gamma_i(m)$ is $(-1)^m$, and as in sector (3) we get stretched exponential decay.

If the sector is such that correlations in the IS imply that $\tilde{\gamma}_i(q)$, the Fourier transform of $\gamma_i(m)$, varies as $|q|^\phi$ for small q, then $D_i(t)$ would vary as $t^{-(1+\phi)/4}$.

It is evident that the HCRWS predictions for $D_i(t)$ accord very well with the Monte Carlo results for $C_i(t)$ shown in Fig. 1. In particular, the predicted difference in qualitative behavior on different sublattices in sector (4) is quite unusual, and it is gratifying to note that it agrees with the results of simulations. The occurrence of stretched exponentials in the [↑↓] sector explains why earlier attempts [3] to fit finite size data to power laws yielded anomalously low values of the dynamical exponent z.

We now show how the DE model may be viewed as a matrix generalization of the KPZ model of interfacial growth [5]. For any configuration $\{S_i\}$ of spins on the line, we define the matrix variable at site i by the equation

$$I_i(t) = \prod_{j=1}^{i} A(S_j(t)), \tag{6}$$

where $A(1)$ and $A(-1)$ are 2×2 complex matrices given by

$$A(1) = \begin{pmatrix} 1 & x \\ 0 & \omega \end{pmatrix} = A(-1)^{\dagger}, \tag{7}$$

where $\omega \equiv \exp(2\pi i/3)$ and x is a real parameter. Then as $A(1)^3 = A(-1)^3 = 1$, it follows that $I_L(t)$ is a constant of motion [4]. The evolution of the variables $\{I_i(t)\}$ is Markovian, governed by local transition rates. If we start with an empty line, $I_i(t = 0)$ take simple values, and get progressively disordered as time increases. For large x, the ratio $\ln \mathrm{Tr} I_r(t)/\ln x$ behaves in qualitatively the same way as $N(r,t)$. At $t = 0$, the variance σ^2 of $N(r,t)$ is zero, and grows to a value proportional to r in the steady state. If $N(r,t)$ obeyed a stochastic evolution equation of the KPZ type [5], this variance would grow as $t^{2/3}$. Our numerical results (Fig. 2) clearly rule out this possibility. From Figs. 1 and 2 we see that the decay of the autocorrelation function is characterized by $\theta \simeq 0.6$, while correlations in the length of irreducible string up to a given point grow anomalously slow, with $\beta \simeq 0.19$. Static correlation functions in this sector are characterized by power-law decays as well. For instance, if we define an indecomposable substring as one which can be completely reduced, but which cannot be written as the concatenation of smaller substrings reducible to the null string [4], the probability of occurrence of an indecomposable substring of length r varies as $r^{-3/2}$ for large r. This is consistent with a random-walk picture of fluctuations in nearest-neighbor distance, and thus a static correlation exponent $\chi = 1/2$. The dynamical exponent $z = \chi/\beta$ is thus $\simeq 2.6$. A numerical diagonalization study [10] of the spectrum of the relaxation operator for finite systems with $L \leq 30$ yields a value 2.55 ± 0.15. This value of z is nonstandard, indicating that the dynamics in this sector is characterized by a new universality class. In this sense, our model is different from an earlier attempt [11] to generalize the interface growth problem by allowing the height function to be an N component vector; it was argued that $z = 3/2$ independent of N in $d = 1$.

To summarize, we have shown that the deposition-evaporation model exhibits great diversity in the manner in which the autocorrelation function decays. There is strong evidence for the conjecture that in sectors with finite IS density ℓ/L the dynamical behavior is essentially the same as that of the spin correlation function of a system of hard core random walkers with the corresponding spin sequence. The diversity of relaxational behaviors in the DE model is linked to differences in correlations in spin patterns in the IS which labels that sector. In the sector composed of completely reducible configurations, the dynamics is characterized by a critical exponent $z \simeq 2.5$, indicative of a new universality class.

M. B. acknowledges the hospitality of the Condensed Matter Group at the International Centre for Theoretical Physics, Trieste where part of this work was done.

[1] B.I. Halperin and P.C. Hohenberg, Rev. Mod. Phys. **49**, 435 (1977).
[2] M. Barma, M.D. Grynberg, and R.B. Stinchcombe, Phys. Rev. Lett. **70**, 1033 (1993).
[3] R.B. Stinchcombe, M.D. Grynberg, and M. Barma, Phys. Rev. E **47**, 4018 (1993).
[4] D. Dhar and M. Barma, Pramana **41**, L193 (1993).
[5] M. Kardar, G. Parisi, and Y. Zhang, Phys. Rev. Lett. **56**, 889 (1986); J. Krug and H. Spohn, in *Solids Far from Equilibrium*, edited by C. Godrèche (Cambridge Univ. Press, Cambridge, 1991).
[6] R. Arratia, Z. Ann. Probab. **11**, 362 (1983); T.M. Liggett, *Interacting Particle Systems* (Springer, New York, 1985).
[7] The reason for the factor 3 in this definition is that particle j can come to a site visited earlier by particle j' only if $j - j'$ is a multiple of 3.
[8] S. Alexander and P. Pincus, Phys. Rev. B **18**, 2011 (1978).
[9] S.N. Majumdar and M. Barma, Physica (Amsterdam) **177A**, 366 (1991).
[10] P.B. Thomas, M.K. Hari, and D. Dhar (to be published).
[11] J.P. Doherty, M.A. Moore, J.M. Kim, and A.J. Bray, Phys. Rev. Lett. **72**, 2041 (1994).

Spontaneous Symmetry Breaking in a One Dimensional Driven Diffusive System

M. R. Evans,[1,2] D. P. Foster,[3] C. Godrèche,[2,3] and D. Mukamel[2,4]

[1]*Laboratoire de Physique Statistique de l'École Normale Supérieure, 24 rue Lhomond, F-75231 Paris 05 Cedex, France**
[2]*Newton Institute for Mathematical Sciences, 20 Clarkson Road, Cambridge CB2 0EH, United Kingdom*
[3]*Service de Physique de l'État Condensé, Centre d'Études de Saclay, F-91191 Gif-sur-Yvette Cedex, France*
[4]*Department of Physics of Complex Systems, Weizmann Institute, Rehovot 76100, Israel*
(Received 6 July 1994)

A simple model of a driven diffusive system which exhibits spontaneous symmetry breaking in one dimension is introduced. The model has short range interactions and unbounded noise. It is characterized by an asymmetric exclusion process of two types of charges moving in opposite directions on an open chain. The model is studied by mean field and Monte Carlo methods. Exact solutions can be found in a restricted region of its parameter space. A simple physical picture for the symmetry breaking mechanism is presented.

PACS numbers: 05.40.+j, 05.60.+w, 05.70.Fh, 05.70.Ln

The question of spontaneous symmetry breaking and long range order in one dimensional (1D) systems with short range interactions and small but unbounded noise is an intriguing one. It is well known that under these conditions no phase transition takes place in thermal equilibrium. This is the case provided the local variable which describes the state of the system can take only a finite number of possible values, such as in the Ising or the Potts models. When the local variable is not restricted to a finite set of values, such as in solid-on-solid models for chain unbinding, phase transitions, and symmetry breaking may take place [1]. Systems far from thermal equilibrium are, on the other hand, less restrictive, and the question of whether they are capable of exhibiting spontaneous symmetry breaking under the above conditions even when the local variable can take only a finite set of values has been open for quite some time. Recently, an example of such a phase transition, in the context of error correcting computation algorithms, has been given [2]. However, this example is rather complicated and not widely understood.

In the present Letter we introduce a simple nonequilibrium one dimensional model with short range interactions and unbounded noise which exhibits spontaneous symmetry breaking in the thermodynamic limit. The local dynamical variable associated with the model is restricted to take only a finite number of possible states. The model belongs to a class of traffic jam models or driven diffusive systems. It may also describe the dynamics of a certain growth process [3]. To be specific we consider a 1D lattice of length N. Each lattice point may be occupied by either a positive (+) or a negative (−) particle, or by a hole (0). The (+) particles move to the right while the (−) particles move to the left. The two kinds of particles may pass each other. The positive (negative) particles are supplied at the left (right) end and removed at the right (left) end of the system. The model possesses a right-left symmetry, and the dynamical rules are invariant under charge conjugation combined with space inversion. Under the conditions where the symmetry of the dynamical process is preserved, one expects the two currents of the positive and the negative charges to be equal. When spontaneous symmetry breaking takes place, the two currents become unequal in the thermodynamic limit (defined below).

We now define more precisely the dynamics of the system. During an infinitesimal time interval dt, the following exchange events take place between two adjacent sites:

$$+0 \to 0+, \quad 0- \to -0, \quad +- \to -+, \qquad (1a)$$

with probabilities dt, dt, and $q\,dt$, respectively. Furthermore, at the two ends, particles may be introduced or removed. At the left boundary ($i = 1$) one has

$$0 \to +, \quad - \to 0, \qquad (1b)$$

with probabilities $\alpha\,dt$ and $\beta\,dt$, respectively. Similarly at the right boundary ($i = N$):

$$0 \to -, \quad + \to 0, \qquad (1c)$$

with probabilities $\alpha\,dt$ and $\beta\,dt$, respectively. The dynamical process in the bulk is conservative: It conserves both the positive and the negative charges. However, these quantities are not conserved at the two ends.

One is interested in the steady state calculating, say, the density profiles of the two charges and the corresponding currents. The model defined above is a generalization of the totally asymmetric exclusion model of a single type of particles [4–7]. The bulk dynamics (1a) of two species but with periodic boundary conditions has been studied in connection with the behavior of shock fronts [8]. Related models in higher dimensions have also been considered recently [9,10].

In the present work we make the observation that the model (1) exhibits spontaneous symmetry breaking for a certain range of the parameters α, β, q which define its dynamics. We find two phases in which the currents

 0031-9007/95/74(2)/208(4)$06.00

corresponding to the positive and negative charges are not equal in the thermodynamic limit. In each of these phases the system may be in either one of two states related to each other by charge conjugation and space inversion $i \rightarrow N - i + 1$. Symmetry breaking does not take place in the single species model.

The model (1) is exactly soluble for $\beta = 1$ and in the limit $\alpha \rightarrow \infty$. The steady state profiles and the currents may be obtained in this case by the recently introduced matrix method [5,8]. However, the phases are symmetric for this set of parameters (for details see [3]).

In order to obtain the qualitative features of the global phase diagram we first study the model within the mean field approximation, where two broken symmetry phases are found for sufficiently small β. We then carry out numerical simulations of the dynamical equations in which the predictions of the mean field approximation, and, in particular, the existence of broken symmetry phases, are substantiated. A simple argument supporting these findings is given. This Letter is concluded by a study of the time scale associated with flipping between the two states of a broken symmetry phase for a finite system.

To study the mean field phase diagram of the model we denote the density of the positive and the negative charges at site i by p_i and m_i, respectively. Within the mean field approximation one neglects density-density correlations and obtains the following equations for the steady state:

$$\begin{aligned} J_+ &= p_i[1 - p_{i+1} - (1 - q)m_{i+1}], \\ J_- &= m_{i+1}[1 - m_i - (1 - q)p_i], \end{aligned} \tag{2a}$$

for $i = 1,\ldots,N - 1$, where J_+ and J_- are the currents of the positive and negative charges, respectively. We have used the fact that, in the steady state, the currents $J_\pm$ are independent of position. In addition to the bulk equations (2a) one has four other equations for the currents at the boundaries,

$$\begin{aligned} J_+ &= \alpha(1 - p_1 - m_1) = \beta p_N, \\ J_- &= \beta m_1 = \alpha(1 - p_N - m_N). \end{aligned} \tag{2b}$$

For simplicity we discuss the resulting phase diagram for the case $q = 1$ (the qualitative results remain unchanged for $q \neq 1$). It is readily seen from Eqs. (2) that when $q = 1$ the two sets of bulk equations decouple. The reason is that, away from the boundaries, a positive particle does not distinguish between a hole and a negative particle and neither does a negative particle distinguish between a hole and a positive particle. However, at the boundaries the two systems of particles are coupled via the boundary equations (2b). Defining

$$\begin{aligned} \alpha^+ &= \alpha h_1/(1 - p_1) = J_+/(J_+/\alpha + J_-/\beta), \\ \alpha^- &= \alpha h_N/(1 - m_N) = J_-/(J_-/\alpha + J_+/\beta), \end{aligned} \tag{3}$$

where $h_i = 1 - p_i - m_i$ is the hole density at site i, the problem is reduced to two one-species totally asymmetric exclusion processes on a lattice of size N. One process corresponds to the (+) particles with boundary parameters (α^+, β) and the other corresponds to the (−) particles with boundary parameters (α^-, β). The only coupling between the two processes is via the boundary equations (3).

The phase diagram for the one-species process with boundary parameters (α^s, β^s) is known [4–6] (here s stands for single species). It exhibits three phases: (a) a power law phase for $\alpha^s \geq \frac{1}{2}, \beta^s \geq \frac{1}{2}$. In this phase the approach to the bulk density ($= \frac{1}{2}$) is algebraic, and the current is maximal ($J^s = \frac{1}{4}$). (b) A low density phase for $\alpha^s < \beta^s$ and $\alpha^s < \frac{1}{2}$. Here the approach to the bulk density ($= \alpha^s$) is exponential, and the current is $J^s = \alpha^s(1 - \alpha^s)$. (c) A high density phase for $\beta^s < \alpha^s$ and $\beta^s < \frac{1}{2}$, in which the approach to the bulk density ($= 1 - \beta^s$) is exponential and the current is $J^s = \beta^s(1 - \beta^s)$. The high and low density phases coexist on the line $\alpha^s = \beta^s < \frac{1}{2}$. Using these results and Eq. (3) to get the effective feeding parameters α^+, α^-, the (α, β) phase diagram of the model (1) and closed expressions for the transition lines may be obtained (for details see [3]). The phase diagram (Fig. 1) exhibits four phases of which two are symmetric and two are nonsymmetric. One of the symmetric phases is characterized by a power law decay of the local density, and the other is a low density phase. The two nonsymmetric phases are characterized by high density–low density (hd-ld) and low density–low density (ld-ld) profiles, and they exist in the low β region of the phase diagram. In these phases the currents of the positive and the negative charges are unequal.

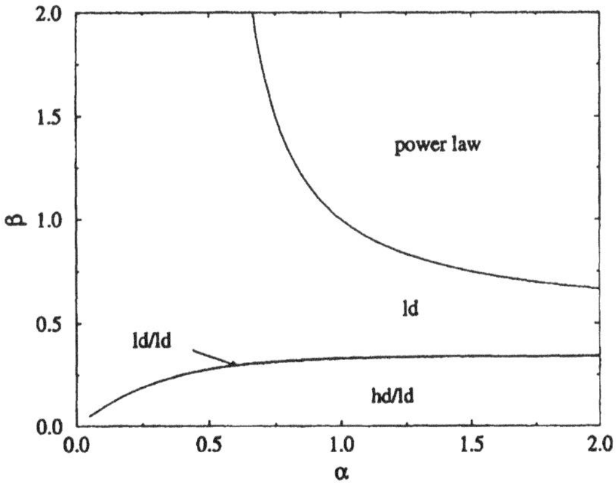

FIG. 1. The (α, β) mean field phase diagram for $q = 1$. It exhibits two symmetric phases: power law and low density (ld), and two nonsymmetric phases: low density–low density (ld-ld) and high density–low density (hd-ld). The ld-ld phase occupies a narrow region which appears as a line on the scale of the figure. The transitions between the various phases are continuous.

To demonstrate that spontaneous symmetry breaking does exist in the stochastic model (1), we carried out extensive Monte Carlo (MC) simulations. Figure 2 shows typical density profiles in the two nonsymmetric phases. They are obtained by averaging the occupations of each site over the simulation. To calculate these profiles one has to run the MC simulations long enough to reach the steady state. However, running time should be smaller than $\tau(N)$, the characteristic flipping time between the two states of the broken symmetry phase. The density profiles given in Fig. 2 are flat in the bulk, with some structure near the ends. In the hd-ld phase, the density of the negative charges is larger than $\frac{1}{2}$, while the density of the positive charges is lower than $\frac{1}{2}$. In the ld-ld phase both densities are smaller than $\frac{1}{2}$.

Consider now the characteristic time $\tau(N)$ between flips. It has to diverge in the thermodynamic limit in order to have a stable broken symmetry phase. Moreover, since the bulk dynamics of this model is conservative, even if $\tau(N)$ grows like N^2 for large N it may not be sufficient to demonstrate spontaneous symmetry breaking. For example, if one considers the 1D Ising model with conserved bulk dynamics but with some nonconserved dynamics at the boundaries, the characteristic time associated with the decay of magnetization grows like N^2. This divergence is a result of the slow conserved dynamics, and does not indicate spontaneous symmetry breaking. The thermodynamic limit is thus taken to be the large t and N limit where $O(N^2) < t < O(\tau(N))$.

In the present model $\tau(N)$ is expected to grow much faster than N^2. This can be seen as follows. Let the broken symmetry phase be characterized by currents j_1, j_2, and bulk densities ρ_1, ρ_2 of the two species. In the two symmetry related states, denoted by A and B, the currents and the bulk densities are given by $J_+ = j_i, \rho_+ = \rho_1$; $J_- = j_2, \rho_- = \rho_2$ in state A, and $J_+ = j_2, \rho_+ = \rho_2$; $J_- = j_1, \rho_- = \rho_1$ in state B. Here $\rho_\pm$ are the bulk densities of the positive and negative charges, respectively. The question is how does a finite system flip from, say, state A to state B? Clearly, a change in the density throughout the lattice has to be induced by fluctuations at the boundaries, since the bulk dynamics is conservative. Suppose such a fluctuation takes place and a droplet of state B is generated near one of the ends of a system of state A. This droplet is separated from the rest of the system by a domain wall. However, unlike a domain wall separating two equivalent states in thermal equilibrium (such as a domain wall between the up and the down states in the Ising model), here the domain wall is not stable [11]. The reason is that the two currents of, say, the positive charges in the two coexisting states, j_1 and j_2, are different from each other and there is a net flux of particles into or out of the domain wall region. As a result, the droplet B is expelled and the system relaxes back to the A state on a time scale which depends on the length of the initial B droplet. The system eventually flips to state B by a mechanism explained below.

To examine this mechanism we consider small β, where the two coexisting states are basically either all positive ($\rho_+ \simeq 1$) or all negative ($\rho_- \simeq 1$). Suppose the system is in the $\rho_+ \simeq 1$ state. In this phase there is a low flux of negative charges and holes which enter the system at the right end and leave at the left end. As the system evolves, a blockage of negative particles (a droplet) may temporarily be formed at the left end due to some fluctuation. When this happens, holes enter the system from both ends resulting in a net influx of holes. They get trapped in between the positive and negative regions. A typical configuration would thus look like $(- - -0000+ + \cdots +)$. Usually, the block at the left end leaves the system after some time and the system relaxes to the all (+) state. However, if the block persists for time of order N, the system will be filled with holes, and thus has a chance of switching to the negatively charged state. Therefore one is interested in the probability of a block persisting for a time of $O(N)$. To estimate this probability we consider the dynamics of the block. The right end of the block, located at a distance $x(t)$ from the left end of the system, performs a diffusive motion biased to the left. The bias is due to the fact that in the blocked state the influx of negative charges at the right end, $j_{\text{in}} = \beta\alpha/(1 + \alpha)$, is smaller than the flux leaving the system at the left end, $j_{\text{out}} = \beta$. The dynamics of $x(t)$ may thus be described by a biased random walk with absorption at $x = 0$. The probability $P(x_0, t)$ that such a walker starting at position x_0 reaches the origin at time t is given by

$$P(x_0, t) \propto \frac{x_0}{t^{3/2}} \exp[-(x_0 - v_t)^2/ct] \qquad (4)$$

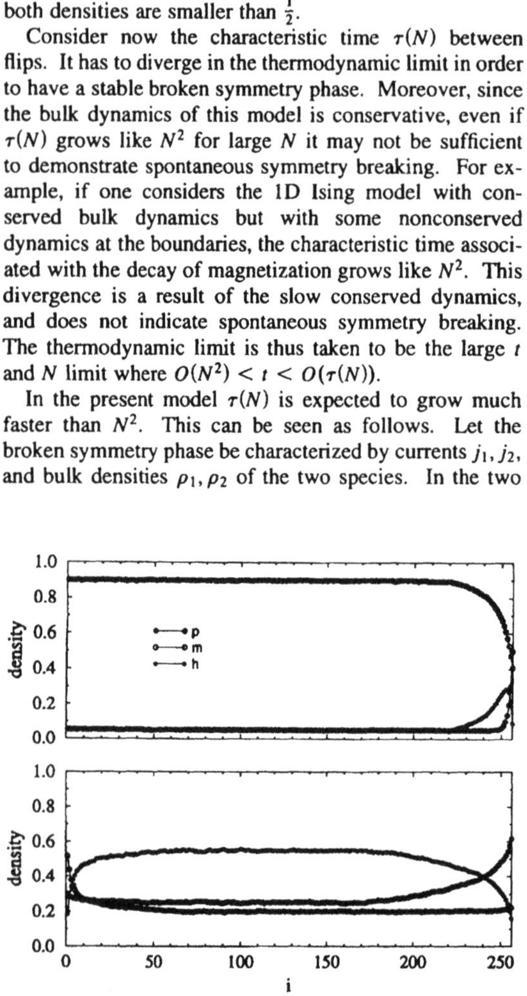

FIG. 2. Density profiles of the positive (p) and negative (m) charges and hole (h) in the hd-ld ($\alpha = 1, \beta = 0.1, q = 1$) and ld-ld ($\alpha = 1, \beta = 0.333, q = 1$) phases.

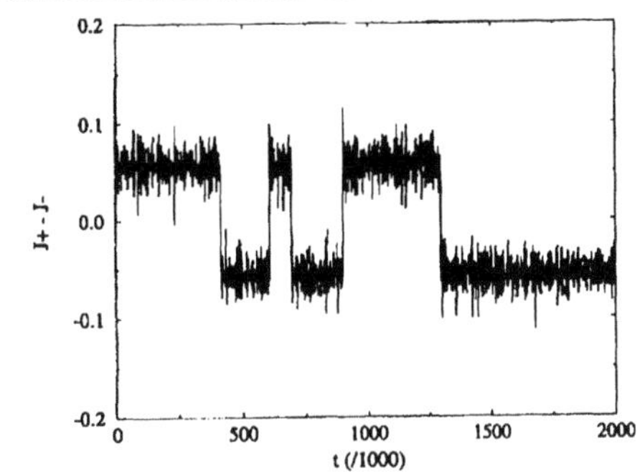

FIG. 3. Time evolution of the current difference in the hd-ld phase ($\alpha = q = 1, \beta = 0.15, N = 80$). Each point represents an average of the current difference over 1000 sweeps. Flips between the two symmetry related states are clearly seen.

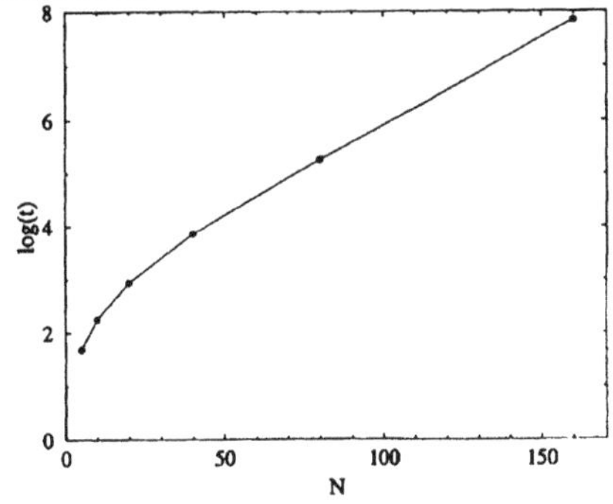

FIG. 4. Time scale $\tau(N)$ as a function of N for $N \leq 160$.

for large t. Here v is the average rate at which the droplet shrinks, and c is a constant. Within this picture the probability that a droplet of initial size of $O(1)$ persists for time of $O(N)$ is exponentially small, $\exp(-bN)$, where b is a constant. The time scale between flips is therefore of order $\exp(bN)$. A detailed discussion of the switching mechanism and the associated time scales will be given elsewhere [12].

This behavior is very different from that of a one species model on the coexistence line between the high and the low density states. In that case the two states have the *same* current and therefore a domain wall between the high density and the low density states is a *stable* object with a vanishing net velocity. It displays a diffusive motion and thus takes a time of order N^2 to traverse and hence flip the system.

To study the flipping process in a finite system we simulated the dynamics of Eqs. (1). In Fig. 3 we present the time evolution of the current difference $J_- - J_+$ for a typical run in the hd-ld broken symmetry phase. Similar behavior is found for the density difference between the two charges. It is clear from Fig. 3 that at any given time the two states do not coexist in the system (except, maybe, during the short time when a flip takes place). This supports the argument given above, namely that droplets of the "wrong" state are expelled from the system unless they are macroscopically large.

To evaluate the time scale $\tau(N)$ we averaged the current difference over many runs, starting from the initial configuration where all sites are occupied by positive charges. This average decays at large time t as $\exp[-t/\tau(N)]$ and thus yields $\tau(N)$. We have measured the time scale for systems of size 5, 10, 20, 40, 80, 160, and the results are given in Fig. 4. At first glance they seem to suggest that $\tau(N)$ grows somewhat slower than exponentially with N, maybe like $\exp(aN^\gamma)$ with $\gamma < 1$. However, in trying to fit the data to a stretched exponential form we find that γ tends to grow with the system size for $N \leq 160$. This may indicate that the stretched exponential form is related to finite size effects, and that, in fact, $\tau(N)$ grows exponentially for large N, as suggested by the droplet dynamics discussed above.

We thank B. Derrida, T. Halpin-Healy, J. Miller, and E. R. Speer for helpful discussions.

*Laboratoire associé au Centre National de la Recherche Scientifique et aux Universités Paris VI et Paris VII.

[1] See, for example, J. M. J. van Leeuven and H. J. Hilhorst, Physica (Amsterdam) **107A**, 319 (1981).
[2] P. Gacs, J. Comput. Sys. Sci. **32**, 15 (1986).
[3] M. R. Evans, D. P. Foster, C. Godrèche, and D. Mukamel, J. Stat. Phys. (to be published).
[4] B. Derrida, E. Domany, and D. Mukamel, J. Stat. Phys. **69**, 667 (1992).
[5] B. Derrida, M. R. Evans, V. Hakim, and V. Pasquier, J. Phys. A: Math. Gen. **26**, 1493 (1993).
[6] G. Schütz and E. Domany, J. Stat. Phys. **72**, 277 (1993).
[7] J. Krug, Phys. Rev. Lett. **67**, 1882 (1991).
[8] B. Derrida, S. A. Janowsky, J. L. Lebowitz, and E. R. Speer, Europhys. Lett. **22**, 651 (1993); J. Stat. Phys. **73**, 813 (1993).
[9] B. Schmittmann, K. Hwang, and R. K. P. Zia, Europhys. Lett. **19**, 19 (1992).
[10] D. P. Foster and C. Godrèche, J. Stat. Phys. (to be published).
[11] Similar behavior was found for walls separating two time periodic states. See G. Grinstein, D. Mukamel, R. Seidin, and C. H. Bennett, Phys. Rev. Lett. **70**, 3607 (1993).
[12] M. R. Evans, D. P. Foster, C. Godrèche, D. Mukamel, and E. R. Speer (to be published).

Authors' note

The following papers were listed as "in preparation" in the two letters reprinted above:

"Asymmetric Exclusion Model with Two Species: Spontaneous Symmetry Breaking," M. R. Evans, D. P. Foster, C. Godreche and D. Mukamel, J. Stat. Phys. 80, 69-102 (1995).

"Jamming Transition in a Homogeneous One-Dimensional System: The Bus Route Model," O. J. O'Loan, M. R. Evans and M. E. Cates, Phys. Rev. E 58, 1404-1418 (1998).

EUROPHYSICS LETTERS 15 April 1998
Europhys. Lett., **42** (2), pp. 137-142 (1998)

Spontaneous jamming in one-dimensional systems

O. J. O'LOAN, M. R. EVANS and M. E. CATES
Department of Physics and Astronomy, University of Edinburgh
Mayfield Road, Edinburgh EH9 3JZ, UK

(received 16 December 1997; accepted 23 February 1998)

PACS. 05.70Ln – Nonequilibrium thermodynamics, irreversible processes.
PACS. 64.60–i – General studies of phase transitions.
PACS. 89.40+k – Transportation.

Abstract. – We study the phenomenon of jamming in driven diffusive systems. We introduce a simple microscopic model in which jamming of a conserved driven species is mediated by the presence of a non-conserved quantity, causing an *effective* long-range interaction of the driven species. We study the model analytically and numerically, providing strong evidence that jamming occurs; however, this proceeds via a strict phase transition (with spontaneous symmetry breaking) only in a prescribed limit. Outside this limit, the nearby transition (characterised by an essential singularity) induces sharp crossovers and transient coarsening phenomena. We discuss the relevance of the model to two physical situations: the clustering of buses, and the clogging of a suspension forced along a pipe.

Many non-equilibrium physical situations can be modelled as driven diffusive systems [1]. An intriguing feature of certain driven systems is their propensity to jam —in traffic flow [2] jamming behaviour is a fact of modern life and in colloid rheology the phenomenon of shear thickening (dilatancy) is widely studied [3].

One-dimensional (1d) driven systems exhibit a wide variety of interesting phenomena, including phase transitions and spontaneous symmetry breaking [4], which are precluded from 1d equilibrium systems (in the absence of long-range interactions). This suggests that the physics of jamming might be captured in simple 1d models. In previous studies of simple 1d non-equilibrium models, jamming arises because of the presence of disorder or inhomogeneities such as defect sites [5]. In contrast, the model we introduce below is homogeneous; the jamming emerges via spontaneous symmetry breaking. Jamming arises through a mechanism in which a non-conserved quantity in the dynamics mediates an *effective* long-range interaction of a conserved quantity (driven species), even though the microscopic dynamics is *local* and stochastic.

We now define the microscopic model we study, which we refer to as the Bus Route Model (BRM) for reasons to become clear. The BRM is defined on a 1d periodic lattice with L sites. Site i has two variables τ_i and ϕ_i associated with it, each of which can be either 1 or 0. When a site is occupied by a "bus", τ_i is 1 and if ϕ_i is 1 the site is said to have "passengers" on it [6]; τ_i and ϕ_i cannot both be 1 simultaneously. There are M buses in total and the bus density

$\rho = M/L$ is a conserved quantity. However, the total number of sites with passengers is *not* conserved.

In updating the system, a site i is chosen at random. If both τ_i and ϕ_i are 0, then $\phi_i \to 1$ with probability λ. If $\tau_i = 1$ and $\tau_{i+1} = 0$, then the bus at site i hops forward with probability $1-(1-\beta)\phi_{i+1}$. If the bus hops, ϕ_{i+1} becomes 0. Thus, a bus hops with probability 1 onto a site without passengers, and probability β onto a site with passengers thereby removing them. The probability that passengers arrive at an empty site is λ. We generally take $\beta < 1$, reflecting the fact that buses are slowed down by having to pick up passengers. Buses are forbidden from overtaking each other but relaxing this condition will have no significant effect [7]. We remark that the dynamics is local and does not satisfy detailed balance.

At this point it is useful to discuss two scenarios which illustrate possible applications of the model and highlight the roles of the conserved and non-conserved quantities. The first and most obvious example is that of buses moving along a bus-route. Clearly, the ideal situation is that the buses are evenly spaced so that they pick up roughly equal numbers of passengers. However, what commonly occurs is that a bus falls behind the one in front and consequently has more passengers awaiting it. Thus the bus becomes further delayed and at the same time, following buses catch up with it, leading to a cluster of buses. The number of passengers awaiting a bus gives an indication of the elapsed time since the last bus went past and in this way communicates information between the two buses, resulting in an effective long-range interaction [8].

We now turn to an alternative interpretation of the model describing a system of driven particles, each of which can exist in two states of mobility. Each time a bus hops to the right in the BRM, a vacancy moves to the left. In the new interpretation of the model, which can be thought of as the dual of the BRM, the vacancies become "particles" and the non-conserved variable is the mobility (hopping probability) of a particle, which is either 1 or β [9]. A possible application of this dual model is to the phenomenon of clogging. A simple scenario is the flow of particles suspended in a fluid being forced through a pipe. The pipe is narrow enough to prevent the particles passing each other and stationary particles may become weakly attached to the pipe (with rate λ), reducing their mobility from 1 to β. At high density, individual particles move more slowly and therefore are more likely to become attached to the pipe, thus impeding the motion of the following particles and encouraging them to attach. Hence clogging ensues. Although set up as a strictly 1d model (requiring the diameters of the particles and the pipe to be comparable), a similar scenario could affect the flow of any heterogeneous material with a tendency to solidify when at rest [10].

From our study of the BRM we provide strong evidence, both numerical and analytical, that a true jamming phase transition does occur, but only in the limit $\lambda \to 0^+$ with $\lambda L \to \infty$. The transition is from a low-density "jammed" phase to a high-density homogeneous phase. When λ is small but finite, we find two strong signatures of the transition. Firstly, the transition is rounded to a crossover; but this is exponentially sharp in $1/\lambda$. Secondly, apparent coarsening occurs where over long time scales, the system separates into jammed regions of finite size with long but finite lifetimes.

We first present some simulation results for the BRM. Figure 1 shows a space-time plot of the system at low density and small λ. As passengers enter the system, one sees the large inter-bus gaps increasing in size until the system comprises several distinct clusters (or "jams") of buses. The system then coarsens via coalescence of the bus clusters until finally, only a single large cluster remains. For high densities, we find that the system is homogeneous —a snapshot of the system as a whole resembles the high-density final cluster in fig. 1. Figure 2 shows a space-time plot for the same system as in fig. 1, with the exception that now $\lambda = 0.1$. While small, transient clusters of buses do appear, the "phase-separation" seen for $\lambda = 0.02$ does not

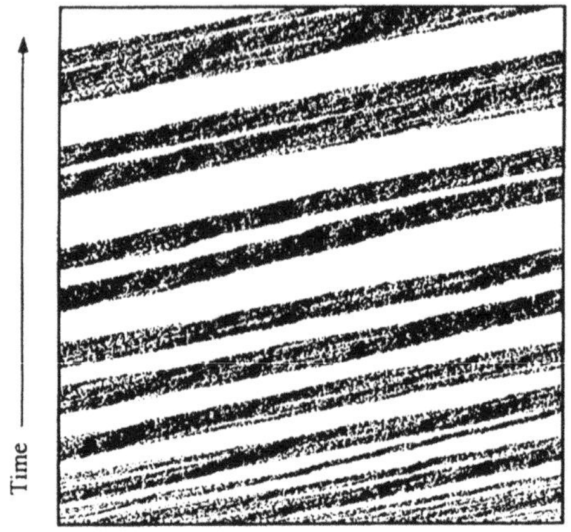

Fig. 1

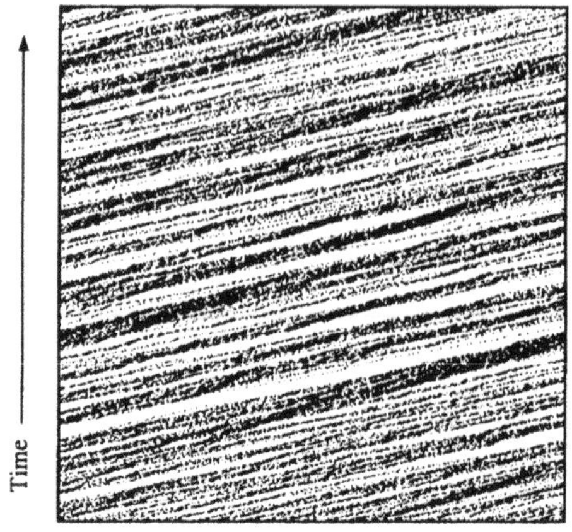

Fig. 2

Fig. 1. – Space-time plot of bus positions for $\lambda = 0.02$, $\rho = 0.2$, $\beta = 0.5$ and $L = 500$. There are 10 time steps between each snapshot on the time axis. Initially, the buses are positioned randomly and there are no passengers.

Fig. 2. – Space-time plot of bus positions for the same parameters as in fig. 1 with the exception that here $\lambda = 0.1$.

occur. Figure 3 shows plots of bus velocity v (average rate of hopping forward) against bus density ρ. For the two larger values of λ, velocity decreases smoothly with increasing density. However, for $\lambda = 0.02$, $v(\rho)$ has an *apparent* cusp at an intermediate value of the density, suggesting the presence of a phase transition.

We now show that the BRM exhibits a phase transition in the limit $\lambda \to 0$ with $\lambda L \to \infty$. To see this, consider a system comprising a single large cluster (as in fig. 1). If $\lambda L \to \infty$, then the site in front of the leading bus has passengers with probability one (because the time since that site was last visited by a bus is $\propto L$). Hence, the leading bus hops forward with probability β. Since all of the gaps *within* the cluster are finite, there are no passengers within the cluster as $\lambda \to 0$; the buses within the cluster hop with probability one into unoccupied sites. The velocity (average rate of hopping forward) of these buses is $1 - \rho_c$ [11], where ρ_c is the density of buses in the cluster. For the cluster to be stable, this velocity must equal that of the leading bus and so we have $\rho_c = 1 - \beta$. For overall bus densities greater than ρ_c, the system becomes homogeneous with all gaps finite. Therefore, we identify ρ_c as the critical density.

This shows that the BRM exhibits a phase transition in the limit of $\lambda \to 0$. We now present a two-particle approximation to the problem which suggests that there is no strict transition for non-zero λ. First, let us approximate the probability that a bus hops into a gap of size x by $u(x) = f(x) + \beta(1 - f(x))$, where $f(x)$ is an estimate of the probability that there are no passengers on the first site of the gap. The average time since a bus last left this site is x/v (where v is the average velocity in the system), so we estimate $f(x) = \exp[-\lambda x/v]$ to give

$$u(x) = \beta + (1 - \beta)\exp[-\lambda x/v] \quad \text{for} \quad x > 0 \tag{1}$$

with exclusion requiring $u(0) = 0$. This is in the spirit of a mean-field approximation for the BRM, the nature of which is to replace the "induced" interaction between buses (which is

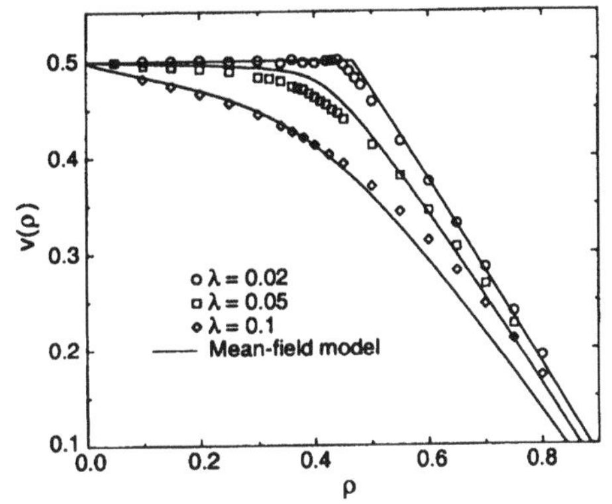

Fig. 3

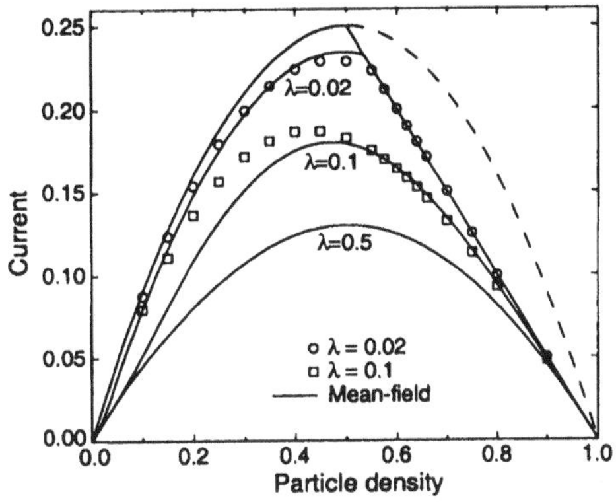

Fig. 4

Fig. 3. – The velocity as a function of bus density for $\beta = 0.5$ and various values of λ. The symbols are simulation results for the BRM with $L = 10\,000$ and the lines are mean-field model (see below) results in the thermodynamic limit.

Fig. 4. – Current as a function of density for the dual model for several values of λ. $L = 10\,000$ for all simulation data and MFM results are in the thermodynamic limit. The uppermost solid curve is the MFM result in the limit $\lambda \to 0$ with $\lambda L \to \infty$. The dashed curve is the exact result when λ is set equal to zero before the thermodynamic limit is taken. The latter two curves are identical for $\rho < 0.5$.

subject to stochastic variation) with a deterministic one.

Now consider a "jammed" system as in fig. 1, with the large gap in front of the leading bus in the cluster having size kL (where k is independent of L). We denote the size of the gap between the leading two buses by x so that, using the mean-field hopping rate in (1), we may write a Langevin equation for the dynamics of this gap size as

$$\dot{x} = u(kL) - u(x) + \eta(t) \equiv -\frac{\mathrm{d}\Phi}{\mathrm{d}x} + \eta(t), \tag{2}$$

where $\eta(t)$ is a noise term (say white noise of unit variance [12]). The gap size x has the dynamics of a particle diffusing in a potential well $\Phi(x)$ given by (1), (2). The potential has a maximum at $x^* = kL$ so that when $x > x^*$, the particle has escaped from the well, or equivalently, the leading bus has left the cluster. We denote the average time for this break-up to occur by τ, which is given by $\exp[\Phi(x^*) - \Phi(0)]$ to a good approximation [13]. In the limit $L \to \infty$, this becomes

$$\tau \sim \exp\left[\frac{\beta(1-\beta)}{\lambda}\right], \tag{3}$$

which is finite for $\lambda > 0$, implying that a jam is not a stable object and will eventually break up. However, when $\lambda \to 0$, the jam becomes stable in agreement with our previous argument. When λ is small but non-zero, τ is exponentially large in $1/\lambda$ and it can appear that a jam is stable when in fact it has a finite lifetime. Thus, we do not expect true phase separation to occur for non-zero λ.

Let us now move beyond the two-particle picture described above. Consider a model of hopping particles where the hopping rate of a particle is a function $u(x)$ of the size of the gap x in front of that particle. By using the mean-field expression for $u(x)$ given in (1), one defines a new model which we call the mean-field model (MFM). The (rigorous) solution and analysis of the steady state [14] of the MFM can be found in [11]; here we present some selected results.

The MFM exhibits no phase transition for non-zero λ in agreement with our two-particle argument but there is indeed a transition in the limit $\lambda \to 0$ with $\lambda L \to \infty$. Figure 3 compares velocity as a function of density in the MFM (solved analytically) and the BRM (simulated); the agreement is quite good. For $\lambda = 0.02$ in both models, $v(\rho)$ has an *apparent* cusp at an intermediate value of the density. We know that for the MFM, $v(\rho)$ is in fact non-singular since there is rigorously no transition for non-zero λ. Since we believe that the MFM captures the essential physics of the BRM, we expect that likewise there is no transition for non-zero λ in the BRM. When λ is small there is, however, a very sharp crossover between a low-density "jammed" regime with $v \simeq \beta$, and a high-density "congested" regime where v decreases roughly linearly with increasing density.

To quantify the sharpness of the crossover for λ close to zero in the MFM, we calculated $\kappa_{\max}$, the maximum curvature of $v(\rho)$. For λ small (less than about 0.02), we found [11] that $\kappa_{\max}$ varies as $\exp[a/\lambda]$, where a depends on β. Therefore, although a strict phase transition occurs only in the limit $\lambda \to 0$, the crossover is exponentially sharp in $1/\lambda$ for small λ.

We now comment on the occurrence of apparent coarsening (see fig. 1) in a system which, according to the above discussion, does not strictly phase-separate. (On the one hand, we have argued that large clusters are ultimately unstable, while on the other hand, fig. 1 appears to show a fully phase-separated system.) We believe [11] that sufficiently large systems coarsen up to some finite length scale which is exponentially large in $1/\lambda$. For the system in fig. 1, this length scale is much larger than the system size.

Let us now return to the dual model defined earlier and interpret our findings in that context. Since an inter-bus gap in the BRM corresponds to a cluster of particles in the dual model, jamming is now a *high-density* phenomenon, characterised by the presence of large clusters of particles. This restores to the word "jamming" a meaning closer to that used in everyday life. In the limit $\lambda \to 0$, a phase transition arises from a low-density homogeneous phase in which the particles move quickly, to a high-density jammed phase which is characterised by macroscopic clusters of particles and a slow flow. An infinitesimal rate λ can result in macroscopic inhomogeneity and decrease in flow. This is illustrated in fig. 4 which shows the current (velocity times density) as a function of particle density for the dual model.

A different interpretation of the dual model is as a model of stop-start traffic flow with the particles representing cars. The longer a car is at rest, the more likely it is that the driver will be slow to react when it is possible to move again. This is related to several "slow-to-start" cellular automaton traffic models studied recently [15].

In conclusion, we have found that the BRM exhibits a jamming transition from a high-density homogeneous phase to a low-density jammed phase. There is a spontaneously broken symmetry in the jammed phase: one bus is selected over all others to head the jam, even though all buses are identical. We have argued, however, that a strict phase transition occurs only in the limit $\lambda \to 0$ with $\lambda L \to \infty$ and that for non-zero λ, one sees crossover behaviour which is exponentially sharp in $1/\lambda$. Thus the model exhibits an essential singularity at $\lambda = 0$ which causes, alongside the dramatic crossover, the transient coarsening behaviour observed (see fig. 1 and [11]) for small, positive λ. If similar phenomena were to arise in other models, this could easily be interpreted as signifying a true phase transition where in fact none exists. Such phenomena may indeed arise in certain cellular automata models of traffic [16].

***

OJO is supported by a University of Edinburgh Postgraduate Research Studentship. MRE is a Royal Society University Research Fellow.

REFERENCES

[1] SCHMITTMANN B. and ZIA R. K. P., in *Phase Transitions and Critical Phenomena*, edited by C. DOMB and J. L. LEBOWITZ, Vol. **17** (Academic Press, London) 1995; HALPIN-HEALY T. and ZHANG Y.-C., *Phys. Rep.*, **254** (1995) 215.

[2] NAGEL K., *Phys. Rev. E*, **53** (1996) 4655.

[3] See, *e.g.*, FARR R. S., MELROSE J. R. and BALL R. C., *Phys. Rev. E*, **55** (1997) 7203.

[4] EVANS M. R., FOSTER D. P., GODRÈCHE C. and MUKAMEL D., *Phys. Rev. Lett.*, **74** (1995) 208.

[5] JANOWSKY S. A. and LEBOWITZ J. L., *Phys. Rev. A*, **45** (1992) 618; SCHÜTZ G., *J. Stat. Phys.*, **71** (1993) 471; MALLICK K., *J. Phys. A*, **29** (1996) 5375; DERRIDA B., in *Statphys19*, edited by B.-L. HAO (World Scientific) 1996; EVANS M. R., *Europhys. Lett.*, **36** (1996) 13; KRUG J. and FERRARI P. A., *J. Phys. A*, **29** (1996) L213; BEN-NAIM E., KRAPIVKSY P. L. and REDNER S., *Phys. Rev. E*, **50** (1994) 822.

[6] While we have taken the passenger variable ϕ_i to be binary, this does not to forbid the presence of more than one passenger at a site; we merely require that the extra passengers have no further effect on the dynamics.

[7] This is because the interchange of a fast-moving bus with a slower-moving bus in front also interchanges their velocities. This contrasts sharply with 1d models where jamming is induced by quenched disorder [5].

[8] The relevance of the BRM to real buses is discussed further in [11].

[9] In an exact mapping from the dual model to the BRM, a particle attempts to hop to the left when the site to its left is updated. However, a more natural dynamics with no significant difference in behaviour is to update only particles [11].

[10] Note that this violates Galilean invariance, requiring a pipe or some other fixed reference frame to be present.

[11] O'LOAN O. J., EVANS M. R. and CATES M. E., cond-mat/9712243 preprint, 1997.

[12] The variance of the noise should strictly depend on β but since we are primarily interested in the effect of λ on the dynamics of the gap, we ignore this dependence.

[13] KRAMERS H., *Physica*, **7** (1940) 284.

[14] The MFM is an example of a zero-range process for which the steady state is given by a product measure.

[15] BENJAMIN S. C., JOHNSON N. F. and HUI P. M., *J. Phys. A*, **29** (1996) 3119; SCHADSCHNEIDER A. and SCHRECKENBERG M., *Ann. Phys. (Leipzig)*, **6** (1997); 541 cond-mat/9709131 preprint, 1997.

[16] SASVÁRI M. and KERTÉSZ J., *Phys. Rev. E*, **56** (1997) 4104; EISENBLÄTTER B., SANTEN L., SCHADSCHNEIDER A. and SCHRECKENBERG M., *Phys. Rev. E*, **57** (1998) 1309; cond-mat/9706041 preprint, 1997.

Authors' note

The following papers were listed as "in preparation" in the two letters reprinted above:

"Asymmetric Exclusion Model with Two Species: Spontaneous Symmetry Breaking," M. R. Evans, D. P. Foster, C. Godreche and D. Mukamel, J. Stat. Phys. 80, 69-102 (1995).

"Jamming Transition in a Homogeneous One-Dimensional System: The Bus Route Model," O. J. O'Loan, M. R. Evans and M. E. Cates, Phys. Rev. E 58, 1404-1418 (1998).

Self-organization and a dynamical transition in traffic-flow models

Ofer Biham and A. Alan Middleton
Department of Physics, Syracuse University, Syracuse, New York 13244

Dov Levine
Department of Physics, Technion, Israel Institute of Technology, Haifa 32000, Israel
(Received 2 June 1992)

A simple model that describes traffic flow in two dimensions is studied. A sharp *jamming transition* is found that separates between the low-density dynamical phase in which all cars move at maximal speed and the high-density jammed phase in which they are all stopped. Self-organization effects in both phases are studied and discussed.

PACS number(s): 02.70.+d, 05.70.Ln, 64.60.Cn, 89.40.+k

Traffic problems have been studied extensively in recent years in order to help in the design of transportation infrastructure and to optimize the allocation of resources. Traffic simulations, based on various hydrodynamic models, have provided much insight and are in good agreement with experiments for simple systems such as a freeway, a tunnel, or a single junction [1]. However, the simulation of traffic flow in a whole city is a formidable task as it involves many degrees of freedom such as local densities and speeds. The availability of powerful supercomputers is likely to make these simulations feasible in the near future, but models that are simpler and more flexible than hydrodynamic models will be needed in order to achieve this task.

Cellular automaton (CA) models [2] are increasingly used in simulations of complex physical systems such as fluid dynamics [3], driven diffusive systems [4], sandpiles [5], and chemical reactions [6]. In some of these systems the cellular automaton models provide only some general qualitative features of the system while in other cases useful quantitative information can be obtained. For some problems involving complex geometries, such as simulations of fluid dynamics in porous media, cellular automata are found to be superior to other methods.

In this paper we present three variants of a simple cellular automaton model that describes traffic flow in two dimensions. The first two variants are three-state CA models on a square lattice. Each site contains either an arrow pointing upwards, an arrow pointing to the right, or is empty. In the first variant (model I) the dynamics is controlled by a traffic light, such that the right arrows move only in even time steps and the up arrows move in odd time steps. On even time steps, each right arrow moves one step to the right unless the site on its right-hand side is occupied by another arrow (which can be either an up or a right arrow). If it is blocked by another arrow it does not move, even if during the same time step the blocking arrow moves out of that site. Similar rules apply to the up arrows, which move upwards. Note that this is a fully deterministic model; randomness enters only through the initial conditions. In this model the traffic problem is reduced to its simplest form while the essential features are maintained. These features include the simultaneous flow in two perpendicular directions of objects that cannot overlap. No attempt is made here to draw a more direct analogy between our model and real traffic problems.

The model is defined on a square lattice of $N \times N$ sites with periodic boundary conditions. Due to the periodic boundary conditions the total number of arrows of each type is conserved. Moreover, the total number of up arrows in each column and the total number of right arrows in each row are conserved, giving rise to $2N$ conservation rules.

The density of right (up) arrows is given by $p_{\rightarrow} = n_{\rightarrow}/N^2$ ($p_{\uparrow} = n_{\uparrow}/N^2$), where $n_{\rightarrow}$ ($n_{\uparrow}$) is the number of right (up) arrows in the system. Here we examine the case where $p_{\rightarrow} = p_{\uparrow} = p/2$. The (average) velocity v of an arrow in a time interval τ is defined to be the number of successful moves it makes in τ divided by the number of attempted moves in τ. It has maximal value $v = 1$, indicating that the arrow was never blocked, while $v = 0$ means that the arrow was stopped for the entire duration τ, and never moved at all. The average velocity $\overline{v}$ for the system is then obtained by averaging v over all the arrows in the system.

We have performed extensive simulations of the model starting with an ensemble of random initial conditions. After a transient period that depends on the system size, on p, and on the random initial condition, the system reaches its asymptotic state. We found two qualitatively different asymptotic states, which are separated by a sharp dynamical transition. Below the transition all the arrows move freely in their turn and the average velocity is $\overline{v} = 1$, while above it they are all stuck and $\overline{v} = 0$ with very high probability. A typical configuration below the transition is shown in Fig. 1, where the system is self organized into separate rows of right and up arrows along the diagonals from the upper-left to the lower-right corners. This arrangement enables the arrows to achieve the maximal speed. When a row of horizontal arrows moves, it makes space for a row of vertical arrows to move in the

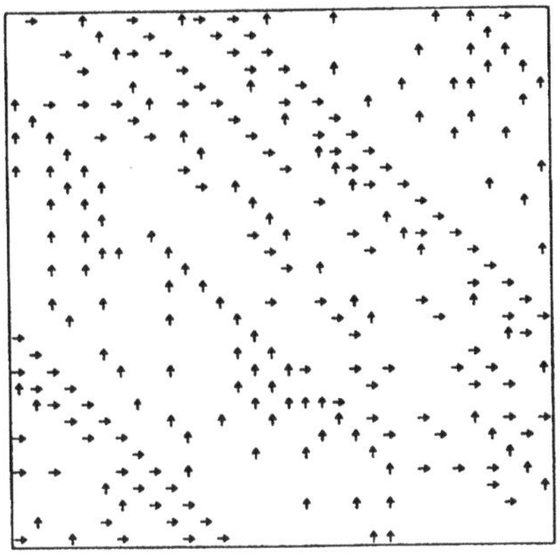

FIG. 1. A typical dynamic configuration in the low-density phase below the transition. The system is self-organized into a pattern of lines of arrows from the upper-left to the lower-right corners and $\bar{v} = 1$. The system size is 32×32 and $p = 0.25$.

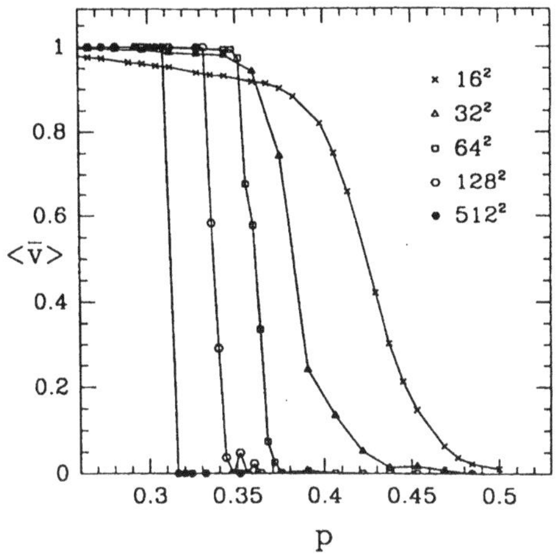

FIG. 3. The ensemble average velocity $\langle \bar{v} \rangle$ as a function of the concentration p for five different system sizes (model I). As the system size increases the transition becomes sharper, and the ensemble-average velocity changes rapidly from $\langle \bar{v} \rangle = 1$ below $p_c(N)$ to $\langle \bar{v} \rangle = 0$ above it.

next step, such that they never collide. Above the transition all the arrows are stopped in a global cluster, shown in Fig. 2 (by global cluster we mean a cluster that connects one side of the system to the other). This global cluster is oriented along the diagonal from the upper-right to the lower-left corners. This way it blocks the paths of all arrows which finally get stopped [7].

These two states are separated by a sharp *jamming transition* in which the ensemble-average velocity changes rapidly from $\langle \bar{v} \rangle = 1$ to $\langle \bar{v} \rangle = 0$ as p is varied (see Fig. 3). Results for five system sizes from 16×16 to 512×512 are presented. Small-size systems (up to 128×128) have been simulated on sequential machines, while the larger systems were simulated on a DECmpp parallel computer with 8k processors. For small-size systems the transition is not sharp but there is a range of densities in which both asymptotically dynamic and asymptotically static states are found with a non-negligible probability (depending on the initial condition). We define $p_c(N)$ to be at the center of this region, which is characterized by very long transients. As the system size increases, $p_c(N)$ tends to decrease while the coexistence region shrinks, giving rise to a sharper transition. From our simulations we have not been able to obtain conclusive results for p_c in the infinite system limit. We find that the transition is very sharp for large systems. However, $p_c(N)$ keeps decreasing as N increases, and we have not been able to determine whether it converges to a finite p_c or to zero in the infinite system limit. The difficulty results from the long equilibration times near the transition (see Fig. 4) and from the slow convergence of $p_c(N)$ as the system size increases.

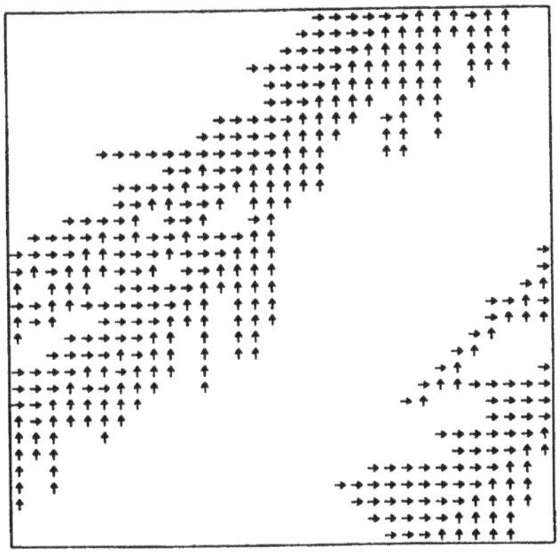

FIG. 2. A typical static configuration in the high-density phase above the transition. Here the global cluster is oriented between the upper-right and the lower-left corners, and blocks the paths of all the arrows until they get stopped. The system size is 32×32 and $p = 418/1024 \approx 0.4082$.

Being a transition as a function of the concentration p, between a state with no global cluster below p_c to a state with a global cluster above p_c, it resembles the percolation transition [8]. However, the percolation transition is a second-order transition and has no dynamics. The jamming transition can also be considered in the context of pinning transitions that occur in extended systems with quenched random impurities such as charge-density waves [9]. In our model there is no quenched component and the two sets of arrows pin each other when the density increases above threshold.

In order to examine the robustness of the jamming transition we have also studied a nondeterministic variant of our model (model II). In model II the traffic light is removed and all arrows move in all time steps (unless

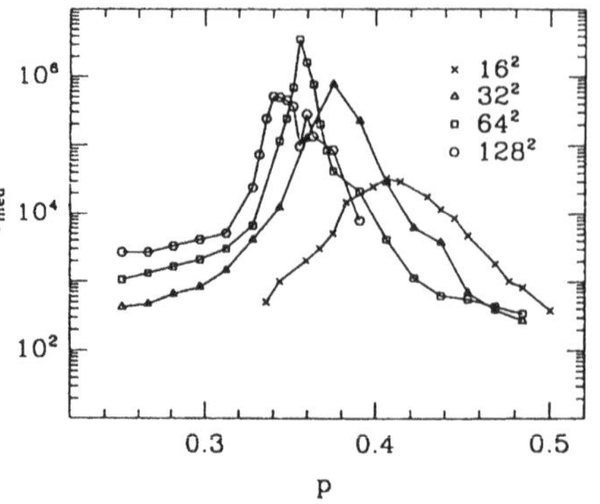

FIG. 4. The median equilibration time T_{med} for model I as a function of p for four different system sizes. The equilibration time is the number of time steps it takes to reach a periodic cycle or to get stuck. The peak around p_c becomes higher and sharper as the system size increases up to 64×64, and then becomes more flat for 128×128. It is not clear how to interpret this behavior for 128×128, although it may be that there is a very narrow peak that we have not been able to resolve.

they are stopped). If both an up and a right arrow try to move to the same site, one of them will be chosen randomly, with equal probabilities [10]. For this model we also find a sharp transition. The values of $p_c(N)$ are smaller than for model I (approximately 0.10 for systems of size 512×512). The value of p_c in the infinite system limit cannot be determined from our data.

By choosing a two-dimensional model that has only right and up arrows, and does not have left and down arrows, we simplify the problem without losing most of its essential features. The essential problem that causes traffic jams is the need of the right and up arrows to cross each other's paths, while each site can be occupied by only one arrow. There is no such problem between the up and down, or between the right and left arrows, as they can move in parallel paths that do not intersect. In models that have both right, up, left, and down arrows one can have a stable finite traffic jam. A simple example is a set of four arrows in which an up arrow blocks a left arrow, which blocks a down arrow, which blocks a right arrow, which blocks the first up arrow. This is the *gridlock* mechanism, which may occur at any density p. In our model gridlocks are not possible, and the jamming transition occurs only when a global cluster forms.

To obtain a better understanding of the model we now describe the one-dimensional analog which can be solved analytically. In one dimension there is only one type of arrows (say right arrows) that move along a closed ring. Every time step each arrow moves to the right unless it is blocked by another arrow [11]. The asymptotic velocity $\overline{v}$ is independent of initial conditions. It is $\overline{v} = 1$ for $p < 1/2$, while for $p > 1/2$ it decreases continuously to zero according to $\overline{v}(p) = (1-p)/p$ [12]. We thus conclude that the sharp first-order transition is indeed a result of the interaction between the horizontal and vertical arrows due to the excluded volume. To clarify this point further we have performed preliminary simulations on a variant (model III) in which a right and an up arrow are allowed to occupy the same site. In this four-state model all arrows try to move at every time step. If both an up arrow and a right arrow try to move to an empty site *at the same time step* they both move in and overlap. On the other hand no arrow can move into a site which is already occupied. This model is designed to have weaker excluded volume effects between arrows of different types. Indeed our simulations show that model III exhibits a continuous transition which is qualitatively similar to the one-dimensional case.

In summary, we have presented a cellular automaton model that describes traffic flow in two dimensions. Our simulations of finite systems up to 512×512 show a sharp transition that separates a low-density dynamical phase in which all cars move at a maximum speed and a high-density static phase in which they are all stuck in a global traffic jam. Such behavior is found both in a deterministic and a nondeterministic variants of the model and we thus believe that it is robust and represents a general feature of traffic flow in two dimensions. We believe that cellular automata provide a useful framework for traffic simulations that should be developed further. These models are especially suitable for simulations on parallel computers, and their flexibility is especially important in the complex geometries of traffic networks.

This work was performed using the computational resources of the Northeast Parallel Architectures Center (NPAC) at Syracuse University.

[1] See, e.g., *Transportation and Traffic Theory*, edited by N.H. Gartner and N.H.M. Wilson (Elsevier, New York, 1987); W. Leutzbach, *Introduction to the Theory of Traffic Flow* (Springer-Verlag, Berlin, 1988); D.L. Gerlough and M.J. Huber, *Traffic Flow Theory* (National Research Council, Washington, D.C., 1975).

[2] S. Wolfram, *Theory and Applications of Cellular Automata* (World Scientific, Singapore, 1986).

[3] See, e.g., *Lattice Gas Methods for PDE's*, Proceedings of the NATO Advanced Research Workshop, edited by G.D. Doolen, [Physica D **47**, 1 (1991)].

[4] B. Schnittman, Int. J. Mod. Phys. B **4**, 2269 (1990).

[5] P. Bak, C. Tang, and K. Wiesenfeld, Phys. Rev. Lett. **59**, 381 (1987).

[6] M. Gerhardt and H. Schuster, Physica D **36**, 209 (1989).

[7] Note that the fact that we find static behavior above p_c and dynamic behavior below does not mean that there are no static configurations with $p < p_c$ or dynamic configurations with $p > p_c$. What it means is that these cases are atypical and have very small basins of attraction in

the ensemble of random initial conditions. In fact, the static configuration with the smallest possible $p > 0$ has $p = 2/N \to 0$ as $N \to \infty$, and the dynamic configuration with $\overline{v} = 1$ and the largest possible p has $p = 2/3$.

[8] D. Stauffer, *Introduction to Percolation Theory* (Taylor & Francis, London, 1985).

[9] See, e.g., *Charge Density Waves in Solids*, edited by G. Hutiray and J. Sólyom (Springer-Verlag, Berlin, 1985); G. Grüner, Rev. Mod. Phys. **60**, 1129 (1988); *Charge Density Waves in Solids*, edited by L. P. Gorkov and G. Grüner (Elsevier, New York, 1989).

[10] Note that a similar situation occurs in lattice-gas cellular automata on the triangular lattice, where some states can have several different outcomes. In this case one can use either a deterministic approach such as using different outcomes in even and odd time steps or the nondeterministic approach. The deterministic approach is more efficient in numerical simulations, while there seems to be no significant difference in the results between the two approaches.

[11] This one-dimensional model is identical to Wolfram's CA rule no. 184, which is asymmetric and thus does not belong to the set of 32 legal rules with a three-site neighborhood. This model was previously studied in the context of surface growth through ballistic deposition [J. Krug and H. Spohn, Phys. Rev. A **38**, 4271 (1988)]. Related stochastic models in two dimensions with one type of arrow have also been studied [S.A. Janowsky and J.L. Lebowitz, Phys. Rev. A **45**, 618 (1992)].

[12] This result is obtained analytically by considering the vacant sites as left-moving arrows, with an exchange dynamics such that the numbers of right arrows and left arrows moving in each time step are the same.

Milton Keynes UK
Ingram Content Group UK Ltd.
UKHW051851071024
449327UK00025B/1911